CO-BIA-880

GRAPHICS

ANALYSIS AND CONCEPTUAL DESIGN

Other Books by A. S. Levens:

Nomography, Second Edition
Also translated into a Japanese Edition

Graphical Methods in Research

A. S. LEVENS Professor of Mechanical Engineering, University of California, Berkeley

SECOND EDITION

GRAPHICS

ANALYSIS AND CONCEPTUAL DESIGN

JOHN WILEY & SONS, INC. NEW YORK LONDON SYDNEY

Library of Congress Catalog Card Number: 67-29941
Printed in the United States of America

PREFACE

Engineering as a profession is concerned primarily with design. The core of engineering is design, in a broad sense, of machines, processes, structures, and circuits, and combinations of these components, into plants and systems. The professional engineer must be capable of predicting the costs and performances of the components, plants, and systems to meet specific requirements. The scientist is primarily concerned with the discovery of new knowledge and the development of new scientific principles, whereas the engineer is concerned with the implementation of these principles. The professional engineer deals with the applications of science, tempered by judgment based on experience, to the solution of engineering problems.

The late Dr. Theodor von Kármán stated it very nicely when he said: "The scientists explore what is, and the engineers create what has never been." In the development of solutions to real engineering problems, the creative art of conceiving a physical means of achieving an objective is the first and most important step. Analysis of the possible solutions is the next step. Stated simply, synthesis in conception precedes the analysis required to refine the conception.

Graphics plays a very significant role in the conceptual design phase; and in many cases is most effective in the analysis stage. Surely, the study of analytical courses is essential because it is difficult, without analysis, to predict performance and costs of a conceptual design while it is in the paper stage.

Designing is a conceptual process which is done largely in the mind, and the making of sketches is a recording process, a reliable memory system, which the engineer uses for self-communication—talking to himself—to help him "think-through" the various aspects of his project. Graphics is an integral part of the conceptual phase because, more often than not, the making of a simple sketch to express a design conception does of itself suggest further ideas of a conceptual nature.

Engineers who have developed the ability to form a visual image of geometrical and physical configurations and to "think graphically" have a tremendous advantage in creating a physical means of achieving a technological objective.

The "thinking-through" process is an exercise of the mental powers of judgment, conception, and reflection for the purpose of reaching a conclusion, i.e., a design which is the "best compromise" solution to a given project.

The technological explosion in the present era has brought about the need for much closer working relations between engineers and scientists. For effective communication both must be proficient in mathematics, physics, chemistry, graphics, and certainly in a common form of oral and written expression.

Engineering education must contribute, significantly, to the development of young, well-qualified persons who can and will face challenging engineering situations with imagination and confidence.

The student who undertakes preparation for the stimulating and exciting profession of engineering must have adequate education in mathematics, physics, chemistry, graphics, the engineering sciences, and a good background in the humanities and social sciences. With respect to graphics this means good facility in freehand sketching; thorough knowledge of (*a*) the fundamental principles of orthogonal projection and experience in applying the principles to the solution of space problems that arise in both engineering and science; (*b*) knowledge and use of graphical methods of computation; and (*c*) the development of the design process and the capability of coping with the "many-solutions" type of problems that are so characteristic of professional engineering.

It is the objective of this book to give the student an up-to-date treatment of graphics and an introduction to design that will help him become "graphically literate," so that he can apply graphics, with confidence, to synthesis, analysis, and solutions of problems that arise in the fields of design, development, and research, whenever appropriate. In contrast to the first edition, this new edition lays much more stress on the design process and includes several industrial examples which serve to provide the student a vicarious design experience. In addition, the very latest standards are used throughout the book. Moreover, many new examples and figures pervade the entire text.

Part 1, Fundamental Principles and Applications of Orthogonal Projection, lays *stress on principles and concepts.* Thorough grasp of the few basic principles greatly enhances the student's ability to analyze and solve various space problems that arise in both engineering and science. Use of *freehand sketches* in recording the student's thought process and in planning a suitable arrangement of views in solving problems is stressed. *The emphasis is on thinking, not on draftsmanship.* A variety of thought-provoking space problems is available in this part of the text and in the accompanying workbooks.

Part 2, Graphical Solutions and Computations, includes: (*a*) graphical presentation of data; (*b*) graphical mathematics, including arithmetic, algebra, and calculus; (*c*) empirical equations—forms frequently used in engineering; (*d*) functional scales; and (*e*) an introduction to nomography. The student will find this part both stimulating and very useful. His experience with the material on graphical calculus will greatly enhance his understanding of integration and differentiation. He will see that, in several cases, it is not convenient to express the relation between two variables algebraically, and that a graphical solution is best suited to the problem. This is very often true in dealing with experimental data.

When the material on graphical, numerical, and grapho-numerical methods of integration and differentiation is presented, it is appropriate to introduce *computer solutions.* For example, a computer solution of the application of Simpson's rule could be demonstrated. Also, Cal-comp or Gerber plot-outs could be shown if equipment is available. Included in this part are several examples of the use of graphical methods in the solution of problems that have arisen in research projects.

The material on nomography will be found most interesting. Nomograms play an important role in the repeated solution of formulas and also in *studying the interrelationships among four or more variables for which an explicit or implicit relation exists.* Further study of the fascinating field of nomography is provided for in the author's book, *Nomography.**

Part 3, Introduction to Design, provides an opportunity for the student "to be on his own" to a large degree. The experience in facing up to projects that have several solutions will enable the student to approach, with much greater confidence, real engineering situations that will confront him in his professional career.

We cannot start too early to give the student the experience of confronting and solving the "open-ended" type of problem. The first year in engineering education is not too soon! We recognize that the student's background is quite limited at this stage of his career, and therefore the proposed projects are relatively simple. Nevertheless, in principle, they are of the same character as some of the more involved projects that arise in engineering practice. The student's experience in coping with open-ended projects will be found to be stimulating, challenging, and most rewarding. He will find a good spirit of competition among his classmates, especially with regard to who has developed the "best" design solution. In this part of the text he will learn to understand the design process and its employment in undertaking a project.

The chapters on pictorial drawing, sections and conventional practices, fasteners, dimensions and specifications for precision and reliability, constitute additional background material for the chapter on conceptual design. It is presumed that the student has already developed a reasonable degree of proficiency in manual dexterity with respect to use of drawing instruments, lettering, geometric constructions, etc., in high school courses. When the student has had such experience in the above areas, including freehand pictorial sketching, more time can be devoted to conceptual design. This text and the two workbooks will also serve the needs of those high school seniors and junior college students who are committed to study engineering in preparation for an engineering career.

It is strongly recommended that some time be

**Nomography*, 2nd ed., John Wiley and Sons.

devoted to the chapter on dimensioning for precision and reliability so that students will be aware of the *important role* of this material in present day design and manufacture.

As the student progresses in his engineering education and enters upon the study of mechanics, strength of materials, design (in the broad sense), and research, he should continue to employ, whenever appropriate, graphical methods to solve problems that arise in these areas.

Graphics, Analysis and Conceptual Design, as presented in this text, reflects today's thinking and our continual effort to develop a meaningful and worthwhile treatment, which, we believe, is consistent with the needs of a scientific and engineering era. The experience with our students continues to be most gratifying. *Students are stimulated to learn and to apply fundamental principles; to "think-through" a problem, rather than depend upon rote learning;* to learn and use graphical methods of computation; and to appreciate, through their own experience with conceptual design, that the qualified engineer must have the necessary education and experience to cope with real engineering situations that arise, and will continue to arise, in an ever-growing, dynamic technological era.

I am deeply appreciative of the cooperation received from the following organizations and editors of a number of journals and periodicals, for permission to use certain photographs, charts, and drawings: Automobile Manufacturing Association, The Bendix Corporation, The Boeing Company, *Experimental Mechanics Engineering,* Falk Corporation, General Motors Corporation, *Industrial Fasteners Institute,* Librascope, Inc., Lockheed-Georgia Company, *Machine Design,* Mattel, Inc., Mobil Oil Corporation, National Safety Council, *Plant Engineering, Product Engineering,* Republic Steel, Sandia Corporation, San Francisco Naval Shipyard, TRW Systems, Inc., and U.S. Weather Bureau.

My thanks, too, to the Expo '67 Publicity Board for the photograph used as the frontispiece and on the cover. Also to the following for the use of photographs on the chapter title pages: Chapter 2, National Museum of Science & Technology, Milan, Italy; Chapter 3, Courtesy of Chemcut; Chapter 4, Weiner Bischof, Magnum Photos; Chapter 5, George Schlosser, The Devilbiss Company; Chapter 6, Charles Rotkin, P.F.I.; Chapter 7, Courtesy of Bethlehem Steel Corporation; Chapter 8, Hedrich-Blessing; Chapter 9, © Ezra Stoller; Chapter 10, Courtesy of Cornell News Bureau; Chapter 11, Courtesy of General Electric; Chapter 13, Courtesy of the Clevite Corporation; Chapter 15, Courtesy of Vecco; Chapter 16, Courtesy of the Gerber Scientific Instrument Company; Chapter 17, Courtesy of TRW Systems; Chapter 18, Courtesy of the Ford Motor Company; Chapter 19, Courtesy of UNIROYAL; Chapter 20, Courtesy of General Electric; Chapter 21, Courtesy of General Motors.

My heartiest thanks to Mr. J. F. Schon of the City College of San Francisco for his most valuable contribution to the chapter on dimensions and specifications for precision and reliability; and for a critical review of the presentation. I am also appreciative of the efforts made by Professor R. W. Reynolds, California State College at San Luis Obispo, and by Mr. Richard Neitzel, Santa Rosa Junior College, for their assistance in developing some problem material.

I am indebted to Mr. Norman Waner, Chief Engineer of Hallikainen Instruments, Inc., and to Mr. Richard J. Leuba, formerly Project Engineer of the same company, for the material on the Osmometer design, and to the latter for his thorough review of the presentation; to Mr. Jack Ryan and Mr. Gilbert Thomas of Mattel, Inc., for the material on the design of the "ride-a-way" toy, and to Mr. R. Glenn Mackenzie of TRW Systems, Inc. for his cooperation in obtaining industrial examples and photographs.

I am grateful to my colleagues at the University of California: Professor H. D. Eberhart for the use of the material from the Prosthetic Devices Research Project; Professors D. M. Cunningham, F. E. Hauser, and H. W. Iversen, for experimental data used in connection with several examples in the chapters on empirical equations and graphical calculus.

And to my wife, Ethel, whose continuous encouragement, patience, and understanding made this work possible, my heartfelt and everlasting appreciation.

A. S. Levens

Berkeley, California

PREFACE TO THE FIRST EDITION

The well-qualified engineer must be proficient in mathematics, graphics, physics, chemistry, and the engineering sciences. In addition, he should have a good background in the humanities and social sciences.

In developing solutions to real engineering problems, the creative art of conceiving a physical means of achieving an objective is the first and most important step. Analysis of the possible solutions is the second step. Stated simply, synthesis in conception precedes the analysis required to refine the conception.

Graphics plays a most important role in the conceptual design phase; and in many cases it is also most effective in the analysis stage. Of course, the study of analytical courses is very necessary because it is difficult without analysis to predict performance of a conceptual design while it is still on paper.

The technological demands of our scientific era have brought about much closer working relations between engineers and scientists. For effective communication both must be proficient, not only in mathematics, physics, and chemistry, but also in engineering graphics; and certainly in a common form of oral and written expression.

Engineering education must contribute, significantly, to the development of young, well-qualified persons who can and will face new and challenging engineering situations with imagination and confidence.

With respect to graphics and its important role in engineering, it is essential to provide programs of study that are pertinent to the students' preparation for the engineering profession. Certainly, development of facility in freehand sketching—both pictorial and orthographic—is most important for both engineering and science students in order to provide an effective *graphic extension of the mind.*

Engineering as a profession is concerned primarily with design. In fact, the core of engineering is design which, in the broadest sense, includes circuits, machines, structures, processes, and combinations of these components into systems and plants. The professional engineer must be capable of predicting the performance and cost of the components, systems, and plants to meet specified requirements. We must not overlook the fact that the theoretical scientist is primarily concerned with the discovery of new knowledge and the development of new scientific principles, whereas the engineer is concerned with the implementation of these principles. The professional engineer deals with the applications of science, tempered by judgment based on experience, to the solution of real engineering problems.

Designing is a conceptual process which is done largely in the mind, and the making of sketches is a recording process, a reliable memory system, which the engineer uses for self-communication to help him "think-through" the various aspects of his project. Graphics is an integral part of the conceptual phase because, more often than not, the making of a simple sketch to express a design conception does of itself suggest further ideas of a conceptual nature. Engineers who have developed their ability to visualize geometrical and physical configurations and to "think graphically" have a decided advantage in creating a physical means of achieving an objective.

The student who undertakes preparation for the exciting and stimulating profession of engineering must have adequate education in graphics. This means not only facility in freehand sketching, but also thorough knowledge of (*a*) the fundamental principles of orthogonal projection and experience in the application of those principles to the solution of space problems that arise in both engineering and science; (*b*) knowledge and use of graphical solutions and methods of computation; and (*c*) the development of capability to cope with the "many-solutions" type of problems that are so characteristic of engineering.

The principal objective of this book is to provide for the student a modern treatment of graphics and an introduction to conceptual design that will help him become "graphically literate," so that with confidence he can employ graphics—a powerful mode of expression—to the synthesis, analysis, and solution of problems that arise in the fields of design, development, and research.

It is presumed that the student has already developed a reasonable degree of proficiency in manual dexterity with respect to lettering, use of drawing instruments, geometric constructions, etc., in high school or, if feasible, in noncredit prerequisite work in a college. The material in Appendix A of this text, and the problems in the appendix of the workbook, number one, provide ample subject matter for this purpose.

Part 1, Fundamental Principles and Applications of Orthogonal Projection, *lays stress on principles. Rote learning is discouraged. Thorough grasp* of the few fundamental principles will greatly enhance the students' ability to "think through" and solve various space problems that arise in both engineering and science. The student is encouraged to use freehand sketches in planning the arrangement of views that may be necessary for the solutions and in recording his thought process in evolving the solutions. *Emphasis is on thinking, not on draftsmanship.* Interesting examples of the application of the fundamental principles of orthogonal projection to problems that have arisen in the field of research are included.

The diligent student will be able to apply the principles to the solution of many problems that "appear" to be different, but are *basically* the same.

The treatment of vector quantities and vector diagrams with applications to three-dimensional force systems affords the student another opportunity to apply the fundamental principles of orthogonal projection to the solution of concurrent, non-coplanar force problems.

Chapter 11, the final chapter of Part 1, is intended to strengthen the students' ability in analysis. It is suggested that reference to this chapter be made whenever appropriate.

Part 2, Graphical Solutions and Computations, includes the graphical presentation of data; graphical mathematics—arithmetic, algebra, calculus; empirical equations (forms most frequently encountered in engineering); functional scales; and an introduction to nomography. The student will find this part most interesting and stimulating. His understanding of integration and differentiation will be enhanced by his experience with the material on graphical calculus. The student will realize that many problems cannot be conveniently expressed in the symbolic language of mathematics and that a graphical (which is also mathematical in the broad sense) method may be better suited to the solution of the problem.

At the time when the material on graphical, numerical, and grapho-numerical methods of integration and differentiation is presented, it has been found appropriate to *introduce computer solutions.* A number of examples are included to demonstrate the employment of graphical methods in the solution of problems that have arisen in research projects at the University of California at Berkeley.

The material on nomography is an adequate introduction to this most useful and fascinating field. The student will dicover that nomograms play an important roll in the repeated solution of formulas and also in analyzing the relations among three or more variables for which an explicit or an implicit relation exists. (In very recent years a nomographic method has been developed to test the validity of a family of experimental data curves.*)

Part 3, Introduction to Conceptual Design, affords the student an opportunity "to be on his own." His background in graphics—based on the material in the first two parts—coupled with his course work in mathematics, chemistry, physics, and perhaps some actual job experience will have a direct bearing on his progress in dealing with *projects that have many solutions.* The experience in facing up to such problems will enable the student to approach, with greater confidence, real engineering situations that will confront him later. *We cannot start too early to give the student the experience of confronting and solving, reasonably well, the "open-ended" type of problem. The freshman year is not too soon!* It is recognized that the student's background is quite limited at this stage of his career, and because of this the proposed problems are relatively simple and limited in scope. *Nevertheless, in principle* they are of the same character as some of the most advanced problems in engineering practice. The student will find his experience in conceptual design stimulating, challenging,

*In the author's 2nd edition of *Nomography,* John Wiley and Sons.

and rewarding. A good spirit of competition is generated among the students—especially with regard to who has the "best" design.

The chapters on pictorial drawing, sections and conventional practices, fasteners, dimensions and specifications, dimensioning for precision and reliability, constitute additional background material for the chapter on conceptual design. Where the student has had good experience in the above areas, more time can be devoted to conceptual design.

An appreciation of some of the problems that confront the engineer is the matter of "dimensioning for precision and reliability." It is strongly recommended that some time be devoted to this chapter so that students will understand the importance of this subject, especially today in connection with the design of various space vehicles.

As the student progresses in his engineering education and enters upon the study of mechanics, strength of materials, design (in the broad sense), and research, he should continue to employ, whenever appropriate, graphical methods to solve problems that arise in these areas.

Graphics, with an introduction to conceptual design, as presented in this text, reflects our continual effort to develop a meaningful and worthwhile treatment which, we believe, is consistent with the needs of a scientific era. The experience with our students continues to be most gratifying. *Students are stimulated to learn and to apply fundamental principles; to "think-through" a problem, rather than depend upon rote learning;* to learn and use graphical methods of computation; and to appreciate, through their own experience with conceptual design, that the qualified engineer must have the necessary education and experience to cope with real engineering situations that arise, and will continue to arise, in an ever-growing, dynamic technological era.

I deeply appreciate the cooperation received from the following organizations and editors of several journals and periodicals, in permitting the use of certain photographs, charts, and drawings: Aerojet-General Corp., American Standards Association, The Bendix Corporation, The Boeing Co., Columbia-Geneva Steel Division of the U.S. Steel Corp., *Chemical Engineering,* Continental Can Co., Inc., Convair Division of General Dynamics Corp., Firestone Tire and Rubber Co. of California, Food Machinery and Chemical Corp., General Motors Corp., Hiller Helicopter Corp., *Industrial Fasteners Institute,* International Business Machines Corp., Leeds and Northrup Co., Librascope, Inc., Lockheed Aircraft Corp., *Machine Design, Materials in Design Engineering, Military Systems Design,* North American Aviation, Inc., *Product Engineering,* San Francisco Naval Shipyard, Society for Automotive Engineers, Toby Enterprises Corp., San Francisco, Calif., U.S. Electrical Motors, Inc., Varian Associates, and Westinghouse Electric Corp.

My thanks to Messrs. A. E. Edstrom and J. F. Schon of the City College of San Francisco—the former for many valuable suggestions, and the latter for his assistance in procuring excellent practical examples for the material on dimensioning for precision and reliability; and for a critical review of the presentation. I am also appreciative of the efforts made by Professor R. W. Reynolds, California State Polytechnic College at San Luis Obispo, California, for his assistance in developing some problem material.

I am grateful to my colleagues at the University of California: Professor H. D. Eberhart, for the use of materials from the Prosthetic Devices Research Project; Professors L. J. Black, J. E. Dorn, F. E. Hauser, R. R. Hultgren, and H. W. Iversen, for experimental data used in connection with several illustrated examples in the chapters on empirical equations and graphical calculus; and to Mr. Paul Urtiew, graduate student, for the photograph and the delineation of the system used in connection with the research project on gas dynamics, shown as an example in the chapter on graphical calculus.

And to my wife, Ethel, whose patience and understanding made possible an environment which was conducive to productive writing, my ever-lasting and heartfelt appreciation.

A. S. Levens

Berkeley, California
April, 1962

CONTENTS

PART 1
FUNDAMENTAL PRINCIPLES AND APPLICATIONS OF ORTHOGONAL PROJECTION

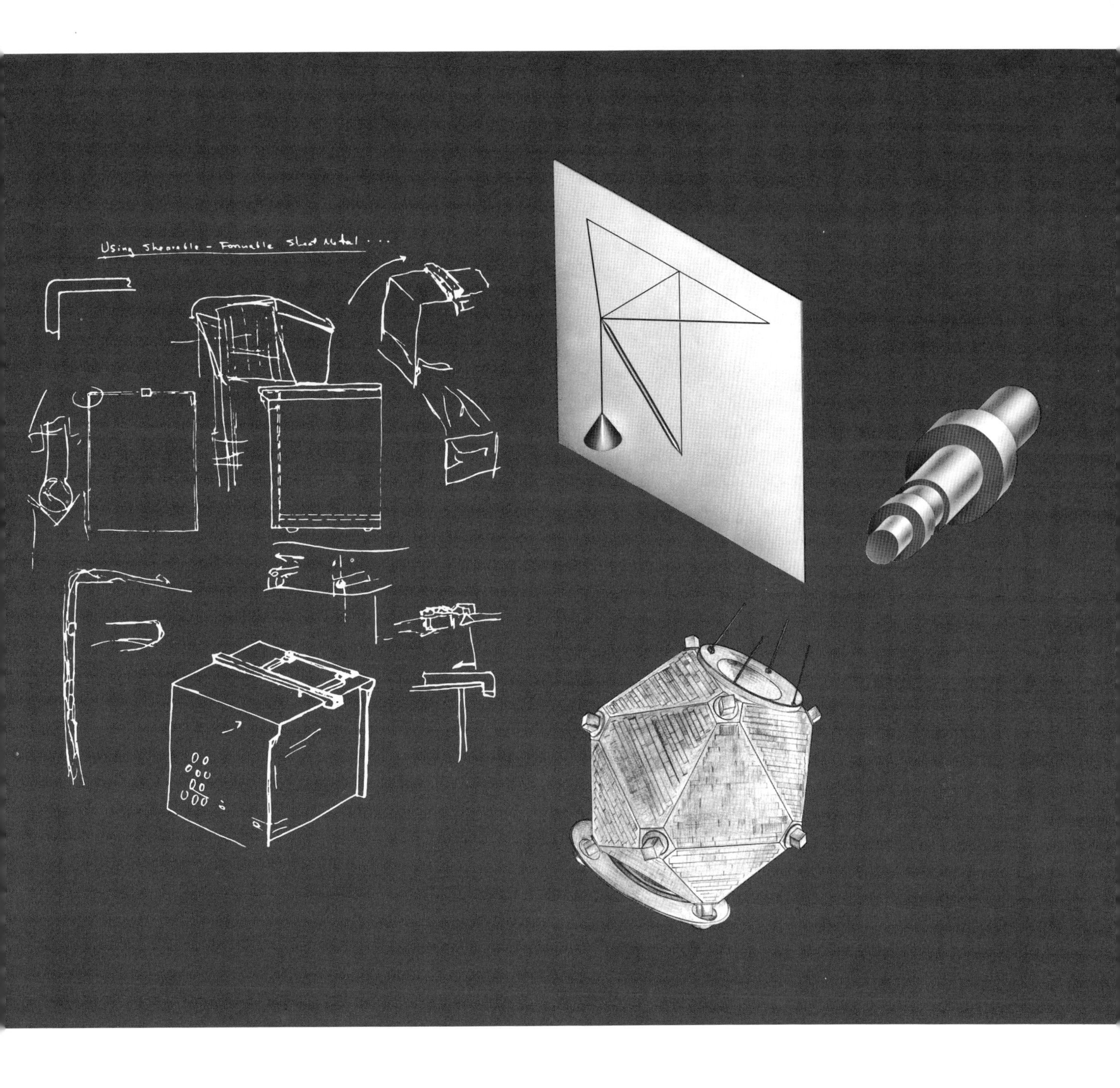

INTRODUCTION 1

The history of the development of engineering education reveals significant changes in engineering curricula. This is as it should be in the dynamic growth of the engineering profession. Changes have been necessary in order to keep pace with the accelerated growth in technology. This is especially true of the period since World War II. The importance of design, research, and development has increased the need for "science-oriented" curricula in engineering. There is now greater emphasis on mathematics, physics, chemistry, and the engineering sciences than on "the art of engineering." In recent years, however, there is the realization that *design* is the core of engineering and that a much greater effort must be made to enhance the engineering student's understanding of the design process and also to provide opportunities for design experiences. In the broadest sense design includes circuits, machines, structures, and processes, and their combinations. It is the responsibility of the professional engineer to use these components in the design of plants and systems; and also to predict their performance and costs to meet specified requirements. Engineers are often guided by the new-knowledge discoveries made by scientists, but the end products —hardware—are not the result of the physical facts alone. Many other factors must be considered. For example, the design of a space vehicle to meet certain environmental and performance specifications, or the design of a device to monitor blood pressure continuously without discomfort to the patient (or, for that matter, other useful products), requires of the engineer effective knowledge of the mathematical and physical sciences, of engineering graphics, of the engineering sciences, coupled with engineering experience which reflects good judgment in determining need, feasibility, reliability, marketability, etc. Engineering education must contribute, significantly, to the development of young men who can face new and challenging situations with imagination and confidence. Meeting such challenges invariably involves both professional and social responsibilities.

In the first stages of a technological project many ideas are presented in written form with accompanying "idea sketches," charts, models, and layouts, to provide means for studying such problems as feasibility, reliability, costs, etc. In some situations it may be necessary to carry on research, development, and design to resolve a number of technical problems. In these areas considerable use is made of mathematics, graphics, physics, chemistry, and the engineering sciences—mechanics, properties of materials, etc.

The role of graphics is very important in the "ideation stage"—the conceptual design phase. Conceptual design is a mental process. It is the creative process of conceiving a physical means of achieving a physical objective. The making of sketches is a very important recording process—a reliable memory system—which enables the design engineer to "talk to himself." The recording of the initial conceptual design sketches serves to suggest further items of a conceptual nature. Today much of the work associated with the preparation of detail drawings of components, assemblies of components, product design, and production illustration is done by well-trained technicians who are supervised by graduate engineers.

As a student who undertakes the study of engineering with the hope of becoming a competent professional engineer, you must have adequate education in the field of engineering graphics. This means thorough knowledge of the fundamental principles of orthogonal projection and experience in the application of these principles to the solution of three-dimensional problems that arise in the various fields of engineering and science; *proficiency in free-hand sketching*—a powerful tool for expressing ideas, for planning solutions, for recording analyses (the "thinking-through process") of space problems; and an understanding and use of graphical mathematics, including graphical calculus, empirical equations, functional scales, and elements of nomography.

In addition, experience in dealing with "open-ended" projects—the type of problems that have several solutions—is essential to your understanding of the "real" problems that continually face the engineer. You will develop, early in your career, a "feel" for the engineering approach to the solutions of project-type situations.

Many problems that arise in projects are best solved graphically. Graphical solutions, in many cases, are much quicker, more vivid, more practical, and less likely to incur accidental errors than solutions obtained algebraically. In some cases the problem cannot be adequately modeled mathematically. Technical problems in many situations are concerned with length, time, temperature, pressure, etc., quantities that can be measured only approximately. The reading of a voltmeter, for example, can be only as accurate as, among other things, the graduated scale (graphical) and the visual acuity of the observer. This is also true of lengths, whose measurement again depends on a graphical scale; or of pressure, whose measurements depend on a gage mechanism and a scale for reading numbers; or, of temperatures, currents, and many other quantities which are measured by devices that indicate values on a scale that is read by an individual. The *data are graphical* in nature. Certainly a graphical solution of a problem that inherently is based on graphical data can be arrived at by graphical methods with sufficient accuracy. The fact that we use numbers, obtained by reading various scales, in a mathematical solution of the problem does not, by any stretch of the imagination, make such a solution more accurate than a graphical one.

We know that thorough training in mathematics is essential to sound engineering and scientific activity. It is, however, important that you as a student do not get a *warped* view, so that your training will lead to *the intelligent use of several methods* and, it is hoped, to the *development of good judgment* in the choice of methods. In many cases a combination of algebraic, graphic, numerical, and electromechanical methods is best suited to the solution of technical problems.

In this treatment of graphics and conceptual design, Part 1 stresses the *fundamental principles* of orthogonal projection and the application of these principles to the solution of a variety of problems that arise in technology. Every effort is made to strengthen your ability to visualize and to analyze (the "thinking-through process") space problems, and then to record the solutions graphically. Many opportunities are given you to develop your *creative thinking* and "imagineering."

Included in Part 1 are problems dealing with the determination of true lengths of members (rods, pipes, cables, etc.); the determination of clearances, as in the case of a bomb-release cable and an aileron-control cable, or for that matter, in any case where specified clearances must be maintained between wires, pipes, or structural members; distance problems such as (*a*) from point to line, (*b*) from point to plane, (*c*) between parallel planes, etc.; angle problems such as (*a*) between lines, (*b*) between planes, (*c*) between line and plane; intersections, developments; vectors, and *graphic statics*, the treatment of which should enable the student to use the fundamental principles effectively in analyzing force problems and in solving such problems graphically. The first portion of graphic statics is concerned with two-dimensional systems, whereas the second portion deals with the determination of forces in three-dimensional frames. These problems afford the student an excellent opportunity to make use of the fundamental principles. Throughout Part 1—Fundamental Principles and Applications of Orthogonal Projection—the emphasis is on principles, analysis, and synthesis. Interesting examples of problems in the areas of design and research are demonstrated throughout Part 1. The development of power to visualize and analyze a space problem and then to solve it by employing the fundamental principles is an essential goal in educating the student to become self-reliant. The engineer who has the ability to "think graphically" has a decided advantage in conceiving a physical means of achieving an objective.

Reasonable proficiency in the use of drawing instruments is desirable, but *greater emphasis should be placed upon the development and use of good freehand techniques. The ability to make good free-hand sketches is an invaluable asset.* Engineers and scientists should be able to use the techniques of freehand sketching effectively in presenting ideas to their co-workers. Engineers can save much time, and thereby do a more economical job, by preparing good freehand sketches *for use by the draftsman* in making finished working drawings. While still an engineering or science student you should develop facility in making good freehand pictorials and orthographic drawings. You should take advantage of every opportunity—not only in graphics courses—to gain experience in technical sketching.

Practice in making freehand sketches should become a hobby. The full realization of the importance of developing skill in *"talking with a pencil"* will become increasingly evident after you have graduated from college and have entered upon the practice of engineering. You need not be an artist or have special talents to produce good freehand sketches. The essential requirement is to draw a reasonably straight line between two points. Of course, some training and practice are necessary. The skill required is, surprisingly, no greater than any student needs for good legible writing. Good habits of accuracy, neatness, etc., are just as essential in freehand work as they are to the preparation of drawings with the use of instruments.

Part 2—Graphical Solutions and Computations— is primarily concerned with graphical analysis and graphic methods of computation. Included are chapters on graphical mathematics; graphical calculus; empirical equations; functional scales; and nomography. You will greatly enhance your ability through the power of graphical methods that give meaning to algebraic-type problems. You will be able to use graphical methods to solve rate, displacement, velocity, and acceleration problems. In addition, you will have an opportunity to learn the theory and design of functional scales—*their importance and use in connection with the representation and interpretation of experimental data.* Again, in the design and construction of both Cartesian charts and alignment charts (nomography), further use of functional scales will be made.

Moreover, as a student you will soon realize the importance of nomography in dealing with the graphical representation of equations that require repeated solutions; and more importantly in design, in analyzing the relationships among several design parameters. It will become apparent that the elementary phases of nomography can be grasped quite readily; that the basic theory is not beyond your understanding. The applications of nomography cut across many fields—engineering, physics, chemistry, psychology, biology, statistics, medicine, business administration, production, research, etc.

The chapter on graphical calculus sets forth, in a simple manner, the meaning of integration and differentiation. Many examples—problems dealing with displacement, velocity, acceleration, areas, volumes, etc.—are included to provide interest for the student and to demonstrate the power of graphical calculus, as well as its usefulness.

Part 3—Introduction to Design—deals with *conceptual design, the creative art of conceiving a physical means of achieving an objective. This is the first and most crucial step in an engineering project.* The chapters on pictorial drawing, sections and conventional practices, fasteners, dimensions, and specifications, constitute additional background material for the important final chapter—Conceptual Design; Developing Creativity. The material of Parts 1 and 2 will be used in connection with Part 3. Projects provide for experience in *synthesis*—the opportunity to tie together fundamental principles, analysis, graphic methods of computation, and some introductory experience in engineering.

It should be quite clear that the material presented in this book is a *very significant part of the education of engineers and scientists—not* the *training* of draftsmen.

It is my strong conviction that effective education in graphics must include the material presented in these three parts. Students so educated in graphics will have a fuller appreciation of the power of graphics and will be better able to integrate graphics with mathematics, physics, chemistry, mechanics, strength of materials, and engineering design. To increase your competence in graphics, take advantage of every opportunity given you to apply what you learn now to course work in the upper division (junior and senior years of college) and at the graduate level to course work and research. Sound education in graphics is of utmost importance for engineers and scientists who can best meet the challenges of our accelerated growth in both engineering and science.

The engineers and scientists who will have had this education in graphics will greatly enhance their effectiveness by using their "graphical thinking" capacity in analyzing and solving the challenging technological problems of our times.

FREEHAND SKETCHING 2

The ability to make good understandable freehand sketches is invaluable to both the engineer and the scientist.

Effectiveness in "talking with a pencil" is most essential to (*a*) the project and preliminary design engineers in their communication with their supporting staffs; (*b*) the design engineers in their communication with design draftsmen; (*c*) the design draftsmen in their communication with technicians and production personnel; and (*d*) the technical illustrator in his preparation of pictorial sketches for instruction and maintenance manuals.

During this twentieth century, particularly since World War II, we have witnessed the ever-increasing interdependence of engineering graphics, research, design, and manufacturing. Engineers and scientists recognize the importance of pictorial sketches in the preparation of "idea sketches" of preliminary design studies, in the interpretation of orthographic drawings, in designating the sequence in which components of a mechanical or an electrical unit fit together, etc. The lines of communication from the "idea engineer," or the "idea scientist," to the research, development, design, production, and sales personnel can be kept open and clear through the effective use of freehand sketches.

The full realization of the importance of developing skill in freehand sketching will become more and more evident after the student has graduated from college and has started his career in engineering.

In school the student will have many opportunities to use his freehand sketching ability in several of his courses—mathematics, physics, chemistry, graphics, various engineering courses, etc.; i.e., in taking usable notes, in preparing design layouts, in writing papers and reports. In Engineering Graphics, freehand sketches will be found very useful *in planning the best arrangement of views in the solution of a variety of three-dimensional problems that arise in the various fields of engineering;* also, in making freehand isometric, oblique and perspective pictorials; in the preparation of freehand orthographic drawings; and in the translation of orthographic design drawings to pictorials when necessary.

A student need not be an artist nor have special talents in order to produce good freehand sketches. So let us observe some useful hints in making freehand sketches.

ELEMENTS OF TECHNIQUE

Pencil, paper, and occasionally an eraser are the simple tools employed in making freehand sketches. These will usually include orthographic, isometric and oblique drawings. Special grid sheets (coordinate paper) are available for isometric and perspective sketches. Every effort should be made to sketch objects in true proportion—not necessarily to a specified scale, although in some cases this may be required. It will be found advantageous to use coordinate paper because it helps one to maintain reasonably accurate proportions. Should it be necessary to prepare freehand sketches on plain sheets, however, it is suggested that transparent paper be used so that a coordinate *underlay* may be employed. Should copies of the sketches be required, it is suggested that paper known as "fade-out blue" be used. This is a transparent sheet with coordinate lines which will not show when blueprints or other types of machine copies are made.

It is recognized that individuals will develop freehand sketching techniques that may vary from recognized practices. No attempt is made in this brief treatment to detail all methods that could be employed in technical sketching. The following suggestions have proved useful; and are best illustrated by the accompanying sketches.

Sketching Vertical Lines, Fig. 2.1. The forearm and elbow should rest on the drawing surface. For trial lines and construction lines, the pencil is held lightly. Note that the pencil point extends about 2 inches beyond the fingers.

Sketching Horizontal Lines, Figs. 2.2 and 2.3. Lightly drawn segments, approximately 3 inches long, are made by using a sliding action of the forearm. Final lines are strengthened by applying more pressure.

Fig. 2.1

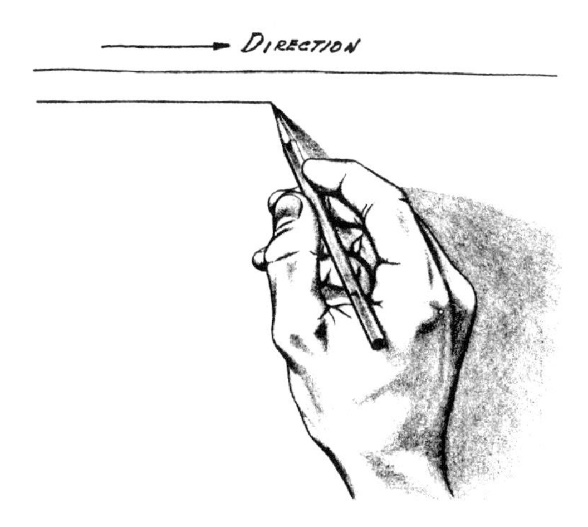

Fig. 2.2

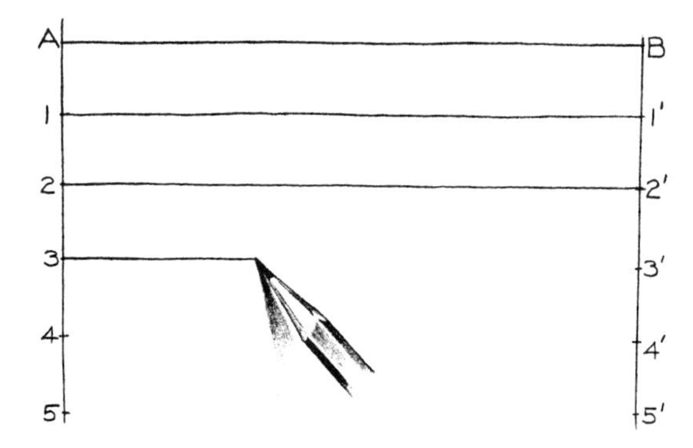

Fig. 2.3

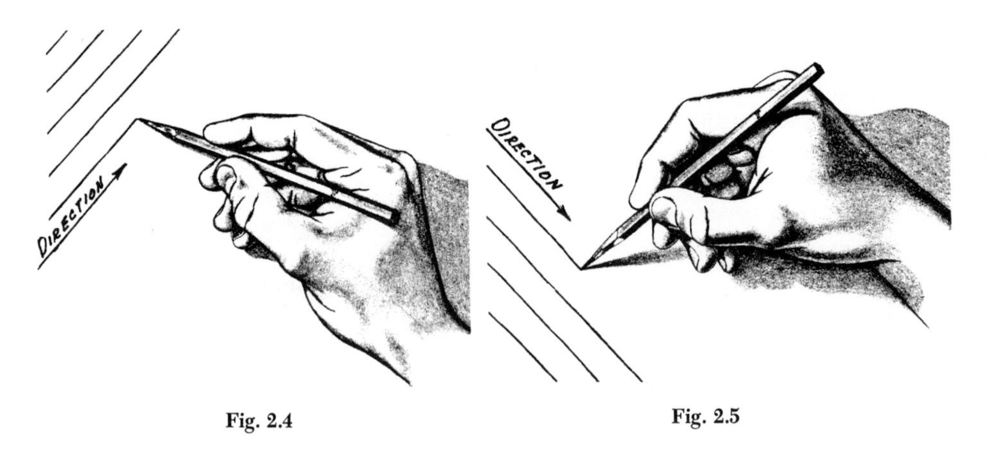

Fig. 2.4

Fig. 2.5

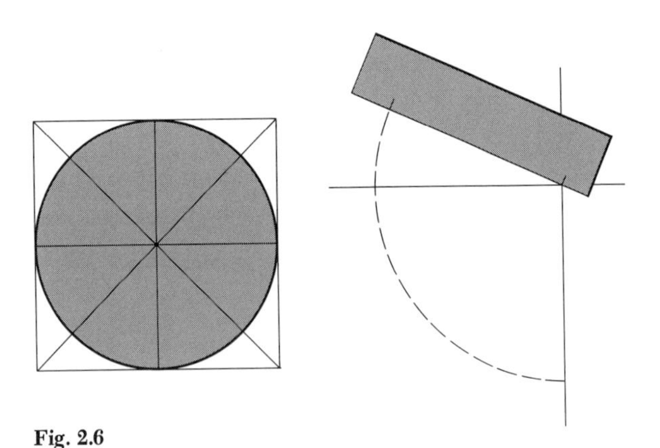

Fig. 2.6

Sketching Sloping Lines, Figs. 2.4 and 2.5. Figure 2.4 shows hand and pencil positions for drawing lines upward and to the right. Figure 2.5 illustrates position for lines drawn downward and to the right. Since most persons find it quite easy to draw horizontals, all lines, provided the size of the sheet makes it possible, may be drawn "as horizontals" by simply rotating the sheet to accommodate this position.

Sketching Circles, Fig. 2.6. A light, freehand square, including the diagonals and the horizontal and vertical mid-lines, is drawn first. Points are located on the diagonals at a distance from the center approximately equal to the radius of the circle. A fine-line circle is sketched through those points and the end points of the mid-lines. The circle is then strengthened by applying more pressure in finishing the sketch. An alternative method for locating points on the circle is shown on the right of Fig. 2.6. A rotation method is shown in Fig. 2.7.

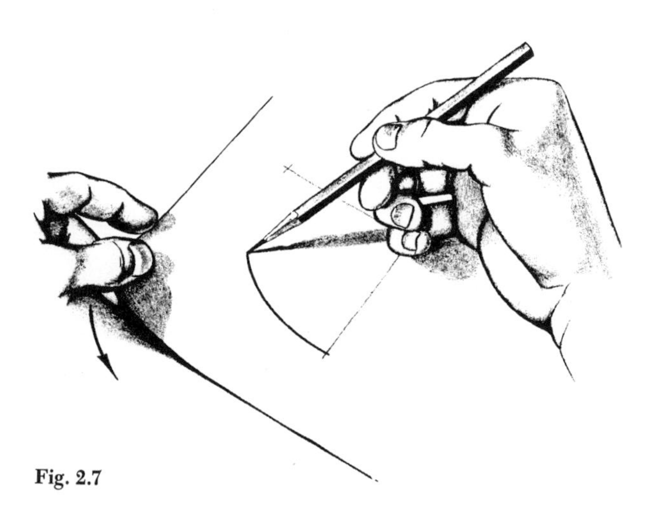

Fig. 2.7

Sketching Ellipses, Fig. 2.8. Points on an ellipse can be located by laying off on a strip of paper, a distance such as AC equal to half of the major axis and distance BC equal to half of the minor axis, and then marking point C as points A and B take different positions on the minor and major axes, respectively. The ellipse is then sketched through these points. (The method used in this case is known as the "trammel" method.)

Fig. 2.8

Pictorial sketches of circles usually appear as ellipses. Figure 2.9 shows circles on three faces of a cube. The construction for locating the centers of arcs that closely approximate portions of the ellipses is clearly indicated.

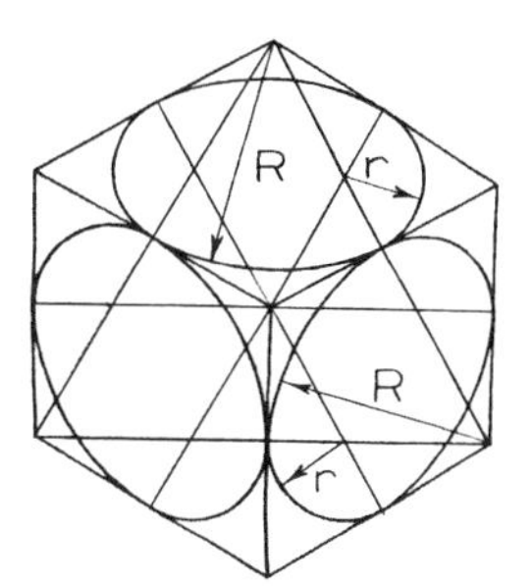

Fig. 2.9

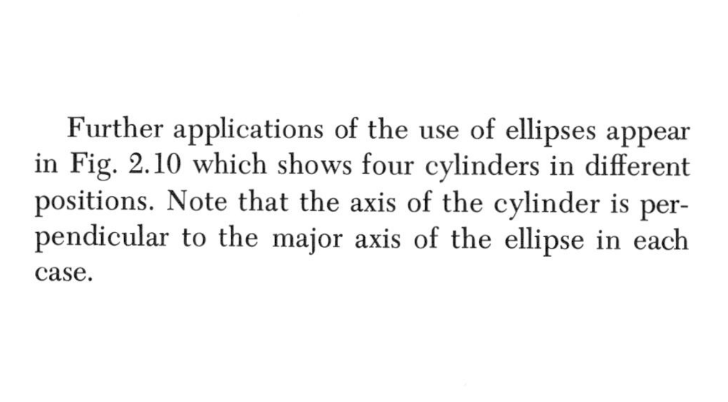

Further applications of the use of ellipses appear in Fig. 2.10 which shows four cylinders in different positions. Note that the axis of the cylinder is perpendicular to the major axis of the ellipse in each case.

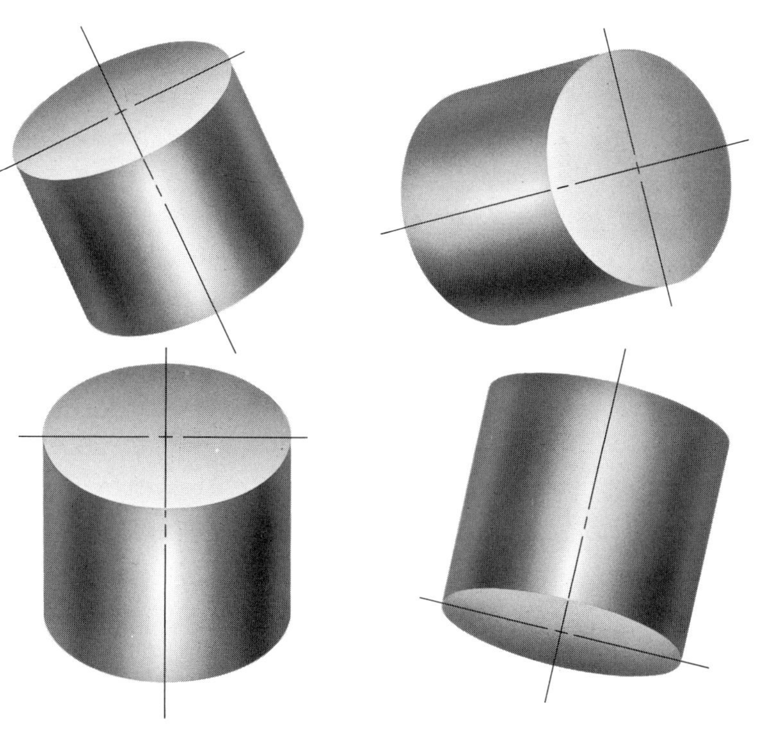

Fig. 2.10

TYPES OF FREEHAND SKETCHES

Freehand sketches may be orthographic or pictorial. Orthographic freehand sketches are often drawn on coordinate grid sheets or on transparent paper with a coordinate-grid underlay. Pictorial sketches include isometric, dimetric, trimetric, oblique, and perspective sketches. Of all these, considerable use is made of orthographic, isometric, and perspective sketches.

A practical example of a freehand orthographic design is shown in Fig. 2.11.

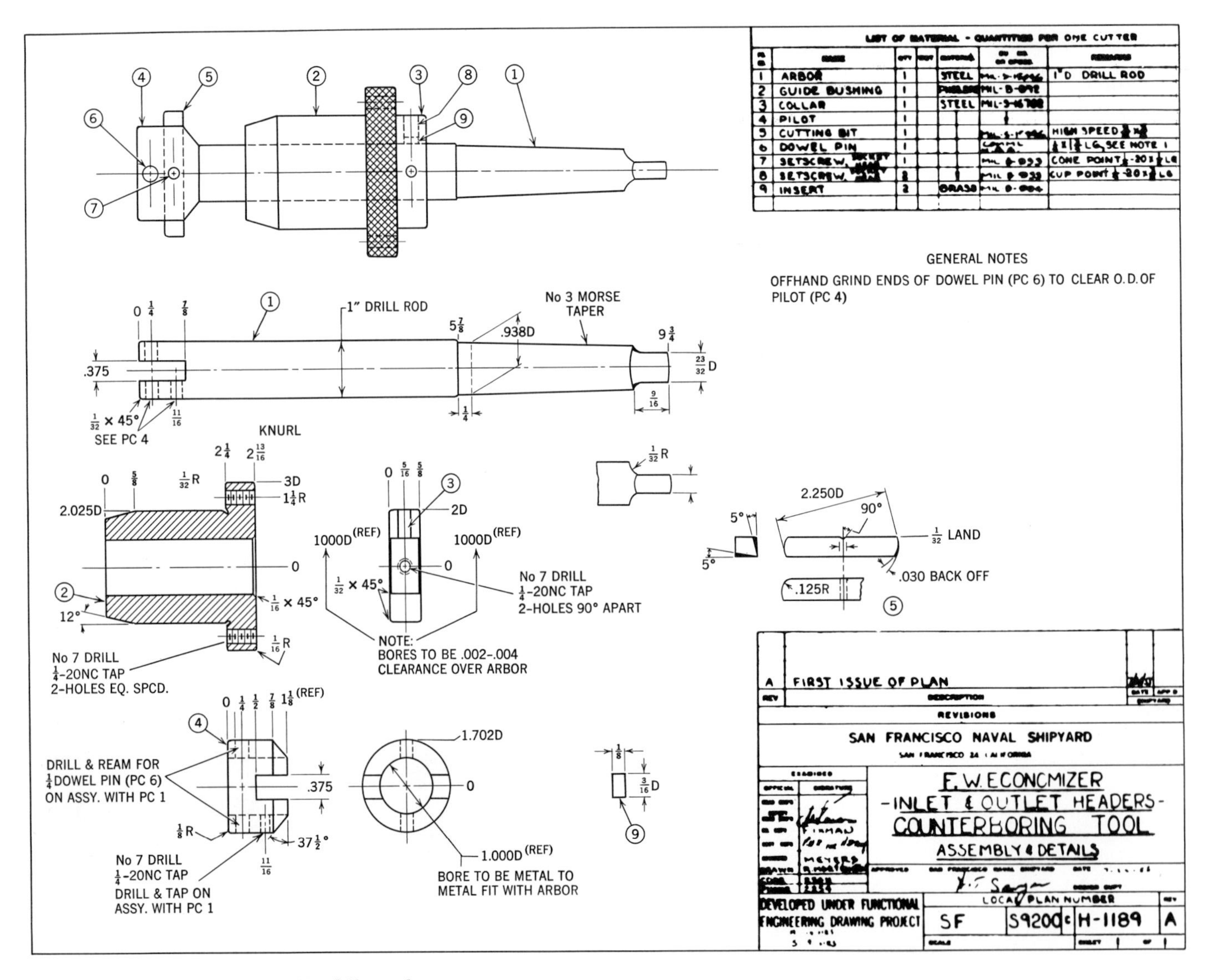

Fig. 2.11 Courtesy San Francisco Naval Shipyard.

Examples of freehand sketches on isometric grids are shown in Figs. 2.12 and 2.13.

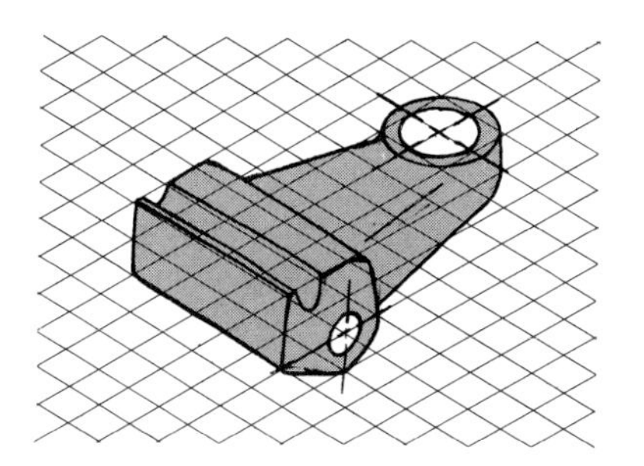

Fig. 2.12 Transmission part.

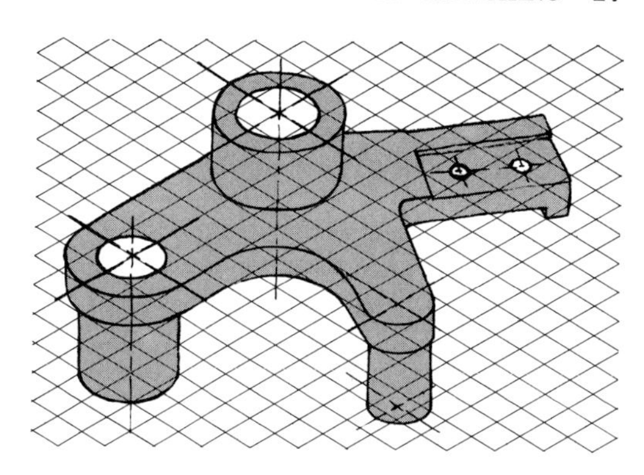

Fig. 2.13 Cam for food-vending machine (Visi-Vend model).

Freehand oblique sketches are shown in Figs. 2.14 and 2.15.

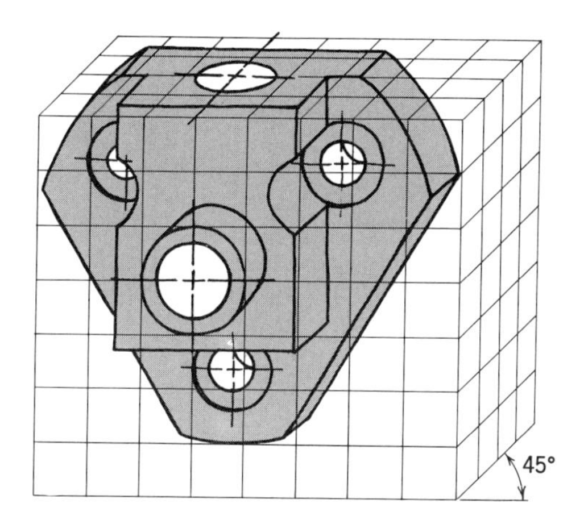

Fig. 2.14 Redundant seal flange. (45° oblique; ½ depth)

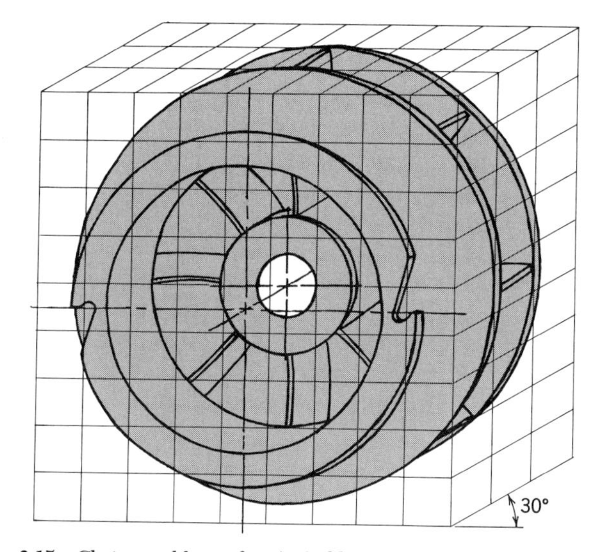

Fig. 2.15 Chain-saw blower fan. (30° oblique)

Two examples of freehand sketches on perspective grids are shown in Figs. 2.16 and 2.17.

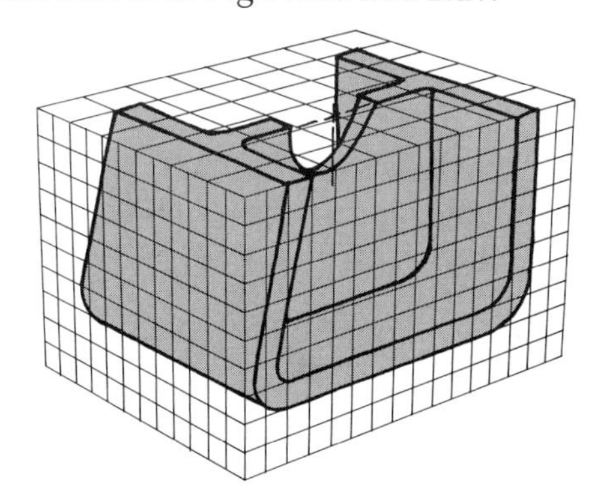

Fig. 2.16 Aircraft Ejection-seat bracket.

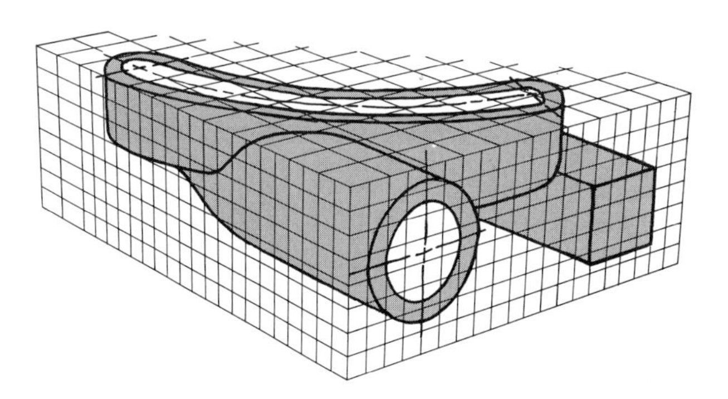

Fig. 2.17 Nozzle for steam turbine.

Conceptual design perspective sketches of space vehicles are shown in Figs. 2.18, 2.19, and 2.20.

Fig. 2.18 Artist's conception of Air Force's newest lifting body design known as the SV-5. Research craft is expected to furnish basic data on aerodynamics, materials, and structural characteristics of maneuverable re-entry vehicles.

Fig. 2.19 The USAF Manned Orbiting Laboratory (MOL) and a modified Gemini capsule are shown in this artist's concept. (Courtesy U.S. Air Force)

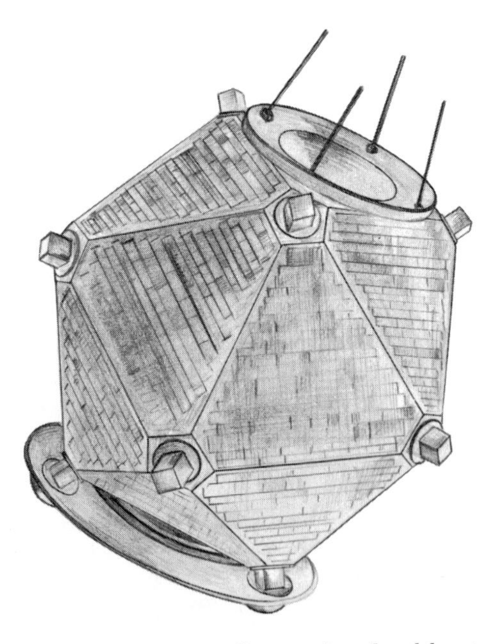

Fig. 2.20 Vela nuclear detection satellites circling the globe at high altitudes keep a lookout for treaty-banned nuclear blasts.

SKETCHING PROCEDURE

When making conceptual design sketches the professional engineer is recording a mental process concerned with the possible solutions to a technological problem. The ability to sketch *rapidly* is very important in order to capture some of the brilliant ideas that originate in flashes of thought. The preparation for rapid sketching is practice, practice, and more practice.

In the learning stage the student would profit from the following suggestions:

1. Use an H- or HB-grade pencil, sharpened to a fairly long conical point for (*a*) the overall outline of the areas to be used; and (*b*) for construction, dimension, and center lines, *applying little pressure to the pencil.*
2. For the finished work apply *more* pressure to produce clean-cut black lines. Suggested line weights are shown in Fig. 2.21.
3. Maintain reasonable proportions of details as related to the major unit—that is, avoid distortions of the elements that make up the part, unit, or system.
4. At this point it is suggested that freehand sketching practice be devoted to drawing (*a*) lines—horizontals, verticals, and at an angle; (*b*) arcs, circles, and ellipses; (*c*) objects such as pyramids, prisms, cylinders, cones, and rectangular forms; (*d*) items such as a book case, a bracket, a lamp, a desk, an antenna, a file cabinet, an engineer's scale, a gasket, etc. These shapes make up the components of a high percentage of engineering-designed "hardware" such as space vehicles, appliances, transportation systems, computer equipment, etc.

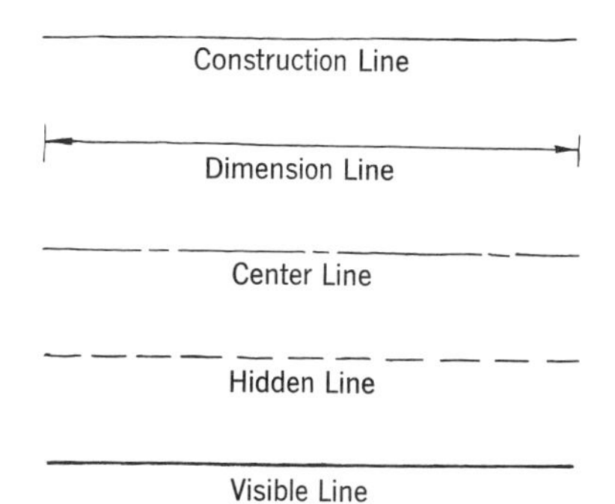

Fig. 2.21

It is interesting to note the geometric shapes in Fig. 2.22 which shows the "Balsa Geometry" to be used in studying complex methods of structuring balsa around the Mars-Impact capsule.

Isometric, perspective, and coordinate grid sheets may be used in making the freehand sketches of the above-mentioned objects and items.

Additional exercises are provided in the workbook which accompanies this textbook.

Students should not hesitate to make freehand sketches when taking class notes and in planning layouts for the graphical solution of three-dimensional problems that arise in subsequent chapters of this book. This will be most helpful in avoiding situations where a portion of the solution falls just off the drawing sheet.

Further use of freehand sketching will be made in translating orthographic drawings to pictorials and vice versa; and in the preparation of conceptual design sketches.

Fig. 2.22 Balsa geometry study for Mars-impact capsule. (Courtesy *Machine Design*)

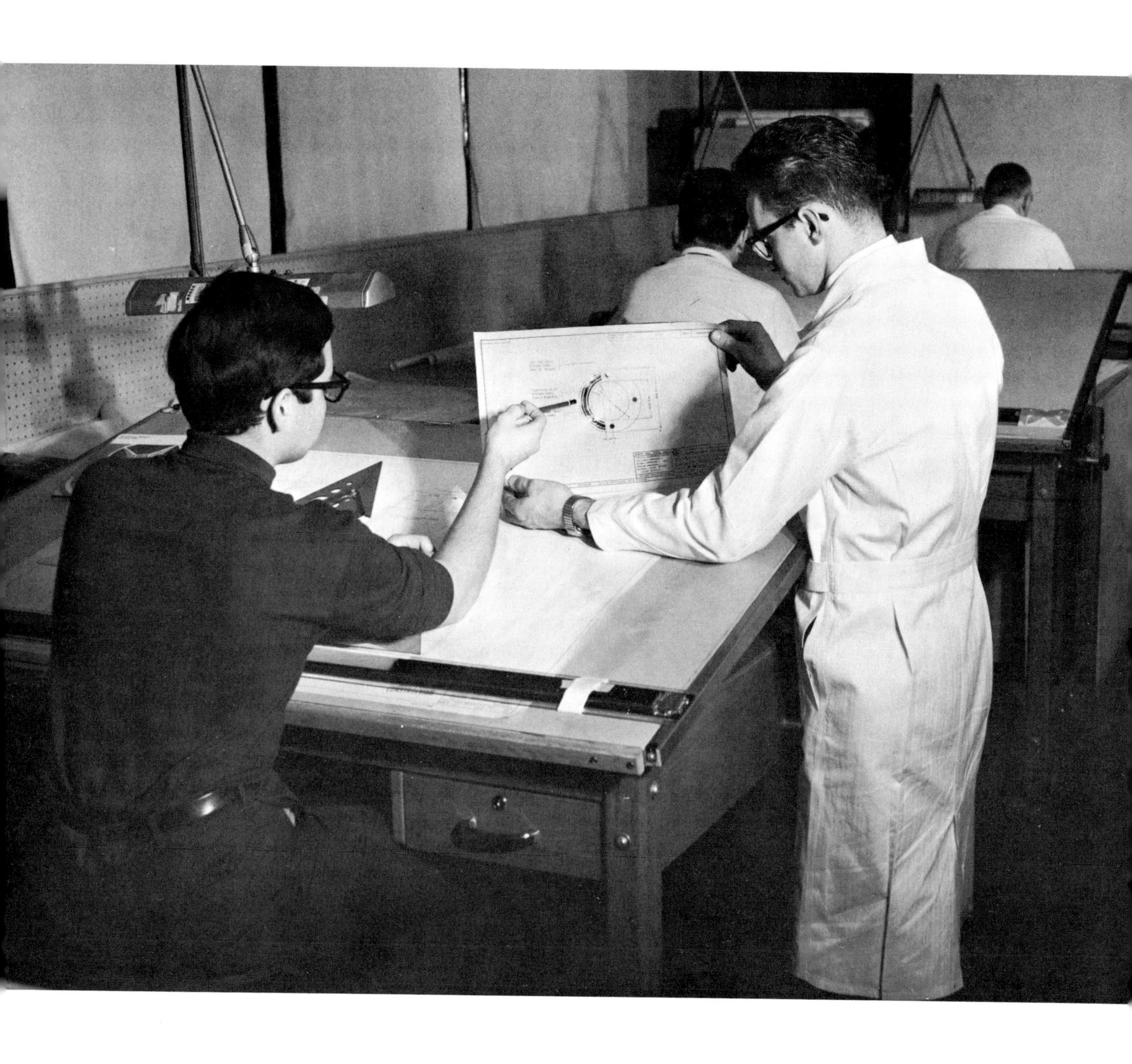

FUNDAMENTAL PRINCIPLES OF PROJECTION 3

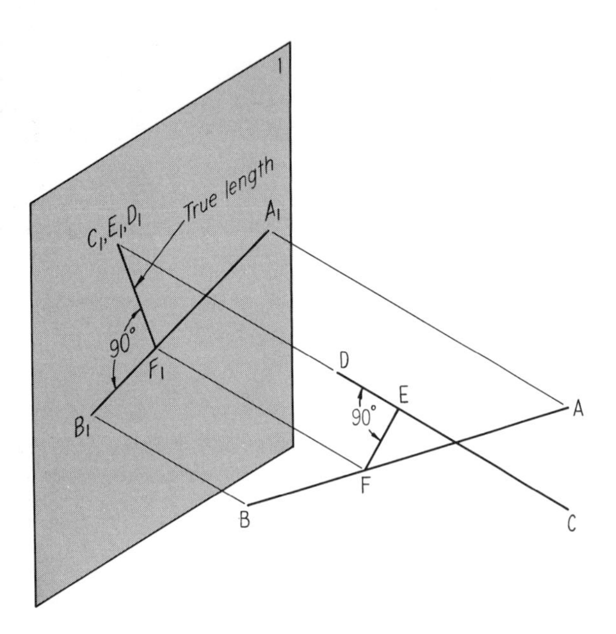

Fig. 3.1

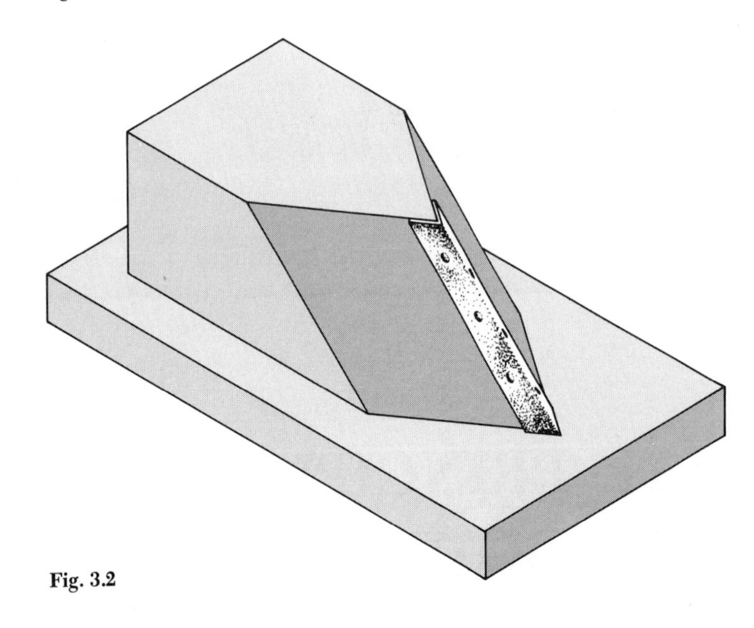

Fig. 3.2

INTRODUCTION

The graphical solutions of many three-dimensional problems that arise in the various fields of engineering and science depend on the abilities of both the engineer and the scientist to analyze the problems, and to apply the fundamental principles of orthogonal projection to their solution. Once the student thoroughly understands the application of the fundamental principles to *find:*

(*a*) the true length of a line segment,
(*b*) the point view of a line,
(*c*) the edge view of a plane surface, and
(*d*) the true shape of a plane surface,

he will be able to analyze and solve more involved space problems without much difficulty.

Analysis Is the "Thinking-through Process." It is the visualization of the problem and its solution. For example, suppose it is necessary to determine the clearance between two high-tension wires, or between an aileron-control cable and a bomb-release cable, or, for that matter, between two structural members, between two pipes, etc. Basically, all these problems belong to one family, namely, skew lines (lines that are neither parallel nor intersecting). *Essentially, then, our problem is to find the perpendicular distance between two skew lines.* Analysis of the problem should lead us to conclude that the distance will be apparent when one of the two skew lines appears as a point. The view of the two lines upon a plane which is perpendicular to one of them will show the perpendicular distance between the two skew lines. See Fig. 3.1.

Now let us consider another problem, i.e., to determine the angle of bend of the edge-reinforcing plate for the concrete pier, shown in Fig. 3.2. Analysis of the problem shows that the angle of bend of the reinforcing plate can be determined, if we know how to find the angle between the two front faces of the pier. *Actually, then, the problem is reduced to "the determination of the angle between two intersecting planes."* The solution of this problem is quite simple. Again, "thinking-through the problem"—analysis—reveals the fact that the angle between the two planes will be seen in a view which shows the line of intersection of the two planes as a point. In this view, the two planes will appear as straight lines. The angle between the lines (edge views of the planes) is the required angle. See Fig. 3.3 which, in general, typifies the solution pictorially. *Observe that, although the last problem appears to be quite different from the clearance problems first cited, the analyses of the problems reveal the fact that their solutions are basically the same.*

In addition to the types of problems discussed above, there are others that occur frequently in engineering and science, i.e., (*a*) distance from a point to a plane; (*b*) distance between parallel planes; (*c*) distance from a point to a line, etc. You should encounter no difficulty in analyzing these problems. *Try them now, just for fun.* Write the analyses and sketch the solutions.

Frequently it is necessary to determine the *true shape of a plane surface.* This can be done by obtaining a view of the surface on a plane which is parallel to the surface; or, by first finding the true lengths of the sides (i.e., a triangle) of the surface and then constructing its true shape. Patterns may be developed by obtaining true shapes of surfaces (Fig. 3.4) or, in the case of "transition pieces," by

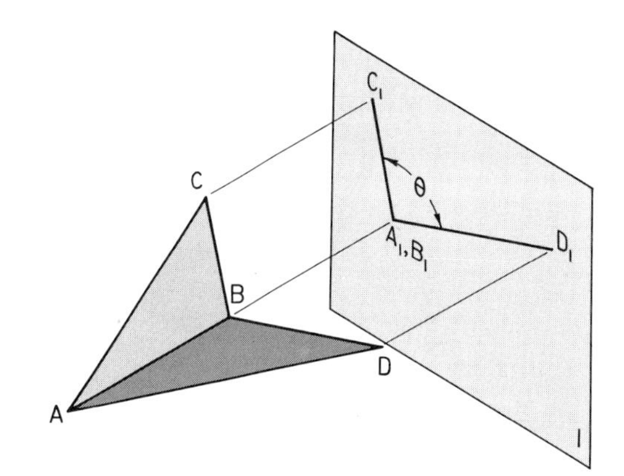

Fig. 3.3

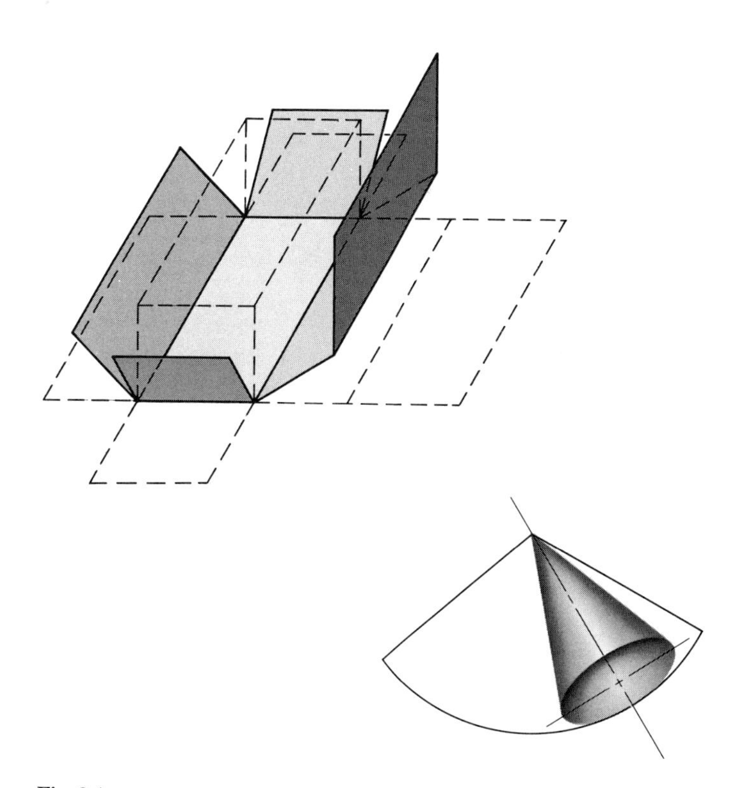

Fig. 3.4

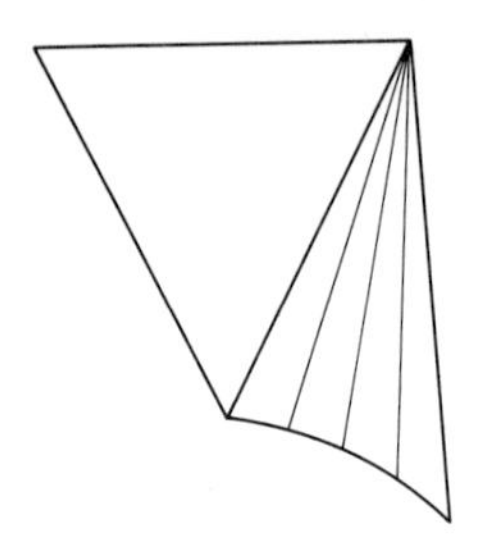

Fig. 3.5

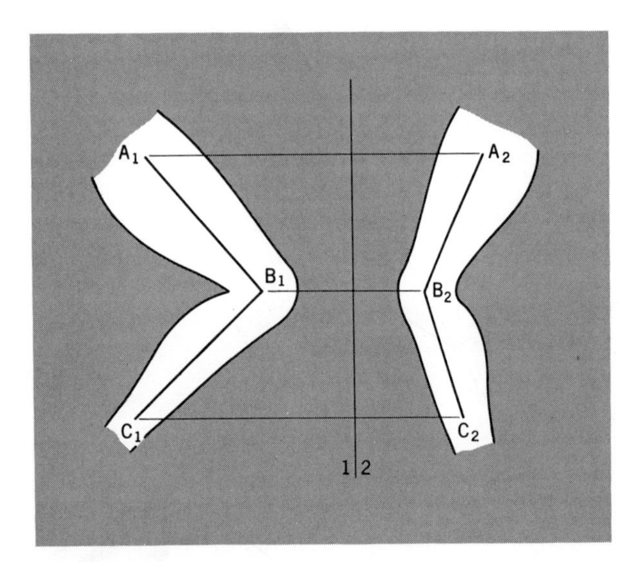

Fig. 3.6

combining true shapes of plane surfaces with curved surfaces that can be developed by the method of "triangulation." See Fig. 3.5.

Many engineering research and design problems include the determination of true lengths, true shapes, intersection of surfaces, angle between surfaces, clearances, forces in concurrent, non-coplanar systems, etc. In other instances, for example, in prosthetic devices research (the improvement in the design of artificial limbs) it has been necessary to employ graphical methods to determine magnitudes of angular motion, velocities, accelerations, etc. In Fig. 3.6 is shown a problem which required the determination of the angle between the femur (thigh bone represented by line *AB*) and the tibia (shin bone represented by line *BC*). The solution of this problem and other related problems will be considered later when we discuss "the angle between lines" and "the angle between a line and plane."

Thorough mastery of the few fundamental principles, development of analytic (thinking) power, and reasonable proficiency in recording the graphical solutions to the various problems that arise in both engineering and science are essential to your development as qualified engineers.

Let us now direct our attention to the fundamental principles of orthogonal projection.

FUNDAMENTAL PRINCIPLES OF ORTHOGONAL PROJECTION

Introduction

The core of engineering is design, which embraces the basic fields of mathematics, graphics, physics, and chemistry; the engineering sciences of mechanics of solids, fluid mechanics, thermodynamics, electronics, heat and mass transfer, properties of materials; economics, aesthetics, production, etc.

Once decisions have been made with respect to a feasible solution to a design problem (i.e., a space vehicle which will satisfy a specification set forth by NASA*, or an automobile that will satisfy the needs and desires of a large portion of the available market), it becomes necessary to develop the designs of major components and of the parts that are needed for the proper functioning of the end product. Size and shape of the various components must be determined by engineers and their supporting staffs. Thorough understanding and use of the fundamental principles of orthogonal projection are necessary for accurate shape determination.

What do we mean by the term orthogonal projection? Before we arrive at an answer to this question, let us first consider the elements that are common to the systems of projection which are identified as (*a*) central or perspective projection; and (*b*) parallel projection. The common elements are the following: a plane of projection (picture plane); a point of sight; and a given object.

Central or Perspective Projection

Let us examine Fig. 3.7, which shows a simple object, the plane of projection, and point of sight or station point, S.

Suppose that a line (projector), r, is drawn from the station point, S, to some point, P, of the object. The intersection, P', of line r with the "picture plane" is the "perspective projection" of point P. If this process of projection were repeated for the other corners of the object, the lines joining the projections of these points would then form the "projection of the object," in this case, *a central or perspective projection.*

The three-dimensional effect obtained in perspective results in a "pictorial" drawing. It lends itself to an easy understanding of the object. It simplifies visualization of the object. On the other hand, it contains distortions of linear and angular magnitudes, the true values of which may be necessary for computation, design, and production.

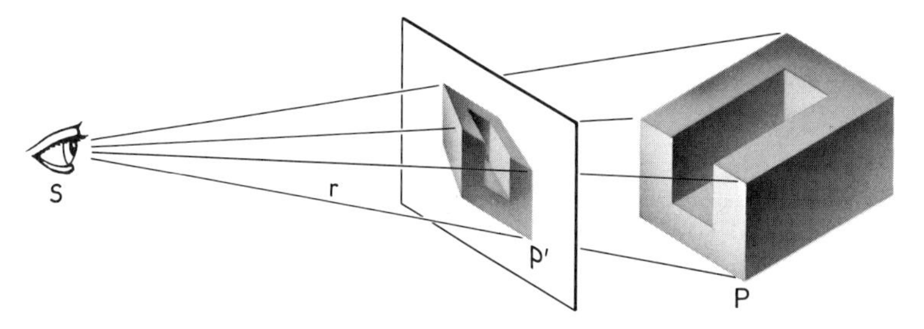

Fig. 3.7

* National Aeronautical Space Administration

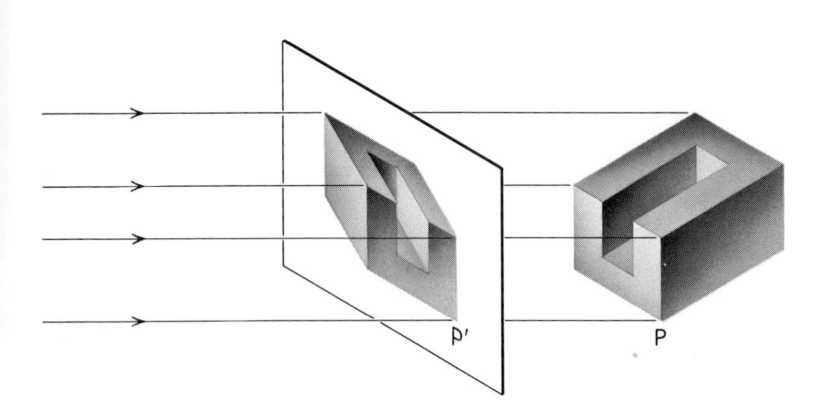

Fig. 3.8

Fig. 3.9

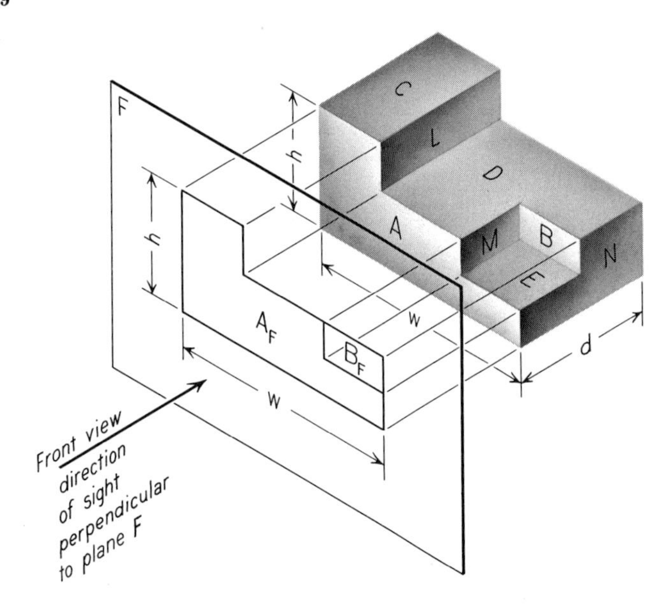

Fig. 3.10

Parallel Projection—General Case

If the station point, S, is an infinite distance from the object, the projectors will be parallel to each other, as shown in Fig. 3.8. Lines joining the points of intersection of these parallels with the plane of projection will form a view having a three-dimensional effect. If the object is so oriented that certain surfaces are parallel to the plane of projection, their true shapes will be revealed. Although this may be advantageous, it should be recognized that certain distortions still exist, i.e., elements not parallel to the plane of projection will not be shown in their true magnitudes.

Parallel Projection—Special Case, Orthogonal

When the parallel projectors are oriented at right angles to the "picture plane" the resulting projection is known as an orthogonal projection. This is the system that is most frequently employed in technical fields because it gives accurate and complete information. It is true that in a number of relatively simple cases complete information can be given in a pictorial drawing; nevertheless, it has been found that in most instances orthographic solutions are best adapted to engineering practice.

Views of an Object

EXAMPLE 1

Let us consider the block shown in Fig. 3.9. Suppose we introduce a reference plane, *F*, known as the *frontal plane,* parallel to surface *A* of the block, Fig. 3.10. Perpendiculars from the various corners of the block will intersect plane *F* in points which are properly connected to form the *front view* of the block. *It is the view obtained by looking at the object in a direction of sight which is perpendicular to the F plane.* Orienting the plane parallel to surface *A* enables us to obtain a front view which includes the true shape of the surface. Note carefully that the height, *h*, and the width, *w*, are available in this view. Depth, *d*, however, cannot be seen in the front view.

Now we will *introduce a horizontal reference plane, H, which is perpendicular to the F-reference plane.* See Fig. 3.11. Perpendiculars from the various corners of the block to the *H*-plane will intersect that plane in points which are properly connected to form the top view of the block. *This view is obtained by looking at the object in a direction of sight which is perpendicular to the H-plane.* Again, note carefully that the width, *w*, and the *depth, d*, are available in this view. In addition, observe that the top view includes the true shapes of surfaces, *C*, *D*, and *E*. Why is this true?

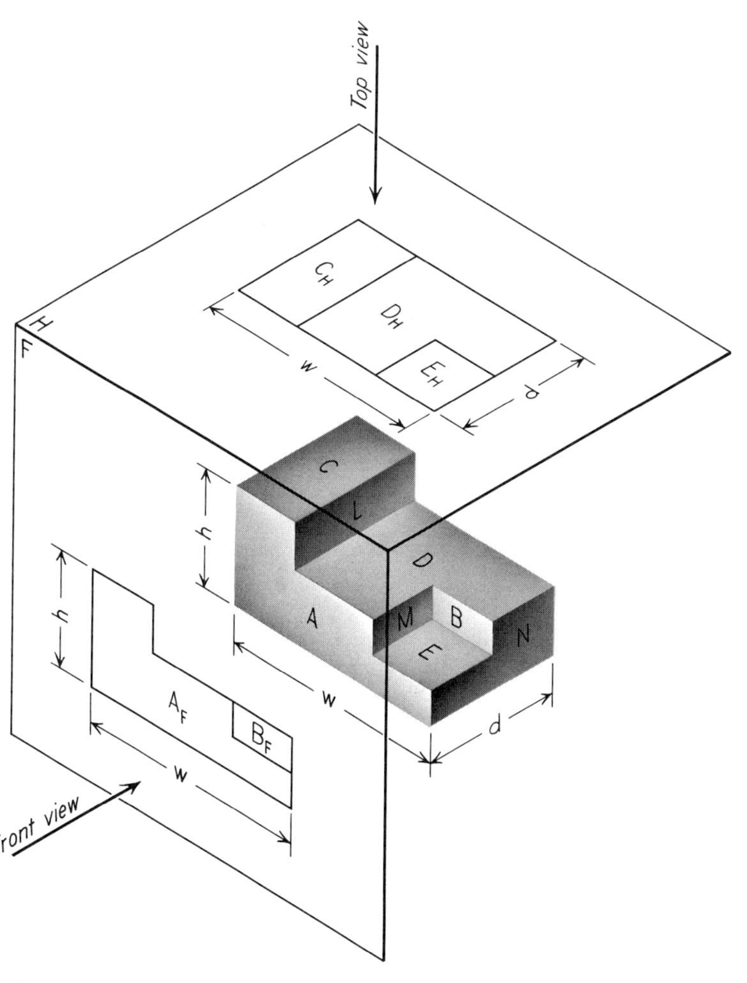

Fig. 3.11

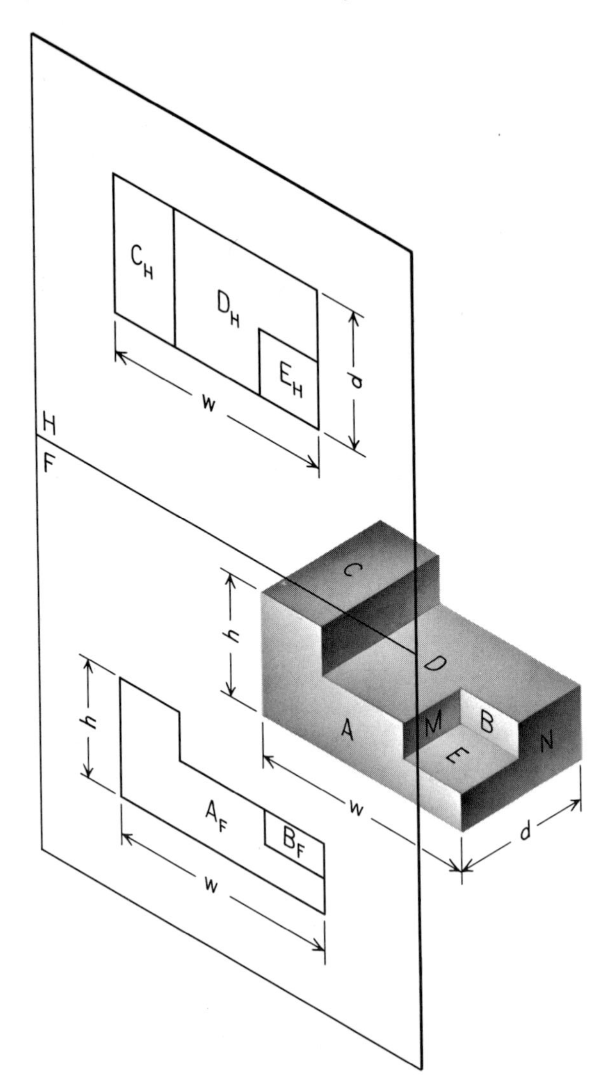

Fig. 3.12

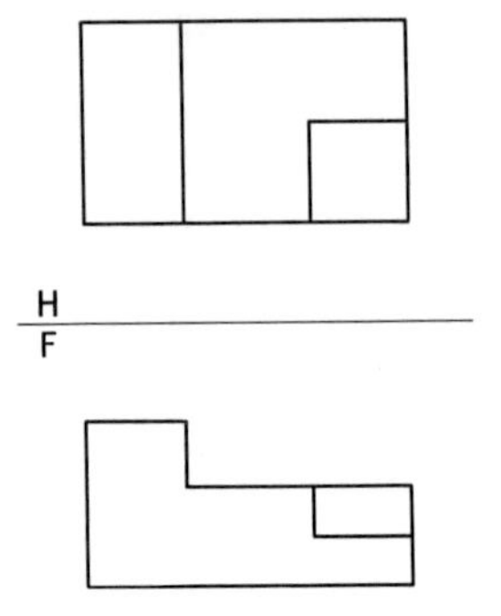

Fig. 3.13

Let us now revolve the *H*-plane, as shown in Fig. 3.12, so that the two views, *H* and *F*, are in one flat sheet. Since the sizes of the reference planes, *H* and *F*, are arbitrary, we can omit their boundary lines and simply show the two views as drawn in Fig. 3.13. Now we have the top (*H*) and front (*F*) views of the block.

If we were given these two views, would we have enough information to describe the shape of the block? Off-hand, we would be inclined to say yes. Yet, it is possible to have other blocks that would have the same two views. For example, one block could look like the one shown in Fig. 3.14; another, like the one shown in Fig. 3.15.

We must conclude, then, that in this case another view is necessary to describe fully the shape of the object. In many cases the additional view is a profile view; *in others it may be necessary to add supplementary views.*

In Fig. 3.16, we have added a *profile reference plane P* on which the profile view is shown. This added information leaves no doubt as to the *designer's intent* of the shape of the block.

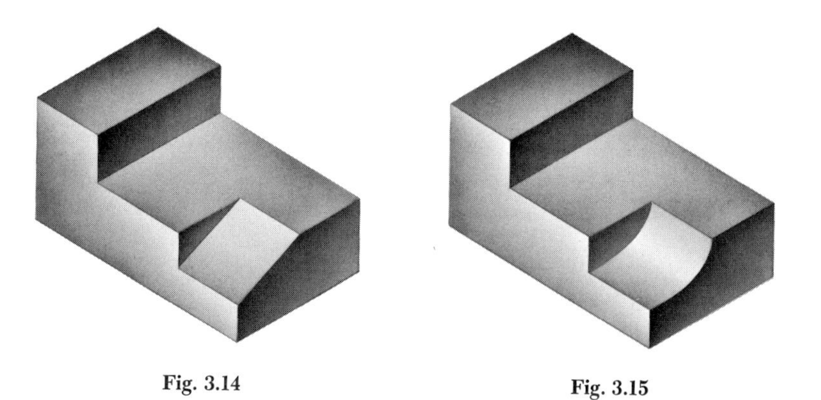

Fig. 3.14

Fig. 3.15

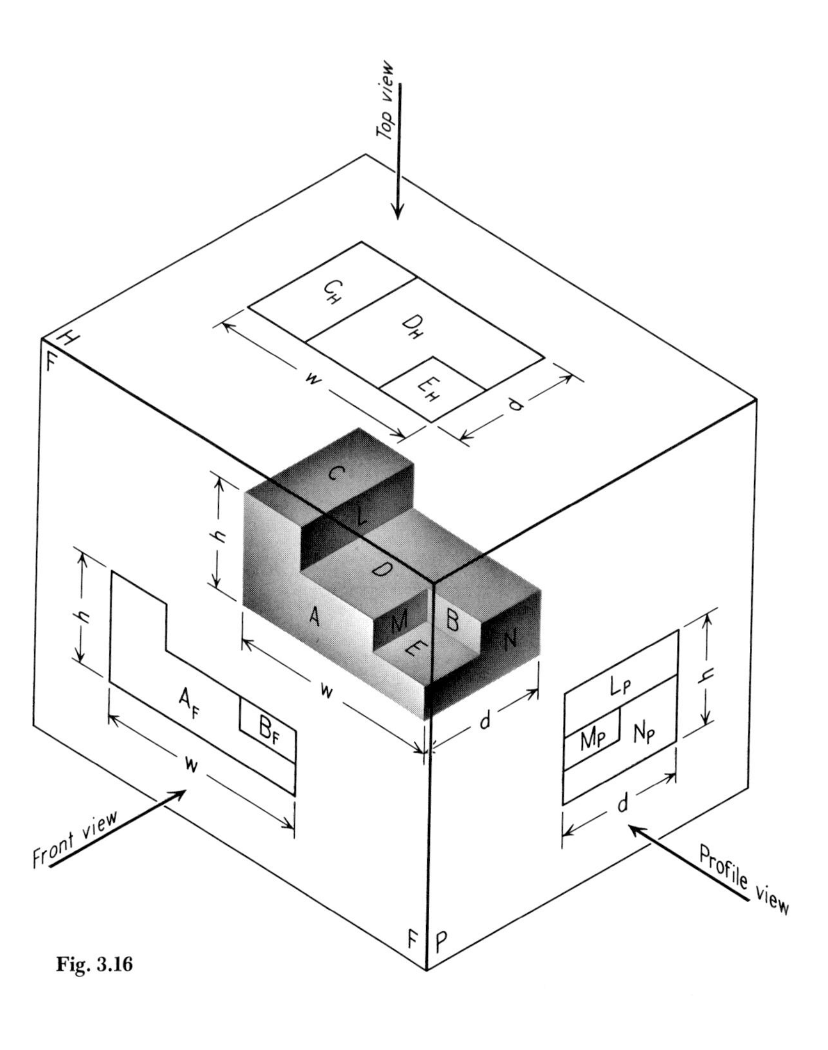

Fig. 3.16

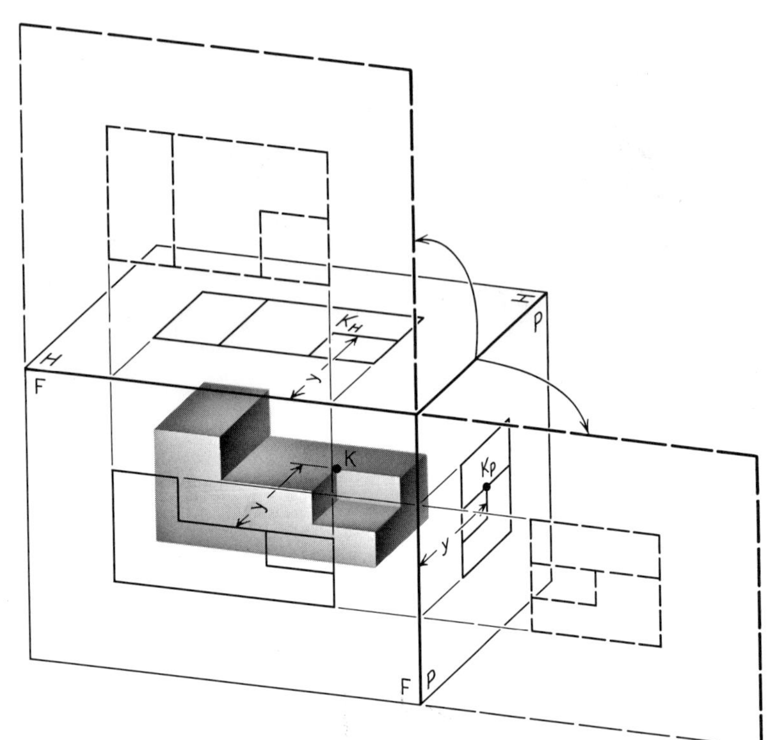

Fig. 3.17

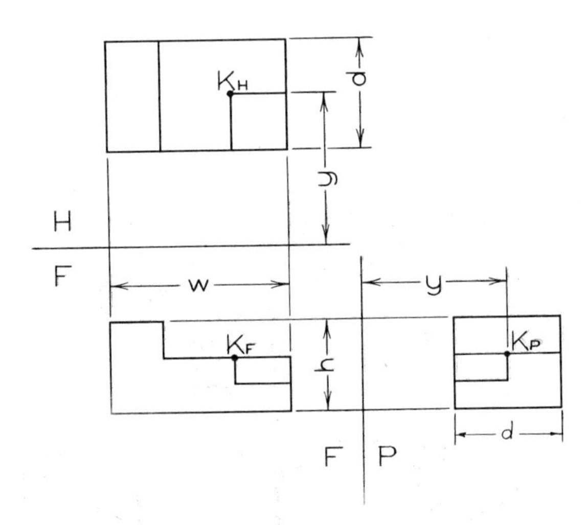

Fig. 3.18

The flat-sheet representation of the three views is shown pictorially in Fig. 3.17, and orthographically in Fig. 3.18.

At this time we should *carefully study* the relationship of the *H*-plane to the *F*-plane; the relationship of the *P*-plane to the *F*-plane; and the relationship of any point of the object, such as *K*, to the three reference planes *H*, *F*, and *P*.

We then reach the following conclusions:

1. Planes *H* and *F* are perpendicular to each other. (These are known as *adjacent* planes.)
2. Planes *P* and *F* are perpendicular to each other. (These are adjacent planes.)
3. Planes *H* and *P* are each perpendicular to plane *F*; therefore,
4. *The distance, y, that point K is behind the F-plane will be seen twice; once in the top view, H, and again in the profile view, P.*
5. The top and front views of point *K* lie on a line which is perpendicular to the intersection of the *H* and *F* planes.
6. The front and profile views of point *K* lie on a line which is perpendicular to the intersection of the *F* and *P* planes.

The full significance of these conclusions will be realized as we consider other problems later. At this point, however, let us return to Fig. 3.18 and consider the meaning of lines marked $\frac{H}{F}$ and $F|P$.

First, with respect to the horizontal line marked $\frac{H}{F}$, notice that:

1. The top view includes both the top view of the block *and the top or edge view of the F-plane.* The latter is represented by the horizontal line marked $\frac{H}{F}$. The distance, *y*, shows how far point *K* is *behind* the *F*-plane.
2. *When we observe the front view, the same horizontal line,* $\frac{H}{F}$, *represents the front or edge view of the H-plane.*

Second, with respect to the vertical line marked $F|P$, observe that:

1. The front view includes both the front view of the block *and the front, or edge view of the P-plane.*

2. *When we look at the profile view the same vertical line represents the profile view of the F-plane.* Since this is true, *the distance, y, again shows how far point K is behind the F-plane.*

Now let us consider the pictorial Fig. 3.19. Carefully observe that the *P*-plane was first rotated into a position coincident with the *H*-plane, and then coincident with the *F*-plane. The flat-sheet representation, without the boundaries of the reference planes, is shown in Fig. 3.20.

Again, let us study a corner of the block, such as point *A*. Note the following relations:

1. Reference planes *H* and *F* are perpendicular to each other. (These are adjacent planes.)
2. Reference planes *H* and *P* are perpendicular to each other. (These are adjacent planes.)
3. *Reference planes F and P are each perpendicular to plane H; therefore,*
4. *The distance Z, that point A is below the H-plane is seen twice: once in the front view and again in the profile view.*

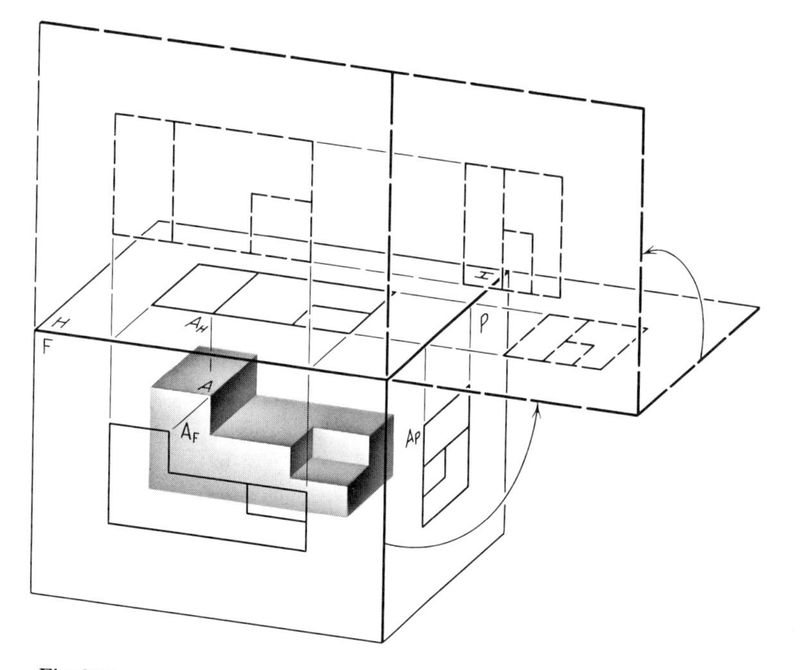

Fig. 3.19

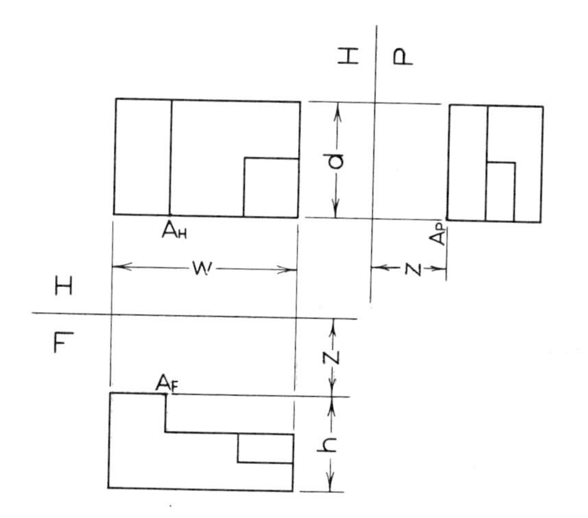

Fig. 3.20

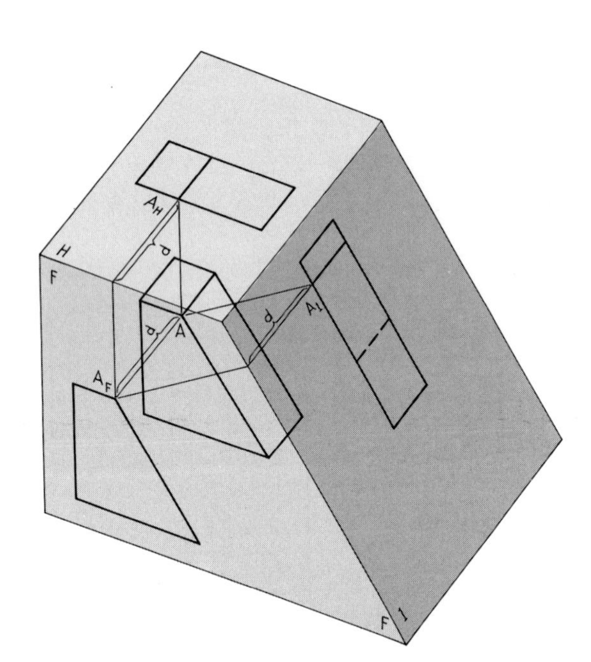

Fig. 3.21

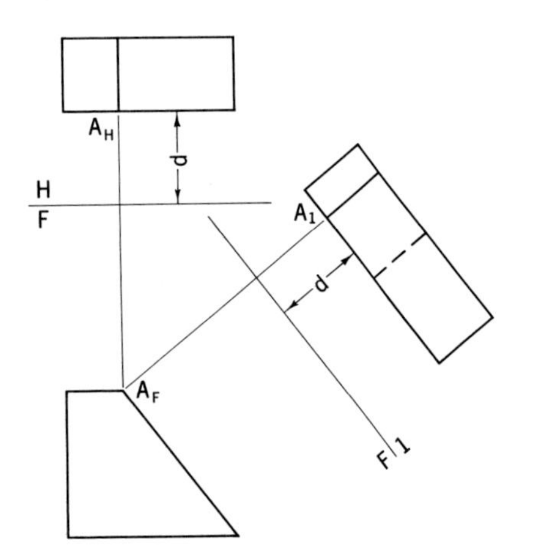

Fig. 3.22

Now let us consider the pictorial Fig. 3.21, and the corresponding flat-sheet representation shown in Fig. 3.22. It should be pointed out that:

1. Planes *H* and *F* are perpendicular to each other. (These are adjacent planes.)
2. Planes *F* and 1 are perpendicular to each other. (These are adjacent planes.)
3. *The H and 1 planes are each perpendicular to the F-plane; therefore,*
4. *The distance, d, that any point such as A is behind the F-plane will be seen twice: once in the top view, H, and again in the supplementary view, 1.*
5. The top and front views of point *A* lie on a line which is perpendicular to the horizontal line which represents the intersection of the *H* and *F* planes.
6. The front and supplementary views of point *A* lie on the perpendicular to the inclined line which represents the intersection of the front (*F*) and supplementary (1) planes.
7. When looking at the front view the inclined line marked *F*\1 *is the front, or edge, view of the supplementary plane 1; and when looking at the supplementary view, 1, the same inclined line is the supplementary, or edge view of the F-plane.*

THE TWO FUNDAMENTAL PRINCIPLES OF ORTHOGONAL PROJECTION

We may now state the two fundamental principles of orthogonal projection:

1. *Adjacent planes are perpendicular to each other and the views of a point on these planes lie on a line which is perpendicular to the line of intersection of the adjacent planes.* See Figs. 3.23 and 3.24.

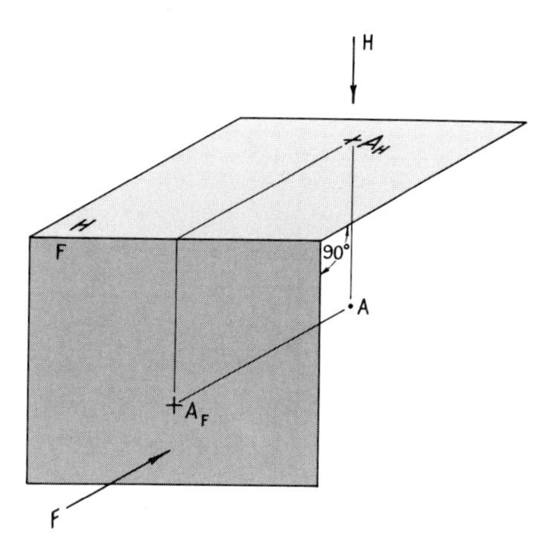

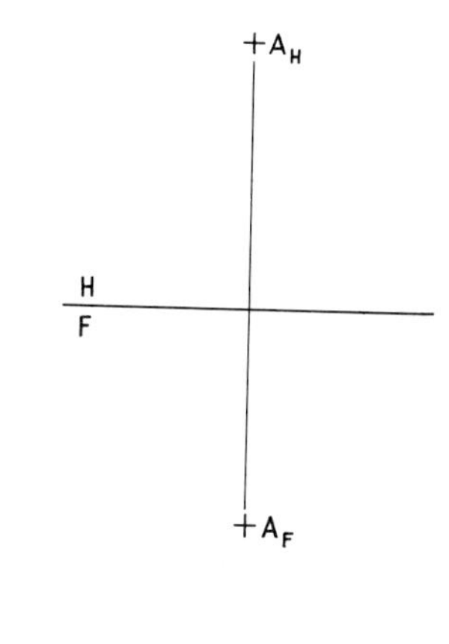

Fig. 3.23

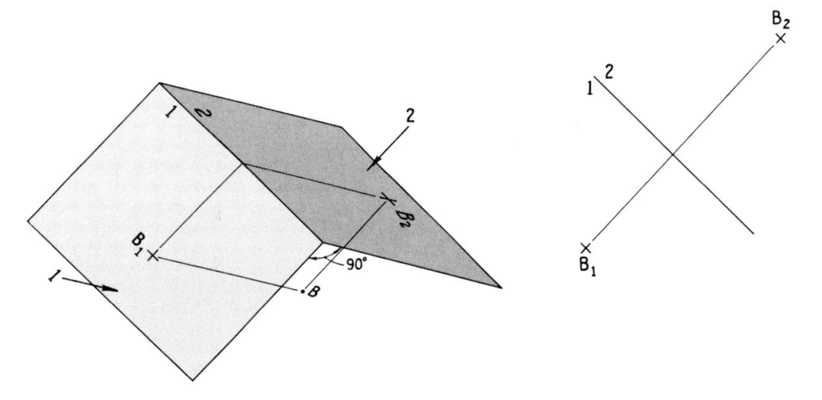

Fig. 3.24

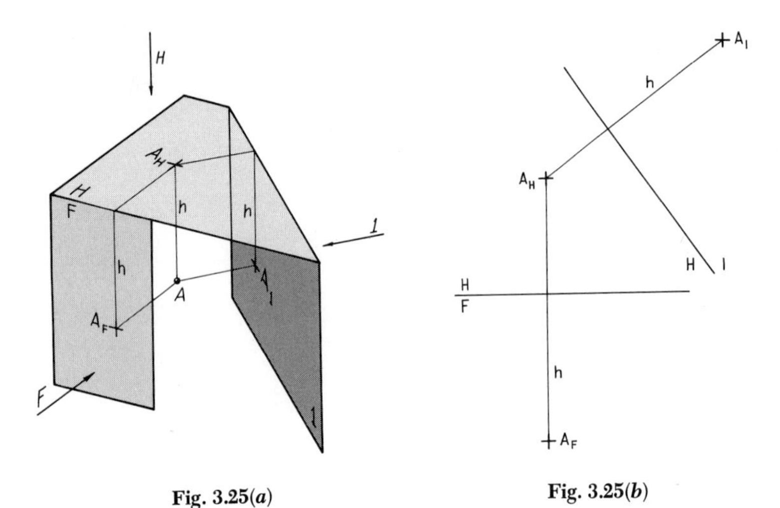

Fig. 3.25(*a*) Fig. 3.25(*b*)

2. *When two planes are perpendicular to a third plane, the distance that a point is from the third plane will be seen twice; once in each of the views upon the other two planes.* See Figs. 3.25 and 3.26.

Note very carefully that in all these examples the relationships among the views are consistent with the two fundamental principles. To further strengthen our understanding and use of the fundamental principles, consider the problems in the following examples.

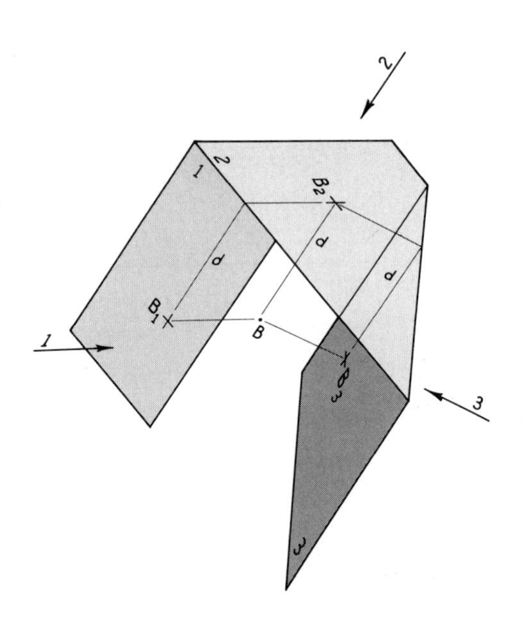

Fig. 3.26(*a*)

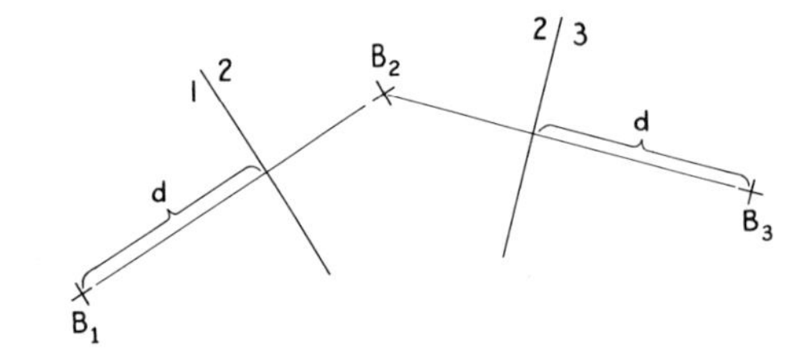

Fig. 3.26(*b*)

EXAMPLE 1

We wish to obtain views of the block shown in Fig. 3.27 on planes 1 and 2 (Fig. 3.28).

Let us think the problem through together. Surely we can understand what the block looks like from the pictorial shown in Fig. 3.27.

First, let us select a point such as *A*, one of the corners of the block, and then locate its top and front views. Now we should have no difficulty in locating the view of point *A* on supplementary plane 1. Since planes *H* and 1 are perpendicular to each other (adjacent planes), we know from the first fundamental principle that A_H and A_1 must lie on the perpendicular to the line $H\backslash 1$. Furthermore, from the second basic principle, distance *d* must be the same in both the front view and the supplementary view 1. This fact enables us to locate view A_1.

In order to locate view A_2, we again use the two fundamental principles. Since planes 1 and 2 are perpendicular to each other (adjacent planes), views A_1 and A_2 must lie on the perpendicular to the line $\frac{1}{2}$. The next step, to locate A_2, is easily taken, if we note carefully that (*a*) planes *H* and 1 are perpendicular to each other; (*b*) planes 1 and 2 are perpendicular to each other; and, therefore, (*c*) planes *H* and 2 are each perpendicular to plane 1. Now, applying fundamental principle two, the distance that point *A* is from plane 1 is seen twice: once in the *H*-view and again in the supplementary view on plane 2. This means that distance *k* must be the same in both views.

Once we know how to obtain the views of point *A* on planes 1 and 2, it is no hardship to obtain the views of additional points in order to complete the views of the entire object (Fig. 3.29).

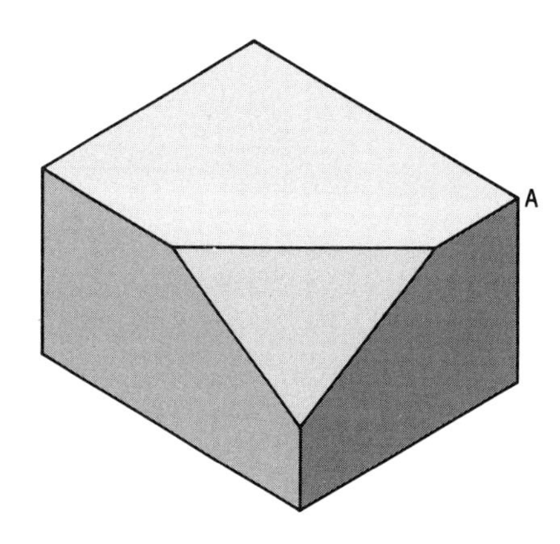

Fig. 3.27

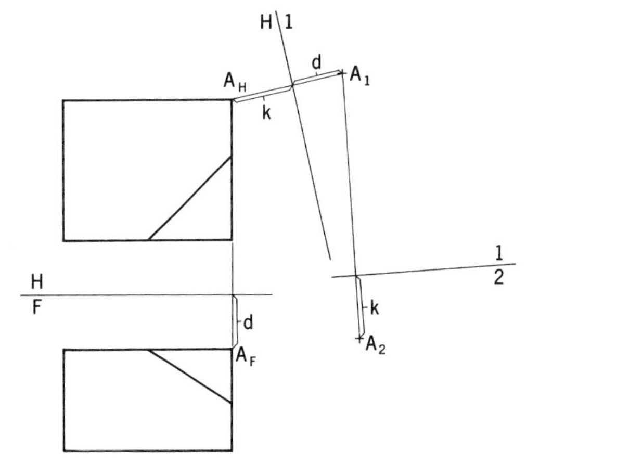

Fig. 3.28

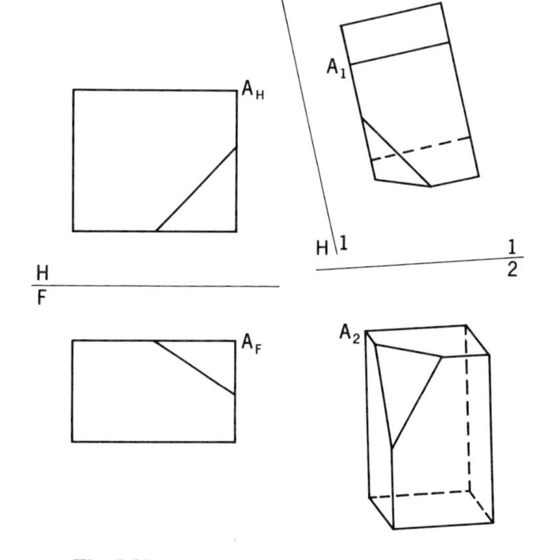

Fig. 3.29

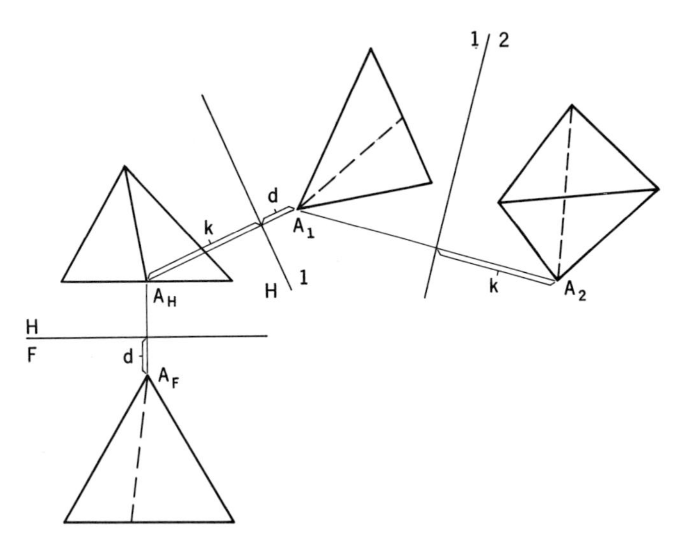

Fig. 3.30

Again we see that in obtaining the view on plane 1 we are concerned with the views on planes *H* and *F*; and that in obtaining the view on plane 2, we are concerned with the views on planes *H* and 1. In each case we are concerned with *three views only*—and that each set of three views is composed of two views that are on planes which are perpendicular to the plane showing the other view.

In simple summary, we see that the basic relationship among the views on planes *H*, *F*, and 1 is the same as the basic relationship among the views on planes *H*, 1, and 2.

Now let us examine another problem.

EXAMPLE 2

Suppose we wish to obtain the views of the block shown in Fig. 3.30 upon supplementary planes 1 and 2. Let us assume that the top and front views have been established. Now let us concentrate on the problem of finding the view on supplementary plane 1.

We observe that:

1. Planes *H* and 1 are perpendicular to each other (adjacent planes).
2. Planes *H* and *F* are also perpendicular to each other; therefore,
3. Planes 1 and *F* are each perpendicular to plane *H*.

Emphasizing the *first fundamental principle,* we know that A_H and A_F must lie on the perpendicular to the line marked $\frac{H}{F}$; and also that A_H and A_1 must lie on the perpendicular to the line marked $H\backslash 1$.

Using the *second fundamental principle,* we know that distance *d* (the distance that point *A* is from plane *H*) will be seen twice: once in the front view and again in the supplementary view on plane 1. Therefore, we can easily locate the view A_1, by laying off distance *d* as shown in the figure.

NOTE VERY CAREFULLY, THAT WE WERE CONCERNED WITH ONLY THREE PLANES AND THE VIEWS ON THESE THREE PLANES: THE TWO THAT WERE GIVEN OR ASSUMED (*H* & *F*) AND THE VIEW ON PLANE 1, THE VIEW WE WISHED TO OBTAIN.

Now let us concentrate on the problem of finding the view on supplementary plane 2. We observe that:

1. Planes H and 1 are perpendicular to each other.
2. Planes 1 and 2 are perpendicular to each other. Therefore,
3. Planes H and 2 are each perpendicular to plane 1.

Employing the *first fundamental principle,* we know that A_1 and A_2 must lie on the perpendicular to the line marked $\frac{1}{2}$. Where on this perpendicular will A_2 be located?

Using the *second fundamental principle,* we know that distance k will be seen twice: once in the top view and again in the supplementary view on plane 2. Therefore, we can easily locate the view A_2 by laying off distance k as shown in the figure.

Again note that we were concerned with only three planes (and the views on them)—the two, H and 1, and the one we wished to obtain, namely, the view on plane 2.

The relationship among the views H, 1, and 2 is fundamentally the same as the relationship among the views F, H, and 1.

To help visualize the relationship among the views on plane H, 1, and 2 we could actually rotate the entire Fig. 3.30 until the line marked $H\backslash 1$ was horizontal, think of the views H and 1 as the given views, and then locate the view on plane 2. Note this arrangement of the views in Fig. 3.31.

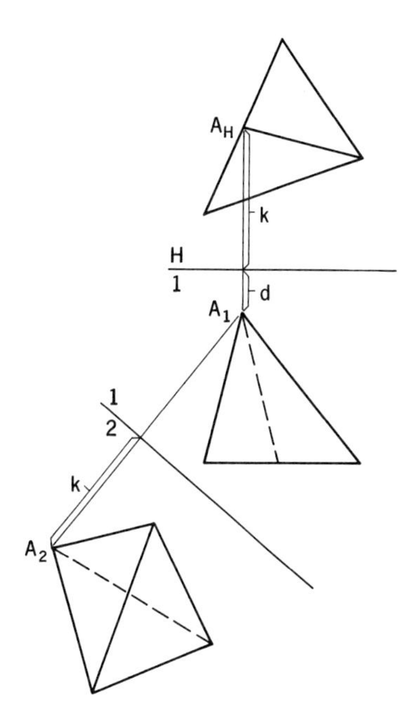

Fig. 3.31

EXERCISES

1. Determine the views of line segment AB upon planes F, 1, and 2. Position the H and F views of the line segments and planes F, 1, and 2 approximately as shown in Fig. E-3.1. Use an $8\frac{1}{2}'' \times 11''$ sheet. The problem may be solved by *freehand* sketches.

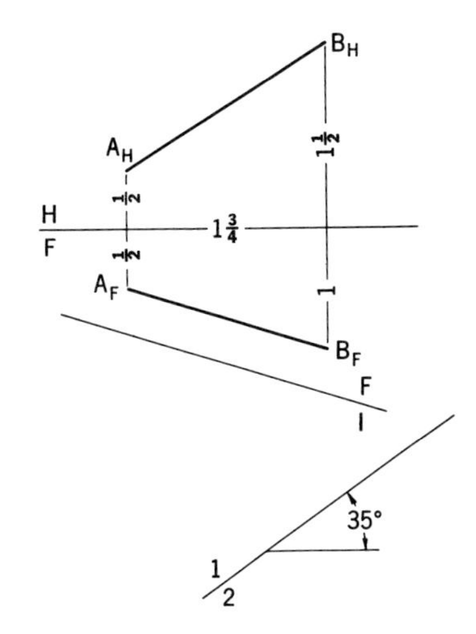

Fig. E-3.1

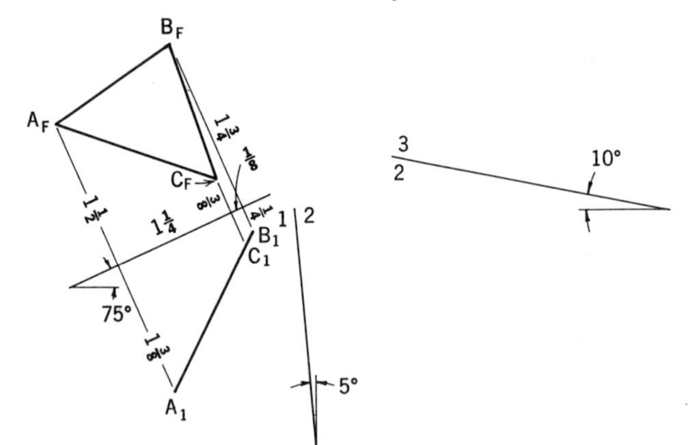

Fig. E-3.2

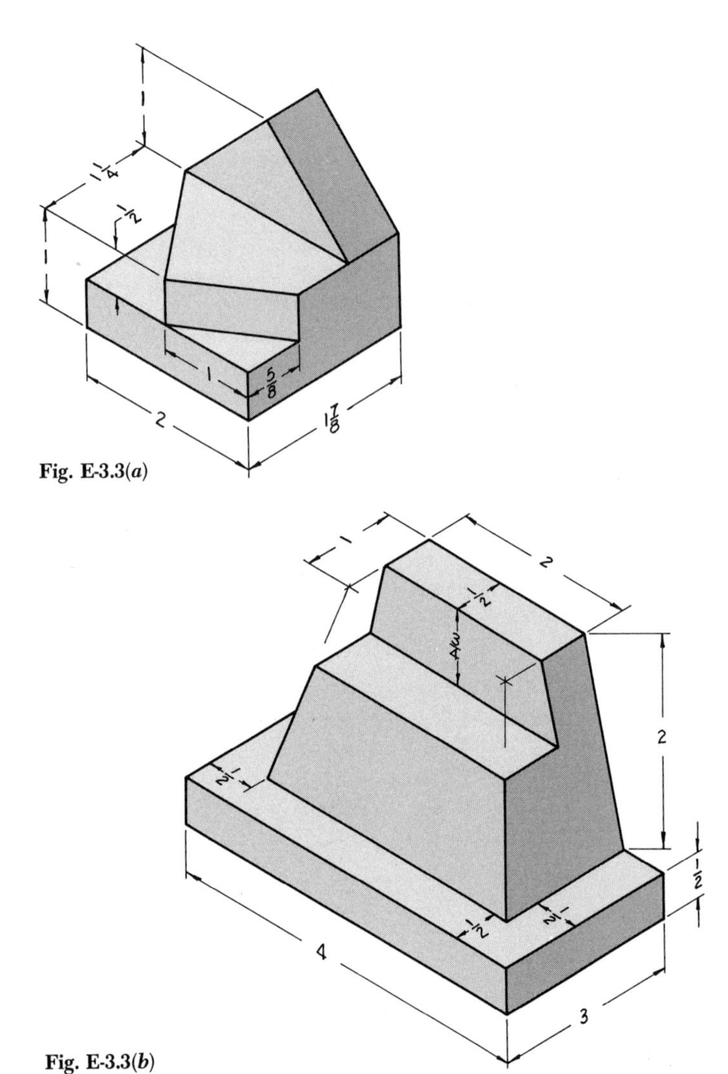

Fig. E-3.3(*a*)

Fig. E-3.3(*b*)

2. Determine the views of surface *ABC* upon planes 2 and 3. Place the *F* and 1 views approximately as shown in Fig. E-3.2. Use an $8\frac{1}{2}'' \times 11''$ sheet.

3. Prepare *freehand* orthographic views of the objects shown in Fig. E-3.3. Use $8\frac{1}{2}'' \times 11''$, $\frac{1}{4}$-inch grid sheets. Do not dimension the views. Select the orthographic views which best describe each object.

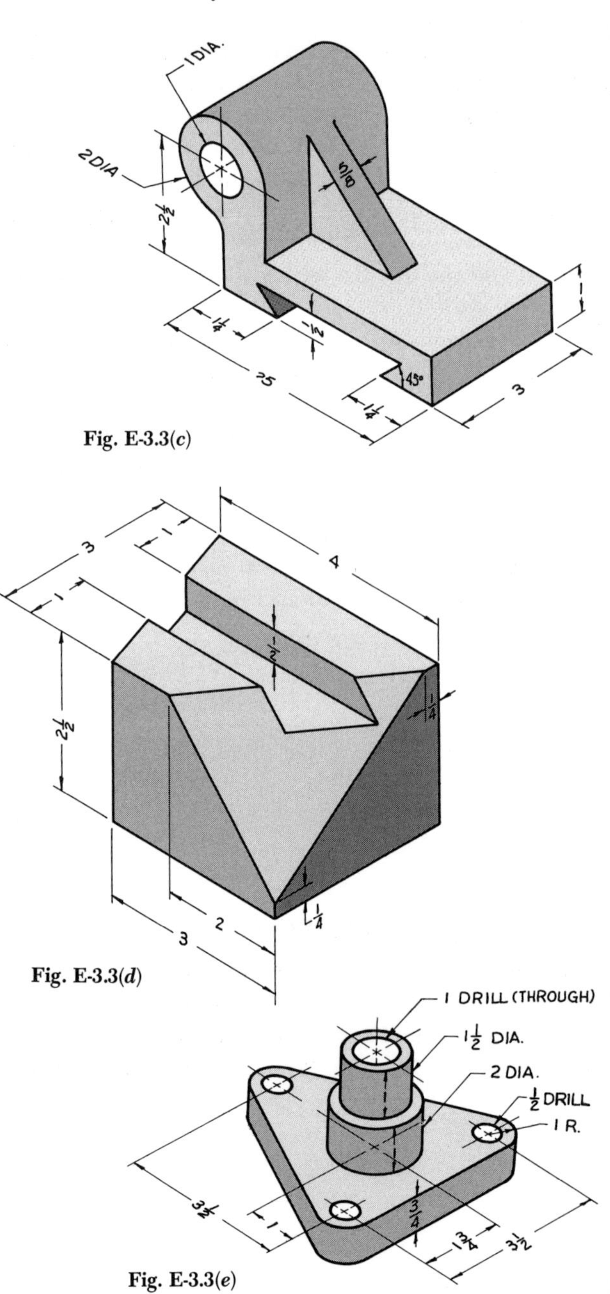

Fig. E-3.3(*c*)

Fig. E-3.3(*d*)

Fig. E-3.3(*e*)

4. Reproduce the given views and then add the supplementary views as indicated in Fig. E-3.4. Prepare preliminary freehand layouts to make certain that the views will fit $8\frac{1}{2}'' \times 11''$ sheets.

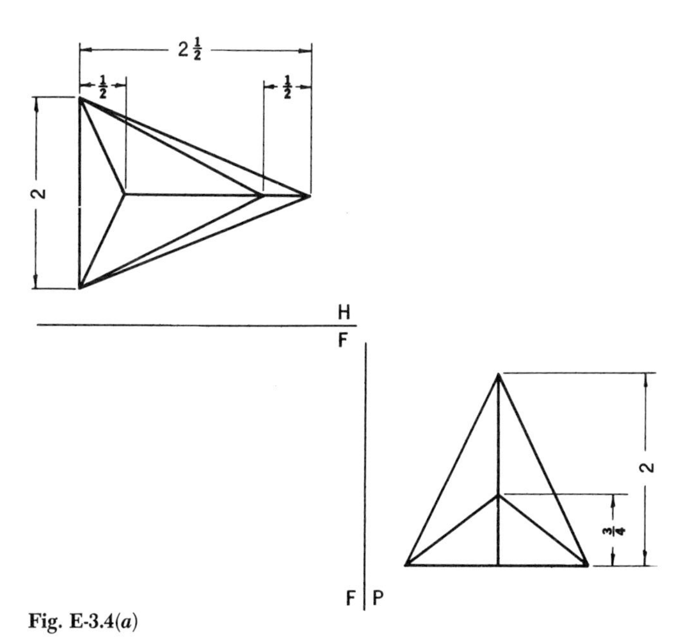

Fig. E-3.4(*a*)

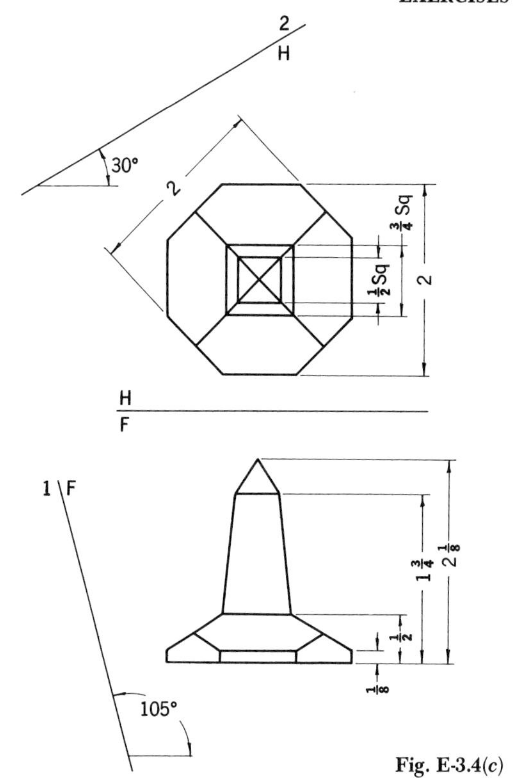

Fig. E-3.4(*c*)

Fig. E-3.4(*b*)

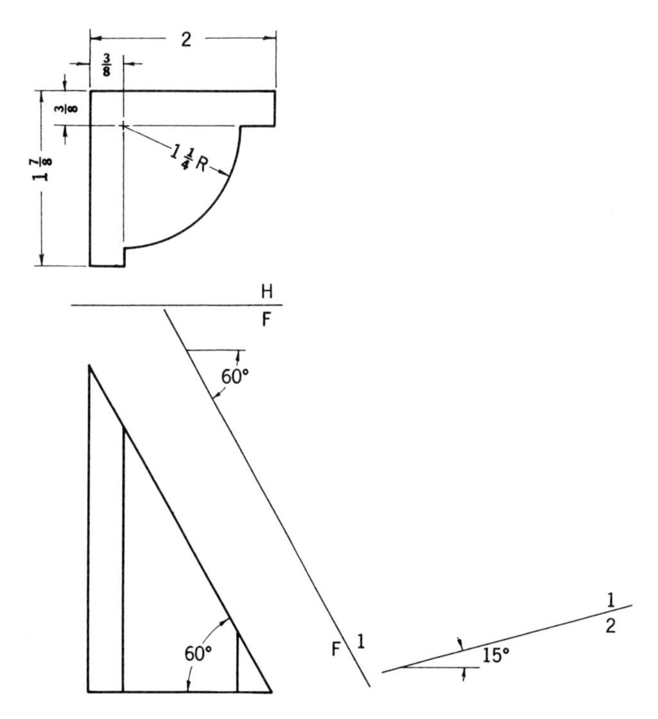

Fig. E-3.4(*d*)

VISIBILITY 4

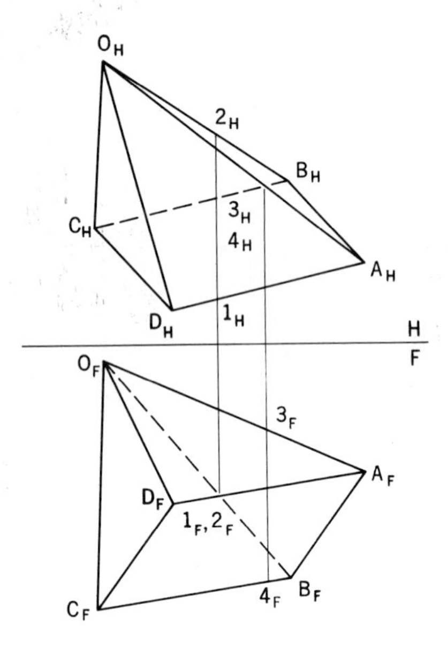

Fig. 4.1

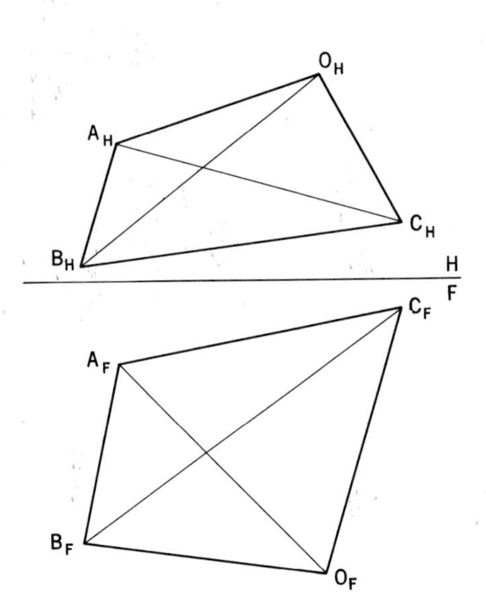

Fig. 4.2

We note that some edges are not visible in the supplementary views of Figs. 3.30 and 3.31, and that these edges are represented by dashed lines. In general, how shall we determine which edges of a solid are visible and which are hidden? This is now discussed in the following examples.

EXAMPLE 1

The top and front views of pyramid *O-ABCD* are shown in Fig. 4.1. There should be no doubt that *the outside edges* of each view of the pyramid *are visible.* Now let us consider the front view of the edges *OB*, *OD*, *AD*, and *CD*. One possibility is that edge *OB* is visible and edges *OD*, *AD*, and *CD* are invisible. The other possibility is that *OB* is invisible and edges *OD*, *AD*, and *CD* are visible. How shall we determine the correct visibility of these edges? Let us study the *apparent* intersection of edges *OB* and *AD*. This apparent intersection is the front view of two points, one on edge *AD*, the other on edge *OB*. Let us designate the front view of these points as 1_F and 2_F, respectively. The top view of these points is easily located. The top view clearly shows that point 1 is in front of point 2; therefore, *when we observe the front view,* point 1 is closer to the observer than is point 2. This means that edge *AD* which contains point 1 is closer to the observer than is edge *OB*; hence, edge *AD* is shown as a solid line and edge *OB* as a dashed line to indicate that edge *OB* is not seen in the front view. Edges *OD* and *CD*, of course, are visible.

In a similar manner let us analyze the *apparent* intersection of edges *OA* and *BC* in the top view. The apparent intersection represents the top view of two points, one (point 3) on edge *OA* and the other (point 4) on edge *BC*. Assume that 3_H and 4_H are the top views of these points. The front views, 3_F and 4_F are easily obtained. The front view shows that point 3 is above point 4; therefore, *when we observe the top view,* point 3 is closer to the observer than is point 4. This means that edge *OA*, which contains point 3, is closer to the observer than is edge *BC*; hence, edge *OA* is shown as a solid line and edge *BC* as a dashed line to indicate that edge *BC* is not seen in the top view. Edge *OD*, of course, is visible.

EXAMPLE 2

The top and front views of pyramid *O-ABC* are shown in Fig. 4.2. The edges in question, as to visibility, are shown as light solid lines. At first glance we might conclude that the object is not a pyramid but, rather, a quadrilateral, since the apparent diagonals in each view intersect in a point whose top and front views are correctly oriented with respect to the *H*- and *F*-reference planes.

Careful analysis of the views, however, will show that the object is a pyramid. Let us examine the front view carefully (Fig. 4.3). The apparent intersection of edges *OA* and *BC* is actually the front view of two points, one on edge *BC* and the other on edge *OA*. Let us designate these points as 1 and 2 respectively. The front view of these two points is shown as 1_F and 2_F. Now, note very carefully that 1_H is on B_HC_H (since point 1 is on edge *BC*) and that 2_H is on O_HA_H. We see that point 1 is in front of point 2; therefore, in observing the front view, point 1 is closer to the observer than is point 2. This means that edge *BC*, which contains point 1, is the visible edge and that edge *OA* is invisible; hence, B_FC_F is shown as a solid line and O_FA_F as a dashed line (Fig. 4.4).

Now let us examine the top view. The apparent intersection of A_HC_H and O_HB_H is actually the top view of two points, 3 and 4, one on edge *AC* and the other on edge *OB*. From the top view we can easily locate 3_F (on A_FC_F, since point 3 is on edge *AC*) and 4_F (on O_FB_F, since point 4 is on edge *OB*). The front view now clearly shows that point 3 is above point 4; therefore, in observing the top view, point 3 is closer to the observer than is point 4. This means that edge *AC*, which contains point 3, is the visible edge and that edge *OB* is the invisible edge; hence, A_HC_H is shown as a solid line and O_HB_H as a dashed line. Figure 4.4 shows the completed two views.

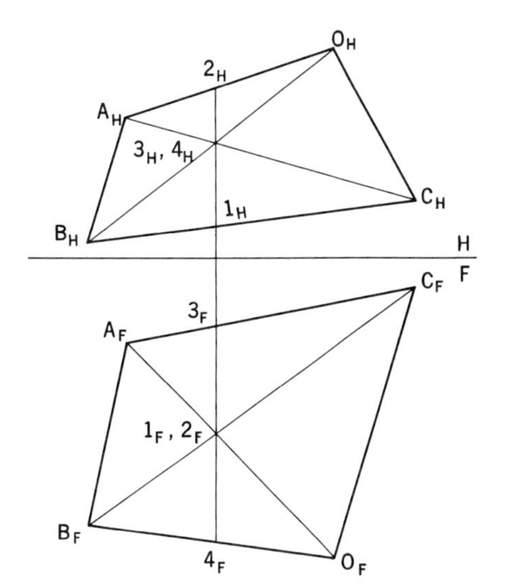

Fig. 4.3

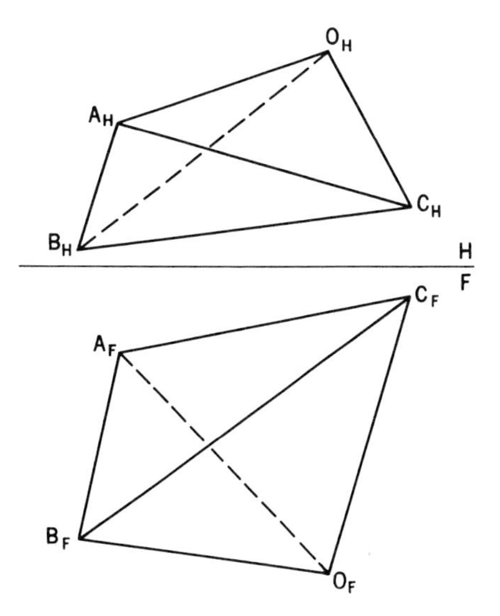

Fig. 4.4

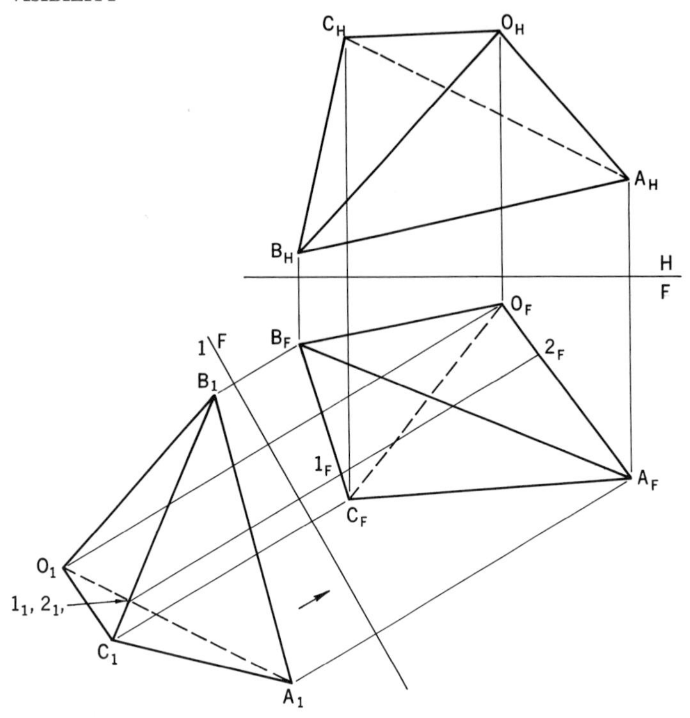

Fig. 4.5 Part 1

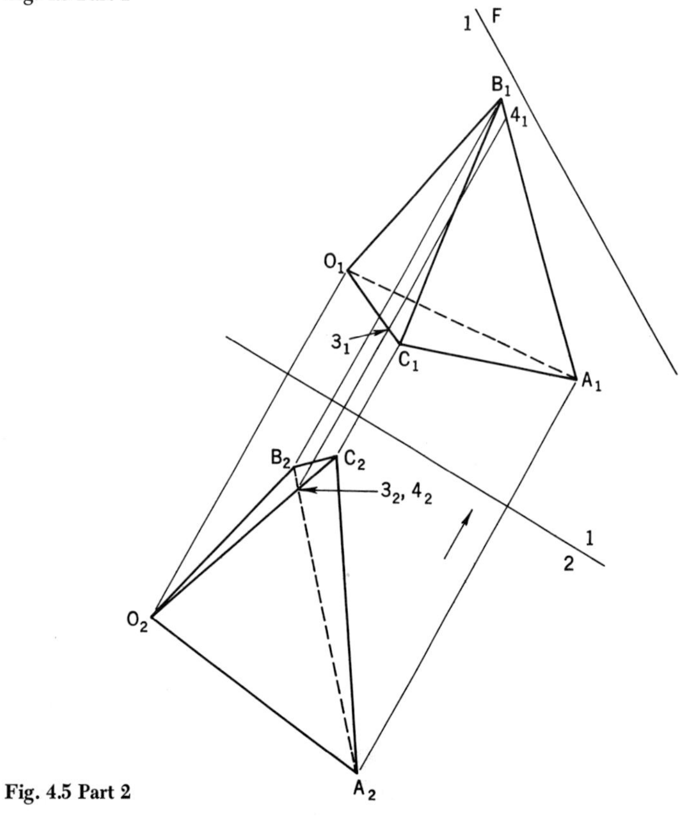

Fig. 4.5 Part 2

EXAMPLE 3

Let us consider the object shown in Fig. 4.5. We will assume that the visibility has been determined in the top and front views. Our problem is to determine the visibility in supplementary views 1 and 2. We will first consider supplementary view 1. The apparent intersection of edges *OA* and *BC* in this view is actually the supplementary view of two points, one on edge *BC* (point 1) and the other (point 2) on edge *OA*. The front view of these two points, 1_F and 2_F, is easily located. When we look at the front view we see supplementary plane 1 as a line and we observe that point 1 is closer to supplementary plane 1 than is point 2. When we look at supplementary plane 1, in the direction of the arrow, we see that edge *BC* which contains point 1 is closer to us than is edge *OA*. This means that edge *BC* will be visible and that edge *OA* will be invisible. We should note that the method of analysis used to determine the visibility in the supplementary view is basically the same as the analysis used in the first two examples.

Now let us consider supplementary view 2. The apparent intersection of edges *OC* and *AB* is actually supplementary view 2 of two points, one (point 3) on edge *OC*, and the other (point 4) on edge *AB*. The views of these two points are shown at 3_2 and 4_2 on plane 2 and at 3_1 and 4_1 on plane 1. When we look at supplementary view 1 we see the edge view of plane 2 and we observe that point 3 is closer to plane 2 than is point 4. This means that when we look at plane 2, in the direction of the arrow, edge *OC*, which contains point 3, is closer to us than is point 4; therefore, edge *OC* is visible and edge *AB* is invisible.

EXAMPLE 4

Let us consider the pyramid O-ABC shown in Fig. 4.6. We know that the outside edges AB, BC, and CA are visible. What about the edges OA, OB, and OC as seen in the top view? Evidently there is no apparent intersection of edges in this view that might be compared with the previous examples. How shall we apply the method of analysis used in the previous problems? Suppose edge OC is extended so that its top view crosses the top view of edge AB. The apparent intersection of these two lines is the top view of two points, 1 and 2. The front view of the two points is shown as 1_F (on O_FC_F extended) and 2_F (on A_FB_F). It is quite evident that the front view shows that point 1 is above point 2; therefore, in looking down on the pyramid, edge OC (extended) which contains point 1 will be visible in the top view, and certainly edges OB and OA will also be visible in the top view.

In a similar manner we may determine the visibility in the front view. You should have no difficulty in doing so.

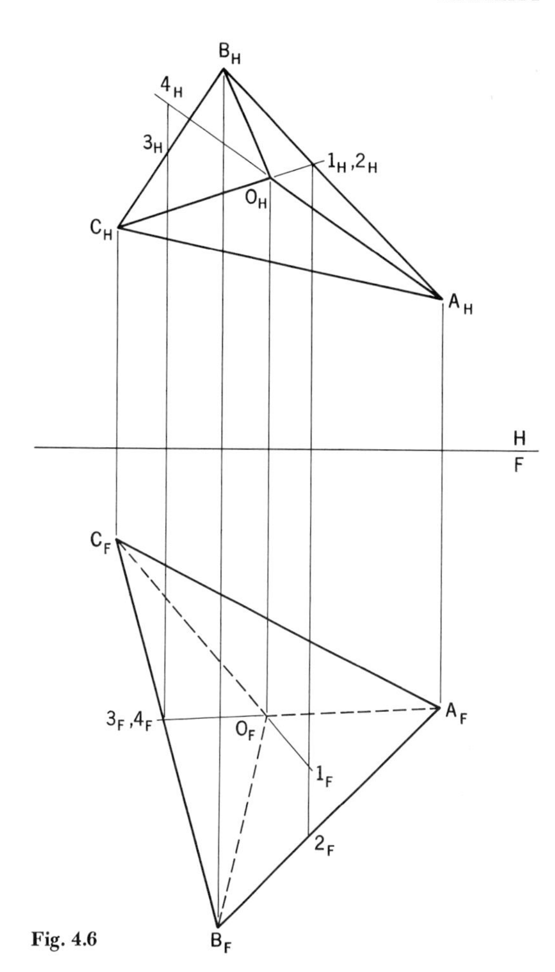

Fig. 4.6

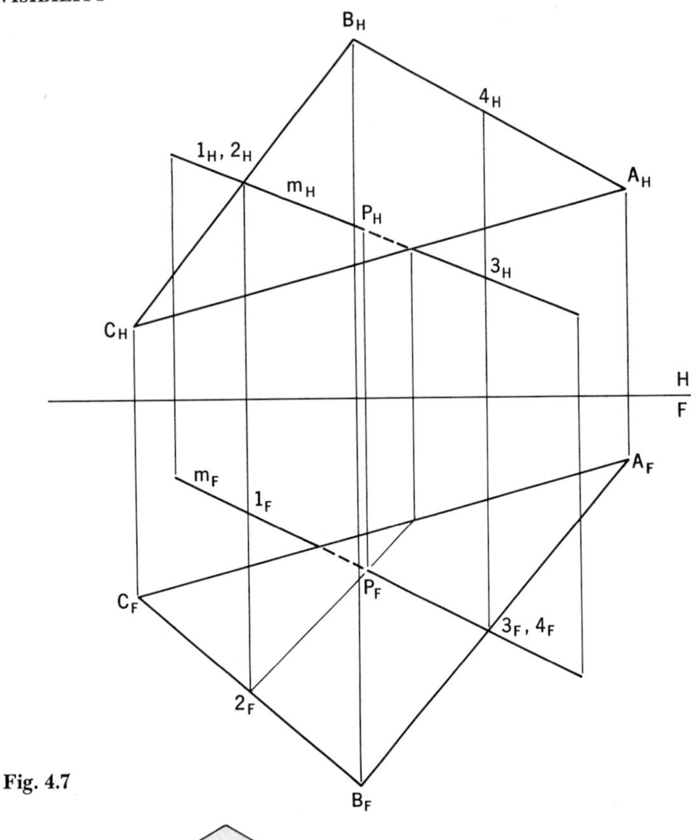

Fig. 4.7

Fig. 4.8

EXAMPLE 5

Let us now consider triangle ABC and line m shown in Fig. 4.7. We will assume that the triangle is opaque (not transparent) and that point P, the intersection of line m with the triangle, is known. Our problem is to determine which portion of m_H is visible and which portion of m_F is visible.

Consider the top view. The apparent intersection of line m with side BC is the top view of two points, one on line m (point 1) and the other on side BC (point 2). The front view of these two points, namely, 1_F and 2_F, clearly shows that point 1 is above 2. Thus, when we look down (top view), line m which contains point 1 is above side BC. Therefore, the portion of m_H from point 1 to point P is visible, and certainly the portion from P to side AC is invisible.

Now let us consider the front view. The apparent intersection of line m and side AB is the front view of two points, one on line m (point 3) and the other on side AB (point 4). The top view of these two points clearly shows that point 3 is in front of point 4. Thus, in looking at the front view, line m which contains point 3 is in front of side AB. Therefore, the portion of m_F from point 3 to point P is visible and the portion from P to side AC is invisible. A pictorial is shown in Fig. 4.8.

EXERCISES

1. Complete the top and front views of pyramid *O-ABC* (Fig. E-4.1). Add supplementary views 1 and 2. Include visibility.

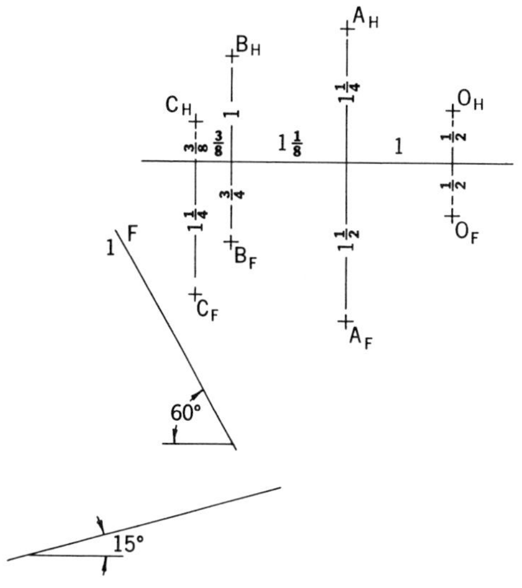

Fig. E-4.1

2. Point *P* is the intersection of line *m* with the opaque (nontransparent) surface *ABC*, as shown in Fig. E-4.2. Determine the solid and dashed-line portion of line *m* in each of the given views. Add supplementary view 1, and then determine the visible portion of line *m*.

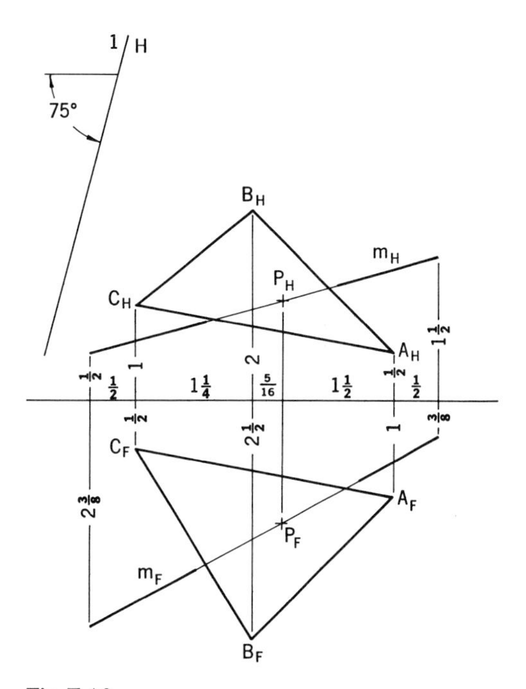

Fig. E-4.2

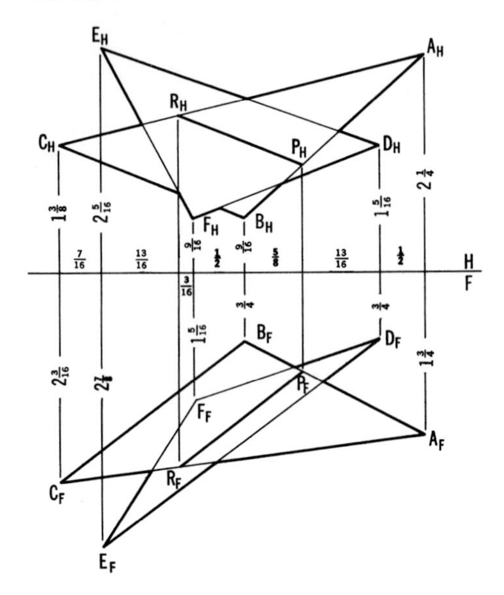

Fig. E-4.3

3. Triangles *ABC* and *DEF* intersect in line *PR*. Assume that the triangles are opaque. Determine the visible and hidden edges in each view (Fig. E-4.3).

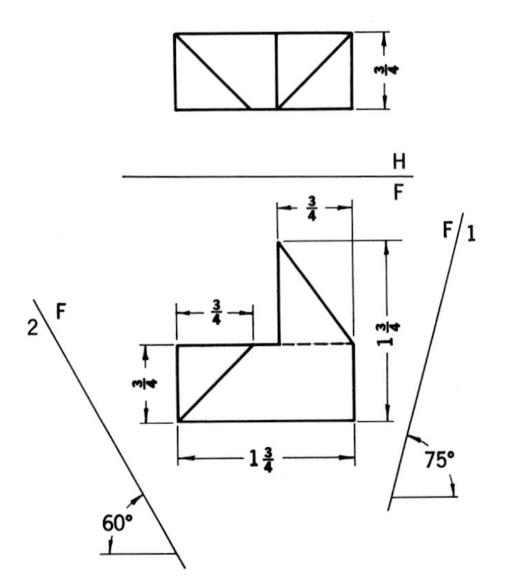

Fig. E-4.4

4. Reproduce the *H* and *F* views shown in Fig. E-4.4. Add the views of the *solid* as seen on planes 1 and 2.

5. Point O is the apex of a pyramid having base ABC. Complete the three views. Show correct visibility (Fig. E-4.5). B_F is on the line joining A_F and C_F.

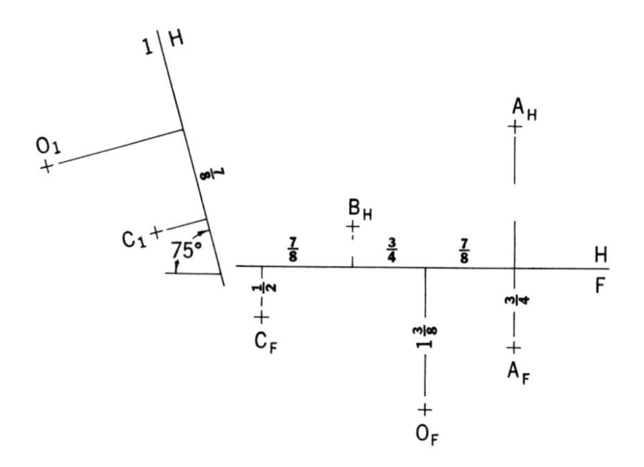

Fig. E-4.5

6. Reproduce the views shown in Fig. E-4.6. Obtain the views on supplementary planes 1 and 2. Show correct visibility.

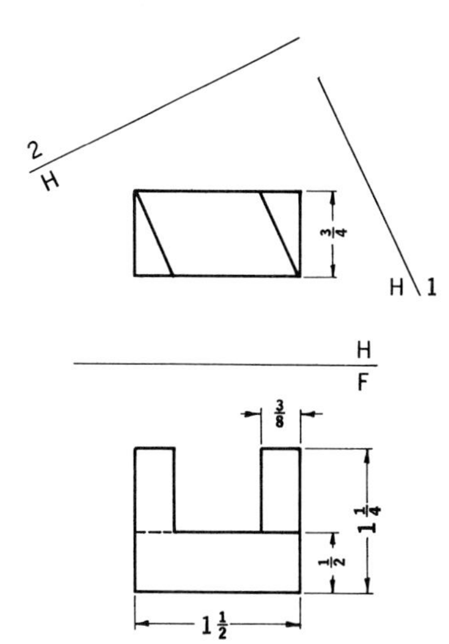

Fig. E-4.6

INTERPRETING ORTHOGRAPHIC DRAWINGS 5

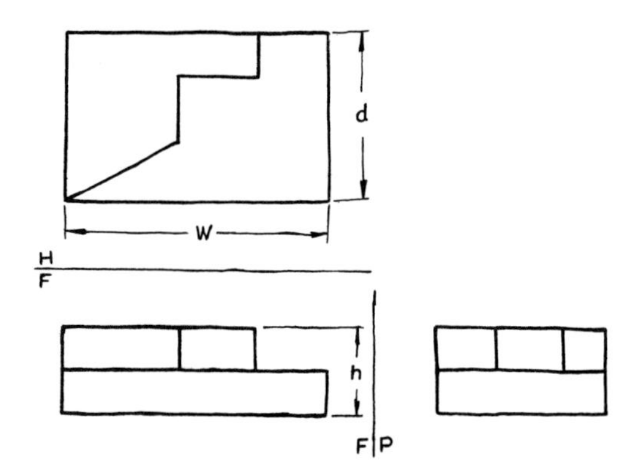

Fig. 5.1

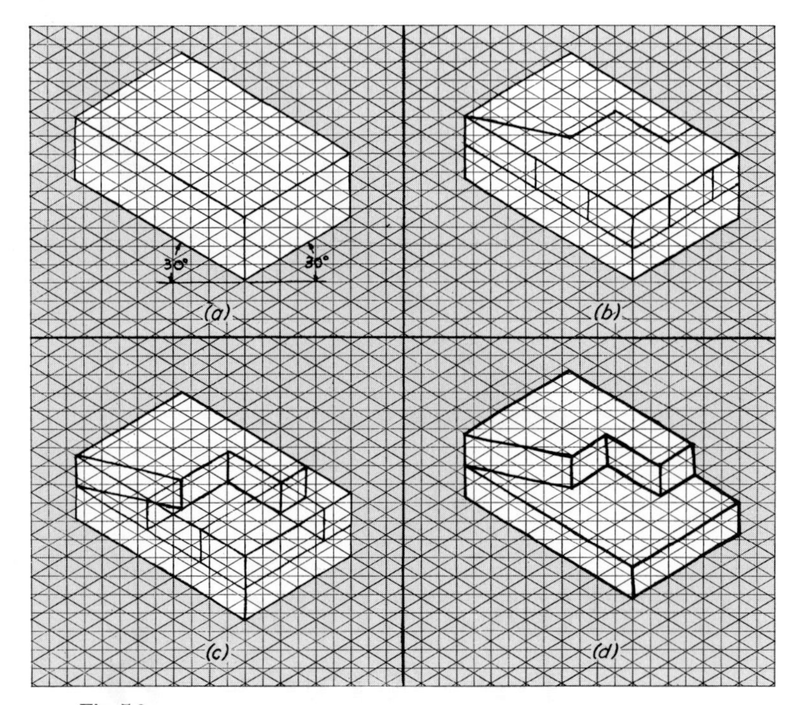

Fig. 5.2

Now that we have had some experience in freehand sketching; in the use of the two fundamental principles of orthogonal projection; and in the determination of visibility, let us apply this knowledge to the interpretation of orthographic representations.

We visualize the shape of an object by "reading" the orthogonal views and forming a mental image of the object. By drawing a freehand pictorial sketch of the mental image we can verify the designer's intent as represented in the orthogonal views.

The pictorial sketches may be (1) axonometric—isometric, dimetric, or trimetric; (2) oblique; or (3) perspective.

In our discussion of freehand sketching (Chapter 2) reference was made to the use of isometric grid sheets and of perspective grid sheets. Examples of orthographic, isometric, oblique, and perspective sketches were included.

EXAMPLE 1

Now let us consider the object shown orthographically in Fig. 5-1. First we should carefully study each of the three views and try to form a mental image of the object. Try to visualize the object. As an initial step in translating the three views it is suggested that a sketch be made of the "total rectangular block." This is known as "blocking-in" the object. Estimate distances *w*, *d*, and *h*, or lengths proportional to these distances, and lay them off on the isometric axes. [See Fig. 5.2(*a*).] This should be done with an H pencil and drawn lightly.

As a second step sketch the lines of each orthographic view on the corresponding surfaces of the total block. [See Fig. 5.2(*b*).]

For the third step sketch the lines of intersection where surfaces meet. These are shown as heavier lines in Fig. 5.2(*c*). Finally, strengthen the remaining edges of the object as shown in Fig. 5.2(*d*) and then check the isometric pictorial with the orthographic views to verify the designer's intent.

The lines may be left in if they do not detract from the clearness of the sketch. Pads, $8\frac{1}{2}'' \times 11''$, with 30° grid lines are commercially available for freehand isometric sketching.

A more complete treatment of Pictorial Drawing is presented in Chapter 17. The theory and methods used for obtaining isometric, dimetric, and trimetric *views;* oblique views; and perspective views are well covered in this chapter.

EXAMPLE 2

Let us consider the object shown in Fig. 5.3. Careful study of the three views reveals the following: (*a*) the left half of the object is a rectangular block except for the sloping triangular surface 1, 2, 3; (*b*) the right half of the object is primarily a triangular wedge shape with a cut-out portion, 4, 5, 6, 7.

The steps to form the isometric pictorial are shown in Figs. 5.4(*a*), and 5.4(*b*). We should recall that true measurements are made on the isometric axes or on lines parallel to the axes.

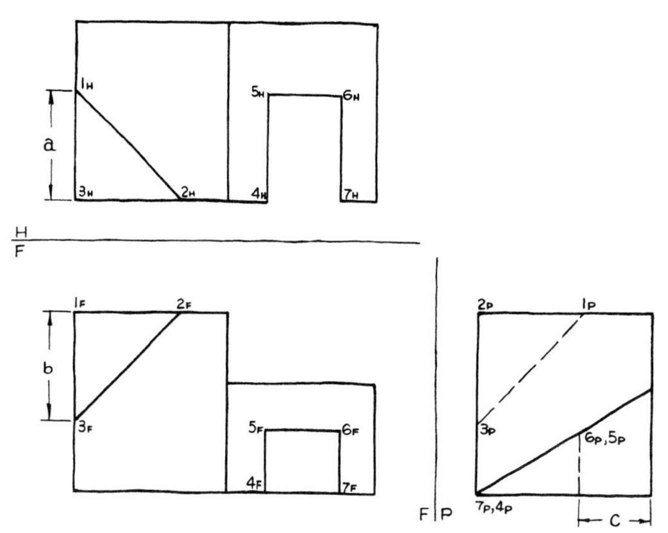

Fig. 5.3

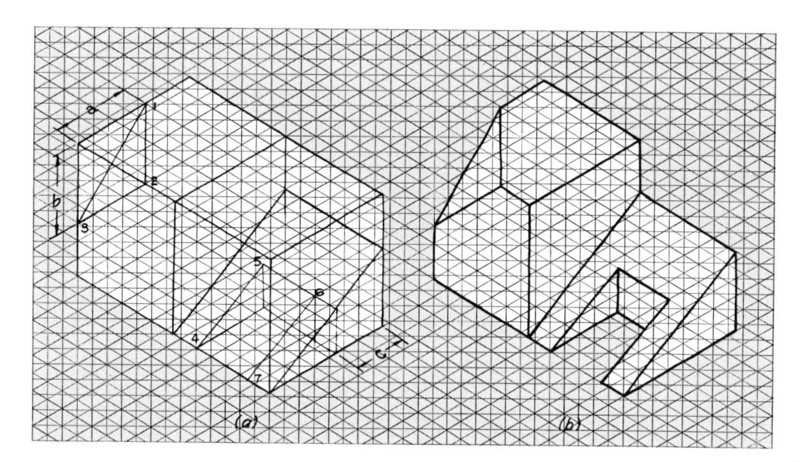

Fig. 5.4

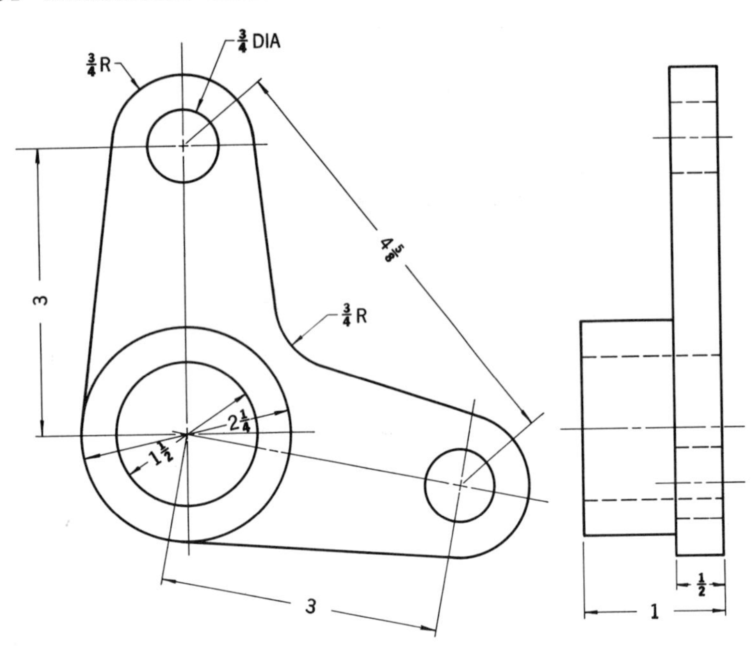

Fig. 5.5 Rocker arm in orthographic.

EXAMPLE 3

The "rocker arm" shown in Fig. 5.5 is best represented pictorially by an *oblique* sketch because the circular arcs and circles shown in the front view can be retained in the sketch. The steps necessary to produce the oblique sketch are shown in Fig. 5.6 and Fig. 5.6(*a*).

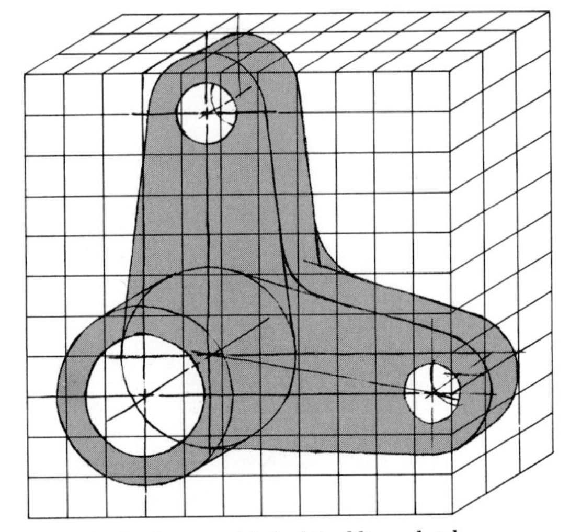

Fig. 5.6 Rocker arm, blocked-in oblique sketch.

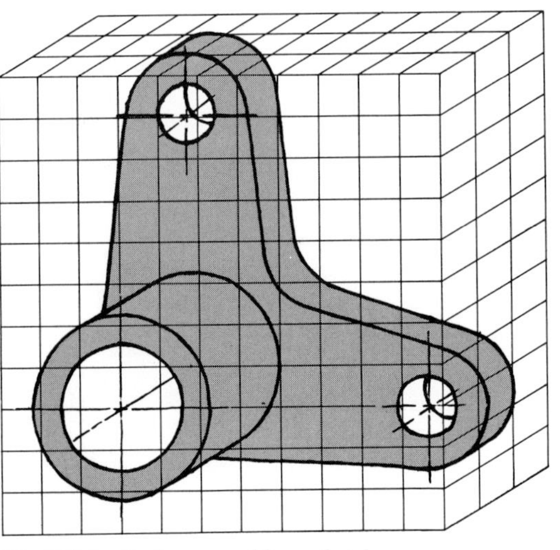

Fig. 5.6(*a*) Rocker arm, oblique sketch.

Perspective sketches present a more natural appearance and are quite useful in making presentations to executive personnel—i.e., members of a board of directors, many of whom may have little technical background.

The orthographic drawing shown in Fig. 5.7 can be presented in perspective by using a suitable perspective grid. The sketches shown in Fig. 5.8 are quite easily understood.

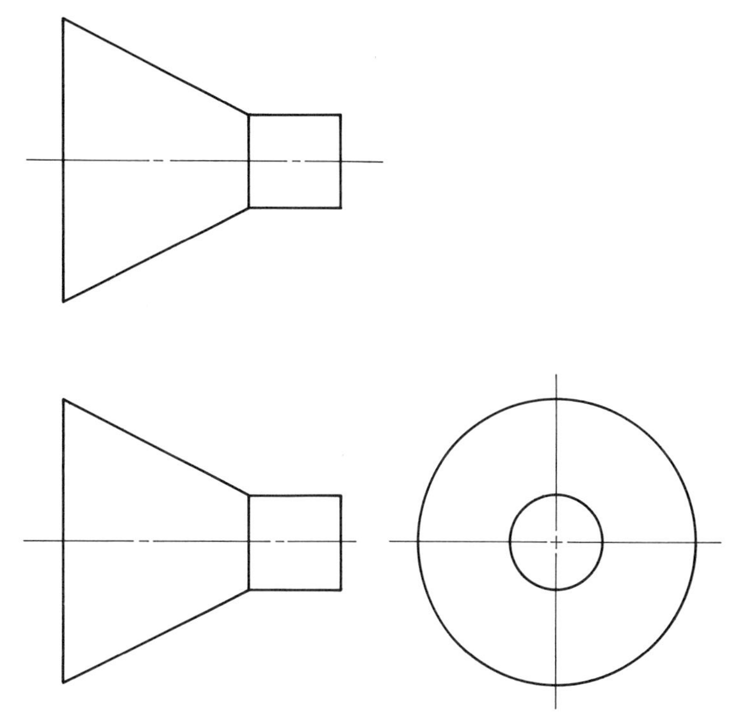

Fig. 5.7 Capsule in orthographic.

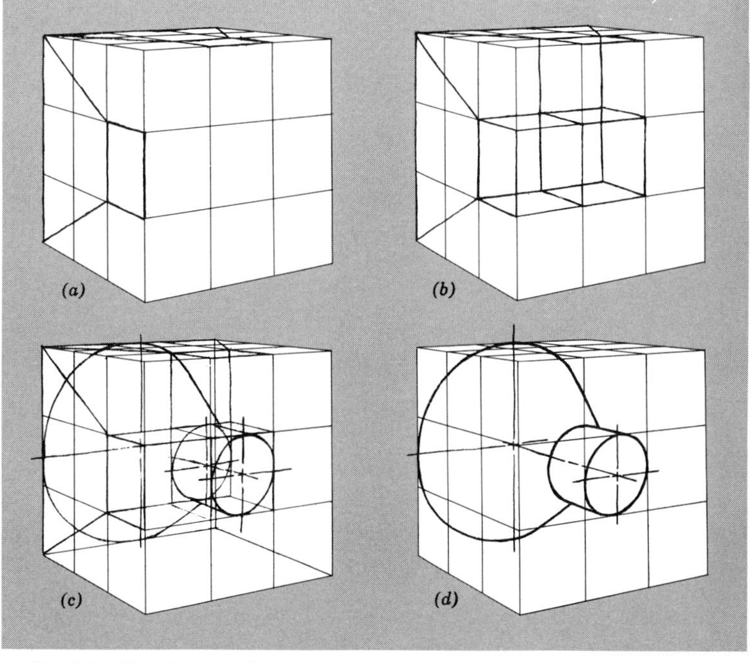

Fig. 5.8 Capsule, steps for perspective sketch.

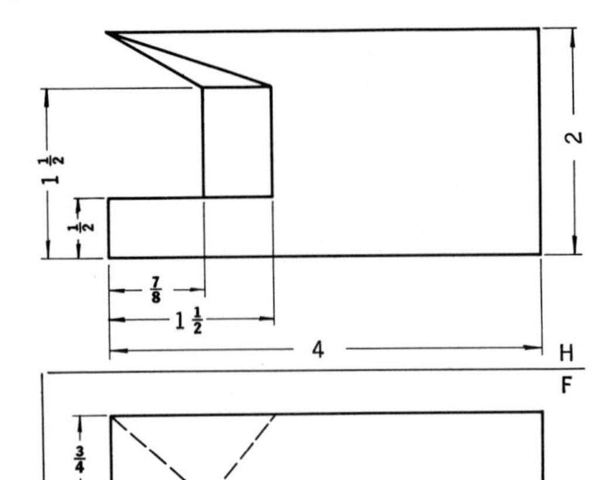

Fig. 5.9

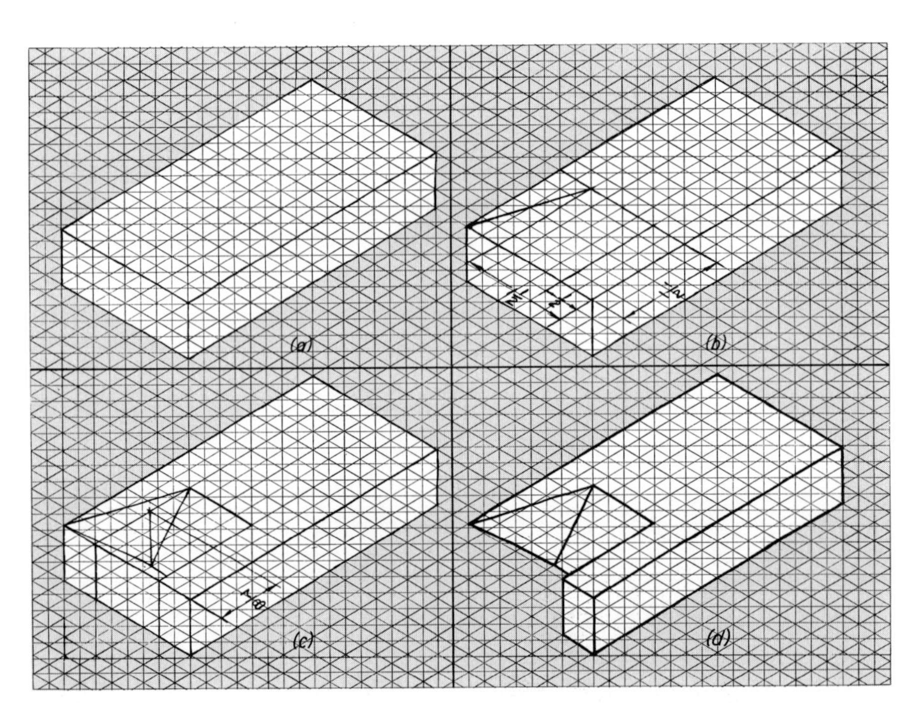

Fig. 5.10

EXAMPLE 5

Let us study the object shown in Fig. 5.9. An interpretation of the object is delineated in the four isometric sketches shown in Fig. 5.10. Part (*a*) shows the "over-all" block that contains the piece (a cutting tool). Parts (*b*) and (*c*) show the location of salient points and edges; and part (*d*) the completed isometric sketch. The latter is then compared with the orthographic views in Fig. 5.9 to make certain the interpretation is correct.

EXAMPLE 6

In some cases it is necessary to locate points of an object by establishing their coordinates in the pictorial sketch. Let us consider the object shown in Fig. 5.11. Suppose we wish to make an isometric sketch of the object. The base is easily constructed. [See Fig. 5.12(*a*).] Points 1, 2, and 3 of the pyramid are located by their coordinates. For example, point 1 is established by the intersection of the line which is parallel to the left top edge of the base at distance $(a + b)$ from that edge, with the line which is parallel to the front top edge of the base at distance r from that edge. In a similar manner, points 2 and 3 are established. [See Fig. 5.12(*b*) and (*c*).]

Point 4 which is the foot of the perpendicular from point 5 to the top surface of the base is also located in the same manner as points 1, 2, and 3. Now, from point 4 we can construct a vertical line and then lay off the length from point 4 to point 5. The length is seen in the front view of the object. Finally, the points are properly connected to form the isometric sketch shown in Fig. 5.12(*d*).

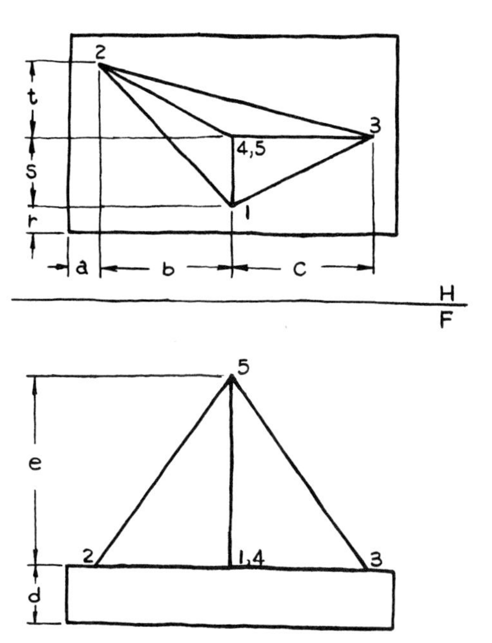

Fig. 5.11

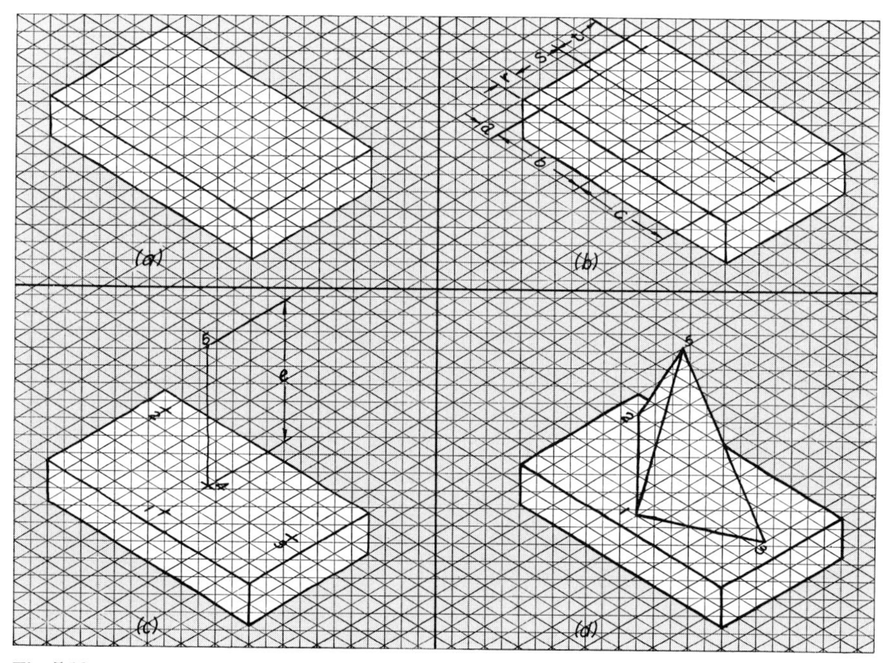

Fig. 5.12

EXERCISES

1. *Sketch* the missing views in Figs. E-5.1, E-5.2, E-5.3, E-5.4, and E-5.5 on separate sheets. Also prepare freehand isometric sketches of each piece. Use $8\frac{1}{2}'' \times 11''$ isometric grid sheets. Be sure to check the isometric sketches to verify the interpretation of the orthographic representations.

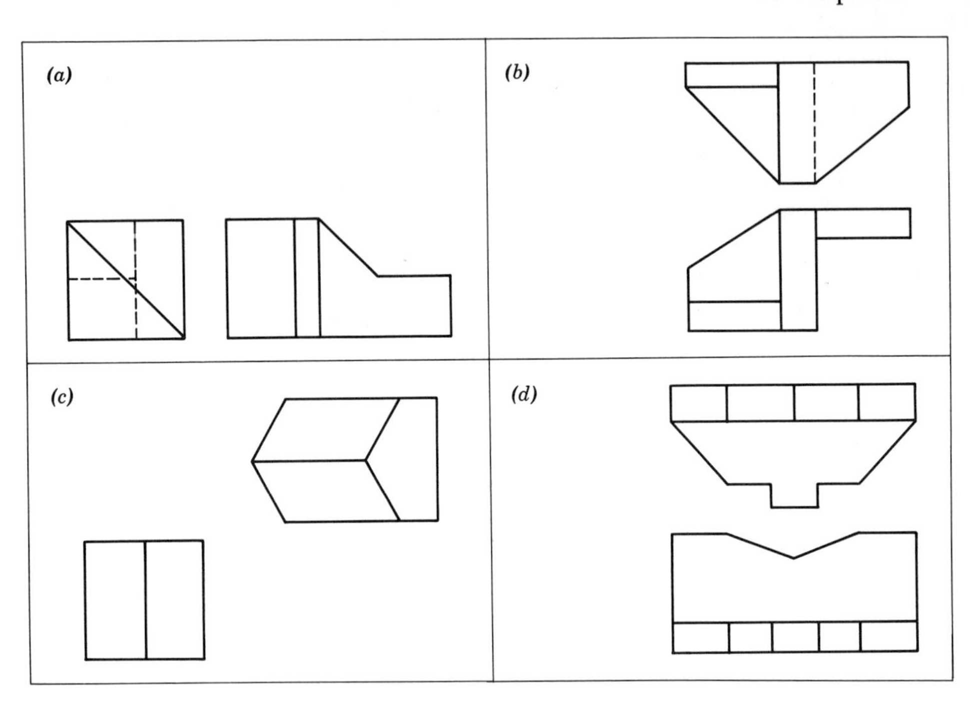

Fig. E-5.1

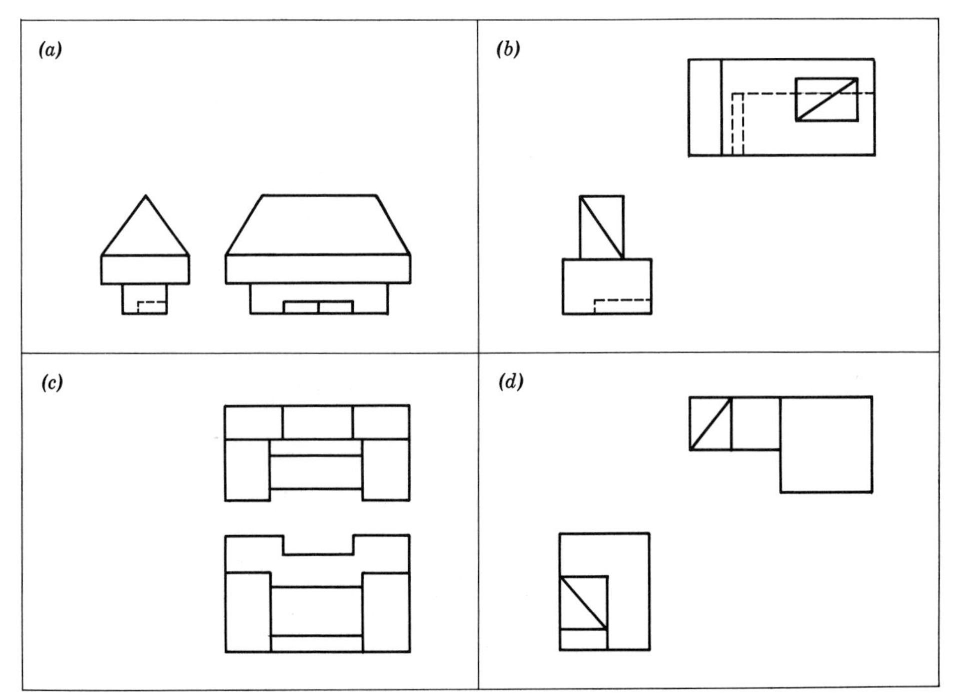

Fig. E-5.2

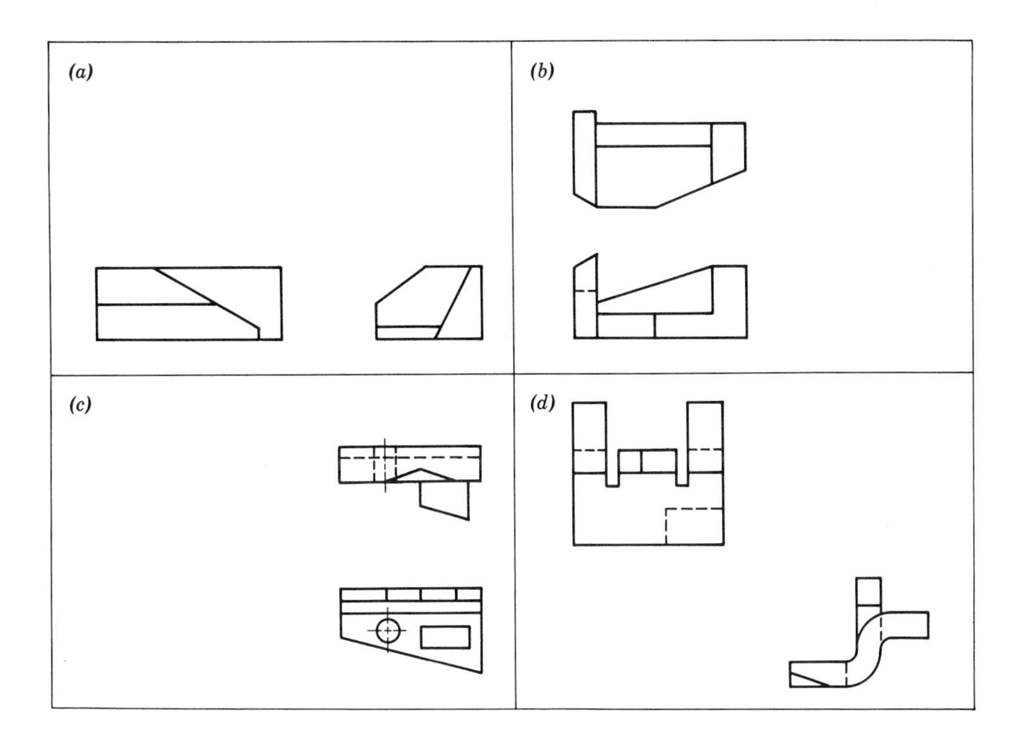

Fig. E-5.3

Fig. E-5.4

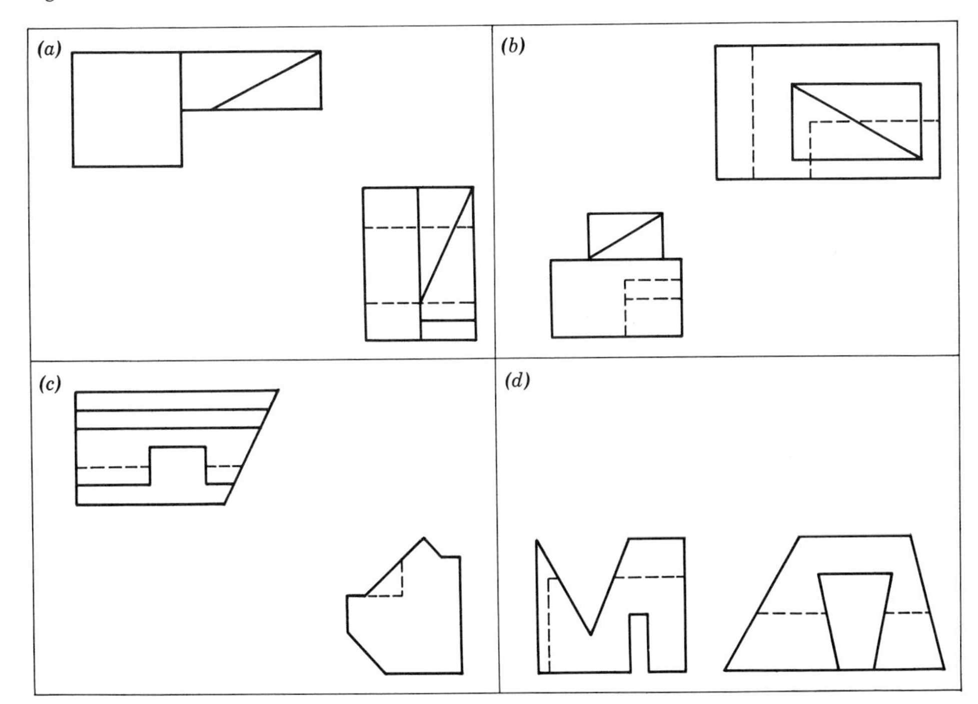

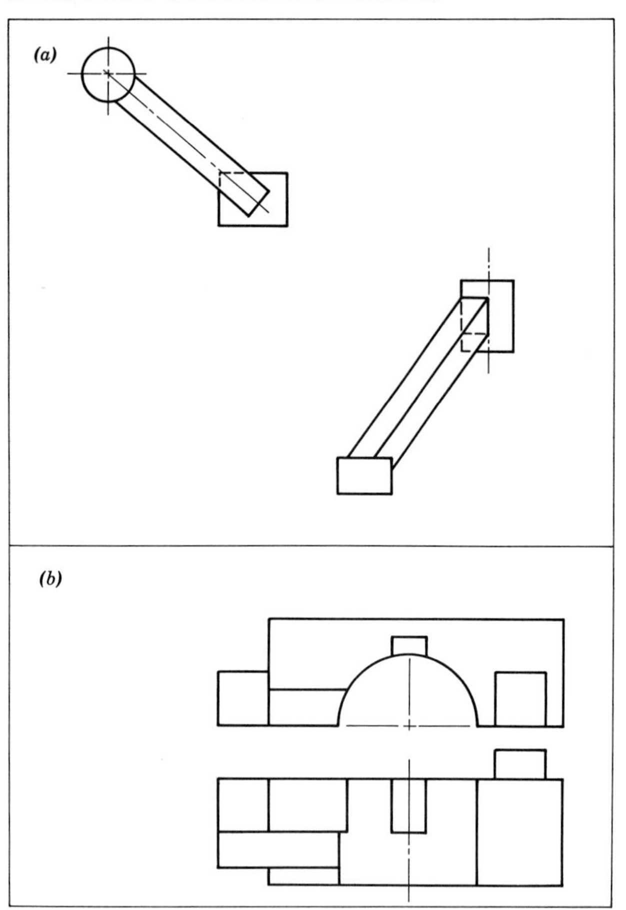

Fig. E-5.5

2. In Figs. E-5.6, E-5.7, and E-5.8 some lines are missing in one or more of the views. *Sketch* the given views of each piece on separate $8\frac{1}{2}'' \times 11''$, $\frac{1}{4}$-inch grid sheets, add the missing lines, and then prepare *freehand* pictorial sketches of the pieces. Use isometric, oblique or perspective grids in accordance with your choice that you believe will best portray each piece. Verify the pictorial interpretation of the orthographic representation.

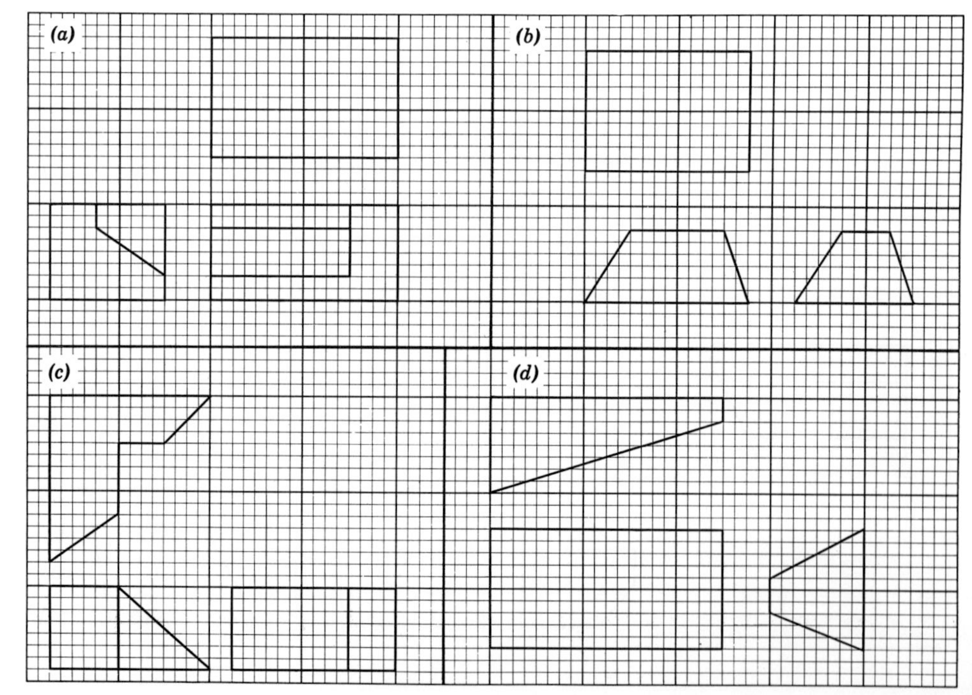

Fig. E-5.6

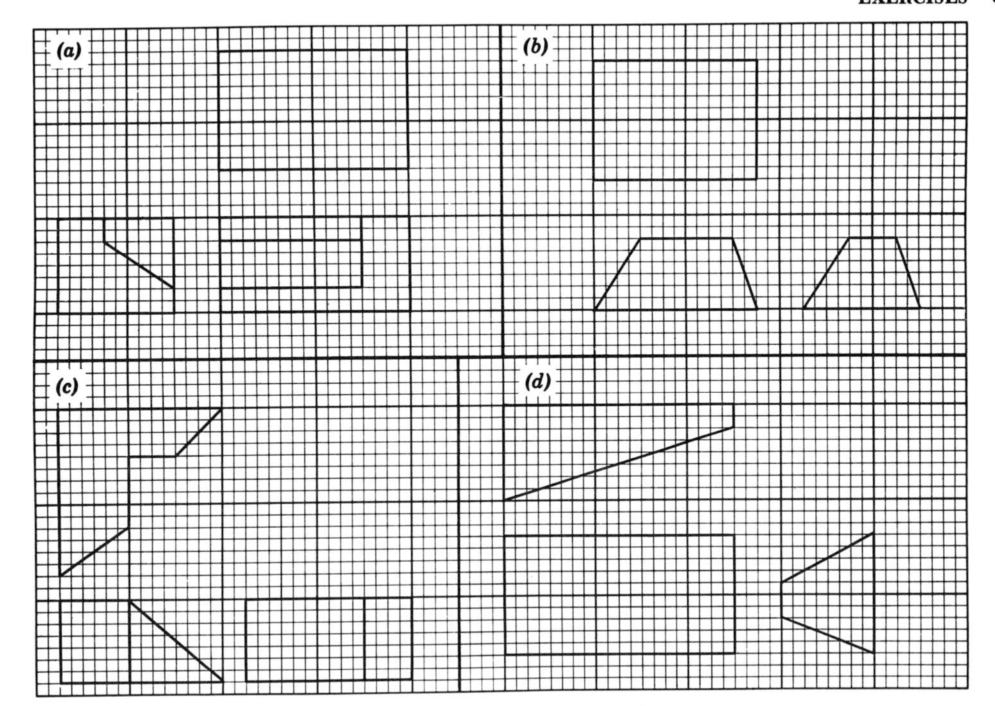

Fig. E-5.7

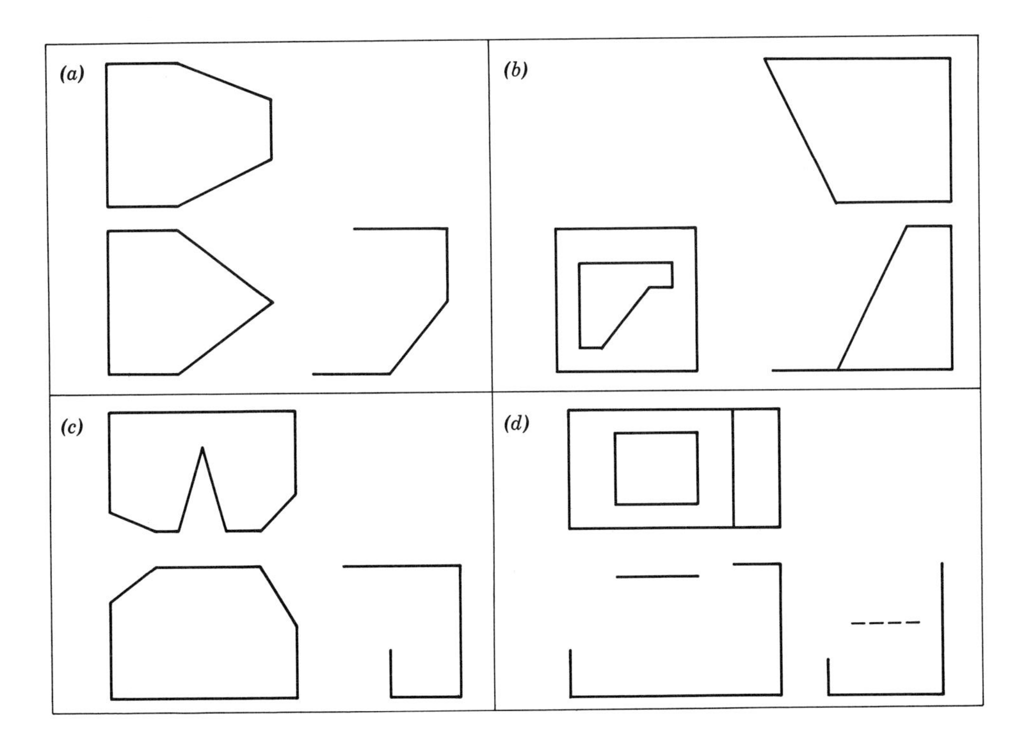

Fig. E-5.8

APPLICATIONS OF THE FUNDAMENTAL PRINCIPLES OF ORTHOGONAL PROJECTION 6

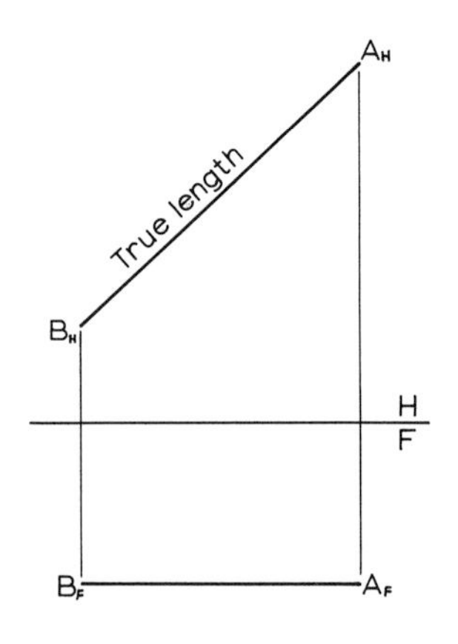

Fig. 6.1

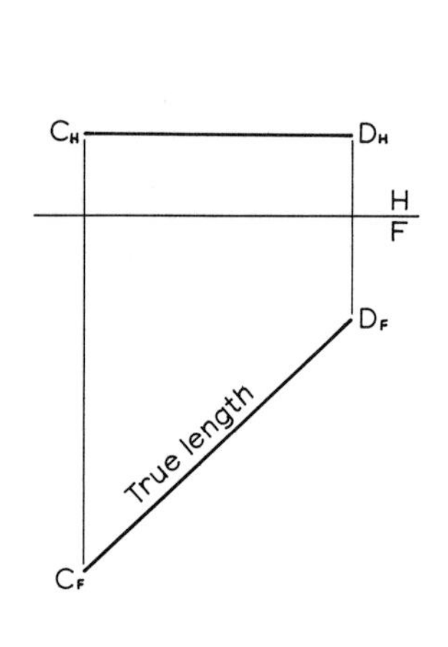

Fig. 6.2

BASIC PROBLEMS

We should not lose sight of the emphasis that has been placed on analysis—"the mental process of thinking through the solution of a problem"—and the application of the two fundamental principles of orthogonal projection to the graphical expression of the solution. We have observed that many engineering design problems include the determination of true lengths of members, the angles between surfaces, the angles between lines, clearances between wires, cables, pipes, rods, etc., true shapes of surfaces, etc.

The analysis and solution of these types of problems are quite simple once we understand thoroughly the solution of the four basic problems:

1. To find the true length of a line segment.
2. To find the point view of a line.
3. To find the edge view of a plane surface.
4. To find the true shape of a plane surface.

Let us consider each of these four basic problems.

1. TRUE LENGTH OF A LINE SEGMENT—FIRST BASIC PROBLEM

The true length of a line segment will be seen in the view on a plane which is parallel to the line segment.

Case 1. The Line Is Parallel to a Reference Plane

In Fig. 6.1, line AB is parallel to the H-reference plane. This is evident from the fact that the front view of line AB is horizontal, indicating that all the points on line AB are the same distance below the H-plane. The top view, A_HB_H, therefore, shows the true length of line AB.

A line which is parallel to the H-reference plane is called a horizontal line.

Now in Fig. 6.2, line CD is parallel to the F-reference plane. The front view of the line shows its true length. Note carefully that the top view of the line is seen as a line which is parallel to the top view (edge view) of the F-plane. The top view of line CD clearly shows that all its points are the same distance behind the F-plane.

A line which is parallel to the F-plane is called a frontal line.

You should experience no difficulty in drawing the H, F, and P views of a line which is parallel to the P-reference plane. *Try it freehand.*

The three lines, horizontal, frontal, and profile, are classified as *principal lines.*

Case 2. The Line Is Not Parallel To a Reference Plane

Supplementary plane method. Suppose the line segment AB is represented by the H and F views shown in Fig. 6.3. It should be quite evident that neither view shows the true length of the line, since the line is not parallel to either the H- or F-reference plane.

It will be necessary to introduce a supplementary plane which will be parallel to the line *and perpendicular* to one of the reference planes. Suppose we introduce plane 1, parallel to line AB and perpendicular to the H-plane. (See Fig. 6.4.) Careful study of this figure shows that the distances from the H-plane to each of the points A and B are seen twice: once in the front view and again in the supplementary view. You will recall the second fundamental principle: *when two planes are perpendicular to a third, the distance that a point is from the third plane will be seen twice, once in each of the views upon the other two planes.* In Fig. 6.4, we see that planes F and 1 are perpendicular to the H-plane; therefore, the distances that points A and B are below the H-plane are seen in both views on planes F and 1. We also see that since plane 1 is parallel to line AB, its true length will be seen in the view A_1B_1.

It should be recognized that we could have introduced a supplementary plane parallel to the line and perpendicular to the F-plane. The view of the line on the supplementary plane would disclose the true length of the line. [See Fig. 6.4(a).]

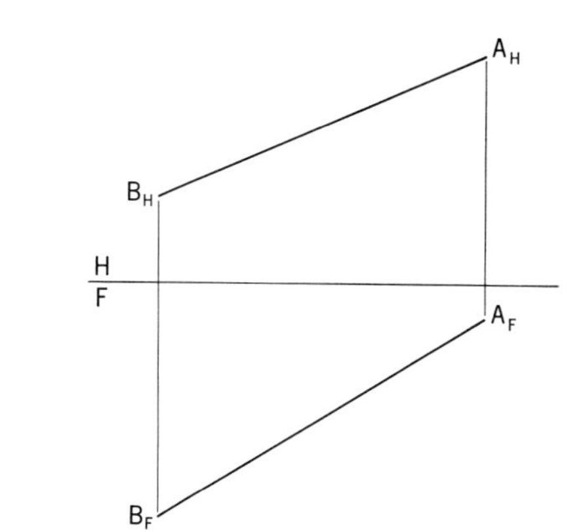

Fig. 6.3

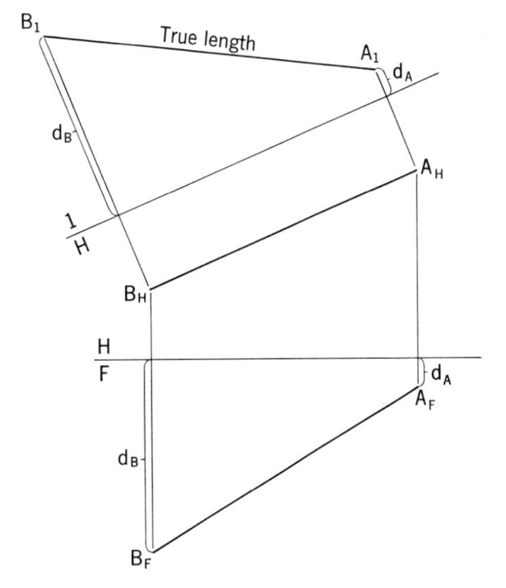

Fig. 6.4

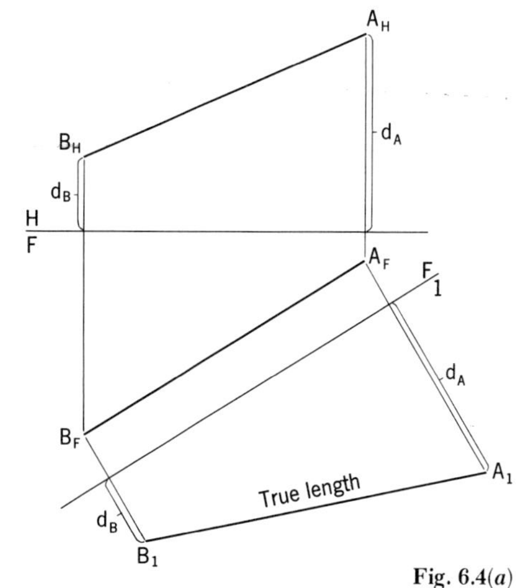

Fig. 6.4(a)

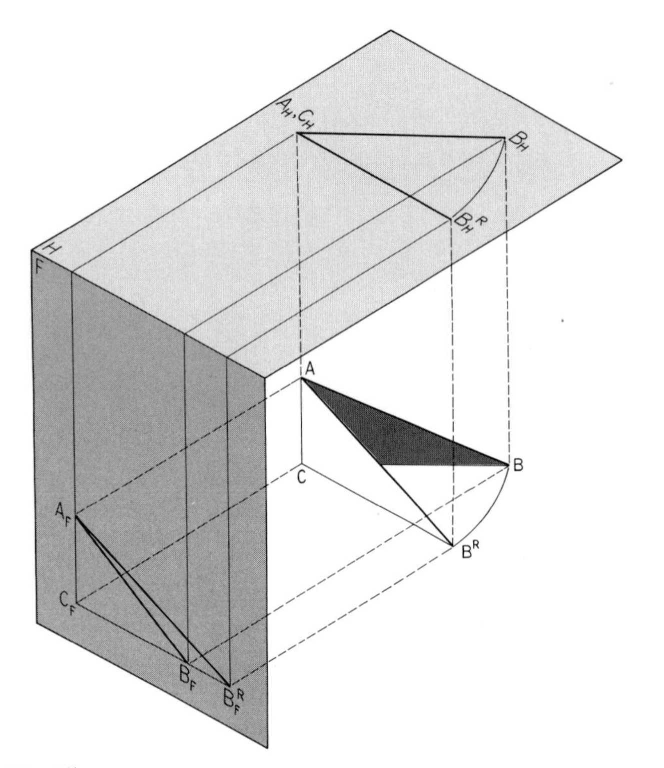

Fig. 6.5

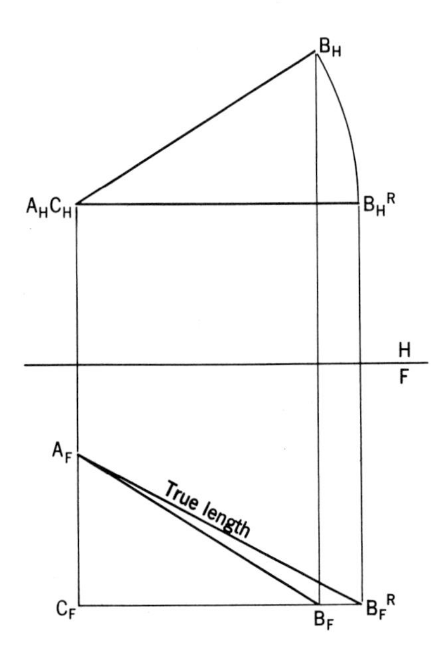

Fig. 6.5(*a*)

Rotation method. Let us consider the pictorial shown in Fig. 6.5. It is assumed that the given line segment is AB. Note that neither the top nor front views of the line shows the true length of the line. Now let us introduce a vertical line through point A and a horizontal line through point B to intersect the vertical line at point C. We now have formed the right triangle ABC. Let us rotate the triangle about line AC until it is parallel to the F-plane. The front view of the rotated triangle will show its true shape, and hence the true length of line segment AB. The orthographic solution is shown in Fig. 6.5(a).

EXAMPLE

Suppose we are given the front view and true length of line segment AB as 2 inches. How shall we determine the top view of segment AB?

Figure 6.6 shows the front view of line segment AB. In Fig. 6.7, $A_FC_FB_F^R$ shows the rotated, or true shape, view of triangle ABC. This is similar to Fig. 6.5(a). The location of A_H is arbitrary. The axis of rotation of triangle ABC is line AC. The true length of the base (CB) of the right triangle ABC is equal to the distance from C_F to B_F^R. Since line CB is horizontal, its top view will show its true length. Therefore, an arc of length C_F to B_F^R and center C_H will cut the vertical through B_F in points B_H and B_H'. Lines A_HB_H and A_HB_H' show the *two* solutions to the problem.

Grade of a line

The grade of a line is the ratio of the vertical displacement or "rise" of two of its points to the horizontal projected length, or "spread," of the line segment. The per cent of grade is this ratio multiplied by 100.

EXAMPLE 1

Let us consider Fig. 6.8 which shows line AB and its views on the H and F planes. Now suppose we draw a vertical line through point A to intersect the horizontal line through point B to form the right triangle ABC.

The per cent of grade of line AB is the ratio of AC (rise) to CB (spread) times 100, or $AC/CB \times 100$, or $A_FC_F/A_HB_H \times 100$.

The grade of line AB is *negative,* since motion from A to B is downward. The grade of line BA, however, is *positive,* since motion from B to A is upward. The magnitude of the grade is the same in both cases.

Again, in Fig. 6.8 we note that the per cent of grade of line AB is $AC/CB \times 100 = A_FC_F/C_HB_H \times 100 = A_FC_F/A_HB_H \times 100 = A_1C_1/A_HC_H \times 100$.

Note carefully that the rise is shown in the views on planes F and 1 which are *both perpendicular to the H-plane.* The view, $A_1B_1C_1$, also shows the true shape of the right triangle ABC and, hence, the distances A_1C_1 (rise) and C_1B_1 (spread); and, of course, the true length of the line segment AB. We should also observe that in Fig. 6.5 we have the true shape of the right triangle ABC. The per cent of grade of line AB is $AC/CB \times 100 = A_FC_F/C_HB_H \times 100 = A_FC_F/C_FB_F^R \times 100$.

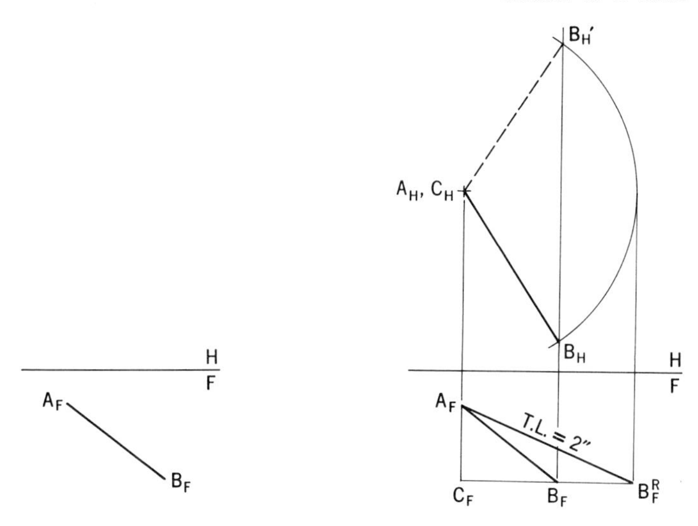

Fig. 6.6

Fig. 6.7

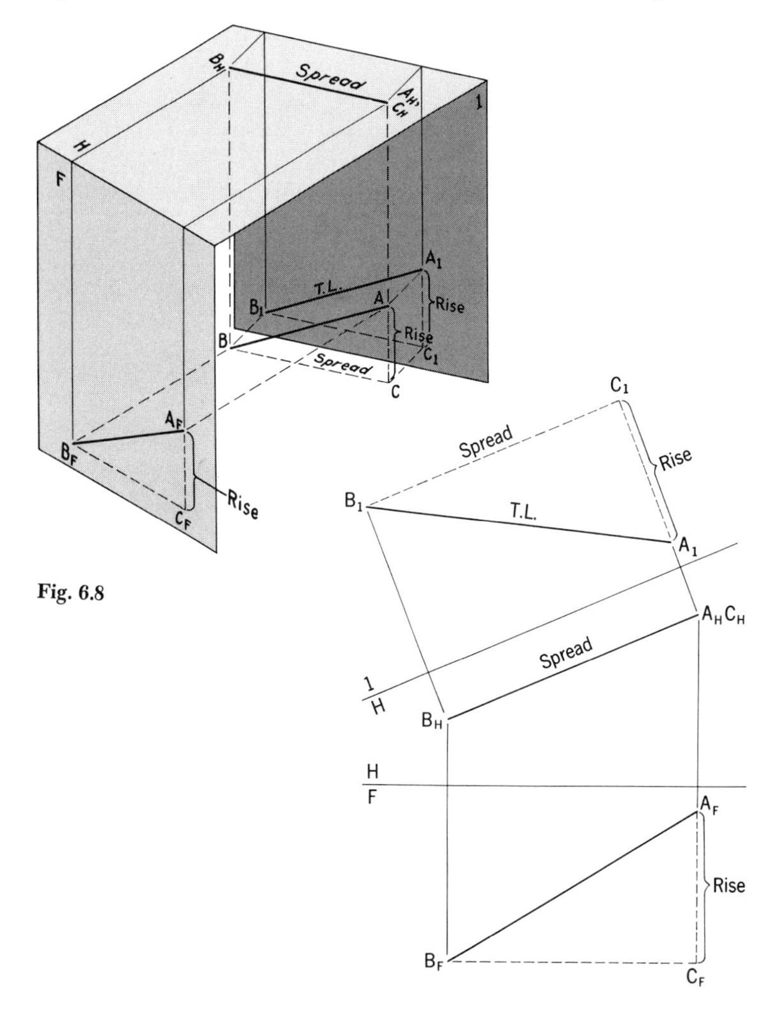

Fig. 6.8

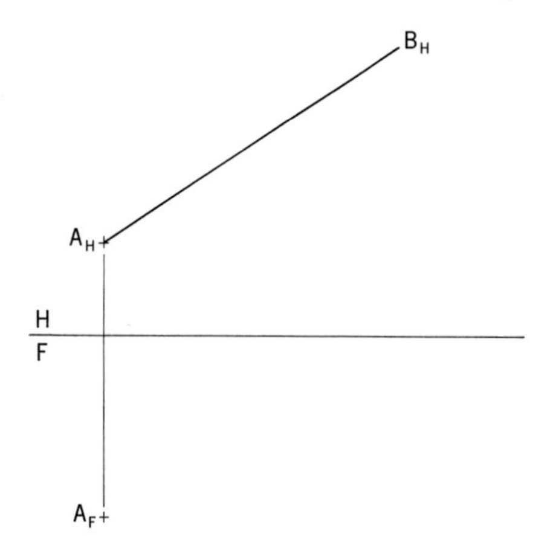

Fig. 6.9

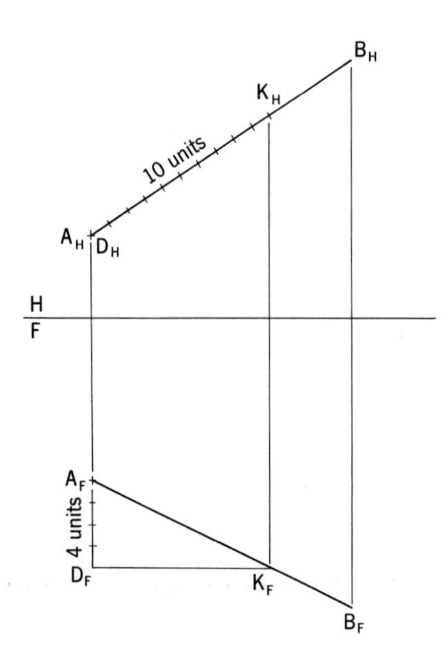

Fig. 6.10

EXAMPLE 2

Suppose we are given the top view of line segment AB, the front view of point A, and the grade of the line as -40%. (See Fig. 6.9.)

We are required to locate the front view of point B.

At the outset we do know that B_F lies on the vertical line drawn through B_H, and that B_F is below A_F, since the grade of line AB is negative. Now, how shall we locate B_F?

Let us refer back to either Fig. 6.5 or Fig. 6.8. We see that the per cent of grade of the line is $AC/CB \times 100 = A_FC_F/C_HB_H \times 100 = A_FC_F/A_HB_H \times 100$.

Now, in Fig. 6.10 we introduce a vertical line $AD = 4$ units (of a convenient length). In the top view, we lay off a distance, $D_HK_H = 10$ units (same unit of measure), where point K is on the segment line AB. Then we can establish the view, K_F; and we can show the top and front views of the right triangle AKD. The ratio AD/DK is the grade of the line. Finally, it is a simple matter to locate B_F. Since K is a point on line AB, A_FK_F is extended to intersect the vertical through B_H, to locate B_F.

Bearing of a Line

The bearing of a line is the angle, less than 90°, that its horizontal view makes with a north-south line. The bearing of a line is determined from its top view only.

EXAMPLE 1

The bearing of line *AB* is N $\theta°$ E; and the bearing of line *CD* is also N $\theta°$ E. See Figs. 6.11 and 6.12.

The bearing of line *EF* is N $\phi°$ W; of line *GK*, S $\phi°$ E. (See Figs. 6.13 and 6.14, respectively.)

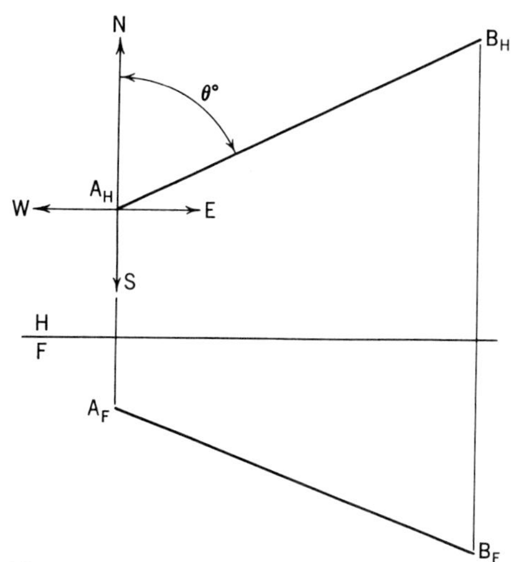

Fig. 6.11

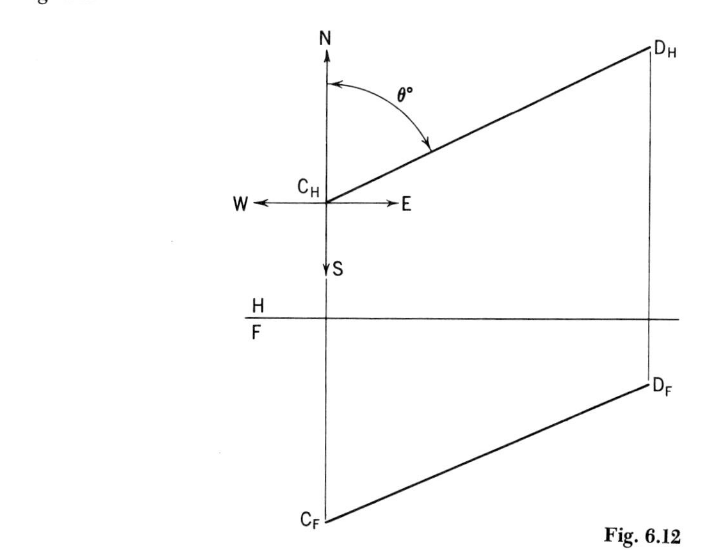

Fig. 6.12

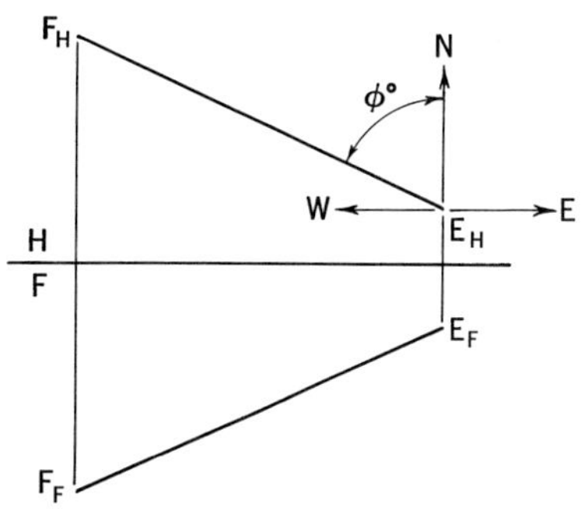

Fig. 6.13

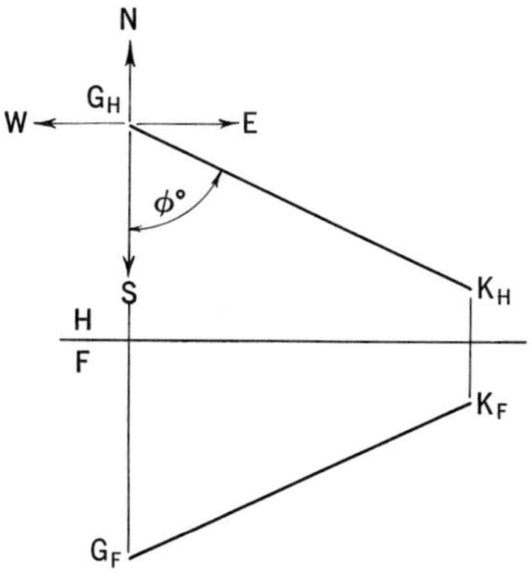

Fig. 6.14

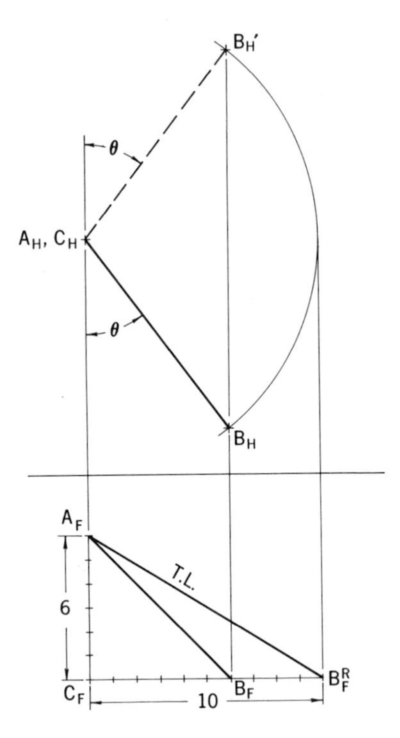

Fig. 6.15

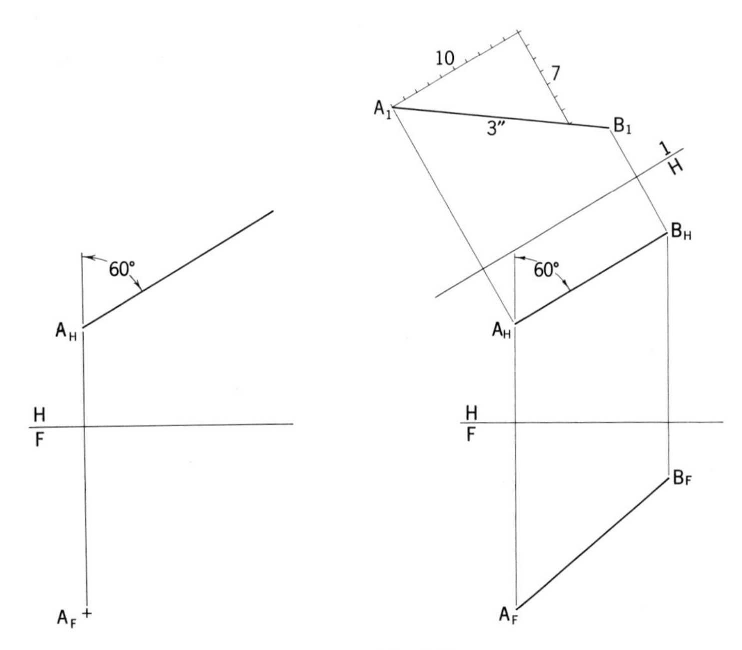

Fig. 6.16

Fig. 6.17

Summary Problems

EXAMPLE 1

The front view and grade (−60%) of line segment AB are given. (See Fig. 6.15.) It is required to find the bearing of line AB. We can readily construct the true shape of right triangle ABC. This is shown as $A_FC_FB_F{}^R$, where $A_FC_F = 6$ units and $C_FB_F{}^R =$ 10 units. The grade, we recall, is rise over spread or 6/10. Therefore, the distance from C_F to $B_F{}^R$ is the length of the spread or the magnitude of A_HB_H or A_HB_H'. There are these two solutions. The bearing is either S $\theta°$ E or N $\theta°$ E.

EXAMPLE 2

Suppose we have the following data: (*a*) Line segment AB has a bearing of N 60° E. (*b*) The grade of the line is 70%. (*c*) The true length of the line is 3 inches. (*d*) Point A is known.

It is required to establish the top and front views of line AB.

Analysis and Solution

1. The bearing of line AB is seen in the top view; therefore, we can draw the top view of line AB. (See Fig. 6.16.)

2. The view of the line on a supplementary plane which is parallel to the line and perpendicular to the H-plane will show both the grade and the true length of the line segment.

Supplementary plane, 1, is introduced parallel to the line and perpendicular to the H-plane. The view on plane 1 shows the grade and true length of line segment AB. See Fig. 6.17. Once we have determined the view A_1B_1, it is a simple matter to locate B_H and B_F. The completed solution is shown in Fig. 6.17.

EXAMPLE 3

Line segment AB has a bearing of N 50° E, a grade of −70% and a true length of $1\frac{3}{4}$ inches. Assume that point A is known. Let us proceed to locate the top and front views of the line segment *employing the method of rotation.*

1. The top view of line AB is easily established, since the bearing is given. The line drawn through A_H, with bearing N 50° E, is the top view of the *line* AB (not the *segment* AB). (See Fig. 6.18.)

2. We recall the definition of grade of a line as "the ratio of rise to spread of a segment of the line." Now, let us lay off 10 units (of convenient length) from A_H to K_H, where K is a point on line AB. The distance from A_H to K_H is the spread of line segment AK.

Now it is a simple matter to locate K_F, since the stated grade is -70%. Note that the "rise" of 7 units establishes K_F.

Line AK has the correct bearing and grade. The length, AK, however, is either longer or shorter than the specified $1\frac{3}{4}$ inches for line segment AB. By employing the rotation method, as discussed earlier, we can find the true length of segment AK. This is shown as length $A_F K_F^R$. Now we can lay off the prescribed length of segment AB on the true length line. This length is shown as $A_F B_F^R$. The locations of B_F and B_H are quite easily established and are shown in Fig. 6.18.

Before we proceed to discuss the point view of a line and associated problems, we should understand, quite clearly, the orthogonal representation of intersecting lines, skew lines, parallel lines, and perpendicular lines.

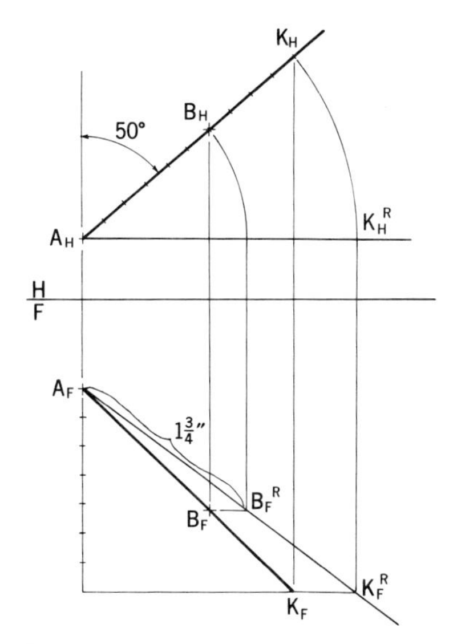

Fig. 6.18

Lines—A. Intersecting. B. Skew. C. Parallel. D. Perpendicular

A. Intersecting Lines. Intersecting lines have a point in common. In Fig. 6.19, the H and F views of lines m and n are shown. Note that their intersection, point P, is represented by the consistent views P_H and P_F.

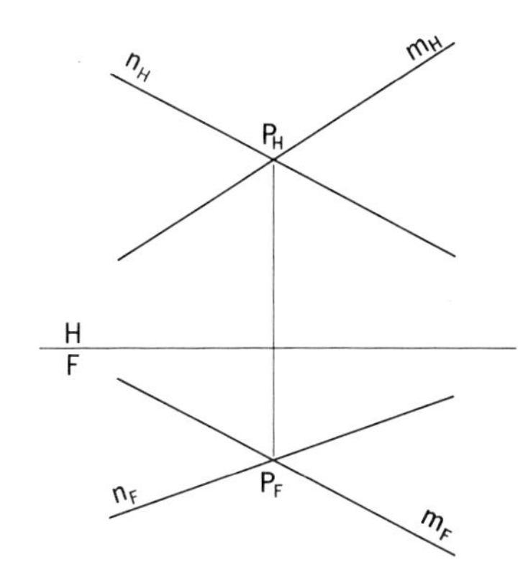

Fig. 6.19

In Fig. 6.20, lines AB and CD appear to intersect. Since line AB is a profile line (parallel to the P-plane), is is possible that the two lines do not intersect. A *profile view* of the two lines would clearly show whether or not there is an intersection. Obviously, there is no intersection, since the F and P views of the "apparent" intersection do not lie on the same horizontal line.

When two lines are general, as shown in Fig 6.19, two adjacent views are sufficient to determine whether or not there is a point of intersection. What use can we make of intersecting lines?

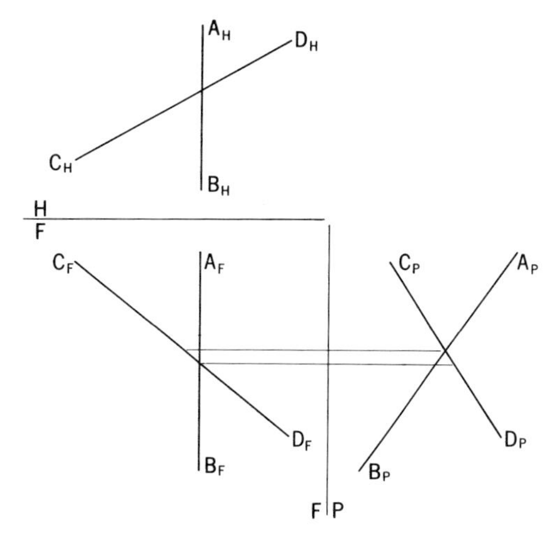

Fig. 6.20

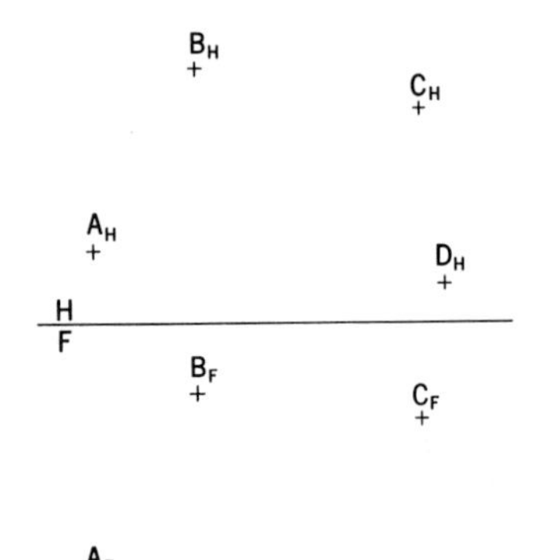

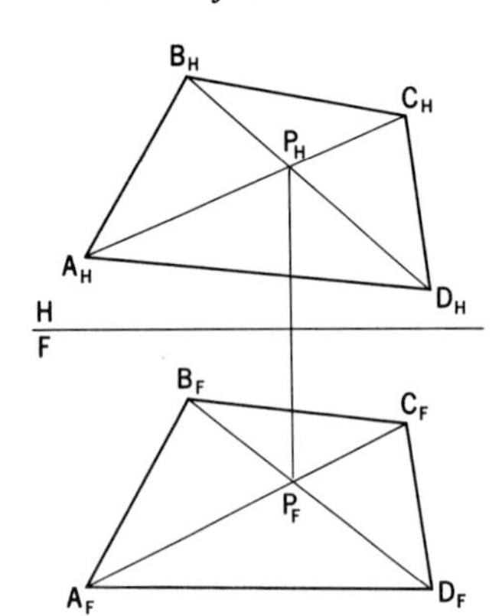

Fig. 6.21

EXAMPLE 1

Suppose we have the top view of four points and the front view of three of these points. Let us assume that the four points are in the same plane. How shall we locate the front view of the fourth point to satisfy the assumption?

In Fig. 6.21, the top view of points *A*, *B*, *C*, and *D* is shown; and the front view of points *A*, *B*, and *C*.

Analysis and Solution

1. Since the four points lie in the same plane, the diagonals *AC* and *DB* must intersect. The top view of the point of intersection, *P*, is readily located.

2. The front view of diagonal *AC* may be drawn, since the front view of points *A* and *C* is known. Now, since point *P* is the intersection of the diagonals, we can easily establish the front view of point *P* on the front view of diagonal *AC*. Points *B*, *P*, and *D* lie on one line; therefore, we can establish the line through B_F and P_F which contains D_F, the location of which is now quite obvious.

Let us consider another example of a similar problem.

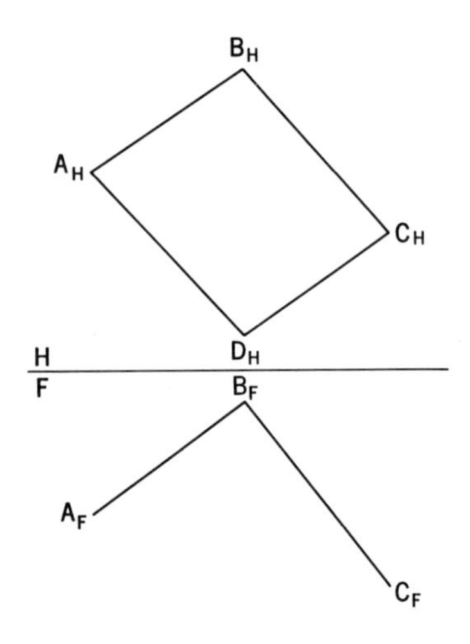

Fig. 6.22

EXAMPLE 2

Figure 6.22 shows a top view of plane *ABCD* and the corresponding front view of a portion of the plane. It is required to locate D_F and complete the front view. The method used in Example 1 is not applicable to the situation presented in Fig. 6.22.

Analysis and Solution

1. It is quite evident that we cannot use the point of intersection of the diagonals because the location of D_F is not unique.

2. We can, however, introduce a new line in the plane, such as *DK*. Once the top view of *DK* is drawn, its intersection, P_H, with diagonal *AC* is readily located. Next P_F and K_F are located. Line *PK* contains point *D*. Therefore, the intersection of line P_FK_F with the vertical passing through D_H uniquely locates D_F. The front view of plane *ABCD* is readily completed.

Let us consider Fig. 6.23. It is assumed that the top view of point K is known and that point K is in plane ABC. Our problem is to determine the front view of point K.

Analysis and Solution

1. Since point K is in plane ABC, there are many lines in plane ABC that pass through point K. One such line is AD.
2. The top view of line AD is easily established by connecting A_H and K_H and locating D_H which is the intersection of line segment AK (extended) and B_HC_H.
3. Since lines AD and BC intersect at point D, we can easily locate D_F and then establish the front view of line AD.
4. Now K_F can be located on A_FD_F, since we know that *"when a point is on a line, the views of that point will lie on the corresponding views of the line."*

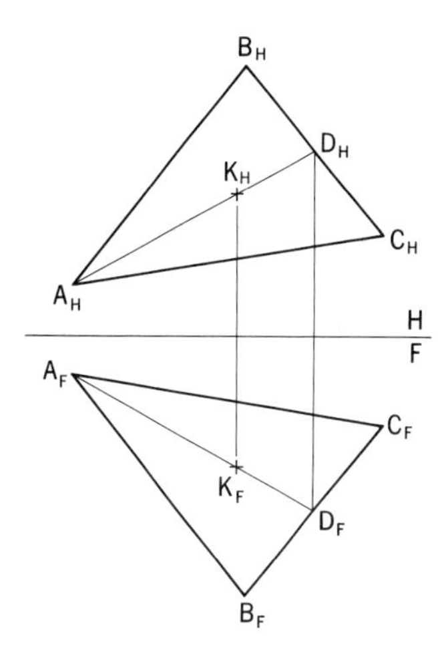

Fig. 6.23

B. Skew Lines. Skew lines do *not* have a common point. Let us consider Fig. 6.24 which shows the H and F views of lines AB and CD. At first glance we might conclude that the lines do intersect. However, careful study will show that the lines actually do *not* intersect. Note that the *apparent* intersection of the line A_HB_H with C_HD_H is the H view of two points, one on line AB and the other on CD. If we label these points E and F, respectively, we can locate both the H and F views. Now, since point F is above point E, we can see that line CD, which contains point F, is above line AB.

In a similar manner the apparent intersection of A_FB_F with C_FD_F is the F view of two points, one on line AB and the other on CD. If we label these points G and K, respectively, we can locate their H and F views. Now, since point G is in front of point K, line AB, which contains point K, is in front of line CD.

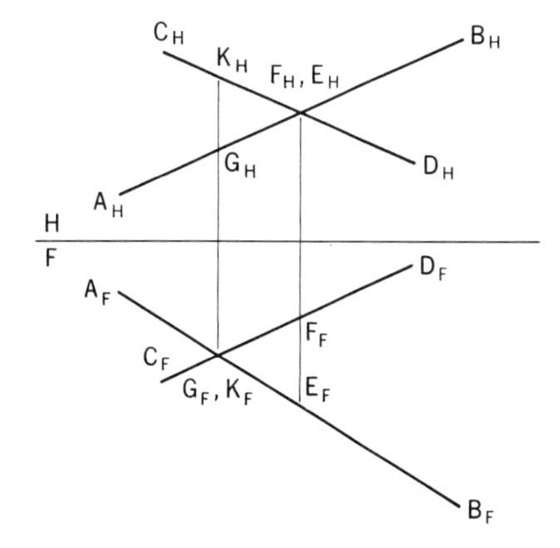

Fig. 6.24

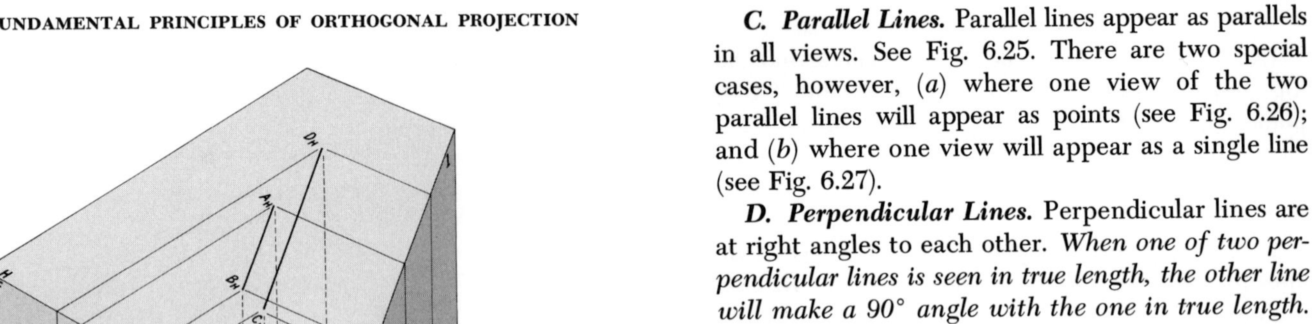

C. *Parallel Lines.* Parallel lines appear as parallels in all views. See Fig. 6.25. There are two special cases, however, (*a*) where one view of the two parallel lines will appear as points (see Fig. 6.26); and (*b*) where one view will appear as a single line (see Fig. 6.27).

D. *Perpendicular Lines.* Perpendicular lines are at right angles to each other. *When one of two perpendicular lines is seen in true length, the other line will make a 90° angle with the one in true length.*

Fig. 6.25(*a*)

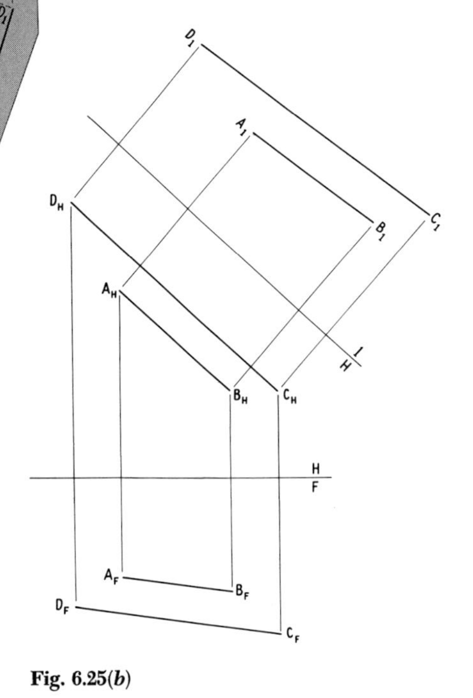

Fig. 6.25(*b*)

Fig. 6.26

Fig. 6.27

EXAMPLE 1

Let us consider Figs. 6.28 and 6.29 which show lines DC and DE perpendicular to line AB. Since line AB is parallel to the F-plane, the true length of AB is seen in the front view as A_FB_F; and the front view of lines DC and DE forms a 90° angle with A_FB_F.

EXAMPLE 2

In Fig 6.30, we observe that both lines DC and DE are perpendicular to line AB. This is evident from the fact that the top view, A_HB_H, of line AB is true length and the angle between C_HD_H and A_HB_H is 90°; and the angle between E_HD_H and A_HB_H is also 90°.

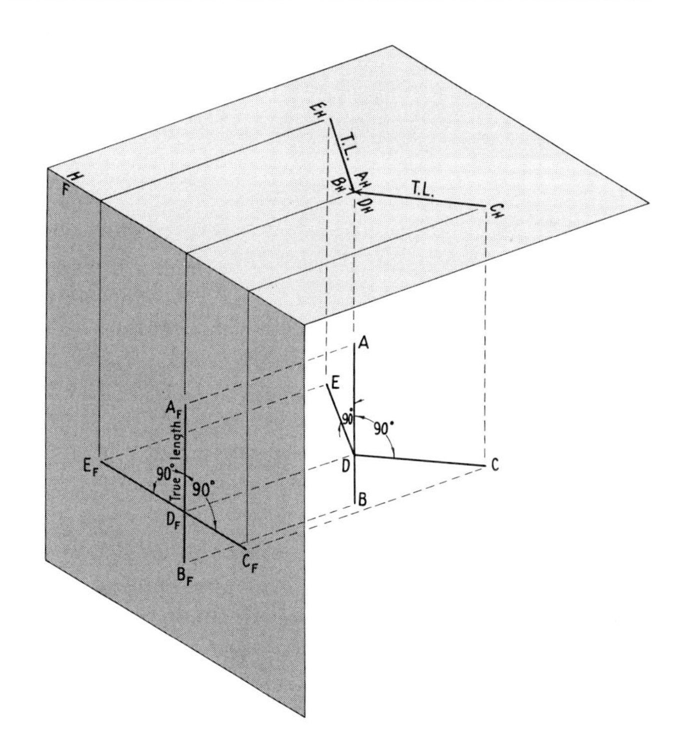

Fig. 6.28

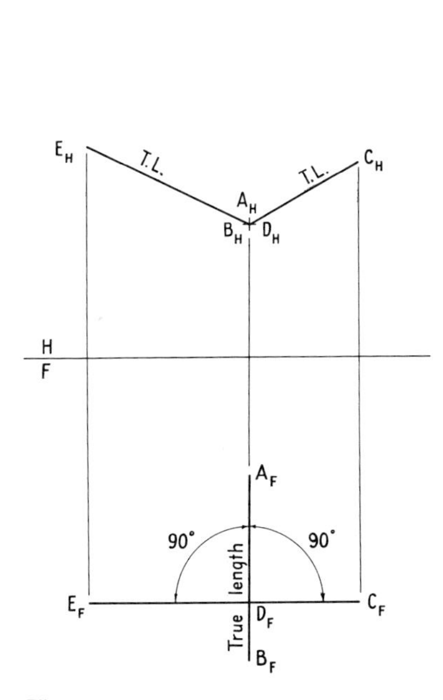

Fig. 6.29

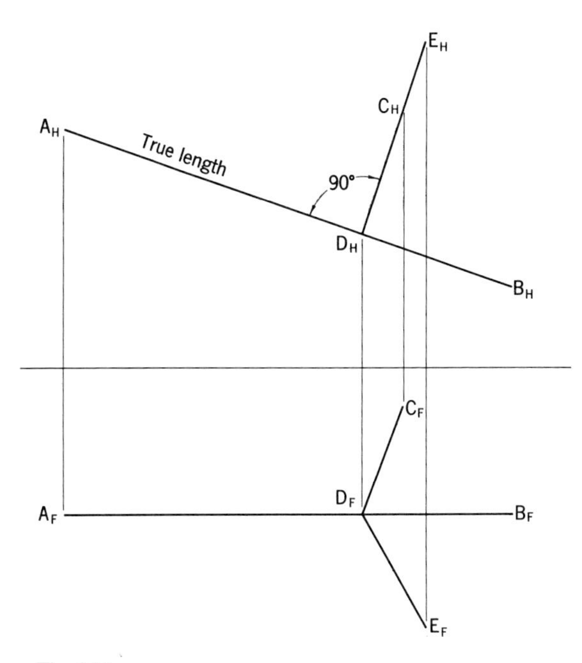

Fig. 6.30

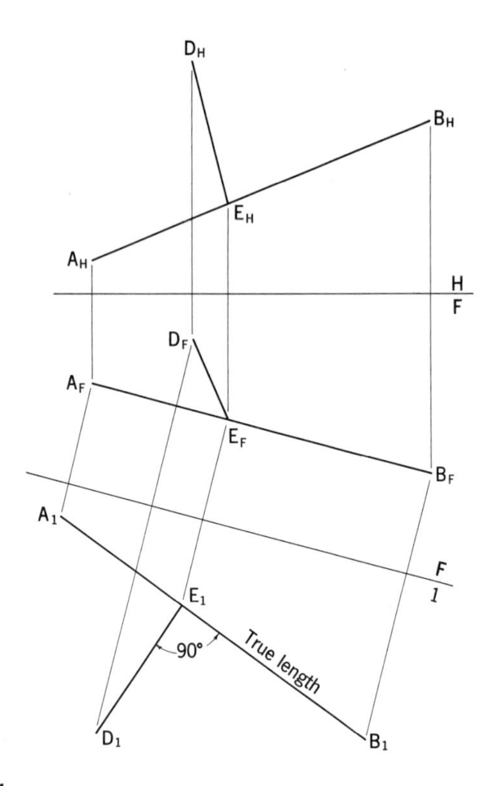

Fig. 6.31

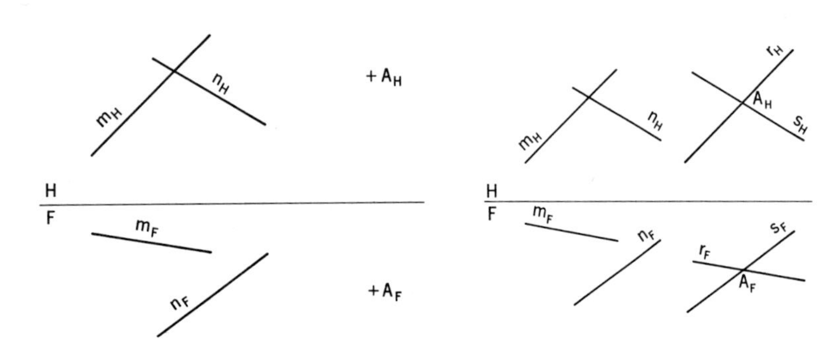

Fig. 6.32

Fig. 6.32(a)

EXAMPLE 3

Suppose we are given line AB and point D (Fig. 6.31) and that we wish to determine the top and front views of line DE which is perpendicular to line AB. It is also assumed that point E is on line AB. It is quite evident that neither the top view nor the front view of line AB shows the true length of line AB. We can determine the true length of line AB, however, by introducing a supplementary plane, 1, parallel to line AB and perpendicular to the F-plane. The view on plane 1 shows the true length of line AB and also the corresponding view of point D (shown as D_1). Now we can draw, through D_1, a line perpendicular to A_1B_1. The point of intersection of this perpendicular with A_1B_1 establishes E_1. Now it is a simple matter to locate E_H and E_F, and the corresponding views of the perpendicular DE.

Problems Dealing with Parallels and Perpendiculars

Let us now consider several problems that can be solved by employing parallels and perpendiculars.

EXAMPLE 1

Suppose it is required to establish a plane through a given point, A, and parallel to two skew lines, m and n. Figure 6.32 is a graphical representation of the problem. Let us now analyze the problem and its solution.

Analysis and Solution

1. A plane may be determined by (a) two intersecting lines; (b) two parallel lines; (c) a point and a line which does not contain the point; or (d) three points that are not on one line.

2. *When a line is parallel to a plane, the line will be parallel to a line of the plane.* We can now, proceed to (a) establish line, r, through point A and parallel to line m; and (b) another line, s, also through point A but parallel to line n. We now have the required plane, determined by intersecting lines r and s, passing through point A and parallel to both skew lines m and n. The solution is shown in Fig. 6.32(a).

EXAMPLE 2

Let us consider the following problem: Establish the H and F views of line RS which fulfills these conditions:

(*a*) The bearing of line RS is N 60° E.
(*b*) Line RS is parallel to plane ABC.
(*c*) Line RS is 3 inches long.

The information that is available is shown in Fig. 6.33. The solution is shown in Fig. 6.34.

Analysis and Solution

1. To satisfy condition (*a*), we can easily establish the bearing by drawing a line m_H through R_H at N 60° E.

2. To satisfy condition (*b*), we know that when a line is parallel to a plane the line (m in this case) is parallel to a line of the plane; therefore, we can readily establish a line, n, in plane ABC and parallel to line m. The top view of line n is drawn through A_H and parallel to m_H (remember, parallel lines are seen as parallels in the respective views). Now, since line n lies in plane ABC, we can establish n_F.

3. Finally, to satisfy condition (*c*), point S is located by first finding the true length of a portion of line m (such as R-1) and then by laying off the specified length 3 in. You will recall that we solved a similar problem in our discussion of true length of a line segment.

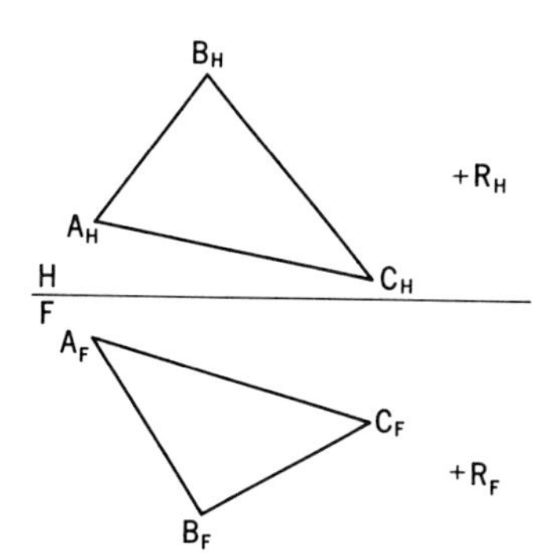

Fig. 6.33

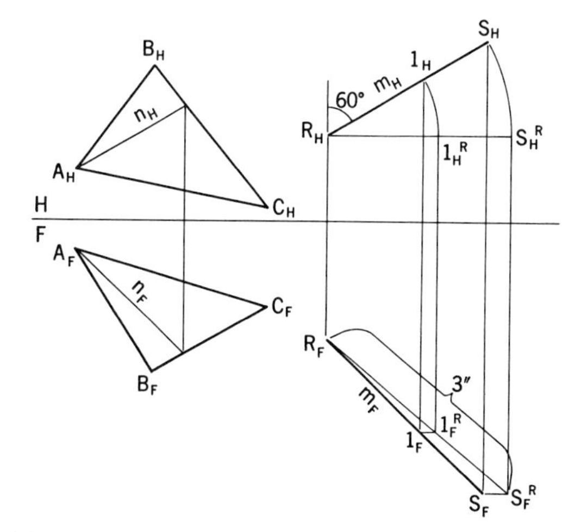

Fig. 6.34

EXAMPLE 3

Suppose we are given the H and F views of angle ABC as shown in Fig. 6.35. Is the angle 90°? How can we determine the answer?

Analysis

One approach could be this: Find the true length of each side of the triangle and then construct a triangle with the true lengths previously found; and finally measure angle ABC (we could virtually determine by visual inspection whether or not angle ABC is 90°). While the above approach yields an answer to the question, it is rather a time-consuming method (hence, costly).

A much simpler solution requires only the recognition of the fact that, "*when two lines are perpendicular to each other, the view of the two lines upon a plane which is parallel to one of them will reveal a right angle.*"

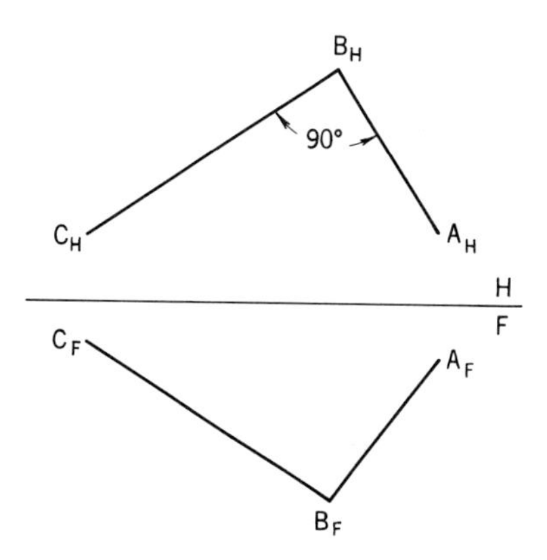

Fig. 6.35

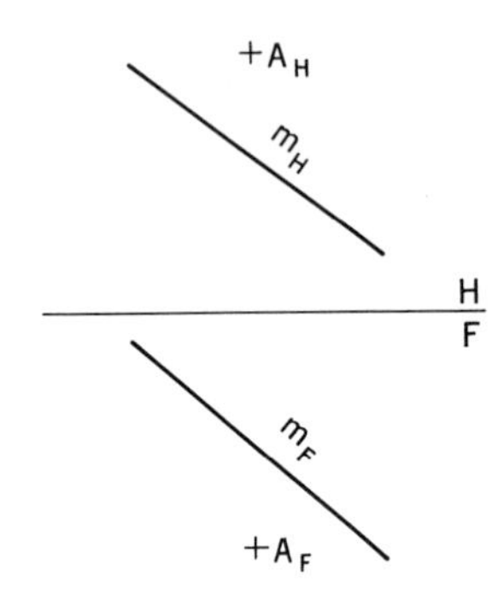

Fig. 6.36

Therefore, we may conclude at once that the angle ABC is *not* 90° since the view marked 90° does not show the true length of either line AB or BC.

Question. If the top view, $A_HB_HC_H$ and the front view of BC remain as shown in Fig. 6.35 where would you place A_F so that the angle $ABC = 90°$?

EXAMPLE 4

Let us try a more sophisticated problem. Suppose we are given line m and point A as shown in Fig. 6.36. We are required to fulfill the following specifications:

(*a*) Establish the plane through point A perpendicular to line m.

(*b*) Represent this plane by two lines: one, AB, which shall be parallel to the H-plane; and the other, AC, which shall be parallel to the P-plane.

(*c*) The lengths of lines AB and AC are arbitrary.

Analysis

1. When we obtain a view of line m showing its true length, we can also see the edge view of any plane that is perpendicular to the line. Therefore, we can easily select the one plane that is both perpendicular to line m and contains point A, since we have the view of point A.

2. The edge view of the plane is also the corresponding view of all the lines that lie in the plane.

Now, our problem is to select the two specified lines AB and AC. Let us concentrate on line AB.

3. Line AB must be parallel to the H-plane—as specified. We should recall that "the front view of a horizontal line is horizontal." Therefore, we can establish a horizontal line through A_F and arbitrarily select B_F on that horizontal. Thus far we have the front view of AB. How shall we obtain the top of point B?

4. Line AB is in the required plane; therefore, we can readily locate B_1. Now, recalling the fundamental principle: "When two planes are perpendicular to a third plane, the distance that a point is from the third plane will be seen twice, once in each of the views upon the other two planes," we can easily locate B_H.

5. To locate line AC which is parallel to the P-plane and in the specified plane, we recall that both the top and front views (H and F) of a profile line

appear as vertical lines. Therefore, we establish the front view of AC by drawing a vertical line through A_F and arbitrarily selecting C_F on this vertical. Once we have chosen C_F it is a simple matter to locate C_1 and finally C_H. The H and F views of the specified lines AB and AC now can be drawn—and our problem is solved. See Fig. 6.37.

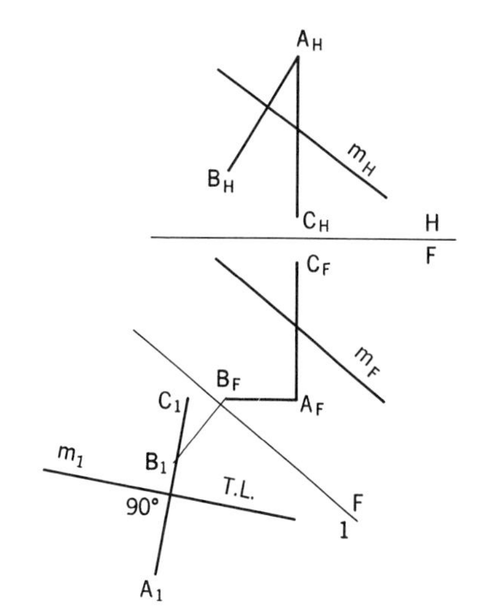

Fig. 6.37

2. POINT VIEW OF A LINE—SECOND BASIC PROBLEM

Once we learn how to obtain the point view of a line, we will then have a powerful graphical method for analyzing and solving such space problems as the following:

1. The distance between two skew lines; i.e., the clearance between two cables, between two steel rods, or between two pipes, etc.
2. The edge view of a plane.
3. The angle between plane surfaces.
4. The distance between parallel lines.
5. The perpendicular distance from a point to a line.

Several practical applications are included in the exercises and in the workbook.

Let us consider Fig. 6.38 which shows line segment AB and its views upon the horizontal and frontal reference planes H and F, respectively. It is quite evident that the line segment AB is parallel to the H-plane and, therefore, its true length is seen in the view, A_HB_H.

Now let us introduce supplementary plane 1, perpendicular to line AB. Plane 1 is, therefore, perpendicular to the H-plane. The view of line AB upon plane 1 is a point, shown as A_1B_1.

If the given line were not parallel to either the H- or F-plane, it would be impossible to introduce a supplementary plane that would be perpendicular to the given line and also to either the H- or F-plane.

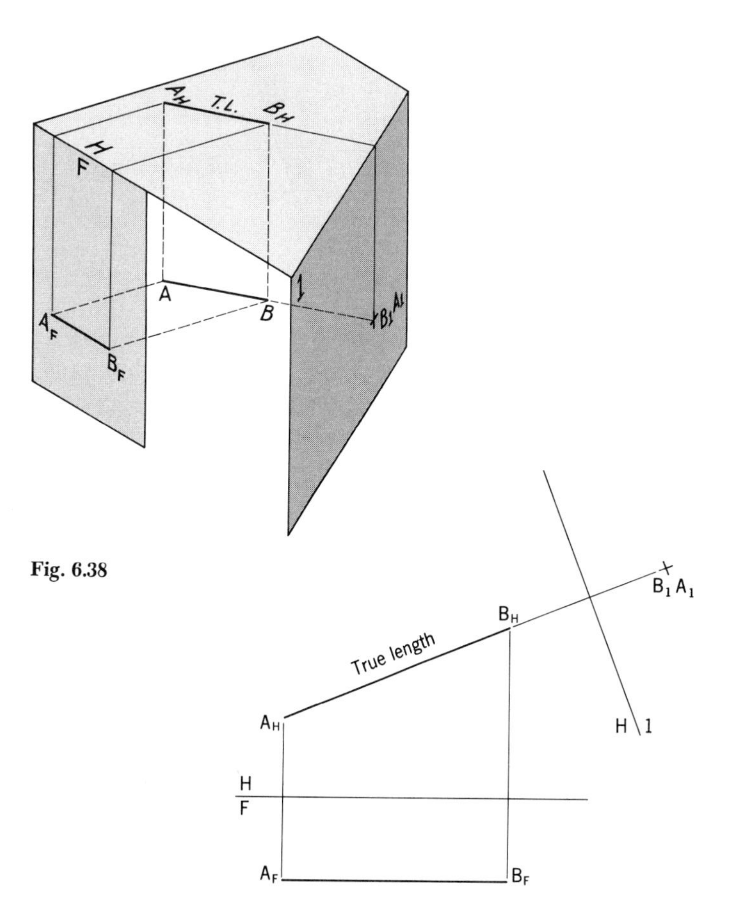

Fig. 6.38

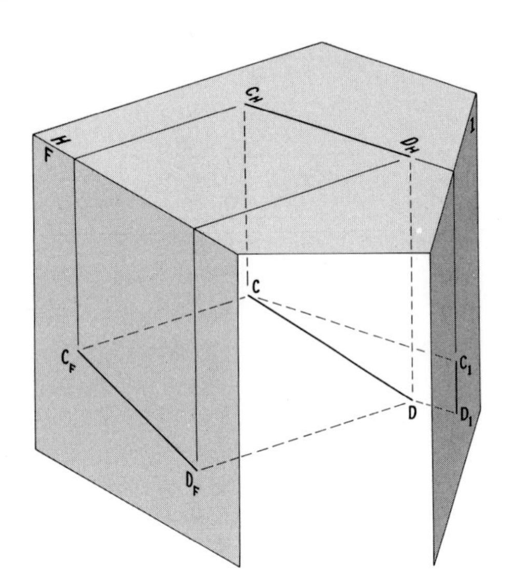

Fig. 6.39

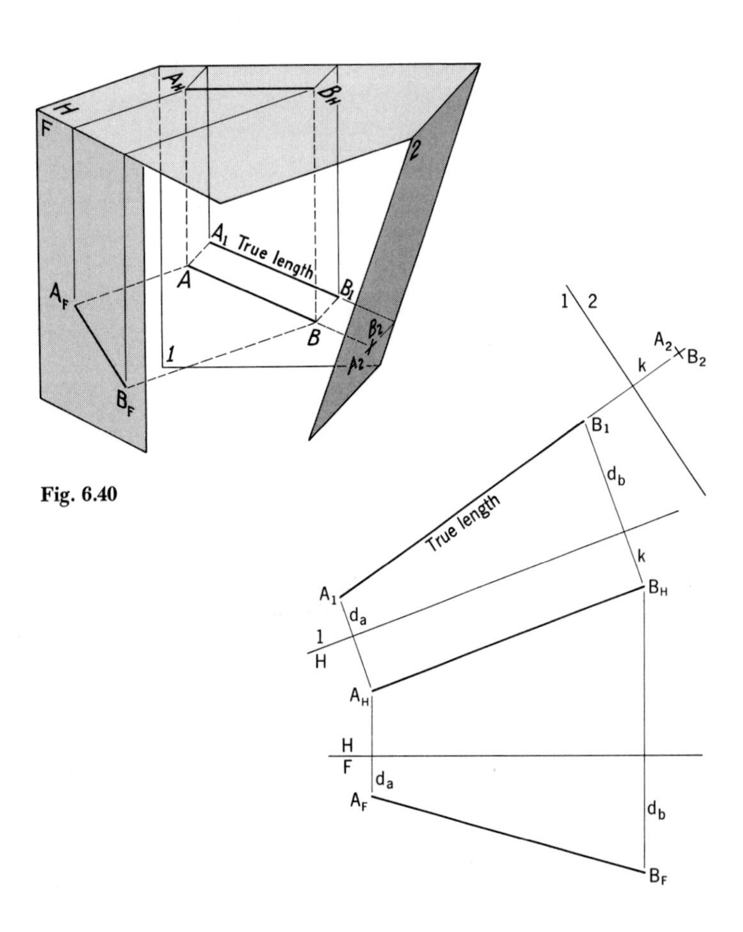

Fig. 6.40

Let us see what would happen if we introduced a supplementary plane perpendicular to the H-plane and *the top view* of the line. (See Fig. 6.39.)

It is quite evident that the view of line CD upon plane 1 is *not* a point. This shows that plane 1 is *not* perpendicular to the line.

We recall that adjacent planes must be at right angles to each other; i.e., H and F are adjacent planes; and H and 1 are also adjacent planes.

How shall we, then, solve our problem? We observed in the first case, Fig. 6.38, that it was a simple matter to introduce plane 1 perpendicular to both the line and the H-plane since line AB was parallel to the H-plane and the true length of line segment AB was seen in the H-view. Therefore, if we can reduce the general problem to the simple case we will encounter no difficulty in formulating the solution. *Actually this means nothing more than first finding the true length view of the line segment and then determining the point view of the line.*

EXAMPLE

Let us consider the line segment AB shown in Fig. 6.40. A true-length view of AB can be obtained, quite readily, by introducing supplementary plane 1 parallel to the line and perpendicular to the H-plane. The view, A_1B_1 on plane 1 shows the true length of line segment AB.

Now, if we consider the views A_HB_H *and* A_1B_1 *as the two given views* (forgetting the existence of the F view), we can readily place supplementary plane 2 perpendicular to both line AB and plane 1. The view, A_2B_2, on plane 2 is the point view of line AB.

To strengthen our understanding of the two fundamental principles, note carefully that:

1. Planes H and F are adjacent planes.
2. Planes H and 1 are adjacent planes.
3. Planes F and 1 are perpendicular to the H-plane. Therefore, *the distances that points A and B are below the H-plane are seen twice, once in the F view and again in the supplementary view 1.*
4. Planes H and 1 are adjacent views.
5. Planes 1 and 2 are adjacent views.
6. Planes H and 2 are perpendicular to the supplementary plane 1. Therefore, *the distances that points A and B are from plane 1 are seen twice, once in the H-view and again in the supplementary view 2.*

As a good exercise in stimulating your thinking and developing your understanding of the fundamental principles it is suggested that you solve the same problem by starting with a supplementary plane that is parallel to line *AB* and perpendicular to the *F*-plane.

3. EDGE VIEW OF A PLANE SURFACE—THIRD BASIC PROBLEM

The second basic problem—"point view of a line"—enables us to obtain the edge view of a plane surface. This is fairly evident from the following: "When a line is perpendicular to a plane, all plane surfaces that contain the line will be perpendicular to the same plane."

For example, suppose line *m* is perpendicular to the *H*-reference plane; then, all plane surfaces that contain line *m* will be perpendicular to the *H*-plane.

If we have a plane surface, say, triangle *ABC*, then when we obtain the point view of a side of the triangle we will also have the edge view of the triangle.

Once we learn how to find the edge view of a plane surface we will know how to find (*a*) the distance between parallel planes, and (*b*) the perpendicular distance from a point to a plane. Now let us solve an elementary problem—to find an edge view of plane *ABC* shown in Fig. 6.41. When we obtain the point view of one of the sides of the triangle, we will have an edge view of the surface *ABC*. Let us choose side *AB* (either of the other two sides could have been selected). First, we must obtain a true length view of side *AB*. The true length view is seen on supplementary plane 1, which is parallel to side *AB* and perpendicular to the *F*-plane. The view on plane 2, which is perpendicular to both plane 1 and side *AB*, shows the point view of side *AB* and, therefore, the edge view of plane *ABC*.

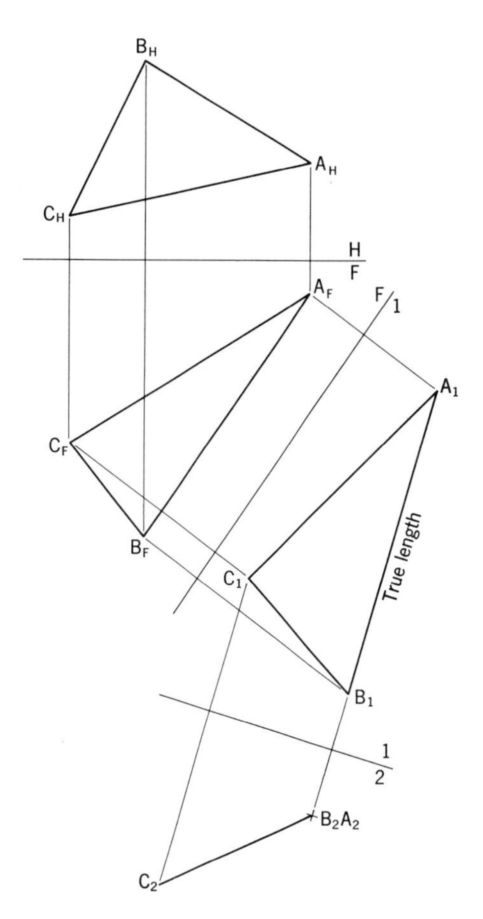

Fig. 6.41

Alternative Solution

We can eliminate the need for the first supplementary plane 1 by establishing a true length line in the surface ABC. Suppose we select a line through point A and parallel to the F-plane. (See Fig. 6.42.) This is the line AD which we should recognize as one of the principal lines; in this case, a *frontal line*. The true length of line AD, therefore, is seen in the front view, shown as $A_F D_F$. Now we can introduce supplementary plane 1 perpendicular to both line AD and the F-reference plane. The view on plane 1 shows line AD as a point, $A_1 D_1$, and the plane surface as a line which represents the edge view of surface ABC; shown as $A_1 B_1 C_1$. A horizontal line could be used. This is shown in Figs. 6.43(a) and 6.43(b).

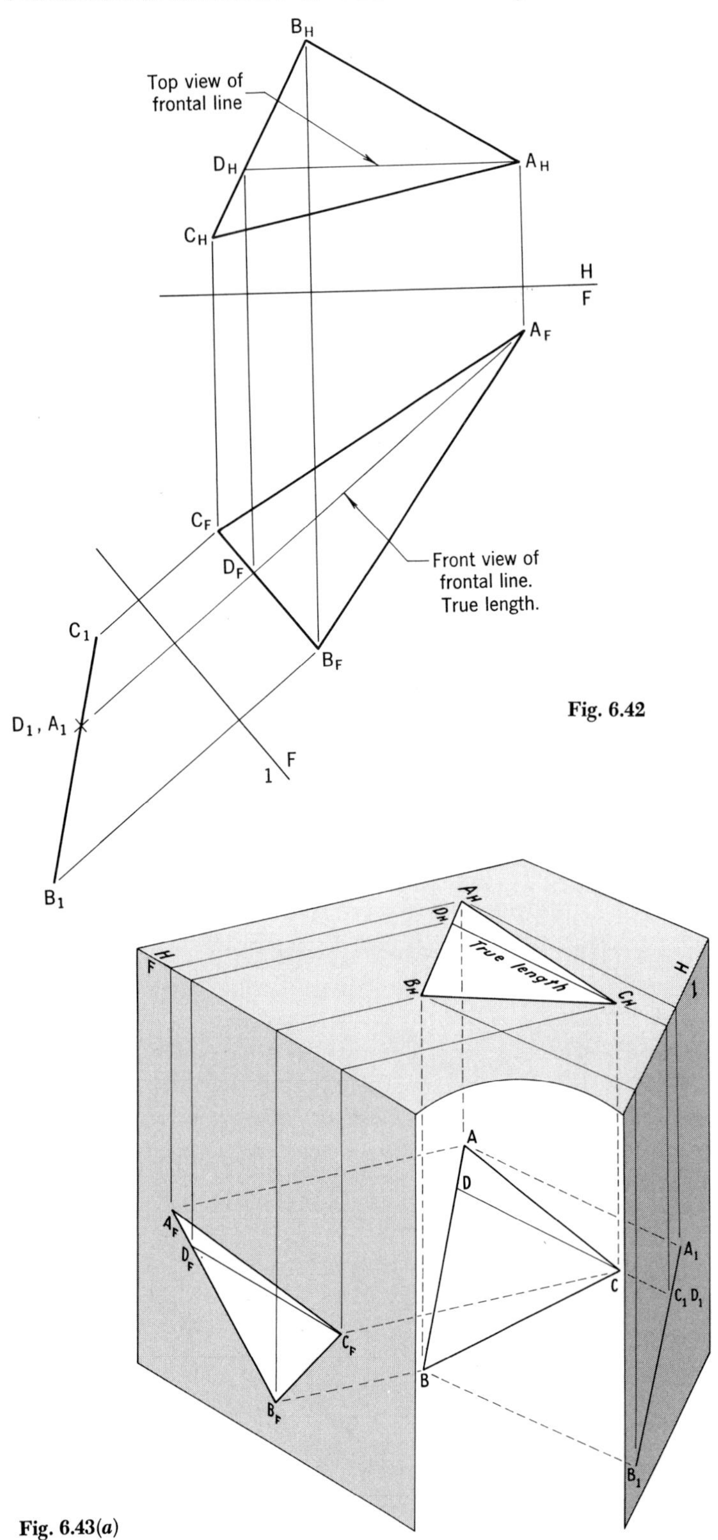

Fig. 6.42

Fig. 6.43(a)

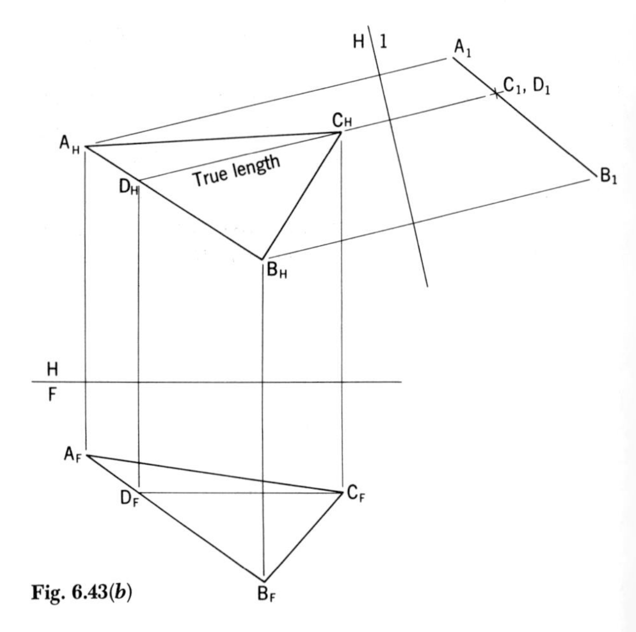

Fig. 6.43(b)

EXAMPLE 1

Now let us consider the problem—"to determine the perpendicular distance between parallel surfaces *ABC* and *DEF*," shown in Fig. 6.44.

Analysis

1. The point view of a line in either surface *ABC* or *DEF* will include the edge views of the parallel surfaces.
2. The distance between the parallel planes can be measured in the same view.

Graphic Solution

Frontal line *AK*, drawn in surface *ABC*, is seen in its true length as A_FK_F. The point view of line *AK* is seen in the view on supplementary plane 1, which is perpendicular to both line *AK* and the *F*-plane. The edge view of surface *ABC* is seen as $A_1B_1C_1$ and the edge view of surface *DEF* as $D_1E_1F_1$. The edge view of both surfaces must appear as parallel lines since the surfaces are parallel.

The perpendicular distance, *d*, between the edge view of the surfaces is the solution to the stated problem.

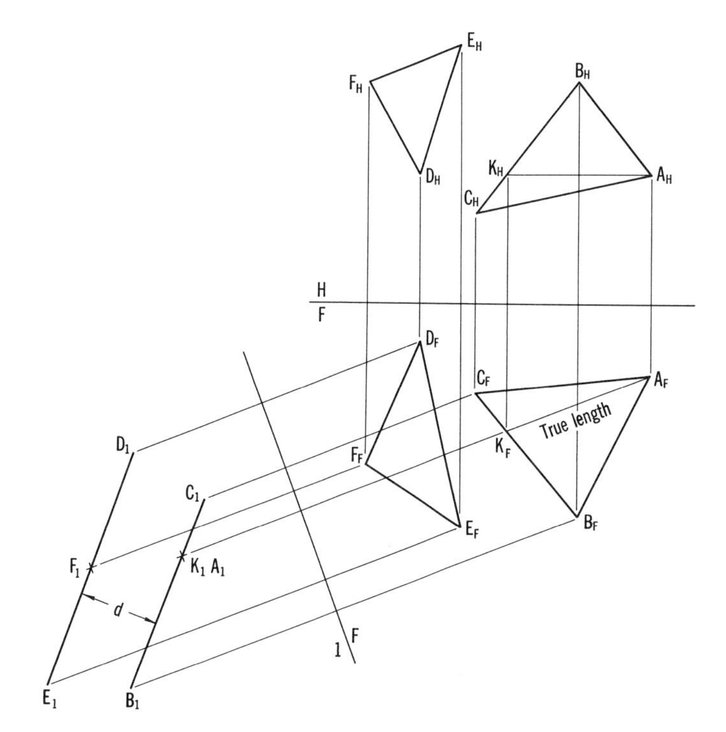

Fig. 6.44

EXAMPLE 2

Suppose we wish to find the perpendicular distance from point *P to the plane surface defined by points A, B, and C;* and also the views of the perpendicular. (See Fig. 6.45.)

Analysis

The perpendicular distance from a point to a plane will be seen in the view which shows the plane surface as a line and also the corresponding view of the point.

Graphic Solution

The triangle formed by joining points *A*, *B*, and *C* is *only a portion* of the plane surface defined by these points; however, we will use the triangle portion to determine the edge view of the plane surface. We observe that line *AB* is parallel to the *H*-plane; therefore, the top view of line *AB* is true length. The point view of line *AB* is seen on supplementary plane 1 which is perpendicular to both line *AB* and the *H*-plane. We also see the edge view of surface *ABC* and the corresponding view of point *P*.

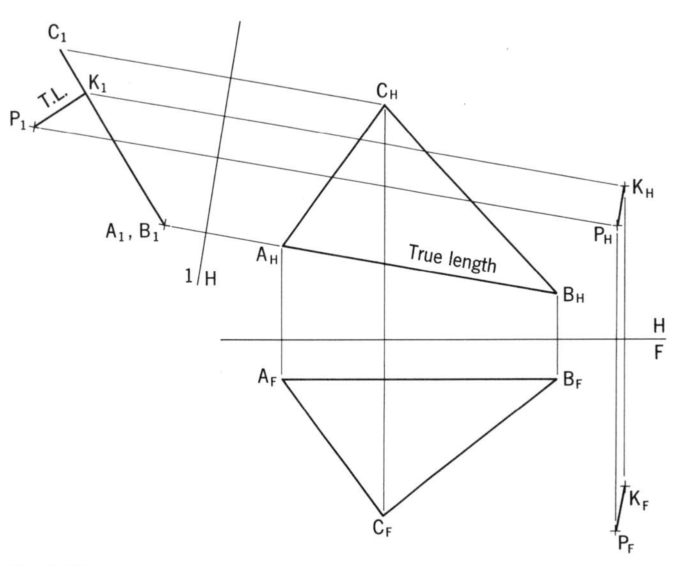

Fig. 6.45

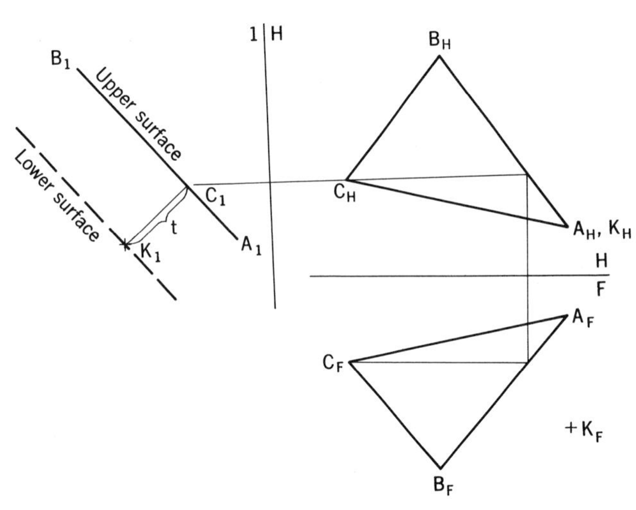

Fig. 6.46

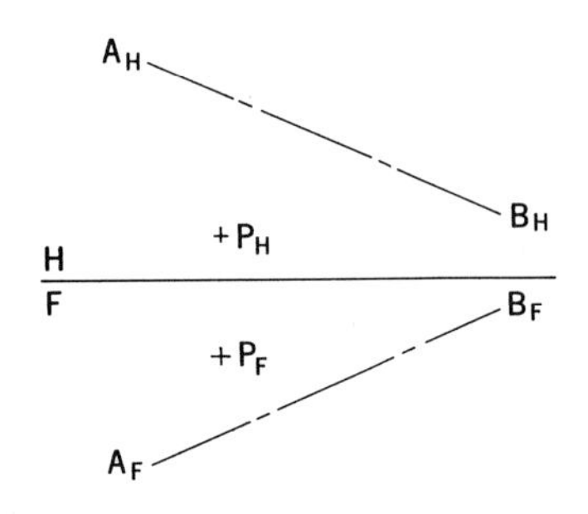

Fig. 6.47

The perpendicular distance from P_1 to $A_1B_1C_1$ is the distance from point P to plane surface ABC. Now let us consider the second part of the problem—"locate the H and F views of the perpendicular."

In supplementary plane 1 we can easily draw the perpendicular; shown as P_1K_1. Now we must determine the location of K_H and K_F. We know from our study of the fundamental principles of orthogonal projection that K_1 and K_H lie on a line which is perpendicular to the intersection of planes H and 1. This means that we can establish the line on which K_H will lie (since we know the location of K_1). Where on this line is K_H? Since P_1K_1 is true length, it must be parallel to plane 1; therefore, P_HK_H must also be parallel to plane 1 in order to show that points P and K (and all points on line PK) are the same distance from plane 1. Once K_H has been located, it is a simple matter to locate K_F.

We should not be disturbed by the solution shown because the perpendicular PK seemingly falls "outside" the triangle ABC. Remember that the points A, B, and C define the plane surface which is unlimited in extent, and that the triangle ABC is only a portion of the plane surface.

EXAMPLE 3

Suppose it is required to determine the thickness of a rock formation which is defined by points A, B, and C on the upper plane surface of the formation, and point K on the lower parallel surface. See Fig. 6.46. (It is presumed that a drilling operation from the surface of the ground to the formation had been performed to locate the points.) We recall from our previous work that the perpendicular distance from a point to a plane surface (the same as distance between parallel surfaces, if the point is in one of the parallel surfaces) will be seen in that view which shows the surface on edge. In Fig. 6.46, view 1 shows the edge view of surface ABC and also K_1 of point K. The perpendicular distance, t, from K_1 to the edge view of surface ABC shown as $A_1B_1C_1$ is a measure of the thickness of the rock formation.

EXAMPLE 4

Line AB is the axis of a right circular cylinder. Point P is on the surface of the cylinder. What is the diameter of the cylinder? (See Fig. 6.47.)

Analysis

1. Let us visualize any circular cylinder in space and a point on its surface. We can readily conclude that the perpendicular distance from the point to the axis of the cylinder is the radius of the circle whose diameter is the diameter of the cylinder.

2. Our problem now is reduced to "finding the perpendicular distance from a point to a line." This distance, r, is seen in the supplementary view that shows the line as a point. Twice this distance, r, is the diameter of the cylinder. (See Fig. 6.48 for the complete solution.)

Shortest Distance Between Two Skew Lines

There are many instances in engineering design where it is necessary to determine the clearance between cables, between structural members, between pipe lines, etc.; or where it is essential to maintain specified distances between skew (nonparallel) members.

The determination of the shortest distance between two skew members is based on the simple concept —"point view of a line." When we obtain the point view of one of the skew lines (which could be taken as the axis of the member) we will be in position to see and measure the shortest (perpendicular) distance between the two skew lines.

EXAMPLE 1

Let us consider the skew lines AB and CD shown by the H and F views in Fig. 6.49. If we obtain a point view of either line, the perpendicular distance between the two lines will be apparent. Let us find the point view of line AB.

We know from our previous study of the basic problem, "to find the point view of a line," that it is first necessary to find the true length of the line. Therefore, we will introduce supplementary plane 1 parallel to line AB and perpendicular to the H-plane. The new view shows line AB in its true length and a foreshortened length of CD. Now the second supplementary plane 2 is introduced, perpendicular to both line AB and plane 1. The final view shows line AB as a point and line CD foreshortened. The shortest (perpendicular) distance between the skew lines can now be measured (shown as E_2F_2).

The other views of the common perpendicular,

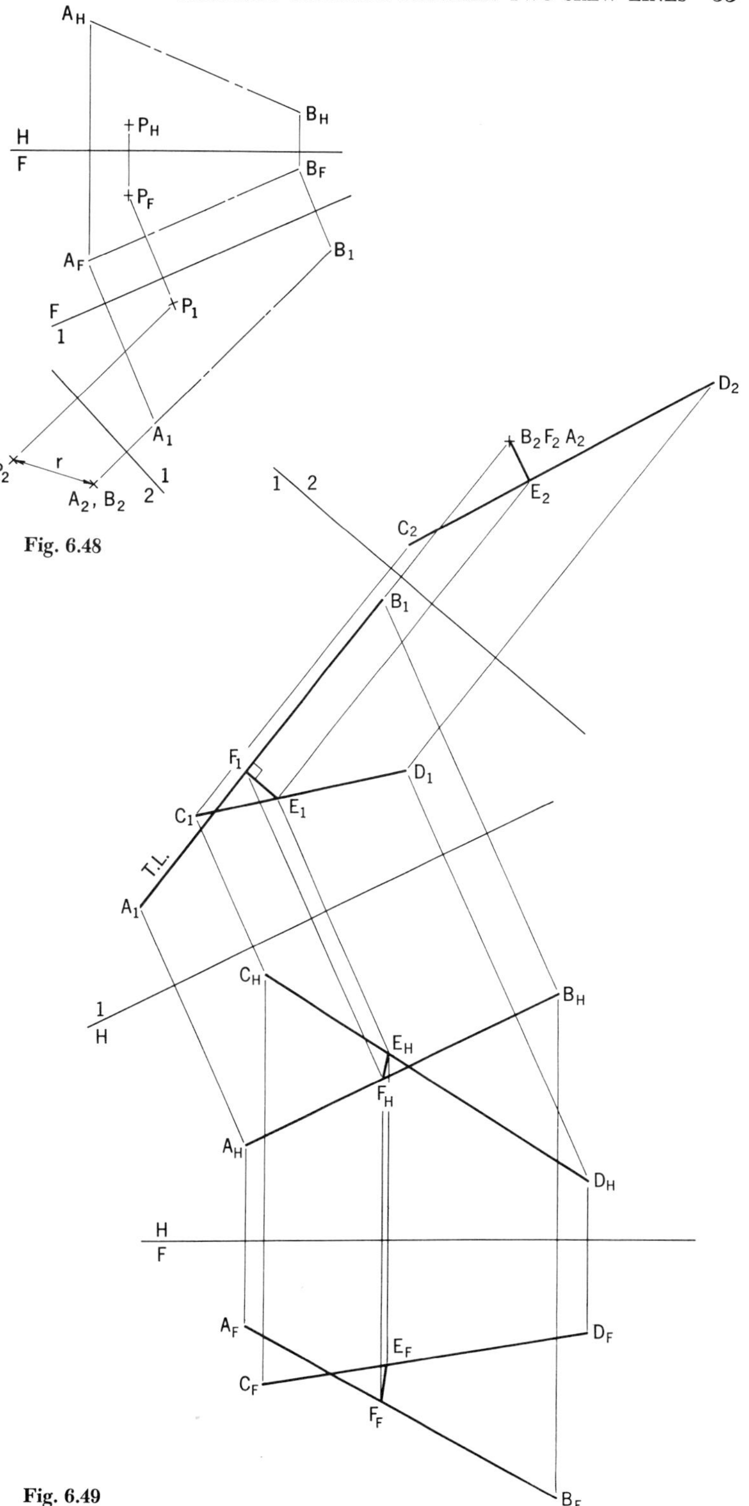

Fig. 6.48

Fig. 6.49

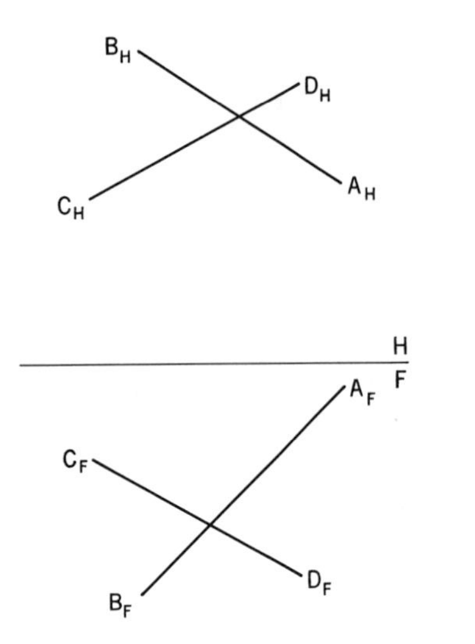

Fig. 6.50

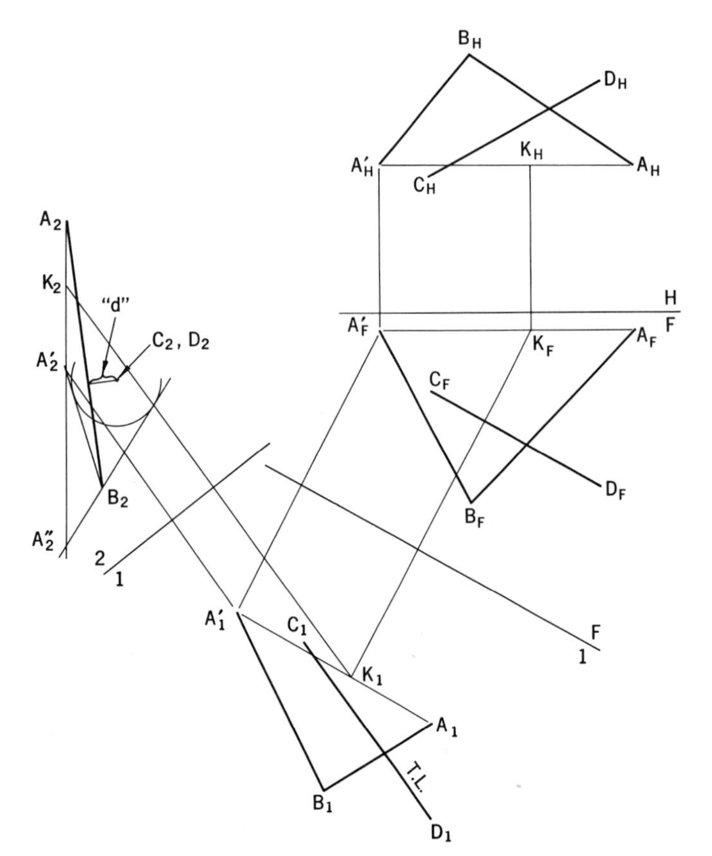

Fig. 6.51

EF, can be readily established. The views of point *E*, which lies on line *CD*, are easily located since the views of the point lie on the corresponding views of the line. Point *F*, however, especially the location of the view, F_1, requires some thought.

We recall that "when two lines are perpendicular to each other, the view of the two lines upon a plane which is parallel to one of them will reveal a right angle"; therefore, the view E_1F_1 will be perpendicular to A_1B_1 (which is the true length of *AB*). Once we have located F_1, it is a simple matter to locate the other views of point *F* and of line *EF*.

EXAMPLE 2

Now let us try to solve a problem which is more challenging. The specifications are:

1. First find the clearance between cables *AB* and *CD*, as shown in Fig. 6.50.
2. Increase the clearance $\frac{1}{4}$ inch.
3. Cable *CD* is to remain fixed.
4. Point *B* of Cable *AB* is also fixed.
5. The new position of point *A* must be on a line which passes through the original position of *A* and parallel to both the *H* and *F* planes. The new position of *A* must be as close to its original position as is possible.

Analysis (the "thinking-through process" of the problem)

1. The clearance between the cables is easily found, since this is the same problem as that in Example 1. (See Fig. 6.51.)

2. The new position of point *A*, represented by *A′*, must lie somewhere on the tangents drawn through B_2 (shown in supplementary view 2) to the circle whose center is at C_2D_2 and whose radius is $(d + \frac{1}{4}'')$. This must be true, if we are to satisfy the condition that the original clearance between the cables has been increased $\frac{1}{4}$ inch.

3. The other condition, that point *A* moves on a line through the original position of *A*, parallel to both the *H* and *F* planes, defines the locus (path of motion) of point *A*. Graphically we can establish this locus by drawing its top and front views (represented by A_HK_H, and A_FK_F, respectively, where point *K* is selected arbitrarily), and then locating line *AK* in both supplementary views.

4. Now, point A' must be somewhere on line AK (extended, if necessary).

5. In order to satisfy both conditions, as specified, point A' must be at the intersection of line AK and the tangents drawn through B. The possible locations of A' are shown in supplementary view 2, as A'_2 and A''_2. Since the new position of A must be as close to its original position as is possible, the correct location is at A'_2. The other views of point A' now are easily obtained. The complete solution is shown in Fig. 6.51.

Question. Is the new cable length longer or shorter than the original length from A to B?

EXAMPLE 3

Let us consider the cable arrangement shown in Fig. 6.52. We are given the following specifications:

1. Cable CD is fixed.
2. Point B of cable AB is fixed.
3. Point A moves in a plane parallel to the F-plane.
4. The length of cable AB does not change.
5. What is the *vertical* clearance between the cables as shown in the Fig. 6.52?
6. Where is the new position of cable AB when the clearance between the cables is reduced to zero?

Analysis

1. The *vertical* clearance between the cables is seen in the front view, shown as distance, d. See Fig. 6.53.

2. Since point A travels in a plane parallel to the F-plane, and since the length of cable AB does not change, the locus of point A is a circular arc whose center is at B_F and whose radius is the length from B_F to A_F.

3. When the clearance between the cables is zero, the cables intersect at point K.

4. The new position of point A, namely, A', therefore, is at the intersection of line BK and the circular arc; and the new position of cable AB is shown in the views $A'_F B_F$ and $A'_H B_H$.

Question. How would you solve this problem if cable AB is fixed; point D is fixed; point C moves in a plane which contains cable CD and is perpendicular to the F-plane; and the length of cable CD does not change? Where is the new position of cable CD?

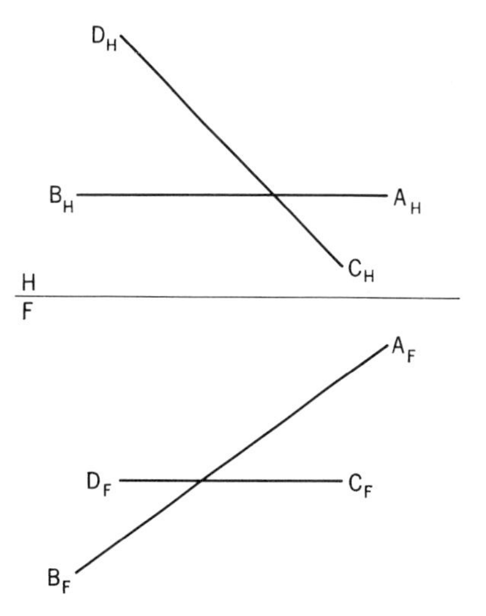

Fig. 6.52

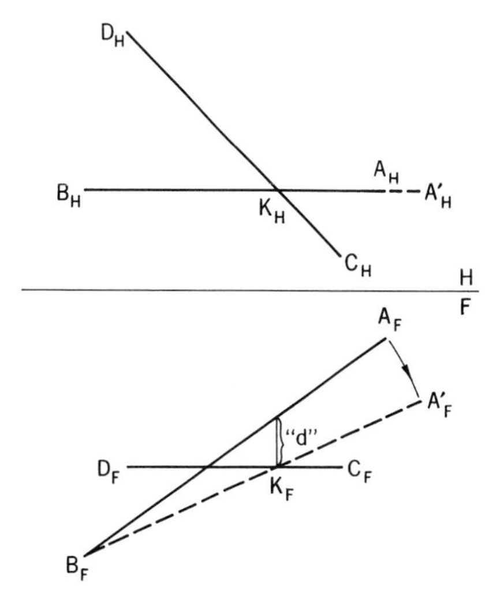

Fig. 6.53

Quite often students will make a "mess" of this type of problem because they plunge into the solution without carefully analyzing ("thinking through") the specifications. Many students immediately jump to the conclusion that a point view of line *CD* is necessary. A little thought about the problem should bring good results.

EXAMPLE 4. SHORTEST DISTANCE BETWEEN SKEW LINES—(A) PERPENDICULAR, (B) HORIZONTAL, AND (C) AT A GIVEN GRADE

A. *The Shortest Perpendicular Between Two Skew Lines*

Previously we found the shortest (perpendicular) distance between two skew lines by finding the point view of one of the lines. Now let us consider an *alternative method* which can also be applied to the determination of the shortest horizontal between two skew lines, and also to the shortest line of a given grade between two skew lines.

Let us consider Fig. 6.54, which shows skew members *AB* and *CD*. Suppose we introduce a plane through one of the lines, *AB*, and parallel to the other, *CD*. Now we can visualize a perpendicular drawn from any point on *CD* to the plane which contains line *AB*. The length of the perpendicular is the shortest distance between the two skew lines. It remains to find the position of the perpendicular that intersects both lines *AB* and *CD*.

In Fig. 6.54 we have introduced line m through point A and parallel to line *CD*. Now lines m and *AB* determine the plane which contains line *AB* and is parallel to line *CD*. We can now find an edge view of the plane by finding the point view of horizontal line *BK*. In supplementary view 1, we see the edge view of the plane and also the view of line *CD* which is parallel to the plane. Now the perpendicular distance, D_1T_1, from any point (i.e., D_1) on line *CD* (as seen in plane 1) to the plane *ABK*, which appears on edge in plane 1, is the shortest distance between the skew lines. The point view of the *common perpendicular* is easily found by introducing supplementary plane 2 perpendicular to the direction D_1T_1. In the view on plane 2 we see the skew lines as A_2B_2 and C_2D_2; and also the point view of the common perpendicular, E_2F_2. The other views of line *EF*, which is the shortest or perpendicular distance between the two skew lines, are easily obtained.

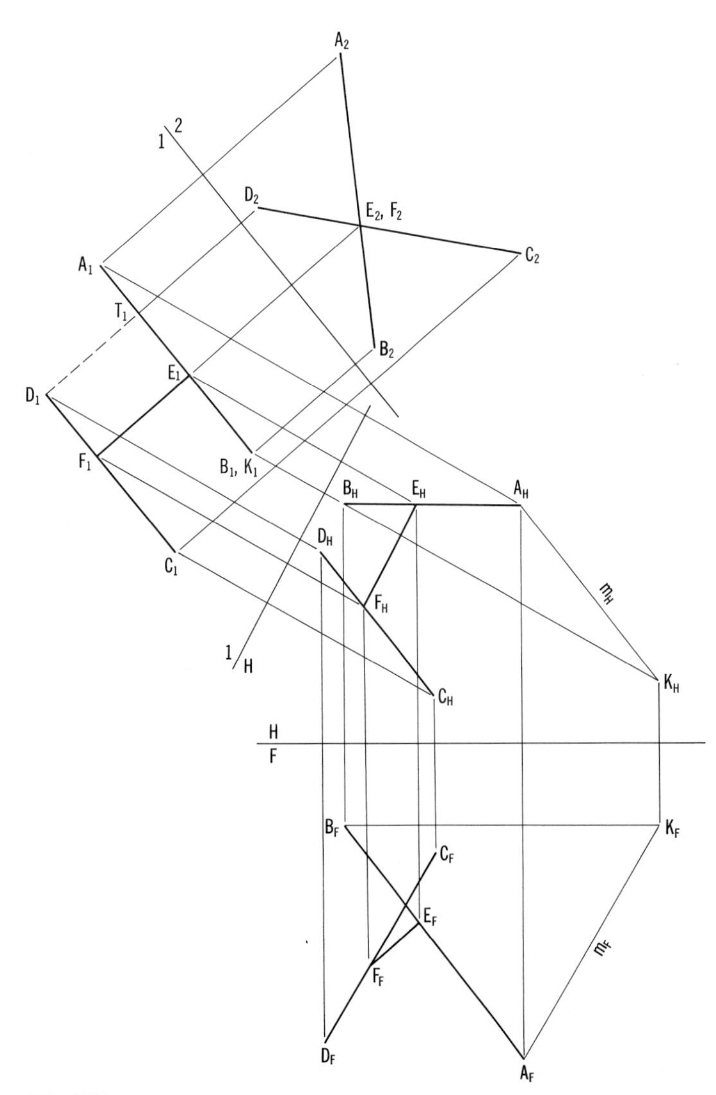

Fig. 6.54

B. The Shortest Horizontal Between Two Skew Lines

Let us consider Fig. 6.55 which shows the same two skew lines, *AB* and *CD*. The steps leading to the view of plane *ABK* and line *CD* on supplementary plane 1 are the same as in the previous problem. Now in plane 1, the true length of the shortest horizontal is shown by the dashed line drawn from any point on line *CD* (i.e., *G*) to the plane *ABK* and parallel to the *H* plane. The point view of the shortest horizontal, *EF*, is seen on supplementary plane 2 which is perpendicular to the direction of line G_1T_1. The other views of line *EF* are easily obtained.

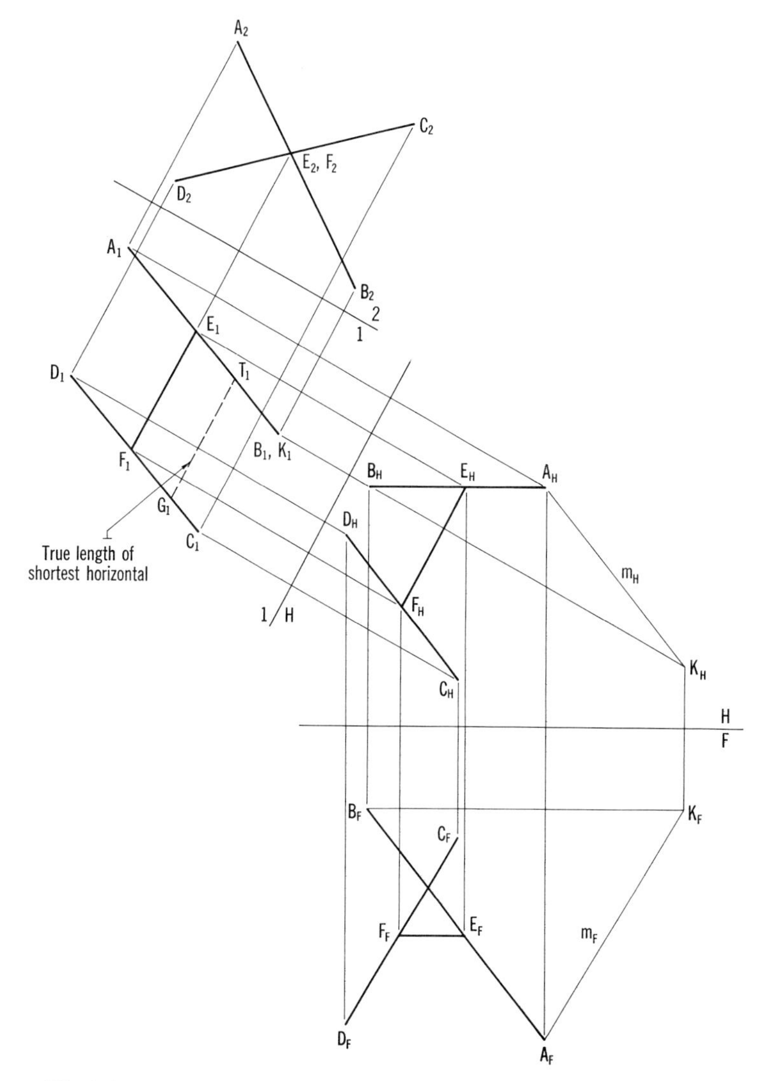

Fig. 6.55

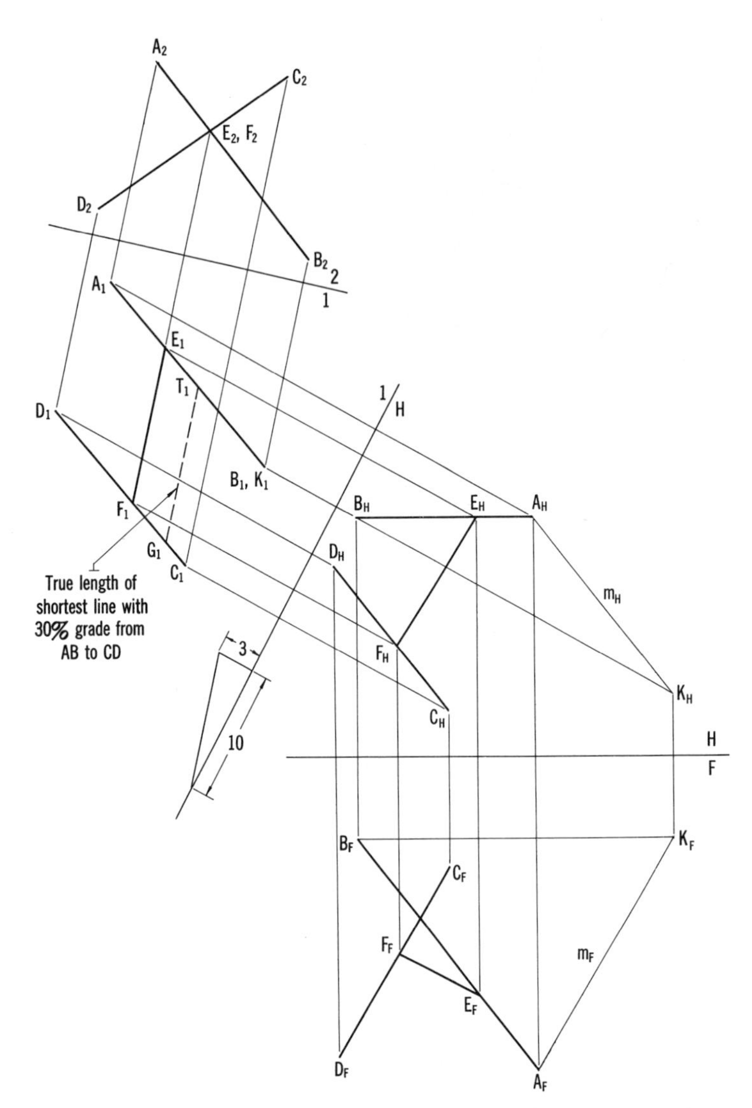

Fig. 6.56

C. The Shortest Line of a Given Grade from One Skew Line to Another

Let us consider Fig. 6.56 which shows the same two skew lines, *AB* and *CD*. Suppose we are to establish the shortest line, at a 30% grade, from line *AB* to *CD*. The steps leading to the view of plane *ABK* and line *CD* on supplementary plane 1 are the same as in the last two cases.

Now the 30% grade can be established in supplementary view 1 since we have an edge view of the *H*-plane. The slope 3:10 is easily established, as shown in the figure. The true length of the shortest line, at the 30% grade, from any point on *CD* to plane *ABK* is shown by the dashed line G_1T_1. The point view of the shortest line, *EF*, at the 30% grade, is seen on supplementary plane 2 which is perpendicular to the direction of line G_1T_1. The other views of line *EF* are readily established.

Question. How would you determine the shortest line, at a 30% grade, from line *CD* to *AB*?

4. TRUE SHAPE OF A PLANE SURFACE—FOURTH BASIC PROBLEM

The true shape of a plane surface, for example, a triangle *ABC*, may be determined in one of several ways—(*a*) by finding the true lengths of each side of the triangle and then constructing the triangle from these true lengths; (*b*) by rotating the triangle about a principal line of its surface parallel to the corresponding reference plane, thus revealing its true shape in the view upon that reference plane; and (*c*) by projecting the triangle upon a plane which is parallel to the triangle. The view of the triangular surface upon that plane would show the true shape of the surface.

Method (*a*) is quite simple but time consuming. Methods (*b*) and (*c*) will be illustrated in the following examples.

EXAMPLE 1. METHOD (b)

Let us consider the triangle *ABC* shown in Fig. 6.57. We will introduce line *CD*, a horizontal line, in surface *ABC*. Now, triangle *ABC* is rotated about horizontal line *CD* until the triangle is parallel to the *H*-plane. How is this accomplished?

Supplementary plane 1 is introduced perpendicular to both line *CD* and the *H*-plane. The view on plane 1 shows the point view of line *CD* (the axis of rotation) and the edge view of the triangle. As we rotate the triangle, about *CD*, parallel to the *H*-plane, points *A* and *B* will move along the circular arcs shown in plane 1. Since the triangle is rotated parallel to the *H*-plane, the rotated position of the triangle appears as the dashed line in the view on plane 1, and as the dashed triangle shown in the *H* view. Triangle $A_H{}^R B_H{}^R C_H$ is the true shape of the triangle *ABC*. It should be noted that a frontal line, such as *AD* shown in Fig. 6.58, could be used to rotate the triangle parallel to the *F*-plane. In this case the dashed triangle $A_F B_F{}^R C_F{}^R$ shows the true shape of the triangle *ABC*.

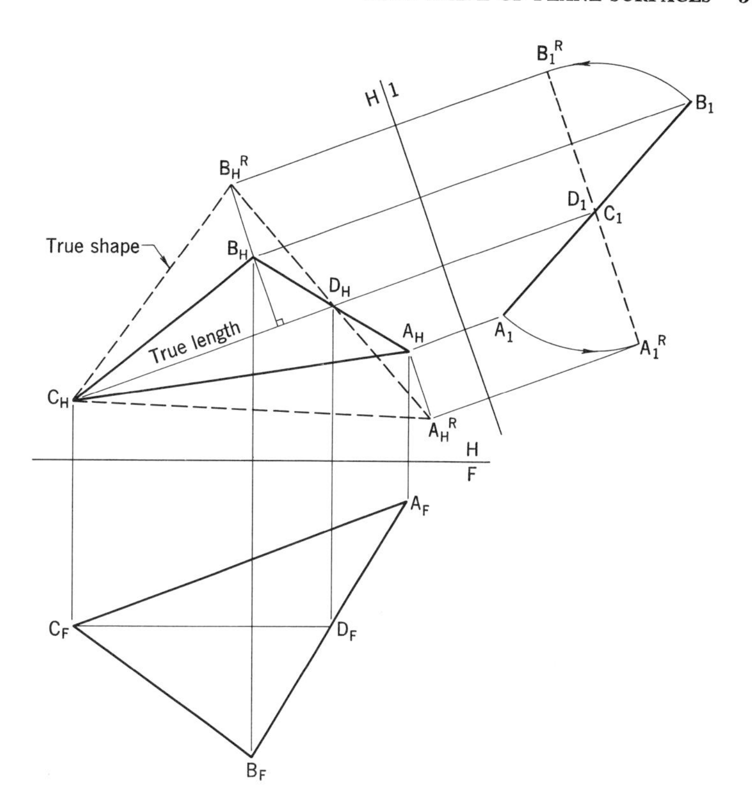

Fig. 6.57

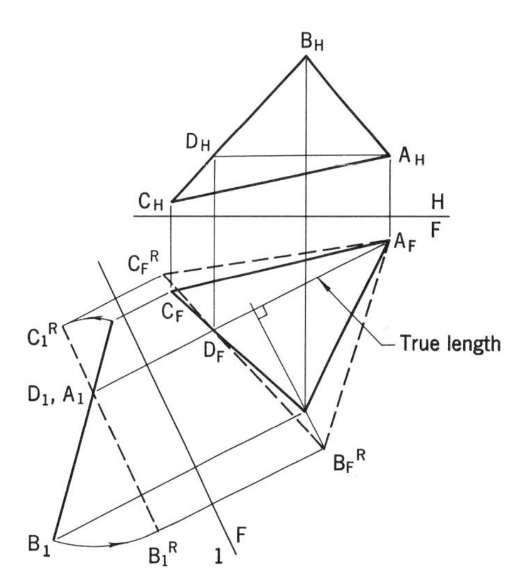

Fig. 6.58

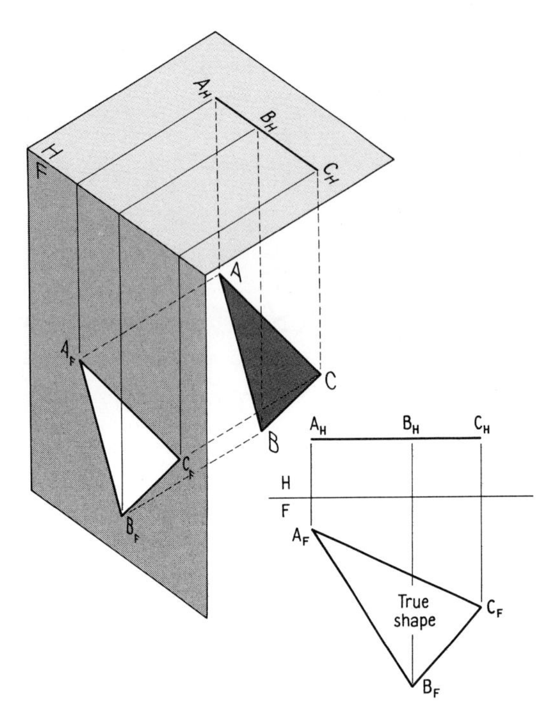

Fig. 6.59

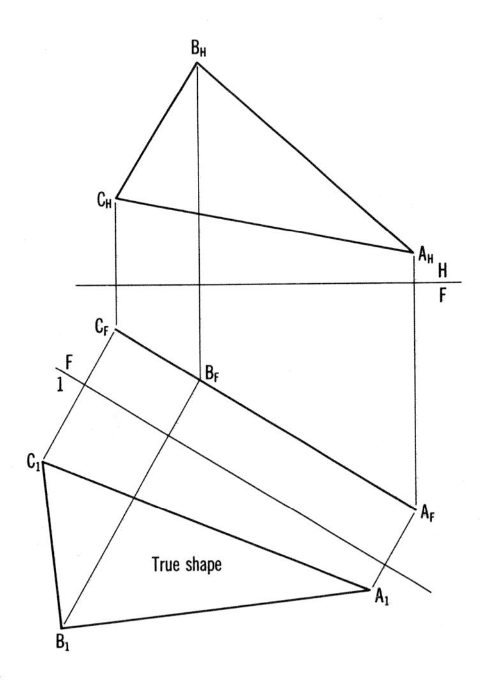

Fig. 6.60

EXAMPLE 2. METHOD (*c*)

This method is most useful when it is necessary to dimension the surface and to locate components that may be attached to the surface. This is so because the true shape view is a completely separate view. Let us now consider three cases.

Case I. The Plane Surface Is Parallel to a Reference Plane. Suppose we are given plane surface *ABC* as shown in Fig. 6.59. Careful study of the pictorial and orthographic drawings leads us to conclude that the surface *ABC* is parallel to the *F*-reference plane. Therefore the front view, $A_FB_FC_F$, shows the true shape of the surface.

Case II. The Plane Surface Is Perpendicular to a Reference Plane, but not Parallel to the Other Reference Planes. Let us consider the plane surface *ABC* shown in Fig. 6.60. We observe that the surface is perpendicular to the *F*-plane since it appears on edge in the front view. Moreover, the evidence is quite clear that the surface is *not* parallel to the *H*-plane.

Since we have an edge view of the surface *ABC*, it is a rather simple matter to introduce a plane parallel to the surface. Supplementary plane 1 has been placed both parallel to the surface and perpendicular to the *F*-plane. The view of surface *ABC* on plane 1 shows the true shape of the surface.

It should be stressed that the relationship between the front view and the supplementary view is basically the same as that between the top and front views shown in the orthographic drawing of Fig. 6.59.

Case III. The Plane Surface Is Not Parallel to Any Reference Plane. The most general case is shown in Fig. 6.61. Study of the top and front views reveals the fact that neither the top view nor the front view shows the true shape of the surface ABC. We can easily reduce this general case to Case II, by finding an edge view of the surface. An edge view of the surface is seen in the view on supplementary plane 1 which is perpendicular to both line AD (a frontal line in surface ABC) and the F-plane. Once we have obtained the edge view, shown as $A_1B_1C_1$, it is a simple matter to introduce supplementary plane 2 parallel to surface ABC and perpendicular to plane 1. The view, $A_2B_2C_2$, on plane 2 shows the true shape of surface ABC.

Again, it should be stressed that the relationship among the front view, the supplementary view on plane 1, and the supplementary view on plane 2, is basically the same as the relationship among the top view, front view, and the supplementary view on plane 1 in Fig. 6.60.

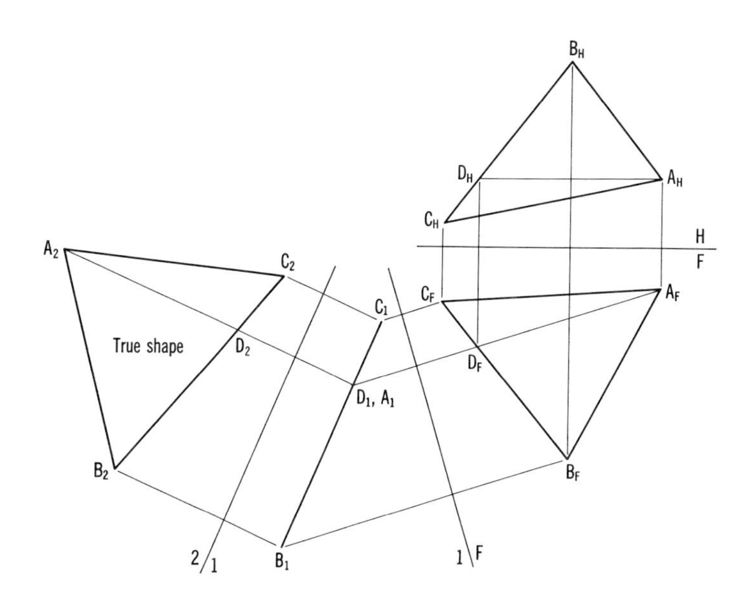

Fig. 6.61

EXAMPLES OF THE APPLICATION OF THE FOUR BASIC PROBLEMS

EXAMPLE 1

Let us consider the following problem: "The upper and lower surfaces of a rock formation are parallel. Points A and B lie in the upper surface; and points C and D lie in the lower surface. (See Fig. 6.62.) What is the thickness of the rock formation?"

Analysis

1. Let us first visualize two parallel planes in space, one above the other. Picture points A and B in the upper plane and points C and D in the lower plane.
2. Now let us join points C and D to form line CD and similarly join points A and B to form line AB.
3. Since line AB is parallel to the lower surface, it is parallel to a line in that surface; therefore, we can easily establish the lower surface by constructing, through point C, a line, m, parallel to line AB.

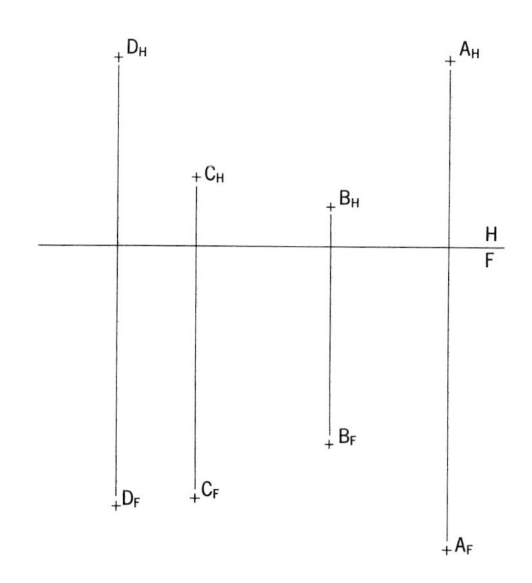

Fig. 6.62

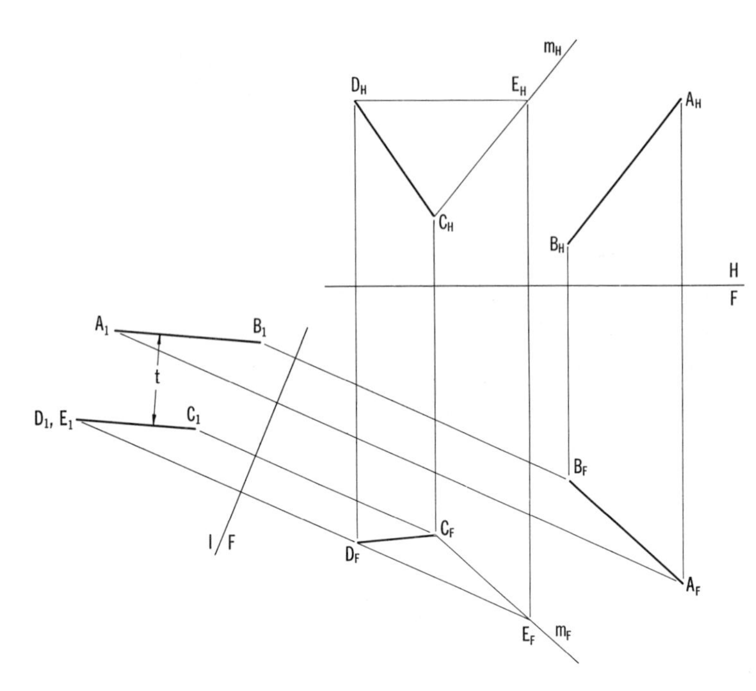

Fig. 6.63

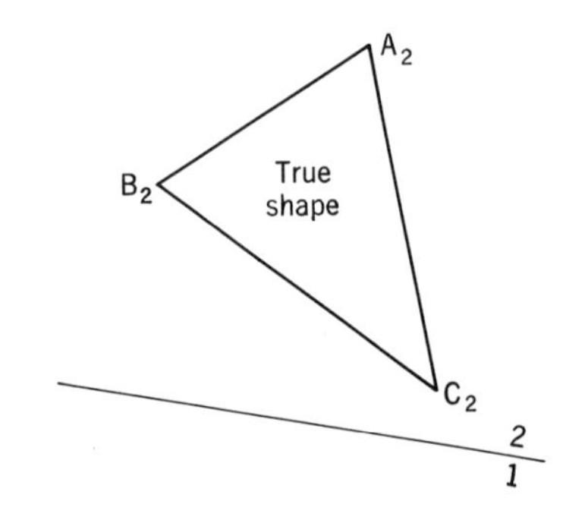

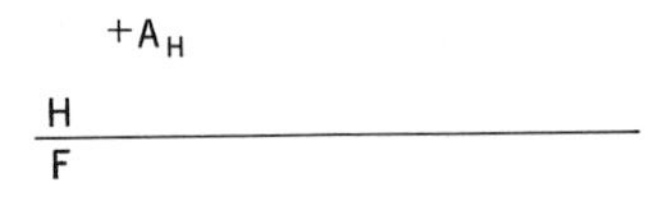

Fig. 6.64

4. We can proceed to obtain an edge view of the lower plane surface and the corresponding view of line *AB*. The perpendicular distance from any point on line *AB* to the edge view of the lower plane is the measure of the thickness of the rock formation. (See Fig. 6.63.)

EXAMPLE 2

Suppose we are presented with Fig. 6.64 which shows the true shape of triangle *ABC*, the *H* and *F* views of point *A* and the line of intersection of supplementary planes 1 and 2. How shall we determine the *H* and *F* views of the triangle?

Analysis

1. We should know from the first fundamental principle of orthogonal projection that views A_2 and A_1 lie on a perpendicular to the line of intersection of planes 1 and 2.

2. Based on the second principle of orthogonal projection we know that the perpendicular distance from A_H to the line of intersection of planes *H* and 1 must be equal to d_1 and that the perpendicular distance from A_1 to the line of intersection of planes *H* and 1 must be equal to distance *d*. Therefore, the distance from A_H to A_1 is equal to $d + d_1$.

3. Using this distance as a radius and point A_H as a center we draw an arc to cut the perpendicular which was drawn through A_2 in a point which uniquely locates A_1.

4. The line which joins A_1 and A_H is perpendicular to the line of intersection of planes H and 1. This line is distance d_1 from A_H or distance d from A_1. The complete solution is shown in Fig. 6.65.

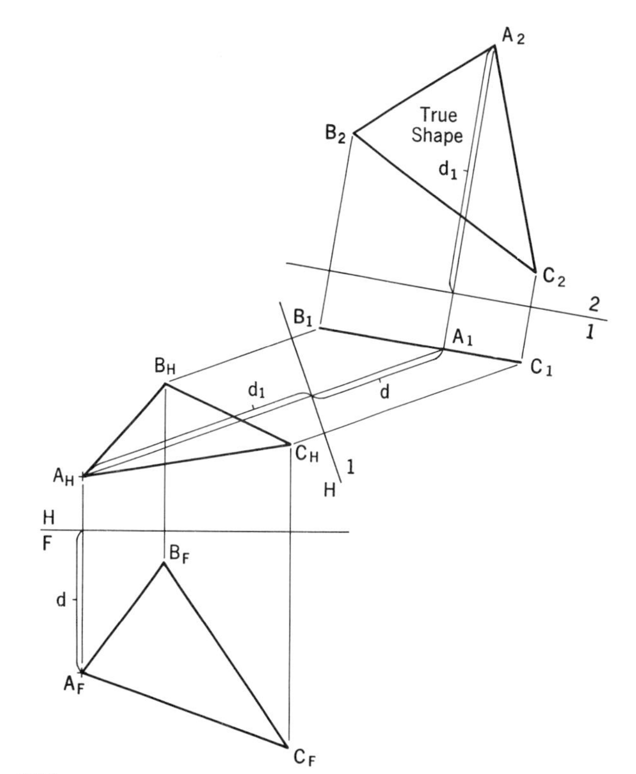

Fig. 6.65

EXAMPLE 3

We are to determine the shortest distance between the center line of the hole and the line of intersection of surfaces $ABCD$ and $BCEF$. (See Fig. 6.66.)

Analysis

1. Careful study of the two given views reveals the fact that line BC is the intersection of the two surfaces $ABCD$ and $BCEF$.

2. Our problem is now reduced to "finding the perpendicular distance between two skew lines BC and the center line of the hole."

3. We know that the distance between two skew lines will be seen in a view which shows one of the lines as a point.

4. Observe that the front view shows the center line of the hole as a point. Therefore, the shortest distance between the center line and line BC will be seen in the front view. The distance is shown as d.

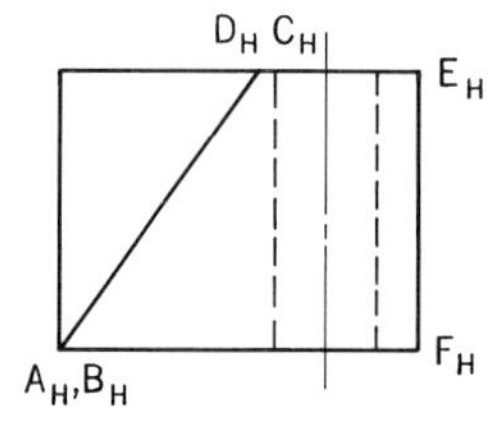

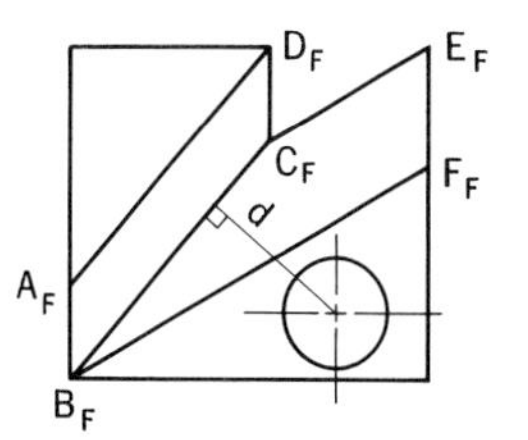

Fig. 6.66

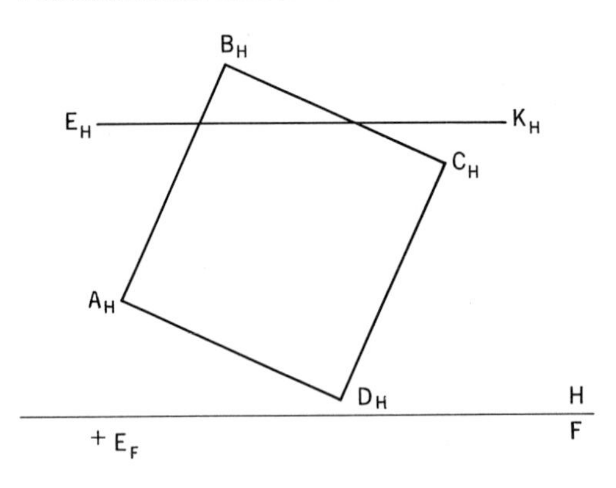

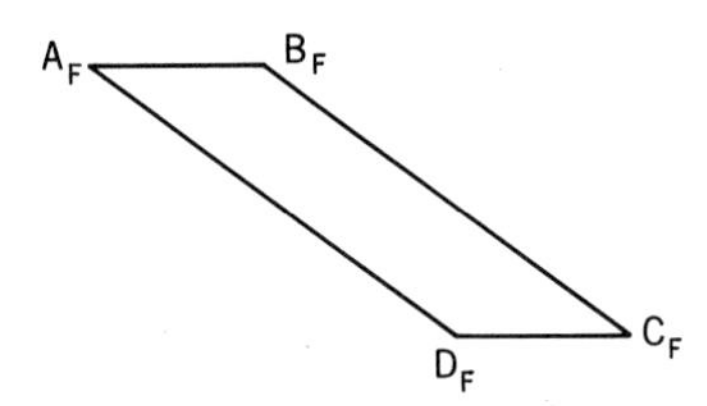

Fig. 6.67(*a*)

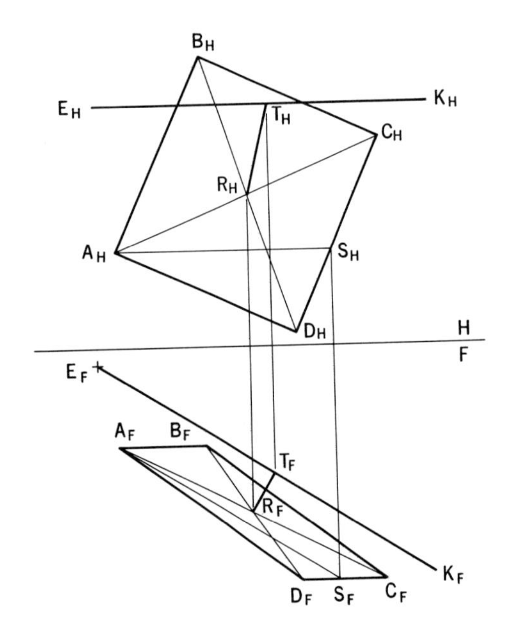

Fig. 6.67(*b*)

EXAMPLE 4

The H and F views of a rectangular plate $ABCD$; the F view of point E; and the H view of cable EK which is parallel to the plate are shown in Fig. 6.67(*a*). It is required to establish the H and F views of the centerline of a supporting arm, RT, which connects the center of the plate with the cable such that the arm is perpendicular to the cable.

Analysis and Solution

1. Since the cable is parallel to the plate it is parallel to a line in the surface of the plate. Such a line is AS as shown in Fig. 6.67(*b*).

2. We know that parallel lines appear as parallels in the respective views; therefore, E_FK_F will be parallel to A_FS_F.

3. We note further that cable EK is parallel to the F-plane; therefore, E_FK_F is in true length.

4. The F view of the arm RT is perpendicular to E_FK_F because E_FK_F is in true length. We recall: "The view of two perpendicular lines upon a plane which is parallel to one of the lines, will reveal a right angle."

5. Having located R_FT_F it is quite simple to establish R_HT_H. (The location of R_F and R_H is quite obvious.)

EXAMPLE 5

In Fig. 6.68(a) is shown plate ABC and point P. It is required to establish through point P two lines, n and t which describe a plane that is perpendicular to plate ABC and parallel to the intersection of the H and F-planes.

Analysis and Solution

1. Let us assume that line t is parallel to the intersection of the H- and F-planes. Since the intersection of planes H and F is both a horizontal line and a frontal line, t_H and t_F will appear as horizontal lines. [See Fig. 6.68(b).]

2. Let us further assume that line n is perpendicular to plate ABC. From our past studies we know that "when a line is perpendicular to a plane it must be perpendicular to two lines that lie in the plane."

3. In Fig. 6.68(b) n_F is perpendicular to A_FD_F and n_H is perpendicular to B_HE_H. Why is this so?

4. Since line n is perpendicular to both lines AD and BE which lie in surface ABC, it is perpendicular to surface ABC.

5. Now we have two lines, n and t, which determine a plane that passes through the given point P and is both perpendicular to plate ABC and parallel to the line of intersection of the H and F planes.

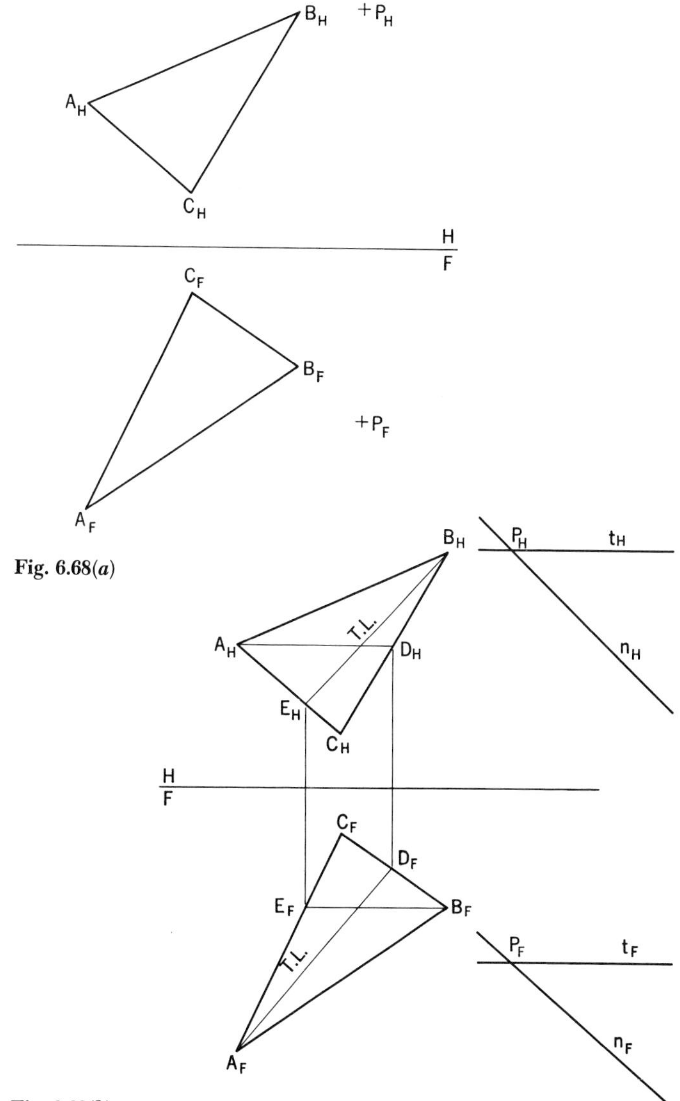

Fig. 6.68(a)

Fig. 6.68(b)

EXERCISES

A. Problems dealing with true length, grade, and bearing of a line

1. Determine the true length, grade and bearing of line segment AB (Fig. E-6.1).

2. Determine the true length, grade and bearing of line segment CD (Fig. E-6.2).

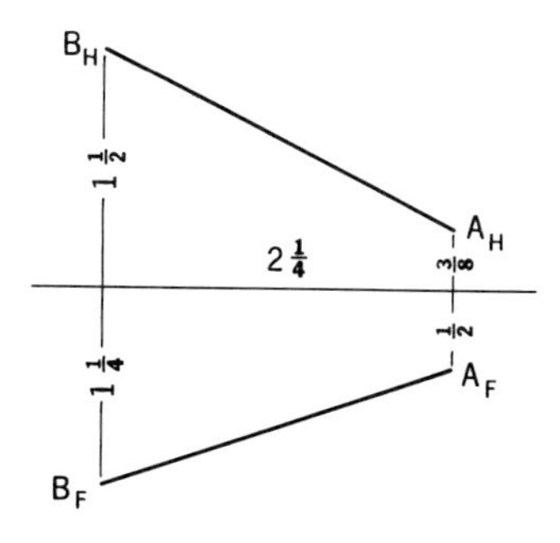

Fig. E-6.1

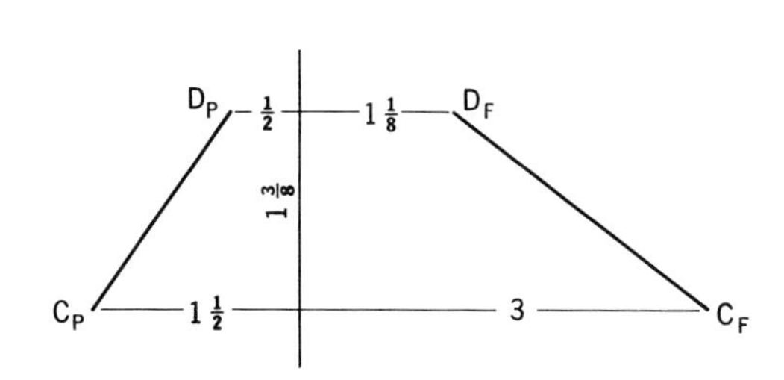

Fig. E-6.2

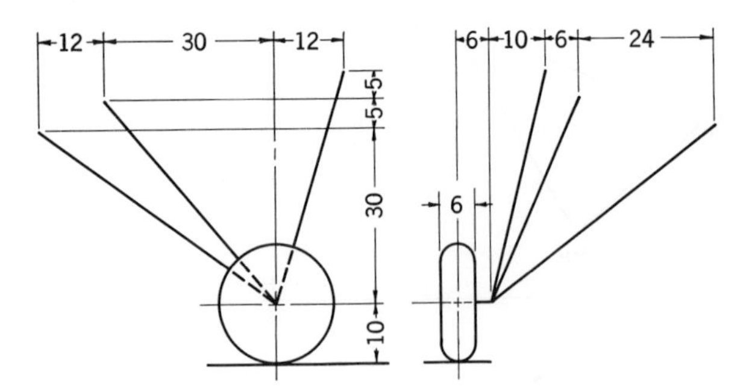

Fig. E-6.3

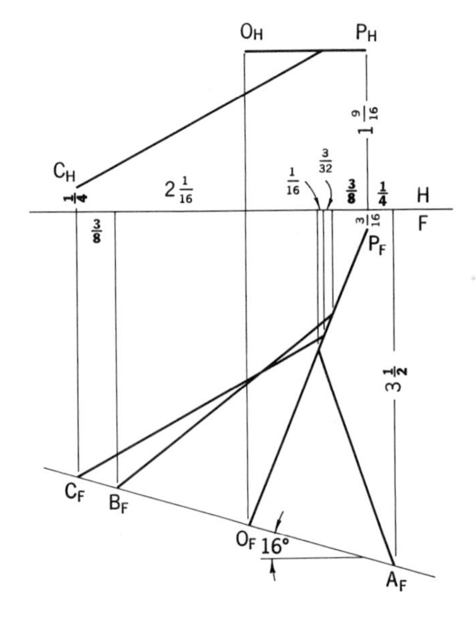

Fig. E-6.4

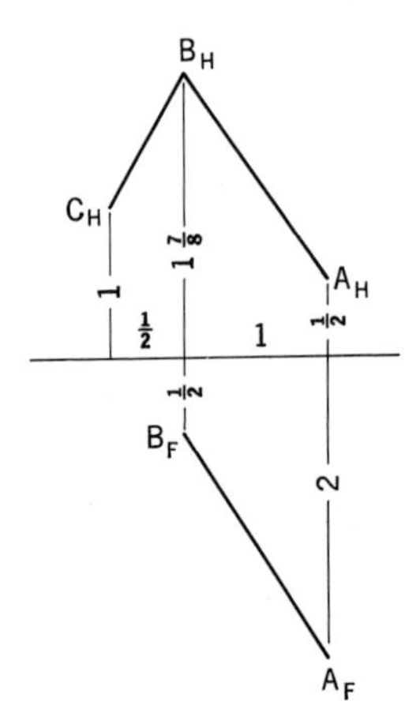

Fig. E-6.5

3. Determine the top and front views of line segment AB which is $2\frac{1}{4}$ inches long, bears N 60° W, and has a grade of −30%. Point A is 1 inch behind the F-plane and $\frac{1}{2}$ inch below the H-plane.

4. The design of a hydroelectric power plant requires a 450-foot tunnel from A to B. The elevations of A and B are 1400 feet and 1250 feet, respectively. Point B is on line AB whose bearing is S 60° E. Assume the H and F views of point A. (*a*) Determine the H and F views of point B. (*b*) Determine the grade of the center line (AB) of the tunnel. Scale: 1″ = 150 feet.

5. Determine the true length and slope (grade) of each strut of the airplane landing gear shown in Fig. E-6.3.

6. The guy wires A, B, and C that support the strut OP are all the same length (Fig. E-6.4). Point A is on the front side of OP, and point B is behind OP. (*a*) Complete the H view (*b*) What is the true length of the guy wires? Scale: 1″ = 8′–0″.

B. Problems dealing with intersecting lines, skew lines, parallel lines, and perpendicular lines

1. Assume the H and F views of four points A, B, C, and D such that the points are *not* on a line. Show that the points are or are not in one plane.

2. Assume the H view of points A, B, C, and D and the F view of points A, B, and C. How would you locate the F view of point D so that the four points will lie in one plane?

3. Assume the H and F views of a triangle ABC and the H and F views of an external point K. Determine the H and F views of a line that passes through point K and parallel to both plane ABC and the H reference plane.

4. Locate the F view of arm BC which is perpendicular to arm AB (Fig. E-6.5).

5. Determine the H and F views of plane nt that passes through point K and is parallel to line m and perpendicular to plane rs (Fig. E-6.6).

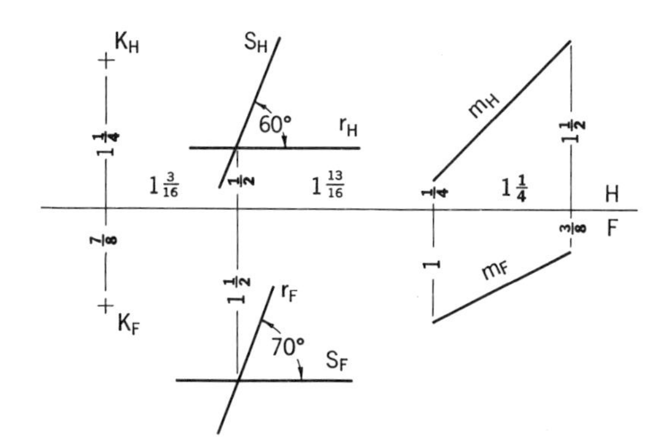

Fig. E-6.6

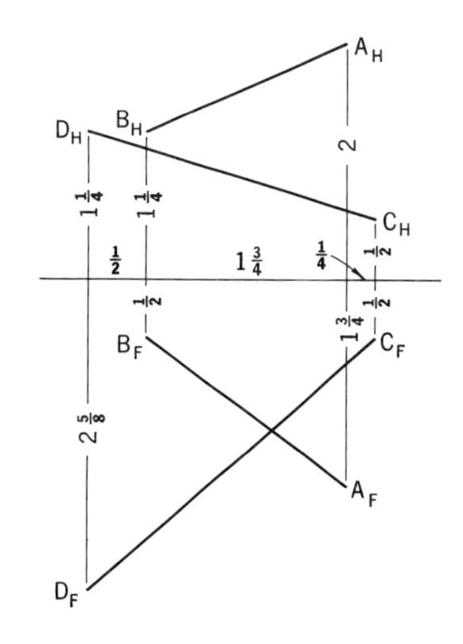

Fig. E-6.7

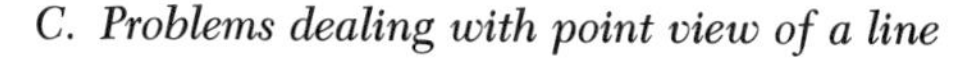

C. Problems dealing with point view of a line

1. Determine the perpendicular distance between the center lines of pipes AB and CD. Also show the H and F views of the center line of the shortest connector pipe EF. Point E is on AB (Fig. E-6.7).

2. Find the minimum clearance between cables AB and CD (Fig. E-6.8).

3. The coordinates of a high-tension power cable AB and a telephone cable CD are given in the table below. The specified minimum clearance between the cables is 30 feet. Check the clearance. Does it satisfy the specification? Select an appropriate scale.

Points	Coordinates		Elevations
	South	East	
A	850′	250′	720′
B	765′	470′	670′
C	800′	270′	645′
D	905′	370′	640′

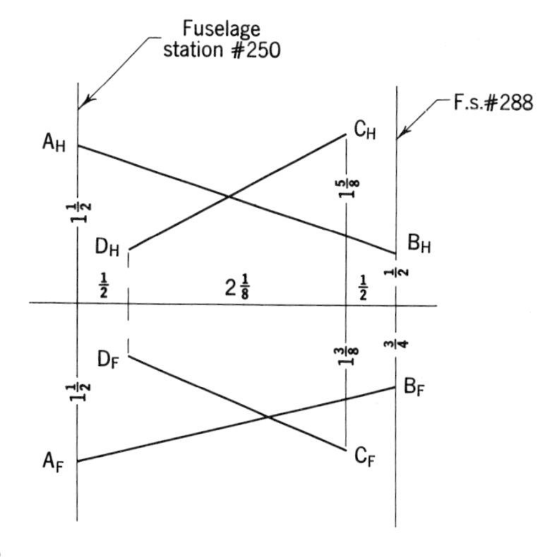

Fig. E-6.8

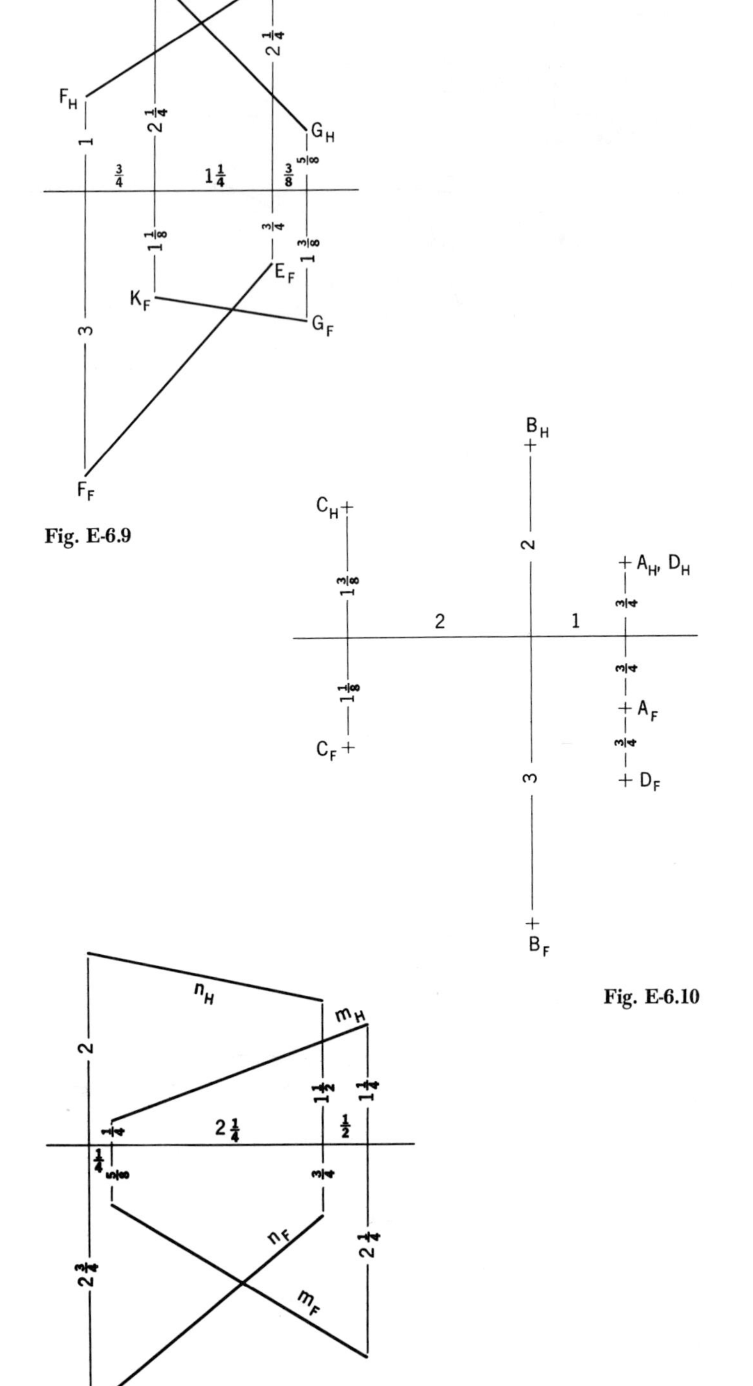

Fig. E-6.9

Fig. E-6.10

Fig. E-6.11

4. The clearance between cables EF and GK is 4 inches. Determine the new position of point E, which moves only on a line perpendicular to the F-plane, so that the clearance between the cables will be 6 inches. How long is cable EF in the new position? Points F, G, and K remain fixed (Fig. E-6.9).

D. Problems dealing with edge view of a plane

1. Points A, B, and C are on the upper surface of a rock formation. Point D is on the lower parallel surface of the formation (Fig. E-6.10). What is the thickness of the formation? Scale: 1 inch = 100 feet.

2. Assume the H and F views of a plane surface (i.e., triangle ABC) and an external point K. Determine the perpendicular distance from point K to the plane. Also show the H and F views of the perpendicular.

3. Assume the H and F views of a plane surface ABC. Determine the H and F views of a plane that is parallel to surface ABC and 1 inch from ABC.

4. Assume the H and F views of line AB. Determine the H and F views of plane mn which is perpendicular to line segment AB and passes through its midpoint.

E. Problems dealing with connectors between skew lines

1. Locate the H and F views of the shortest *horizontal connector* shaft, s, between shafts m and n (Fig. E-6.11). How long is shaft s? Scale 1 inch = 50 feet.

2. Locate the H and F views of the shortest connector shaft, s, which has a 30% grade from shaft m to shaft n (Fig. E-6.12). How long is shaft s? Scale: 1 inch = 100 feet.

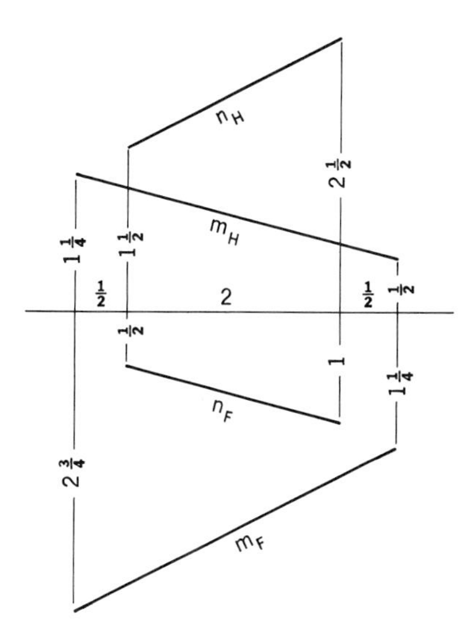

Fig. E-6.12

3. Determine the views of the shortest connector between AB and CD. The connector is parallel to plane 1 (Fig. E-6.13).

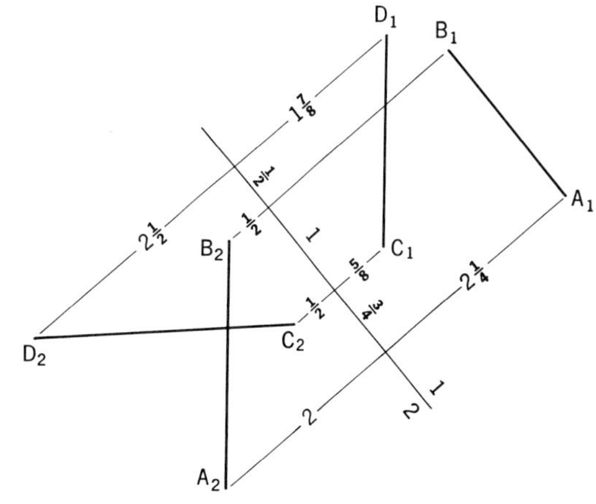

Fig. E-6.13

F. Problems dealing with true shape of plane surfaces

1. Determine the true shapes of roof surfaces A and B shown in Fig. E-6.14.

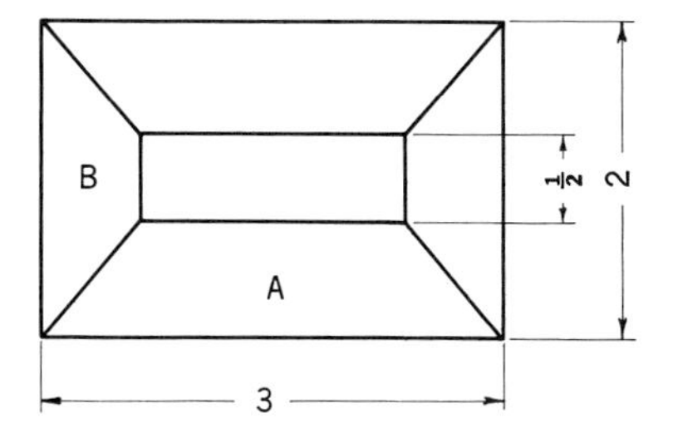

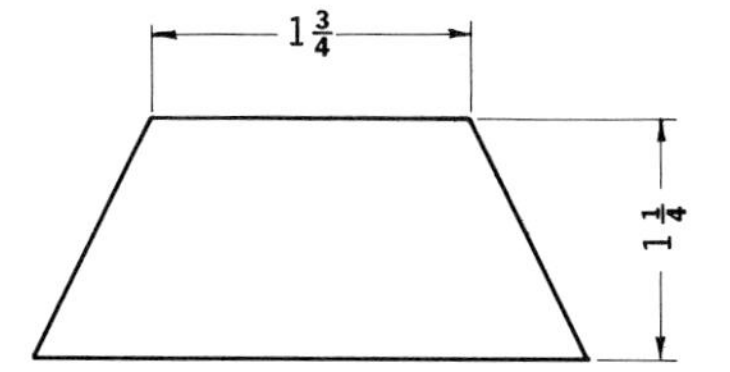

Fig. E-6.14

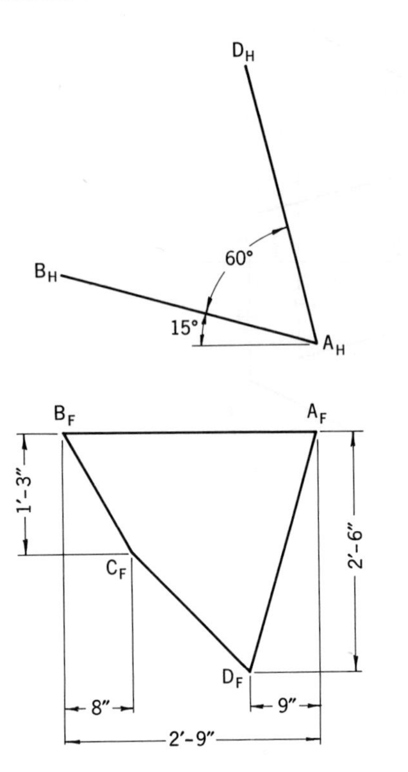

Fig. E-6.15

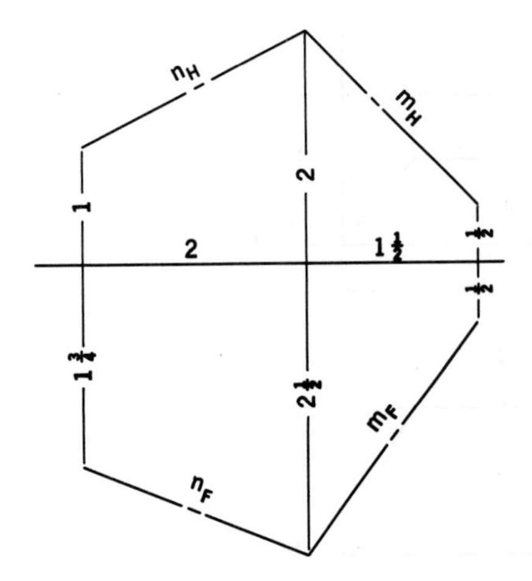

Fig. E-6.16

2. Determine the true shape of plate *ABCD* shown in Fig. E-6.15.

3. Determine the *H* and *F* views of the circular arc (radius 1 inch) which is tangent to *m* and *n* (Fig. E-6.16).

4. A partial front view and a complete profile view of a flat glass plate are shown in Fig. E-6.17. Determine (*a*) the true shape of the plate; (*b*) the complete front view.

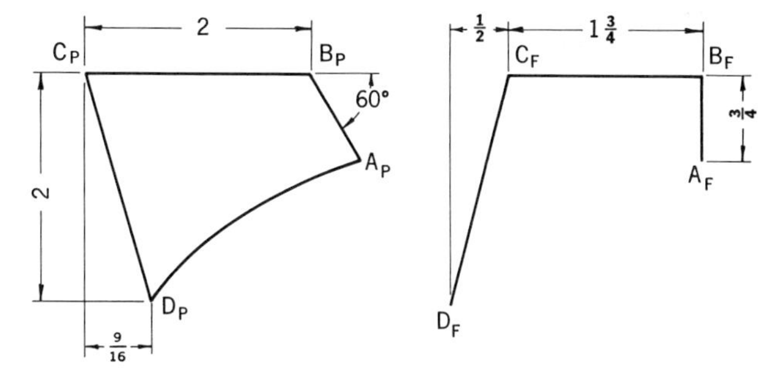

Fig. E-6.17

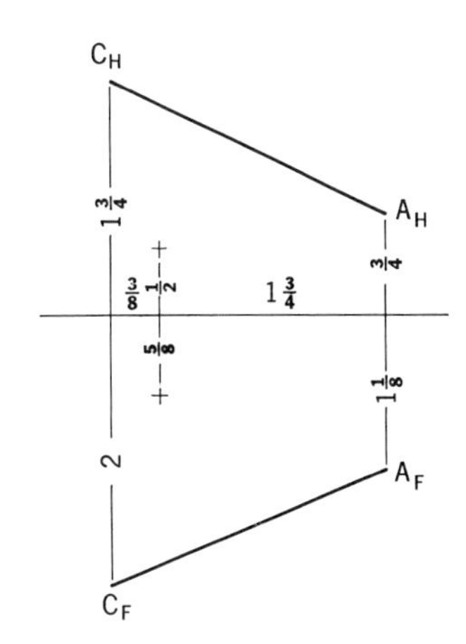

Fig. E-6.18

5. Lines *AC* and *BD* are diagonals of plane figure *ABCD*. The diagonals intersect at right angles and are of equal length (Fig. E-6.18). Determine: (*a*) the true shape of the plane figure *ABCD*; (*b*) the *H* and *F* views of the figure; and (*c*) the true length of the diagonals.

6. Assume a point *A* and a line *m*. Determine the *H* and *F* views of a line from point *A* and intersecting line *m* at an angle of 60°.

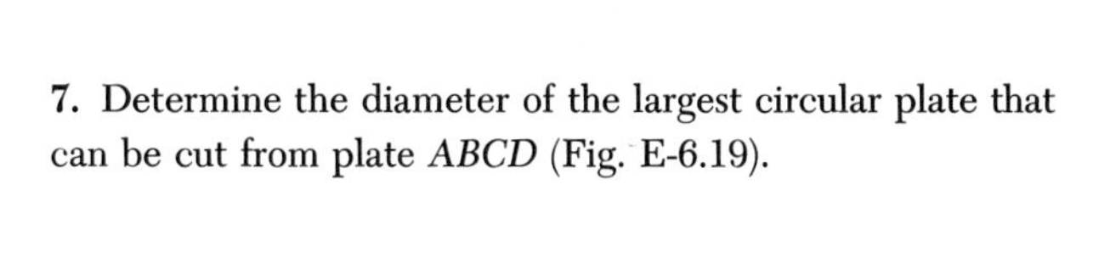
7. Determine the diameter of the largest circular plate that can be cut from plate *ABCD* (Fig. E-6.19).

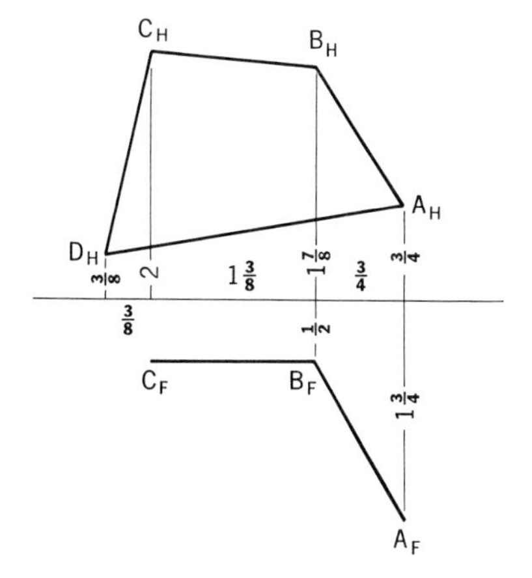

Fig. E-6.19

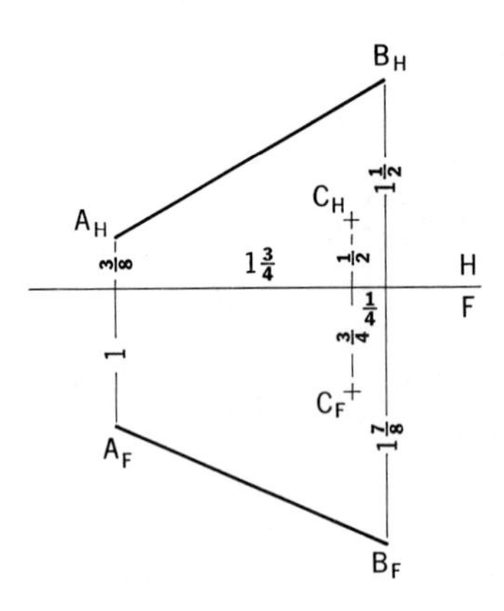

Fig. E-6.20

8. An equilateral plate $A'B'C$ joins point C to member AB. Determine the H and F views of the plate. (See Fig. E-6.20.)

ANGLE PROBLEMS 7

ANGLE BETWEEN PLANE SURFACES

Many engineering design problems often include the need to determine the angle between plane surfaces. For example, in the design of connection plates between structural steel members (see Fig. 7.1) it is necessary to determine the angle of bend of the plates so that the intended positioning of the steel members is accomplished.

Another example is shown in Fig. 7.2 where the angles θ and ϕ must be maintained in the design of a chute. In Fig. 7.3 the supporting members of the pilot's windshield must be properly designed to receive the glass plates. Here again the angles of bend of the supports must be determined.

How shall we proceed to determine the angle between two plane surfaces?

Analysis. *The angle between two plane surfaces will be seen in a view which shows the line of intersection* (the common line) *of the two planes as a point.* Since three points (not on one line) determine a plane, it is only necessary to consider two points on the common line of the two planes, and one additional point in each of the given planes, in the solution of the problem.

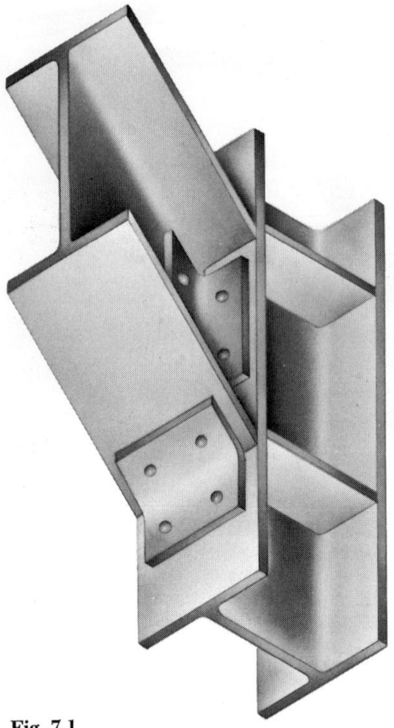

Fig. 7.1

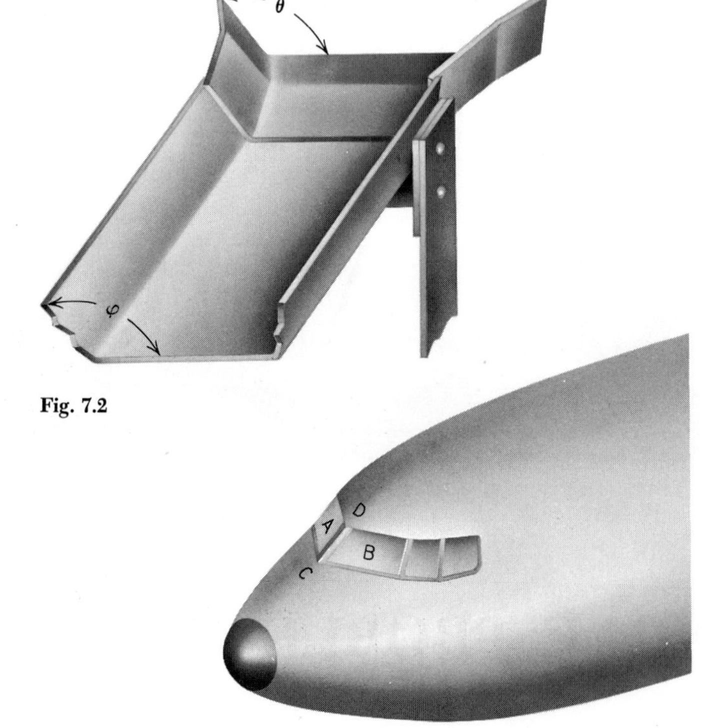

Fig. 7.2

Fig. 7.3

EXAMPLE 1

Let us consider the hopper shown in Fig. 7.4. We will suppose that plates *A* and *B* are to be reinforced by an angle iron along edge *CD*. The angle to which the reinforcing member is to be bent is the true angle between surfaces *A* and *B*.

The solution to our problem is quite simple. A view which will show edge *CD* as a point will also show surfaces *A* and *B* as lines; the angle between these "lines" (edge views of the surfaces *A* and *B*) is the required angle. We know from our previous study that we must first obtain a true length view of a given line before we can introduce a plane perpendicular to the line to show its point view.

In Fig. 7.5 supplementary plane 1 has been introduced parallel to edge *CD* and perpendicular to the *F*-reference plane. The view on plane 1 shows *the true length of edge CD* and the view of points *E* and *F* which lie in surfaces *A* and *B*, respectively. (We recognize that points *CDE* determine plane surface *A*; and that points *CDF* determine plane surface *B*.)

Supplementary plane 2 which has been introduced perpendicular to edge *CD* shows the point view of edge *CD* and the views of points *E* and *F*. The angle θ between the surfaces *A* and *B* is now apparent in the second supplementary view, since edge *CD* appears as a point and the two surfaces appear as lines. The reinforcing angle iron would be bent to angle θ.

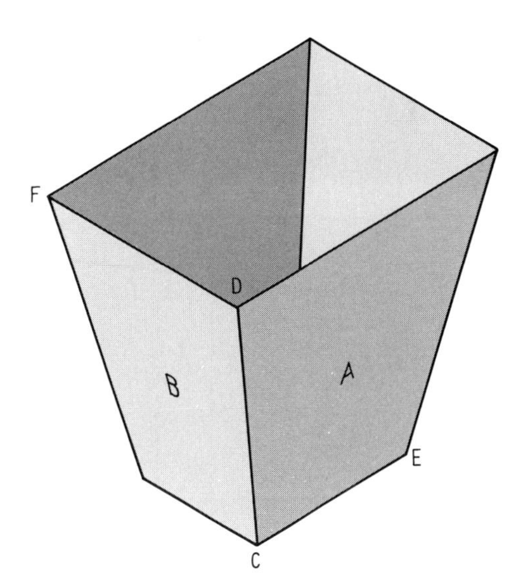

Fig. 7.4

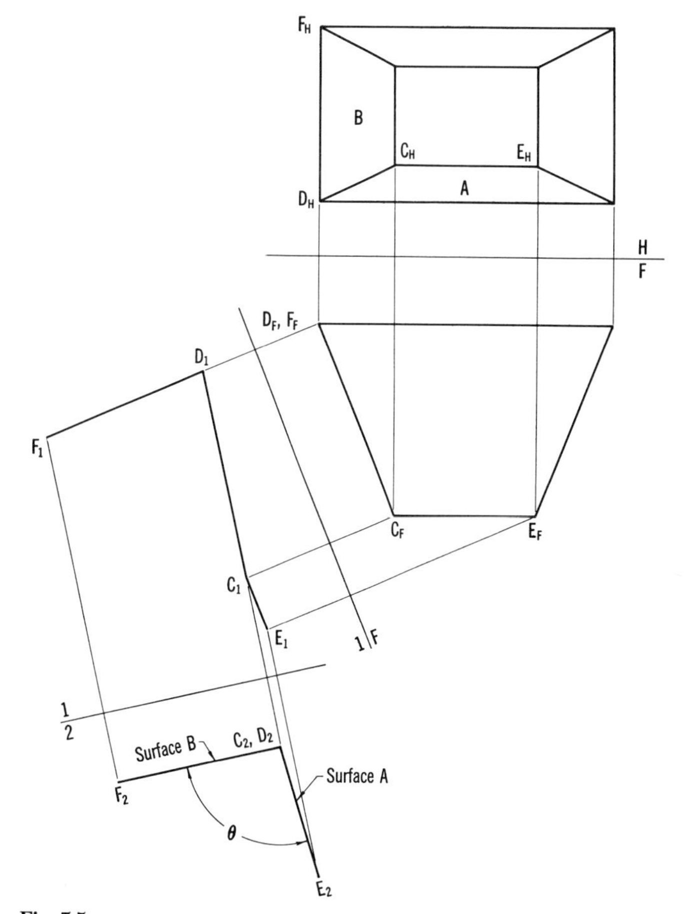

Fig. 7.5

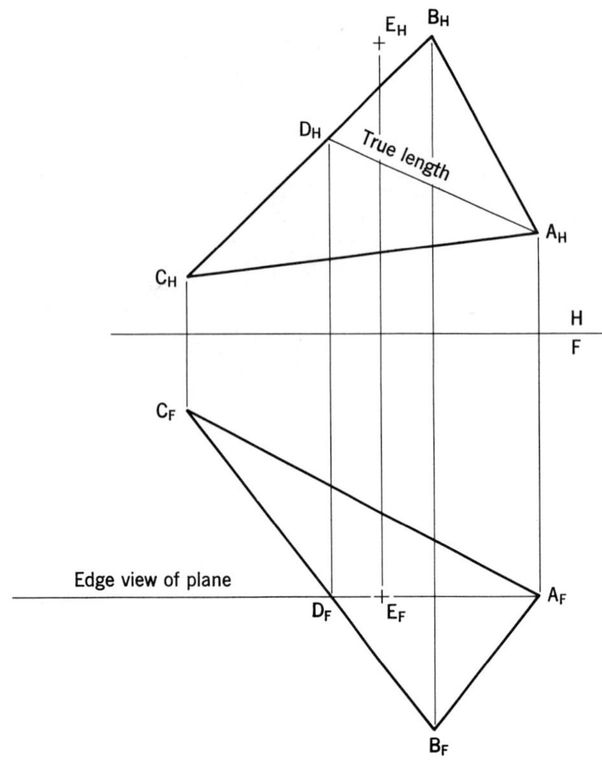

Fig. 7.6

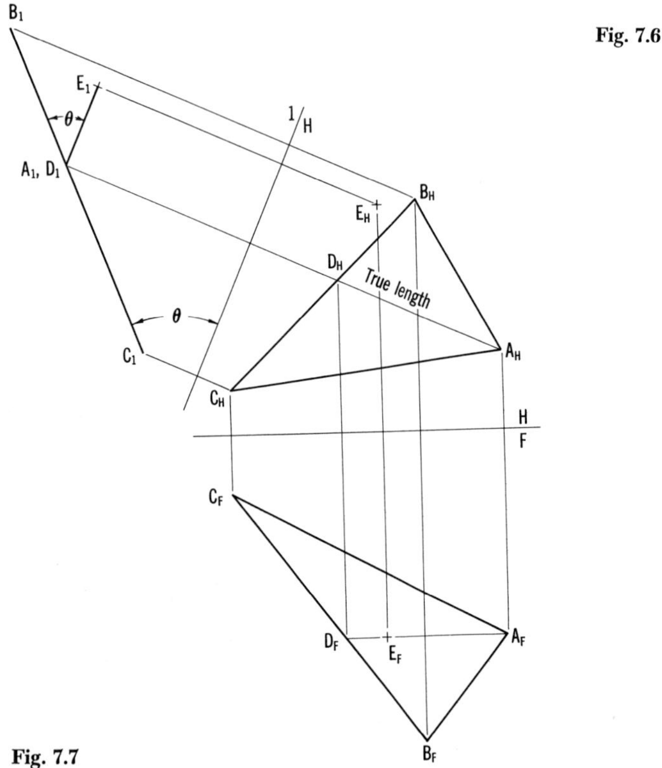

Fig. 7.7

EXAMPLE 2

Now let us consider another problem. Suppose we wish to determine the angle between plane *ABC* and the *H*-reference plane. (See Fig. 7.6.) We know that it is necessary to obtain a point view of the common line. But where is the common line? Actually, we can establish a plane parallel to the *H*-reference plane. The parallel plane could be used since the angle between surface *ABC* and the parallel plane would be the same as the angle with the *H*-reference plane. Let us introduce plane *ADE* which is parallel to the *H*-reference plane and contains line *AD* of surface *ABC*. Point *E* is selected arbitrarily in horizontal plane *ADE*.

Careful study of the figure shows that plane *ADE* is horizontal and, moreover, that line *AD* is common to both surfaces *ABC* and *ADE*. (See Fig. 7.7.)

The angle between these two surfaces is easily obtained. *A point view of the common line AD is seen on supplementary plane 1* which is perpendicular to common line AD. In addition, the edge views of surfaces ABC and ADE are seen and, of course, the angle θ between the surfaces. Further study of the figure shows that, since the H-reference plane is seen on edge in the supplementary view 1, the angle θ is easily identified without the need for the establishment of the surface ADE in view 1. This is shown in Fig. 7.8(a) and in the accompanying pictorial sketch, Fig. 7.8(b).

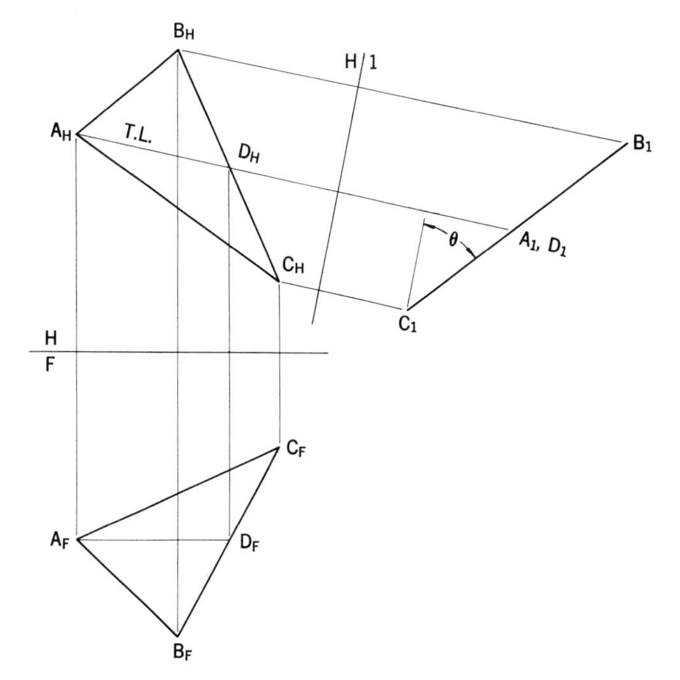

Fig. 7.8(a)

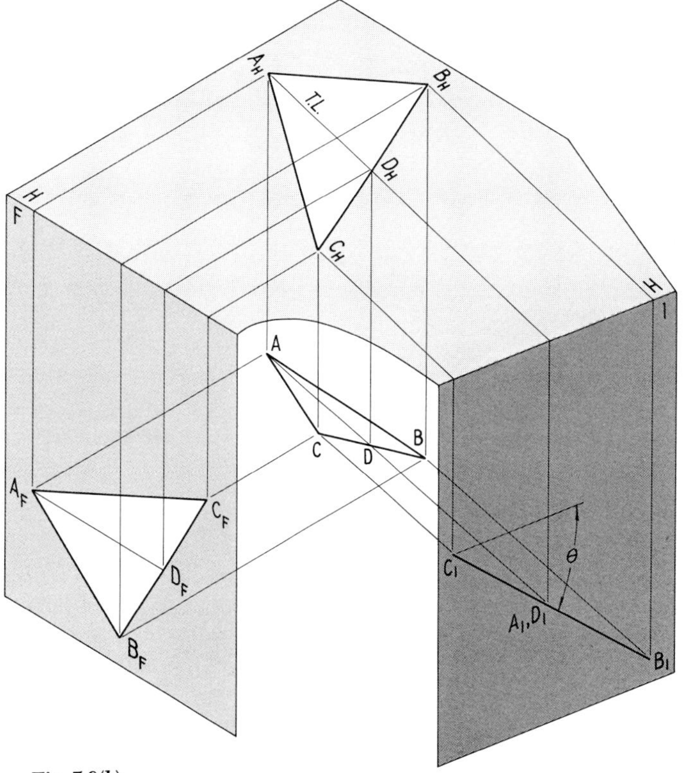

Fig. 7.8(b)

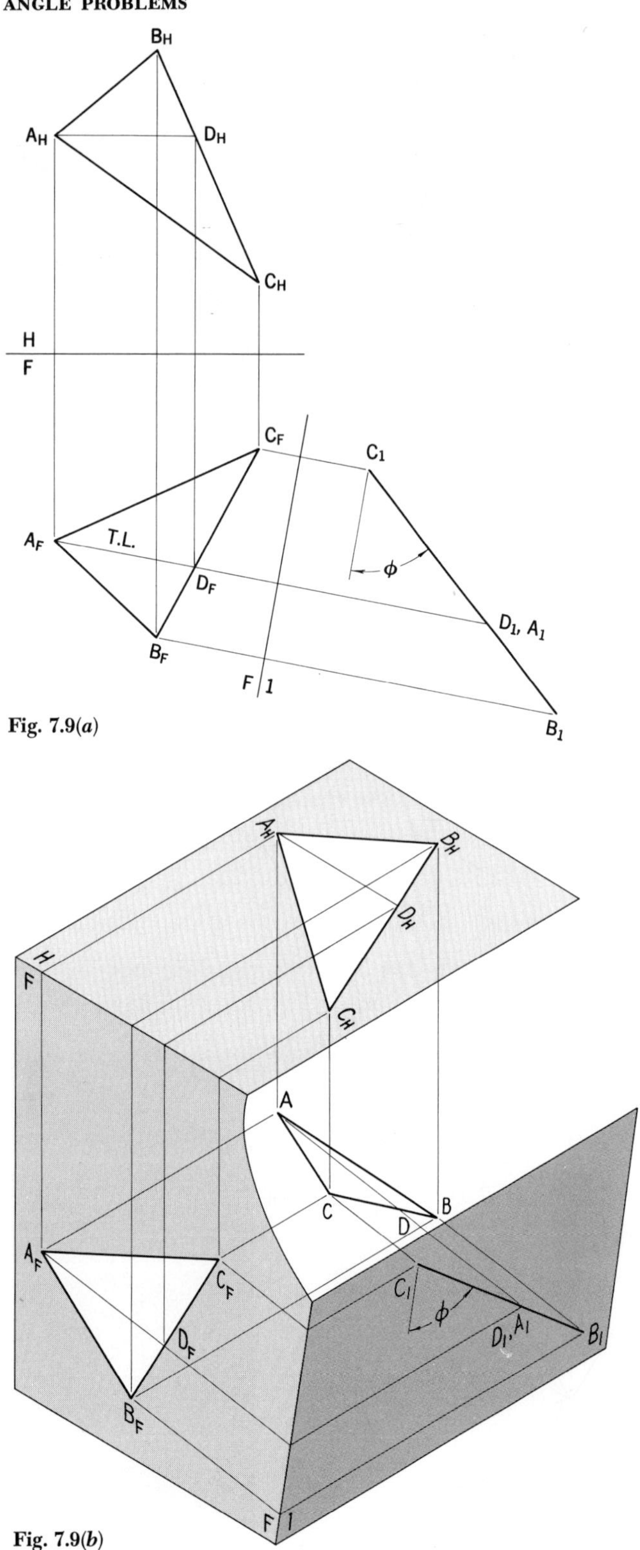

Fig. 7.9(*a*)

Fig. 7.9(*b*)

The angle between a given plane surface and the *F* reference plane can be determined in a similar manner. In Fig. 7.9(*a*), line *AD*, a frontal line, has been established in surface *ABC*. Since line *AD* is parallel to the *F*-plane its true length will be seen as $A_F D_F$. Supplementary plane 1 is positioned perpendicular to line *AD*. The view of surface *ABC* on plane 1 is, therefore, a line. Plane *F* is also seen on edge in the same view. Angle ϕ is the required angle.

A pictorial representation is shown in Fig. 7.9(*b*).

STRIKE LINE AND DIP ANGLE OF A PLANE SURFACE

The terms *strike line* and *dip angle* of a plane surface are commonly used in the fields of geology and mining.

A strike line is a horizontal line that lies in the plane surface. It is identified by its bearing.

The dip angle is the inclination of the plane surface to the horizontal. The magnitude of the inclination is measured in degrees, and below the horizontal plane.

In Fig. 7.10 we see that line *AD* is a horizontal line in plane surface *ABC*. *Line AD is a strike line.* It is identified by its bearing N θ° W.

In our previous example we observed that line *AD* was common to plane surface *ABC* and horizontal surface *ADE*. Now we should recognize the fact that any horizontal line (strike line) in surface *ABC* is common to both surface *ABC* and the horizontal plane which contains the horizontal line. Since the angle between two planes will be seen in the view that shows the common line as a point, it is only necessary to obtain the point view of the strike line *AD* in order to see the angle between plane surface *ABC* and the horizontal. This angle, ϕ, is the *dip angle.* (See Fig. 7.11.)

Now let us consider three additional angle problems: (*A*) the angle between two intersecting lines; (*B*) the angle between two nonintersecting lines; and (*C*) the angle between a line and a plane. In research, development, and design many problems arise, including the angle of bend of pipe members, reinforcing steel, rods; i.e., such as are used to connect an accelerator pedal to a carburetor, the clutch pedal to the clutch (rods will have bends to clear other parts); or cable sleeves for passing cable through a bulkhead (partition). Also, for example, in the field of biomechanics it may be necessary at times to determine the angle between bone members of the human body. An illustration of such a problem will be presented later in this chapter.

Now let us consider the three angle problems stated above.

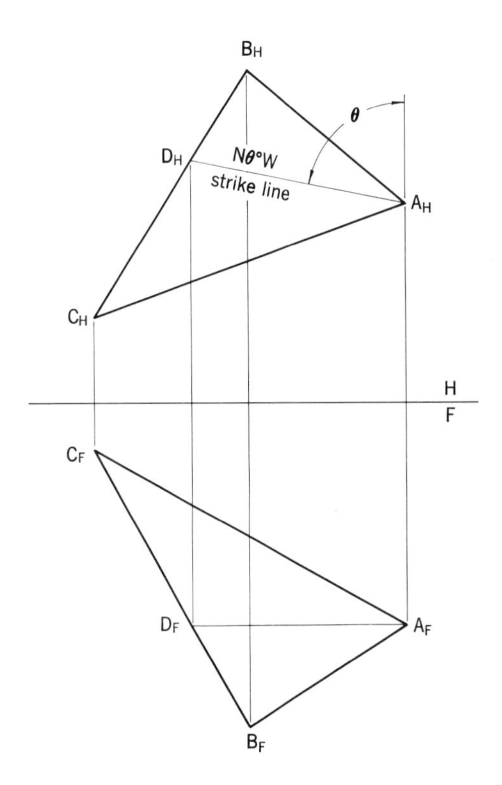

Fig. 7.10

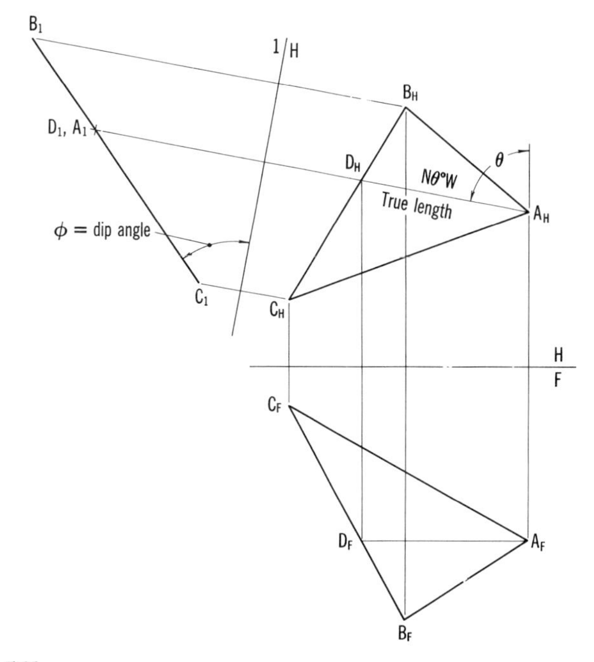

Fig. 7.11

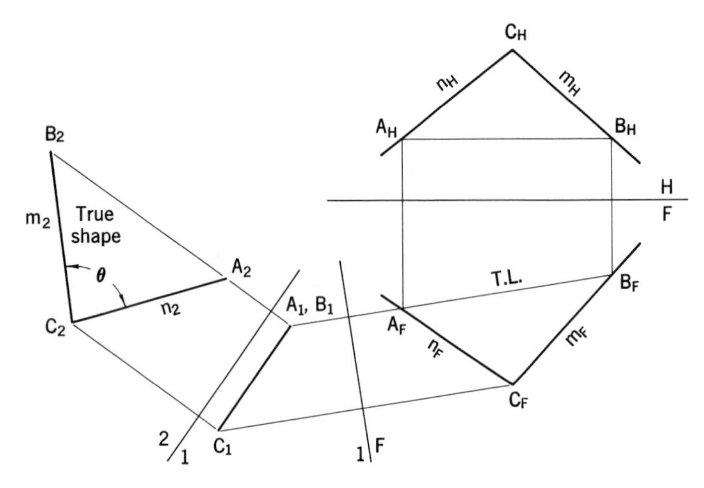

Fig. 7.12

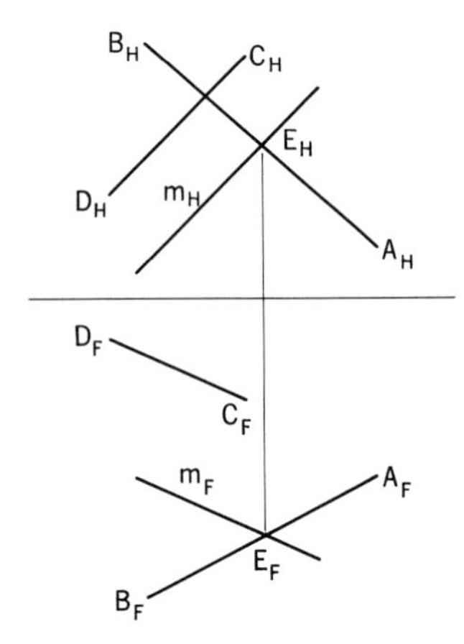

Fig. 7.13

A. Angle Between Two Intersecting Lines

Let us consider the two intersecting lines, m and n, shown in Fig. 7.12. Suppose we introduce frontal line AB, across lines m and n, to form triangle ABC. Now, if we proceed to find the true shape of triangle ABC, we will then obtain the true angle between lines m and n. From our previous discussion on true shape we may proceed to use one of the methods employed in finding the true shape of a plane surface. If we choose method (C), then we can easily obtain a point view of side AB and the corresponding edge view of triangle ABC, shown as $A_1B_1C_1$. Once we have this edge view it is easy to introduce supplementary plane 2 parallel to the surface ABC. The view, $A_2B_2C_2$ on plane 2 shows the true shape of the triangle and the true angle, θ, between lines m and n.

B. The Angle Between Two Nonintersecting Lines (Skew Lines)

The angle between two nonintersecting lines is the same as the angle between intersecting lines that are, respectively, parallel to the nonintersecting lines. For example, consider the skew lines AB and CD shown in Fig. 7.13. Suppose we select point E on line AB and then introduce a new line m through point E and parallel to line CD. The angle between lines m and AB is the required angle between the two given skew lines AB and CD.

The graphical solution is quite simple since the problem has been reduced to the previous case—the angle between intersecting lines.

C. Angle Between a Line and a Plane

The angle between a line and a plane is defined as "the angle between the line and its orthogonal projection upon the plane." The orthogonal projection can be obtained by determining the line which joins the points in which two perpendiculars (drawn through two points of the given line) intersect the plane. The angle between the given line and its projection, then, is the required angle. *It is well to recognize that this angle is complementary to the angle between the line and either one of the perpendiculars to the plane.*

Therefore, one approach to the solution is (a) first introduce a perpendicular to the plane from a point on the line, and then (b) find the angle between the given line and the perpendicular. This angle is complementary to the required angle. Therefore, the required angle, θ, is equal to $(90° - \alpha)$ where α is the angle between the line and the perpendicular to the plane.

Before we proceed to the general case of the angle between a line and any plane surface, it will be helpful to know how to establish the views of a perpendicular from a given point to a plane surface.

Suppose we are given point K and a plane, part of which is triangle ABC, as shown in Fig. 7.14. Let us first introduce horizontal line AD. Its true length is seen in the view $A_H D_H$. The perpendicular, n, through point K has a top view, n_H, that passes through K_H and perpendicular to $A_H D_H$. We recall that, "when two lines are perpendicular to each other, the view upon a plane that is parallel to one of the lines will reveal a right angle."

Now we introduce frontal line AE. Its true length is seen in the view $A_F E_F$. The perpendicular n through K has a front view, n_F, that passes through K_F and perpendicular to $A_F E_F$. Line n is perpendicular to plane ABC because the line is perpendicular to two lines, AD and AE, that lie in the plane.

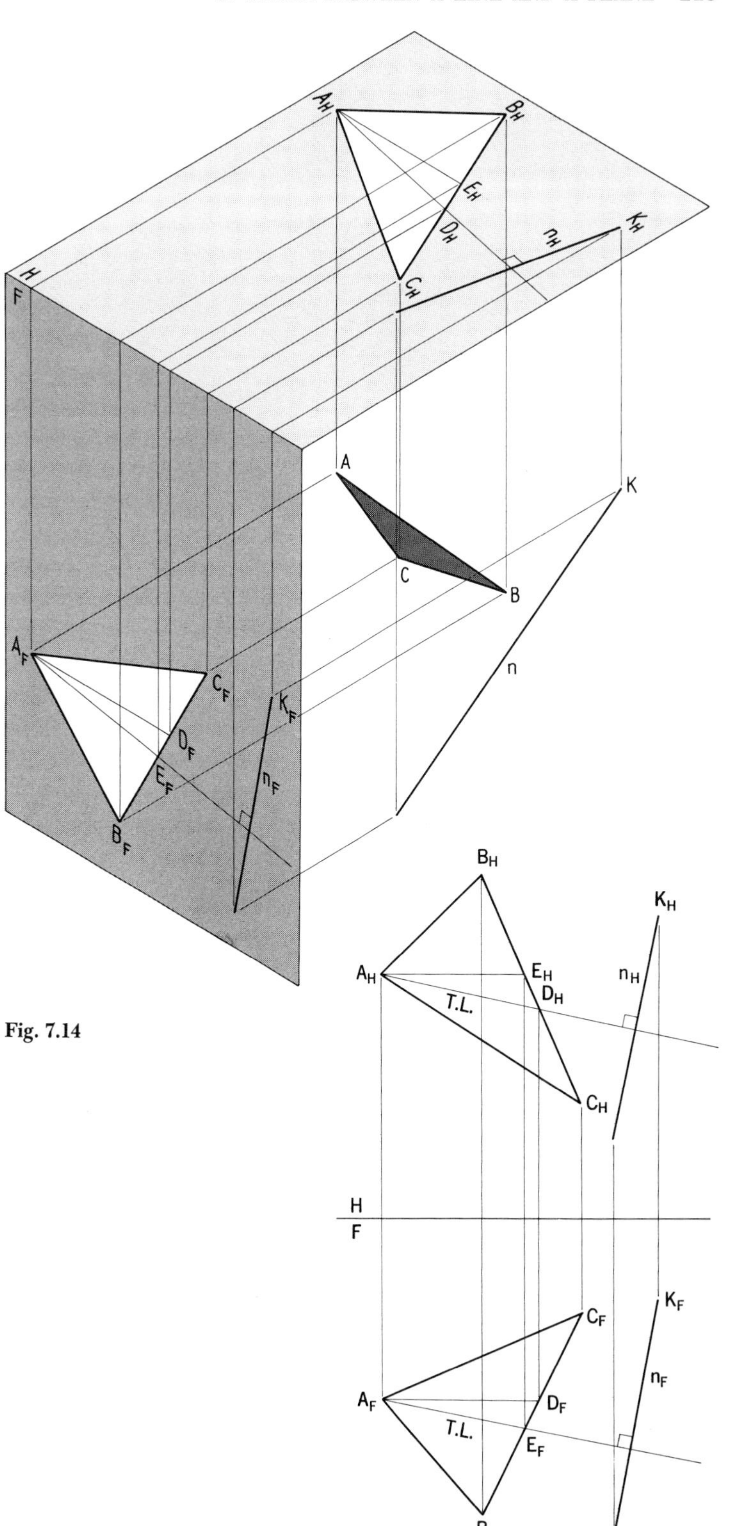

Fig. 7.14

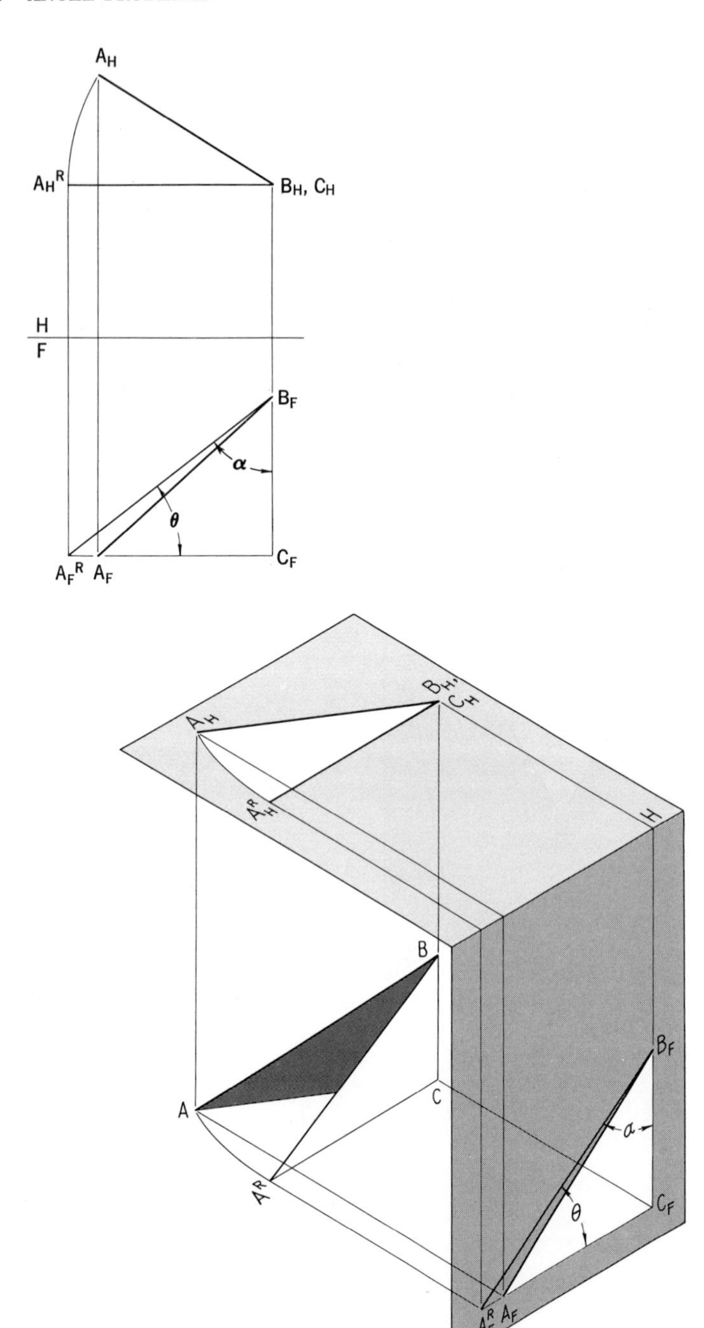

Fig. 7.15

EXAMPLE 1

Let us determine the angles θ, ϕ, and ψ that line AB makes with the H-, V-, and P-planes, respectively. Let us consider the determination of angle θ (Figure 7-15). Through point B a perpendicular BC is drawn to the H-plane. Line AC is horizontal. The true shape of right triangle ABC will show the angle, α, between lines AB and BC. The complement to this angle is θ which is the angle between line AB and its projection upon the H-plane. The true shape of triangle ABC is shown as $A_F{}^R B_F C_F$.

In a similar manner we can determine angle ϕ. (See Fig. 7.16.) Through point B a perpendicular BD is drawn to the F-plane. Line AD is a frontal line (parallel to the F-plane). The true shape of right triangle ABD will show the angle, β, between lines AB and BD. The complement to this angle is ϕ, the required angle with the F-plane.

The determination of angle ψ is found in a like manner. The reader should be able to do this without much difficulty.

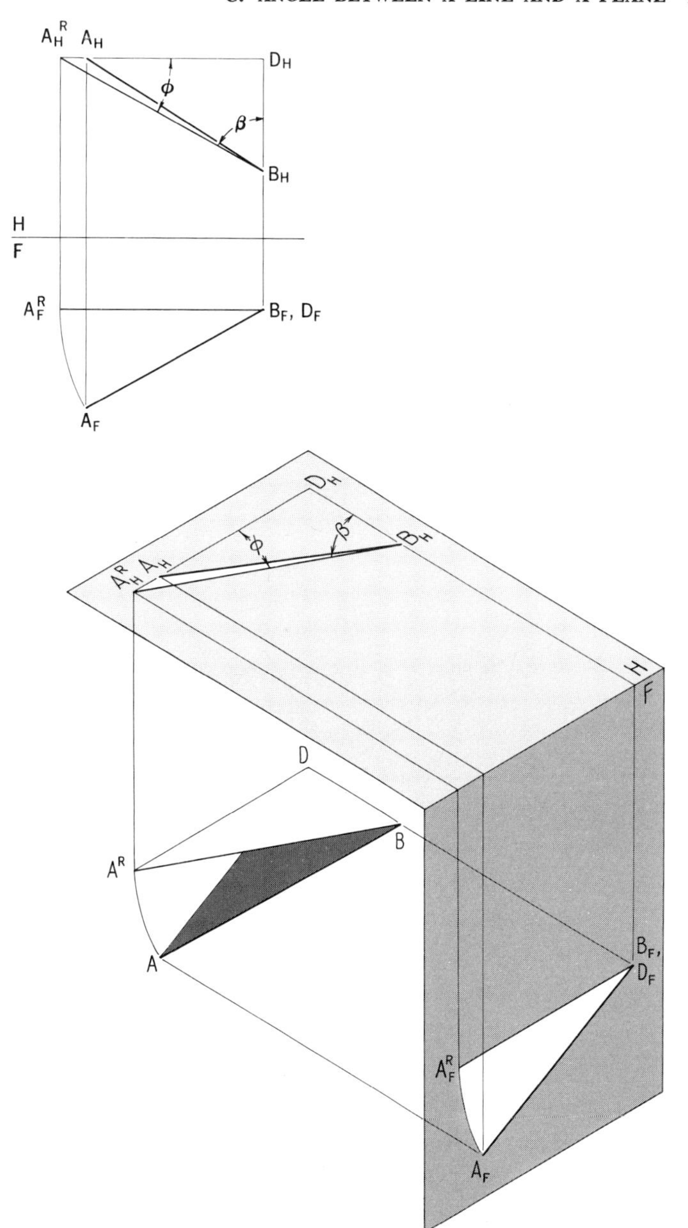

Fig. 7.16

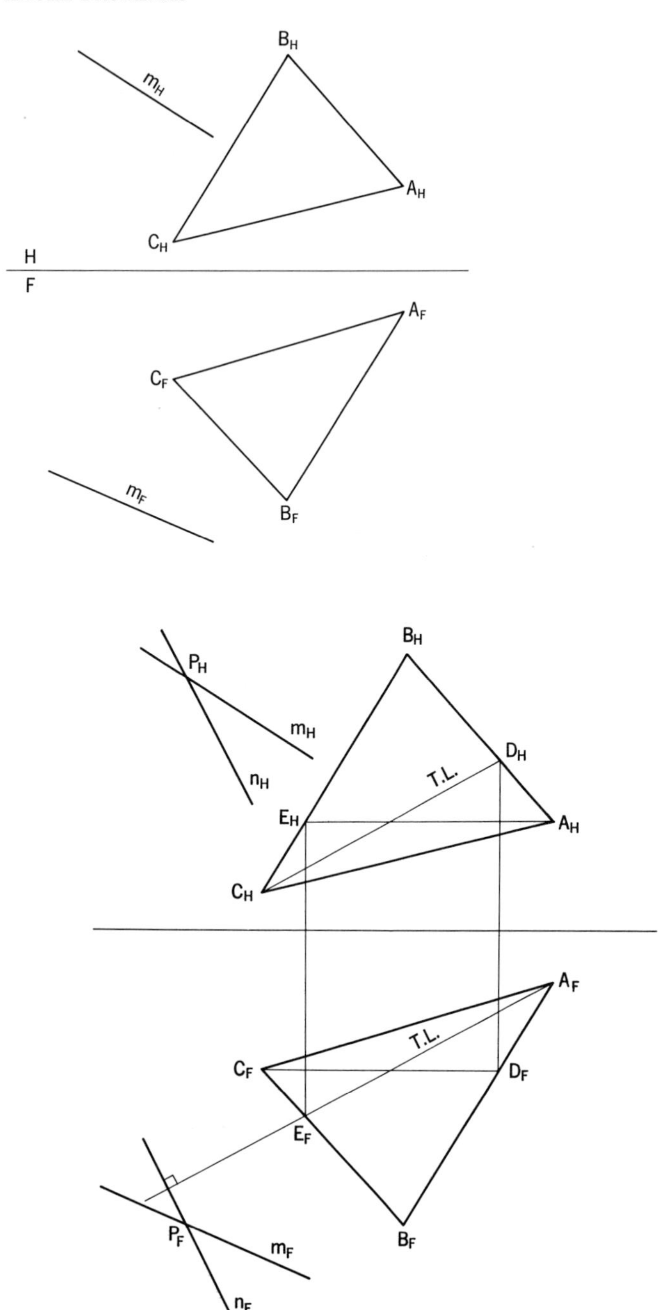

Fig. 7.17

EXAMPLE 2

Let us now consider Fig. 7.17 which shows line m and plane ABC. We are to determine the angle between line m and the plane by first finding the angle between line m and a perpendicular, n, to the plane; and then recording the complement to this angle.

By employing the method shown in Fig. 7.14 we can readily establish the H and F views of line n through point P, an arbitrary point on line m. Once this is done the solution is quite simple since we need only find the angle between lines m and n. The complement of this angle is the required angle between line m and plane ABC.

Alternative Solution. Let us now consider a method which employs supplementary views. *The angle between a line and a plane is seen in a view which shows the line in true length and the plane on edge.*

Case I. Let us consider Fig. 7.18, which shows line *DE* in true length in the front view and also the edge view of plane *ABC* in the *same* view. The angle, θ, between line *DE* and plane *ABC*, is also seen in its true magnitude.

How do the two views, *H* and *F*—one showing the true shape of the plane figure *ABC* and the other showing both the edge view of the plane *ABC* and the true length of line *DE*—satisfy the definition: "the angle between a line and a plane is the angle between the line and its orthogonal projection upon the plane"?

Where is the orthogonal projection of the line upon the given plane? Suppose we select any two points, such as 1 and 2 on line *DE*. Perpendiculars drawn through these points to the plane *ABC* intersect that plane in points 3 and 4, respectively. The line joining points 3 and 4 is the projection of line 1–2 (a portion of line *DE*) upon plane *ABC*. Since the front view shows the true lengths of both lines *DE* and 3–4, the angle θ between them appears in true magnitude.

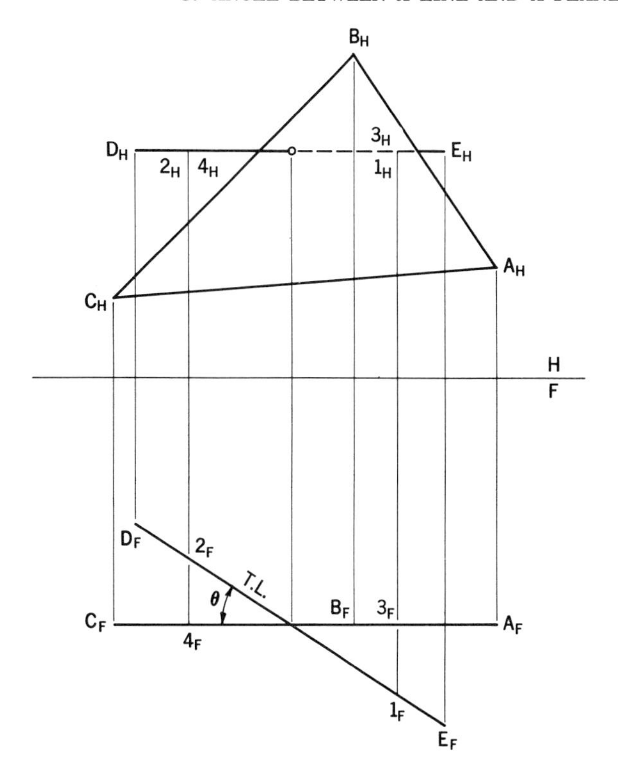

Fig. 7.18

Case II. Let us next consider Fig. 7.19. We observe that the front view shows plane *ABC* as a line, since the plane is perpendicular to the frontal plane, and shows line *DE* in true length. The angle between line *DE* and plane *ABC* is apparent in the front view. Although the top view of plane *ABC* is not true shape, nevertheless the line *DE* and its projection (line 1–2) upon the plane *ABC* are shown in true length in the front view; therefore, the angle θ between line *DE* and plane *ABC* appears in its true magnitude in the front view.

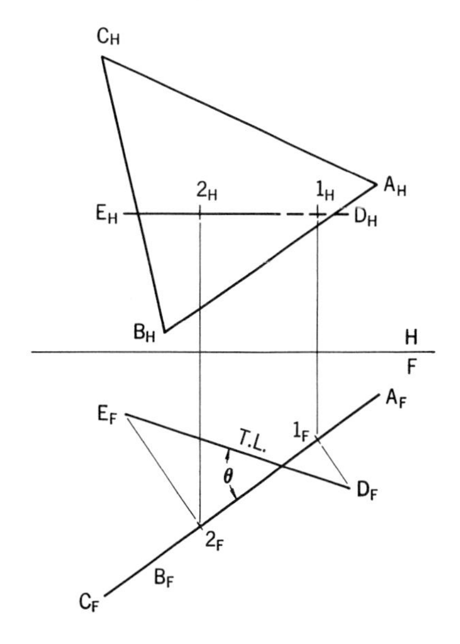

Fig. 7.19

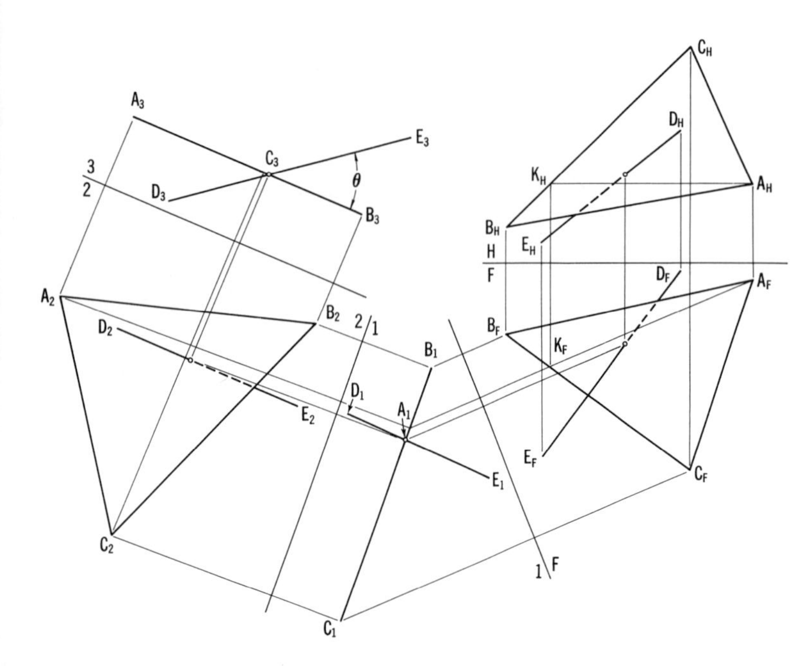

Fig. 7.20

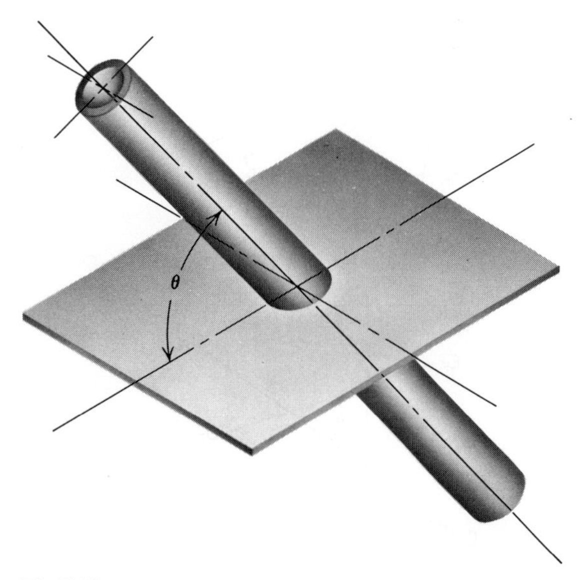

Fig. 7.21

Case III. Now let us consider the general case shown in Fig. 7.20. We can reduce this general case to Case I in the following manner.

Since Case I includes a true shape view of the plane, we can readily establish: (*i*) supplementary plane 1 perpendicular to both frontal line AK and the F-plane. The view on plane 1 then shows the edge view of the plane as $A_1B_1C_1$, and the line as D_1E_1; and (*ii*) supplementary plane 2 which is parallel to surface ABC and perpendicular to plane 1. The view on plane 2 shows the true shape of plane ABC as $A_2B_2C_2$, and the line as D_2E_2; and (*iii*) supplementary plane 3 which is parallel to line DE and perpendicular to plane 2. The view on plane 3 now shows the edge view of surface ABC as $A_3B_3C_3$ and the *true length* of line DE as D_3E_3, and the required angle, θ, between the line and the plane surface.

It should be pointed out that the above solution is ideally suited to the preparation of a working drawing that would be necessary for the production, for example, of a plate (ABC) and a connecting rod (DE), since the views on supplementary planes 2 and 3 disclose the true shape of the surface ABC, the true length of DE, and the true angle, θ.

Another application is shown in Fig. 7.21 which represents a portion of a bulkhead (partition) and a cylindrical sleeve to guide a control cable. Here again it is necessary to produce orthographic views which will tell the whole story—the size of the bulkhead, the size of the cylindrical sleeve and the angle θ.

Alternative Solution 2. *Example 1.* In some cases it is desirable to determine only the angle, θ. When this is so, we can determine the angle in the following manner.

Let us consider Fig. 7.22. We observe that the front view shows the edge view of surface ABC. The true length of line DE, however, is *not* shown in the front view; therefore, the true angle, θ, is not seen in the front view. We can, however, rotate line DE parallel to the F-plane so as to show its true length. The angle, θ, is now seen as the angle between $D_F E_F{}^R$ and the edge view of the plane, shown as $A_F B_F C_F$.

It is very important to observe that point E moves in a plane parallel to the surface ABC, to establish the position $E_F{}^R$.

Example 2. Let us consider Fig. 7.23, which shows the H and F views of plane surface ABC and line DE. We can easily reduce this problem to that

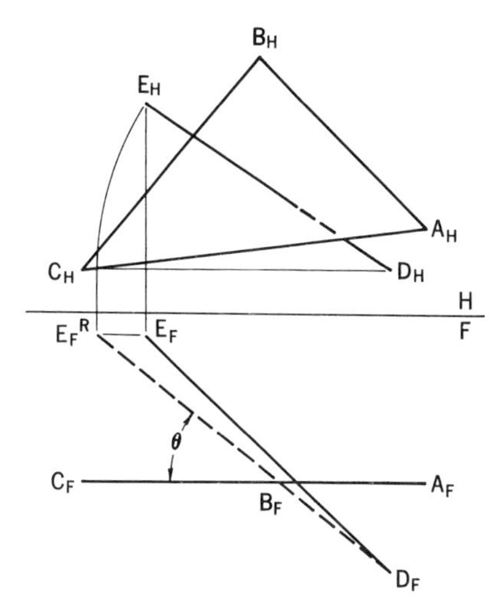

Fig. 7.22

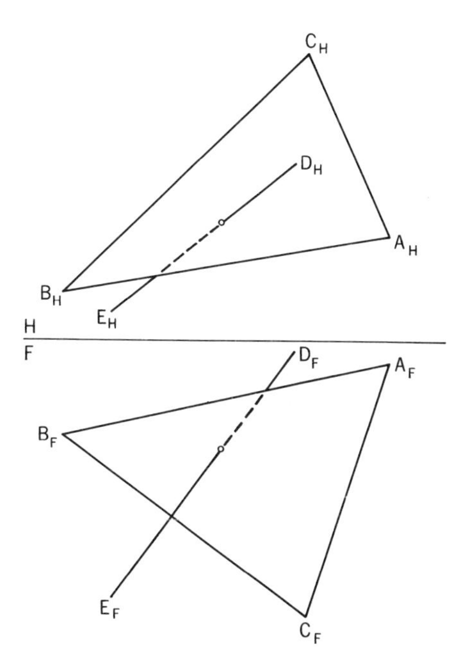

Fig. 7.23

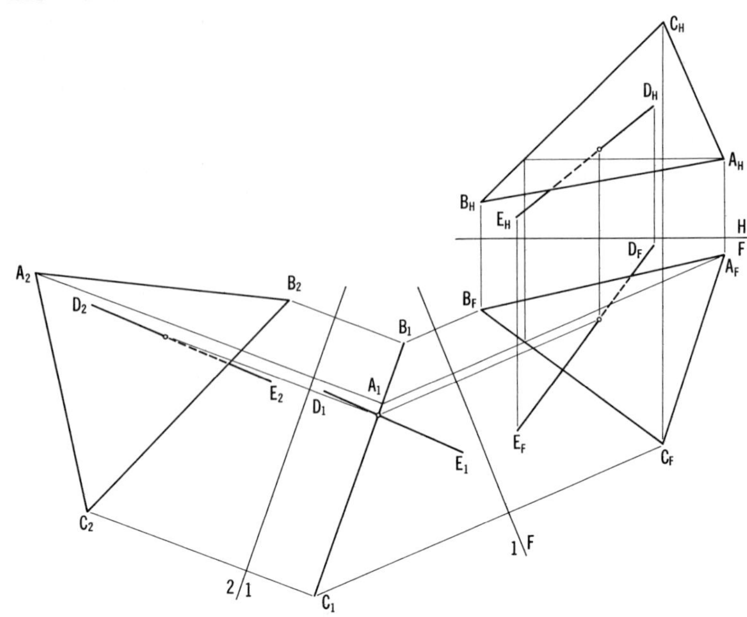

Fig. 7.24

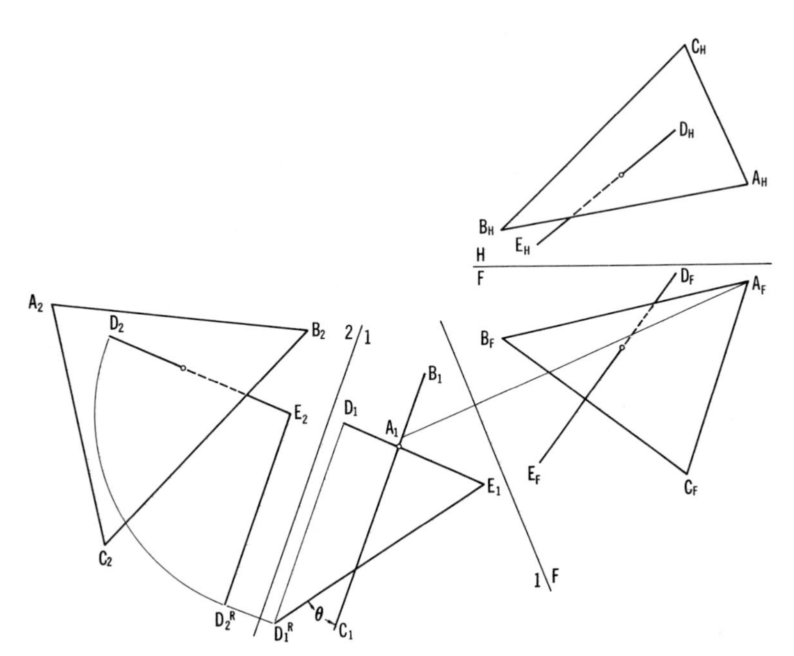

Fig. 7.25

of Case II by the introduction of two supplementary views (Fig. 7.24), one showing the edge view of the surface, such as the view on plane 1, the other showing the true shape of the surface ABC on plane 2; and, of course, the corresponding views of line DE. Now we can treat the views on planes 1 and 2 as the given views and proceed to solve the problem in the manner shown in Fig. 7.22. This means we can rotate line ED parallel to plane 1 and obtain its true length in the view on plane 1—shown in Fig. 7.25 as $E_1D_1^R$ (*note again that D_1 moves in a plane parallel to the surface which is shown as $A_1B_1C_1$*). The angle, θ, between line DE and surface ABC, is seen in the view on plane 1.

Remark. It is possible to find angle θ without using the second supplementary plane.

Application of "the Angle Between a Line and Plane" to a Problem in Prosthetic Devices Research

In 1945 the Department of Engineering of the University of California at Berkeley undertook a research project (supervised by Professor H. D. Eberhart, Project Director) in the field of prosthetic devices, dealing with problems related to the improvement and design of artificial limbs. This project, which is still active, embraces the following:

1. Studies of human locomotion—the mechanics of motion of the legs, measurements of the ranges of motion in space, including rotations of the major segments of the legs during locomotion, and, in particular, the study of the action in the major joints of the leg, i.e., the hip, knee, and ankle joints.
2. Studies and analysis of the phase and action of the musculature.
3. Studies relating to factors contributing to the comfort of the amputee.
4. Muscle energy output in relation to locomotion and energy characteristics for design of artificial limbs.
5. Structural analysis, design, and accelerated testing of various types of leg prostheses.

Several problems that arose in the study of human locomotion were best solved graphically. Among these was one dealing with the determination of the "tibiofemoral" angle; namely, the angle between the femur (thigh bone) and the plane determined by the tibia (shin bone) and a steel pin screwed into the tibia (under sterile conditions and anesthesia, of course).

The graphical representation and solution of the problem is shown in Fig. 7.26.

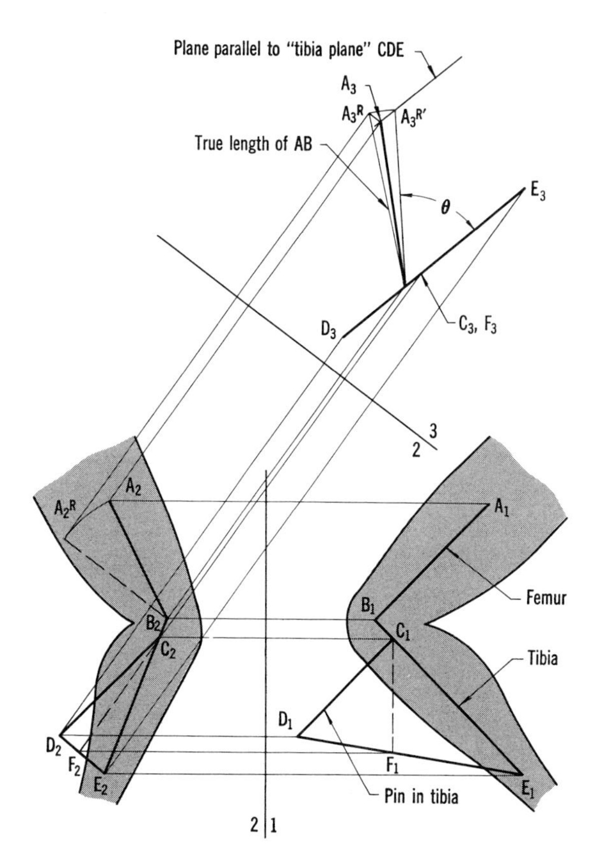

Fig. 7.26

EXERCISES

1. Determine the angle between lines m and n shown in Fig. E-7.1.

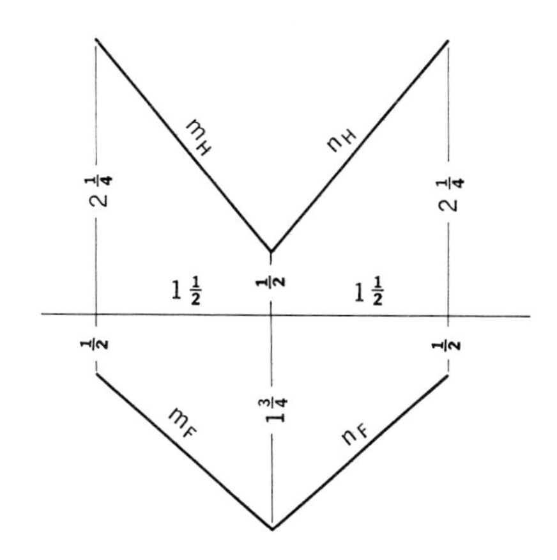

Fig. E-7.1

2. Determine the angle between skew lines m and n shown in Fig. E-7.2.

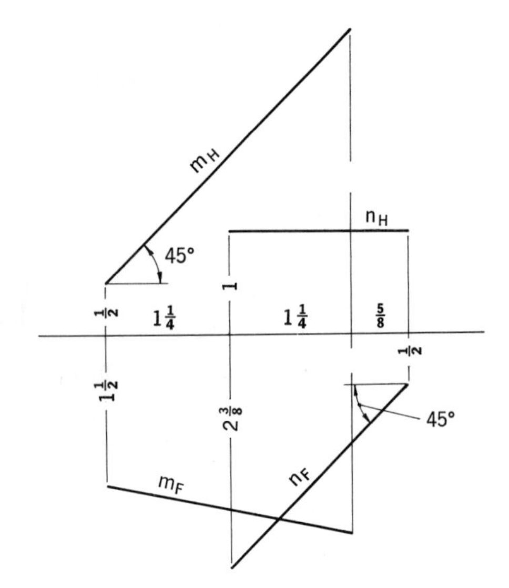

Fig. E-7.2

3. Determine the angle between control cable EF and bulkhead (plane surface) $ABCD$ shown in Fig. E-7.3.

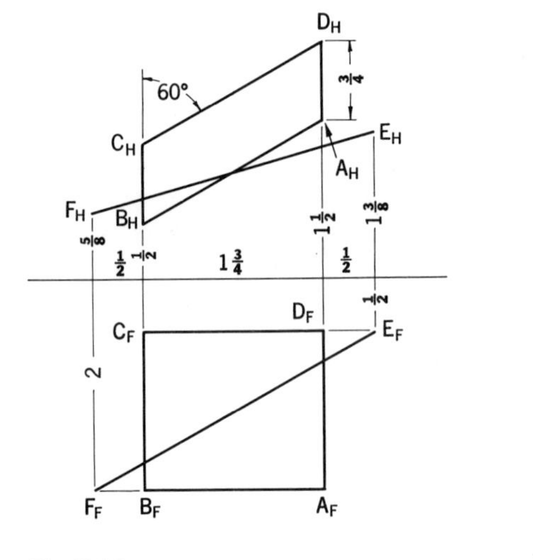

Fig. E-7.3

4. Determine the angles between guy wires OA, OB, and OC, and the roof surface ABC shown in Fig. E-7.4.

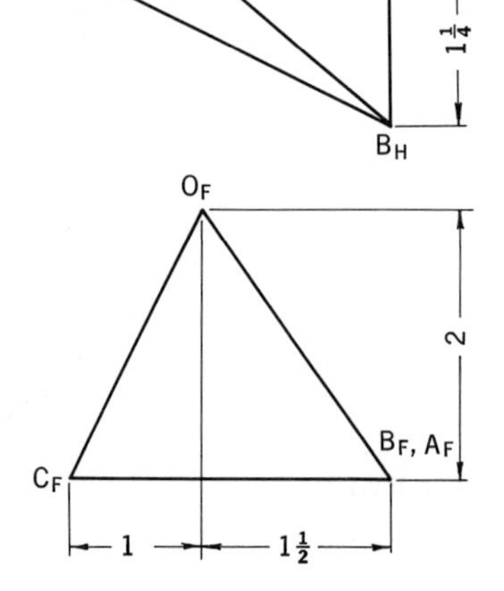

Fig. E-7.4

5. What is the magnitude of the angle between member *OA* and plate *ABC* shown in Fig. E-7.5?

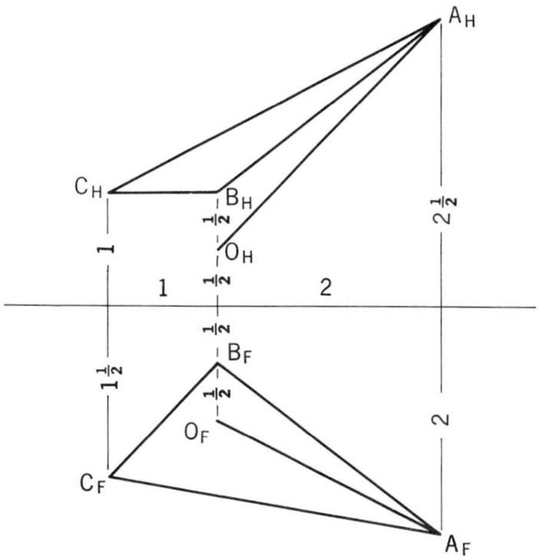

Fig. E-7.5

6. What is the magnitude of the angle between member *OA* and the plane determined by members *OB* and *OC* shown in Fig. E-7.6?

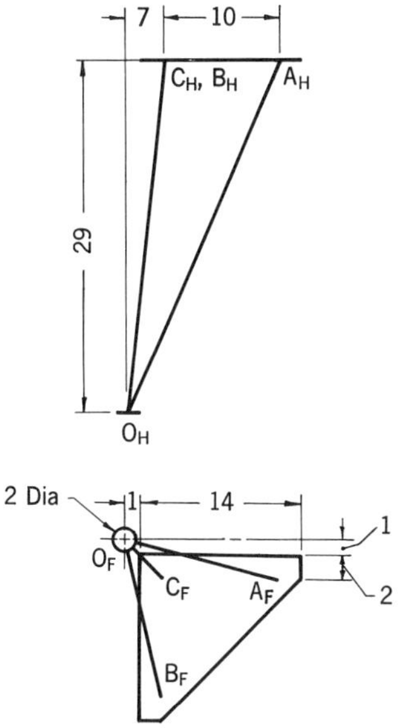

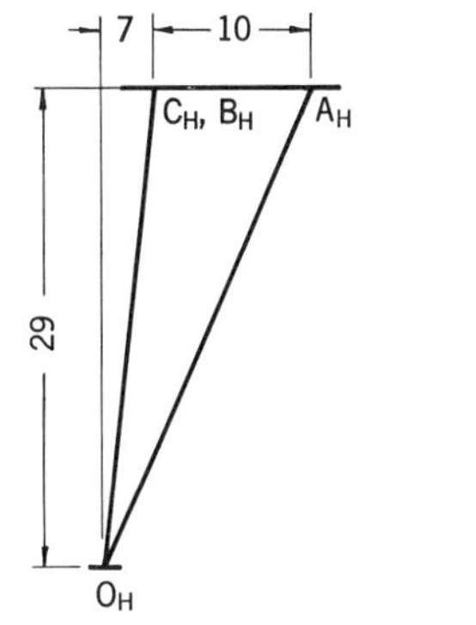

Fig. E-7.6

7. Determine the magnitude of the angles between member *AB* and the surfaces shown in Fig. E-7.7.

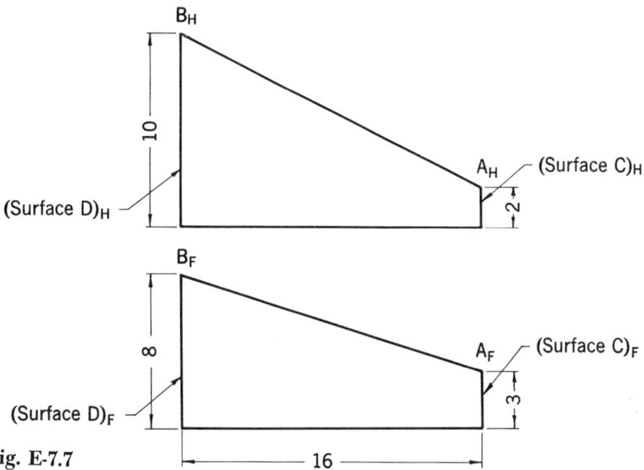

Fig. E-7.7

8. Determine the angle between plates *ABC* and *ADB*. Point *D* is 1½ inches beneath point *A* (Fig. E-7.8).

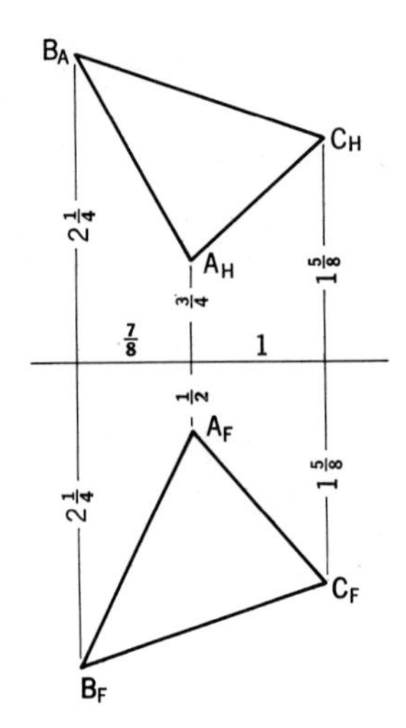

Fig. E-7.8

9. Determine the strike line and dip angle of surface *ABC* (Fig. E-7.9).

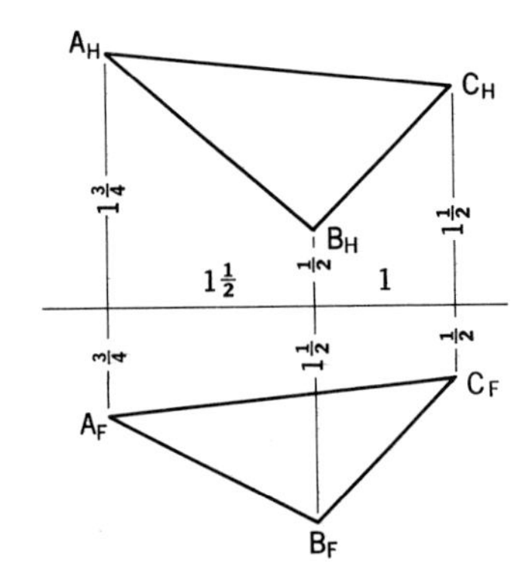

Fig. E-7.9

10. The bearing of strike line *CD* of plane *ABC* is N 60° E. Determine the dip angle of the surface; also complete the *H* view (Fig. E-7.10).

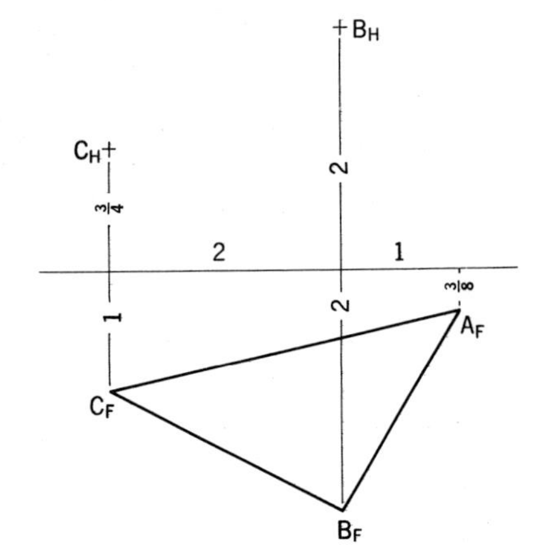

Fig. E-7.10

DEVELOPMENTS 8

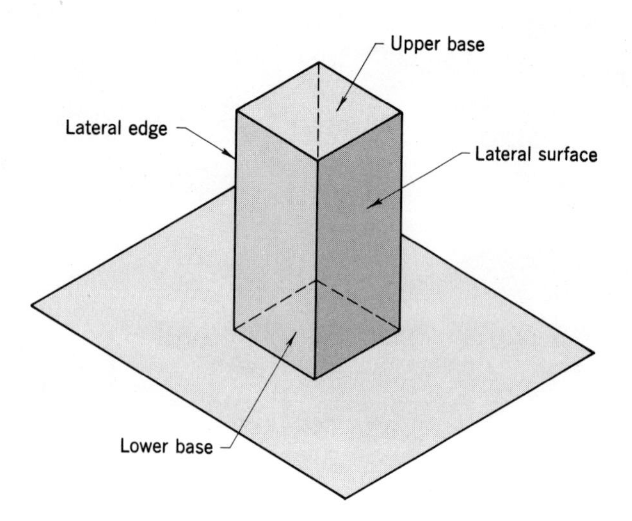

Fig. 8.1

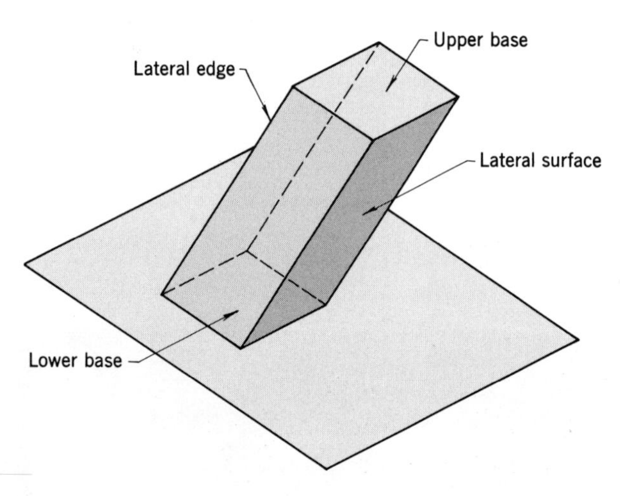

Fig. 8.2

In many engineering designs, parts may consist of plane surfaces, of cones, of cylinders, and of other single-curved surfaces that can be developed accurately. Many of the space vehicles are combinations of the above mentioned surfaces. Developments are flat-sheet layouts of the surfaces. These layouts provide patterns for cutting material that can be properly bent or shaped to form the desired geometry.

In some designs there are parts that include double-curved or warped surfaces. The double-curved surfaces can be developed approximately by assuming that they consist of small portions of developable surfaces, such as cones and cylinders; and the warped surfaces by assuming that they consist of relatively small triangles which, of course, are developable.

The theory employed in making a development from orthographic views of the part is very simple. *Essentially it consists of finding true lengths of the elements of the given surface* (or of the substituted surface) and the correct *placement of one element with respect to an adjacent element.* Let us first consider the development of prisms.

DEVELOPMENT OF PRISMS

Prisms are solids bounded by lateral faces that are parallelograms and by bases that lie in parallel planes. (See Fig. 8.1.) When the lateral edges are perpendicular to the bases, the prism is a right prism and the lateral surfaces are rectangles. When the bases do not lie in parallel planes, the prism is a truncated prism. An oblique prism is shown in Fig. 8.2. In this case the lateral edges are not perpendicular to the bases.

EXAMPLE 1

As a very simple example we will consider the development of the right triangular prism shown in Fig. 8.3. The true lengths of edges AB, CD, and EG are shown in the front view. *Since the top view shows the edges as points, the perpendicular distances between them are readily available.* These distances are shown as k, m, and n in the top view. The development, or flat-sheet layout, of the lateral surfaces of the prism can now be easily constructed. Edge EG (we could have used one of the other edges) is laid off a convenient distance from E_FG_F, equal in length to E_FG_F and parallel to it, as shown in the figure. On a line perpendicular to EG, distances k, m, and n are laid off. These are the perpendicular distances between the adjacent lateral edges.

Since the lateral edges are the same length, we can readily complete the development which is shown in the figure. The development can now be used as a pattern from which the material could be cut; and lines CD and AB as the bend lines to form the lateral surfaces of the prism. Two triangular shapes, $A_HC_HE_H$, would form the top and bottom bases of the prism.

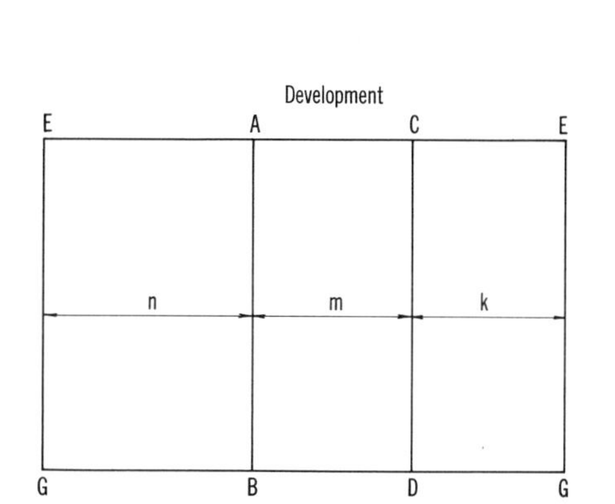

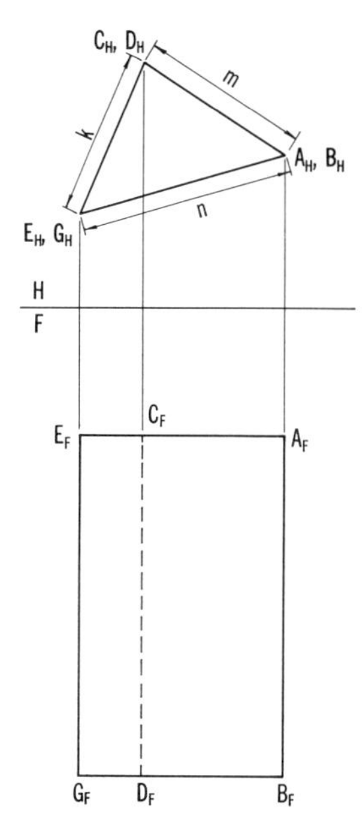

Fig. 8.3

EXAMPLE 2

Case I. Truncated Triangular Prism—Edges Perpendicular to a Reference Plane. Let us consider Fig. 8.4, which shows a truncated triangular prism. The true lengths of the edges AB, CD, and EG are shown in the front view. Since the top view shows the edges as points, the perpendicular distances between them are readily available. These are shown as k, m, and n. The development, or pattern, of the lateral surface of the prism can now be easily constructed. Edge EG is laid off a convenient distance from E_FG_F and parallel to it as shown in the figure. On a line perpendicular to EG, distances k, m, and n are laid off. They locate the positions of the edges. Terminal points of lines CD, AB, and EG may be obtained by direct projection from the front view as illustrated in the development.

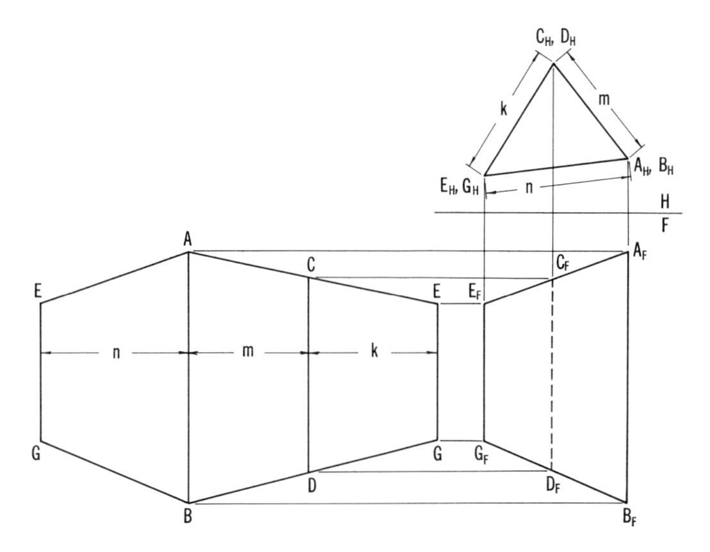

Fig. 8.4

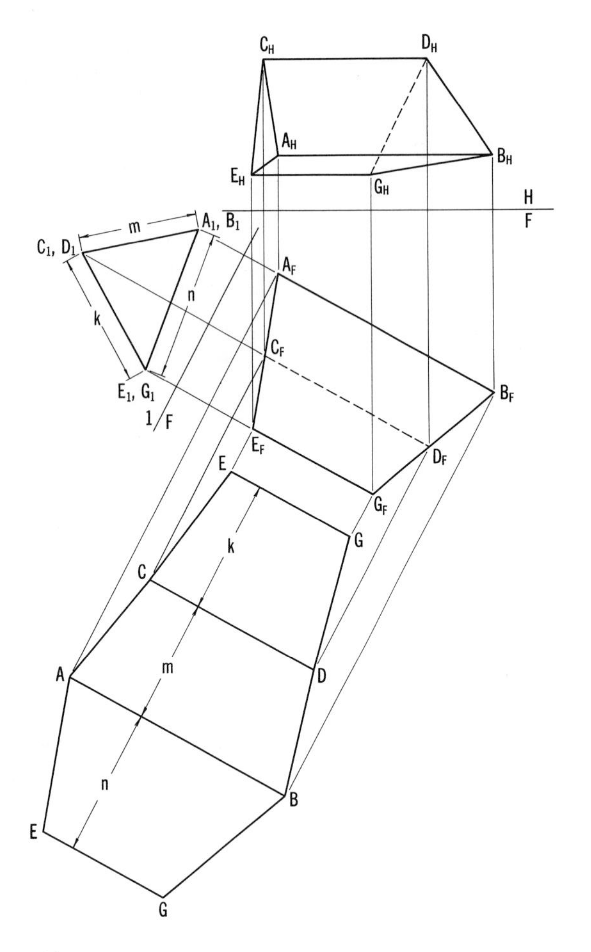

Fig. 8.5

Case II. Edges Parallel to One Reference Plane. Now suppose that the given views of the prism are the top and front views, as shown in Fig. 8.5. Careful study of the top and front views leads to the conclusion that the edges are parallel to the frontal plane and, therefore, appear in true length in the front view. Supplementary plane 1 can be easily set up perpendicular to the edges of the prism, thus yielding point views of the edges. The distances between the edges are available in this view since the edges appear as points.

The problem has now been reduced to Case I (see Fig. 8.4), if we regard the front and supplementary views as the given views. The development is established in the manner previously discussed.

Case III. Edges of Prism Not Parallel to a Reference Plane. Now let us consider the general case shown in Fig. 8.6. It should be quite clear that neither the top view nor the front view discloses the true lengths of the edges of the prism. We can, however, reduce this problem to Case I, previously discussed (Fig. 8.4), by the introduction of (*a*) supplementary plane 1 which is parallel to the edges of the prism, thereby yielding a view which shows the true lengths of the edges; and (*b*) supplementary plane 2, perpendicular to the edges, thus enabling us to obtain a point view of the edges. The relationship between the supplementary views 1 and 2 is basically the same as the relationship between the top and front views shown in Fig. 8.4. The development is established in the manner discussed in the previous two examples.

DEVELOPMENT OF CYLINDERS

The development of the lateral surface of a cylinder requires only the determination of "the true lengths of the elements and the correct placement of one element with respect to an adjacent element." Since we cannot, from a practical point of view, use all (the infinite number) the elements, we choose a sufficient number to define the development within the desired accuracy. By doing so we have, in effect, replaced the cylinder with a prism, so that the method employed in developing a cylinder is, *basically,* the same as that used in developing a prism.

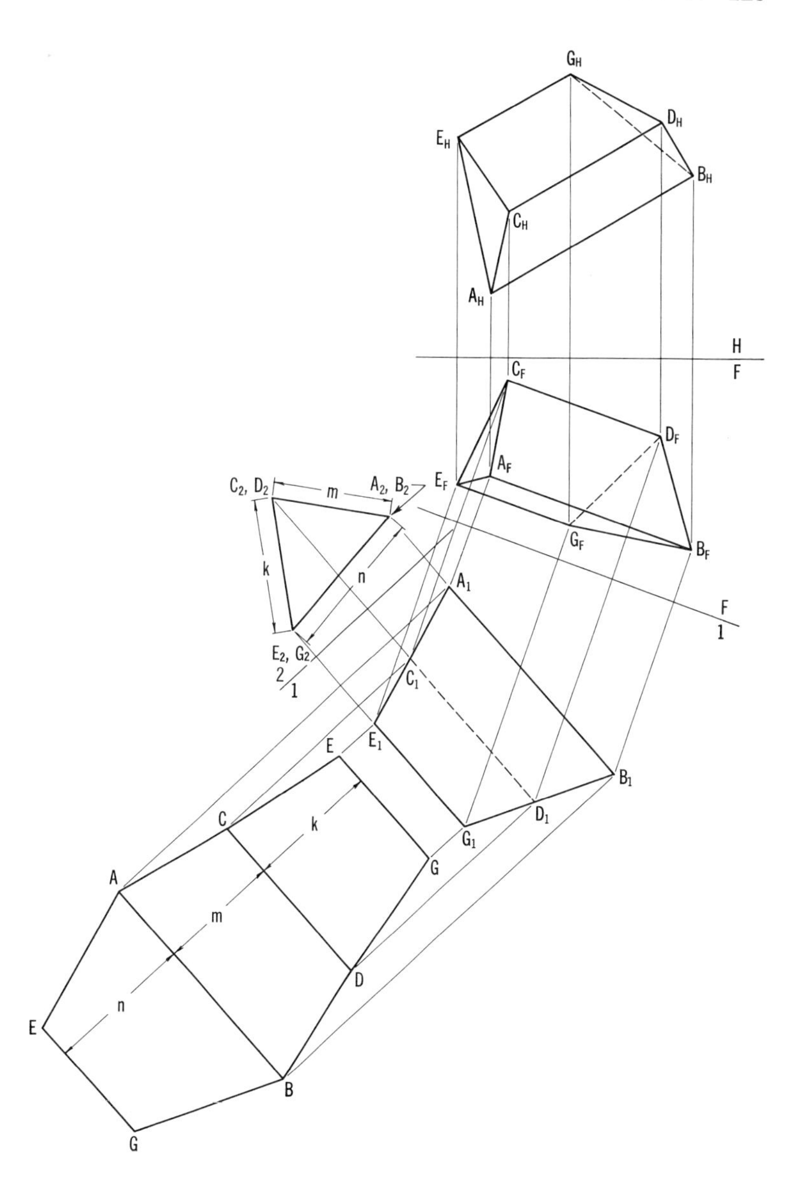

Fig. 8.6

EXAMPLE 1

Let us consider the circular cylinder shown in Fig. 8.7. The top view shows the point views of the elements and, therefore, the distances between them. All the elements are shown in true length in the front view. Now we have all the information necessary for the development of the lateral surface of the cylinder. We note that the method is the same as that used in developing the lateral surface of the prism shown in Fig. 8.4. It should be pointed out that the end points of the elements, shown in the development, are joined by a smooth curve.

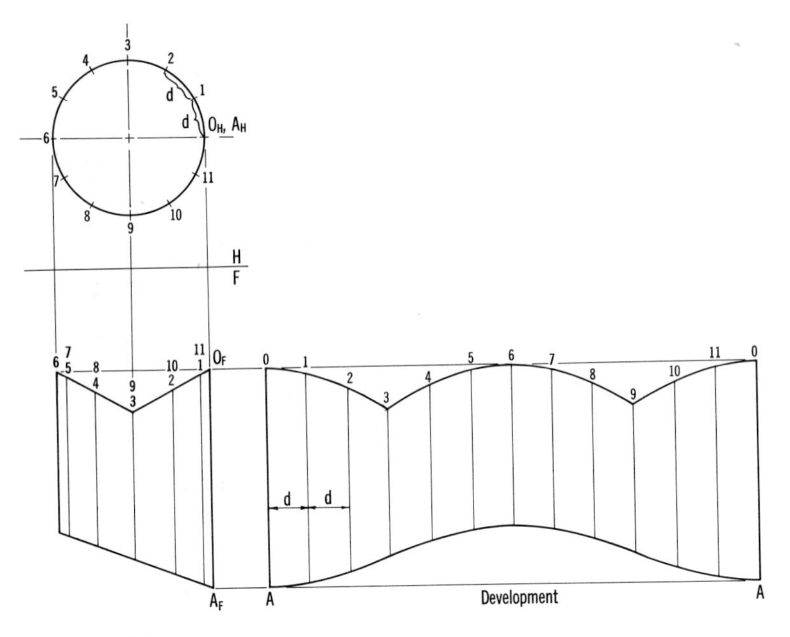

Fig. 8.7

EXAMPLE 2

Let us consider the elliptical cylinder shown in Fig. 8.8. The top view shows the point views of the elements and, therefore, the distances between them. In this example it should be observed that the elements are *not* equally spaced since the curvature varies. The elements are closer together where the curvature is sharper, and further apart where the curvature is flatter. The development, however, is made in the same manner employed in Example 1.

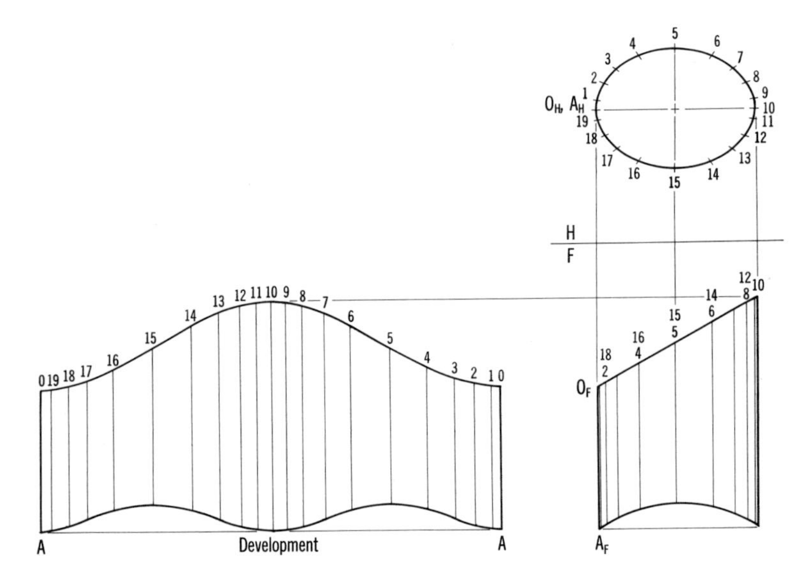

Fig. 8.8

EXAMPLE 3

Now let us consider the general case shown in Fig. 8.9. We see that neither the top view nor the front view shows the true lengths of the elements of the cylinder. We can, however, obtain a view which does show the true lengths of the elements and also a view which shows the distances between the elements. The view on plane 1 shows the true lengths and the view on plane 2 shows the point view of the elements and, therefore, the distances between them. The development of the lateral surface is easily constructed since the problem has been reduced to the simple case shown in Fig. 8.7.

Actually *the relationship among the views on planes 1 and 2 and the development is basically the same as the relationship among the top and front views, and the development shown in Fig. 8.7.*

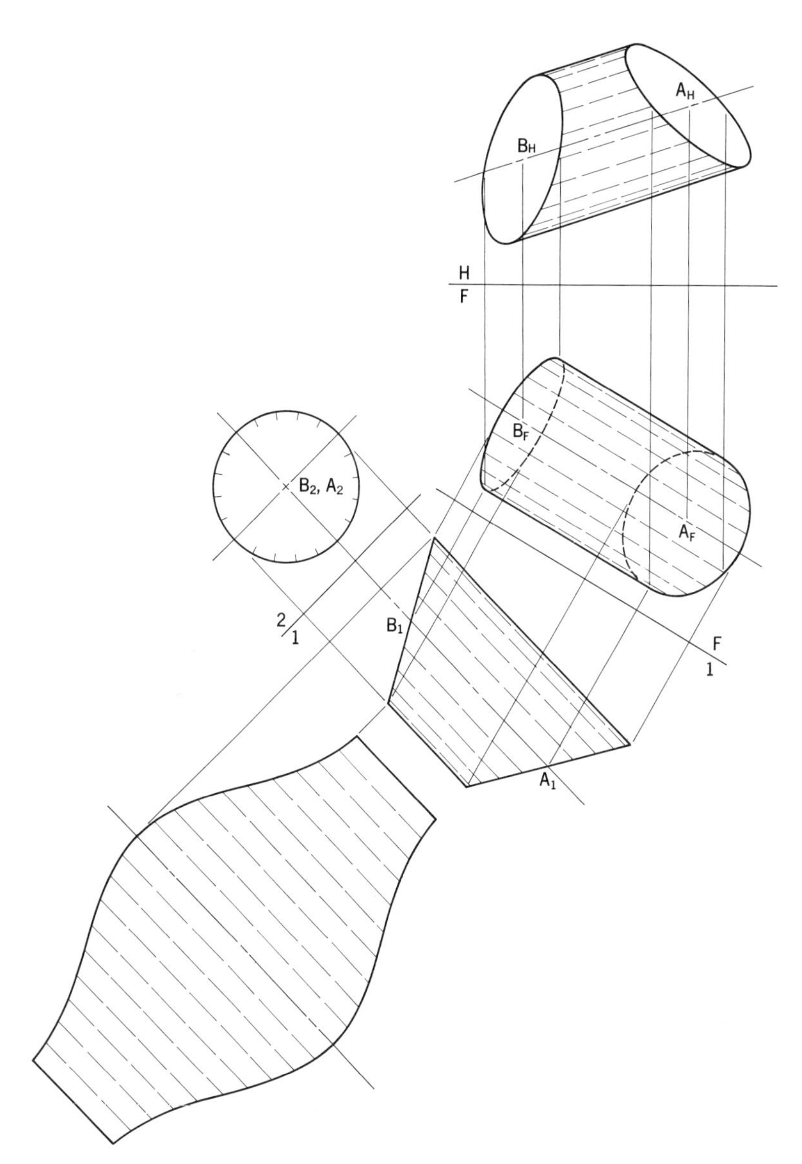

Fig. 8.9

DEVELOPMENT OF PYRAMIDS

EXAMPLE 1

Let us consider the triangular pyramid shown in Fig. 8.10. A pattern or development of the three lateral surfaces of the pyramid can be formed by joining the true shapes of triangles *OAB*, *OBC*, and *OCA*. The true shape of each triangle can be constructed, once the lengths of the sides are known.

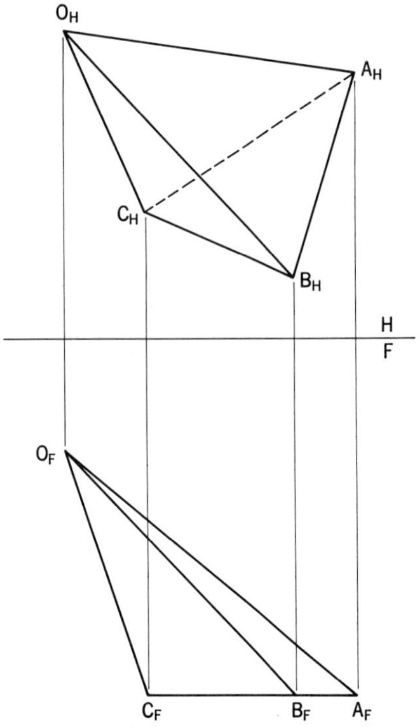

Fig. 8.10

The true lengths of edges *OA*, *OB*, and *OC* are determined by rotating them into positions parallel to the frontal plane, as shown in Fig. 8.11. Edges *AB*, *BC*, and *CA* are shown in true length in the top view.

The pattern can now be laid out. True length *OA* is drawn in a convenient position. With *O* and *A* as centers and radii, respectively, equal to the true lengths of *OB* and *AB*, arcs are drawn to intersect in point *B*. Triangle *OBA* can now be constructed.

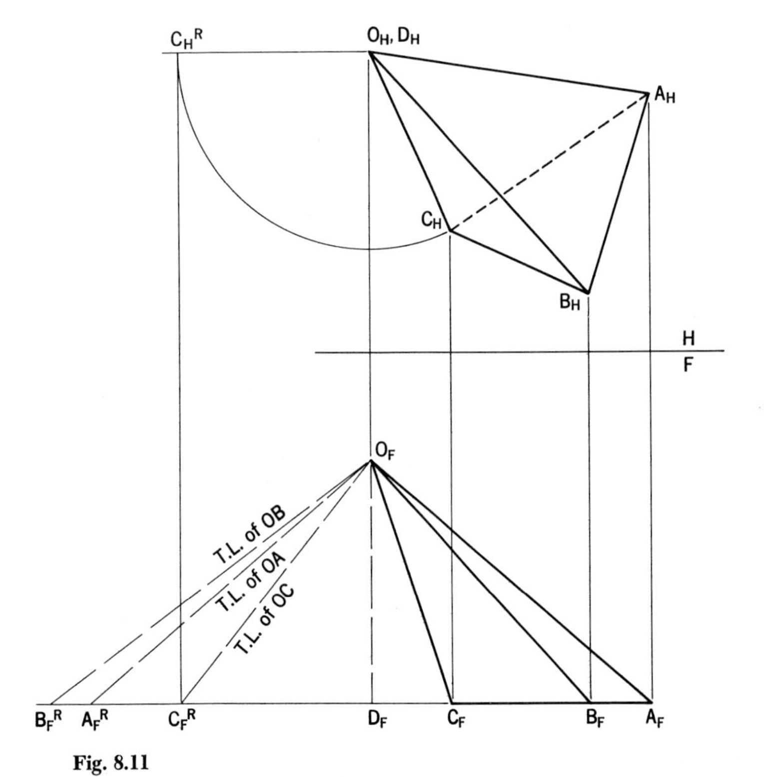

Fig. 8.11

With O and B as centers and radii, respectively, equal to the true lengths of OC and BC, arcs are drawn to intersect in point C. Lines CO and CB complete triangle OBC. In a similar manner, triangle OCA can be established, thus completing the pattern or development of the three lateral surfaces, shown in Fig. 8.12.

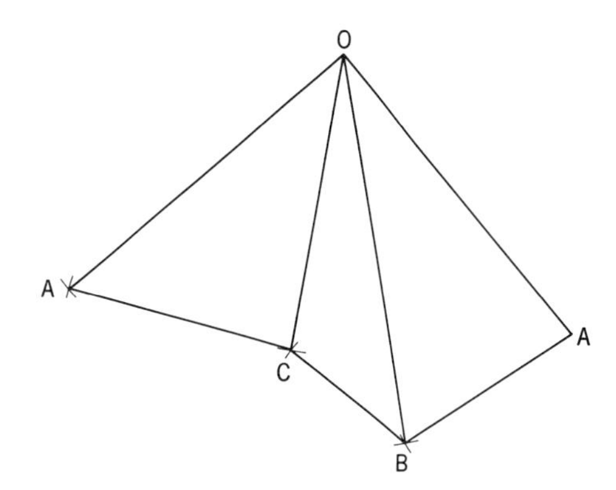

Fig. 8.12

EXAMPLE 2

Suppose we are given the truncated pyramid shown in Fig. 8.13. The true lengths of OA, OB, OC, OD, OK, OG, OE, and OF are easily obtained by rotating these lines parallel to the F-plane. The true lengths are shown in the front view. Now the development of the lateral surface is easily constructed. Point O is selected arbitrarily. An arc with center at O and with radius equal to the true length of OA (OB, OC, and OD are of the same length) is drawn. Point B is conveniently located on the arc. With B as center and radius equal to the true length of AB (shown as A_HB_H), an arc is drawn to intersect the large arc at point A. In a similar manner points D, C, and B can be located. The rest of the construction of the development is quite evident in the solution shown. If it is desirable to add the base, we note that its true shape is shown in the top view. If the sloping face $EFGK$ is to be added, its true shape is easily obtained from a supplementary view on a plane which would be introduced parallel to the face. This is easily done since we have an edge view of the face as shown in the front view.

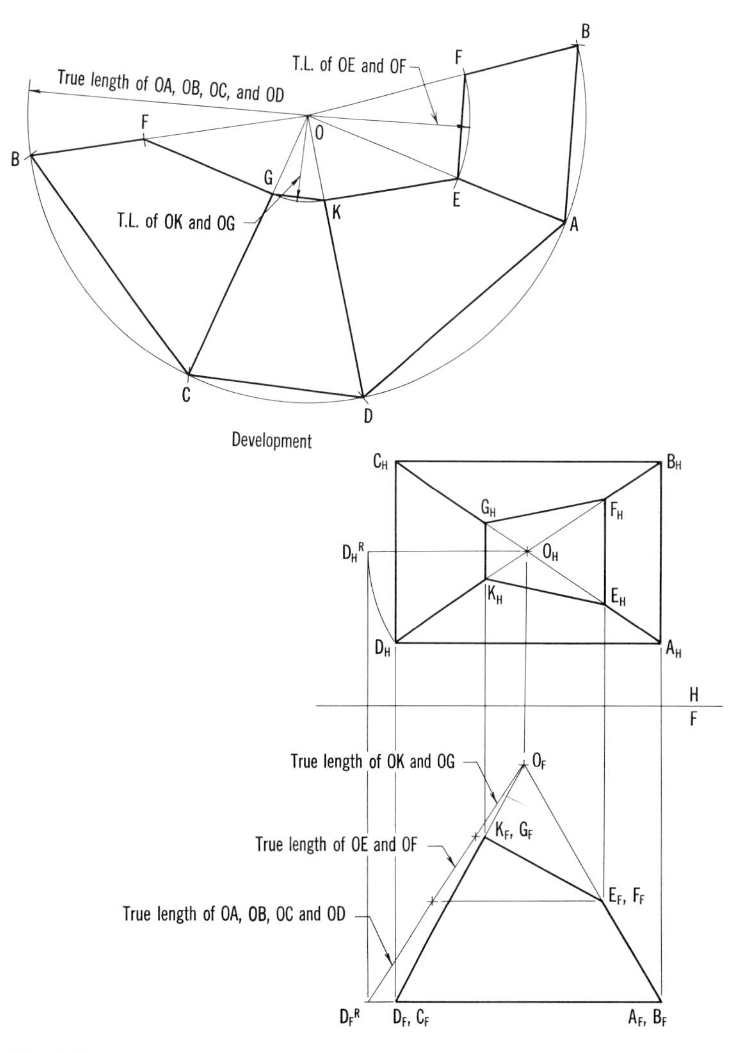

Fig. 8.13

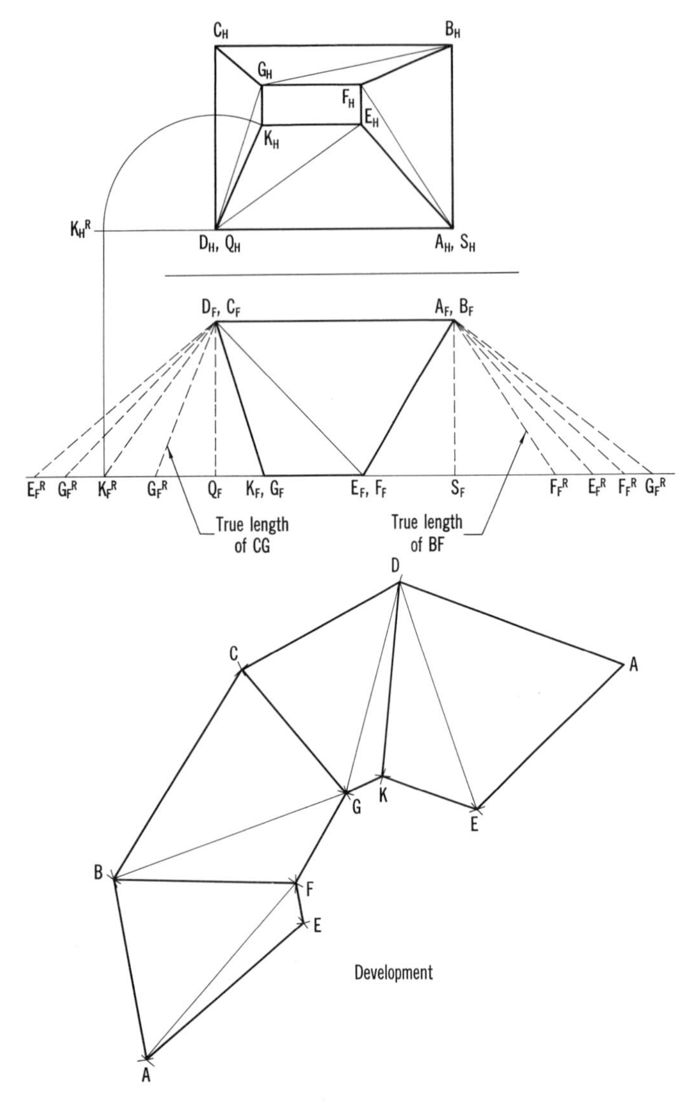

Fig. 8.14

Development of a Hopper

Let us now consider the hopper shown in Fig. 8.14. First, let us determine the true lengths of the elements which, in this case, are regarded as the edges of the hopper. This can be done by the method of supplementary views or by the method of rotation.

The rotation method is more advantageous because the construction can be reduced to a minimum. Suppose we find the true length of edge DK. If we introduce a vertical line through point D and a horizontal line through point K intersecting the vertical line at point Q, we will have formed the right triangle DQK. If we rotate this triangle about line DQ until it is parallel to the F-plane and then obtain a new front view, we will see the true length of DK as $D_FK_F^R$.

Now, if we observe that the altitudes of the right triangles formed in a similar manner for edges AE, BF, and CG will be the same length—equal to D_FQ_F or A_FS_F—we can lay off the top view lengths of AE and BF at right angles to A_FS_F through S_F and draw the hypotenuse for each right triangle. In a similar manner C_HG_H is laid off at right angles to C_FQ_F from Q_F. The hypotenuse of triangle $C_FQ_FG_F^R$ is then drawn. The hypotenuse lengths will be the true lengths of the edges. In this example they are shown as $A_FE_F^R$, $B_FF_F^R$, and $C_FG_F^R$, respectively.

The development can now be started. Suppose we lay off a length equal to AD (true length is available in both the top and front views). The true length of EK is also available in the top and front views. The location of edge EK with respect to edge AD cannot be established until we fix the location of either point E or point K. Let us agree to locate the position of point E. This can be done by drawing *a temporary line* from D to E. The true length of DE can be easily determined. It is shown as $D_FE_F^R$. Now, if we use A and D as centers and use radii respectively equal to the true lengths of AE and DE, we can draw arcs which intersect at point E. Line EK is now laid off parallel to AD, and line DK is drawn to complete the true shape of surface $ADKE$. If we introduce temporary line DG and find its true length, we can proceed in a similar manner to lay out the true shape of surface $DCGK$. The other two surfaces can be found similarly and laid out in proper sequence to form the complete development or pattern.

DEVELOPMENT OF CONES

EXAMPLE 1

Let us consider the *right circular cone* shown in Fig. 8.15. As in the previous examples we must determine the true lengths of the elements of the cone and their positions in the development.

It is fairly obvious that the elements, in this case, are all the same length. Moreover, since element O–8 is parallel to the F-plane, its true length is seen in the front view. Now we can construct the development by drawing an arc with center at point O (conveniently located) and radius equal to the true length of element O–8. On the arc we can locate points 0, 1, 2, . . . , 15, 0 such that the distances between 0 and 1, 1 and 2, etc., are equal to the distances between these points as seen in the top view. It should be pointed out, however, that a simple calculation will give us the magnitude of angle θ which can be used directly in making the development without the need for locating the elements. Angle θ is computed from the relation,

$$\frac{\theta°}{360°} = \frac{2\pi r}{2\pi S} \text{ or } \theta = \frac{r}{S} \times 360°$$

where r = the radius of the base circle and S = the slant height of the cone.

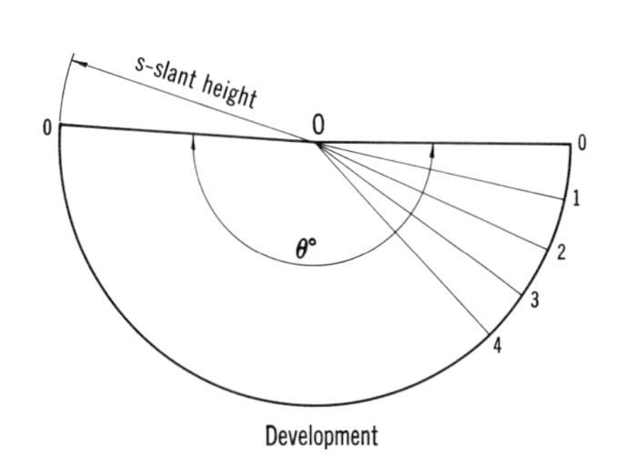

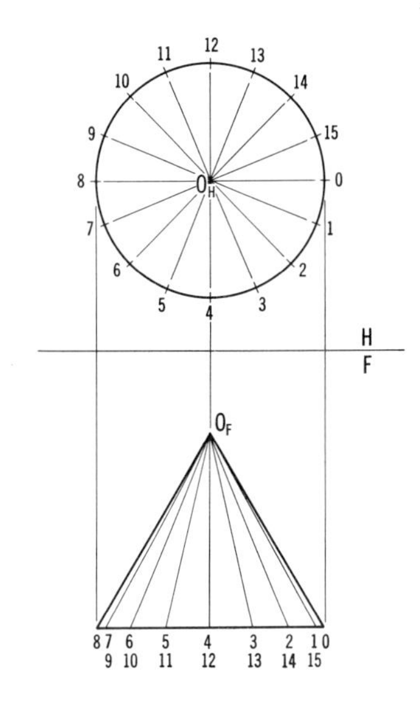

Fig. 8.15

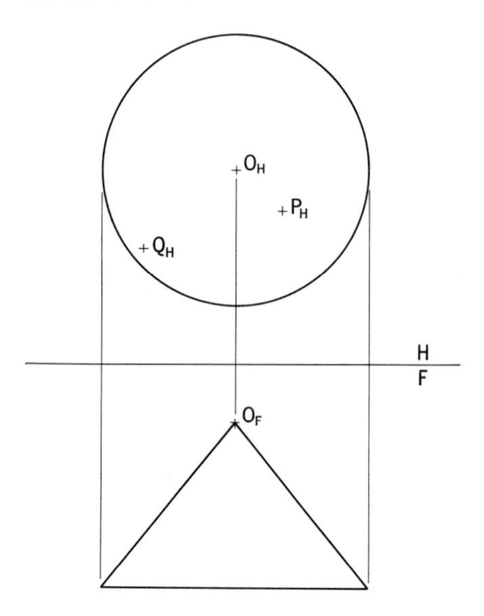

Fig. 8.16(*a*)

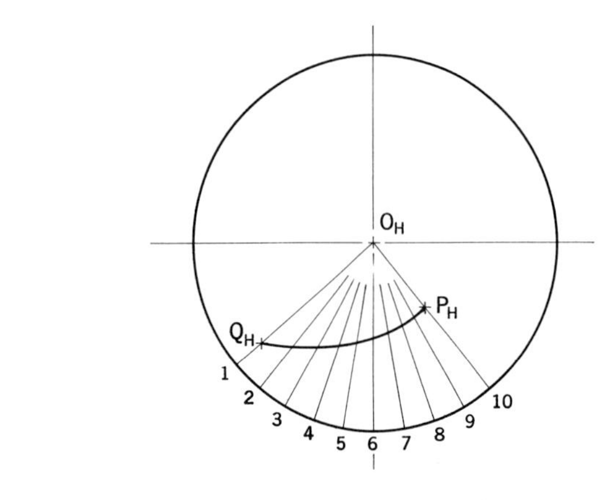

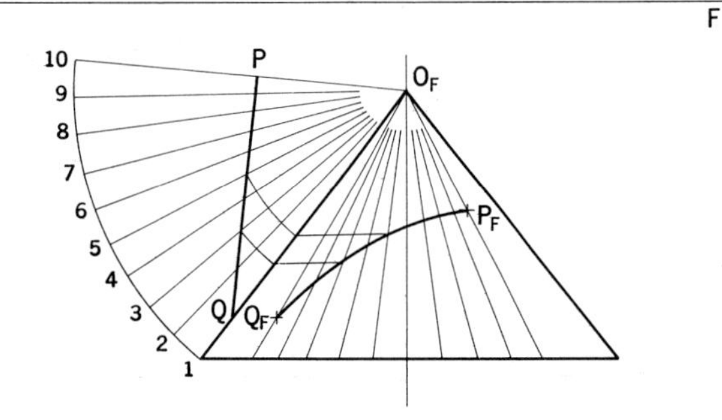

Fig. 8.16(*b*)

EXAMPLE 2

It is required to locate the shortest path *on the surface* of the right circular cone, between points *P* and *Q*. (See Fig. 8.16(*a*).

Analysis

1. At first glance it would appear that the shortest path, as seen in the top view, is the straight line joining P_H and Q_H. If this were true, then the implication is that the shortest path is part of the hyperbola formed by the intersection of the vertical plane (containing points *P* and *Q*) with the cone. This cannot be correct because it is just as logical to conclude that the shortest path, as seen in the front view, is the straight line joining P_F and Q_F. This implies that the shortest path is part of the ellipse formed by the intersection of the plane that contains *P* and *Q* and is perpendicular to the *F*-plane, with the cone. We see that neither path as described above is correct.

2. How shall we find the shortest path? Suppose we develop the portion of the surface of the cone between elements *O*–1 and *O*–10; and then locate points *P* and *Q* in the development. The line joining points *P* and *Q* is the shortest path. It remains, then, to establish the top and front views of a number of points on line *PQ*. The curve which joins the top view of the points is the top view of the shortest path and, similarly, the curve which joins the front view of the points is the front view of the shortest path. The complete solution is shown in Fig. 8.16(*b*).

EXAMPLE 3

Now let us consider the development of the oblique cone shown in Fig. 8.17. The surface of the cone can be approximated by an inscribed pyramid whose apex is point A and whose base is the polygon 0, 1, 2, 3, . . . , 34, 35, 0. The true lengths of the lateral edges of the pyramid are found in the manner described in the previous example. These are shown as $A_F O_F{}^R$, $A_F 1_F{}^R$, . . . , $A_F 18_F{}^R$. The sides of the base are shown in true length in the top view. Therefore, the true lengths of the sides of the triangles, A–0–1, A–1–2, A–2–3, . . . , are known and, hence, their true shapes can be constructed. When the triangles are joined in the manner shown in Fig. 8.17(a) and a smooth curve is drawn through points 0, 1, 2, 3, . . . , the pattern or development of the lateral surface of the cone is completed.

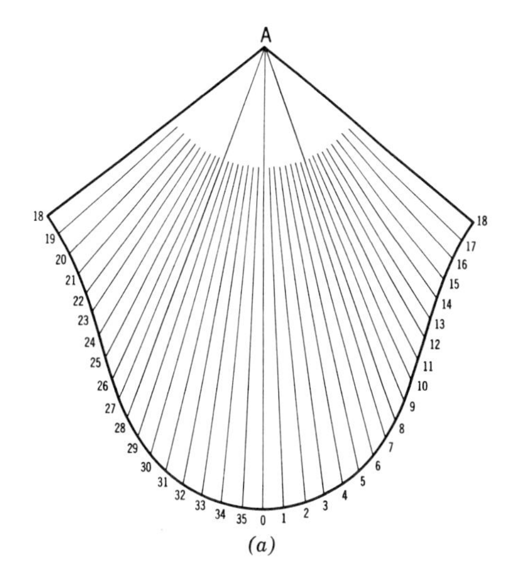

(a)

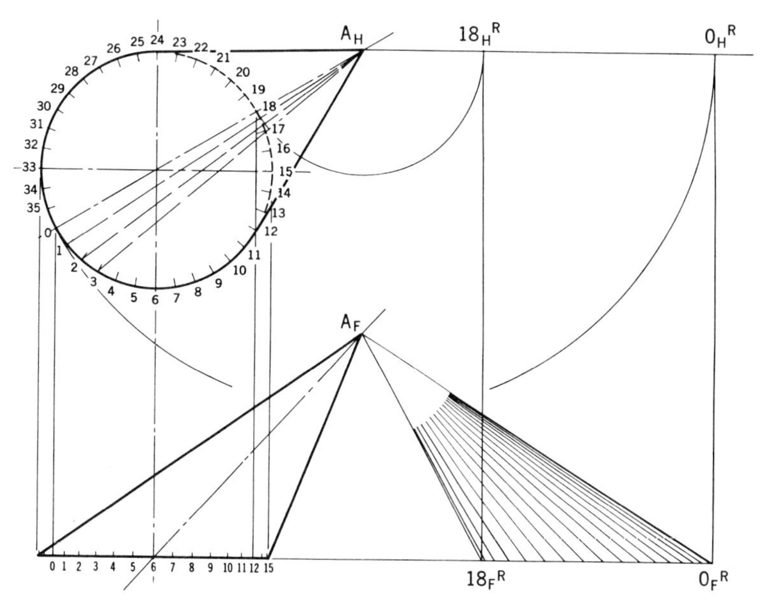

Fig. 8.17

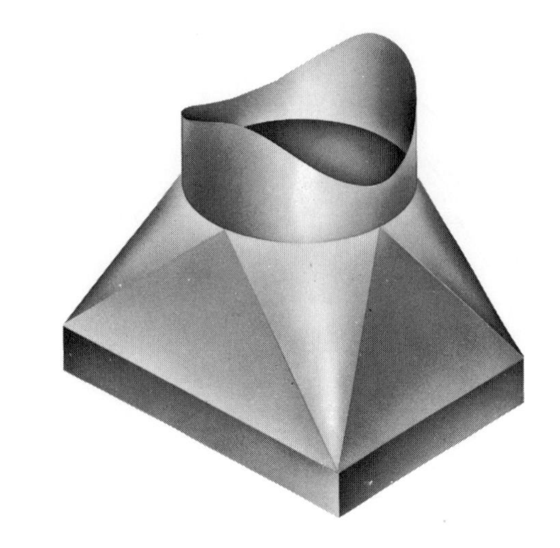

Fig. 8.18

DEVELOPMENT OF TRANSITION PIECES

A transition piece is generally used to provide for a change in cross section. For example, in Fig. 8.18 the transition piece is used to accommodate a change from the rectangular collar to the cylindrical duct of a ventilating system. The front and top views of the transition piece are shown in Fig. 8.19. The development of the surface presents no serious difficulty. The transition piece is made up of four triangular surfaces and four conical surfaces. True lengths of elements of the cones and also of the sides of the triangles can be easily determined by the method of rotation, which is convenient to use in this type of problem. The development consists of the true shapes of the triangles *AED*, *DHC*, *CGB*, and *BFA*, and the true shapes of smaller triangles such as *DEK*, joined together in sequence to form the pattern shown in Fig. 8.19. Transition pieces may be made up from a variety of shapes resulting from a combination of plane surfaces and curved surfaces. The latter may be conical, cylindrical, spherical, or warped.

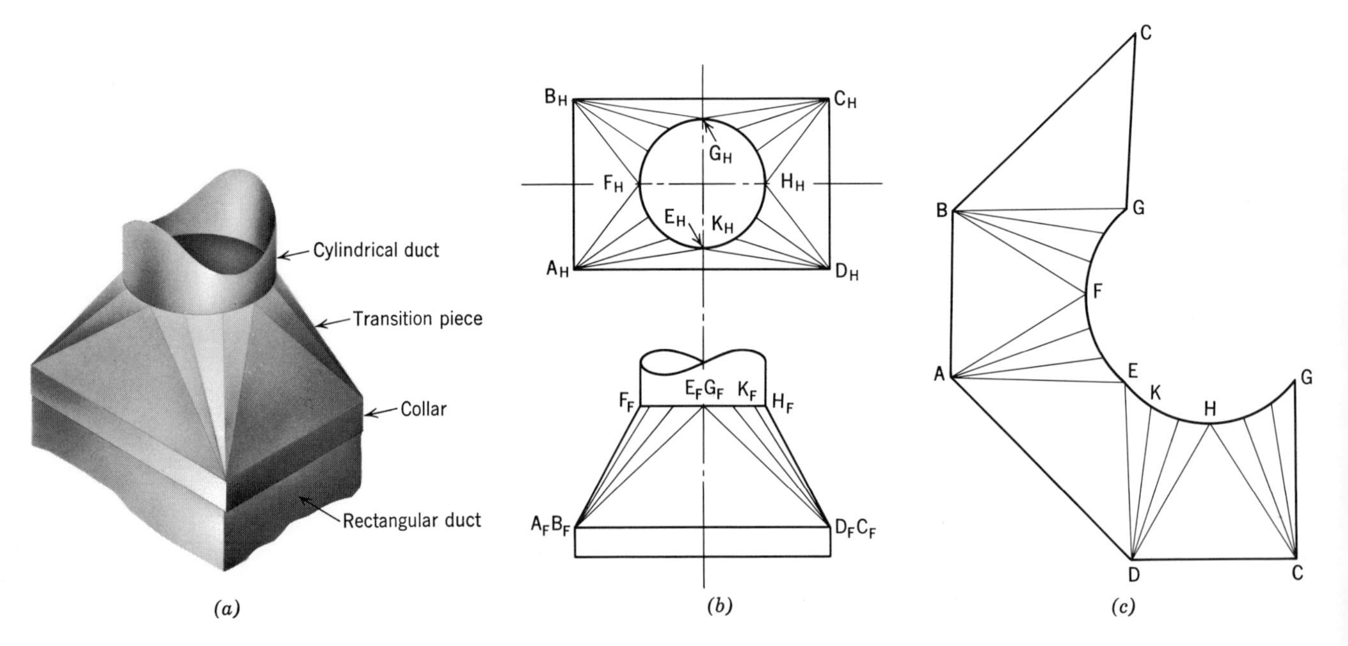

Fig. 8.19

Warped Surfaces

There are several warped surfaces, such as the *helicoid, conoid, cylindroid,* and *hyperbolic paraboloid,* that may be generated to form useful shapes.

The *helicoid surface* is generated by a straight line (the generatrix) which moves along two concentric helices (directrices) while maintaining a constant angle with the axis of the helices. In a number of applications the generatrix intersects the axis. When the elements of the surface make an angle of 90° with the axis, the surface is identified as a right helicoid, otherwise as an oblique helicoid. Screw conveyors and chutes typify the right-helicoid surface, Fig. 8.20, whereas the sloping faces of thread forms are typical of the oblique-helicoid surface.

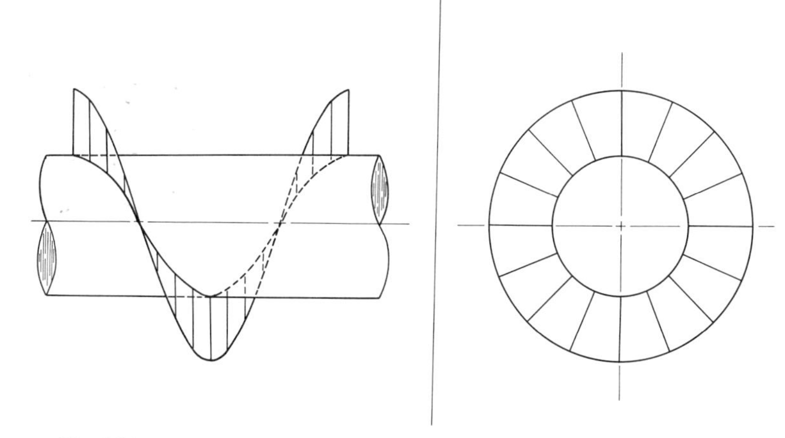

Fig. 8.20

The *conoid surface* is generated by a straight line (generatrix) which remains parallel to a plane (known as the plane director) while intersecting a straight line (directrix) and a curved line (directrix). This surface is quite useful in fairing (smoothing) a roof surface from a curved shape such as a parabola to the edge of a supporting wall. See Figs. 8.21(*a*) and 8.21(*b*).

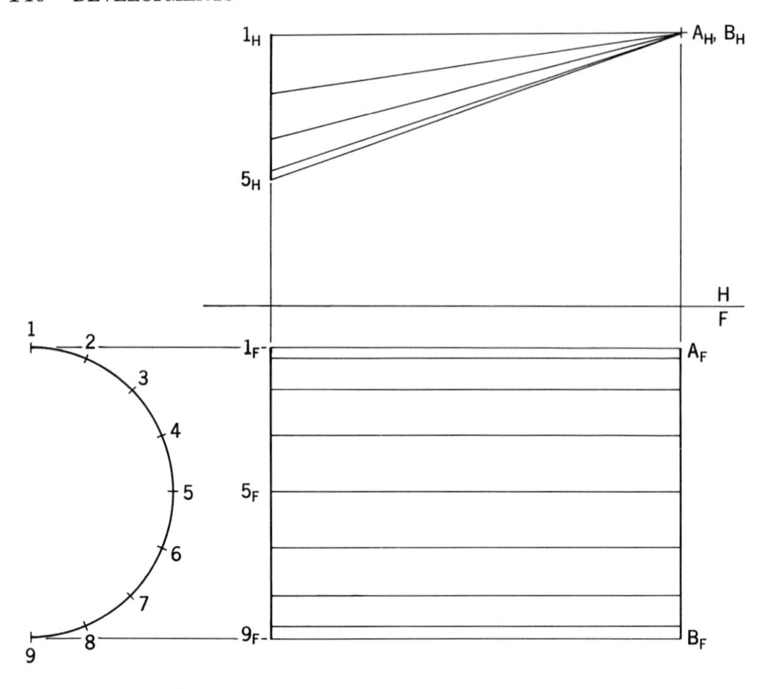

Fig. 8.21(*a*)

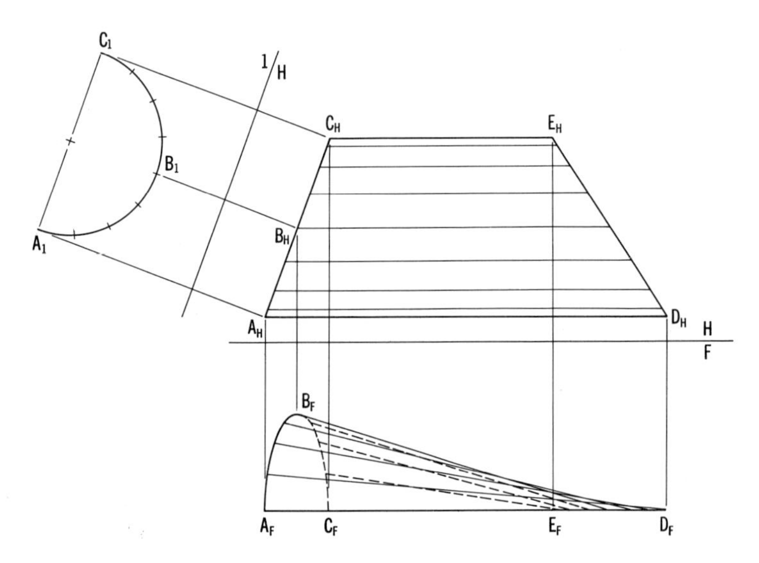

Fig. 8.21(*b*)

The *cylindroid surface* is generated by a line that remains parallel to a plane director while intersecting two curved-line directrices which usually lie in nonparallel planes. (See Fig. 8.22.)

The *hyperbolic paraboloid* surface is quite useful in the design of modern roof structures, ducts, concrete walks, culverts, etc. The hyperbolic paraboloid surface is generated by a straight line (generatrix) which intersects two skew lines (directrices) and remains parallel to a plane.

Fig. 8.22

EXAMPLE 1

Suppose the given skew lines are *AB* and *CD*, and that the *F*-plane is the director plane, as shown in Fig. 8.23. The top view of elements of the surface is easily established since the generatrix is parallel to the *F*-plane. We may introduce a sufficient number of elements to delineate the surface. Eight are shown in Fig. 8.23(*a*).

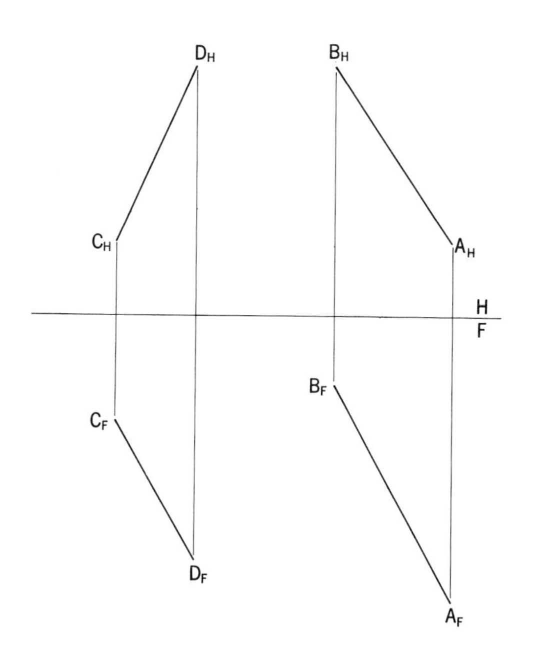

Fig. 8.23

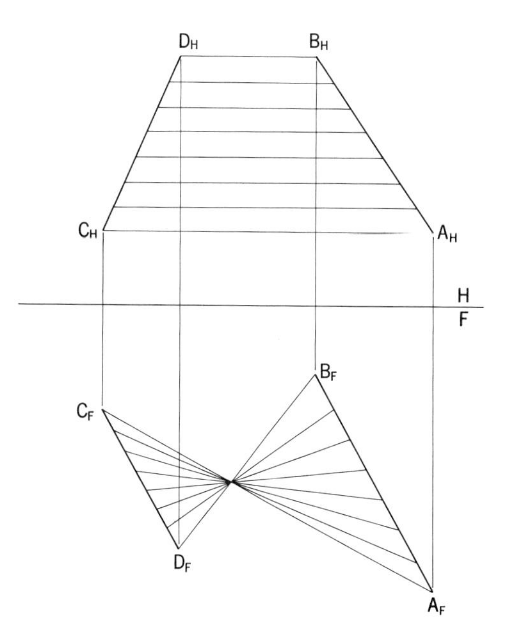

Fig. 8.23(*a*)

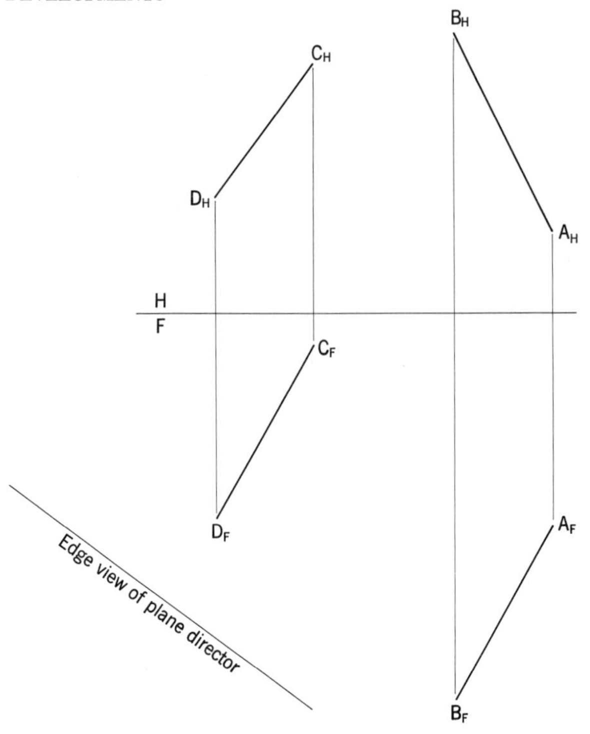

Fig. 8.24

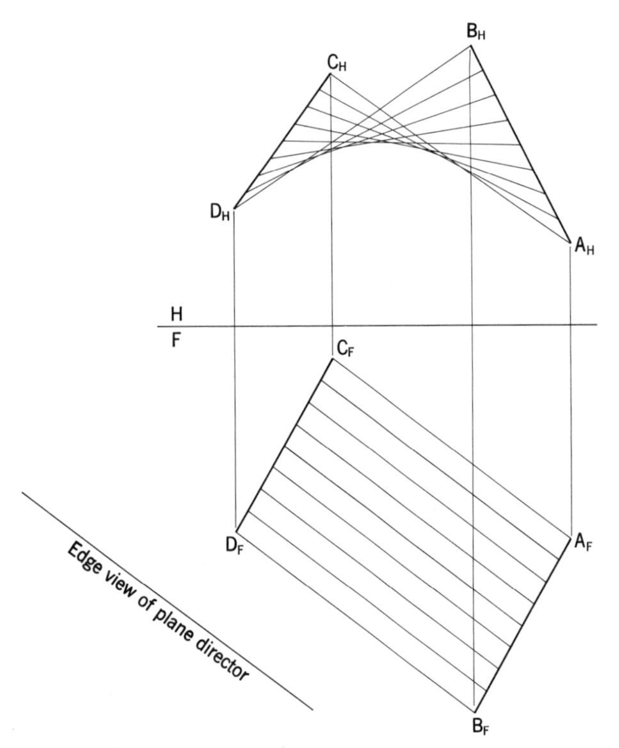

Fig. 8.24(*a*)

EXAMPLE 2

Consider the case where the plane director is perpendicular to the *F*-plane, as shown in Fig. 8.24. All the elements in the front view will be parallel to the edge view (front view) of the plane director. The corresponding top view of the elements is easily established. [See Fig. 8.24(*a*).]

In Fig. 8.25, the plane director appears as a triangle in the top and front views. It is a simple matter to reduce this figure to the first two examples by finding an edge view of the plane director. Supplementary plane 1 is perpendicular to line *GK* which lies in surface *EFG*.

The supplementary view shows the edge view, $E_1F_1G_1$, and the corresponding views of *AB* and *CD*. The elements of the hyperbolic paraboloid as seen in plane 1 will be parallel to the edge view $E_1F_1G_1$. The top and front views are easily established. See Fig. 8.25(*a*) for the solution.

It should be carefully observed that the elements divide each generatrix (*AB* and *CD*) into the same number of parts in each view. Recognizing this fact simplifies the work of locating the elements in each view and, moreover, enhances the accuracy.

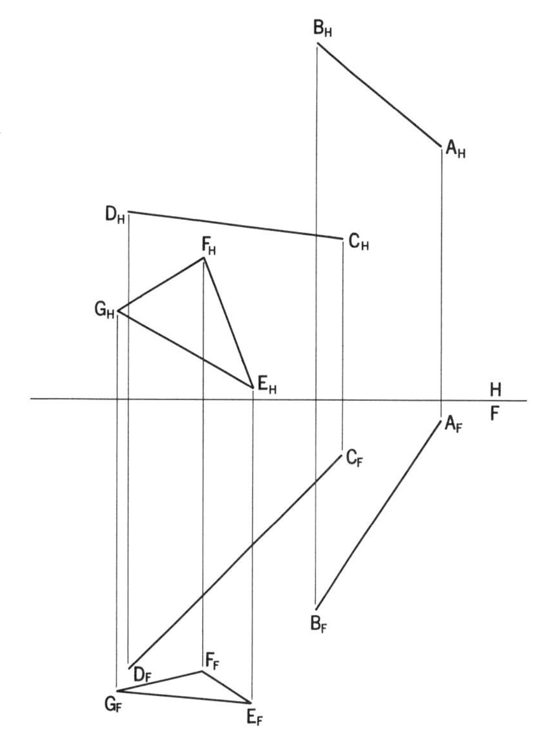

Fig. 8.25

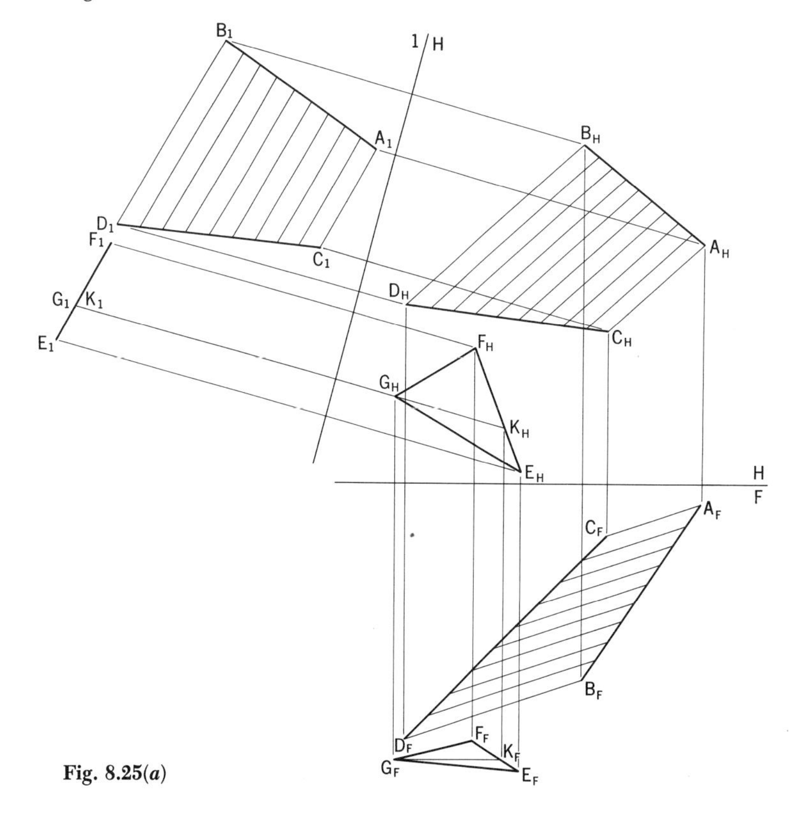

Fig. 8.25(*a*)

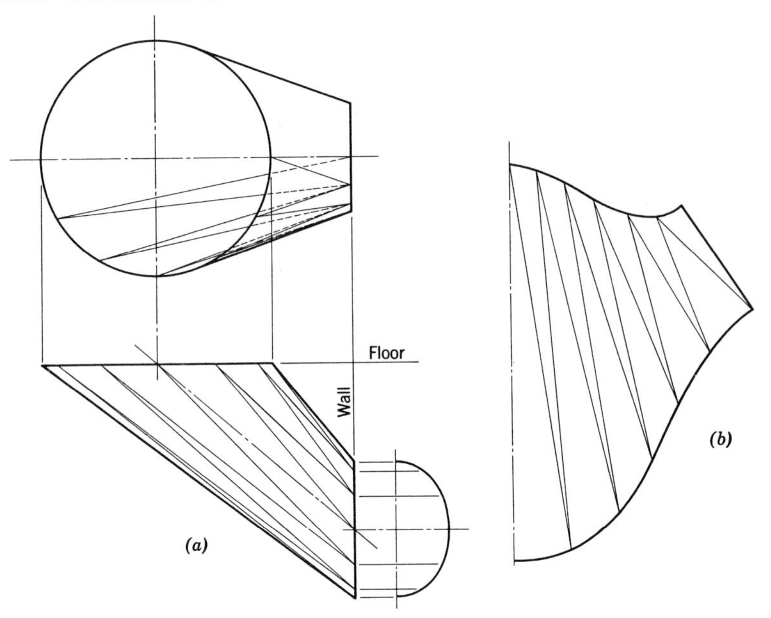

Fig. 8.26

EXAMPLE 4

Let us consider the warped transition piece shown in Fig. 8.26(*a*). This piece is designed to connect two openings that lie in intersecting planes. The opening in the floor is circular and the one in the wall is elliptic. The surface is approximated by the triangles shown in the given views. The method employed for finding the true lengths of the sides of the triangle is rotation. Again, the development consists of the true shapes of the triangles arranged in sequence to form the pattern shown in Fig. 8.26(*b*).

EXAMPLE 5

Let us consider the warped transition piece shown in Fig. 8.27. The warped surface is approximated by a number of small triangles. The upper and lower circular openings are divided into the same number of divisions (i.e., upper circle divisions are marked 0, 1, 2, etc., and the corresponding divisions on the larger circle are 0′, 1′, 2′, etc.). Areas such as

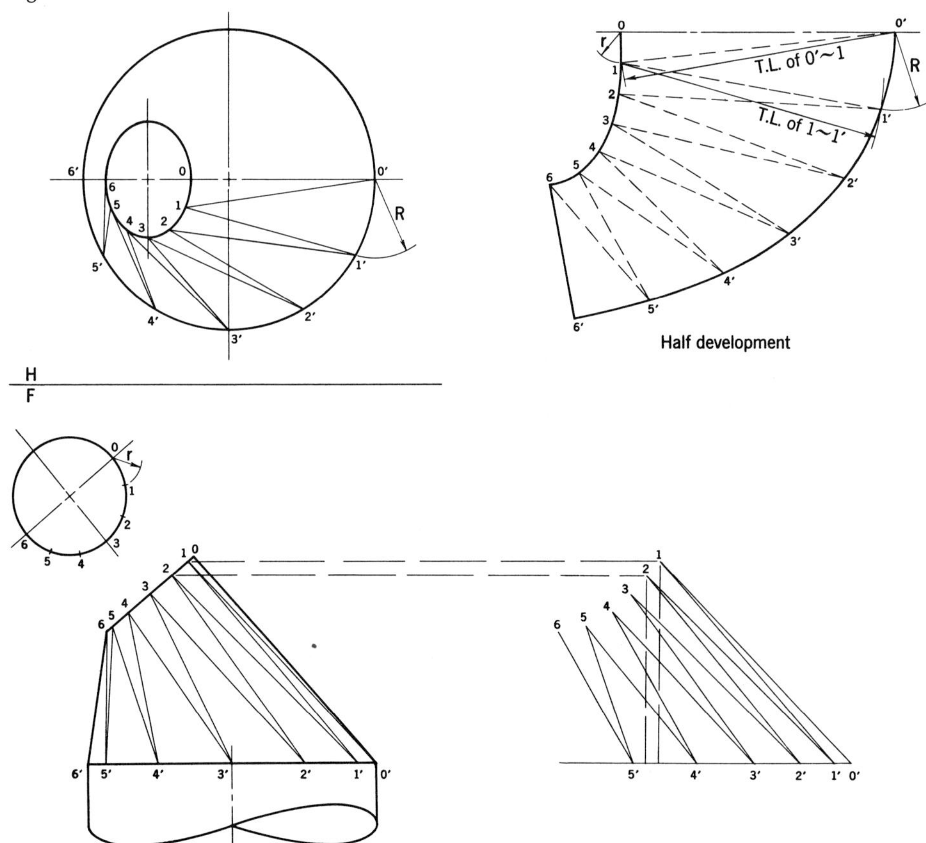

Fig. 8.27

0–0′–1′–1 are divided into two triangles by the insertion of line 0′–1. The true shapes of the triangles are constructed by using the true lengths of the sides of the triangles.

For example, let us consider triangle 0–0′–1. The true length of side 0–0′ is available in the front view, since side 0–0′ is parallel to the *F*-plane. The true length of side 0′–1 is equal to the hypotenuse of the right triangle whose altitude is equal to the *vertical* distance between points 0′ and 1 (this distance is available in the front view and is shown on the right in the true length diagram); and whose base is equal to the length of the top view of 0′–1. You will recall that we used this method in the development of an oblique cone.

In a similar manner the true length of side 1–1′ of the adjacent triangle is obtained. The true-length diagram shows only a few sides. This has been done in the interest of clarity. Of course, the true lengths of the sides of all of the triangles must be determined in order to construct the adjacent triangles. Now the true length, *r*, of side 0–1 is seen in the supplementary view which shows the true shape of the smaller circle. Since we now have the true lengths of the sides of triangle 0–0′–1 we can construct its true shape. The construction is shown in the partial development. The adjacent triangle, 0′–1–1′, is constructed in a similar manner. Note that the true length, *R*, of side 0′–1′ is available in the top view of the large circle. After points 0, 1, 2, etc., and 0′, 1′, 2′, etc., are located in the development, smooth curves are drawn through them as shown. These curves compensate, fairly well, for the differences between the chord lengths and the arc lengths (i.e., chord 0′–1′ and arc 0′–1′).

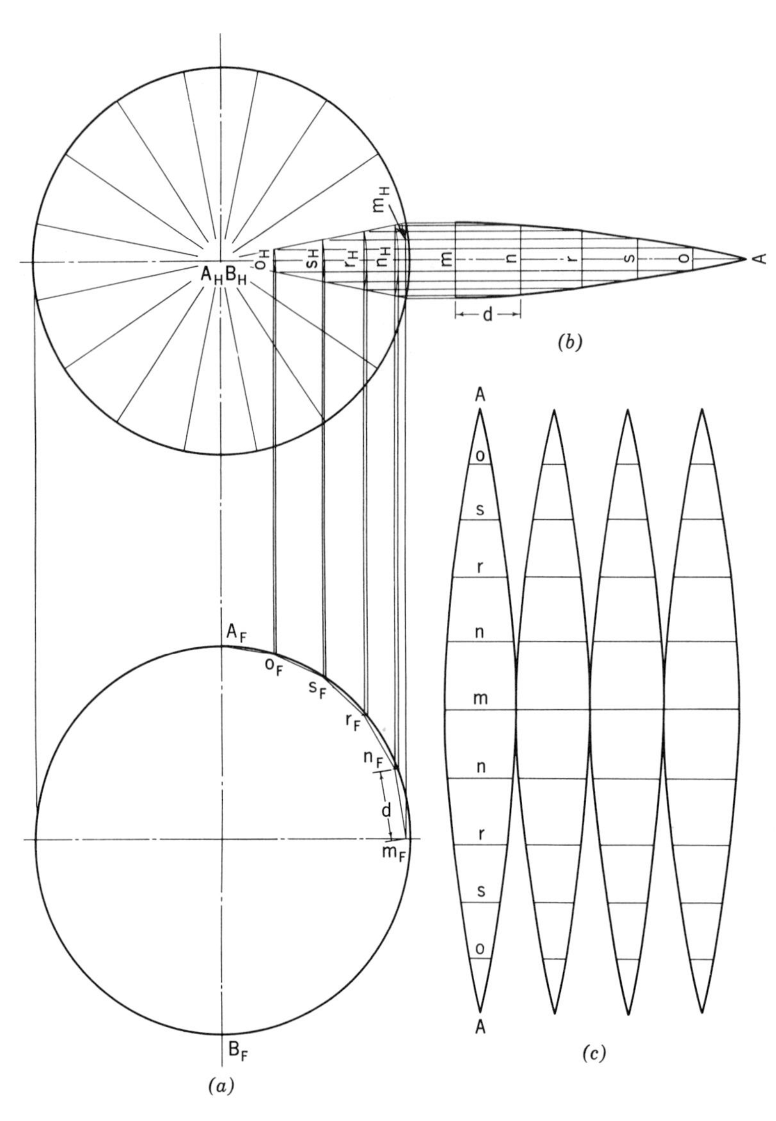

Fig. 8.28

DEVELOPMENT OF A SPHERICAL SURFACE

Let us consider the sphere shown in Fig. 8.28(*a*). Suppose the sphere is divided into an equal number of wedge-like segments formed by passing planes through the vertical axis *AB*. Now the surface of each segment can be approximated by a portion of a cylindrical surface. Since all the segments are the same, we will deal with the single segment *m* . . . *A*. The surface of this segment is approximated by the inscribed cylindrical surface with elements *m*, *n*, *r*, *s*, *o*, and *A* (more elements could be used). Careful study of the top and front views leads us to conclude that (*i*) these elements are true length in the top view and (*ii*) the distances between the elements are available in the front view since they appear as points. The pattern or development can now be constructed. Element *m* can remain in its original position or be placed at a convenient distance from this position as shown in Fig. 8.28(*b*). Element *n* is parallel to *m* and at a distance *d* from *m*. In a similar manner, elements *r*, *s*, *o*, and *A* can be located. In fact, the development of the cylindrical surface, in this problem, is basically the same as in the example (Fig. 8.7) which we studied earlier. The complete development of the sphere would consist of sixteen areas, four of which are shown in Fig. 8.28(*c*).

Conical surfaces may also be used to approximate the spherical surface. Figure 8.29(*a*) shows the sphere divided into a number of horizontal zones. The surfaces of these zones can be approximated by frustums of cones. For example, the surface of zone *A* can be approximated by the surface of the frustum of the cone with apex *V*. The bases of the frustum are shown. The development of zone *A*, which is included in Fig. 8.29(*b*), is basically the same as the development of the conical surface discussed previously and shown in Fig. 8.15.

It should be quite clear that throughout our discussion of developments we have used a very simple method, i.e., the determination of true lengths of elements and the correct placement of one element with respect to an adjacent element. It should be emphasized that any development problem must be carefully analyzed with respect to the use of simple geometric shapes which closely approximate the given surface. The layout of the simple geometric shapes, in proper sequence, will form the pattern or development.

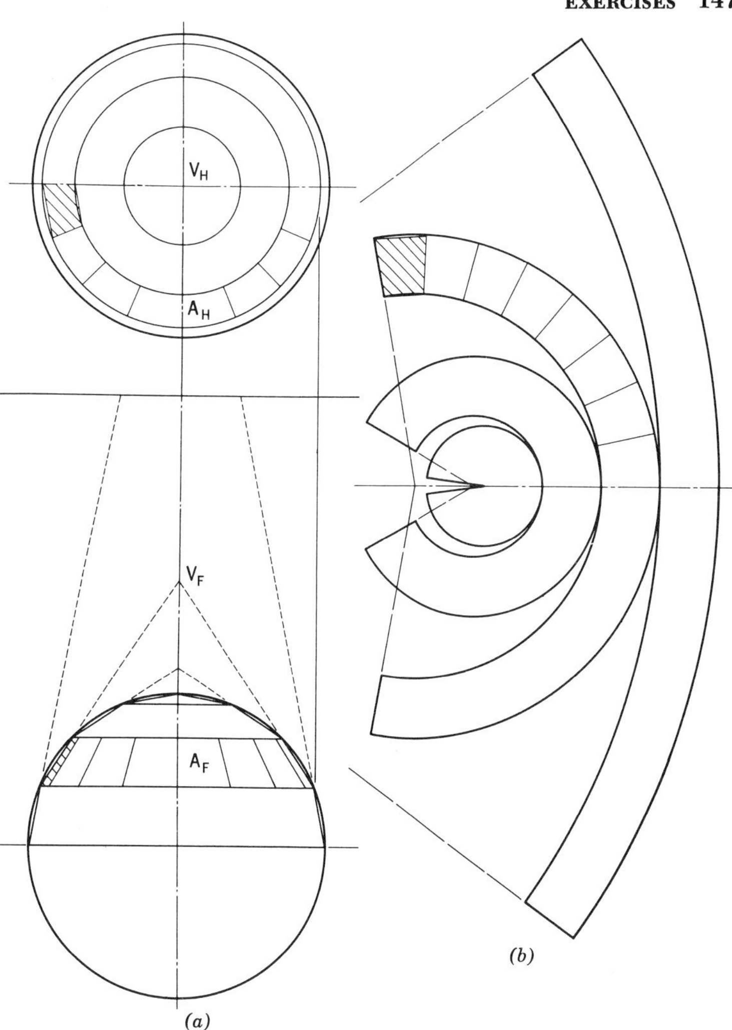

Fig. 8.29

EXERCISES

1. Develop the lateral surface of the prism shown in Fig. E-8.1.

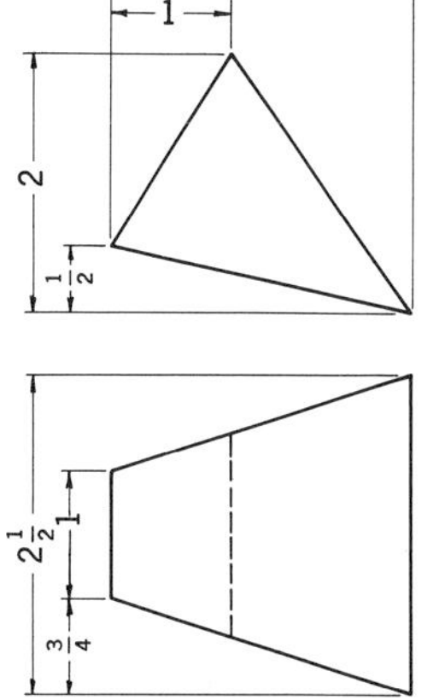

Fig. E-8.1

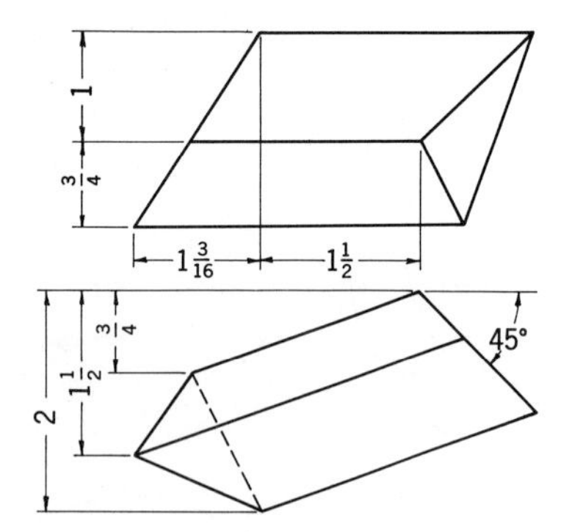

Fig. E-8.2

2. Develop the lateral surface of the prism shown in Fig. E-8.2.

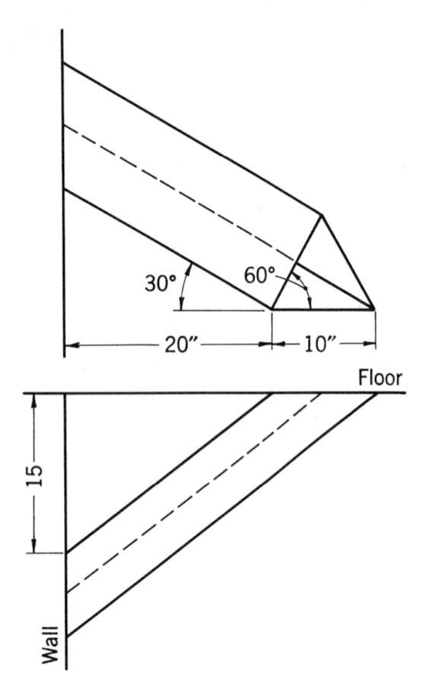

Fig. E-8.3

3. Develop the surface of the chute shown in Fig. E-8.3.

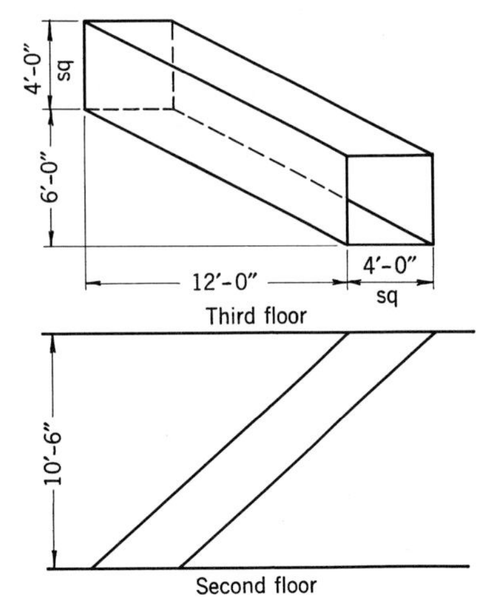

Fig. E-8.4

4. Develop the surface of the chute shown in Fig. E-8.4.

5. Develop the surface of the hopper shown in Fig. E-8.5.

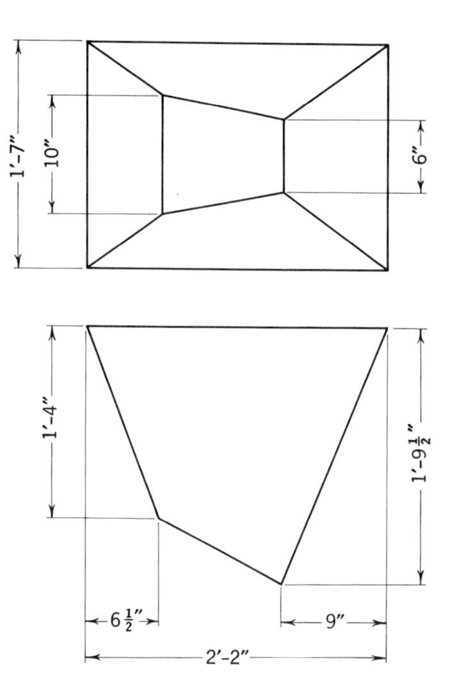

Fig. E-8.5

6. Develop the surface of the cone shown in Fig. E-8.6. Show the *H* and *F* views of the shortest path *on the surface of the cone* from point *A* to point *B*.

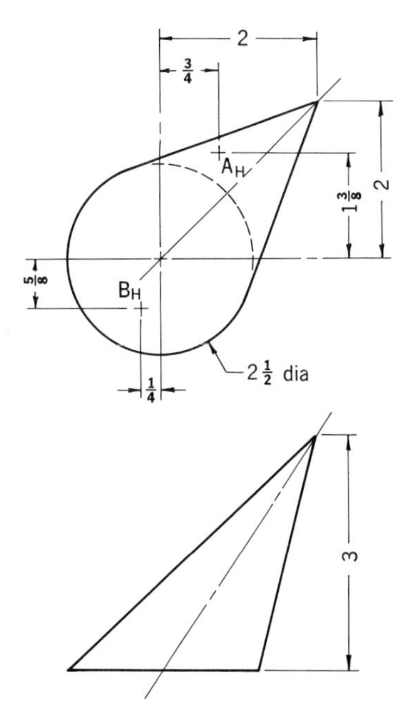

Fig. E-8.6

7. Develop the surface of the conoid shown in Fig. E-8.7.

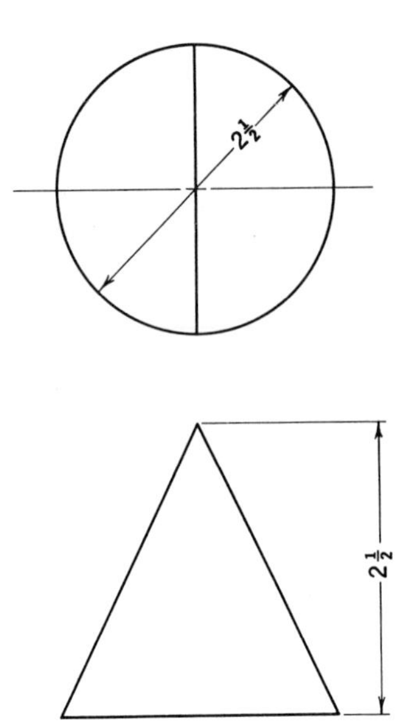

Fig. E-8.7

8. Develop the surface of the transition piece shown in Fig. E-8.8.

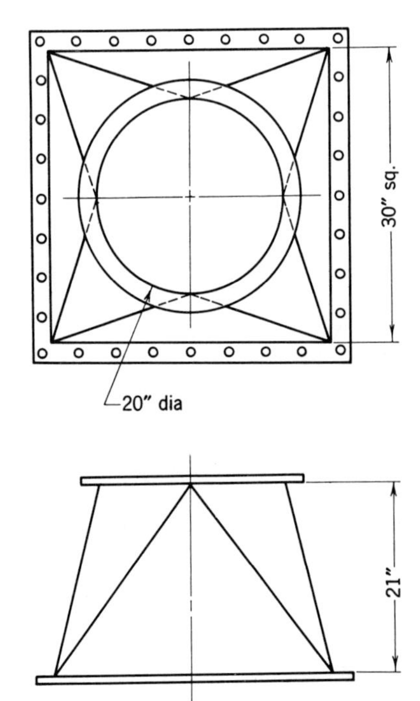

Fig. E-8.8

9. Design a transition piece to connect the rectangular openings shown in Fig. E-8.9. Develop the transition piece.

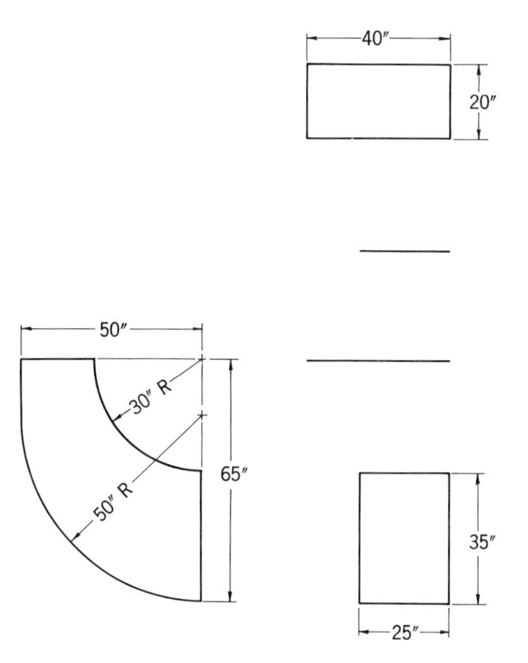

Fig. E-8.9

10. Design a transition piece to connect the rectangular opening with the circular collar shown in Fig. E-8.10. Develop the transition piece.

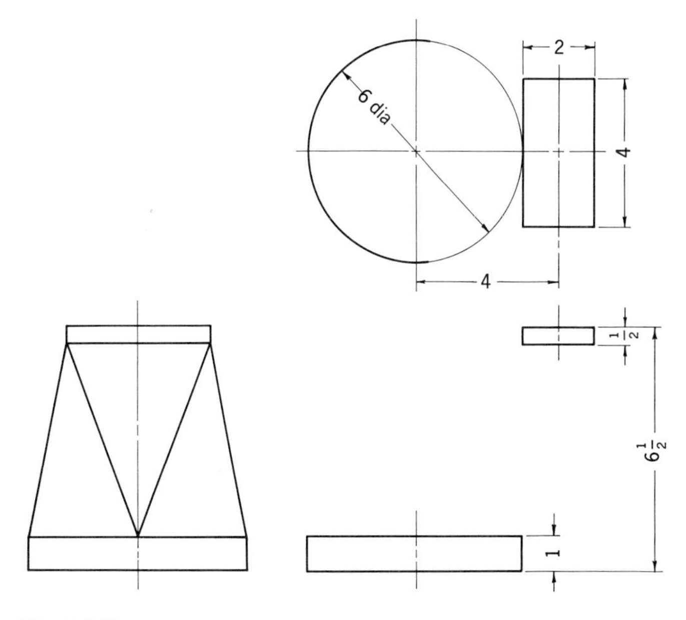

Fig. E-8.10

INTERSECTIONS 9

The design of airplanes, space vehicles, automobiles, appliances, ventilating systems, electrical and mechanical components, etc., very often includes intersecting surfaces. The intersection of plane surfaces, for example, is shown in Fig. 9.1; of cylindrical surfaces, in Fig. 9.2; of spherical and cylindrical surfaces, in Fig. 9.3; and of curved surfaces, in Fig. 9.4.

Fig. 9.1 Intersection of plane surfaces. McGregor Memorial Hall, Wayne University, Detroit, Mich. Minoru Yamasaki & Associates, Architect. (Photo Hedrich-Blessing)

Fig. 9.2 Intersection of cylindrical surfaces. Pumping station and surge tanks. (Courtesy Mobil Oil Corp.)

Fig. 9.3 Intersection of spherical and cylindrical surfaces. Oil refinery storage tanks. (Courtesy Mobil Oil Corp.)

Fig. 9.4 Intersection of curved surfaces. (Courtesy Lockheed-Georgia Co.)

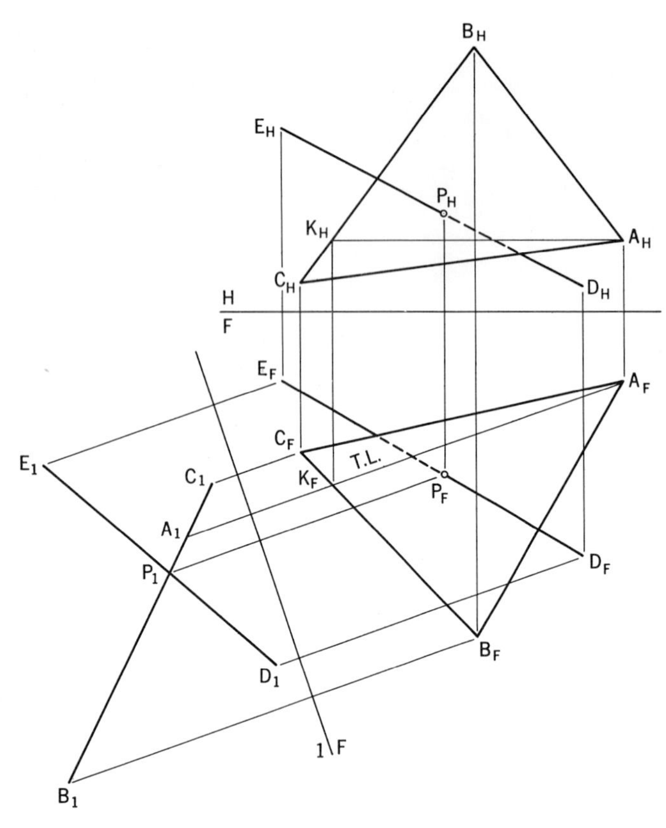

Fig. 9.5

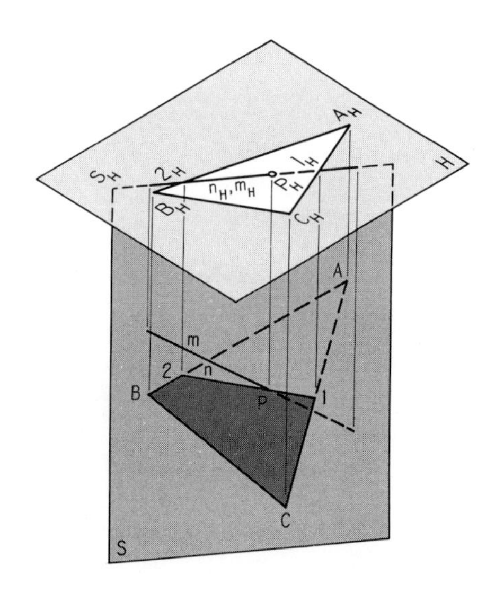

Fig. 9.6

BASIC PROBLEMS

A fundamental and direct approach to the solution of intersection problems requires an ability to analyze and solve the following *basic* problems:

1. The intersection of a line and a plane surface.
2. The intersection of a line and a conical surface.
3. The intersection of a line and a cylindrical surface.
4. The intersection of a line and a spherical surface.

Let us now solve these basic problems.

1. Intersection of a Line and a Plane

In our discussion of the problem "to determine the perpendicular distance from a point to a plane" we actually found the intersection, point *K*, of the perpendicular with the plane and then measured the distance from the given point, *P*, to point *K*. The method used was based on the fact that "the point of intersection of a line with a plane is apparent in the view which shows the plane as an edge."

EXAMPLE

Suppose we are given the *H* and *F* views of plane surface *ABC* and line *DE* (Fig. 9.5). The intersection (point *P*) of line *DE* with the plane *ABC* is quite easily established. We can obtain an edge view of surface *ABC once we have a point view of a line in the surface.* In this example a frontal line, *AK*, is drawn through point *A* and in the surface *ABC*. We know, from our previous study, that the true length of line *AK* is seen as A_FK_F. Supplementary plane 1, which is perpendicular to both line *AK* and the *F*-plane, is also perpendicular to surface *ABC*. The view on plane 1 shows the edge view of the surface *ABC*, the view of line *DE*, and of the point of intersection, *P*.

The locations of P_F and P_H are readily obtained.

Alternative Solution. The point of intersection of line *m* with plane *ABC* can be determined *without resorting to the use of a supplementary plane.*

Let us consider Fig. 9.6, which is a pictorial drawing of triangle *ABC* and line *m*. Point *P*, the intersection of line *m* with triangle *ABC*, is located in the following manner:

1. Pass plane S through line *m* perpendicular to the horizontal plane, *H*. (Plane S could be passed through line *m* in any one of many directions. It is

most convenient to orient the plane perpendicular to either the H- or F-plane.)

2. Find the intersection of plane S and triangle ABC. This is line n. *Note carefully that the top view of lines m and n and plane S appears as one line.*

3. Locate the intersection of lines m and n; namely, point P, which is the required point.

EXAMPLE 1

Intersection of Line and Plane Surface—Alternative Solution. Now suppose we are given the top and front views of triangle ABC and line m as shown in Fig. 9.7. On the basis of our study of Fig. 9.6 and the three steps outlined above, we can proceed to locate the top and front views of the point of intersection P of line m and plane ABC; therefore, we will:

1. Pass a plane S through line m perpendicular to the horizontal plane. The top view of this plane is a line, S_H, which coincides with the top view, m_H, of the given line m.

2. Find the line of intersection, n, of plane S with surface ABC. The top view n_H of line n also coincides with the top view of line m and of plane S; that is, n_H, m_H, and S_H are coincident since S_H is the top view of all lines that lie in plane S.

Now it is very important to bear in mind that line n lies *in plane ABC*. This being true, we can easily establish the front view, n_F, of line n simply by locating the front view of points 1 and 2, which are on line n and also on sides AC and AB, respectively.

Since 1_H and 2_H are available, it is easy to establish 1_F and 2_F. The line joining them is n_F.

3. Find the intersection of lines m and n. In the orthographic drawing this means nothing more than locating P_F at the intersection of the front views of lines m and n, and then locating P_H on m_H. This is a very simple step, if we have not forgotten the two fundamental relations:

(*a*) If a point is on a line, views of that point lie on corresponding views of the line.

(*b*) Adjacent views of a point lie on a line which is perpendicular to the line representing the edge views of the adjacent planes; thus, in Fig. 9.7, the line joining P_F and P_H is perpendicular to the horizontal line marked H–F.

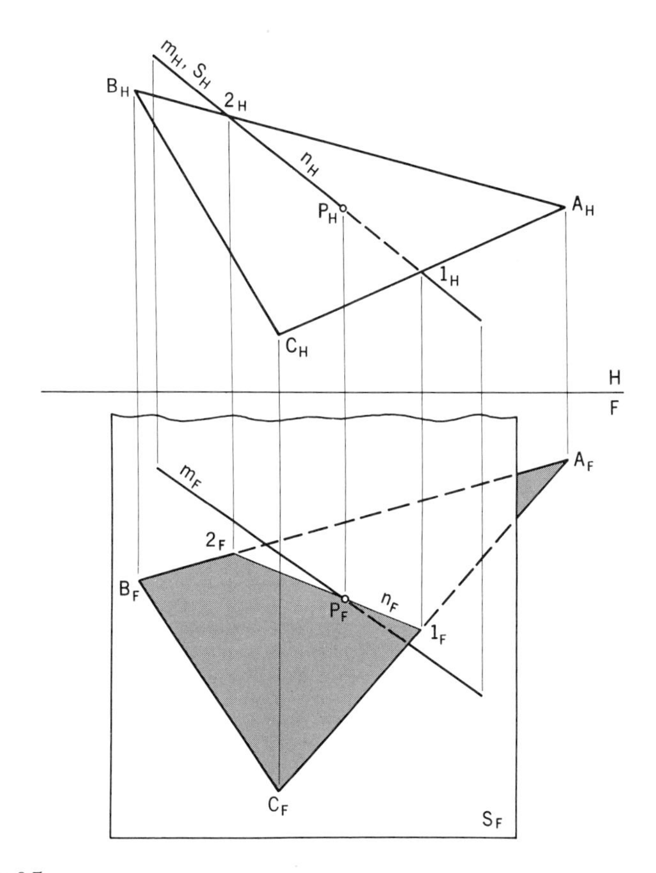

Fig. 9.7

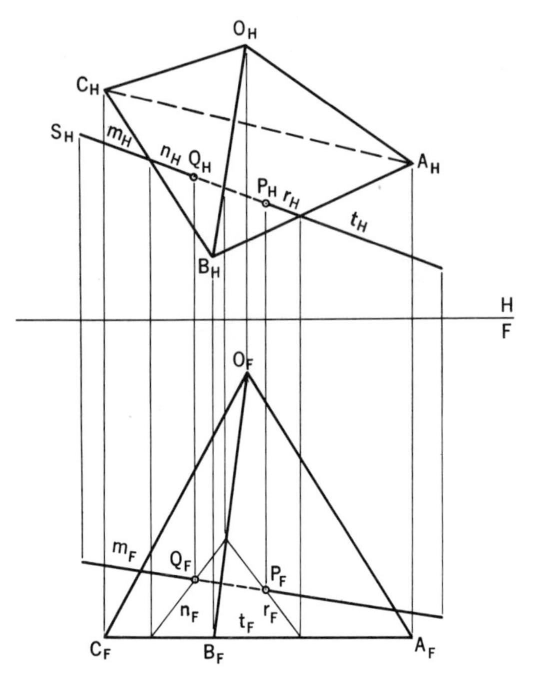

Fig. 9.8(*a*)

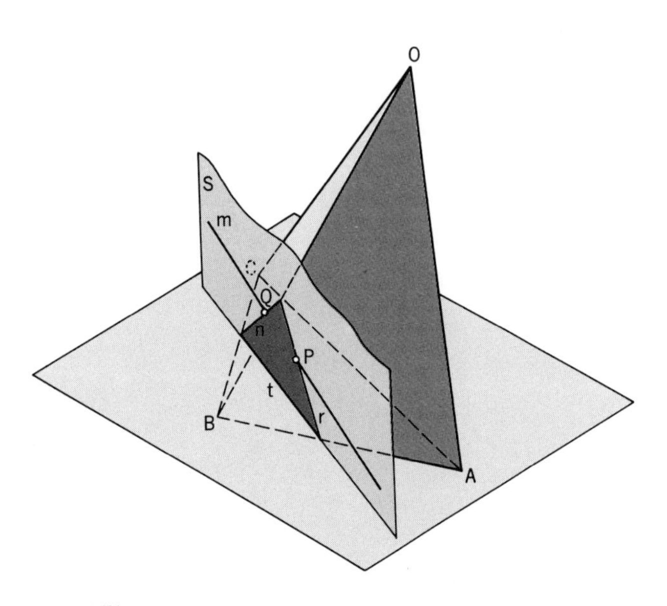

Fig. 9.8(*b*)

It should be quite clear that *the second method is the simpler of the two and, moreover, it will be found to be much more convenient when we consider problems dealing with the intersection of solids.*

EXAMPLE 2

Intersection of Line and Pyramid. Now let us consider the problem of locating the points in which a line intersects a pyramid.

Suppose we are given line *m* and pyramid *O–ABC* as shown in Fig. 9.8. The points in which line *m* intersects the pyramid can be easily determined in the following manner.

First, let us pass plane S through line *m* perpendicular to the *H*-plane. Plane S intersects surface *OBC* of the pyramid in line *n*, surface *OAB* in line *r*, and surface *ABC* in line *t*. The three lines *n*, *r*, and *t* are the sides of the triangle formed by the intersection of plane S with the pyramid. We should observe that the method used in locating lines *n*, *r*, and *t* is the same as that employed in the previous problem —intersection of a line and plane surface. In fact, thus far we have already made use of the first two steps, i.e., (1) *Pass a plane through the line.* (2) *Find the intersection of the plane with the given surface* (*or surfaces, in this case*).

Finally, the points of intersection, *P* and *Q*, are located by finding the intersections of line *m* with lines *n* and *r*. This is the same as the third step we discussed in the previous example. It should be quite evident that lines *m* and *t* do not intersect within the limits of the pyramid; hence, there are only two points of intersection, *P* and *Q*.

2. Intersection of a Line and a Cone

To emphasize further the employment of the three steps taken in determining the intersection of a line and a surface, let us consider the problem of locating the points in which a line intersects a cone (Fig. 9.9).

Suppose that, as the first step, we pass a plane through the line and perpendicular to the H-plane (as we did in the previous Examples 1 and 2). Now the second step is to find the intersection of the plane with the surface of the cone. This would be an hyperbola. Finally, the intersection of the line with the hyperbola would locate the points in which the line intersects the cone. This solution, while a correct one, does require the construction of a portion of the hyperbola. This is time consuming.

Let us try another approach. Suppose, as the first step, we pass a plane through the line and perpendicular to the F-plane. Now, the second step is to find the intersection of the plane with the surface of the cone. This would be an ellipse. Finally, the intersection of the line with the ellipse would locate the points in which the line intersects the cone. Again, this solution is also correct; however, it is necessary to construct a portion of the ellipse.

How can we pass a plane through the line so as to intersect the cone in a simple geometric shape? A little reflection on this question leads to the conclusion that a plane which passes through the line and the apex of the cone will intersect the cone in a triangle, two sides of which are elements of the cone and the third side is a chord of the base of the cone.

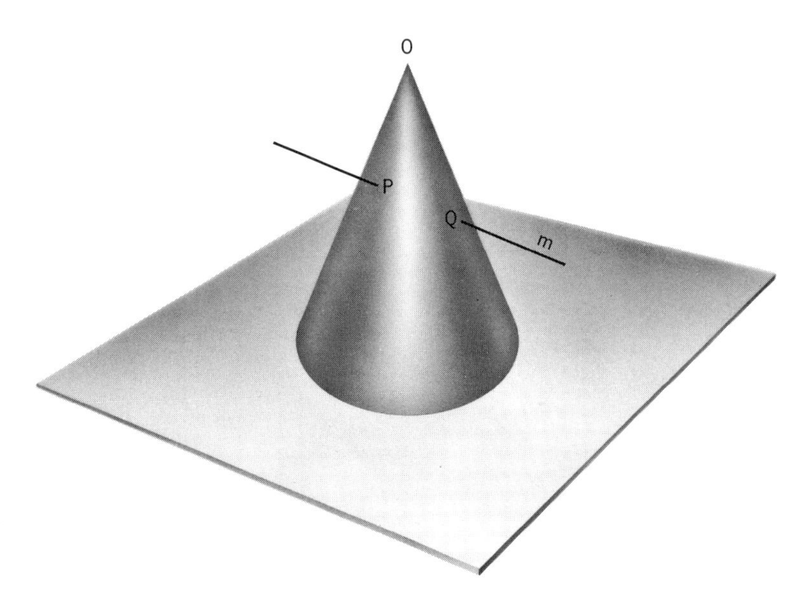

Fig. 9.9

EXAMPLE 1

In Fig. 9.10, the given line is m. Now, the plane which contains line m and apex O intersects the cone in triangle O–1–2. Note carefully that the location of points 1 and 2 is easily established, once we find line BC. This line is determined by locating point C which is the intersection of line m *with the plane that contains the base of the cone;* and by locating point B which is the intersection of line n *with the plane that contains the base of the cone.* Line n is any convenient line that lies in the plane determined by apex O and line m. Line n is the *line OA* where A is chosen, arbitrarily, on line m.

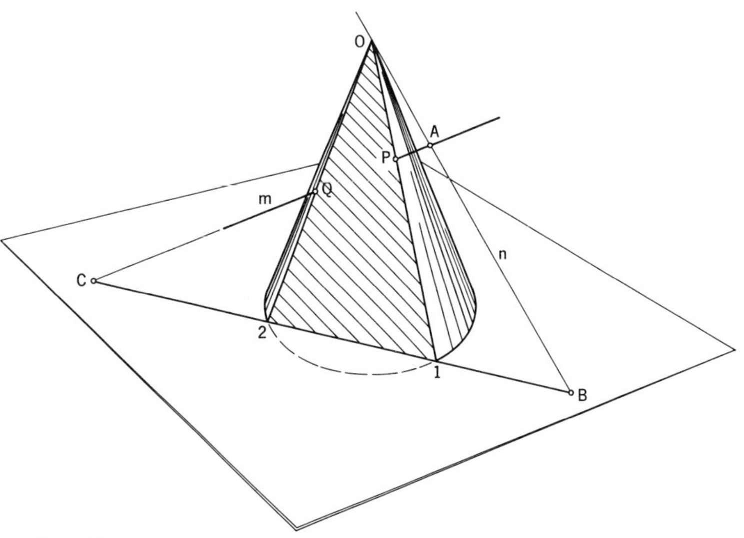

Fig. 9.10

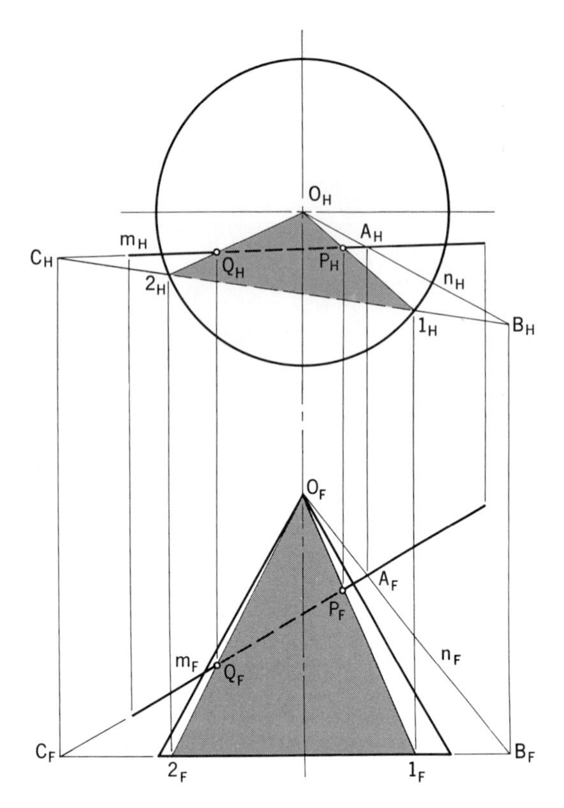

Fig. 9.11

The orthographic solution to the problem is shown in Fig. 9.11. The plane which contains line m and apex O cuts the base of the cone in line 1–2. This line is established in the following manner. We must first select point A on line m. This is a simple step since we should know that *"the views of the point lie on the corresponding views of the line."* Now we can construct line n by joining points O and A. The views of n are n_H and n_F.

How shall we locate points B and C? Consider point B, which is the intersection of line n with the plane containing the base of the cone. Since this plane appears on edge in the front view, we can locate B_F, the intersection of n_F with the base plane of the cone. Once B_F is determined we can locate B_H. In a similar manner the top and front views of point C are determined.

The line joining points B and C cuts the base of the cone in chord 1–2, which together with elements O–1 and O–2 forms triangle O–1–2. The top and front views of this triangle are shown in Fig. 9.11.

The views of points P and Q, the required points of intersection, are uniquely located by finding the points in which the respective views of line m intersect the corresponding views of the sides of the triangle.

If we analyze this problem carefully, it should be quite clear that the three steps used in solving the simple case—intersection of line and plane surface—were also followed in the line and cone problem.

Line and Triangle Problem	*Line and Cone Problem*
Step 1. Plane S passed through line m.	Step 1. Plane m–n passed through line m.
Step 2. The intersection (line n) of plane S and triangle ABC determined.	Step 2. The intersection (triangle O–1–2) of plane m–n and cone determined.
Step 3. Point P, the required point of intersection, determined by the intersection of lines m and n.	Step 3. Points P and Q, the required points of intersection, determined by the intersection of line m with lines O–1 and O–2.

EXAMPLE 2

Let us consider line m and the cone shown in Fig. 9.12.

Our first step in locating the points in which line m intersects the cone is to pass a plane through apex O and line m. The second step is to find the intersection of that plane with the surface of the cone, i.e., triangle O–1–2. The final step is to find the points in which line m intersects the triangle. The plane passed through the apex O and line m intersects the *plane which contains the base of the cone* in line BC. To locate point B, we find the intersection of line n with the plane containing the base of the cone. Note that an edge view of that plane is seen in the front view; therefore, it is a simple matter to locate B_F from which we can readily locate B_H on n_H. In a similar manner we can locate C_F and C_H. Now the intersection of line BC with the base of the cone locates points 1 and 2. Triangle O–1–2 is now constructed. The points of intersection, P and Q, result from the intersection of line m with the triangle O–1–2.

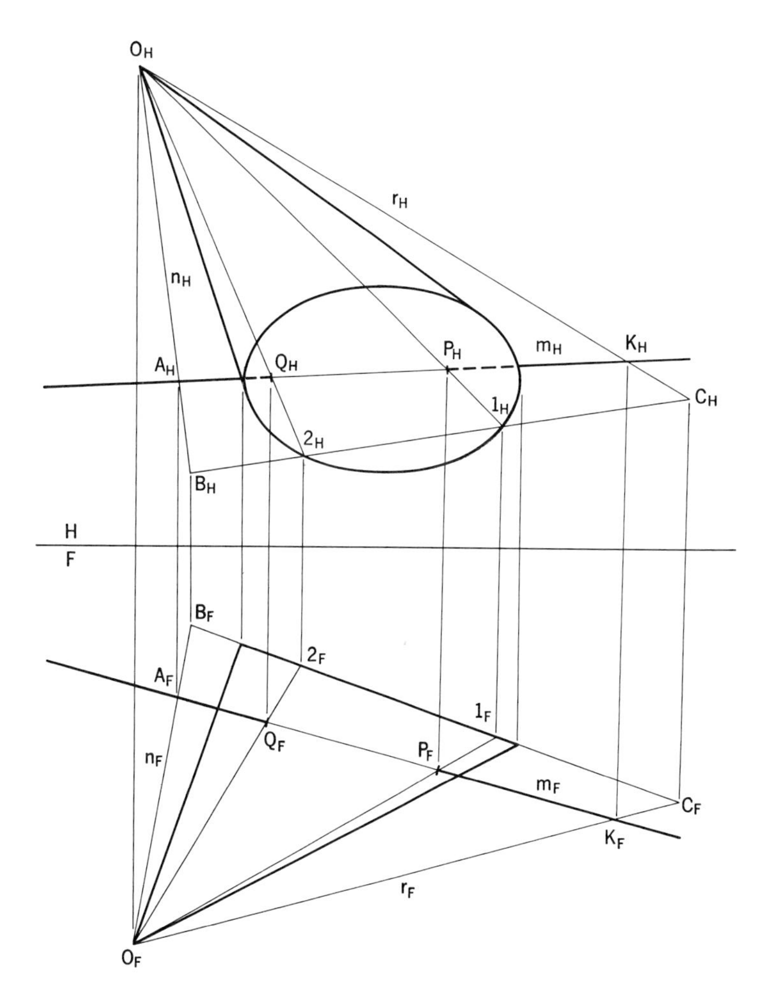

Fig. 9.12

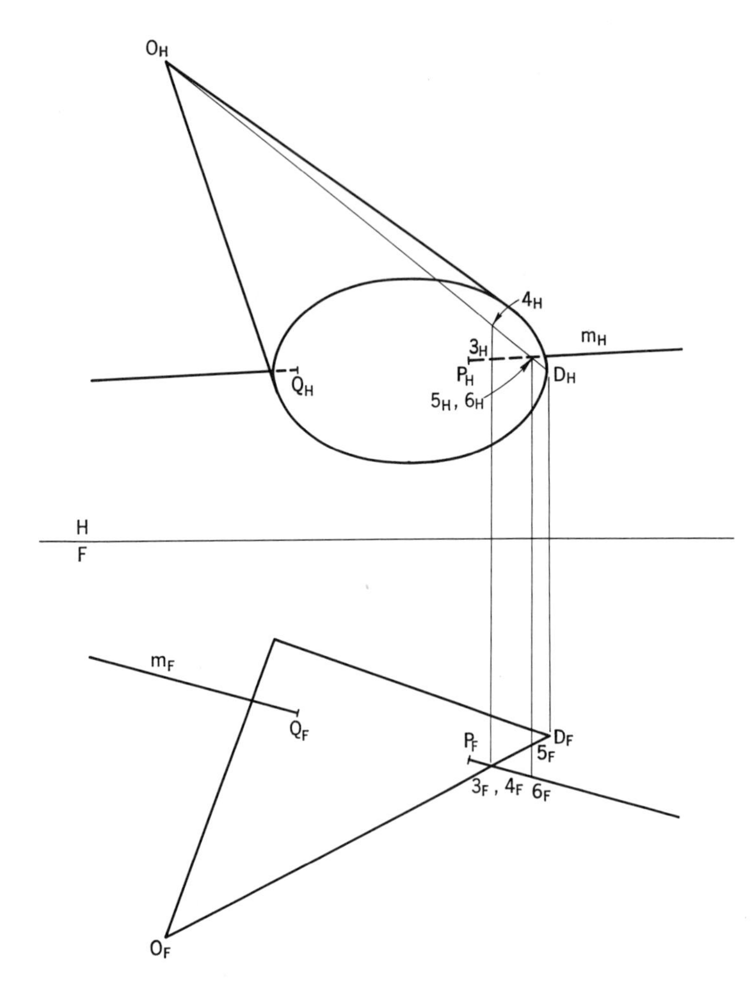

Fig. 9.13

Now let us assume that the cone is opaque. How shall we determine the visibility of line m? We will employ the same analysis we used previously in Chapter 4 when discussing visibility. To see this clearly, let us examine Fig. 9.13, which is the same as Fig. 9.12 except for the omission of the lines used to locate points P and Q. Points 3 and 4 are located on line m and element OD, respectively. Of these two points, 3 is in front of 4; therefore, line m, which contains point 3, is in front of element OD. This means that the portion of m_F between 3_F and P_F is visible and, therefore, is shown as a solid line. Now let points 5 and 6 be located on element OD and line m, respectively, as shown. Of these two points, 5 is above 6 (seen in the front view); therefore, element OD is above line m, and the hidden portion of m_H to P_H is shown as a dashed line. In a similar manner you could analyze the visibility of the portion of line m to the left of point Q. Try it.

Questions. Suppose that neither the H nor the F views of the cone shows an edge view of the base of the cone. How shall we proceed to solve this problem? Can we introduce a supplementary view that will show an edge view of the base and thereby reduce the problem to the former examples? How else could we obtain an edge view of the base or better perhaps introduce a new base which would appear on edge in the H or F views?

3. Intersection of a Line and a Cylinder

The problem of locating the points of intersection of a line with a given cylinder can be solved in a manner similar to that used in the line and cone problem.

If we *pass a plane through the line and parallel to the axis of the cylinder,* the resulting intersection will include elements of the cylinder. This is shown in Fig. 9.14.

The pictorial drawing shows the introduction of a plane through line *m* and parallel to the axis, S, of the cylinder. This was accomplished by establishing a line *n* through point *A*, any convenient point on line *m*, parallel to axis S. The plane determined by intersecting lines *m* and *n* cuts the cylinder in elements 1–4 and 2–3. These elements are easily established, since we can locate line *BC*, which is the intersection of plane *mn* and the plane containing the base of the cylinder. The intersection of line *BC* and the base of the cylinder uniquely locates points 3 and 4, which lie on the elements. The intersections of line *m* with these elements determine points *P* and *Q*, the points in which line *m* intersects the cylinder.

The orthographic solution of this problem is shown in Fig. 9.15. It should be recognized that if point *B* is inaccessible, we can introduce another line, *r*, through point *D* (an arbitrary point on line *m*) and parallel to the axis S. Line *r* intersects the plane which contains the lower base of the cylinder at point *E*. Now line *CE* can be used to locate points 3 and 4.

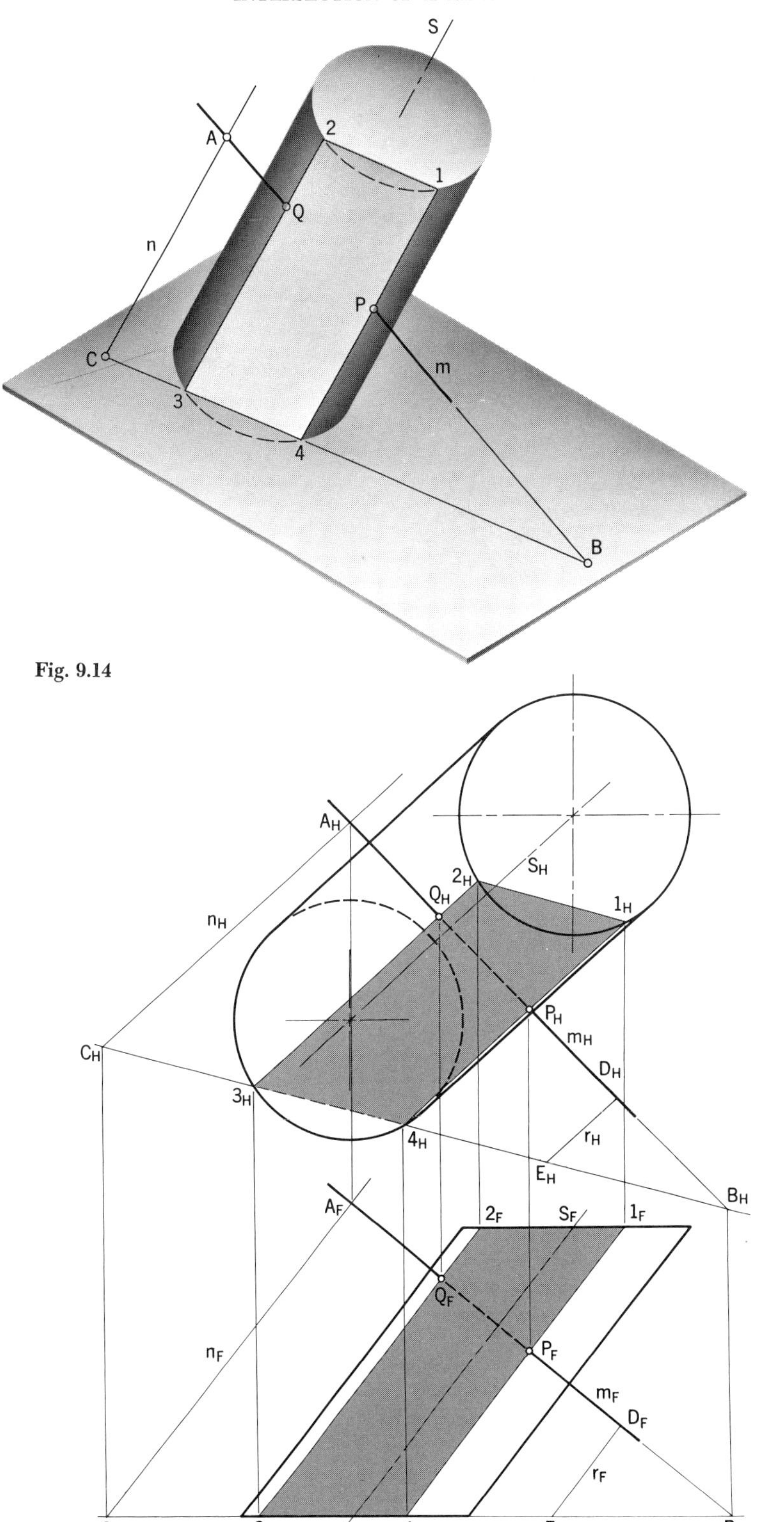

Fig. 9.14

Fig. 9.15

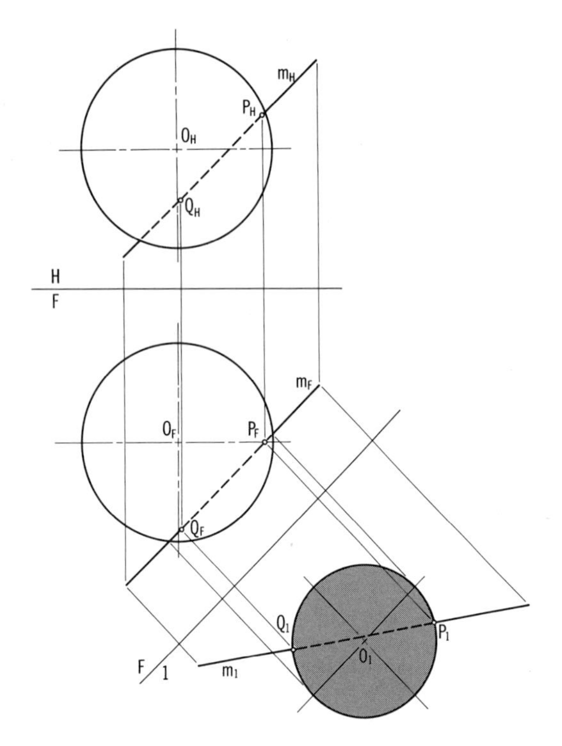

Fig. 9.16

4. Intersection of a Line and a Sphere

Suppose we are given line m and sphere O (Fig. 9.16). A plane containing line m will intersect the sphere in a circle. In general, the top and front views of the circle will be ellipses. The accuracy of the solution of the problem will depend upon the accuracy in drawing the ellipses. We can avoid drawing the ellipses, if we proceed in the following manner:

Let us pass a plane through line m and perpendicular to the F-plane (we could have passed a plane through m and perpendicular to the H-plane). This plane intersects the sphere in a circle, the front view of which appears as a line. Now we can introduce a supplementary plane parallel to the circle, thus showing the circle as a circle. The supplementary view of the points of intersection, P and Q, are seen as the points in which the supplementary view of line m intersects the circle. There is no difficulty, now, in locating the front and top views of points P and Q.

As to the determination of the visibility of line m, let us consider Fig. 9.17, which only shows the H and F views of the solution. Points 1 and 2 are, respectively, on line m and the great circle which is parallel to the F-plane. Point 1 is in front of point 2; therefore, line m, which contains point 1, is in front of the great circle. This means that the portion of m_F between 1_F and Q_F is visible. Points 3 and 4 are, respectively, on the great circle and line m. Point 3 is in front of point 4; therefore, the great circle, which contains point 3, is in front of the line. This means that the portion of m_F between 4_F and P_F is hidden and is, therefore, represented by a dashed line. The analysis for the visibility of m_H is the same. For example, points 5 and 6 are, respectively, on line m and the great circle which is parallel to the H-plane. Point 5 is below point 6; therefore, line m, which contains point 5, is below the great circle. This means that the portion of m_H between 5_H and Q_H is hidden and is, therefore, represented by a dashed line. You should experience no difficulty in verifying the visibility of the other portion of m_H.

Thorough review of the above intersection problems (relating to a line and a surface) should lead us to conclude that *basically* both the analysis and the method of solution are the same.

We now proceed to study intersection problems dealing with (a) two plane surfaces (b) two solids bounded by plane surfaces, (c) two solids bounded by plane and curved surfaces, and (d) two surfaces, one of which is plane, the other topographic.

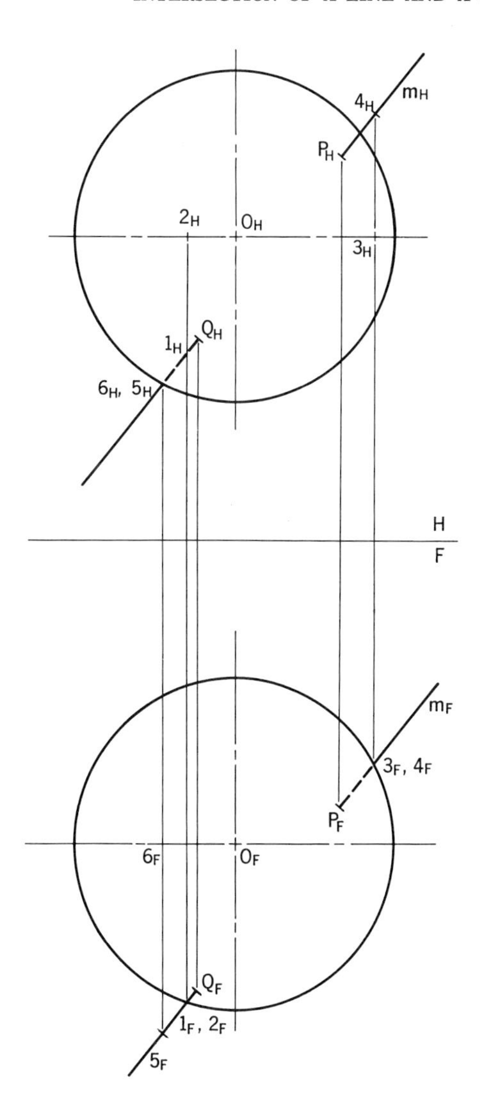

Fig. 9.17

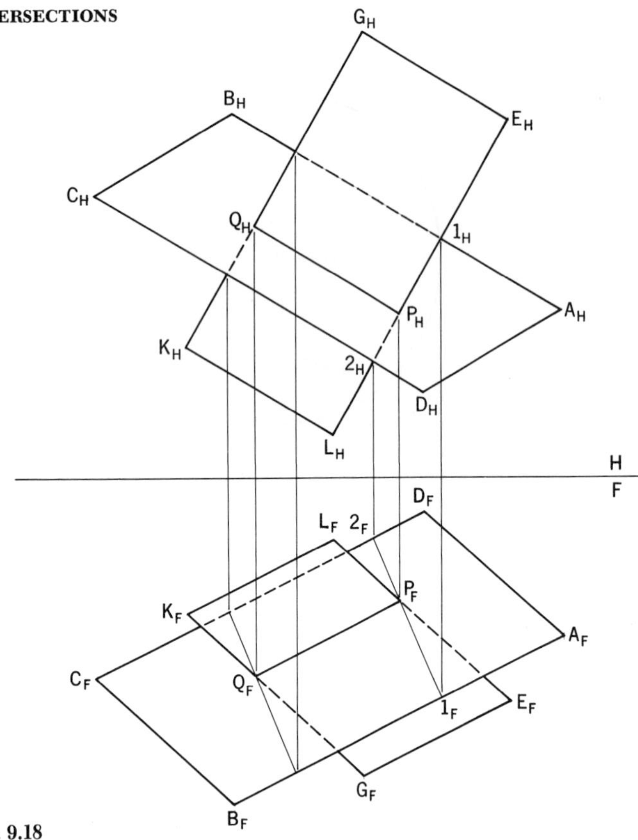

Fig. 9.18

Intersection of Plane Surfaces

EXAMPLE 1

First consider the two parallelograms shown in Fig. 9.18. The intersection, point *P*, of side *EL* with parallelogram *ABCD* is found by using the three steps stated previously. In this example, a plane has been passed through side *EL* and perpendicular to the *H*-plane. The intersection of that plane with the parallelogram *ABCD* is line 1–2; and the intersection of line 1–2 with side *EL* is the point of intersection, *P*. In a similar manner, point *Q*, the intersection of side *GK* with parallelogram *ABCD*, has been located. Line *PQ* is the intersection of the two parallelograms.

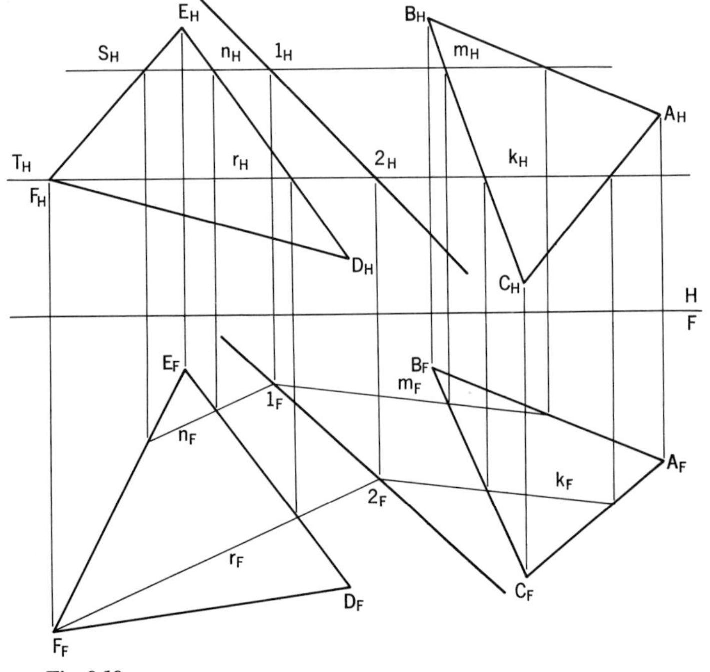

Fig. 9.19

EXAMPLE 2

Consider next the limited surfaces *ABC* and *DEF* shown in Fig. 9.19. It is quite obvious that there is no line of intersection of the limited surfaces. Now suppose we wish to determine the line of intersection of the unlimited surfaces which are defined by *ABC* and *DEF*. We first introduce a plane, such as S, parallel to the *F*-plane (it could be placed in any other convenient position) and crossing surfaces *ABC* and *DEF*. Plane S intersects surface *ABC* in line *m*, and surfaces *DEF* in line *n*. Point 1, the intersection of lines *m* and *n*, is common to all three planes and, therefore, is a point on the intersection of unlimited surfaces *ABC* and *DEF*. To locate a second point, we introduce plane *T*, which is parallel to plane S. Plane *T* intersects surface *ABC* in line *k*, and surface *DEF* in line *r*. Point 2, the intersection of lines *k* and *r*, is common to the three planes, *ABC*, *DEF*, and *T*. The line which joins points 1 and 2 is the line of intersection of the unlimited surfaces *ABC* and *DEF*.

Intersection of Solids Bounded by Plane Surfaces

Basically the problem of determining the lines of intersection of two solids bounded by plane surfaces is the same as finding the intersection of plane surfaces (which actually reduces to the problem of finding the intersection of a line with a plane).

Let us consider the solids shown in Fig. 9.20. First, we will obtain the intersections of edges *m*, *n*, and *v*, of the triangular piece, with the faces of the larger block. This is very easy to do because the lateral surfaces of the larger block appear on edge in the top view. Points *P* and *Q* are the points in which edges *n* and *m*, respectively, intersect surface *ABCD*. *R* is the point in which edge *v* intersects surface *BCEF*. Second, let us determine the points in which edges of the larger block intersect the faces of the triangular piece. The orthographic views of the two solids (treated as a single casting) show that only edge *BC* intersects faces of the triangular piece. Our problem, now is to find the intersection, point S, of edge *BC* with the plane determined by parallel edges *n* and *v*; and also the intersection, point *T*, with the plane determined by parallel edges *m* and *v*.

To locate point S, let us employ the three steps for finding the intersection of a line with a plane. *Step 1*—pass a plane through line *BC*. The plane is surface *ABCD* extended to cut across the plane determined by edges *n* and *v*. *Step 2*—the intersection of the two planes is line *P*–1. *Step 3*—the intersection of line *P*–1 with line *BC* is the required point S.

To locate point *T*, which is the intersection of line *BC* with the plane determined by parallel edges *m* and *v*, we pass the same plane, *ABCD* extended, through line *BC*. This plane intersects the plane determined by edges *m* and *v* in line *Q*–1. Finally, the intersection of line *Q*–1 with line *BC* is point *T*, the intersection of line *BC* and surface *mv*.

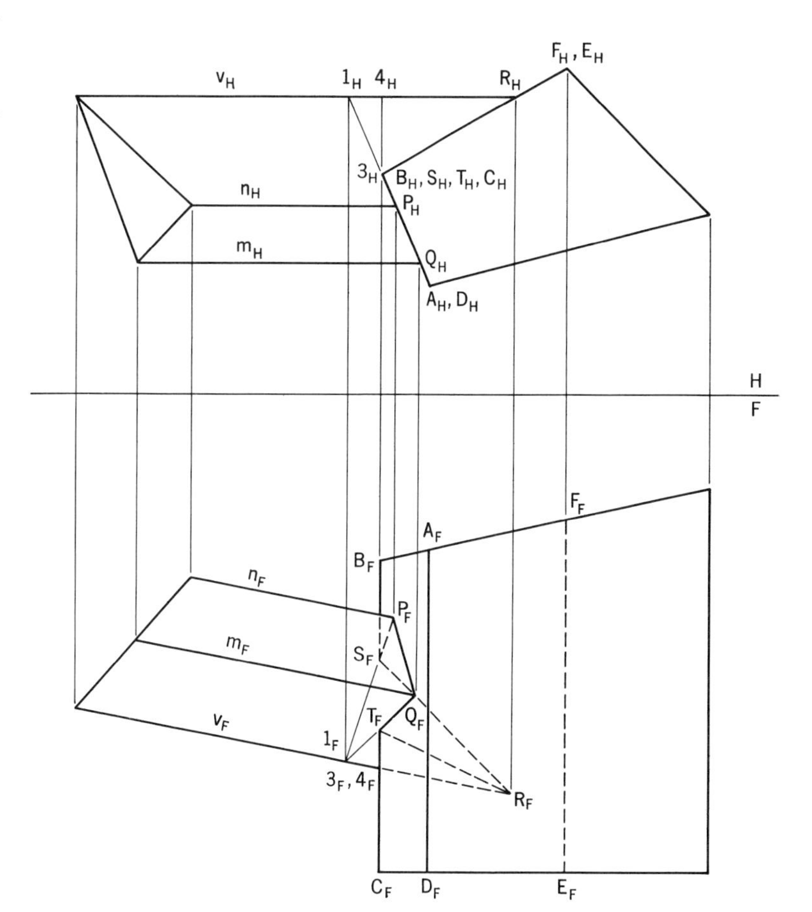

Fig. 9.20

Alternative Solution. An alternative solution for the location of points S and *T* is shown in Fig. 9.21. The supplementary view shows the lateral surface of the triangular piece on edge. The intersections, S and *T*, of line *BC* with the lateral surfaces are easily located in the supplementary view, and from which it is a simple matter to locate the other views of points S and *T*.

Now that we have all of the possible points of intersection, how shall we connect them?

The line which joins two points must be common to two surfaces. For example, the line which joins points *P* and *Q* is common to surfaces *ABCD* and *mn*. Also, the line which joins points *P* and S is common to surfaces *ABCD* and *nv*, etc. On the other hand, we observe that a line such as *PR* or *QR* is *not* common to two surfaces. The determination of the hidden lines is easily analyzed. For example, the front view of points 3 and 4 (point 3 on line *TC*

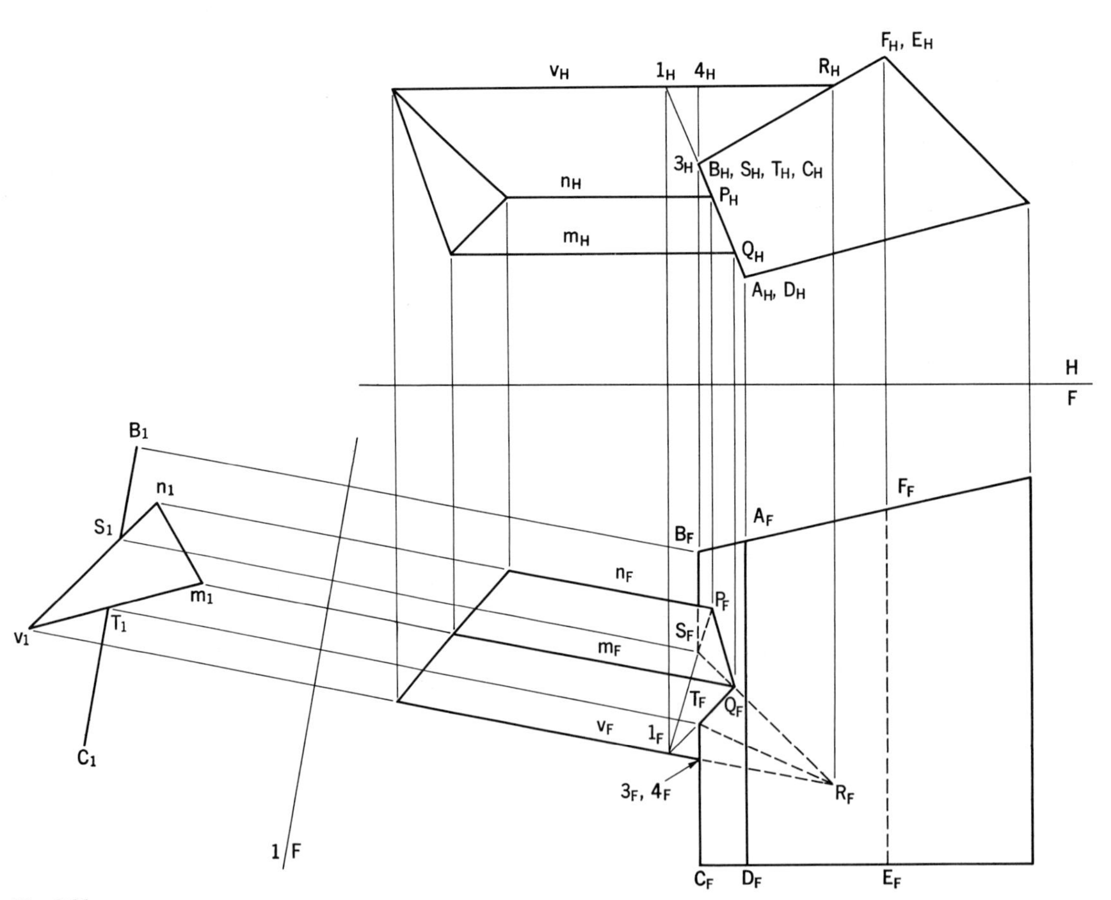

Fig. 9.21

and point 4 on line v) appears as one point. Which of these two points is in front of the other? The top shows, quite clearly, that point 3 is in front of point 4. Since line TC contains point 3, line TC is in front of line v and, therefore, the portion of line v from 4_F to R_F is hidden and is represented by the dashed line. In a similar manner we can verify the visibility of the other lines.

Intersection of a Cone and a Cylinder

In our discussion of "the intersection of a line and a cone" it was pointed out that a plane passed through the line and the apex of the cone would cut a triangle out of the cone, if the plane intersected the base of the cone.

In the case of the "intersection of a line and a cylinder" we found that a plane passed through the line and parallel to the axis of the cylinder would cut elements from the cylinder, if the plane intersected the cylinder.

Therefore, to cut the simplest shapes out of both the cone and the cylinder, it is necessary to introduce planes through the apex of the cone and parallel to the axis of the cylinder.

This can be done by first introducing a line through the apex of the cone and parallel to the axis of the cylinder. Now all the planes that contain this line will cut elements from both the cone and the cylinder, if the plane intersects both surfaces.

Suppose we are given the cone and cylinder shown in Fig. 9.22. Let us determine the location of points on the curve of intersection of the two solids.

Our first step is to pass a plane through the apex of the cone and parallel to the axis of the cylinder. In order to accomplish this we shall first introduce line n through point C parallel to the axis of the cylinder. The top and front views of line n are easily determined since we know (a) that the views of a line pass through the corresponding views of the point which lies on the line, and (b) views of parallel lines appear as parallel lines; in this case n_H is parallel to the top view of the axis of the cylinder and n_F is parallel to the front view of the axis of the cylinder.

All planes containing line n will intersect the cone and cylinder in elements, if the planes cut the two solids.

Let us select one plane, S, for example, which contains line n and cuts both solids. How shall we

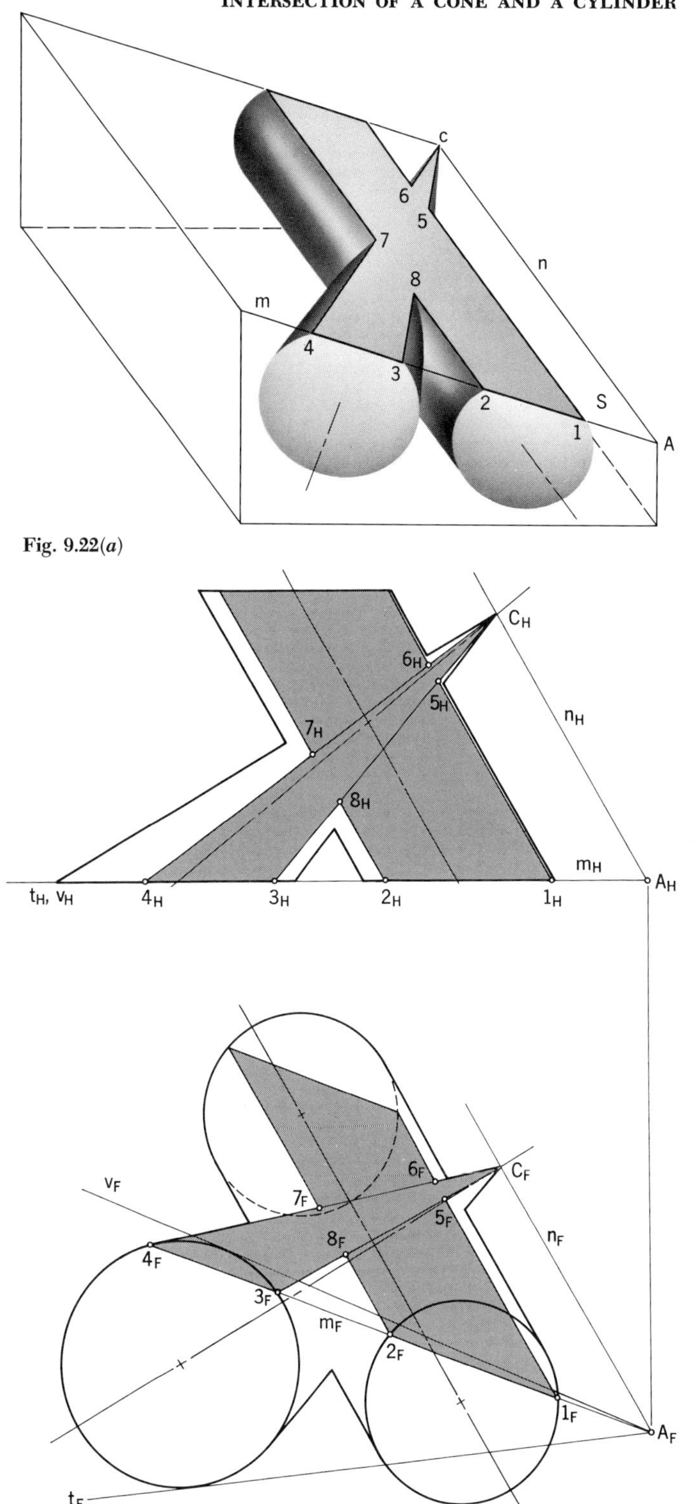

Fig. 9.22(a)

Fig. 9.22(b)

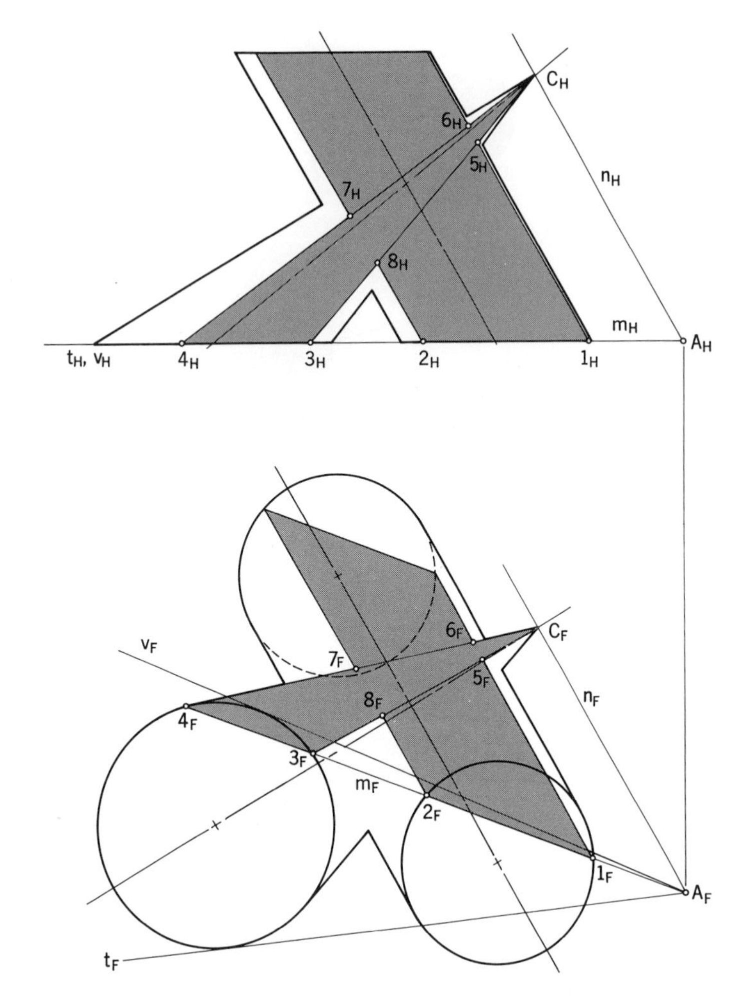

Fig. 9.22 (*b*) (Repeated)

proceed to do this? Before we attempt to answer this question, we should first investigate another one: How can we establish any plane through line n? This question is easily answered. We should know that a plane may be determined by (*a*) two intersecting lines, (*b*) two parallel lines, (*c*) three noncollinear points, or (*d*) a line and an external point. Therefore, any plane through line n may be determined, for example, by line n and the intersecting line, m.

If we select the intersecting line m in the plane of the bases of the cone and cylinder, the point on n through which line m passes must also lie in the plane of the bases of the two solids. Thus, we see that this point A must be the point in which line n intersects the plane of the bases. This point is readily located since an edge view of the plane of the bases is available in the top view.

Now plane S, determined by lines n and m, intersects the cylinder in elements 1 and 2, and intersects the cone in elements 3 and 4. Since all four elements lie in plane S, the points in which the elements intersect will be common to the two surfaces. These four points are 5, 6, 7, and 8. In a similar manner we can establish other planes through line n and then determine additional points common to the two surfaces. The required intersection (curves in this case) is drawn through the points which lie in both surfaces.

It should be carefully noted that the planes determined by lines n and t, and by lines n and v, are the "limiting planes"; i.e., the intersection of the cone and cylinder will lie between the limiting planes.

Alternative Solution. It should be pointed out that *this problem* (Fig. 9.22) *could be solved by passing planes parallel to the bases of the cone and cylinder.* For example, a plane parallel to the bases would intersect the cone in a circle and would also intersect the cylinder in a circle. Since both circles would lie in the same plane, the points of intersection of the circles would be points on the curve of intersection of the cone and cylinder.

Intersection of Two Cones

We have previously noted that any plane which intersects a cone and passes through its apex will cut out elements of the cone. If we connect the vertices of the given cones, planes containing the connecting line will intersect both cones in elements provided that the planes intersect both solids.

The common points of the elements of the two cones are points on the line of intersection of the two solids. A sufficient number of planes should be passed through the line joining the vertices of the cones to determine an adequate number of points on the curve of intersection.

EXAMPLE 1

Let us consider the two cones shown in Fig. 9.23. The planes which contain the line *n* (line *n* connects the apex *O* with apex *B*) and intersect the cones will cut triangles from the cones; i.e., plane *mn* cuts triangle *OCD* from the cone with apex *O* and triangle *BGK* from the other cone. The points 1, 2, 3, 4, of intersection of the sides (elements of the cones) of the triangles are points on the curve of intersection of the two cones. In a similar manner additional points can be located by passing several planes through line *n*. It should be pointed out that the curve of intersection of the two cones will lie between the "limiting planes *nt* and *nr*."

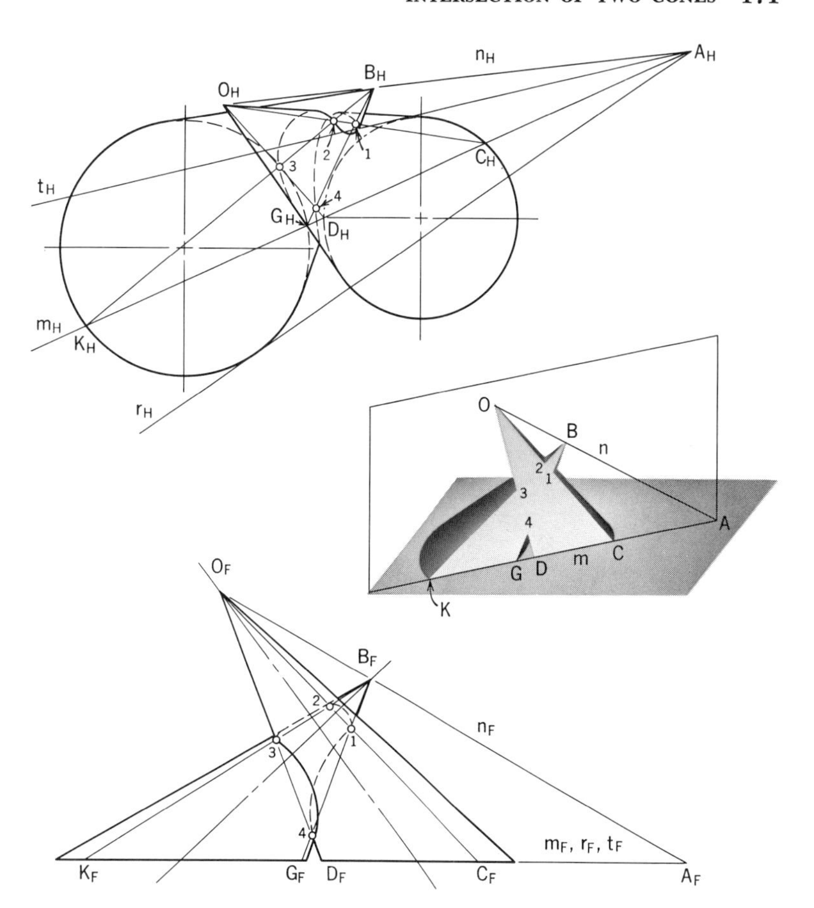

Fig. 9.23

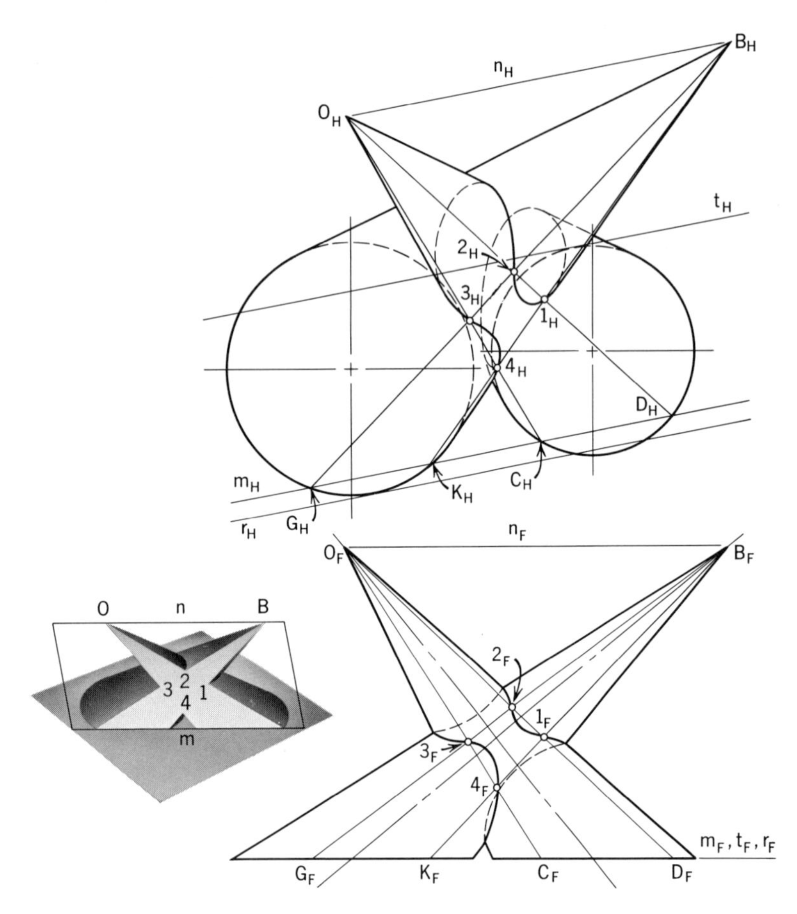

Fig. 9.24

EXAMPLE 2

Let us consider a unique case where the line joining the apices is parallel to the plane which contains the bases of the cones. In Fig. 9.24 we observe that line n which joins the apices of the cones is parallel to the plane which contains the bases of the cones and, therefore, does not intersect this plane. How shall we identify planes that contain line n?

Any plane that contains line n will intersect the plane which contains the bases of the cones in a line which is parallel to n. We know that the views of parallel lines will appear as parallels; therefore, we can easily establish planes through line n. For instance, line m is drawn in the plane, which contains the bases, parallel to line n. This plane, mn, obviously contains line n and intersects the plane of the bases in line m. Moreover, plane mn intersects the cones in triangles OCD and BGK. The intersections of the sides (elements of the cones) of these triangles are points on the curve of intersection of the two cones. In a similar manner additional planes may be passed through line n to determine more points on the curve of intersection of the cones. Again, it should be observed that the curve of intersection will lie between the limiting planes nt and nr.

How else could you solve this problem? How would you pass planes across the two cones to cut simple geometric shapes from each cone?

Intersection of Any Two Solids

Recall the problem of finding the intersection of a cone and cylinder (Fig. 9.22). The alternative solution suggested that planes could be introduced parallel to the bases of the cone and cylinder in order to locate points on the curve of intersection. The curve of intersection of any two solids can be determined, in general, by passing a number of planes across the two solids. Each plane will cut from the solids curves which lie in that plane. The points in which these curves intersect are points on the intersection of the two solids. The choice of the cutting planes is most important. The selection should be based upon the objective that *the simplest geometric shapes result from the intersection of the selected plane with the two solids.*

Let us consider the two surfaces of revolution shown in Fig. 9.25. We could find the intersection of the two surfaces by first finding the points in which elements of the cone intersect the ellipsoid and then joining the points of intersection by a smooth curve. This method is laborious because the intersection of an element of the cone with the ellipsoid involves the three steps: (*a*) a plane through the element; (*b*) the intersection of the plane with the ellipsoid surface; and (*c*) the intersection of the element with the curve cut from the ellipsoid.

A simpler solution is possible by using the *sphere method.* This method is very useful in finding the intersection of surfaces of revolution, *if their axes intersect and are seen in true length in one view.* The sphere method is based on the fact that a sphere with its center at the intersection of the axes will cut circles from each surface of revolution. The circles will appear as lines in the view which shows the axes in true length. The points of intersection of the edge views of the circles will be points on the curve of intersection of the surfaces of revolution.

In Fig. 9.25, sphere 1 intersects the cone in circle *A* and the ellipsoid in circle *B*. The circles appear as straight lines in the front view. The intersections of the circles are points 1 and 2, shown in the front view as 1_F and 2_F. The top view of the points are shown as 1_H and 2_H, which are on the top view of circle *A*. Additional spheres may be introduced to obtain more points which, when properly connected, will establish the intersection of the cone and ellipsoid. The limiting spheres are shown in Fig. 9.25. Other spheres would lie between the limiting spheres.

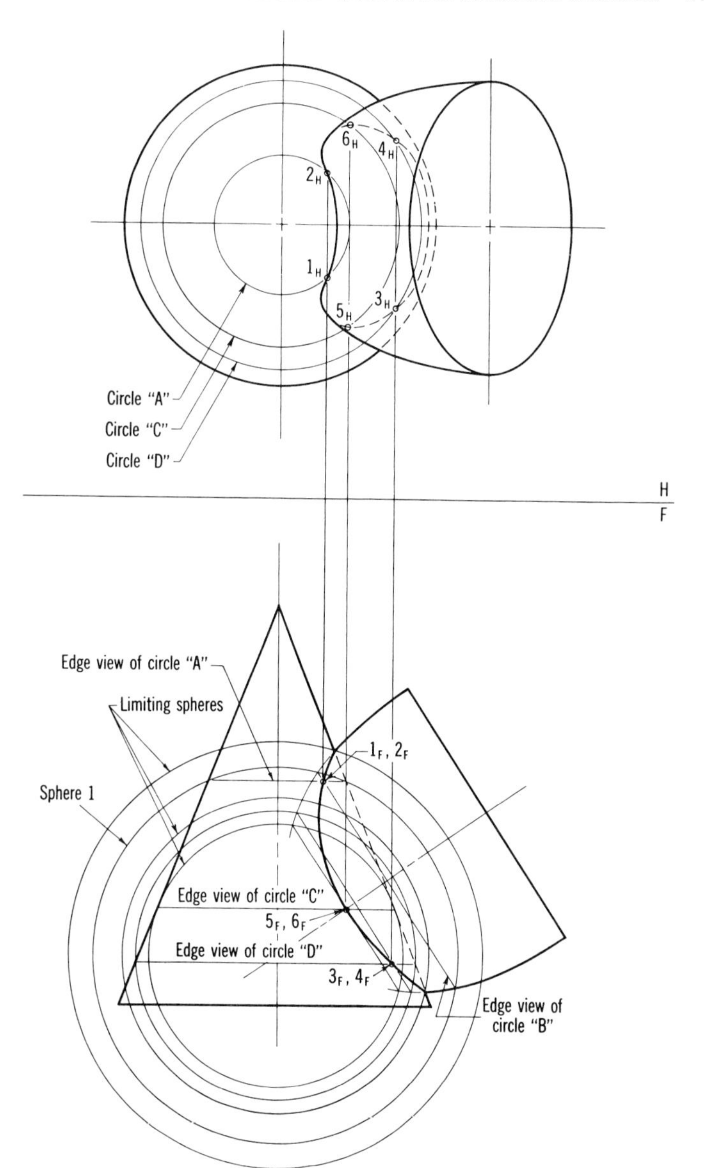

Fig. 9.25

Intersection of a Plane with a Topographic Surface

An interesting intersection problem arises in the determination of (*a*) the widths of road "right-of-way" and (*b*) earth quantities as they pertain to highway design.

EXAMPLE

Let us assume that line AB is the center line of a proposed 50-foot highway having a 10% grade, Fig. 9.26. At station 1 + 00* the elevation of the finished subgrade is 40 feet. Earth cuts are to be made at a slope of 1:1 (i.e., one unit horizontally to one unit vertically, as shown in Fig. 9.27). Earth fills are to be made at a slope of $1\frac{1}{2}:1$ (see Fig. 9.28).

Our problem is to determine the lines in which the sloping surfaces of the proposed highway inter-

* Notation used in surveying. Station 2 + 00 is 100 feet from station 1 + 00. A point half way between these two stations would be marked 1 + 50.

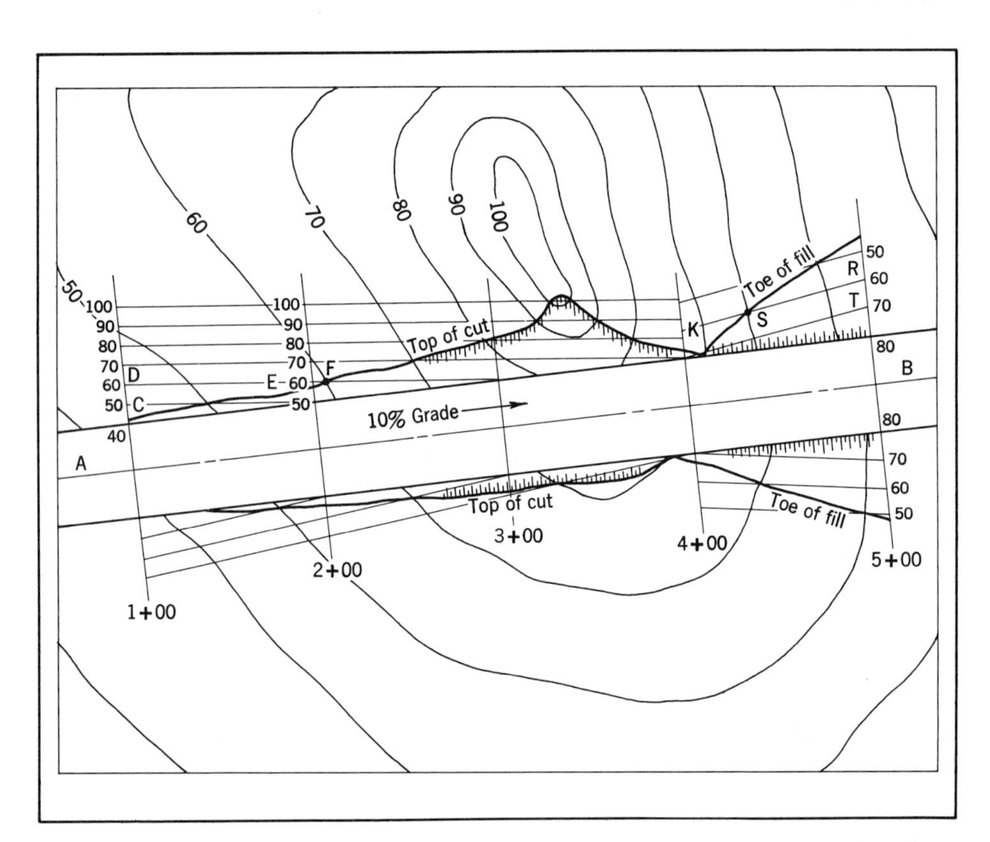

Fig. 9.26

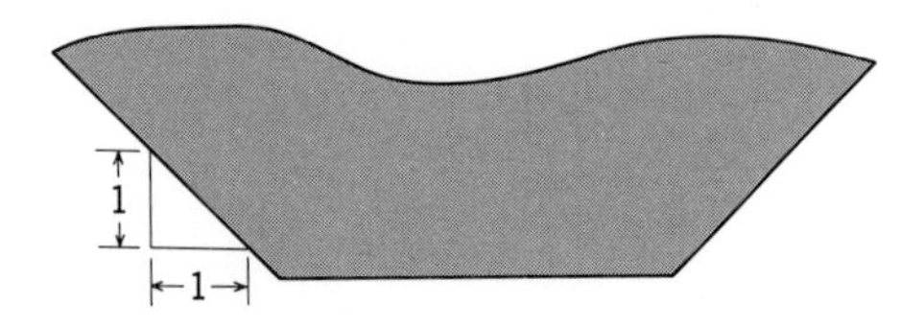

Fig. 9.27

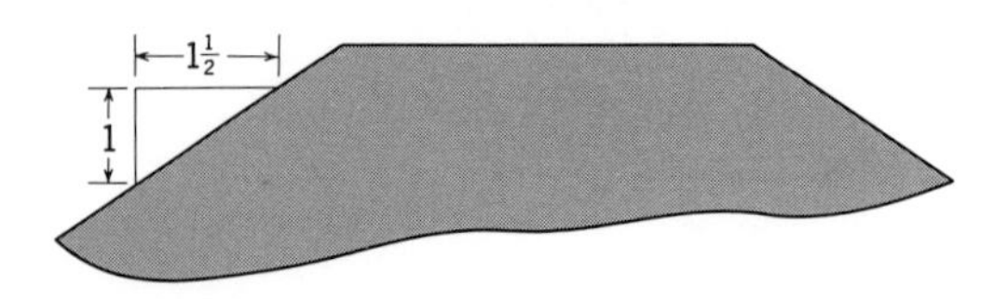

Fig. 9.28

sect the topographic surface (earth surface). The line in which the 1:1 surface intersects the earth surface is called the "top-of-the-cut." The line in which the $1\frac{1}{2}$:1 surface intersects the earth surface is called the "toe-of-the-fill."

Let us start at station 1 + 00. If we lay off a distance of 10 feet from the upper edge of the road, the elevation of the end point, *C*, will be 50 feet. This is true because the contour* lines in the vicinity of stations 1 + 00, 2 + 00, . . . indicate that the ground is at a higher elevation than the elevation of the proposed subgrade; hence, cuts will be required. Since the slope of the cut is 1:1, a horizontal distance of 10 feet will reflect an increase in elevation of 10 feet. If we lay off an additional 10 feet from *C* to *D*, the elevation of *D* will be 60 feet, and so on for additional points.

Now let us move to station 2 + 00. The elevation of the subgrade at this station is 50 feet. Why is this true? If we lay off a distance of 10 feet from the upper edge of the road, the end point, *E*, will be at elevation 60 feet, and so on for additional 10-foot increments. We should observe that points *E* and *D* are at the *same* elevation, 60 feet. *The line joining these points is a contour line on the sloping 1:1 surface. This contour line intersects the earth contour* (elevation 60) *at point F, which is a point common to the earth surface and the plane surface, 1:1.*

Now let us see how the three steps used in finding the intersection of a line with a plane were employed in this problem. What is our problem? Actually, we wish to find the intersection of the contour line at, for example, elevation 60 feet with the plane surface having the 1:1 slope.

You will recall that the first step in finding the intersection of a line with a plane is to pass a plane through the line. In this case the only plane that contains contour 60 is the horizontal plane in which the contour lies. The second step is to find the line in which the horizontal plane intersects the 1:1 surface. The line of intersection is *DE*. The final step is to locate the point (or points) in which line *DE* intersects the contour 60. That point is *F*. In a similar manner we locate additional points. The line (usually irregular) which joins these points is the "top-of-the-cut."

As we approach station 4 + 00, we observe that

* Lines all of whose points are at the same elevation.

the cut has nearly run out and that fill will be necessary for the continuation of the highway. At this station the subgrade elevation is 70 feet. If we lay off a distance of 15 feet from the upper edge of the roadway (remember the slope in a fill is $1\frac{1}{2}:1$), the elevation of the end point K, will be 60 feet. If we lay off another 15 feet from point K, the elevation will be 50 feet, and so on.

Now let us move to station 5 + 00. The subgrade elevation is 80 feet. Again let us lay off a distance of 15 feet from the upper edge of the roadway. The elevation of the end point, T, is 70 feet. If we lay off another 15 feet from point T to point R, the elevation of the latter will be 60 feet. The line joining points R and K is the 60-foot contour on the sloping surface, $1\frac{1}{2}:1$. The intersection of this contour with the 60-foot natural ground contour is the common point S. Additional points are located in a similar manner. The line which joins the common points is the "toe-of-the-fill."

If the same procedure is followed in the determination of the top-of-the-cut and the toe-of-the-fill

Fig. 9.29

for the other side of the highway, it will then be possible to draw cross sections of the road. The areas of the cross sections can be determined (usually by the use of a planimeter—shown on p. 300. If the distances between the sections are known, the volumes of fill and cut can be calculated. The widths of right-of-way now can be established on the contour drawing since the top-of-the-cut and toe-of-the-fill lines have been located. Figure 9.29 shows a typical pictorial of the top-of-the-cut and toe-of-the-fill lines.

Another example is shown in Fig. 9.30. The same basic method used in the previous example is applied to the solution of this problem.

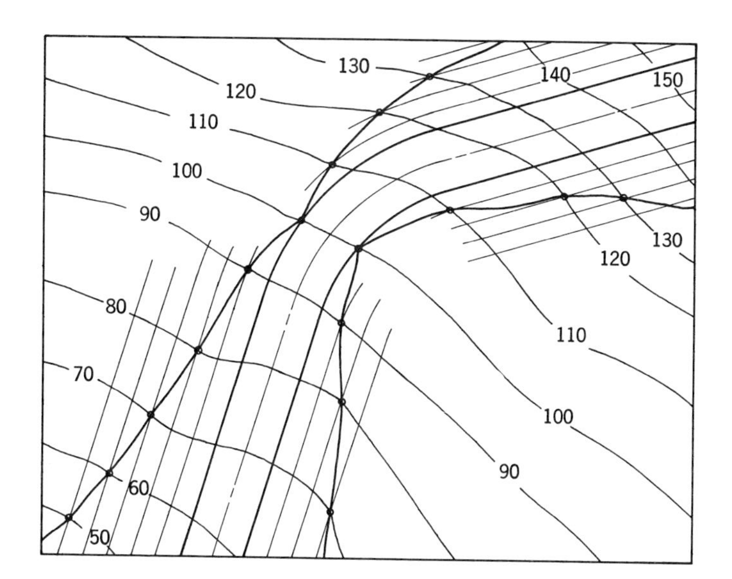

Fig. 9.30

STIMULATING THINKING—REVIEW OF BASIC MATERIAL

Thus far we have developed some experience in (*a*) the employment of the two fundamental principles of orthogonal projection; (*b*) the application of the basic problems—true length of a line segment, point view of a line, edge view of a plane, and true shape of a plane surface; (*c*) the basic intersection problems; and (*d*) the development of surfaces. Now let us test our knowledge of a few thought-provoking problems.

EXAMPLE 1

Let us locate the *H* and *F* views of line *n* which passes through point *K*, is parallel to plane *ABC*, and intersects line *m*. (See Fig. 9.31.)

Analysis

1. Any line through point *K* will satisfy the first condition. Among the lines that pass through point *K*, some will intersect line *m*; and of these, one will be parallel to plane *ABC*. How shall we find that one line?

2. We can establish a plane through point *K* parallel to plane *ABC*; then all the lines in the new plane will be parallel to plane *ABC*. We can easily establish the new plane by introducing lines *s* and *t* through point *K* and respectively parallel to *BC* and *AB*. Of all the lines in the new plane we can find the one line which intersects line *m*. Now, since the one line, *r*, we are looking for is in plane *st*, and since the point of intersection of line *r* with line *m* must also lie in plane *st*, we need only find point *P*, the intersection of line *m* with plane *st*.

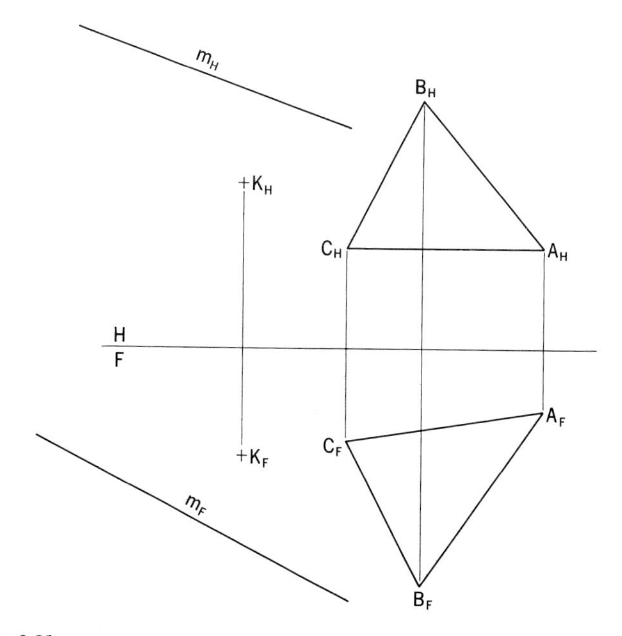

Fig. 9.31

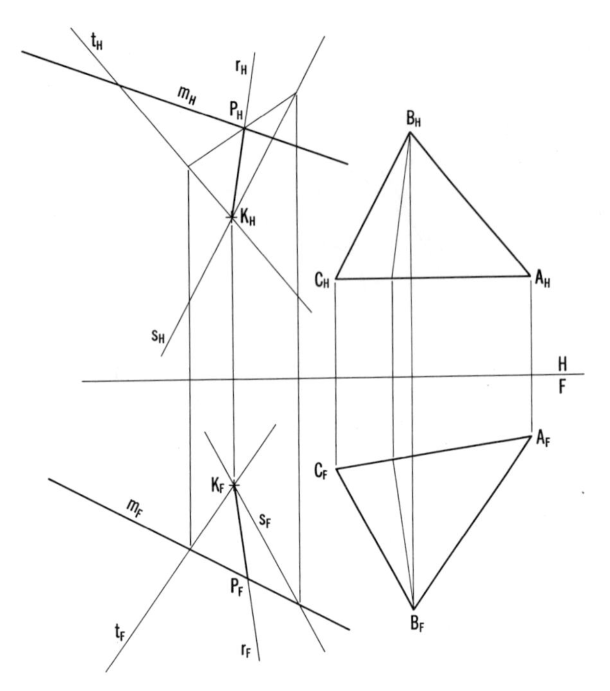

Fig. 9.32

3. The line joining points K and P is the required line r. The solution is shown in Fig. 9.32.

In *summary,* the solution of this problem included the following problems that we solved earlier:

(*a*) To pass a plane through a point parallel to a given plane.
(*b*) The intersection of a line with a plane.

Question. Could you analyze and solve the problem by employing a supplementary view? Try it.

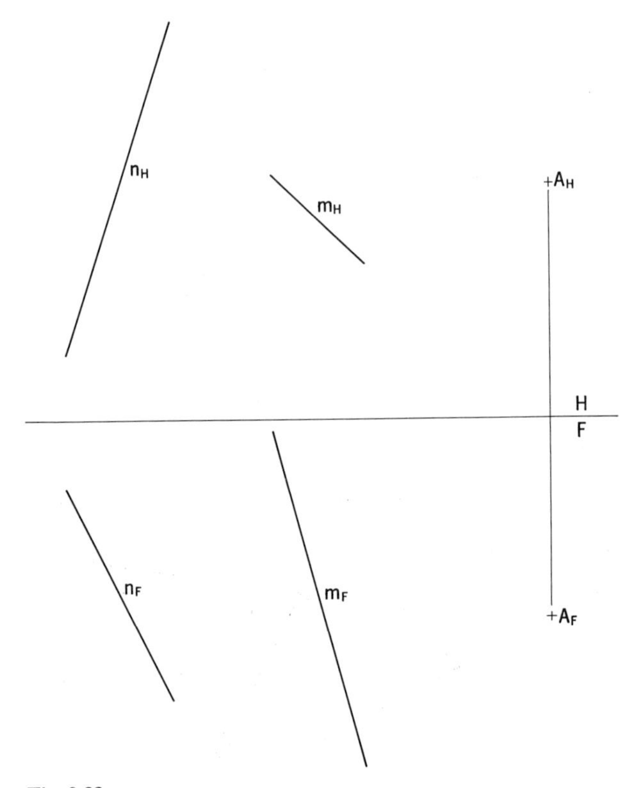

Fig. 9.33

EXAMPLE 2

Consider the problem shown in Fig. 9.33. Lines m and n represent two tunnels. It is proposed to locate another tunnel, starting at point A and intersecting tunnels m and n. We are to determine the H and F views of the proposed tunnel AB, where point B is on tunnel n.

Analysis

1. Tunnel AB must lie in the plane determined by point A and tunnel n.
2. Tunnel AB must also lie in the plane determined by point A and tunnel m.
3. Therefore, tunnel AB is the line of intersection of the two planes described in steps 1 and 2.
4. To determine the line of intersection of the two planes, we only need to find the point, P, in which tunnel m intersects the plane determined by point A and tunnel n. The line joining points A and P will intersect tunnel n at point B. Tunnel AB, then, connects points A and B. The solution is shown in Fig. 9.34.

Question. Could you analyze and solve the problem by employing a supplementary view?

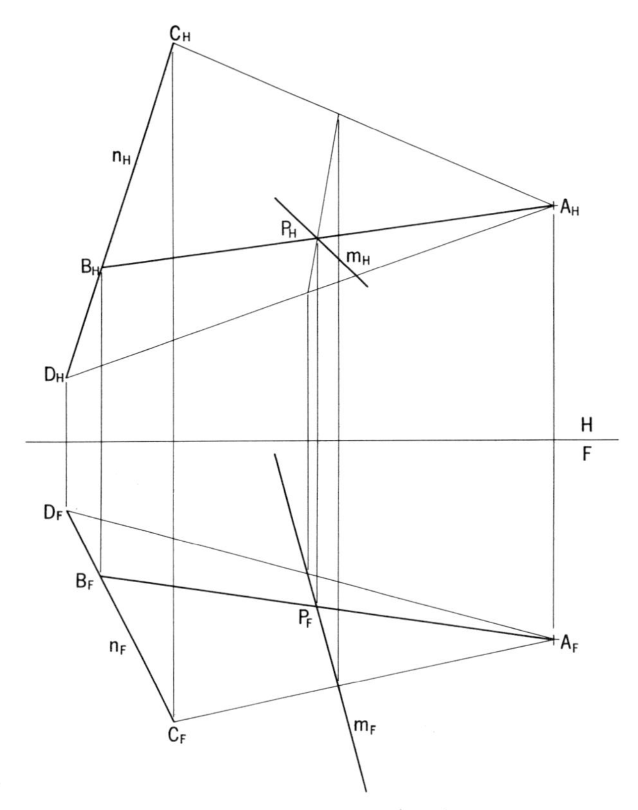

Fig. 9.34

EXAMPLE 3

Shown in Fig. 9.35 are the H and F views of point O and line m. We are to locate all lines that pass through point O, intersect line m, and make an angle of 60° with the F-plane.

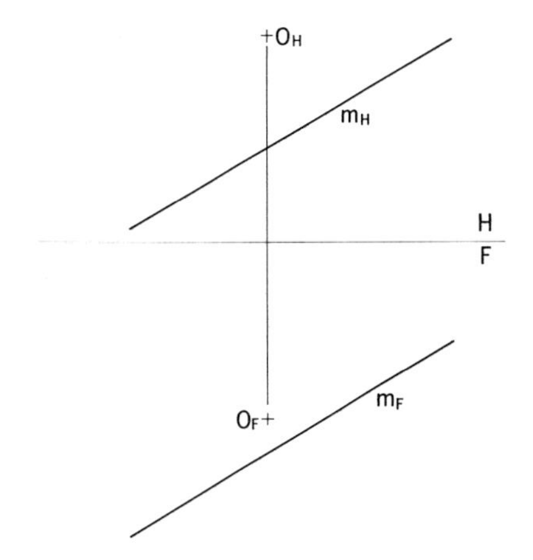

Fig. 9.35

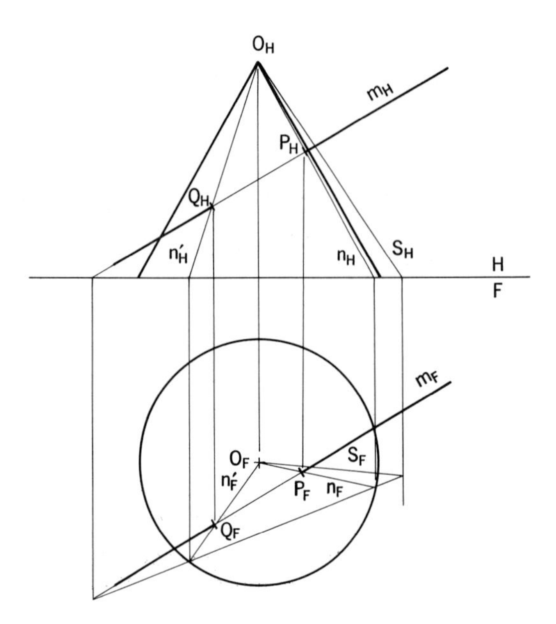

Fig. 9.36

Analysis

1. We can construct any number of lines through point O and intersecting line m. Only some of these lines make an angle of 60° with the F-plane. How shall we determine those that do? Let us consider the following attack on the problem. Suppose we *first visualize lines through point A and at angle of 60° with the F-plane.* All such lines will be elements of a right-circular cone, with apex O and base angle, 60°. Now, which of these elements intersects line m?

2. The intersection of line m with the cone will identify the elements of the cone that intersect line m. The solution is shown in Fig. 9.36.

Question. Under what conditions might there be (*a*) only one solution; (*b*) no solution?

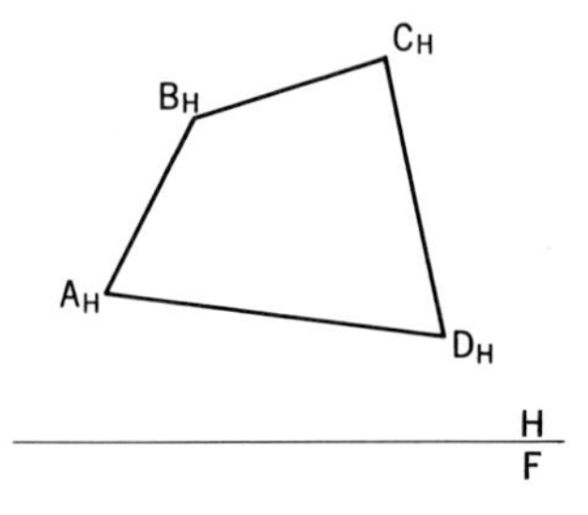

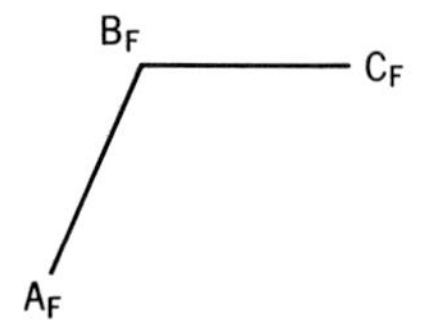

Fig. 9.37

EXAMPLE 4

Determine the diameter of the largest circular plate that can be cut from the flat plate $ABCD$. (See Fig. 9.37.)

Analysis

1. First, obtain a view which shows the true shape of plate $ABCD$. This can be done, quite readily, by obtaining an edge view and then a view upon a plane which is parallel to the surface $ABCD$. The edge view is obtained by introducing supplementary plane 1 perpendicular to side BC which appears in true length in the top view. The view on supplementary plane 2, which is parallel to surface $ABCD$, reveals the true shape of the plate. (See Fig. 9.38.)

2. Second, locate the center of the largest circle. This is a plane geometry problem, the solution of which is left to the student.

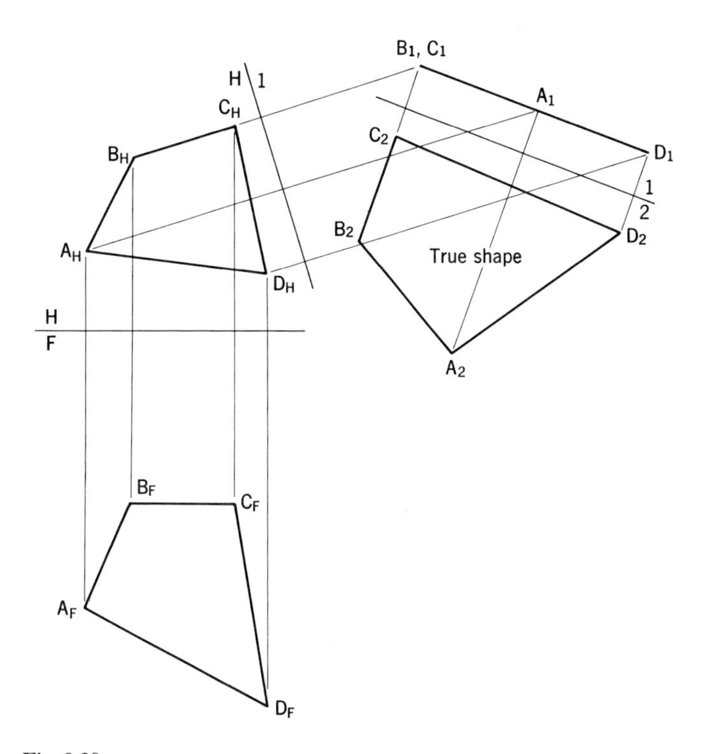

Fig. 9.38

EXAMPLE 5

Consider the plate ABC and point P as shown in Fig. 9.39. It is required to establish a plane defined by two lines m and n which pass through point P. It is specified that line m is perpendicular to plate ABC and that line n has a bearing of N 60° W and a grade of 50%. In addition, the angle between plate ABC and plane mn is to be determined.

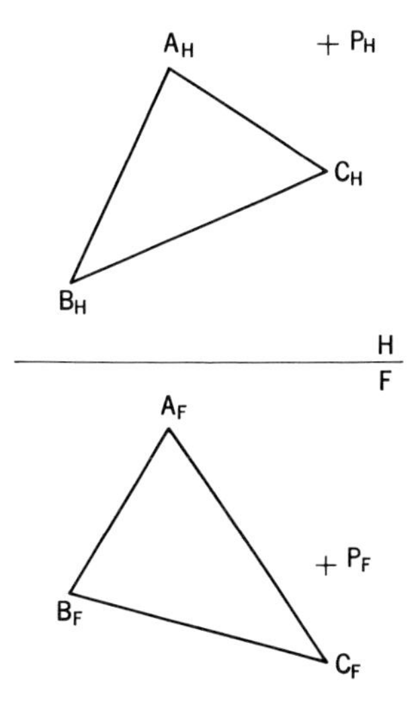

Fig. 9.39

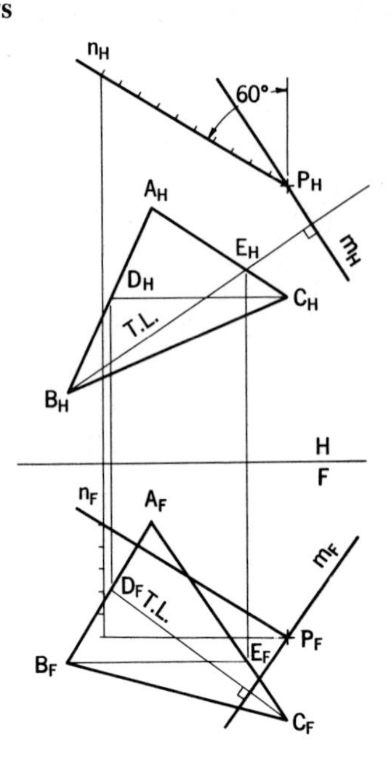

Fig. 9.40

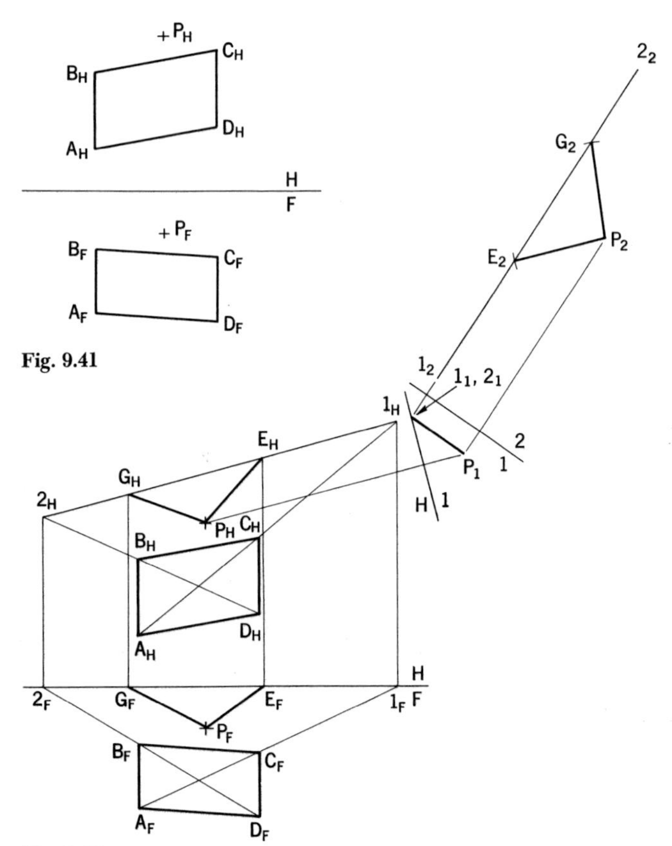

Fig. 9.41

Fig. 9.42

Analysis

1. The *H* and *F* views of line *m* can be easily located if we recall that "the view of two perpendicular lines upon a plane that is parallel to one of the lines will reveal a right angle between the two lines." Therefore, m_H is perpendicular to B_HE_H. Similarly, m_F is perpendicular to C_FD_F. Line *m* is perpendicular to plate *ABC* because it is at right angles to two lines *BE* and *CD* that lie in surface *ABC*.

2. Line *n* has a bearing of N 60° W. This information enables us to establish n_H. Knowing that the grade is 50% we can establish n_F. Lines *m* and *n*, as shown in Fig. 9.40, satisfy the specification.

3. Now the angle between plane *mn* and surface *ABC* is 90° because line *m* is perpendicular to surface *ABC*. This is true because "all planes that contain line *m* are perpendicular to surface *ABC*."

EXAMPLE 6

Shown in Fig. 9.41 are plate *ABCD* and point *P*. It is specified that (*a*) the plate is to be extended to the *H*-plane; and (*b*) that two connectors, *PE* and *PG*, are to be established from point *P* to the edge of the plate that lies in the *H*-plane. The connectors are of a length which is represented by a 1-inch line on the drawing. We are to determine the *H* and *F* views of the connectors.

Analysis

1. The first step is to establish the line in which plane *ABCD* intersects the *H*-plane. We can do this by finding point 1, the intersection of diagonal *AC* with the *H*-plane. The front view of point 1 is easily located since we have an edge view of the *H*-plane. This is shown as 1_F (see Fig. 9.42). Having located 1_F it is a simple matter to locate 1_H. In a similar manner we can establish point 2 which is the intersection of diagonal *BD* with the *H*-plane. The line which joins 1_H and 2_H is the intersection of plane *ABCD* with the *H*-plane.

2. Now we know that point *P* and line 1–2 determine a plane. The true shape of a portion of this plane is easily obtained. This is shown in the view upon supplementary plane 2. In this view, points *E* and *G* can be located since we know that the distance from *P* to these points is represented by a 1-inch line.

3. It is now a simple task to locate the H and F views of points E and G, and then to connect these points to point P, thus satisfying the specification.

EXAMPLE 7

In Fig. 9.43, point P, the front view of rod AB, and the top view of point A are known. We are required to locate the H and F views of member PC which connects P to rod AB, makes an angle of 60° with H, and is 2 inches long.

Analysis

1. All of the 2-inch lines that pass through point P at an angle of 60° to the H-plane are elements of a right-circular cone whose axis is vertical and whose base angle is 60°. The H and F views of the cone are shown in Fig. 9.44.

2. Point C must be on the circumference of the base of the cone and on AB. Therefore, the front view of point C, namely, C_F, must be at the intersection of $A_F B_F$ with the front view of the base of the cone.

3. Once C_F is located it is a simple matter to locate C_H and then B_H, thus determining the top view of rod AB.

Question. Is another solution possible?

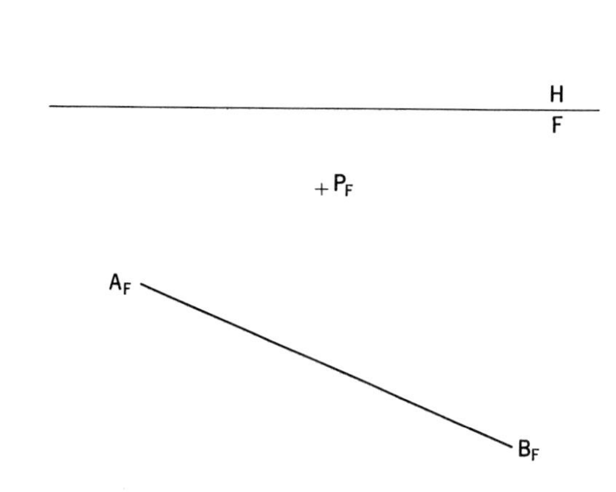

Fig. 9.43

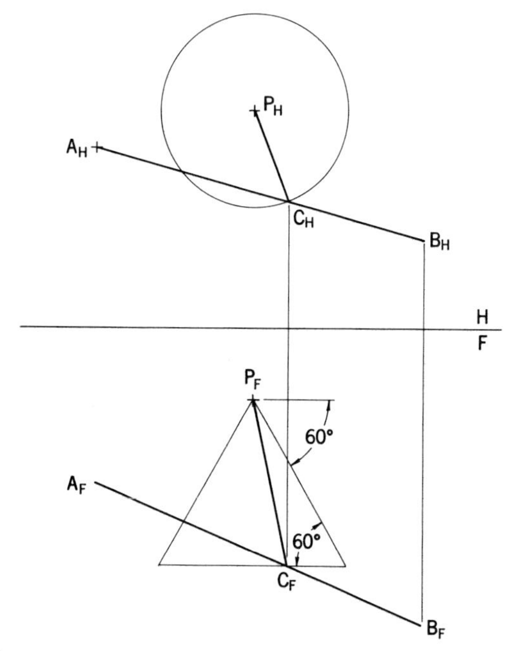

Fig. 9.44

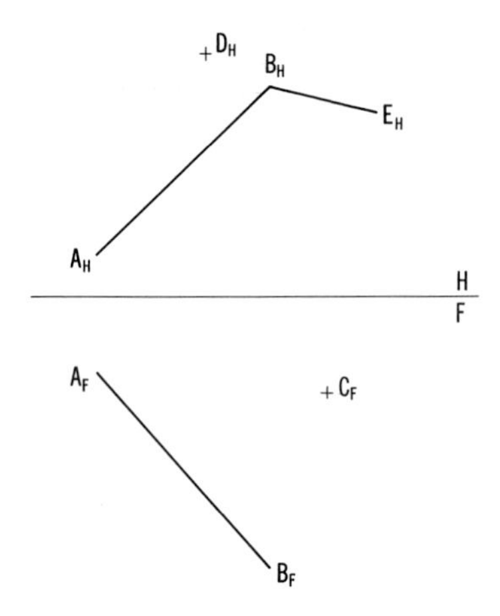

Fig. 9.45

EXAMPLE 8

Shown in Fig. 9.45 are the *H*-view of cables *AB* and *BE*, and of point *D*; and the *F* view of cable *AB* and of point *C*. It is required to establish the *H* and *F* views of cable *CD*, and the *F* view of cable *BE* to meet the following conditions:

1. Cables *CD* and *BE* are at right angles to cable *AB*.
2. The *H* view of cables *AB* and *CD* shows a right angle.

Make a freehand sketch of Fig. 9.45 and then try to analyze and solve the problem.

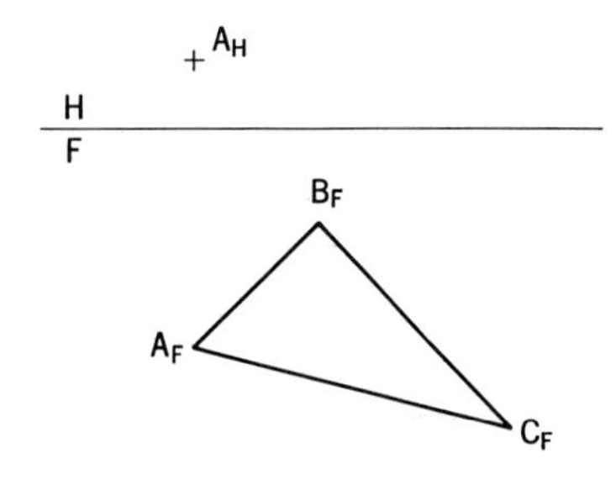

Fig. 9.46

EXAMPLE 9

In Fig. 9.46 are shown the *F* view of plate *ABC* and the *H* view of point *A*. It is required to complete the *H* view of the plate *ABC* to satisfy the specification that (1) edge *AB* has a grade of 50%; and (2) the true length of the strike line from *A* to edge *BC* is $1\frac{3}{4}$ inches long. After the *H* view is completed find the angle between surface *ABC* and the *F* plane. *Make a freehand sketch of Fig. 9.46, analyze and solve the problem.*

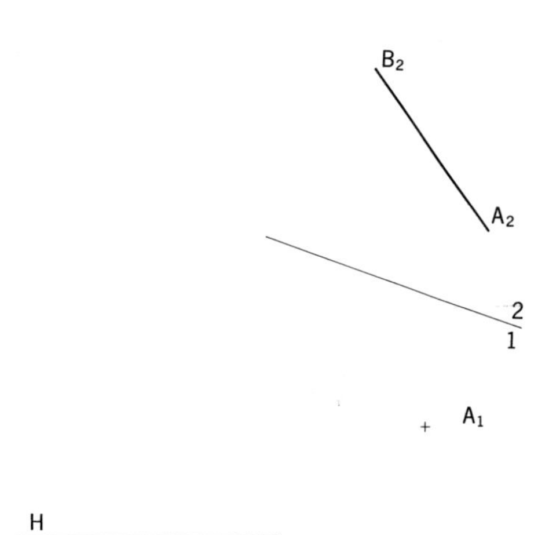

Fig. 9.47

EXAMPLE 10

Figure 9.47 shows the *F* view of point *A*; the view of *A* on supplementary plane 1; and the view of line segment *AB* on supplementary plane 2. It is required to locate the *H*, *F*, and 1 views of line segment *AB*, knowing its bearing is N 60° W. Planes *H* and 1 are perpendicular to each other. *Make a freehand sketch of Fig. 9.47, analyze, and solve the problem.*

EXERCISES

1. Find the intersection of line m and surface ABC, in each case as shown in Fig. E-9.1. Regard the surfaces as opaque and show correct visibility.

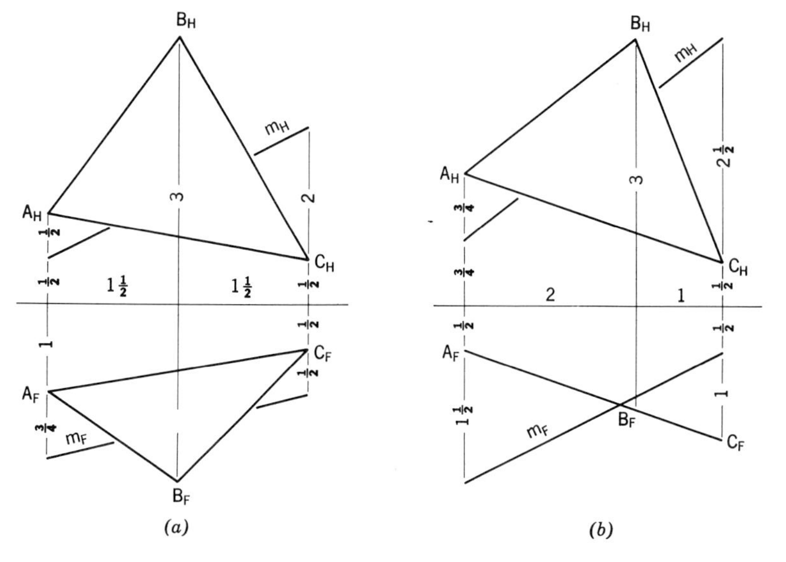

Fig. E-9.1(*a*) and (*b*)

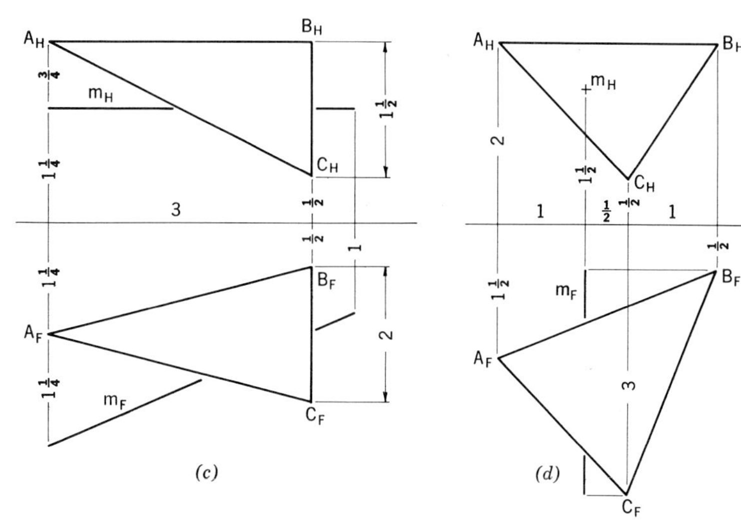

Fig. E-9.1(*c*) and (*d*)

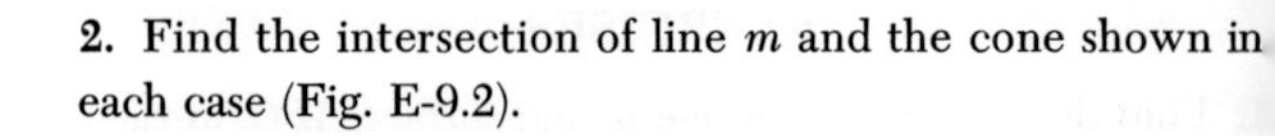

2. Find the intersection of line m and the cone shown in each case (Fig. E-9.2).

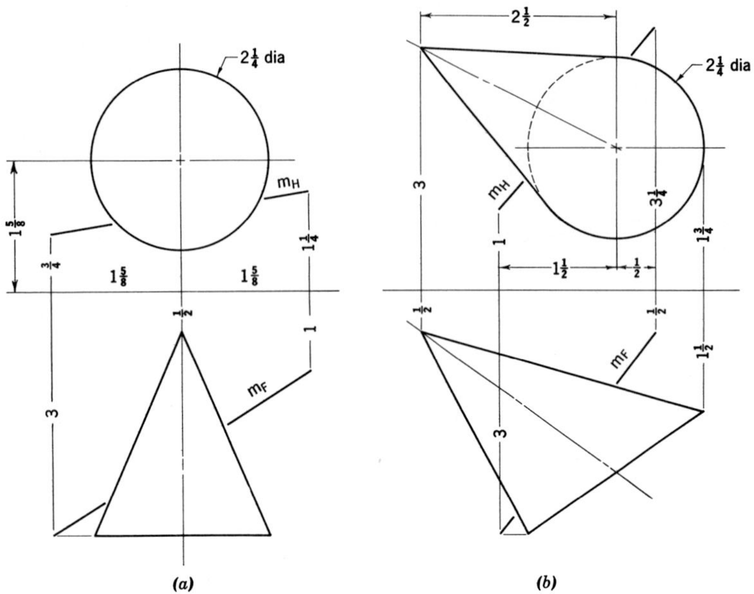

Fig. E-9.2(*a*) and (*b*)

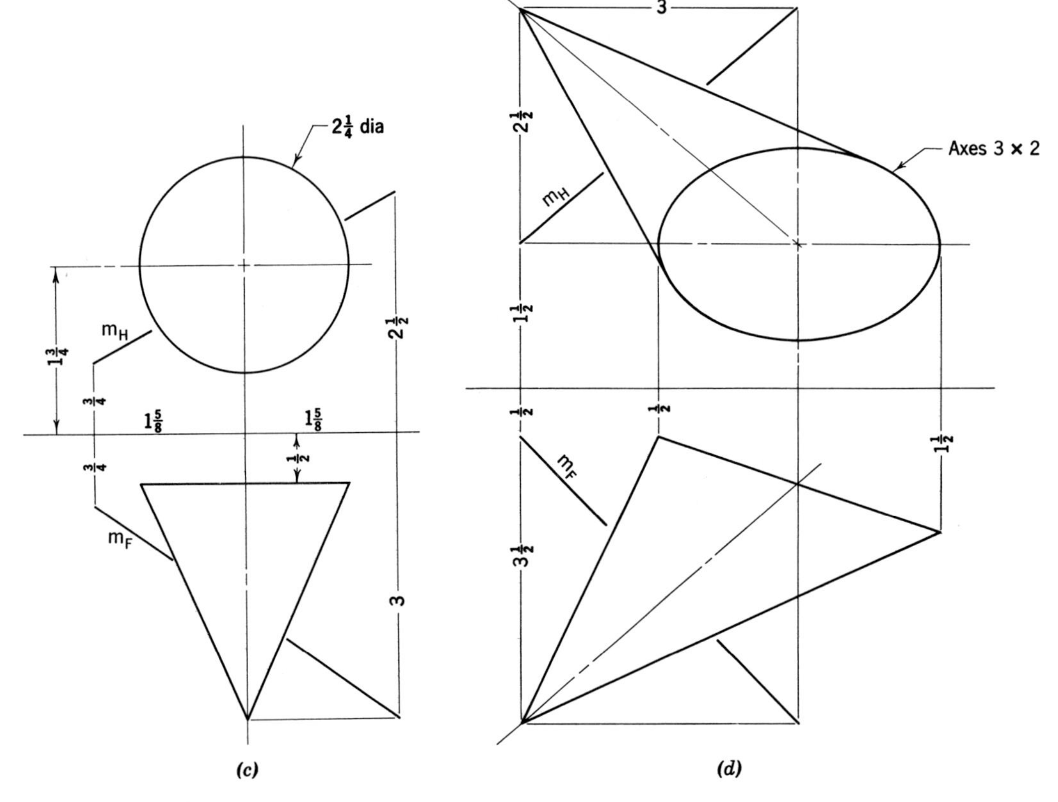

Fig. E-9.2(*c*)

Fig. E-9.2(*d*)

3. Find the intersection of line m and the cylinder as shown in Fig. E-9.3.

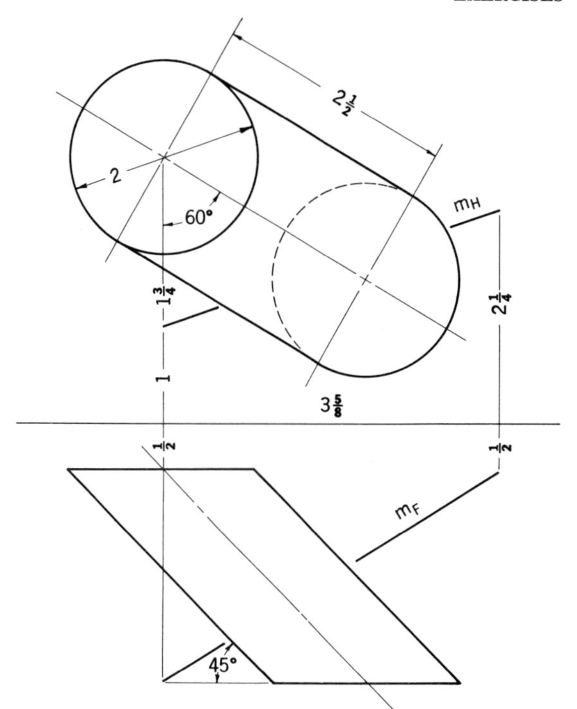

Fig. E-9.3(*a*)

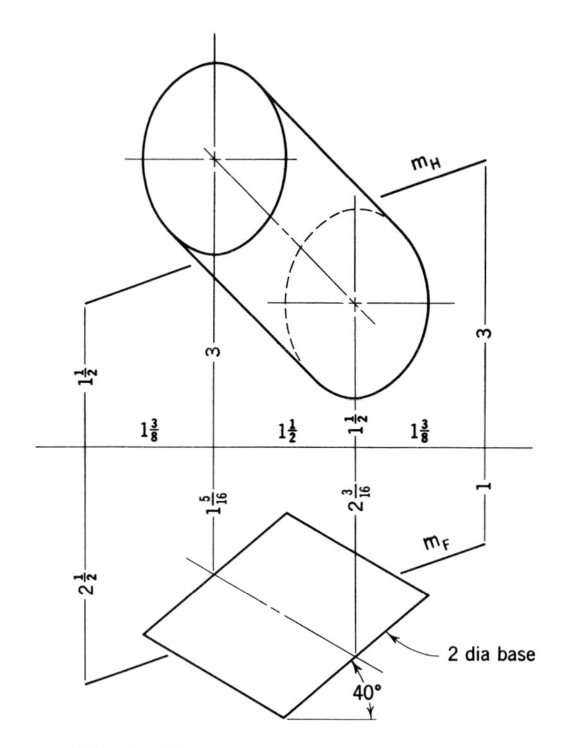

Fig. E-9.3(*b*)

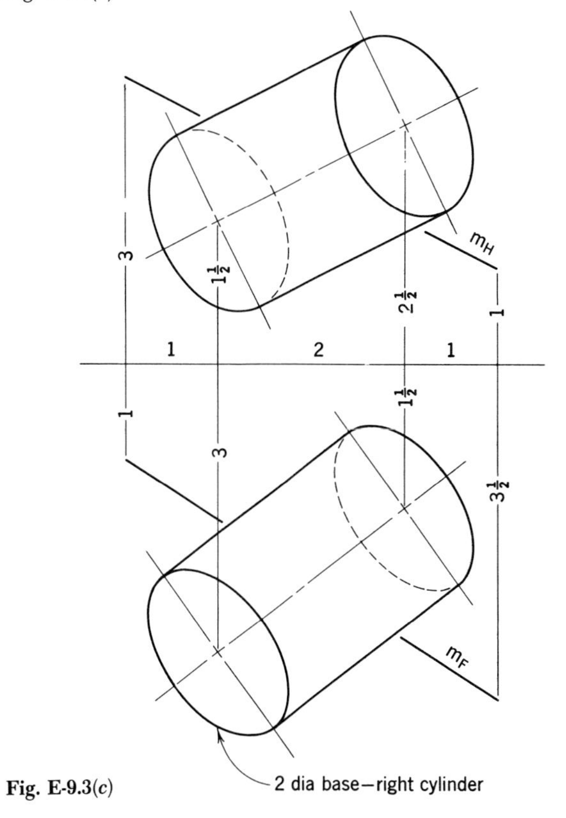

Fig. E-9.3(*c*)

4. Find the intersection of line m and the sphere (Fig. E-9.4).

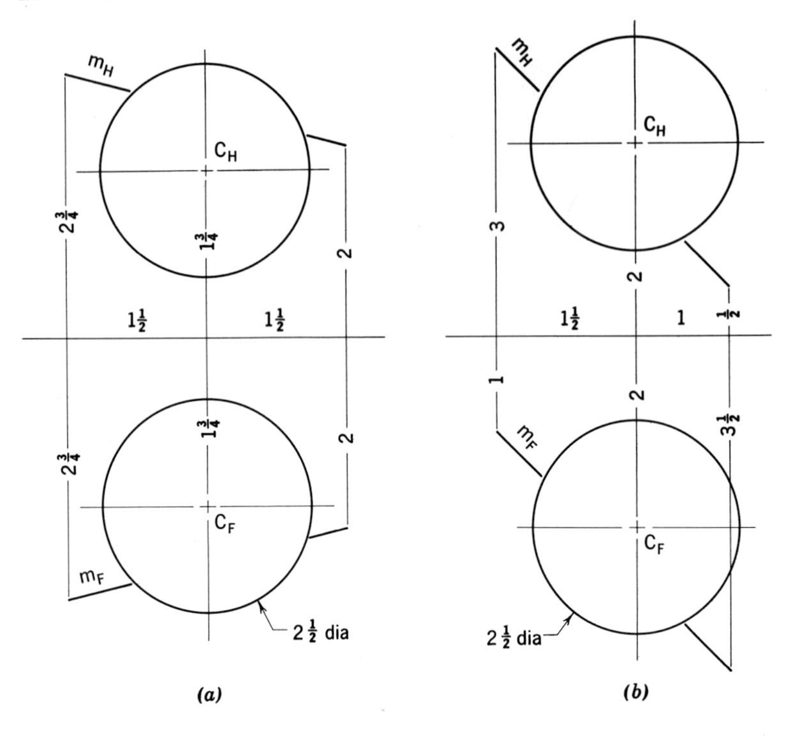

Fig. E-9.4

5. Find the intersection of the plane surfaces shown in Fig. E-9.5.

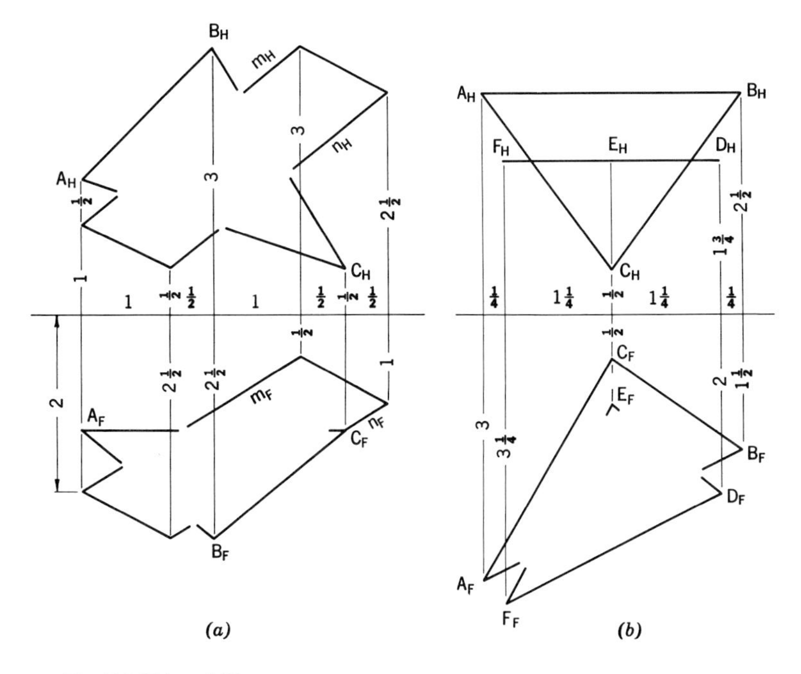

Fig. E-9.5(*a*) and (*b*)

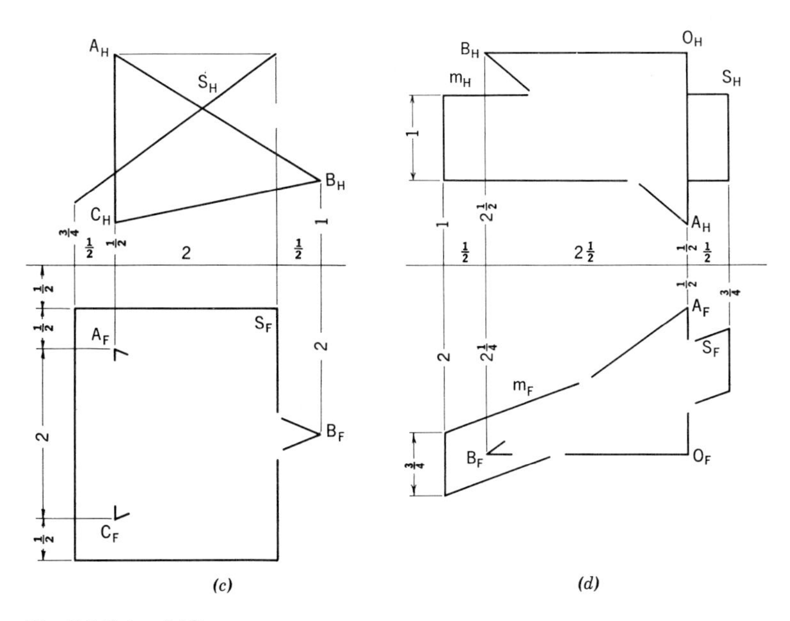

Fig. E-9.5(*c*) and (*d*)

6. Find the intersection of the planes that are determined by shapes shown in Fig. E-9.6.

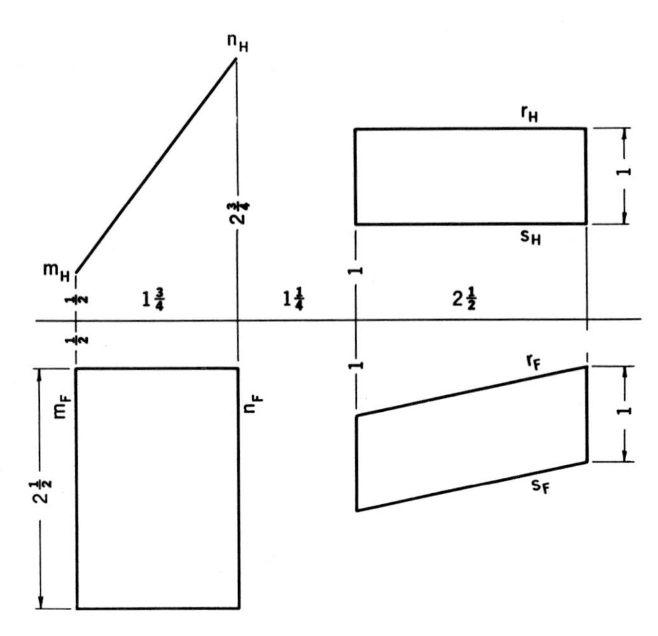

Fig. E-9.6(*a*)

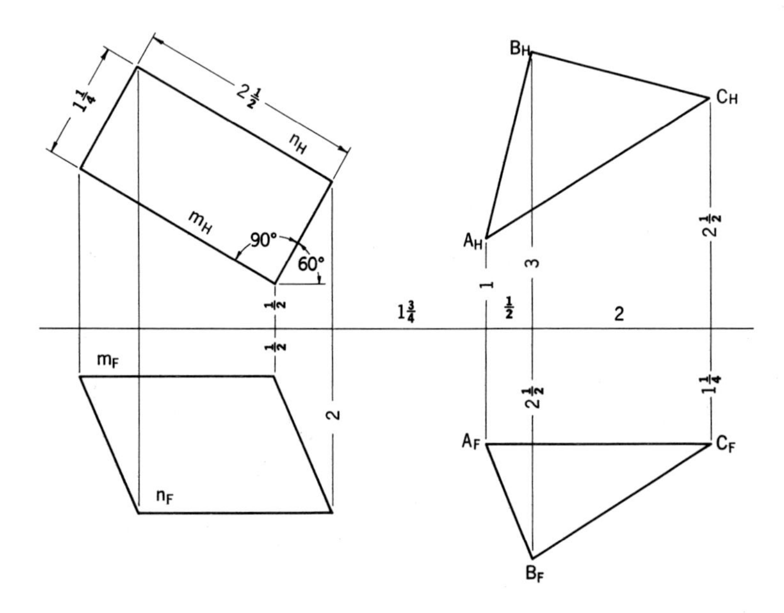

Fig. E-9.6(*b*)

7. Find the intersection of the plane and cone shown in Fig. E-9.7.

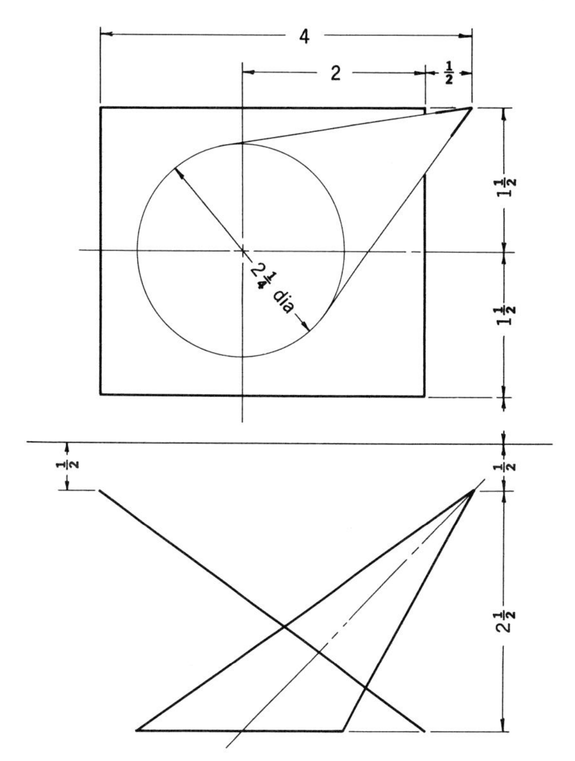

Fig. E-9.7(*a*)

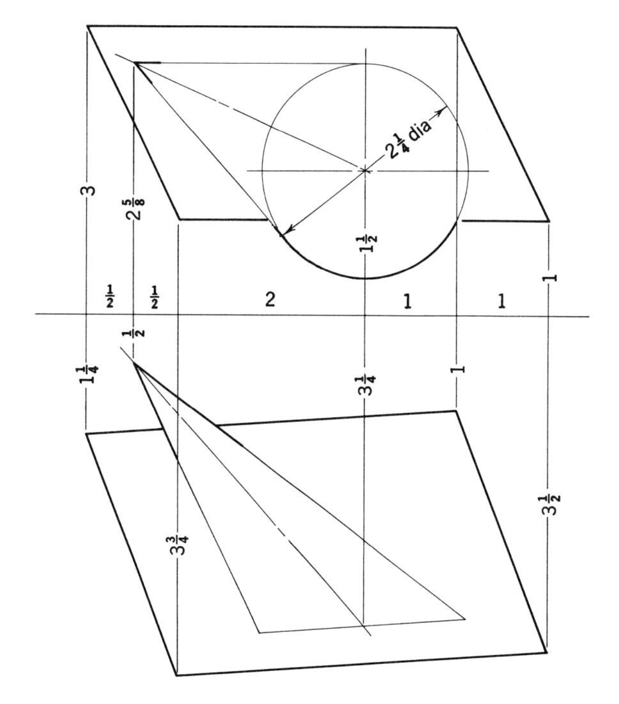

Fig. E-9.7(*b*)

8. Find the intersection of the solids shown in Fig. E-9.8.

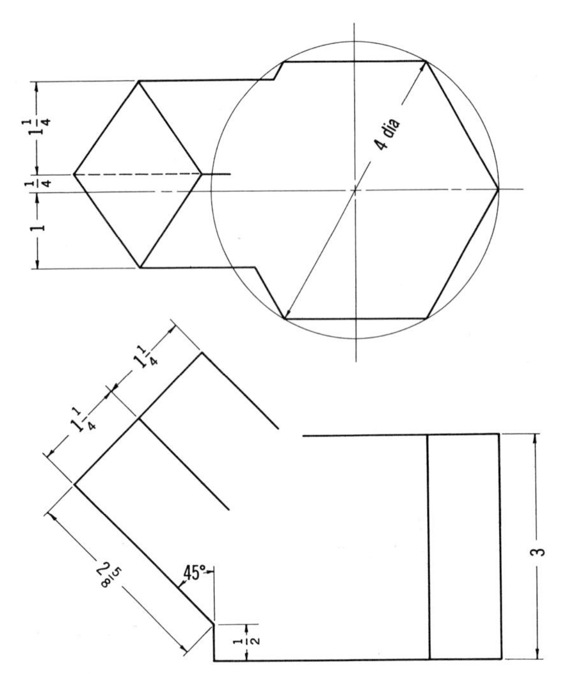

Fig. E-9.8(*a*)

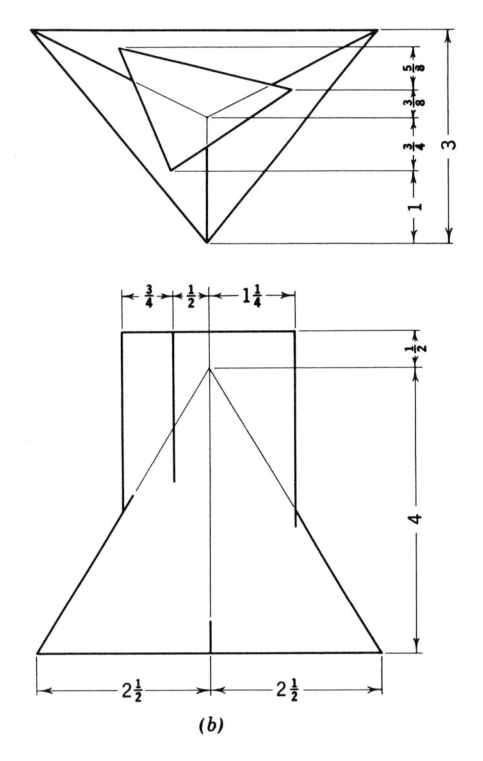

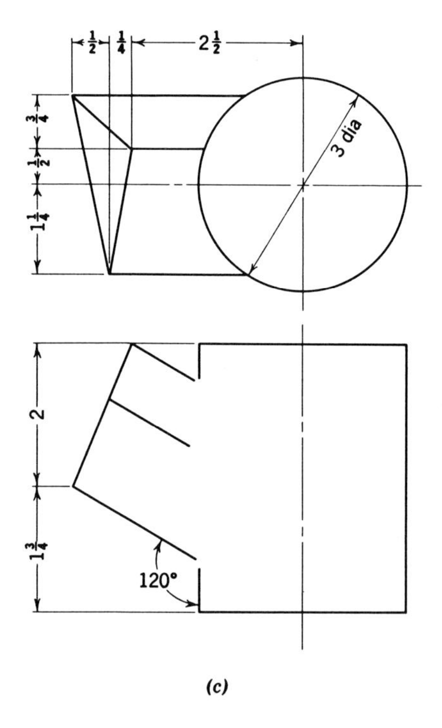

Fig. E-9.8(*b*) and (*c*)

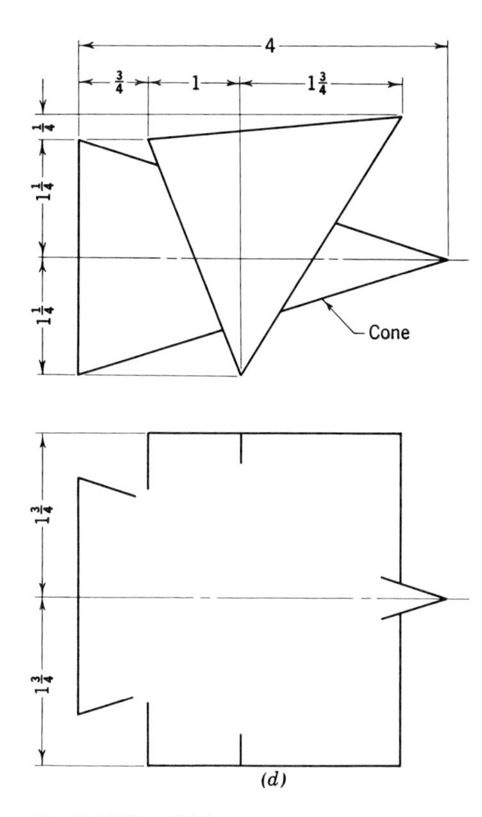

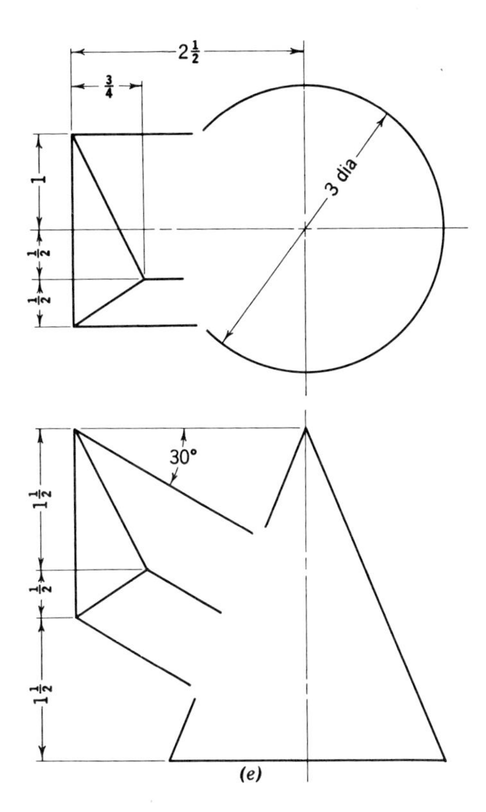

Fig. E-9.8(*d*) and (*e*)

9. Find the intersection of the cones shown in Fig. E-9.9.

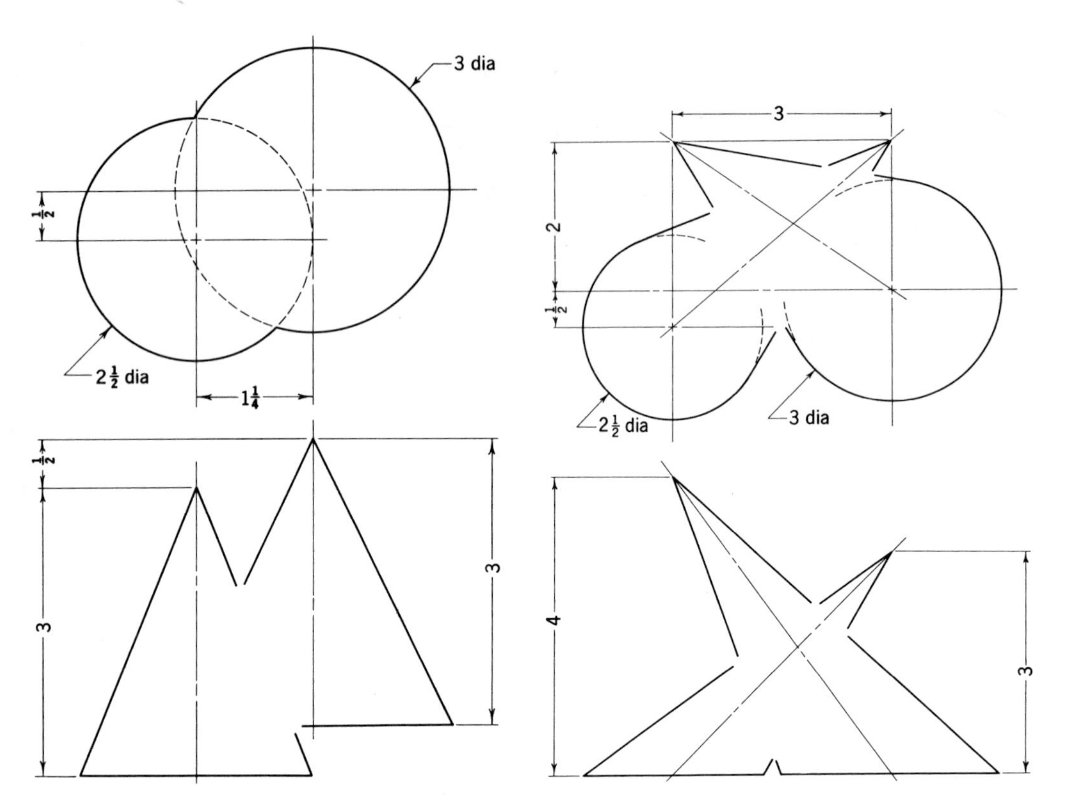

Fig. E-9.9(*a*) and (*b*)

10. Find the intersection of the cone and cylinder shown in Fig. E-9.10.

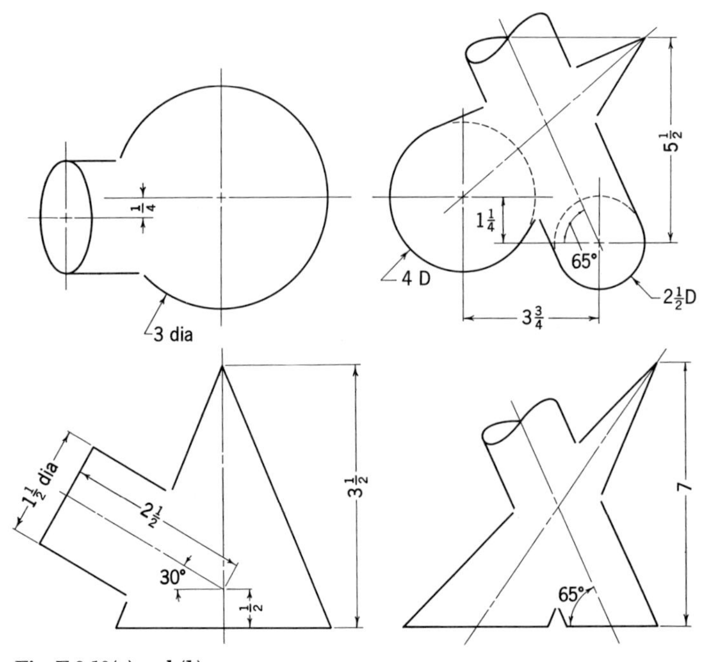

Fig. E-9.10(*a*) and (*b*)

11. Find the intersection of the center line of a culvert with the highway embankments shown in Fig. E-9.11.

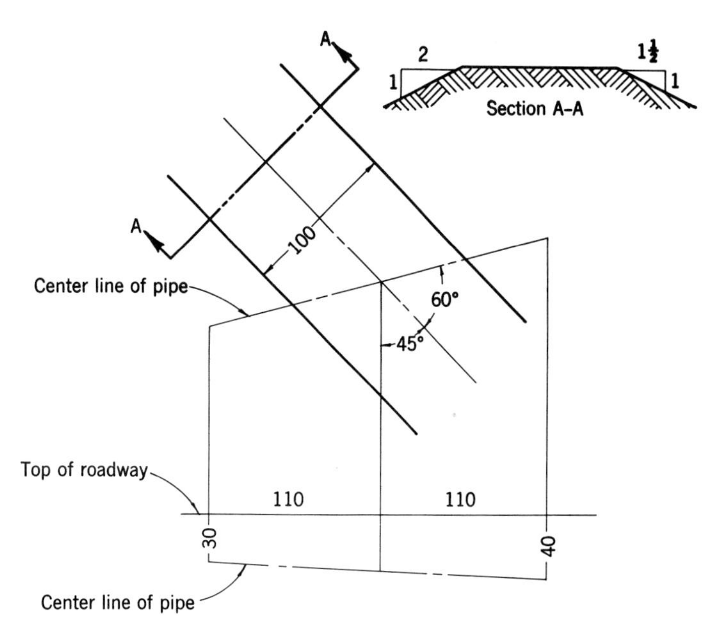

Fig. E-9.11

12. Points A, B, and C are on the upper surface of a rock formation (Fig. E-9.12). Determine the intersection of the surface with the natural terrain.

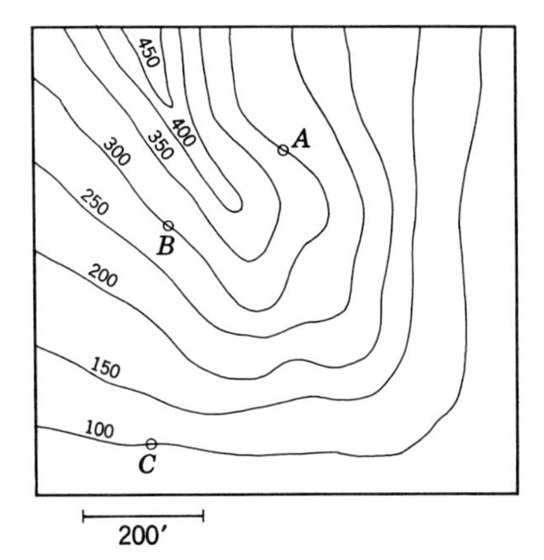

Fig. E-9.12

13. Locate the "top-of-the-cut" (1:1) and the "toe-of-the-fill" ($1\frac{1}{2}$:1) of a proposed highway to be specified by the instructor (Fig. E-9.13).

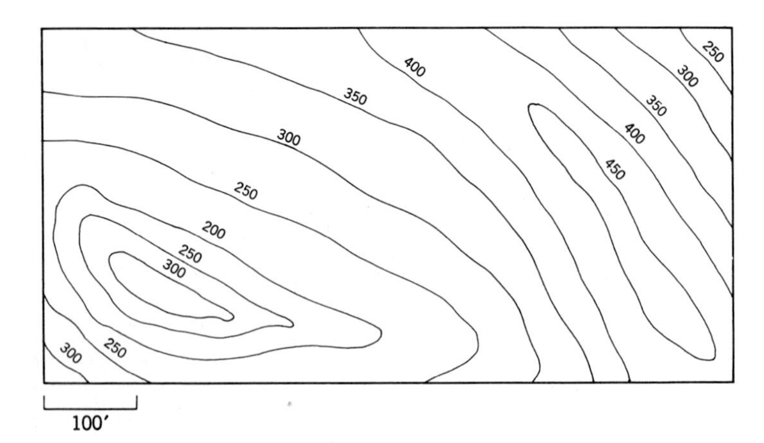

Fig. E-9.13

14. Find the intersection of the flat windshield with the portion of an airplane fuselage shown in Fig. E-9.14.

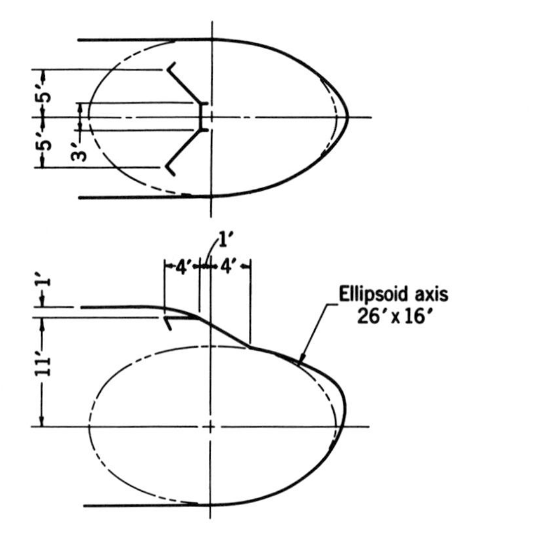

Fig. E-9.14

15. Find the intersection (*a*) between the fuselage and the engine shroud, and (*b*) between the engine shroud and the stabilizer fin. (See Fig. E-9.15.)

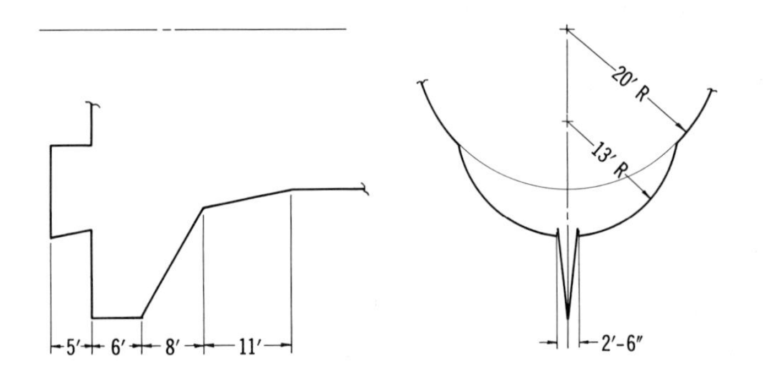

Fig. E-9.15

VECTOR QUANTITIES AND VECTOR DIAGRAMS 10

VECTOR QUANTITIES

In both engineering and science there are many problems that involve such quantities as (*a*) displacement—change of position; (*b*) velocity—rate of change of displacement with respect to time; (*c*) acceleration—rate of change of velocity with respect to time; and (*d*) force—the action of one body on another, or the product of mass and acceleration. *These quantities have magnitude, direction, and line of action (position).*

Vector Diagrams

A vector quantity is represented graphically by a line segment, called a vector, which has a definite length and is provided with an arrowhead at one end.

The length of the line segment represents the magnitude of the vector quantity, to some scale; and the direction along the line segment from tail end to arrow end gives the direction of the vector quantity. The line of action is shown on the drawing.

EXAMPLE 1

Case 1. Suppose that a point moves from A to B (Fig. 10.1). When this has happened we say that the point has received a displacement, the magnitude of which is the length (to some convenient scale) of line AB and the direction of which is the direction of AB as shown by the arrow.

Case 2. Now suppose that the point which moved from A to B did not travel directly from A to B, but took a path from A to C and then from C to B, as shown in Fig. 10.2. *The final displacement of the point is the same as in Case 1, but the final displacement is the sum of the separate displacements AC and CB. Vector AB is the resultant of vectors AC and CB, whereas vectors AC and CB are the components of vector AB.*

Case 3. Again suppose that the point which moved from A to B did not travel directly from A to B, but took a path through points C, D, and E, as shown in Fig. 10.3. *The final displacement of the point, however, is the same as in the first movement, but the final displacement is the sum of the separate displacements, AC, CD, DE, and EB.* In simple terms, vector AB is the *resultant* or sum of the vectors AC, CD, DE, and EB.

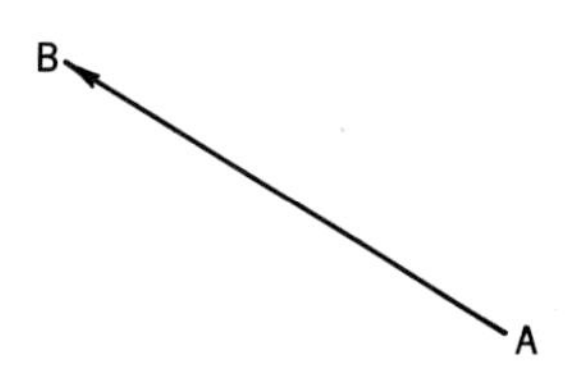

Fig. 10.1

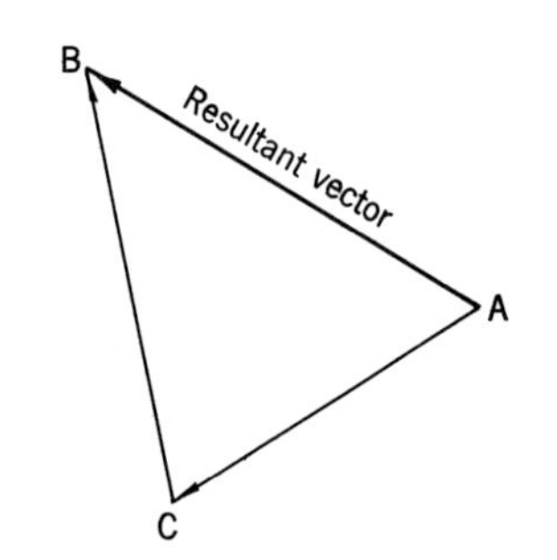

Fig. 10.2

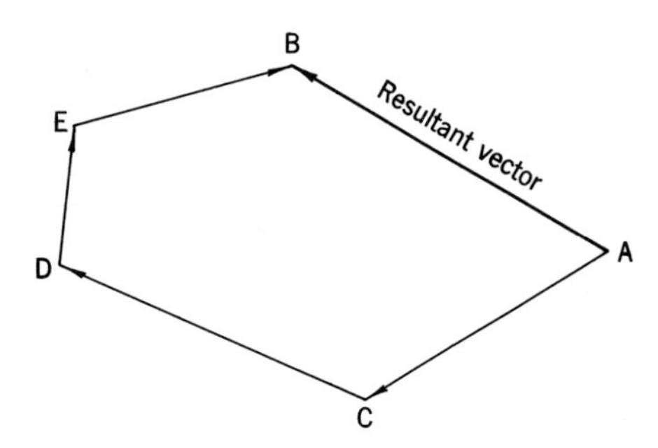

Fig. 10.3

EXAMPLE 2

Case 1. Assume that vectors OA and OB, as shown in Fig. 10.4, represent the magnitude and direction of forces acting at point O. The diagonal, OR, of the parallelogram $OARB$ (rectangle in this case) represents the magnitude and direction of the single force that produces the same effect on point O as do the two forces. *Vector OR is the resultant of vectors OA and OB.*

Case 2. Now let us consider the forces represented by vectors OA and OB, as shown in Fig. 10.5. The diagonal, OR, of the parallelogram $OARB$ represents the magnitude and direction of the single force that has the same effect on point O as do the two forces. It should be noted that the magnitude of the vector OR is greater than the magnitude of either vector OA or OB. This is not always true. It is possible to obtain a resultant vector whose magnitude is less than one or both of the given forces. In Fig. 10.6(a) we observe that the magnitude of resultant vector OR is less than that of vector OA. In Fig. 10.6(b) the resultant vector OR is less than either vector OA or OB.

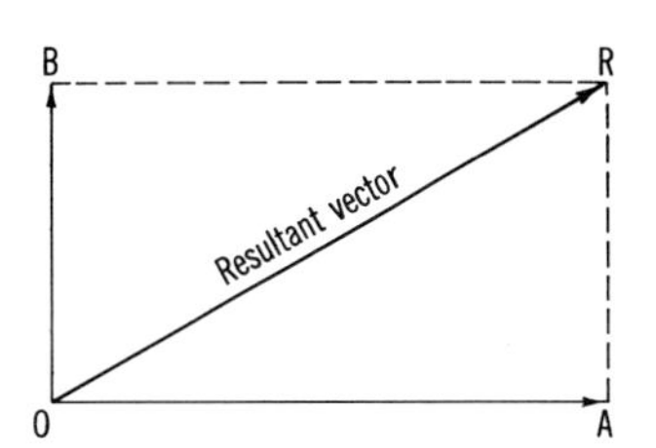

Fig. 10.4

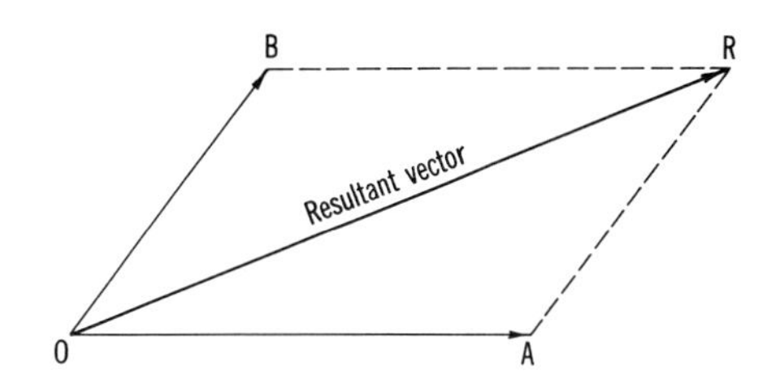

Fig. 10.5

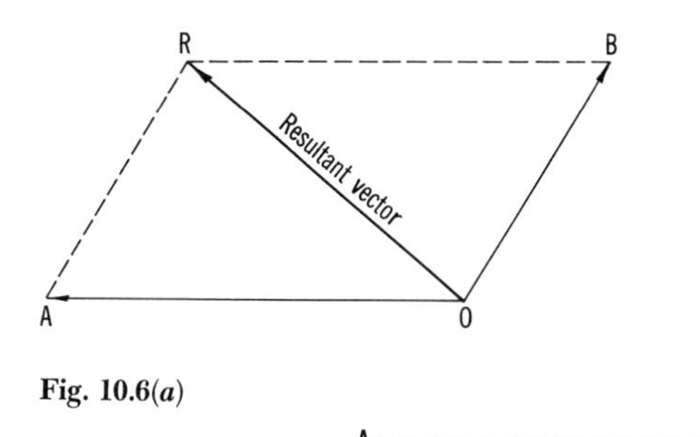

Fig. 10.6(a)

Fig. 10.6(b)

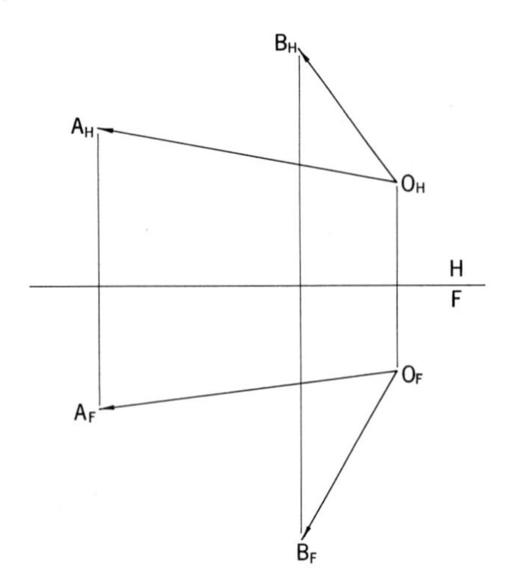

Fig. 10.7

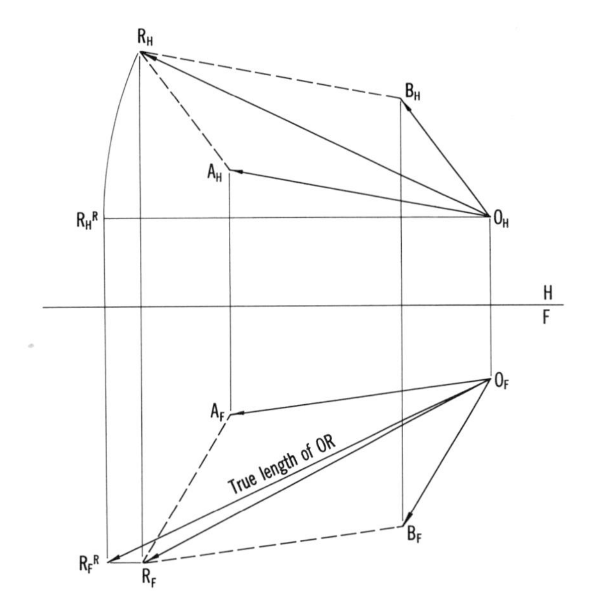

Fig. 10.8

EXAMPLE 3

Now let us consider Fig. 10.7, which shows the *H* and *F* views of vectors *OA* and *OB*. Suppose we wish to find the magnitude and direction of the resultant vector *OR*. The *H* and *F* views of the parallelogram *OARB* are easily constructed since we know that "parallel lines appear as parallels in the respective views." The top and front views of resultant vector *OR* can now be drawn. To obtain the magnitude of vector *OR* is a simple matter. This merely means that we need only find the true length of line *OR*. The solution is shown in Fig. 10.8.

GRAPHIC STATICS

Our knowledge of vectors, of the fundamental principles of orthogonal projection, and of the solutions of the basic problems may now be nicely applied to the analysis and solution of problems that arise in statics.

Graphic statics, as treated in this chapter, deals in the main with the graphical solution of elementary two-dimensional and three-dimensional force problems. However, some additional problems are included to illustrate slightly more advanced applications of the simple principles, and it is hoped they will prove thought-provoking and will encourage further study by the interested student.

Most problems in statics, like those in many other subjects, can be solved by either graphical or algebraic methods. In some cases the graphical solution is quicker and, therefore, more economical. In other cases the algebraic solution is more readily obtained. It is important for the student to become sufficiently familiar with both methods so that he can make an intelligent choice of the method which is best adapted to attack a particular problem. Regardless of which method is used, however, the basic conditions to be satisfied in the solution are the principles of statics. It is essential, therefore, that these few and simple principles be clearly understood and strictly applied to the solution of the problem at hand. Consequently, this chapter attempts to emphasize, first, the principles which are applied and, second, the means of expressing them in the language of graphics.

Statics is that part of mechanics which deals with balanced force systems (systems of forces in equilibrium). The term *statics* implies a static state or state of rest for the bodies on which the forces act, as in the case of roof trusses, walls, floors, and columns of a building, the structural members of a highway or railroad bridge, storage tanks, etc. It should be pointed out, however, that *all* bodies in equilibrium, including those in motion with constant velocity, e.g., constant-speed motors, shafts, pulleys, gears, and conveyor belts, may be analyzed by using the principles of statics. Dynamics, on the other hand, is the branch of mechanics which deals in general with bodies in motion and the relations between their motions and the forces causing them. Whenever bodies move with varying velocities, their motions and the applied forces must be analyzed by using the principles of dynamics. The design and analysis of airplane structures, variable-speed mechanisms, and vibrating systems, for example, require careful study of their dynamic behavior.

Force is the action of one body on another; it changes, or tends to change, the state of rest or motion of the body acted upon. Force is also equal to the product of mass and acceleration. Force is a vector quantity having magnitude and direction. In addition, every force has a line of action upon which its effect partly depends. Suppose, for example, that two forces, identical in magnitude and direction, are both applied perpendicular to the axis of a flywheel. Let us further suppose, however, that the line of action of one force intersects the flywheel axis, while the other force acts tangentially to the rim. The first force has no turning effect, but the second force tends to rotate the flywheel.

We repeat that force is a vector quantity and that it is represented graphically by a line segment, called a *vector*, which has a definite length and is provided with an arrowhead at one end. The length of the line segment represents the magnitude of the force to some scale, and the direction along the line segment from tail end to arrow end gives the direction of the force. The line of action of the force is shown on a drawing of the body to which the force is applied.

Addition of Forces

EXAMPLE 1

Consider forces F_1 and F_2 shown in Fig. 10.9. These forces are added graphically by completing the parallelogram of which two sides are the given forces, and then drawing the diagonal, R. Vector R represents the sum of vectors F_1 and F_2. *The single force R is called the resultant of the two given forces.*

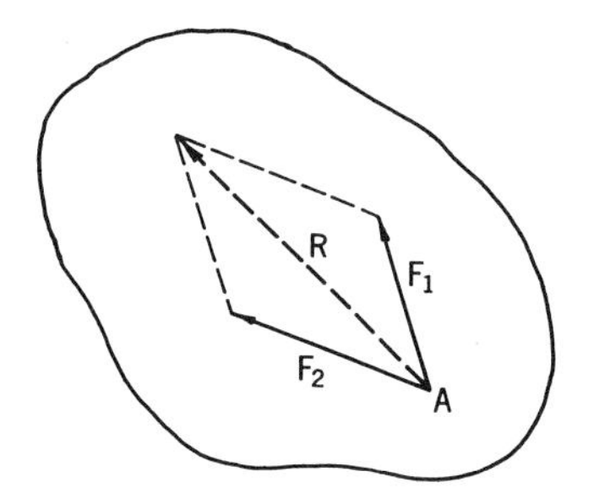

Fig. 10.9

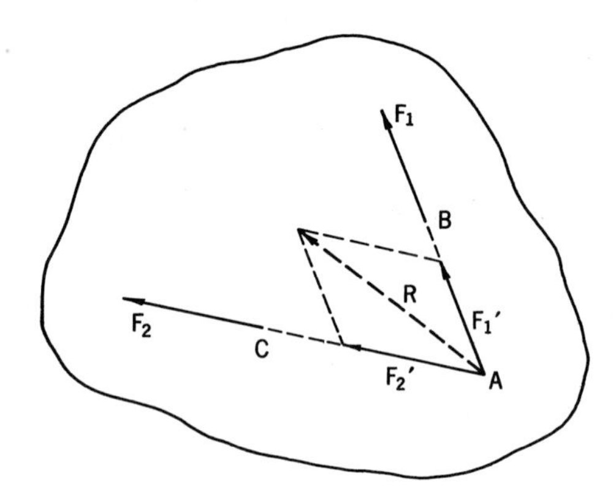

Fig. 10.10

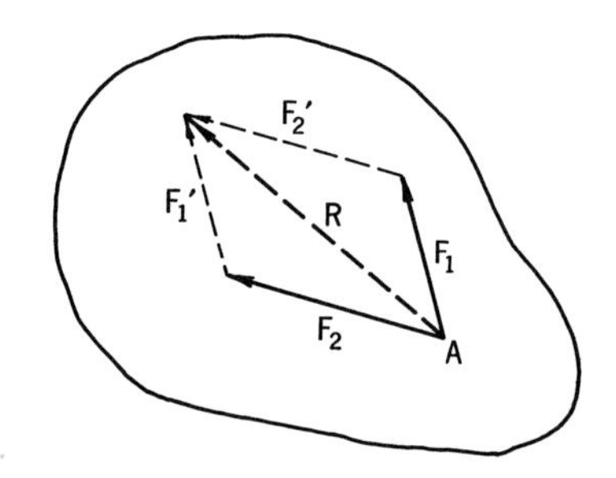

Fig. 10.11

Its effect on the body is the same as the combined effect of forces F_1 and F_2.

EXAMPLE 2

Now suppose force F_1 is applied at point B and force F_2 is applied at point C (forces F_1 and F_2 are in the same plane). *The resultant R is found by sliding the vectors along their respective lines of action to their intersection point A,* completing the parallelogram, and drawing the diagonal as in Example 1. See Fig. 10.10 for the graphical solution. The vectors F_1' and F_2' represent the forces F_1 and F_2 in their new positions.

EXAMPLE 3

In the use of the parallelogram method illustrated in Fig. 10.9 it may be noted that the resultant R divides the parallelogram into two triangles. Thus it appears that R may also be found by constructing either of these triangles, as shown in Fig. 10.11. The vector F_1' is drawn with its tail end coinciding with the arrow end of F_2, and R is drawn from the tail end of F_2 to the arrow end of F_1', thus completing one of the two possible triangles (the other would consist of F_1 as the first vector, F_2', as the seecond, and R as their resultant). *We now recognize that, in the use of this so-called triangle method of vector addition, the magnitude and direction of the sum of two vectors are independent of the order in which they are added.*

Unlike the parallelogram method, however, the triangle method does not necessarily give the correct location of the line of action of the resultant. Consider, for example, a triangle addition of the forces F_1 and F_2 in Fig. 10.10. If F_2 is left in its original position with its point of application at C, and if F_1 is placed with its tail end coinciding with the arrow end of F_2, the resultant R so determined passes through point C instead of the point A on its correct line of action. If, on the other hand, F_2 is added to F_1 without moving the latter, the resultant R appears to act through B instead of A. *Thus the use of the triangle method requires that a point on the line of action of the resultant be known in advance, as in Example 4, or else that it be found by some other means, as in Example 5.*

EXAMPLE 4

Let us assume a body acted upon by *three forces which have one point in common* (*concurrent forces*) *and which lie in one plane* (*coplanar forces*). (See Fig. 10.12.) If we use the parallelogram construction, we can first obtain R_1, the resultant of forces F_1 and F_2, and then combine R_1 with force F_3 to form a second parallelogram, the diagonal of which is R, the resultant of the system of forces F_1, F_2, and F_3. Less construction is necessary if we use the triangle method. This is shown in Fig. 10.13, where F_2' is parallel and equal to F_2; and where F_3' is parallel and equal to F_3.

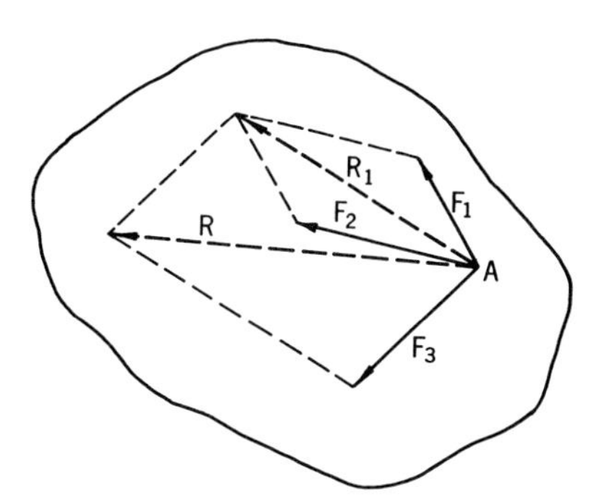

Fig. 10.12

EXAMPLE 5

Consider the body shown in Fig. 10.14, acted upon by forces F_1 and F_2 at points A and B, respectively. How shall we proceed to determine the line of action, magnitude, and direction of the resultant R since the intersection of the lines of action of the given forces is inaccessible? Along the line AB let us introduce force F_3 at point A and an equal, opposite, and collinear force F_3' at point B. The original force system is not affected by forces F_3 and F_3' since they cancel each other. Now two resultants, R_1 for forces F_1 and F_3, and R_2 for forces F_2 and F_3', are easily constructed. Point P, the intersection of the lines of action of R_1 and R_2, is on the line of action of the resultant R of the forces F_1 and F_2. The direction and magnitude of R can now be established by drawing the diagonal of the parallelogram whose sides are R_1' and R_2'. Vector R_1' is obtained by sliding vector R_1 to point P, and similarly vector R_2' is obtained by sliding vector R_2 to point P. *It should be observed that the magnitude of forces F_3 and F_3' will affect the location of point P without changing the line of action of the resultant force R.*

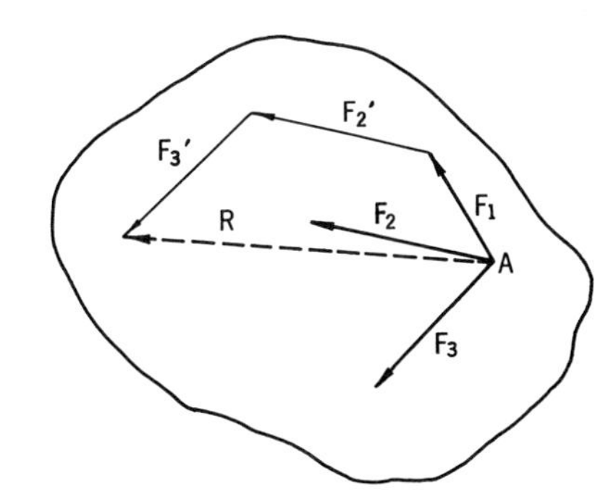

Fig. 10.13

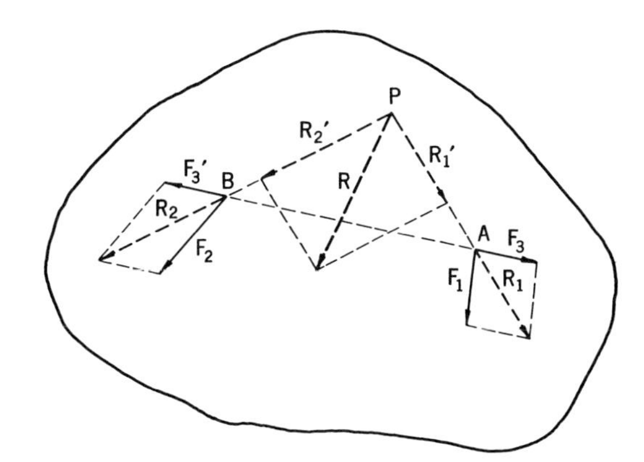

Fig. 10.14

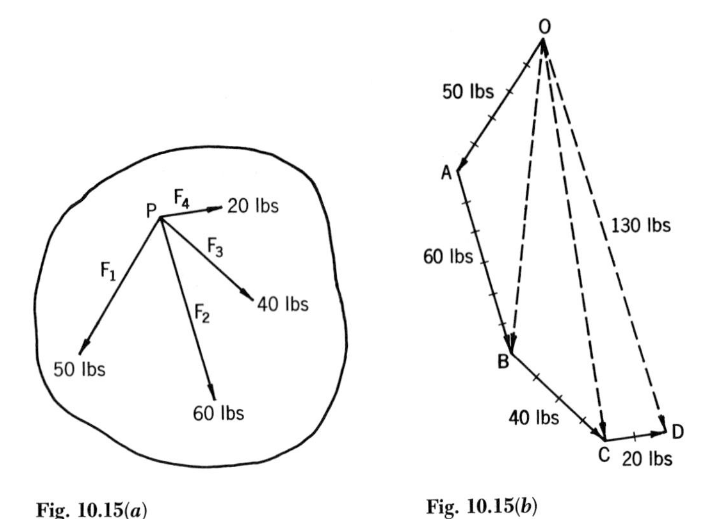

Fig. 10.15(a)

Fig. 10.15(b)

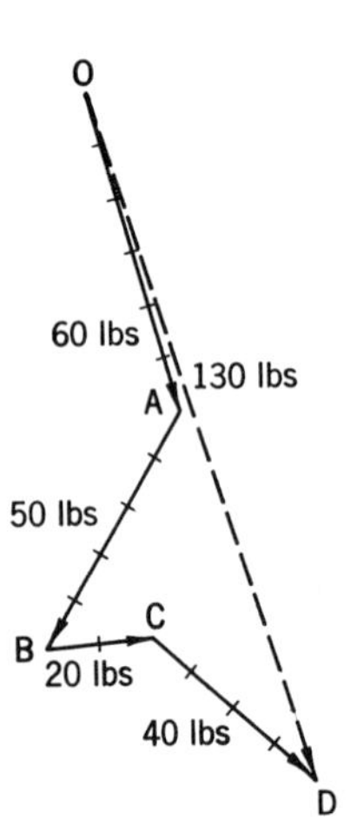

Fig. 10.16

EXAMPLE 6

Consider the *coplanar, concurrent* force system shown in Fig. 10.15(a). A force polygon can be constructed in the following manner to determine the direction and magnitude of the resultant.

1. Through point O in Fig. 10.15(b) line OA is drawn parallel to force F_1 and is laid off to a length which represents 50 pounds.
2. Through point A line AB is drawn parallel to force F_2 to a length which represents 60 pounds to the *same* scale used in laying off OA; similarly for lines BC and CD.
3. The vector OD determines the direction and magnitude of the resultant. It should be observed that the force polygon $OABCD$ is actually a combination of force triangles OAB (in which OB is the resultant of forces F_1 and F_2), OBC (in which OC is the resultant of force F_3 and the resultant of forces F_1 and F_2), and OCD (in which OD is the resultant of force F_4 and the resultant of forces F_1, F_2, and F_3).

Since the resultant is known to act through the point of concurrency of the original forces, a line drawn parallel to OD, through P in Fig. 10.15(a) would locate the resultant of the given force system. In addition, it should be pointed out that *the sequence of drawing the vectors of the force polygon need not follow the order* F_1, F_2, F_3, *and* F_4. Any sequence may be used without affecting the direction and magnitude of the resultant. This is true because the order of addition of any two vectors is arbitrary, as pointed out in Example 3. For instance, in Fig. 10.16 the sequence is F_2, F_1, F_4, and F_3. In summary, we should note:

(a) *The arrow end of one vector is the beginning of the following vector.*
(b) *The arrow end of the resultant vector touches the arrow end of the last vector.*

Forces in Equilibrium

If a body is in a state of equilibrium, the resultant of all forces applied to the body is zero. Suppose we have a system of concurrent forces, F_1, F_2, and F_3, as shown in Fig. 10.17(*a*). The force triangle shown in Fig. 10.17(*b*) reveals the fact that the resultant is zero, since the arrow end of the last force coincides with the beginning of the first one. Hence, *the system is in balance, i.e., it is in equilibrium.* We should note that *when the force system is in equilibrium, the arrowheads of the sides of the force polygon follow each other around the polygon.*

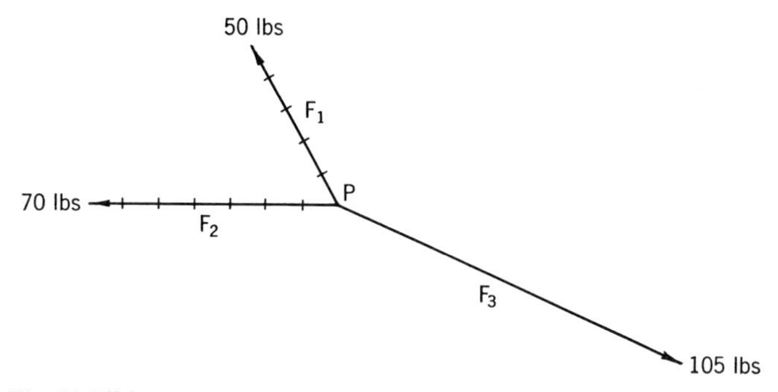

Fig. 10.17(*a*)

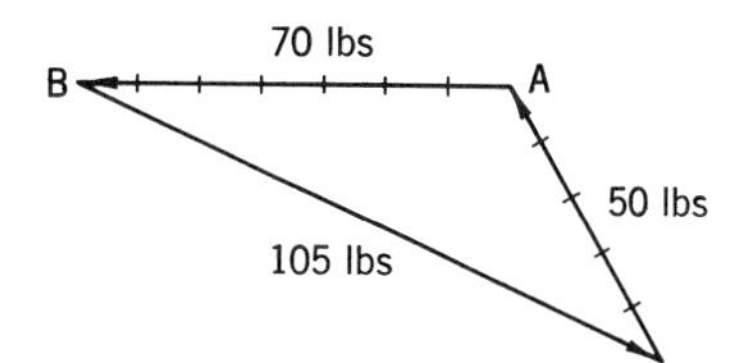

Fig. 10.17(*b*)

EXAMPLE 1

Consider Fig. 10.18, which shows forces F_1 and F_2 and the lines of action of forces F_3 and F_4 which are concurrent with F_1 and F_2. If the system of forces is in equilibrium, what are the magnitudes of F_3 and F_4?

To simplify and shorten our further discussion we introduce a system of lettering known as Bow's notation. In Fig. 10.18, which shows the lines of action of the forces acting at a point, we write the lower-case letters a, b, c, d in the spaces between the lines of action so that we can designate any line of action by the letters in the adjacent spaces; for example, the line of action of force F_2 is designated by the letters ab, and those of F_4, F_3, and F_1 are designated by bc, cd, and da, respectively. (*In concurrent force systems, the forces are conventionally designated by the letters as read in a clockwise progression around the point of concurrency.*) Now in the force polygon (Fig. 10.19) capital letters are used to specify the corresponding vectors, the first letter denoting the tail end and the second denoting the arrow end; for example, force F_2 is lettered AB, F_4 lettered BC, etc.

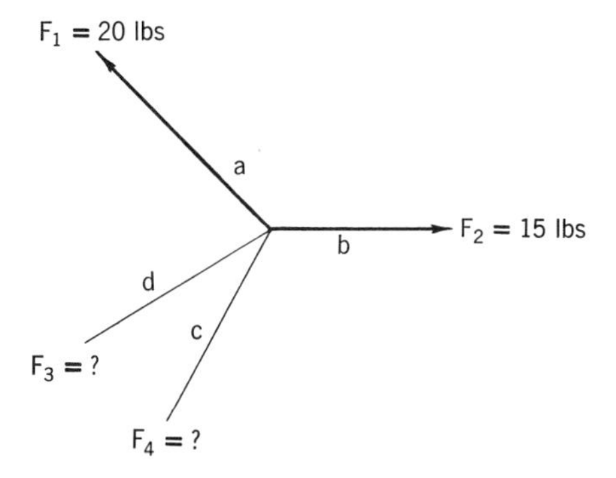

Fig. 10.18

With Bow's notation in mind, the analysis of the problem now proceeds as follows. Since the system is in equilibrium, the vector sum of the four forces is zero, so that the arrow end D of force F_3 must coincide with the beginning (also D) of F_1 when forces F_1, F_2, F_4, and F_3 are added head-to-tail in that order. Since forces F_1 and F_2 are known, the corresponding sides DA and AB may be constructed immediately. The direction and one end point of each of the remaining two sides are now known, so that the polygon can be completed and the unknown magnitudes determined.

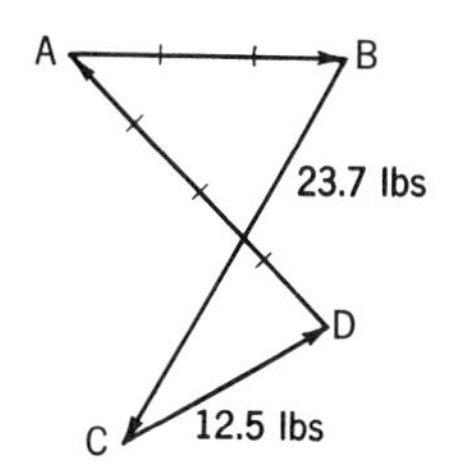

Fig. 10.19

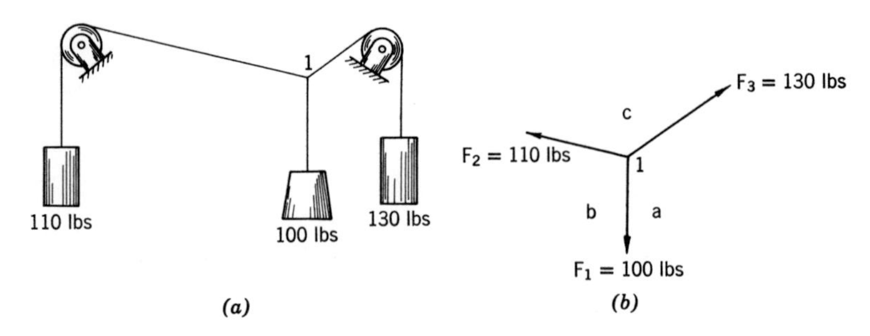

Fig. 10.20

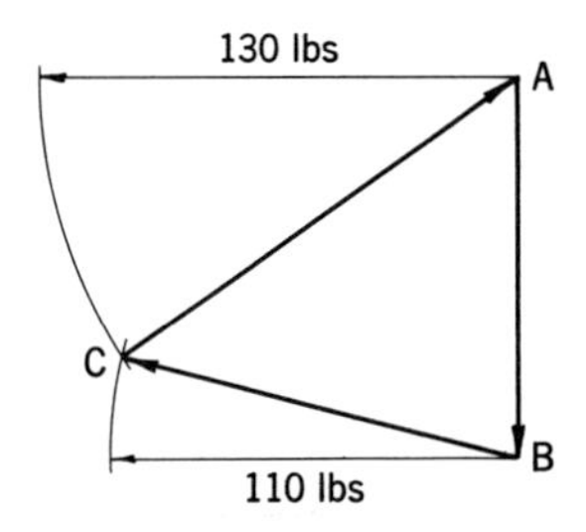

Fig. 10.21

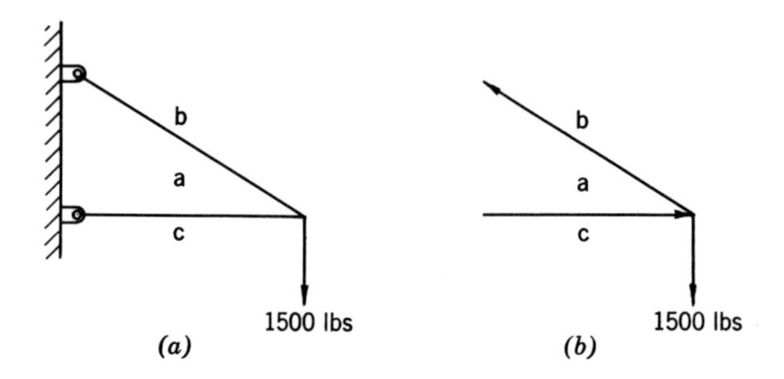

Fig. 10.22

EXAMPLE 2

Figure 10.20(*a*) shows a system of three weights held in equilibrium by the tensions in the connecting ropes. It is desired to determine the equilibrium position of point 1 and thus of the entire system. This may be done by finding the directions of the two inclined ropes. In Fig. 10.20(*b*), point 1 is *isolated* and is shown in equilibrium under the actions of forces F_1, F_2, and F_3. F_2 and F_3 are known in magnitude but have the unknown directions of the inclined ropes.

Equilibrium requires that the polygon of the three forces must close so that their resultant is equal to zero. In Fig. 10.21, we first lay off the vector AB representing F_1, the only force which is completely known. Point C, the arrow end of the vector BC (force F_2), must now lie on a circular arc whose center is B and whose radius, to scale, is 120 pounds, the known magnitude of force F_2. Since point C is also the tail end of the closing vector CA (force F_3), C lies also on an arc whose center is A and whose radius, to scale, is 100 pounds. The intersection of these two arcs locates point C and, when the arrowheads are placed in sequence, gives the directions of the unknown forces. (It may be noted that the two arcs have a second intersection. What does this second result mean physically?)

Determination of Forces in Two-Dimensional Trusses*

EXAMPLE 1

A simple truss is shown in Fig. 10.22(*a*). What are the forces acting in members *ca* and *ab*? We first *isolate* the joint to which the known load is applied and show all forces acting on this joint [see Fig. 10.22(*b*)]. This isolation serves two very important purposes. First, it enables us to account, clearly and completely, for *all* forces which act on the joint. These forces, and only these, hold the joint in equilibrium. Second, it gives meaning to the distinction between the forces acting *on* the joint and those exerted *by* the joint on the members in contact with it. For example, the force which the member *ab*

* A truss is a structural framework which consists of members arranged and connected to form a system of triangles. The forces applied to any member are usually assumed to act in the direction of that member.

exerts on the joint is accompanied by an equal, opposite, and collinear force having the same point of application but exerted by the joint on that member. *The isolation of the joint is an essential step in understanding the action of the forces.* The more complicated the structure and the more complex the force system, the more vital becomes this process of isolating the body considered, since it is truly the key step in the solution of the problem.

Why was the loaded joint isolated instead of one of the left-hand joints? The answer is clearly that the loaded joint is the only one at which a known force acts, and is therefore the only one for which the force polygon can be constructed.

The magnitude and direction of the load *BC* are known; therefore, we can start the graphical solution by drawing a vertical line, *BC* (Fig. 10.23), to represent 1500 pounds. The lines of action of the forces in members *ca* and *ab* are known. Hence, we may draw through point *C* a line parallel to member *ca*, and through point *B* another line parallel to member *ab*. The intersection of these lines locates point *A*. The magnitudes of the forces in members *ca* and *ab* are obtained by measuring line segments *CA* and *AB* with the unit of measure used in laying off *BC*. The directions of the forces in these members are established by placing arrowheads in sequence: at *A* (for *CA*) and at *B* (for *AB*). This we recall from our discussion of forces in equilibrium (Figs. 10.17 and 10.19). We now observe that force *AB*, exerted by member *ab* on the isolated joint in Fig. 10.22(*b*), pulls away from the point of concurrency. This means that member *ab* is in *tension.* On the other hand, force *CA*, exerted by member *ca* on the isolated joint, pushes toward the joint. This means that member *ca* is in *compression.*

EXAMPLE 2

Now let us consider the truss shown in Fig. 10.24. It is assumed that the reactions (the 1250-lb forces at the supports) have already been determined. What are the forces acting in the members of the truss?

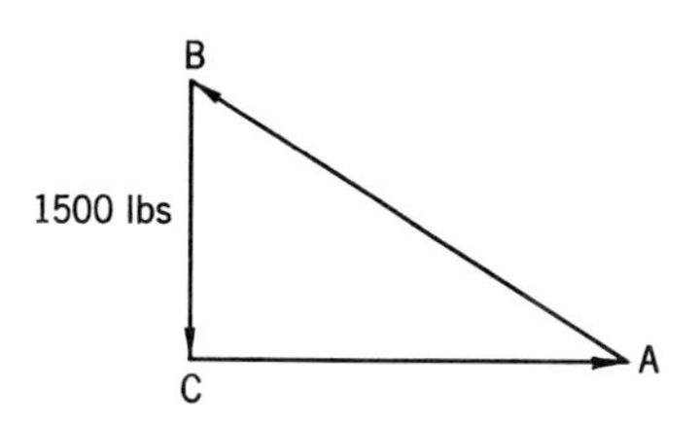

Fig. 10.23

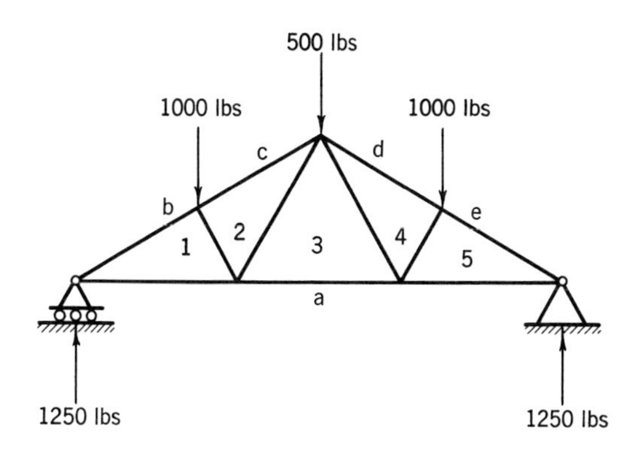

Fig. 10.24

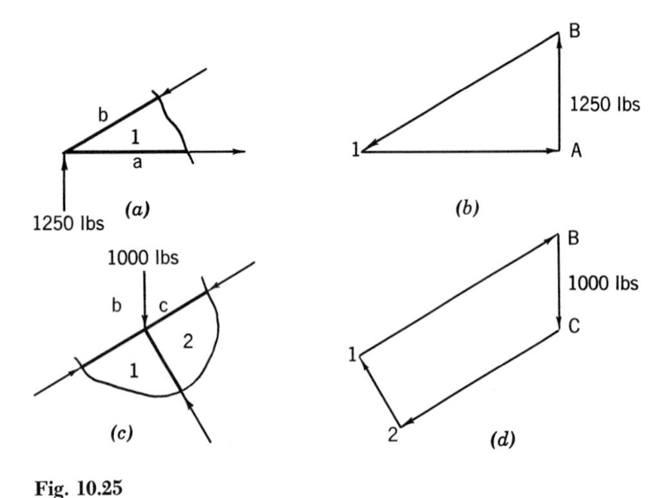

Fig. 10.25

We shall first consider the isolated portion of the truss shown in Fig. 10.25(*a*). The determination of the forces in members *b*–1 and 1–*a* is essentially the same as the problem just discussed. The force polygon is shown in Fig. 10.25(*b*). The magnitudes of vectors *B*–1 and 1–*A* are the forces in members *b*–1 and 1–*a*, respectively. Moreover, we observe that member *b*–1 is in compression and that member 1–*a* is in tension. Now let us analyze the forces acting at the joint which is common to members *c*–2, 2–1, and 1–*b* [Fig. 10.25(*c*)]. *We know the force in member b–1 from Fig. 10.25(b).* The force polygon for the forces acting at the joint is shown in Fig. 10.25(*d*). *BC* is laid off as a vertical line to represent 1000 pounds. Through point *C*, a line is drawn parallel to member *c*–2, and through point 1 (known because we have both the magnitude and direction of the force in member 1–*b*) a line is drawn parallel to member 2–1. The intersection of these lines locates point 2. The magnitudes of the forces in members *c*–2 and 2–1 can be measured in the force polygon. The combined use of Figs. 10.25(*c*) and 10.25(*d*) shows that both members *c*–2 and 2–1 are in compression. *It should be pointed out that the lines of action of all forces at a joint, including forces in the members, are identified by reading clockwise about the joint.* For example, at the peak joint (Fig. 10.24) the reading would be *c*–*d*, *d*–4, 4–3, 3–2, and 2–*c*. We shall see that this same sequence, when applied to the force polygon, provides a simple method for determining the members that are in compression and also those that are in tension.

The analysis of the forces in the members of the peak joint and at the remaining joints can be made in a manner similar to that used with the first two joints. It should be noted, particularly with reference to the peak joint, that the force in member 2–*c* is known from the analysis of the forces in the members of the previous joint; and that likewise the force in member *d*–4 is known since the truss is symmetrical, and is loaded symmetrically, so that the forces in both members 2–*c* and *d*–4 are the same. If this were not true, we could first determine the force in member 3–2 from an analysis of the forces acting at the joint common to members *a*–1, 1–2, 2–3, and 3–*a*, and then proceed to construct a force polygon which would contain the magnitudes of the forces in members 2–3 and 3–*a*. This will be done in the solution which follows.

Rather than construct separate force polygons for all joints, it is most convenient to combine them into a single force diagram for the entire structure. Such a diagram is known as a *Maxwell diagram,* in recognition of the work of Clerk Maxwell, who presented this method in 1864.

Figure 10.26 shows the combined force diagram for the determination of the forces in the members of the truss. First the external loads—*BC*, *CD*, *DE*, *EA*, and *AB*—are laid off to a convenient scale. Now, if we consider the first joint at the left [same as shown in Fig. 10.25(*a*)], we can locate point 1 by drawing through *B* a line parallel to member *b*–1, and another line through point *A*, parallel to member *a*–1. The intersection of these lines locates point 1. Point 2 is located at the intersection of the line drawn through point *C* parallel to member *c*–2, with the line drawn through point 1 parallel to member 1–2. Point 3 is at the intersection of the line drawn through point *A* parallel to member *a*–3, with the line drawn through point 2 parallel to member 2–3, and so forth for the other points 4 and 5.

Once the force diagram is completed, the magnitudes of all forces can be scaled directly. The determination of the kind of force—compressive or tensile—can also be made. For example, consider the peak joint. Starting with the known external load, *c*–*d*, and reading clockwise, the sequence is *c*–*d*, *d*–4, 4–3, 3–2, and 2–*c*. Now, using this sequence in the force diagram, we observe that force *D*–4 (arrow if placed would have been at 4) applied by the member *d*–4 to the isolated peak joint would push toward the joint; hence member *d*–4 is in compression. Reading force 4–3 (from point 4 to point 3) and applying it to the isolated joint shows that the force is pulling away from the joint; hence, member 4–3 is in tension. Now, proceeding from point 3 to point 2 in the force diagram, applying force 3–2 to the isolated joint shows that the force is pulling away from the joint and, therefore, the member 3–2 is in tension. Finally, force 2–*C* on the peak joint acts toward the joint; hence, member 2–*c* is in compression.

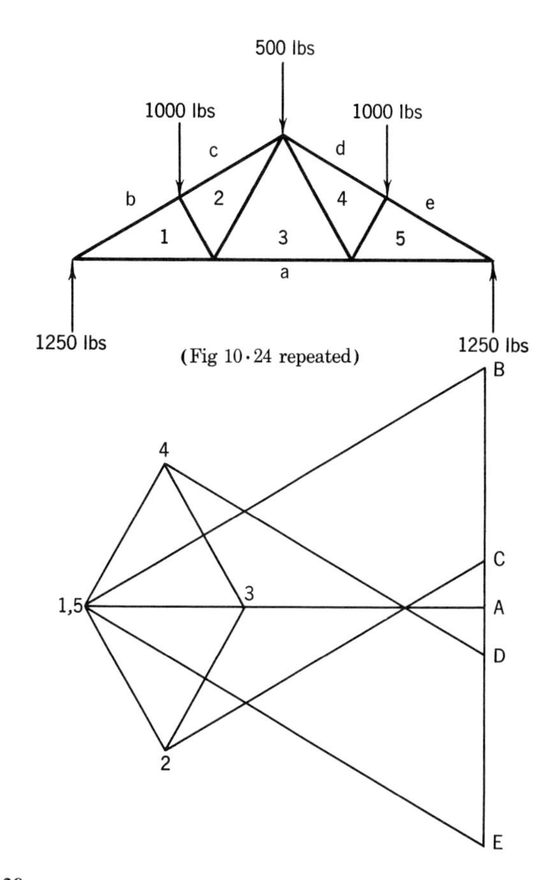

Fig. 10.26

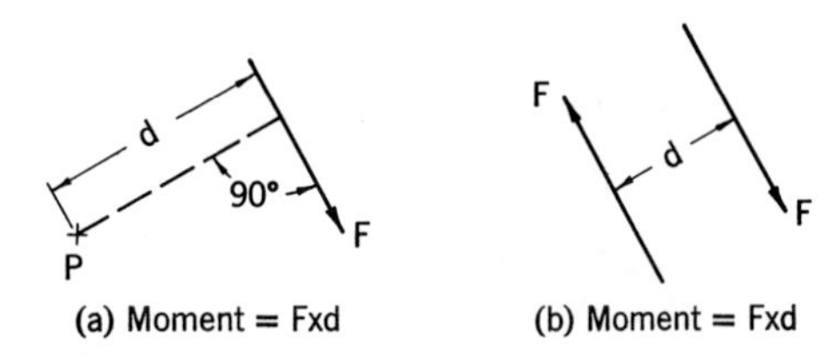

(a) Moment = Fxd (b) Moment = Fxd

Fig. 10.27

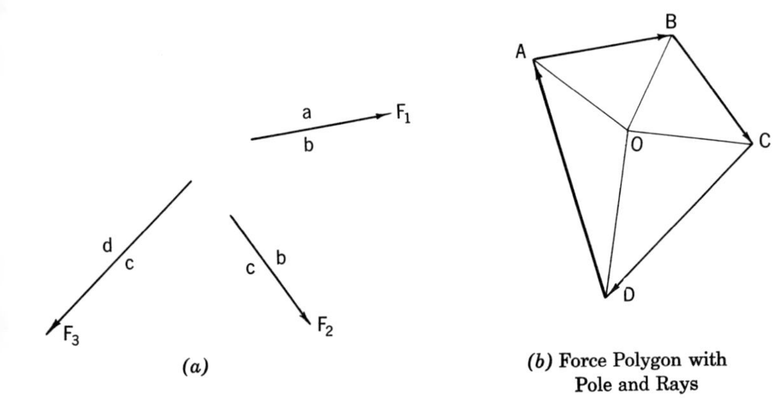

(a)

(b) Force Polygon with Pole and Rays

Fig. 10.28

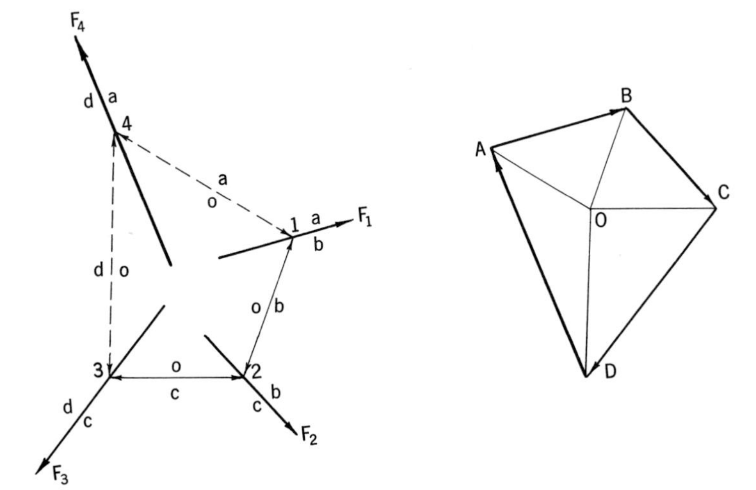

Fig. 10.29

Nonconcurrent Forces in a Plane

Before considering this subject, we should point out what is meant by the *moment of a force.* The *moment of a force* such as F [Fig. 10.27(a)] *about a point,* such as P, *is the product of the force and the perpendicular distance from the point to the line of action of the force.* The distance is known as the *moment arm.* If the length of the arm is in inches and the force is given in pounds, then the moment of the force about the point is given in inch-pounds. If we have *two equal and opposite forces whose lines of action are parallel* [Fig. 10.27(b)], *the moment of the forces about any point is equal to Fd. This system of forces is known as a couple.* A system of nonconcurrent forces in a plane is in equilibrium, provided that *both the resultant force and resultant moment are zero.*

Now let us consider the force system shown in Fig. 10.28(a). It is required to determine the magnitude, direction, and line of action of the single additional force which will produce equilibrium. Again let us use Bow's notation, so that the line of action of any force is designated by adjacent lowercase letters and the vector representing the force in magnitude and direction is designated by the corresponding capital letters placed at its ends. The magnitude and direction of the required force can be easily determined from the line DA which closes the vector polygon [Fig. 10.28(b)] and satisfies the requirement that the resultant force must be equal to zero.

It remains to determine a point on the line of action da of the required force DA. Let us assume a pole O in (or outside) the vector polygon, and then draw lines OA, OB, OC, and OD. We may regard each of the four triangles OAB, OBC, OCD, and ODA as a vector triangle, one side of which in each case in the resultant of the forces represented by the other two sides. For example, consider triangle OAB. If vector AB is treated as the resultant of vectors AO and OB, then the given force F_1 (AB) may be replaced by forces AO and OB.

Hence, if we select an arbitrary point 1 on the line of action ab of force AB (see Fig. 10.29) and then draw lines ao and ob through point 1 parallel, respectively, to vectors AO and OB, we shall have represented the lines of action of the two forces AO and OB by which force AB may be replaced. Now let us consider triangle OBC, in which force BC is the resultant of forces BO and OC. We observe that force BO is equal and opposite to OB, one of the two forces by which AB was replaced at point 1. If we extend the line of action ob until it intersects bc at point 2, and if we then replace BC by the two forces BO and OC at this point, we see that the equal and opposite forces OB and BO have the common line of action ob and therefore cancel each other. The remaining forces AO and OC have now completely replaced the two forces AB and BC.

Similarly, we see that force CD may be replaced by the two forces CO and OD. If this replacement is made at point 3 on the line of action oc of force OC, the equal and opposite forces OC and CO will be collinear and will cancel each other. The forces AO and OD which remain are now completely equivalent to the original forces AB, BC, and CD. Finally, considering triangle DOA, we observe that the required force DA may be replaced by the two forces DO and OA. If this replacement is made at the point, 4, common to the lines of action of forces AO and OD, all forces cancel and the requirements for equilibrium are satisfied. Therefore, point 4 is a point on the line of action da of the required force DA, and this line of action may be drawn through point 4 parallel to vector DA.

The polygon 1–2–3–4 is known as a *string polygon or funicular polygon.* Each side of this polygon represents the common line of action of two equal and opposite forces and is quite analogous in this respect to a tension or compression member of a truss or, in the case of tension, to a string pulled tight under the action of the two forces.

We now see that *the force system—F_1, F_2, F_3, F_4 —is in balance,* i.e., it is in equilibrium, *since both the force polygon and the string polygon are closed.* It is possible to have a closed force polygon and yet not have the system in equilibrium, if the string polygon does not close. This is clearly seen if we arbitrarily shift the line of action of the force F_4, and so replace it by the parallel and equal force F_4' (see Fig. 10.30). We see now that the system of forces F_1, F_2, F_3, F_4' is equivalent to the equal and opposite, but noncollinear, forces DO and OD, which are separated by the moment arm d and thus form a counterclockwise couple, which is the resultant of forces F_1, F_2, F_3, F_4'. Hence, these forces are *not* in equilibrium. It should be quite clear that *a*

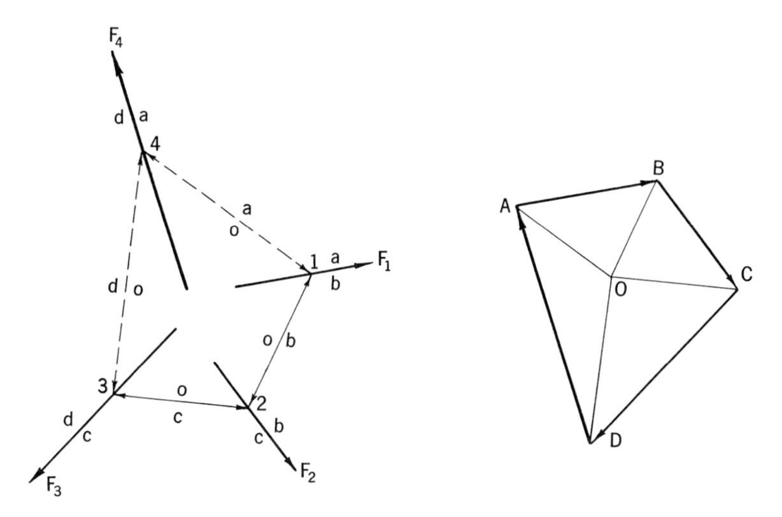

Fig. 10.29 (Repeated)

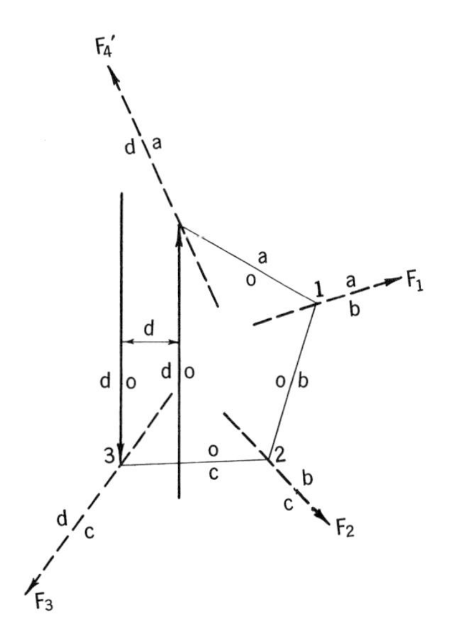

Fig. 10.30

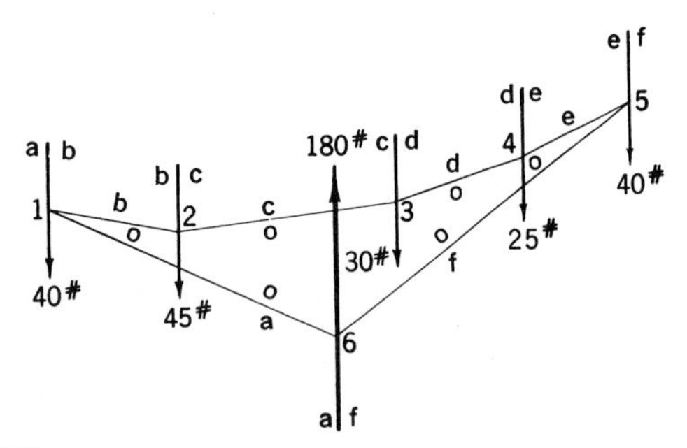

Fig. 10.31

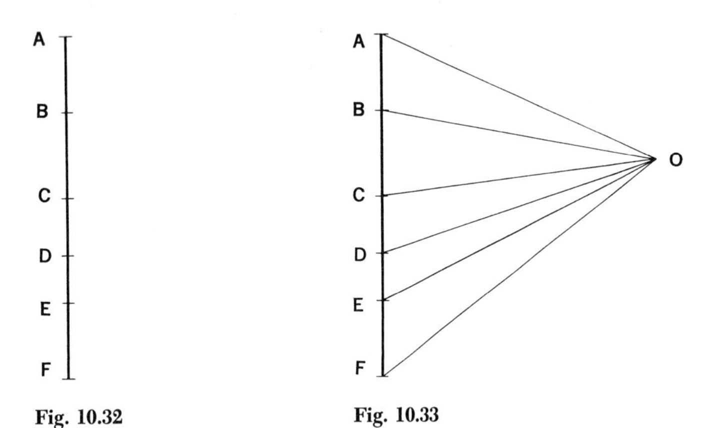

Fig. 10.32

Fig. 10.33

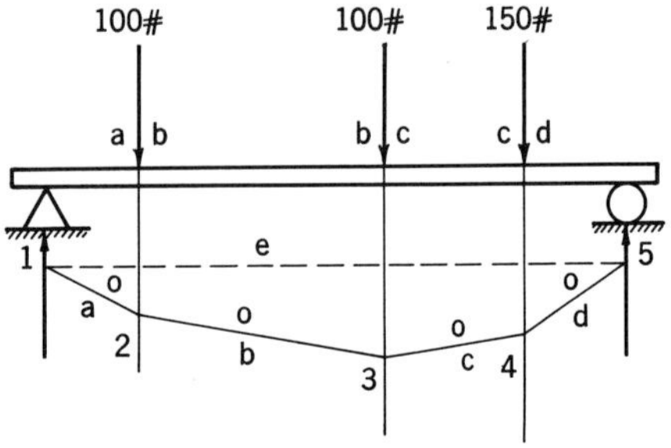

Fig. 10.34

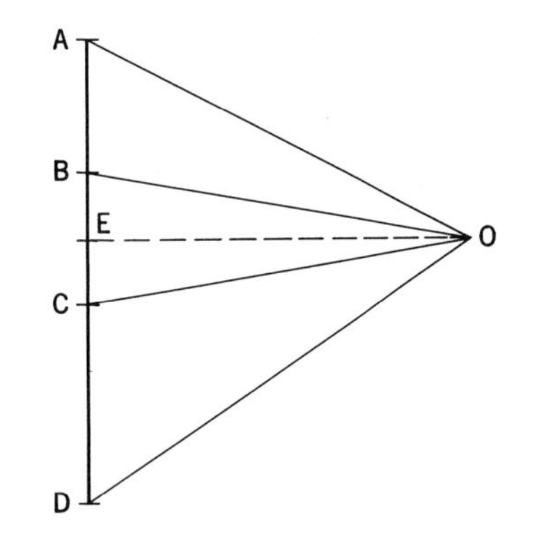

Fig. 10.35

nonconcurrent coplanar force system is in equilibrium only when both *the resultant moment and the resultant force of the system are zero; that the closed force polygon is the necessary condition for zero resultant force; and that the closed string polygon is the necessary condition for zero resultant moment.*

EXAMPLE 1

Consider the force system shown in Fig. 10.31. It is required to determine the magnitude, direction, and line of action of the single additional force which will produce equilibrium.

Let us start by constructing the force polygon, *ABCDEF*, as shown in Fig. 10.32. The magnitude, 180 pounds in this case, and direction of the required force are represented by vector *FA*. To locate a point on the line of action of the required force we shall construct a string polygon as shown in Fig. 10.31. First we assume a pole *O* (Fig. 10.33) and then draw rays *OA*, *OB*, . . . , *OF*. Force *AB* is replaced by two forces *AO* and *OB*, which have the lines of action *ao* and *ob*, respectively (Fig. 10.31). Through point 1, an arbitrarily selected point on the line of action of force *AB*, these two lines *ao* and *ob* are drawn respectively parallel to *AO* and *OB*. The intersection of line *ob* with line *bc* locates point 2, through which *oc* is drawn parallel to *OC*. The intersection of line *oc* with line *cd* locates point 3, through which line *od* is drawn parallel to *OD*. In a similar manner points 4 and 5 are located. Finally, a line *of* is drawn through point 5 parallel to *OF*. The intersection of lines *oa* and *of* locates point 6, a point on the line of action of the required force, which acts along a vertical line (since *FA* of the force polygon is vertical). The magnitude and direction of force *FA* are included.

EXAMPLE 2

Suppose we have a beam loaded as shown in Fig. 10.34. It is required to determine the magnitudes of the support reactions (forces *DE* and *EA*). In this problem we recognize that the direction of the forces *DE* and *EA* is known to be vertical, since the roller on the right will support only a vertical force. The force polygon is first drawn and then a pole *O* is located arbitrarily (Fig. 10.35). Rays *OA*, *OB*, *OC*, and *OD* are then established. We know the lines of action of the reactions *EA* and *DE*; however, their magnitudes can be determined only by locating point E in the force polygon. In order to do this we must determine the direction of ray OE. This ray is parallel to string oe of the funicular polygon, whose direction may be established from the requirement that the funicular polygon must close. Starting from point 1 on the line of action ea of reaction EA, we construct the successive strings ao, bo, co, and do parallel respectively to the corresponding rays in the force polygon. Thus the points 2, 3, 4, and 5 are successively located. The string polygon is closed by the string eo which joins points 1 and 5. Ray OE is drawn parallel to string oe. The magnitude of vector DE determines the force (right-hand reaction) whose line of action is de; and the magnitude of vector EA is the left-hand force, whose line of action is ea. It should be clear from a study of Figs. 10.34 and 10.35 that both the vector polygon, $ABCDE$, and the string polygon, 1–2–3–4–5, are closed polygons and, therefore, the force system is in equilibrium.

EXAMPLE 3

Consider the truss shown in Fig. 10.36. It is required to determine the magnitude of force CD and both the magnitude and direction of force DA.

A portion of the force polygon can be drawn by laying off vectors AB and BC as shown in Fig. 10.37. A vertical line may then be drawn through point C because we know the direction of force CD, which must be vertical since the right support is on rollers. At this stage we do not know the location of point D because the magnitude of force CD is unknown. We shall employ a string polygon to locate point D.

In Fig. 10.38, pole O and rays OA, OB, and OC have been added to Fig. 10.37. The string polygon is started at point 1, the only known point on the line of action da of force DA. Lines oa, ob, and oc are respectively parallel to OA, OB, and OC. Closing line od (1–4) establishes the direction of ray OD. Now that point D is available, the force polygon is completed by drawing the closing line DA, thus establishing both the magnitude and the direction of force DA. The magnitude of force CD is the magnitude of vector CD. The supporting forces CD and DA are known as reactions.

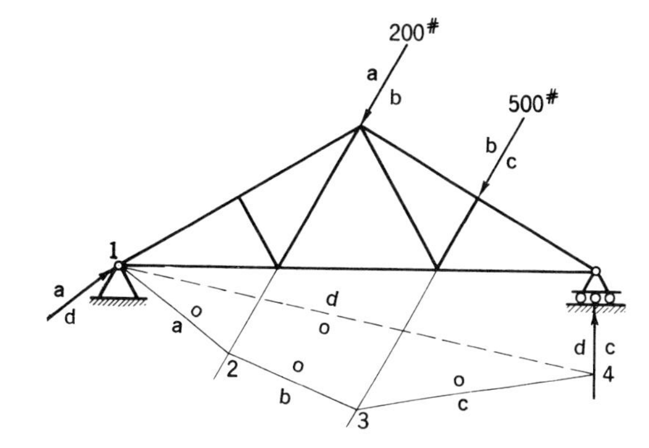

Fig. 10.36

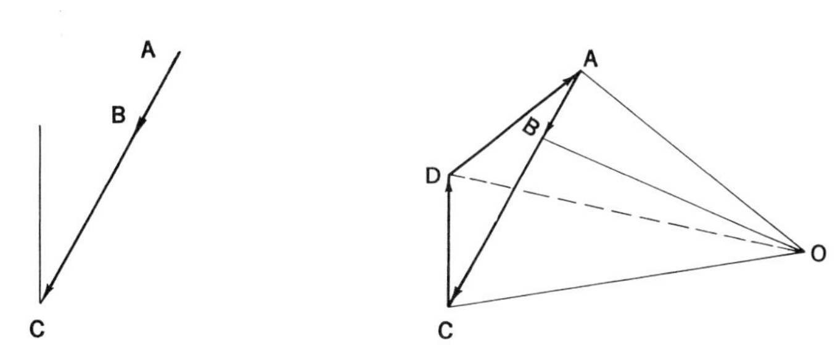

Fig. 10.37

Fig. 10.38

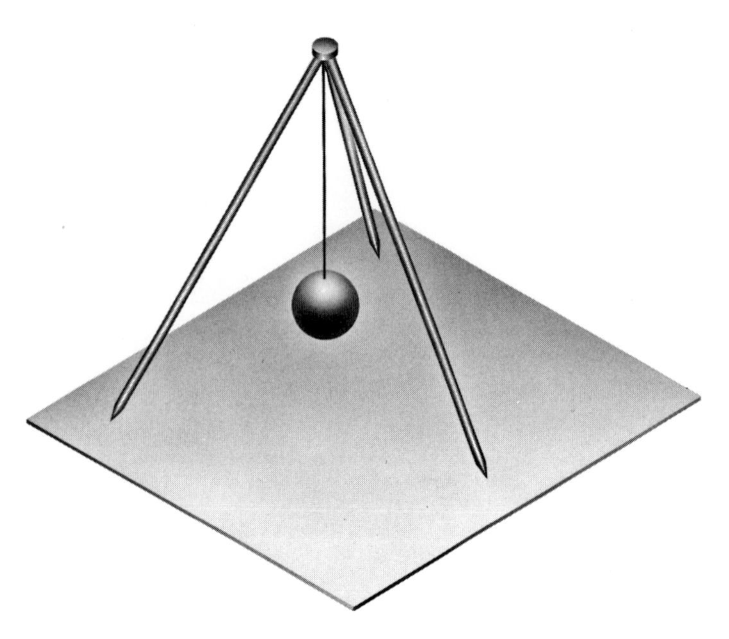

Fig. 10.39

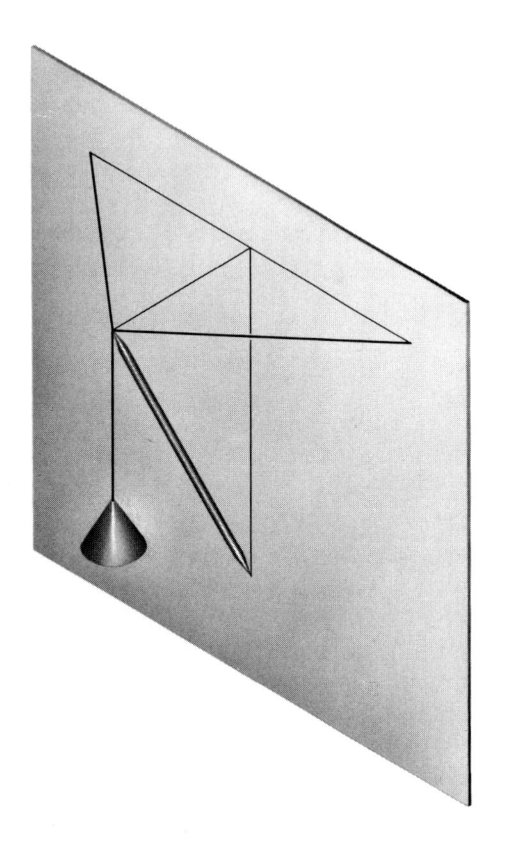

Fig. 10.40

Concurrent Noncoplanar Force Systems

In the design of three-dimensional frames (space frames) it is necessary to determine the forces acting in the members of the frame. Simple space frames may consist of three concurrent noncoplanar members which support a load, such as the ones shown in Figs. 10.39 and 10.40.

Before we consider problems dealing with the determination of forces acting in the members of space frames let us see what is meant by a *space force polygon.*

In Fig. 10.41 there are shown three concurrent noncoplanar forces *OA*, *OB*, and *OC*. Let us assume that we wish to find the resultant of these forces. To do this, we may employ the "parallelogram of forces" method. In Fig. 10.42 we observe that force *OD* is the resultant of forces *OA* and *OB*. Now, when we combine force *OD* with force *OC*, we will determine the resultant *OE* of the three forces *OA*, *OB*, and *OC*. *We note that OE is the body diagonal of the parallelepiped having edges OA, OB, and OC.*

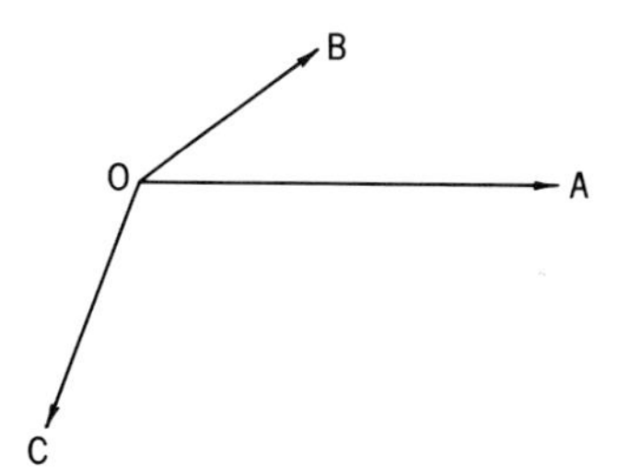

Fig. 10.41

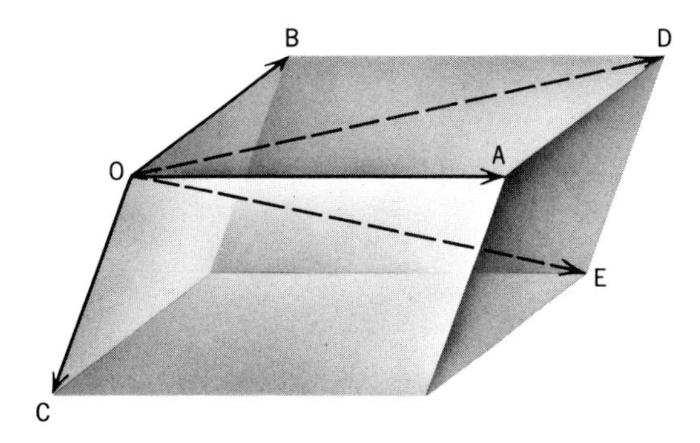

Fig. 10.42

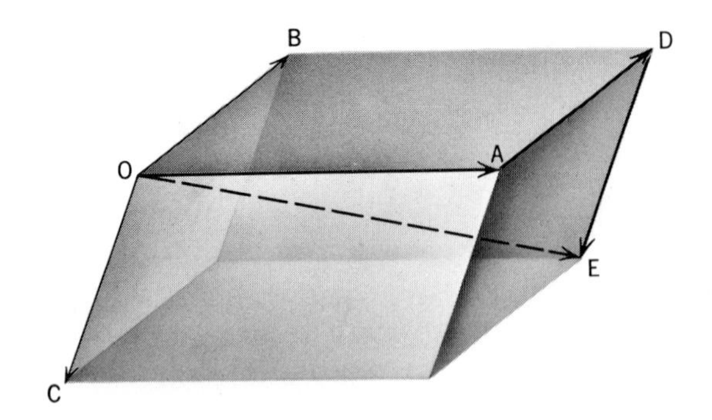

Fig. 10.43

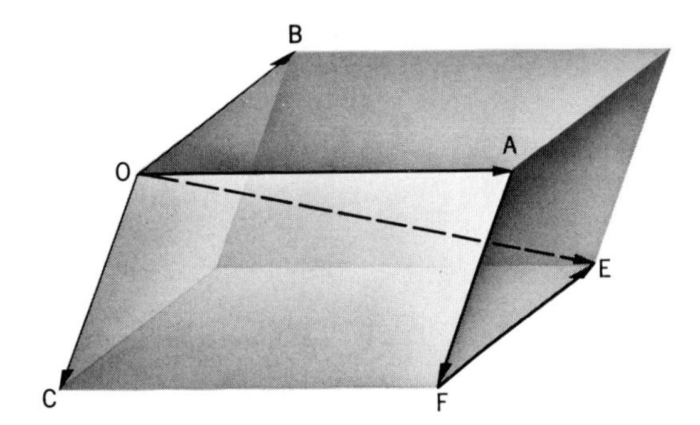

Fig. 10.44

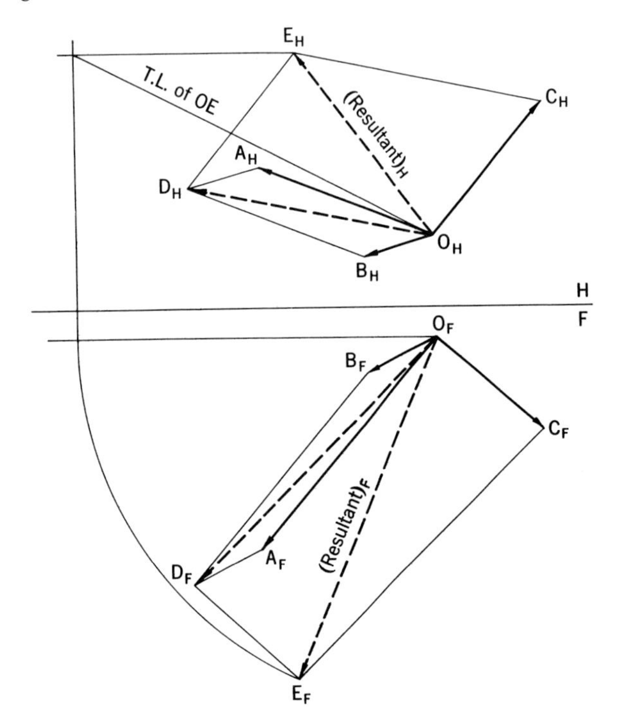

Fig. 10.45

It is quite evident now that it is not necessary to construct the entire parallelepiped to obtain the resultant force *OE*. Actually, it is sufficient to construct the "space force polygon" *OADEO* or the space force polygon *OAFEO*. These are shown in Figs. 10.43 and 10.44. An orthographic solution is shown in Fig. 10.45. In this figure the space force polygon *OADEO* is included. The magnitude of the resultant force *OE* can be determined by measuring the true length of *OE* in terms of the scale adopted in setting up the original data.

Figure 10.46 shows four concurrent forces, *OE*, *AO*, *BO*, and *CO*, which are in equilibrium. Therefore, their force polygon *OEDAO* closes. This figure suggests a solution to the following problem. If three concurrent members of a space frame support a known load, what are the forces* in the members? If we regard *OE* as the known load and *m*, *n*, and *s* as the concurrent members, we can determine the force polygon, *OEDAO*. Since point *E* is known, we may introduce a line *t* through this point and parallel to member *n*. Line *t* intersects the plane of members *m* and *s* in point *D*. Now through point *D* a line is drawn parallel to member *m* and intersecting member *s* in point *A*. The force polygon is easily constructed since it consists of sides *OE*, *ED*, *DA*, and *AO*. The true lengths of sides *AO*, *DA*, and *ED* determine the magnitudes of the forces in members *s*, *m*, and *n*, respectively. We now see that when force *ED* is applied to member *n*, *the force acts toward point O and, hence, the member is in compression.* Similarly, when force *DA* is applied to member *m* this force also acts toward point *O* and, therefore, member *m* is in compression. Finally, we observe that the force *AO* in member *s* also acts toward point *O*, so that member *s* is also in compression.

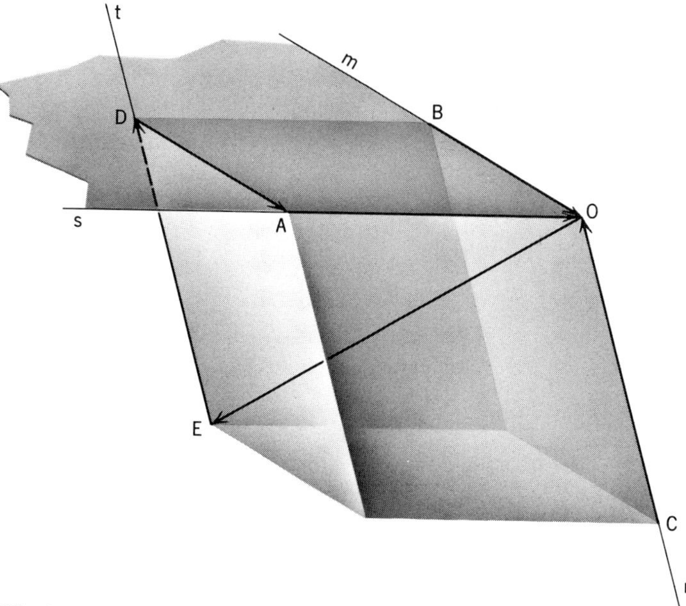

Fig. 10.46

* It is assumed that the forces applied to any one of the members *m*, *n*, and *s* act only in the direction of that member. In other words, forces applied perpendicular to any member are neglected in comparison with the forces applied along that member. We should recognize that the same assumption is basic to the analysis of the two-dimensional truss; hence, the frame just analyzed is a simple example of a particular type of space frame, namely, the space truss. No other type of space frame will be considered in this chapter.

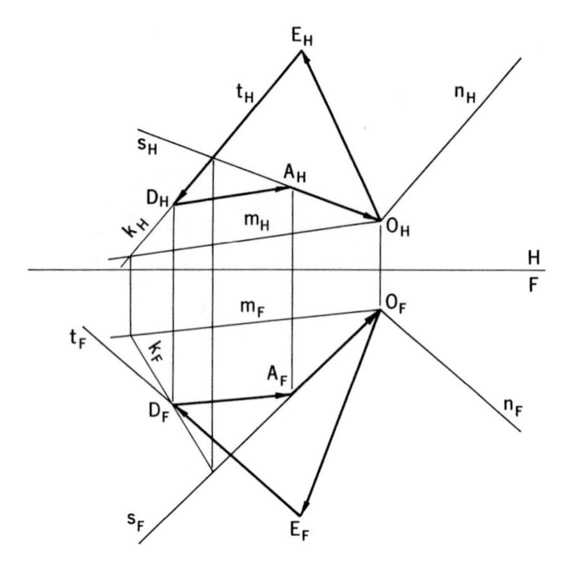

Fig. 10.47

EXAMPLE 1

Now let us proceed to solve this type of problem by means of an orthographic drawing such as the one shown in Fig. 10.47. Members m, n, and s support a load which is represented by vector OE. Through point E, line t is introduced parallel to member n. The intersection of line t with the plane determined by members m and s is point D. *We may recall the three basic steps that are employed in locating the point in which a line intersects a plane:*

1. Pass a plane through the line. (In this example the plane containing line t is perpendicular to the horizontal reference plane.)
2. Find the intersection of this plane with the given plane. (This is line k in the example.)
3. Find the intersection of lines k and t. This is point D. Now the line drawn through point D parallel to member m intersects member s in point A. The force polygon is $OEDAO$. The magnitudes of the forces in members m, n, and s are determined by measuring the true lengths (not shown) of vectors DA, ED, and AO, respectively. If the forces are applied to the members, it is evident that the members are in compression.

EXAMPLE 2

Consider the space frame and the load shown in Fig. 10.48. Carefully observe that *the plane of members m and n is shown on edge in the front view.* Advantage should be taken of this because it enables us to locate very easily the intersection, point D, of line t with the plane of members m and n. Then we introduce a line through point D parallel to member m and intersecting member n in point A (we could use a line through D parallel to member n and intersecting m). The force polygon is $OEDAO$. The true lengths of the vectors are readily established, since the vector ED is in true length in the front view and a single auxiliary view shows the true lengths of vectors DA and AO. The true lengths, when measured in accordance with the scale shown in Fig. 10.48, result in the values given in the table. The minus signs are one way of indicating compression, since tensions in truss members are ordinarily considered positive.

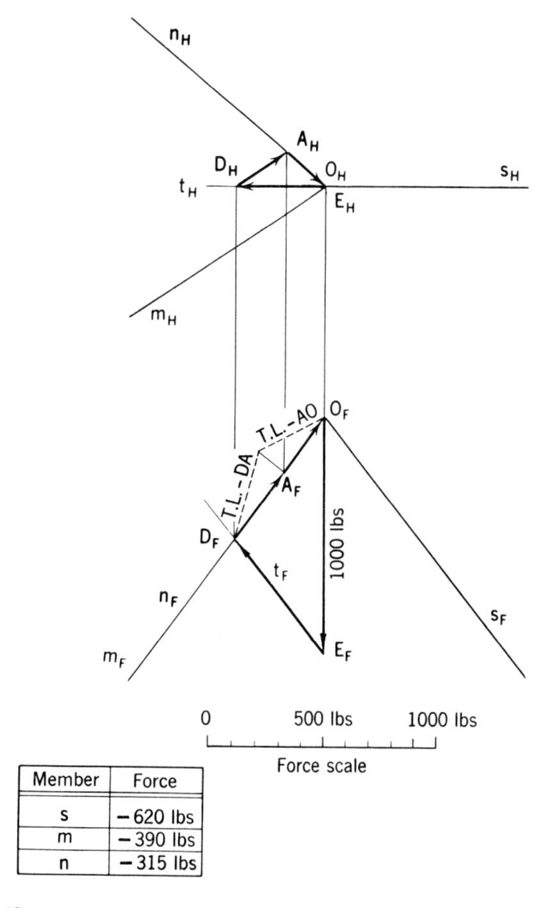

Member	Force
s	−620 lbs
m	−390 lbs
n	−315 lbs

Fig. 10.48

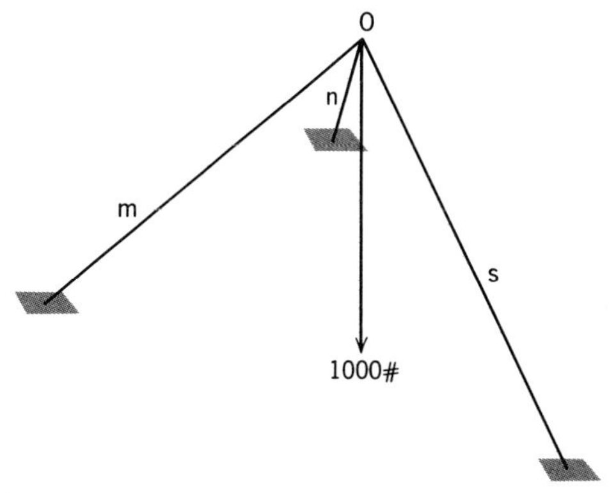

Fig. 10.48(*a*)

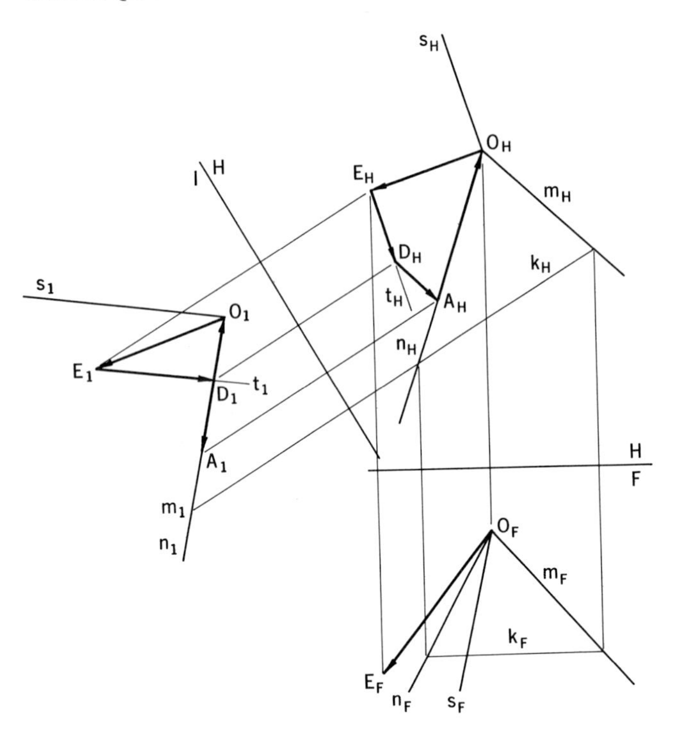

Fig. 10.49

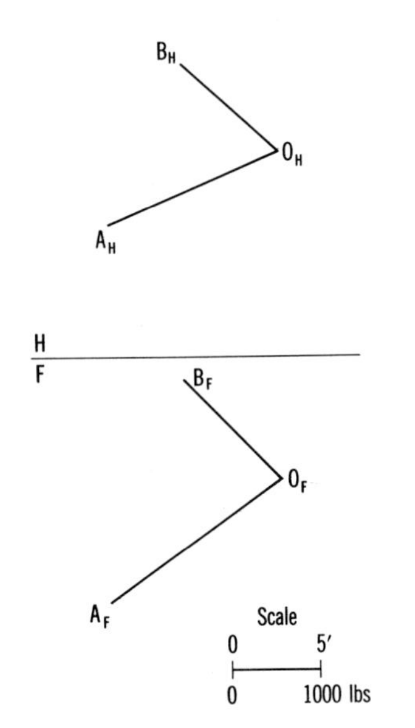

Fig. 10.50

EXAMPLE 3

Let us now consider the space frame and the load shown in Fig. 10.49. The method employed in Example 2 may be applied to this problem. It is necessary only to obtain a view which shows the plane of members m and n as a line. An edge view of plane mn is easily obtained by first establishing a point view of a line that lies in the plane. The horizontal line k is such a line. Supplementary plane 1 is perpendicular to the plane of members m and n. Once the supplementary view of the space frame and the load is established, it is necessary only to regard the top and supplementary views as given, and then to proceed as in Example 2 to determine the views of the force polygon.

In determining the forces in the members of the space frames just discussed, we have regarded the single joint as apart from the rest of the structure. We have thus isolated the joint *mentally* without actually drawing the views of the isolated joint and the forces acting upon it. Exactly the same thing was done in constructing and using the Maxwell diagram for the two-dimensional truss. We can safely do this only because we are dealing with frames of the truss type, so that the unknown forces act along the members whose directions are already shown in the views of the frame. For almost all other types of structures and machines this is usually *not* true and, therefore, the portion to be analyzed should be isolated by means of a clear diagram which shows all the forces that act on the body.

EXAMPLE 4

Carefully examine Fig. 10.50. Members OA, OB, and OC of this space frame support a vertical load of 2000 lb. Member OC is horizontal and is 10 ft long. The force acting in member OA is 1000 lb. Determine the views of member OC and the forces acting in members OB and OC.

The solution is shown in Fig. 10.51. We know the force acting in member OA is 1000 lb.; therefore, we can obtain a true length view of OA, shown as O_1A_1, and then lay off the vector $E_1O_1 = 1000$ lb. The H and F views of point E are readily established. Through point E a parallel to member OB is drawn. The intersection of this parallel with the horizontal drawn through W_F establishes D_F (remember that member OC is horizontal). Once we have located D_F, the F view of the force polygon is fixed—as $O_FW_FD_FE_FO_F$. The H view of the force polygon is readily established. The H and F views of member OC are now drawn—we know that OC is horizontal and is 10 ft long (2″ on the drawing). The forces in members OB and OC can be found by first obtaining the magnitudes of vectors DE and WD, respectively. Study of the force polygon shows that member OA is in compression, and that members OB and OC are in tension.

The two examples which follow deal with the determination of the forces in space frames which are more complex than those previously discussed. It will be observed that these examples afford an opportunity to make more extensive use of the principles which have been employed in solving the problem of the simple tripod frame (Fig. 10.39). The tripod problem is easily analyzed because there is only a single joint involved. In the case of the more complex space frames, however, we again regard each joint as apart from the rest of the structure (isolated mentally) in analyzing the system of forces which acts at that joint. The force polygon for the force system at each joint will be drawn separately.

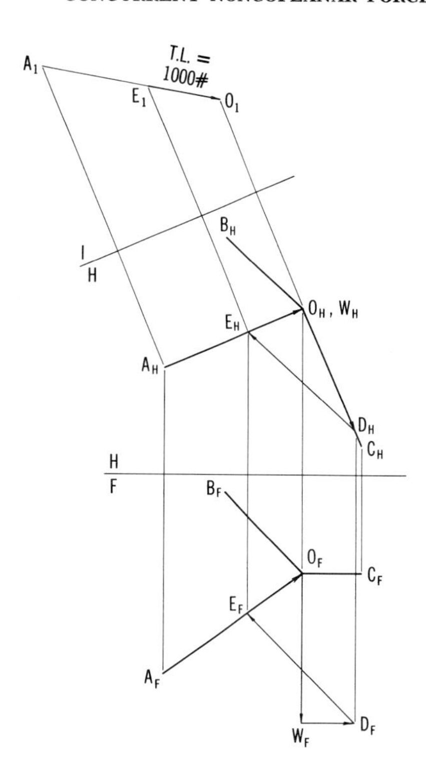

Fig. 10.51

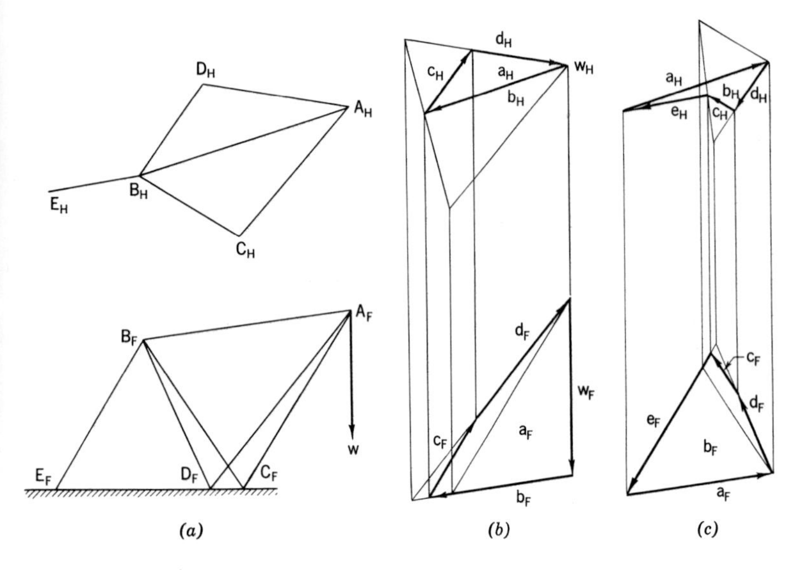

Fig. 10.52

EXAMPLE 5

Figure 10.52(a) shows a space frame which is slightly more complex than those so far considered. Before proceeding with the analysis, we should recognize clearly that, for a balanced system of concurrent forces in space, the unknown magnitudes of three forces can be found, provided that all the other forces are known. Thus, for the frame shown in Fig. 10.52(a), we consider first the joint A, where the load W is known and the three forces in members AB, AC, and AD are unknown. We then complete the analysis by considering joint B, where the force in member AB is now known and the forces in members BC, BD, and BE are the only unknowns.

Figures 10.52(b) and 10.52(c) are the force polygons for joints A and B, respectively. The polygons are labeled with a modified Bow's notation so that the letter inside the polygon designates the joint on which the forces act, and the letter outside the polygon and adjacent to any force designates the other end of the member which exerts that force. Thus, in Fig. 10.52(b), the letter a inside the polygon identifies the forces as acting on joint A, and the adjacent letter b designates the force exerted on joint A by member AB. For example, the letters a_F and b_F identify the front view of the force AB which acts in member AB.

For each polygon the views of the vector representing the known force are first laid off to scale; then a line is drawn through one end of that vector parallel to one of the unknown forces, and a plane is determined by lines drawn through the other end parallel, respectively, to the remaining two unknown forces. The intersection of the line and the plane is found by the method reviewed in Example 1, and the views of the force polygon are completed. The true lengths of the vectors are not shown.

The force polygon for joint A shows that member AB is in tension. Hence, in the views of the force polygon for joint B [Fig. 10.52(c)], the direction of force AB is such that it would pull away from that joint when applied by member AB. The completed force polygon clearly shows that members BD and BC are in compression, while member BE is in tension.

EXAMPLE 6

Let us now consider the loaded frame shown in Fig. 10.53(*a*). Is the frame stable, or will it collapse? If it is not stable, there is no problem to solve; on the other hand, if it can be shown that the frame is stable, we can proceed to determine the forces in the members. We first observe that the location of point *C* is established by the members *CD*, *CE*, and *CF*, which are attached to the foundation at their lower ends. Point *B* is then uniquely located by the members connecting it to the fixed points *C*, *E*, and *D*. Finally, point *A* is fixed by members *AB*, *AC*, and *AD*.

Next, the forces *AB*, *AC*, and *AD* are found from the force polygon for joint *A* [Fig. 10.53(*b*)] in the same manner as previously discussed in Example 2. It is seen that the plane of forces *AB* and *AC* is shown as a line in the front view. This enables us to locate immediately the front view of the point of intersection of the line of force *AD* with the plane of forces *AB* and *AC*. The views of the force polygon are then completed. Since force *AB* is now known, only three unknown forces remain at joint *B*. The force polygon for joint *B* is shown in Fig. 10.53(*c*). The two known forces, *AB* and *T*, are first drawn. Since member *BC* appears as a point in the front view of the frame, two planes of unknown forces will appear as lines in the front view of the force polygon, namely, the plane of forces *BC* and *BE*, and the plane of forces *BC* and *BD*. We may thus regard the completion of the front view of the force polygon as locating the front view of either of two points, i.e., either the intersection of force *BD* with the plane of forces *BC* and *BE*, or the intersection of force *BE* with the plane of forces *BC* and *BD*. We then complete the top view of the force polygon, which shows the sequence *BE*, *BC*, and *BD* for the three unknown forces. It should be recognized that this sequence is only one of three for which the front view would have the same shape, although the labeling *and the top view* would be different. The other two are *BC*, *BE*, and *BD*; and *BE*, *BD*, and *BC*.

We may now construct the views of the force polygon for joint *C* [Fig. 10.53(*d*)] where forces *AC* and *BC* are now known and the three forces *CD*, *CE*, and *CF* are unknown. The plane of forces *CE* and *CF* appears as a line in the front view, and the intersection of the line of force *CD* with this plane is easily found in that view.

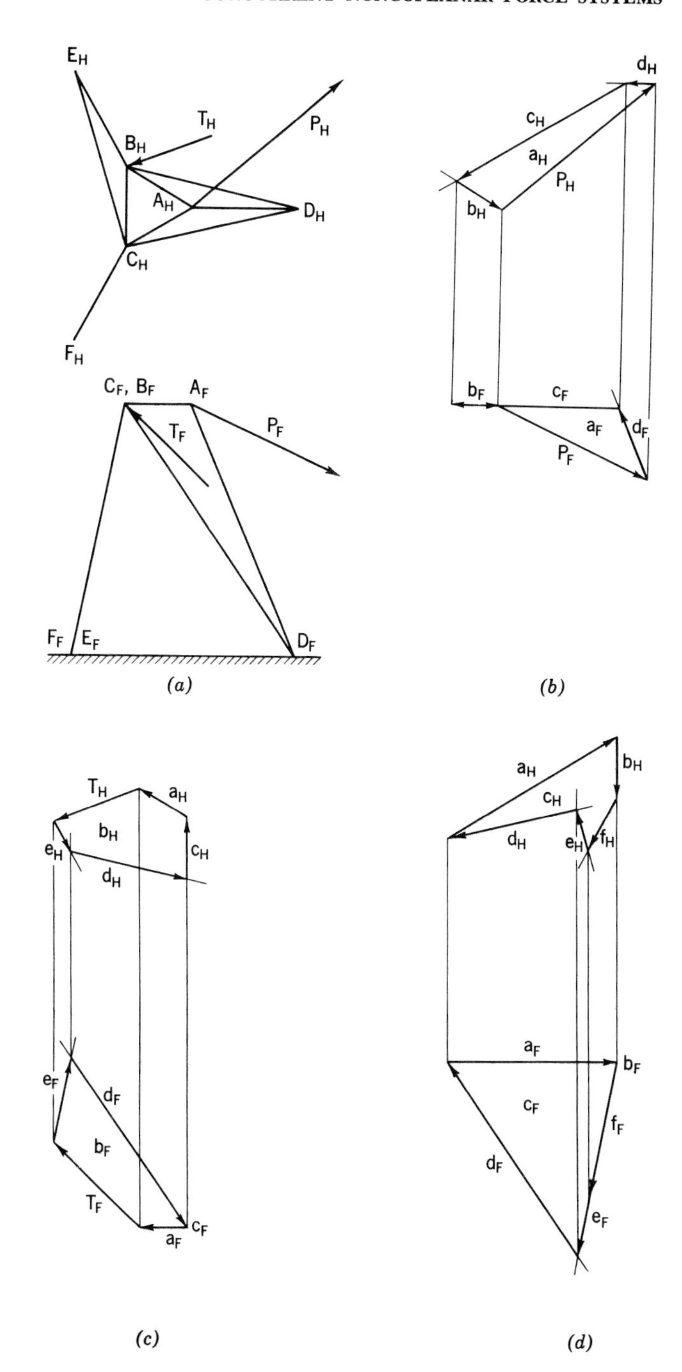

Fig. 10.53 Orthographic solution of forces in space frame supporting two loads.

It should be noted that, in this example, several planes appear as lines in the front view. This is a convenience in solving that part of the problem which deals with the intersection of a line and a plane. It should be recognized, however, that if this were not true we could still find the intersection by employing the three basic steps which have been referred to several times. Do you still recall what they are?

EXERCISES

Solve all problems graphically. Select convenient scales and sheet sizes.

1. A 400-lb force, inclined at an angle of 30° with the horizontal, acts on a body which rests on the floor. What is the magnitude of the force which tends to move the body along the floor, and of the force which tends to lift the body?

2. Two forces, $F_1 = 85$ lb and $F_2 = 120$ lb, in the same vertical plane, act on a body. F_1 is inclined at an angle of 20° with the horizontal, and F_2 at an angle of 40° with the horizontal. Both forces act in the same general direction. Determine the magnitude of the force which tends to lift the body.

3. Determine the resultant force of the concurrent force system shown in Fig. E-10.1.

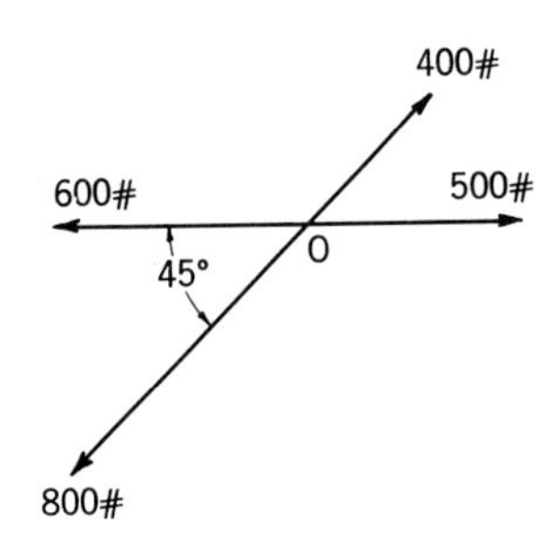

Fig. E-10.1

4. Determine the resultant force of the concurrent force system shown in Fig. E-10.2.

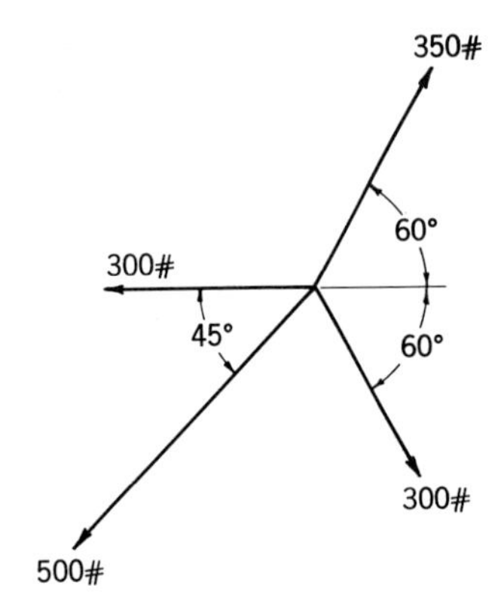

Fig. E-10.2

5. Determine the resultant force of the concurrent force system shown in Fig. E-10.3.

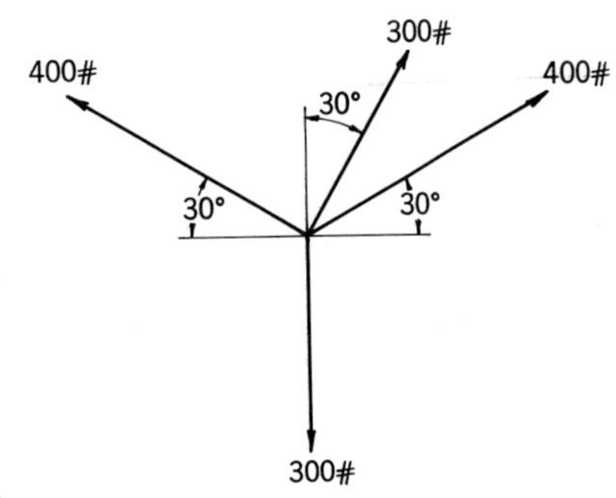

Fig. E-10.3

6. Determine the magnitude and direction of force F of the concurrent force system shown in Fig. E-10.4.

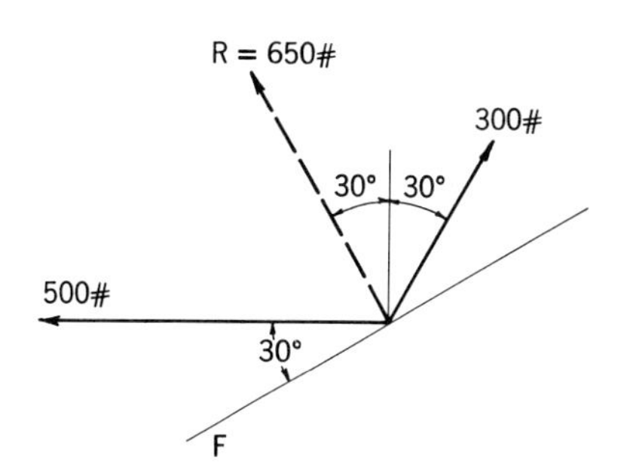

Fig. E-10.4

7. Determine the magnitudes of the vertical resultant R, and of the force, F, of the force system shown in Fig. E-10.5.

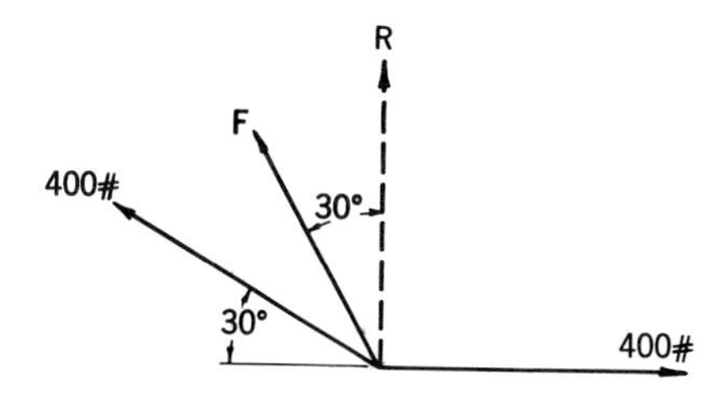

Fig. E-10.5

8. Determine the magnitude and sense (direction) of each of the forces F_1 and F_2 of the balanced force system shown in Fig. E-10.6.

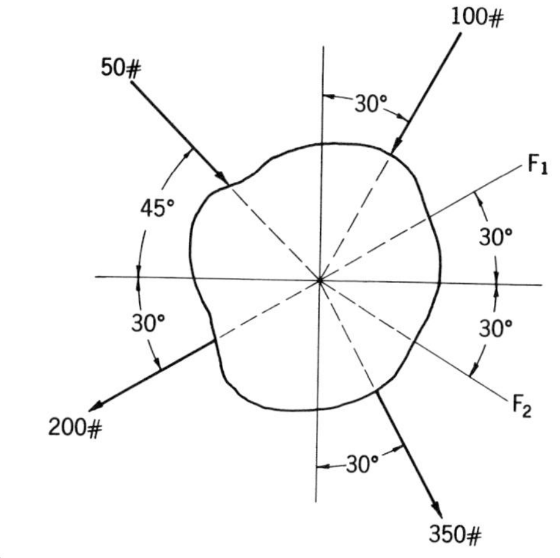

Fig. E-10.6

9. Particle O is in equilibrium when acted upon by the forces shown in Fig. E-10.7. Determine the magnitude and sense of each of the unknown forces.

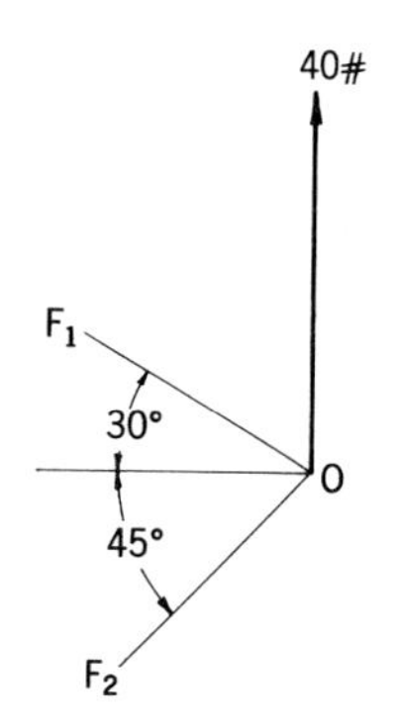

Fig. E-10.7

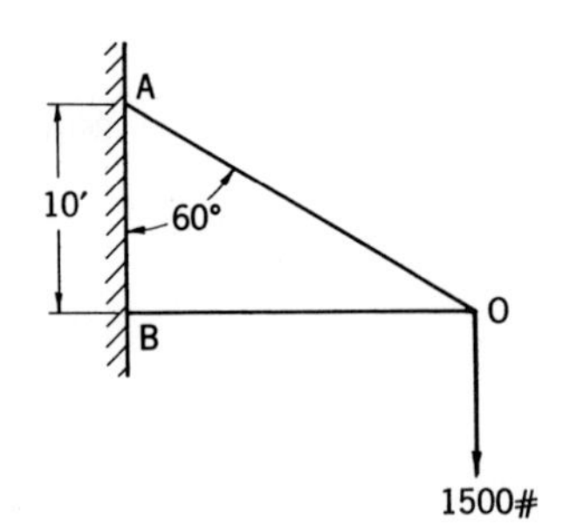

Fig. E-10.8

10. Determine the forces acting in members *OA* and *OB* of the frame shown in Fig. E-10.8. Indicate whether each force is in compression or tension.

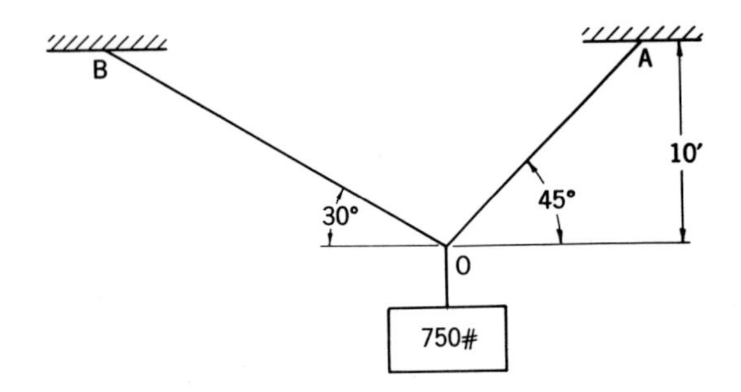

Fig. E-10.9

11. A 750-lb weight is supported by cables *OA* and *OB* as shown in Fig. E-10.9. The force acting in cable *OA* is 500 lb. Determine the force acting in cable *OB* and the location of point *O* with respect to point *B*.

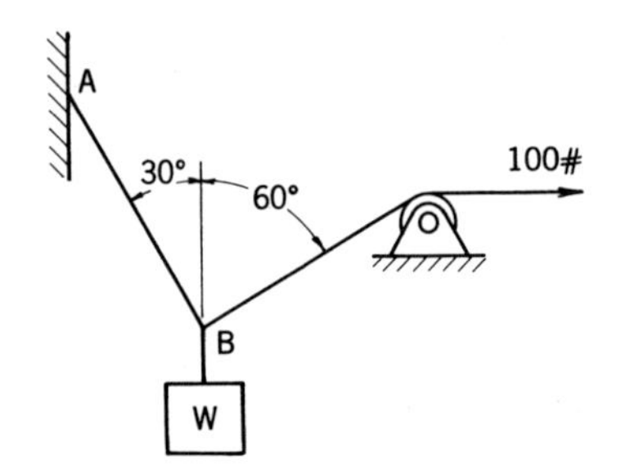

Fig. E-10.10

12. Determine load *W* and the magnitude of the tension in cable *AB* of the system shown in Fig. E-10.10. The system as shown is in equilibrium.

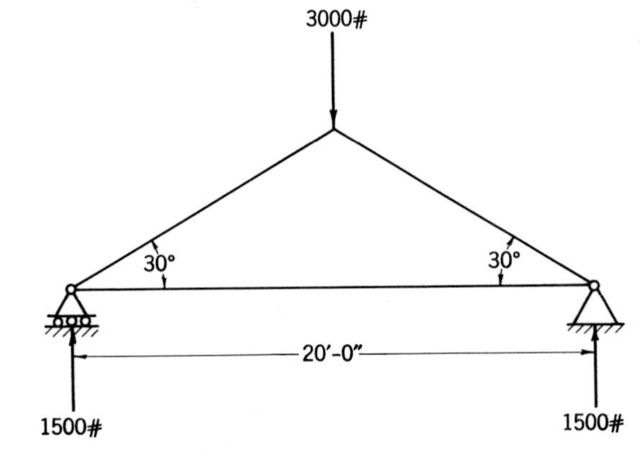

Fig. E-10.11

13. Determine the forces acting in the members of the truss shown in Fig. E-10.11. Designate compression by a minus sign.

14. Determine the forces acting in the members of the truss shown in Fig. E-10.12. Use Bow's notation and the Maxwell diagram.

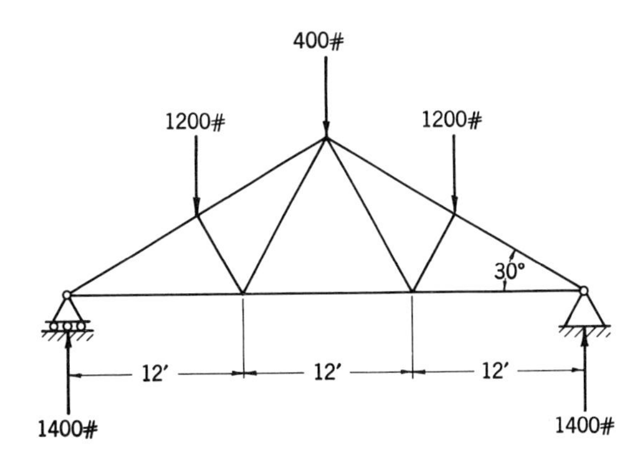

Fig. E-10.12

15. Determine the forces acting in the members of the truss shown in Fig. E-10.13.

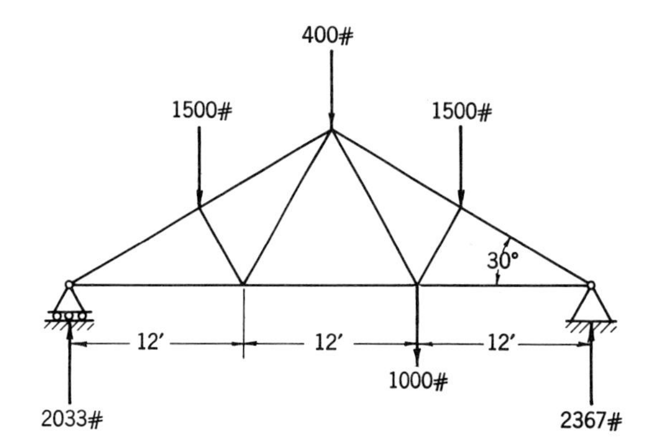

Fig. E-10.13

16. Determine the magnitude, sense, and line of action of the resultant of the force system shown in Fig. E-10.14.

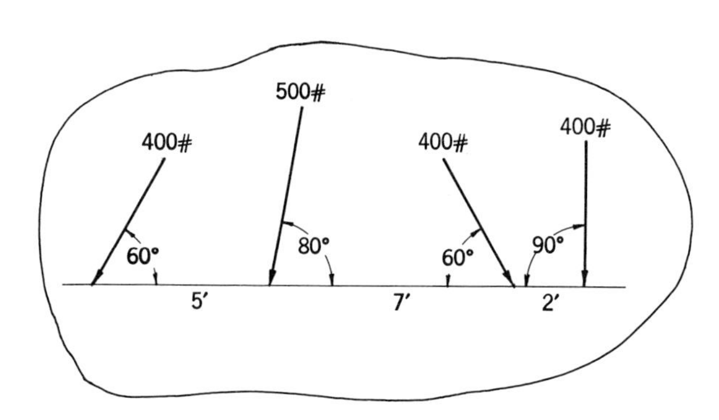

Fig. E-10.14

17. Determine the magnitude, sense, and line of action of the resultant of the force system shown in Fig. E-10.15.

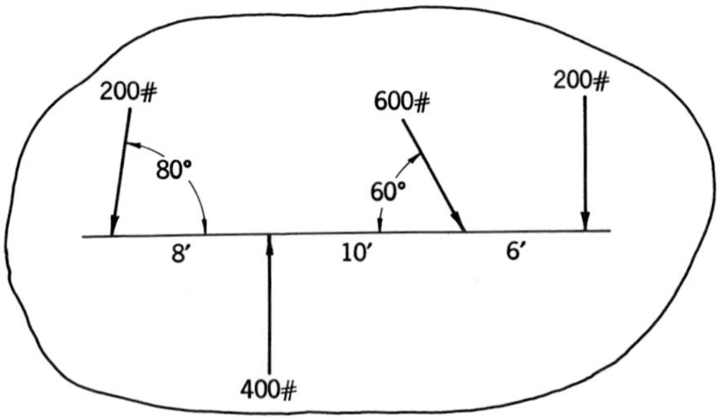

Fig. E-10.15

18. Determine the magnitudes of the reactions R_1 and R_2 on the beam shown in Fig. E-10.16.

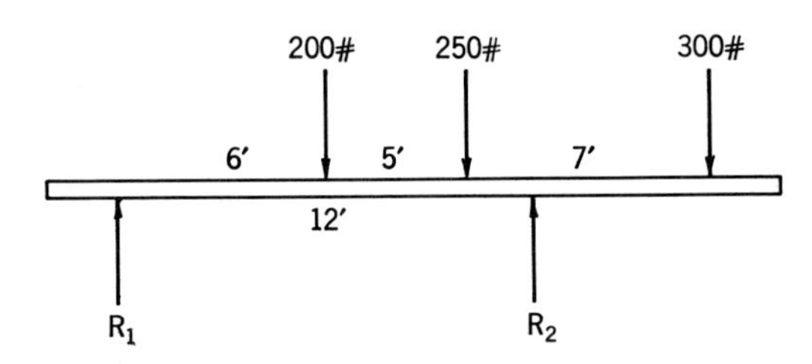

Fig. E-10.16

19. Determine the magnitude of the force *BC* acting on the beam shown in Fig. E-10.17. (*Hint:* Start the funicular polygon at point *A*.)

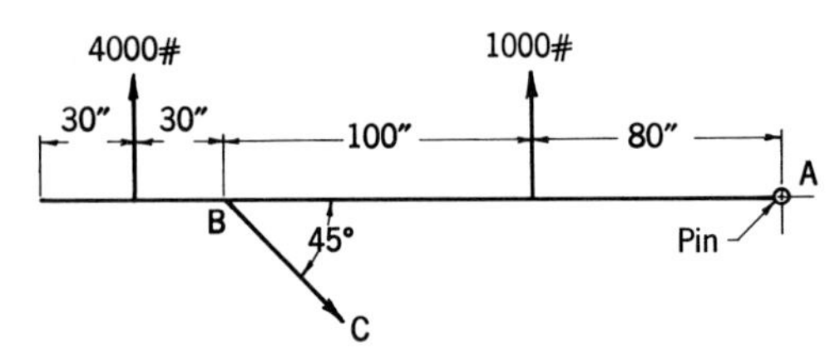

Fig. E-10.17

20. Determine the reactions and the forces in the members of the truss shown in Fig. E-10.18.

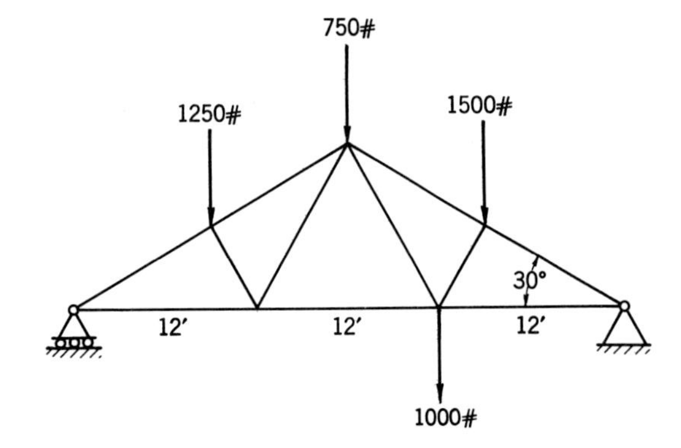

Fig. E-10.18

21. Determine the reactions and the forces in the members of the truss shown in Fig. E-10.19.

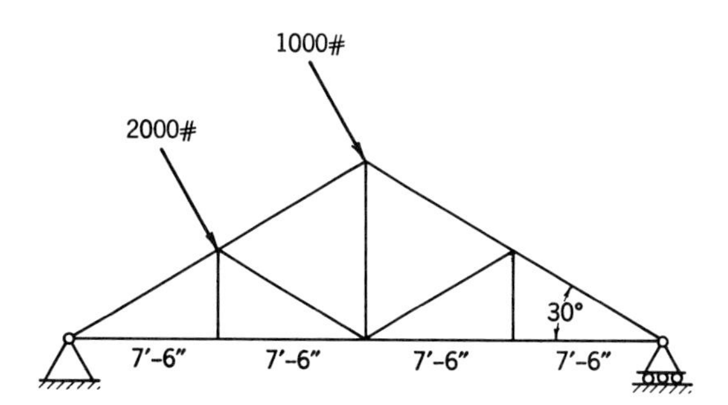

Fig. E-10.19

22. Determine the reactions and the forces in the members of the truss shown in Fig. E-10.20.

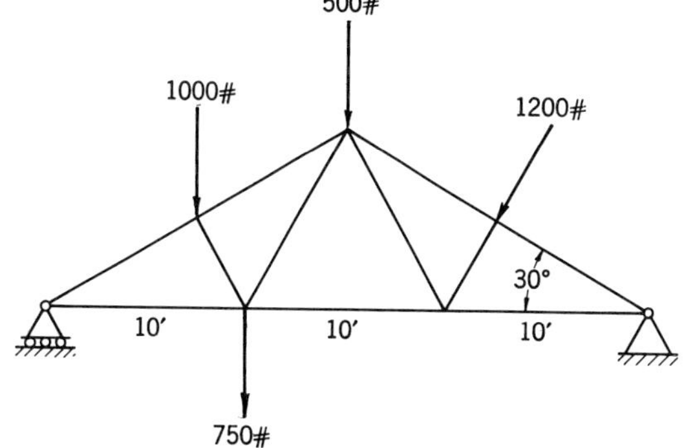
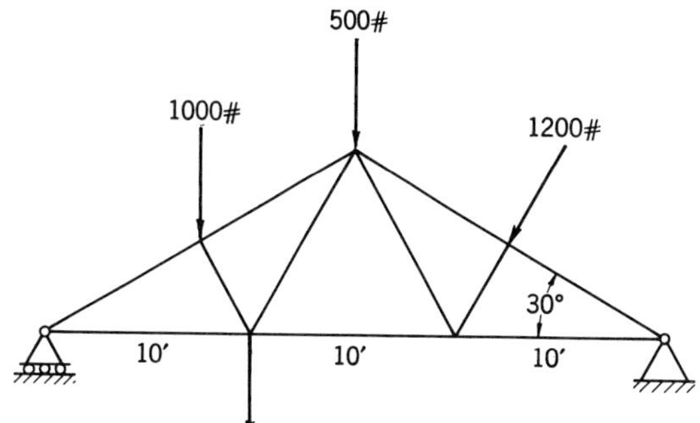

Fig. E-10.20

23. Determine the reactions and the forces in the members of the truss shown in either Fig. E-10.21(*a*) or Fig. E-10.21(*b*).

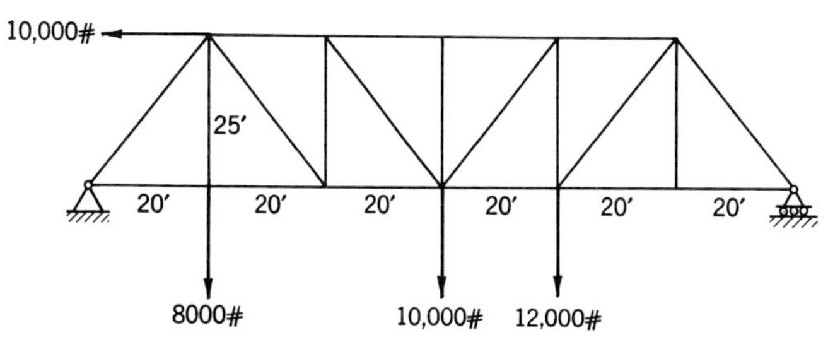

Fig. E-10.21(*a*)

Fig. E-10.21(*b*)

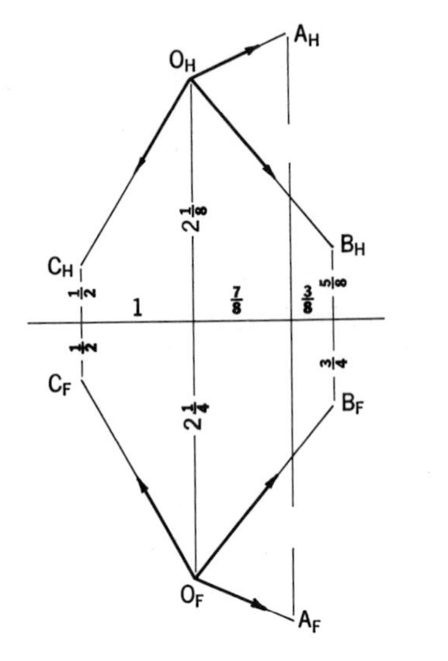

Fig. E-10.22

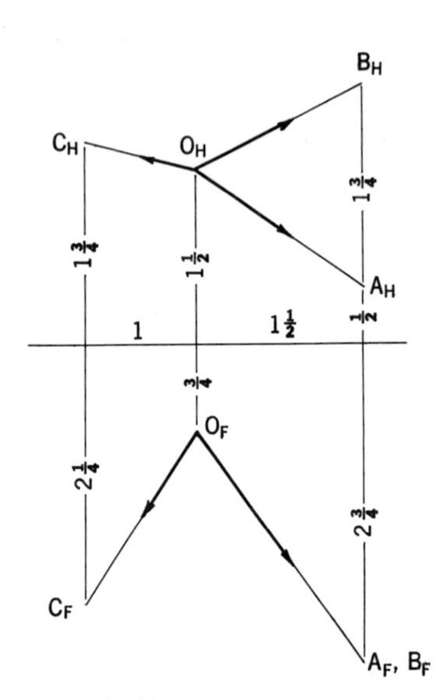

Fig. E-10.23

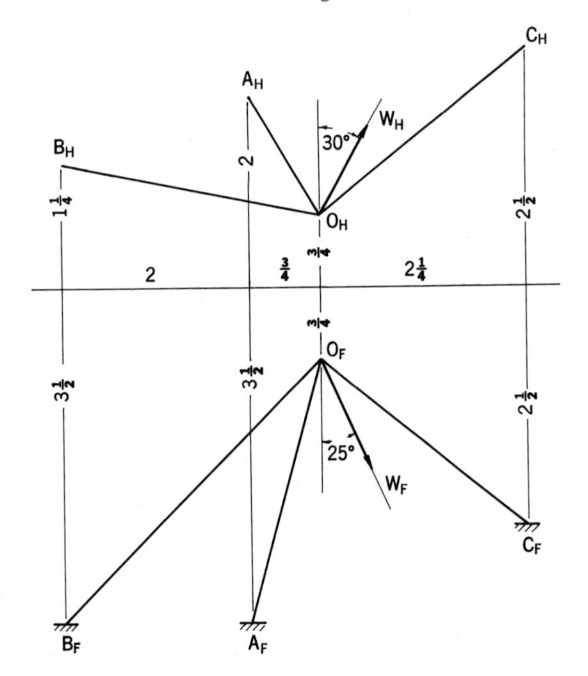

Fig. E-10.24

24. Determine the resultant of the concurrent noncoplanar force system shown in Fig. E-10.22. Scale: 1 in. = 50 lb. Assume the magnitudes of the forces.

25. Determine the resultant of the concurrent noncoplanar force system shown in Fig. E-10.23. Scale: 1 in. = 50 lb.

26. Find the forces acting in the members of the space frame shown in Fig. E-10.24. The load $OW = 1000$ lb. Scale: 1 in. = 500 lb.

27. Find the forces acting in the members of the space frame shown in Fig. E-10.25. The load $OW = 1000$ lb. Scale: 1 in. = 500 lb.

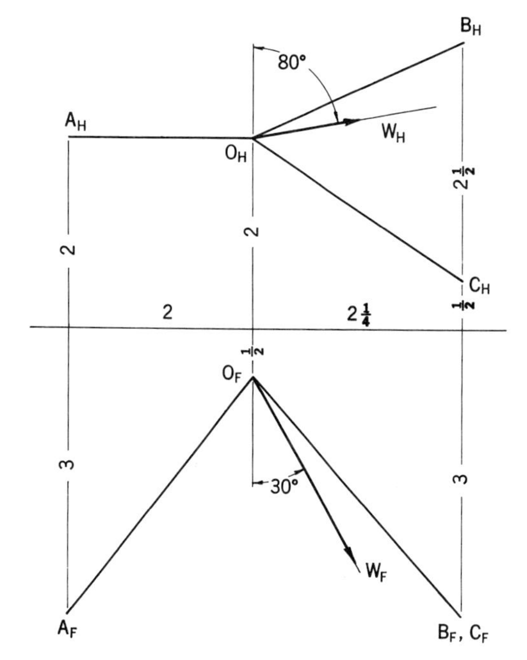

Fig. E-10.25

28. Find the forces acting in the members of the space frame shown in Fig. E-10.26. A vertical load of 3000 lb is suspended from point O. Scale: 1 in. = 1500 lb.

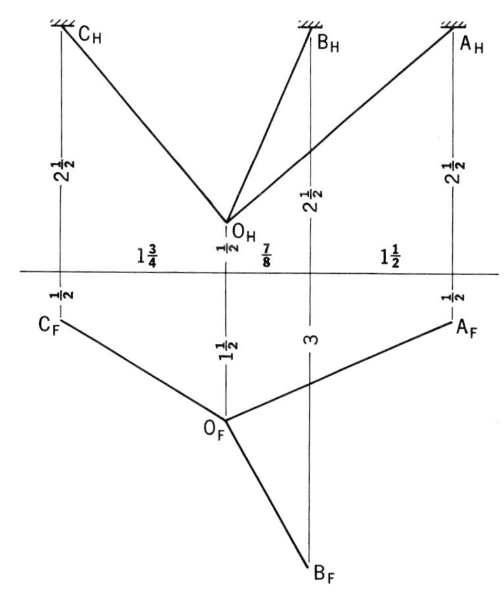

Fig. E-10.26

29. Find the forces acting in the members of the space frame shown in Fig. E-10.27. Load $OW = 3000$ lb. Scale: 1 in. $=$ 1000 lb.

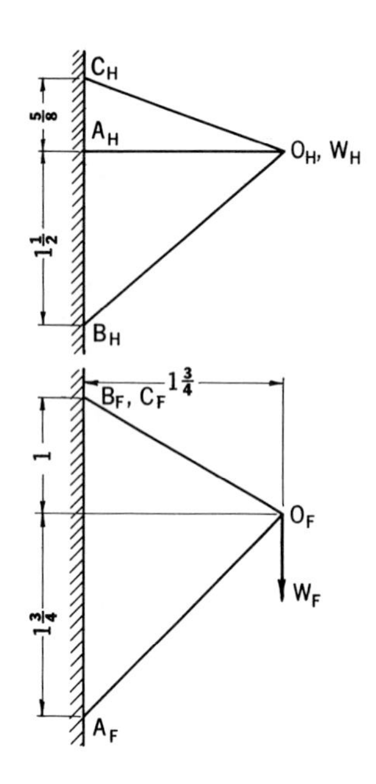

Fig. E-10.27

30. Find the forces acting in the members of the space frame shown in Fig. E-10.28. Load $OW = 1500$ lb. Scale: 1 in. $=$ 500 lb.

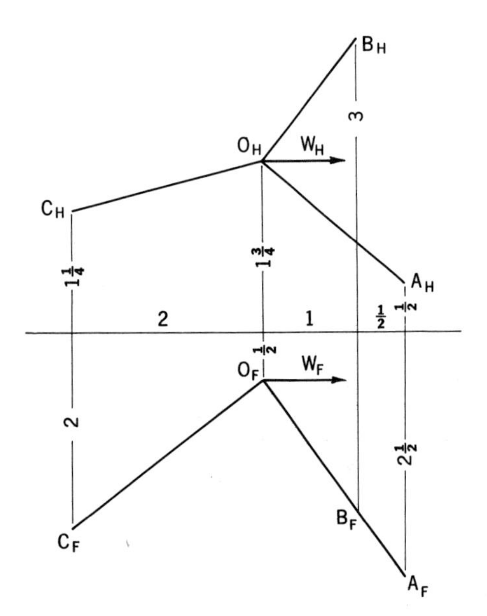

Fig. E-10.28

31. Find the forces acting in the members of the space frame shown in Fig. E-10.29. Load $OW = 4000$ lb. Scale: 1 in. = 2000 lb.

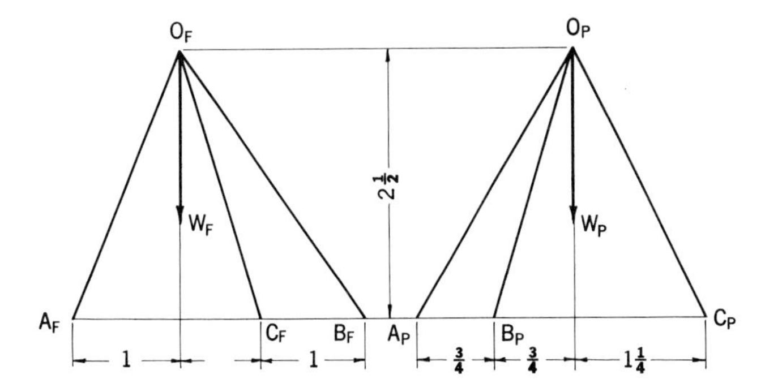

Fig. E-10.29

32. Find the forces acting in the members of the space frame shown in Fig. E-10.30. Load $OW = 1500$ lb. Scale: 1 in. = 500 lb.

33. Find the forces acting in the members of the space frame shown in Fig. E-10.30. Change load OW to 2000 lb perpendicular to the frontal plane and acting away from the observer.

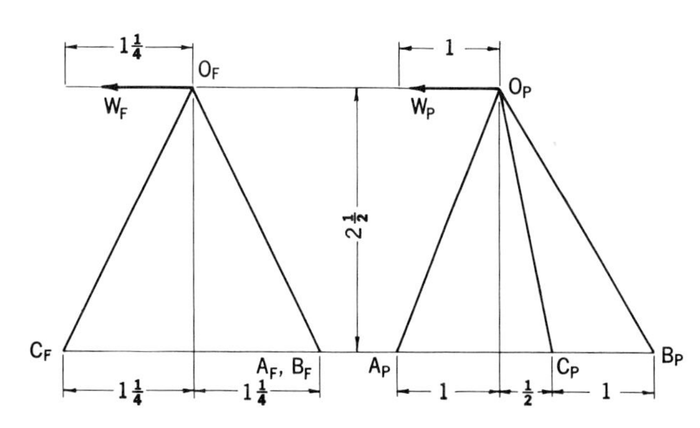

Fig. E-10.30

34. Find the forces acting in the members of the frame shown in Fig. E-10.31. The load $L = 10$ tons.

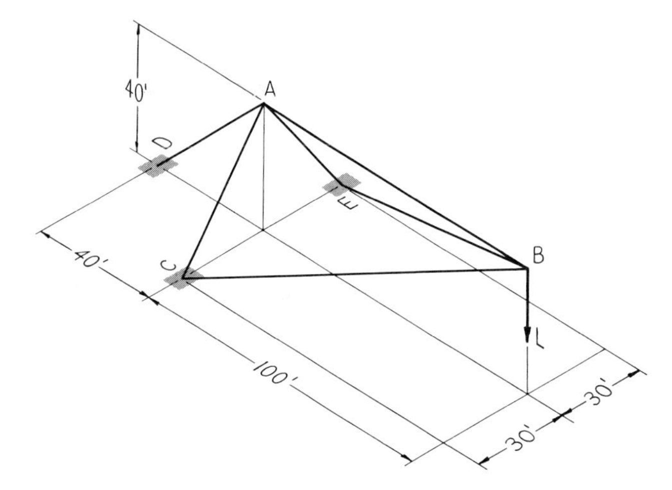

Fig. E-10.31

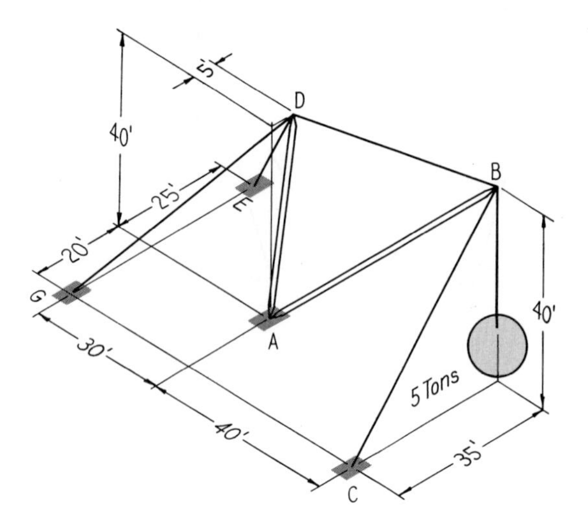

Fig. E-10.32

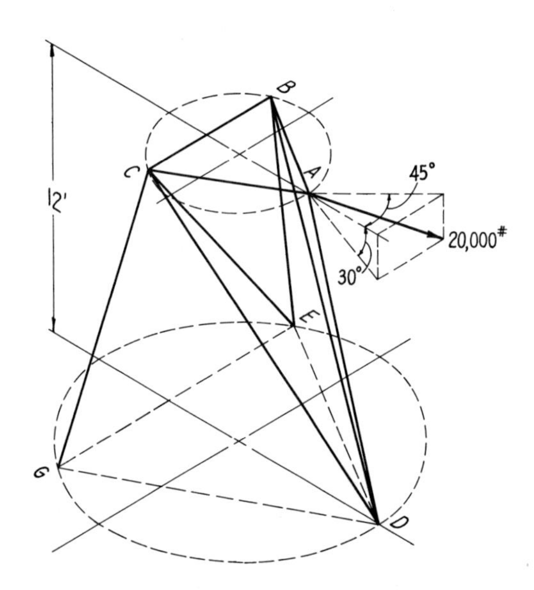

Fig. E-10.33

35. The derrick shown in Fig. 10.32 is the type that handles cargo aboard ships. Mast AD is slightly inclined so that the boom tends to swing toward its lowest position when loaded. Therefore, cable BC must be used to overcome this tendency and to position the boom. Determine the forces in the mast, the boom, cable BC, and the cables DE and DG that provide support for the mast.

36. The frame in Fig. E-10.33 resembles that in Fig. 10.51. Points A, B, C, and D, E, G are the corners of two horizontal equilateral triangles whose centers lie on the same vertical line and whose sides are respectively parallel. Length $\overline{AB} = 6$ ft, and length $\overline{DE} = 15$ ft. Determine the forces induced in all members of the frame by the 20,000-lb load. Attempt to explain any unusual or unexpected feature of your results.

37. Determine the forces acting in the members of the frame shown in Fig. E-10.34. Joints O, K, F, E, and D are pin connected.

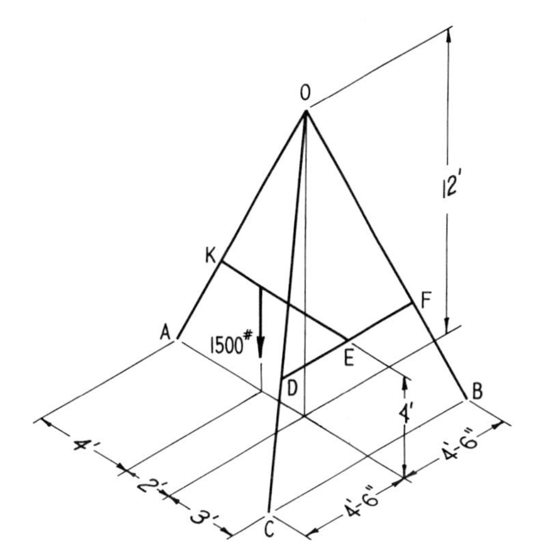

Fig. E-10.34

38. Determine the forces acting in the members of the frame shown in Fig. E-10.35. The joints are pin connected.

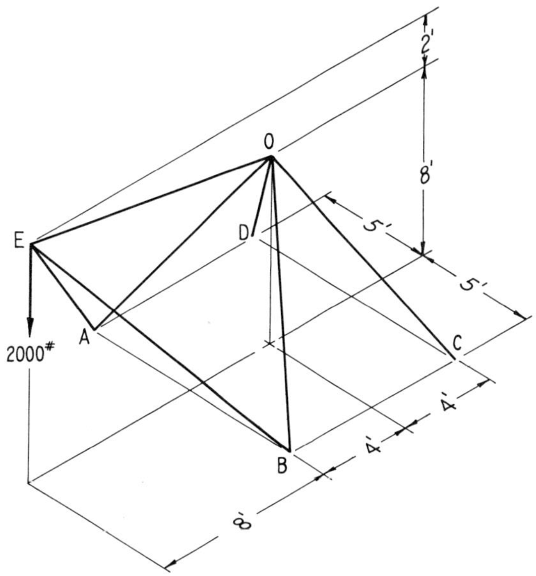

Fig. E-10.35

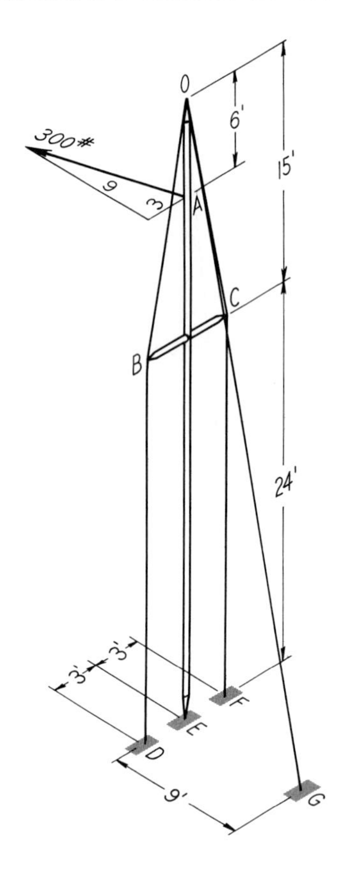

Fig. E-10.36

39. Find the forces acting in the members of the tower frame shown in Fig. E-10.36. The joints are pin connected.

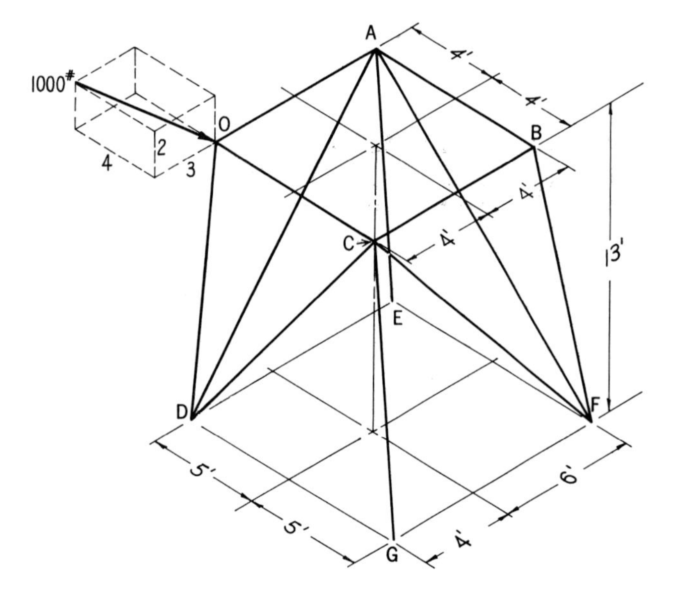

Fig. E-10.37

40. Find the forces acting in the members of the frame shown in Fig. E-10.37. The joints are pin connected.

PART 2
GRAPHICAL SOLUTIONS AND COMPUTATIONS

GRAPHICAL PRESENTATION OF DATA 11

INTRODUCTION

The purpose of Part 2 is to help the student develop his ability to understand and use graphical mathematics and to make him aware of graphical solutions wherever appropriate. There are many instances where mathematics in its symbolic form cannot be used advantageously, and in some cases it is impossible to set up suitable equations to provide design criteria with sufficient accuracy to satisfy the *intent of the designer.* In both research and design it has been found desirable to employ graphical mathematics in determining displacements, velocities, accelerations, etc., for those cases where the data are experimental and for which a mathematical expression is not easily obtained. In other cases, the problem can be programmed and, where costs are justified, computer systems may be used to yield the desired information.

Moreover, it is believed that the symbolic form of mathematics will be more meaningful to the student when he *sees* that the graphical method portrays the problem and its solution in a manner which can be readily understood.

Also included in this part of the book are graphic methods as applied to (*a*) graphical presentation of data; (*b*) arithmetic and algebra; (*c*) calculus—integration and differentiation; (*d*) the representation of experimental data and the determination of simple forms of *empirical equations;* (*e*) functional scales; and (*f*) an introduction to nomography, including both (1) concurrency charts and (2) alignment charts.

In both engineering and science the application of mathematical methods of calculation—*both symbolic and graphical*—is very important. For the solution of many technological problems the graphical method is sufficiently accurate and is preferred to algebraic methods, being much quicker (hence, more economical) and much less susceptible to error. In fact, it is quite easy to detect a graphical error, but in the case of an algebraic solution considerable time may be wasted in hunting for an error (signs, addition, etc.). There are instances, however, where an exact answer must be obtained, i.e., computation of interest on a principal sum of money. In such cases, a graphical solution would only be a close approximation.

In technology, the very nature of the problems precludes exact answers. Data that enter many problems most often depend on readings of instruments that are provided with scales of various types. These readings are approximate since they, in turn, depend on the *graphic accuracy of the scales* and the *visual acuity of the observer.* Despite the fact that high-precision instruments are employed in certain cases, the solutions are still close approximations. This is true because the human element is involved in both the manufacturing and the reading of the instruments.

Assuming that a person has normal vision and works under favorable conditions, the accuracy of graphical solutions will depend on:

1. The degree of accuracy that can be tolerated (i.e., forces computed to the closest 100 lb may be sufficient; areas to the closest square yard; or volume to the closest 10 cu ft, etc.).
2. The accuracy of the instruments used.
3. The scale selected for drawing the graphical solution.
4. The materials used.
5. The care exercised by the person solving the problem.

Both the engineer and the scientist should be capable of using the symbolic and graphic forms of mathematics whenever appropriate. Education and experience will help develop good *judgment* in selecting the form best suited to the problem. In some cases, a combination of the symbolic and graphic forms may yield the best results. Often it is difficult to work with only one form or method. We may believe that a certain solution is entirely algebraic, but in analyzing the solution we may find that helpful sketches are employed in formulating the basis for writing the equations.

So let us first consider the *graphical presentation of data.*

Graphical presentation of both technical and nontechnical data has proved to be a very effective means of conveying facts and their significance to business and professional groups, to manufacturing personnel, and to engineers and scientists. All about us, in fact, we see various types of graphs and charts in our daily newspapers, in magazines, and in technical publications.

In general, we find *two classes* of graphs and charts: (1) those employed primarily for popular appeal, as in advertising and in conveying information of general interest—population growth, use of the tax dollar, etc.; and (2) those usually employed by the engineer and scientist who may be primarily concerned with the meaning of experimental data—the interrelationship among several variables, the use of charts (nomograms) for making calculations quickly and quite accurately, etc. In some cases, both classes of charts and graphs may appear in annual reports of the activity of a company.

The engineer, while more concerned with the technical and scientific areas, should nonetheless be acquainted with the preparation and use of graphs and charts in both classes.

GRAPHICAL PRESENTATION OF DATA

Types of Charts

Frequent use is made of the following types of charts: (*a*) bar charts, (*b*) pie charts, (*c*) pictorial charts, (*d*) maps, and (*e*) organization charts. These charts are typical of class 1. In the technical and scientific fields the following types are typical: (*a*) polar charts, (*b*) trilinear charts, (*c*) flow charts, (*d*) Cartesian coordinate charts, and (*e*) nomograms. Cartesian charts may consist of the plot of data on a grid sheet that has uniform scales or functional scales (i.e., log-log or semi-log) for values of the independent (usually the x-axis) and dependent (usually the y-axis) variables.

Now let us consider the types of charts as mentioned above.

Bar Charts. Bar charts (column charts) are generally employed in presenting facts considered to be of interest to the layman. For example, the one shown in Fig. 11.1 shows how the federal budget for 1966–67 compares with the budgets of others. In some cases single bars are used to show two or more variables. For example, in Fig. 11.2 each bar is divided into two parts, the upper portion to show federal excise taxes and the shaded part to show state and local taxes for motor vehicles in 1964. Another interesting bar chart is shown in Fig. 11.3 which shows a comparison of unemployment rates between high-school dropouts and high-school graduates.

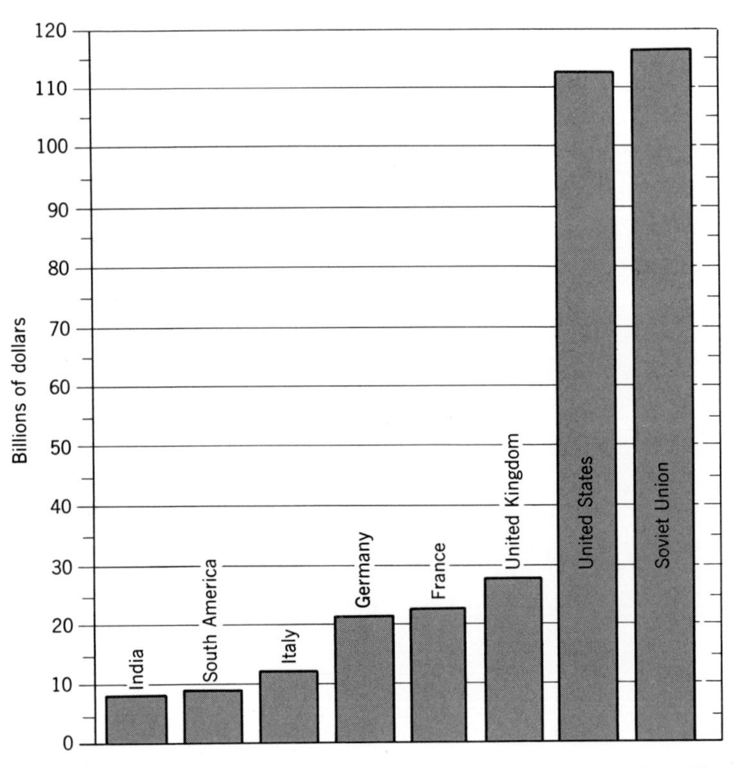

Fig. 11.1 Bar chart showing world budgets, 1966–67. (From AP Wirephoto Chart, 1966)

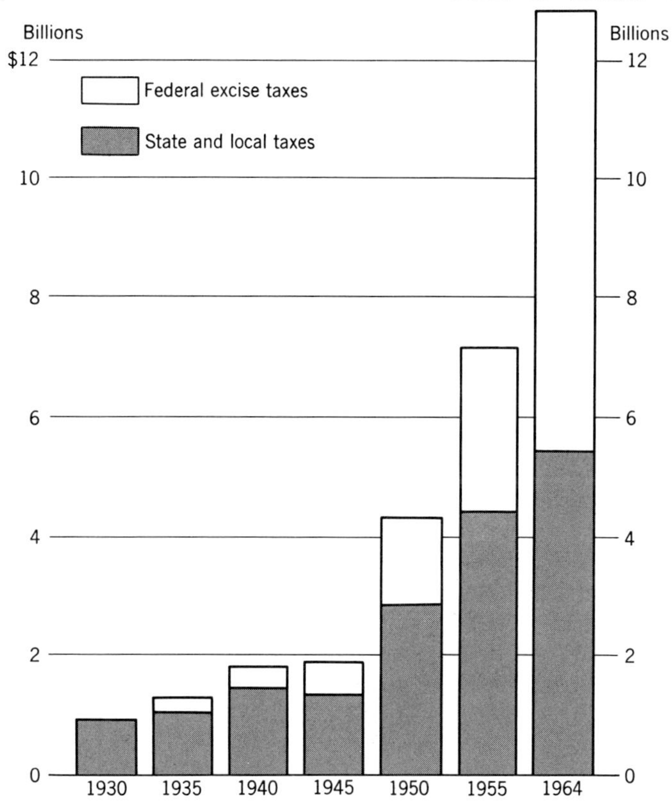

Fig. 11.2 Bar chart showing special motor vehicle taxes, 1930–1964. (From *Automobile Facts and Figures,* 1965)

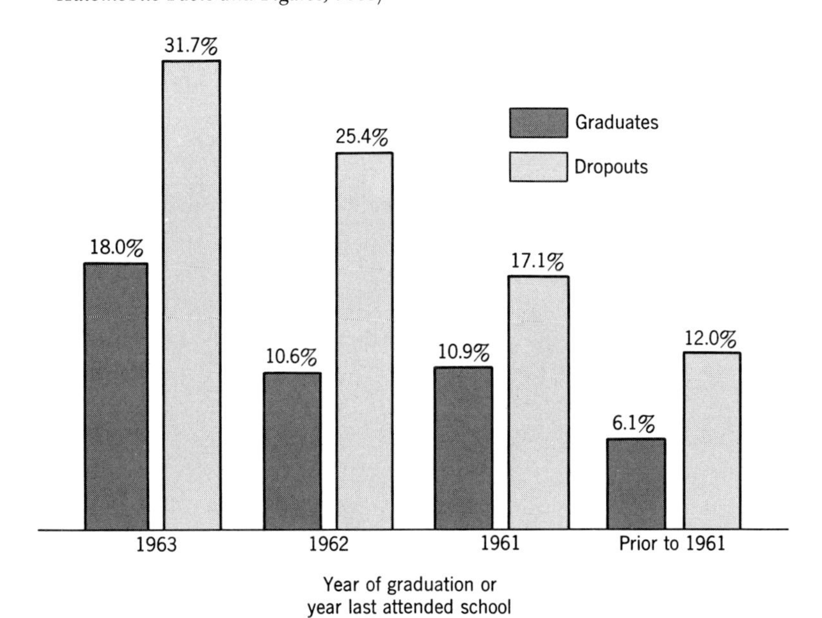

Fig. 11.3 Comparison of unemployment rate between dropouts and high school graduates. (Source: *PG & E Progress,* Jan. 1966)

Pie Charts. These are circular charts used to show the distribution of related components of a unit. For example, the federal budget dollar, estimated for 1967, is shown quite clearly in the pie charts of Fig. 11.4.

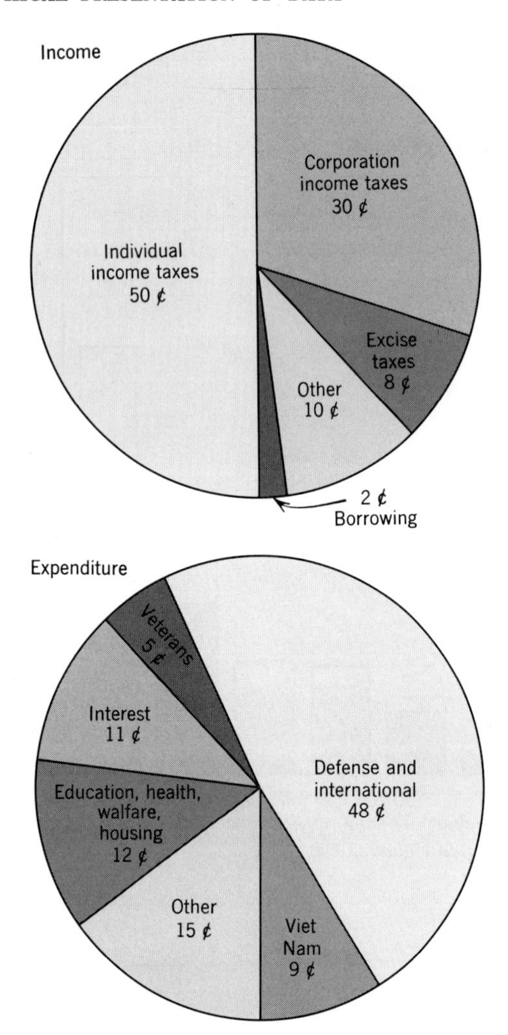

Fig. 11.4 Pie charts showing U.S. budget dollar, fiscal year 1967 estimate. (Source: Bureau of the Budget)

Pictorial Charts. An example of a pictorial chart is shown in Fig. 11.5 which shows car ownership by regions in the United States.

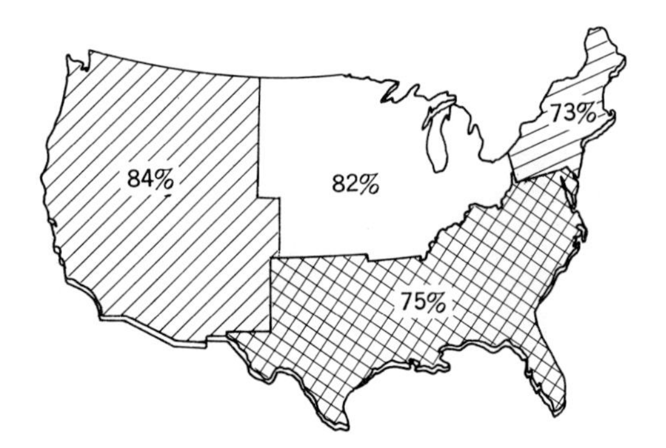

Fig. 11.5 Pictorial chart showing car ownership by regions in the U.S. (From *Automobile Facts and Figures*, 1965)

Map-Type Charts. These charts are used in showing distribution systems—water lines, power systems, etc.; providing information concerning city or county zoning—areas planned for various types of dwellings, for manufacturing facilities, etc.; and population densities, mineral deposits, etc. An example with which most of us are familiar is shown in Fig. 11.6.

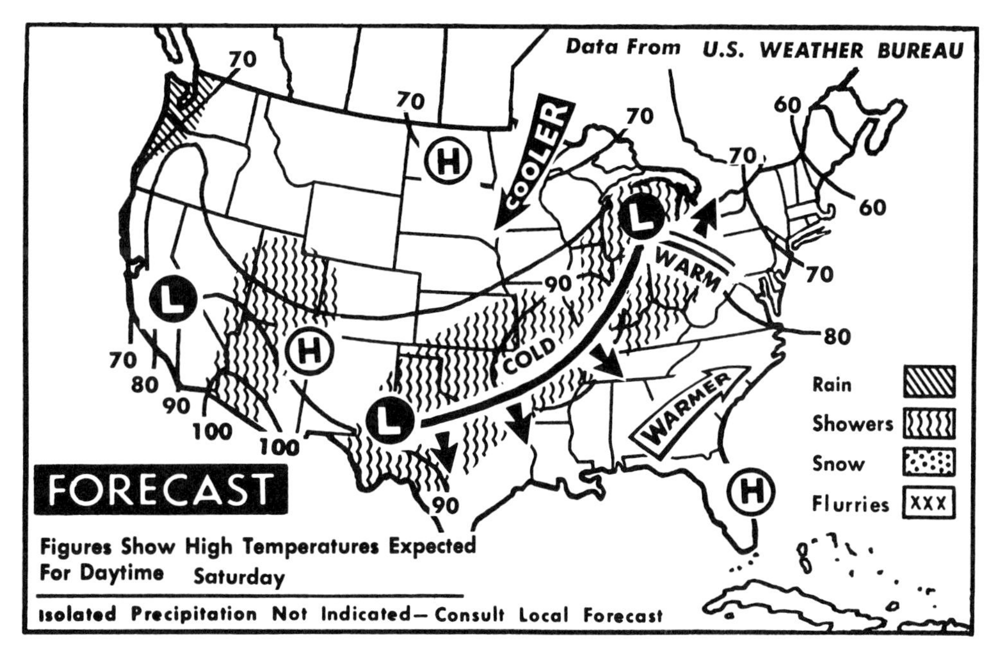

Fig. 11.6 Map-type chart showing weather bureau data. (Courtesy U.S. Weather Bureau)

Flow Charts. Included in this category are organization charts, process or operations charts, and progress charts. *Organization charts* show the structure of a company and the relationship among its components. An example is shown in Fig. 11.7. *Process* or *operations charts* show the steps, for example, in converting materials to a desired end product; or in solving a problem by a computer for which the problem has been programmed. (See Figs. 11.8 and 11.9.)

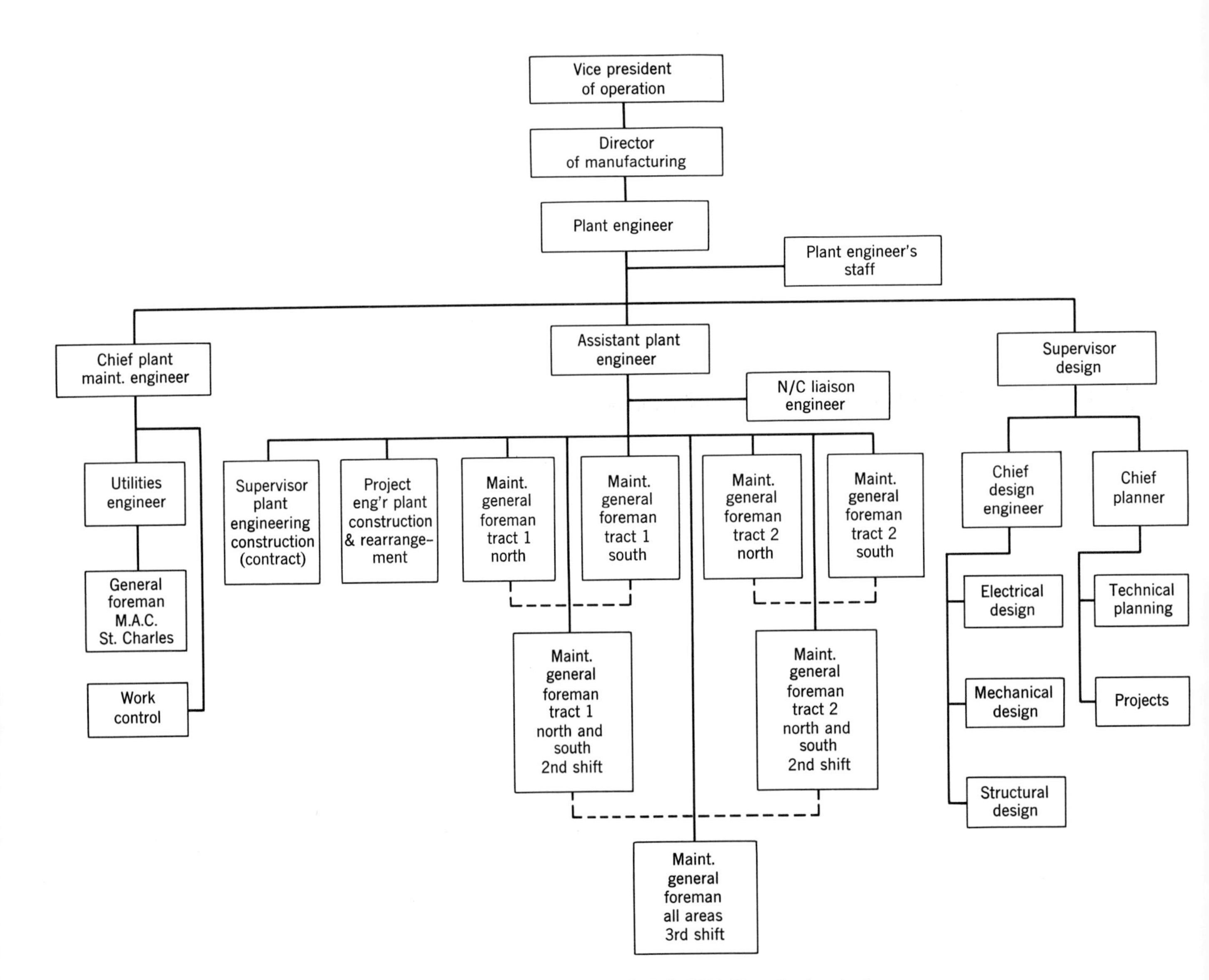

Fig. 11.7 Organization chart. Ref.: Feb. 1966, *Plant Engineering*)

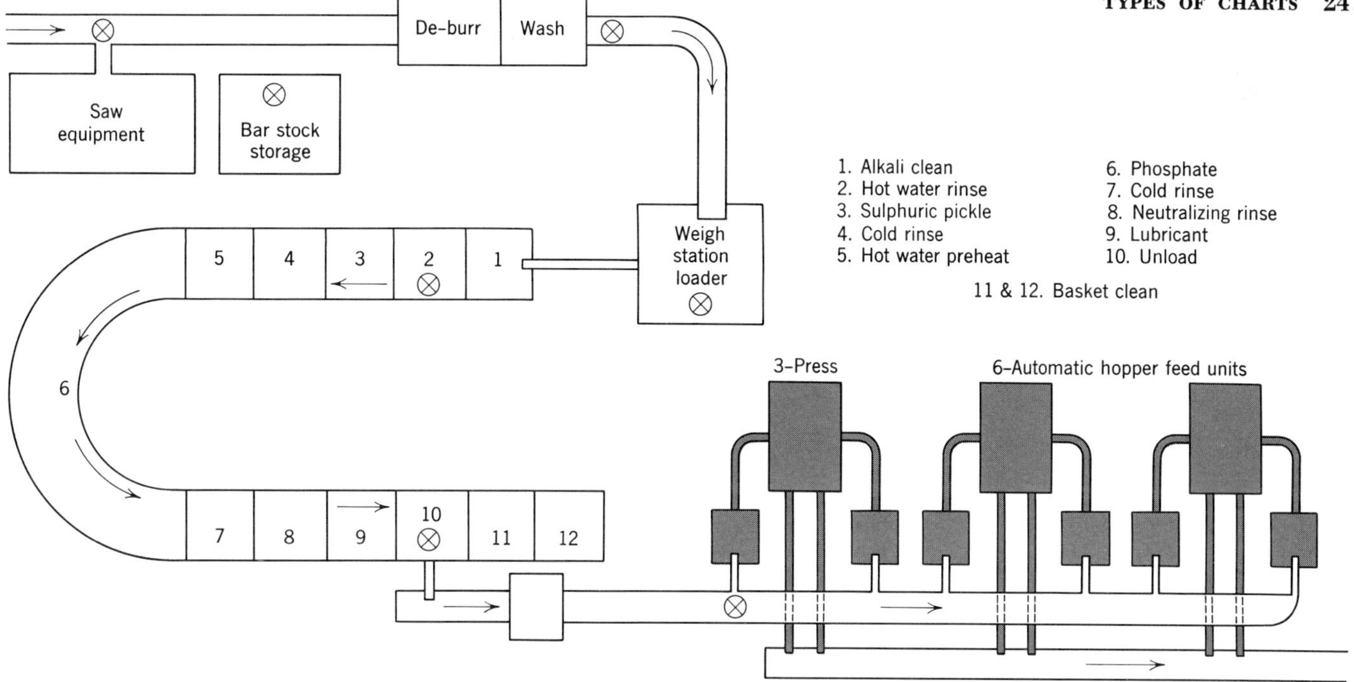

Fig. 11.8 Process flowsheet. (Source: Republic Steel, 1965)

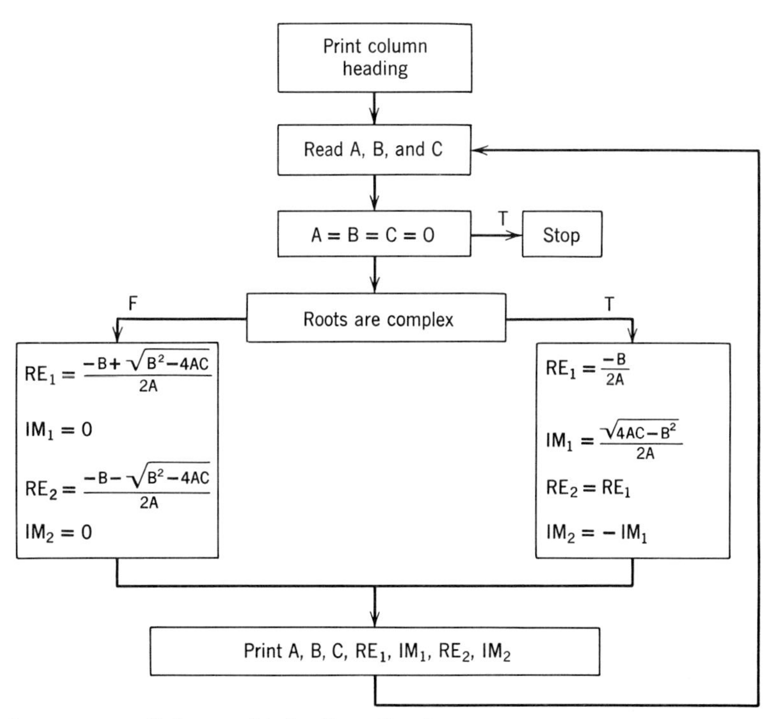

Fig. 11.9 Flow chart for program to find roots of $Ax^2 + Bx + C = O$.

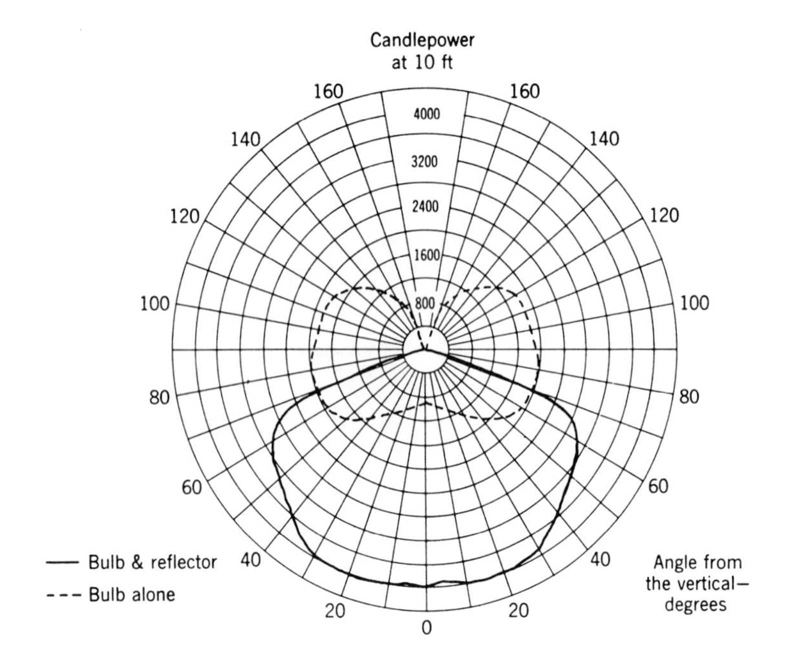

Fig. 11.10

Polar Charts. These charts are drawn on polar coordinate paper which usually consists of equally spaced concentric circles and radii passing through the pole or center of the circles. Recording devices make possible the plot of continuous curves, each point of which results from the simultaneous measurement of a linear distance from the pole and the corresponding angle expressed in degrees or radians. An example of a polar coordinate chart is shown in Fig. 11.10.

Trilinear Charts. These charts are in the shape of an equilateral triangle. It can be shown that the sum of the perpendiculars, drawn from a point within the triangle, to the sides of the triangle is equal to the altitude of the triangle. Now, when we graduate each side of the triangle in equal divisions ranging from 0% to 100%, as shown in Fig. 11.11, and regard each side as representing a variable, we have a means for determining the percentages of each of the three variables that will make their sum equal 100%.

Note carefully that for any point, such as *P*, the perpendicular distance, *PD*, represents 30% of substance *A*; the perpendicular distance, *PE*, represents 20% of substance *C*; and the perpendicular distance, *PF*, represents 50% of substance *B*; and that their sum equals 100%, which is represented by an altitude of the equilateral triangle.

When certain combinations of the three substances, resulting from percentages of each (totaling 100%), exhibit a common characteristic—for example, that they weigh the same, say, 1000 lb—we could establish the contour, 1000 lb, on the chart. Perhaps a simpler method would be to take an arbitrary percentage of each substance, assume that their total is 100%, then weigh the combination and plot the point which represents the percentage of each substance. This point would be marked with the weight of the combination of the three substances. In a similar manner additional points could be located. Finally, contours would be drawn to establish desired weight increments. This process is virtually the same as that used in establishing contours on a topographic map, or in determining the location of "isobars" from barometric pressure data.

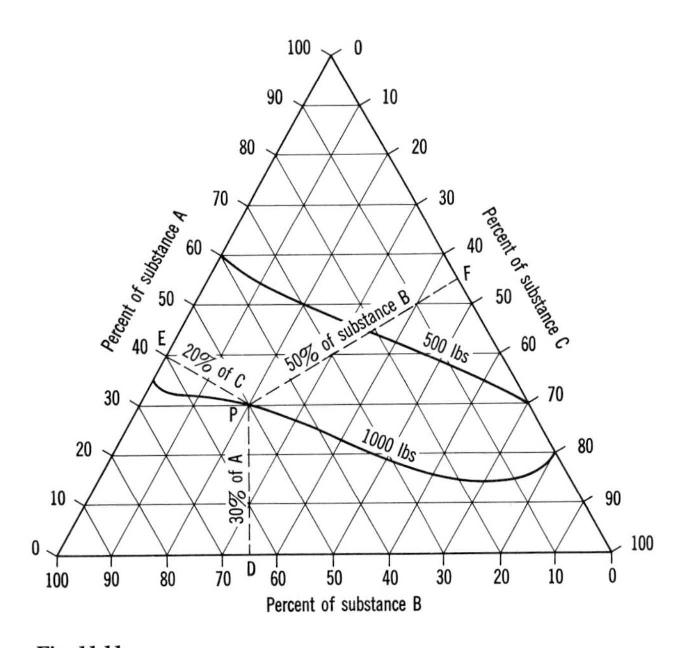

Fig. 11.11

Rectangular Coordinate Charts. Engineers and scientists make frequent use of rectangular coordinate charts. Laboratory experiments quite often deal with the simultaneous measurement of two related quantities. Study and analysis of the relation between the quantities are greatly facilitated by a graphical plot of the data. Also, the meaning of mathematical expressions of the form $y = f(x)$ is often enhanced by a graphical plot of the equation. In many instances the rectangular coordinate-type chart is used to present information of general interest. Much time can be saved by taking advantage of printed coordinate sheets that are commercially available. The sheets that are commonly used consist of (*a*) equally spaced horizontal and vertical lines; (*b*) logarithmic grids in which the spacing of the horizontals and verticals is logarithmic (proportional to the logarithms of numbers in one or more cycles; i.e., 1 to 10 in both x and y directions; or, in some cases, two cycles in the y-direction and one in the x-direction); (*c*) semi-logarithmic grids, where the verticals are equally spaced and the horizontals are spaced logarithmically; and (*d*) several special types of grids in which the spacing of the horizontals and verticals follow a prescribed function; i.e., x^2, or $x^{1/2}$, etc.

Preparation of Graphs. The basic steps in preparing graphs are:

1. *Examine the tabular data* and determine therefrom the ranges of each variable.
2. *Select a suitable commercial grid sheet* or, if necessary, prepare a grid to accommodate the ranges of each variable (including the zero lines if desired).
3. *Select a scale* that will convey the information to the best advantage.

   "The choice of scales is the most important and at the same time the most difficult step in chart construction.

   "Scale selection does more to shape the picture than all other steps combined. It can make the difference between an accurate picture; between a revealing presentation and a misleading one.

   "Scale selection is a subtle problem: the proper scale under one set of conditions may not be proper under other conditions; the proper scale for a chart shown alone may be improper for the same chart shown with other charts. Scales should never be chosen carelessly or left to chance: only rarely will you get a satisfactory scale by accident."*

4. *Place the origin* in the lower left portion of the sheet when the ranges of the variables are positive. Otherwise, locate the origin to accommodate both positive and negative values. It is customary to use the x-axis for the independent variable. There are, however, some exceptions; e.g., when a dependent variable, such as time, is usually plotted on the x-axis.
5. *The plotted data points should be identified by a symbol,* such as a small open circle. When more than one set of data is plotted on the same grid sheet, different symbols† should be used to identify each set.
6. *The drawn curve* connects the data points by straight lines when the data are discontinuous or do not follow an implicit relationship; otherwise, a smooth curve is drawn to balance the plotted points, so that some points are on the curve and others are near the curve. Do not draw the curve through the symbols, but stop at the boundary.
7. *The curve, or curves, should be properly identified by* suitable labels placed to advantage—in some cases on the curve; in others, near the curve.
8. *Appropriate titles and notes* should accompany the charts. The lettering should be placed so it is readable from the bottom and right side of the chart.
9. *If the chart is to be presented in a technical publication or used as a slide* in the oral presentation of a technical paper, it is suggested that reference be made to ASA Y15.1–1959, *Illustrations for Publication and Projection,* for many valuable suggestions concerning the preparation of charts for each purpose.

Several examples of rectangular coordinate charts are shown in Figs. 11.12 to 11.18. Further applications of the rectangular coordinate-type chart will be discussed in Chapter 14.

* Quoted from Section 3, "Scale Selection," ASA Y15.2–1960. This standard, *Time-Series Charts,* is highly recommended for good practice in the design of various types of charts.

† See p. 50, Fig. 5.2.5, ASA Y15.2–1960.

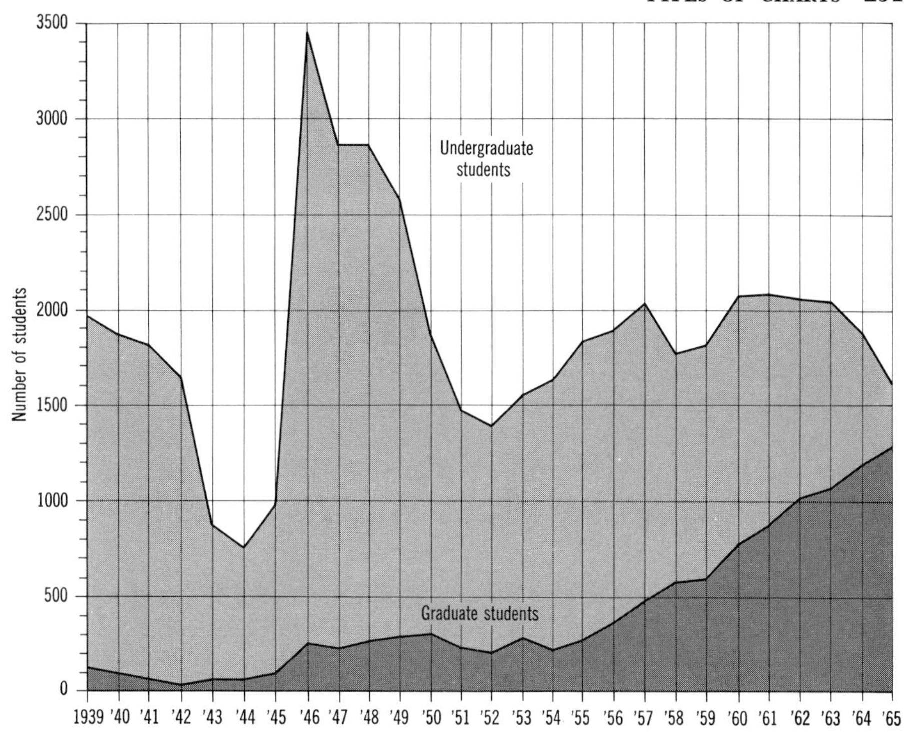

Fig. 11.12 Fall registration, College of Engineering, University of California at Berkeley.

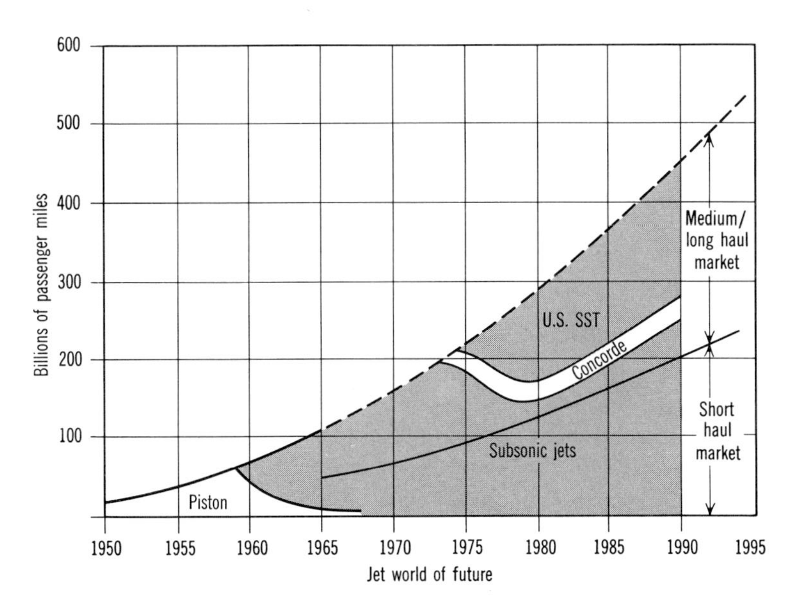

Fig. 11.13 (Source: *Los Angeles Times*)

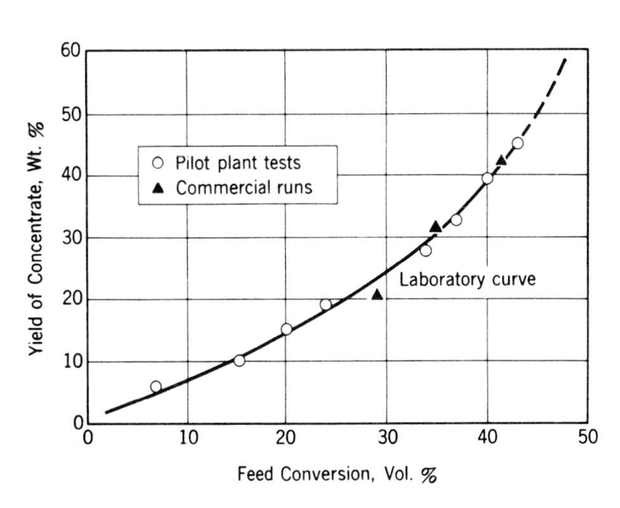

Fig. 11-14 Rectangular Cartesian chart. (From ASA *Standard* Y-15.1, 1959)

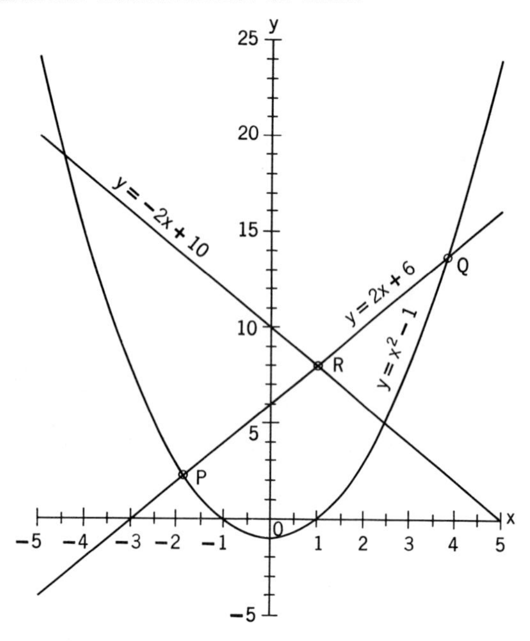

Fig. 11.15 Cartesian coordinate representation of equations: $y = x^2 - 1$; $y = 2x + 6$; and $y = -2x + 10$.

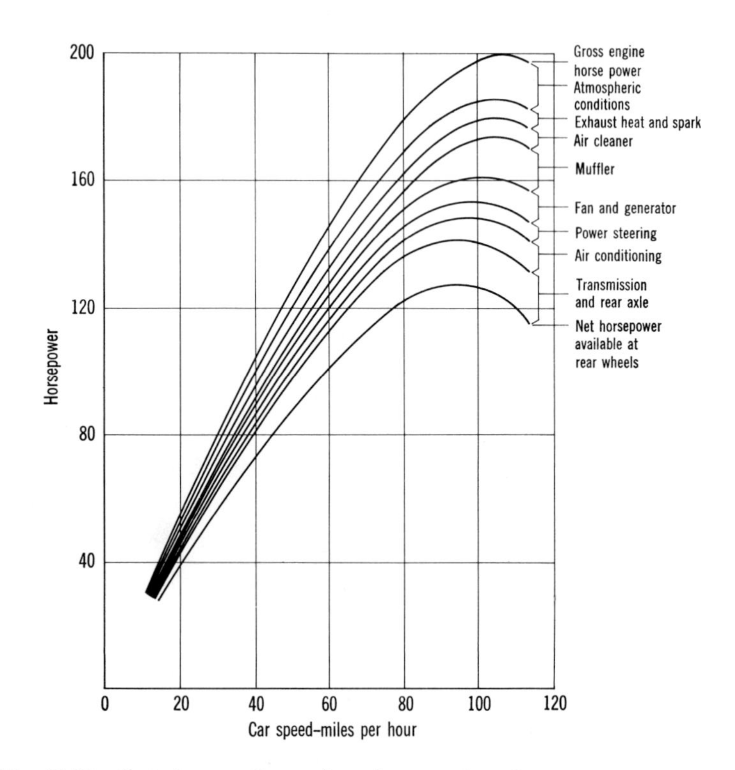

Fig. 11.16 Cartesian coordinate chart showing where horsepower goes. (Source: "Design for Safety," General Motors Corp., 1965)

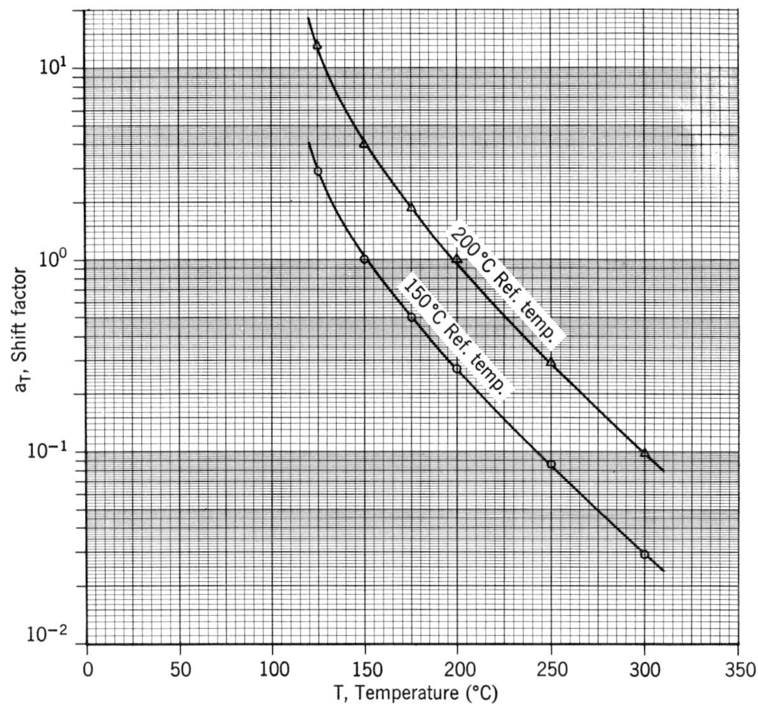

Fig. 11.18 Temperature-shear rate superposition shift factors versus temperature for both 150°C and 200°C reference temperatures. (Source: *SPE Transactionst*, Jan. 1965, p. 38)

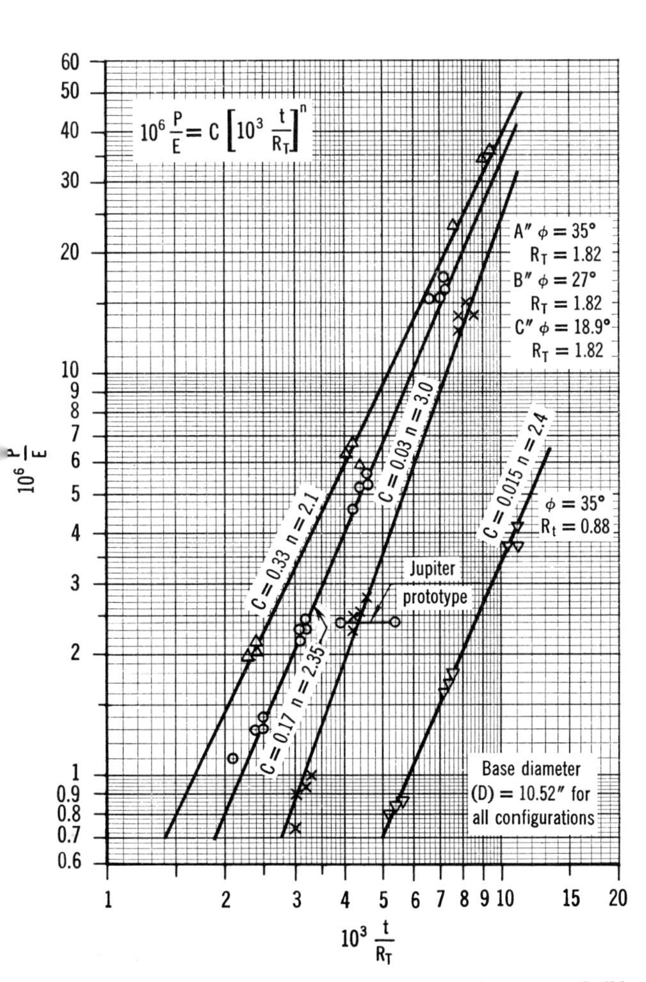

Fig. 11.17 Buckling pressure versus thickness for various bulkhead configurations. (Source: *Experimental Mechanics*, Aug. 1964)

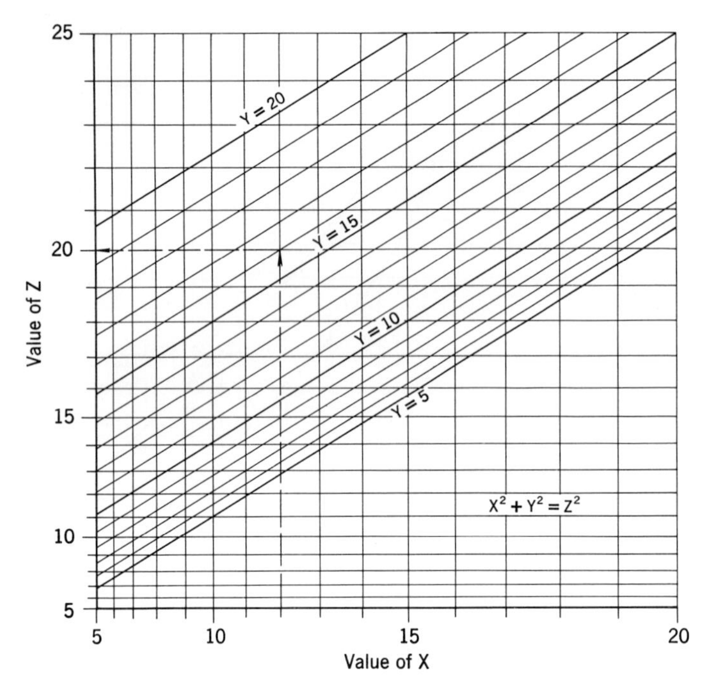

Fig. 11.19 Concurrency chart for $x^2 + y^2 = z^2$. Example: When $x = 12$ and $y = 16$, then $z = 20$.

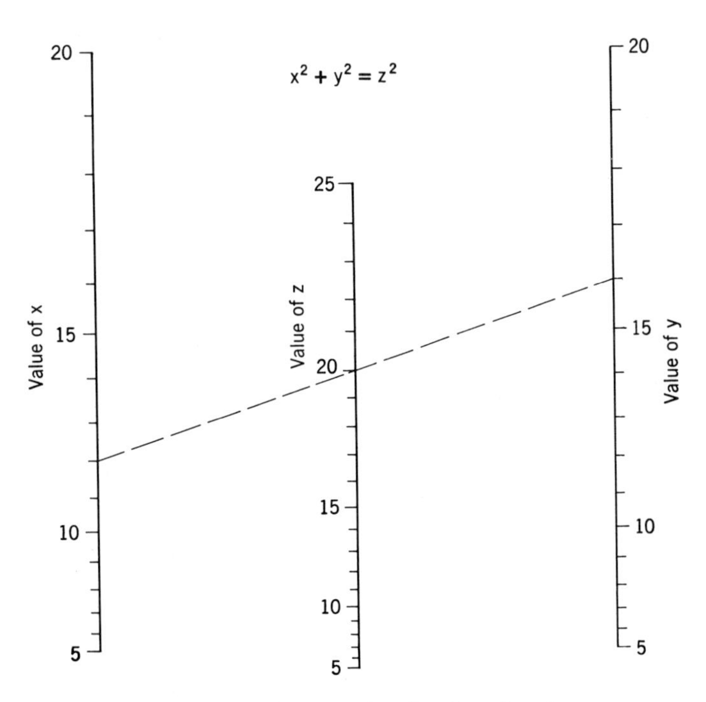

Fig. 11.20 Alignment chart for $x^2 + y^2 = z^2$.

Nomograms. Nomography is most useful in making repeated calculations of equations having three or more variables *and in studying the interrelationships among the variables.* Nomograms may be designed as rectangular coordinate charts (concurrency) or as alignment charts, examples of which are shown in Figs. 11.19 to 11.23.

Suffice it to say at this point that the importance of nomography has increased tremendously during the last decade. Applications of nomography have been made in all areas of engineering, the physical and biological sciences, medicine, ballistics, biomechanics, food technology, and business. Development of the theory and design of nomograms is presented in Chapter 16, but only an introductory coverage of the field of nomography* is intended there.

* A much fuller coverage of nomography is available in the author's textbook, *Nomography*, 2nd ed., John Wiley and Sons, New York, 1959.

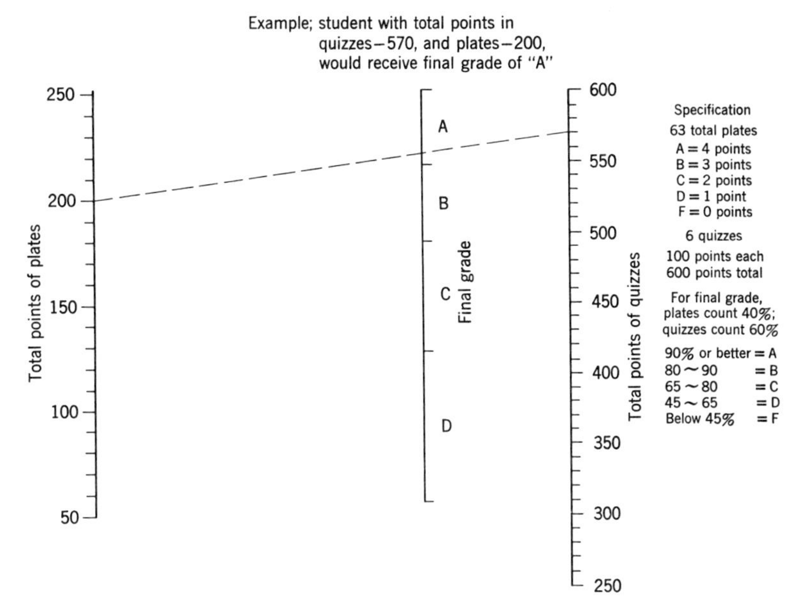

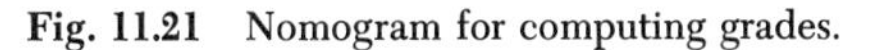
Fig. 11.21 Nomogram for computing grades.

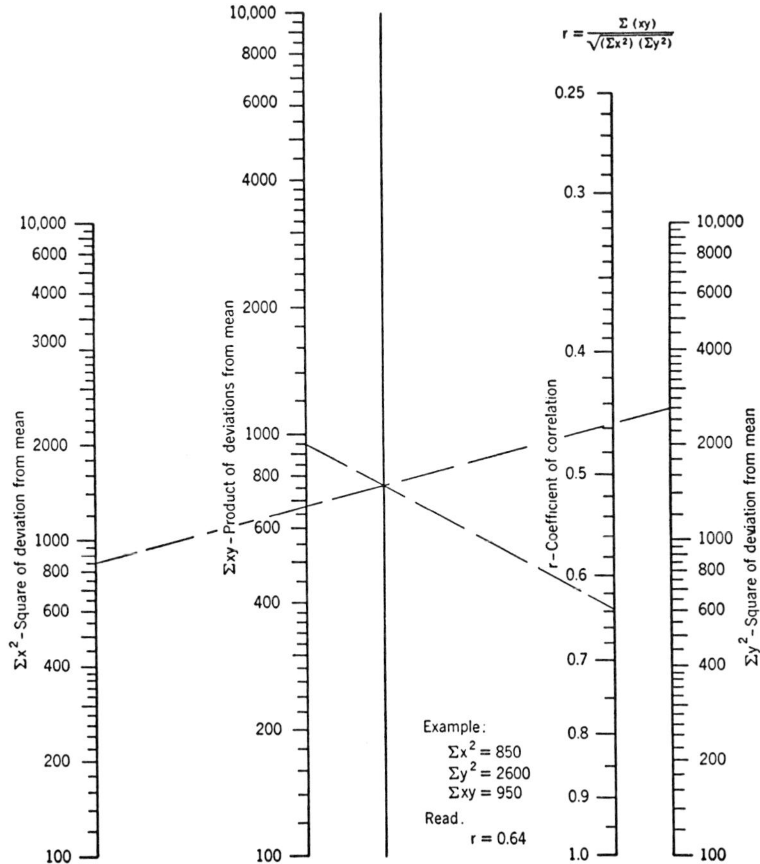

Fig. 11.22 Nomogram for coefficient of correlation. (See p. 221, Levens, *Nomography.*)

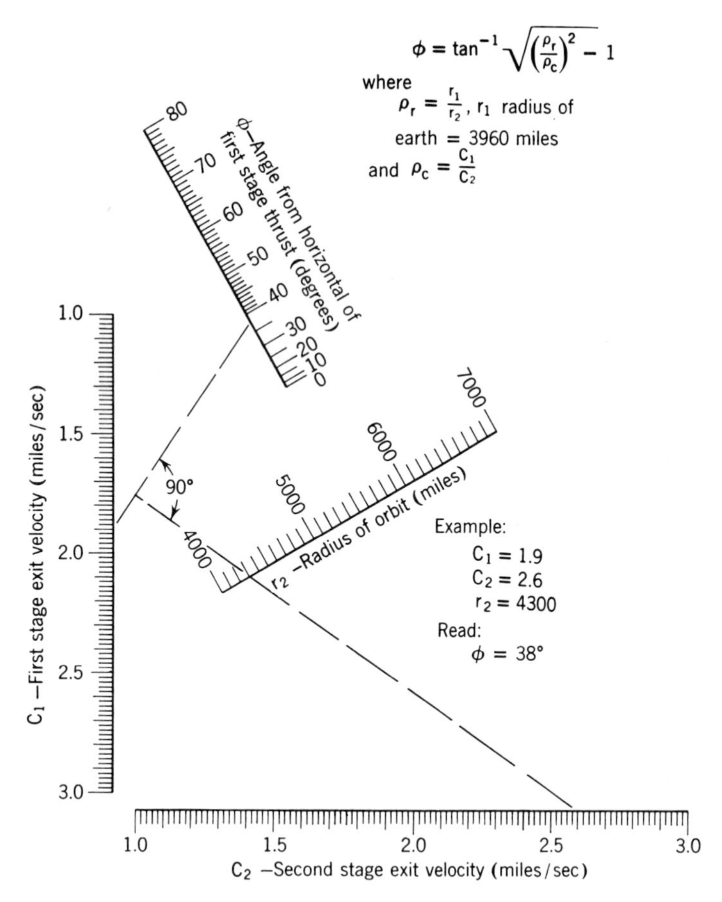

Fig. 11.23 Nomogram for optimum direction angle, ϕ, from horizontal of first-stage thrust (launch direction); for the case $\rho_c = \rho_r$. (See p. 275, Levens, *Nomography*.)

EXERCISES

1. The following data* show Motor Vehicle Factory Sales in U.S. plants for the years 1945 to 1964. Present the data graphically. Make a judgment as to the "best" form for presentation.

* *Source:* Automobile Manufacturers Association.

Motor Vehicle Factory Sales, U.S. Plants
(Value in thousands of dollars)

	Passenger Cars		Motor Trucks and Buses		Total	
	Number	Value	Number	Value	Number	Value
1900	4,192	$ 4,899	...	...	4,192	$ 4,899
1901	7,000	8,183	...	...	7,000	8,183
1902	9,000	10,395	...	...	9,000	10,395
1903	11,235	13,000	...	...	11,235	13,000
1904	22,130	23,358	700	$ 1,273	22,830	24,630
1905	24,250	38,670	750	1,330	25,000	40,000
1906	33,200	61,460	800	1,440	34,000	62,900
1907	43,000	91,620	1,000	1,780	44,000	93,400
1908	63,500	135,250	1,500	2,550	65,000	137,800
1909	123,990	159,766	3,297	5,334	127,287	165,099
1910	181,000	215,340	6,000	9,660	187,000	225,000
1911	199,319	225,000	10,681	21,000	210,000	246,000
1912	356,000	335,000	22,000	43,000	378,000	378,000
1913	461,500	399,902	23,500	44,000	485,000	443,902
1914	548,139	420,838	24,900	44,219	573,039	465,057
1915	895,930	575,978	74,000	125,800	969,930	701,778
1916	1,525,578	921,378	92,130	161,000	1,617,708	1,082,378
1917	1,745,792	1,053,506	128,157	220,983	1,873,949	1,274,488
1918	943,436	801,938	227,250	434,169	1,170,686	1,236,107
1919	1,651,625	1,365,395	224,731	371,423	1,876,356	1,736,818
1920	1,905,560	1,809,171	321,789	423,249	2,227,349	2,232,420
1921	1,468,067	1,038,191	148,052	166,071	1,616,119	1,204,262
1922	2,274,185	1,494,514	269,991	226,050	2,544,176	1,720,564
1923	3,624,717	2,196,272	409,295	308,538	4,034,012	2,504,810
1924	3,185,881	1,970,097	416,659	318,581	3,602,540	2,288,677
1925	3,735,171	2,458,370	530,659	458,400	4,265,830	2,916,770
1926	3,692,317	2,607,365	608,617	484,823	4,300,934	3,092,188
1927	2,936,533	2,164,671	464,793	420,131	3,401,326	2,584,802
1928	3,775,417	2,572,599	583,342	460,109	4,358,759	3,032,708
1929	4,455,178	2,790,614	881,909	622,534	5,337,087	3,413,148
1930	2,787,456	1,644,083	575,364	390,752	3,362,820	2,034,835
1931	1,948,164	1,108,247	432,262	265,445	2,380,426	1,373,691
1932	1,103,557	616,860	228,303	137,624	1,331,860	754,485
1933	1,560,599	773,425	329,218	175,381	1,889,817	948,806
1934	2,160,865	1,140,478	576,205	326,782	2,737,070	1,467,260
1935	3,273,874	1,707,836	697,367	380,997	3,971,241	2,088,834
1936	3,679,242	2,014,747	782,220	463,719	4,461,462	2,478,467
1937	3,929,203	2,240,913	891,016	537,315	4,820,219	2,778,227
1938	2,019,566	1,241,032	488,841	329,918	2,508,407	1,570,950
1939	2,888,512	1,770,232	700,377	489,787	3,588,889	2,260,018
1940	3,717,385	2,370,654	754,901	567,820	4,472,286	2,938,474
1941	3,779,682	2,567,206	1,060,820	1,069,800	4,840,502	3,637,006
1942	222,862	163,814	818,662	1,427,457	1,041,524	1,591,270
1943	139	102	699,689	1,451,794	699,828	1,451,896
1944	610	447	737,524	1,700,929	738,134	1,701,376
1945	69,532	57,255	655,683	1,181,956	725,215	1,239,210
1946	2,148,699	1,979,781	940,963	1,043,247	3,089,662	3,023,028
1947	3,558,178	3,936,017	1,239,443	1,731,713	4,797,621	5,667,730
1948	3,909,270	4,870,423	1,376,274	1,880,415	5,285,544	6,750,898
1949	5,119,466	6,650,857	1,134,185	1,394,035	6,253,651	8,044,892
1950	6,665,863	8,468,137	1,337,193	1,707,748	8,003,056	10,175,885
1951	5,338,435	7,241,275	1,426,828	2,323,859	6,765,263	9,565,134
1952	4,320,794	6,455,114	1,218,165	2,319,789	5,538,959	8,774,903
1953	6,116,948	9,002,580	1,206,266	2,089,060	7,323,214	11,091,640
1954	5,558,897	8,218,094	1,042,174	1,660,019	6,601,071	9,878,113
1955	7,920,186	12,452,871	1,249,106	2,020,973	9,169,292	14,473,844
1956	5,816,109	9,754,971	1,104,481	2,077,432	6,920,590	11,832,403
1957	6,113,344	11,198,379	1,107,176	2,082,723	7,220,520	13,281,102
1958	4,257,812	8,010,366	877,294	1,730,027	5,135,106	9,740,393
1959	5,591,243	10,534,421	1,137,386	2,338,719	6,728,629	12,873,140
1960	6,674,796	12,164,234	1,194,475	2,350,680	7,869,271	14,514,914
1961	5,542,707	10,285,777	1,133,804	2,155,753	6,676,511	12,441,530
1962	6,933,240	13,071,709	1,240,168	2,581,756	8,173,408	15,653,465
1963	7,637,728	14,427,077	1,462,708	3,090,345	9,100,436	17,517,422
1964	7,751,822	14,860,000	1,540,453	3,095,000	9,292,275	17,955,000

NOTE: A substantial proportion of the trucks and buses consists of chassis only; therefore the value of the bodies for these chassis is not included. Value is based on vehicles with standard equipment. Prior to July 1, 1964, certain firms included tactical vehicles in factory sales data. After July 1, 1964, all tactical vehicles are excluded. Federal excise taxes are excluded.

2. The following data show Monthly Motor Vehicle Production for the period 1958 to 1964. Present the data graphically. Make a judgment as to the "best" form for presentation.

Monthly Motor Vehicle Production

	1958	1959	1960	1961	1962	1963	1964
Jan.	489,841	546,319	688,770	416,111	628,706	688,074	745,835
Feb.	392,644	479,086	660,091	364,889	536,314	601,702	675,581
Mar.	357,443	576,409	655,127	408,536	603,358	648,058	723,811
Apr.	317,005	579,339	583,660	447,345	617,708	691,839	786,824
May	349,946	547,397	612,127	542,867	674,109	715,415	726,007
June	337,656	558,306	613,754	560,000	564,828	689,996	777,595
July	321,228	555,420	434,743	399,587	589,769	655,669	587,292
Aug.	180,584	239,152	306,023	195,866	196,300	156,976	190,159
Sept.	130,337	258,220	408,305	355,202	471,177	505,232	573,420
Oct.	261,945	508,297	618,539	557,808	724,239	799,375	411,501
Nov.	514,535	255,352	598,068	646,507	687,971	747,103	680,835
Dec.	594,263	496,195	523,901	627,301	648,855	744,938	866,632
Total	4,247,427	5,599,492	6,703,108	5,522,019	6,943,334	7,644,377	7,745,492

3. The following data show New Motor Vehicle Registrations during the years 1958 to 1964. Present the data graphically. Make a judgment as to the "best" form for presentation.

New Motor Vehicle Registration

	1958	1959	1960	1961	1962	1963	1964
Jan.	382,240	420,751	430,116	413,563	491,683	553,852	612,032
Feb.	333,818	425,095	494,178	374,877	475,365	497,978	551,844
Mar.	400,763	497,651	596,669	480,067	611,072	624,158	636,880
Apr.	418,598	574,922	647,287	496,059	640,449	758,755	812,327
May	423,753	583,459	647,055	543,975	657,839	714,688	780,576
June	411,017	585,932	599,864	571,953	625,772	691,624	754,321
July	406,265	566,453	546,535	500,534	613,599	705,992	724,208
Aug.	375,373	533,636	525,400	470,646	540,180	552,861	648,682
Sept.	321,223	458,434	458,765	370,505	373,943	403,624	565,439
Oct.	324,942	534,847	547,461	549,624	677,673	714,680	658,457
Nov.	338,688	428,306	543,042	557,894	637,475	640,171	563,523
Dec.	517,834	430,830	544,278	525,690	644,442	711,970	756,828
Total	4,654,514	6,041,275	6,576,650	5,854,747	6,938,863	7,556,717	8,065,150

4. The following data show Motor Vehicle Registrations in several states. Present the data graphically. Make a judgment as to the "best" form for presentation.

Motor Vehicle Registrations

State	1963	Prelim. 1964	State	1963	Prelim. 1964
Alabama	1,187,582	1,264,000	Minnesota	1,417,382	1,461,000
California	7,749,114	8,085,000	Missouri	1,539,804	1,549,000
Connecticut	1,132,891	1,199,000	New Jersey	2,439,864	2,532,000
Florida	2,381,452	2,484,000	New York	4,942,780	5,107,000
Georgia	1,442,735	1,529,000	North Carolina	1,546,384	1,628,000
Illinois	3,606,672	3,748,000	Ohio	3,982,192	4,158,000
Indiana	1,843,139	1,890,000	Pennsylvania	4,034,588	4,193,000
Iowa	1,152,577	1,193,000	Tennessee	1,231,600	1,284,000
Kentucky	1,073,847	1,128,000	Texas	4,010,925	4,216,000
Louisiana	1,041,831	1,089,000	Virginia	1,389,481	1,446,000
Maryland	1,151,909	1,211,000	Washington	1,231,238	1,267,000
Massachusetts	1,773,713	1,824,000	Wisconsin	1,438,988	1,486,000
Michigan	3,160,610	3,318,000			
Totals				57,903,298	60,289,000

5. The following data show Average Age of Passenger Cars and Trucks in U.S. Present the data graphically. Make a judgment as to the "best" form for presentation.

Average Age of Passenger Cars and Trucks

	Passenger Cars	Motor Trucks		Passenger Cars	Motor Trucks
1938	5.5 years	. . .	1953	6.5 years	6.6 years
1939	5.6	. . .	1954	6.2	6.6
1940	5.7	5.8 years	1955	5.9	6.7
1941	5.5	5.6	1956	5.6	6.8
1944	7.3	7.6	1957	5.5	7.0
1946	9.0	8.6	1958	5.6	7.2
1947	8.9	8.1	1959	5.8	7.5
1948	8.8	7.8	1960	5.9	7.7
1949	8.5	7.4	1961	6.0	7.9
1950	7.8	7.0	1962	6.0	8.0
1951	7.1	6.6	1963	6.0	8.1
1952	6.8	6.6	1964		

6. The following data show Traffic Fatalities and Vehicle Miles, 1955 to 1964. Present the data graphically. Make a judgment as to the "best" form for presentation.

Traffic Fatalities and Vehicle Miles*

		Deaths from Collision with:							Total Death Rates		
	All Deaths†	Deaths from Noncollision Accidents	Pedestrians	Other Motor Vehicles	Railroad Trains	Street Cars	Bicycles and Horsedrawn Vehicles	Fixed Objects	Per 100,000 Population	Per 10,000 Motor Vehicles	Per 100,000,000 Vehicle Miles
1955	38,426	12,100	8,200	14,500	1,490	15	500	1,600	23.4	6.1	6.4
1956	39,628	13,000	7,900	15,200	1,377	11	540	1,600	23.7	6.1	6.3
1957	38,702	11,800	7,850	15,400	1,376	13	540	1,700	22.7	5.8	6.0
1958	36,981	11,600	7,650	14,200	1,316	9	530	1,650	21.3	5.4	5.6
1959	37,910	11,800	7,850	14,900	1,202	6	550	1,600	21.5	5.3	5.4
1960	38,137	11,900	7,850	14,800	1,368	5	540	1,700	21.2	5.2	5.3
1961	38,091	12,200	7,650	14,700	1,267	5	570	1,700	20.8	5.0	5.2
1962	40,804	12,900	7,900	16,400	1,245	0	590	1,800	22.0	5.1	5.3
1963	43,564	13,900	8,200	17,600	1,340	10	650	1,900	23.1	5.3	5.4
1964 (Est.)	47,800	15,000	8,900	19,600	1,500	5	800	2,000	25.0	5.5	5.7

* *Source:* "Accident Facts," National Safety Council.

† Totals do not quite equal the sum of the various types because the estimates were generally made only to the nearest 10 deaths, and to the nearest 50 deaths of certain types.

7. Present the following land-use data in graphical form:

Streets	25.9%	Commerce	3.9%
Public	14.3%	Residence	48.1%
Industry	5.6%	Vacant Land	2.2%

8. Present the following temperature data in graphical form:

Month	Mean Temperature	Month	Mean Temperature
Jan.	48°F	July	64°F
Feb.	54°F	Aug.	64°F
Mar.	58°F	Sept.	67°F
Apr.	59°F	Oct.	63°F
May	60°F	Nov.	58°F
June	62°F	Dec.	51°F

9. Present the following "cost-of-industrial-gas" data* in graphical form:

Interruptible Service			Firm Service		
Monthly Delivery Therms†	Average Per Therm	Average Per MCF	Monthly Delivery Therms†	Average Per Therm	Average Per MCF
500,000	3.7¢	42.5¢	500,000‡	5.1¢	57.1¢
100,000	4.4¢	49.4¢	100,000‡	5.2¢	57.5¢
50,000	4.6¢	50.7¢	50,000‡	5.2¢	58.0¢
25,000	4.7¢	52.2¢	25,000	5.3¢	58.8¢
5,000	4.9¢	54.6¢	5,000	5.4¢	59.8¢

* *California State Chamber of Commerce Standard Industrial Survey Summary Report,* March 1965.

† 1 Therm—100,000 B.t.u. B.t.u. content per cubic foot: 1,100

‡ Subject to availability.

10. Present the following "cost-of-electric-power" data in graphical form:

Maximum Demand	Monthly Usage KWH	Net Mo. Elec. Bill*	Av. Cost Per KWH	Monthly Usage KWH	Net Mo. Elec. Bill*	Av. Cost Per KWH
5000 KW	1,000,000	$12,844.60	1.284¢	2,000,000	$19,919.20	0.996¢
2500 KW	500,000	6,657.10	1.331¢	1,000,000	10,572.70	1.057¢
1000 KW	200,000	2,944.60	1.472¢	400,000	4,520.20	1.130¢
500 KW	100,000	1,592.40	1.592¢	200,000	2,410.20	1.205¢
300 KW	60,000	986.40	1.644¢	120,000	1,492.20	1.244¢
150 KW	30,000	531.90	1.773¢	60,000	803.70	1.340¢

* Subject to discount for primary service.

11. Visit the Gas and Electric Company of your city. Obtain information about (*a*) kilowatt-hour output each year over the past decade; (*b*) gas and electric rates for householders over the same decade; and (*c*) number of employees each year of the same decade. Present the information in the "best" graphical form.

12. Visit your City Engineer's office and obtain information about (*a*) the location of parks; (*b*) the location of libraries; and (*c*) the location of schools. Prepare a map presentation of (*a*), (*b*), and (*c*).

13. Plot the following experimental data and draw the curve which best represents the plotted points:

L, load in pounds	8,000	12,000	14,000	16,500	19,000	20,500	20,000	17,000
E, elongation in inches	0	0.29	0.39	0.58	0.74	0.88	1.18	1.47

Plot E along the x-axis. Choose suitable scales. Graduate and label each scale.

14. Plot the following test data and draw the curve which is most representative of the plotted points:

I, indicated horsepower	1	2	4	6	8	10	12	14	20
P, pounds of steam per hour	44	60	91	119	153	188	221	265	308

I is the independent variable (plot along the x-axis). Choose suitable scales for I and P. Graduate and label each scale.

15. Plot the following data and draw the "best" curve to represent the data:

E, in feet	0	950	2800	4800	6500	10,600
P, inches of mercury	29	28	26	24	22	19

P represents barometric pressure and E the elevation in feet above sea level. Graduate and label the scales.

16. Plot the following equations:

(a) $y = 2X^2 - 2$ $\quad$ X varies from 0 to 6

(b) $y = \dfrac{X^3 + 3}{X - 1}$ $\quad$ X varies from 2 to 5

17. Plot the equation $V = \frac{4}{3}\pi r^3$ where V = the volume of a sphere whose radius varies from 0 to 10 inches.

18. Plot the equation $y = 2 \log X$, where X varies from 0.1 to 20.

19. Plot the equation $h = v^2/2g$, where h = velocity head; v = velocity in feet per second (10 to 50), and g = 32.2 ft/sec^2.

20. Plot the equation $S = 4\pi r^2$, where r = the radius of a sphere (0 to 10 inches) and S = the area of the surface of the sphere in square inches.

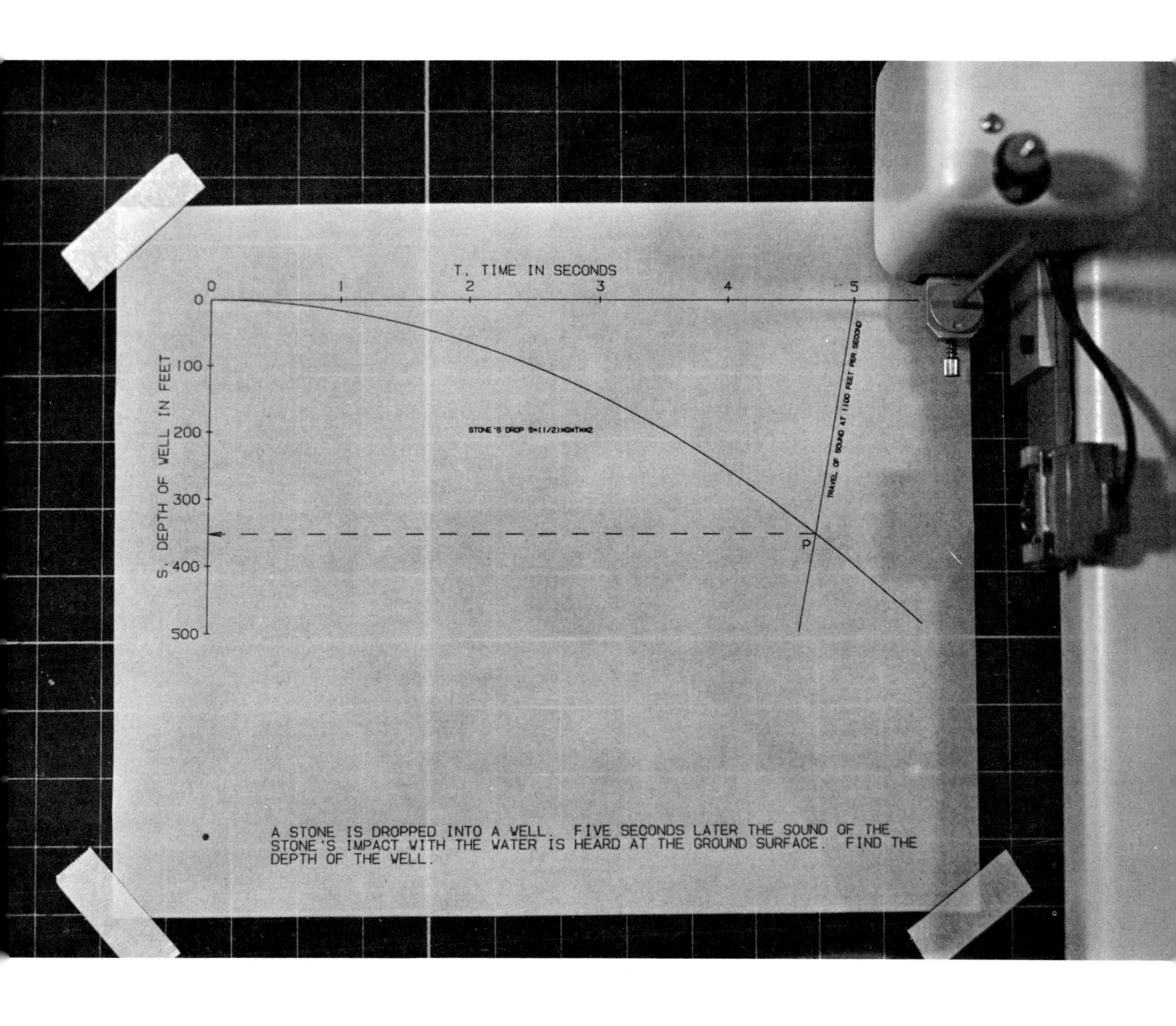

GRAPHICAL MATHEMATICS— ARITHMETIC AND ALGEBRA 12

Let us now consider graphical methods that may be employed in the solution of arithmetic and algebraic problems.

GRAPHICAL ARITHMETIC

A number or quantity can be easily represented by a geometrical magnitude, such as a length laid off to an appropriate scale, or by an angle which can be laid off by a protractor or by the *chord method* which will be described in Example 3.

Graphical Addition and Subtraction

EXAMPLE 1

Suppose we wish to find the sum of the numbers 7, 3, 8, and 10. First we draw a fine line and then lay off, to a convenient scale, distances AB, BC, CD, and DE, respectively equal to the given numbers. The distance from A to E, measured in terms of the selected scale, is the sum of the numbers. Figure 12.1 shows the method used in laying off the segments $AB, \ldots, DE$.

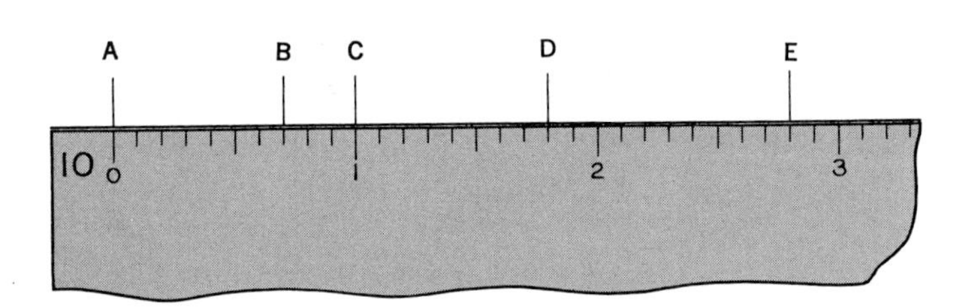

Fig. 12.1

A sharp-pointed pencil, or a needle point fitted to a convenient holder, should be used in marking points A, B, C, D, and E.

Figure 12.2 shows the use of a needle point and holder in laying off distances.

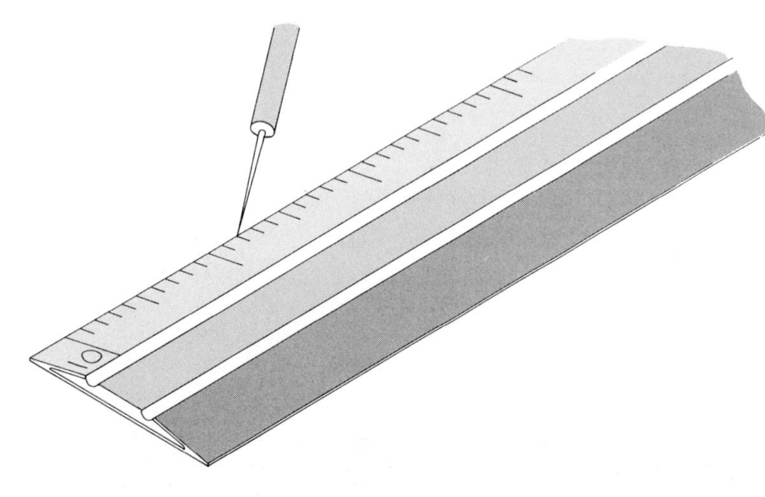

Fig. 12.2

Distances that are set off with a fine metal point can be read to the nearest 0.01 in. without difficulty by using the 50-scale and estimating hundredths. The accuracy of a graphical solution will generally increase with an increase of the scale used. Good, reliable results can be obtained if good equipment is used and if care is exercised in laying off distances.

It should be recognized that *graphical subtraction* is actually the same process; namely, the addition of negative quantities. Graphically, this simply means that attention must be given to *direction.* If we agree to lay off positive values to the right, then negative quantities are laid off to the left. For example, the sum of 5, 3, -4, and 6 can be determined graphically in the following manner. Lay off, to a convenient scale, segments AB, BC, CD, and DE, respectively equal to the given numbers (Fig. 12.3). Note carefully that segments AB and BC were first laid off to the right, that segment CD was laid off to the left (negatively for subtraction), and finally seg-

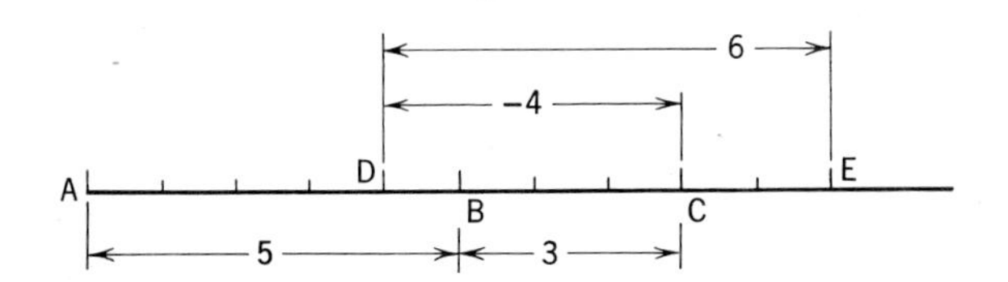

Fig. 12.3

ment DE to the right again. Segment AE, in this case 10 units long, represents the sum of the given numbers.

EXAMPLE 2

Suppose we wish to solve the equation, $a + b = c$. For given values of a and b we can find the value of c in a manner similar to that used in Example 1. A simpler method is shown in Fig. 12.4. We can construct a "slide rule" with values of a represented by the scale on the upper portion of the slide rule, and values of b on the movable scale. Assume that $a = 5$ and $b = 10$. We place the zero reading of the b scale opposite the 5 on the a scale. Now we add $b = 10$ to $a = 5$ and read $c = 15$. Note that the distance from 0 to 5 on the a scale represents $a = 5$, and that the distance from 0 to 10 on the b scale represents $b = 10$. Therefore, the distance from 0 to 15 represents $c = 15$. Values of a and c appear on the same scale.

If $a = 3$ and $b = 6$, we simply move the movable scale (b) to the left until the zero on the b scale is under $a = 3$. Then locate the value $b = 6$, and read $c = 9$, directly above the value $b = 6$.

Question. Could you use the slide rule to subtract? For example, try $a = 10$ and $b = 7$ in the equation $a - b = c$. Later, when we discuss functional scales, you will have sufficient knowledge to construct simple slide rules for such equations as $a^2 + b^2 = c^2$, and $ab = c$.

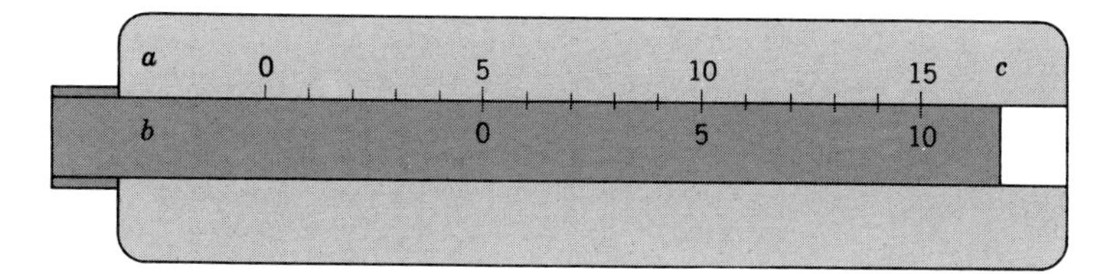

Fig. 12.4

EXAMPLE 3

Suppose we wish to lay off an angle of 20°. We could use a protractor to do so; or, better, we could use the *chord method*. For the latter purpose we use the table of chords shown in the Appendix on page 645. In Fig. 12.5 we lay off on line AB a distance of, say, 3 in.—distance AC. The tabular value opposite 20° is 0.3473. This value is multiplied by 3, yielding 1.0419. Now, with C as center and radius 1.04 inches, an arc is drawn. A second arc, with center A and radius 3 inches, is drawn to intersect the first arc at point D. Line AD is the line which forms an angle of 20° with line AB.

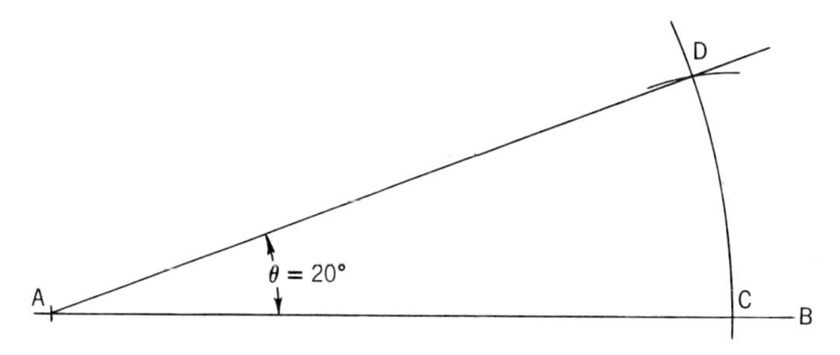

Fig. 12.5

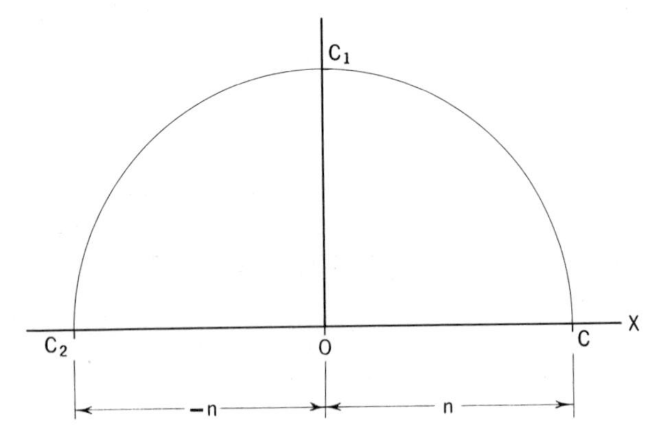

Fig. 12.6

Graphic Representation of Complex Numbers

First consider the graphic representation of complex (imaginary) numbers. In Fig. 12.6, point C is taken n units to the right of the origin. Line segment OC represents the number n. Segment OC_2, laid off in the negative direction, represents $-n$. We observe that line OC can be brought into position OC_2 by a counterclockwise rotation through an angle of 180° about the origin, O, as center. We also observe that, algebraically, n has been transformed into $-n$ by multiplying by -1. We conclude that *the multiplication of a real number by -1 is shown graphically by a counterclockwise rotation through an angle of 180°.*

EXAMPLE 1

Now let us consider the algebraic meaning of a rotation through an angle of 90°, where the rotation represents the multiplication by an unknown number p.

Therefore,	$OC_1 = n \cdot p$
and	$OC_2 = OC_1 \cdot p = n \cdot p^2$
or	$-n = np^2$
and hence,	$p = \sqrt{-1}$
also,	$OC_1 = n\sqrt{-1}$

It is now apparent that multiplication by $\sqrt{-1}$ is represented by a counterclockwise rotation through an angle of 90°.

EXAMPLE 2

Let us consider (Fig. 12.7) the algebraic meaning of a rotation through an angle of 45°, where the rotation represents the multiplication by an unknown number p.

Therefore,	$OC_1 = n \cdot p$
and	$OC_2 = OC_1 \cdot p = n \cdot p^2$
and	$OC_3 = OC_2 \cdot p = n \cdot p^3$
and	$OC_4 = OC_3 \cdot p = n \cdot p^4$
or	$-n = n \cdot p^4$
from which	$p = \sqrt[4]{-1}$

We now see that multiplication by $\sqrt[4]{-1}$ is represented by a counterclockwise rotation through an angle of 45°.

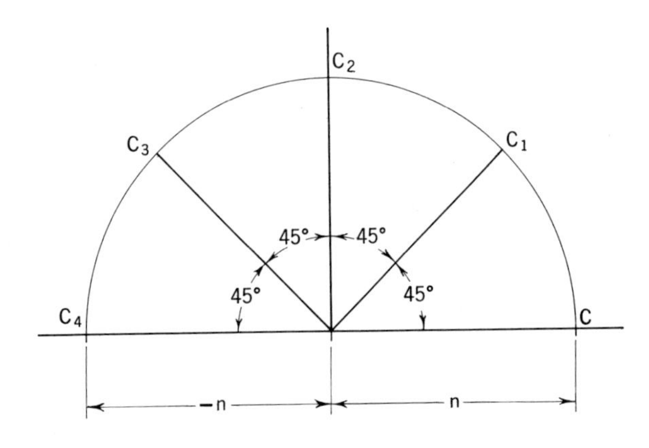

Fig. 12.7

EXAMPLE 3

Let us determine, graphically, the different values of $\sqrt[6]{1}$.

In Fig. 12.8, lines OC, OC_1, OC_2, OC_3, OC_4, and OC_5 have been drawn to form angles of 60° between adjacent lines. When OC lies on the axis of real numbers, lines OC, OC_1, OC_2, OC_3, OC_4, and OC_5 represent the six values of $\sqrt[6]{1}$. It is readily seen that the six values are:

$$
\begin{aligned}
OC &= 1\\
OC_1 &= \tfrac{1}{2} + \tfrac{1}{2}\sqrt{3}\,i\\
OC_2 &= -\tfrac{1}{2} + \tfrac{1}{2}\sqrt{3}\,i\\
OC_3 &= -1\\
OC_4 &= -\tfrac{1}{2} - \tfrac{1}{2}\sqrt{3}\,i\\
OC_5 &= \tfrac{1}{2} - \tfrac{1}{2}\sqrt{3}\,i
\end{aligned}
$$

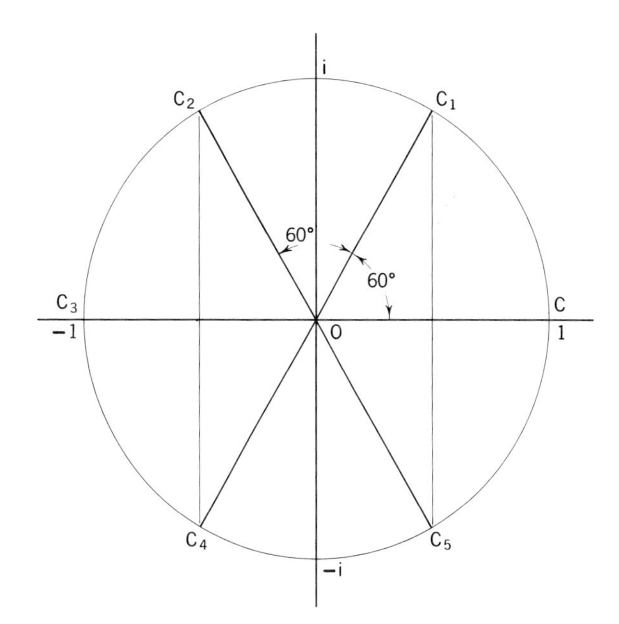

Fig. 12.8

Graphically we can now represent complex numbers by setting up a coordinate system in which the x-axis is graduated to represent values of the real numbers, and the y-axis, values of the imaginary part.

EXAMPLE

Let us represent the complex numbers $(5 + 4i)$ and $(-3 - 3i)$ graphically. In Fig. 12.9 we lay off $OC = 5$ and $OC_1 = 4i$. The horizontal line through C_1 intersects the vertical line through C at point A. OA represents the sum $5 + 4i$, both in length and direction. In a similar manner we locate point B. Line OB represents the sum $-3 - 3i$.

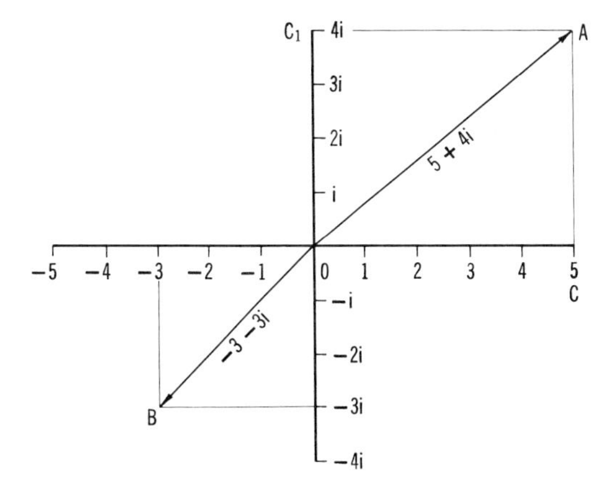

Fig. 12.9

Graphical Addition and Subtraction of Complex Numbers. Consider the graphical addition of complex numbers $(2 + 3i)$ and $(4 + i)$. These numbers are represented by OA and OB, as shown in Fig. 12.10. The sum of the complex numbers is represented by OC. Point C is located by drawing BC parallel and equal to OA (or AC parallel and equal to OB). It is a simple matter to show that OC represents the sum of the two complex numbers. Let us draw a vertical through C to intersect the horizontal drawn through B at D. Now triangle BCD is congruent to triangle OAE; therefore, $BD = OE = 2$, and $CD = AE = 3i$; also, $BF = DG = i$ and $CG = 4i$. Hence, $OC = OE + OF + (EA + FB)i$, or $OC = 2 + 4 + (3 + 1)i$ which is the sum of complex numbers $(2 + 3i)$ and $(4 + i)$.

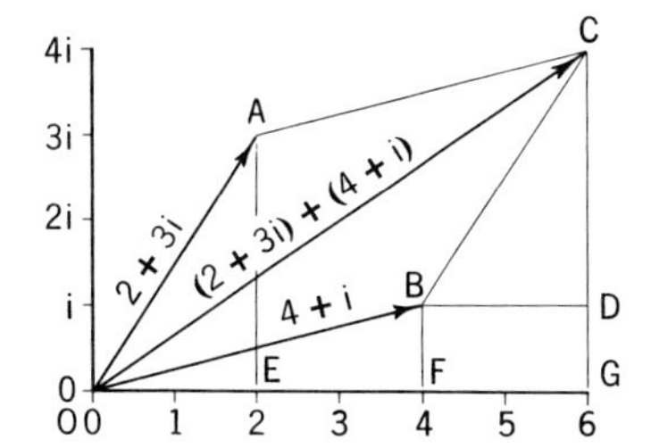

Fig. 12.10

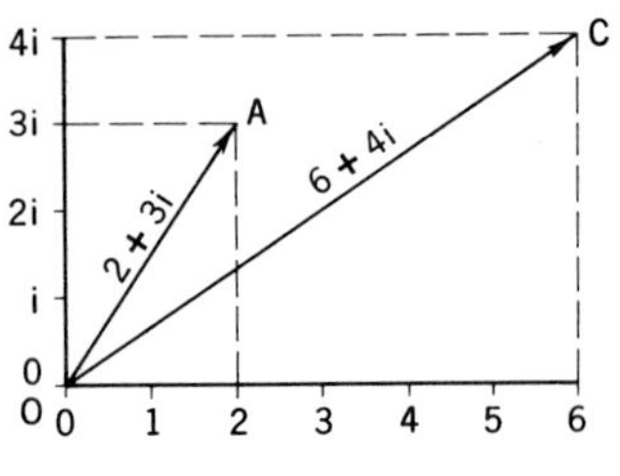

Fig. 12.11

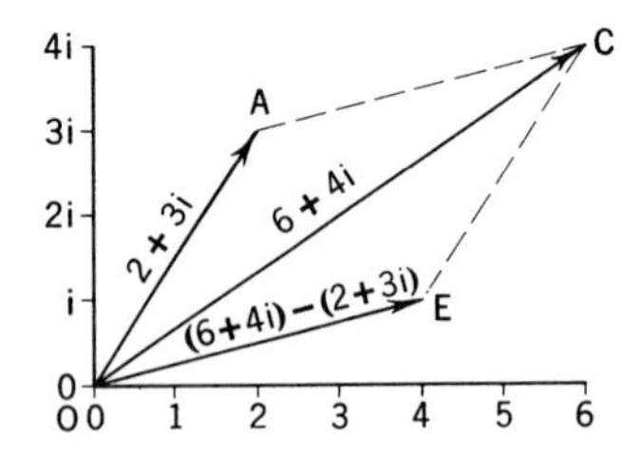

Fig. 12.12

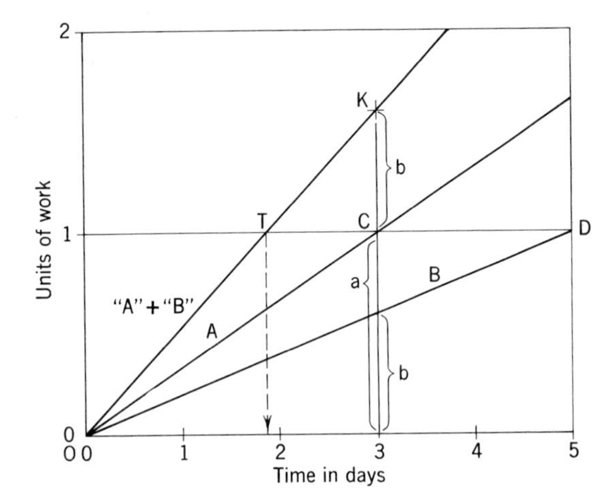

Fig. 12.13 Graphical solution of work problem (uniform rate of work).

Now suppose we want to subtract complex number $(2 + 3i)$ from complex number $(6 + 4i)$. The graphic representation of these numbers is shown in Fig. 12.11 and the solution is shown in Fig. 12.12. Through point C a line, CE, is drawn parallel and equal to OA in a direction opposite to OA. $OE = (6 + 4i) - (2 + 3i) = (4 + i)$.

APPLICATIONS

Graphical Addition

EXAMPLE 1

Let us consider the problem in which "A" does a unit of work in 3 days and "B" does a similar unit of work in 5 days. If they work together on a like unit, how long will it take? This is a typical rate problem usually encountered in elementary algebra classes. After some discussion by the instructor, the student discovers that the solution can be found by evaluating x in the equation $1/3 + 1/5 = 1/x$. Many students do not comprehend the meaning of the equation.

A graphic approach to the solution greatly enhances the meaning of the equation. Consider Fig. 12.13. The graphic representation of the performances by "A" and "B" is shown by the lines marked A and B. Line A is located by joining the origin, O, with point C, and line B by joining point O and D. Now at any time, say, three days, "A" and "B" together complete $(a + b)$ units of work. If we connect points O and K (K was established by graphically adding b to a) we will have established line OK, which shows the rate at which "A" and "B" work together. The intersection of line OK with the horizontal line drawn through point 1 on the "units-of-work" scale locates point T. The x-value of this point is $1\frac{7}{8}$ days, which is the required time to complete a unit of work if "A" and "B" work together.

It is true that, in this case, the "mechanics of the solution" is quite easily effected by solving the equation algebraically; nevertheless, the graphic solution simplifies the "visualization and the analysis" of the problem.

EXAMPLE 2

The solution in Example 1 is based on the assumption that the rates of work are uniform. In actual practice, however, the rates of work vary.

Suppose now that "A" and "B" work in accordance with the work patterns shown in Fig. 12.14 and that we wish to know how long it would take to accomplish a unit of work when "A" and "B" work together, assuming the work patterns so shown.

An algebraic solution in this case is quite difficult. We can, however, use the method of graphical addition to solve this problem quite easily in the same manner used in Example 1. It is only necessary to introduce a number of vertical lines, and on each of them graphically to add segments such as a to the corresponding segments representing the work performance of "B." The curve drawn through the newly established points, such as K, intersects the horizontal line drawn through point 1 of the "units-of-work" scale in point T. The vertical line through T intersects the time scale at 2.75 days, the required time.

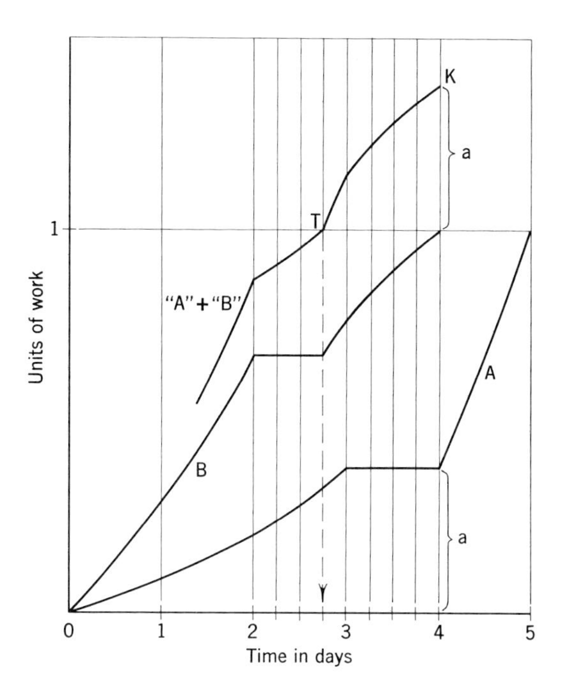

Fig. 12.14

EXAMPLE 3

A vehicle travels at a speed of s feet per second at any time, t seconds. The data are:

s	0	15.5	30.0	48.0	57.2	56.1	42.5
t	0	5	10	15	20	25	30

Determine the number of feet traveled in 30 seconds.

First, the data are plotted and then a "fair" curve is drawn through the points as shown in Fig. 12.15.

The area under the curve represents the distance traveled in 30 seconds. The area will be determined by a combination of graphical and numerical methods. We start by dividing the area into strips of equal width, since the change in curvature is not great. Now, horizontal lines, such as the one through point A, are introduced across each strip so as to balance the areas above and below the curve. Positioning the horizontals is easily judged by eye. The sum of the areas of the rectangle is a very close approximation of the area under the curve.

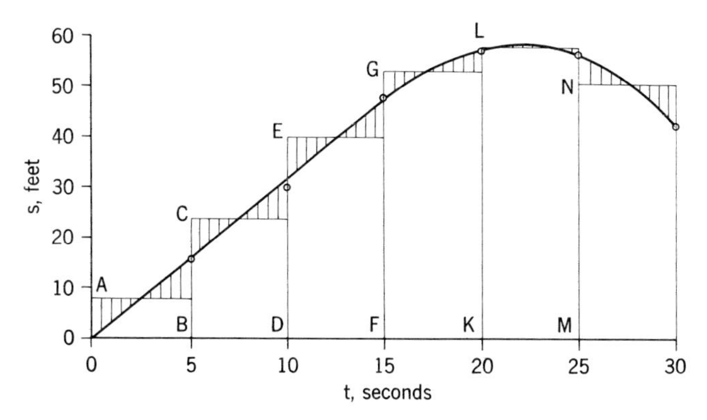

Fig. 12.15

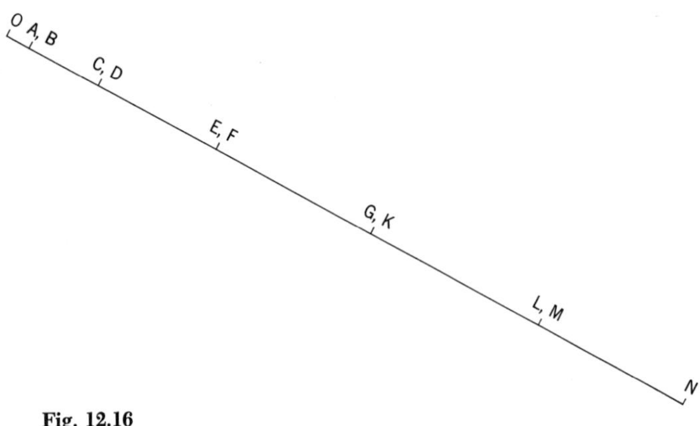

Fig. 12.16

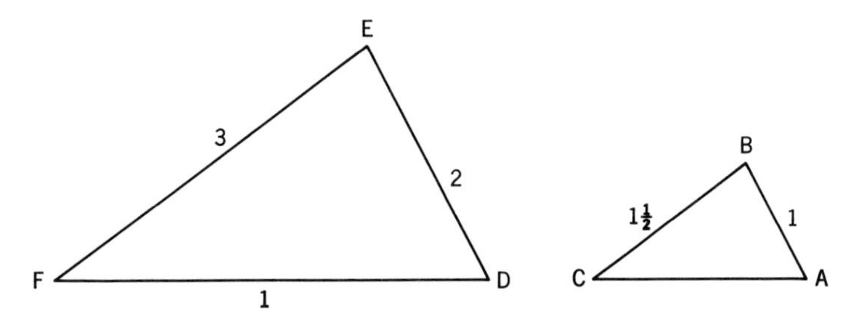

Fig. 12.17

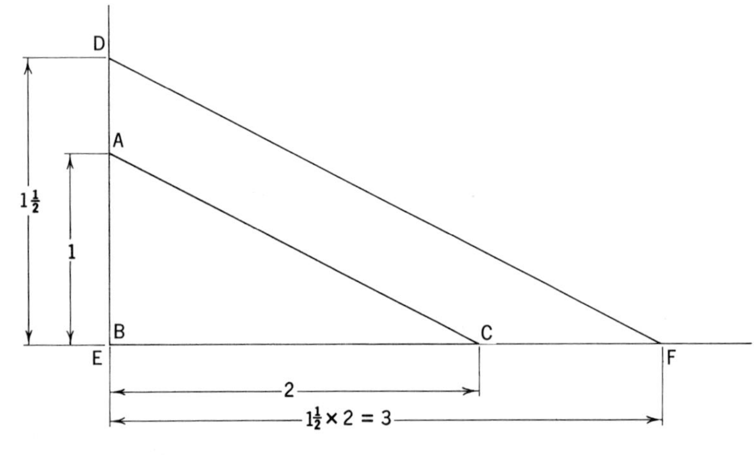

Fig. 12.18

Next, the sum of the segments OA, BC, DE, FG, KL, and MN is obtained as shown in Fig. 12.16. The total length, ON, is 232 feet, measured in terms of the s scale. This length multiplied by 5 (the width of each strip) is 1160 feet, the total distance traveled in 30 seconds.

The above examples stress the point that both the engineer and the engineer-scientist should be familiar with both the algebraic and the graphic methods and should then exercise good judgment with regard to the choice of the method that is best suited to the solution of the problem. It is through education and experience that each of us develops a "feel" for the most effective use of both the symbolic and graphic methods.

Graphical Multiplication

Note the similar triangles ABC and DEF in Fig. 12.17. We know that corresponding sides of similar triangles are in the same ratio. This means that $AB/DE = BC/EF = AC/DF$.

EXAMPLE 1

Now, when $AB = 1$ unit, $BC = 1\frac{1}{2}$ units, and $DE = 2$ units, then $\frac{1}{2} = 1\frac{1}{2}/EF$, or $EF = 3$ units.

The problem can be easily solved graphically by arranging similar triangles in a convenient and compact manner, as shown in Fig. 12.18. Right triangles are used to facilitate the construction.

EXAMPLE 2

Suppose it is desired to determine graphically the product of a, b, c, and d.

Let $$a \times b = R \tag{1}$$
and $$R \times c = S \tag{2}$$
and $$S \times d = T \tag{3}$$

where T is the result.

The first equation may be rewritten as

$$\frac{1}{a} = \frac{b}{R} \tag{4}$$

The second equation may be rewritten in a similar manner as

$$\frac{1}{c} = \frac{R}{S} \tag{5}$$

and, finally, the third equation as

$$\frac{1}{d} = \frac{S}{T} \tag{6}$$

Graphically, equation (4) can be solved as in the previous case. This is shown in Fig. 12.19.

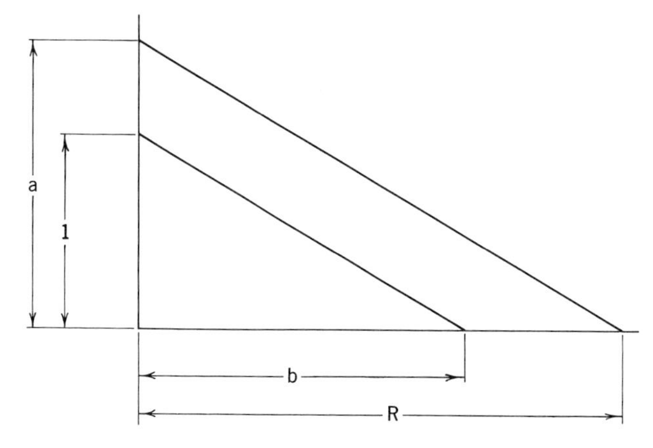

Fig. 12.19

In a similar manner, equations (5) and (6) may be solved. Figure 12.20 shows the solution of equation (5), and Fig. 12.21 the solution of equation (6).

Now, instead of drawing three separate figures, we may combine them as shown in Fig. 12.22.

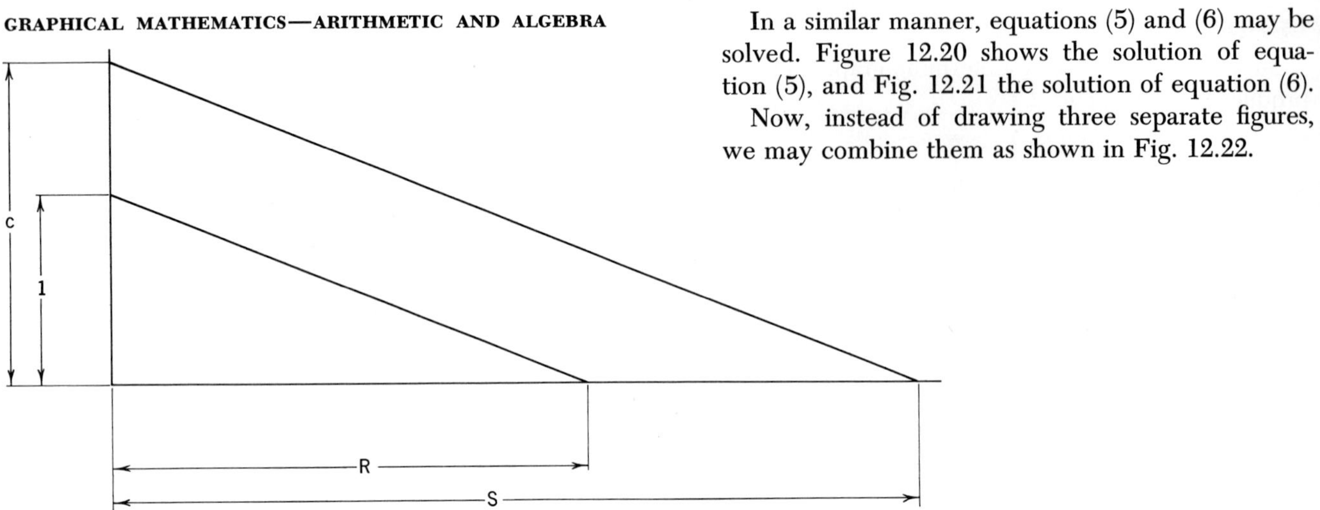

Fig. 12.20

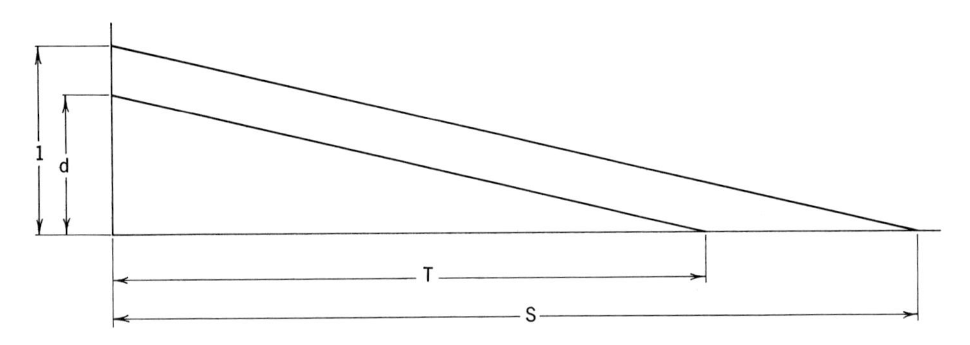

Fig. 12.21

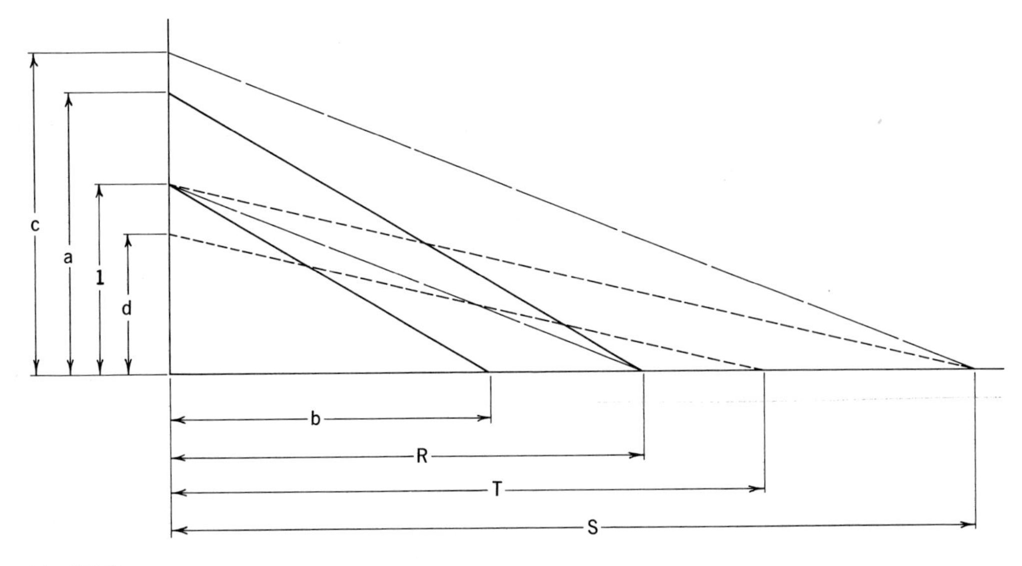

Fig. 12.22

Graphical Division

Graphical division can be performed in a similar manner. Consider the similar triangles shown in Fig. 12.23. It is quite evident that $AB/DE = BC/EF$. When $DE = 1$ unit; $BC = 3$ units; and $EF = 2$ units, then $AB = DE \times BC/EF = 1 \times 3/2 = 1\frac{1}{2}$ units. Graphically, we can easily obtain the length of AB by drawing through point C a line parallel to FD, thereby locating point A. Length AB is then measured to the same scale that was used in laying off the other lengths.

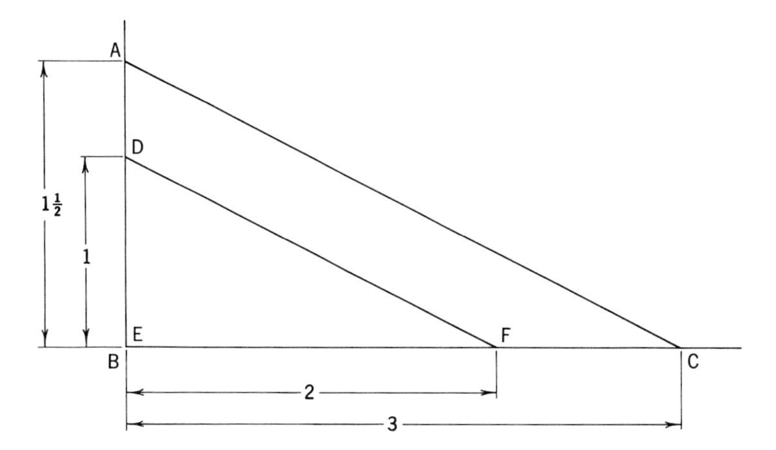

Fig. 12.23

Graphical Addition and Multiplication

Let us consider the relation, $b + ac = d$. We can replace this equation with two equations:

$$T = ac \tag{1}$$

and

$$b + T = d \tag{2}$$

Equation (1) can be written

$$\frac{T}{c} = \frac{a}{1} \tag{3}$$

Graphically the solution for equation (3) is based on the fact that corresponding sides of similar triangles are in the same ratio. In Fig. 12.24 we observe that $T/c = a/1$. Now, when we add b to T we obtain d, thus solving equation (2). Distance d, measured to the same scale used in laying off distances a, b, and c, is the solution to the expression $b + ac = d$.

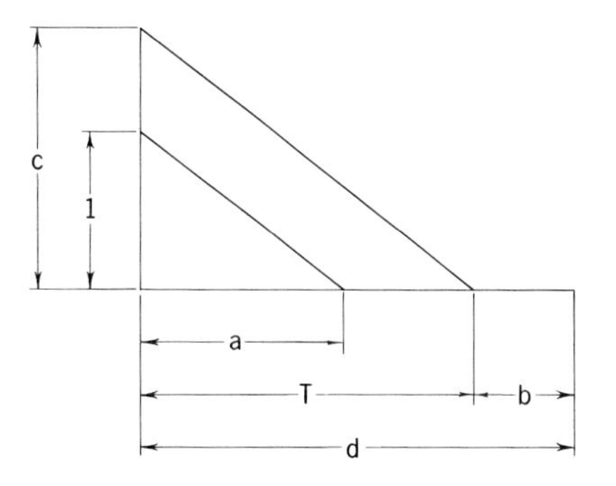

Fig. 12.24

EXAMPLE

In equation $b + ac = d$, let $a = 1\frac{1}{2}$ units; $b = 1\frac{1}{2}$ units; and $c = 2$ units. The construction for the solution is shown in Fig. 12.24(a).

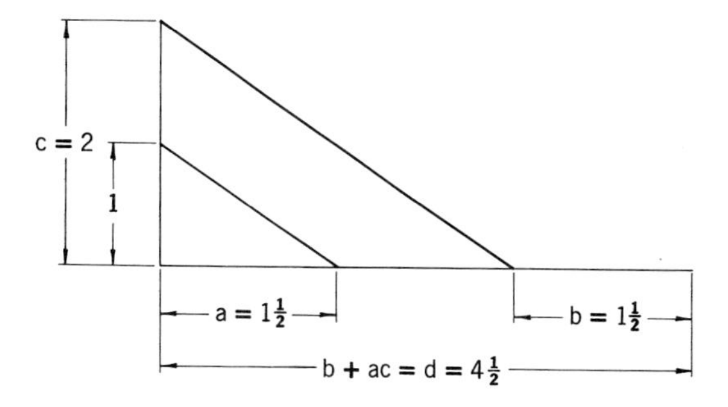

Fig. 12.24(a)

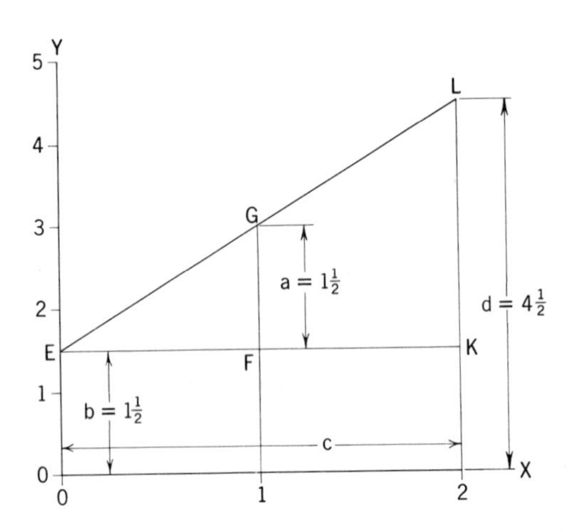

Fig. 12.25 d = ac + b where a = $1\frac{1}{2}$, b = $1\frac{1}{2}$, c = 2, and d = $4\frac{1}{2}$.

Alternative Solution. In Fig. 12.25, scales are laid off on the X and Y coordinate axes to include the values stated in the problem. Note that the "unit-of-measure" used in laying off the scale on the Y-axis differs from the unit-of-measure used on the X-axis. The choice made in preparing Fig. 12.25, although arbitrary, provided a very satisfactory graphical solution. Had the same unit-of-measure been used on both axes, the slope of line EL would have been increased. Distances $a = 1.5$, $b = 1.5$, and $c = 2$ have been laid off as shown in the figure. The value of $d = b + ac$ is easily obtained. It is clearly seen that triangles EFG and EKL are similar; therefore,

$$\frac{EF}{FG} = \frac{EK}{KL}$$

or

$$\frac{1}{1.5} = \frac{2}{KL};$$

$$KL = 3 = ac$$

Now $ac + b = d = 4.5$.

In terms of X and Y, we see that $Y = mX + b$, where m is the slope of line EGL ($FG/FE = 1.5/1$) and b is the Y-intercept (1.5) of the line EGL.

The line EGL is represented mathematically by the expression,

$$Y = 1.5X + 1.5$$

Now, when $X = 2$ (where 2 is the value of c), $Y = 1.5 \times 2 + 1.5 = 4.5$, which is the value of d in the stated problem.

Thus we have both the graphical solution of the problem and the symbolic (equational) solution which was derived from the graphical. *Both solutions are mathematical—one employs graphics, the other symbols that are expressed in equational form.*

Addition, Multiplication, and Division

1. Some problems involve addition, multiplication, and division. For example, two resistances, R_1 and R_2, in parallel are equivalent to R, where

$$R = \frac{R_1R_2}{R_1 + R_2}$$

This equation can be solved by combining the graphical solutions for parts of the equation.

If we let $X = R_1R_2$

and $Y = R_1 + R_2$

and $R = \frac{X}{Y}$

it is seen that each of the above equations can be solved by the methods previously discussed. A combination of these can be arranged to effect a solution of the given equation.

It should be quite obvious that the proposed solution is a bit cumbersome.

A much simpler solution is shown in Fig. 12.26. It is necessary only to lay off distances OA and BC to represent the magnitudes of R_1 and R_2, and then to locate point D, which is the intersection of lines OC and AB. The vertical distance, DE, from point D to base line OB (any convenient length), measured to the same scale used in laying off R_1 and R_2, is the required value of R.

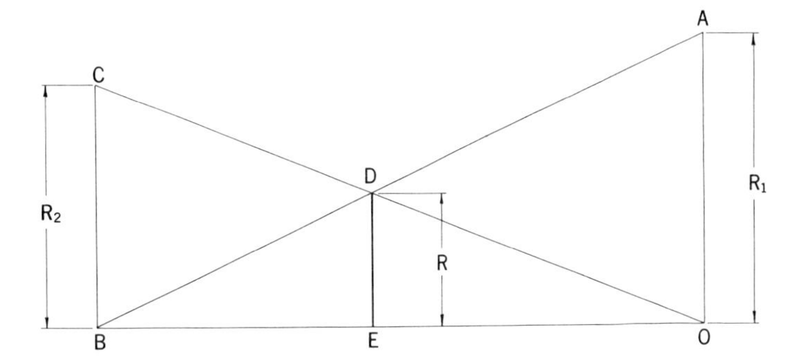

Fig. 12.26

That the above construction is correct can be easily shown. Consider Fig. 12.27, which is essentially the same as Fig. 12.26. Let us draw a horizontal line through point D to intersect lines AO and BC in points F and G, respectively. Now, since triangles AFD and DEB are similar,

$$\frac{AF}{DE} = \frac{FD}{EB}$$

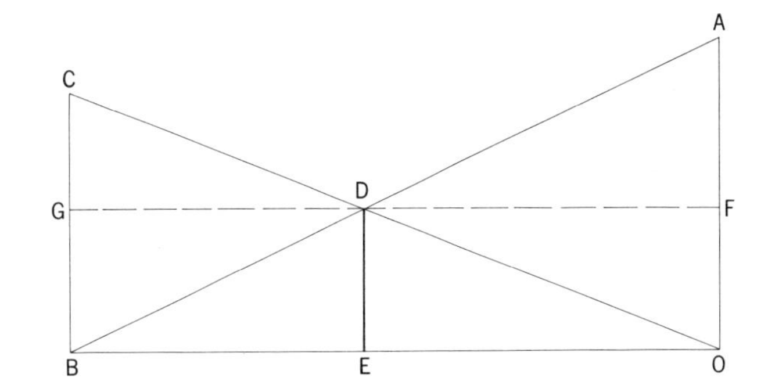

Fig. 12.27

Likewise, triangles DEO and CGD are similar. Hence,

$$\frac{DE}{CG} = \frac{EO}{DG} = \frac{FD}{EB}$$

Therefore,

$$\frac{AF}{DE} = \frac{DE}{CG} \text{ or } \overline{DE^2} = AF \times CG$$

Now, if we compare Figs. 12.26 and 12.27, and keep in mind the above relation, $\overline{DE^2} = AF \times CG$, we see that

$$R^2 = (R_1 - R)(R_2 - R)$$

or

$$R^2 = R_1R_2 - R_1R - RR_2 + R^2 \text{ or } R = \frac{R_1R_2}{R_1 + R_2}$$

or

$$R = \frac{1}{\left(\frac{1}{R_1} + \frac{1}{R_2}\right)}$$

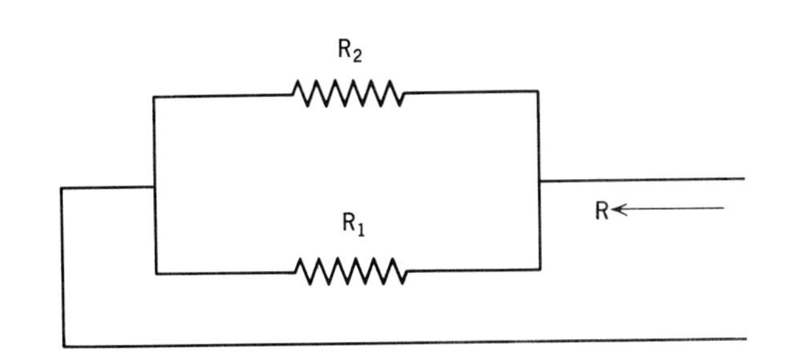

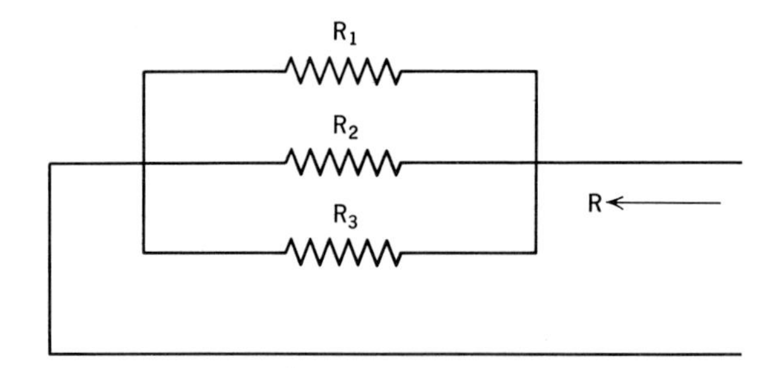

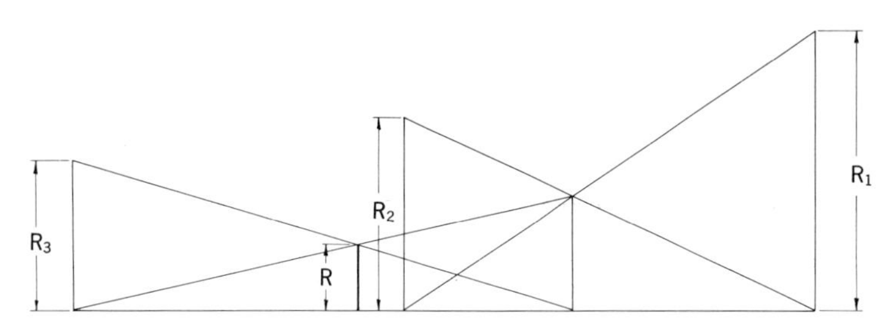

Fig. 12.28

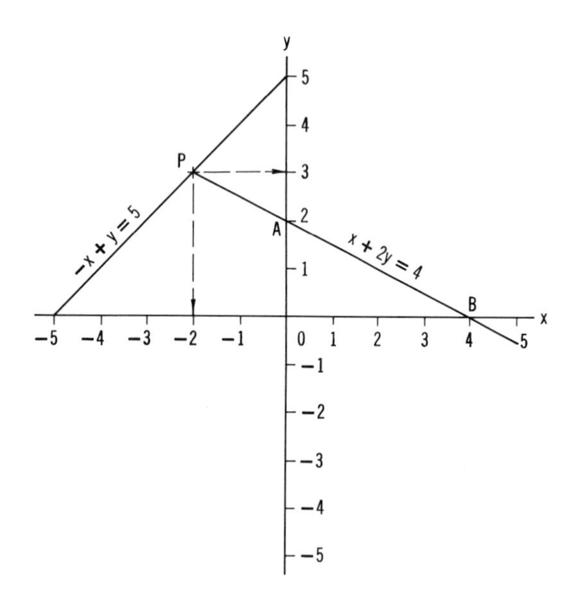

Fig. 12.29

2. Should three resistances in parallel be given, then

$$R = \frac{1}{\left(\frac{1}{R_1} + \frac{1}{R_2} + \frac{1}{R_3}\right)}$$

Graphically, this is solved quite simply. (See Fig. 12.28.)

There are many equations that can be solved by the simple operations of graphical mathematics. In fact, the use of a single basic relationship will enable anybody who has some spark of ingenuity to solve a variety of problems graphically. This relationship is "corresponding sides of similar triangles are in the same ratio." This will be further evident in some of the problems dealing with graphical algebra.

Graphical Algebra

Let us now consider several problems which (*a*) are expressed in algebraic form and (*b*) others that are stated in word form.

EXAMPLE

Suppose we are given the equations

$$x + 2y = 4 \tag{1}$$

$$-x + y = 5 \tag{2}$$

Each equation is shown graphically in Fig. 12.29. Consider equation (1). When $x = 0$, $y = 2$. The point represented by these coordinates is *A*. Now, when $y = 0$, $x = 4$. The point which has these coordinates is *B*. The line which joins points *A* and *B* is the graphical representation of the equation $x + 2y = 4$. In a similar manner we can draw the line which represents the equation $-x + y = 5$.

Observe that the distance from the origin to point B is the x-intercept of the line which represents equation (1), and that the distance from the origin to point A is the y-intercept of the line.

When the equations are written in the *intercept form* it is a simple matter to draw the lines which represent equations (1) and (2) graphically.

We can write the equations in the intercept form by rewriting the equations so that the right-hand side of the equations is 1. Thus,

$$x + 2y = 4 \text{ becomes } \frac{x}{4} + \frac{y}{2} = 1 \tag{1$'$}$$

and $\quad -x + y = 5$ becomes $\dfrac{x}{-5} + \dfrac{y}{5} = 1 \qquad (2)'$

Note that the denominator values in equations (1)′ and (2)′ are the intercepts in each case. Now the intercept distances can be laid off and the lines can then be drawn to represent the equations.

The intersection of the two lines is point P whose coordinates are $x = -2$ and $y = 3$. These are the coordinates of the point which is common to both lines. If we were to solve equations (1) and (2) simultaneously we would obtain the same values of x and y.

Now let us find the equation of a line from its graphical representation.

EXAMPLE 1

Consider the line shown in Fig. 12.30. The x and y intercepts are 4 and 15 respectively; therefore, the equation of the line, in intercept form, is

$$\frac{x}{4} + \frac{y}{15} = 1$$

from which we obtain

$$15x + 4y = 60$$

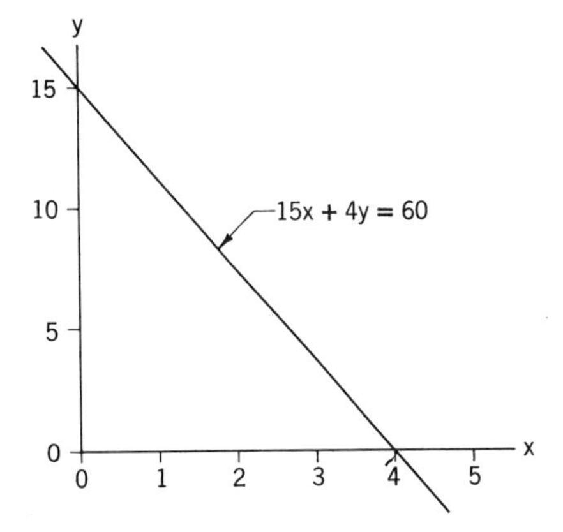

Fig. 12.30

EXAMPLE 2

Let us consider lines m, n, and s shown in Fig. 12.31. We observe that the lines are parallel to each other. The x and y intercepts of line m are 10 and 3, respectively. The equation of the line is

$$\frac{x}{10} + \frac{y}{3} = 1 \qquad \text{or} \qquad 3x + 10y = 30 \qquad (1)$$

Similarly, the equation of line n is

$$\frac{x}{20} + \frac{y}{6} = 1 \qquad \text{or} \qquad 3x + 10y = 60 \qquad (2)$$

And the equation of line s is

$$\frac{x}{-20} + \frac{y}{-6} = 1 \qquad \text{or} \qquad 3x + 10y = -60 \qquad (3)$$

Note that equations (1), (2), and (3) are the same except for the constants 30, 60, and -60. A family of parallel lines, therefore, can be easily recognized from their equations.

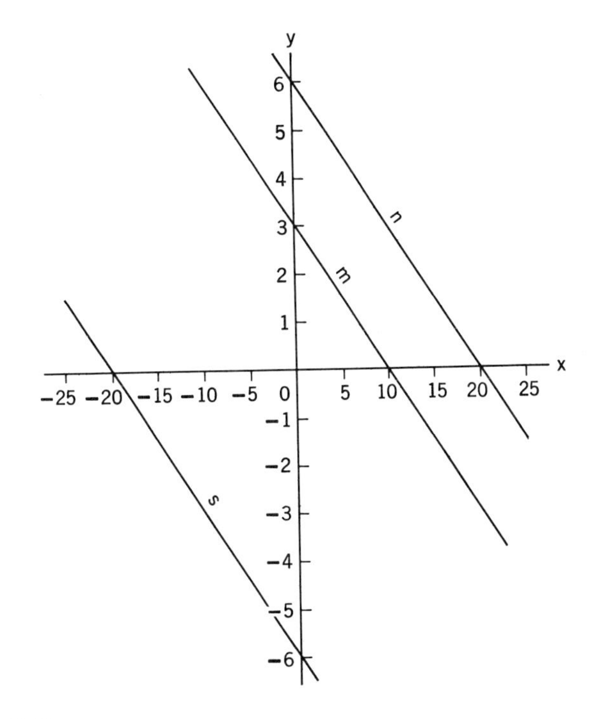

Fig. 12.31

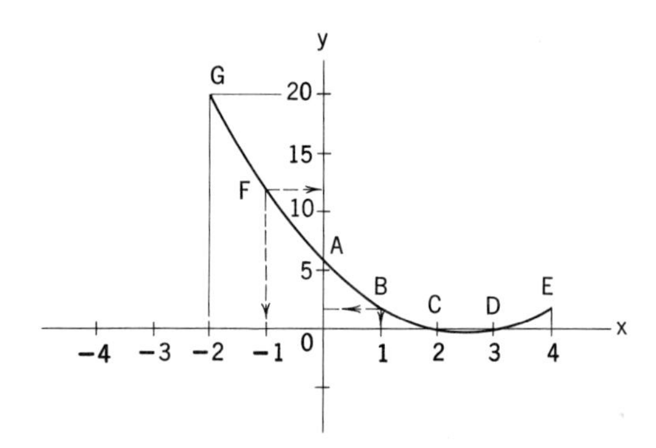

Fig. 12.32

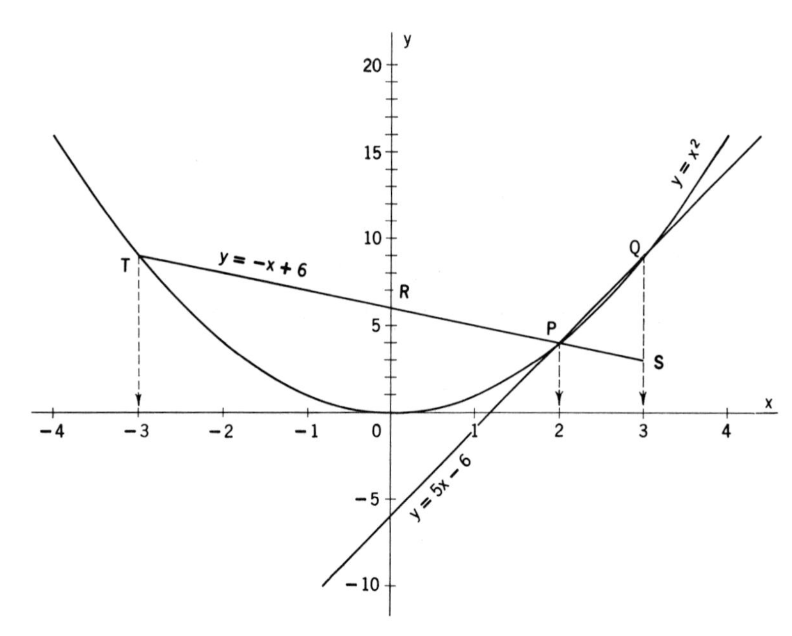

Fig. 12.33

Roots of a Quadratic Equation. The roots of a quadratic equation, such as $x^2 - 5x + 6 = 0$, are the values of x which satisfy the equation. How shall we obtain these values graphically?

One approach to the solution would be to let $y = x^2 - 5x + 6$ and then plot the curve. This means, simply, the determination of the coordinates of points which lie on the curve.

For example,

when $x = 0, y = 6$

when $x = 1, y = 2$

when $x = 2, y = 0$

etc.

We could arrange a convenient table of the values of y for the different values of x, thus:

Point	A	B	C	—	F	G	—
x	0	1	2	—	−1	−2	—
y	6	2	0	—	12	20	—

The smooth curve (Fig. 12.32) drawn through the points is the graphical representation of the given equation.

It is clearly seen that when $y = 0$, $x = 2$ and $x = 3$ (points C and D). These values of x satisfy the given equation.

A second approach makes the solution possible in a simple manner, not only for the given equation, but also, in general, for equations of the form $Ax^2 + Bx + C = 0$. In fact, the method to be presented is basically the same as the one used previously in solving for the values of x and y for the two equations $x + 2y = 4$ and $-x + y = 5$.

The equation $x^2 - 5x + 6 = 0$ may be rewritten as

$$y = x^2 \quad (1)$$

and

$$y = 5x - 6 \quad (2)$$

The graphical simultaneous solution of these two equations is easily obtained. Plot first the equation $y = x^2$, and then the equation $y = 5x - 6$. These are shown in Fig. 12.33. The intersections of the curve and line, namely, points P and Q, have x coordinates 2 and 3, respectively. These are the values of x which satisfy the given equation

$x^2 - 5x + 6 = 0$. If solutions are desired for a number of different equations of the form $Ax^2 + Bx + C = 0$, the advantage in using the latter method of solution should be quite evident, since the fixed curve $y = x^2$ need be drawn only *once*. The plotting of the line is a simple matter.

Consider the equation

$$x^2 + x - 6 = 0$$

Let $\quad y = x^2$

and $\quad y = -x + 6$

The curve for $y = x^2$ is already drawn. It is necessary only to locate the line $y = -x + 6$ (added to Fig. 12.33). This is easily done. When $x = 0$, $y = 6$ (point R), and when $x = 3$, $y = 3$ (point S). Point S was used because the line $y = -x + 6$ intersects the x-axis at a point ($y = 0$, $x = 6$) which is beyond the limit of the graduated x-axis. Line RS intersects the curve in points P and T, the x-values of which are 2 and -3, respectively, the values which satisfy the given equation $x^2 + x - 6 = 0$.

Question 1. What is the meaning of a case in which the straight line is tangential to the curve, $y = x^2$?

Question 2. What is the meaning of a case in which the straight line does *not* intersect the curve? How would you proceed with a graphical solution to the problem?

WORD PROBLEMS

Graphical solutions to word problems most frequently enhance the understanding of the problem and, moreover, are much more effective than algebraic solutions. In many cases it is not easy to translate the word problem to equational form in a reasonable amount of time. Where an exact mathematical answer is required, a graphical presentation of the problem is most useful in the development of the necessary equations.

EXAMPLE 1

An airplane, A, leaves Los Angeles at 3:00 P.M., traveling toward San Francisco at an average speed of 200 mph until 4:00 P.M., when it lands because of engine trouble. One hour later it continues its flight at an average speed of 275 mph. Another airplane, B, leaves San Francisco at 4:30 P.M., traveling toward Los Angeles (both planes travel the same route) at

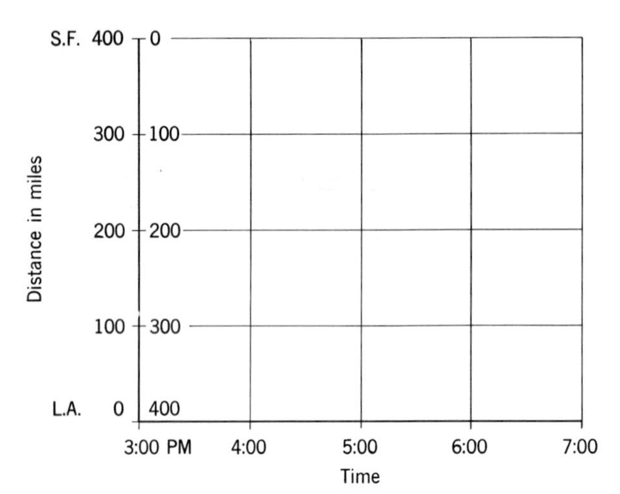

Fig. 12.34

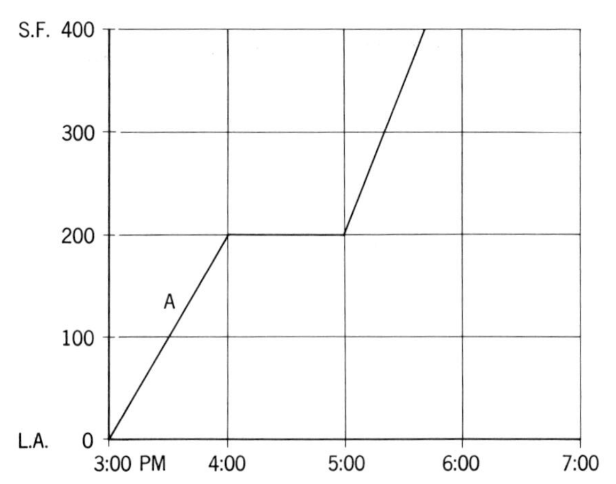

Fig. 12.35

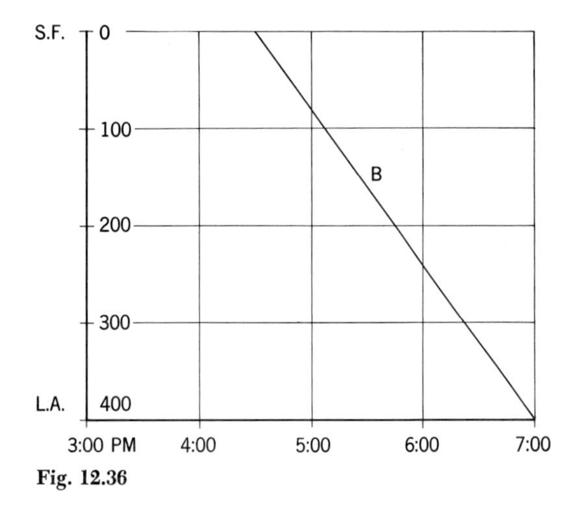

Fig. 12.36

an average speed of 165 mph. Assuming that the distance between the cities is 400 air miles, at what time will the planes pass each other, and at what distance from San Francisco will this occur?

Graphical Solution

1. First, let us prepare coordinate axes that show time and distance. (See Fig. 12.34.)

2. Now the described travel of plane *A* is plotted. This is shown in Fig. 12.35.

3. The described travel of plane *B* is shown in Fig. 12.36.

4. The complete problem and its solution are shown in Fig. 12.37. The coordinates of point *P* are 5:15 P.M. (the time at which the planes pass each other) and 122 miles (the distance from San Francisco).

Question. How would you solve this problem algebraically?

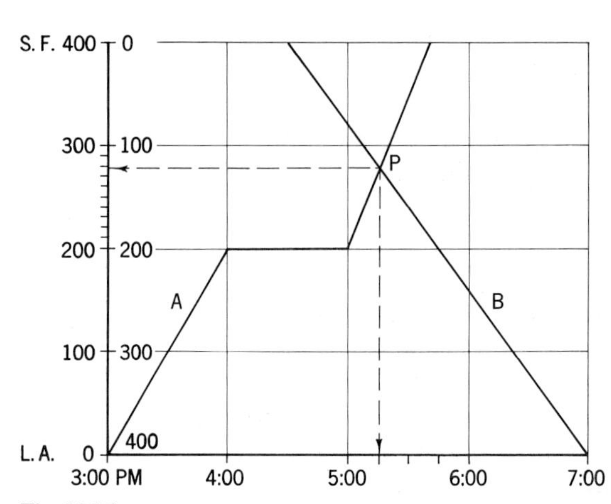

Fig. 12.37

EXAMPLE 2

One of two grades of zinc ore contains 60% zinc and the other contains 30% zinc. How many pounds of each should be used to make a 2500-pound mixture containing 40% zinc?

Graphical Solution

1. First, we prepare coordinate axes showing pounds of zinc ore and pounds of zinc. This is shown in Fig. 12.38.
2. Now we can show the 30%, 40% and 60% zinc lines. (See Fig. 12.38.)
3. Through point A, Fig. 12.39, a line is drawn, parallel to the 30% line and intersecting the 60% line at point C. The horizontal through point C intersects the vertical scale at 833, the required number of pounds of zinc ore containing 60% zinc.
4. In a similar manner a line is drawn through point A, parallel to the 60% line and intersecting the 30% line at point B. The horizontal through point B intersects the vertical scale at 1667, the required number of pounds of zinc ore containing 30% zinc.

It is easy to understand the graphical solution because our problem is reduced to the location of point B on the 30% line so that a line through B, parallel to the 60% line, will pass through point A. The construction for locating point B is self-evident. Study of the parallelogram $ACDB$ verifies the fact that the zinc coordinate of point B added to the zinc coordinate of point C is equal to 1000 pounds, which is the required zinc content of the mixture.

Question. How would you solve the problem algebraically?

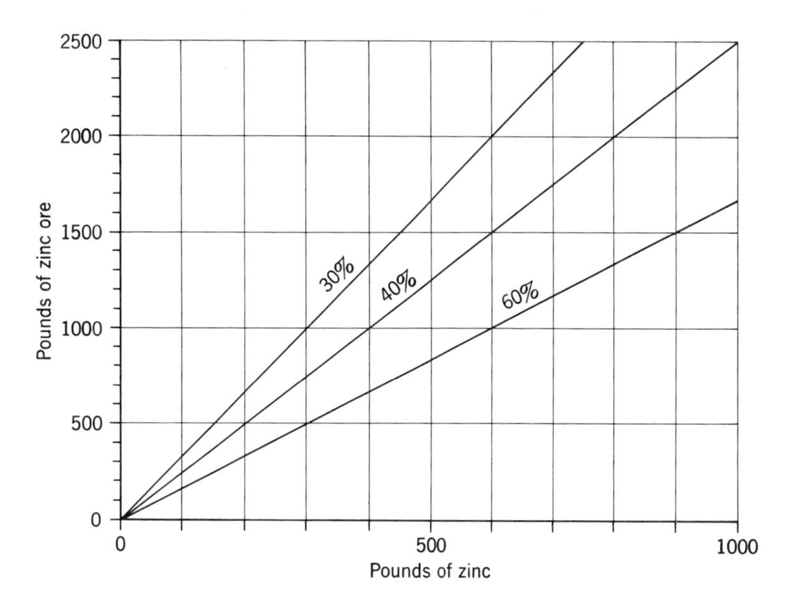

Fig. 12.38

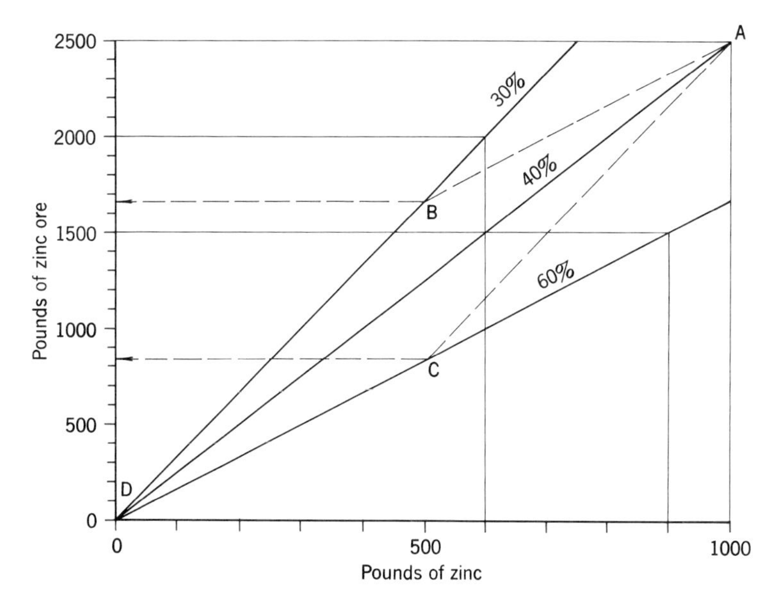

Fig. 12.39

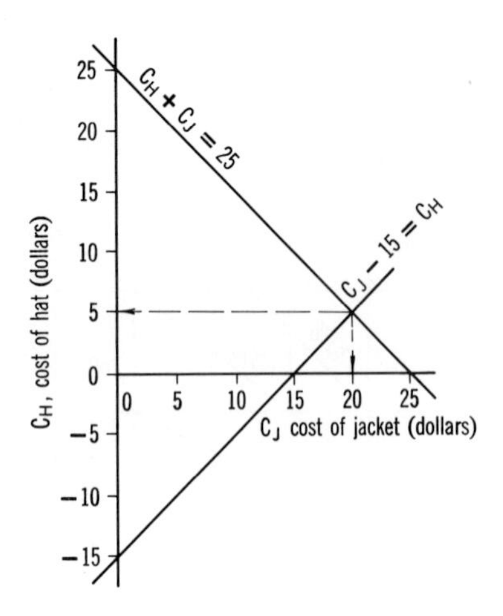

Fig. 12.40

EXAMPLE 3

A hat and a jacket cost \$25. The cost of the hat is \$15 less than the cost of the jacket. What is the cost of each? In Fig. 12.40, the cost of the hat and jacket is shown by the line $C_H + C_J = 25$; and the statement, "the cost of the hat is \$15 less than the cost of the jacket," by the line $C_J - 15 = C_H$. The intersection of the two lines, point A, has coordinates 20 and 5 which represent the costs of the jacket and hat, respectively.

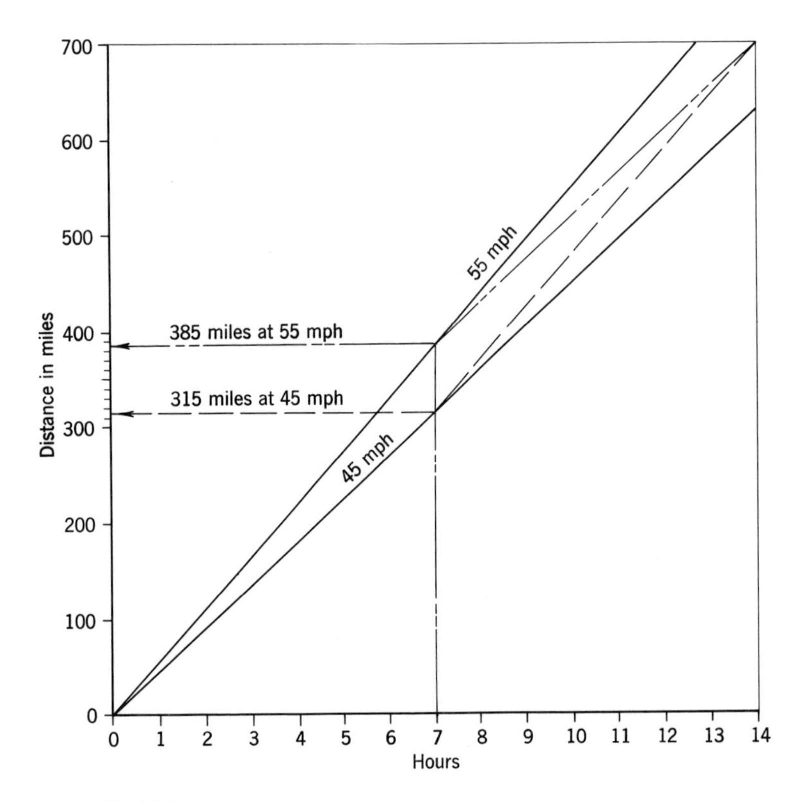

Fig. 12.41

EXAMPLE 4

George made a trip of 700 miles. Part of the trip was made at an average rate of 45 miles per hour, and part at an average rate of 55 miles per hour. The trip took him 14 hours. How far did he travel at each rate? The solution is shown in Fig. 12.41 which is similar to Fig. 12.39.

EXAMPLE 5

The circumference of the rear wheel of a vehicle is 4 feet more than the circumference of the front wheel. The rear wheel makes 100 revolutions less than the front wheel while traveling 5000 feet. What is the diameter of each wheel? The solution is shown in Fig. 12.42.

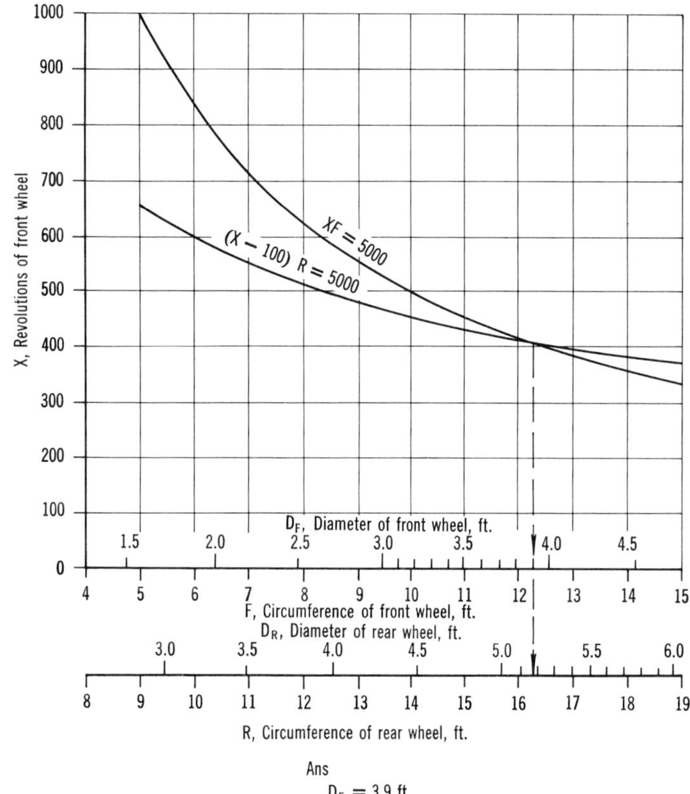

Fig. 12.42

EXERCISES

Solve all problems **graphically.**

1. What is the value of X in the equation $\frac{1}{2} + \frac{1}{3} = 1/X$?

2. What is the value of X in the equation $\frac{2}{5} + \frac{3}{7} = 1/X$?

3. Divide $\sin\theta$ by $\cos\theta$. θ varies from 0° to 90°.

4. Construct $2 + 2i$, and then multiply by $\sqrt{-1}$.

5. Add $(3 + 2i)$ to $(-3 - 2i)$.

6. Present the roots of $X^3 = 1$.

7. Determine the roots of the equations:
(*a*) $X^2 + 2X = 8$
(*b*) $X^2 - 5X = -15$

8. Determine the roots of the equations:
(*a*) $X^3 - 12X = 8$
(*b*) $X^3 + 4X = 3$
(*c*) $X^3 - 3X = -1$

9. Determine the root between $X = 4$ and $X = 8$, to three significant figures, of the equation $\tan X = X - X^2$.

10. Plot the line $Y = 2X + 2$, and then establish a line through point A (3, 3) and parallel to the original line. What is the equation of the new line?

11. Plot the line $Y = -X + 7$, and then establish a line through point B (2, 4) and perpendicular to the original line. What is the equation of the new line?

12. Mr. Smith traveled from Chicago to Detroit (270 miles) by automobile. Part of the trip was made at an average rate of 50 mph, the other part at 40 mph. How far did he travel at each rate? The trip required $6\frac{1}{2}$ hours.

13. A freight train travels at the rate of 30 mph. Two and one half hours later a passenger train leaves the same station, traveling at the rate of 50 mph on a parallel track and in the same direction as the freight train. Find the time required to overtake the freight train.

14. A tank can be filled in 7 hours by water flowing through pipes A and B. This can also be done by pipes B and C in 5 hours, and by pipes A and C in 3 hours. How long would it take to fill the tank if each pipe were operated separately?

15. A freight train leaves San Francisco at 10:00 A.M., traveling south at an average speed of 35 mph. At 1:00 P.M. it is delayed for 35 minutes, and then proceeds at an average speed of 30 mph. At 2:30 P.M. an express train leaves from San Jose (48 miles south of San Francisco), traveling south on a parallel track at an average speed of 60 mph. How far from San Francisco and at what time does the express train overtake the freight train?

16. An airplane usually travels between cities A and B at the average of 160 mph. On one of its trips it leaves city A an hour late, but travels at 210 mph to arrive at city B on schedule. What is the distance between the cities?

17. An airplane, A, leaves Los Angeles at 3:00 P.M. headed toward San Francisco, traveling at an average speed of 225 mph until 4:00 P.M., when it lands because of engine trouble. Two hours later it continues its flight at an average speed of 280 mph. Another airplane, B, leaves San Francisco at 4:30 P.M. headed toward Los Angeles (both planes travel the same route), traveling at an average speed of 165 mph. Assuming that the distance between the two cities is 400 air miles, at what time will the planes pass each other, and at what distance from San Francisco will this occur?

18. A 120-gallon tank, initially empty, has a hole in it through which water can discharge at the constant rate of $2\frac{1}{2}$ gallons per minute. A faucet is turned on so that water enters the tank at the rate of 5 gallons per minute, and operates for 10 minutes before a second faucet is turned on through which water enters the tank at the rate of 3 gallons per minute. Discharge from both faucets continues until the tank is full. Draw a graph to show the amount of water that is in the tank at any instant, and determine the time required to fill the tank.

19. One of the grades of zinc ore contains 70% zinc and the other contains 20% zinc. How many pounds of each should be used to make a 1500-pound mixture containing 50% zinc?

20. A stone is dropped into a well. Five seconds later the sound of the stone's impact with the water is heard at the ground surface. How deep is the well?

21. An airplane that is at elevation 1000 ft and traveling at 120 mph in an easterly direction drops a shatterproof instrument. How far east of the starting point will the instrument strike the ground (assumed to be elevation 100 ft)?

22. An open tank with square base is to be constructed from 100 square feet of sheet metal. Determine the dimensions of the tank to accommodate the maximum capacity possible. *Hint*: First set up the equation for volume, and plot this equation. Then determine the value of X (where X = length of side of the base) which corresponds to the largest volume.

23. A sheet-metal open tank contains 300 gallons of water. The base of the tank is a square. Determine the dimensions of the tank, assuming that it is made from the smallest area of metal.

24. What are the cross-sectional dimensions of the largest rectangular piece that can be cut from a circular log which has a 20-inch diameter?

25. A circular steel tank, open at the top, holds 10,000 gallons of water when the water level is 3 inches below the top of the tank. Determine the dimensions of the tank so that the least area of steel plate is used.

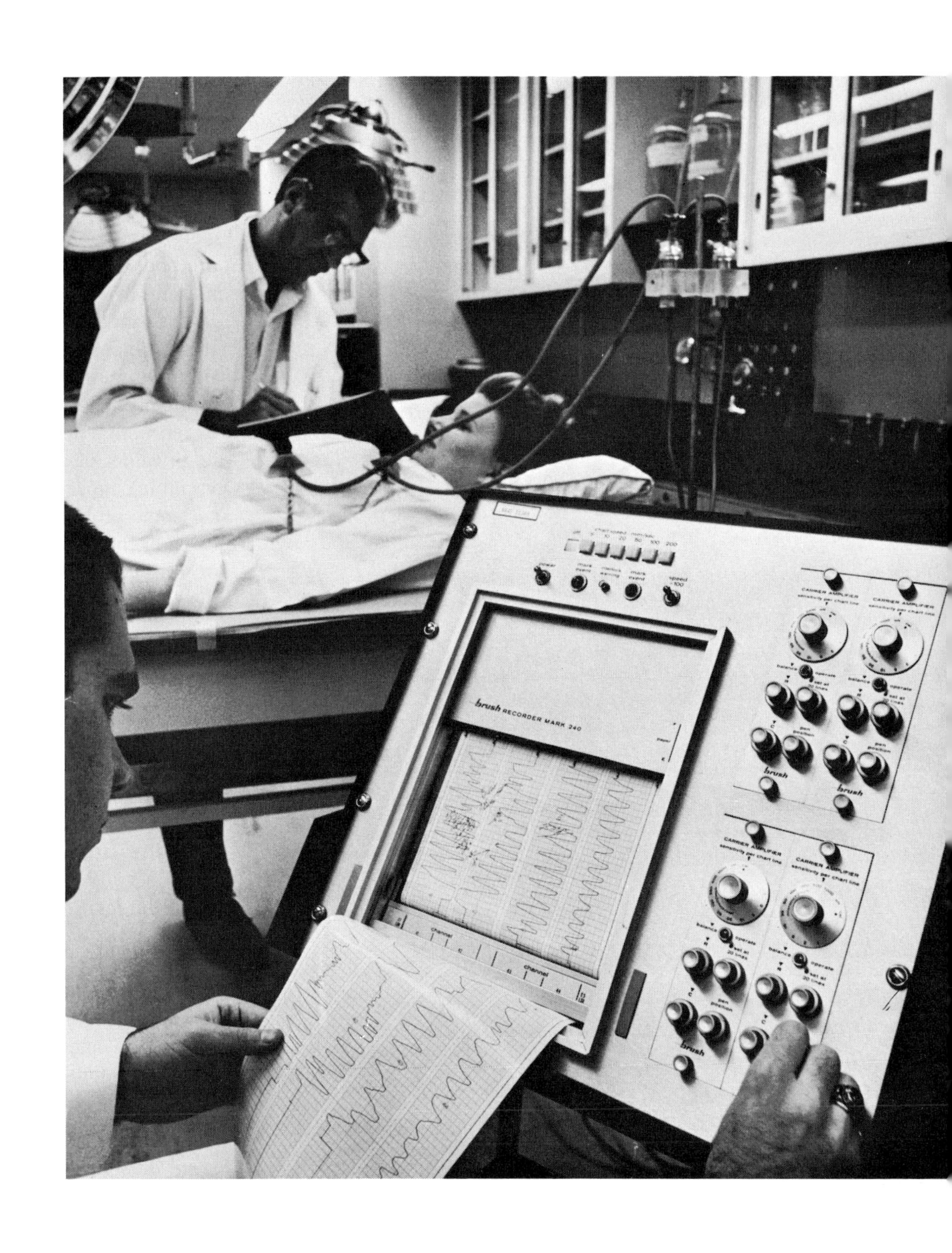

GRAPHICAL CALCULUS 13

In both engineering and science many problems arise that deal with the determination of areas, volumes, centroids, moments of inertia, displacements, velocities, accelerations, etc. Such problems may be solved graphically, numerically, mechanically, mathematically, or by a combination of these methods. The method selected will depend on the manner in which the problem can be best presented. If, for example, it is found most convenient to state the problem by means of an algebraic expression, then we may proceed to solve the problem mathematically. If, on the other hand, the given data are experimental or have been collected by measurement (e.g., values of one variable with respect to a related variable), most likely it will be more economical to solve the problem graphically, numerically, mechanically, or by a combination of graphical and numerical or graphical and mechanical methods. These will be discussed in this chapter.

GRAPHICAL INTEGRATION

Integration is a summation process. In a physical sense it is the summation of many small quantities. In the algebraic form of the calculus the quantities are regarded as infinitely small, whereas in practical problems that *cannot be conveniently expressed algebraically,* and that arise in engineering, the quantities are necessarily taken sufficiently large to effect a good solution. The volume of a sphere, for example, can be determined by formal calculus (integration) since it is possible to write the equation of the spherical surface. On the other hand, it would not be feasible to find the volume of an irregularly shaped reservoir by the formal calculus methods of integration since a precise algebraic expression could not conveniently be written to describe the exact shape of the reservoir. Solutions of such problems, and of many more in this category, are best solved by graphical, mechanical, or numerical methods of integration, or by combinations of these methods.

The significance of integration can be further demonstrated by considering the following example.

Suppose we have an irregularly shaped ground depression which could be used for a reservoir. We will assume that field data have been obtained to determine horizontal cross-sectional areas and the perpendicular distances between the sections (see table below). It is desired to compute the volume of the reservoir.

A, area in sq ft	10	20	35	50	75	90	85	75	60
S, distance in ft	0	5	8	10	15	18	22	25	30

A plot of these data is shown in Fig. 13.1. Consider the very thin slice represented by the area shown shaded. This slice has a surface area, A, and has a width dS (a little bit of S). The volume of this slice is, then, $A\ (dS)$. If all such slices are summed, we will obtain the total volume, or $V = \int A\ (dS)$, where the symbol $\int$ is used for "summing-up" or integrating.

In general terms, $V = \int y\, dx$ if the curve is defined as $y = f(x)$, where y corresponds to the area A and x to the distance S. If the algebraic expression for the curve is available, formal calculus methods can be used for integrating the slices; if not, the methods referred to above will be quite satisfactory.

Graphical integration, in effect, *is a process for finding the area under a curve.*

Let us now consider several methods that may be used to find the area under a curve. Among these is the "pole-and-ray" method which is the best one to use when a plot of the "integral curve" is necessary or desirable. Once this curve is plotted, areas under portions of the original curve can be readily determined.

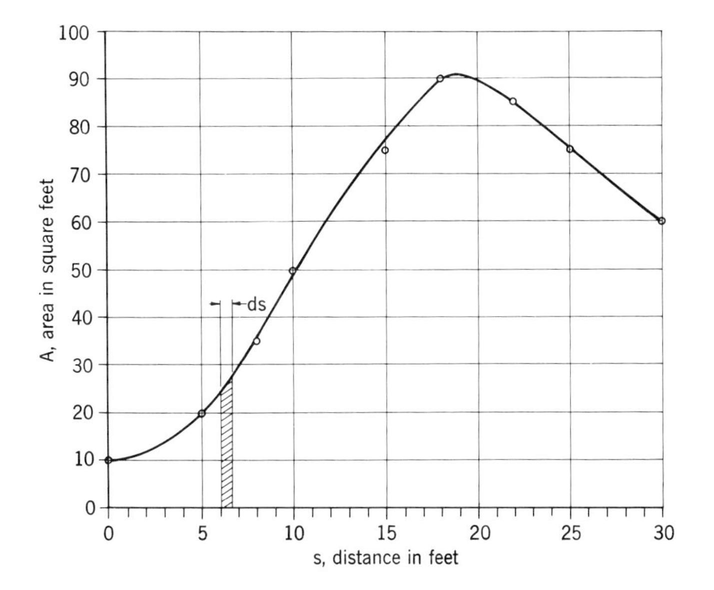

Fig. 13.1

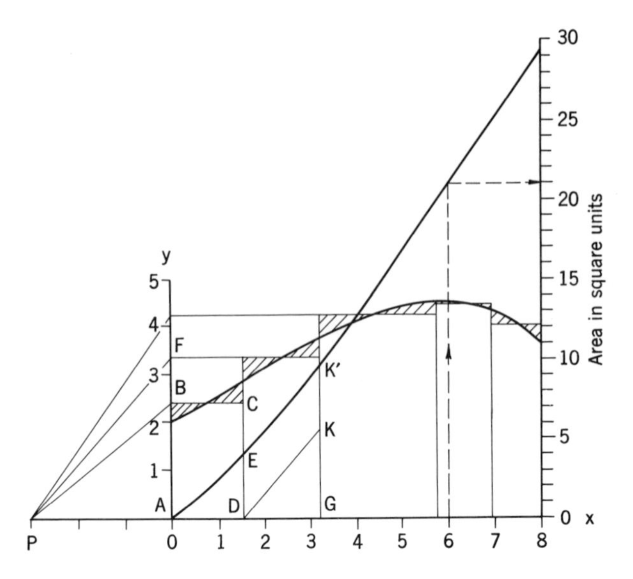

Fig. 13.2

The Pole-and-Ray Method

This method of integration is based on the well-known geometric relationship: "corresponding sides of similar triangles are in the same ratio."

Let us consider Fig. 13.2. The area under the curve is first divided into strips of varying widths which depend on the sharpness or flatness of the curve. The area of the first strip is closely approximated by that of the rectangle $ABCD$ where line BC is placed so that the shaded areas are in balance. This is done quite accurately by eye.

We now lay off a convenient "pole distance," AP, measured in terms of the unit on the horizontal scale. In Fig. 13.2 we show that $AP = 3$. A larger value would *decrease* the length of the area scale shown on the right while a smaller value would *increase* the length of the area scale.

Through point P, rays such as PB are drawn. Through point A, line AE is drawn parallel to ray PB. The area of the first strip is equal to $\overline{DE} \times \overline{AP}$. This is easily shown in the following manner:

1. Triangles BAP and EDA are similar by construction.
2. Therefore, $ED/DA = BA/AP$ (corresponding sides of similar triangles are in the same ratio).
3. $ED \times AP = DA \times BA$.
4. $DA \times BA$ is the area of the rectangle $ABCD$ which is equivalent to the area of the first strip.
5. Therefore, $ED \times AP$ is equal to the area of the first strip.

The area of the second strip is found in a similar manner. Line DK is drawn parallel to ray PF. The area of this strip is equal to $\overline{KG} \times \overline{AP}$. Now, if a line is drawn through point E parallel to DK, distance $K'G$ equals $\overline{DE} + \overline{KG}$ and, therefore, the sum of the areas of the first two strips is equal to $\overline{K'G} \times \overline{AP}$. It is quite evident that it is not necessary to draw DK, since EK' can be drawn directly. When this process is repeated for the other strips and a smooth curve is drawn through points A, E, K', etc., we shall have established the *integral curve*—the summation curve—which will enable us to determine the total area under the given curve or the area of any portion of the total area. It is readily seen that the area under the given curve between $x = 0$ and $x = 6$ is found by following the vertical dashed line drawn through $x = 6$ to the *integral curve* and then the horizontal dashed line to the area scale shown on the right. *It should be observed that the values of the graduations on the area scale are equal to the corresponding y-scale values multiplied by the pole distance.* In Fig. 13.2 the pole distance is 3; therefore, the location of a graduation, say, 15, on the area scale is on the horizontal passing through $y = 5$. Once a convenient graduation, such as the one marked 15, has been located, other graduations are readily established without further calculation since the distance from the 0 to the 15 is known.

EXAMPLE

Let us consider the area under the curve shown in Fig. 13.3. We have assumed a pole distance $= 3$ units of x. The integral curve has been established in the same manner as previously described. A graduation on the area scale, say, 60, lies on the horizontal which would pass through $y = 20$. Other graduations are readily obtained. Had we selected the pole distance $= 2$ units of x, the area scale would be longer. In this case, with the pole distance $= 2$, the 60 on the area scale would lie on the horizontal through $y = 30$. *It should now be clear that we can select a pole distance that is compatible with a desired length of area scale.* To do this we would first estimate the area under the given curve. Although the estimate may be off 5% to 10%, it may be used to closely approximate the pole distance, since the pole distance is equal to the estimated area divided by the y-scale value corresponding to the desired length of the area scale.

In this example, the long-dash horizontal line has been drawn (by eye) so that the area below the horizontal closely balances the area above the horizontal. The area of the rectangle, formed by the horizontal and the x-axis from $x = 0$ to $x = 5$ (60 square units) is a close approximation of the area under the given curve. The pole distance, then, is $60/20 = 3$, when the distance from $y = 0$ to $y = 20$ is the desired length of the area scale.

We should not forget that the pole distance is always laid off in terms of the units on the x-axis. *If, for example, the units on the x-axis had been in tenths, then the corresponding pole distance would be three tenths.*

The pole-and-ray method as used in Figs. 13.2 and 13.3 establishes chords of the integral curve.

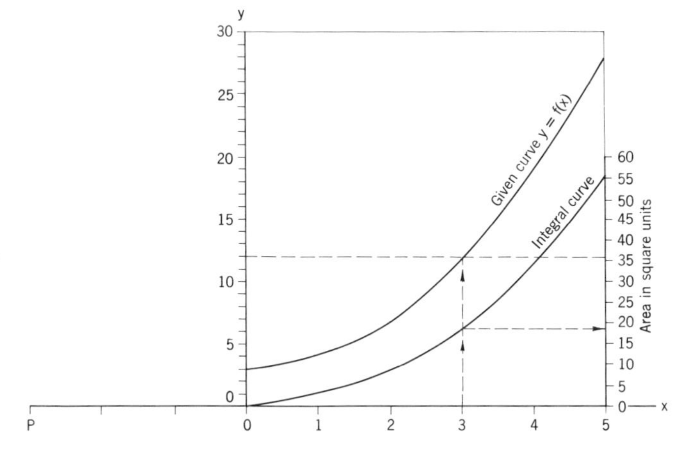

Fig. 13.3

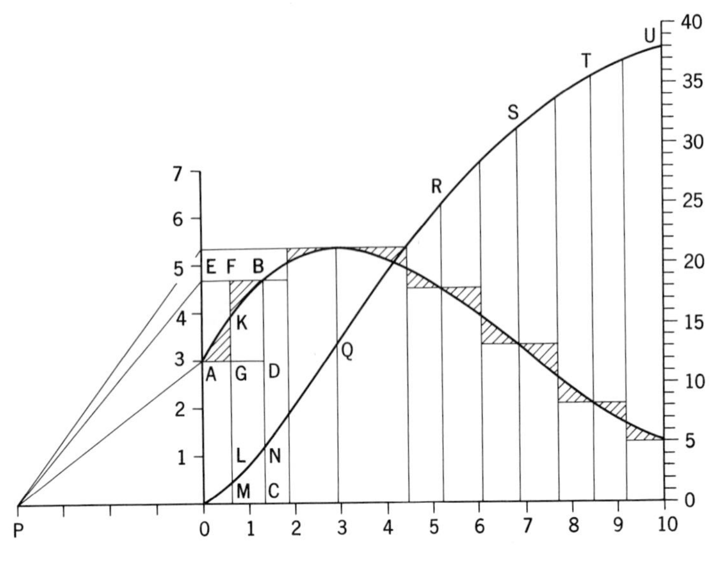

Fig. 13.4

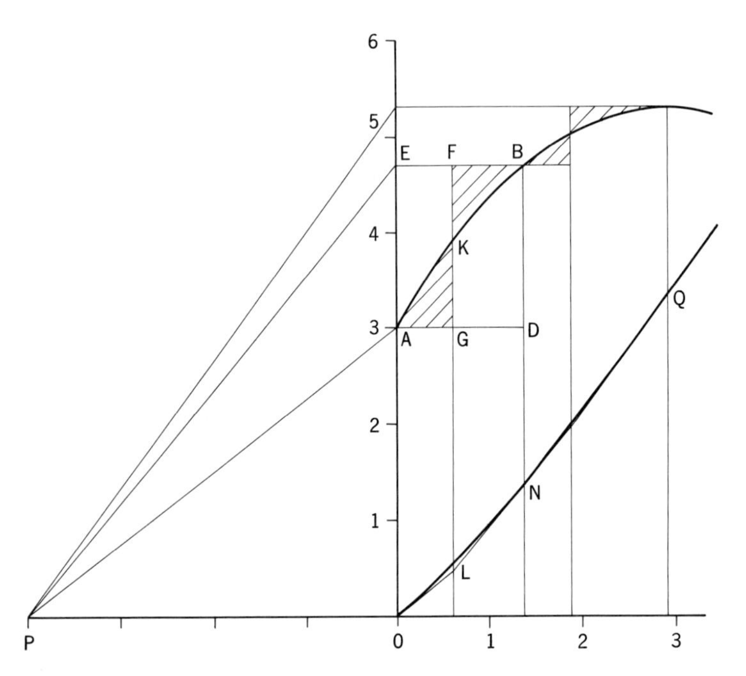

Fig. 13.5

This may invite some inaccuracy in the determination of the path of the integral curve. An alternative construction which provides better control of the path of the integral curve will now be described.

Pole-and-Ray Method; Alternative Construction

Let us consider vertical strip *OABC*, shown in Fig. 13.4. The width of the strip depends on the sharpness or flatness of the curve. Now, through points *A* and *B* we draw horizontals *AD* and *BE*, respectively. Vertical line *FG* is drawn so that areas *KGA* and *KFB* are approximately equal. Line *OL* is drawn parallel to ray *PA*. We know from our previous discussion that area $OAGM = \overline{LM} \times \overline{OP}$. Now line *LN* is drawn parallel to ray *PE*. It follows that $\overline{NC} \times \overline{OP} =$ area $\overline{OAKBC}$. The integral curve, it is seen, passes through points *O* and *N* tangentially to *OL* and *NL*. In a similar manner additional points and tangents are established. The integral curve passes through points *O*, *N*, *Q*, *R*, *S*, *T*, and *U* tangentially to the line segments adjacent to these points. *Q* is the inflection point. An enlargement of a portion of Fig. 13.4 is shown in Fig. 13.5, which clearly shows the integral curve passing through points *O* and *N* tangentially to line segments *OL* and *LN*.

Several examples employing the pole-and-ray method of integration follow.

EXAMPLE 1

Let us consider the subdivision of the plot shown in Fig. 13.6. The specification calls for two lots—the left one is to have an area equal to one-half of the right lot. The common line of the two lots is to be perpendicular to the property line OA.

First, we find the area of the plot by employing the pole-and-ray method. Now, distance AB, which is a measure of the area of the plot, is divided into parts AC and CB where $AC/CB = \frac{1}{2}$. The horizontal line drawn through point C to the integral curve establishes point 1. The vertical line, DE, through point 1 establishes the common line of the two lots.

EXAMPLE 2

An automobile starting from rest attains a speed of 12 mph during the first 5 seconds (low gear), remains at that speed for 1 second (shifting from low to second), attains a speed of 20 mph during the next 4 seconds (driving in second), remains at that speed for 1 second (shifting from second to high), attains a speed of 35 mph in the next 9 seconds, and continues to operate at that speed for 5 seconds more. What is the distance traveled?

A graphical representation of the above information is shown in Fig. 13.7 where the given data are satisfied and the curve appears reasonable for the acceleration periods.

If we integrate the curve graphically, using the alternative construction of the pole-and-ray method, we obtain the integral curve which enables us to read not only the distance, 805 feet, traveled at the end of 25 seconds, but also the distance traveled at the end of any time between zero and 25 seconds. Here we have an example which clearly shows the advantages of using a graphical solution because it portrays the important relationships in an understandable manner. Moreover, it is much faster and, therefore, is more economical than an algebraic solution.

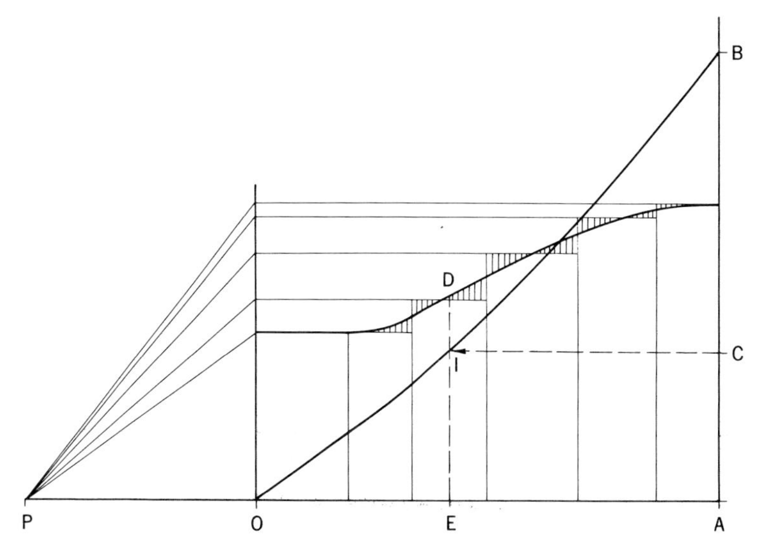

Fig. 13.6

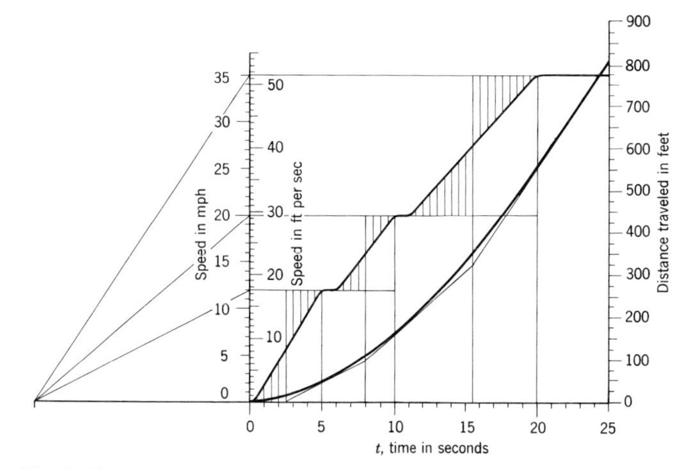

Fig. 13.7

EXAMPLE 3

A *research project*[*] dealing with problems related to the improvement and design of artificial limbs included studies of human locomotion. These studies covered the mechanics of motion of the legs, measurements of the ranges of motion in space, including rotations of the major segments of the legs during locomotion; and, in particular, the study of the action in the major joints of the leg—the hip, knee, and ankle joints. Other problems dealt with the determination of the vertical components of velocity and displacement of the center of gravity of a normal subject during level walking. A study of power supply by muscles and energy level of the shank during level walking as a function of time is shown in Fig. 13.8. The solid curve shows the relation between power supplied by the muscles and time.

[*] Prosthetic Devices Research Project conducted by the College of Engineering, University of California at Berkeley.

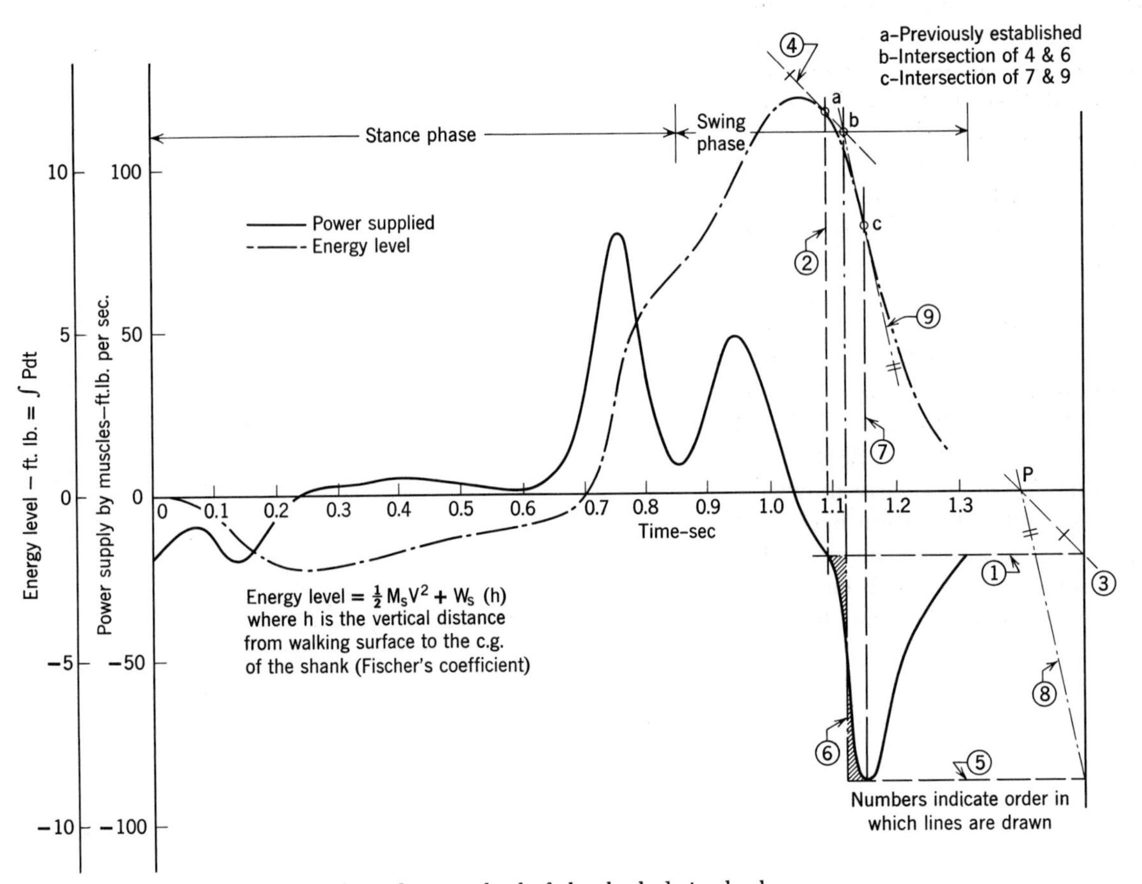

Fig. 13.8 Power supply by muscles and energy level of the shank during level walking versus time.

This curve was integrated by the pole-and-ray method to establish the dashed curve which shows the relation between energy level and time.

EXAMPLE 4 CALIBRATION OF ACCELEROMETER

The following experimental data show the relation between accelerometer output (in volts) and time after impact in milliseconds.

t, sec $\times 10^{-3}$	0	0.25	0.50	0.75	1.00	1.25	1.50	2.00	2.50
e, volts	0	−0.30	−0.15	−0.94	−0.94	−0.87	−0.83	−0.82	−0.81
t, sec $\times 10^{-3}$	3.00	3.50	4.00	4.50	5.00	5.50	6.00	6.50	7.00
e, volts	−0.77	−0.72	−0.66	−0.59	−0.51	−0.42	−0.30	−0.19	−0.13
t, sec $\times 10^{-3}$	7.50	8.00	8.50	9.00	10.00				
e, volts	−0.08	−0.05	−0.03	−0.01	0.00				

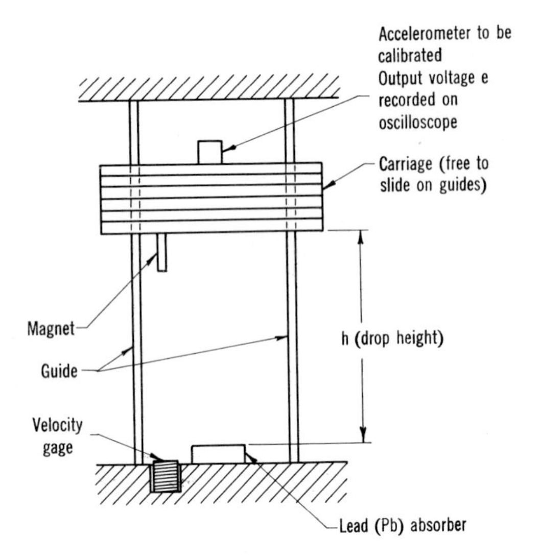

Fig. 13.9

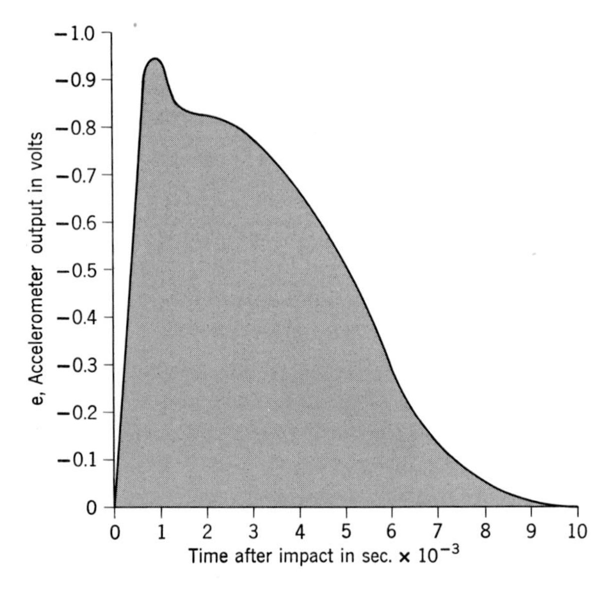

Fig. 13.10

A sketch of the experimental set-up is shown in Fig. 13.9; and a plot of the data curve in Fig. 13.10.

Our problem is to calibrate the accelerometer. In the accelerometer, $a = Ke$,

where $e =$ output voltage from a piezoelectric crystal in the accelerometer

$K =$ accelerometer constant to be determined experimentally.

Solution

1. When the free-falling carriage comes to a stop after striking the lead absorber we see that:

$$\Delta v = \int a\,dt$$

or

$$\Delta v = K\int e\,dt$$

and

$$\Delta v = KA$$

where A is the area under the plotted curve shown in Fig. 13.10.

2. For the free-falling carriage, released from height h,

$$v = \sqrt{2gh}$$

since $v = \Delta v$ when the carriage comes to a stop after dropping the distance, h.

3. Therefore, when $h = 3$ ft,

$$K = \frac{\sqrt{2gh}}{A} = \frac{13.9\ fps}{A}$$

By integrating the curve—either by the pole-and-ray method (or by using a planimeter which is discussed later)—we obtain the value of $A = 0.0044$. Now,

$$K = \frac{13.9}{0.0044} = 3158.9$$

Other Methods of Integration

In addition to the pole-and-ray method there are others that give good results, especially when the integral curve is not needed. Among these are (1) the method of trapezoids; (2) the method of rectangles; (3) the method of parabolas; (4) a numerical method which employs Simpson's Rule; and (5) a mechanical method.

Each of these methods will now be discussed.

1. Method of Trapezoids. Let us consider the problem of finding the area under the curve shown in Fig. 13.11. The area is divided into vertical strips, the widths of which depend on the sharpness or flatness of the curve. Narrower strips are employed for the sharper portions. When the portion of the curve AD is very nearly a straight line, the area of strip S_1 is closely approximated by the area of the trapezoid $OADE$. The area of this trapezoid is equal to $\frac{1}{2}(\overline{OA} + \overline{DE}) \times \overline{OE}$. In a similar manner the areas of strips S_2, S_3, S_4, S_5, and S_6 can be determined. The area under the curve is $A = S_1 + S_2 + S_3 + S_4 + S_5 + S_6$.

2. Method of Rectangles. Consider the area $OABC$, shown in Fig. 13.12. The area is divided into strips whose widths depend upon the degree of curvature of the curve. Consider strip S_1. A horizontal line m is introduced so that the shaded areas are approximately equal. This is done quite accurately by eye. The area of strip S_1 is now equal to the area of the rectangle whose base is OE and whose altitude is DE. This area is, of course, $\overline{OE} \times \overline{DE}$. In a similar manner the areas of the other strips may be found. The total area under the curve is the sum of the areas of the rectangles that are equivalent in area to the various strips.

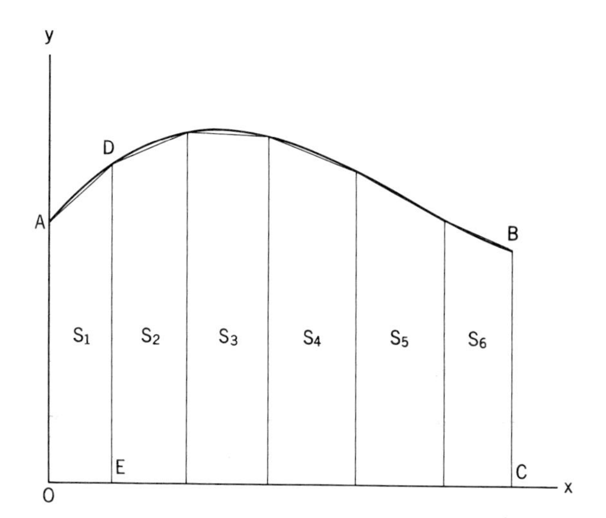

Fig. 13.11

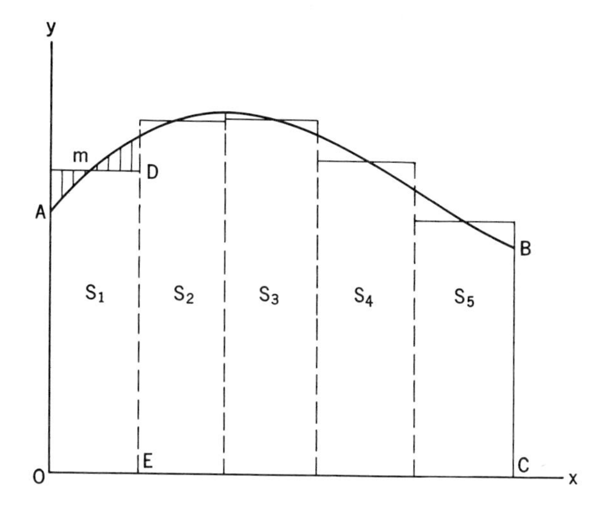

Fig. 13.12

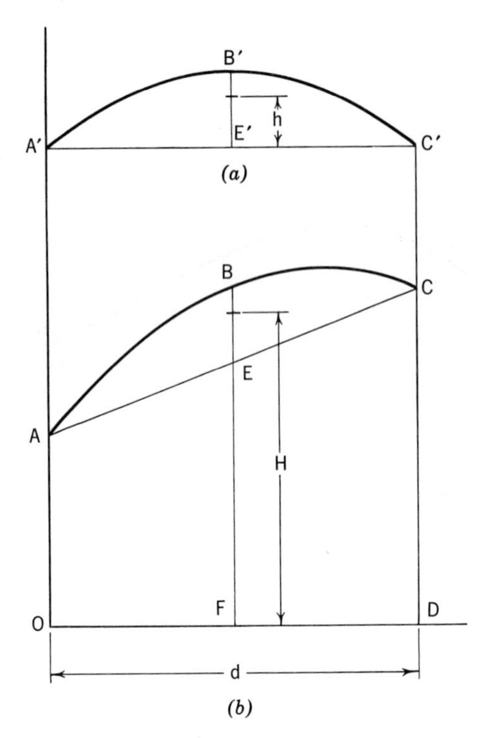

Fig. 13.13

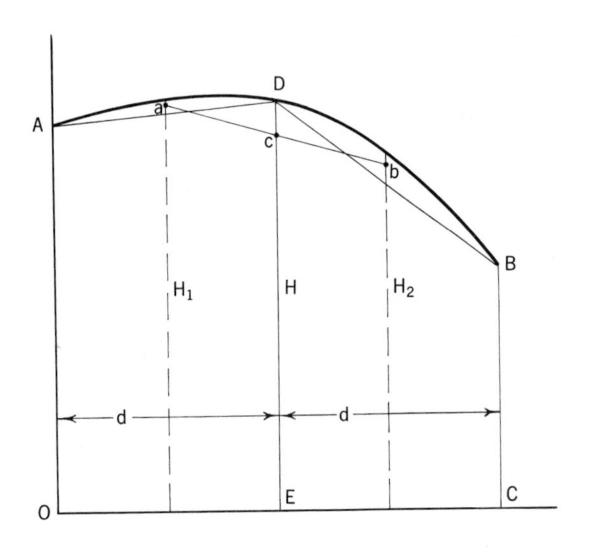

Fig. 13.14

3. ***Method of Parabolas.*** * This method is rapid and gives close approximations of the areas. For example, let us find the area *OABCD* (Fig. 13.13).

First, consider part (*a*) of this figure. Assume that the curve $A'B'C'$ is a parabola and that $\overline{B'E'}$ is equal to $\overline{BE}$. Now area $A'B'C' = hd$, where $h = (\frac{2}{3})B'E'$. If all the vertical lines, such as $B'E'$, of area $A'B'C'$ are approximately equal to corresponding lines, such as BE, of area ABC, then the two areas are approximately equal to each other.

Therefore, area $OABCD$ is equal to Hd, where $H = EF + (\frac{2}{3})BE$. This is true since the total area is made up of trapezoid $OACD$, the area of which is $EF \times d$, and area ABC, the area of which is $(\frac{2}{3})BE \times d$.

Now let us apply this method to a larger area, such as that shown in Fig. 13.14. Suppose we divide area $OABC$ into two strips of equal width, and then determine H_1 and H_2 for the two strips in a manner similar to that used in Fig. 13.13. The line which joins points a and b (the upper end points of H_1 and H_2 respectively) intersects line DE at point c.

The total area is equal to $H \times 2d$, where $H = cE$. Obviously, this method may be extended to include larger areas.

* "Fast Method for Finding Areas or Mean Ordinates of Curves," by A. D. Moore, *Journal of Engineering Education*, Vol. 31, No. 7, March 1941.

4. Method Employing Simpson's Rule. This is a *numerical* method for finding the area under a curve. Let us consider the area *OABCD*, shown in Fig. 13.15. The area is divided into two strips of equal width. If we regard the curve *ABC* as part of a parabola, it will be shown that the total area can be computed from the equation

$$A = \frac{d}{3}(y_0 + 4y_1 + y_2)$$

where y_0, y_1, and y_2 are the ordinates of points *A*, *B*, and *C*, respectively.

If we employ the method of parabolas, it follows that the area under the curve *ABC* may be expressed as

$$A = \frac{2}{3}\left(y_1 - \frac{y_0 + y_2}{2}\right)2d + \left(\frac{y_0 + y_2}{2}\right)2d$$
$$= \frac{d}{3}(y_0 + 4y_1 + y_2)$$

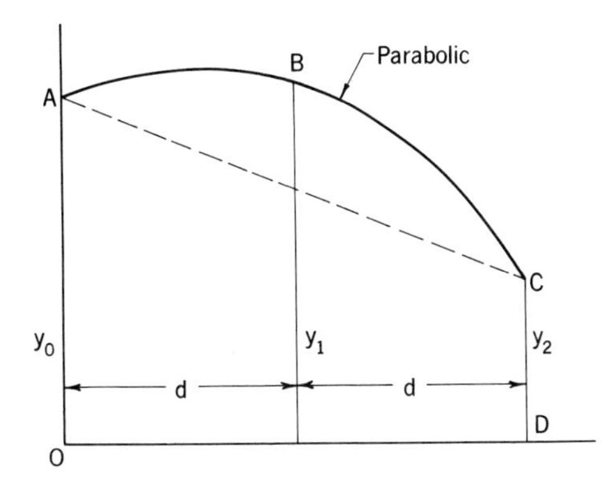

Fig. 13.15

EXAMPLE

Suppose we are to determine the magnitude of the area *OABC*, shown in Fig. 13.16. First, we divide the area into an *even* number of strips (ten strips in this case). Now we measure the distances $y_0, y_1, \ldots, y_{10}$, which are recorded in the table.

y_0	y_1	y_2	y_3	y_4	y_5	y_6	y_7	y_8	y_9	y_{10}
5	6.5	7.1	7.1	6.8	6.5	6.5	6.8	7.1	8	9.1

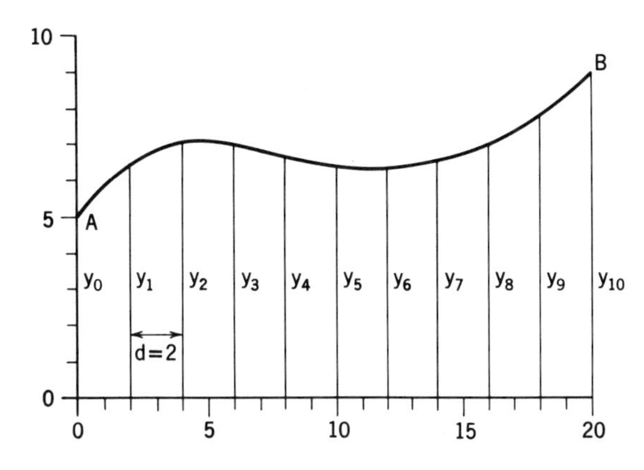

Fig. 13.16

From our previous discussion we recall that the area of the first two strips can be determined from the expression $(d/3)(y_0 + 4y_1 + y_2)$.

Now the area of the third and fourth strips can be determined in a similar manner from the expression $(d/3)(y_2 + 4y_3 + y_4)$. Therefore, the sum of the areas of the first four strips is equal to $(d/3)(y_0 + 4y_1 + 2y_2 + 4y_3 + y_4)$.

It should be clear that the single expression, $(d/3)(y_0 + 4y_1 + 2y_2 + 4y_3 + 2y_4 + 4y_5 + 2y_6 + 4y_7 + 2y_8 + 4y_9 + y_{10})$, sums the areas of all of the strips. The total, therefore, is $(\frac{2}{3})(5 + 26 + 14.2 + 28.4 + 13.6 + 26 + 13 + 27.2 + 14.2 + 32 + 9.1) = 139.1$ square units.

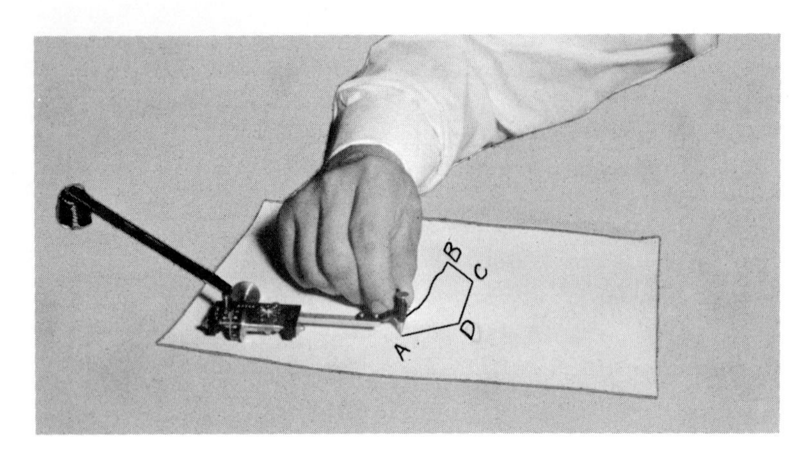

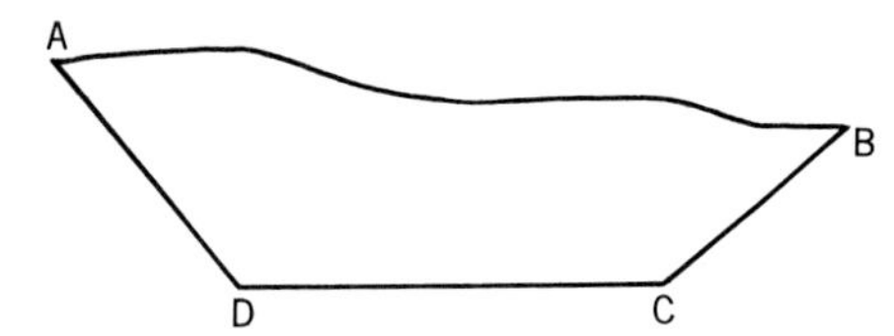

Fig. 13.17

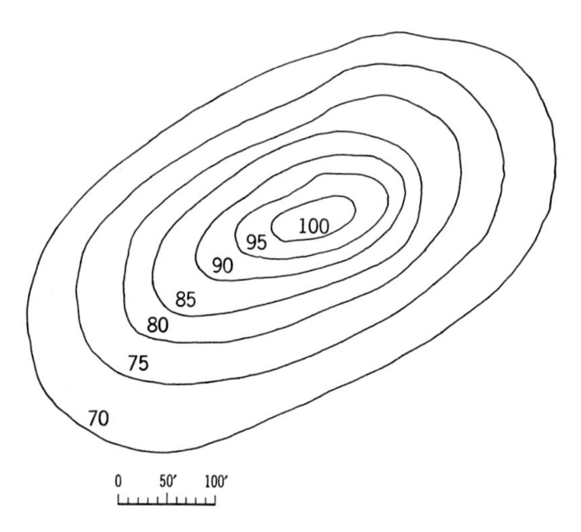

Fig. 13.18

5. *Method Employing the Use of a Planimeter.*[*] Another method which is very practical employs the use of an instrument known as a *planimeter* which, in effect, is a mechanical integrator. Figure 13.17 is a photograph of a polar planimeter set up for the determination of the area of a road cross section. The tracer point is started at point A (any other point would do) and is then moved to follow along the irregular (topographic) curve to point B, then to point C along line BC, then along line CD, and finally back to point A along line DA. The number of square units in the cross-sectional area is read on the graduated wheel, whose axis is horizontal.

If the areas of consecutive cross sections are determined, the number of cubic yards of earth required for fills and the number of cubic yards that must be removed can be computed, if the distances between the cross sections are known. Such information is necessary when considering costs of highway construction.

Other examples follow.

EXAMPLE 1

Let us consider the hill which is shown topographically in Fig. 13.18. It is proposed to remove the earth above contour 70 ft. How many cubic yards will be removed?

The area bounded by each contour is first determined by using a planimeter. The values thus obtained are shown in this table:

Contour, feet	100	95	90	85	80	75	70
Area, sq ft	2,600	9,400	18,800	36,200	59,000	95,000	164,500

[*] Most calculus books include a mathematical development of the principle on which the design of a planimeter is based.

A plot of these values is shown in Fig. 13.19. The volume, in cubic feet, is obtained by integrating (pole-and-ray method) the area under the curve which passes through the plotted points. The volume scale is shown both in cubic feet and cubic yards. The amount of earth to be removed is 55,000 cubic yards. How many cubic yards would be removed if it had been proposed to level off at contour 85 ft?

EXAMPLE 2

The ampere-time record of a power station is shown in the table. What was the total number of ampere-hours supplied by this power station?

t, time	12 noon	1	2	3	4	5	5:30	6
A, amperes	4700	3100	3000	2700	3100	7900	15000	19000
t, time	6:30	7	8	9	10	11	12	
A, amperes	22500	23000	20000	17000	11000	9500	8200	

A plot of the data is shown in Fig. 13.20. The area under the curve which best fits the data represents the total number of ampere-hours supplied by the station. The area is determined, quite easily, by using a planimeter. Try to check the result by employing *Simpson's Rule.*

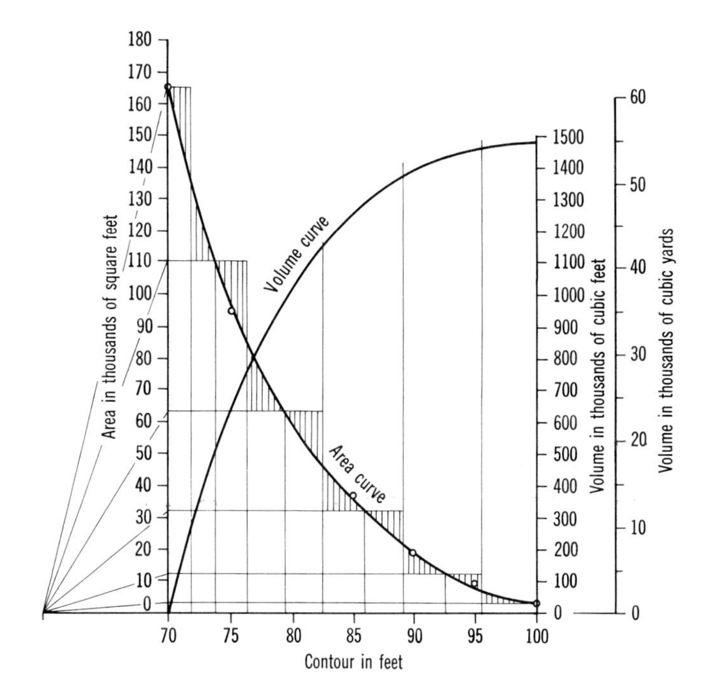

Fig. 13.19

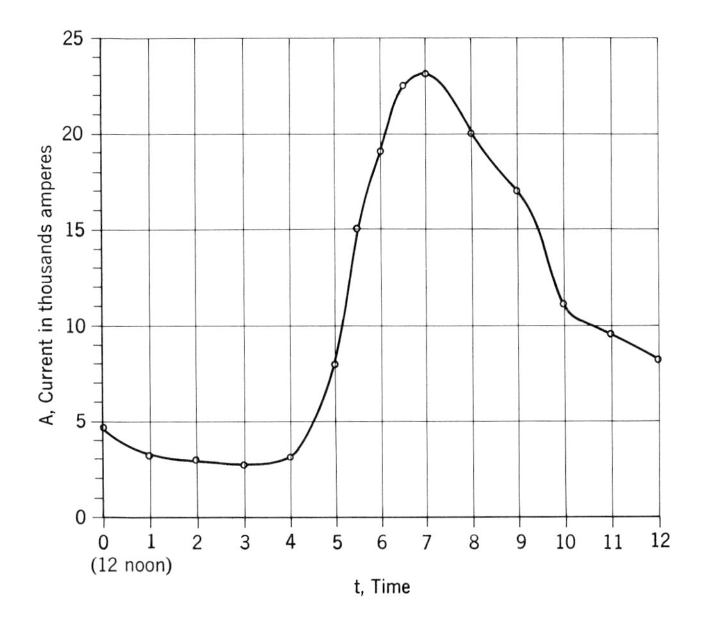

Fig. 13.20

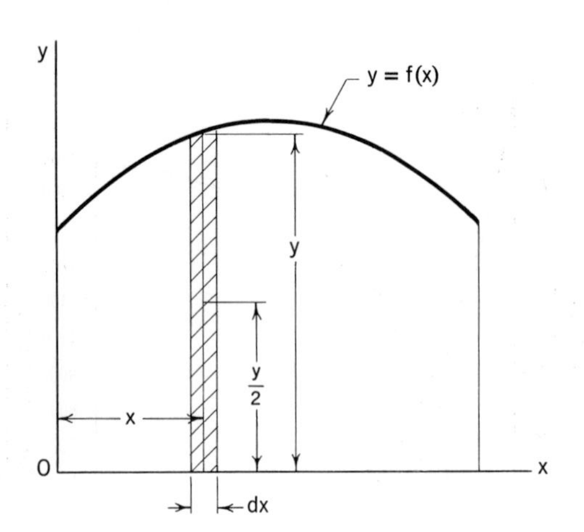

Fig. 13.21

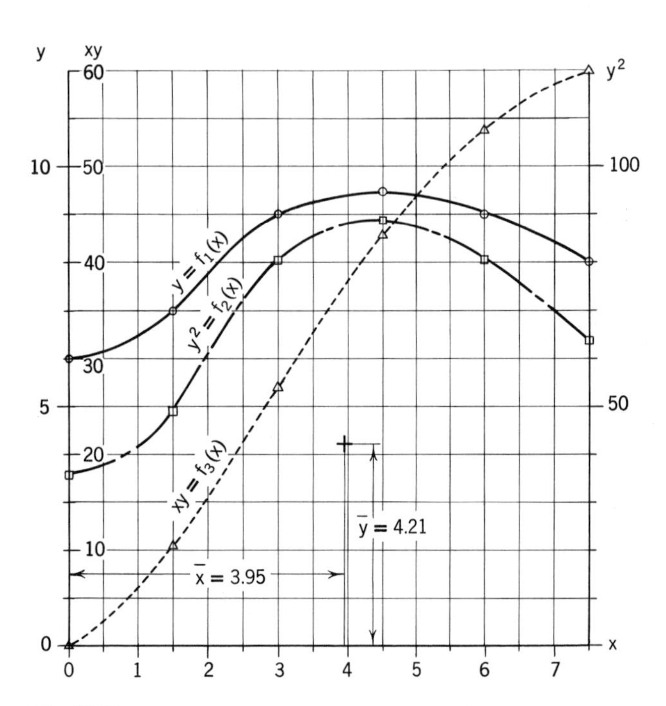

Fig. 13.22

EXAMPLE 3 CENTROIDS OF AREAS

Let us consider the area under the curve shown in Fig. 13.21. The centroid of the area is that point about which the algebraic sum of all the moments of infinitesimal areas of the surface is zero. It is required to determine the coordinates $\bar{x}$ and $\bar{y}$ of the centroid, C.

The coordinate $\bar{x}$ is determined by taking moments about the y-axis and then dividing the sum of the moments of all strips (width dx) by the total area.

The moment about the y-axis of a typical strip, such as $y\,dx$, is equal to $xy\,dx$. The summation of the moments of all strips is $\int xy\,dx$. Therefore, the x-coordinate of the centroid is

$$\bar{x} = \frac{\int xy\,dx}{\int y\,dx}, \qquad \text{where } \int y\,dx \text{ is the total area.}$$

By taking moments about the x-axis, we find that the moment of strip $y\,dx$ is equal to $(y\,dx)(y/2) = \frac{1}{2}y^2\,dx$, where $y/2$ is the moment arm. Therefore, the $\bar{y}$-coordinate of the centroid is

$$\bar{y} = \frac{\frac{1}{2}\int y^2\,dx}{\int y\,dx}.$$

The above equations are useful, provided we know the equation of the curve. In many cases the data are experimental and the equation is unknown and cannot be economically determined. We can, however, solve the problem graphically.

Let us determine the centroid of the area under the curve shown in Fig. 13.22. This curve is based upon the data in the first table.

x	0	1.5	3	4.5	6	7.5
y	6	7	9	9.5	9	8

Since $\bar{x} = \dfrac{\int xy\,dx}{\int y\,dx}$, we can plot a curve whose ordinates are (xy) and whose abscissas are values of x, as shown in the second table.

x	0	1.5	3	4.5	6	7.5
xy	0	10.5	27	43	54	60

Now the area under each curve can be determined by the use of a planimeter or by graphical integration. Once this is done

$$\bar{x} = \frac{\text{area under the curve } xy = f_3(x)}{\text{area under the curve } y = f_1(x)} = 3.95$$

In a similar manner, if we plot the curve $y^2 = f_2(x)$, we can determine the area under this curve and then evaluate

$$\bar{y} = \frac{\frac{1}{2}\text{ area under the curve } y^2 = f_2(x)}{\text{area under the curve } y = f_1(x)} = 4.21$$

Figure 13.22 also includes the curve $y^2 = f_2(x)$, which is based upon the coordinates in the third table.

x	0	1.5	3	4.5	6	7.5
y^2	36	49	81	89	81	64

A very useful grapho-mechanical solution to centroid and moment of inertia problems is applied to the next example.

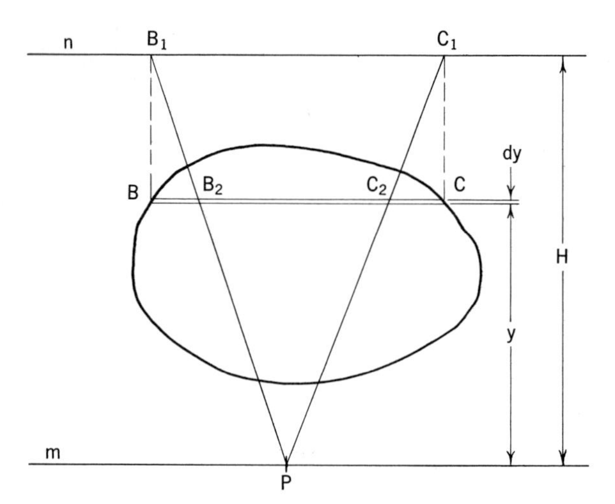

Fig. 13.23

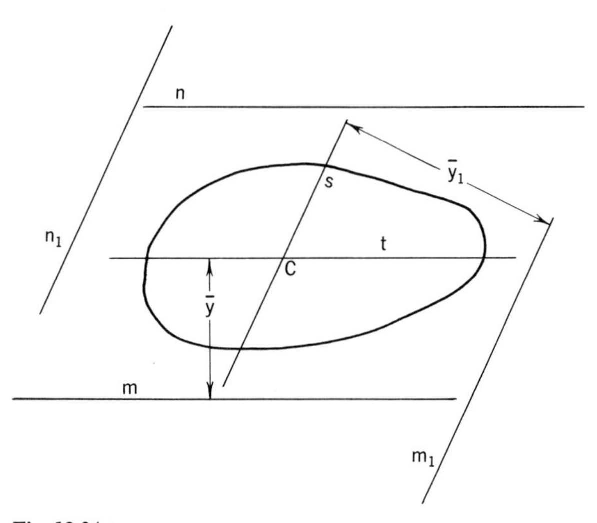

Fig. 13.24

EXAMPLE 4

Let us consider the area bounded by the curve shown in Fig. 13.23. It is required to determine the location of its centroid.

First, draw parallel lines m and n a convenient distance (H) apart. Now introduce line BC parallel to m and project points B and C to line n by perpendiculars to line n. These points are shown as B_1 and C_1, respectively. Lines PB_1 and PC_1 (where P is any point on line m) intersect line BC in points B_2 and C_2. From similar triangles, PB_1C_1 and PB_2C_2, it follows that

$$\frac{B_2C_2}{B_1C_1} = \frac{y}{H}, \quad \text{or} \quad \frac{dA_1}{dA} = \frac{y}{H},$$

if we let $dA_1 = B_2C_2 \times dy$, and $dA = BC \times dy$. Therefore,

$$A_1 = \int dA_1 = \frac{1}{H}\int y\, dA \quad \text{and} \quad \bar{y} = \frac{A_1 H}{A},$$

since

$$\bar{y} = \frac{\int y\, dA}{A}$$

If the construction shown in Fig. 13.23 is repeated for additional parallels to BC, the area bounding the curve which passes through such points as B_2C_2 will be area A_1, which can be determined by the planimeter method or one of the methods of graphical integration. This is also true for area A.

It should be clear that parallel lines m and n could be drawn in any direction. If the construction is repeated for a new orientation of parallels m and n, another centroidal distance ($\bar{y}_1$) can be found with respect to the new position of line m. Then the centroid is the point in which the $\bar{y}$-line intersects the $\bar{y}_1$-line.

For example, assume that $\bar{y}$ for the area shown in Fig. 13.24 has been found with respect to the horizontal position of lines m and n. Now suppose that $\bar{y}_1$ has been determined for the new positions of lines m and n as indicated by m_1 and n_1. The intersection of lines s and t is the centroid, C.

Let us consider the area, A, shown in Fig. 13.25. The moment of inertia of this area about axis m is the summation of all the products formed by multiplying every elementary area by the square of its distance from the axis.

First, let us repeat the construction shown in Fig. 13.23 to locate points B_2 and C_2. Now project these points to line n to locate points B_3 and C_3, respectively. Lines PB_3 and PC_3 will intersect line BC in points B_4 and C_4, which are on the boundary of area A_2.

From similar triangles PB_3C_3 and PB_4C_4 it follows that

$$\frac{B_4C_4}{B_3C_3} = \frac{y}{H}, \quad \text{or} \quad \frac{dA_2}{dA_1} = \frac{y}{H}$$

$$A_2 = \frac{1}{H}\int y\,dA_1 = \frac{1}{H^2}\int y^2\,dA,$$

since $$dA_1 = \frac{y\,dA}{H} \text{ (see Example 4)}$$

Therefore, the moment of inertia about axis m is A_2H^2. Area A_2 can be determined by the planimeter method or by one of the methods of graphical integration.

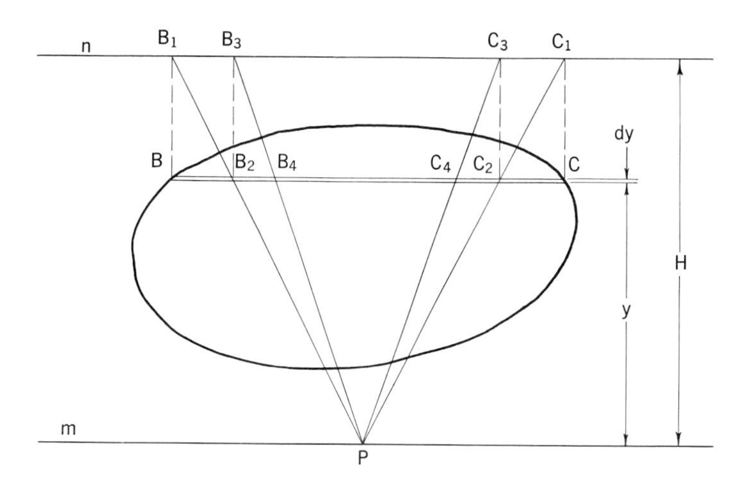

Fig. 13.25

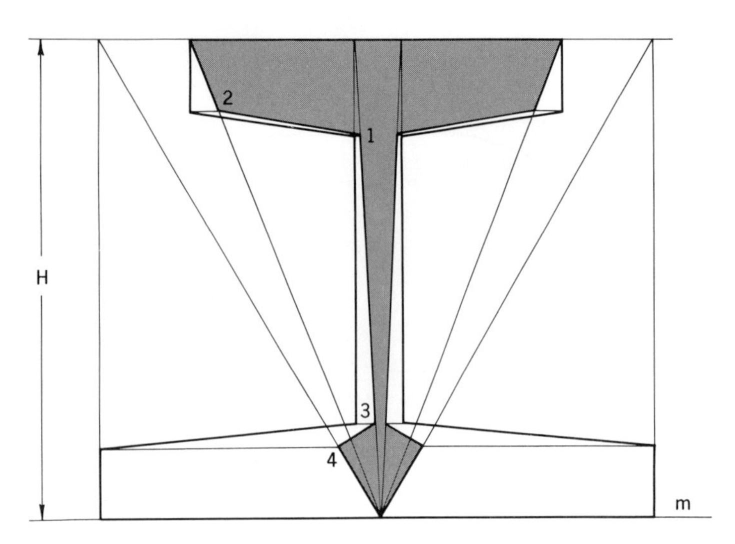

Fig. 13.26

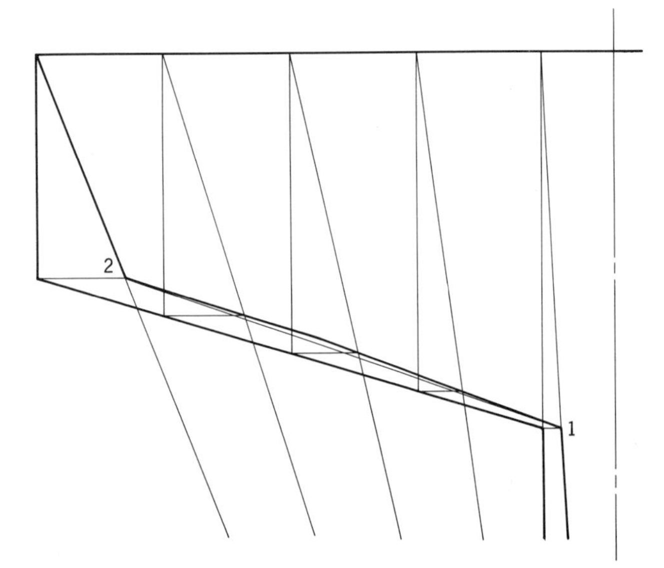

Fig. 13.27

Now let us consider the section shown in Fig. 13.26. It is required to determine its centroid. The construction used in Example 5 yields the shaded area A_1, which can be evaluated by the planimeter method. The $\bar{y}$ value for the centroid is $\dfrac{A_1 H}{A}$, and $\bar{x}$ is zero, if the y-axis is coincident with the axis of symmetry.

If area A_1 is regarded as the given section and the process repeated to determine area A_2, the moment of inertia about axis m is $A_2 H^2$.

It should be noted that actually line 1–2 and line 3–4 are not straight lines, but in the above example are drawn as straight lines since they do not deviate appreciably from the correct curve. Figure 13.27 shows an enlarged drawing of the portion of the area that affects line 1–2, to illustrate the difference between the line 1–2 and the correct curve 1–2.

GRAPHICAL DIFFERENTIATION

In a physical sense, we may regard differentiation as meaning *"the determination of the rate of change of one variable with respect to a related variable."* Let us study this statement carefully. Suppose the velocity, V, of a body as measured at various distances, S, from a known location is given as in the table.

S, in feet	0	0.25	0.50	0.75	1.00	1.25	1.50
V, in feet per second	0	0.10	0.18	0.42	0.95	1.70	3.22

What are the values of the rate of change of velocity, V, with respect to distance, S, for the intervals shown in the table?

The change in velocity, V, corresponding to the displacement from 0 to 0.25 feet is expressed as

$$\frac{\text{Change in } V, (\Delta_1 V)}{\text{Change in } S, (\Delta_1 S)} = \frac{0.10 - 0}{0.25 - 0} = 0.40 \text{ (ft/sec)/ft.}$$

In the first interval, therefore, the *average* rate of change of velocity with respect to the corresponding distance is 0.40 (ft/sec)/ft.

In the second interval it is

$$\frac{\Delta_2 V}{\Delta_2 S} = \frac{0.18 - 0.10}{0.50 - 0.25} = 0.32 \text{ (ft/sec)/ft.}$$

In the third interval it is

$$\frac{\Delta_3 V}{\Delta_3 S} = \frac{0.42 - 0.18}{0.75 - 0.50} = 0.96 \text{ (ft/sec)/ft.}$$

In a similar manner we can compute the values for the other intervals. We should note carefully that each of the values is the *average* value for an interval.

If the distance interval, ΔS, had been smaller, we still would have obtained an *average* value for each interval. If, however, ΔS were made infinitely small (ΔS approaching zero as a limit), the *average* values would become *exact* values at the respective values of S. The *exact* values are denoted by the expression dV/dS, where this ratio is the rate of change of V with respect to S. In the algebraic form of the calculus the relation between the variables is usually expressed algebraically (i.e., $V = S^3$) and then differentiated. (That is, $dV/dS = 3S^2$ in this case. The rate of change of V with respect to S can be obtained for any value of S from this equation.)

If the above data are plotted and a curve is drawn through the points, the average rates of change, $\Delta V/\Delta S$, would be recognized as slopes of chords (formed by joining the points) of the curve; and it would be recognized that, as ΔS approaches zero, each chord approaches a tangent to the curve at a particular point. Thus, graphically, differentiation is merely the determination of the slope of a tangent to a curve. For example, if $y = f(x)$, then dy/dx is the slope of the tangent to this curve at any point of the curve. It is clearly seen in Fig. 13.28 that, at point P, dy/dx for the curve is equal to the slope, $\Delta y/\Delta x$, of tangent t at point P. It should be observed that Δy *is measured in terms of the scale on the y-axis* and, similarly, that Δx *is measured in terms of the scale on the x-axis.*

Methods for determining the tangent at any point of a given curve and the procedures used in graphical differentiation will now be considered.

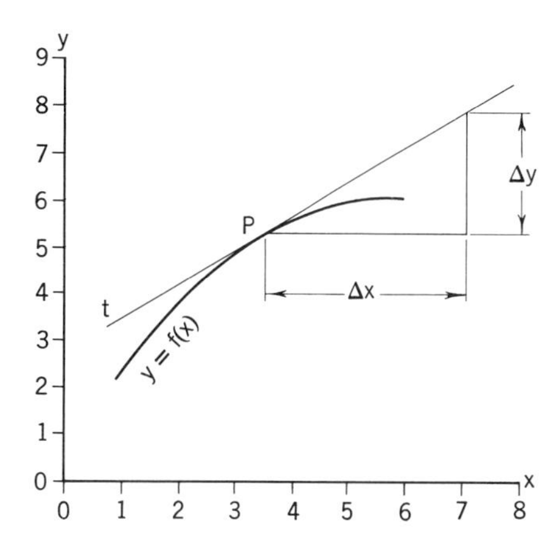

Fig. 13.28

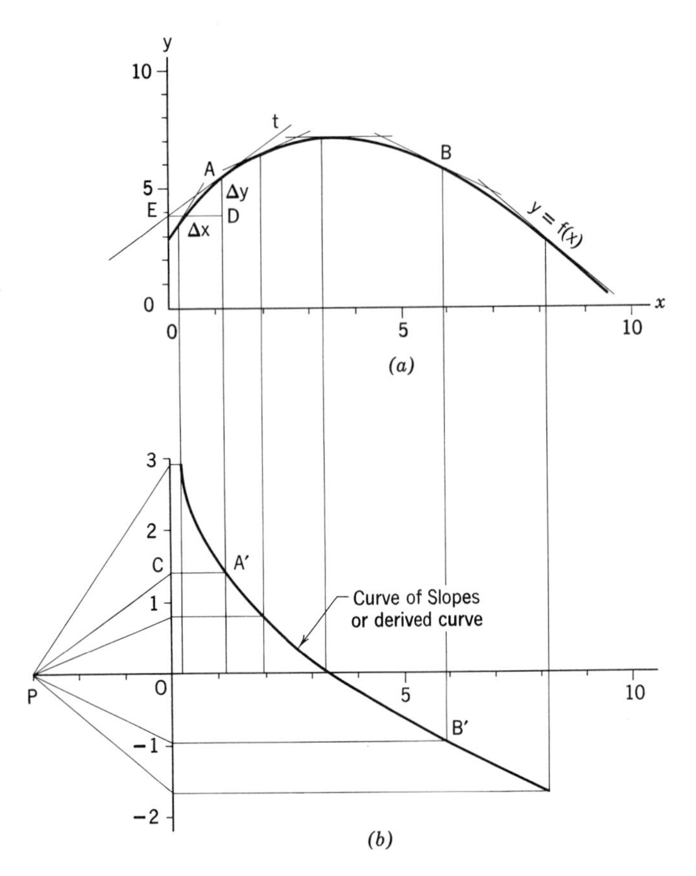

Fig. 13.29

Suppose we are given the curve $y = f(x)$ as shown in Fig. 13.29(a). Now let us draw line t tangentially to the curve at point A (any point on the curve). If we plot the magnitude of the slope of line t, and of additional tangents to other points of the given curve, as ordinate values corresponding to the abscissa values of the points, the curve passing through these ordinate values will be the *curve of slopes* or the *derived curve.*

A very simple graphical method will now be used to determine this curve. In Fig. 13.29(b) a distance OP, known as *the pole distance,* is laid off to the left of the origin. This distance is, for convenience, chosen as an integral number of the units used along the x-axis. In this case $OP = 3$ units of x.

Now through point P a line is drawn parallel to the tangent t [Fig. 13.29(a)] to intersect the y-axis at point C. The point of intersection, A', of the vertical line drawn through point A [Fig. 13.29(a)], with the horizontal line drawn through point C, is a point on the curve of slopes or the derived curve.

If this is true, the ordinate value of A' must be the measure of the magnitude of the slope of line t. In Fig. 13.29(a) we observe that the ratio $AD/DE = (\Delta y)/(\Delta x)$ is a measure of the slope of line t. It should be pointed out again, however, that the *measurement of distance AD must be in terms of the vertical scale,* and that the *measurement of distance DE must be in terms of the horizontal scale.* Now, since triangles ADE and COP are similar (by construction), it follows that $\Delta y/\Delta x = AD/DE = CO/OP$. It should be clear that a change in the location of point P affects the length of CO.

Now, if distance CO, which is a measure of the ordinate of point A', is to represent the correct magnitude of the slope of tangent t, the y-axis scale must be determined accordingly. Since the x-axis graduations are the same for both figures (usual and convenient practice), it should be fairly evident that the values on the y-axis of Fig. 13.29(b) are equal to the values on the y-axis of Fig. 13.29(a) *divided by the pole distance* (measured in terms of the horizontal scale). In this case the location of point 1 on the vertical axis of the derived curve is the same distance from the origin as point 3 of the given curve, because the pole distance is 3.

The curve of slopes, or the derived curve, passes through points such as A'. In graphical terms, differentiation is simply a process for determining the magnitude of the slope at any point of the given curve. The location of a tangent to the curve can be closely approximated by first drawing two parallel chords, and then locating the point in which the line joining the midpoints of the chords intersects the curve. The tangent to the curve at the point of intersection is approximately parallel to the chords. (See Fig. 13.30.)

A more precise method can be used to locate the tangent. It consists of placing a polished surface or a mirror perpendicular to the paper and across the given curve. If the reflected portion of the curve shows no break in the curve, i.e., if it appears to be a smooth continuation of the portion of the curve in front of the polished surface, the line of intersection of the planes of the paper and of the polished surface is the normal (perpendicular) to the curve at the point of intersection. The required tangent is perpendicular to this normal. (See Fig. 13.31.)

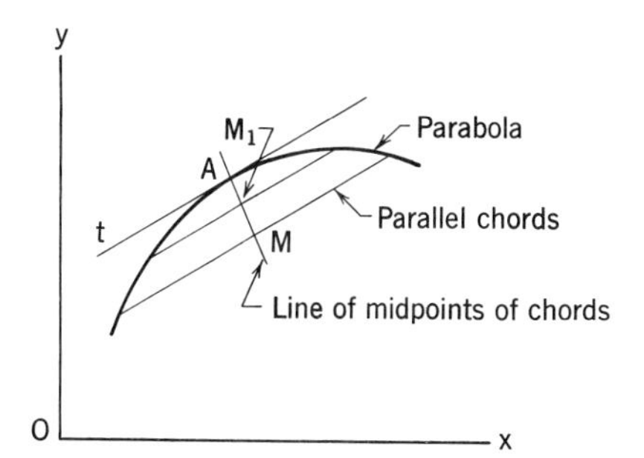

Fig. 13.30

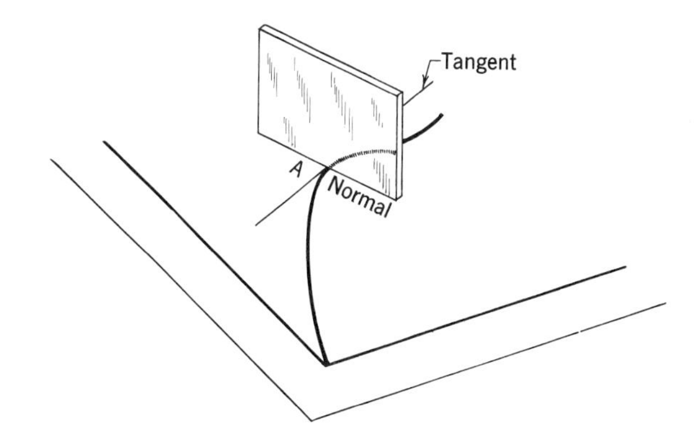

Fig. 13.31

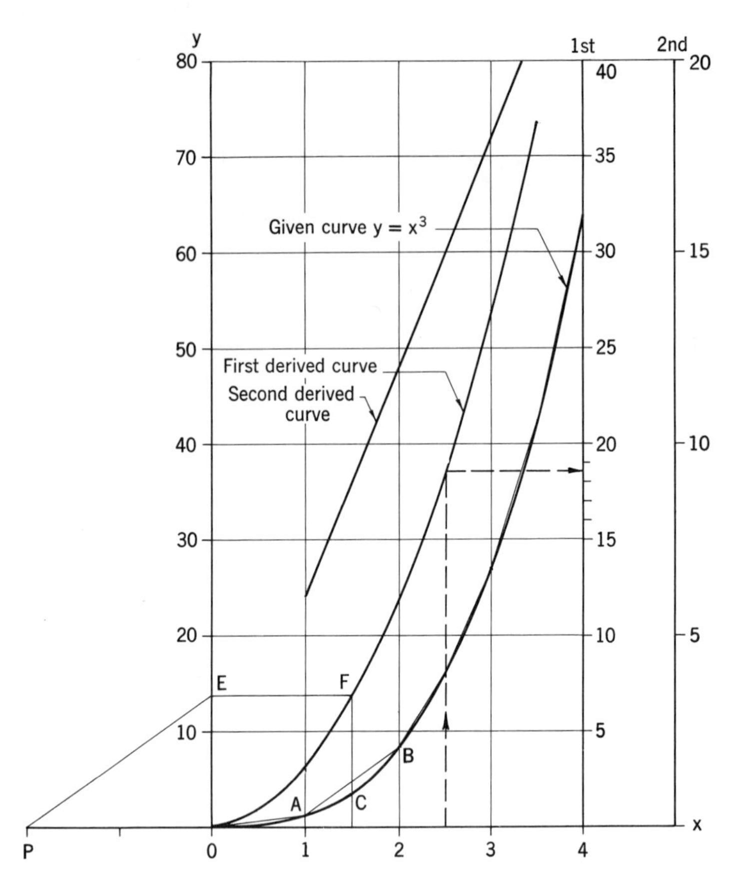

Fig. 13.32

EXAMPLE 1

Let us consider the equation $y = X^3$. A portion of the curve described by the equation is shown in Fig. 13.32. We are to differentiate the plotted curve twice. We will assume a pole distance of 2 units of X. Now let us introduce chords such as AB. The chords are approximately parallel to the tangents at points such as C which lie on the vertical lines that bisect those chords. It should be carefully noted that the selected chord lengths will depend on the curvature of the given curve, shorter chords being used for the sharper portions of the curve. A line (ray) is drawn through point P parallel to chord AB to intersect the y-axis at point E. A horizontal line drawn through point E intersects the vertical line drawn through the midpoint of chord AB at point F which is a point on the *first derived curve.* Other points are located in a similar manner. The scale on the right marked *1st* is used to read the magnitude of the slope at any point of the given curve. This is done by drawing a vertical line through the selected value, for example, $X = 2\frac{1}{2}$, to intersect the first derived curve, and then drawing through that point a horizontal line to the scale marked *1st.* In this case the reading is 18.75 which is the slope when $X = 2\frac{1}{2}$. *We should not forget that the values on the scale marked 1st are equal to the values on the y-axis divided by the pole distance.*

The second derived curve is obtained in a similar manner, *by considering the first derived curve as the given curve.* In the example, Fig. 13.32, the same pole distance, 2, is used; however, a different value could be chosen. Again, we should note that the values on the scale marked *2nd* are equal to the values on the scale marked *1st* divided by the pole distance.

EXAMPLE 2

The *prosthetic devices research project,** dealing with problems of human locomotion, included a study of the determination of the variation of ankle angle, θ_A, and of angular velocity with time in seconds. A plot of the data relating ankle angle to time is shown in Fig. 13.33. The curve which was drawn through the data points was differentiated graphically to establish the velocity-time curve. The advantage in using graphical differentiation in this case is quite evident. It would be quite uneconomical to solve the problem algebraically.

* This project, conducted at the University of California by faculty members of the College of Engineering and of the Medical School, was started in 1945 on the Berkeley campus and is still active.

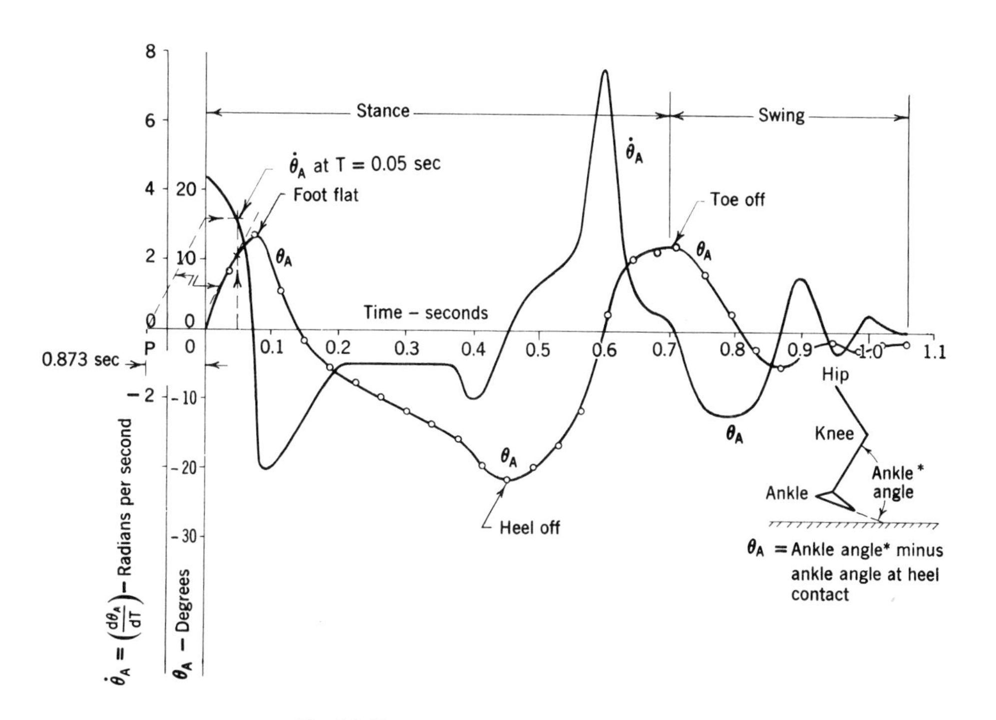

Fig. 13.33

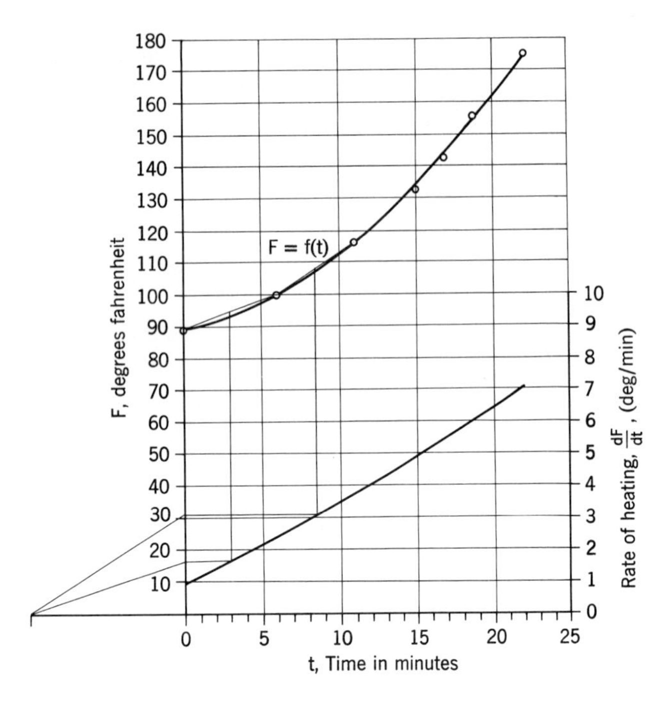

Fig. 13.34

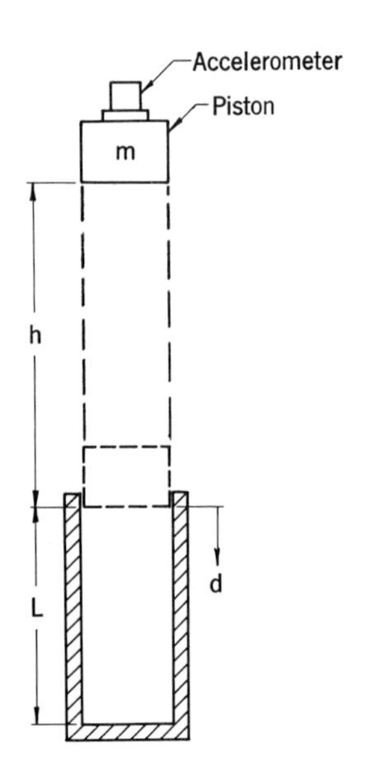

Fig. 13.35

EXAMPLE 3

The relation of temperature, F, in degrees Fahrenheit, of heating water to time, t, is shown in this table:

t, minutes	0	6	11	15	17	19	22
F, deg. Fahrenheit	89.6	99.5	116.9	131.9	142.7	156.2	176.0

We are to determine the rate of heating at any time, t.

A plot of the data and a "fair" curve drawn through the plotted points are shown in Fig. 13.34. To solve our problem we differentiate the curve by using the pole-and-ray method. The scale for the "rate-of-heating" curve is easily obtained by *dividing* the F-scale by the pole distance.

EXAMPLE 4

An accelerometer calibrator consists of a piston, of mass m, which is dropped freely in the earth's gravitational field a distance, h. The piston then engages an air-filled cylinder of length, L. (See Fig. 13.35.) The piston decelerates because of the increase in the air pressure, comes to a stop, and then finally bounces out. Our problem is to find the acceleration-time relationship for a small accelerometer mounted on the piston. The data for the related variables t, time in milliseconds, and displacement d, in inches, are shown in this table:

t	0	2	4	6	8	10	12	14	16	18
d	0	0.556	1.112	1.665	2.213	2.752	3.279	3.791	4.283	4.748
t	20	22	24	26	28	30	32	34	36	38
d	5.181	5.574	5.919	6.208	6.433	6.586	6.659	6.648	6.556	6.389

The curve through the plotted points is differentiated twice (Fig. 13.36). The second derived curve shows the acceleration-time relationship.

EXAMPLE 5 ANALYSIS OF A CONTROL SYSTEM

Our problem is to determine the storage-level relationship for a cylindrical tank when the supply and demand patterns are known. (See Fig. 13.37.) Now we integrate the "supply-demand" curve to obtain the storage curves as a function of time. (This is shown in Fig. 13.38.) The left portion of this figure shows the horizontal scale, storage in gallons, and the vertical scale level in inches. The 45° line is a transfer line and the curve just above this line is the "characteristic curve" for a horizontal circular tank (in this example, the tank is 20 inches in diameter and 60 inches long).

To determine the level in the tank at any time, say, 1 minute, a horizontal line is drawn through point *A* to intersect the 45° line at point *B*, through which a vertical line is drawn to intersect the characteristic curve at point *C*. The horizontal drawn through point *C* intersects the vertical drawn through *A* at point *D*, which is a point on the "level" curve. Additional points on the "level" curve, can be established in a similar manner.

In this example the tank was assumed to be empty at time zero minutes. Now, let us assume an initial

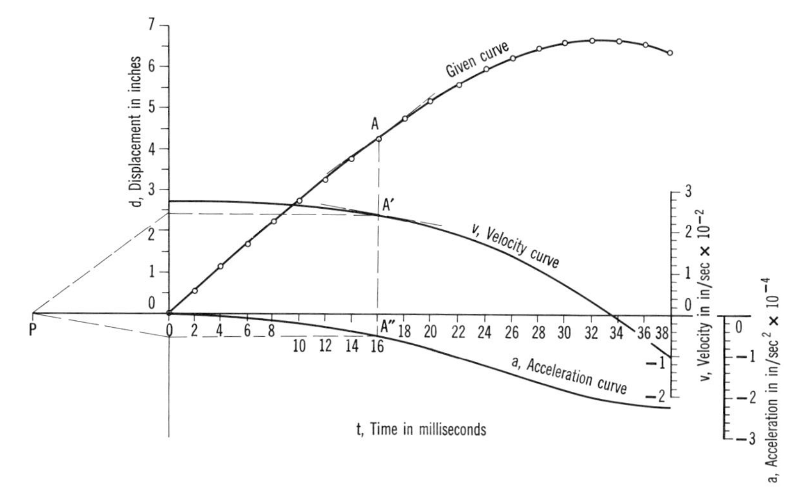

Fig. 13.36

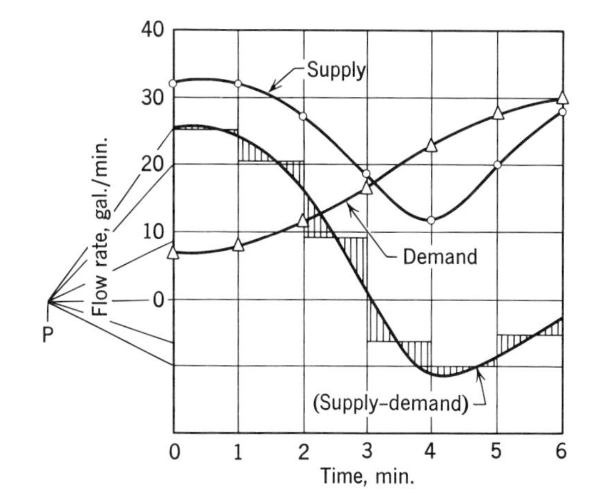

Fig. 13.37

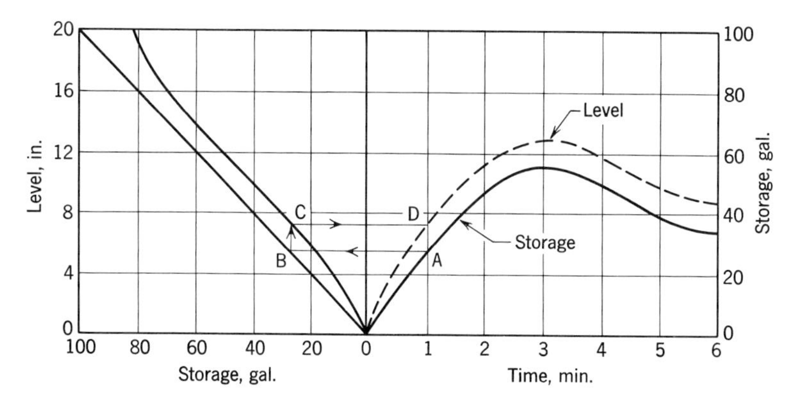

Fig. 13.38

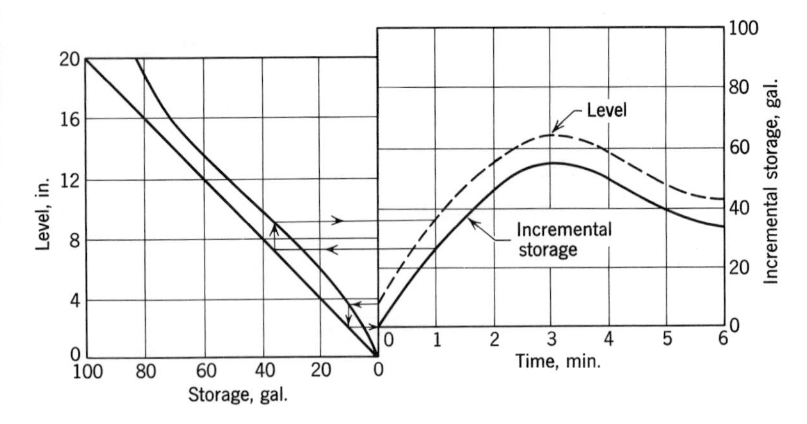

Fig. 13.39

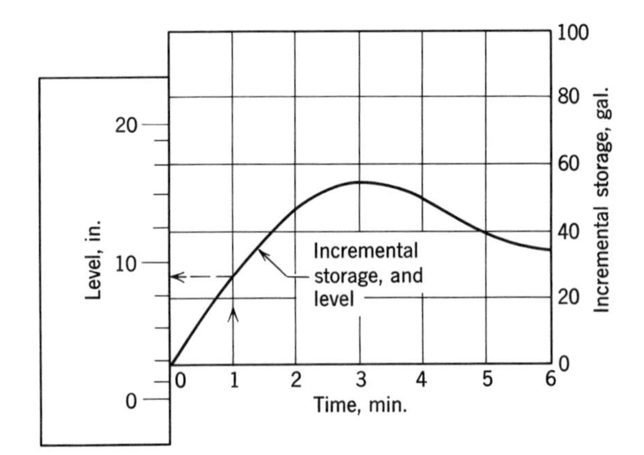

Fig. 13.40

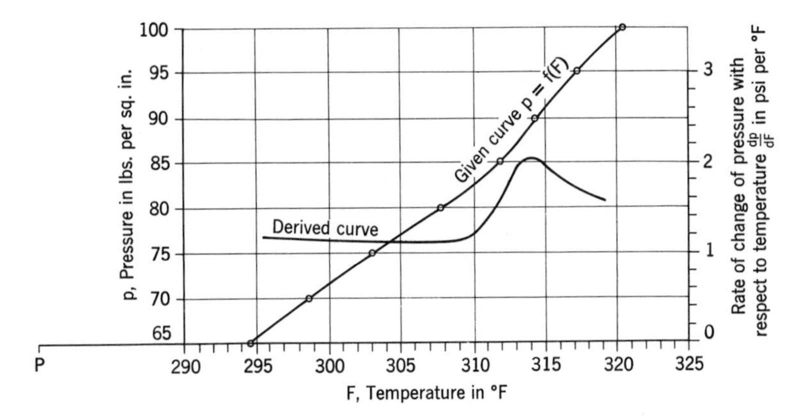

Fig. 13.41 Rate of change of pressure with respect to temperature.

storage of 10 gallons, which corresponds to level 3.6 inches. Figure 13.39 shows the method that can be used to determine storage levels when the above initial condition obtains. The "incremental storage" curve is the same as that shown in Fig. 13.38. It should be noted, however, that the belt portion of the chart has been positioned to account for the initial storage condition. Once this is done the method for determining the "level" curve is the same as that used previously in Fig. 13.38.

It is possible to eliminate the 45° transfer line and the characteristic curve by the use of a functional scale. (This is shown in Fig. 13.40.) The "level" scale is graduated as a functional scale. If this scale were moved upward until its graduation points were on the horizontal axis (time-scale), the values, incremental storage in gallons, would correspond to readings on the "level" scale. The position of this scale, as shown, indicates that there is an initial depth of 3.6 inches in the tank. Moreover, it is now possible to read "levels," at any time, directly; for example, at 1 minute, the "level" reading is 9 inches. In this manner we can first prepare a number of functional "level" scales based upon the geometry of the storage vessel, and use such scales to determine levels similar to the method employed in the case shown on Fig. 13.40.

EXAMPLE 6

The data shown below relate pressure, p, in pounds per square inch to corresponding values of temperature, F, in degrees Fahrenheit. We are to determine the rate of change of pressure with respect to temperature.

p, lb/sq in.	65	70	75	80	85	90	95	100
F, degrees	294.6	298.7	303.2	307.7	311.9	314.2	317.3	320.5

A plot of the data is shown in Fig. 13.41. The curve of the rate of change of p with respect to F is obtained by differentiating the curve of the data points.

Grapho-Numerical Method of Differentiation

In a number of cases it has been found that relatively large errors occur in the second derived curve as a result of the errors in the first derived curve. This was observed in several problems that arose in the research project* dealing with studies of human locomotion. To overcome this difficulty a *grapho-numerical* method was developed and successfully used in the project.

Essentially, the method consists of (*a*) plotting the data; (*b*) drawing a "best" smooth curve to represent the data graphically; and (*c*) reading the ordinate values of points *on the curve* corresponding to equal increments, h, along the horizontal axis. The ordinates of these points are then used in formulas that yield ordinate values for the first and second derived curves. The formulas can be developed in the following manner.

Let us assume that a portion of a parabola passes through points A, B, and C whose ordinates are y_{n-1}, y_n, and y_{n+1}, respectively, as shown in Fig. 13.42. When the increment, h, is made small, the slope of chord AC closely approximates the slope of the tangent to the curve at point B. This slope may be expressed as

$$\frac{y_{n+1} - y_{n-1}}{2h} = \text{Slope at point } B$$

Now, when we replace n with $n + 1$, the slope of the tangent to the curve at point C is

$$\frac{y_{n+2} - y_n}{2h} = \text{Slope at point } C$$

Similarly, when we replace n with $n - 1$, the slope at point A is

$$\frac{y_n - y_{n-2}}{2h} = \text{Slope at point } A$$

When, for example, experimental data express the relationship between displacement and time, we now have available, in the above expressions, a simple means for computing velocities (first derived curve) at points on the curve corresponding to y_{n+1} and y_{n-1}. But, more importantly, since acceleration is the rate of change of velocity with respect to time, we can easily develop a formula for acceleration from the above expressions.

Since,

$$V_{n+1} = \frac{y_{n+2} - y_n}{2h} \qquad \text{(velocity at point } C\text{)}$$

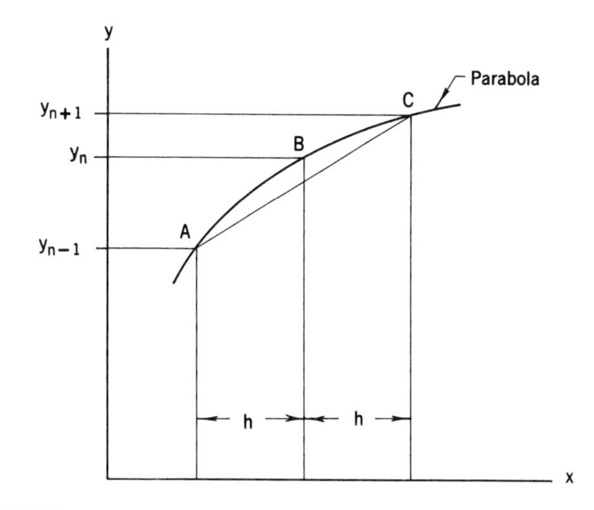

Fig. 13.42

* Prosthetic Devices Research Project, University of California at Berkeley.

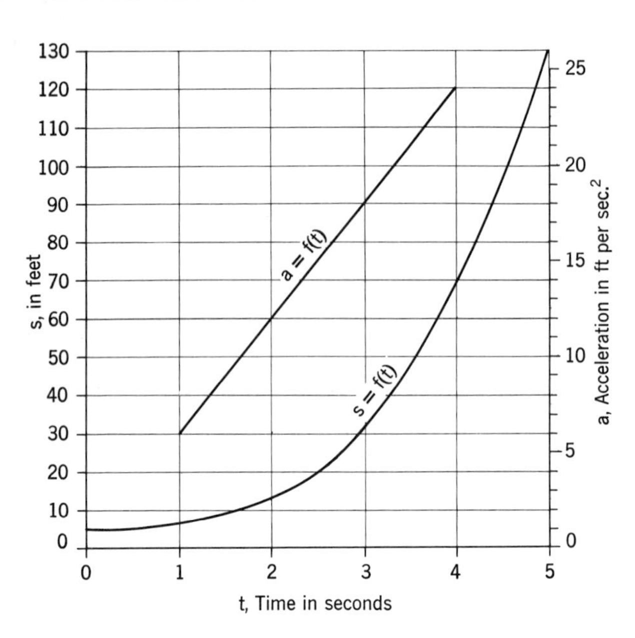

Fig. 13.43

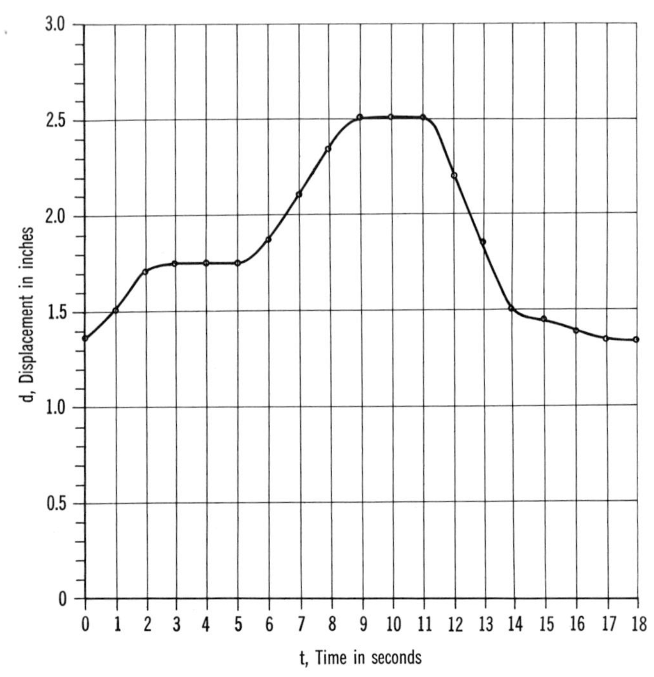

Fig. 13.44

and,

$$V_{n-1} = \frac{y_n - y_{n-2}}{2h} \qquad \text{(velocity at point } A)$$

then,

$$a_n = \frac{(y_{n+2} - y_n) - (y_n - y_{n-2})}{4h^2}$$

(acceleration at point B)

or

$$a_n = \frac{y_{n+2} + y_{n-2} - 2y_n}{4h^2}$$

Now we see that, by using the last equation, we may compute accelerations (second derived curve) *without actually first determining the velocity curve.*

EXAMPLE 1

Assume that the following data are given, where t represents time in seconds and S displacement in feet. It is desired to determine the acceleration curve using the grapho-numerical method.

t, in seconds	0	1	2	3	4	5
S, in feet	5	6	13	32	69	130

A plot of the above data is shown in Fig. 13.43. The increment along the t-axis is taken as $h = 0.5$ sec. Now we measure the ordinates, to the smooth curve that passes through the data points, at the end of each interval. A convenient table is prepared for the corresponding values of S and t, and for the necessary computations of the accelerations. This is shown in the accompanying table.

t, sec	S, ft	$2S_n$	S_{n+2}	S_{n-2}	$a_n = \frac{S_{n+2} + S_{n-2} - 2S_n}{4h^2}$; $h = 0.5$
0.0	5.0	10.0	6.0		
0.5	5.125	10.25	8.375		
1.0	6.0	12.0	13.0	5.0	$a = 13.0 + 5.0 - 12.0 = 6$ ft/sec²
1.5	8.375	16.75	20.625	5.125	$a = 20.625 + 5.125 - 16.75 = 9$ ft/sec²
2.0	13.0	26.0	32.0	6.0	$a = 32.0 + 6.0 - 26.0 = 12$ ft/sec²
2.5	20.625	41.25	47.875	8.375	$a = 47.875 + 8.375 - 41.25 = 15$ ft/sec²
3.0	32.0	64.0	69.0	13.0	$a = 69.0 + 13.0 - 64.0 = 18$ ft/sec²
3.5	47.875	95.75	96.125	20.625	$a = 96.125 + 20.625 - 95.75 = 21$ ft/sec²
4.0	69.0	138.0	130.0	32.0	$a = 130.0 + 32.0 - 138.0 = 24$ ft/sec²
4.5	96.125	192.25		47.875	
5.0	130.0	260.0		69.0	

When we plot the acceleration values (Fig. 13.43), we observe that the curve through these points is a straight line, showing that, in this case, the acceleration varies uniformly. The length of the acceleration curve can be increased by assuming a smaller value for the increment, h. For an experiment that involves a large amount of data points, the calculations may be performed by an automatic computing machine, thus greatly reducing calculation time.

The data for the above example were based on the formula $S = t^3 + 5$. Students acquainted with elementary calculus will, of course, know that since $V = ds/dt$ and $a = dV/dt$, the corresponding relations are $V = 3t^2$ and $a = 6t$.

The grapho-numerical method is, indeed, very well suited to those cases where the experimental data curve contains no severe changes of slope or curvature.

EXAMPLE 2

The following data show the relation of time, t, in seconds to displacements, d, in inches of the center of a cam follower from the center of a rotating cam.

t, sec	0	1	2	3	4	5	6	7	8	9
d, inches	1.37	1.51	1.72	1.75	1.75	1.75	1.87	2.11	2.35	2.51
t, sec	10	11	12	13	14	15	16	17	18	
d, inches	2.51	2.51	2.22	1.87	1.52	1.46	1.40	1.37	1.37	

We first plot the data points and the curve through these points, as shown in Fig. 13.44 on page 316. Now we will use the grapho-numerical method to establish the *acceleration-time* curve. We will assume the interval $h = 0.5$. The calculations of accelerations are shown in the accompanying table.

t, sec	d, in.	$2d_n$	d_{n+2}	d_{n-2}	$a_n = \dfrac{d_{n+2} + d_{n-2} - 2d_n}{4h^2}$
0.0	1.37	2.74	1.51		
0.5	1.43	2.86	1.61		
1.0	1.51	3.02	1.72	1.37	$a = 1.72 + 1.37 - 3.02 = \ 0.07$ in./sec^2
1.5	1.61	3.22	1.73	1.43	$a = 1.73 + 1.43 - 3.22 = -0.06$
2.0	1.72	3.44	1.75	1.51	$a = 1.75 + 1.51 - 3.44 = -0.18$
2.5	1.73	3.46	1.75	1.61	$a = 1.75 + 1.61 - 3.46 = -0.10$
3.0	1.75	3.50	1.75	1.72	$a = 1.75 + 1.72 - 3.50 = -0.03$
3.5	1.75	3.50	1.75	1.73	$a = 1.75 + 1.73 - 3.50 = -0.02$
4.0	1.75	3.50	1.75	1.75	$a = 1.75 + 1.75 - 3.50 = \ 0$
4.5	1.75	3.50	1.78	1.75	$a = 1.78 + 1.75 - 3.50 = \ 0.03$
5.0	1.75	3.50	1.87	1.75	$a = 1.87 + 1.75 - 3.50 = \ 0.12$
5.5	1.78	3.56	1.99	1.75	$a = 1.99 + 1.75 - 3.56 = \ 0.18$
6.0	1.87	3.74	2.11	1.75	$a = 2.11 + 1.75 - 3.74 = \ 0.12$
6.5	1.99	3.98	2.24	1.78	$a = 2.24 + 1.78 - 3.98 = \ 0.04$
7.0	2.11	4.22	2.35	1.87	$a = 2.35 + 1.87 - 4.22 = \ 0$
7.5	2.24	4.48	2.46	1.99	$a = 2.46 + 1.99 - 4.48 = -0.03$
8.0	2.35	4.70	2.51	2.11	$a = 2.51 + 2.11 - 4.79 = -0.08$
8.5	2.46	4.92	2.51	2.24	$a = 2.51 + 2.24 - 4.92 = -0.17$
9.0	2.51	5.02	2.51	2.35	$a = 2.51 + 2.35 - 5.02 = -0.16$
9.5	2.51	5.02	2.51	2.46	$a = 2.51 + 2.46 - 5.02 = -0.05$
10.0	2.51	5.02	2.51	2.51	$a = 2.51 + 2.51 - 5.02 = \ 0$
10.5	2.51	5.02	2.41	2.51	$a = 2.41 + 2.51 - 5.02 = -0.10$
11.0	2.51	5.02	2.22	2.51	$a = 2.22 + 2.51 - 5.02 = -0.29$
11.5	2.41	4.82	2.00	2.51	$a = 2.00 + 2.51 - 4.82 = -0.31$
12.0	2.22	4.44	1.87	2.51	$a = 1.87 + 2.51 - 4.44 = -0.06$
12.5	2.00	4.00	1.65	2.41	$a = 1.65 + 2.41 - 4.00 = \ 0.06$
13.0	1.87	3.74	1.52	2.22	$a = 1.52 + 2.22 - 3.74 = \ 0$
13.5	1.65	3.30	1.47	2.00	$a = 1.47 + 2.00 - 3.30 = \ 0.17$
14.0	1.52	3.04	1.46	1.87	$a = 1.46 + 1.87 - 3.04 = \ 0.29$
14.5	1.47	2.94	1.42	1.65	$a = 1.42 + 1.65 - 2.94 = \ 0.13$
15.0	1.46	2.92	1.40	1.52	$a = 1.40 + 1.52 - 2.92 = \ 0$
15.5	1.42	2.84	1.38	1.47	$a = 1.38 + 1.47 - 2.84 = \ 0.01$
16.0	1.40	2.80	1.37	1.46	$a = 1.37 + 1.46 - 2.80 = \ 0.03$
16.5	1.38	2.76	1.37	1.42	$a = 1.37 + 1.42 - 2.76 = \ 0.03$
17.0	1.37	2.74	1.37	1.40	$a = 1.37 + 1.40 - 2.74 = \ 0.03$
17.5	1.37	2.74		1.38	
18.0	1.37	2.74		1.37	

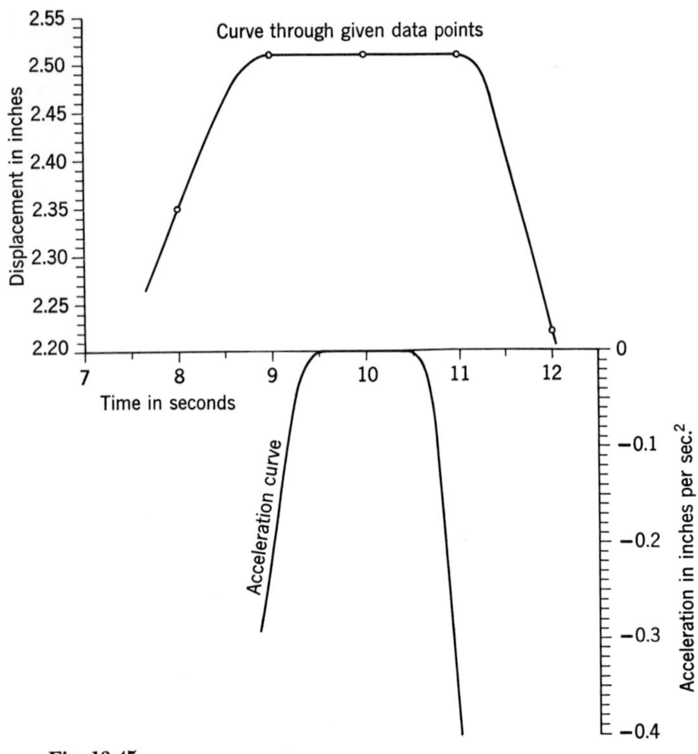

Fig. 13.45

Observe that in regions such as $t = 9.00$ to $t = 11.00$ the acceleration should be zero. The values shown in the above table include values near zero. We can, however, reduce the interval h to $\frac{1}{4}$ and obtain the values shown in the table below. Now the results are much more accurate. See Fig. 13.45 for the graphical representation.

t	d_n	$2d_n$	d_{n+2}	d_{n-2}	$a_n = \dfrac{d_{n+2}+d_{n-2}-2d_n}{4h^2}$
8.50	2.46	4.92	2.51	2.35	$a_n=(2.51+2.35-4.92)4=-0.24'/\text{sec}^2$
8.75	2.50	5.00	2.51	2.41	$a_n=(2.51+2.41-5.00)4=-0.32'/\text{sec}^2$
9.00	2.51	5.02	2.51	2.46	$a_n=(2.51+2.46-5.02)4=-0.20'/\text{sec}^2$
9.25	2.51	5.02	2.51	2.50	$a_n=(2.51+2.50-5.02)4=-0.04'/\text{sec}^2$
9.50	2.51	5.02	2.51	2.51	$a_n=(2.51+2.51-5.02)4=\ 0$
9.75	2.51	5.02	2.51	2.51	$a_n=(2.51+2.51-5.02)4=\ 0$
10.00	2.51	5.02	2.51	2.51	$a_n=(2.51+2.51-5.02)4=\ 0$
10.25	2.51	5.02	2.51	2.51	$a_n=(2.51+2.51-5.02)4=\ 0$
10.50	2.51	5.02	2.51	2.51	$a_n=(2.51+2.51-5.02)4=\ 0$
10.75	2.51	5.02	2.49	2.51	$a_n=(2.49+2.51-5.02)4=-0.08'/\text{sec}^2$
11.00	2.51	5.02	2.41	2.51	$a_n=(2.41+2.51-5.02)4=-0.40'/\text{sec}^2$
11.25	2.49	4.98	2.31	2.51	$a_n=(2.31+2.51-4.98)4=-0.64'/\text{sec}^2$

EXERCISES

1. The relation between volume and pressure is given by the following data:

V	3	4	5	6	7	8	9	10	11	12
P	10	70	56	42	33	28	24	21	18	17

V represents volume in cubic feet per pound of steam, and P steam pressure in pounds per square inch. Find the work done by the piston.

2. Determine graphically the area under the curve. Use the "pole-and-ray" method. Include an area scale so that the area of any portion of the total area can be read. Now determine the area by (*a*) the use of a planimeter (if available); (*b*) by the parabola method; and (*c*) by Simpson's rule. Tabulate results and compare. The curve is defined by the following data:

x	0	1	2	3	4	5	6	7	8
y	30	38	41	43	41	38	31	25	22

3. A relation between t, time in minutes, and S, speed in miles per hour, is shown in the table below. Graduate values of t on the x-axis and of S on the y-axis. Plot the curve

$S = f(t)$; use the "pole-and-ray" method to integrate this curve; and include a scale to read distance traveled in miles during any interval of time. What is total distance traveled?

t, minutes	0	0.5	1	2	3	4	5	6	7	8	9	10
S, mph.	0	12	20	25	30	33	40	45	42	35	30	20

4. Assume a semicircular plot of ground. Subdivide the plot so that the area of one lot is three times the area of the remaining lot. The dividing line between the two lots is perpendicular to the diameter line of the plot.

5. Determine the capacity, in cubic feet, of a reservoir for which the following data are available:

H, height above sea level, feet	220	210	200	190	185	182	180
A, area of horizontal section, square feet	27,300	21,000	11,200	9500	7200	3200	0

6. Plot the river cross section from the following data, and determine the flow in cubic feet per minute when the average velocity is 5.6 feet per second. (Use the "pole-and-ray" method to integrate the curve.)

S, distance in feet	0	10	20	30	40	50	60	70	80	90
d, depth in feet	0	14	15.7	16.5	16.2	15.2	14.0	12.6	11.2	0

7. Determine the area of the midship section of a vessel, using the following data:

W, half width° in feet	7.2	13.1	18.1	22.3	25.1	25.6	26.1
H, height above keel	0	3	7	11	15	18	21

° Symmetrical about a vertical line.

8. The following data show the values of velocities, v, in feet per second and the corresponding values of accelerations, a, in feet per second squared. It is required to establish the integral curve to determine the time, $t = \int \frac{1}{a}\, dv$.

v	0	1	2	3	4	5	6
a	0.455	0.410	0.333	0.229	0.119	0.063	0.040

9. The approximate speeds of a freight truck during a 30-minute run are shown in the table. Plot a curve of these data and then (*a*) determine the acceleration curve; and (*b*) include an acceleration scale. Read values of acceleration that correspond to $t = 9$ and $t = 24$.

t, minutes	0	0.5	1	3	5	7	10	13	14	20	22	23	25	26	27	30
S, mph.	0	10	20	30	35	38	42	45	45	46	53	55	55	52	42	0

10. Plot the data shown below. Draw a "fair" curve to represent these data and then determine (*a*) the velocity-time curve and (*b*) the acceleration-time curve. (Use the "pole-and-ray" method.) Include velocity and acceleration scales. Plot values of S along the y-axis and values of t along the x-axis.

t, seconds	0	0.1	0.2	0.4	0.6	0.8	1.0	1.2	1.4	1.5
S, feet	0	2.5	4.5	7.0	7.5	7.2	6.5	6.0	6.5	7.5

11. Plot the curve for $y = \sin\theta$ (θ varies from 0 to π radians). Determine the curve $y = \cos\theta$ by graphical differentiation.

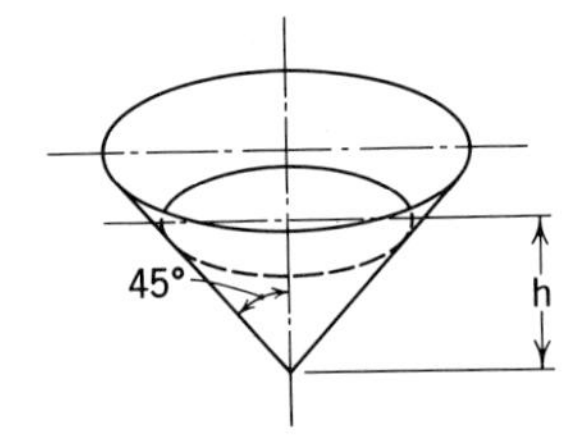

12. Water flows into an inverted right circular cone (base angle 45°) at the rate of 15 cubic inches per minute. At what rate is the surface of the water rising? What is the rate of increase when $h = 20$ inches? (Hint: Plot the equation $V = \pi h^3/3$ for the range of h from 0 to 30 inches. Use the x-axis for values of h and the y-axis for values of volume, V.)

13. A 20-foot beam supported at both ends and uniformly loaded has the following bending moments at 2-foot intervals along the beam:

l, distance in feet	0	2	4	6	8	10	12	14	16	18	20
M, bending moment in foot-pounds	0	2000	3500	4400	4600	4800	4600	4400	3500	2000	0

Draw the bending moment curve, and then determine (*a*) the shear curve (shear is measured by the rate of change of bending moment) and (*b*) the load curve (load is measured by the rate of change of shear). What is the load per foot of beam?

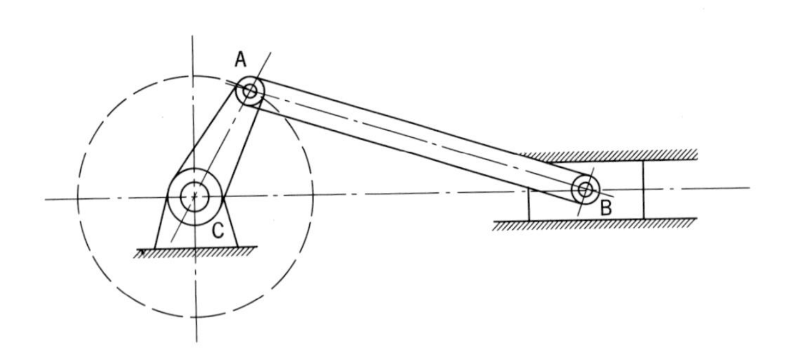

14. The slider-crank mechanism shown below is composed of crank $CA = 5$ inches, and arm $AB = 15$ inches, which is attached to the crank at A and to the sliding unit at B. The crank rotates about C at a speed of 500 rpm. Plot the displacement-time curve for point B and then determine the velocity-time and acceleration-time curves.

15. A rectangular plate 6 ft wide and 8 ft long is immersed in fresh water. One 6-ft edge is in contact with the water surface. Thc plate is in a vertical position. Locate the center of pressure (C.P.) on the plate, and determine the total force on one side of the plate. The C.P. is the point at which the total force may be assumed to act.

16. A plate in the form of a 10-in. equilateral triangle is immersed in fresh water in a vertical position. A 10-in. side is in contact with the water surface. Locate the C.P. on the plate; and determine the total force on one side of the plate.

17. Find the moment of inertia of a 6-in. × 8-in. rectangle about an axis which contains a 6-in. edge. Use the grapho-mechanical method.

18. Find the moment of inertia of an 8-in. equilateral triangle about an axis which contains a side of the triangle.

19. Find the moment of inertia of the section shown on the right about axis *AB*. Use the grapho-mechanical method.

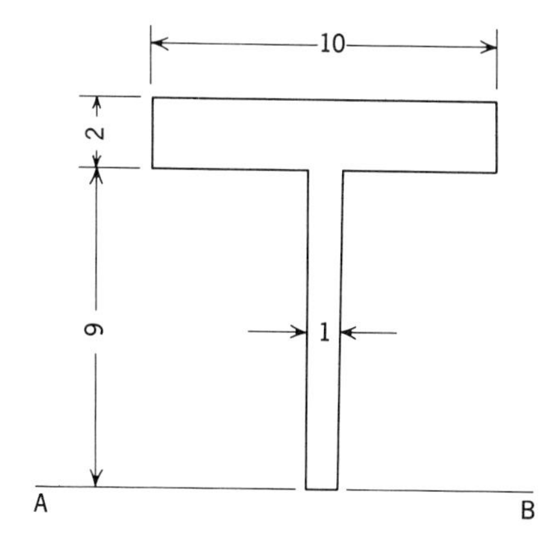

20. Draw a 3-inch diameter circle. Find the moment of inertia about a line which is tangent to the circle.

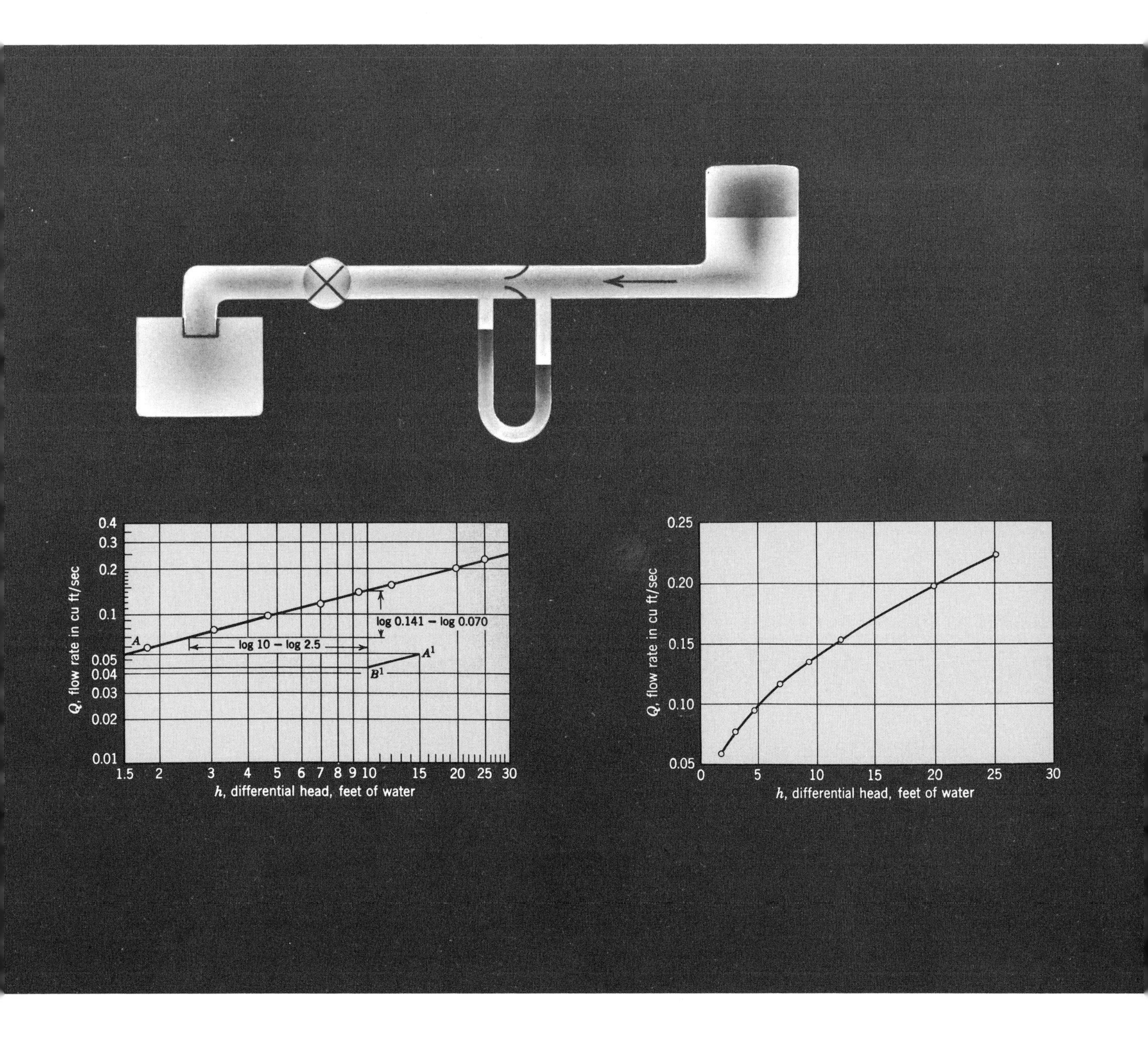

EMPIRICAL EQUATIONS 14

In research, development, and design, experimental work is often necessary to determine the relationship among related variables. Experiments deal with the simultaneous measurement of two quantities and observation of their behavior in relationship to each other. For example, experiments may be concerned with the rate of flow of water over a weir in relationship to the height of water over the crest of the weir; or with the relationship between pressure and volume of saturated steam; or in the study of the relationship between loads applied to metal bars and the corresponding elongation of the bars; such information may contribute to our knowledge of the elastic and plastic properties of the material.

Graphical Representation of Experimental Data

The relationship between the quantities measured is portrayed graphically by plotting the data on a coordinate grid whose x-axis is generally used for values of the independent variable and the y-axis for the corresponding values of the dependent variable. The data should be plotted on a grid sheet large enough to ensure good accuracy.

EXAMPLE 1

Let us suppose that the relation between load, L, and the effort, E, required to lift the load is expressed by the following data:

L, pounds	10	30	40	60	80	100	120	140
E, pounds	1.2	2.5	3.0	4.2	5.4	6.6	7.6	9.0

First, let us plot the above data and then draw the "best" curve to represent the relationship between the two quantities L and E. (See Fig. 14.1.) It is quite evident that the data points fall very nearly on a straight line, which shows that a linear relationship exists between the two variable quantities.

It should be pointed out that, in most cases, the observations of the magnitudes of the quantities are based on readings of instruments—gages, voltmeters, ammeters, thermometers, etc.—which have graduated scales and, therefore, *the data are graphical and approximate.* It is imperative to note this, since it will help us to understand that the equation which symbolizes the relation between the two quantities is also an approximation. Such an equation is known as an *empirical equation.*

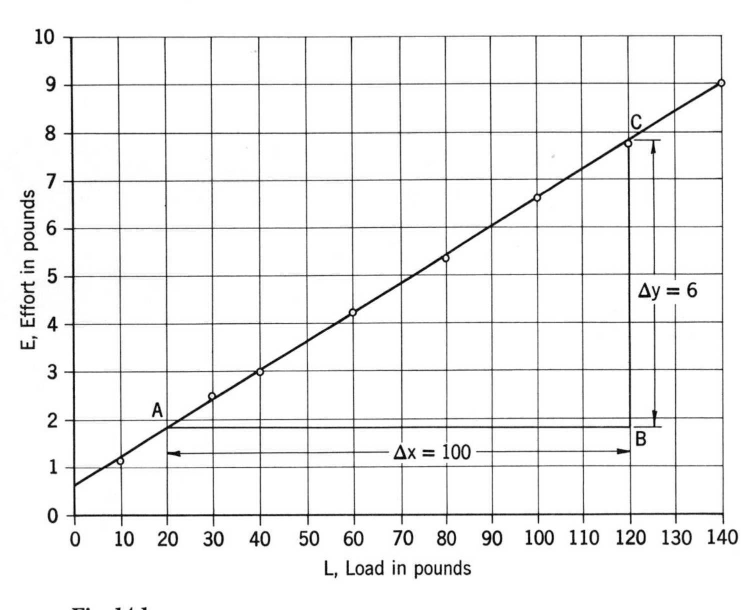

Fig. 14.1

OBTAINING THE EMPIRICAL EQUATION

The equation of a straight line, plotted on uniform coordinates, is $y = mx + b$, where m is the slope of the line and b is the *y-intercept* (the value of y when $x = 0$). The values of the constants m and b can be determined by one of the following methods: (*a*) *graphical;* (*b*) *selected points;* (*c*) *averages;* (*d*) *least squares. Methods* (*a*) *and* (*b*) *are adequate for data accurate to two significant figures, and in some cases for data accurate to three significant figures.* Method (*c*) is quite accurate for most cases, and method (*d*) is more precise than the others, but is also more time-consuming. We will apply the first two methods to the given example. The last two methods are presented in smaller type for reference only. In subsequent examples we will consider methods (*a*) and (*b*) exclusively.

(a) Graphical Method

First, we assume that the "best" straight line has been drawn to represent the relation between the two quantities. This means that the line may pass through some of the plotted points and approximately balance the others so that nearly the same number of points are on each side of the line.

The slope, m, of the line is determined in the following manner:

1. Select two points such as A and C *on the line* and at a reasonable distance apart.
2. Form the right triangle ABC. Now $m = \Delta y/\Delta x = 6/100 = 0.06$. Note carefully that Δx is the length of the horizontal side AB *measured in terms of the units of the horizontal* (x) *scale,* and that Δy is the length of the vertical side BC *measured in terms of the units of the vertical* (y) *scale.*
3. The y-intercept, b, is the value of y when $x = 0$. In this case $b = 0.65$, which is read directly on the y-axis.
4. The empirical equation is $y = 0.06x + 0.65$. Let us now check the accuracy of the equation by computing y-values corresponding to the given x-values and then comparing the computed values with the observed values of y. These are shown in the table on page 326.

From these values and the percentage deviations we note good agreement between the observed and the calculated values of the y quantities.

x	$y_{(obs)}$	$y_{(comp)}$	$y_o - y_c$	% Deviation
10	1.2	1.25	−0.05	4.2
30	2.5	2.45	0.05	2.0
40	3.0	3.05	−0.05	1.7
60	4.2	4.25	−0.05	1.2
80	5.4	5.45	−0.05	0.9
100	6.6	6.65	−0.05	0.8
120	7.6	7.85	−0.25	3.0
140	9.0	9.05	−0.05	0.6

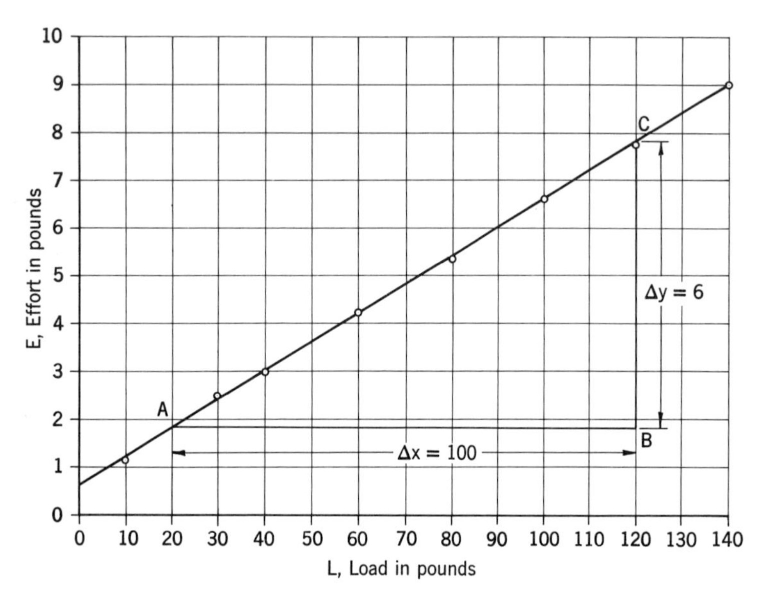

Fig. 14.1 (Repeated)

(b) Method of Selected Points

In this method we first select two points *on the line*. Let us use points A and C. The coordinates of point A are $x = 20$ and $y = 1.8$; and of point C, $x = 120$ and $y = 7.8$. Two points are selected because there are two constants, m and b, that must be determined. Since points A and C are on the line, the coordinates of these points must satisfy the equation $y = mx + b$; therefore,

$$1.8 = 20m + b$$

and

$$7.8 = 120m + b$$

The simultaneous solution of these equations yields

$$m = 0.06$$

and

$$b = 0.60$$

Therefore, the empirical equation of the line is

$$y = 0.06x + 0.60$$

(c) Method of Averages

The best line is based upon the assumption that its location is such as to make the algebraic sum of the differences of the observed and calculated values equal to zero. Algebraically, this means

$$\Sigma(y - mx - b) = 0$$

where y represents any observed value and $(mx + b)$ the corresponding calculated value.

We know that two constants, m and b, must be determined; therefore, the given data are divided into two nearly equal groups. The sum of the differences (r) of the observed and calculated values of each group is placed equal to zero, i.e.,

$$\begin{array}{l} 1.2 - 10m - b = r_1 \\ 2.5 - 30m - b = r_2 \\ 3.0 - 40m - b = r_3 \\ 4.2 - 60m - b = r_4 \\ \hline 10.9 - 140m - 4b = 0 (\Sigma r = 0) \end{array} \qquad \begin{array}{l} 5.4 - 80m - b = r_5 \\ 6.6 - 100m - b = r_6 \\ 7.6 - 120m - b = r_7 \\ 9.0 - 140m - b = r_8 \\ \hline 28.6 - 440m - 4b = 0 (\Sigma r = 0) \end{array}$$

The simultaneous solution of the equations

$$10.9 = 140m + 4b$$

and

$$28.6 = 440m + 4b$$

results in

$$m = 0.059 \text{ and } b = 0.66$$

Therefore, the empirical equation is $y = 0.059x + 0.66$.

(d) Method of Least Squares

The best line is based on the assumption that its location is such as to make the sum of the squares of the differences of the observed and calculated values a *minimum*. Algebraically, this means that $\Sigma(y - mx - b)^2 =$ minimum. Therefore, the derivatives of the above expression with respect to m and b must equal zero. Hence,

$$\Sigma[2(y - mx - b)(-x)] = 0$$

and

$$\Sigma[2(y - mx - b)(-1)] = 0$$

or

$$\Sigma xy = b\Sigma x + m\Sigma x^2$$

and

$$\Sigma y = bn + m\Sigma x$$

where n represents the number of observations.

The simultaneous solution of these two equations will determine the values of constants b and m.

First, let us compute the values of the terms in the above equations. This is conveniently done in the accompanying tabular form.

x	y	xy	x^2
10	1.2	12.0	100
30	2.5	75.0	900
40	3.0	120.0	1,600
60	4.2	252.0	3,600
80	5.4	432.0	6,400
100	6.6	660.0	10,000
120	7.6	912.0	14,400
140	9.0	1,260.0	19,600
$\Sigma =$ 580	39.5	3,723.0	56,600

The two equations are:

$$3{,}723.0 = 580b + 56{,}600m$$

$$39.5 = 8b + 580m$$

The simultaneous solution of the equations yields

$$b = 0.701 \text{ and } m = 0.0586$$

Therefore, the empirical equation is $y = 0.0586x + 0.701$.

EXAMPLE 2

Let us consider the following data:

x	0	1	2	3	4	5	6	7	8
y	32.5	29.7	26.0	23.2	19.9	17.2	13.8	10.3	8.0

A plot of these data is shown in Fig. 14.2. It is readily apparent that the relation between x and y is linear; therefore, the empirical equation is of the form $y = mx + b$. Let us determine the values of m and b by employing (*a*) the graphical method and, as a check, (*b*) the method of selected points.

(*a*) *Graphical Method.* The intercept, b, is easily obtained, since $y = b$ when $x = 0$. In this case, $b = 32.5$. The slope $m = \Delta y/\Delta x = -18.6/6 = -3.1$. It should be observed that the slope is negative, because $\Delta y = (y_1 - y_2)$, $\Delta x = (x_1 - x_2)$ which is negative and, therefore,

$$m = \frac{(y_1 - y_2)}{(x_1 - x_2)}.$$

The empirical equation is $y = -3.1x + 32.5$.

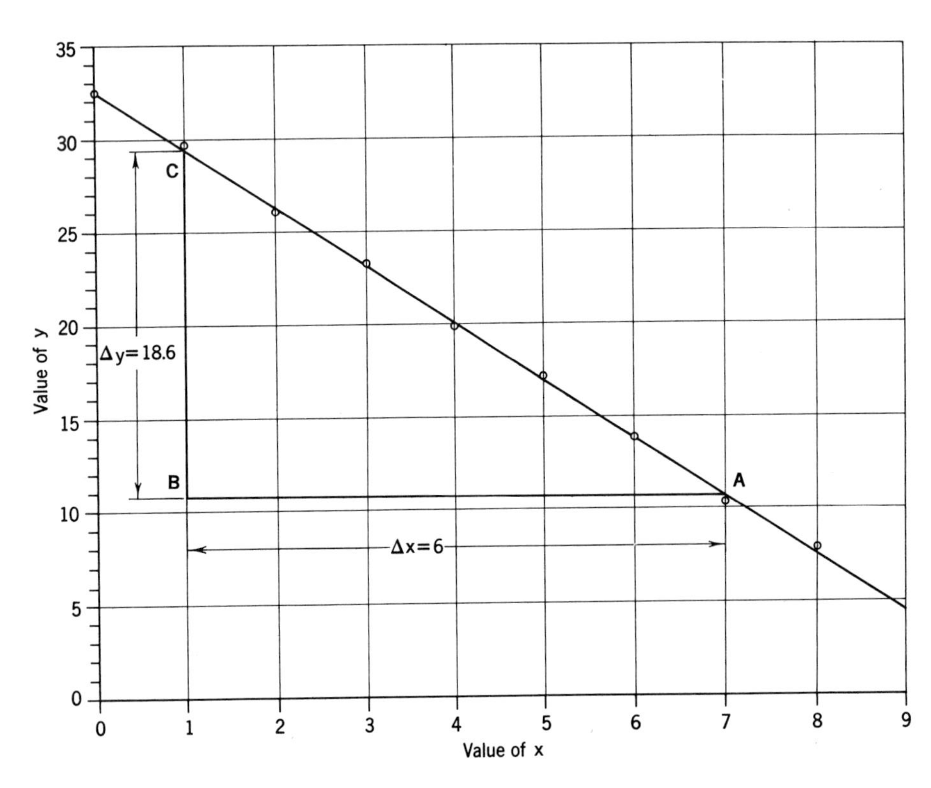

Fig. 14.2

(*b*) *Method of Selected Points.* Points A (7, 10.7) and C (1, 29.3) are on the line; therefore, their coordinates must satisfy the equation of the line. The equation, $y = mx + b$ can be expressed as $y = (y_1 - y_2/x_1 - x_2)\,x + b$, since $m = y_1 - y_2/x_1 - x_2$. Therefore,

$$y = \left(\frac{29.3 - 10.7}{1 - 7}\right)x + b$$

or

$$y = \frac{18.6}{-6}x + b = -3.1x + b.$$

Now, since $y = -3.1x + b$, we can substitute the x and y coordinates of either point A or point C to evaluate b. When we use the x and y coordinates of point A, we obtain

$$10.7 = (-3.1)(7) + b$$

from which

$$b = 32.4$$

When we use the x and y coordinates of point B, we obtain

$$29.3 = (-3.1)(1) + b$$

from which

$$b = 32.4$$

The empirical equation is $y = -3.1x + 32.4$ which checks very closely with the equation previously obtained from the graphical method.

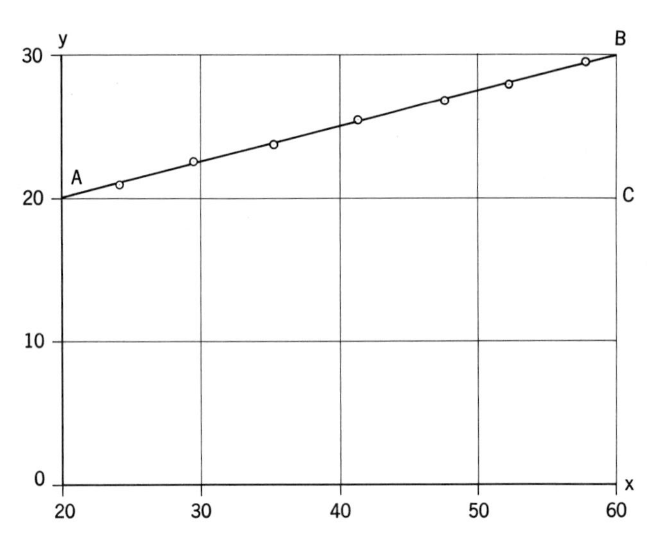

Fig. 14.3

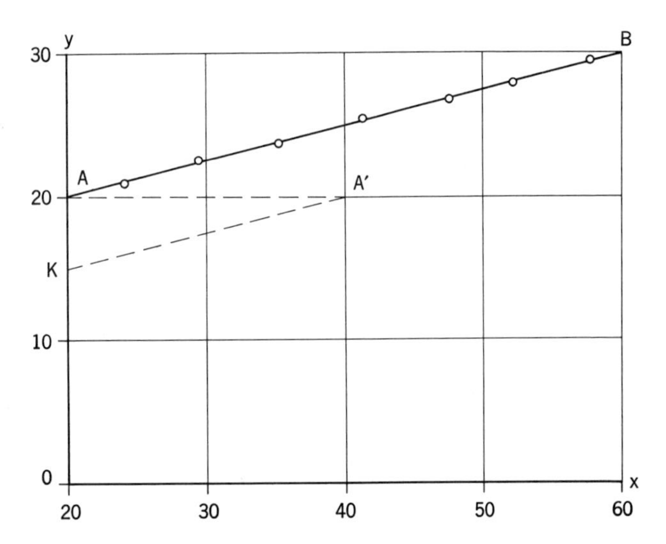

Fig. 14.4

EXAMPLE 3

Suppose that the plot of given data and the "best" straight-line representation of those data is as shown in Fig. 14.3. Let us first use the graphical method. The slope, m, is easily obtained since

$$m = \frac{\Delta y}{\Delta x} = \frac{BC}{CA} = \frac{30 - 20}{60 - 20} = \frac{10}{40} = \frac{1}{4}$$

The intercept b, however, is *not* the y value of point A. We must remember that the value of b is seen only when $x = 0$. In this example, x varies from 20 to 60; therefore, it is necessary to extend the line AB to $x = 0$. This is easily done without actually extending the x-scale. Figure 14.4 shows that it is only necessary to shift point A to A' (a distance of 20 units of x) and then draw $A'K$ parallel to line AB. The y-value of K is the intercept, b. (This is readily seen since point K is on the line $x = 0$.) The value of the intercept b is 15. The empirical equation is $y = x/4 + 15$ or $y = 0.25x + 15$.

As a check on this equation, we will now employ the selected point method. The coordinates of point A are $x = 20$ and $y = 20$; and of point B, $x = 60$ and $y = 30$. The coordinates of these points must satisfy the equation $y = mx + b$. Therefore,

$$20 = 20m + b$$

and

$$30 = 60m + b$$

The simultaneous solution of these two equations yields $m = 0.25$ and $b = 15$, and the empirical equation is $y = 0.25x + 15$.

EXAMPLE 4

Let us determine the relation between I, the indicated horsepower, and P, pounds of steam per hour, from the data shown in this table:

I	1	2	4	6	8	10	12	14
P	44	60	91	119	153	188	221	265

A plot of these data shows that there is a linear relationship between the two quantities (Fig. 14.5). The empirical equation, therefore, is of the form $y = mx + b$ or $P = mI + b$. Employing the graphical method, we obtain

$$m = \Delta P/\Delta I = \frac{225 - 55}{12 - 2} = 17.$$

The intercept $b = 23$. The empirical equation is $P = 17I + 23$.

In many cases the experimental data, when plotted, appear to fall on a curve. We will now consider several types that occur frequently in engineering.

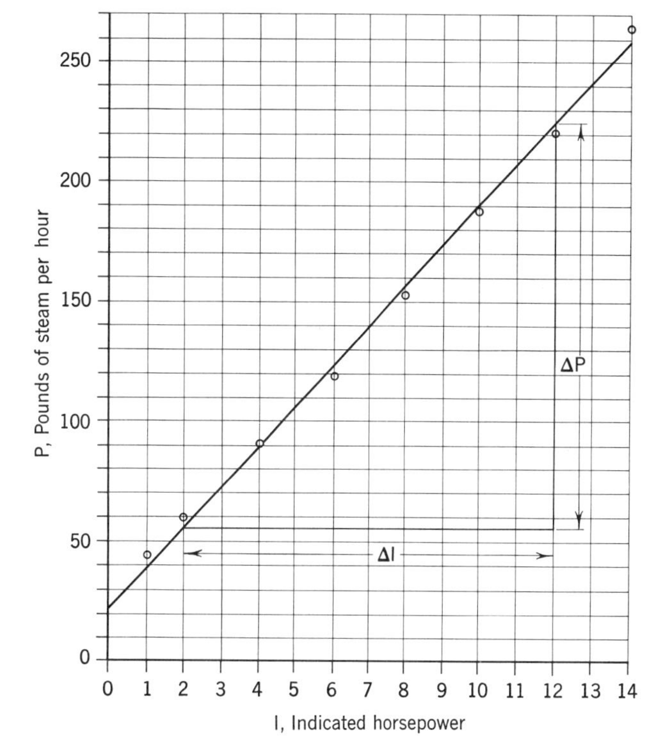

Fig. 14.5

Equations of the form $y = bx^m$

Let us consider the equation $y = x^2$, where $b = 1$ and $m = 2$, in the above form of equation. Points on the curve which represent this equation can be determined by assigning values to x and then computing the corresponding values of y. A portion of the curve is shown in Fig. 14.6.

Additional curves for equations $y = x^3$, $y = x^{0.5}$, $y = x^{-0.5}$, $y = x^{-1}$ are also included in this figure. The *positive exponents* identify the curves known as *parabolas;* the *negative exponents,* curves known as *hyperbolas.* If the plot of experimental data, upon a Cartesian chart having uniform scales, reveals a curve that might possibly belong to the family of curves represented by the form $y = bx^m$, verification of this fact can be made by plotting the data on a chart having logarithmic scales for the x and y axes, in which case the plot of the data should lie approximately on a straight line. This is true because the equation $y = bx^m$, when expressed logarithmically, becomes $\log y = m \log x + \log b$, which is of the form $y_1 = mx_1 + b_1$ (straight line), where $y_1 = \log y$, $x_1 = \log x$, and $b_1 = \log b$.

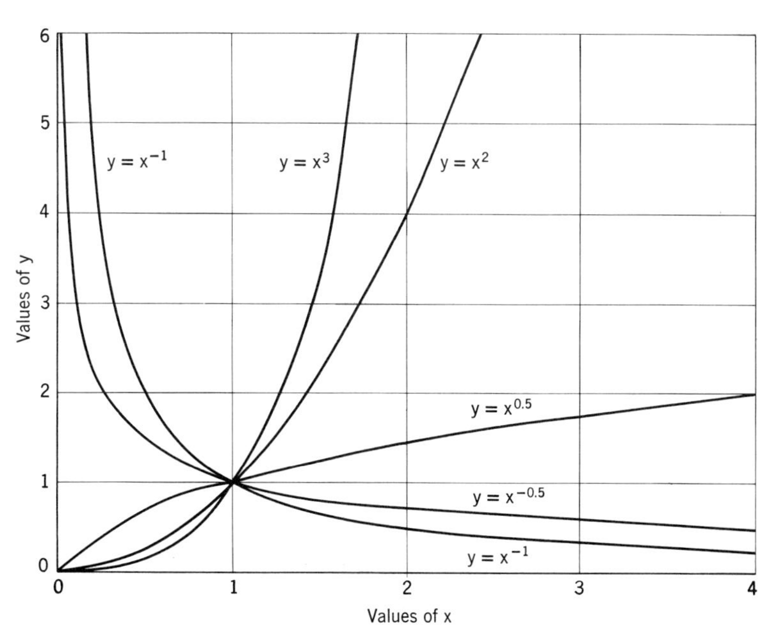

Fig. 14.6

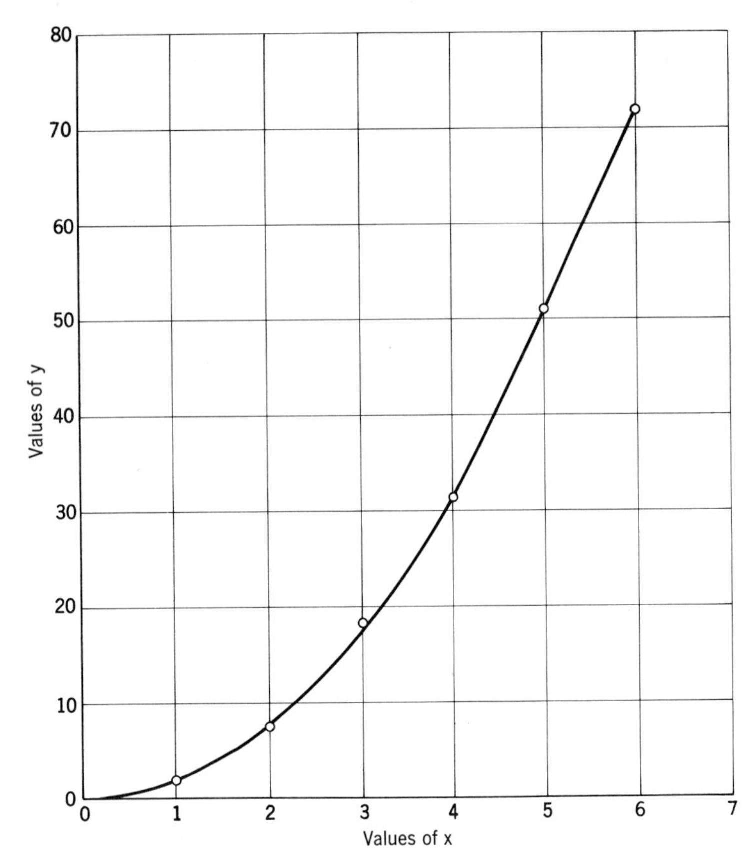

Fig. 14.7

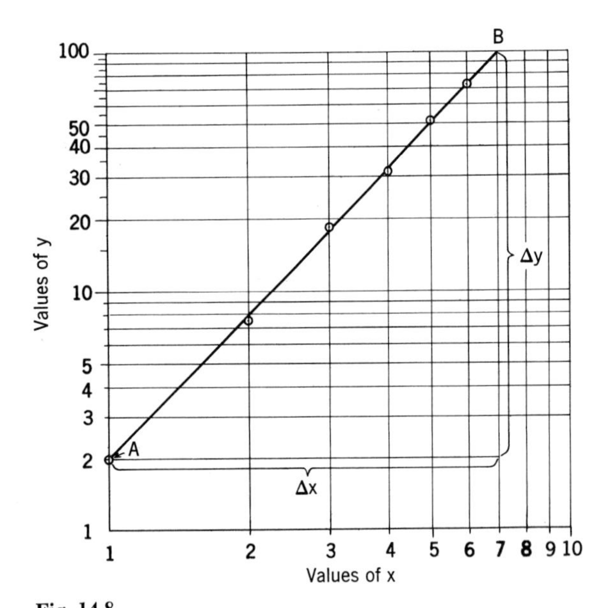

Fig. 14.8

EXAMPLE 1

Let us consider the following data:

x	1	2	3	4	5	6
y	2	7.5	18.4	31.6	51	71.8

A plot of these data and the "best" curve which approximately passes through the points is shown in Fig. 14.7. This curve appears to belong to the family $y = bx^m$.

Now let us plot the same data on a chart having logarithmic scales. We observe that the points lie very nearly on a straight line (Fig. 14.8).

When we employ the graphical method we can readily determine the values of m and b. The value of

$$m = \frac{\Delta y}{\Delta x} = \frac{\log 100 - \log 2}{\log 7 - \log 1} = 2.01$$

In the equation $\log y = m \log x + \log b$ we see that when $x = 1$, $\log 1 = 0$ and, therefore, $\log y = \log b$ or $y = b$.

Hence, the value b is 2. The empirical equation is $y = 2x^{2.01}$.

As a check on this equation let us use the method of selected points. Suppose we select points A (1, 2) and B (7, 100). The coordinates of these points must satisfy the equation $\log y = m \log x + \log b$; therefore,

$$\log 100 = m \log 7 + \log b \tag{1}$$

and

$$\log 2 = m \log 1 + \log b \tag{2}$$

From equation (2) $b = 2$.

Substituting this value in equation (1), we obtain,

$$\log 100 = m \log 7 + \log 2$$

and

$$m = \frac{\log 100 - \log 2}{\log 7} = 2.01$$

The empirical equation is $y = 2x^{2.01}$.

When the lengths of the log cycles are the *same* lengths Δy and Δx can be measured directly in inches. For example, in Fig. 14.9,

$$m = \frac{\Delta y}{\Delta x} = \frac{2.250}{3.750} = 0.6$$

which is the same as

$$m = \frac{\Delta y}{\Delta x} = \frac{\log 80 - \log 20}{\log 10 - \log 1} = 0.6$$

When the lengths of the log cycles are different, the ratio of the lengths of the cycles is used to determine m graphically. For example, in Figure 14.10 the ratio of lengths of the cycles is 1:2. Therefore,

$$m = \frac{\Delta y}{\Delta x} = \left(\frac{2.250}{1.875}\right)\left(\frac{1}{2}\right) = 0.6$$

which is the same as

$$m = \frac{\Delta y}{\Delta x} = \frac{\log 8 - \log 2}{\log 10 - \log 1} = 0.6$$

EXAMPLE 2

Let us determine the empirical equation for the relation between Q, the flow rate in cubic feet per second, of water through a 1-inch nozzle and h, the differential head in feet of water. The test data are shown in the table.

h	1.81	3.06	4.64	6.94	9.34	12.09	19.99	25.10
Q	0.0690	0.0770	0.0949	0.1163	0.1346	0.1528	0.1965	0.2217

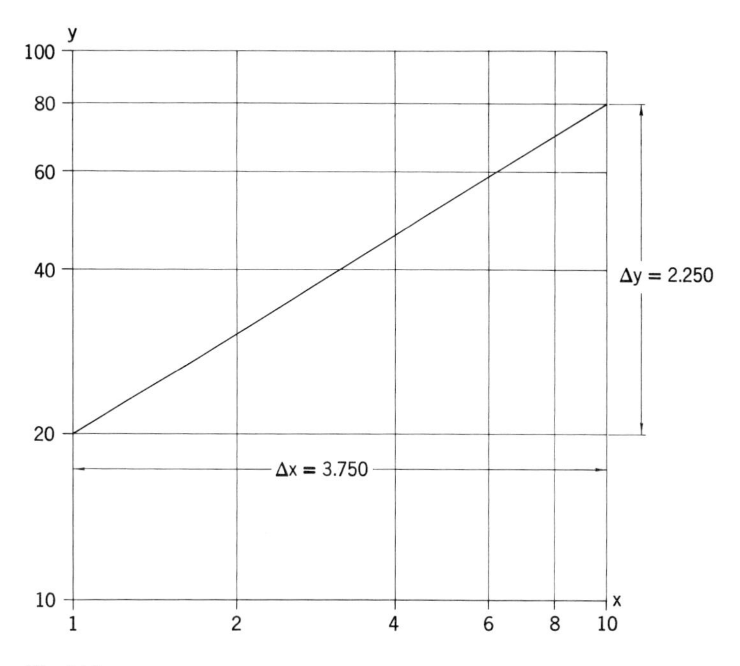

Fig. 14.9

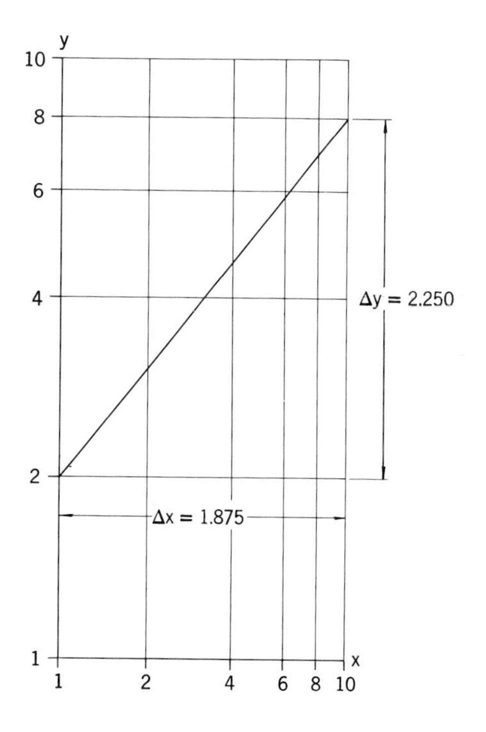

Fig. 14.10

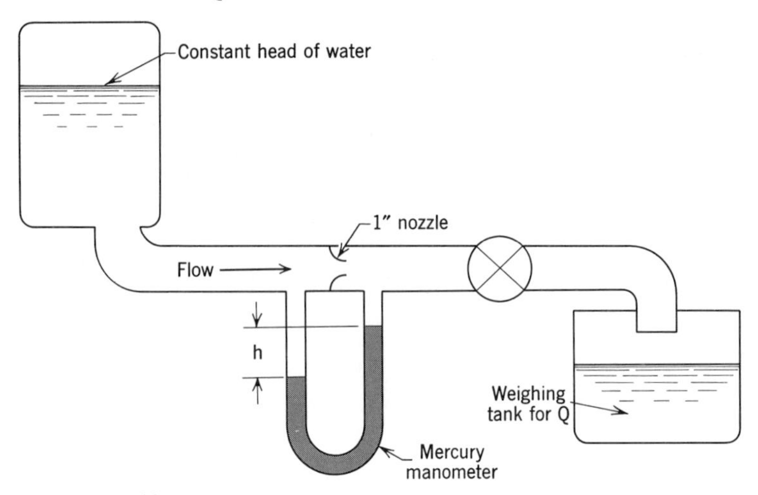

Fig. 14.11

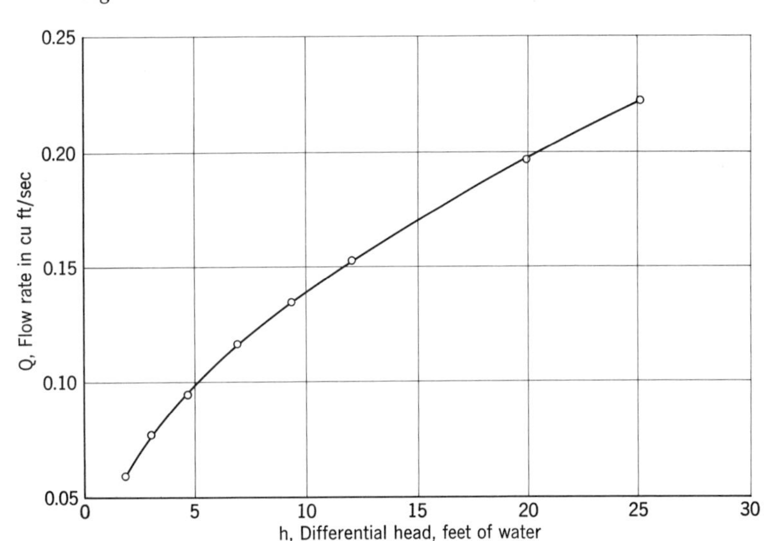

Fig. 14.12

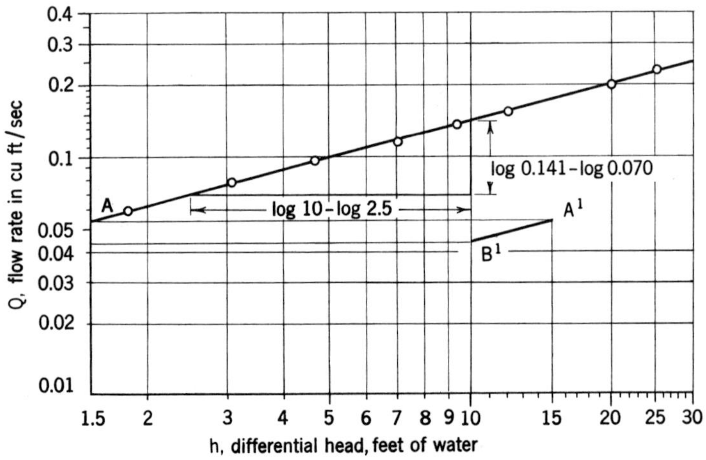

The schematic sketch, Fig. 14.11, shows the test apparatus. A plot of the data is shown in Fig. 14.12.

It is reasonable to assume that the form of empirical equation is $y = bx^m$ since the curve closely resembles $y = x^{0.5}$. Verification of the assumption is seen in Fig. 14.13 which shows that the data, when plotted on a log-log coordinate sheet, lie approximately on a straight line.

In terms of the variables given, the equation is $Q = bh^m$. The constants b and m can readily be determined by using the graphical method. The slope

$$m = \frac{\Delta Q}{\Delta h} = \frac{\log 0.141 - \log 0.070}{\log 10 - \log 2.5} = 0.505.$$

Constant b is equal to the intercept value which occurs at $h = 1$. This, we recognize is true, since

$$\log Q = m \log h + \log b$$

and $$\log Q = \log b, \text{ (when } h = 1)$$

Therefore, $$Q = b.$$

We note that $h = 1$ is not available on the h scale. We can, however, transfer point A to A' and then draw $A'B'$ parallel to the original line. This, in effect, is the same as extending the original line to $h = 1$ and then translating the extended portion to the position $A'B'$. The h-value of A' is then taken as 1.5 and the h-value of B' as 1.0. The Q coordinate of point B' is 0.043, which is the value of the constant b in the equation $Q = bh^m$. Now that we have the values of b and m, the empirical equation is $Q = 0.043h^{0.505}$.

Try to check this result by employing the selected point method.

Fig. 14.13

Equations of the Form $y = be^{mx}$

Exponential or logarithmic curves often approximate a large number of experimental results. The mathematical expression for such curves is $y = be^{mx}$, where $e = 2.718$ (the base of natural logarithms, known as the Naperian system). The form $y = bm^x$ is also used. Figure 14.14 shows several curves of the form $y = be^{mx}$ for $b = 2$ and $m = 1, 2, -1$, and -2. Figure 14.15 shows these curves of the form $y = bm^x$ for $y = 2 \times 3^x$; $y = 2 \times 3^{-x}$; and $y = 2 \times 3^{-2x}$.

Let us now concentrate on the form, $y = be^{mx}$. When the plot of experimental data upon a Cartesian chart having uniform scales reveals a curve similar to the type shown in Fig. 14.14, verification of the form can be made by plotting the data points on a chart having a *logarithmic y-scale* and a *uniform x-scale,* in which case the plot of the data should lie approximately on a straight line.

This is quite evident since the equation $y = be^{mx}$, when expressed logarithmically, becomes

$$\log y = mx \log e + \log b$$

or

$$\log y = (m \log e)x + \log b$$

which is of the form $y_1 = m_1x + b_1$ (straight line), where $y_1 = \log y$; $m_1 = m \log e$; and $b_1 = \log b$.

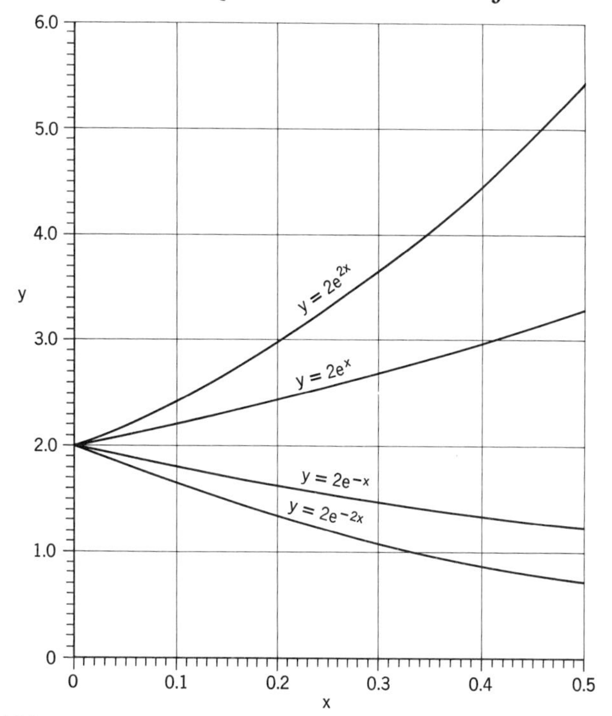

Fig. 14.14

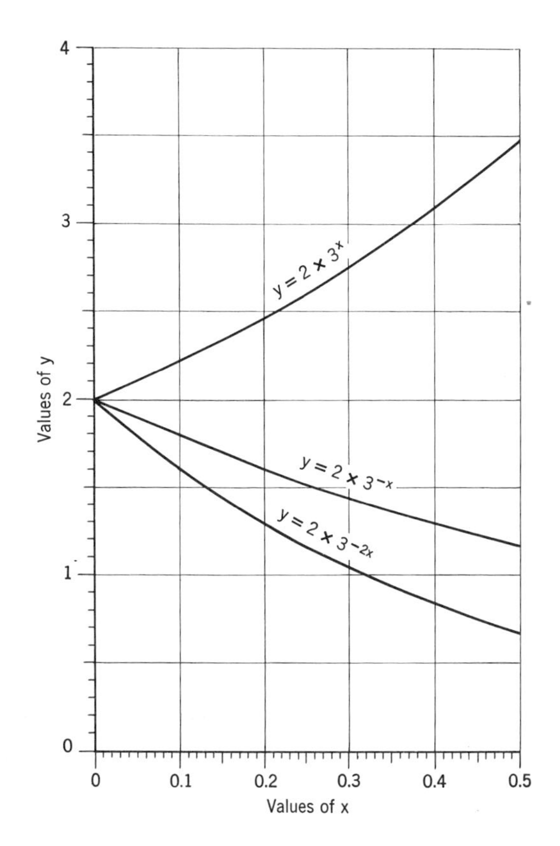

Fig . 14.15

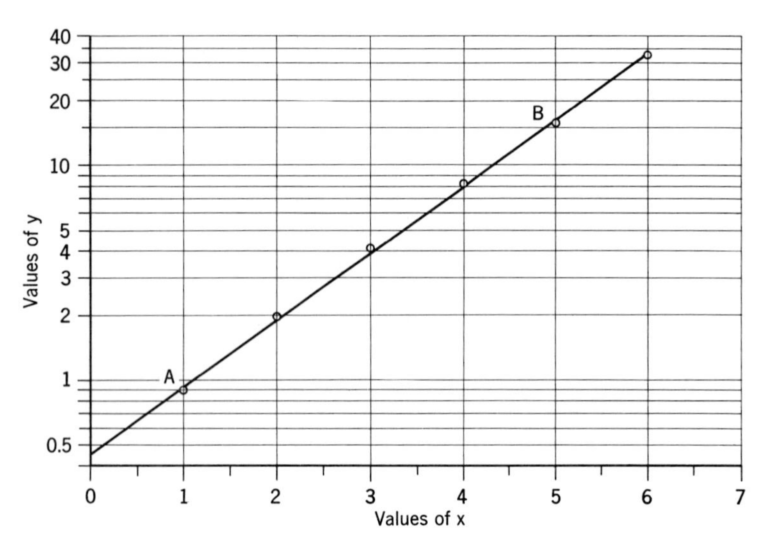

Fig. 14.16

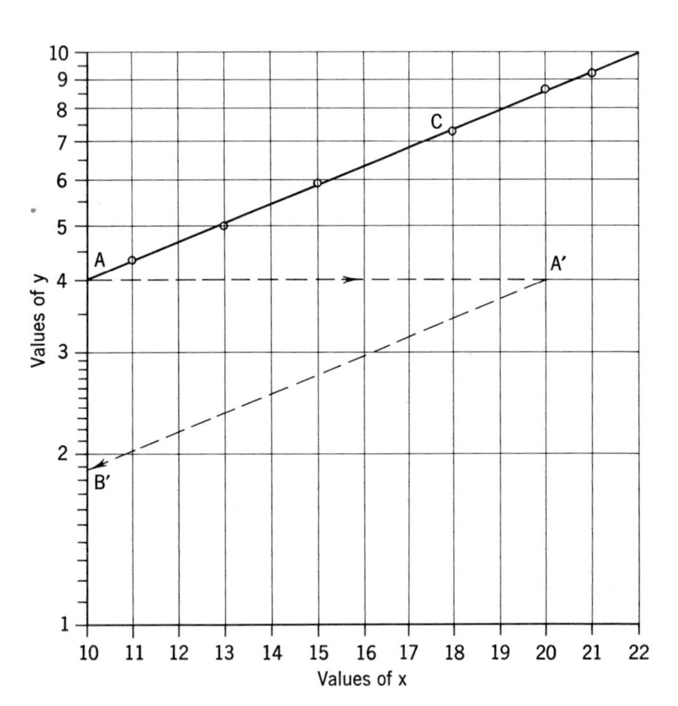

Fig. 14.17

EXAMPLE 1

Let us consider the following data:

x	1	2	3	4	5	6
y	0.9	1.9	4.1	8.2	15.9	32.3

We will assume that a plot of these data on a Cartesian coordinate sheet that has uniform x and y scales shows that the "best" curve, which approximately passed through the points, belongs to the form $y = be^{mx}$. A replot of the data on a semi-logarithmic grid shows that the points lie very nearly on a straight line. This is seen in Fig. 14.16. Now, since $\log y = (m \log e)x + \log b$, we see that *when* $x = 0$, $\log y = \log b$ or $y = b$. Graphically, $b = 0.46$. We also note that

$$m = \frac{\log y - \log b}{(0.434)(x)}; \text{ (where } \log e = 0.434)$$

Let us choose $x = 5$ and $y = 16$ since this point is on the line (Fig. 14.16). Then

$$m = \frac{\log 16 - \log 0.46}{(0.434)(5)} = \frac{\log 34.783}{2.17} = 0.710$$

The empirical equation is, therefore, $y = 0.46e^{0.710x}$.

The value of m could be determined in another way. Instead of using one point, we can select two points *on the line,* such as A (1, 0.93) and B (5, 16).

Now,

$$m \log e = \frac{\Delta y}{\Delta x} = \frac{\log 16 - \log 0.93}{5 - 1} = 0.309$$

and
$$m = \frac{0.309}{\log e} = \frac{0.309}{0.434} = 0.712$$

which is slightly different from the value obtained previously.

EXAMPLE 2

Consider the following data:

x	11	13	15	18	20	21
y	4.3	5.0	5.9	7.3	8.6	9.2

A plot of these data on a semi-logarithmic grid shows that the points lie very nearly on a straight line. This is seen in Fig. 14.17. The empirical equation is of the form $y = be^{mx}$. Expressed logarithmically, we get $\log y = (m \log e)x + \log b$. When $x = 0$, $\log y = \log b$ or $y = b$. We note that $x = 0$ is not available in the original plot of the data. We can, however, shift point A a distance of 10 units to point A'. Now, through A' we can easily draw a parallel to the original line and locate point B'. It should be quite obvious that the y-value, 1.88, of B' corresponds to $x = 0$. Therefore, the value of b in the equation $y = be^{mx}$ is 1.88. We also recall that

$$m = \frac{\log y - \log b}{(0.434)(x)}$$

Let us choose $x = 18$ and $y = 7.3$ which are the coordinates of point C.

Now,

$$m = \frac{\log 7.3 - \log 1.88}{(0.434)(18)} = 0.0754$$

The empirical equation is $y = 1.88e^{0.0754x}$.

EXAMPLE 3

Let us consider the data shown below for P, absolute pressure in pounds per square foot, and corresponding values of H, altitude in thousands of feet.

H, in ft $\times 10^3$	0	1	2	3	4	5	6	7	8	9	10
P, psf	2116	2041	1968	1896	1828	1761	1696	1633	1572	1512	1456

A plot of the data on a semi-logarithmic grid shows that the points lie very nearly on a straight line. (See Fig. 14.18.) The equation is of the form $y = be^{mx}$. The value of b is the y-intercept of the line, or $b = 2116$ (same as the value given in the table, in this case, since the point is on the line).

Now, since

$$m = \frac{\log y - \log b}{(0.434)(x)}$$

let us take point A (10, 1456), then,

$$m = \frac{\log 1456 - \log 2116}{(0.434)(10)} = \frac{-0.1623}{4.34} = -0.0374.$$

The equation is $P = 2116e^{-0.0374H}$.

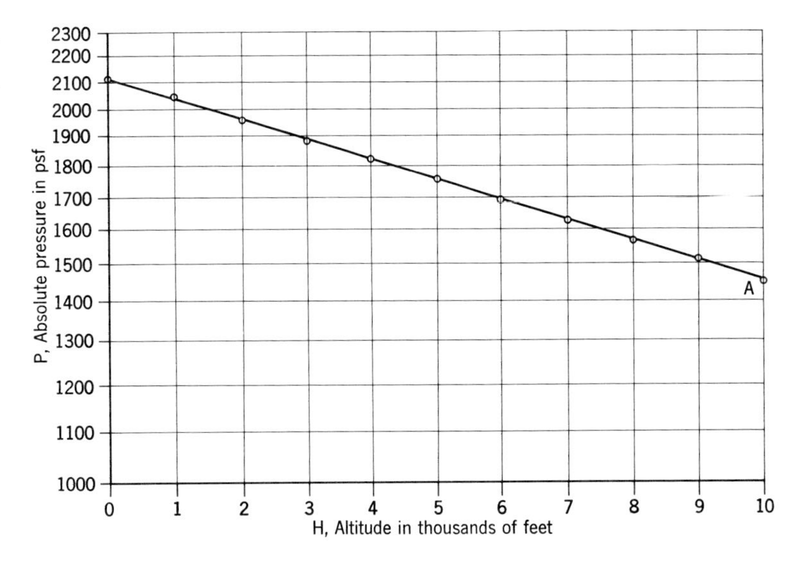

Fig. 14.18

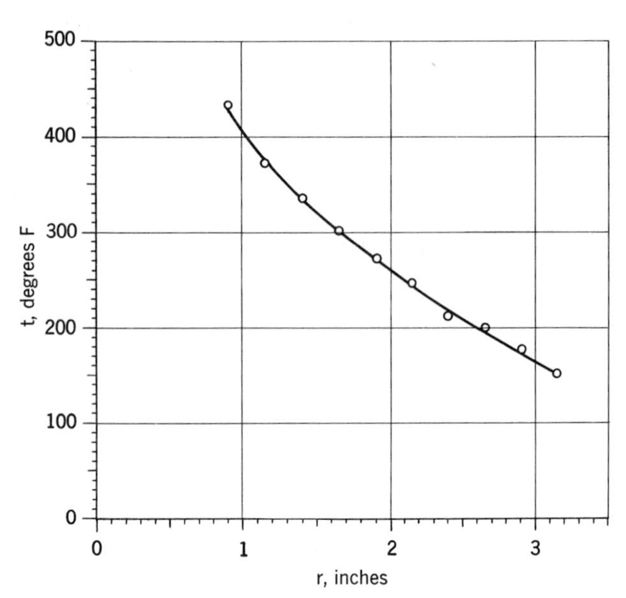

Fig. 14.19

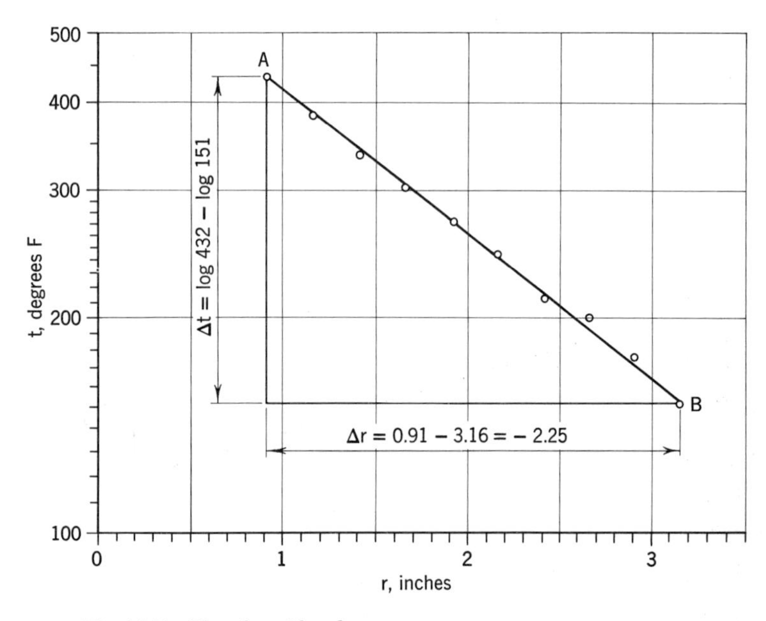

Fig. 14.20 Plot of semi-log sheet.

EXAMPLE 4

An experiment dealing with the variation of temperature in degrees F with radius r in inches for a 3-inch thick insulation around a 1-inch nominal diameter wrought-iron pipe resulted in the following data:

r, in inches	0.91	1.16	1.41	1.66	1.91	2.16	2.41	2.66	2.91	3.16
t, in degrees	432	371	334	302	272	245	211	200	176	151

A plot of these data is shown in Fig. 14.19. An inspection of the curve which best represents the data does not clearly indicate the type of equation which would express the relation between the variables. If the data were plotted on a chart having logarithmic scales, we would find that the points lie very nearly on a curve. If, however, the data were plotted on a chart having a logarithmic scale for one axis and a uniform scale for the other, it would be evident that the points do lie very nearly on a straight line. This is shown in Fig. 14.20. Let us select points A (0.91, 432) and B (3.16, 151) on the line. Now the value of m is determined from

$$m \log e = \frac{\Delta t}{\Delta r}$$

or $$m \log e = \frac{\log 432 - \log 151}{0.91 - 3.16} = -0.203$$

and $$m = \frac{-0.203}{0.434} = -0.467$$

Since the equation is of the form $y = be^{mx}$, it follows that $t = be^{mr}$. We also know that $\log t = (m \log e)r + \log b$.

Let us now evaluate b. Using point A (0.91, 432), we obtain

$$\log 432 = (-0.203)(0.91) + \log b$$

or $$\log b = 2.6355 + 0.1847 = 2.8202$$

Therefore, $b = 661$. The equation is $t = 661e^{-0.467r}$.

Other useful forms of empirical equations include the following:

1. $y = bx^m + c$

   A plot of $\log (y - c)$ and $\log x$ approximates a straight line. The value of c is obtained by selecting two points x_1y_1 and x_2y_2 on the plotted curve, and then measuring the value of y_3 which corresponds to $x_3 = \sqrt{x_1x_2}$. It can be shown that $c = \dfrac{y_1y_2 - y_3^2}{y_1 + y_2 - 2y_3}$.

2. $y = bm^x + c$

Log $(y - c) = x \log m + \log b$. A plot of log $(y - c)$ against x approximates a straight line. To determine a value of c, select points x_1y_1 and x_2y_2 on the curve (the points should be quite far apart), and then find y_3 for $x_3 = \dfrac{x_1 + x_2}{2}$. It can be shown that c has the same value as in the above form.

3. $y = nx^2 + mx + b$

Select a point x_1y_1 on the "best" curve of the plotted data. Then $y_1 = nx_1^2 + mx_1 + b$ and $(y - y_1) = n(x^2 - x_1^2) + m(x - x_1)$ or $\dfrac{y - y_1}{x - x_1} = n(x + x_1) + m$. Therefore, a plot of $\dfrac{y - y_1}{x - x_1}$ against x approximates a straight line since $\dfrac{y - y_1}{x - x_1} = nx + (nx_1 + m)$.

4. $y = px^3 + nx^2 + mx + b$

Select four points x_1y_1, x_2y_2, x_3y_3, and x_4y_4 on the curve and then solve simultaneously the four equations:

$$y_1 = px_1^3 + nx_1^2 + mx_1 + b$$
$$y_2 = px_2^3 + nx_2^2 + mx_2 + b$$
$$y_3 = px_3^3 + nx_3^2 + mx_3 + b$$
$$y_4 = px_4^3 + nx_4^2 + mx_4 + b$$

for the values of the constants p, n, m, and b.

EXERCISES

Use appropriate coordinate sheets—uniform, log-log, or semi-log.

1. Determine the equations of lines A and B as shown in Fig. E-14.1.

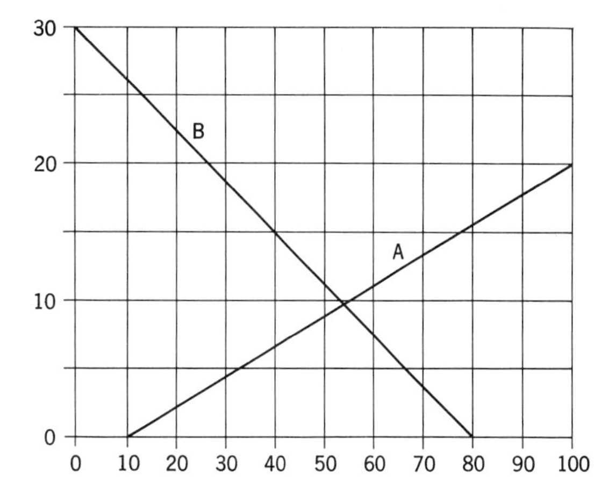

Fig. E-14.1

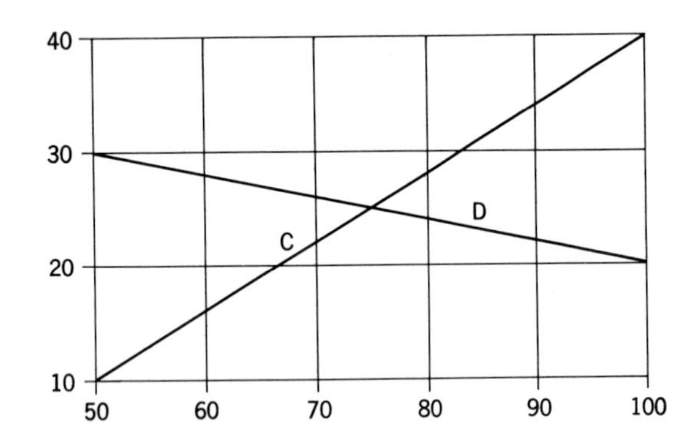

Fig. E-14.2

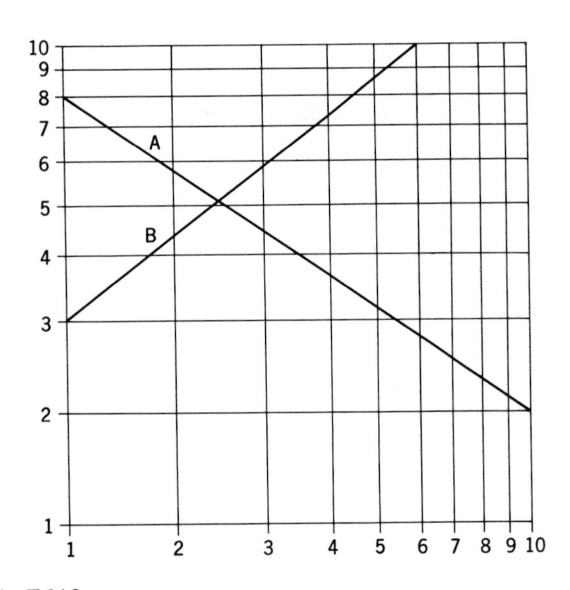

Fig. E-14.3

2. Determine the equations of lines C and D as shown in Fig. E-14.2.

3. Determine the relation between R, resistance, and V, voltage, resulting from a lamp test. The data are shown in the table. Solve (a) graphically and (b) by the method of selected points. (Form: $y = mx + b$.)

V	70	75	80	85	90	95	100	105	110
R	142.5	146	150	154.5	157	162	165	169	173

4. Determine the relation between time, t, in seconds and S, distance in feet, from the data shown below. Solve (a) graphically and (b) by the method of selected points. (Form: $y = mx + b$.)

t, sec	1	2	3	4	5	6	7	8	9
S, ft	8.0	10.3	13.8	17.2	19.9	23.2	26.0	29.7	32.5

5. Determine the equations of lines A and B as shown in Fig. E-14.3.

6. Determine the equations of lines C and D as shown in Fig. E-14.4.

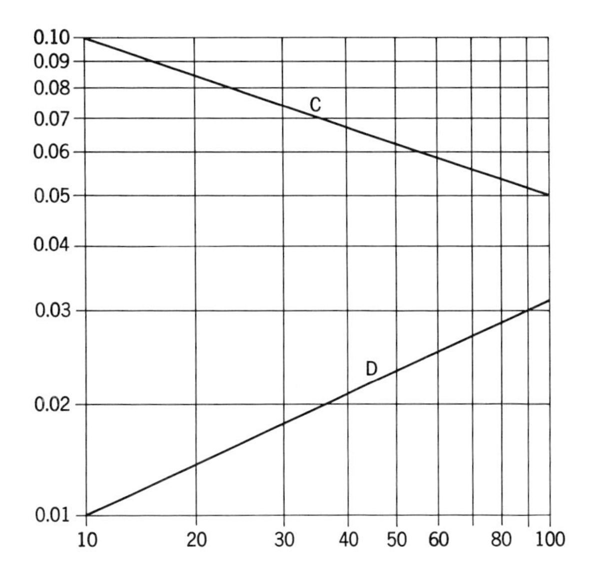

Fig. E-14.4

7. Determine the relation between P, pressure in pounds per square inch, and V, volume in cubic feet of 1 pound of saturated steam, from the data shown in the table. Solve (*a*) graphically and (*b*) by the method of selected points.

V	22.4	19.0	16.3	14.0	12.1	10.5
P	17.5	20.8	24.5	28.8	33.7	39.3

8. The water displacement V, in cubic feet, of a vessel varied with h, the distance in feet from the keel to the water level, as shown in the table.

h, in feet	9.5	10.9	12.8	17.7
V, in cubic feet	41,000	50,900	64,750	106,800

Plot the data as a straight line on the appropriate coordinate grid, and determine the relation between the variables. Solve (*a*) graphically and (*b*) by the method of selected points.

9. Determine the relation between H, the head (in feet) of water over a weir, and Q, the quantity of water (in pounds per minute) flowing over the weir. The data are shown in the table. Solve (*a*) graphically and (*b*) by the method of averages.

H, in feet	0.190	0.241	0.272	0.291	0.312	0.341	0.410
Q, in lb/min	142.3	251.7	325.4	415.3	486.5	610.5	886.5

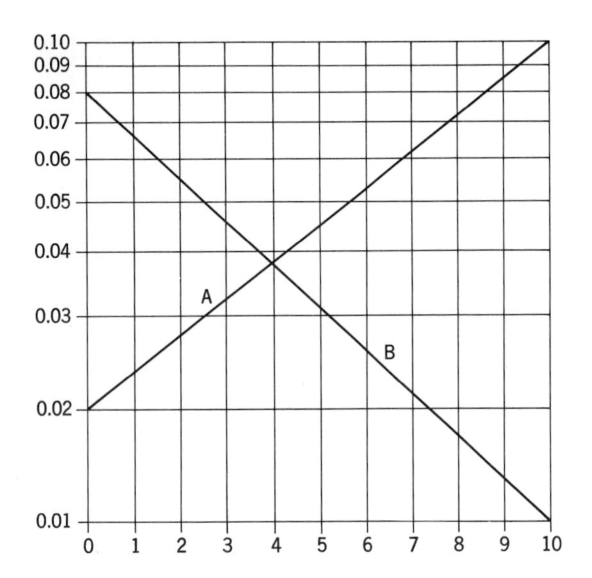

Fig. E-14.5

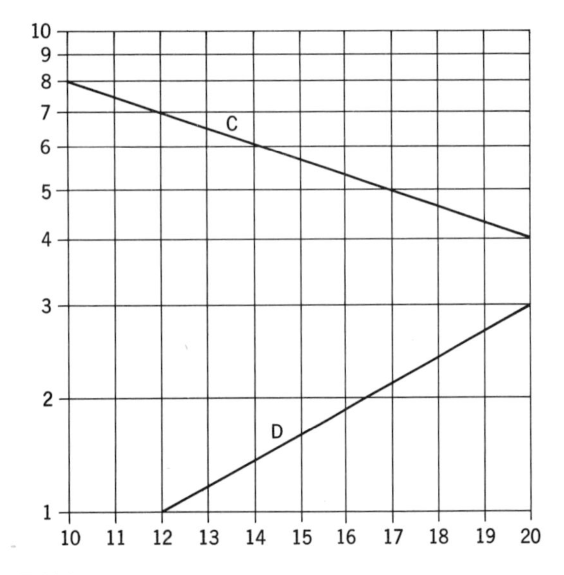

Fig. E-14.6

10. Determine the equations of lines A and B as shown in Fig. E-14.5.

11. Determine the equations of lines C and D as shown in Fig. E-14.6.

12. Determine the relation between μ, the coefficient of friction, and T, temperature of bearings operating at a constant speed (Beauchamp Tower's experiment on friction of bearings). The following data were obtained:

T	120	110	100	90	80	70	60
μ	0.0051	0.0059	0.0071	0.0085	0.0102	0.0214	0.0148

Solve (a) graphically and (b) by the method of averages.

13. Determine the relation between A, the amplitude in inches of a long pendulum, and t, the swinging time.

t	0	1	2	3	4	5	6	7	8
A	10.3	5.12	2.51	1.27	0.67	0.41	0.35	0.24	0.17

14. Determine the relation between barometric pressure, P, and the elevation, E, above sea level from the data shown in the table. Solve (*a*) graphically and (*b*) by the method of averages.

E, in feet	0	900	2750	4760	6940	10,600
P, inches mercury	30	29	27	25	23	20

15. Determine the relation between torque, T, and the twist angle, θ, from the data shown below. Solve (*a*) graphically and (*b*) by the method of averages.

T, in pounds	500	550	600	650	700	750	800	850
θ, degrees	15.2	17.3	20.2	24.0	28.5	34.1	39.4	47.3

FUNCTIONAL SCALES 15

The graphical solutions of the algebraic and calculus problems presented in the previous chapters employed scales whose graduations were spaced uniformly for equal increments of the variables. Uniform scales are quite common, i.e., on measuring devices such as ammeters, speedometers, voltmeters, engineer's and architect's scales, etc. The scales on the slide rule, however, are not uniform. Most of them are logarithmic, i.e., the distances between 1 and 2, between 2 and 3, between 3 and 4, etc., are proportional to the logarithms of the numbers. Logarithmic scales are functional scales.

What do we mean by a scale? *A graphical scale is a curved or straight line with graduations which correspond to a set of numbers arranged in order of increasing magnitude.*

A functional scale is one on which the graduations are marked with THE VALUES OF THE VARIABLE *and on which the distances to the graduations are laid off in proportion to the corresponding* VALUES OF THE FUNCTION *of the variable.*

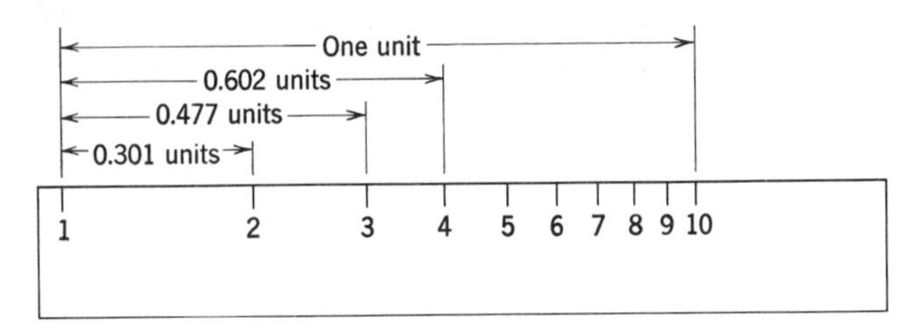

Fig. 15.1

EXAMPLE 1

Let us consider a logarithmic scale of the ordinary slide rule. Figure 15.1 shows a logarithmic scale having a range from 1 to 10. At the outset we see that the graduations are marked 1, 2, 3, . . . , 10, and that the distances between consecutive values of the variable are not uniform. Actually, the distances from the value 1 to each of the values 2, 3, 4, etc. *are proportional to the logarithms of these numbers.* These distances are the *values of the function of the variable.*

If we regard the distances from graduation 1 to the other graduations of the scale as X and the values of the other graduations as u, we may write the equation $X = \log u$ as the expression which we can use to determine the distances from graduation 1 to the other graduations of the scale.

Now, suppose that we want to make the scale 10 units long for the range of the variable from 1 to 10. Suppose also that we decide to use the inch as the unit. Then the logarithmic scale will be 10 inches long for the specified range.

Obviously, the equation $X = \log u$ does not take into account the 10-inch length of the scale.

We know that when we substitute 10 for u in the equation, $X = \log u$, then $X = 1$. Therefore, we must introduce a scale factor, or *scale modulus*, so that $X = 10$. This is easily done by writing the equation $X = m \log u$, where m is the scale factor, or what we commonly refer to as the *scale modulus.* In this example, it is easily seen that when $m = 10$, $X = 10 \log u$.

Now when we substitute values of u in this equation we will obtain the distances from $u = 1$ to the other values of u in inches. (See Fig. 15.2.)

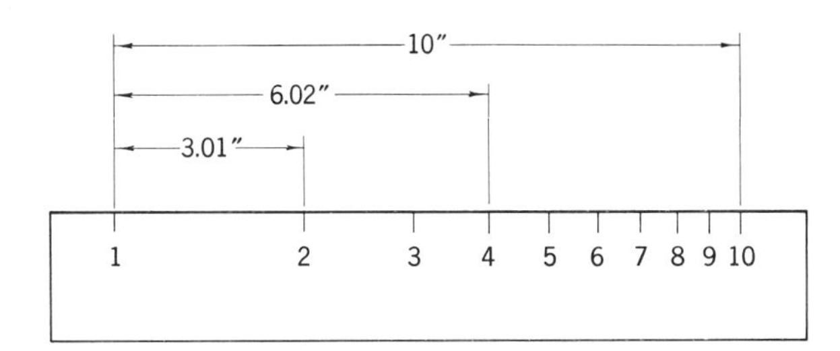

Fig. 15.2

EXAMPLE 2

Now suppose that the $f(u) = \log u$; that u varies from 2 to 8; and that the length, L, of the scale is to be approximately 10 inches.

Can we use the equation $X = m \log u$? Does this equation take into account the range of u from 2 to 8? A little thought on these two questions should lead us to conclude that the answer to these questions is *no.* Can we develop an equation which will satisfy the specifications set forth? Let's try.

We know the specified length, L, of the scale is approximately 10 inches; that one end of the scale must be the value $u = 2$, and the other end $u = 8$.

Now we can set up an expression to satisfy the above conditions by writing

$$m = \frac{L}{f(u)_{\max} - f(u)_{\min}}$$

or

$$m = \frac{10''\pm}{\log 8 - \log 2} = \frac{10''\pm}{\log 4} = 16.6\pm$$

Since the length of the scale is approximately 10 inches, we can let $m = 16$. This value of m will make the scale length 9.63 inches [from (16)(0.602)]. The scale equation, $X = 16(\log u - \log 2)$ is used to locate the graduations from $u = 2$ to $u = 8$. (See Fig. 15.3.)

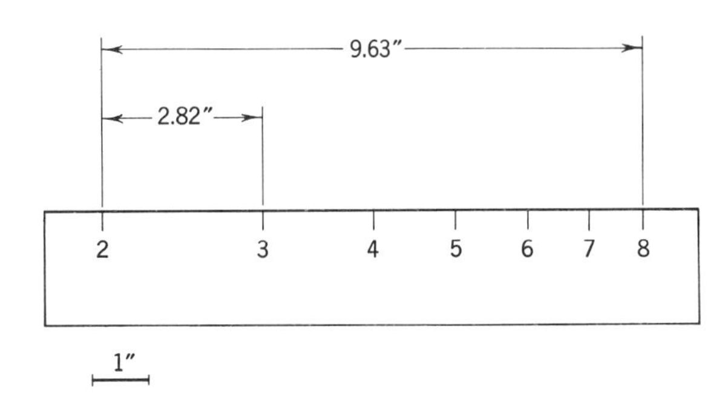

Fig. 15.3 Scale equation, $X = 16 (\log u - \log 2)$.

In general, the *scale equation* is written as

$$X = m[f(u) - f(u_1)]$$

where $f(u_1)$ is the minimum value of the function.

In the above example note that since logarithmic scales are available (printed forms and scales on slide rules) it is not necessary to compute the values of the distances from the 2 to the other graduations of the scale. We can make use of the fact that "corresponding sides of similar triangles are in the same ratio" to locate graduations between 2 and 8 or, for that matter, between any two points. To

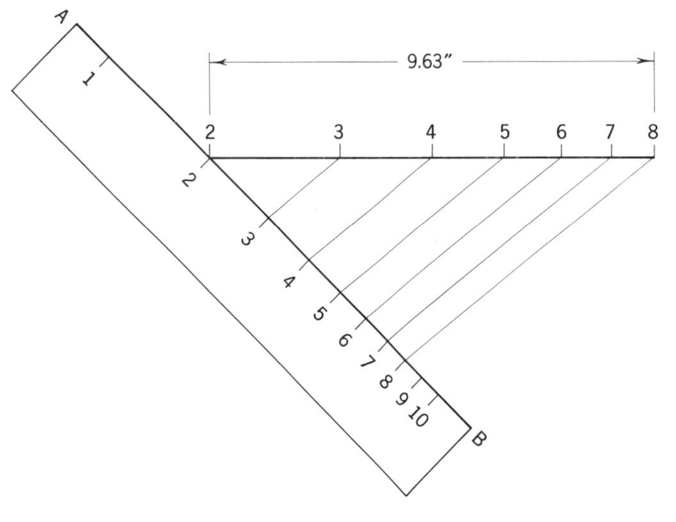

Fig. 15.4

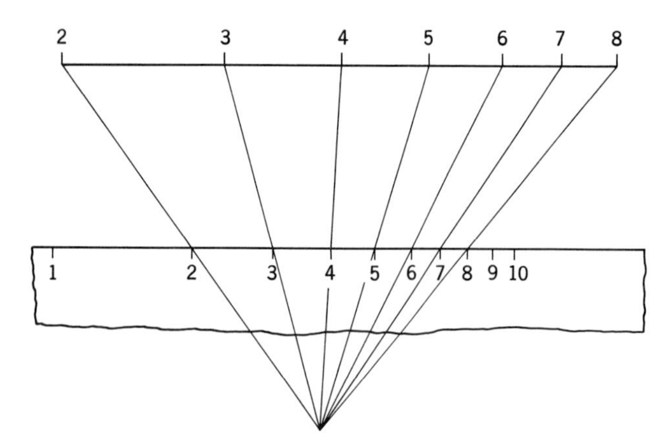

Fig. 15.5

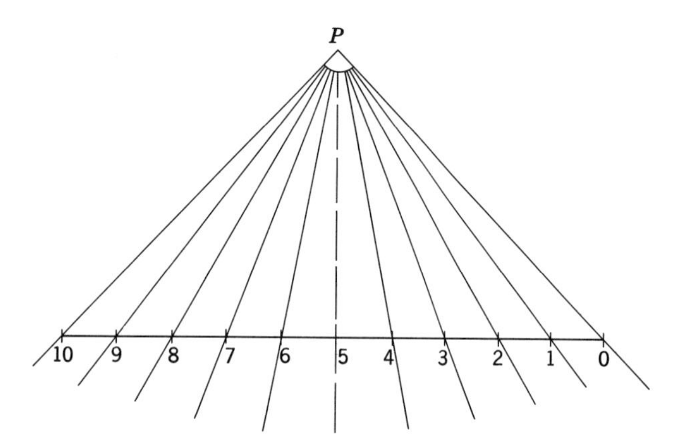

Fig. 15.5(a) Projective pencil.

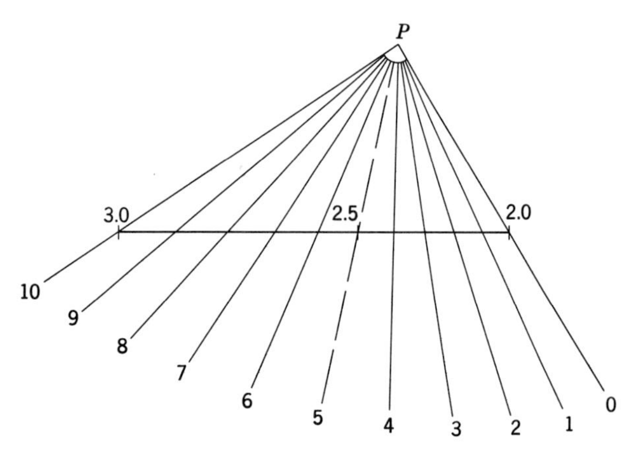

Fig. 15.5(b) Subdivision scale using the projective pencil method.

illustrate this, we lay off a line 9.63 inches long and mark the ends 2 and 8. Now through point 2 we introduce a line, AB, making a convenient angle with the first line. On line AB we mark points 2, 3, 4, 5, 6, 7, and 8 from a printed log scale. Lines drawn through these points, parallel to the line which joins the 8's, will intersect the 9.63-inch scale in points that have corresponding values. (See Fig. 15.4 on page 347.)

An alternative method which is more accurate is shown in Fig. 15.5. This is often preferred to the one shown in Fig. 15.4 because it usually overcomes the difficulty of locating, quite accurately, the points of intersection with the given line. This is especially true when the length of the printed scale is much shorter than the line to be graduated. In such a case the angle of intersection of the parallels with the given line is very small, thus making it difficult to locate the points of intersection.

Often the functional scales are not logarithmic. In such cases it is necessary to compute distances from the end values, using the scale equation. Whenever printed forms (for functions other than logarithmic) are available, it is recommended that they be used if appropriate.

In a number of cases however, subdividing a non-uniform scale can be accomplished reasonably well by using the projective-ray method. First, prepare a "projective pencil" (concurrent rays) as shown in Fig. 15.5(a). This should be done on a transparent sheet.

Now, suppose we wish to subdivide the scale shown in Fig. 15.5(b) to include all of the tenths graduations between 2.0 and 3.0. Place the projective pencil over the scale so that rays P-0, P-5, P-10 pass through 2.0, 2.5, and 3.0 respectively. Next, locate the intersections of the other rays with the scale (use a needle point or compass point to prick through the overlay). This method is based on the assumption that the scale is homographic.* If it is, the locations of the subdivisions are quite accurate; if not, they are close approximations.

* A homographic function is of the form $y = \dfrac{A \cdot f(x) + B}{C \cdot f(x) + D}$, where $AD - BC \neq 0$. The distances to the scale graduations (from an initial point) are proportional to $\dfrac{A \cdot f(x) + B}{C \cdot f(x) + D}$.

EXAMPLE 3

Suppose the $f(u) = u^3$, that u varies from 2 to 4, and that the scale length, L, is approximately 6 inches. The scale equation is $X = m(u^3 - 2^3)$. The scale modulus, m, is computed from the relation

$$m = \frac{L}{f(u)_{\max} - f(u)_{\min}}$$

or

$$m = \frac{6''\pm}{4^3 - 2^3} = \frac{6''\pm}{64 - 8} = \frac{6''\pm}{56} = 0.1$$

This means that the scale length is actually 5.6 inches. Had we decided to use $m = 0.11$, the scale length would be $(0.11)(56) = 6.16$ inches. In either case, *the method* for locating the graduations would be the same. Let us agree to use $m = 0.1$; then, the scale equation is $X = 0.1(u^3 - 2^3)$. Distances from the end point, $u = 2$, to several other values of u are shown in the table. The scale is shown in Fig. 15.6.

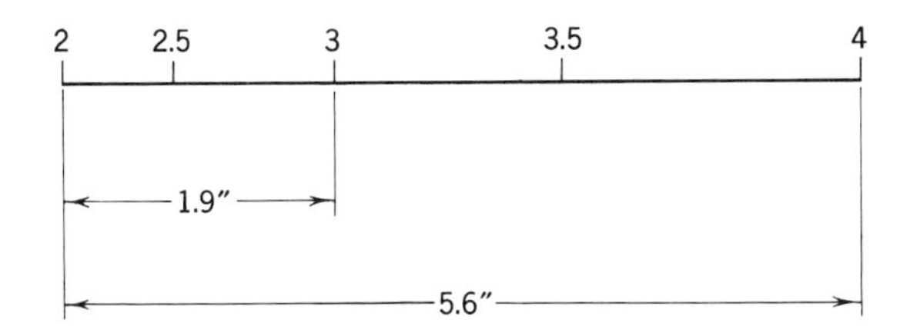

Fig. 15.6

u	2	2.5	3.0	3.5	4
$f(u) = u^3$	8	15.63	27	42.88	64
$X = 0.1(u^3 - 2^3)$	0	0.76	1.9	3.49	5.6

Observe that:

1. The distance between any two graduations, u_1 and u_2, is equal to $X = m[f(u_2) - f(u_1)]$.
2. Any unit of length other than inches could be adopted as the unit of measure.
3. Whereas the distance between any two graduations is equal to the product of the scale modulus and the difference in the *values of the function of the variable,* the *graduations* are marked with the *values of the variable.*

EXAMPLE 4

Let us consider the $f(u) = 2u^3$. We will assume that u varies from 2 to 4 and that the scale length is approximately 6 inches. We know that the total length of the scale is equal to

$$L = m[2 \times 4^3 - 2 \times 2^3]$$

or

$$m = \frac{6''\pm}{2 \times 64 - 2 \times 8} = \frac{6''\pm}{112} = 0.05$$

This means that the total length of the scale is $(0.05)(112) = 5.6$ inches.

Now the scale equation is

$$X = 0.05(2u^3 - 2 \times 2^3) = 0.1(u^3 - 2^3)$$

This last expression is the same as the scale equation in the previous example. This means that the scales would be the same. The logical question is, "What is the significant difference, if any?"

Note that the *actual modulus* in this example is $m = 0.05$ and that the *effective modulus* is 0.1. *The effective modulus is used in graduating the scale. The actual modulus is used in locating the position of scales that occur in the design and construction of alignment charts for the solution of equations containing three or more variables.* The distinction between the actual modulus and the effective modulus will be quite evident when we discuss nomography and the design of nomograms in Chapter 16.

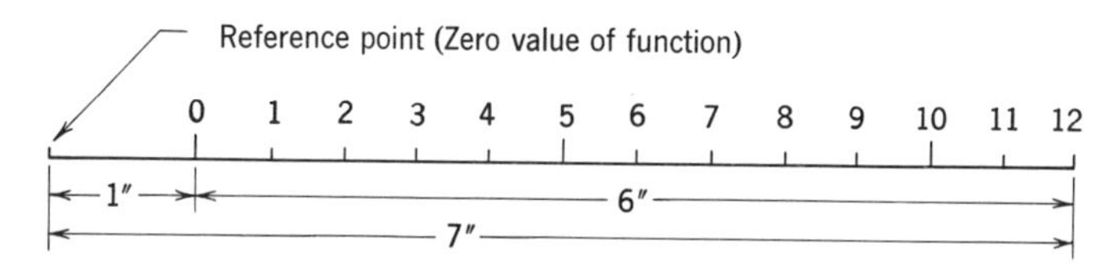

Fig. 15.7

EXAMPLE 5

Suppose the $f(u) = (u + 2)$, that u varies from 0 to 12, and that the scale length is 6 inches. (See Fig. 15.7.)

Now, we know that

$$m = \frac{6}{(12 + 2) - (0 + 2)} = 0.5$$

and that

$$X = 0.5[(u + 2) - (0 + 2)] = 0.5u$$

The scale equation, $X = 0.5u$, would be the same if the $f(u) = u$; however, *with respect to* $f(u) = 0$ {that is, when $u = -2, f(u) = [(-2 + 2) = 0]$}, the zero graduation of the scale for $f(u) = (u + 2)$ is shifted an amount equal to $\frac{1}{2}(0 + 2) = 1$ inch.

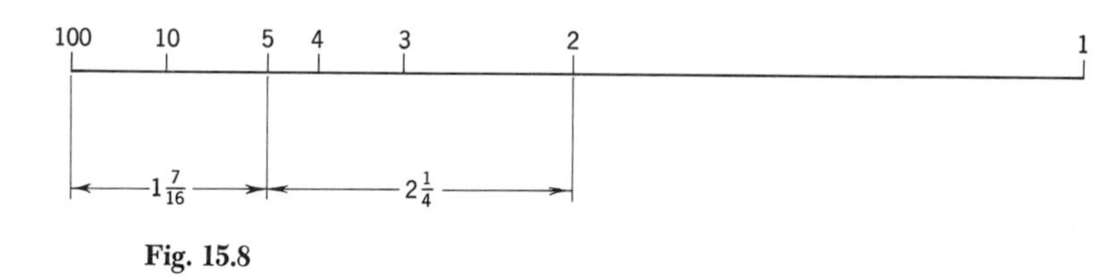

Fig. 15.8

EXAMPLE 6

Let us consider the $f(u) = 1/u$, where u varies from 1 to 5, and where the scale length is 6 inches. (See Fig. 15.8.) The scale modulus is

$$m = \frac{6}{(1/1 - 1/5)} \quad \text{or} \quad \frac{6}{4/5} = 7.5 \qquad (1)$$

Note the value in the denominator, $(1/1 - 1/5)$. Although 5 is numerically greater than 1, we should not forget that the function is $1/u$, and that

$$m = \frac{L}{[f(u)_{\max} - f(u)_{\min}]}$$

Therefore, the correct evaluation of m is as shown in equation (1).

The scale equation is

$$X = 7.5\left(\frac{1}{u} - \frac{1}{5}\right) \tag{2}$$

We observe, in equation (2), that when $u = 5$, $X = 0$. Values of X for other values of u are shown in the table.

u	1	2	3	4	5	—	10	100
$f(u) = 1/u$	1	$\frac{1}{2}$	$\frac{1}{3}$	$\frac{1}{4}$	$\frac{1}{5}$	—	$\frac{1}{10}$	$\frac{1}{100}$
$X = 7.5(1/u - 1/5)$	6	$2\frac{1}{4}$	1	$\frac{3}{8}$	0	—	$-\frac{3}{4}$	$-1\frac{7}{16}$

When we arbitrarily increase u to 10 and then to 100, note that the distances are negative. These distances are laid off to the left of *the starting point* $u = 5$. Moreover, observe the congestion of values of u between 10 and 100. As we increase the values of u, the congestion will become more severe. Therefore, *it is strongly recommended that scales for reciprocal functions be limited to short ranges of the variable.*

Another observation is of interest. Had we elected to graduate the scale with reference to $u = 1$ *as the starting value*, the scale equation would be

$$X = 7.5\left(\frac{1}{u} - \frac{1}{1}\right)$$

In this case the values of X would be those shown in the table below.

u	1	2	3	4	5	—	10	100
$f(u) = 1/u$	1	$\frac{1}{2}$	$\frac{1}{3}$	$\frac{1}{4}$	$\frac{1}{5}$	—	$\frac{1}{10}$	$\frac{1}{100}$
$X = 7.5(1/u - 1/1)$	0	$-3\frac{3}{4}$	-5	$-5\frac{5}{8}$	-6	—	$-6\frac{3}{4}$	$-7\frac{7}{16}$

Obviously, the scale would be the same as that shown in Fig. 15.8.

Adjacent Scales for Equations of the Form $f_1(u) = f_2(v)$

Equations of this form are easily solved by graduating both sides of one line in such a manner that any point on the line will yield values which satisfy the given equation.

The scale equations for the functions are:

$$X_u = m_u f_1(u) \qquad \text{(from } f(u) = 0\text{)}$$

and

$$X_v = m_v f_2(v) \qquad \text{(from } f(v) = 0\text{)}$$

(Note that the subscripts are used to identify the functions.)

Now, for any point on the scale $X_u = X_v$, or $m_u f_1(u) = m_v f_2(v)$. *Since* $f_1(u) = f_2(v)$, *it follows that* $m_u = m_v$.

EXAMPLE 1

Consider the relation, $u = v^2$, where u varies from 0 to 10 in a scale length of 5 inches.

The scale equation for $f(u) = u$ is

$$X_u = m_u u$$

from which

$$m_u = \frac{5}{10} = 0.5$$

Therefore,

$$X_u = 0.5u \tag{1}$$

Since $m_u = m_v$, the scale equation for $f(v) = v^2$ is

$$X_v = 0.5v^2 \tag{2}$$

The graphical representation of $u = v^2$ is shown in Fig. 15.9. Additional graduation can be obtained from the scale equations, if desired.

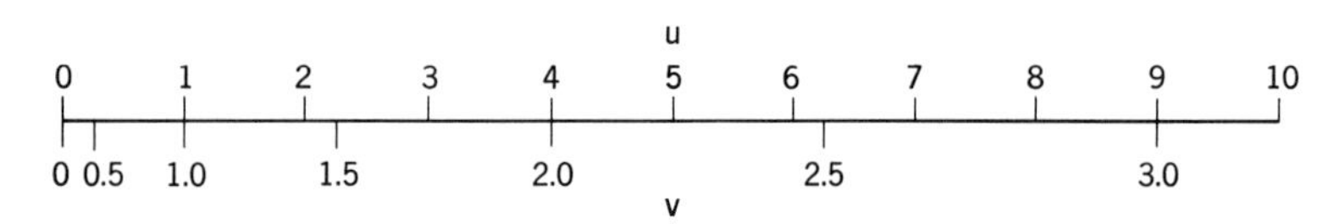

Fig. 15.9 Adjacent scales for the relation $u = v^2$.

EXAMPLE 2

Consider the relation between horsepower and foot-pounds per second,

$$H = \frac{F}{550}$$

Where H = horsepower (0 to 50) and F = foot-pounds per second. We will assume a scale length of 10 units.

The scale equation for horsepower is,

$$X_H = m_H H$$

from which

$$m_H = \frac{10}{50} = 0.2$$

Therefore,

$$X_H = 0.2H \tag{1}$$

The scale equation for the $f(F)$ is

$$X_F = 0.2\frac{F}{550} \tag{2}$$

The modulus must be the same for both functions.

Figure 15.10 shows the graphical representation of the relation $H = F/550$.

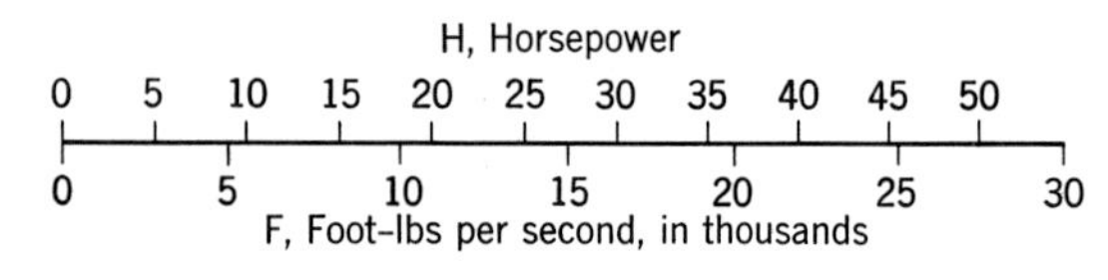

Fig. 15.10

EXAMPLE 3

Let us consider the relation,

$$A = \pi r^2,$$

where r (3 to 10 inches) is the radius of a circle and A is the area in square inches. Scale length approximately 10 units.

As a matter of convenience, it is suggested that the equation be written as $r^2 = A/\pi$ since the range is specified for the variable, r, thus freeing variable r of the constant π.

Now, the modulus, m, for the $f(r) = r^2$ is

$$m = \frac{10\pm}{10^2 - 3^2} = \frac{10\pm}{91} = 0.11$$

The scale equation is

$$X_r = 0.11(r^2 - 3^2) \tag{1}$$

It should be noted that there is an advantage in using $m = 0.11$ rather than the likely choice $m = 0.1$. This becomes quite clear when we write the scale equation for the $f(A) = A/\pi$, i.e.,

$$X_A = \frac{0.11}{\pi}(A - A_1) = 0.035(A - A_1) \tag{2}$$

The actual modulus is 0.11. The *effective* modulus is

$$\frac{0.11}{\pi} = \frac{0.11}{22/7} = 0.035$$

Scale equation (1) enables us to graduate the r scale, which is shown in Fig. 15.11.

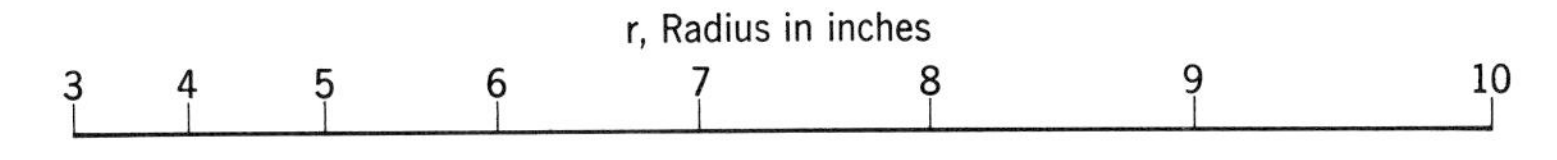

Fig. 15.11

Scale equation (2) cannot be used to graduate the A scale until we obtain a value for A_1. The latter can be any value which corresponds to the value of r, consistent with the relation $A = \pi r^2$. For example, we can let $A = 100$ and compute the corresponding value of r from $100 = \pi r^2$. In this case $r = 5.64$. The distance from $r = 3$ to $r = 5.64$ is easily computed from the scale equation,

$$X_r = 0.11[(5.64)^2 - 3^2] = 2.51 \text{ units}$$

The graduation opposite $r = 5.64$ (which need not be marked) is $A = 100$. (See Fig. 15.12.) Now the scale equation for $f(A) = A/\pi$, is

$$X_A = 0.035(A - 100) \tag{3}$$

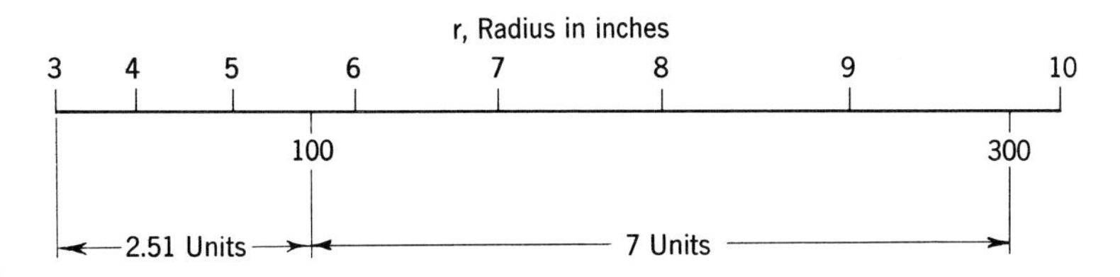

Fig. 15.12

The A scale is now graduated with respect to $A = 100$ *as the starting point.* For example, the dis-

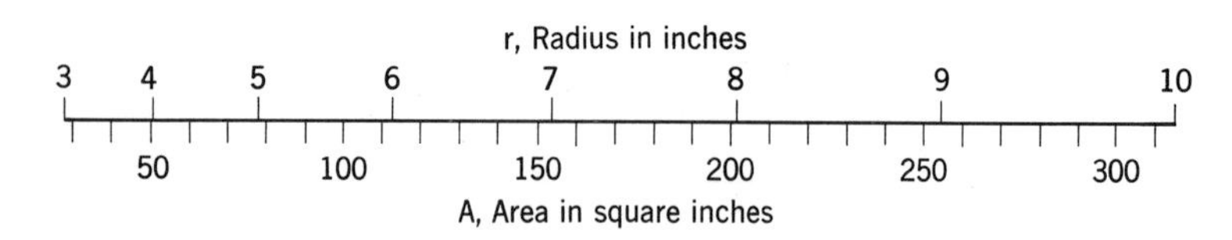

Fig. 15.13

I, Inches
0 1 2 3 4 5 6 7 8 9 10 11 12 13 14 15 16 17 18 19 20
0 5 10 15 20 25 30 35 40 45 50
C, Centimeters

Fig. 15.14

tance from $A = 100$ to $A = 300$ is

$$X_A = 0.035(300 - 100) = 7 \text{ units}$$

Since the scale is uniform (because the function is linear) graduations between $A = 100$ and $A = 300$ can be easily located without further use of equation (3). For that matter, graduations of values below $A = 100$ can also be readily located. As a check on the location of, say, $A = 50$, the scale equation (3) shows that $X_A = 0.035(50 - 100) = -1.75$ units, measured to the left of $A = 100$. The complete solution is shown in Fig. 15.13.

EXAMPLE 4

Let us consider the relation between inches, I, and centimeters, C. We know there are 2.54 centimeters in one inch. When we let I represent the *number* of inches and C the corresponding number of centimeters, then the relation is

$$(2.54)I = C$$

or

$$I = \frac{C}{2.54}$$

Suppose I varies from 0 to 20 inches, and that the scale length is 10 units. Now,

$$X_I = m_I I$$

from which

$$m_I = \frac{10}{20} = 0.5$$

The scale equation is

$$X_I = 0.5I \tag{1}$$

The scale equation for the $f(C) = \dfrac{C}{2.54}$ is

$$X_C = 0.5\left(\frac{C}{2.54}\right) = \frac{C}{5.08} \tag{2}$$

Again we should note that the *actual* modulus is 0.5 and that the *effective* modulus is 1/5.08. It is the scale equation with the effective modulus that is used in graduating the C scale. The solution is shown in Fig. 15.14.

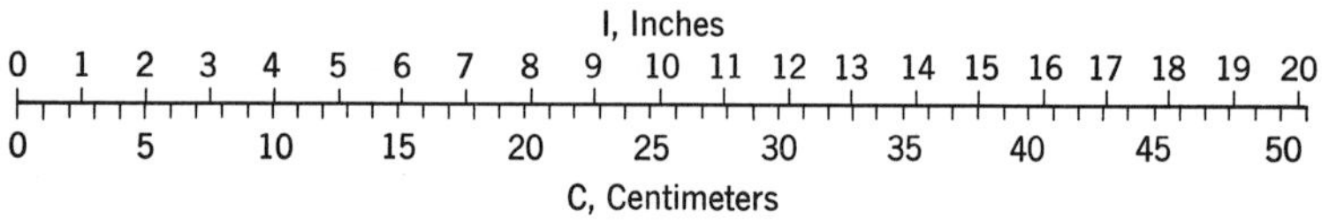

EXAMPLE 5

Suppose the given relation is $u = \sin v$; that v varies from 0° to 90°; and that the scale length is 10 units. The scale equations are:

$$X_u = m_u(u - 0) \quad (1)$$

and

$$X_v = m_v(\sin v^\circ - \sin 0^\circ) \quad (2)$$

Since the range for the variable v is specified we will compute m_v from equation (2),

$$m_v = \frac{10}{\sin 90^\circ - \sin 0^\circ} = 10$$

Therefore,

$$X_v = 10(\sin v^\circ - \sin 0^\circ)$$

and

$$X_u = 10(u - 0) \text{ (remember } m_u = m_v)$$

The solution is shown in Fig. 15.15.

If it is desired to make the v-scale more nearly uniform, we may proceed as follows:

Let $u^2 = \sin^2 v$.

Now,

$$m_v = \frac{10}{\sin^2 90^\circ - \sin^2 0^\circ} = 10$$

and

$$X_v = 10(\sin^2 v - \sin^2 0^\circ)$$

and

$$X_u = 10(u^2 - 0^2) = 10u^2.$$

The solution is shown in Fig. 15.16.

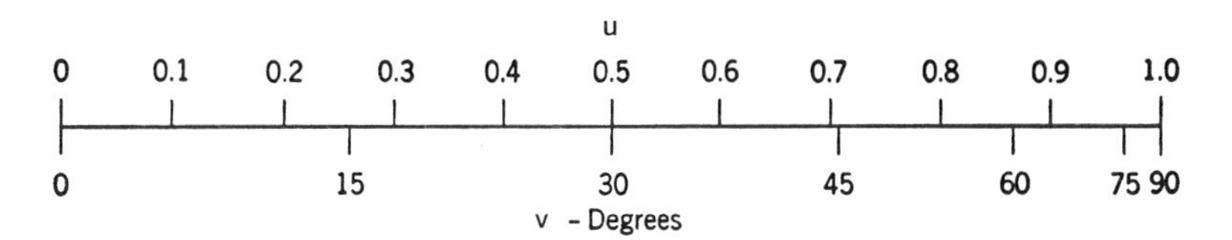

Fig. 15.15 Adjacent scales for equation $u = \sin v$.

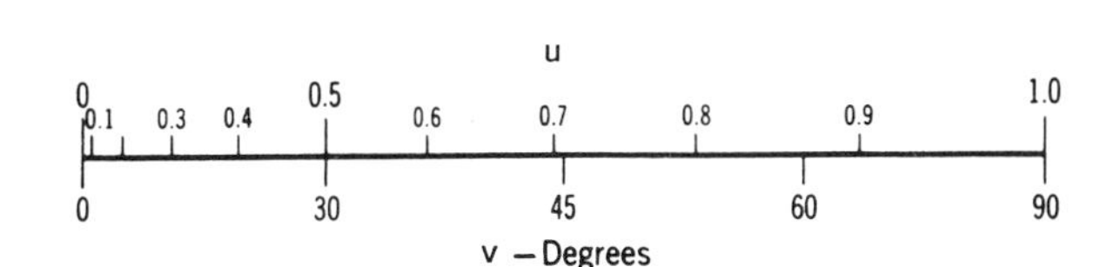

Fig. 15.16

Nonadjacent Scales for Equations of the Form $f_1(u) = f_2(v)$

In our discussion of adjacent scales we pointed out that the *same* modulus must be used in *both* scale equations. Recall that the modulus was based on the specified range of *one* of the variables and the desired length of scale. As a consequence, the adjacent scale for the other variable was fixed. This means that the scale increment between successive values of the variable could be too small for the desired accuracy in reading. To help overcome this difficulty, the scales for the variables can be separated so that *different* moduli can be used. In this manner we can select scale lengths that are consistent with the desired accuracy of reading the values of each variable.

Let us now proceed to the design of separated (nonadjacent) scales for equations of the form $f_1(u) = f_2(v)$.

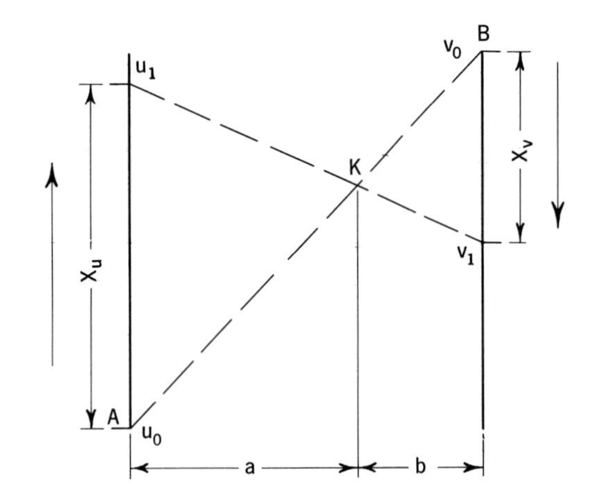

Fig. 15.17

In Fig. 15.17 we observe the following:

(a) The scale equations are

$$X_u = m_u f_1(u) \tag{1}$$

and

$$X_v = m_v f_2(v) \tag{2}$$

since u_0 and v_0 are values of u and v, respectively, which make the *functions* of u and v equal to zero.

(b) Point K is located on line AB so that any line passing through K and a selected value on the u- or v-scale will intersect the other scale in a value which satisfies the given equation. The u and v scales are placed a convenient distance apart.

From the similar triangles Au_1K and Bv_1K,

$$\frac{X_u}{X_v} = \frac{AK}{KB} = \frac{a}{b}$$

Therefore,

$$\frac{m_u f_1(u)}{m_v f_2(v)} = \frac{a}{b} \quad \text{(from equations (1) and (2) above)}$$

Since $f_1(u) = f_2(v)$,

$$\frac{a}{b} = \frac{m_u}{m_v}$$

Hence, point K can be located on the diagonal AB by dividing it into the ratio

$$\frac{AK}{KB} = \frac{a}{b} = \frac{m_u}{m_v}$$

EXAMPLE 1

Let us consider the equation $C = \pi D$, where D, diameter, varies from 0 to 10 inches in a scale length of 10 units. We will rewrite the equation as $D = C/\pi$, since the range of D is specified, and to simplify the expression.

Now the scale equation for variable D is

$$X_D = m_D D$$

where

$$m_D = \frac{10}{10} = 1$$

Therefore,

$$X_D = D \tag{1}$$

The scale equation for variable C is

$$X_C = m_C \frac{C}{\pi}$$

where

$$m_C = \frac{20}{(C/\pi)} = \frac{20}{(10\pi/\pi)} = 2$$

where 20 units is the desired length of the C scale and 10π is the maximum value of C (from $C = \pi D$). Therefore,

$$X_C = \frac{2C}{\pi} = \frac{7}{11}C \quad \left(\pi \text{ was taken equal to } \frac{22}{7}\right) \qquad (2)$$

In Fig. 15.18, scales D and C have been graduated from scale equations (1) and (2). Obviously, it is only necessary to establish the 0 and the 10 on the D-scale since the scale is uniform. Intermediate points can be easily located graphically. As to the C-scale, it is only necessary to locate the 0 and one additional point, such as the 30. Since the C-scale is also uniform, other values of C can be located graphically without any difficulty. Point K, which lies on the line joining the zero values of the functions of C and D, is located by dividing the diagonal into the ratio

$$\frac{a}{b} = \frac{m_D}{m_C} = \frac{1}{2}$$

Point K can be located in another way. We can let $D = 10$ (or any other value) and solve for C from the equation $C = \pi D$. In this case, $C = 10\pi = 31.42$. The line which joins $D = 10$ with $C = 31.42$ intersects the diagonal at point K. All lines through point K intersect the C and D scales in values which satisfy the equation $C = \pi D$.

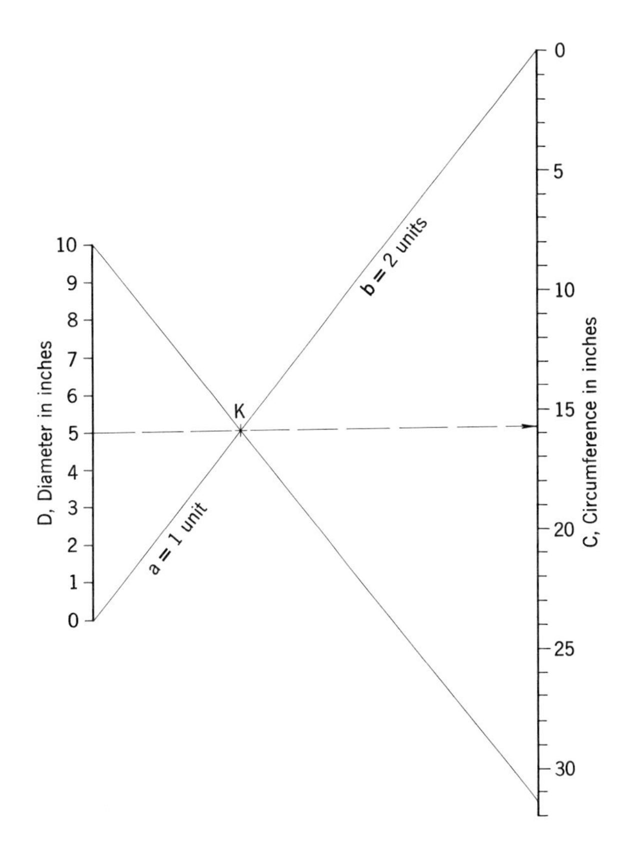

Fig. 15.18

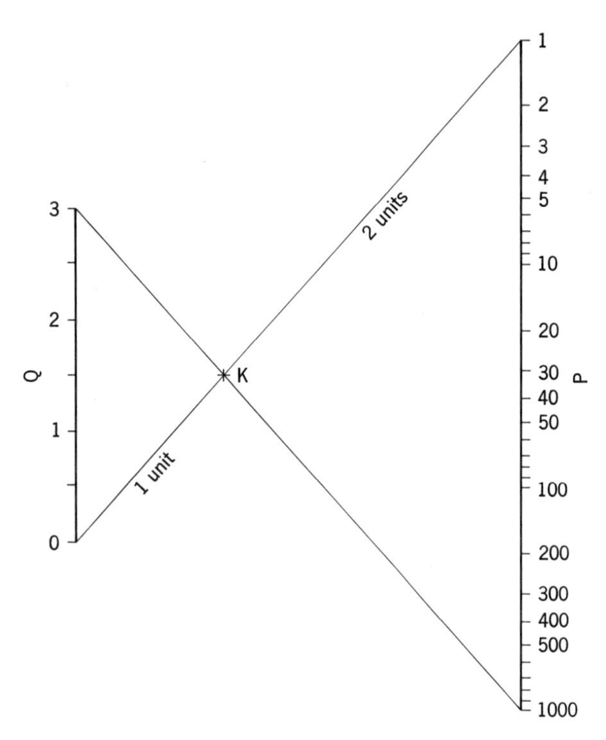

Fig. 15.19 Separated scales for the relation $Q = \log P$.

EXAMPLE 2

Let us consider the relation, $Q = \log P$, where P varies from 1 to 1000 in a scale length of 10 units.

The scale equation for the function of P is

$$X_P = m_P(\log P - \log 1)$$

where

$$m_P = \frac{10}{\log 1000 - \log 1} = \frac{10}{3}$$

Now,

$$X_P = \frac{10}{3}(\log P - \log 1) \qquad (1)$$

The scale equation for Q is

$$X_Q = m_Q Q$$

where

$$m_Q = \frac{5}{3}$$

and where the scale length is 5 units and Q varies from 0 to 3, since $Q = \log P$.

The scale equation is

$$X_Q = \frac{5}{3}(Q - 0) \qquad (2)$$

Figure 15.19 shows the graphical solution. Observe that:

(*a*) The Q scale is graduated from the scale equation (2). It is only necessary to locate the 0 and 3 values of Q since the scale is uniform. Intermediate graduations can be located graphically.

(*b*) The P scale is graduated from the scale equation (1). It is only necessary to locate the graduations 1, 10, 100, and 1000. Graduations between 1 and 10, 10 and 100, and between 100 and 1000 can be located graphically by projection from a log scale.

(*c*) The diagonal line on which point K is located joins the 0 on the Q scale and 1 on the P scale. It should be stressed that the function of P is zero when $P = 1$, since the log $1 = 0$.

(*d*) Point K divides the diagonal in the ratio

$$\frac{a}{b} = \frac{m_Q}{m_P} = \frac{5/3}{10/3} = \frac{1}{2}$$

The length of unit on the diagonal is independent of the unit used to graduate the Q and P scales.

EXAMPLE 3

Let us consider the relation $H = 2.3P$, where P (10 to 110) represents pressure in pounds per square inch, and H pressure head in feet. The equation may be rewritten as

$$P = \frac{H}{2.3}$$

The scale modulus for the P-scale is

$$m_P = \frac{5}{110 - 10} = 0.05$$

(Assumes a scale length of 5 units)

and the scale equation is

$$X_P = 0.05(P - 10) \qquad (1)$$

Next, let us assume a scale length of 10 units for the H-scale. The scale modulus is

$$m_H = \frac{10}{(253 - 23)/2.3} = \frac{23}{230} = 0.1$$

The scale equation is

$$X_H = m_H\left(\frac{H}{2.3} - \frac{23^*}{2.3}\right) = \frac{m_H}{2.3}(H - 23)$$

Now

$$X_H = 0.1\left(\frac{H}{2.3} - \frac{23}{2.3}\right) = \frac{1}{23}(H - 23) \qquad (2)$$

Note carefully that *the actual modulus is 0.1,* whereas the *effective modulus* is 1/23.

The ratio a/b which is used in locating point K is

$$\frac{a}{b} = \frac{m_P}{m_H} = \frac{0.05}{0.1} = \frac{1}{2}$$

Figure 15.20 shows the solution.

Observe that:

(*a*) The P-scale is graduated from the scale equation (1). It is only necessary to locate the values of 10 and 110. Intermediate graduations are easily located graphically, since the scale is uniform.

(*b*) The H-scale is graduated from the scale equation (2). Here again it is only necessary to locate the values of 23 and 253. Intermediate graduations are readily located since this scale is also uniform.

(*c*) The diagonal on which point K is located *must join the zero value of the function of P and the*

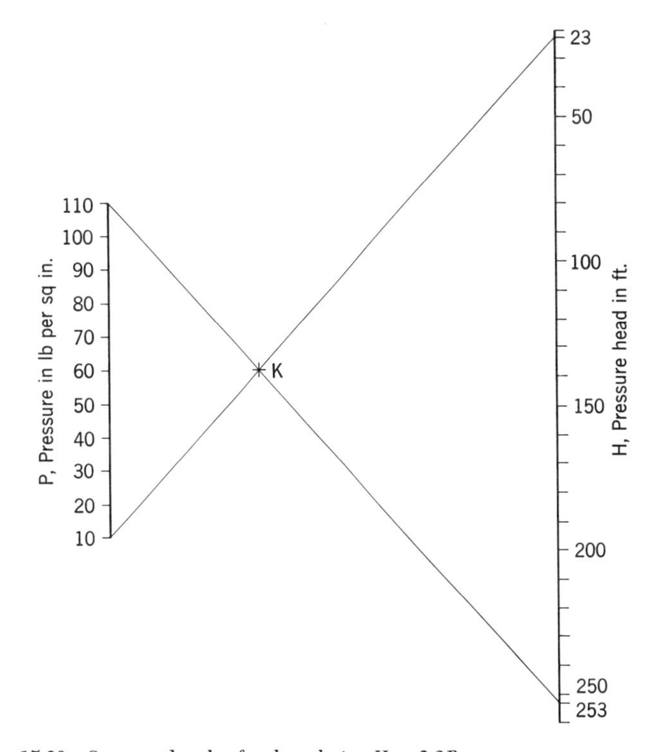

Fig. 15.20 Separated scales for the relation $H = 2.3P$.

* From $H = 2.3P$ when $P = 10$.

zero value of the function of H. While, in this example, it is a simple matter to extend the scales to include the zero values, we should not overlook those cases where it is not convenient to do so. A simple way to overcome this difficulty is as follows:

(1) Select a value of P, say, 10, and compute H from the equation $H = 2.3P$. H then equals 23. Draw a light line joining $P = 10$ and $H = 23$.

(2) Let $P = 110$, then $H = 253$. Draw a light line joining $P = 110$ and $H = 253$.

(3) The intersection of the light lines is point K. Other lines through point K will intersect the P and H scales in values which satisfy the given equation, $H = 2.3P$. As a check on the location of point K, the ratio in which K divides any line passing through K and terminating at the scales P and Q, must be equal to $m_P/m_H = 1/2$.

EXERCISES

Functional Scales

1. Construct a scale for the function $f(u) = u^2$. u varies from 0 to 10. Scale length approximately 6 inches.

2. Construct a scale for the function $f(u) = (u^2 + 2u + 1)$. u varies from 2 to 7. Scale length about 8 inches.

3. Construct a scale for the function $f(u) = 2 \log u$. u varies from 10 to 400. Scale length about 6 inches.

4. Construct a scale for the function $f(u) = 1/u$. u varies from 1/2 to 12. Scale length about 6 inches.

5. Construct a scale for the function $f(u) = (2 - u)$. u varies from 0 to 8. Scale length about 6 inches.

6. Construct a scale for the function $f(u) = 1/\log u$. u varies from 10 to 20. Scale length about 6 inches.

7. Construct a scale for the function $f(u) = \sin u$. u varies from 0° to 180°. Select suitable scale length.

8. Construct a scale for the function $f(u) = \cos u$. u varies from 0° to 90°. Select suitable scale length.

9. Construct a scale for the function $f(u) = \log u$. u varies from 2 to 200.

10. Construct a scale for the function $f(u) = \dfrac{u^2 + 3}{u^2 + 5}$. u varies from 12 to 20.

Adjacent and Nonadjacent Scales

1. Construct adjacent scales for the expression, $A = \pi R^2$; R varies from 2 to 12 inches.

2. Construct adjacent scales for converting Fahrenheit readings of temperature to Centigrade. $C = 5/9(F - 32)$; C varies from $-40°$ to $100°$.

3. Construct adjacent scales for the expression, $H = 2.3P$; P varies from 100 to 500 psi.

4. Construct adjacent scales for the expression $V = 4/3\pi R^3$; where V = volume of a sphere, and R = radius (10 to 25 inches).

5. Construct adjacent scales for the expression, $V = \sqrt{2gh}$, where V = velocity in feet per second, and h (5 to 15 feet) = velocity head in feet.

6. Construct *nonadjacent* scales for the expression, $F = 9/5C + 32$; F varies from $-40°$ to $212°$.

7. Construct *nonadjacent* scales for the expression, $S = 4\pi R^2$, where S = area of a sphere, and R (7 to 20 inches) = radius of sphere.

8. Construct *nonadjacent* scales for the expression, $A = \pi d^2/576$, where A = area of circle in square feet, and d (5 to 20 inches) = diameter of circle.

9. Construct *nonadjacent* scales for the expression, $L = gt^2/4\pi^2$; L varies from 4 to 15 feet; g = 32.2 ft/sec^2.

10. Construct *nonadjacent* scales for the expression $P = A(1 + R)^n$, where P = principal sum in dollars, R = 8%; and n = number of times compounded. Let A = $100 and n (0 to 30).

NOMOGRAPHY 16

Nomography deals with the graphical representation and solution of mathematical relations that are either explicit or implicit. Not only are nomograms very useful as time-saving devices in the repetitive solution of mathematical expressions, but they are also, and quite often more importantly, a means for quickly analyzing the interrelationship among the variables. At best, mathematical studies of the interrelationship among four or more variables of an algebraic expression are both time-consuming and cumbersome.

Among the fields to which nomography can be applied are the following: ballistics, biomechanics, electronics, the various branches of engineering, food science and technology, heat transfer, medicine, physical and biological sciences, production, radioactivity, and statistics.

Engineers and scientists should be educated to understand the mathematical theory and the design of nomograms. Engineering and science students will profit considerably from this introduction to nomography. They will find opportunities to use nomograms in other areas of study and most certainly later on in professional practice. It is hoped that many students will be stimulated to enhance their knowledge of this fascinating and most useful field by additional study provided for in nomography courses.

In general, there are two types of nomograms: one commonly recognized as a rectangular Cartesian coordinate chart (concurrency charts) and the other as an alignment chart. The discussion in this chapter will deal with both types of nomograms.

CONCURRENCY CHARTS

A concurrency chart is a nomogram that presents, in a Cartesian coordinate system, the graphical solution of a relation among three or more variables.

Addition Charts

EXAMPLE 1

Let us consider the relation $a + b = c$. We will assume that a and b each vary from 0 to 10. When we let $x = a$ and $y = c$, we obtain the expression

$$y = x + b$$

This equation is one of the straight-line family,

$$y = mx + b$$

where m is the slope of the line and b is the value of the y-intercept. When we assign values to b, i.e., $b = 0, 1, 2, 3, \ldots, 10$, we will obtain a family of parallel lines, all having a slope of $m = 1$ and intercept values equal to the respective values of b. The concurrency chart is shown in Fig. 16.1. This chart serves as an addition chart since it solves the equation $a + b = c$. The sample problem is solved as indicated by the dashed lines and arrows. How would you use the chart to solve the equations $b = c - a$ and $a = c - b$?

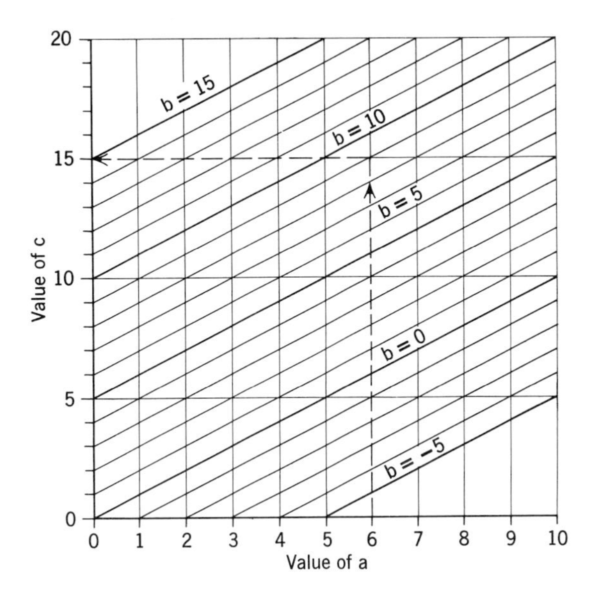

Fig. 16.1 Addition chart graphical solution of the equation $a + b = c$. Example: $a = 6$, $b = 9$. Read $c = 15$.

EXAMPLE 2

Here consider the equation $c = \sqrt{a^2 + b^2}$, where a and c each vary from 0 to 5.

The equation may be written as

$$a^2 + b^2 = c^2$$

When we let $x = a^2$ and $y = c^2$, we obtain $y = x + b^2$, which is the equation of a straight line for any particular value of b.

We should note, especially, that the a-scale will be a functional scale, since $x = a^2$. Similarly, the c-scale will be a functional scale, since $y = c^2$.

Suppose we plot the straight line $c^2 = a^2 + b^2$, when $a = 3$ and $b = 4$. In Fig. 16.2 we observe that the slope of the line is 1 and that the intercept is 4. This is evident from the relation $y = x + 4^2$, since $y = c^2$ and $x = a^2$.

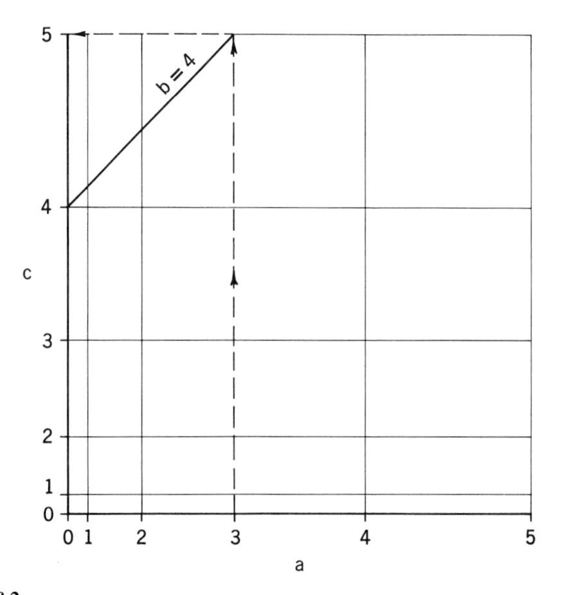

Fig. 16.2

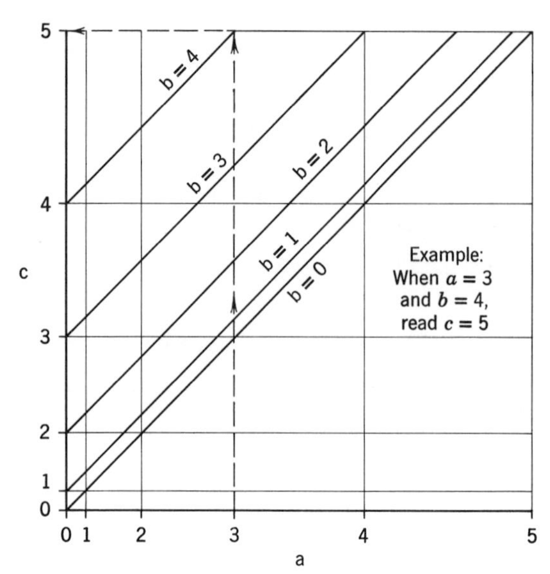

Fig. 16.3

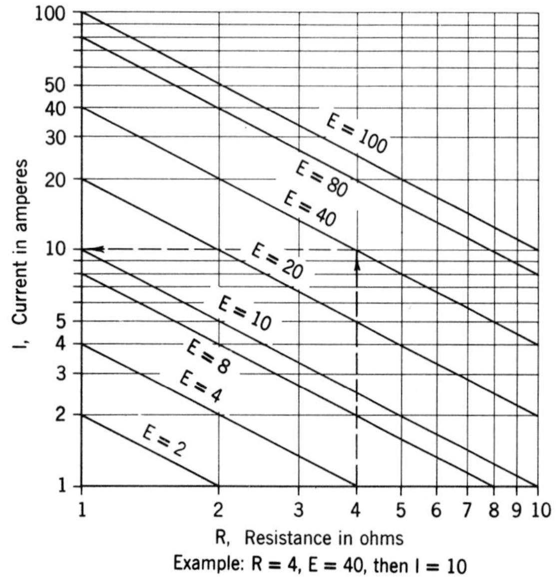

Fig. 16.4

For other values of b, we will obtain straight lines that are parallel to each other (since the slope is 1 in all cases). The completed concurrency chart is shown in Fig. 16.3.

EXAMPLE 3

Now consider the equation $I = E/R$ (Ohm's law) where I(1 to 100) represents current in amperes, R (1 to 10) represents resistance in ohms, and E represents electromotive force in volts.

In order to reduce the equation to the straight-line form $y = mx + b$, we will first write

$$\log I = -\log R + \log E$$

Now, when we let $y = \log I$ and $x = \log R$, we will obtain the expression

$$y = -x + \log E$$

For each value of E we will obtain a straight line having a *negative* slope of 1 and an intercept value corresponding to the value of E.

Figure 16.4 shows the graphical solution of the equation $I = E/R$.

EXAMPLE 4

Let us consider the expression $a + b + c = d$. Let a, b, and c vary as shown in the figures. The given expression can easily be replaced by two equations,

$$a + b = T \qquad (1)$$

and

$$T + c = d \qquad (2)$$

Both equations are of the straight-line form,

$$y = mx + b$$

In equation (1) we will let $y = T$ and $x = a$. Then $y = x + b$. The solution is shown in Fig. 16.5.

Now, in equation (2), we will let $y = T$, and $x = d$. Then $y = x - c$. The solution is shown in Fig. 16.6.

The two charts may now be combined to form Fig. 16.7. Note carefully that the T-scales of Figs. 16.5 and 16.6 *must be the same* in order to have a common scale. It is fairly obvious that this scale need not be graduated.

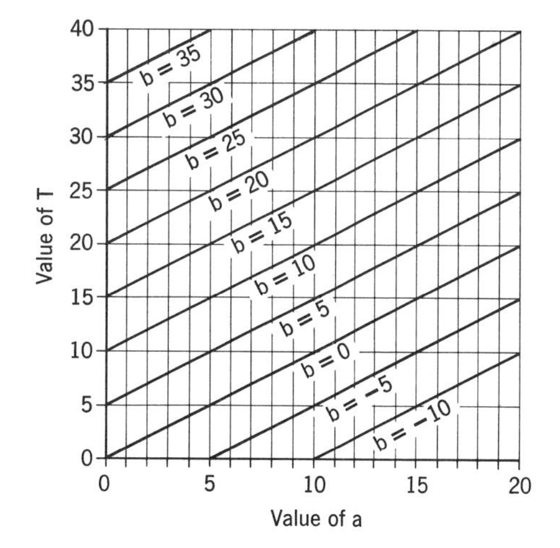

Fig. 16.5

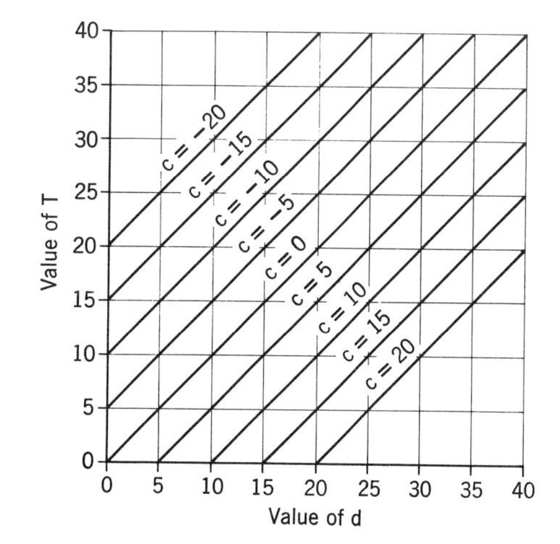

Fig. 16.6

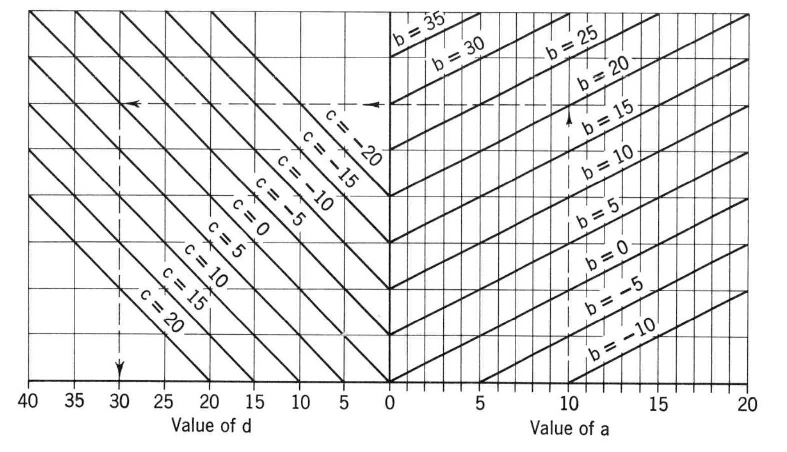

Fig. 16.7

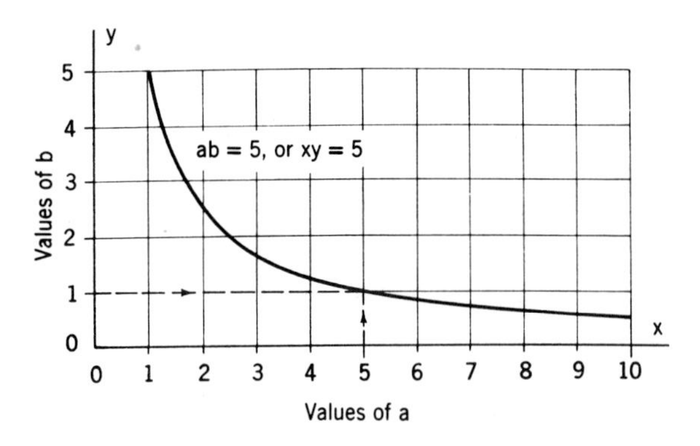

Fig. 16.8 Graphic representation of the equation $ab = 5$. Example: $a = 5$, $b = 1$. Read $ab = 5$.

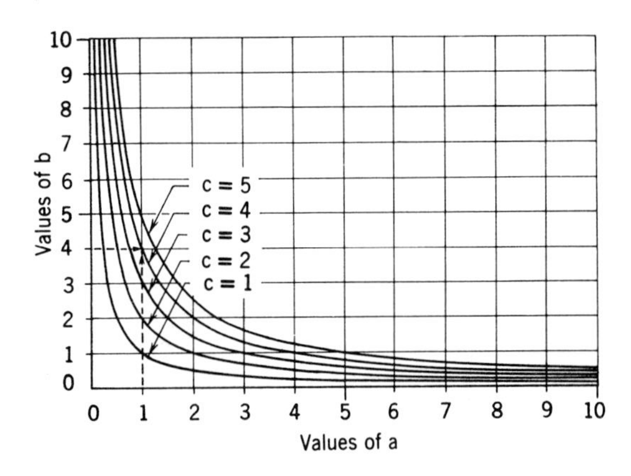

Fig. 16.9 Multiplication chart $ab = c$. Example: $a = 1$, $b = 4$. Read $c = 4$.

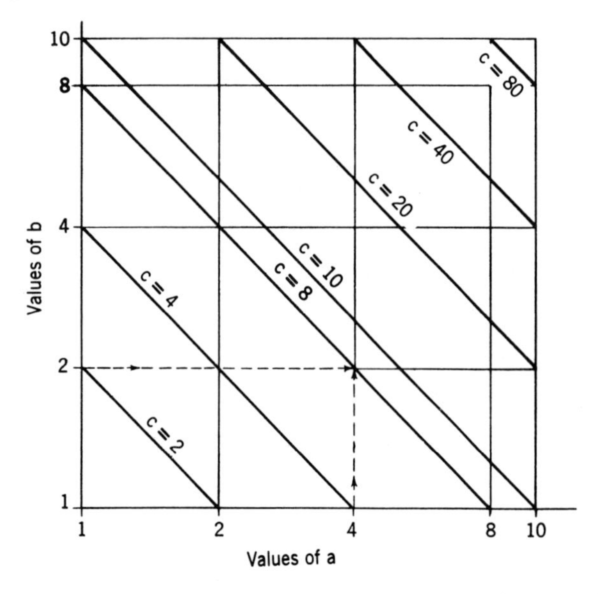

Fig. 16.10 Multiplication chart $ab = c$. Example: $a = 4$, $b = 2$. Read $c = 8$.

Multiplication Charts

EXAMPLE 1

Now let us consider *multiplication charts*. Suppose the given equation is $ab = c$. If we let $x = a$ and $y = b$, then $xy = c$. If $c = 5$, then $xy = 5$. We can plot the curve for this expression by assigning values to x and computing the corresponding values of y. The accompanying table contains several such values.

x	0	1	2	3	4	5	—	10
y	∞	5	5/2	5/3	5/4	1	—	1/2

Figure 16.8 shows the curve $xy = 5$.

If we let $c = 1, 2, 3, 4, 5, \ldots, 10$, we will obtain a family of curves (rectangular hyperbolas) that can be used to solve the equation $ab = c$ (Fig. 16.9).

Evidently this is not a convenient chart for multiplication since interpolation between the curves is not very accurate, and, moreover, the time consumed in drawing the curves may not be justified. ***Alternative Solutions.*** If we expand the equation $ab = c$ logarithmically, we obtain the expression

$$\log a + \log b = \log c$$

If we let $x = \log a$ (a functional scale)

and $y = \log b$

then $x + y = \log c$

If definite values are assigned to c and if $k = \log c$, it follows that $x + y = k$ or $y = -x + k$. Figure 16.10 shows logarithmic scales for a and b and a family of diagonal lines with slope equal to -1 and intercepts equal to the various values of k.

It is clearly seen that this multiplication chart (Fig. 16.10) has some advantage over the one shown in Fig. 16.9 in that only straight lines are involved. Since logarithmic scales are used, interpolation may be inaccurate. Now, if we can combine the best features of both charts, i.e., the uniform scales of Fig. 16.9 with the straight lines of Fig. 16.10, we will have the advantages of both. This can be done in the following manner:

Let $x = a$

and $y = c$

Now $y = xb$

Figure 16.11 shows the solution. It should be observed that, as values are assigned to b, we obtain a family of straight lines from the equation $y = xb$. All these lines pass through the origin since the intercept is zero, and their slopes are equal to the corresponding values of b.

EXAMPLE 2

Now suppose the given equation is $abc = d$. The expression may be divided into two parts:

$$ab = T \qquad (1)$$

and

$$Tc = d \qquad (2)$$

Each of these equations is of the form $ab = c$.

Suppose the ranges of a, b, and c are as shown in Fig. 16.12 and Fig. 16.13. For the graphic solution of equation (1), let

$$x = a$$

and

$$y = T$$

Then

$$y = xb \qquad \text{(Fig. 16.12)}$$

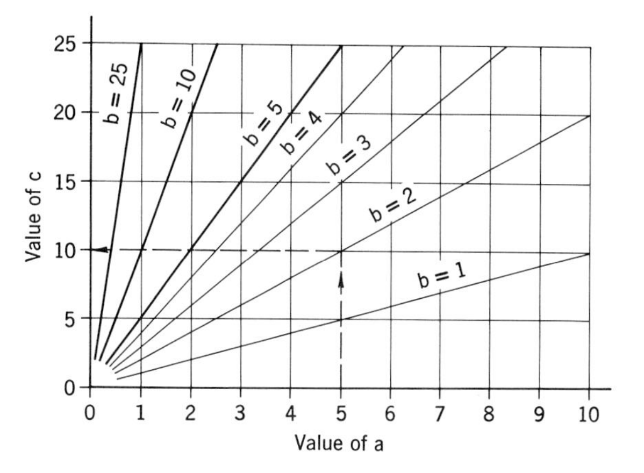

Fig. 16.11 Multiplication chart, $ab = c$. Example: $a = 5$, $b = 2$. Read $c = 10$.

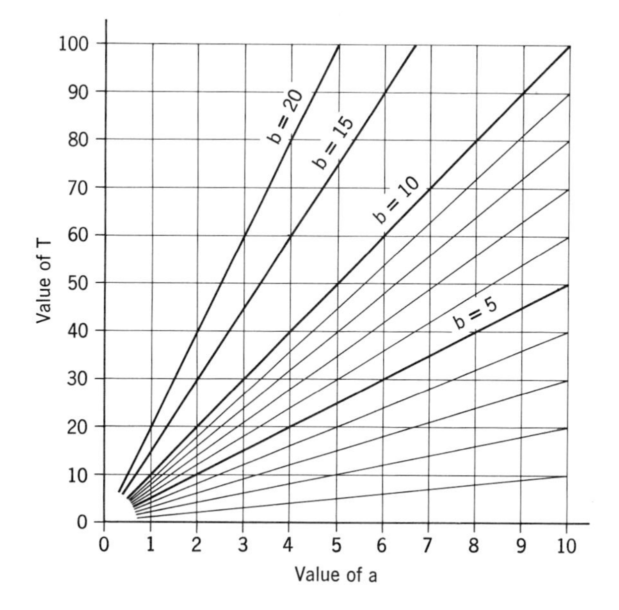

Fig. 16.12 Multiplication chart, $ab = T$.

The second equation, $Tc = d$, is represented graphically in Fig. 16.13. It should be pointed out that the T-scales must be the same in both Figs. 16.12 and 16.13, if we combine the two charts. We will obtain Fig. 16.14, which includes an example of the use of the chart in solving the equation $abc = d$.

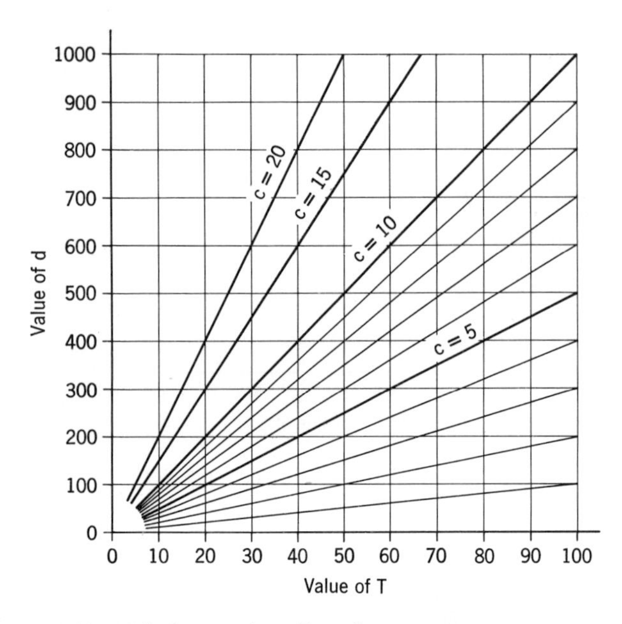

Fig. 16.13 Multiplication chart, $Tc = d$.

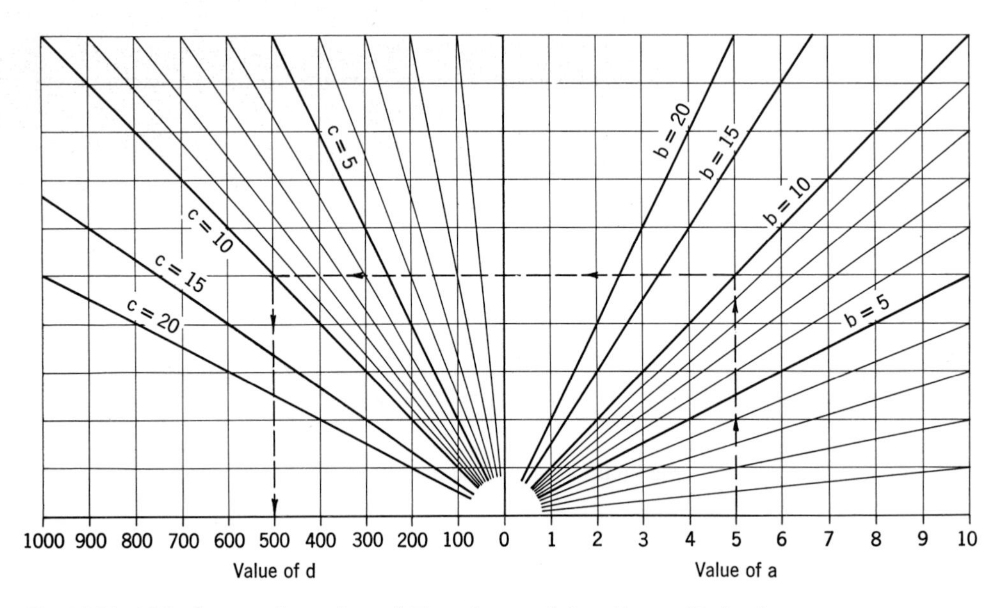

Fig. 16.14 Multiplication chart, $abc = d$. Example: $a = 5$, $b = 10$, $c = 10$. Read $d = 500$.

EXAMPLE 3

Now let us consider the equation

$$Rpm = \frac{336RS}{D}$$

where Rpm = engine revolutions per minute, varying from 0 to 5000 rpm,
R = gear ratio (from 2:1 to 7:1),
S = road speed of vehicle (5 to 60 mph),
D = tire size, nominal outside diameter (28 to 40 inches).

Solution

Let $$Rpm \times D = T \tag{1}$$
and $$336RS = T \tag{2}$$

Each of these equations is of form $ab = c$. The graphic representation of equation (1) is shown in Fig. 16.15.

It should be pointed out that a single calculation, i.e., when $Rpm = 5000$ and $D = 40$, $T = 200{,}000$, locates point A which is joined with zero to establish the line $D = 40$. If the line $D = 30$ (the line joining point B with zero) is established in a similar manner, the other lines for D can be located without further calculation simply by dividing the segment AB into five equal parts, thus locating points on $D = 32$, 34, 36, and 38. The line $D = 28$, of course, can be established by first laying off a distance below point B equal to one-fifth of segment AB, thus locating point C which is then joined with zero to locate the line $D = 28$.

Now the second equation, $336RS = T$, can be represented graphically in a similar manner (Fig. 16.16). One point which should be stressed, however, is that the T-scale must be the same as that used in Fig. 16.15.

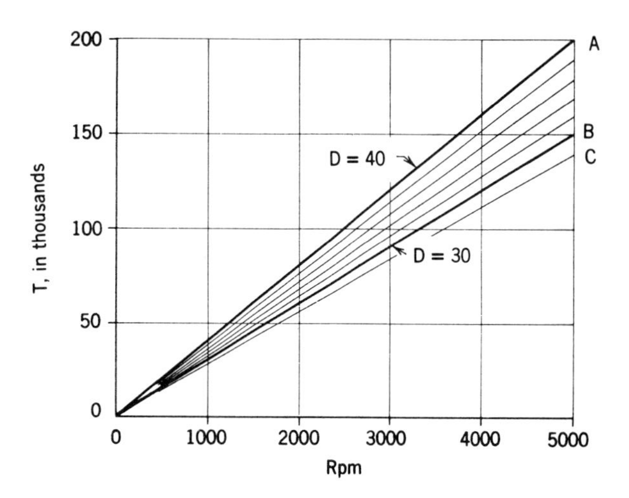

Fig. 16.15 Chart for equation Rpm $\times$ $D = T$.

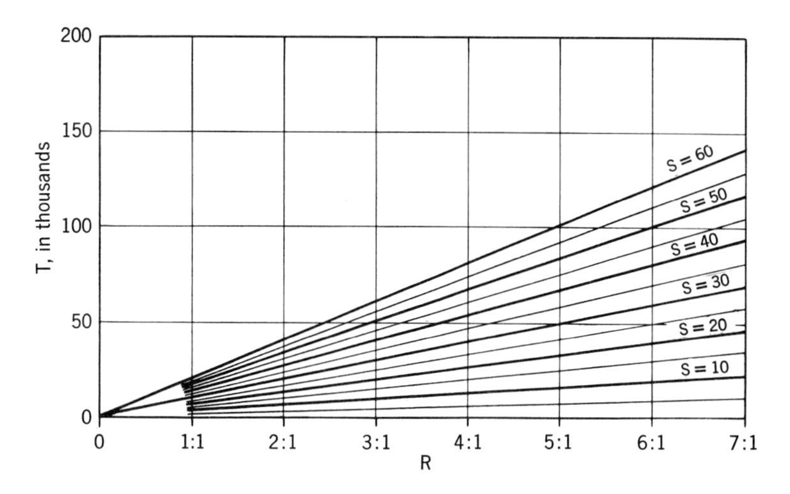

Fig. 16.16 Chart for equation $336RS = T$.

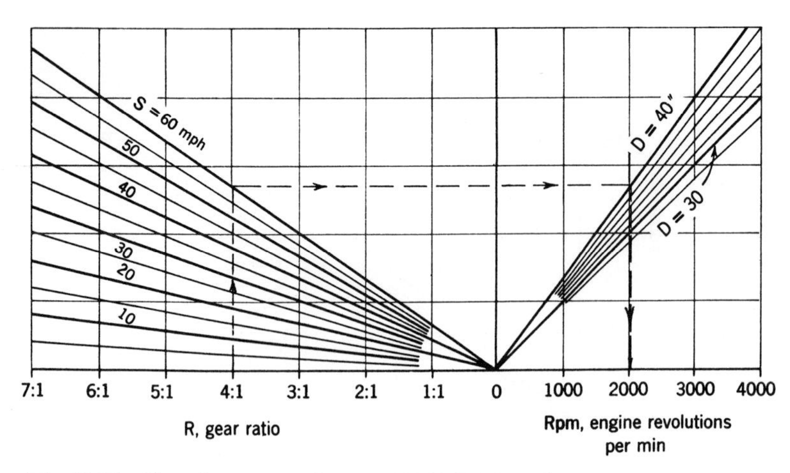

Fig. 16.17 Chart for equation Rpm = $336RS/D$. Example: $R = 4:1$, $S = 60$ mph (road speed of vehicle), $D = 40''$ (nominal outside diameter of tire). Read Rpm = 2000+.

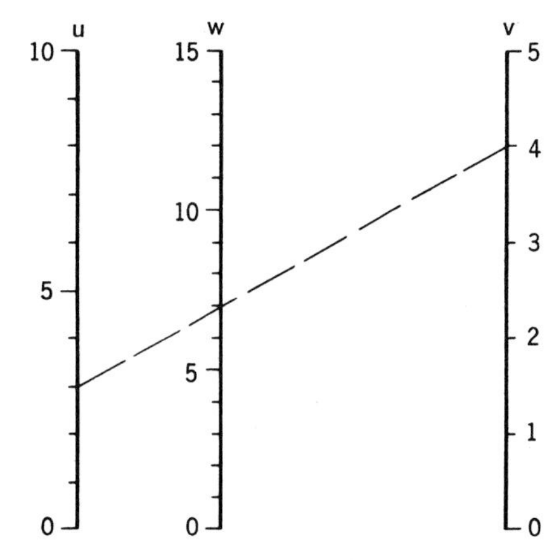

Fig. 16.18 Alignment chart for equation $u + v = w$. Example: $u = 3$, $v = 4$. Read $w = 7$.

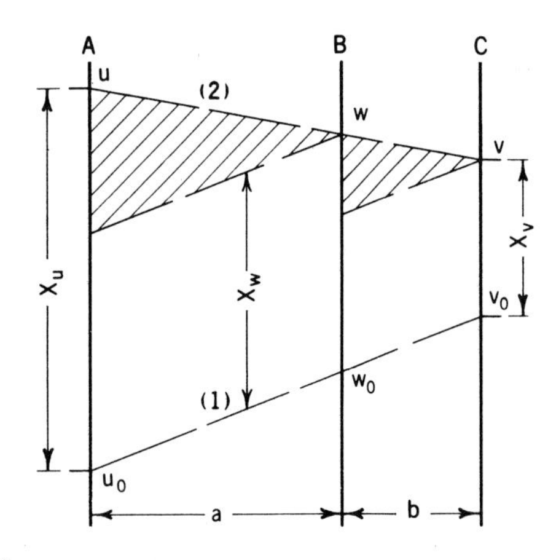

Fig. 16.19

Now the two charts may be combined to form the combination shown in Fig. 16.17. The example shows that when $R = 4:1$, $S = 60$, and $D = 40$, then $Rpm = 2000+$.

Alignment Charts for the Form

$$f_1(u) + f_2(v) = f_3(w)$$

In its simplest form an alignment chart consists of three parallel scales so graduated that a line which joins values on two of the scales will intersect the third scale in a value which satisfies the given equation. For example, in Fig. 16.18, observe that the line which joins $u = 3$ with $v = 4$ passes through the value $w = 7$, which satisfies the equation $u + v = w$.

In order to design alignment charts for equations of the form $f_1(u) + f_2(v) = f_3(w)$, we must know (*a*) how to graduate the scales and (*b*) what determines the spacing of the parallel scales. Let us now proceed to do this.

Suppose we have three parallel scales (Fig. 16.19), *A*, *B*, and *C*, so graduated that lines (isopleths) (1) and (2) cut the scales in values which satisfy the equation $f_1(u) + f_2(v) = f_3(w)$.

Now,

$$X_u = m_u[f_1(u) - f_1(u_0)]$$
$$X_v = m_v[f_2(v) - f_2(v_0)]$$
$$X_w = m_w[f_3(w) - f_3(w_0)]$$

If u_0, v_0, and w_0 respectively represent values of u, v, and w for which the corresponding functions are zero, we may write simply

$$X_u = m_u f_1(u) \quad (1)$$
$$X_v = m_v f_2(v) \quad (2)$$
$$X_w = m_w f_3(w) \quad (3)$$

Let us agree further that the spacing of the scales is in the ratio $a:b$. If we graduate the scales for $f_1(u)$ and $f_2(v)$ in accordance with their scale equations (1) and (2), respectively, what will be the modulus for the scale equation of $f_3(w)$ and what will be the ratio $a:b$, if the chart satisfies this relation $f_1(u) + f_2(v) = f_3(w)$?

In Fig. 16.19, draw lines through points w and v parallel to line u_0v_0. The shaded triangles are similar by construction; hence,

$$\frac{X_u - X_w}{X_w - X_v} = \frac{a}{b}$$

$$X_u b + X_v a = X_w(a + b)$$

$$\frac{X_u b}{ab} + \frac{X_v a}{ab} = X_w \frac{a + b}{ab}$$

or

$$\frac{X_u}{a} + \frac{X_v}{b} = \frac{X_w}{ab/(a + b)}$$

Since

$$X_u = m_u f_1(u)$$

$$X_v = m_v f_2(v)$$

$$X_w = m_w f_3(w)$$

then

$$\frac{m_u f_1(u)}{a} + \frac{m_v f_2(v)}{b} = \frac{m_w f_3(w)}{ab/(a + b)}$$

When $m_u = ka$ and $m_v = kb$, then

$$f_1(u) + f_2(v) = \frac{m_w f_3(w)}{m_u m_v/(m_u + m_v)}$$

Let $m_w = m_u m_v/(m_u + m_v)$. It follows that $f_1(u) + f_2(v) = f_3(w)$. Since $m_u/m_v = ka/kb$, it is evident that $m_u/m_v = a/b$.

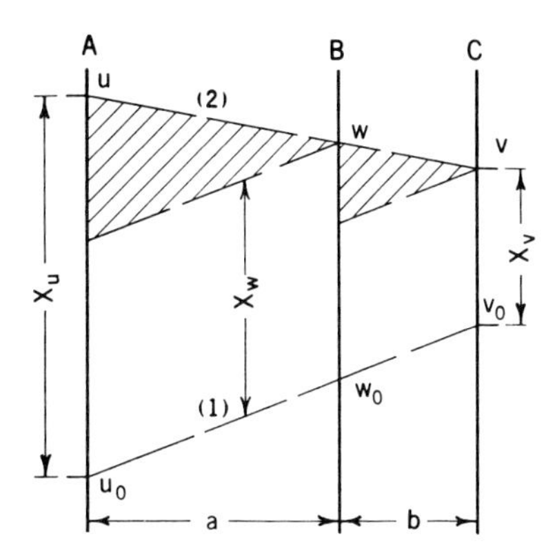

Fig. 16.19 (Repeated)

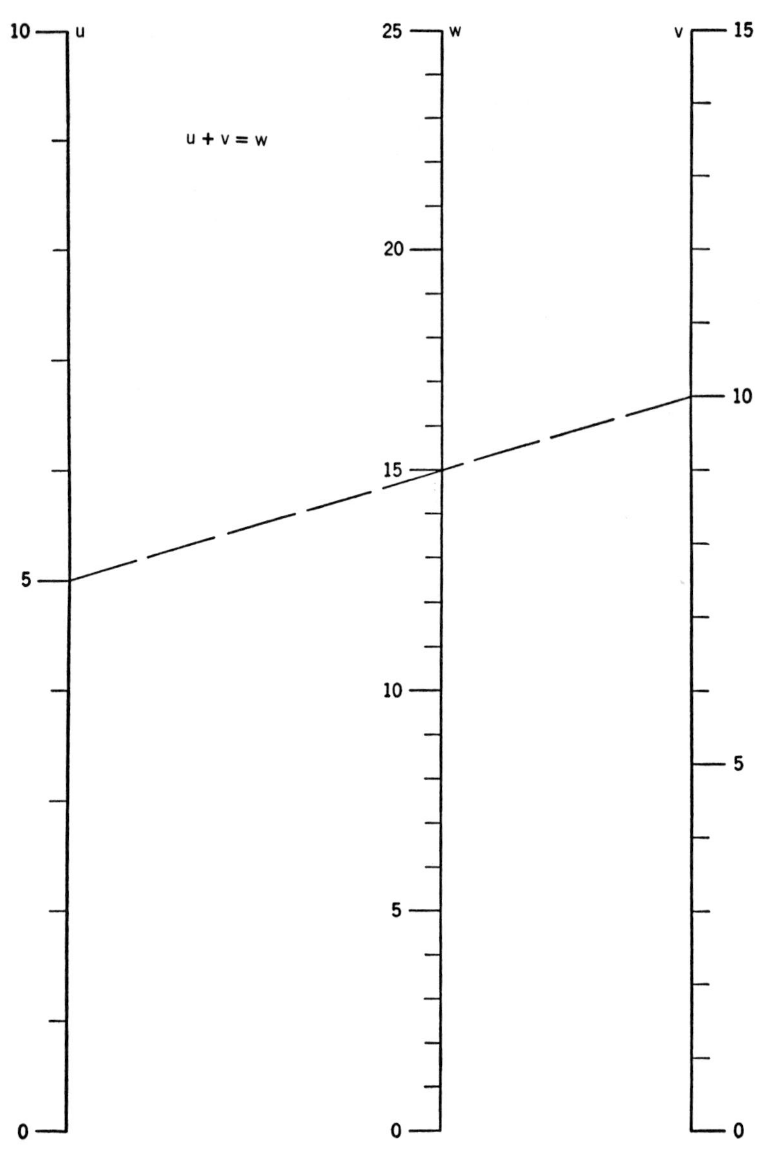

Fig. 16.20 Alignment chart for equation $u + v = w$.

Design Summary

To construct an alignment chart for an equation of the form $f_1(u) + f_2(v) = f_3(w)$,

1. Place the parallel scales for $f_1(u)$ and $f_2(v)$ a convenient distance apart.
2. Graduate these scales in accordance with their scale equations.
3. Locate the scale for $f_3(w)$ so that its distance from the $f_1(u)$ scale is to its distance from the $f_2(v)$ scale as $m_u/m_v = a/b$.
4. Graduate the $f_3(w)$ scale in accordance with its scale equation.

EXAMPLE 1

$u + v = w$ (Fig. 16.20). Let u vary from 0 to 10 and let v vary from 0 to 15. Suppose that the scale lengths are to be 6 inches.

Now

$$m_u = \frac{6}{10} = 0.6; \; X_u = 0.6u$$

and $$m_v = \frac{6}{15} = 0.4; \; X_v = 0.4v$$

and $$m_w = \frac{0.6 \times 0.4}{0.6 + 0.4} = 0.24; \; X_w = 0.24w$$

$$\frac{m_u}{m_v} = \frac{0.6}{0.4} = \frac{3}{2} = \frac{a}{b}$$

Now suppose that it is desirable to cut off the chart at the line (1) (Fig. 16.21), eliminating the portion below line (1). The scale equations will then be [using line (1) as the base line]:

$$X_u = m_u[f_1(u) - f_1(u_1)]$$
$$X_v = m_v[f_2(v) - f_2(v_1)]$$
$$X_w = m_w[f_3(w) - f_3(w_1)]$$

where u_1, v_1, and w_1 satisfy the equation $f_1(u) + f_2(v) = f_3(w)$.

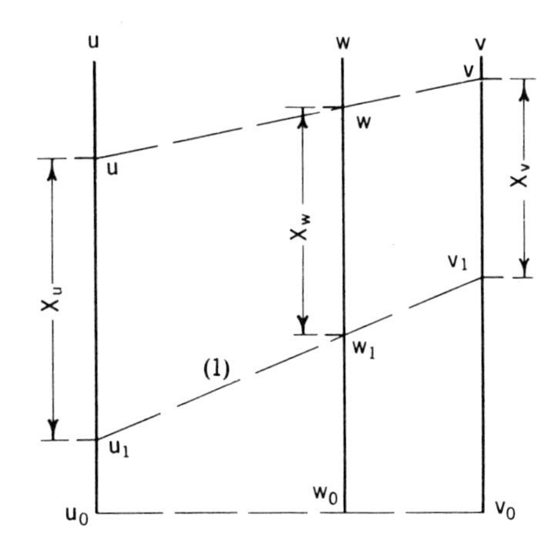

Fig. 16.21

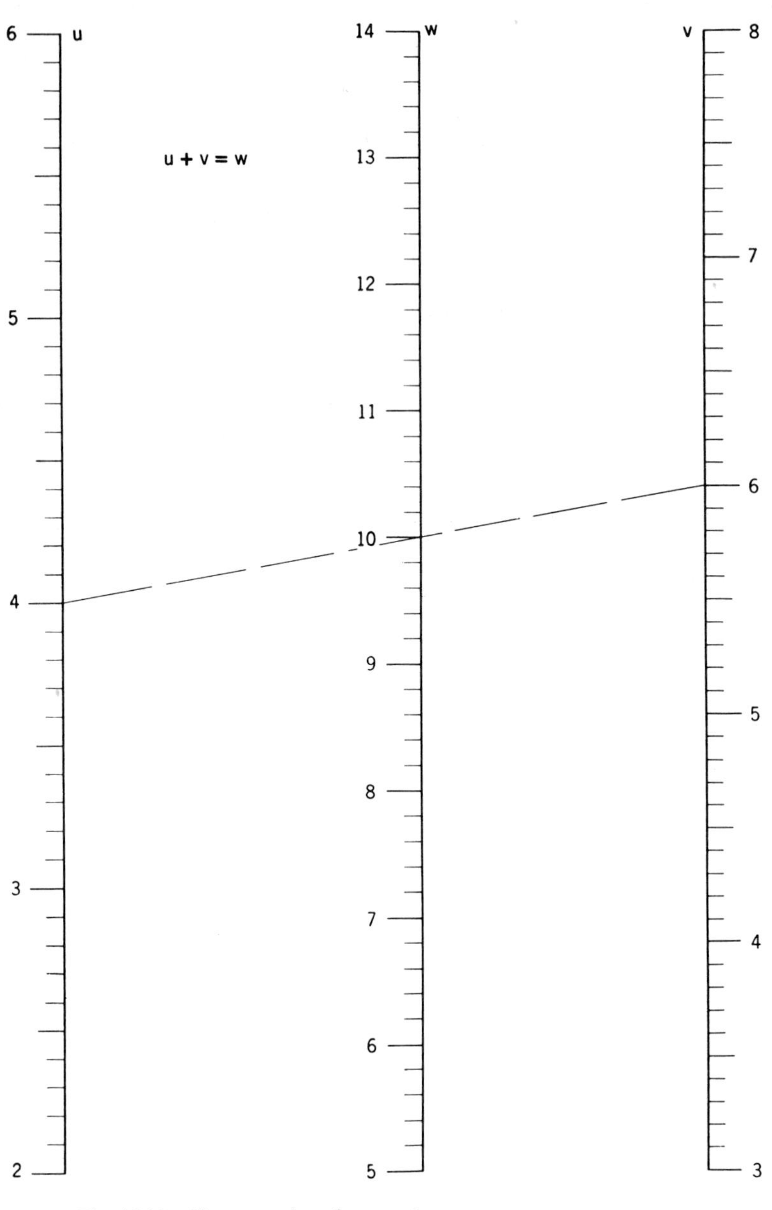

Fig. 16.22 Alignment chart for equation $u + v = w$.

EXAMPLE 2

$u + v = w$ (Fig. 16.22). Let u vary from 2 to 6 and let v vary from 3 to 8. Length of scales, 6 inches.

Now

$$m_u = \frac{6}{6-2} = \frac{3}{2};\ X_u = \frac{3}{2}(u-2)$$

$$m_v = \frac{6}{8-3} = \frac{6}{5};\ X_v = \frac{6}{5}(v-3)$$

Since

$$m_w = \frac{m_u m_v}{m_u + m_v}$$

$$m_w = \frac{(3/2)(6/5)}{3/2 + 6/5} = \frac{18/10}{27/10} = \frac{2}{3}$$

Since $u_1 = 2$ and $v_1 = 3$, therefore, $w_1 = 5$ (*from the original equation* $u + v = w$). Hence,

$$X_w = \frac{2}{3}(w-5)$$

$$\frac{a}{b} = \frac{m_u}{m_v} = \frac{3/2}{6/5} = \frac{5}{4}$$

Form the following tables:

u	2	3	4	5	6
$f_1(u) = u$	2	3	4	5	6
$X_u = 3/2(u-2)$	0	1.5 in.	3.0 in.	4.5 in.	6.0 in.

v	3	4	5	6	7	8
$f_2(v) = v$	3	4	5	6	7	8
$X_v = 6/5(v-3)$	0	1.2 in.	2.4 in.	3.6 in.	4.8 in.	6.0 in.

w	5	6	7	—	14
$f_3(w) = w$	5	6	7	—	14
$X_w = 2/3(w-5)$	0	2/3 in.	4/3 in.	—	6 in.

Much of the calculation set forth in the above tables can be eliminated if we compute the location of the end points for each scale and then project the other points geometrically.

EXAMPLE 3

$I = 1/12bd^3$ (Fig. 16.23), where I is the moment of inertia of a rectangle about its axis parallel to b, b is the width of the rectangle, and d is its height.

Let b and d vary from 1 to 10 inches. Length of scales, 6 inches. The equation, which may be written $bd^3 = 12I$, is put in the type form by taking logarithms; thus we obtain

$$\log b + 3 \log d = \log I + \log 12$$

Now the moduli m_b and m_d are computed as follows:

$$m_b = \frac{6}{\log 10 - \log 1} = 6;$$

$$X_b = 6(\log b - \log 1) = 6 \log b$$

$$m_d = \frac{6}{3 \log 10 - 3 \log 1} = 2;$$

$$X_d = 2(3 \log d - 3 \log 1) = 6 \log d$$

It should be pointed out that the function of d is $3 \log d$, the modulus 2 is the *actual modulus* which is used in locating the I-scale, and the coefficient 6 is the *effective modulus* which is used in graduating the d-scale.

$$m_I = \frac{6 \times 2}{6 + 2} = \frac{12}{8} = \frac{3}{2};$$

$$X_I = \frac{3}{2}[(\log I + \log 12) - (\log I_1 + \log 12)]$$

where I_1 is a value on the I-scale from which the I-scale is graduated.

Note carefully that the actual moduli of b and d are used in computing m_I. Form the following table:

b	1	2	3	—	10
$f(b) = \log b$	0	0.301	0.477	—	1.000
$X_b = 6 \log b$	0	1.81	2.86	—	6.00

The table for d will be the same as the above since $X_d = 6 \log d$.

Now we can graduate the scales for b and d in accordance with the scale equations $X_b = 6 \log b$ and $X_d = 6 \log d$. The scales are placed a convenient distance apart. The position of the I-scale is determined from the ratio $m_b : m_d = 6:2 = 3:1$. Our next step is to locate one point on the I-scale, i.e., point 1. (This is I_1.) Suppose we let $d = 2$. Then

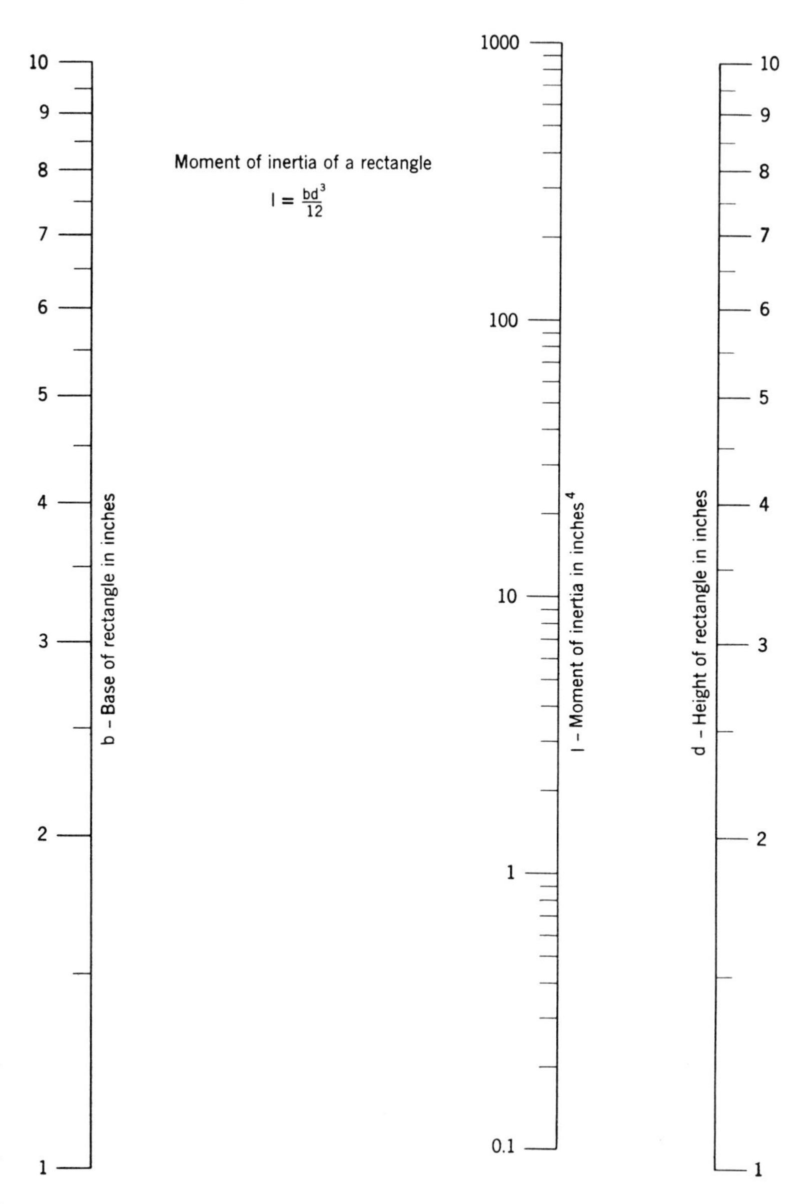

Fig. 16.23 Moment of inertia of a rectangle about its axis parallel to the base.

$b = 12I/d^3$ or $b = (12 \times 1)/8 = 1.5$. The line joining $b = 1.5$ and $d = 2$ cuts the I-scale at point 1. Now we can locate other points on the I-scale from the scale equation

$$X_I = \frac{6 \times 2}{6 + 2}(\log I - \log 1)$$

or

$$X_I = 1.5 \log I$$

This means that points on the I-scale are laid off *from point 1*. If the selected point on the I-scale were 10, graduations would be laid off from this point in accordance with the scale equation

$$X_I = 1.5(\log I - \log 10)$$

or

$$X_I = 1.5(\log I - 1)$$

If the equation were $f_2(u) - f_2(v) = f_3(w)$, the scale equations would be

$$X_u = m_u f_1(u) \tag{1}$$

$$X_v = m_v[-f_2(v)] \tag{2}$$

$$X_w = \frac{m_u m_v}{m_u + m_v} f_3(w) \tag{3}$$

The negative sign in equation (2) implies that positive values of $f_2(v)$ are laid off downward, if we agree to lay off positive values of $f_1(u)$ upward.

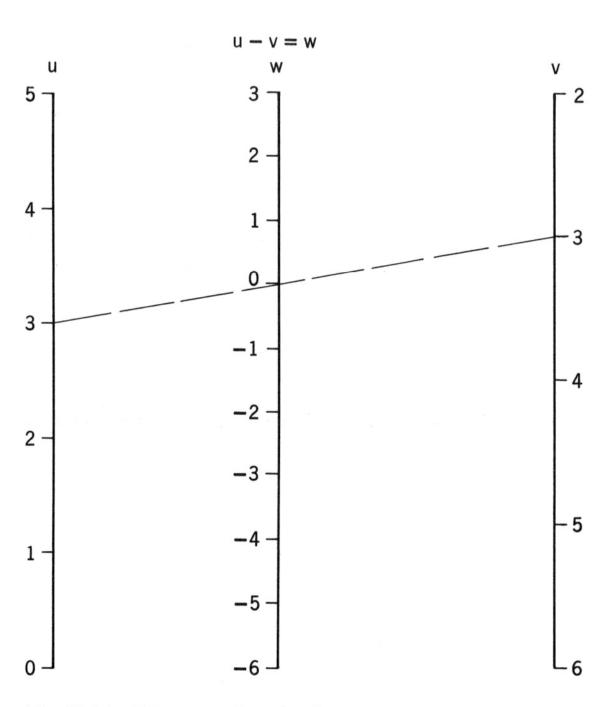

Fig. 16.24 Alignment chart for the equation $u - v = w$.

EXAMPLE 4

$u - v = w$ (Fig. 16.24). Suppose u varies from 0 to 5, and v varies from 2 to 6. Scale lengths, 6 units (inches, centimeters, or any convenient length).

Now,
$$m_u = \frac{6}{5};\ X_u = \frac{6}{5}u$$

$$m_v = \frac{6}{4} = \frac{3}{2};\ X_v = -\frac{3}{2}(v - 2)$$

$$m_w = \frac{(6/5)(3/2)}{6/5 + 3/2} = \frac{9/5}{27/10} = \frac{2}{3};$$

$$X_w = \frac{2}{3}w \text{ (from } w = 0)$$

$$\frac{m_u}{m_v} = \frac{6/5}{3/2} = \frac{4}{5}$$

Scales u and v are placed a convenient distance apart. Scale u is graduated in accordance with the scale equation $X_u = 6/5u$. Scale v is graduated from the equation $X_v = -3/2(v - 2)$. This is done by locating point 2 on the upper end of the v-scale, and laying off distances equal to 3/2 units for each point 3, 4, 5, and 6.

NOTE: POSITIVE VALUES OF "V" ARE LAID OFF DOWNWARD!

A point on the w-scale can be located by solving the original equation, $u - v = w$.

Example. Let $u = 3$ and $v = 3$; then $w = 0$. With one point located on w, other points can be located from the scale equation $X_w = 2/3(w - w_1)$, where w_1 is the value of the located point. In this case, then, $X_w = 2/3(w + 0)$ or simply $X_w = 2/3w$.

Practical Short-Cut Method

If the designer is thoroughly grounded in the theory of alignment charts and fully understands the mathematical methods employed in changing a given equation to a type form, it is frequently possible to short-cut the actual construction of the chart.

EXAMPLE

Suppose that the given equation is $M = wl^2/8$ (bending moment in foot-pounds, simple beam, uniform load), where the ranges are w (10 to 300 lb/ft) and l (5 to 30 ft).

If a chart consisting of parallel scales is desired, the designer recognizes the fact that the equation can be converted to the form:

$$\log w + 2 \log l = \log M + \log 8$$

The chart can now be constructed without making any further calculations. The following procedure is suggested:

1. Draw two parallel lines any convenient distance apart.
2. Graduate the left-hand scale for w by simply marking the lower point 10 and the upper point 300. Other points on the scale may be located by projecting from a log scale (two-cycle slide-rule scale or commercial log sheets having two cycles).
3. Mark the lower point of the l-scale 5 and the upper point 30. Again, locate additional graduations by projecting from a log scale.
4. Now calculate two points for M, i.e.:
   (*a*) Let $w = 40$ and $l = 10$. This yields $M = 500$.
   (*b*) Let $w = 160$ and $l = 5$. This yields $M = 500$.

The point in which the line joining 40 and 10 intersects the line joining 160 and 5 is point $M = 500$. The vertical line through this point locates the

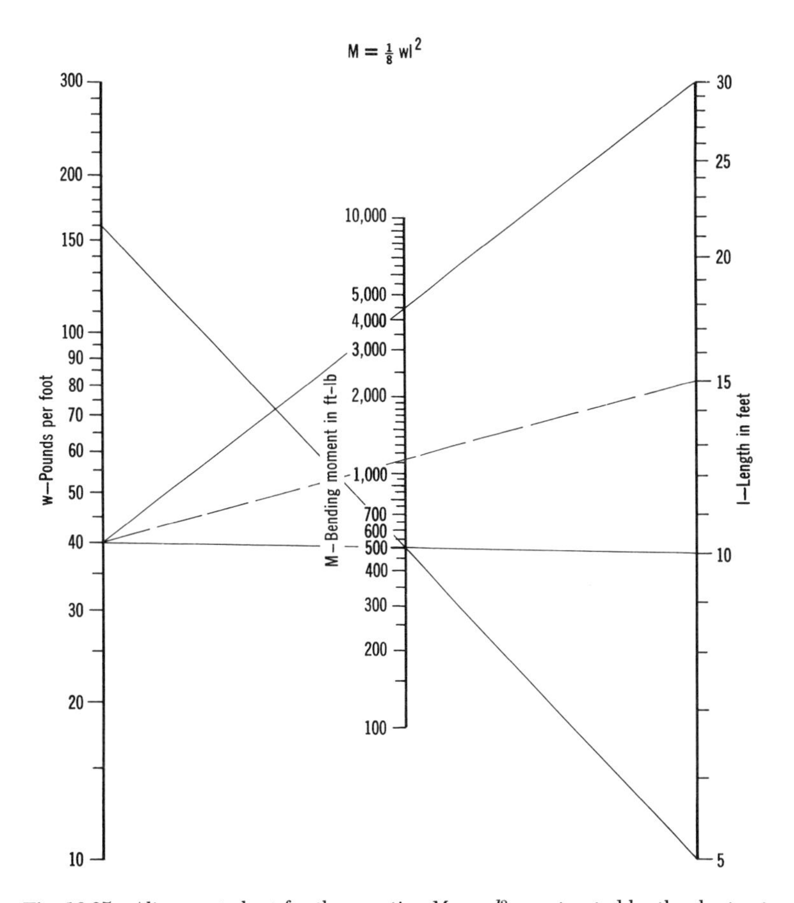

Fig. 16.25 Alignment chart for the equation $M = wl^2$, constructed by the short-cut method.

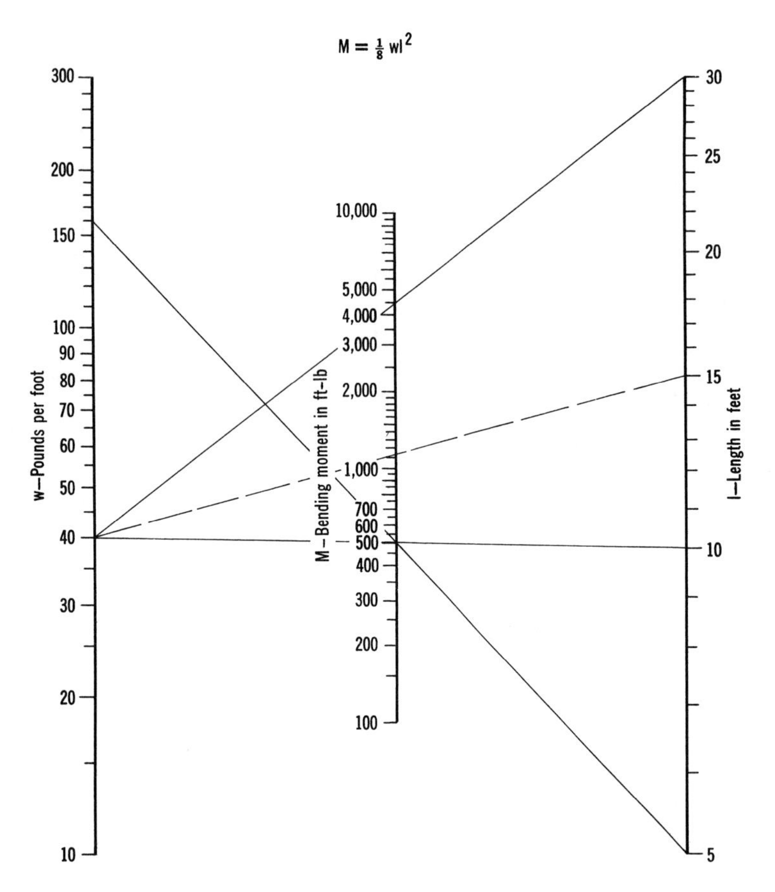

Fig. 16.25 (*Repeated*)

M scale. A second point on this scale can be now located by letting $w = 40$ and $l = 30$, which yield $M = 4500$. The line joining $w = 40$ with $l = 30$, then cuts the M scale in point 4500. Other points may be obtained by projecting from a log scale. The completed chart is shown in Fig. 16.25.

Alignment Charts for Equations of the Form $f_1(u) + f_2(v) + f_3(w) + \cdots = f_4(q)$

EXAMPLE 1

Let us consider the relation

$$u + 2v + 3w = 4t$$

Let
$$u + 2v = Q \tag{1}$$
then
$$Q + 3w = 4t \tag{2}$$

These two equations are of the form just discussed.

Suppose [equation (1)] that $m_u = 1$ and $m_v = 1/2$; then

$$X_u = u \text{ and } X_v = \frac{1}{2}(2v) = v$$

Now

$$\frac{m_u}{m_v} = \frac{1}{1/2} = \frac{2}{1}; \; m_Q = \frac{1(1/2)}{1 + 1/2} = \frac{1}{3}$$

If [equation (2)]

$$m_w = \frac{1}{3}; \; X_w = \frac{1}{3}(3w) = w$$

$$\frac{m_Q}{m_w} = \frac{1/3}{1/3} = \frac{1}{1}; \; m_t = \frac{(1/3)(1/3)}{1/3 + 1/3} = \frac{1/9}{2/3} = \frac{1}{6}$$

Therefore,

$$X_t = \frac{1}{6}(4t) = \frac{2}{3}t$$

From the above calculations we may now proceed to construct the chart (Fig. 16.26).

Now consider equation (1), $u + 2v = Q$. Scales u and v are placed a convenient distance apart. Scale u is graduated from its scale equation, $X_u = u$; and scale v is graduated from its scale equation $X_v = \frac{1}{2}(2v) = v$. The Q-scale is located in accordance with the ratio, $m_u : m_v = 1 : \frac{1}{2} = 2 : 1$. This scale, Q, is *not* graduated.

Now consider equation (2), $Q + 3w = 4t$. Scale w is placed a convenient distance from the Q-scale. Graduations on the w-scale are located in accordance with its scale equation, $X_w = \frac{1}{3}(3w) = w$. The t-scale is located from the ratio, $m_Q : m_w = \frac{1}{3} : \frac{1}{3} = 1 : 1$. The t-scale is then graduated from the scale equation, $X_t = \frac{1}{6}(4t) = \frac{2}{3}t$.

It should be carefully noted that in most of the practical applications of this form, it is necessary to locate a point on the fourth scale (by a computation from the given equation) before graduating that scale.

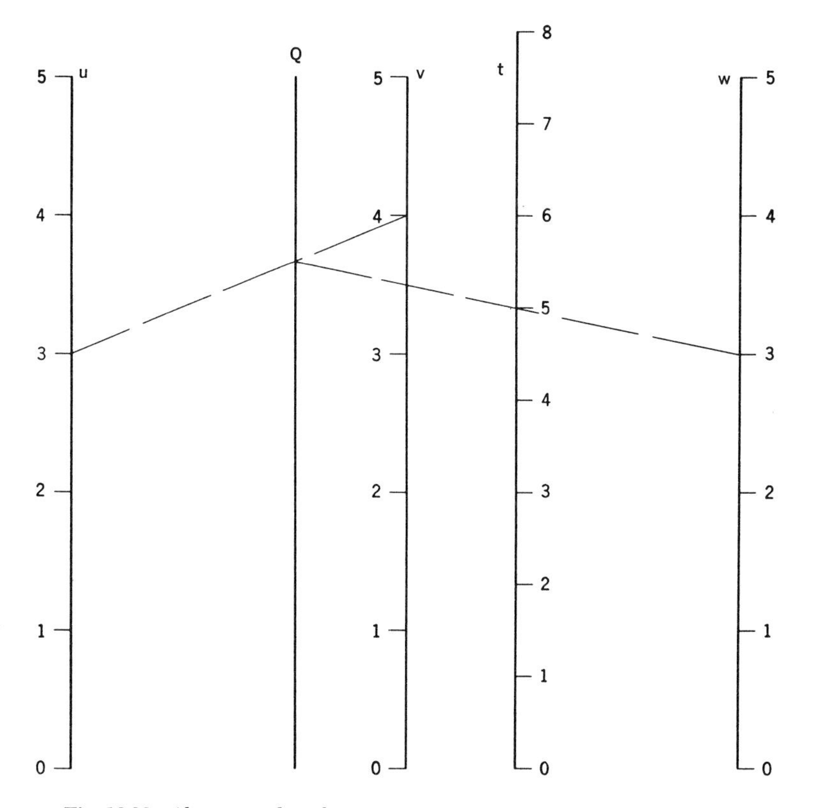

Fig. 16.26 Alignment chart for equation $u + 2v + 3w = 4t$.

EXAMPLE 2

Let us consider the Chezy formula

$$V = C\sqrt{RS}$$

Where V is the velocity of flow in feet per second (0.01 to 100); C, a coefficient related to the channel condition (10 to 200); R, the hydraulic radius (0.2 to 30 feet); and S, the water-surface slope (0.00004 to 0.02).

The equation is written in type form as,

$$\log V = \log C + \tfrac{1}{2}\log R + \tfrac{1}{2}\log S$$

Now we let $\quad \tfrac{1}{2}\log R + \tfrac{1}{2}\log S = T \qquad (1)$

and $\quad T + \log C = \log V \qquad (2)$

Equations (1) and (2) are of the form,

$$f_1(u) + f_2(v) = f_3(w).$$

Let us design a nomogram for equation (1). Now,

$$m_R = \frac{10\pm}{\tfrac{1}{2}(\log 30 - \log 0.2)} = \frac{10\pm}{\tfrac{1}{2}\log 150} = 10$$

Therefore,

$$X_R = 10[\tfrac{1}{2}(\log R - \log 0.2)] = 5\log 5R \qquad (3)$$

$$m_S = \frac{10\pm}{\tfrac{1}{2}(\log 0.02 - \log 0.00004)} = \frac{10\pm}{\tfrac{1}{2}\log 500} = 7.5$$

$$X_S = 7.5[\tfrac{1}{2}(\log S - \log 0.00004)]$$
$$= 3.75\log 25000S \qquad (4)$$

The parallel scales for R and S are graduated by employing scale equations (3) and (4), respectively. The T-scale is located from the ratio

$$\frac{m_R}{m_S} = \frac{10}{7.5}.$$

The modulus of the T-scale is,

$$m_T = \frac{(10)(7.5)}{17.5} = \frac{30}{7}$$

Figure 16.27 shows the nomographic solution of equation (1).

Now, with respect to equation (2),

$$m_C = \frac{10\pm}{\log 200 - \log 10} = 7.5$$

$$X_C = 7.5(\log C - \log 10) = 7.5\log\frac{C}{10}^{*} \qquad (5)$$

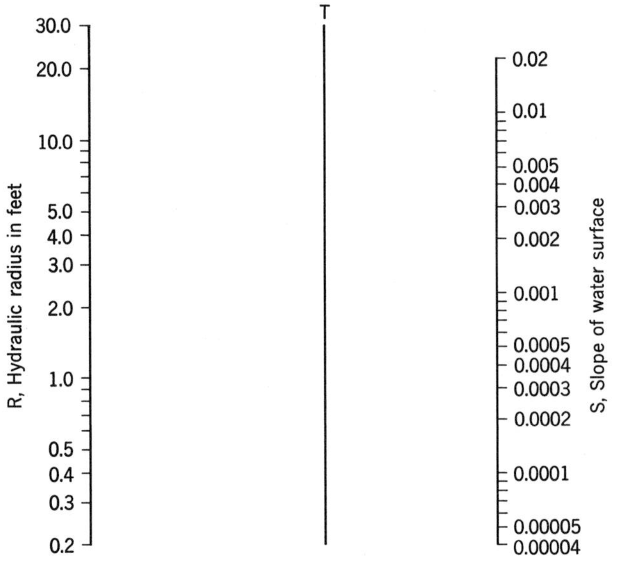

Fig. 16.27

* This equation is correct for graduations from $C = 10$. It is necessary, however, first to locate the graduation $C = 10$. This can be done, quite readily, by letting, for example, $R = 1$; $S = .001$, and $C = 10$; and then determining the value of V from the equation, $V = C\sqrt{RS}$. V, then, is equal to 0.333. Line one of Fig. 16.29 joins $R = 1$ and $S = 0.001$; and line two joins point A and $V = 0.333$. Line two extended intersects the C-scale at $C = 10$.

Having the values of m_C and m_T, we obtain

$$m_V = \frac{(30/7)(15/2)}{30/7 + 15/2} = \frac{30}{11}$$

and $X_V = \dfrac{30}{11}(\log V - \log 0.01) = \dfrac{30}{11}\log 100\,V$ (6)

The scales for C and V are graduated by employing scale equations (5) and (6), respectively. The V-scale is located from the ratio

$$\frac{m_T}{m_C} = \frac{30/7}{15/2} = \frac{4}{7}.$$

Figure 16.28 shows the solution of equation (2). The combination of the two Figs. 16.27 and 16.28 is shown in Fig. 16.29 which is the nomographic solution of the given equation, $V = C\sqrt{RS}$.

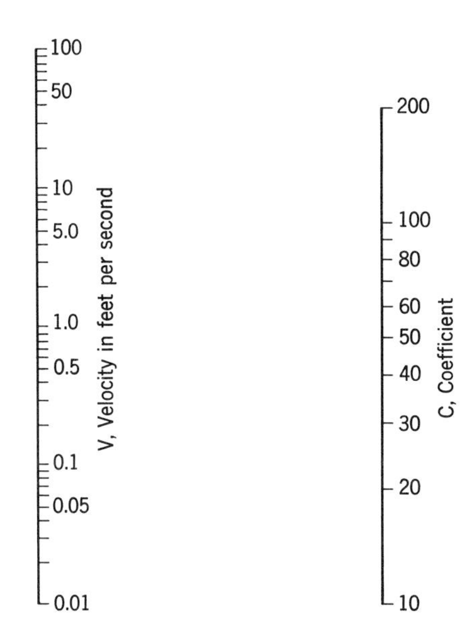

Fig. 16.28

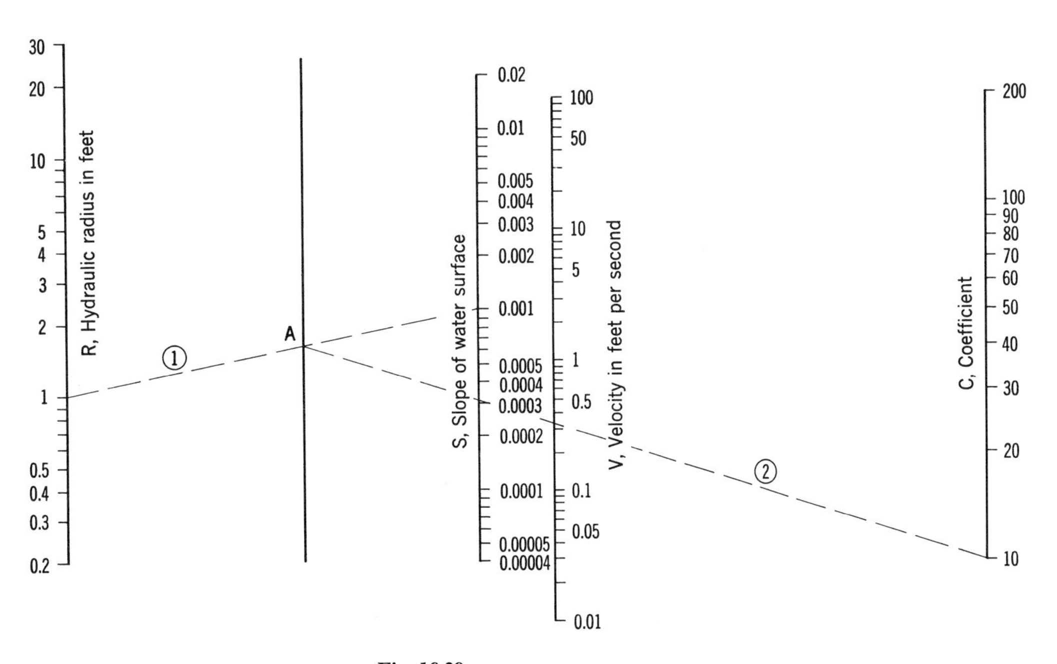

Fig. 16.29

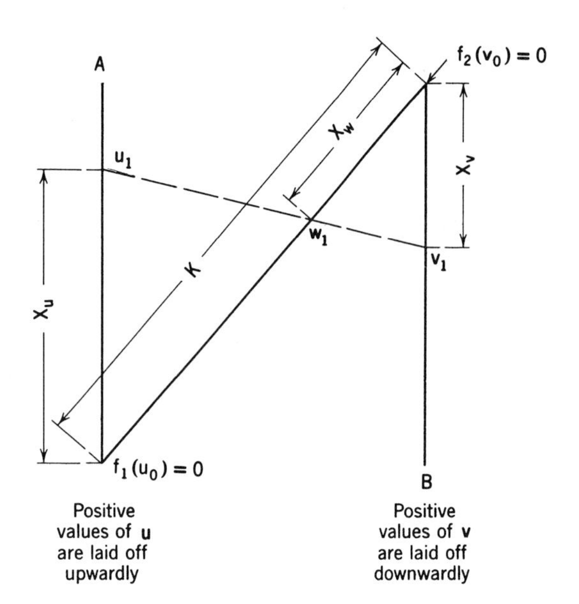

Fig. 16.30

Z-Charts for Equations of the Form $f_1(u) = f_2(v) \cdot f_3(w)$

There are equations of the form $f_1(u) = f_2(v) \cdot f_3(w)$ that can be best solved, nomographically, by three parallel scales. This can be done by expressing the above equation in the form $\log f_2(v) + \log f_3(w) = \log f_1(u)$. In many cases, however, a better solution is the Z-type nomogram. This type is advantageous when one or more of the functions are linear and when the ranges of the variables are compatible with the desired accuracy of scale readings. Let us now consider the geometry of the Z-type nomogram and the development of the scale equation for graduating the diagonal scale.

Suppose that the parallel scales (Fig. 16.30), A and B, are graduated in accordance with their scale equations $X_u = m_u f_1(u)$ and $X_v = m_v f_2(v)$, respectively. *The diagonal scale for* $f_3(w)$ *joins* $f_1(u_0)$ *and* $f_2(v_0)$, *i.e., the zero values of the functions of u and v.*

Let us further suppose that a straight line joining points u_1 and v_1 cuts the diagonal scale in point w_1 so that the equation $f_1(u) = f_2(v) \cdot f_3(w)$ is satisfied. What will be the scale equation for $f_3(w)$?

From the similar triangles $u_0u_1w_1$ and $v_0v_1w_1$,

$$\frac{X_u}{X_v} = \frac{K - X_w}{X_w} \quad \text{or} \quad X_u = X_v \frac{(K - X_w)}{X_w}$$

since

$$X_u = m_u f_1(u) \quad \text{and} \quad X_v = m_v f_2(v)$$

Then

$$m_u f_1(u) = m_v f_2(v) \frac{(K - X_w)}{X_w}$$

When

$$f_1(u) = f_2(v) \cdot f_3(w)$$

then

$$\frac{K - X_w}{X_w} = \frac{m_u}{m_v} f_3(w)$$

from which

$$X_w = \frac{K m_v}{m_u f_3(w) + m_v}$$

or

$$X_w = \frac{K}{\dfrac{m_u}{m_v} f_3(w) + 1} = \frac{K}{r f_3(w) + 1}$$

where

$$r = \frac{m_u}{m_v}$$

If it is desired to graduate the w-scale from u_0 instead of v_0, it can be shown that the distance from u_0 to w_1 is equal to $\frac{Km_u f_3(w)}{m_u f_3(w) + m_v}$. (This should be verified by the reader.)

Design Summary

From the above one can construct this type of chart in the following manner:

1. Draw scales for the variables, u and v, parallel to each other.
2. Graduate the u-scale in accordance with its scale equation $X_u = m_u f_1(u)$.
3. Graduate the v-scale in accordance with its scale equation $X_v = m_v f_2(v)$ (plotting positive values of v downwardly if positive values of u were plotted upwardly).
4. Graduate the w-scale from the upper end of the scale in accordance with its scale equation

$$X_w = \frac{K}{\frac{m_u}{m_v} f_3(w) + 1},$$

or from the lower end in accordance with the scale equation

$$X_w = \frac{Km_u f_3(w)}{m_u f_3(w) + m_v}.$$

EXAMPLE 1

Consider the equation $P = I^2R$, where P = power in watts, I = current in amperes (0 to 10), and R = resistance in ohms (0 to 200). The parallel scales are to be approximately 10 units long.

Now

$$m_P = \frac{10}{20{,}000} = 0.0005$$

(It should be observed that the value of $P = 10^2 \times 200 = 20{,}000$.)

and $$X_P = 0.0005P \tag{1}$$

also $$m_R = \frac{10}{200} = 0.05$$

and $$X_R = 0.05R \tag{2}$$

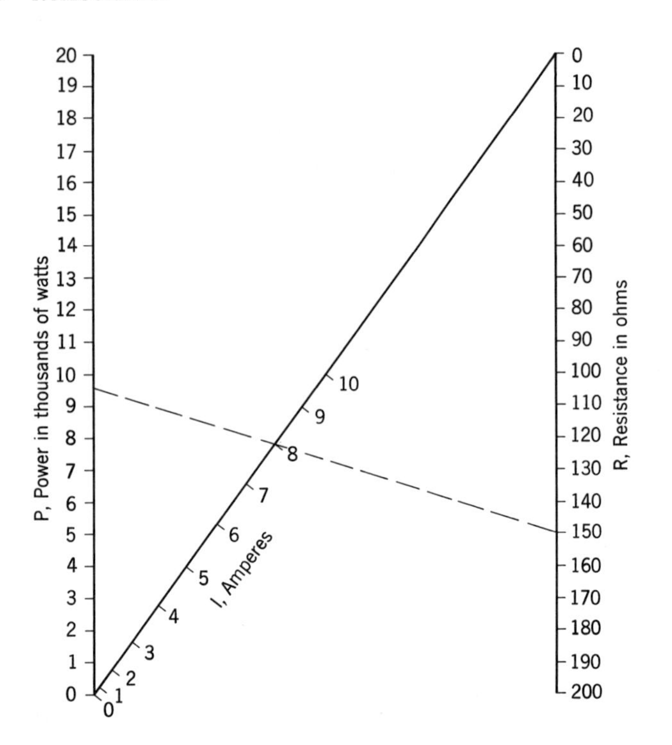

Fig. 16.31 $P = I^2R$. Example: $I = 8$, $R = 150\ \Omega$, then $P = 9600$ watts.

The diagonal scale for the variable I will be graduated by employing the scale equation,

$$X_I = \frac{K}{\frac{0.0005}{0.05}I^2 + 1} = \frac{K}{(0.01)I^2 + 1}$$

The length of the diagonal scale is K which can be assigned any convenient value. Suppose we let $K = 1$.

The scale equation is

$$X_I = \frac{1}{(0.01)I^2 + 1} \tag{3}$$

We compute values of X_I for various values of I from equation (3).

The P and R scales are graduated by employing scale equations (1) and (2). Since the functions of P and R are linear, it is only necessary to locate the end values. The other graduations can be easily located without using the scale equations. The I-scale graduations are located by using the scale equation (3). The nomogram is shown in Fig. 16.31.

EXAMPLE 2

Consider the equation $u + 2 = v^2w$ (Fig. 16.32). Suppose that u varies from 0 to 10, and v from 0 to 5. The scale lengths are to be approximately 6 inches.

$$m_u = \frac{6}{(10 + 2) - (0 + 2)} = 0.6$$

$m_v = 6/25 = 0.24$. (We shall use 0.25, which merely lengthens the scale from 6 to 6.25 in.) Hence $X_u = 0.6(u + 2)$; $X_v = 0.25v^2$. The u and v scales are graduated from these scale equations, and the w scale from

$$X_w = \frac{10}{\dfrac{0.6w}{0.25} + 1} = \frac{10}{\dfrac{12w}{5} + 1}$$

where $K = 10$ (ten of any convenient unit).

Form the following table:

w	0	0.5	1	2	3
X_w	10	50/11	50/17	50/29	50/41

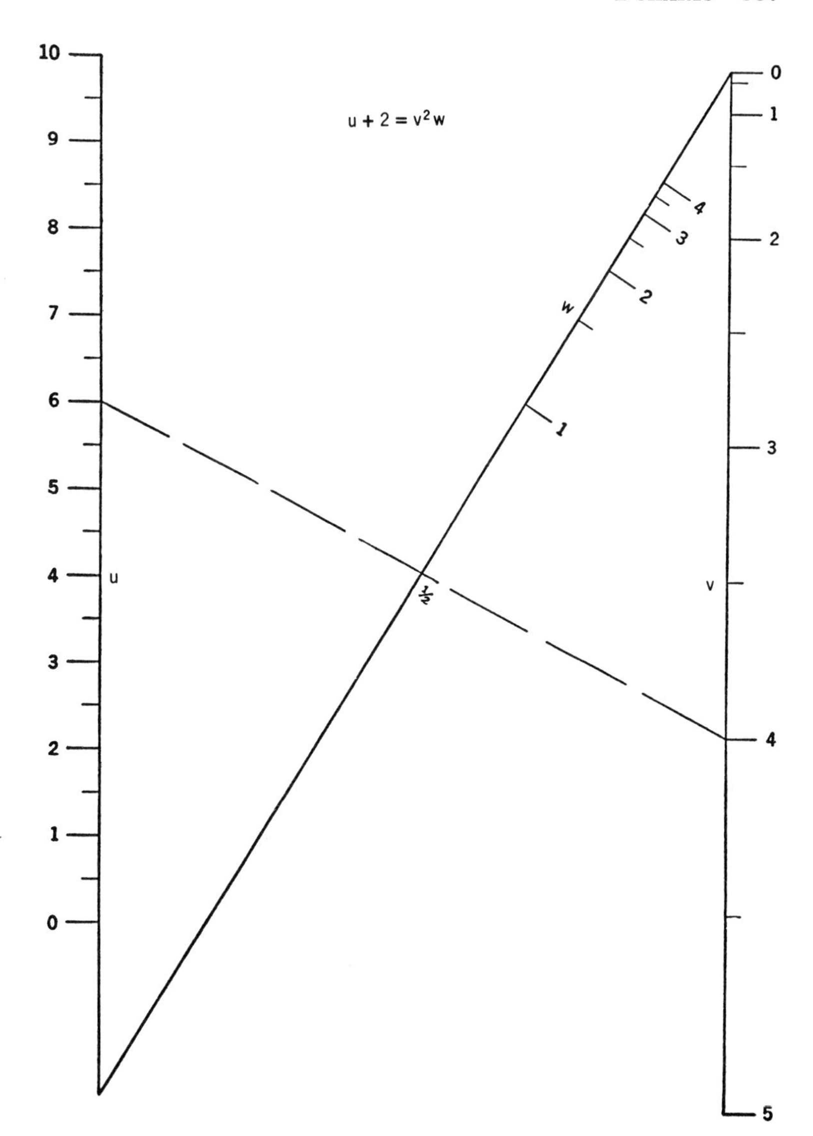

Fig. 16.32 Z-type chart for equation $(u + 2) = v^2w$.

EXAMPLE 3

Let us now consider the equation for the volume of a right circular cylinder, $V = \pi r^2h/144$, where V is the volume in cubic feet, r is the radius of the base circle, in inches (4 to 12), and h is the height of the cylinder in feet (4 to 15). We may write the equation,

$$KV = r^2h, \qquad \text{where } K = 144/\pi$$

The range of V is determined from the ranges of r and h. Simple calculations will show that V varies from 1.40 (or $4\pi/9$) cu ft to 47.15 (or 15π) cu ft.

Now

$$m_V = \frac{10\pm}{\dfrac{144}{\pi}\left(15\pi - \dfrac{4\pi}{9}\right)} = 0.005,$$

where scale length = $10\pm$ and

$$X_V = 0.005\left[\frac{144}{\pi}\left(V - \frac{4\pi}{9}\right)\right]$$

$$m_r = \frac{10\pm}{144 - 16} = 0.08$$

$$X_r = 0.08(r^2 - 4^2)$$

From the above scale equations, we can graduate the V and r scales. It will be observed that it is necessary only to compute the total length of the V-scale; i.e., $X_V = 0.005\left[\frac{144}{\pi}\left(15\pi - \frac{4\pi}{9}\right)\right] = 10.48$. Then we know that the lower point of the scale will be marked 1.40 and the upper point will be marked 47.15. Additional graduations can be obtained by proportion. Since the function is linear, the scale is uniform.

In the case of the r-scale, it should be noted that the function is r^2, and therefore distances between consecutive points are proportional to the square of r.

The location of the diagonal scale must be determined next. Many students make the typical error of connecting point 4 on the r-scale with point 1.40 on the V-scale. Remember that *the diagonal line joins the zero value of the function of r with the zero value of the function of V*. These points would be zero on the r-scale and zero on the V-scale. In this case, it would be possible to include these points on the respective scales. However, often the zero values of the functions are not accessible within the limits of the drawing. Let us assume this to be the case in our problem.

The position of the h-scale can be established by a very simple method. Let us locate points 6 and 12 on the h-scale. If we let $r = 10$, then $V = 13.1$ when $h = 6$. The line joining $r = 10$ with $V = 13.1$ contains $h = 6$. Again, if we let $r = 12$, then $V = 18.9$ when $h = 6$. The line joining $r = 12$ with $V = 18.9$ contains $h = 6$. Therefore, the intersection of these two lines is $h = 6$ and, in addition, is a point on the diagonal. This method can be repeated for another point such as $h = 12$. Other points on the h-scale can then be located from point 12, by properly using the scale equation

$$X_h = \frac{K}{\frac{m_V}{m_r}h + 1} = \frac{K}{\frac{0.005}{0.08}h + 1}$$

It is evident that K must be determined. This can be done, since the distance between points 6 and 12 can be measured. Hence,

$$\frac{K}{\frac{0.005}{0.08} \times 6 + 1} - \frac{K}{\frac{0.005}{0.08} \times 12 + 1} = 2.37$$

from which $K = 15.23$.

The distances from $r = 0$, along the diagonal, can now be computed from

$$X_h = \frac{15.23}{\frac{0.005}{0.08}h + 1} = \frac{243.68}{h + 16}$$

h	4	5	6	7	8	9	10	11	12	13	14	15
X_h	12.18	11.59	11.07	10.58	10.13	9.73	9.36	9.01	8.70	8.40	8.12	7.86
X_h from $h = 12$	3.48	2.89	2.37	1.88	1.43	1.03	0.66	0.31	0	0.30	0.58	0.84
	← Distances below $h = 12$ →									← Above $h = 12$ →		

The alignment chart is shown in Fig. 16.33.

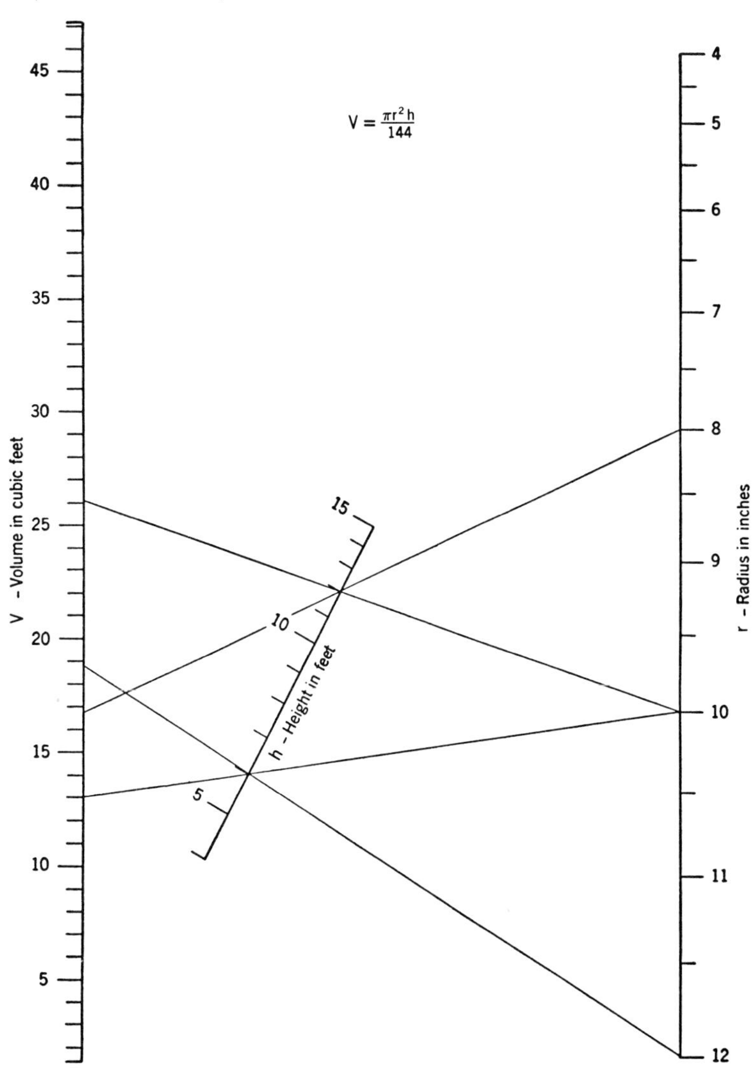

Fig. 16.33 Z-type chart for the relation $V = \pi r^2 h/144$.

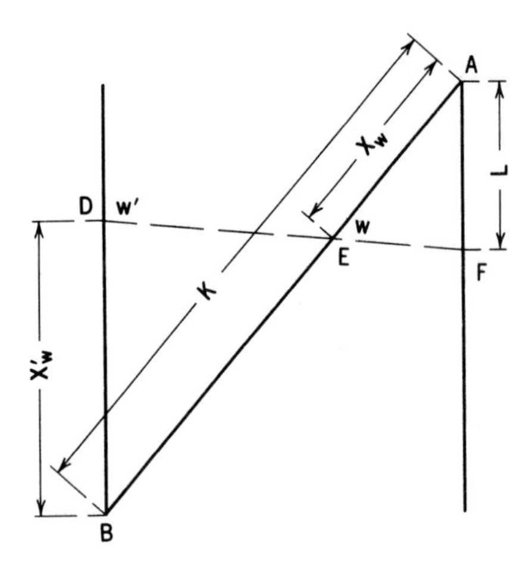

Fig. 16.34

Simplified Method

A method which may simplify the work of graduating the diagonal scale can be developed in the following manner (Fig. 16.34). Let F be a fixed point on the right vertical scale. Let the distance from point A to the fixed point F be L (inches, centimeters, or any other convenient number of units). Suppose that the *right-hand* side of the left vertical scale carries a temporary w-scale. From the similar triangles, BDE and AFE,

$$\frac{X_w'}{L} = \frac{K - X_w}{X_w}$$

$$X_w' = L\left(\frac{K - X_w}{X_w}\right)$$

Previously it had been shown that

$$\frac{K - X_w}{X_w} = \frac{m_u}{m_v}[f_3(w)]$$

Hence,

$$X_w' = L\frac{m_u}{m_v}[f_3(w)]$$

This equation enables us to graduate the temporary w-scale. Lines joining the fixed point, F, with the graduations on the temporary scale will intersect the diagonal in points having the same values of w.

This method has two advantages over the one of locating points on the diagonal from the equation,

$$X_w = \frac{K}{\frac{m_u}{m_v}f_3(w) + 1}.$$

First, if the function of w is linear, a uniform scale can be graduated on the temporary scale; second, the length, K, of the diagonal scale need not be known.

EXAMPLE

Let us consider the equation $Q = 3.33bH^{3/2}$ (Francis' weir formula). The ranges of the variables are: b (3 ft to 20 ft); and H (0.5 ft to 1.5 ft). Scale lengths for Q and H are approximately 6 inches. The preliminary design of the Z-chart will be as shown in Fig. 16.35.

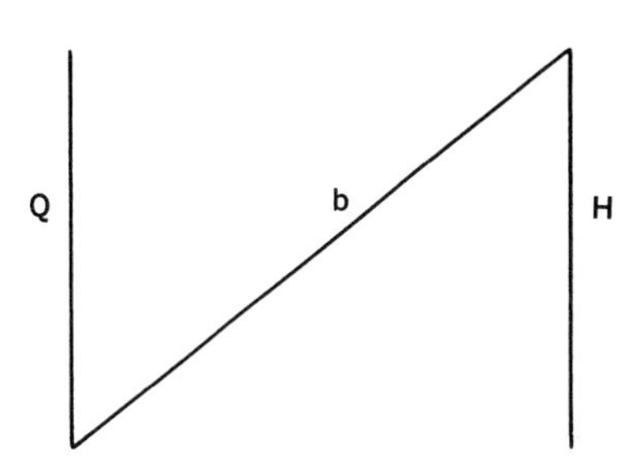

Fig. 16.35

Note carefully that by placing the b-variable on the diagonal scale, the *temporary b-scale will be uniform.* This is a definite advantage over the placement of the H-variable on the diagonal since the temporary scale for H would not be uniform.

Calculations of the scale moduli are:

$$m_Q = \frac{6\pm}{120} = 0.05 \quad (Q_{\max} = 120+)$$

$$m_H = \frac{6\pm}{3.33(1.5^{3/2} - 0.5^{3/2})} = \frac{6\pm}{4.94} = 1.2$$

The scale equations are:

$X_Q = 0.05Q$ (graduated from $Q = 0$)

$X_H = 1.2(3.33H^{3/2}) = 4H^{3/2}$ (graduated from $H = 0$)

or

$X_H = 4(H^{3/2} - 0.5^{3/2})$ (graduated from $H = 0.5$)

$$X_b' = L\frac{0.05}{1.2}b = \frac{5}{120}Lb = \frac{1}{24}Lb$$

Let $L = 6$; then $X_b' = b/4$. The completed nomogram is shown in Fig. 16.36.

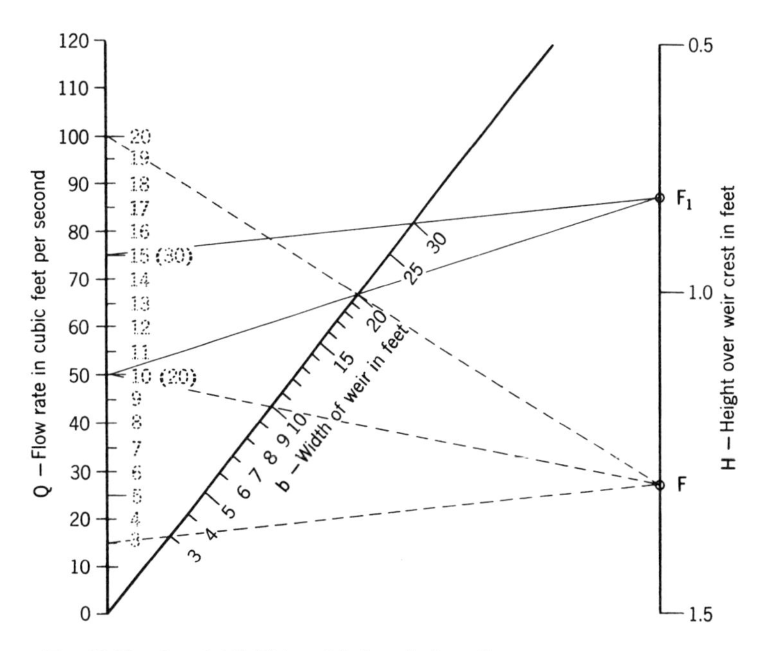

Fig. 16.36 $Q = 3.33bH^{\frac{3}{2}}$ (simplified method used).

Now suppose that b has a maximum value of 30. It is clearly seen that the location of $b = 30$ *on the temporary scale* would fall beyond the maximum length of the present scale. In order to locate $b = 30$ on a temporary scale we shall shift the location of the fixed point F to F_1 which can be determined in the following manner:

Since $X_b' = (1/24)Lb$, let us choose $L = 3$. Then $X_b' = b/8$. When $b = 30$, $X_b' = 30/8 = 3.75$ in. [Note $b = 30$ is shown as (30) on the new temporary scale, $X_b' = b/8$.] The line connecting F_1 and (30) intersects the diagonal scale at $b = 30$.

Locating the Diagonal Scale

A little reflection on the use of two F positions should lead us to conclude that the location of the diagonal scale can be fixed very simply.

The diagonal scale passes through the zero values

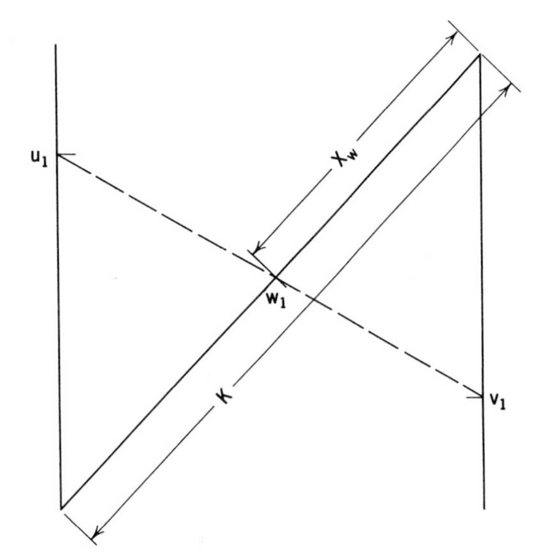

Fig. 16.37

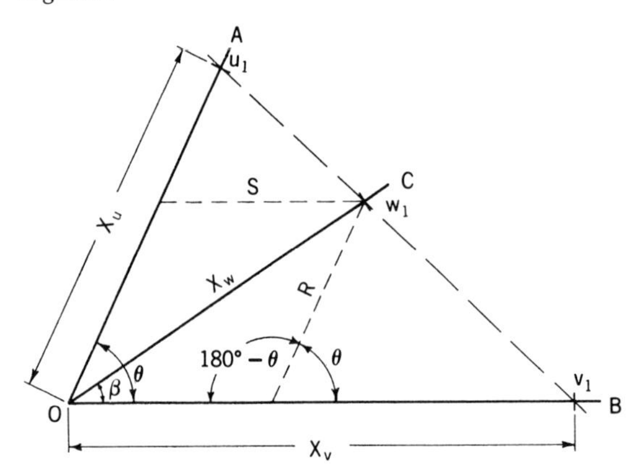

Fig. 16.38

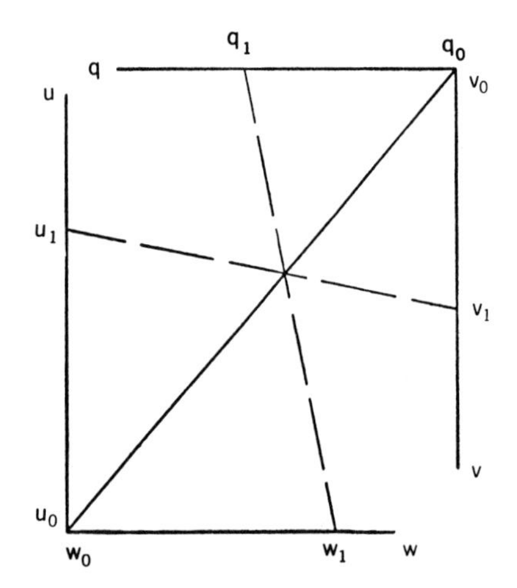

Fig. 16.39

of the functions of Q and H. This means that the diagonal scale passes through $Q = 0$ and $H = 0$. The value $Q = 0$ is available; however, $H = 0$ is not. Therefore, it is only necessary to locate one point on the diagonal, i.e., $b = 20$.

Using F. The line joining F and 20 on the temporary b-scale contains $b = 20$.

Using F_1. The line joining F_1 and (20) also contains $b = 20$.

Therefore, the intersection of the two lines, described above, is $b = 20$. The line joining this point with $Q = 0$ is the diagonal line which carries the graduations of b. Do not forget that L is measured from $f(H) = 0$, which in this case is $H = 0$. When $H = 0$ is not accessible how will you locate F or F_1?

There are several additional forms of nomograms that are used to solve problems that arise in research, design, production, statistics, etc. The forms shown in Figs. 16.37 to 16.41 are employed quite frequently.

1. $f_1(u) + f_2(v) = \dfrac{f_2(v)}{f_3(w)}$. (Fig. 16.37)

Scale equations:

$$X_u = m_u f_1(u)$$
$$X_v = m_v f_2(v)$$
$$X_w = K f_3(w)$$

Note: $m_u = m_v$

2. $\dfrac{1}{f_1(u)} + \dfrac{1}{f_2(v)} = \dfrac{1}{f_3(w)}$. (Fig. 16.38)

Scale equations:

$$X_u = m_u f_1(u)$$
$$X_v = m_v f_2(v)$$

Location of w-scale:

$$\frac{R}{S} = \frac{m_u}{m_v}$$

Graduate w-scale by

(1) $R = m_u f_3(w)$ and parallel to the B-axis, or

(2) $S = m_v f_3(w)$ and parallel to the A-axis; or

(3) $X_w = [m_u^2 + m_v^2 + 2m_u m_v \cos\theta]^{1/2} f_3(w)$

3. $\dfrac{f_1(u)}{f_2(v)} = \dfrac{f_3(w)}{f_4(q)}$. (Fig. 16.39)

Scale equations:

$$X_u = m_u f_1(u)$$
$$X_v = m_v f_2(v)$$
$$X_w = m_w f_3(w)$$
$$X_q = m_q f_4(q)$$

Where, $\dfrac{m_u}{m_v} = \dfrac{m_w}{m_q}$

4. $f_1(u) + f_2(v) = \dfrac{f_3(w)}{f_4(q)}$. (Fig. 16.40)

Scale equations:

$$X_u = m_u f_1(u)$$
$$X_v = m_v f_2(v)$$
$$X_w = m_w f_3(w)$$
$$X_q = m_q f_4(q)$$

Where $m_u = m_v$ and $m_q = Km_w/m_u$
K = length of the diagonal

5. $f_1(u) + f_2(v) \cdot f_3(w) = f_4(w)$. (Fig. 16.41)

Scale equations:

$$X_u = m_u f_1(u)$$
$$X_v = m_v f_2(v)$$
$$X_w = \frac{Km_u f_3(w)}{m_u f_3(w) + m_v}$$
$$Y_w = \frac{m_u m_v f_4(w)}{m_u f_3(w) + m_v}$$

Examples of the above type forms are shown in Figs. 16.42 to 16.47.

These and several other forms are discussed in a fuller treatment of nomography in the author's textbook, *Nomography*, 2nd ed., published by John Wiley and Sons, New York.

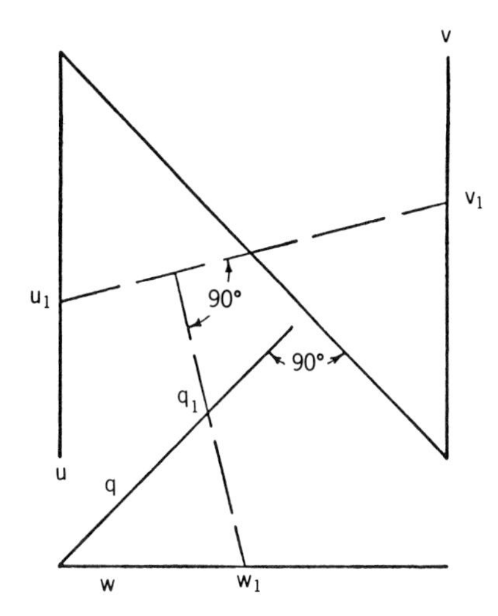

Fig. 16.40

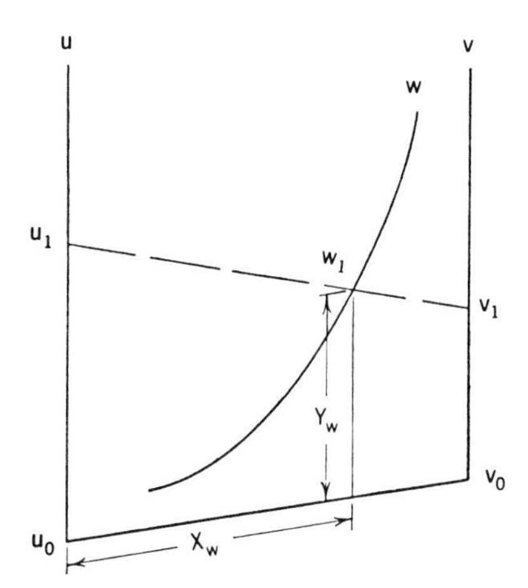

Fig. 16.41

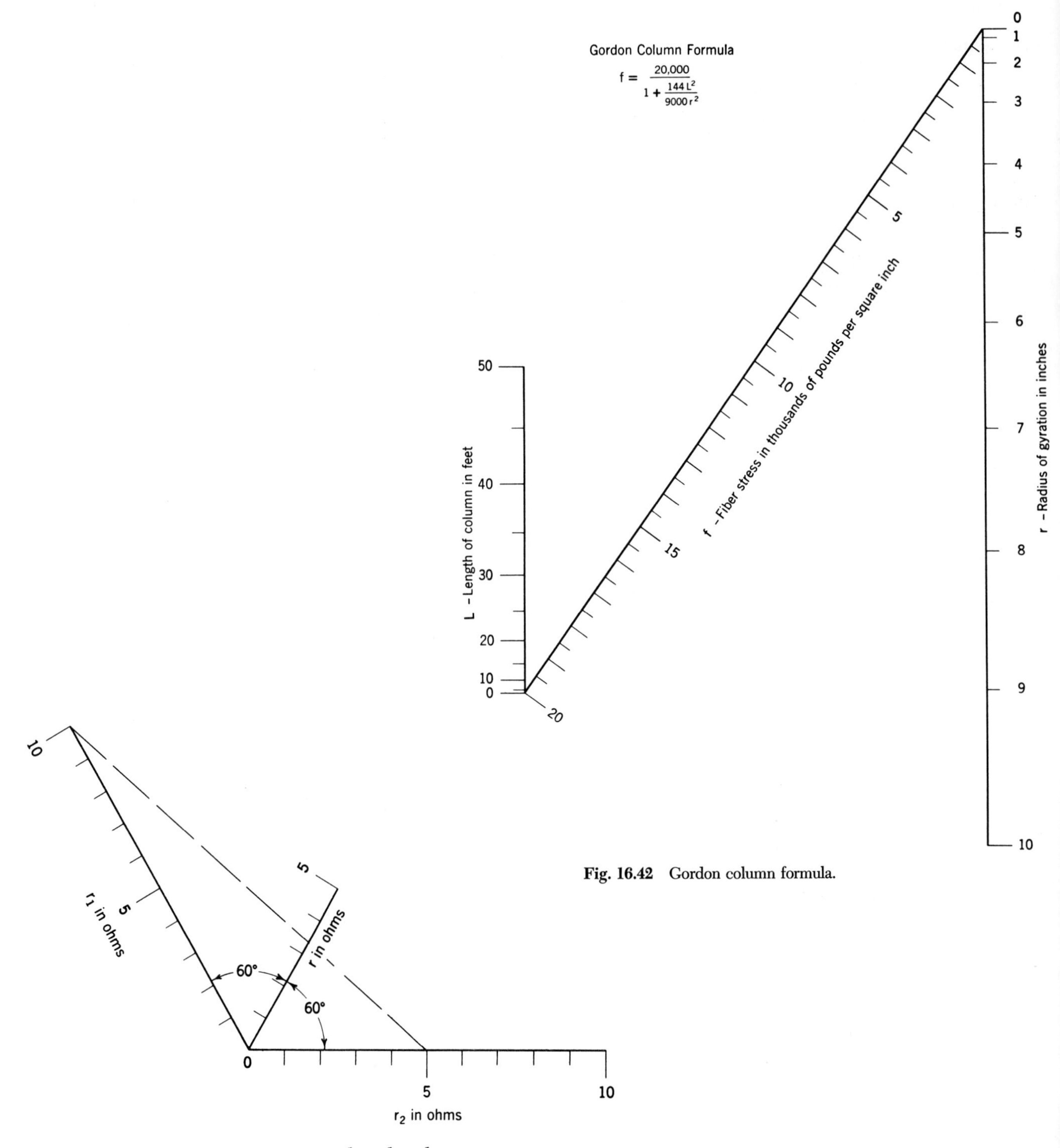

Fig. 16.42 Gordon column formula.

Fig. 16.43 Alignment chart for the equation $\frac{1}{r_1} + \frac{1}{r_2} = \frac{1}{r}$.

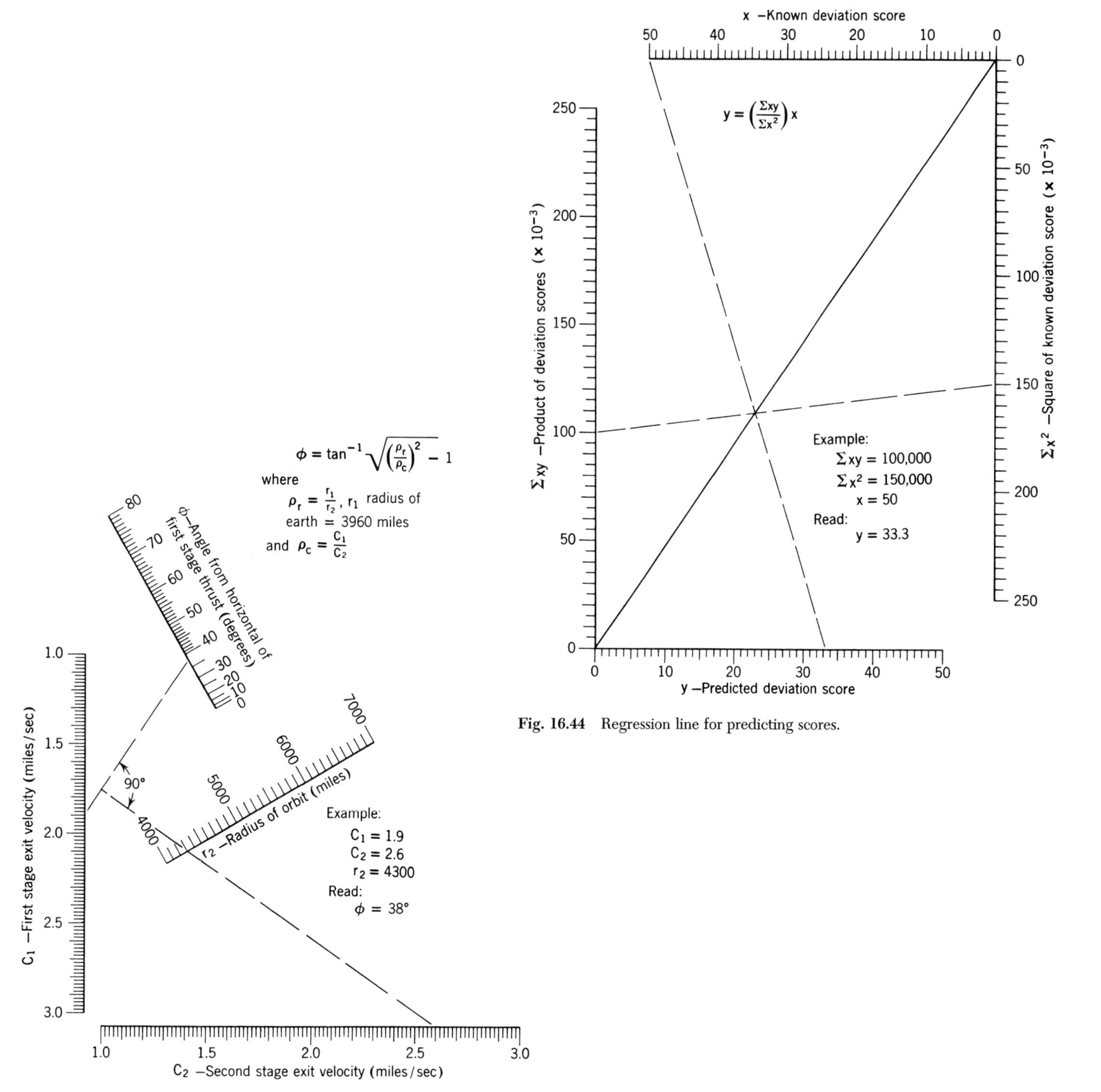

Fig. 16.44 Regression line for predicting scores.

Fig. 16.45 Nomogram for optimum direction angle, ϕ, from horizontal on first-stage thrust (launch direction); for the case $\rho_c = \rho_r$.

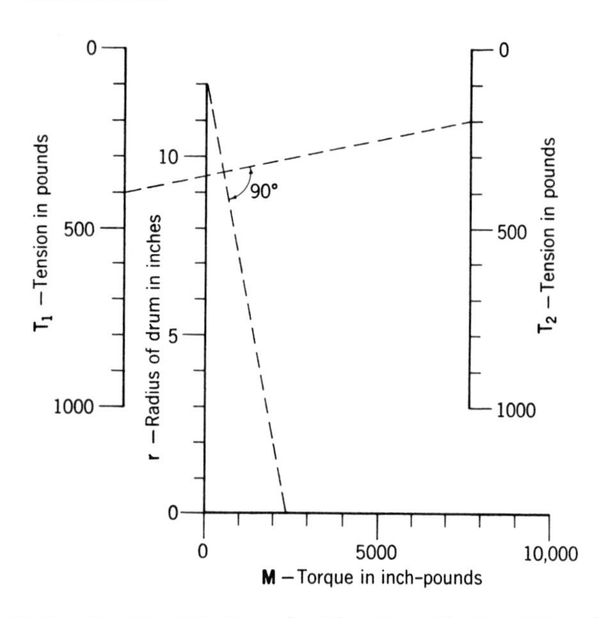

Fig. 16.46 $M = (T_1 - T_2)r$. Example: When $T_1 = 400$, $T_2 = 200$, and $r = 12$, then $M = 2400$.

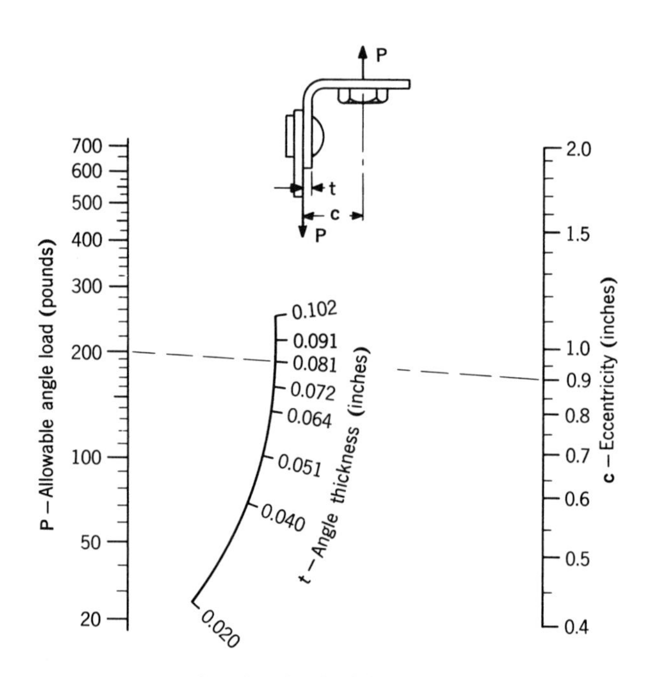

Fig. 16.47 Nomogram for allowable load for sheet-metal angle. (Lockheed Stress Memo 89.)

Computer-Aided Design of Nomograms

During the past few years we have developed computer programs for the construction of nomograms of the forms: (1) $f_1(u) + f_2(v) = f_3(w)$; (2) $f_1(u) = f_2(v) \cdot f_3(w)$; and (3) $f_1(u) + f_2(v) \cdot f_3(w) = f_4(w)$. Examples are shown in Figs. 16.48, 16.49 and 16.50. FORTRAN 4 programs were developed for each of the above forms. Magnetic and paper-punched tapes were produced to obtain either Cal-Comp or Gerber plots of the nomograms.

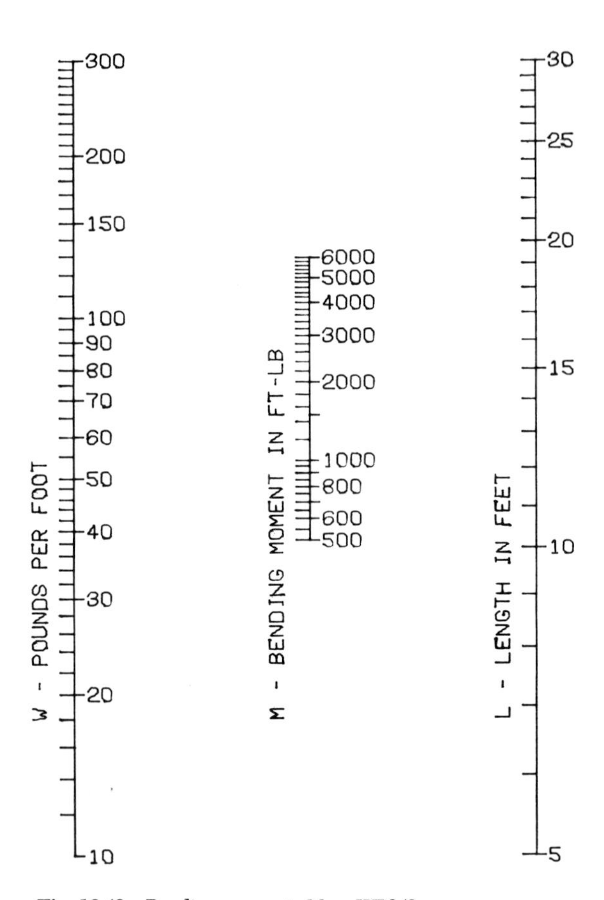

Fig. 16.48 Bending moment, $M = WL^2/8$.

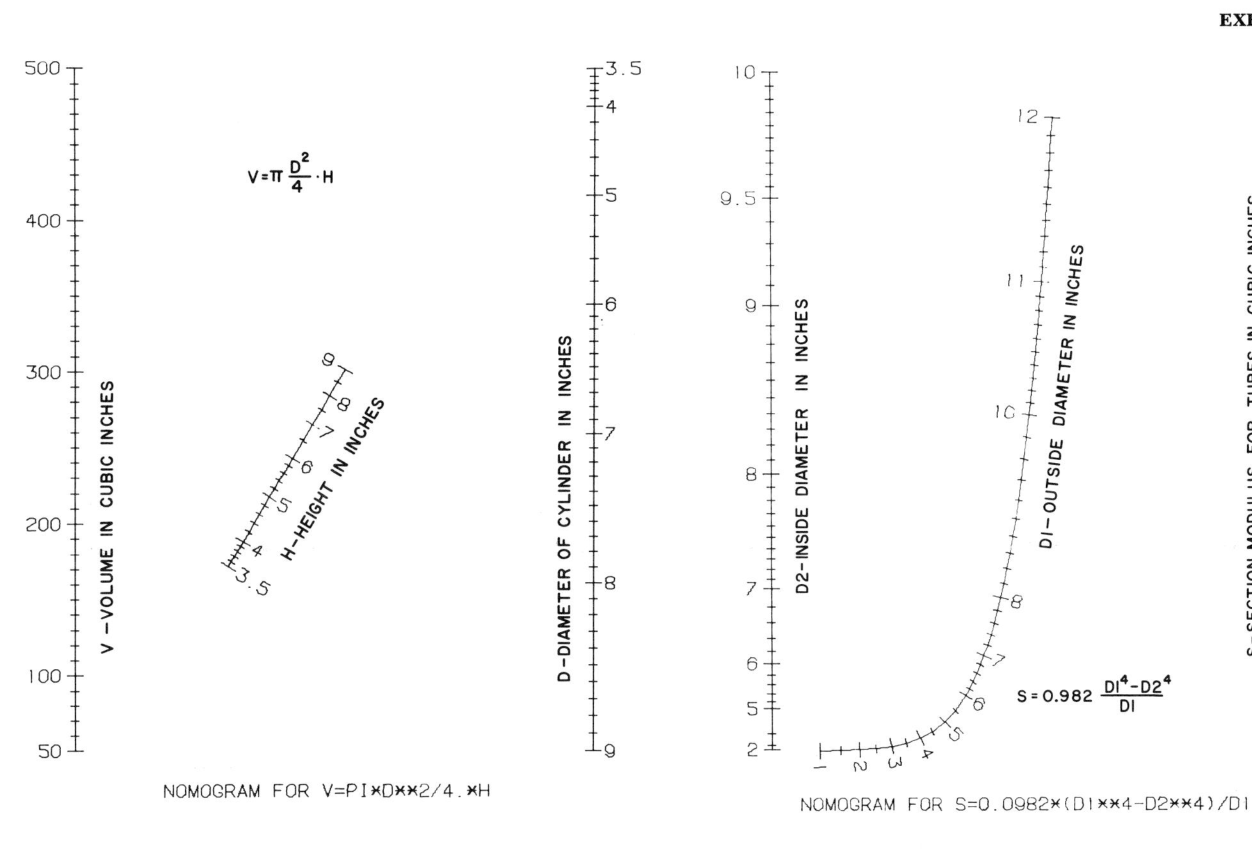

Fig. 16.49 Nomogram for the equation $V = \left(\frac{\pi D^2}{4}\right)H$.

Fig. 16.50

EXERCISES

1. Construct a concurrency chart for the equation $Q = 3.33bH^{3/2}$, where Q = discharge in cubic feet per second, b = width of weir (2 to 25 feet), and H = head above crest (0.5 to 2.0 feet). Include a computation sheet.

2. Construct a concurrency chart for the equation $x^2 + y^2 = r^2$; x and y vary from 0 to 20.

3. Construct a concurrency chart for the equation $I = E/R$, where E = voltage (0 to 400 volts); R = resistance (0 to 20 ohms); and I = amperes.

4. Design an alignment chart for the equation $r = \sqrt{x^2 + y^2}$; x and y vary from 0 to 60.

5. Design an alignment chart for the equation $I = E/R$; E varies from 10 to 1000 volts; R varies from 10 to 60 ohms.

6. Design an alignment chart for the equation $Q = 3.33bH^{3/2}$; Q = discharge in cubic feet per second; b = width of weir (5 to 25 feet); H = head above crest (0.5 to 2.0 feet).

7. Design an alignment chart for the equation $S = \pi DN/12$; S = cutting speed in feet per minute; D = diameter of work (0.50 to 18 inches); N = (100 to 2000 rpm).

8. Design an alignment chart for the equation $P = E^2/R$, where P = power in watts; E = voltage (10 to 220 volts); R = resistance (10 to 1000 ohms).

9. Design an alignment chart for the equation $I = bd^3/12$, where I = moment of inertia; b = width of rectangle (5 to 20 inches); d = depth of rectangle (8 to 30 inches).

10. Design an alignment chart for the equation $I = M/(1 - r)$, where M = yield on tax exempt investments (1% to 10%); r = marginal income tax rate (22% to 90%); I = yield on ordinary investment (3% to 40%).

11. Design an alignment chart for the equation $V = C\sqrt{RS}$, where R = hydraulic radius (5 to 50 feet); S = slope of channel (0.0001 to 0.01); C = (60 to 150).

12. Design an alignment chart for the equation $P = 3EI \cos \theta$, where P = power in watts; E = line voltage (10 to 220 volts); I = line current (5 to 50 amperes).

PART 3
INTRODUCTION TO DESIGN

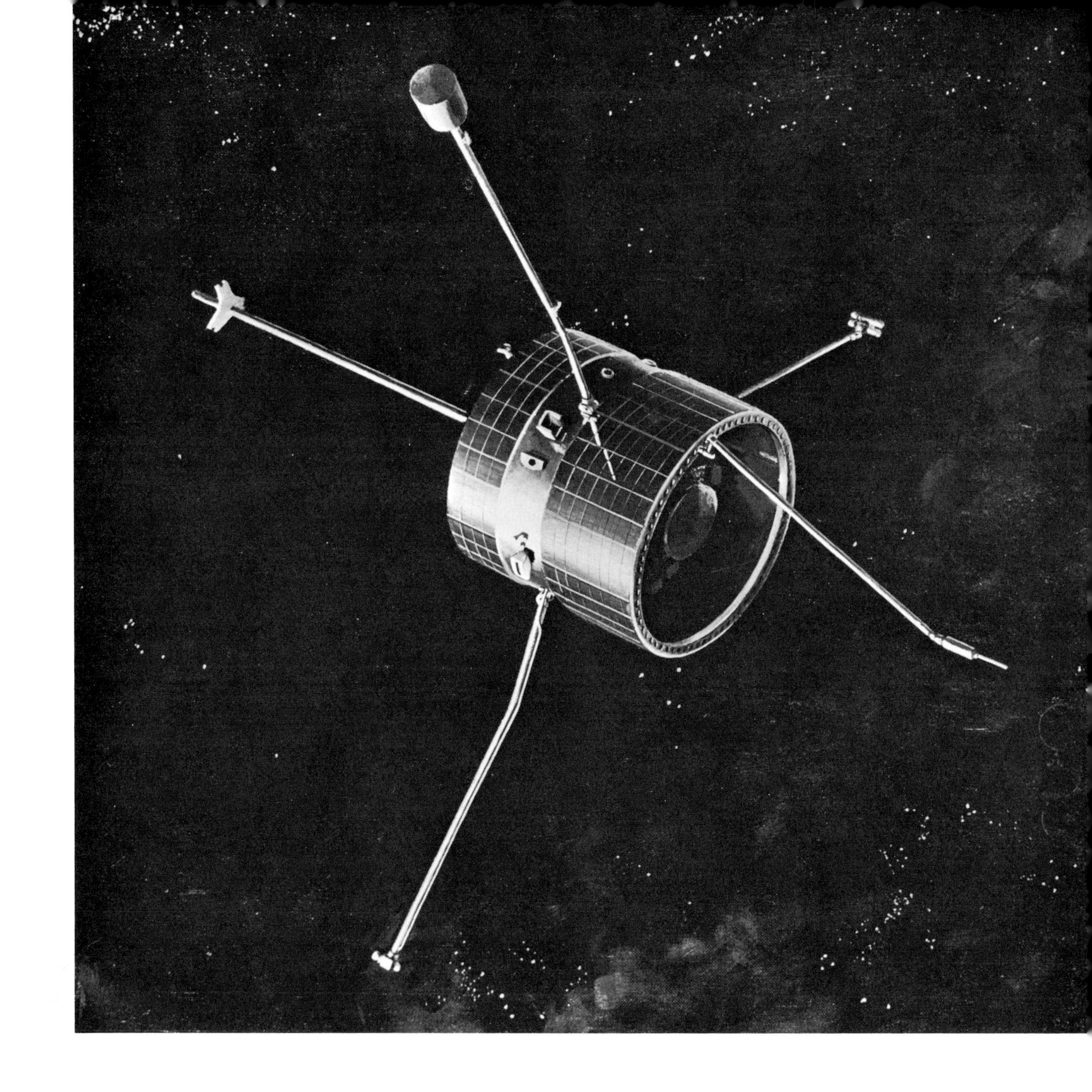

PICTORIAL DRAWING 17

In Chapter 2, emphasis was placed on freehand sketching—"talking with a pencil." In Chapter 5, freehand sketching was used again in making isometrics, obliques, and perspectives from orthographic representations. The importance of freehand sketches in preparing layouts, in recording the "thinking-through-process" in solving three-dimensional problems, and in the interpretation of orthographic drawings has been stressed a number of times. It is hoped that you have achieved reasonable proficiency in making freehand sketches and that you have taken full advantage of the opportunities to employ freehand sketching, not only in engineering courses but also in other areas as well.

The pictorial language (graphics) is the source of two other languages—one which uses symbols (i.e., in mathematics, physics, and chemistry) and the other which uses alphabets. William Mason, in *A History of the Art of Writing*, points out that:

> The printed words that we read, which are made up of the present, purely arbitrary characters which we know as alphabetic signs, convey to the mind through their abstract symbols, word-pictures comparable to the primitive ideographs—idea pictures—which were the ancient prototypes of these symbols. These printed or written words of the classical parchments, the medieval manuscripts, and the modern books, it must be remembered, stand for the phonetic names of the intended things, representing them to the mind, but bearing no resemblance to them in form. By long association, however, these word-symbols, despite their apparently arbitrary character, call up in the mind of the reader the corporate things themselves; not alone their names, but their forms, qualities or attributes that the thousands of words in the language have come to represent. They are idea-pictures as well as sound-pictures. In fact they were idea-pictures first, ages before the primitive culture of advancing man arrived at the conception of using these pictures as signs to represent phonetic sounds.

To this day we still retain many of the symbolic signs, those denoting celestial and terrestrial features, such as the sun, moon, the planets, mountains, flowing water, and trees. The signs of the zodiac, seen in most almanacs, are among the oldest symbols. It is believed that their origin antedates the classic period in Rome. In Babylonia, during the time of Nebuchadnezzar I (1130–1115 B.C.), "boundary stones were carved with the representations of the heavenly bodies and stellar deities."

The North American Indian already had a system of picture writing at the time of the European settlement of this country. Pictorial notices and inscriptions were frequently observed on tree trunks. Many of these were warning notices. Most of the writing was of a pictographic nature, drawing the picture of the object intended to "recall it to the mind of the reader."

The first positive evidence of the use of drawings to guide construction work is found in the writings of Vitruvius (63 B.C.–A.D. 14). He states, "The architect must be skillful with the pencil and have knowledge of drawing so that he readily can make the drawings required to show the appearance of the work he proposes to construct." He speaks of the assistance of geometry in teaching the use of rule and compasses. He defines the ground plan, the elevation, and the perspective, "where the lines meet at the center."

About 100 B.C. Frontinus, who became "Curator Aquarum," tells about plans made of the Roman aqueducts.

So much for a brief review of the period prior to 1450.

More is known of the drawings of Leonardo da Vinci (1452–1519). It is clearly evident that he understood and practiced representations made by what came to be known as orthogonal projection. Da Vinci was not only a great artist, but also a great engineer. His vast collection of sketch books is filled with sketches of machine details and assemblies.

Today, engineers and scientists recognize the importance of pictorial drawing in recording "design ideas" and in conveying their design concepts to those in the research, design, development, and production fields. The effective teacher makes frequent blackboard sketches in his class presentations, and good sketches contribute a good deal to the students' comprehension of the material presented.

To further our knowledge of pictorial drawing let us now consider the following:

1. Isometric *views* and isometric *drawings*.
2. Dimetric *views* and dimetric *drawings*.
3. Oblique *views* and oblique *drawings*.
4. Trimetric *views* and trimetric *drawings*.
5. Perspective *views* and perspective *drawings*.

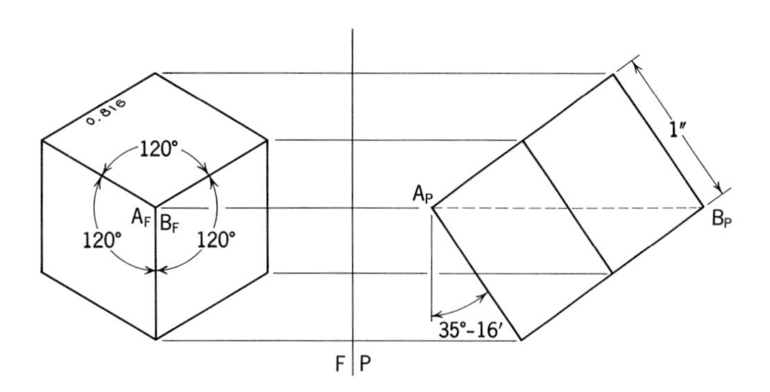

Fig. 17.1

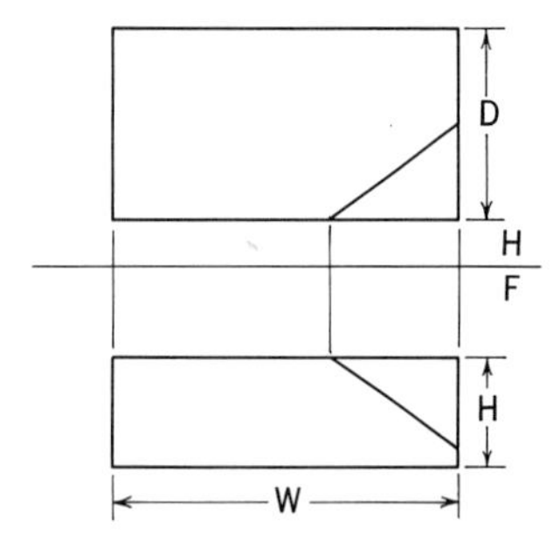

Fig. 17.2

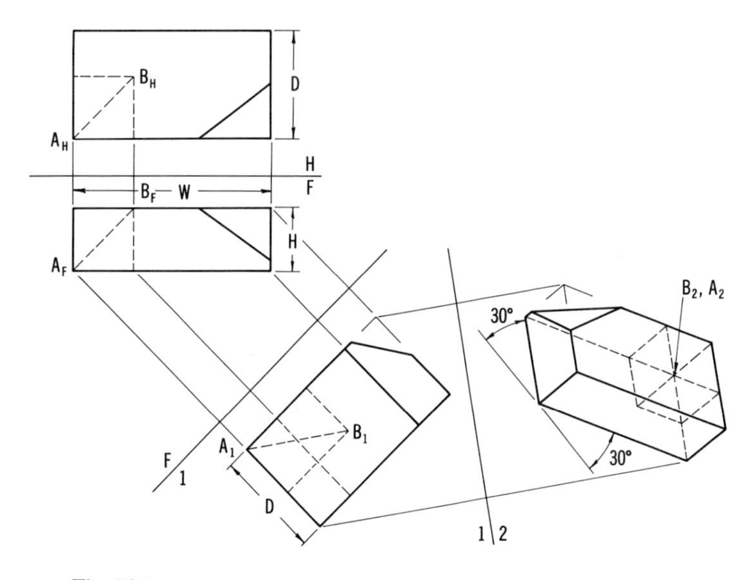

Fig. 17.3

Isometric Views

An orthographic view may be a pictorial view, if the object is so oriented that the resulting view reveals a three-dimensional effect. For example, consider the cube shown in Fig. 17.1. The cube is so oriented that its edges make equal angles with the frontal plane. The front view of the body diagonal AB is a point. The edges as seen in this view are equally foreshortened by an amount equal to $\sqrt{2}/\sqrt{3}$* times the original length. In the *front view* the edges that intersect at point A make angles of 120° with each other, whereas the edges of the cube actually make an angle of approximately 35° 16′ with the frontal plane. The three edges that pass through point A are regarded as the *isometric axes;* lines parallel to the axes are identified as isometric lines. This front view is the *isometric view* of the cube.

EXAMPLE 1

How would we proceed to obtain an isometric view from the top and front views of the block shown in Fig. 17.2?

An isometric view of the box will occur when we obtain the point view of a body diagonal, such as AB, *of the inscribed cube* which is shown dotted in Fig. 17.3. This can be done by introducing supplementary plane 1 parallel to AB, thus revealing its true length A_1B_1, and then introducing supplementary plane 2 perpendicular to line AB, which now appears as a point, A_2B_2. The *isometric view* is apparent in the second supplementary view.

* The student should have no difficulty in verifying this value.

EXAMPLE 2

When the given figure is irregular in shape, a small cube can be introduced adjacent to the figure. The orientation of the supplementary planes 1 and 2 can be controlled by the particular body diagonal selected in the cube.

Let us consider the solid shown in Fig. 17.4. An isometric view of the solid will be seen on supplementary plane 2 which shows the point view of body diagonal *AB* of the cube.

We observe that the method of attack in obtaining an isometric view is simply *an application of the principles employed previously in finding the point view of a line.*

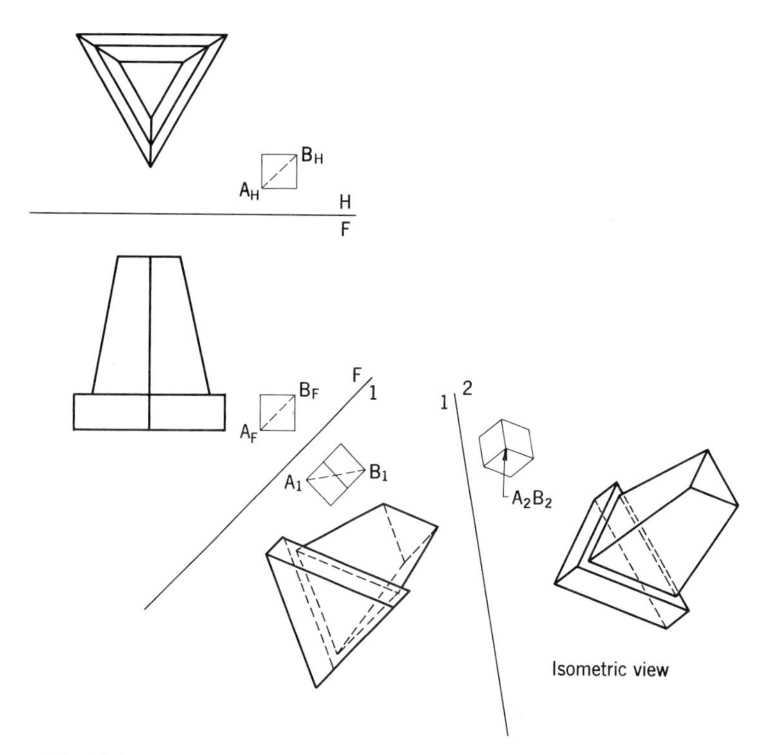

Fig. 17.4

Isometric Drawing

In Chapters 2 and 5 we employed isometric grid sheets to make freehand isometric sketches. Note that an *isometric drawing* is similar to an isometric view. In our discussion of the cube we observed that the edges were foreshortened an amount equal to $\sqrt{2}/\sqrt{3}$ times the true length of the edges. In an isometric drawing, however, *the foreshortening is not considered.* This simply means that in the case of a 1-inch cube, the edges would be drawn 1 inch long (or a length that represents 1 inch). This is a convenience. In most cases an isometric drawing can be made without resorting to supplementary orthogonal views, as was done in Figs. 17.3 and 17.4.

EXAMPLE 1

In making an isometric *drawing* of a 1-inch cube (Fig. 17.5), edge *AB* is laid off as a vertical line 1 inch long; edges *AC* and *AD*, each 1 inch long, will form 30° angles with the horizontal drawn through point *A*. The three edges *AB*, *AC*, and *AD* are regarded as the *isometric axes.* Measurements that are parallel to the axes will be true length. The completion of the drawing of the isometric cube is quite evident. We have previously observed that isometric grid sheets greatly simplify the construction of isometric drawings.

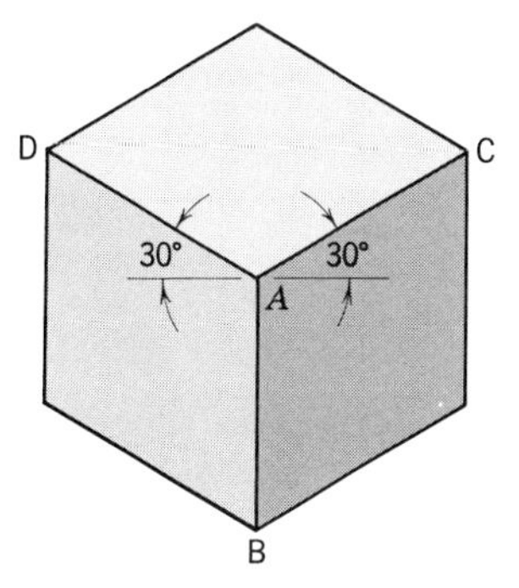

Fig. 17.5

In Chapter 5, we also solved several problems involving the translation from an orthographic representation to an isometric sketch. We noted that true measurements were made on the isometric axes and on lines parallel to the axes. In several examples, the coordinate method was used to locate points in the isometric sketches.

Now let us examine problems that may involve circles or irregular curves.

EXAMPLE 2

Consider the circle shown in Fig. 17.6(a). An isometric drawing can be made by first constructing the circumscribed isometric square and then locating the points of intersection of the circle with lines drawn parallel to diameter cd. The isometric drawing shown in Fig. 17.6(b) is then easily prepared by establishing the points a, b, c, d, 1, 2, etc., on the corresponding lines in the isometric square. The smooth curve drawn through these points is the isometric drawing of the circle.

A close approximation to the true isometric drawing of the circle can be made by using circular arcs. This is shown in Fig. 17.6(c). The centers for the small arcs are found by locating the intersections of the perpendiculars drawn from points o and o' to the opposite sides of the isometric square. The isometric circle is then easily constructed by drawing the two small arcs with radius r and the larger arcs with radius R and centers o and o'. The necessary condition that the approximate ellipse be tangent to the sides of the isometric square at their midpoints is satisfied by the construction described above.

Templates are available for drawing isometric circles.

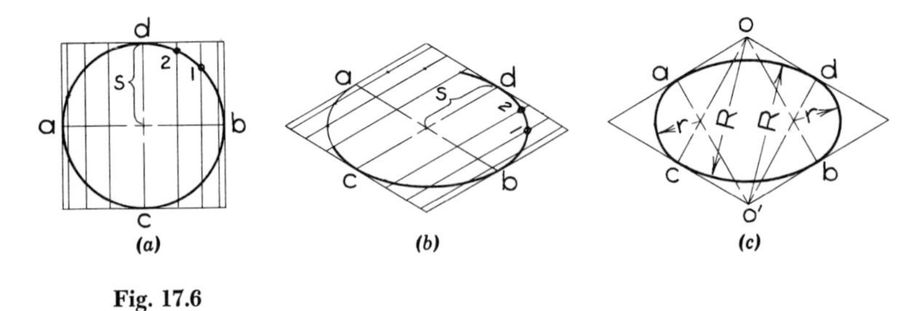

Fig. 17.6

EXAMPLE 3

Now let us consider the object shown in Fig. 17.7(a). The construction for making the isometric drawing of the irregular-shaped surface is clearly shown in Fig. 17.7(b).

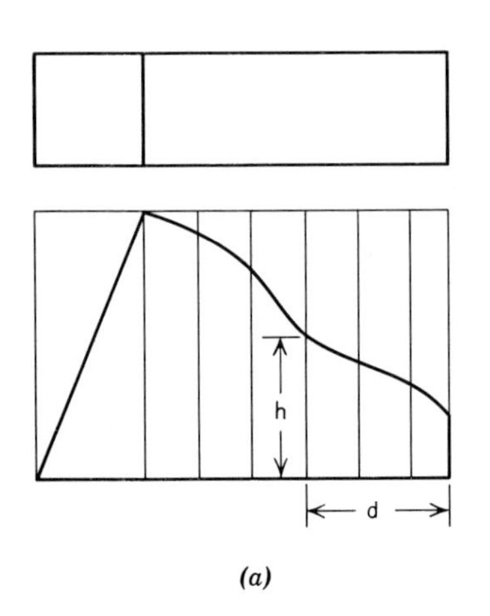

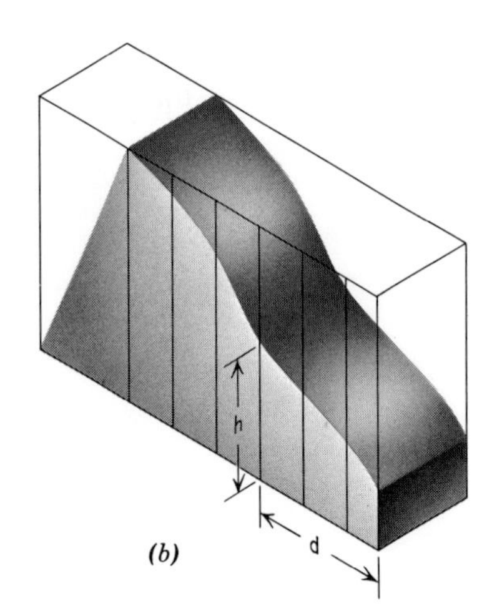

Fig. 17.7

Dimetric Views

When a cube is oriented to reveal an orthographic view that shows equal foreshortening of only two of three concurrent edges, it is known as a *dimetric view*.

For example, consider the top and front views of the cube shown in Fig. 17.8. Select body diagonal AB (we could use one of the other body diagonals) and obtain a view of the cube showing the true length of AB. This is apparent in supplementary view 1. Now establish adjacent supplementary plane 2 in *any desired position* that shows diagonal AB *as a line*. The new *view, 2,* of the cube is a *dimetric view*. Edges EC and ED are equally foreshortened. Edge EA, however, is *not* foreshortened the same amount. By varying the position of plane 2, angle θ can be changed, thus changing the ratio of the lengths of the edges EC (or ED) and EA. The three concurrent edges EA, EC, and ED, as seen in the dimetric view, may be regarded as the *dimetric axes*.

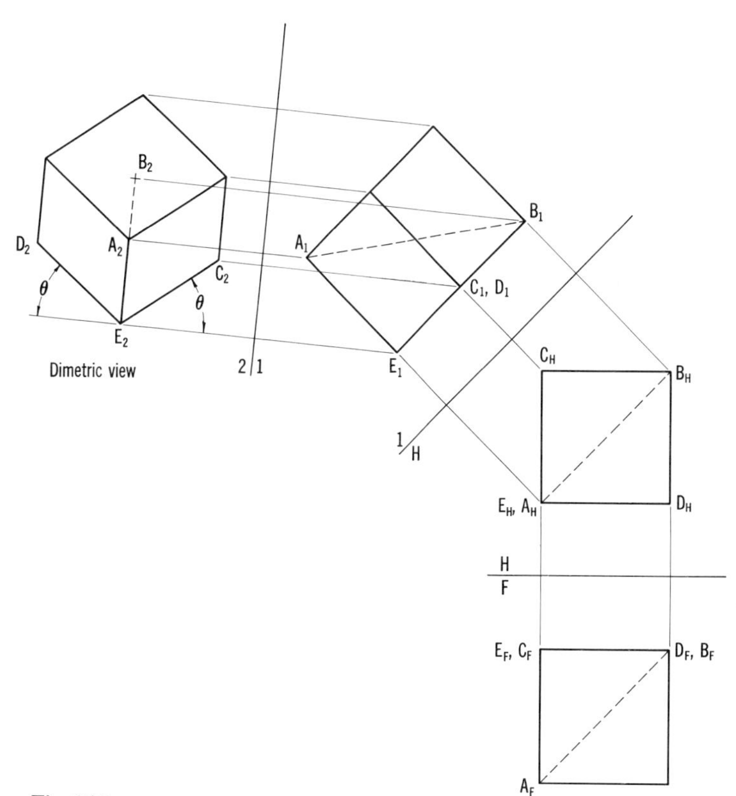

Fig. 17.8

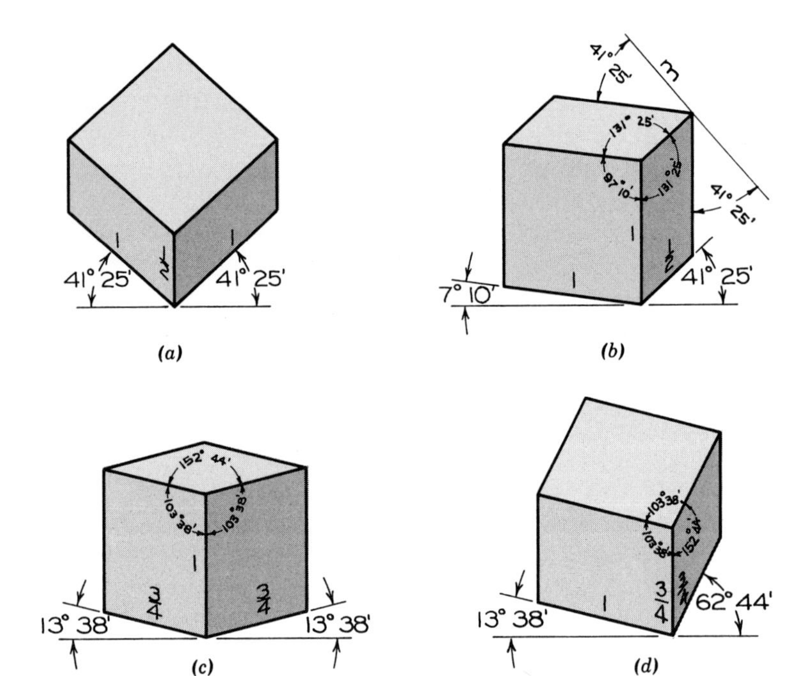

Fig. 17.9

Dimetric Drawing

A *dimetric drawing* can be made as easily as an isometric drawing. Once the axes and the scale ratios for the dimetric drawing have been established, the procedure is the same as that employed in making an isometric drawing. *Measurements parallel to the axes are true length, though measured to different scales.* Points may be located by the coordinate method. It should be pointed out that the angles between the oblique axes and a horizontal line may vary for different positions of the cube. For example, four different combinations of angles and scale ratios are shown in Fig. 17.9. The one shown in part (*b*) is used quite frequently. It should be observed that (*b*) is the same as (*a*) when line *m* is oriented in a horizontal position. Are (*c*) and (*d*) basically the same?

Suppose we were to make a dimetric drawing of a rectangular block 4 inches long, 2 inches high, and 3 inches deep. If we use the scale ratios $1:1:\frac{1}{2}$, as in part (*b*) of Fig. 17.9, the 4-inch and 2-inch dimensions would be laid off full size, whereas the 3-inch depth dimension would be laid off as $1\frac{1}{2}$ inches. If the drawing were made other than full size, the *scale ratios* would be maintained.

A dimetric *drawing* may produce a more pleasing effect than an isometric drawing. See Figs. 17.10 and 17.11 which show, respectively, an isometric and a dimetric drawing of the "Shaft Guide" represented by the orthographic views in Fig. 17.12.

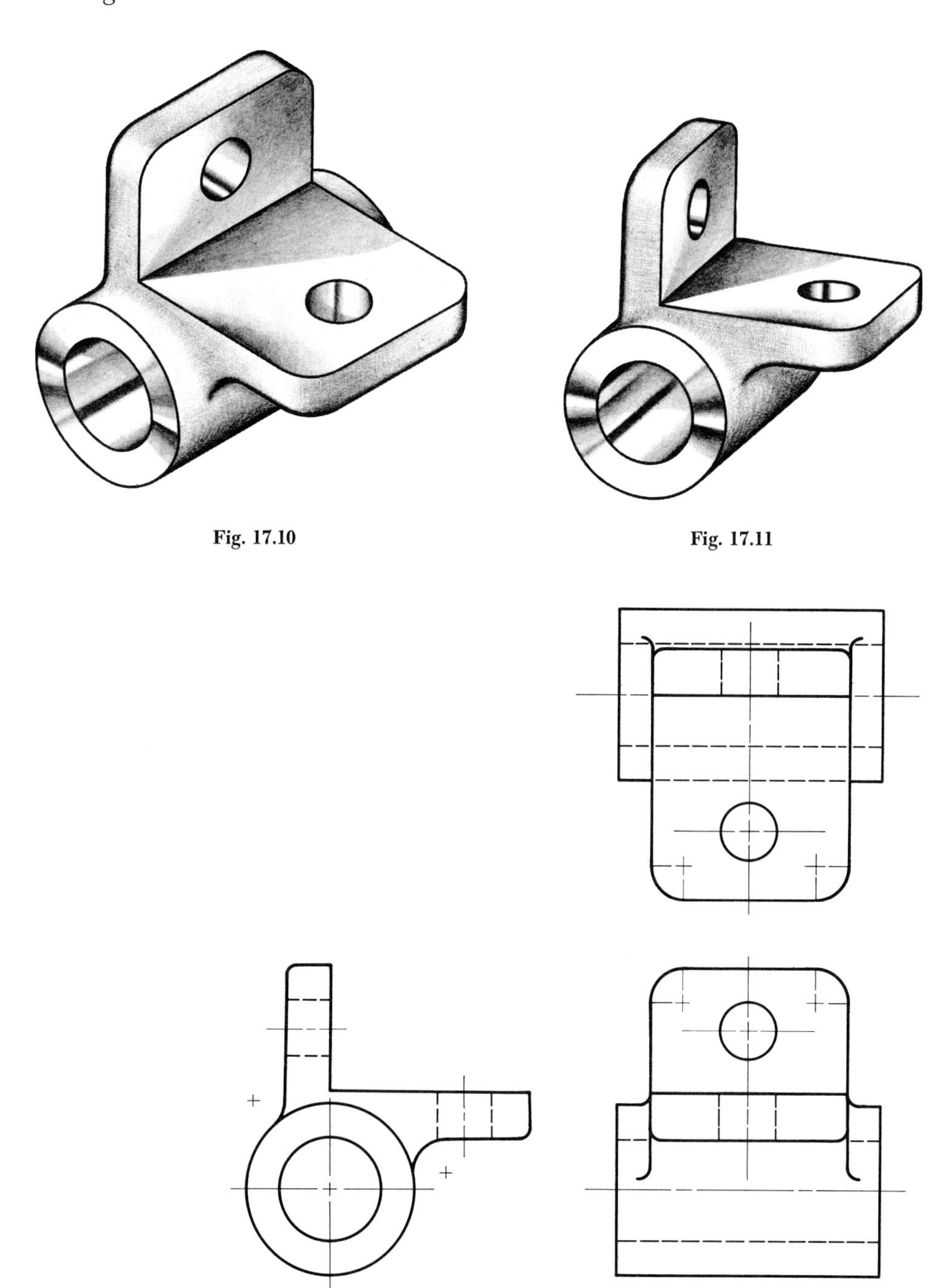

Fig. 17.10

Fig. 17.11

Fig. 17.12

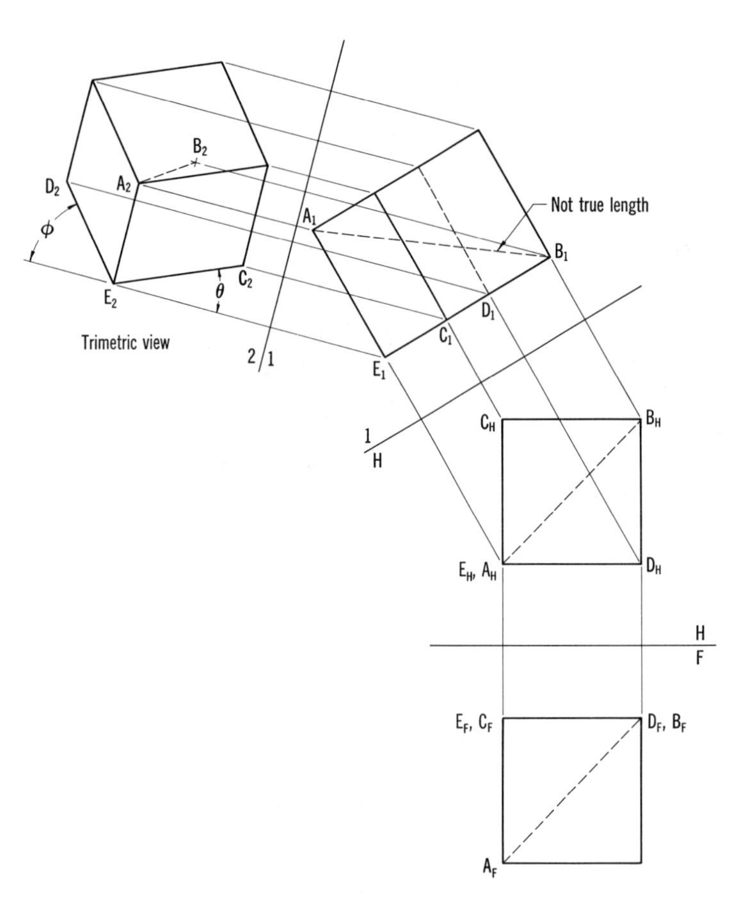

Fig. 17.13

Trimetric Views

When a cube is oriented to reveal an orthographic view that shows *unequal foreshortening of three concurrent edges,* the view is known as *a trimetric.*

For example, suppose we consider the cube shown by the top and front views in Fig. 17.13. Select body diagonal *AB*, or any other body diagonal, and obtain a view of the cube, *not* showing the true length of *AB*. Supplemental view 1 is such a view. Now we can introduce adjacent supplemental plane 2 in any desired position. The new view, 2, of the cube is a *trimetric view.* The concurrent edges *EA*, *EC*, and *ED* are foreshortened *unequally.* They are regarded as the *trimetric axes.* A variety of trimetric views, obviously, is possible by changing the position of supplementary planes 1 and 2.

Trimetric Drawing

A trimetric *drawing* can be made in the same manner as an isometric drawing, once the axes and the scale ratios have been established. As in the isometric and dimetric drawings, measurements parallel to the axes are true length, though *not to the same scale.* Points may be located by using the coordinate method. Again, note that the angles between the oblique axes and a horizontal line will vary for different positions of the cube. For example, four different combinations of angles and scale ratios are shown in Fig. 17.14. Again, note that (*a*) and (*c*) are basically the same. Is this true also of (*b*) and (*d*)?

Suppose we were to make a trimetric drawing of a rectangular block 3 inches long, 2 inches high, and $1\frac{1}{2}$ inches deep. If for the inclined axes we use angles and scale ratios shown in part (*a*) of Fig. 17.14, the 3-inch length would be laid off as 3 inches, the 2-inch dimension would be laid off as $1\frac{1}{3}$ inches, and the $1\frac{1}{2}$-inch dimension as $1\frac{5}{16}$ inches. If the drawing were made to a scale other than full size, the *scale ratios,* $1:\frac{2}{3}:\frac{7}{8}$, would be maintained.

The foregoing discussions on isometric, dimetric, and trimetric *drawing* have pointed out that we can prepare such drawings *directly* by using the proper combinations of *axes angles* and *scale ratios.* Observe further that these angles and ratios are based on the orientation of a cube* with respect to the plane which shows the isometric, dimetric, or trimetric *view;* and that the fundamental principles of orthogonal projection are employed in obtaining the *views.*

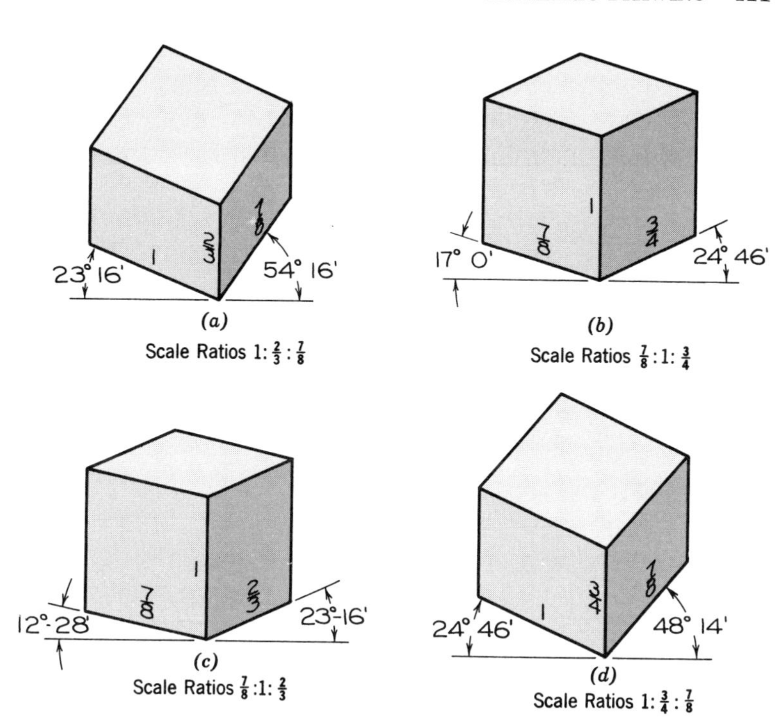

Fig. 17.14 Trimetric drawings of a cube.

* A cube is used because it enables us to deal with the "axes" —three concurrent edges of the cube—and to determine the scale ratios. A mathematical treatment is given in Appendix D, p. 738.

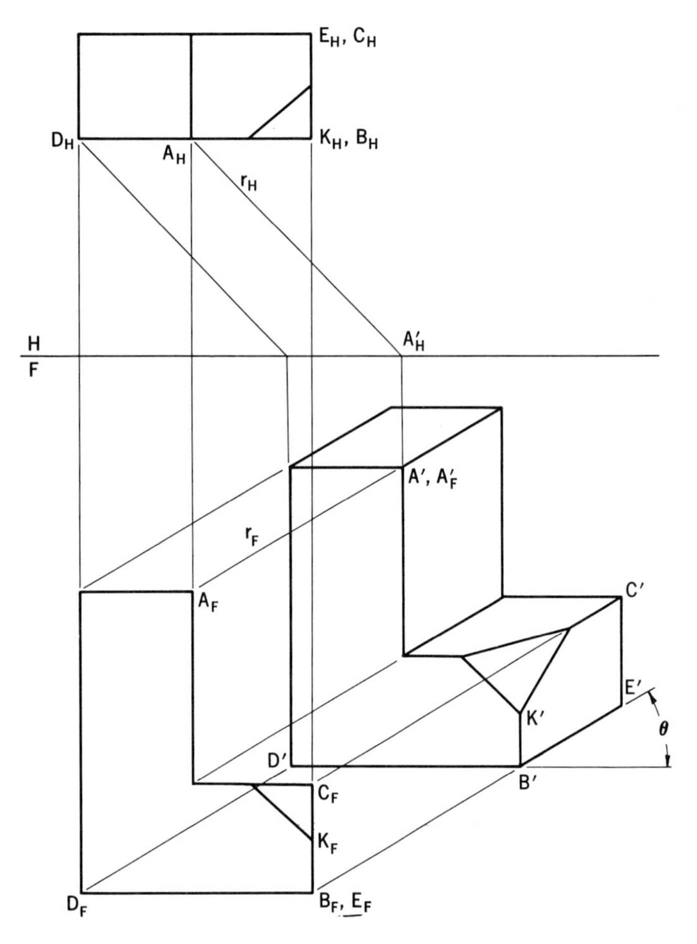

Fig. 17.15

Oblique Views

An L-block is shown in Fig. 17.15. Suppose we introduce ray r through point A and oblique to the frontal plane. Now, let us find the intersection, A', of ray r with the frontal plane. The top view of the point A' is easily determined because we have an edge view of the frontal plane; that is, A'_H will be the point in which r_H intersects the top view of the frontal plane. We also know that *if a point is on a line, the views of that point are on the corresponding views of the line;* hence, A'_F can be readily located. *Point A' is the oblique view of point A.* In a similar manner we may introduce additional *parallel* rays through other salient points of the object and locate their oblique views. The lines joining these points will determine the oblique view of the object.

It should be quite evident that the angle θ will vary with changes in the direction of the parallel rays. The three concurrent edges $B'K'$, $B'D'$, and $B'E'$ may be regarded as the *oblique axes.*

Oblique Drawing

An oblique drawing is easily constructed, once the three axes are established. It is common practice to set up two of the axes at right angles to each other, with one of them vertical, and to use an angle, θ, of 30° or 45° for the third axis. In addition, the ratio of scales used for the horizontal, vertical, and oblique axes is usually either 1:1:1, respectively, or $1:1:\frac{1}{2}$, respectively. The latter ratio tends to overcome the pictorial distortion due to the nonconvergence of the lines parallel to the oblique axis. Drawings made with scale ratios of 1:1:1 and with the oblique axis at 45° are known as "Cavalier" oblique drawings; those made with a scale ratio of $1:1:\frac{1}{2}$ and with oblique axis at 45° are known as "Cabinet" oblique drawings.

It is important to point out that an oblique drawing may have certain advantages over the isometric, dimetric, and trimetric drawings. For example, circular holes and irregularly shaped surfaces will appear true if these shapes lie in a plane parallel to the plane of the horizontal and vertical axes, whereas in the isometric, dimetric, and trimetric drawings ellipses would have to be drawn to represent the circular holes. It is good practice, therefore, to *place the object in a position which will orient irregularly shaped plane surfaces or contours in the plane of the horizontal and vertical axes* (i.e., frontal plane).

EXAMPLE 1

Consider the object shown in Fig. 17.16. The front face in the oblique drawing will be the same as the front view shown in the orthographic drawing. Length AB in the oblique drawing is equal to the length A_HB_H, the true length of edge AB. In a similar manner edge CD is determined. Since curves FD and EC are the same, no difficulty should be experienced in completing the oblique drawing. The center of the arc contained in FD is located on the oblique center line of the hole and at a distance CD from the front face. Figure 17.17 shows the same object drawn to the ratios $1:1:\frac{1}{2}$ with the oblique axis at 30°. In this case, edges AB and CD are one-half of the true lengths shown in the top view of the orthographic drawing.

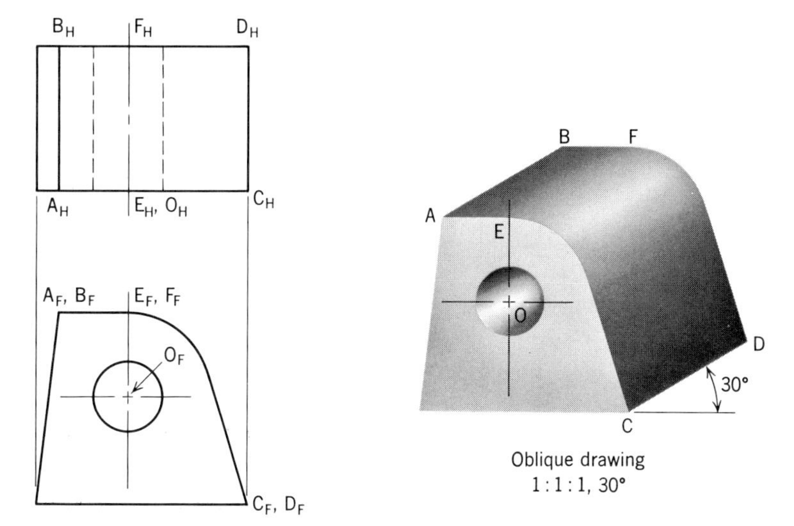

Fig. 17.16

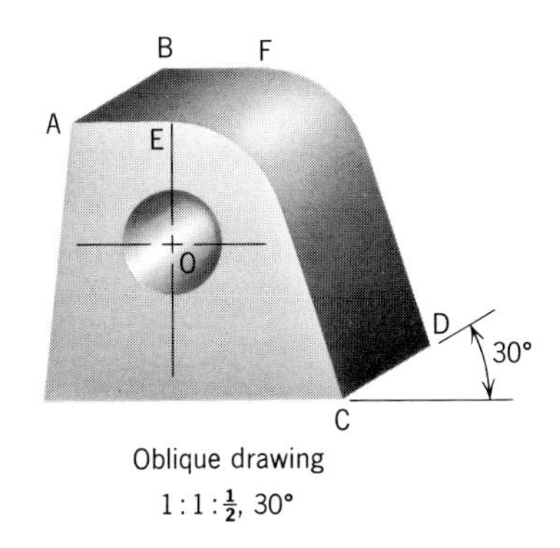

Fig. 17.17

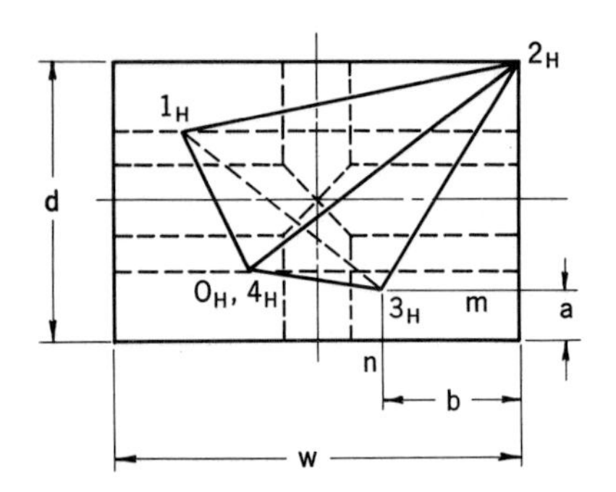

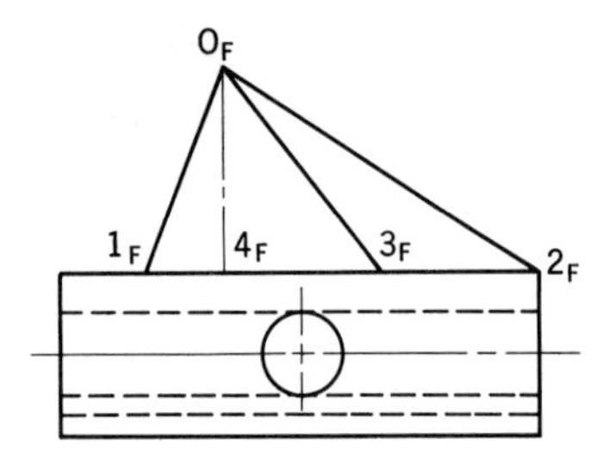

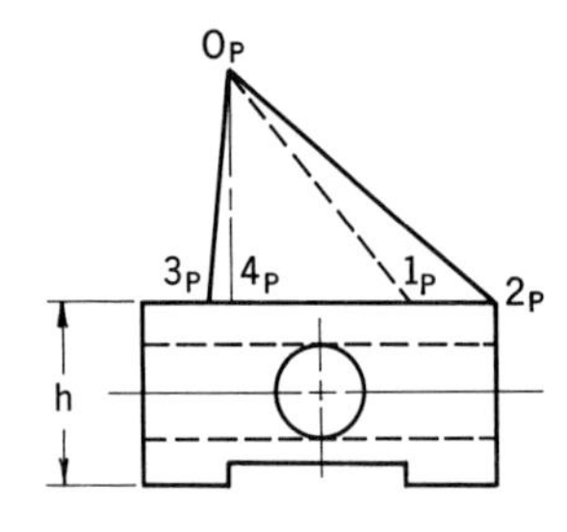

Fig. 17.18

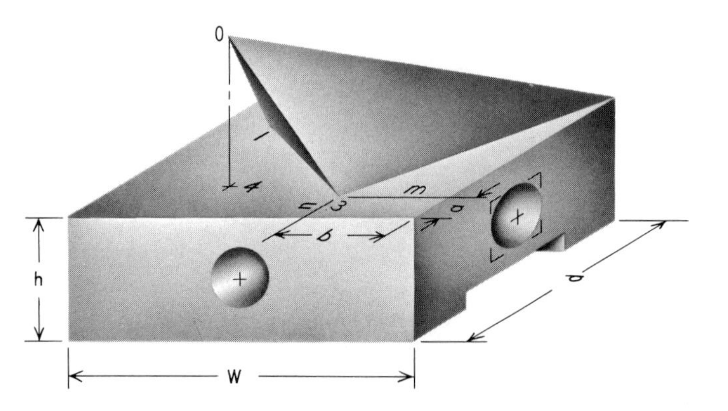

Fig. 17.19

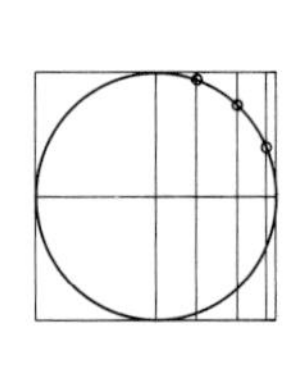

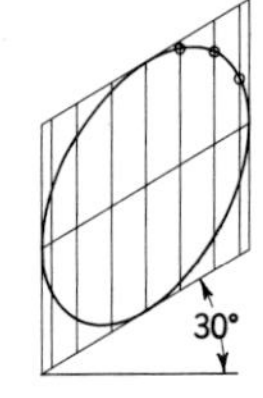

(a) Coordinate Method

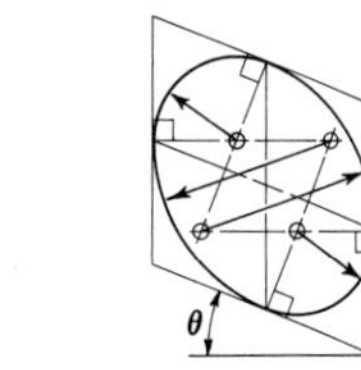

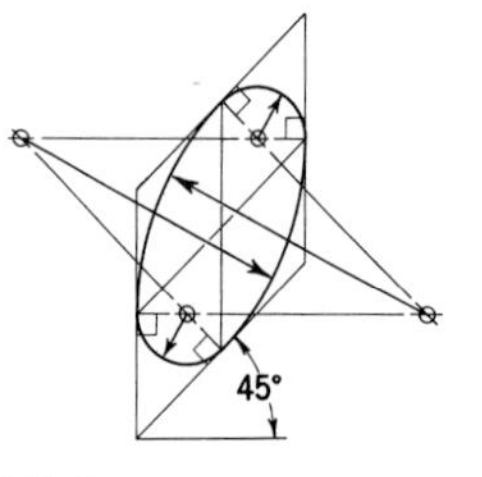

(b) Circle Arc Method
Centers of arcs are at intersection of perpendicular bisectors of the sides of the rhombus.

Fig. 17.20

EXAMPLE 2

In many cases some lines of a given object are not parallel to the axes. It will be necessary to locate points on these lines by the coordinate method. Let us consider the object shown in Fig. 17.18.

We shall have no difficulty in making the oblique drawing, Fig. 17.19, of the base member except for the representation of the hole. Methods for drawing oblique circles are shown in Figs. 17.20(*a*) and (*b*). These should be carefully studied before proceeding with the completion of the oblique drawing.

Now for the location of points 0, 1, 2, and 3, shown in Figs. 17.18 and 17.19, the coordinate method previously discussed in the section on isometric drawing can be employed. For example, point 3 in Fig. 17.19 is the intersection of lines *m* and *n* which are parallel to the horizontal and oblique axes, respectively. Point 0 is determined by first locating point 4, which is the foot of the perpendicular drawn from 0 to the top surface of the base. The vertical distance from point 4 to 0 is, of course, equal to 0_F–4_F, shown in Fig. 17.18. Edges 0–1, 0–2, 0–3, etc., are drawn to complete the oblique drawing of the pyramid. Edge 1–2 is not shown by a dashed line since it is customary to omit hidden edges in pictorial drawings.

The center of the circle can be located in a manner similar to that used in positioning points 1, 2, and 3. Two opposite sides of the rhombus which encloses the oblique circle will be vertical, and the other two will be parallel to the oblique axis. The length of the sides of the rhombus is equal to the diameter of the circle.

In general, it is best to place the long dimension of an object in the horizontal position so as to minimize the pictorial distortion. *Preference, however, should be given to placing the contour surfaces, arcs, and circles parallel to the frontal plane.* (See Fig. 17.21.)

Pictorial drawings, such as those shown in Fig. 17.22, clearly show the frequent use of the ellipse to represent circular shapes.

Fig. 17.21

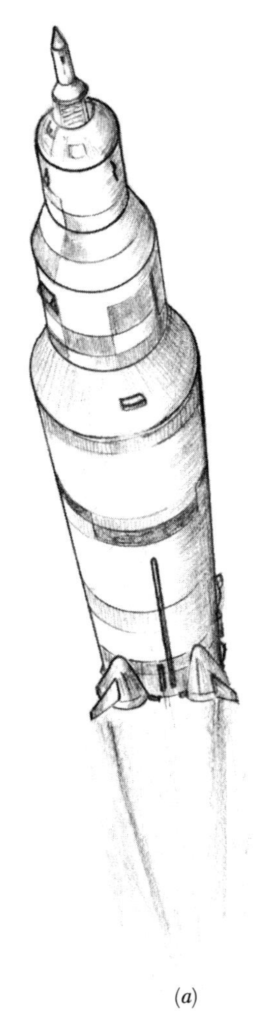

(*a*)

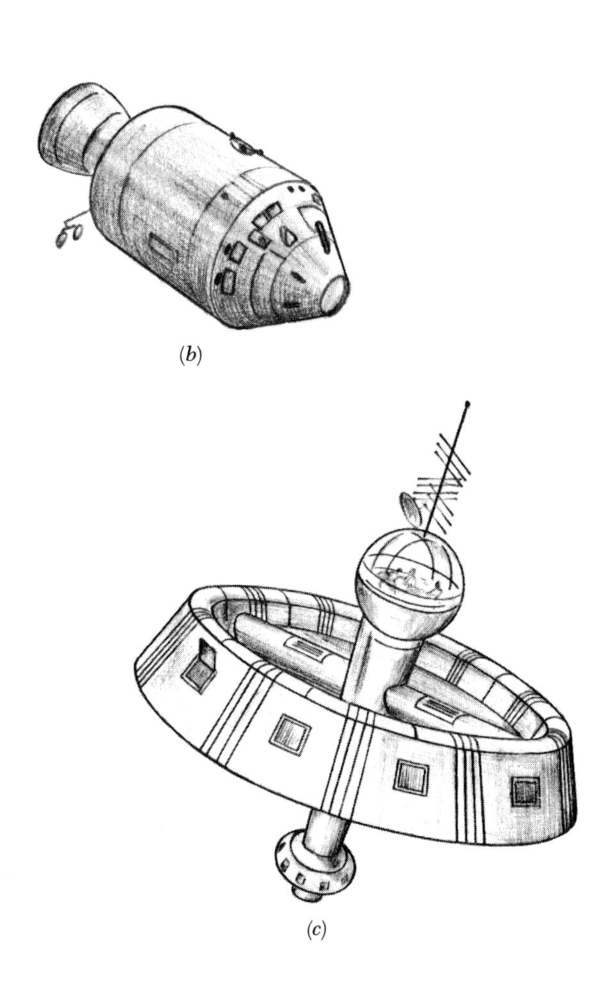

(*b*)

(*c*)

Fig. 17.22

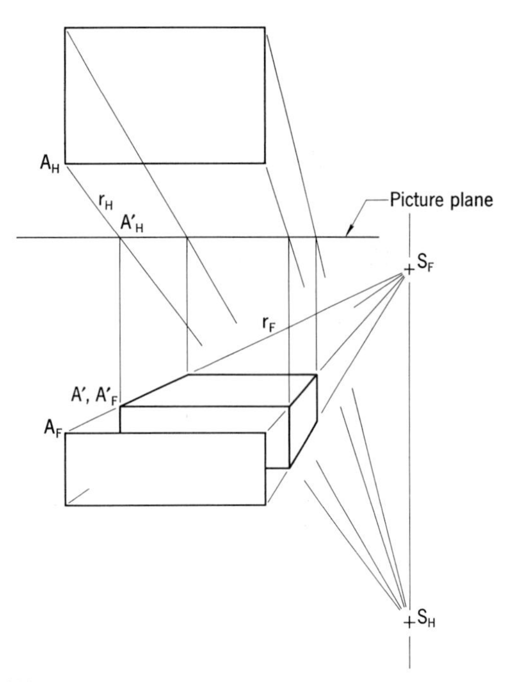

Fig. 17.23

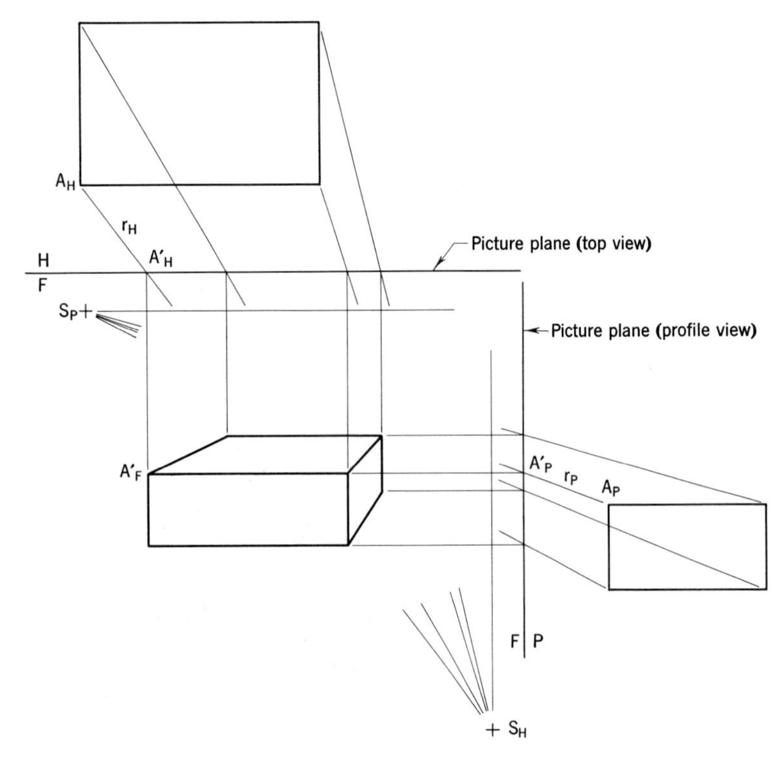

Fig. 17.24

Perspective Views—Ray Method

Figure 17.23 shows the top and front views of a rectangular block. Suppose we assume *point S as a station point,* or point of sight, through which rays are drawn to the corners of the block. If we can find the points in which these rays intersect the frontal plane (commonly described as the "picture plane"), we will obtain the perspective views of these points. The lines which join the points will form the perspective view of the block. How shall we proceed to locate the intersection of the rays with the picture plane? This is not a new problem since we previously solved a similar one in dealing with oblique views.

Let us consider ray r, which joins point A of the block with the station point S. Since we have an edge view of the picture plane, the top view of the point of intersection of ray r with the picture plane is readily determined. It is shown as A'_H. The front view is, of course, on r_F and is shown as A'_F, which is the perspective view of point A. In a similar manner, we can establish the perspective views of the other corners of the block. The lines which join these points form the perspective view of the block.

In order to avoid overlapping of a portion of the perspective view with the front view, we can make use of the top and *profile* views of the block. For example, in Fig. 17.24, the perspective view of point A is determined by locating the point of intersection of ray r with the frontal plane, using the top and profile views of ray r. The top view, A'_H, of the point of intersection is located in the same manner, previously discussed. The profile view, A'_P, is located at the intersection of r_P with the profile view of the frontal plane (picture plane). Since the top and profile views are now available it is a very simple matter to locate the front view, A'_F, because we know that "adjacent views of a point lie on a line perpendicular to the line which represents the edge views of the adjacent planes." In a similar manner, additional corner points are located. These are joined to form the perspective view.

Perspective Views—Line Method

Let us consider the rectangular block shown in Fig. 17.25. The top view is oriented so that a vertical side of the block makes an angle θ with the frontal plane. This angle is quite commonly taken as 30°. It could be any desired angle. The view to the right establishes the vertical distances between points in the upper horizontal surface of the block and those in the lower surface. It may be regarded as a profile view taken when the angle $\theta = 0°$. In addition, its position relative to the station point, S, shows that the observer is above the upper horizontal surface.

If we carefully study the top view, we observe that edge AB is in the picture plane. The perspective view of edge AB, therefore, is the same as the front view of AB.

Now let us determine the location of the perspective view of point C by first locating the perspective views of edges AC and CD. Their intersection will uniquely locate point C. Assume that edge AC is extended from the picture plane to infinity. The perspective view of the point at infinity could be found by the ray method. If ray r were drawn through station point S, and parallel to edge AC, it would connect point S and the point at infinity. *Since parallel lines have parallel projections,* r_H would pass through S_H parallel to A_HC_H, and r_F

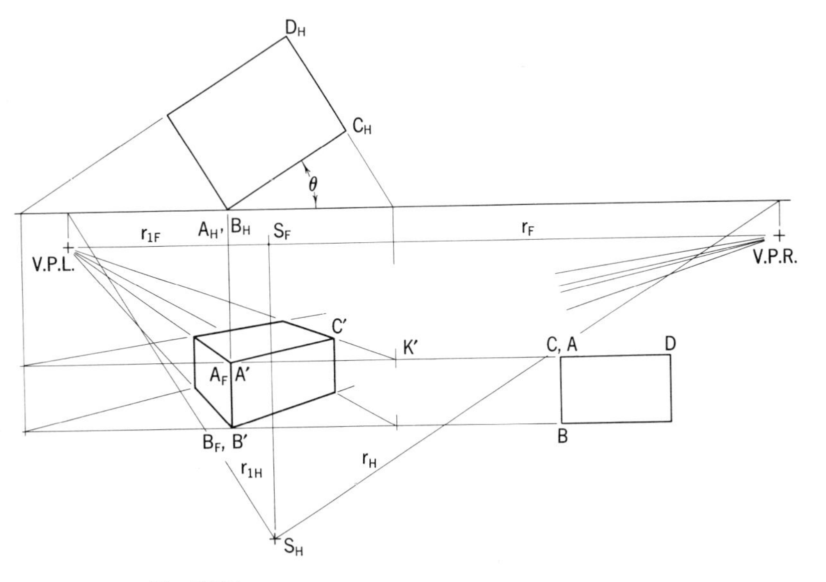

Fig. 17.25

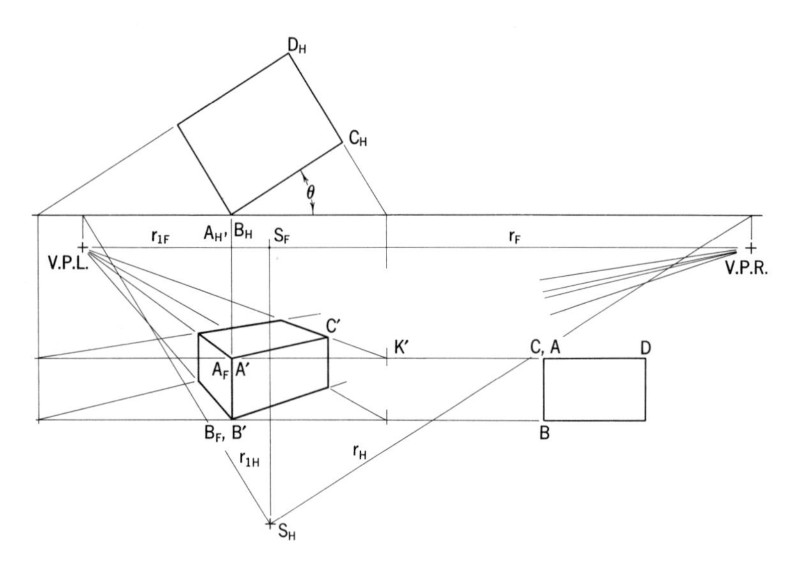

Fig. 17.25 (*Repeated*)

would pass through S_F in a horizontal direction because edge AC is a horizontal line. Ray r intersects the picture plane at the point marked V.P.R. This point is known as the vanishing point. The R simply indicates the vanishing point to the right. *All lines parallel to edge AC will have this same vanishing point.* The reader should satisfy himself that this statement is true. Now the perspective view of edge AC extended to infinity can be drawn. It joins the points A' and V.P.R.

In a similar manner we can locate the perspective view of edge CD extended from the picture plane to infinity. Edge DC, extended to the picture plane, intersects it at point K'. Again, the perspective view of the point at infinity on CD extended can be found by the ray method. Ray r_1, drawn through station point S parallel to edge CD, is the ray which connects point S and the point at infinity. The top view r_{1H} of ray r_1 is parallel to C_HD_H, and the front view r_{1F} passes through S_F in a horizontal direction, since line CD is horizontal. Ray r_1 intersects the picture plane at the point marked V.P.L., which is the vanishing point for all lines parallel to edge CD. Now the perspective view of the edge DC extended to the picture plane and to infinity can be drawn. It is the line joining points K' and V.P.L. The intersection of the perspective views of *lines* AC (infinitely long) and CD (infinitely long) is the perspective view of point C, shown as C'. The advantages of the line method over the ray method will become quite evident after a few problems have been solved. *Again, we must point out that, basically, the method used for determining an oblique view or a perspective view is nothing more than locating the point in which a line intersects a plane.* Printed "perspective-grids" greatly facilitate the preparation of both freehand and mechanical perspective drawings.

Engineers and scientists should thoroughly understand the fundamental principles used in making both pictorial views and pictorial drawings and be capable of making intelligible pictorial sketches. This ability will greatly enhance their power to express themselves graphically. Certainly, it will help them clear the lines of communication between themselves and their co-workers—scientists, engineers, technicians, and production personnel.

EXERCISES

1. Make isometric, dimetric, and oblique drawings of each of the parts shown in Fig. E-17.1. Choose the direction of axes that will best show the parts.

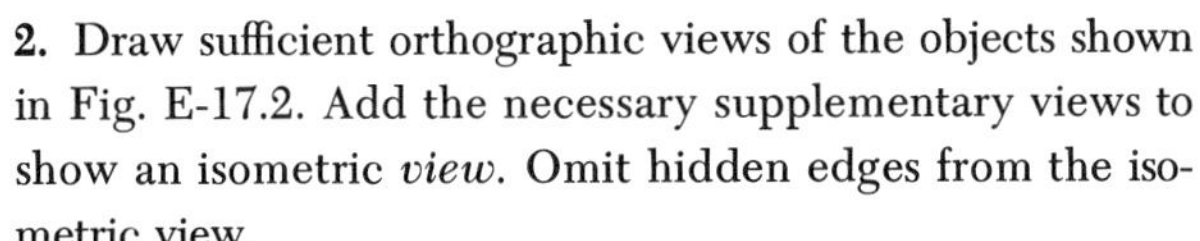

2. Draw sufficient orthographic views of the objects shown in Fig. E-17.2. Add the necessary supplementary views to show an isometric *view*. Omit hidden edges from the isometric view.

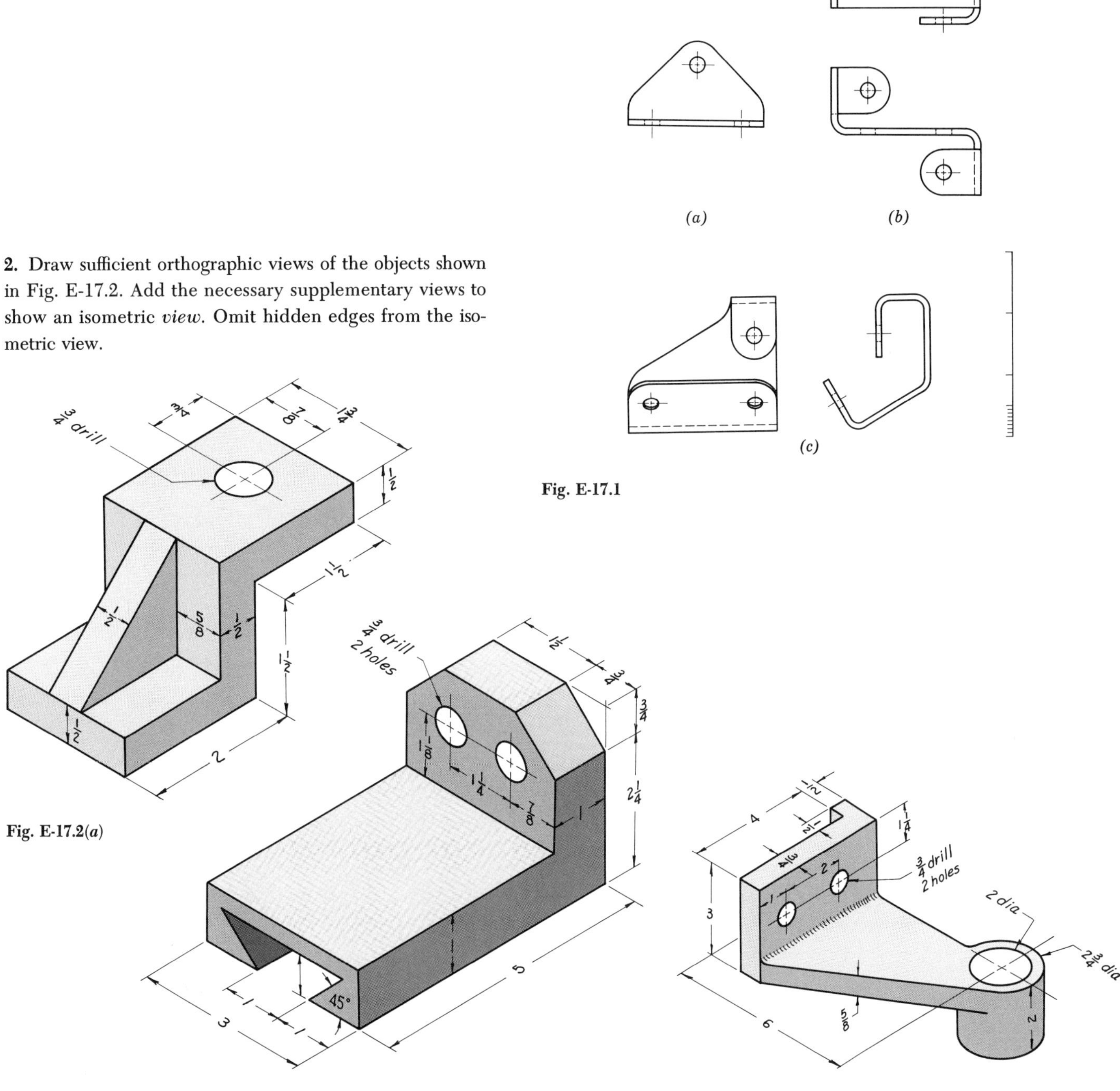

Fig. E-17.1

Fig. E-17.2(*a*)

Fig. E-17.2(*b*)

Fig. E-17.2(*c*)

3. Make a *freehand* isometric *drawing* of each of the objects shown in Fig. E-17.3.

4. Make a *freehand* dimetric *drawing* of each object shown in Fig. E-17.3. Use suitable scale ratios.

5. Make a *freehand* trimetric *drawing* of each object shown in Fig. E-17.3. Use suitable scale ratios.

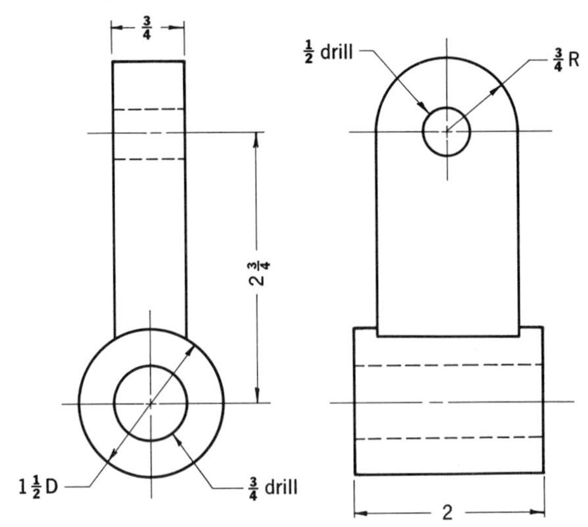

Fig. E-17.3(*a*)

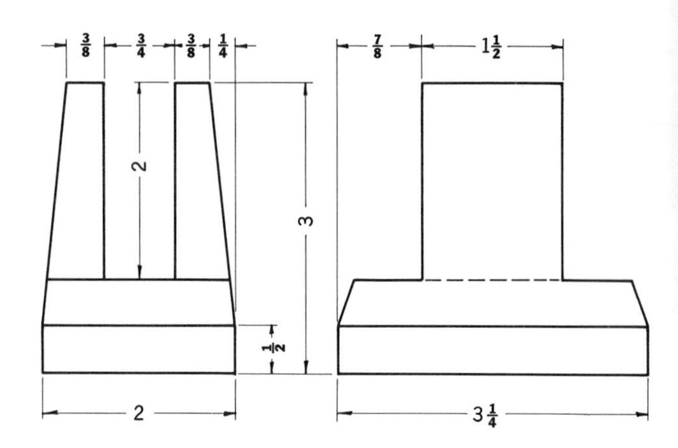

Fig. E-17.3(*b*)

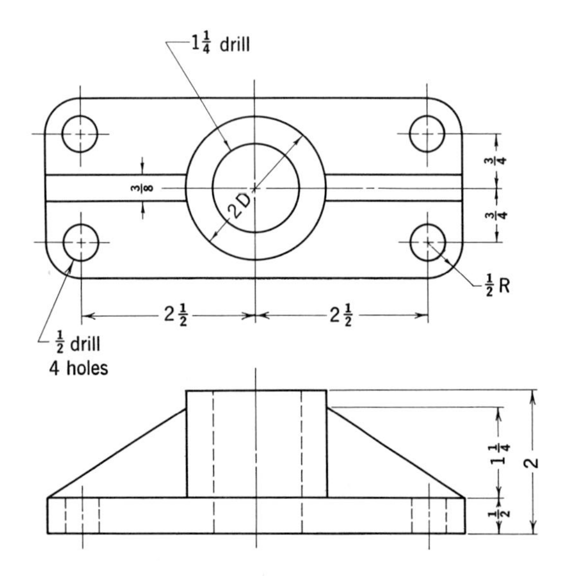

Fig. E-17.3(*c*)

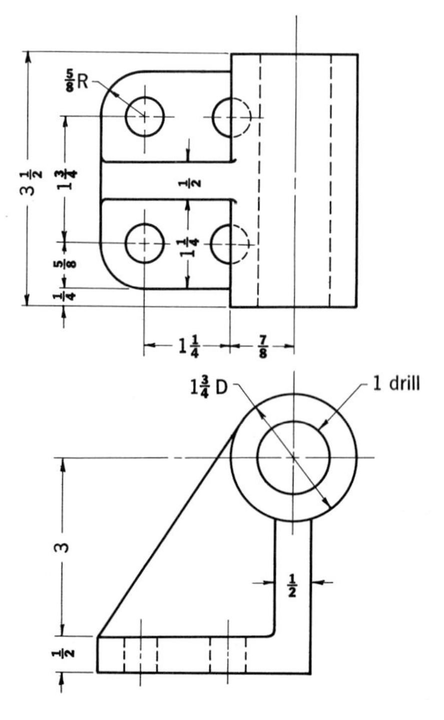

Fig. E-17.3(*d*)

6. Draw sufficient orthographic views of the objects shown in Fig. E-17.2. Add an oblique *view*. Select appropriate scale and angle for the oblique view. Omit hidden edges from the oblique view.

7. Make a *freehand* oblique *drawing* of each object shown in Fig. E-17.4. Select suitable angle and scale ratios.

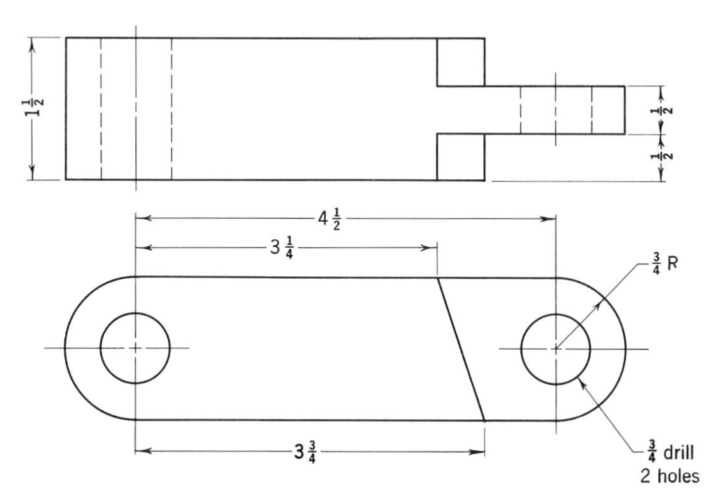

Fig. E-17.4(*a*)

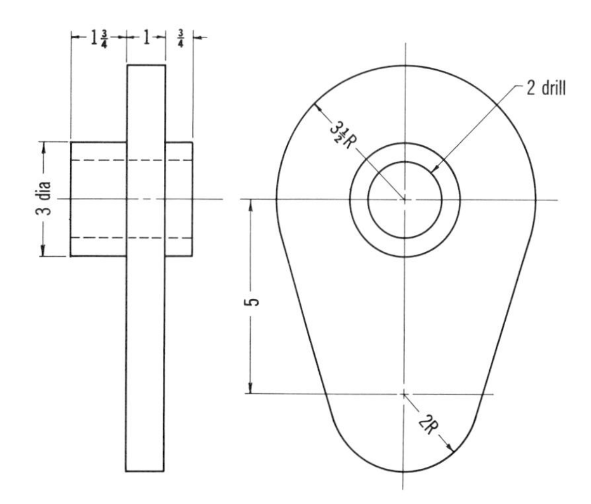

Fig. E-17.4(*b*)

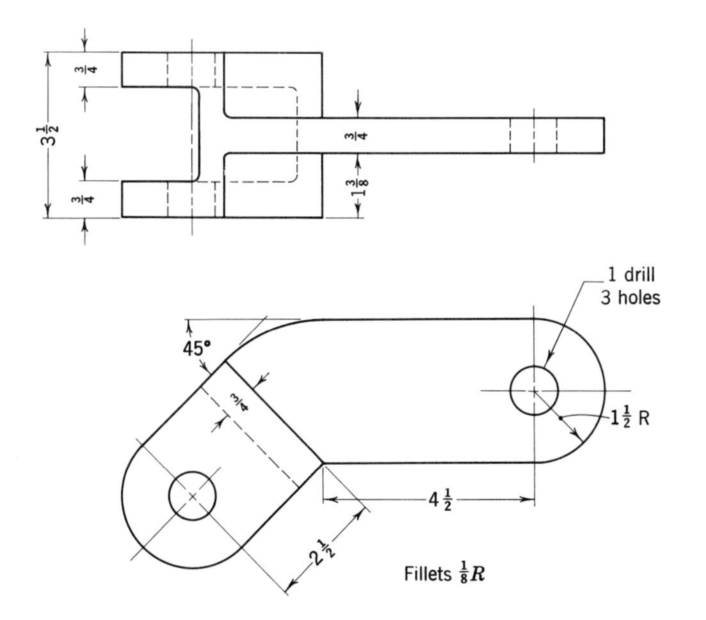

Fig. E-17.4(*c*)

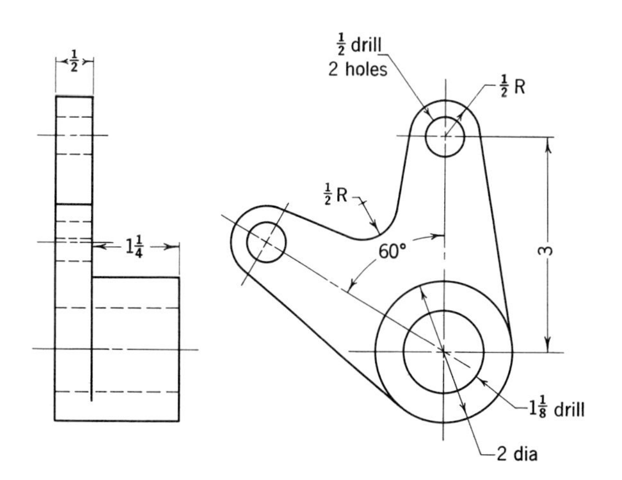

Fig. E-17.4(*d*)

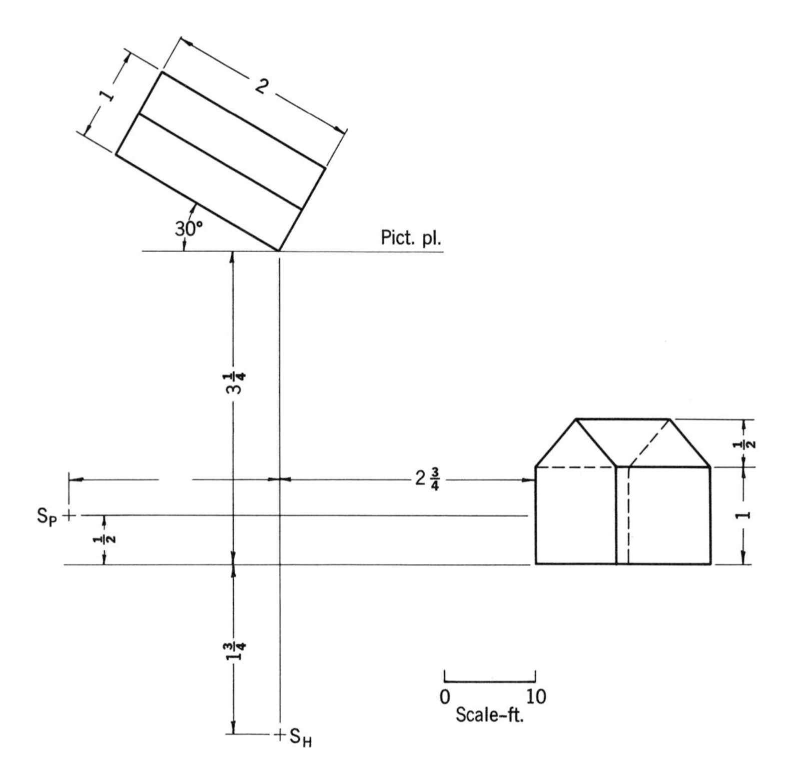

Fig. E-17.5(*a*)

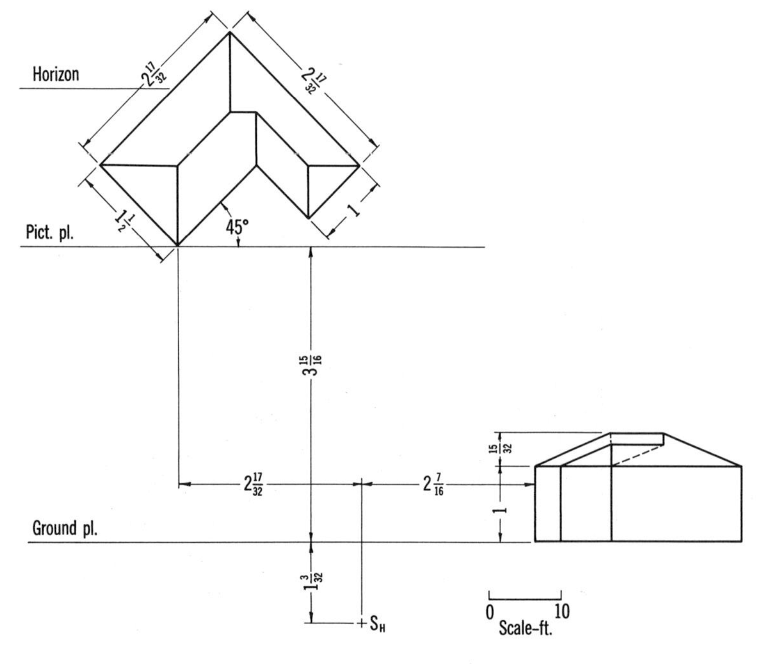

Fig. E-17.5(*b*)

8. Add a perspective *view* of the objects shown in Fig. E-17.5. Use an 8½″ × 11″ sheet for each object.

9. Add a perspective *view* of the shaft support shown in Fig. E-17.6(*a*). Use an $8\frac{1}{2}'' \times 11''$ or a $11'' \times 17''$ sheet.

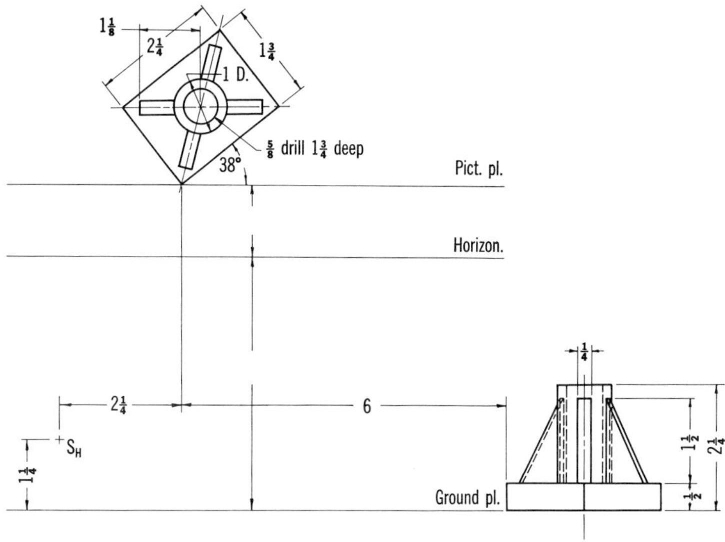

Fig. E-17.6(*a*)

10. Add a perspective *view* of the structure shown in Fig. E-17.6(*b*). Use an $8\frac{1}{2}'' \times 11''$ or a $11'' \times 17''$ sheet.

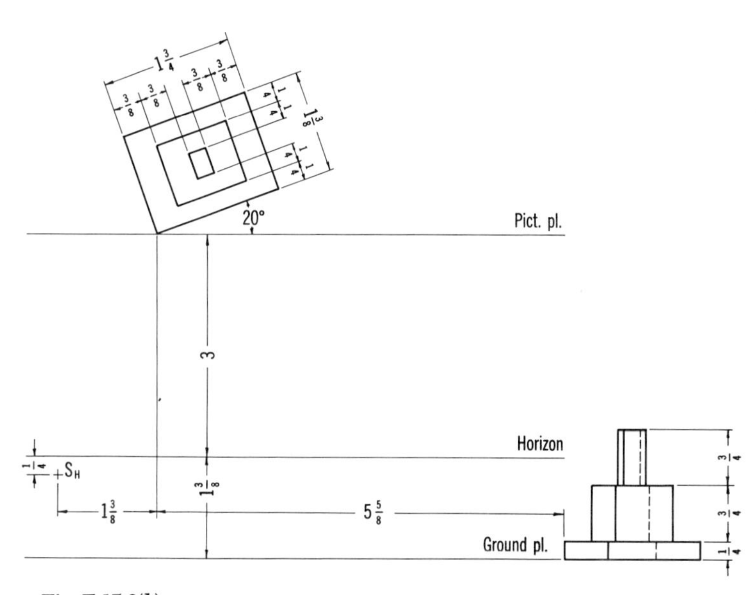

Fig. E-17.6(*b*)

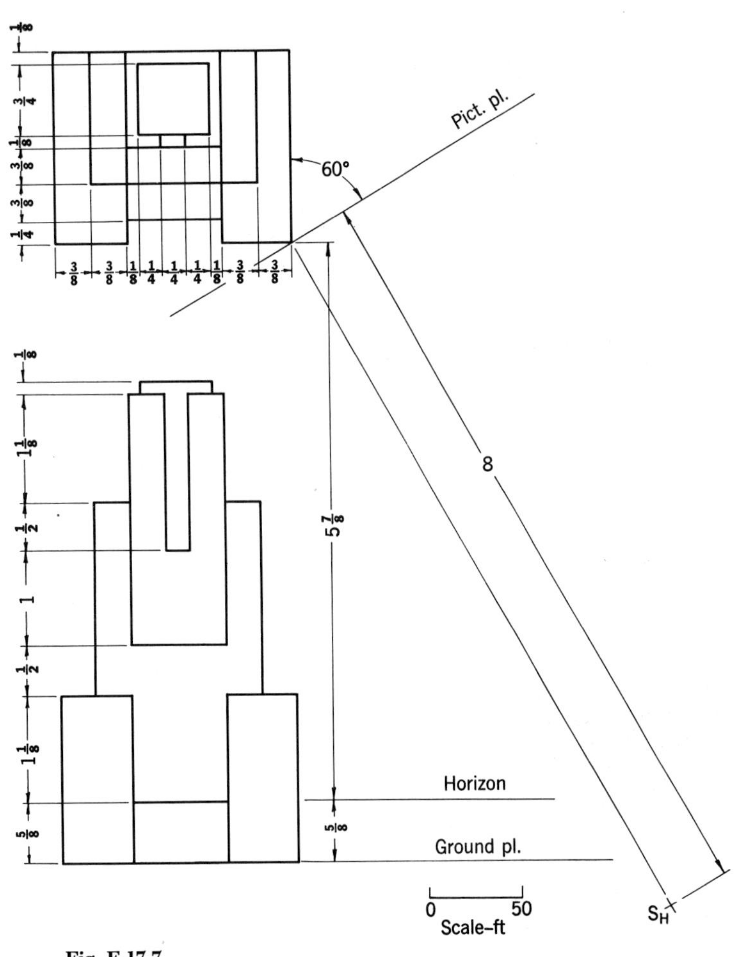

Fig. E-17.7

11. Make a perspective view of the "office building" shown in Fig. E-17.7.

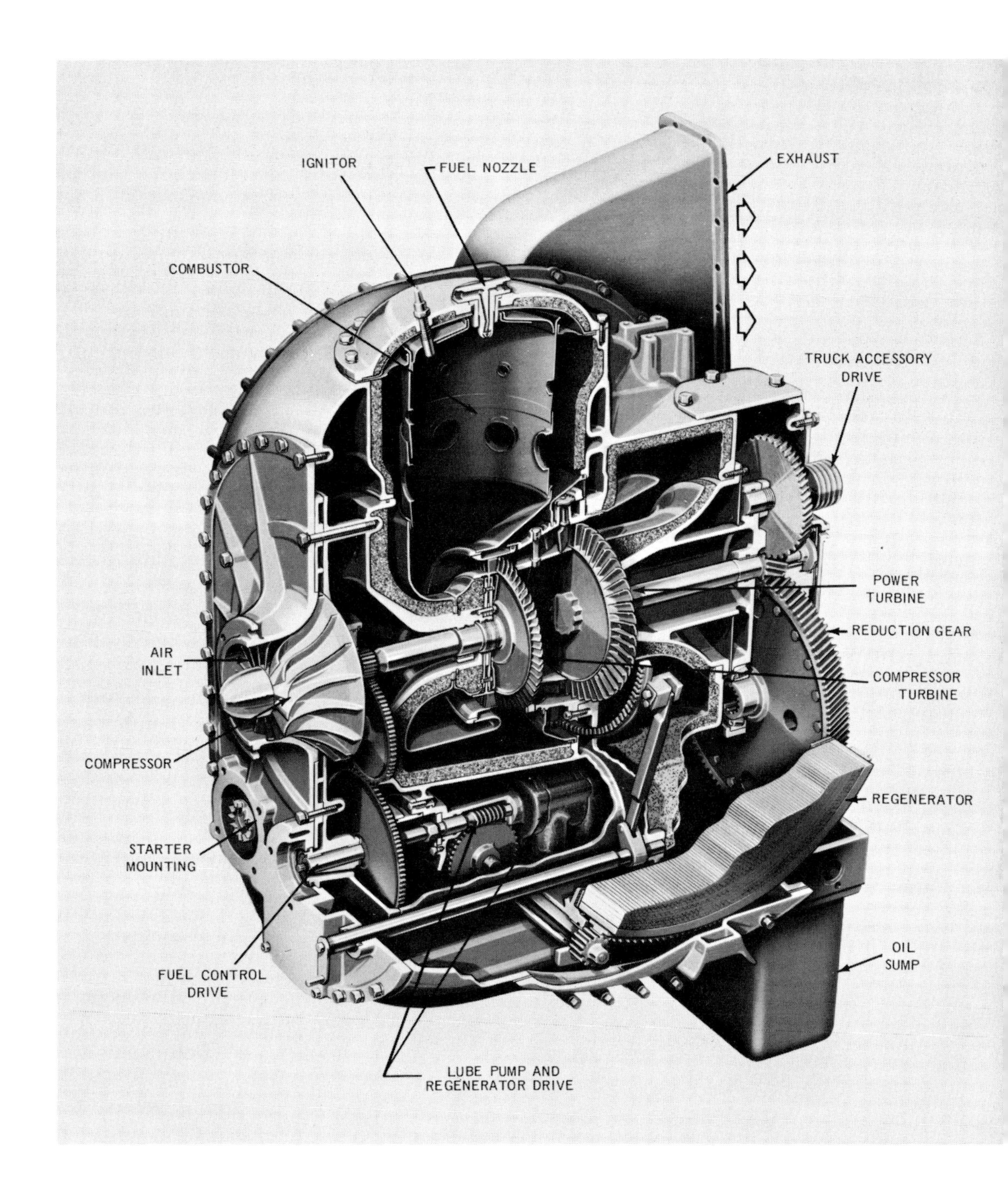

SECTIONS AND CONVENTIONAL PRACTICES 18

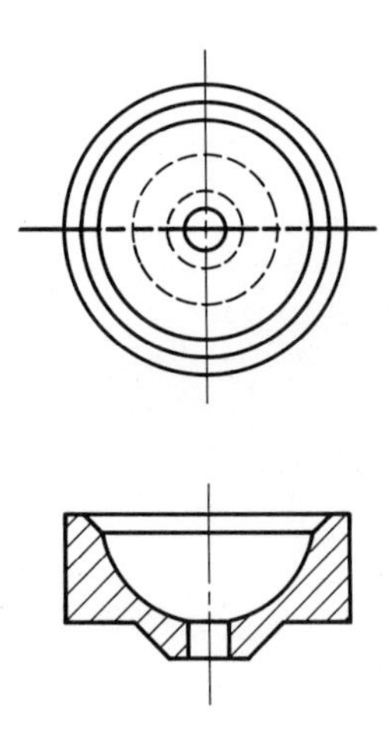

Fig. 18.1 Full section, orthographic.

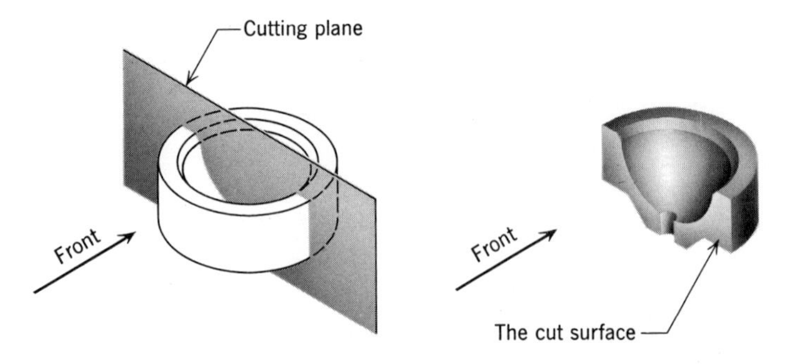

Fig. 18.2 Full section, pictorial.

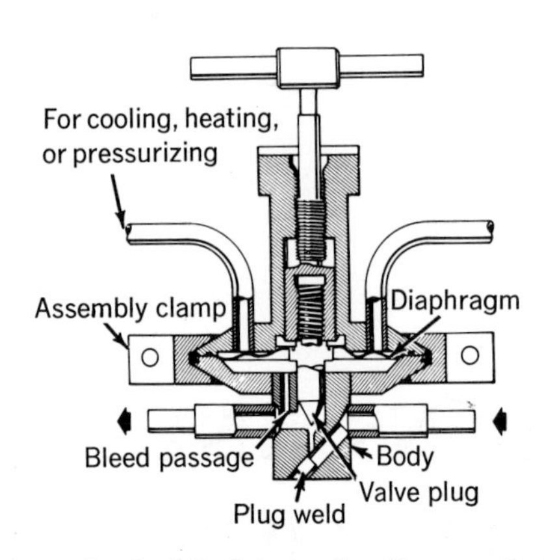

Fig. 18.3 Full section sketch of diaphragm valve. (Courtesy *Product Engineering*)

SECTIONS

Quite often the orthographic views of a device, mechanism, structure, etc., do not clearly reveal the internal parts or elements. In order to simplify the task of interpreting the intent of the engineer's design we use "sectional views." In some cases pictorial drawings require sectional views to clarify the "reading" of the design.

"A section is drawn to show how the object would appear if an imaginary cutting plane were passed through the object perpendicular to the direction of sight and the portion of the object between the observer and the cutting plane were removed or broken away. The exposed cut surface of the material is indicated by section lining or *cross-hatching*."*

It is customary to omit hidden lines behind the imaginary cutting plane. Although specific symbols† for section lining may be used for various materials, it is recommended that the general-purpose symbol (same as that used for cast iron) be used, because material specifications are necessarily more detailed than the identification by name and section-lining symbol. The use of the general-purpose symbol on the design drawing is economical and, moreover, it obviates the necessity of redrawing portions of the design when the material is changed. The use of the general-purpose symbol, which consists of equally spaced, fine, full lines, is evident in the illustrations which follow.

Types of Sections

Various types of sections are used to simplify the interpretation of the engineer's design. The following are typical:

(a) **Full Sections.** The cutting plane extends through the entire object, as in Fig. 18.1. A full section in pictorial is shown in Fig. 18.2. Commercial applications are shown in Figs. 18.3 and 18.4(*b*).

* From ASA Y14.2-1957. Section 2.

† ASA Y14.2-1957. Section 2, p. 9.

Fig. 18.4(*a*) Photograph of moment of inertia device. (Courtesy *Product Engineering*)

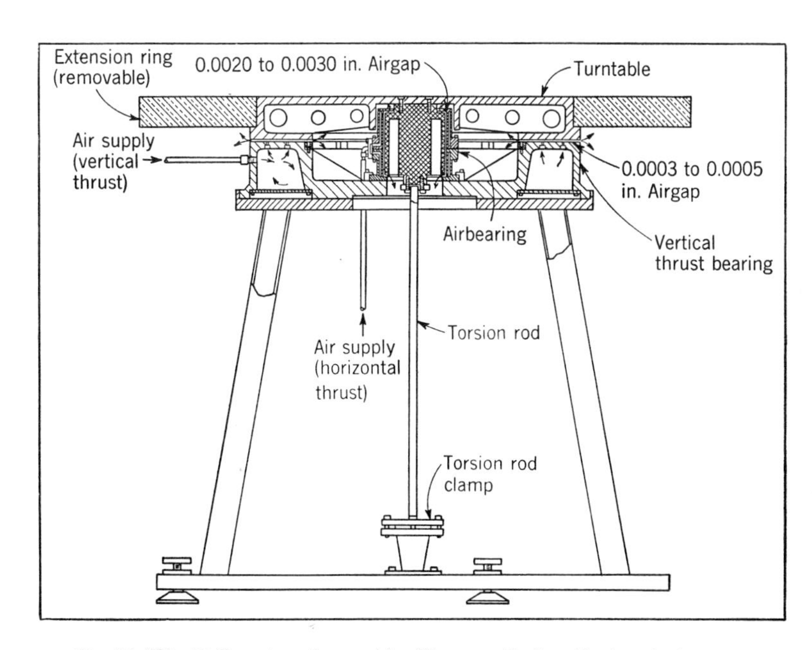

Fig. 18.4(*b*) Full section of turntable. (Courtesy *Product Engineering*)

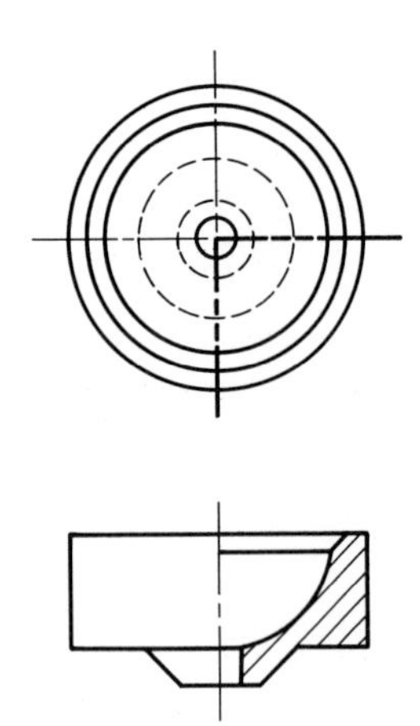

Fig. 18.5 Half section.

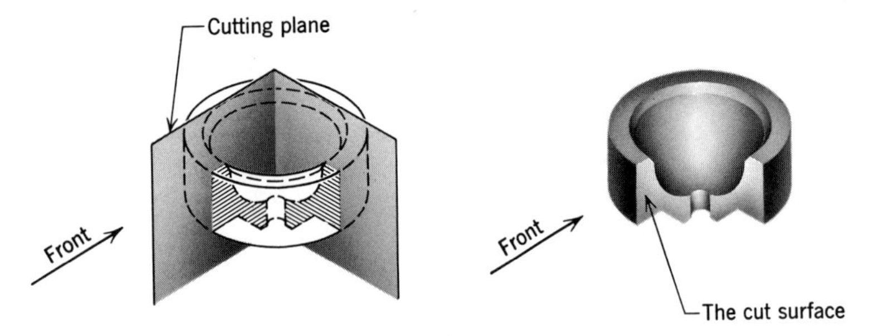

Fig. 18.6 Half section in pictorial.

(b) Half Sections. The cutting plane extends halfway through the object, as in Fig. 18.5. A half-section in pictorial is shown in Fig. 18.6. The section lines are drawn in such directions so that they would appear to coincide, if the planes were folded together.

(c) **Broken-out Sections.** A portion of the object is cut away to expose an interior detail, as in Figs. 18.7 and 18.8.

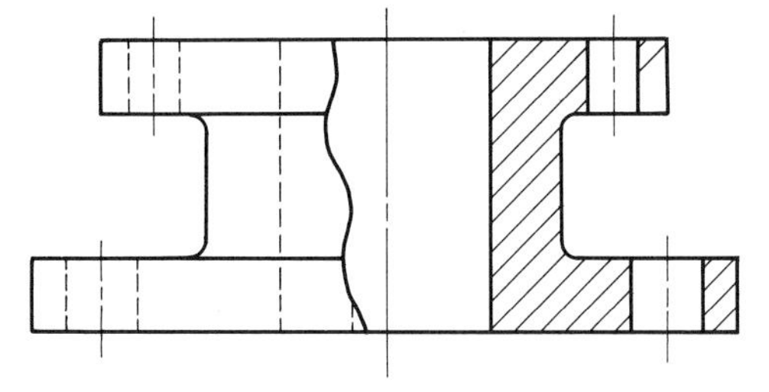

Fig. 18.7 Broken-out section.

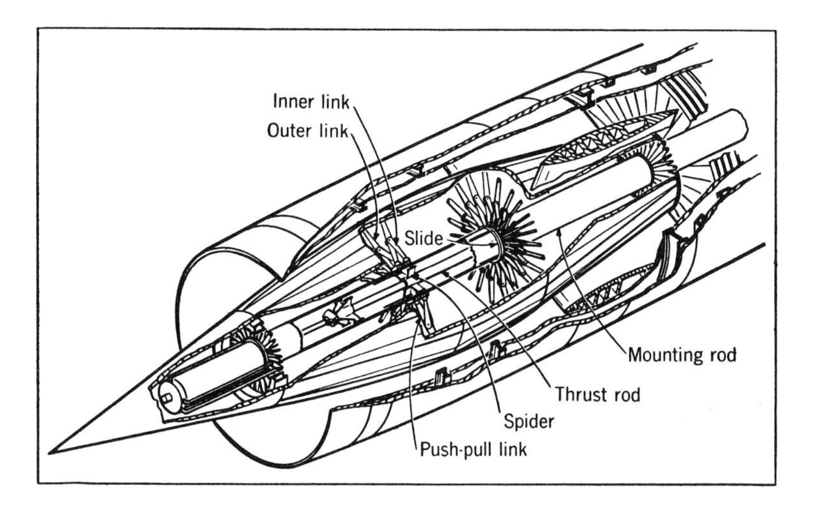

Fig. 18.8 Broken-out section in pictorial. Schematic of SST variable diameter air inlet for engine. (Courtesy *Product Engineering*)

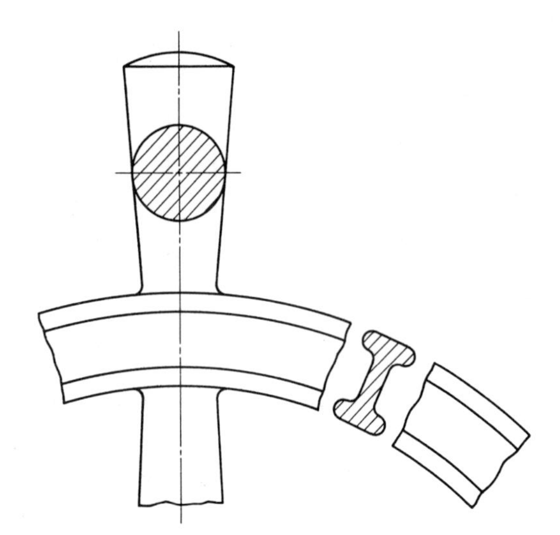

Fig. 18.9 Revolved section.

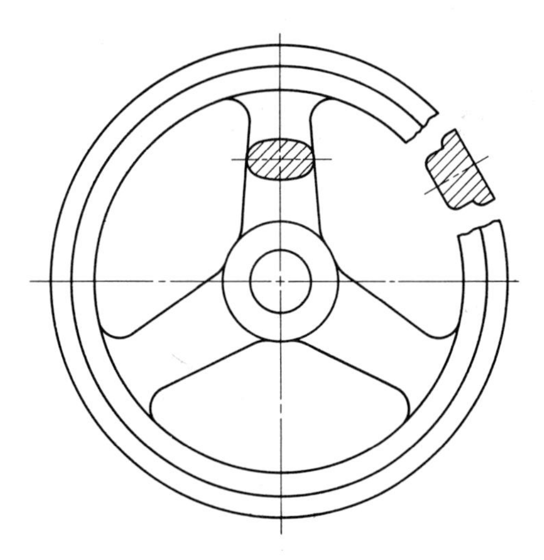

Fig. 18.10 Revolved section.

(d) **Revolved Sections.** The cutting plane is perpendicular to the axis of the member, and the section thus formed is revolved 90° about a line which is perpendicular to the axis. (See Figs. 18.9 and 18.10.)

(e) **Removed Sections.** Usually the same as revolved sections except that they are placed outside of the view. (See Figs. 18.11 and 18.12.)

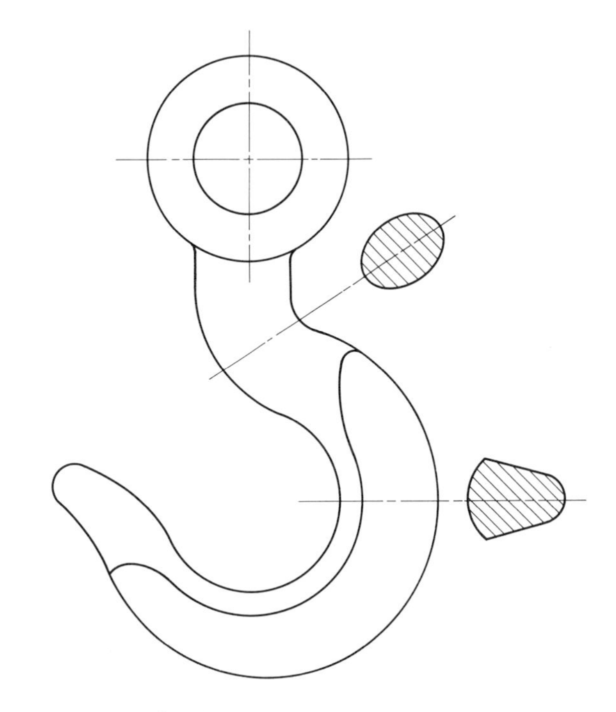

Fig. 18.11 Removed sections.

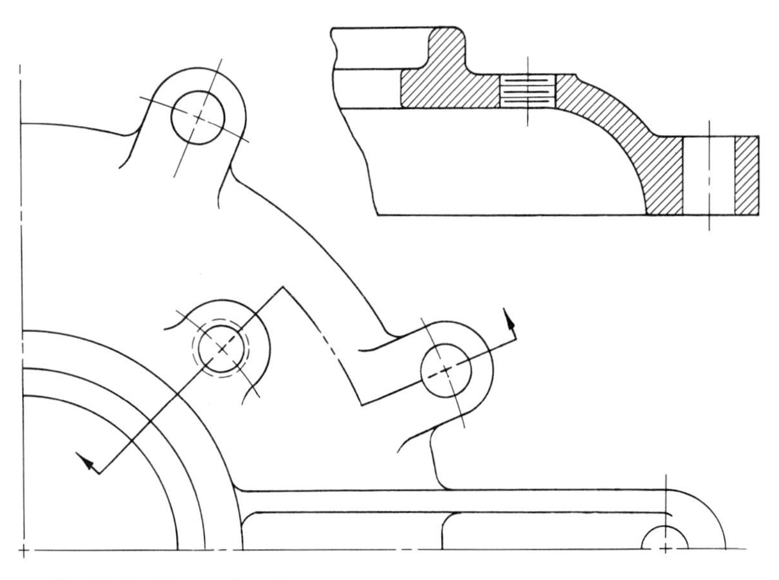

Fig. 18.12 Removed section.

If two or more removed sections are on the same sheet, they should, if possible, be arranged in a consistent sequence as shown in Fig. 18.13.

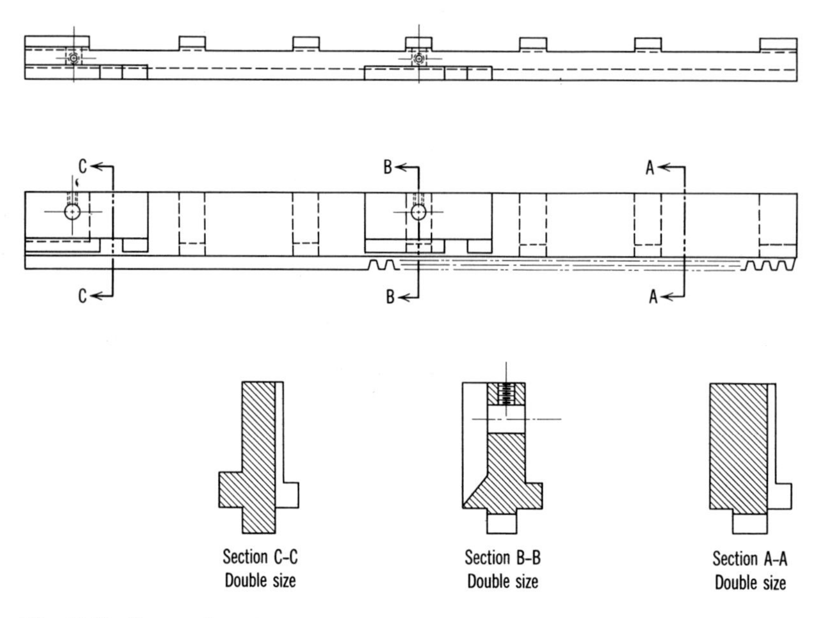

Fig. 18.13 Removed sections.

(f) **Offset Sections.** The imaginary cutting-plane may be stepped, or offset, to include elements that are not in a line. The section, however, is drawn as though the elements were in one plane. (See Fig. 18.14.) Pictorials are shown in Figs. 18.14(*a*) and (*b*).

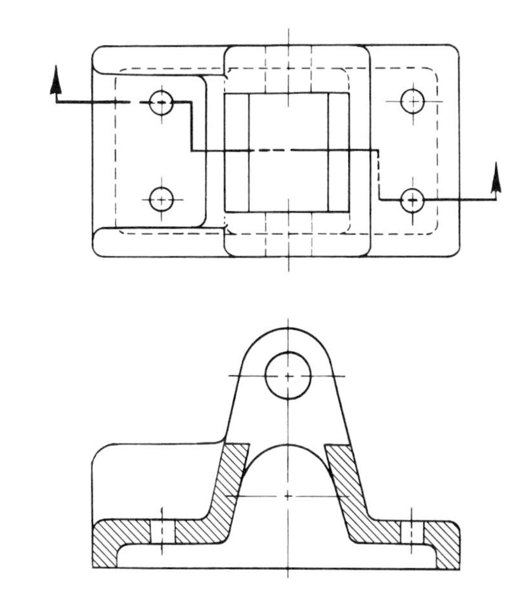

Fig. 18.14 Offset section.

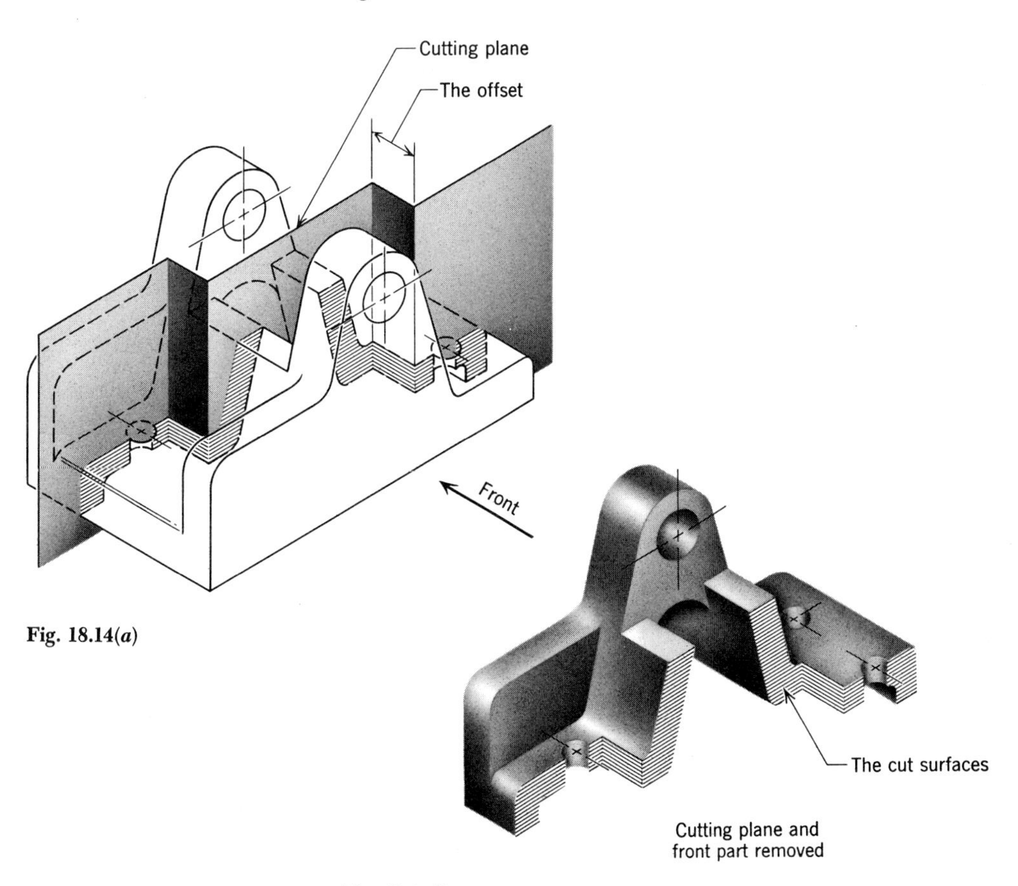

Fig. 18.14(*a*)

Fig. 18.14(*b*)

(g) **Supplementary Sections.** A supplementary view in section is a *supplementary section.* An example is shown in Fig. 18.15.

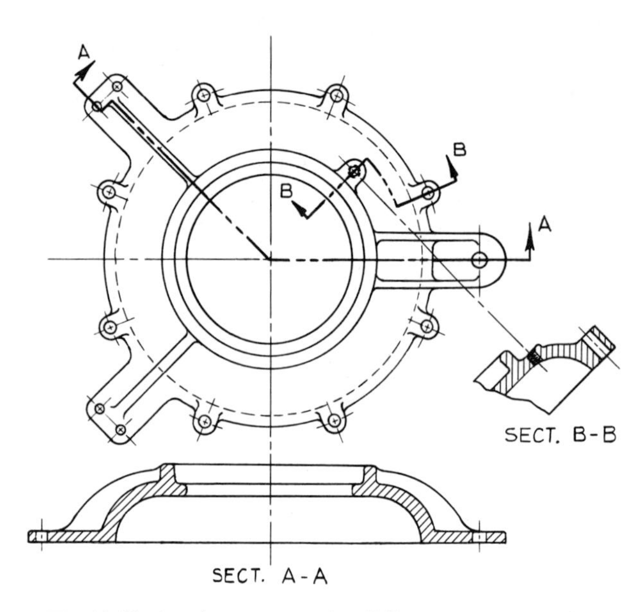

Fig. 18.15 Supplementary section, B-B.

(h) **Thin Sections.** When the material in section is too thin for the effective use of section lining, the material is shown solid, as in Fig. 18.16.

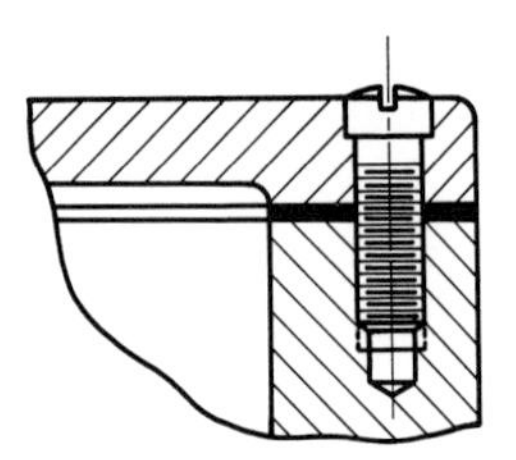

Fig. 18.16 Thin sections, shown solid.

Conventional Practices

In order to simplify further the interpretation of technical design drawings, certain conventional practices have been recommended by the American Standards Association (ASA). These practices, though justified, often violate some of the basic principles of orthogonal projection.

Intersections. While it is true that the intersections of two unfinished surfaces theoretically show no line, it is considered good practice to include the line of intersection. Its location is determined by the theoretical intersection. Several examples are shown in Figs. 18.17 and 18.18.

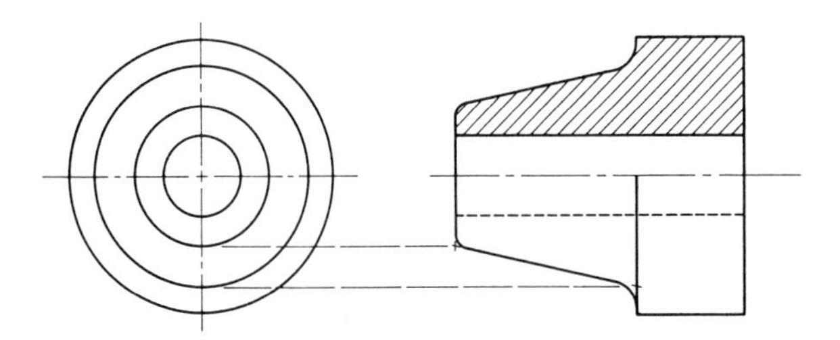

Fig. 18.17 Representation of rounded edges and fillets.

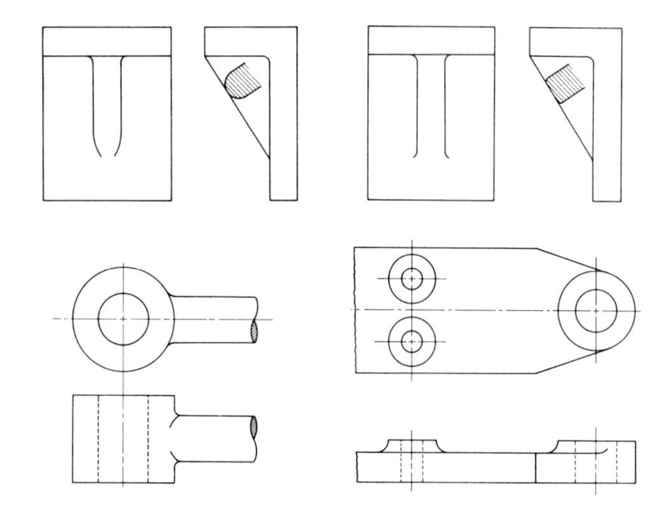

Fig. 18.18 Filleted intersections and runouts.

The intersections of small circular and rectangular shapes are conventionalized in the manner shown in Figs. 18.19 and 18.20.

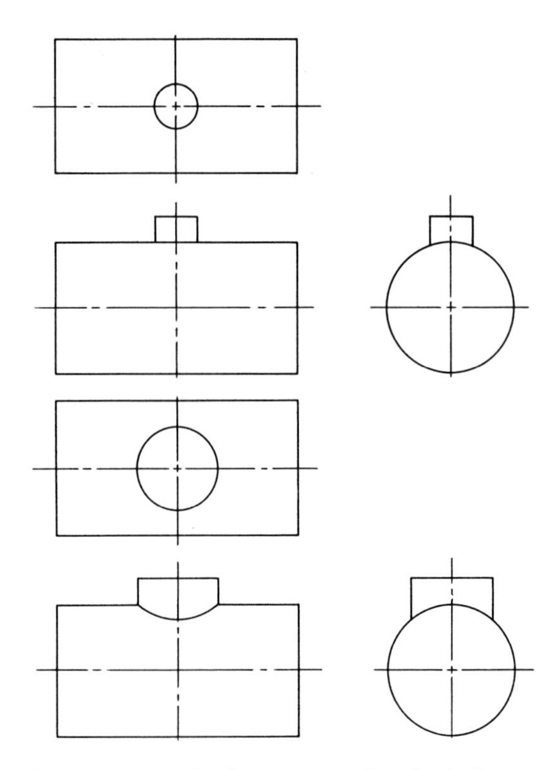

Fig. 18.19 Conventionalized intersections of small cylinders.

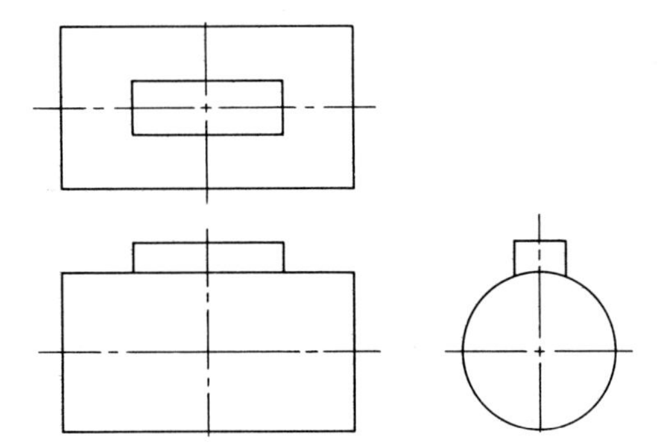

Fig. 18.20 Conventionalized intersection of small cylinders and rectangular shapes.

When a section is drawn through an intersection in which the exact figure or curve of intersection is small, or of little significance, the figure or curve may be shown in simplified form, as in Fig. 18.21. Larger figures of intersections may be shown as in Fig. 18.22.

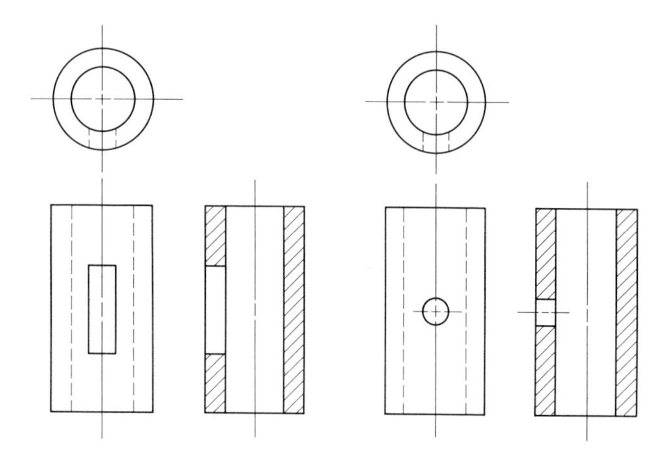

Fig. 18.21 Intersections (small).

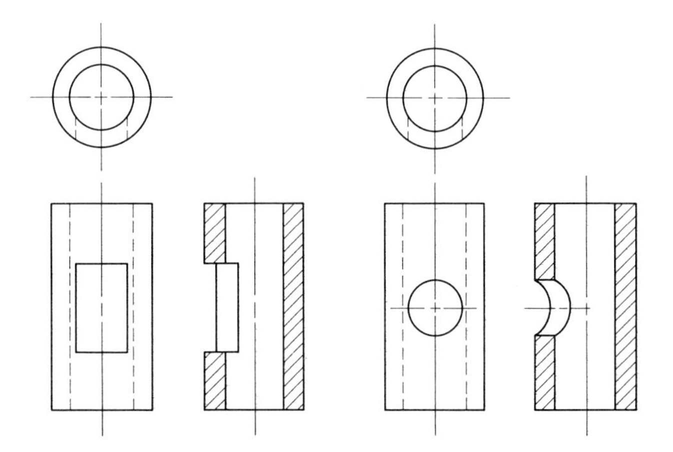

Fig. 18.22 Intersections (large).

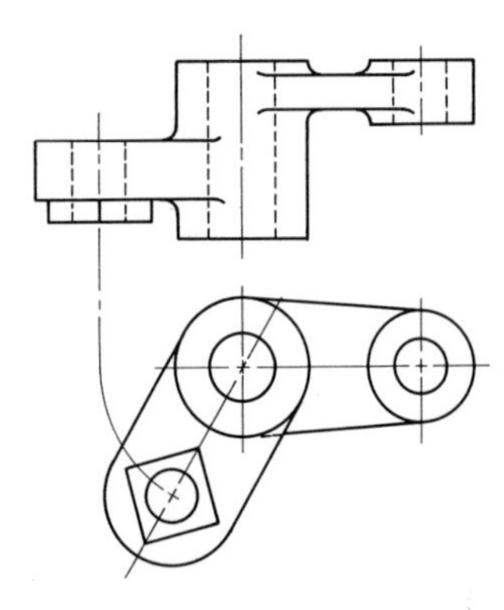

Fig. 18.23 Bell-crank—arm revolved to show relationship of parts. (This is an example of violation of orthogonal projection theory.) (Courtesy ASA)

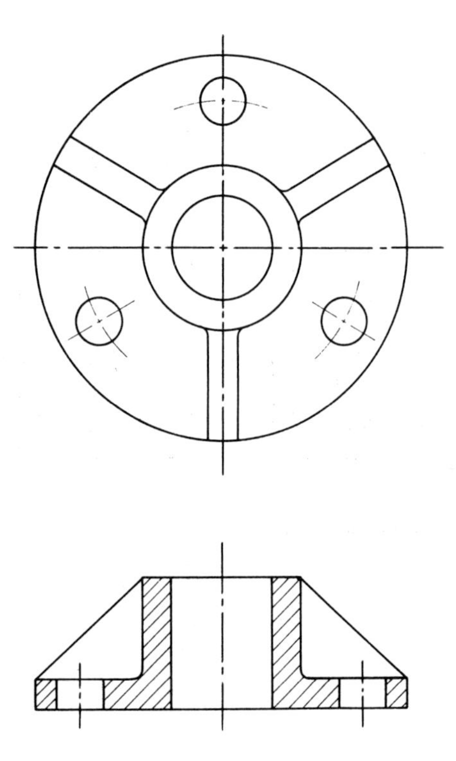

Fig. 18.24 Full section, showing treatment of holes and ribs.

Violations of Orthogonal Projection Theory. In the interest of clarity in interpreting the designer's intent, strict adherence to theory may not be justified. A few selected examples of the violation of orthogonal projection theory are shown in Figs. 18.15, 18.23, and 18.24.

In Fig. 18.15, the treatment of section *A–A* leads to a drawing which is easy to understand, whereas a true front view would be confusing. In Fig. 18.23, one arm of the bell crank has been revolved to show the true relationship of the portions of the piece. A theoretically correct top view would not add to the clarity of the drawing. In most cases it would, in fact, make the reading of the drawing more difficult. In Fig. 18.24, the section shows the ribs and the holes as though they were revolved into the cutting plane. Note the sectional treatment of the ribs. A true orthogonal projection certainly would hamper the ease of interpreting the views.

Treatment of Shafts, Bolts, Nuts, Webs, Keys, etc., in Section. Section lines are omitted when the section passes through shafts, bolts, nuts, webs, keys, rivets, and similar parts whose axes lie in the cutting plane. Figures 18.25 and 18.26 are typical of this treatment.

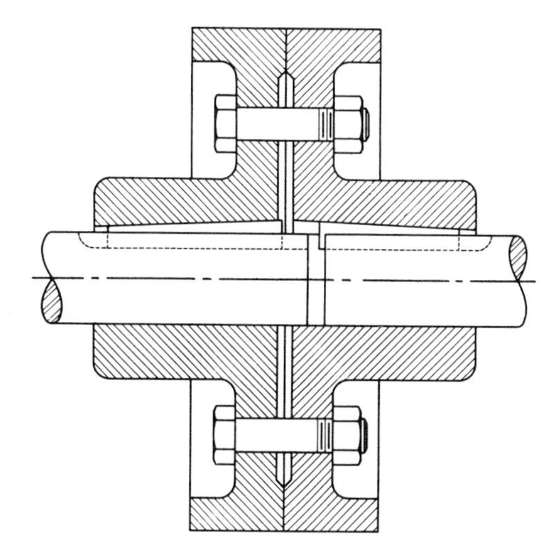

Fig. 18.25 Shafts, keys, bolts, and nuts in a sectional view. (Courtesy ASA)

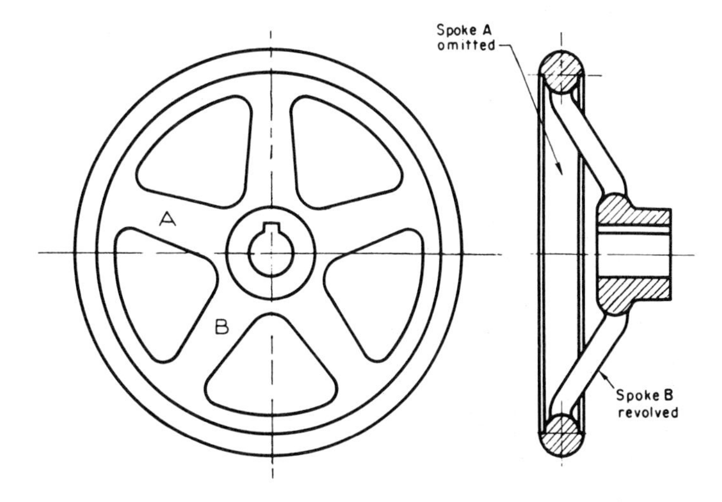

Fig. 18.26 Spokes in section. (Courtesy ASA)

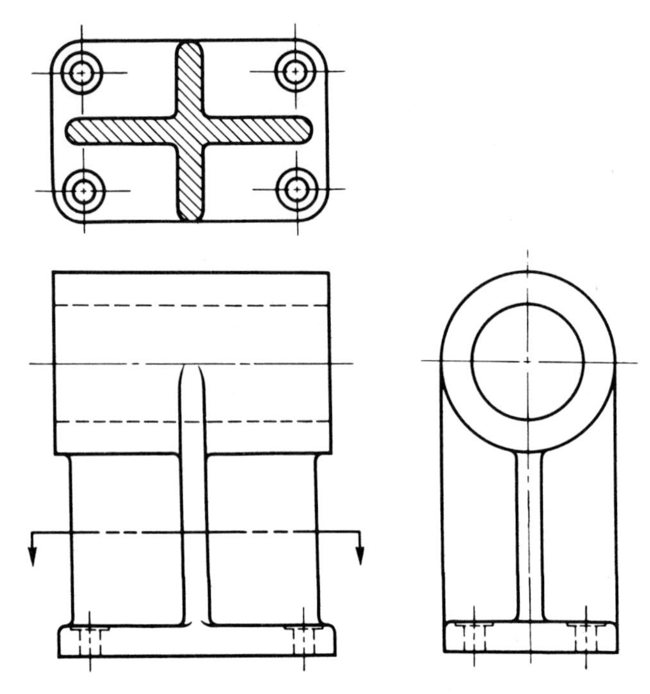

Fig. 18.27 Ribs in section; cutting plane across ribs. (Courtesy ASA)

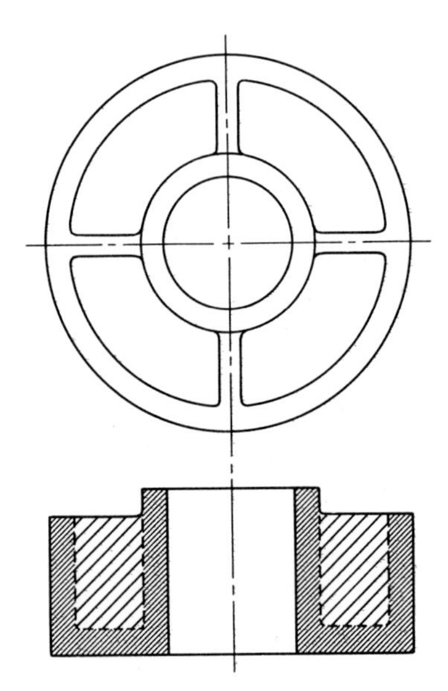

Fig. 18.28

When the cutting plane cuts across the elements, then they should be section lined. (See Fig. 18.27.)

In cases where the presence of a flat element is not clear without section lining, alternate section lines may be used to clarify the engineer's intent. Figure 18.28 shows the treatment which clearly indicates the presence of the ribs.

A pictorial section through an assembly is shown in Fig. 18.29. Additional conventions are included in ASA Y14.2–1957, Section 2, Line Conventions, Sectioning and Lettering; (Figure A–1, Width and Character of Lines; Fig. A-2, Hidden Line Technique; Fig. A-3, Section Lining of Adjacent Parts; Fig. A-4, Vertical Letters; and Fig. A-5, Inclined Letters are included in Appendix A).

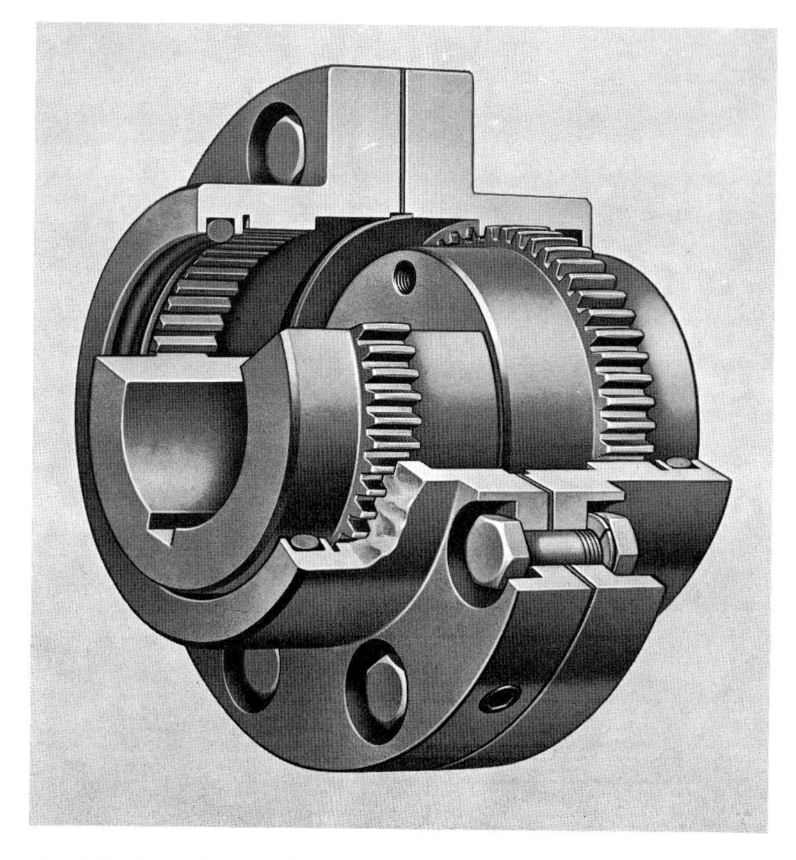

Fig. 18.29 Pictorial section of gear coupling. (Courtesy Falk Corp., Milwaukee, Wisc.)

EXERCISES

1. Draw the right side view of the mounting bracket as a *full section.* Re-align one bolt hole to show in the section. (See Fig. E-18.1.)

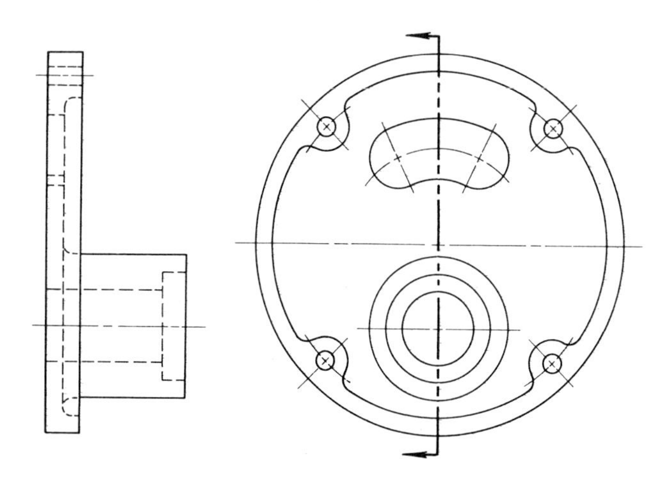

Fig. E-18.1 Mounting bracket.

2. Draw the casting shown in Fig. E-18.2, using the circular view and a full section view on cutting plane *A–A*. *Scale:* Full size, or as directed by the instructor.

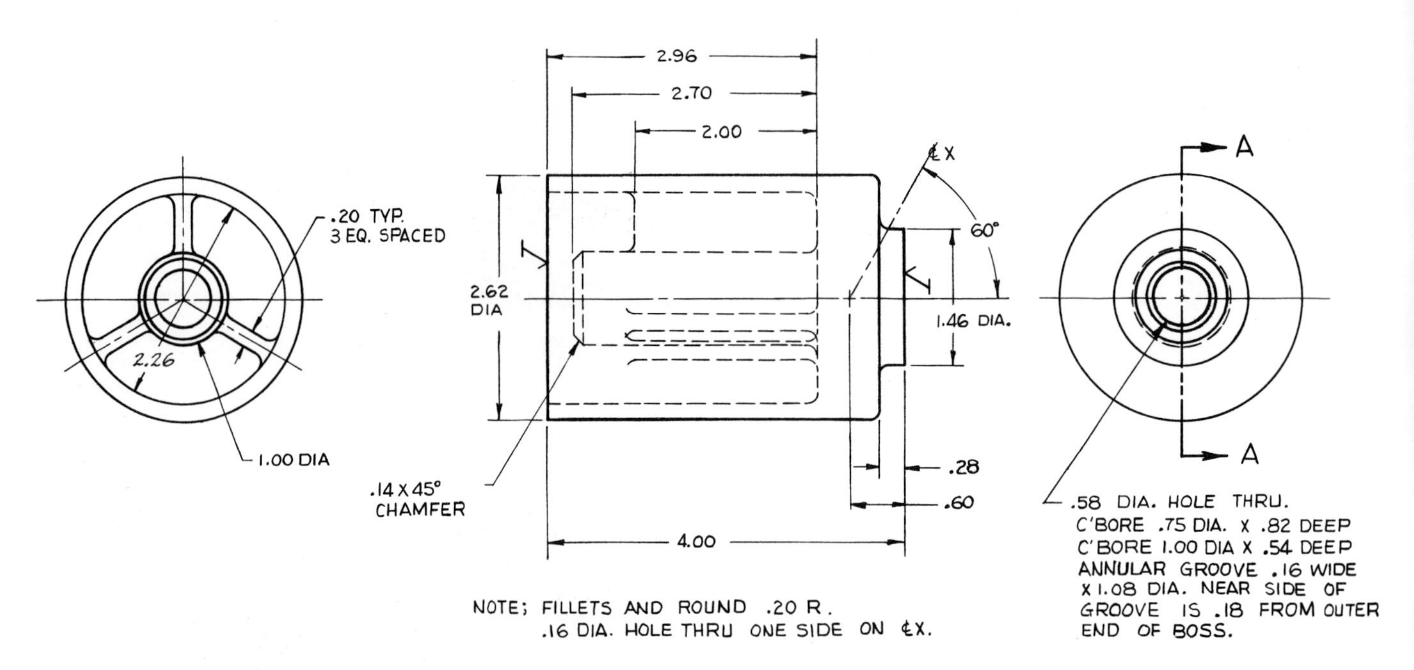

Fig. E-18.2 Single-flange roller and plunger die casting. (Adapted from a drawing by Hewlett-Packard.)

3. Draw the front view, a full section view of cutting plane A–A; a removed section view on cutting plane B–B; and a revolved section on cutting plane C–C. *Scale:* $1\frac{1}{2}'' = 1'\text{-}0''$ or $3'' = 1'\text{-}0''$, as directed by the instructor. (See Fig. E-18.3.)

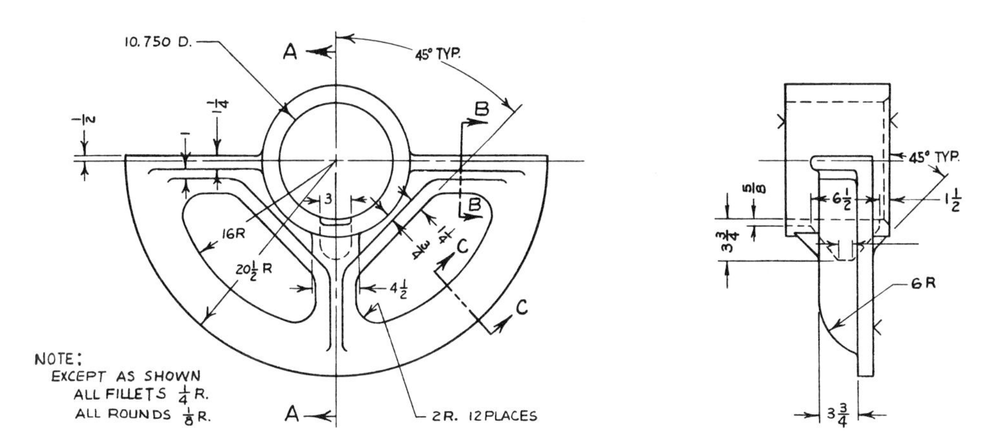

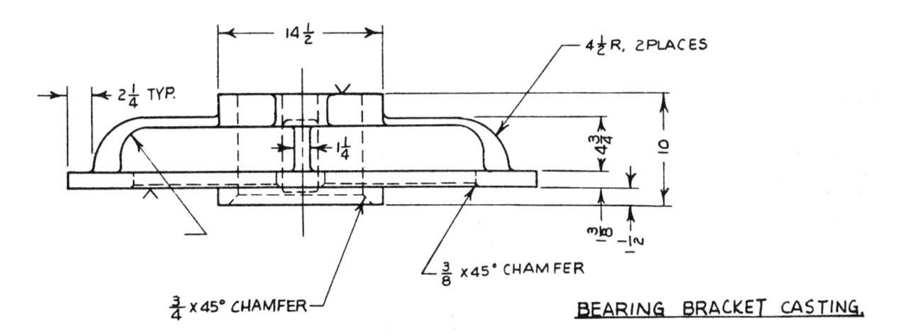

Fig. E-18.3 Bearing bracket casting. (Adapted from a drawing by Westinghouse Electric Corp.)

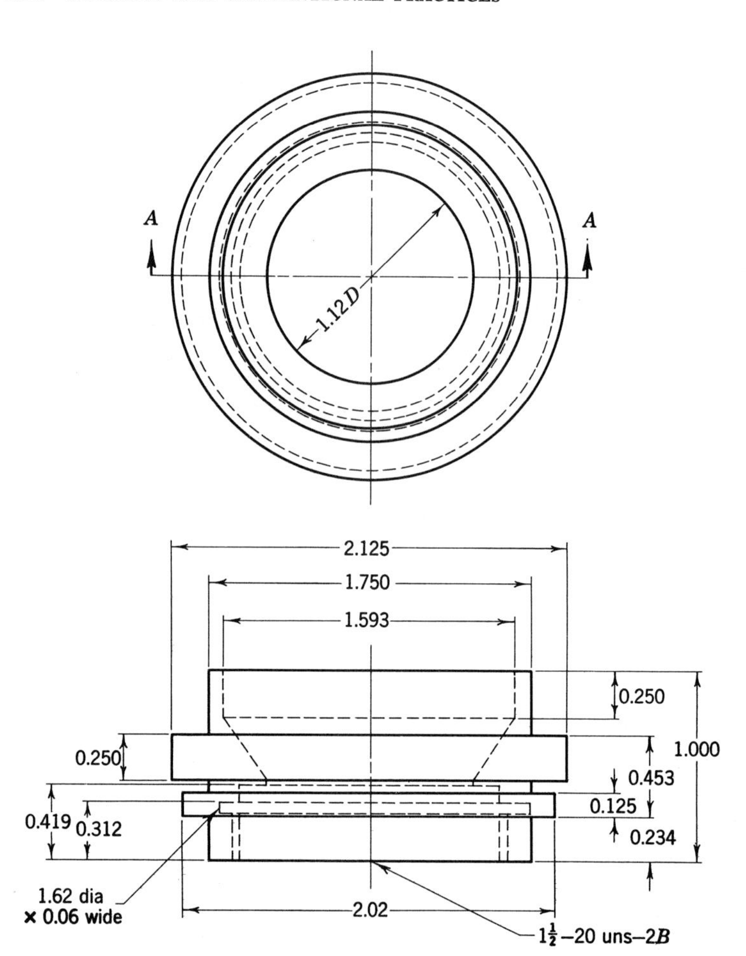

Fig. E-18.4 Base, reticle rotating. (Adapted from a drawing by Librascope, Inc.)

4. Draw a full section view on cutting plane *A–A*. *Scale:* twice size. (See Fig. E-18.4.)

5. Redraw the part shown in Fig. E-18.5, using the circular view plus a half section view on cutting plane *A–A*. *Scale:* 4″ = 1″, or as directed by the instructor.

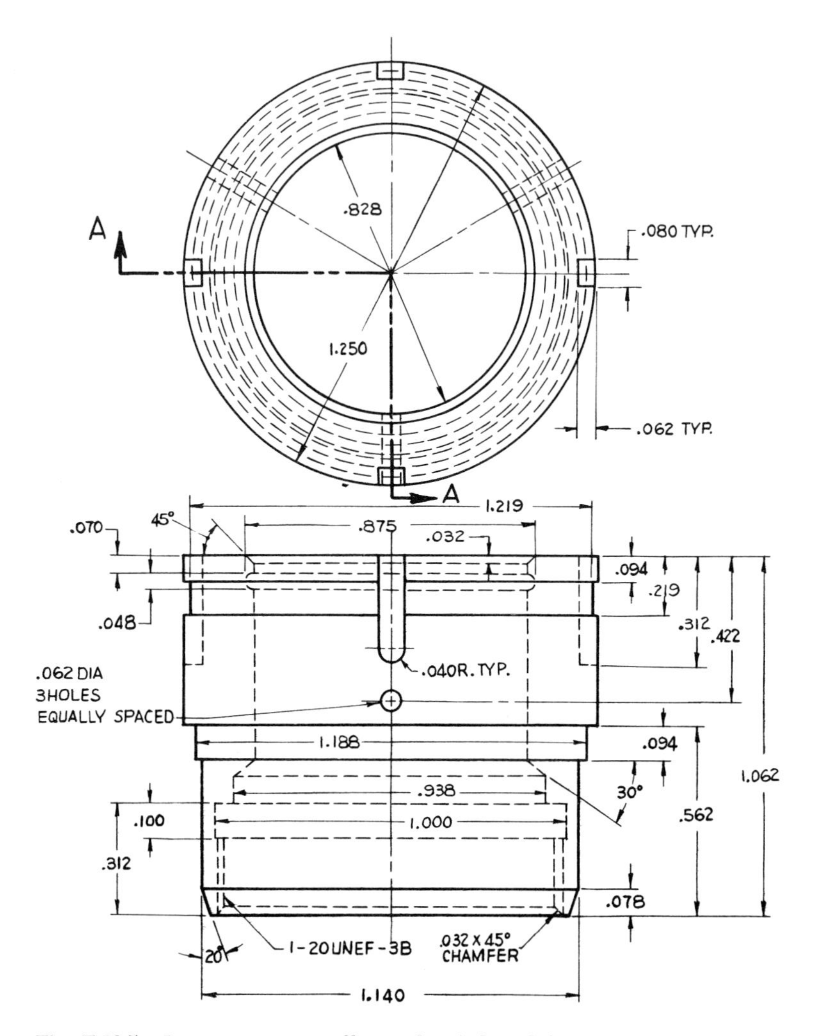

Fig. E-18.5 Lens mount, microfilm reader. (Adapted from a drawing by Data Processing Systems Division of Smith-Corona-Marchant, Inc.)

6. Make a new drawing of the casting consisting of (1) the front view; (2) a full section view on cutting plane *A–A*; (3) a full section view on cutting plane *B–B*; and (4) a removed section on cutting plane *C–C*. *Scale:* Full size, or as directed by the instructor. (See Fig. E-18.6.)

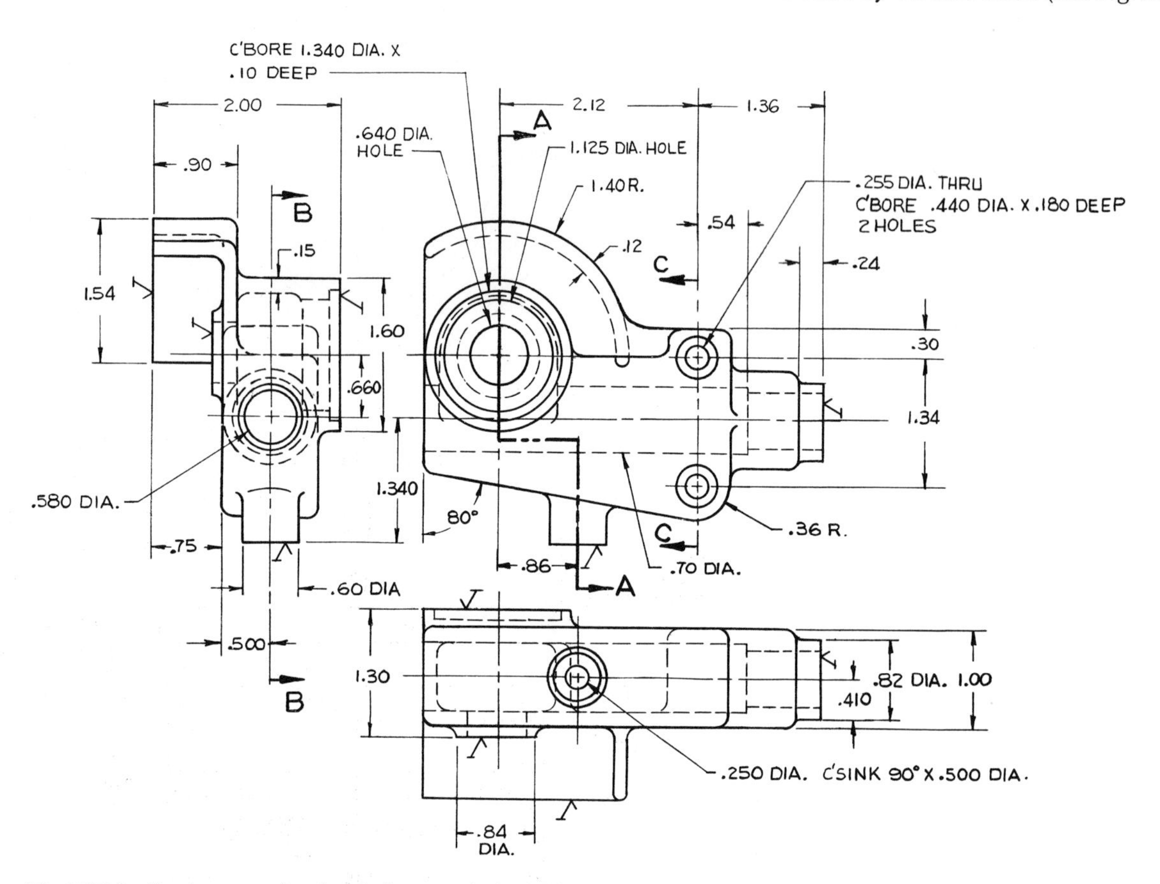

Fig. E-18.6 Gearbox control surface lock casting. (Adapted from a drawing by Convair Division of General Dynamics Corp.)

7. Redraw the casting, utilizing broken-out section views on cutting planes *A–A*, *B–B*, and *D–D*; a removed section view on cutting plane *C–C*; and a revolved section view on axis *X–X*. *Scale:* Full size, or as directed by the instructor. (See Fig. E-18.7.)

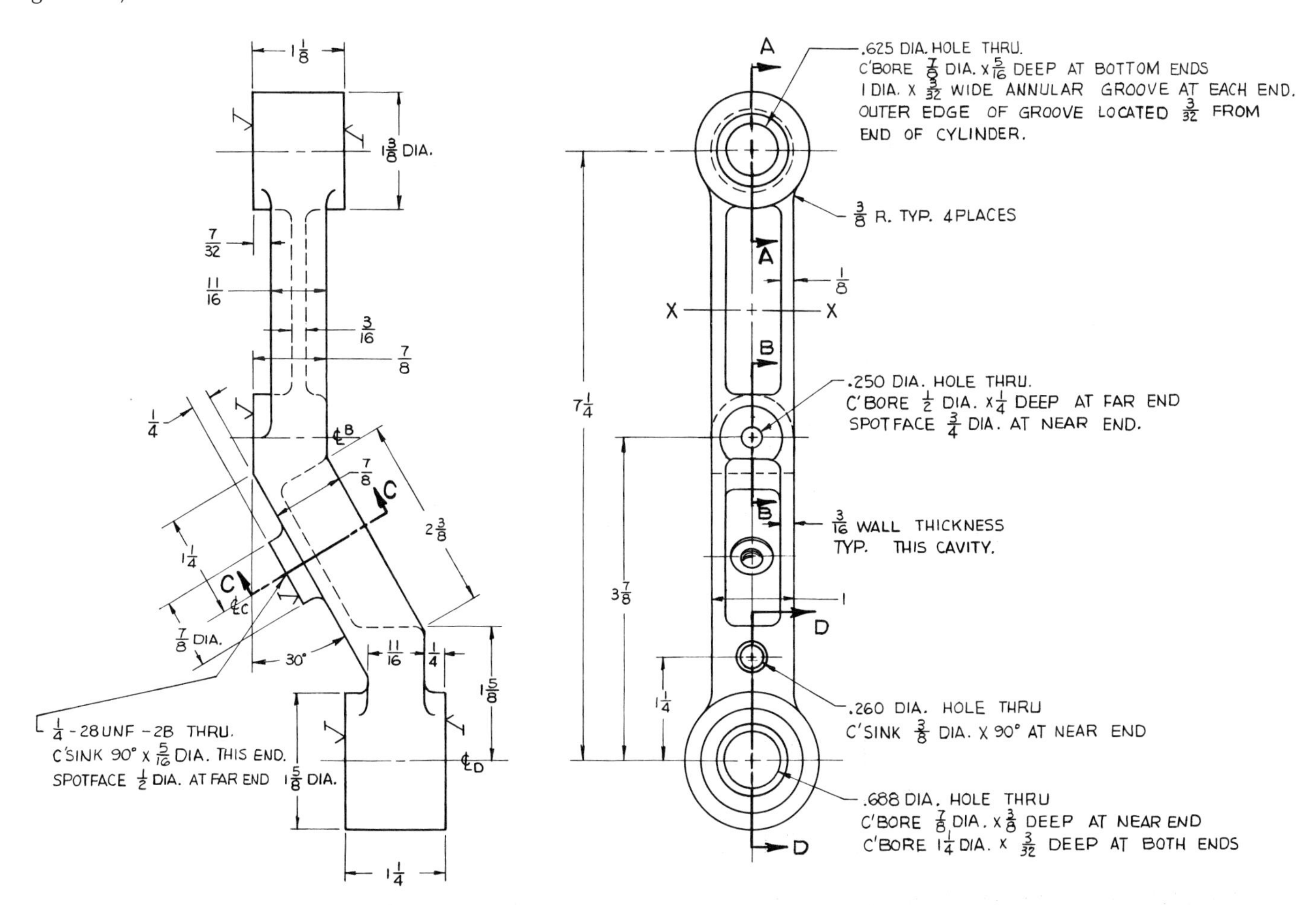

Fig. E-18.7 Arm-servo valve control. (Adapted from a drawing by Northrop Aircraft, Inc.)

8. Redraw the part shown in Fig. E-18.8, using the circular view and a sectional view on cutting plane *A–A*. *Scale:* 2″ = 1″, or as directed by the instructor.

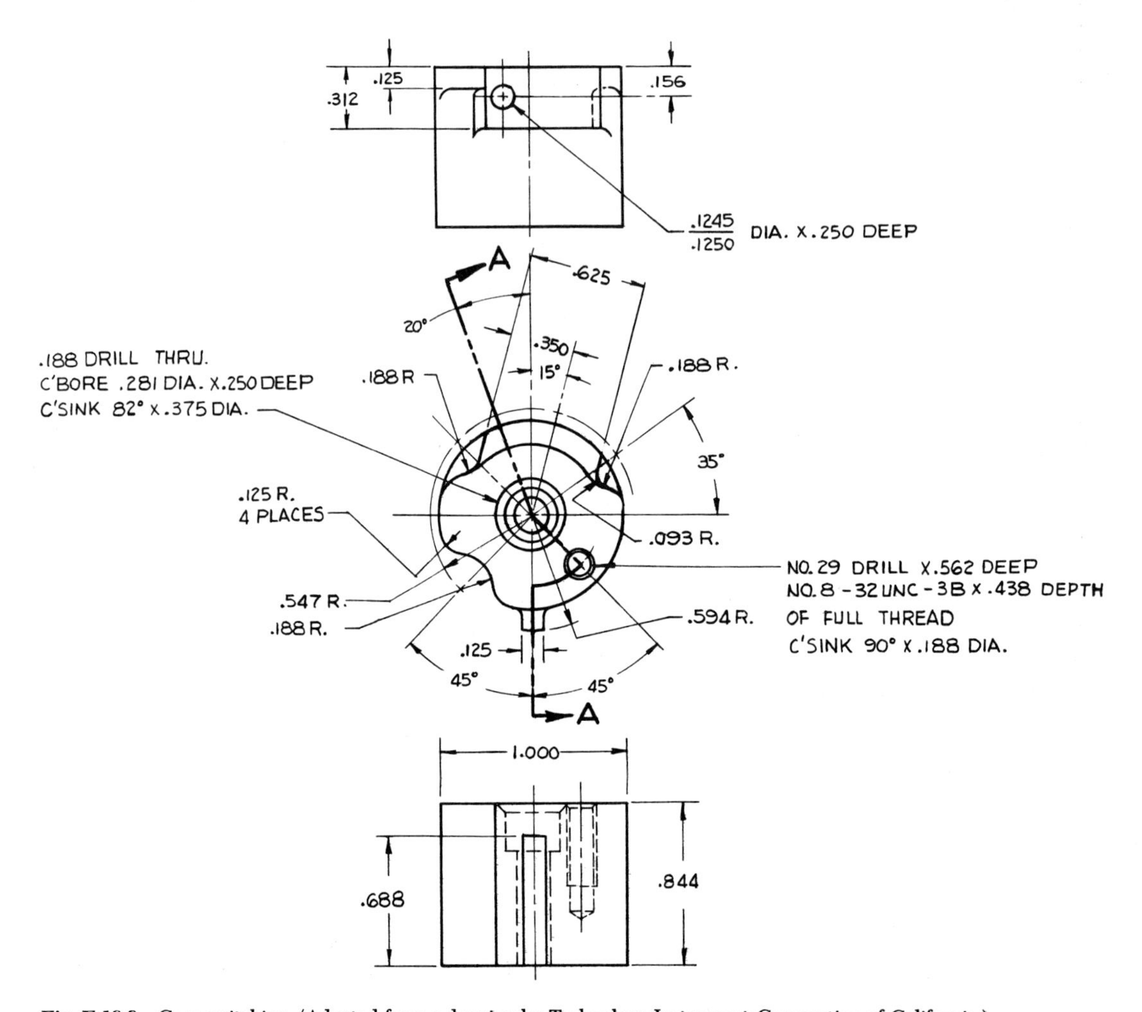

Fig. E-18.8 Cam-switching. (Adapted from a drawing by Technology Instrument Corporation of California.)

9. Make a new drawing of the cam, consisting of the views at the left and right; a full section view on cutting plane *A–A*; and a partial section view on cutting plane *B–B*. *Scale:* 5″ = 1″, or as directed by the instructor. (See Fig. E-18.9.)

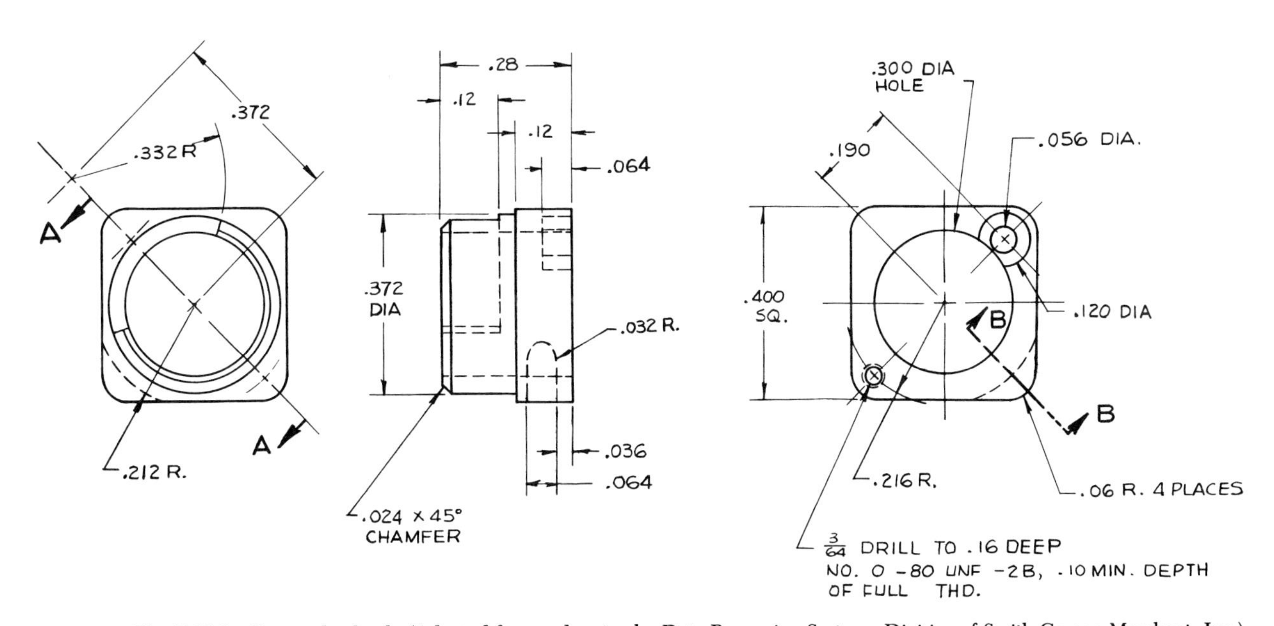

Fig. E-18.9 Cam-code clutch. (Adapted from a drawing by Data Processing Systems Division of Smith-Corona-Marchant, Inc.)

10. Make a new drawing of the part shown in Fig. E-18.10, consisting of the following: (1) the top view; (2) a full section view on cutting plane *A–A*; (3) a broken-out section view on cutting plane *B–B*; and (4) the right-side view. *Scale:* 2″ = 1″, or as directed by the instructor.

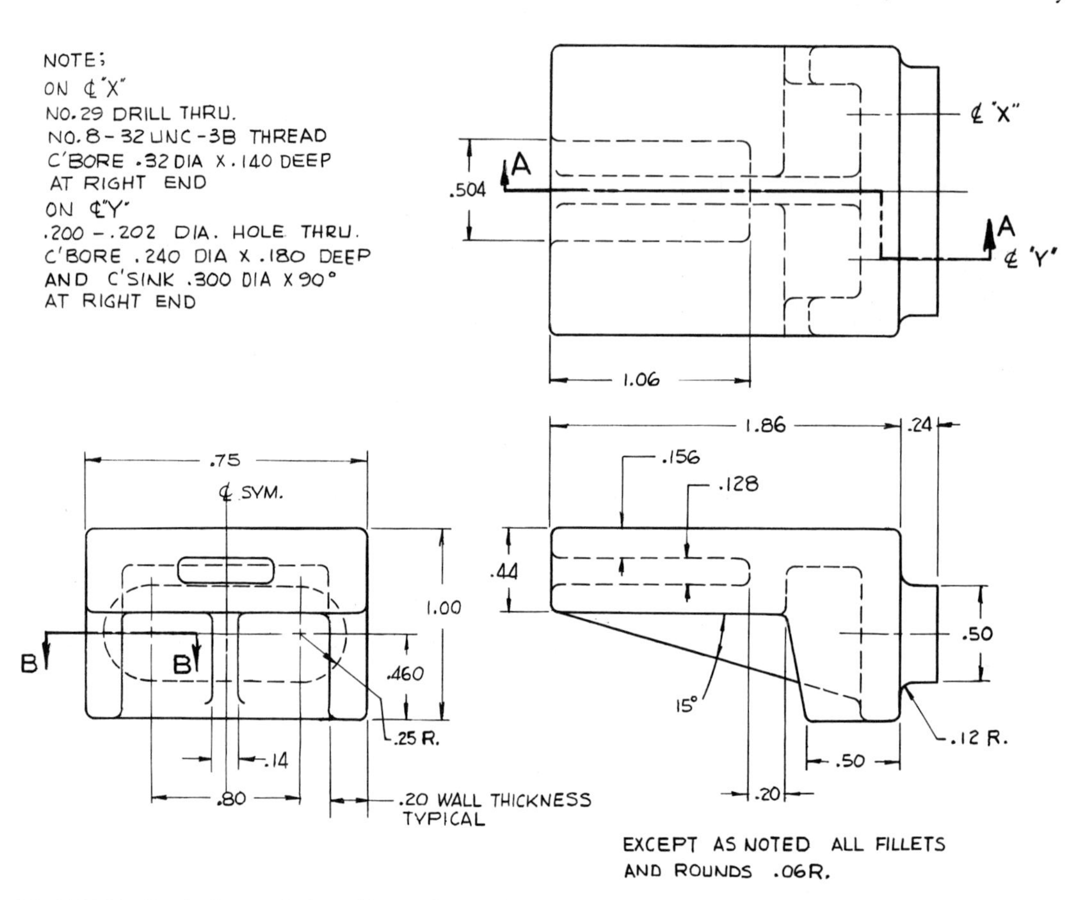

Fig. E-18.10 Reader button. (Adapted from a drawing by Data Processing Systems Division of Smith-Corona-Marchant, Inc.)

11. Make a drawing of the HOOK consisting of a front view (Fig. E-18.11), a full section view on cutting plane *D–D*, a removed section on cutting plane *B–B*, a revolved section on line *A–A*, and revolved section on line *C–C*. (The revolved section on *C–C* may be shown as a removed section moved out along the centerline.) *Note:* the radius contour at the outer edge of the HOOK as shown in the side view is uniform along the HOOK except for the eye at the upper end.

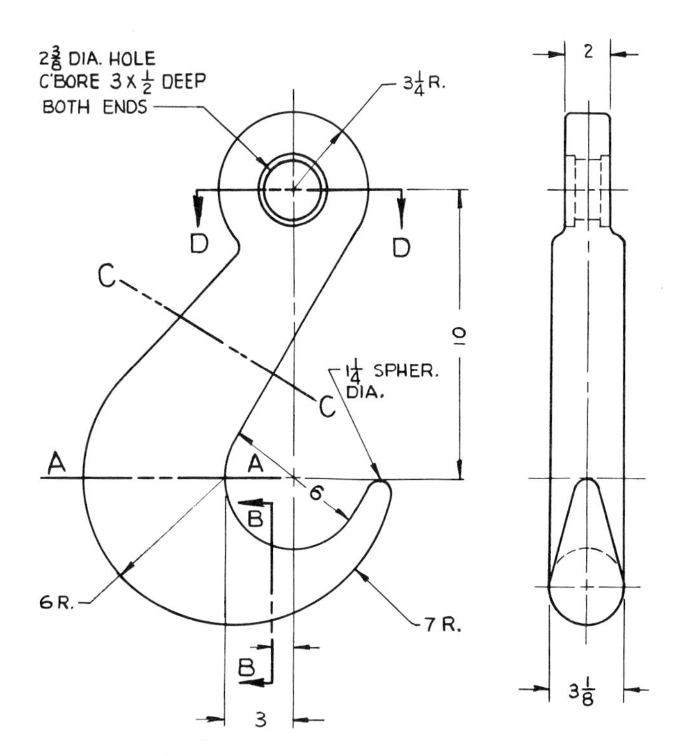

Fig. E-18.11 Hook.

12. Make a new drawing of the part shown in Fig. E-18.12, consisting of the following: (1) the circular view as shown; (2) a full section view on cutting plane *A–A*; (3) a partial section view on cutting plane *B–B*; and (4) a broken-out section view on *C–C* or a removed section on *D–D*. *Scale:* 5″ = 1″, or as directed by the instructor.

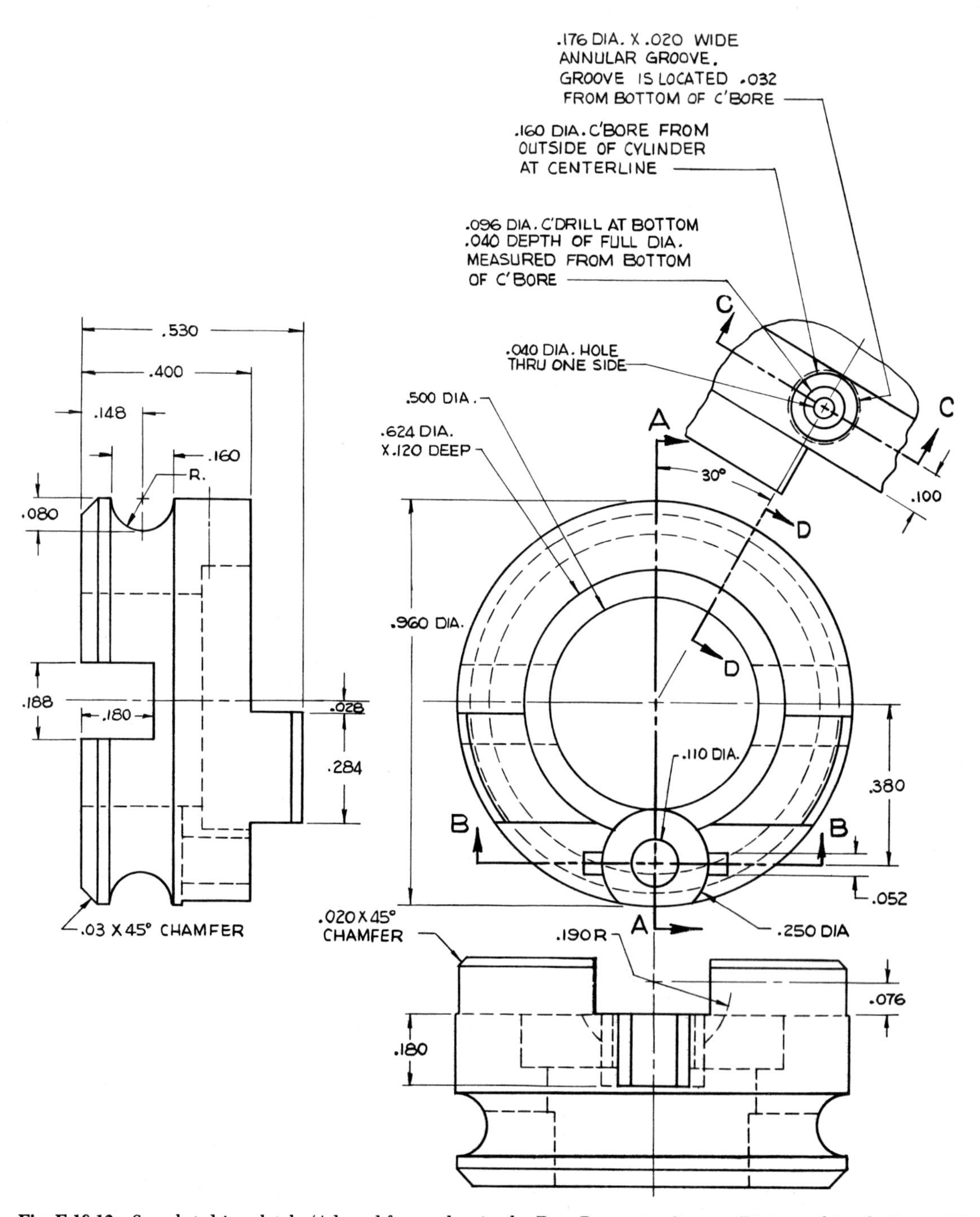

Fig. E-18.12 Sprocket drive clutch. (Adapted from a drawing by Data Processing Systems Division of Smith-Corona-Marchant, Inc.)

13. Make a new drawing of the casting shown in Fig. E-18.13, consisting of (1) the view on the right; (2) a full section view on cutting plane *A–A*; (3) a half view of the projection on the left; and (4) removed sections on cutting planes *B–B* and *C–C*. *Scale:* Half size, or as directed by the instructor.

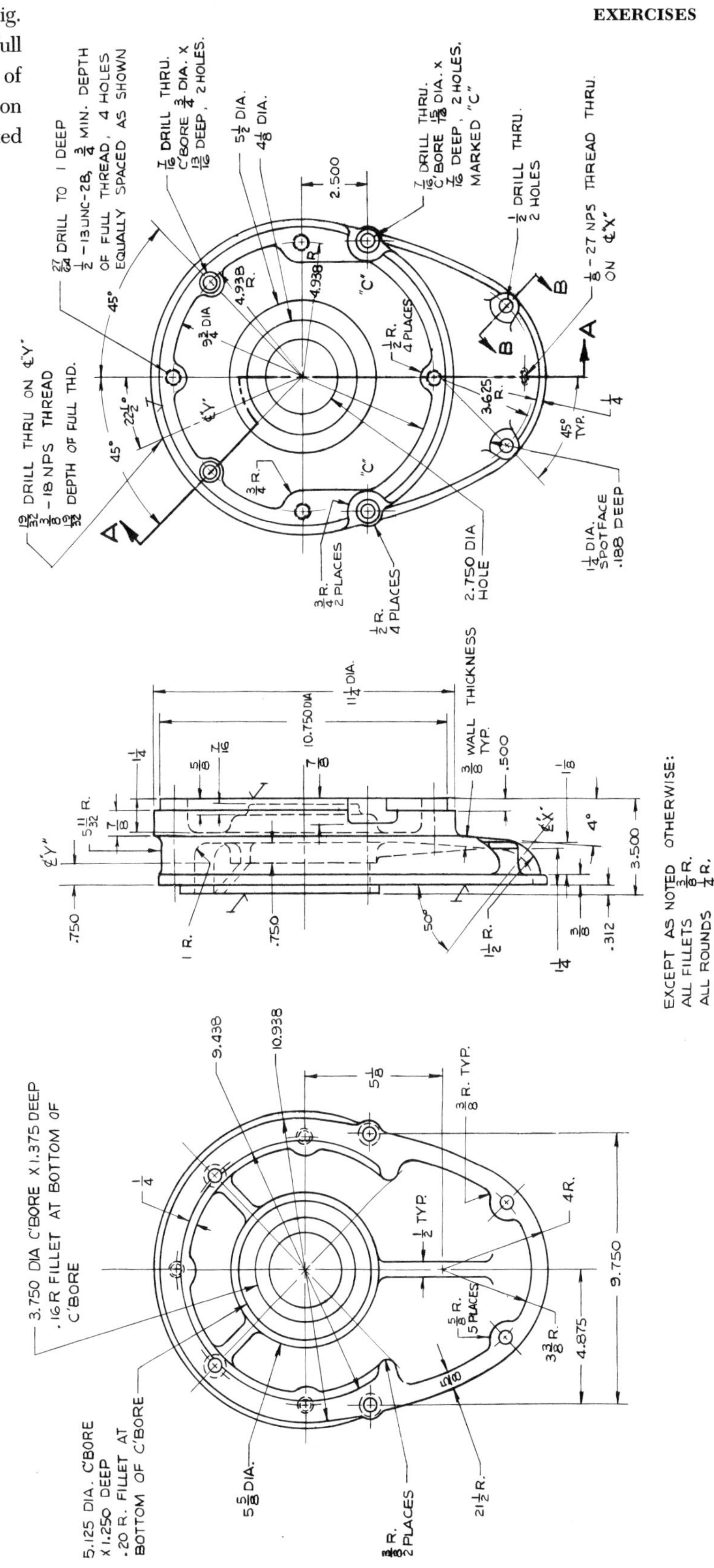

Fig. E-18.13 Pulley end bracket. (Adapted from a drawing by U.S. Electrical Motors, Inc.)

14. Redraw the coupler base shown in Fig. E-18.14 by converting the middle view to a full section on cutting plane *A–A*; and by adding other sectional views to enhance the description of the part. *Scale:* Full size, or as directed by the instructor.

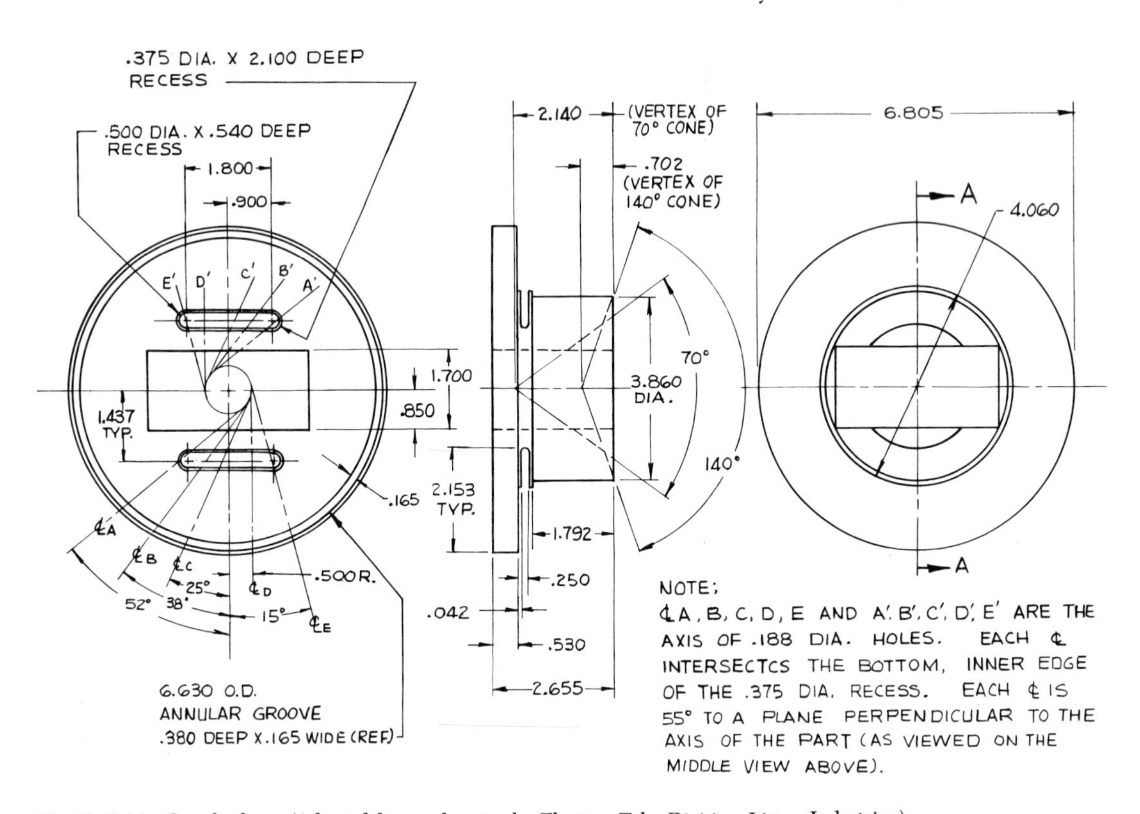

Fig. E-18.14 Coupler base. (Adapted from a drawing by Electron Tube Division, Litton Industries.)

FASTENERS AND SPRINGS 19

Engineering designers must have up-to-date knowledge of the various types of fasteners that are available for the assembly of the component parts of mechanisms, devices, machines, and structures. Good designers will accumulate "data sheets" of fasteners and their uses, especially those that are pertinent to the design areas of their interests.

Many designs require the use of both *permanent fastenings* such as *rivets* or *welds,* and *removable types* such as *screws*, *bolts*, *keys*, and *pins*. Examples of permanent fastenings are such riveted or welded structures as steel bridges, high-rise office buildings, space vehicles, tanks, boilers, automobile frames, etc. The removable-type fastener is used in the assembly of components of refrigerators, household appliances, switches, electric bulbs and sockets, shafts and pulley systems, automobile components, etc.

Engineering students should also be familiar with the common types of fasteners and their application to the assembly of various components. Moreover, students should know how to use recognized standards for the graphic representation and identification of fasteners and screw threads. Appropriate to the students' use are the American Standard Association publications and the material which is made available by the Industrial Fasteners Institute, Cleveland, Ohio, through the periodical, *Fasteners,* which contains the latest information on new developments and new uses of fasteners. The most frequently used fasteners involve screw threads, hence it is incumbent upon us to learn nomenclature and definitions; to study thread profiles and their applications; and to understand and to properly use graphic representations of threads.

Great strides have been made in the unification of screw-thread standards which received their "impetus from the need for interchangeability among the billions of fasteners used in the complex equipment of modern technology and made in different countries. Equally important, however, are international trade in mechanisms of all kinds and servicing of transportation equipment that moves from country to country. These have made unification not only highly advantageous, but practically essential."*

* From ASA B1.1-1960.

DEFINITIONS AND NOMENCLATURE —SCREW THREADS

Abstracted from the American Standard Association's ASA B1.7 are the following definitions correlated with the graphical representation shown in Fig. 19.1.

Screw Thread. A screw thread (hereinafter referred to as a thread) is a ridge of uniform section in the form of a helix on the external or internal surface of a cylinder, or in the form of a conical spiral on the external or internal surface of a cone or frustum of a cone. A thread formed on a cylinder is known as a "straight" or "parallel" thread, to distinguish it from a "taper" thread which is formed on a cone or frustum of a cone.

External Thread. An external thread is a thread on the external surface of a cylinder or cone.

Internal Thread. An internal thread is a thread on the internal surface of a hollow cylinder or cone.

Major Diameter. The major diameter is the largest diameter of a screw thread (Fig. 19.1).

Pitch Diameter. Pitch diameter is the diameter of the imaginary coaxial cylinder whose surface would cut the threads so as to make the widths of the threads equal to the widths of the spaces between the threads (Fig. 19.1).

Minor Diameter. The minor diameter is the smallest diameter of a screw thread (Fig. 19.1).

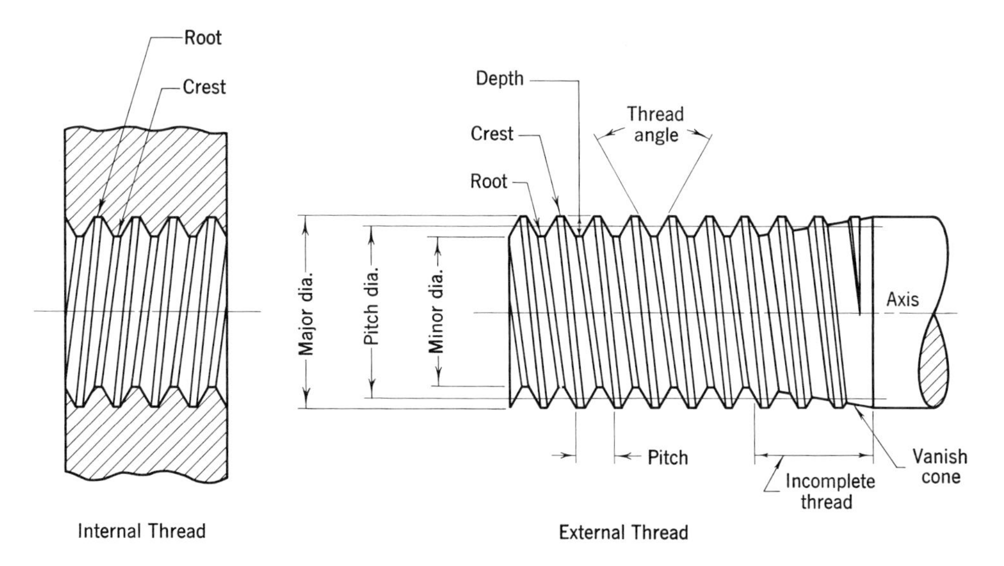

Fig. 19.1 Screw thread nomenclature.

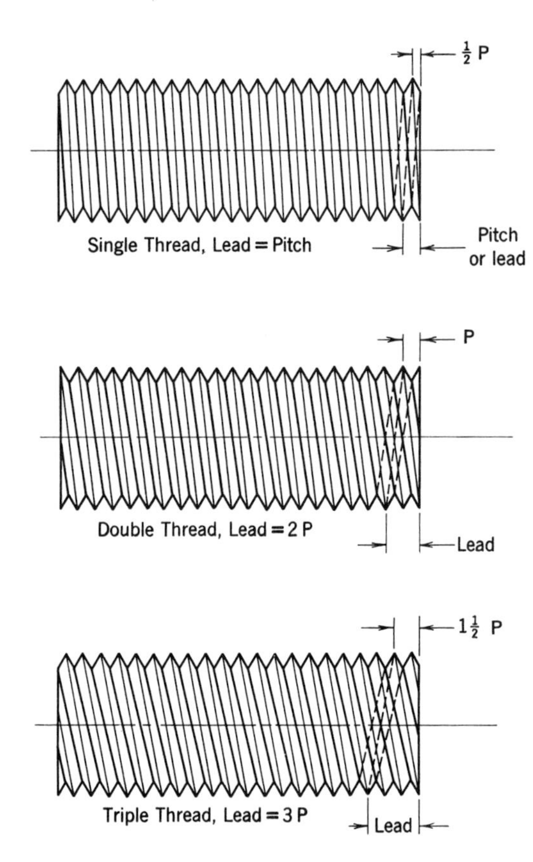

Fig. 19.2 Relationship between lead and pitch.

Pitch. Pitch is the distance measured parallel to the axis from a point on one thread to the corresponding point on an adjacent thread (Fig. 19.1).

Lead. The lead is the distance a threaded part moves axially, with respect to a fixed mating part, in one complete rotation.

Crest. The crest is the top surface which joins the sides of a thread (Fig. 19.1).

Root. The root is the bottom surface which joins the sides of a thread (Fig. 19.1).

Depth of Thread. The depth of thread is the distance, measured perpendicular to the axis, between the crest and root surfaces.

Thread Angle. The thread angle is the angle between the flanks (sides) of the thread measured in an axial plane (Fig. 19.1).

Single Thread. A single (single-start) thread is one having *lead equal to pitch* (Fig. 19.2).

Double Thread. A double (double-start) thread is one in which the *lead is twice the pitch* (Fig. 19.2).

Multiple Thread. A multiple (*multiple-start*) thread is one in which the *lead is an integral multiple of the pitch* (Fig. 19.2).

Right-hand Thread. A thread is a right-hand thread if, when viewed axially, it *recedes when turned clockwise.*

Left-hand Thread. A thread is a left-hand thread if, when viewed axially, it *recedes when turned counterclockwise.* All left-hand threads are designated "L.H."

Thread Profiles—Designer's Choice

The designer's choice of a thread profile (cross section shape) is directly related to the function that is to be served. Several thread forms have been produced for specific purposes, i.e., for making adjustments and for transmitting motion or power. Among these are such threads as the sharp V,

American National, acme, square, buttress, worm, and knuckle threads. (See Fig. 19.3.)

The *sharp V* is used to some extent for adjustment. Usually the sharp V thread is employed in the small diameter range.

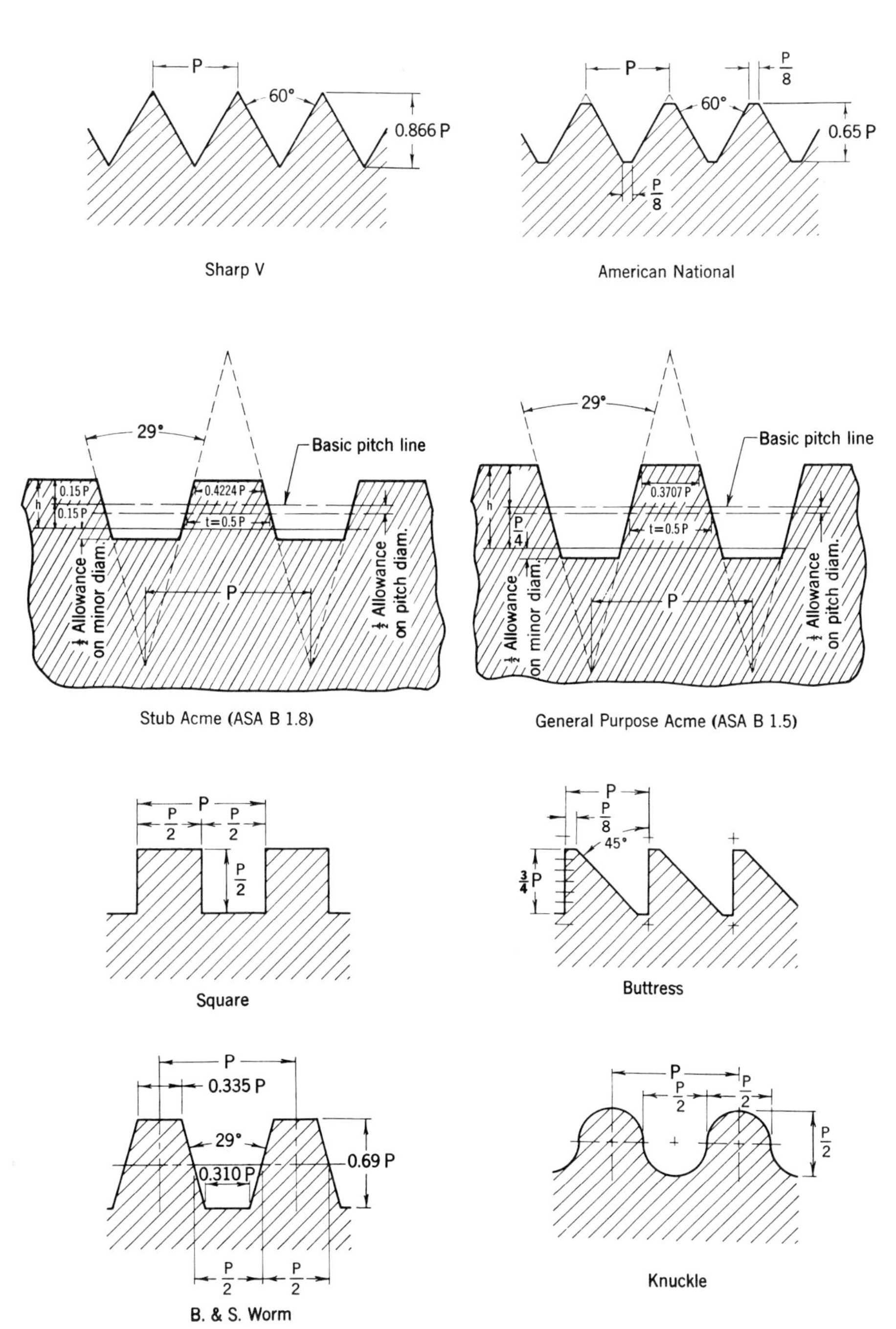

Fig. 19.3

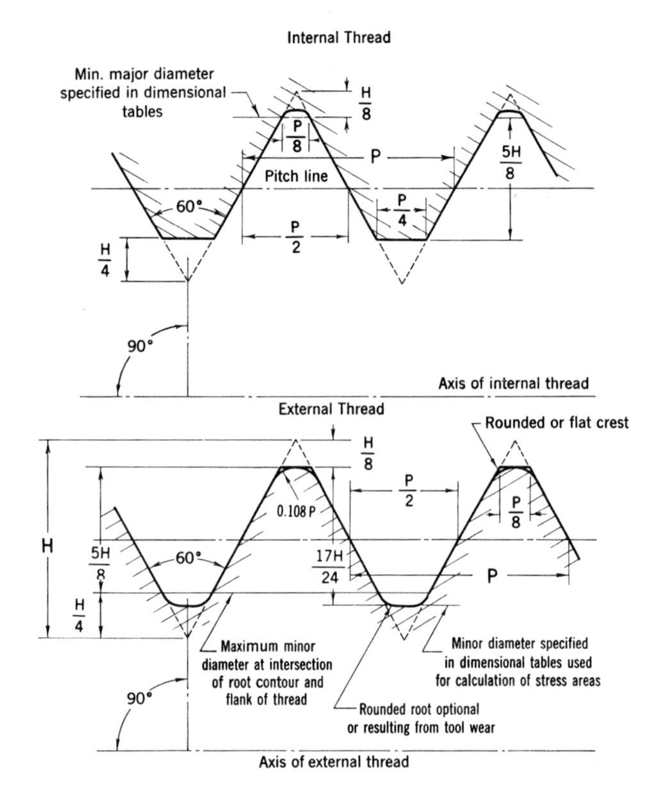

Fig. 19.4 Unified internal and external screw thread design forms. (Courtesy ASA)

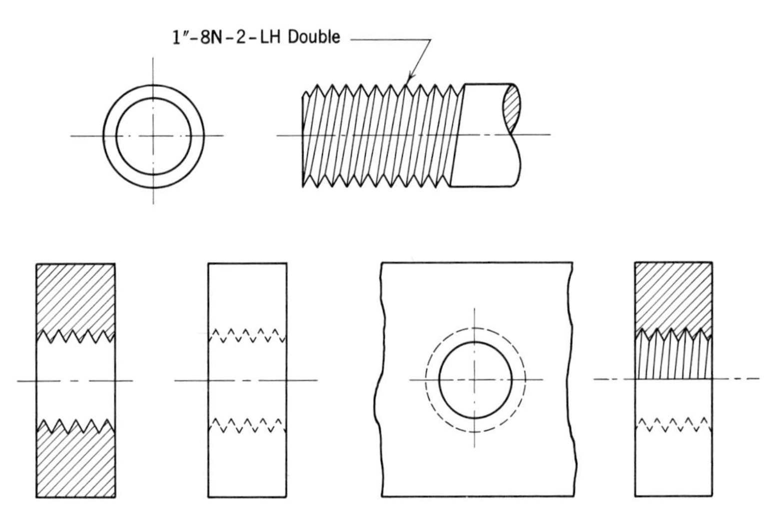

Fig. 19.5 Detailed representation (semiconventions).

The *American National* thread, which is a modification of the sharp V, is used in much of the screw-thread work in this country.

The *acme* and *square* thread forms are used for transmitting power.

The *buttress* thread is also used for transmitting power, but in one direction only.

The *worm* thread is another form for transmitting power, used in mechanisms that involve the transmission of power to worm wheels.

Knuckle thread forms are commonly used for fuses, electric bulbs, and lamps.

Unified Screw Threads

As a consequence of agreements reached by the Standards Associations of the United States, Great Britain, and Canada on November 18, 1948, a new Unified and American Screw Threads Standards, ASA B1.1-1949, was made available through the sponsorship of the Society of Automotive Engineers and the American Society of Mechanical Engineers. Further developments and revisions since that time have culminated in the most recent standard, Unified Screw Threads, ASA B1.1-1960.

The *Unified* thread form is shown in Fig. 19.4.

GRAPHIC REPRESENTATION OF THREADS

Screw thread representation has been greatly simplified by the use of standards* which eliminate the laborious and time-consuming task of drawing true helical curves.

There are three conventions generally used for the representation of threads on design drawings:

(*a*) The detailed representation. (See Fig. 19.5.)

* ASA Y14.6-1957.

(*b*) The schematic representation. (See Fig. 19.6.)
(*c*) The simplified representation. (See Fig. 19.7.)

The detailed representation is a good approximation of the actual appearance of screw threads. The helices are conventionalized as slanting straight lines, and the thread contour is shown as a sharp *V* at 60°.

In the schematic representation the staggered lines, symbolic of the thread crests and roots, are usually drawn perpendicular to the axis. The short lines are usually drawn heavier than the long lines. The spacing of the lines is independent of the actual pitch of the thread, so long as the distances appear reasonable. The construction shown should not be used for external threads in section or for hidden internal threads. *In the interest of economy*, the simplified representation is quite justified and is recommended for extensive use. In some cases, clarity may dictate the use of all three conventions. (See Fig. 19.8.)

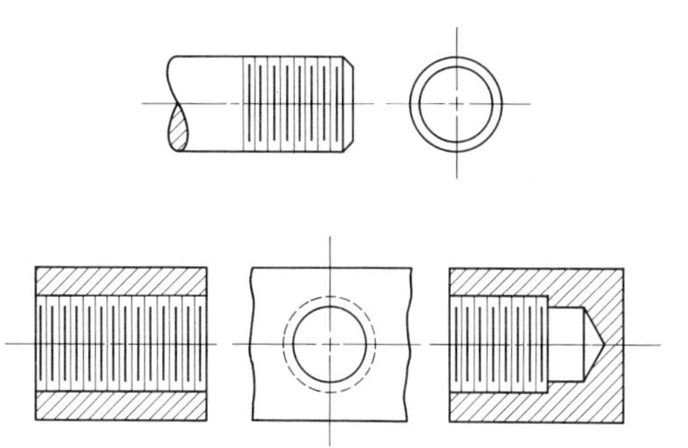

Fig. 19.6 Schematic representation.

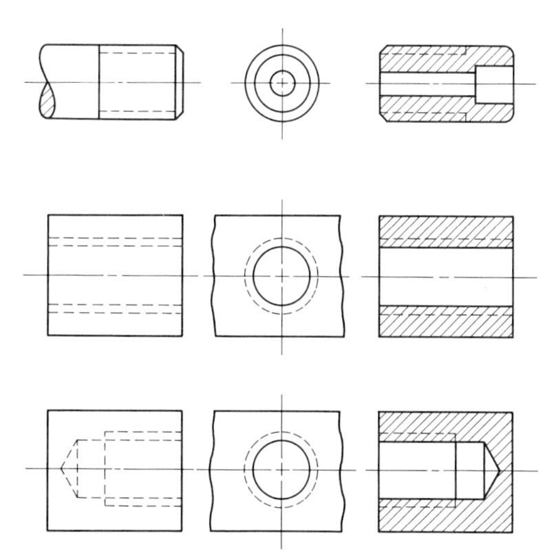

Fig. 19.7 Simplified representation.

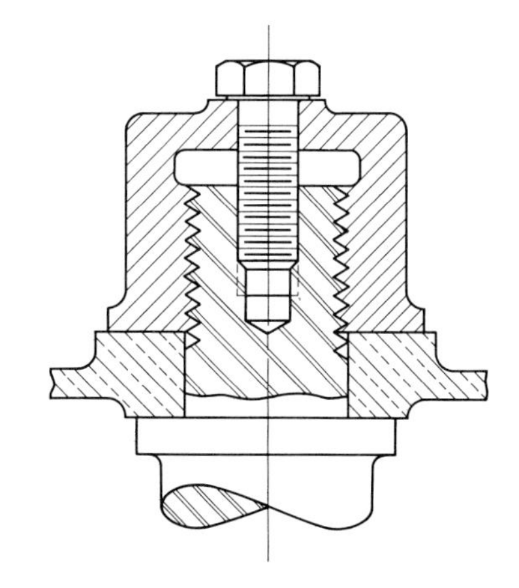

Fig. 19.8 Threads in section assembly.

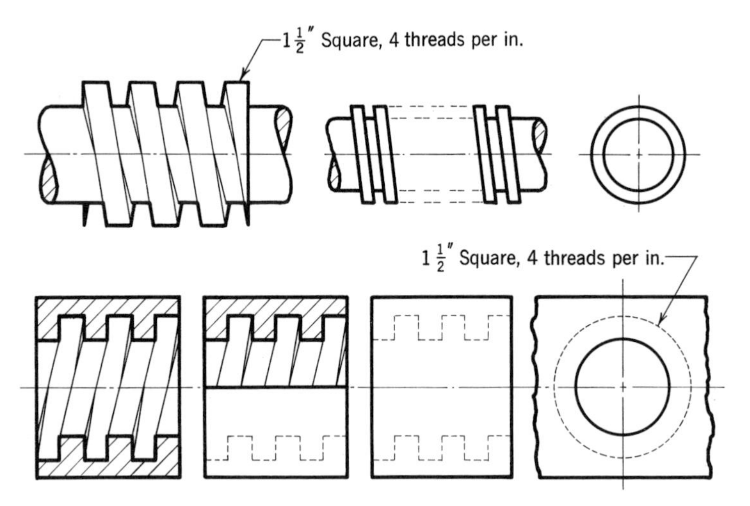

Fig. 19.9 Square thread convention.

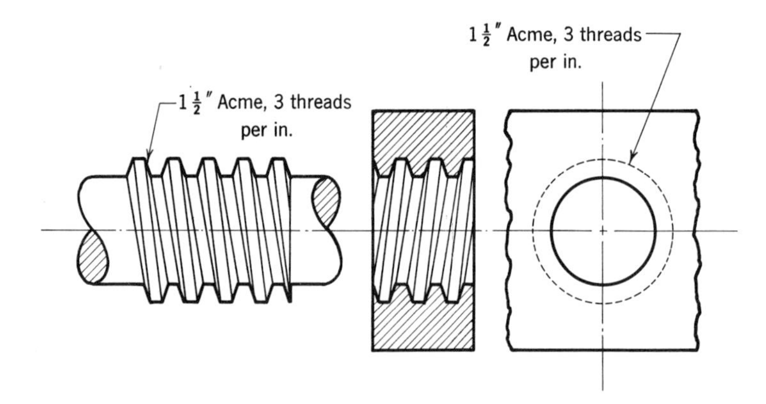

Fig. 19.10 Acme thread convention.

Square Thread Representation. Square threads are presented as shown in Fig. 19.9. A note specifying the type of thread and the pitch should accompany the drawing.

Acme Thread representation is shown in Fig. 19.10. A note specifying the type of thread and the pitch should accompany the drawing.

Thread Series* and Suggested Applications

"Thread series are groups of diameter-pitch combinations which are distinguished from each other by the number of threads per inch applied to a specific diameter." See Table 7, p. 660. The following thread series includes:

1. **Coarse-Thread Series.** This series is utilized for the bulk production of bolts, screws, and nuts. It is used in general applications for threading into lower tensile strength materials—cast iron, mild steel, bronze, brass, plastics, etc.—to obtain the optimum resistance to stripping of the internal thread. It is applicable to rapid assembly and disassembly. Coarse-thread series are designated *UNC* (Unified Coarse) or *NC* (National Coarse).

2. **Fine-Thread Series.** This series is used where the length of engagement is short, or where the wall thickness demands a fine pitch. It is recommended for general use in automotive and aircraft work, and where special conditions require a fine thread. The designation is *UNF* (Unified Fine) or *NF* (National Fine).

3. **Extra-Fine-Thread Series.** This series is applicable where even finer pitches of threads are needed for short lengths of engagements and for thin-walled tubes, nuts, ferrules, or couplings. It is used particularly in space vehicles and auxiliary equipment. The designations for this series are *UNEF* (Unified Extra Fine) or *NEF* (National Extra Fine).

4. **8-Thread Series.** The 8-thread series is a uniform-pitch series for large diameters or as a compromise between coarse- and fine-thread series. It is widely used as a substitute for the coarse-thread series for diameters larger than 1 inch. The designation is 8 *UN*.

5. **12-Thread Series.** The 12-thread series is a

* The American National standards for screw threads are available in condensed form in ASA B1.1-1960 for continued use in existing design. The transition, however, from these standards to the Unified standards is virtually completed.

uniform pitch series for large diameters requiring threads of medium-fine pitch. It is used as a continuation of the fine-thread series for diameters larger than $1\frac{1}{2}$ inches. Designated as 12 *UN*.

6. 16-Thread Series. The 16-thread series is a uniform pitch series for large diameters requiring fine-pitch threads. It is used as a continuation of the extra-fine-thread series for diameters over $1\frac{11}{16}$ inches. It is also used for adjusting collars and retaining nuts.

Although there are additional constant-pitch series with 4, 6, 20, 28, and 32 threads per inch that may be used when the threads in the Coarse, Fine, and Extra-Fine series do not meet *design requirements*, preference should be given wherever possible to the 8, 12, or 16 thread series. In some cases design requirements may dictate the use of Selected Combinations—designated *UNS*.

Screw Thread Classes* and Their Uses—Unified and American

"Classes of thread are distinguished from each other by the amount of tolerance or the amount of tolerance and allowance as applied to pitch diameter."*

The *tolerance*† on a dimension is the total permissible variation in its size.

An *allowance*† is an intentional difference in correlated dimensions of mating parts. It is the minimum clearance (positive allowance) or maximum interference (negative allowance) between such parts.

The *fit*† between two mating parts is the relationship existing between them with respect to the amount of clearance or interference which is present when they are assembled.

*Basic size** is the theoretical size from which the limits of size for that dimension are derived by the application of the allowances and tolerances.

Classes 1*A*, 2*A*, and 3*A* apply to *external* threads only. Classes 1*B*, 2*B*, and 3*B* apply to *internal* threads only.

Classes 1A and 1B are intended *for ordnance and other special uses.* They are used on components that require easy and quick assembly. These classes replace American National Class 1 for new designs.

Classes 2A and 2B are the *most commonly used* standards for general applications, including production of bolts, nuts, screws, and similar fasteners.

* ASA Y14.6-1957. † ASA B1.7-1949.

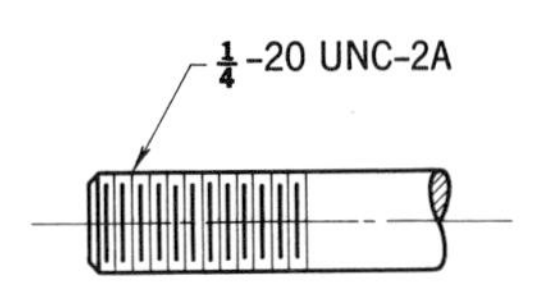

Fig. 19.11

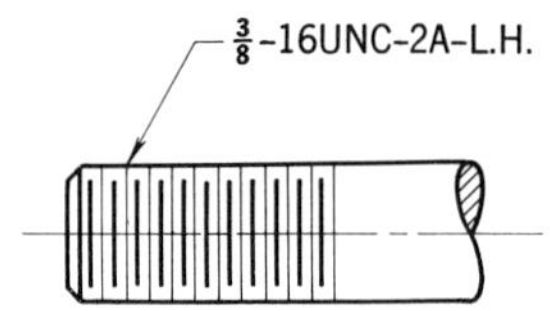

Fig. 19.12

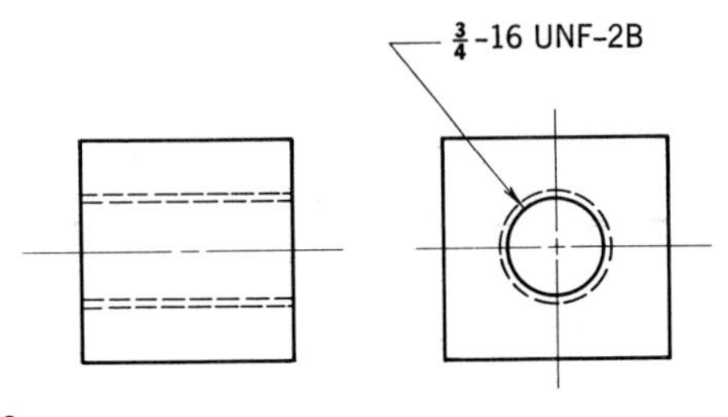

Fig. 19.13

Classes 3A and 3B are used when tolerances closer than those provided by *2A* and *2B* are required.

Classes 2 and 3, because of their long-established use, are still retained in the American Standard. They apply to both internal and external threads. Unified classes, however, are rapidly replacing the American Standard classes in new designs.

Specification and Designation of Screw Threads

"A screw thread is designated on a drawing by a note with leader and arrow pointing to the thread. The minimum of information required in all notes is the specification, in sequence, of the nominal size (or screw number), number of threads per inch, thread series symbol, and the thread class number or symbol, supplemented optionally by the pitch diameter limits. Unless otherwise specified, threads are right-hand and single lead; left-hand threads are designated by the letters *LH* following the class symbol; and double- or triple-lead threads are designated by the words *DOUBLE* or *TRIPLE* preceding the pitch diameter limits."‡

The following examples serve to demonstrate applications of the specification stated above.

EXAMPLE 1

$\frac{1}{4}$–20 *UNC–2A*. (See Fig. 19.11.)
PD. 0.2164–0.2127 (Optional).
$\frac{1}{4}$ = nominal diameter; 20 = number of threads per inch. *UNC* = Unified coarse series; 2A = class of fit (external thread); *PD* = pitch diameter limits, optional.

EXAMPLE 2

$\frac{3}{8}$–16 *UNC*–2A–*LH*. (See Fig. 19.12.)
$\frac{3}{8}$ = nominal diameter; 16 = number of threads per inch.
UNC = unified coarse series; 2*A* = class of fit.
LH = left hand.

EXAMPLE 3

$\frac{3}{4}$–16 *UNF–2B*. (See Fig. 19.13.)
$\frac{3}{4}$ = nominal diameter; 16 = number of threads per inch.
UNF = unified fine series; 2*B* = class of fit.

‡ ASA Y14.6-1957.

Question. Which of the following specifications* are correct?
$\frac{7}{8}$–9 *NC*–2
2–8 *N*–3
$1\frac{1}{8}$–7 *UNC*–3*B*.
$\frac{3}{4}$–9 *NC*–2.

BOLTS AND SCREWS

Billions of bolts are used annually in a variety of applications. Bolts are partially threaded cylindrical metal or plastic pieces. The primary function of bolts is to hold parts or units together. The most commonly used bolts are (*a*) through bolts and (*b*) stud bolts.

A *bolt* is defined "as an externally threaded fastener *designed* for insertion through holes in assembled parts, and is normally intended to be tightened or released by torquing a nut."*

A *screw* is defined "as an externally threaded fastener capable of being inserted into holes in assembled parts, of mating with a preformed internal thread or forming its own thread, and of being tightened or released by torquing the head."†

A *Through bolt* holds two pieces together by passing through holes in the pieces that are clamped by the nut which is screwed on the threaded end of the bolt. (See Fig. 19.14.)

A *Stud bolt* is used to fasten two pieces, one of which has a clear hole and the other a threaded hole. The bolt passes through the clear hole, is screwed into the threaded hole, and the pieces are then clamped by means of a nut screwed on the threaded free end of the bolt. (See Fig. 19.15.)

A *Cap screw* is practically the same as a through bolt, except for a greater length of thread, and is used without a nut. The shank (cylindrical portion) passes through the hole of one piece and is screwed into the threaded hole of the second piece. See Fig. 19.16. Cap screws are used where removal of the pieces is infrequent. They are available in a variety of heads, i.e., flat, fillister, hexagonal, round, etc. (See Appendix B, Table 8, p. 662, for details.)

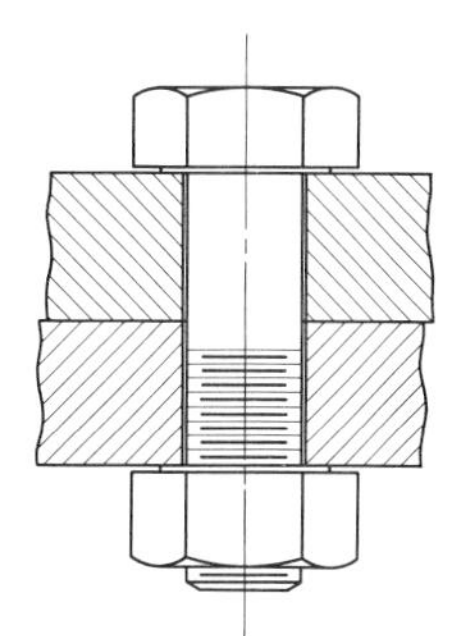

Fig. 19.14 Through bolt.

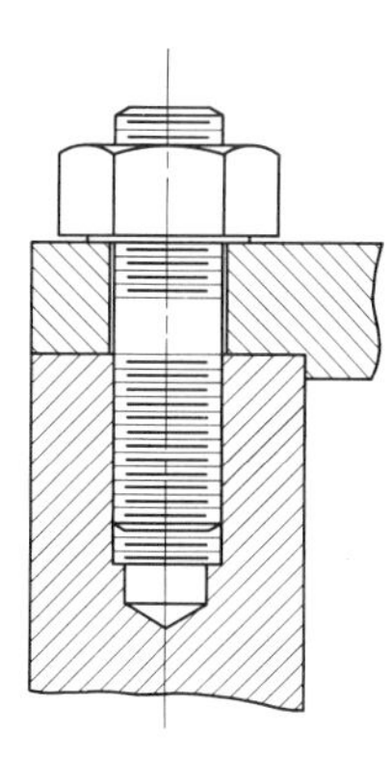

Fig. 19.15 Stud bolt.

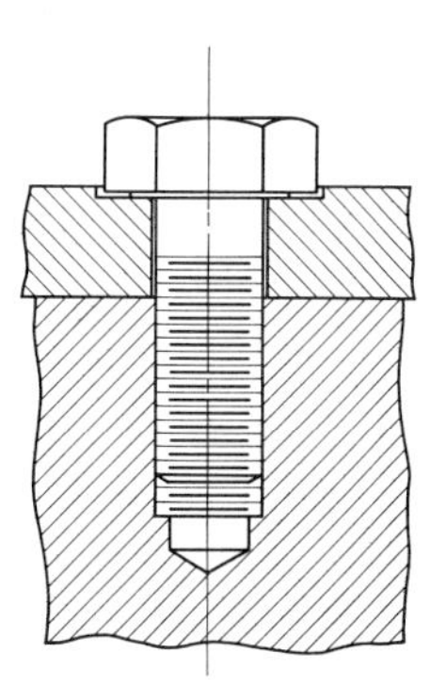

Fig. 19.16 Hexagonal head cap screw.

* See Table 7 in the Appendix, p. 660.
† ASA B18.2.1-1965.

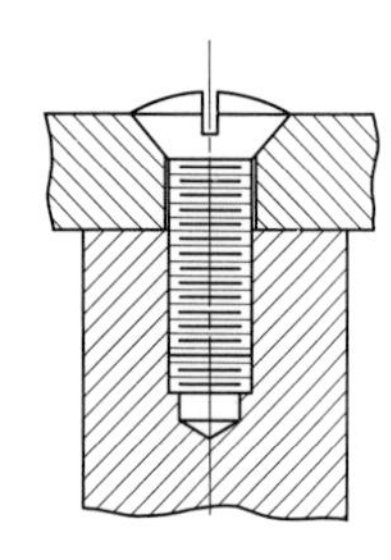

Fig. 19.17 Oval head machine screw.

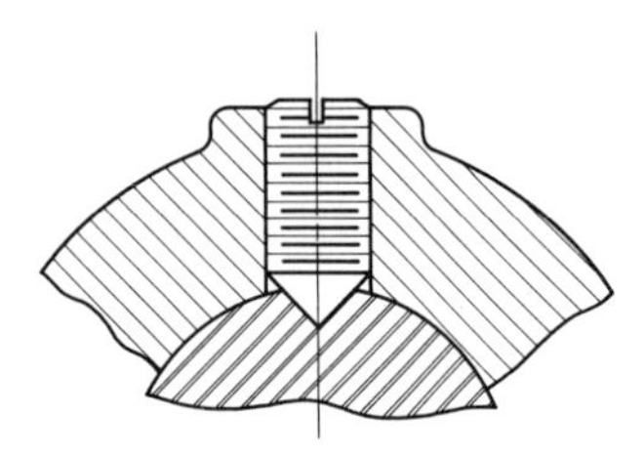

Fig. 19.18 Set screw, cone point.

A *Machine screw* is similar to cap screws except for size, being smaller and more convenient in fastening relatively thin parts. (See Fig. 19.17.) There are available several standard heads, i.e., round, flat, fillister, oval, hexagon, etc. Details of some of these are shown in the Appendix B, Table 9, p. 663.

Set screws are used to prevent relative motion between parts by entering the threaded hole of one part and setting the point against the other part, i.e., a pulley hub fastened to a shaft. (See Fig. 19.18.) There are several standard set screws with a variety of points, i.e., cup, flat, oval, cone, full-dog, and half-dog. Dimensions are given in the Appendix B, Table 10, p. 666

Miscellaneous Screws and Bolts. Many types of screws and bolts other than those presented are used in commercial practice. It would be futile to include all of them. A few of the more common ones, however, are shown in Fig. 19.19.

Graphic Representation of Bolts, Nuts, and Screws

The American Standards, ASA B18.2.1-1965, "Square and Hex Bolts and Screws," and ASA B18.2.2-1965, "Square and Hex Nuts," cover the latest information on:

1. The complete general and dimensional data for the various types of square and hexagon bolts, nuts, and screws recognized as "American Standard."
2. Wrench openings for bolts and screws; grade markings for steel bolts, nuts, and screws; specifications for identification of bolts and screws; and formulas on which dimensional data are based.

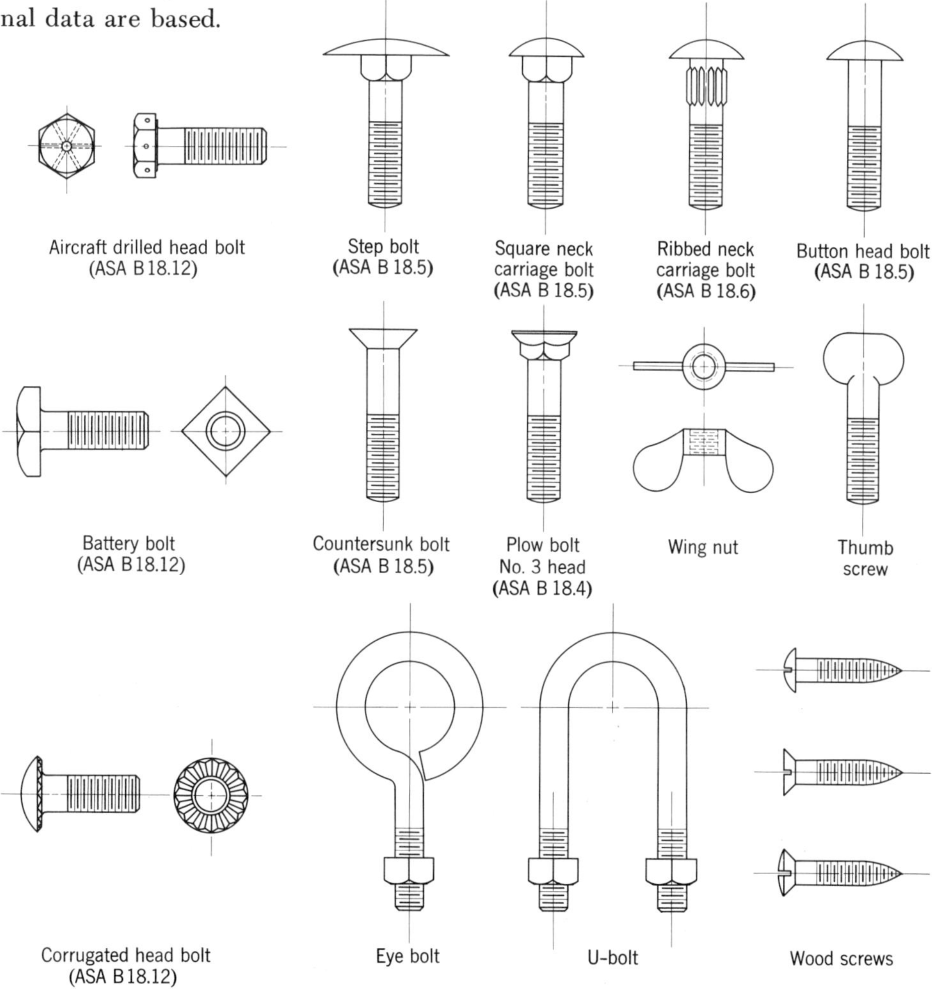

Fig. 19.19 Miscellaneous screws and bolts.

In addition the following definitions are included in these standards:

Washer Face. The washer face is a circular boss on the bearing surface of a nut, bolt, or screw head.

Height of Head. The height of head is the over-all distance from the top of the head to the bearing surface, including the thickness of the washer face where provided.

Thread Length. For purposes of this standard, thread length is the distance from the extreme end of the bolt or screw to and including the last complete (full form) thread.

Bolt or Screw Length. Bolt or screw length is the distance from the bearing surface of the head to the extreme end of the bolt or screw, including point, if product is pointed.

Thickness of Nut. The thickness of nut is the over-all distance from the top of the nut to the bearing surface, including the thickness of the washer face where provided.

Typical details of a modern bolt are shown in Fig. 19.20.

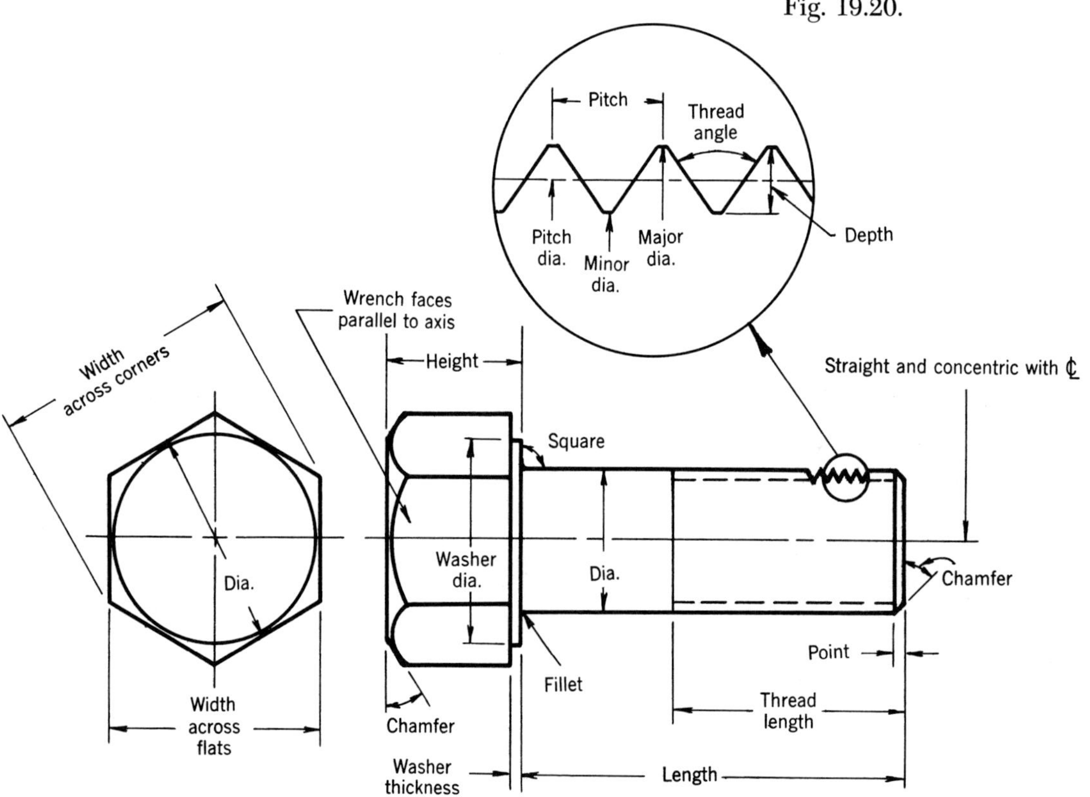

Fig. 19.20 Details that make up the modern bolt. (Courtesy Industrial Fasteners Institute)

Dimensions for hexagonal and square-head bolts and nuts are given in Tables 11–20, pp. 670–679 of the Appendix B. Formulas for widths across flats and heights of head are given in the American Standards. For example, in the case of a square bolt, the width across flats, W, is one and one-half times the nominal bolt size, D; the height of the head is two-thirds D; and the height of the nut is seven-eighths D. Variations from these dimensions for different bolt sizes, semifinished, and finished bolts and nuts are given in the standards. Typical examples are shown in Figs. 19.21, 19.22, and 19.23.

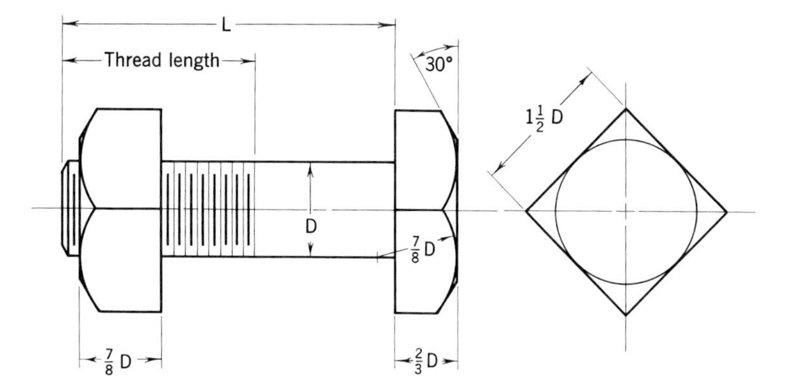

Fig. 19.21 Square bolt and nut.

Bolt Specifications

Detail drawings of standard bolts and nuts are not necessary. Commercial templates are available for drawing bolt heads and nuts. Bolts and nuts are specified by giving the following information: nominal size (diameter of bolt); length (from bearing surface of head to end of shank); thread specification; material (if other than steel); series and finish; shape of head and nut (if different from head); and name (i.e., bolt).

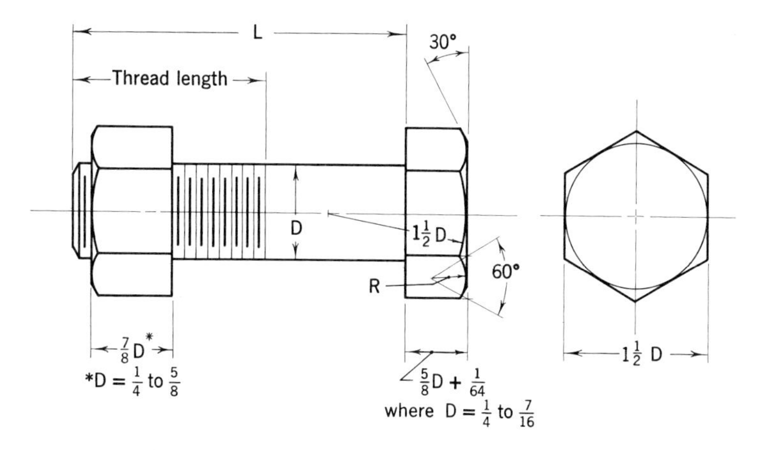

Fig. 19.22 Hexagon bolt and nut.

EXAMPLES

(*a*) $\frac{3}{4} \times 2$, 10 *UNC–2A*, semifinish hex head bolt.
(*b*) $\frac{7}{8} \times 2\frac{1}{2}$, 14 *UNF–2A*, brass fin. hex head bolt.
(*c*) $\frac{3}{4} \times 2$, 10 *NC–2*, heavy unfin. sq. head bolt.
(*d*) $1\frac{5}{8}$–11 *UNC–2B* semifin. hex nut.

Locking Devices

There are several types of standard locking devices, such as *jam nuts,** which tighten against the gripping nut; *slotted** and *castle** nuts which are prevented from turning by the use of cotter pins that pass through a hole in the bolt and are placed in slots of the nuts; and cotter pins that pass through bolt holes just above the nuts.

Standard spring lock washers† are intended for automotive and general industrial application. These devices compensate for developed looseness and the loss of tension between component parts of an assembly.

Today there are locknuts available for nearly all known applications. It is primarily a matter of proper selection to meet given and anticipated conditions.

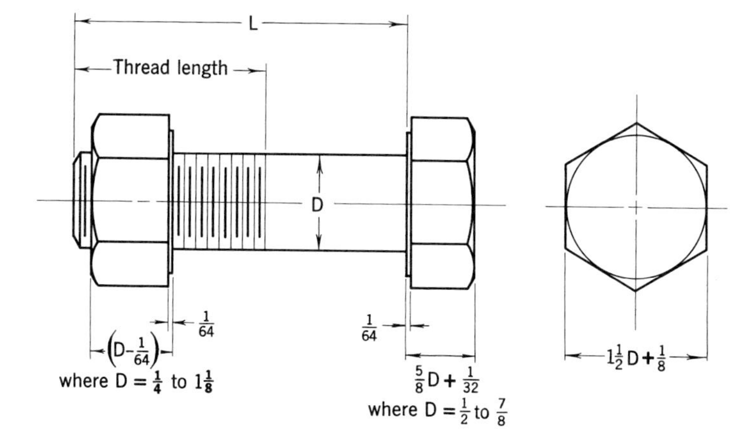

Fig. 19.23 Heavy hexagon bolt and nut.

* ASA B18.2.2-1965.
† See Tables 29 and 30, pp. 690–691.

Locknuts which depend upon pressure contact against the piece to maintain the locking effect are very efficient on bolts in shear; however, where some bolt stretch occurs, as in the case of bolts in tension, pressure on the piece may relax so that the locking effect is lost.

Temperatures affect the operation of locknuts. Some are excellent for use at low and medium temperatures, but lose their grip at high temperatures. Others are most efficient in sizes ranging upwards from $\frac{1}{2}$ inch, but would be quite useless in smaller sizes because of production difficulties which often result in the manufacture of locknuts of nonuniform holding power.

It is difficult to estimate the man-hours that have been saved in the speeding up of assembly operations and in the reduction of maintenance services as a result of using modern locknuts.

While cotters and drilled bolts, lock washers, off-angle and off-lead threads, and the peening of threads, as methods for locking a nut on a bolt, are entirely satisfactory for some purposes, the fact remains that *for a multitude of applications* these locking devices are as out of date as the automobiles that were introduced at the turn of the century. Today's assembly and production procedures call for the use of modern fasteners and locking devices. A few of the modern designs are shown in Fig. 19.24.

KEYS AND KEYWAYS

This type of fastener is used to eliminate relative motion between shafts and wheels, pulleys, etc. There are several commonly used keys such as: (*a*) *square and flat* plain parallel stock keys; (*b*) Woodruff keys which are semicircular; (*c*) *taper* stock keys, both square and flat; (*d*) Gib-head taper stock keys, also both square and flat shapes; and Pratt and Whitney keys.

Square, flat, taper, and gib-head taper stock keys are specified by giving the width, height, and length. Sizes should, whenever possible, conform to ASA B17.1. American standard Woodruff keys, keyslots, and cutters are generally specified by key number. Tables 34 and 35, pp. 695–696, give key details and dimensions.

STANDARD PINS: TAPERS AND SPLINES

In designs for relatively light loading, pins or tapers may be used for fasteners. Detailed dimen-

BETHLEHEM ANCO LOCK NUT

Description: A precision-made, single unit, self-locking nut in which the locking element is a special alloy steel locking pin which has extra high tensile strength. The nut can be reused repeatedly without appreciable loss of locking effectiveness. In sizes from $\frac{1}{4}$-in. to 3-in. diameter, the ANCO Lock Nut is made from a full range of metals, including carbon steel, stainless steel, aluminum, brass, bronze, and silicon-bronze.

Manufacturer: Bethlehem Steel Company, Bethlehem, Pennsylvania.

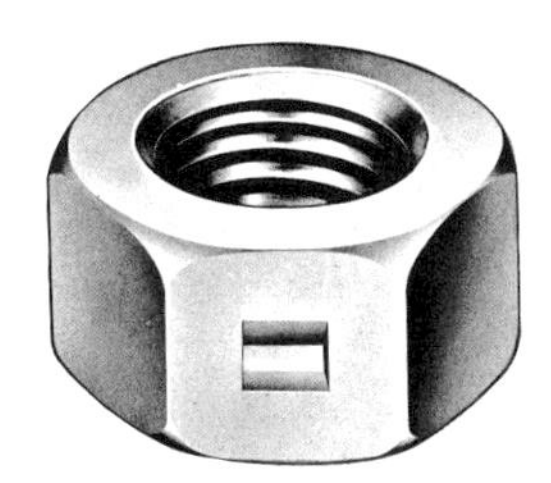

BI-WAY LOKUT

Description: An all-metal, prevailing-torque type self-locking nut designed for free starting from either end. Center threads are altered into an oval configuration to give a stiff spring locking action. Furnished in American Standard Finished Hexagon Series with American Coarse and Fine Threads of Class 2B tolerance.

Manufacturer: Shakeproof Division, Illinois Tool Works Inc., Elgin, Illinois.

CONELOK

Description: The Conelok is a one-piece, reusable, prevailing-torque locknut with the locking characteristics obtained by accurately preforming the threads in the locking section.

Manufacturers: Automatic Products Company, Detroit, Michigan; The National Screw & Manufacturing Co., Cleveland, Ohio.

LAMSON LOCK NUT

Description: The Lamson Lock Nut is a one-piece, spring-action collar style prevailing-torque type lock nut. It is characterized by ability to maintain locking action. The Lamson Lock Nut is available in a light (regular) and a heavy series. Both plain and plated nuts are standard.

Manufacturer: The Lamson & Sessions Co., Cleveland, Ohio.

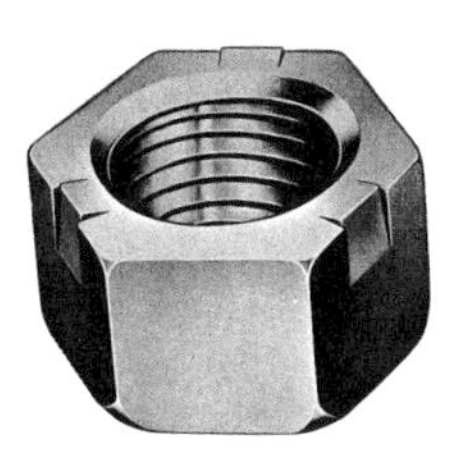

LOKUT

Description: An all-metal, prevailing-torque type, self-locking nut designed for free starting. Unique three point shear depression assures positive locking action. Several top threads are projected inward providing efficient distribution of the locking load. Manufactured in Machine Screw American Standard Light and Regular Series.

Manufacturer: Shakeproof Division, Illinois Tool Works Inc., Elgin, Illinois.

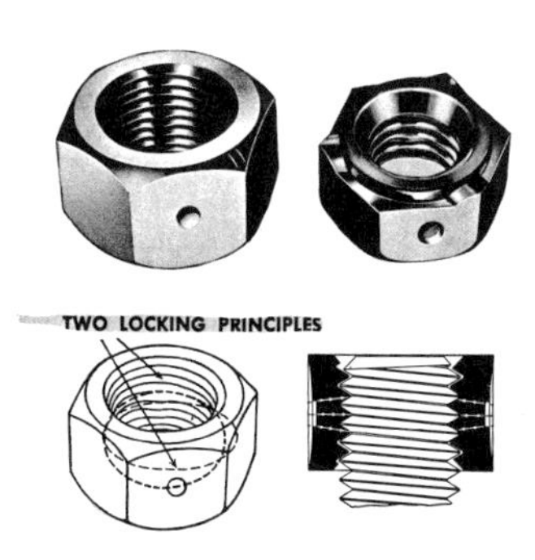

'M-F' TWO-WAY LOCK NUT

Description: The 'M-F' Two-Way Lock Nut is an automatable, reusable, one-piece, all-metal, prevailing-torque type lock nut available in low and high-carbon steels as well as non-ferrous materials. The Two-Way locking feature can be added to other fasteners including cap nuts and weld nuts.

Manufacturers: MacLean-Fogg Lock Nut Co., Chicago, Illinois; Russell, Burdsall & Ward Bolt and Nut Co., Port Chester, New York.

Fig. 19.24 Modern locking designs. (Courtesy Industrial Fasteners Institute)

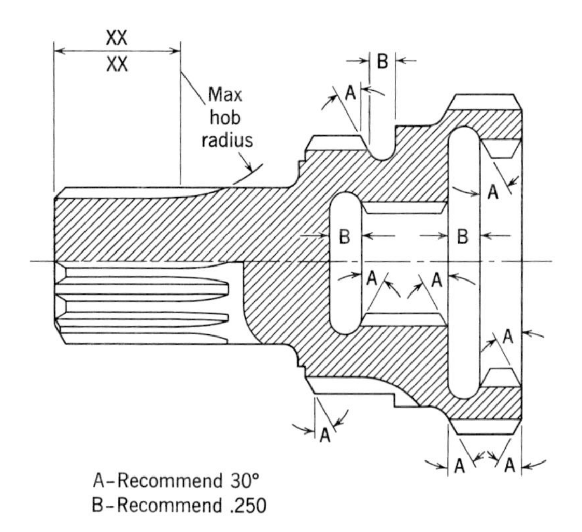

Fig. 19.25 Representation of splines. (Courtesy ASA)

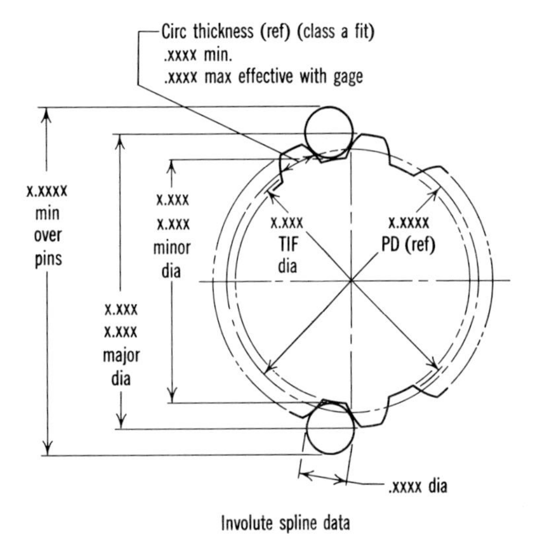

Fig. 19.26 External spline dimensions. (Courtesy ASA)

sions for various pins and tapers are given in ASA B5.20 and ASA B5.10 respectively; however, Tables 39 and 41, pp. 699, 701, contain dimensions for the commonly used taper pin. Spline shafts provide a positive means for the transmission of relatively heavy loads. Information of value to the design engineer is available in ASA B5.15-1960. Figures 19.25 and 19.26 show "representation of splines" and "external spline dimensioning."

RIVETS

Rivets are most commonly used in permanently connecting members of structural frames* and in the fabrication of tanks, boilers, etc. Rivets consist of a cylindrical portion known as the body, or "shank," and a "head" that is integral with the body. Rivets are designated by giving the diameter, length, and type of head, i.e., $\frac{1}{2}$ inch × $2\frac{1}{2}$ inches, button-

* In recent years high-strength bolts have replaced rivets in steel building construction, e.g., Time and Life Building, New York City.

head rivet. Figure 19.27 shows the nomenclature applied to a button-head rivet as manufactured, and after being driven. The length of the rivet depends on the grip dimension, which is the thickness of the metal parts held together by the rivet. There are a number of head shapes, such as, acorn head, cone head, flat-top, etc. A few are shown in Fig. 19.28.

Riveted joints may be either butt joints or lap joints, depending on the type of structure. Butt joints are commonly used in "girder splices," whereas lap joints are used in tank fabrication and in the construction of ship hulls. Figure 19.29 shows examples of both a butt joint and a lap joint. Special rivets (i.e., "blind rivets") can also be used in such cases where only one side of the work is exposed.

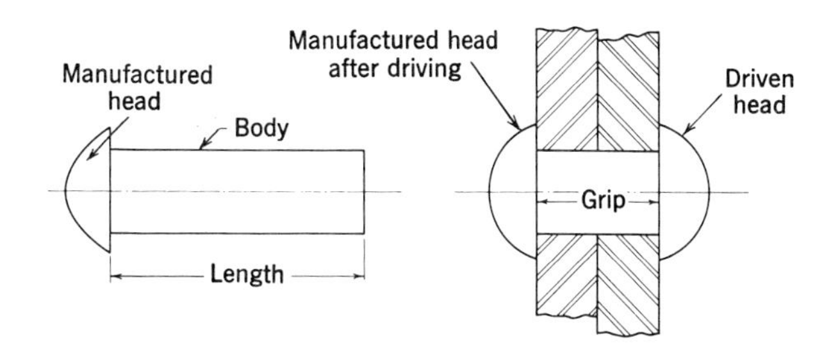

Fig. 19.27 Rivet nomenclature—button-head rivet.

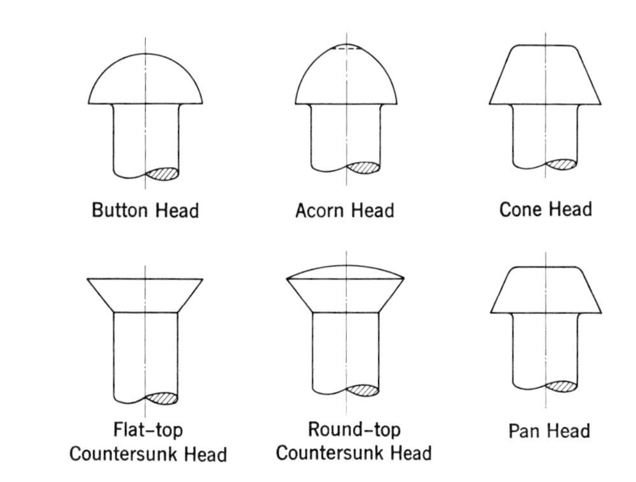

Fig. 19.28 Rivet head shapes.

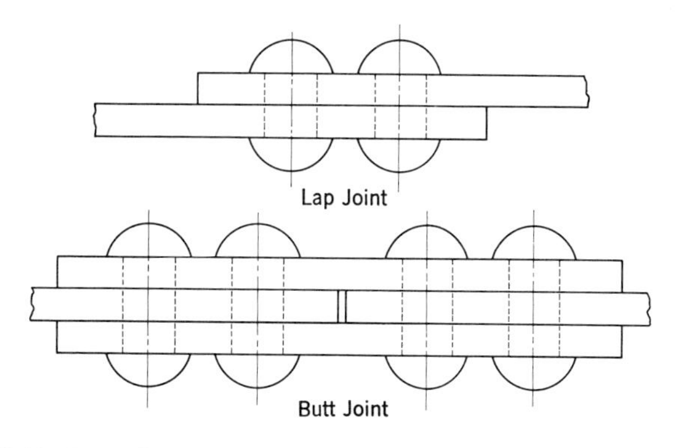

Fig. 19.29 Riveted joints.

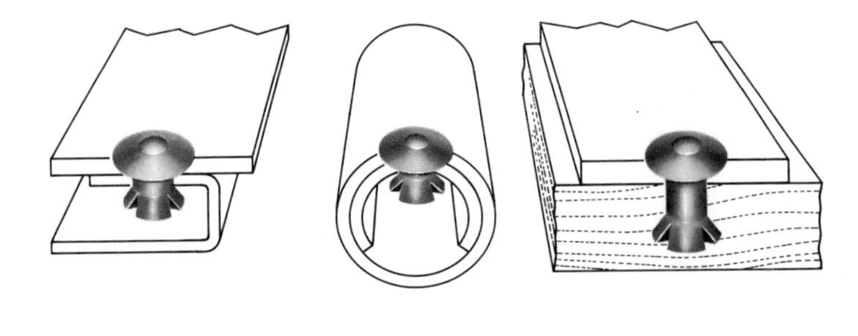

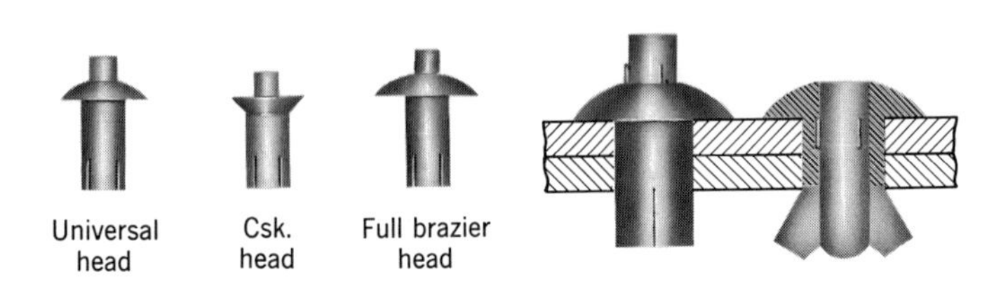

Fig. 19.30 Special purpose rivets.

Head shapes and a few typical examples of "blind rivets" are shown in Fig. 19.30.

WELDING

Welding, an excellent permanent fastener, is a most important method used in manufacturing metal components and in building construction. Welding design is beyond the scope of this text. Appendix C, p. 726, however, contains the commonly used graphical symbols for the representation of type of weld and associated dimensions. In addition, there are examples of the use of resistance welding symbols; of the use of arc and gas welding; and the use of welding symbols on a structural design drawing.

SPRINGS

Springs are elastic units, made from wire or strip material, which stretch, compress, or twist under applied forces. Wire springs may be helical or spiral and may be made from round, square, or special-section wire. They are classified as (*a*) compression, (*b*) extension, or (*c*) torsion types.

Compression springs are open-coiled helical springs that resist compression. The most common form has the same diameter throughout its entire length, and is known as a straight spring. Extensive use, however, is made of tapered and cone-shaped compression springs.

Extension springs are close-coiled springs that resist pulling forces. They are close wound, in contact with each other, and made from round or square wire. The coils may be wound so tight as to require an effort to pull them apart. This coiling load is known as the initial tension.

Torsion springs exert force along a circular path, thus providing a twist or a torque. Although compression and extension springs are subjected to torsional forces, a "torsion" spring is subjected to bending forces. The ordinary spring hinge is typical of one of the common uses of torsion springs.

Spring specifications usually include: material; outside diameter, inside diameter; free length, number of coils; type of ends; position of loops; winding, left or right; and length of coils. Figure 19.31 shows the graphical representation and dimensioning of compression, extension (tension), and torsion springs.

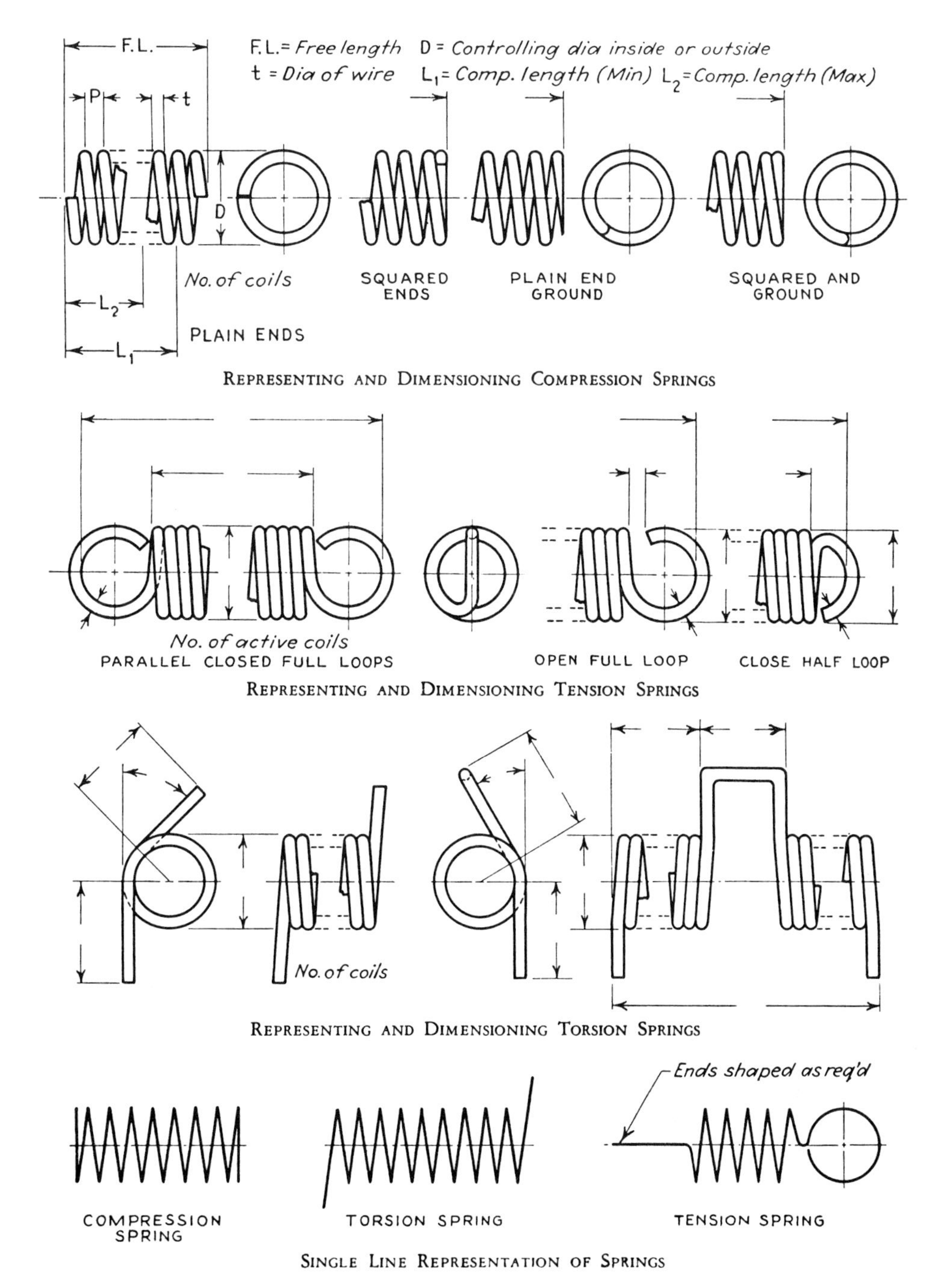

Fig. 19.31 Representation and dimensioning of springs. (Courtesy ASA)

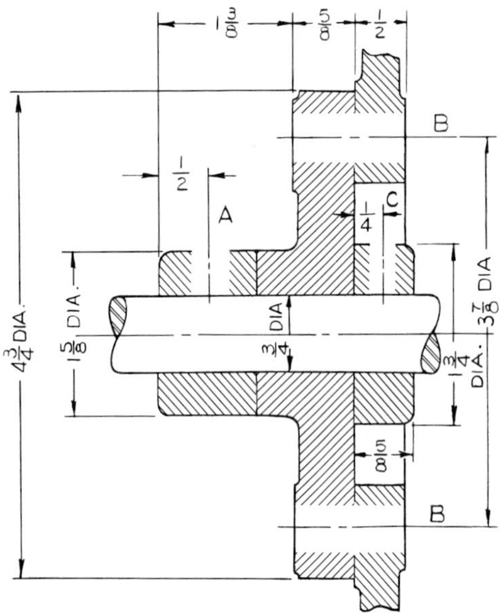

Fig. E-19.1

EXERCISES

1. On center lines *B*, show appropriate size* hexagon-head bolts and nuts (four required). On center line *A*, place a safety setscrew,* and on center line *C*, a hexagon-socket setscrew.* Use cone pointed setscrews. Dimension and specify the fasteners only. (See Fig. E-19.1.)

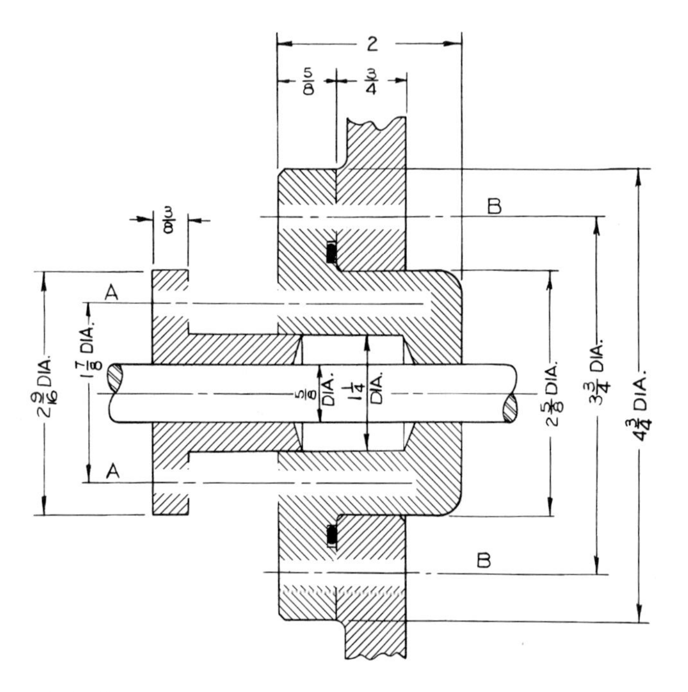

Fig. E-19.2

2. On center lines *A*, show studs* and regular hexagonal nuts.* On center lines *B*, show hexagon-head cap screws* (four required). Dimension and specify the fasteners only. (See Fig. E-19.2.)

* Instructor may specify diameters.

Note: Many problems in Chapters 20 and 21 also include the use of fasteners.

3. Prepare a *freehand* section assembly of the Hanger shown in Fig. E-19.3. The vertical strap is fastened with two $\frac{1}{2}$-13NC-3 hex bolt and nut. *Assume needed dimensions.* Dimension and specify the fasteners.

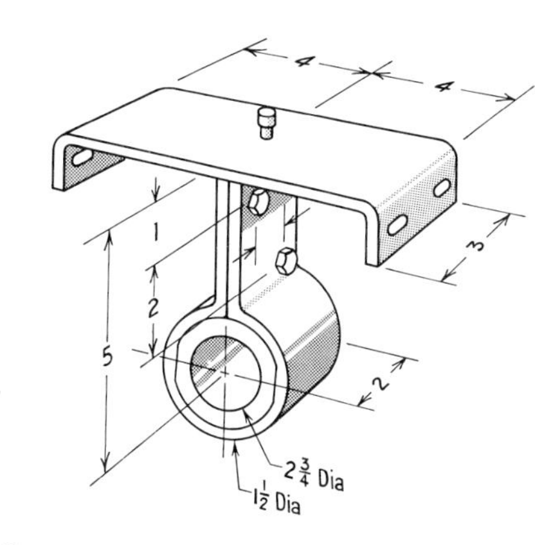

Fig. E-19.3 Hanger.

4. Prepare a *freehand* section assembly of the cast-iron *split pillow block* shown in Fig. E-19.4. The cover is fastened to the base by $\frac{3}{8}$-16NC-3 $\times$ $1\frac{1}{4}$ long stud bolt. Dimension fasteners only.

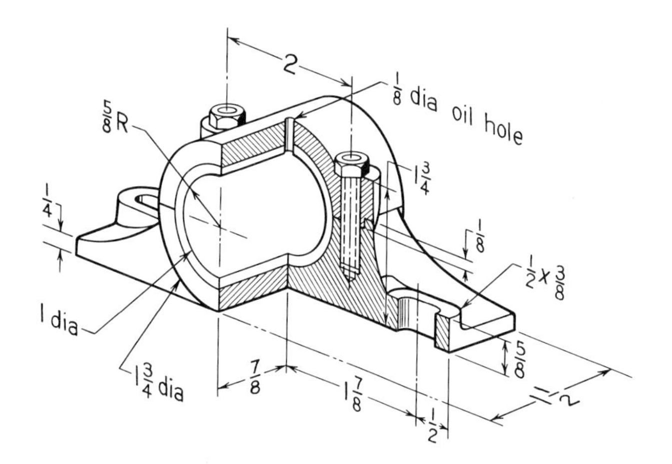

Fig. E-19.4 Split pillow block.

5. Prepare a *freehand* section assembly of the shuttle valve shown in Fig. E-19.5. Assume needed dimensions and dimension the fasteners only.

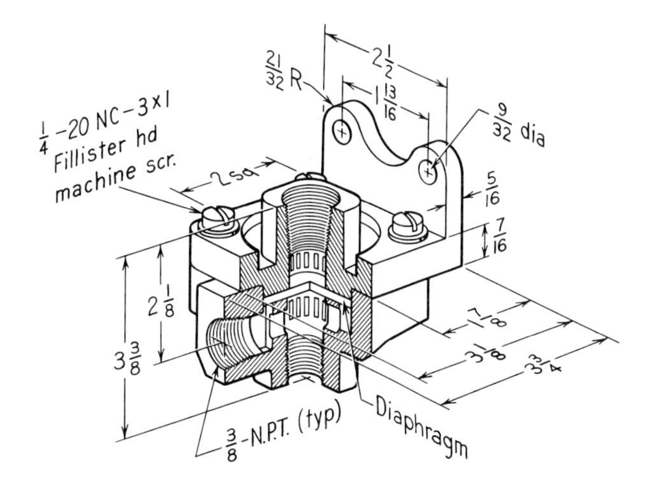

Fig. E-19.5 Shuttle valve.

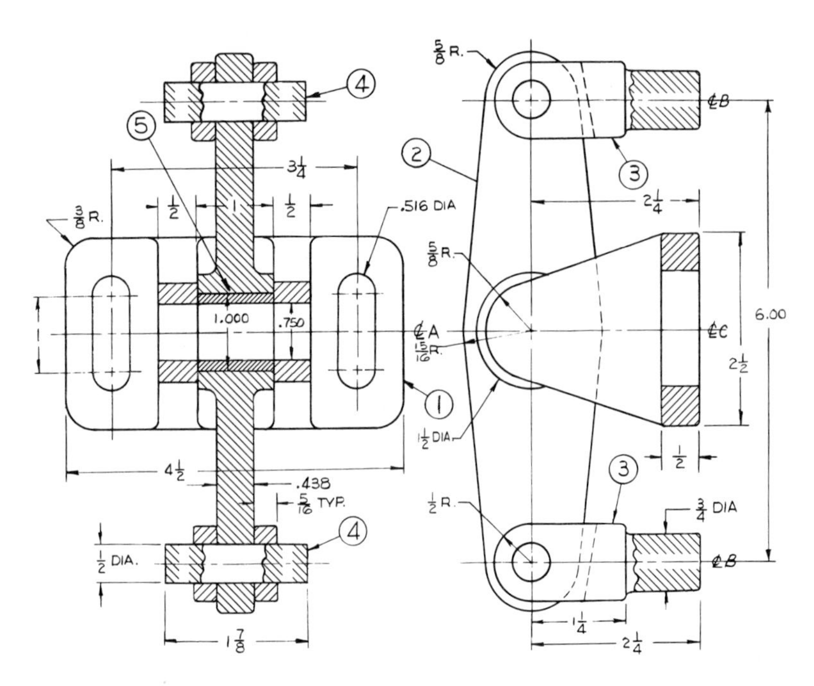

Fig. E-19.6 Rocker arm assembly.

6. Shown in Fig. E-19.6 is a Rocker Arm Assembly.

Instructions:

Piece 1 and piece 2 are held together with a shoulder screw of the following dimensions:

Shank diameter—0.7492 to 0.7484

Length of shank—1.625

Thread—$\frac{1}{2}$-13UNC-2A

Thread length—1.00

Hex head to have same dimensions as standard $\frac{3}{4}''$ bolt.

Bolt is drilled with $\frac{1}{8}$ dia drill to a depth of $1\frac{3}{8}$ at head end. This hole is then counter-drilled and threaded with a $\frac{1}{8}$-NPT thread to a depth needed to accommodate a standard lubrication fitting. A $1\frac{1}{8}$ hole, located $\frac{27}{32}$ from the under side of the head, is drilled across the bolt.

The bolt is secured with a standard wide series plain washer, a semifinished hex slotted nut and standard cotter pin.

Show the bolt, washer, nut and lubrication fitting on center line *A*.

Piece 4 is held in place in pieces 2 and 3 by means of cotter pins or snap rings of a suitable size. Show these drawn in place as required.

On center line *B* show eyebolts of the following dimensions:

Shank—$\frac{3}{8}$-16UNC-2A × $\frac{3}{4}$ long

Inside diameter of eye—$\frac{3}{4}''$

Length from threaded end to center of eye—$2\frac{1}{2}$.

Show the eyebolts screwed into a standard depth of engagement. All parts are made of steel.

On center line *C* show a regular series square head bolt and nut with a type B plain washer, heavy helical spring lock washer, and a regular semifinished hex nut.

7. Shown in Fig. E-19.7 is a Control Cable Pulley Assembly.

Instructions:

On center line *A* show a $\frac{7}{16}$-20UNF-2B fillister head machine screw, $3\frac{1}{2}$ long; lightweight plain washers under the screw head and nut, and a heavy series elastic stop nut.

On center line *B* show a bolt of the following dimensions:

Shank 0.4998 to 0.4981 dia. × 0.875 long

Head dimensions are the same as for a $\frac{9}{16}$ bolt

Threaded portion $\frac{3}{8}$-16UNC-2B × 0.500 long

Heavyweight plain washer, standard semifinished thick slotted nut and cotter key.

On center line *C* show $\frac{5}{16}$-18UNC-2B flat head cap screws, $1\frac{1}{4}$ long hex nut and spring lock washer.

Piece 4 is retained in piece 2 by suitable snap rings. *Scale*—Full size, or as directed by the instructor.

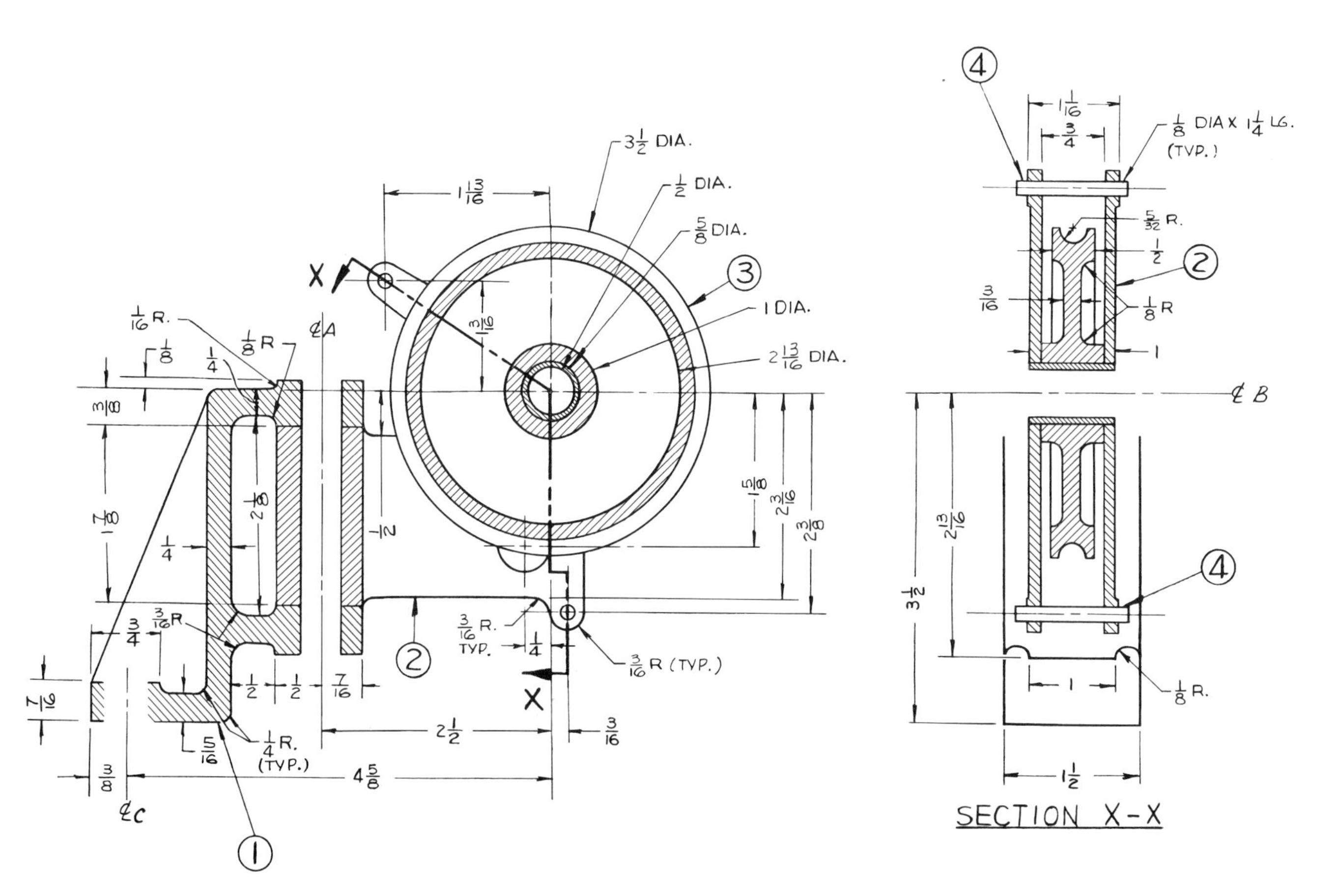

Fig. E-19.7 Control cable pulley assembly.

8. Shown in Fig. E-19.8 are views of a Jet Engine Removal Installation.

A serious accident resulting from the dropping of a jet engine during removal from a plane was narrowly averted when the slipping track roller was noticed in time. Failure of the roller was caused by incorrect assembly of the bolt as shown in the sketches.

Redesign this bolted assembly to prevent this error in installing the bolt.

Redesigning should be accomplished so that the existing rollers and brackets can be used, with the thread size on the bolt unchanged. It would also be desirable to utilize standard bolts.

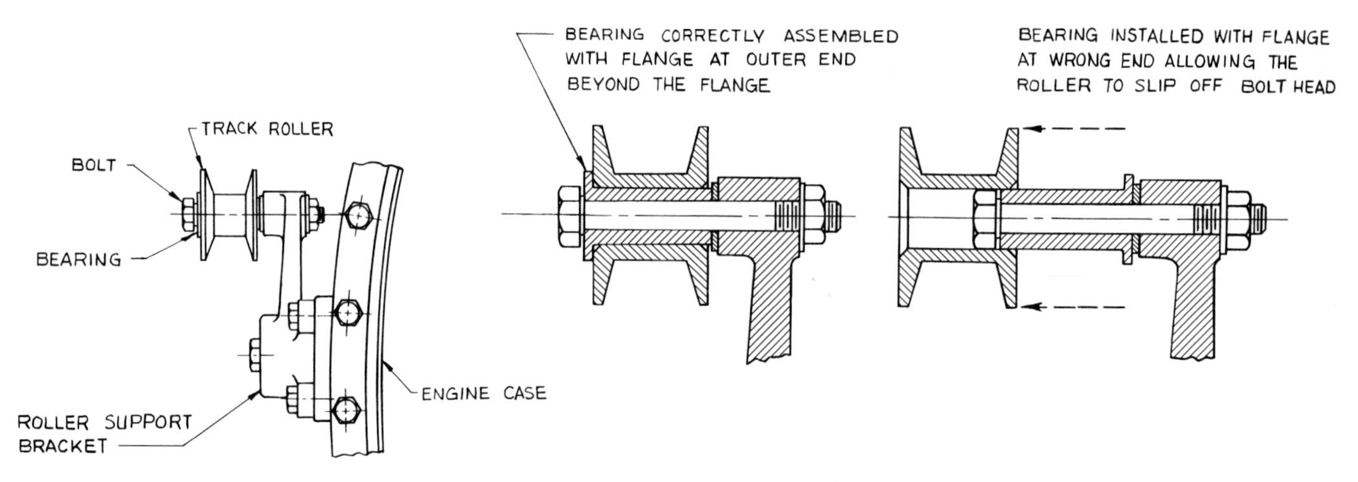

Fig. E-19.8

9. A portion of the Elevator Control System installation on a modern bomber is shown in Fig. E-19.9.

A near crash occurred when the elevator locked in a nose-up attitude shortly after takeoff. Investigation after landing showed the malfunction was due to a pivot bolt which had been installed improperly.

Make a sketch of the bolt installation showing your version of a *new design* which would make this improper assembly procedure impossible. Redesigning should be accomplished with a new fastener arrangement and without major changes in the castings involved.

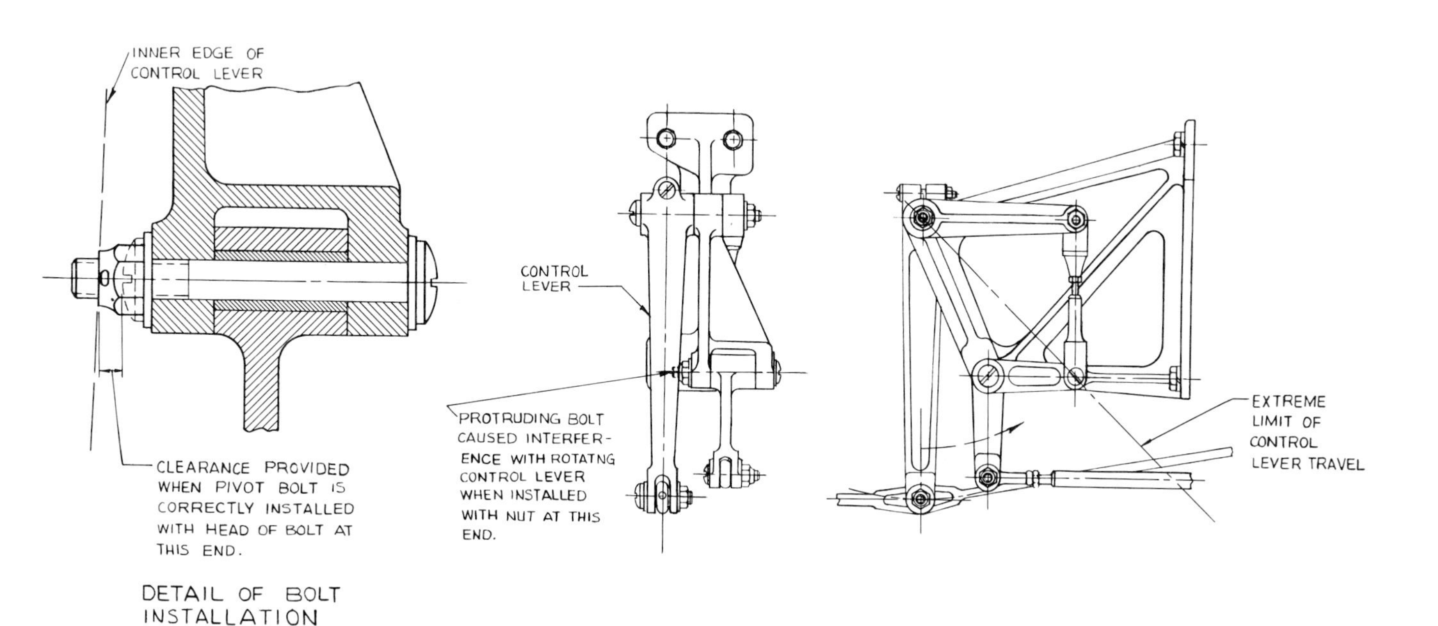

Fig. E-19.9 Elevator control system.

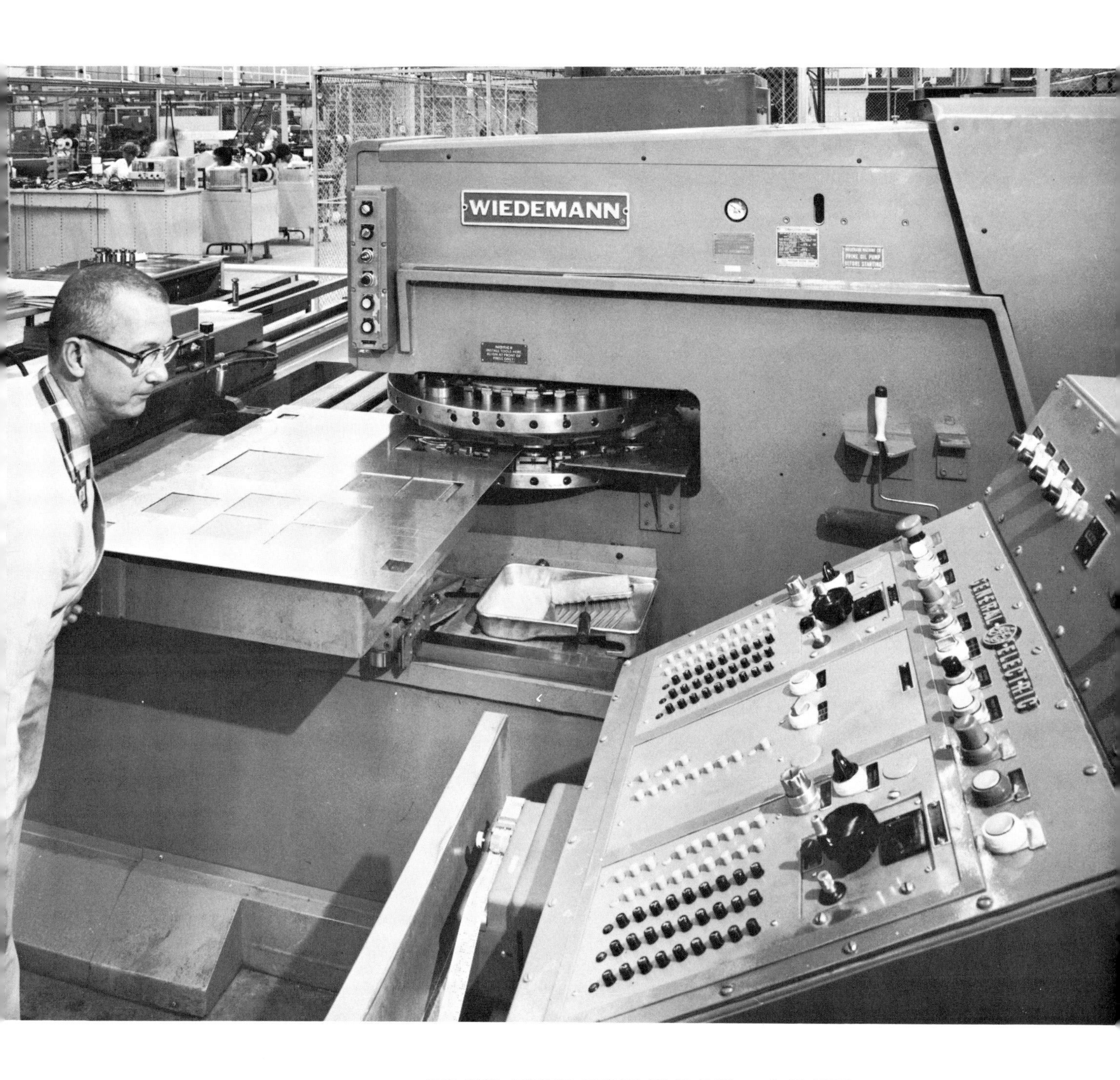

DIMENSIONS AND SPECIFICATIONS FOR PRECISION AND RELIABILITY 20

The production of devices, machines, structures, etc., requires much more than "shape descriptions" of their components. The orthographic (or pictorial) views of a part, or unit, including necessary sections, tell the *shape story* of the object. To specify size, however, dimensions must be added to the views. In addition, efficient production requires information regarding the kind of materials to be used, treatment and finish of the materials, and special instructions, *if necessary,* concerning methods of manufacture and verification, shipping of subassembly units, methods of assembly at the site, etc. *Without dimensions and specifications, the design drawing is useless for practical purposes.*

The well-qualified engineer must have a good knowledge of materials, manufacturing processes, inspection methods, construction methods, and product design in order to develop devices, machines, and structures that are functional, reliable, and economical with the added plus of eye appeal.

As an engineering student you should make every effort to learn as much as possible about materials, manufacturing processes, inspection procedures, and construction methods. Much can be gained from course work, from plant visitations, and from the latest technical periodicals and journals. Summer employment, especially at the end of the freshman and sophomore years, can contribute much to your knowledge and understanding of manufacturing processes and construction methods.

The precision parts that are needed in modern engineering designs require that complete information, *with only one possible interpretation,* be made available in the form of carefully dimensioned and specified drawings. Accurate dimensioning reflects serious and careful thinking. It demonstrates that the engineer has clearly visualized the finished product and the functions of its parts. Moreover, the well-thought-through design is evidence of the engineer's knowledge of feasibility, of availability of materials, of plant facilities for production and verification, and of personnel required to produce the product as planned.

A properly prepared engineering drawing must tell the true and complete story about the design. There is no room for ambiguity that could lead to costly errors.

The physical process of recording dimensions on

the drawing is usually an activity performed by the drafting technician; however, it should be obvious that the design engineer must have a complete understanding of dimensioning so that he can direct and check the work of the drafting technician. *Drawings must be correctly dimensioned to properly communicate the engineer's intent.* This chapter will provide the student with an opportunity to become acquainted with some problems in adequate communication of design specifications and possible solutions which will better ensure the correct interpretation of the design intent.

In dimensioning, as in any form of communication, the message must be carefully thought out. If the communique is to be effective it must be clear and understandable, complete, and free from ambiguities. To achieve a properly dimensioned drawing the designer must anticipate possible misinterpretations and finally establish his specifications in such a way that the reader must accept the information *as it is intended.*

To develop some fundamental concepts let us consider the geometric shapes of engineering structures. We find that the geometry may be resolved into (1) straight lines and plane surfaces, (2) regular curved shapes, and (3) nonregular (free form) shapes. Moreover, we should note that the great bulk of the shapes can be resolved into the simple geometry of the first two forms. In our preliminary discussion we will deal only with straight line, plane, and regular curved shapes and treat the free-form shapes later as a special topic.

Consider the design shown in Fig. 20.1. The plane geometric shapes include straight lines (rectangular and triangular shapes) and regular curved lines (arcs and circles). Consider each "shape element" and decide upon the dimensions which will define the size of the element. In doing this we will omit the (″) symbol because all dimensions are in inches. This is an accepted practice.

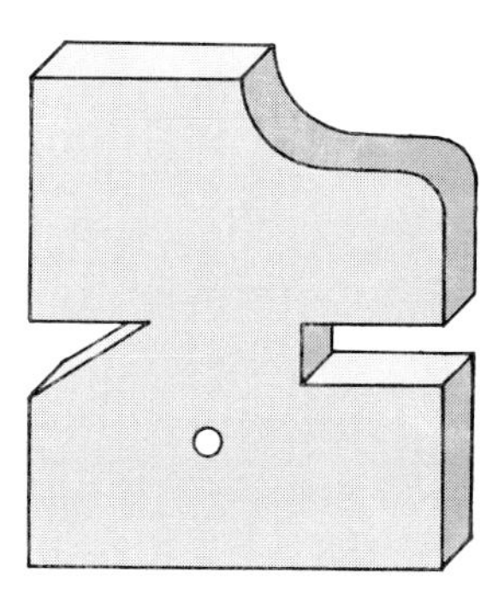

Fig. 20-1

The elements are shown in Fig. 20.2.

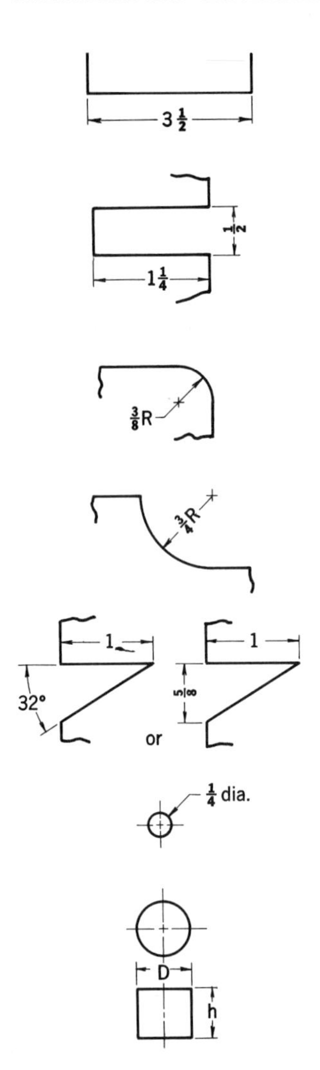

Fig. 20.2 Element shapes.

A *line element* (edge view of a plane surface bounded by two surfaces)

A *rectangular element*—negative (a rectangular slot) NOTE: A positive rectangular element (a rectangular tongue as in Fig. 20.17 would be dimensioned in a similar manner.

A *curved element*—positive arc

A *curved element*—negative arc

A *triangular element*—negative (a triangular slot) NOTE: A positive triangular element, not shown, would be dimensioned in a similar manner.

A *cylindrical element*—negative (a round hole). The depth would be given on an adjacent view.

A *positive cylindrical element* (not included as a feature of the design part) would be dimensioned by giving the diameter and height on the rectangular view.

Now when the elements are dimensioned on the drawing of the design part, Fig. 20.3, it is quite apparent that additional information is required to establish the dimensional specifications of the part. By adding dimensions which show the locational relationship of the elements, we establish the drawing as shown in Fig. 20.4. Notice that the dimensions could have been arranged in several different ways. The choice of arrangement depends on the *functional* requirements of the part and the use to which the drawing will be put. These considerations are discussed in detail later.

Now, to test the adequacy of the dimensional specifications in Fig. 20.4 we ask this question: "Can the part be made from the given information?" The answer is an emphatic, "NO!" To make the part we must know the material, surface treatment, and other such specifications, but aside from these—please recognize that *the basic dimensions given define a perfect part.* Unfortunately, the human being can achieve perfection only by accident. To deal with reality, we must assign a tolerance to each basic dimension. By tolerance we mean the permissible variation that can (and will) be allowed to occur in any dimension without adversely affecting the functionability of the part. In actual practice it is necessary to carefully examine and then assign the tolerance for each individual dimension. In this case we are not aware of the function of the part; therefore, we cannot properly assign the tolerances. For the sake of illustration only, we may arbitrarily assign a tolerance of $\pm\frac{1}{64}$ to each linear dimension and $\pm 1°$ to the angular dimension. Now, if "reasonable care" is exercised in the manufacturing processes, and if the tolerances assigned are functionally proper, the part should be acceptable. We will see, as we develop our knowledge of dimensioning, that a typical precision part cannot afford the above "ifs" and a considerably greater amount of information must be given to provide complete and proper dimensional specifications.

From the foregoing example we may establish the procedural concept of dimensioning; namely: (1) determine the basic (perfect) dimensions which define the size of each element, (2) add the dimensions to locate the elements, (3) assign the tolerance for each basic dimension, (4) establish the verbal specifications (i.e., notes describing the material requirements,

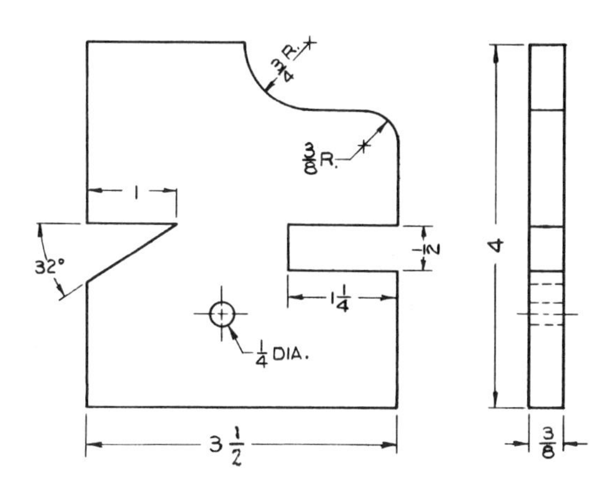

Fig. 20.3

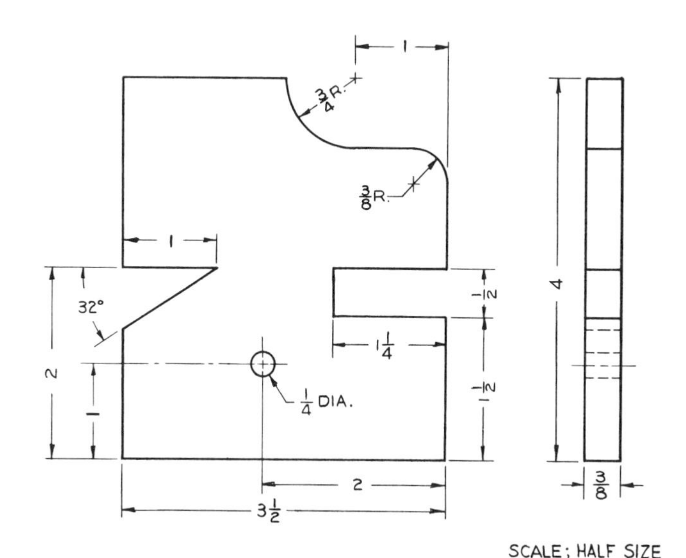

Fig. 20.4

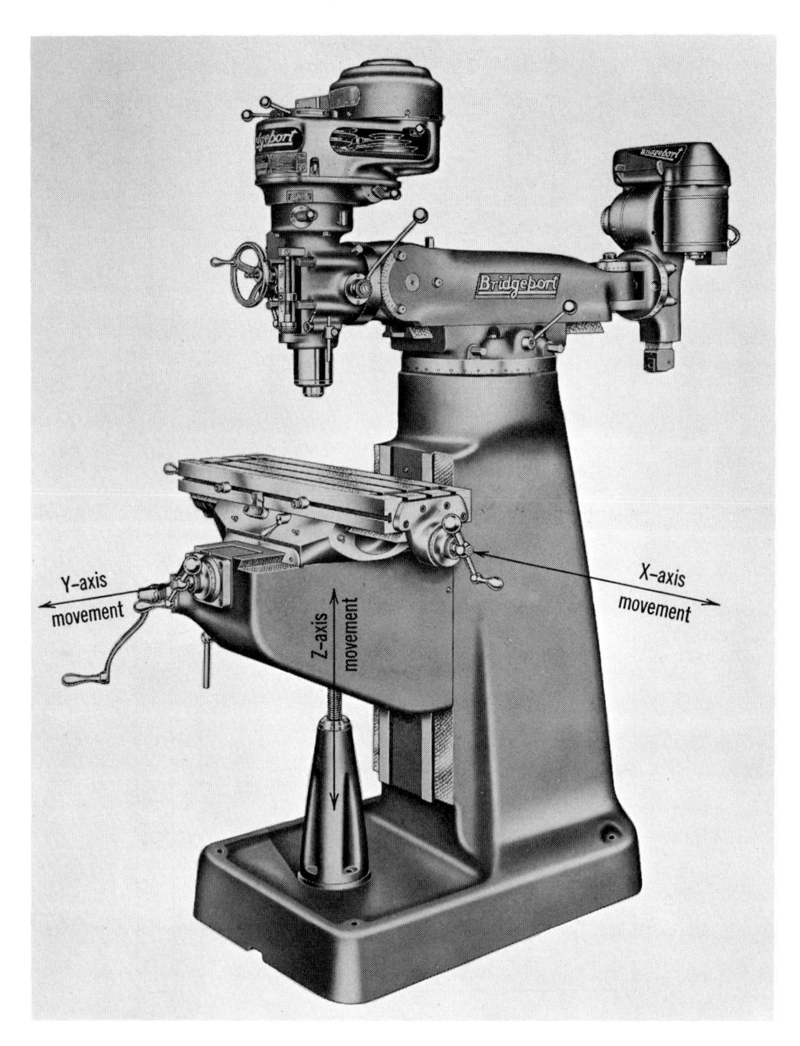

Fig. 20.5

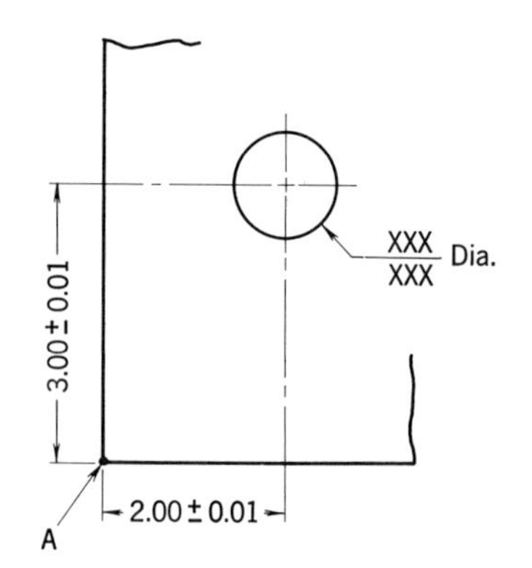

Fig. 20.6

surface treatment, etc.), and (5) CHECK the dimensional drawing by considering the possibility of making the part from the given information.

TOLERANCE

We previously defined tolerance as the permissible variation that can be allowed to occur in any dimension without adversely affecting the functionability of the part. It is axiomatic that all measurements are imperfect. We cannot expect perfection so we must determine the degree of imperfection which can be permitted and make this fact known. In dimensioning we must assign tolerances to size, to location, and to form. The tolerances may be expressed according to the older "coordinate method" or the more recent "true position method." We will first consider the coordinate method of assigning tolerances, and then investigate some of the advantages of the true position method.

The coordinate method owes a large measure of its popularity to its compatibility with machine-tool operation. That is to say, the adjustment of position of many machine tools is by handwheels which position the work along two (or three) axes, Fig. 20.5. For example, in an operation to form a hole as specified in Fig. 20.6, the perfect position of the hole center (the bull's-eye) is located at $X = 2.00$ and $Y = 3.00$ from A. The tolerance ± 0.01 permits the operator to overrun or fall short of the bull's-eye by 0.01 along each axis. Taking a closer look to understand the full significance of the specification, Fig. 20.7, we see that the dimensions identify a 0.02×0.02 square tolerance zone in which the hole center must be located if the part is to agree with the specifications. And so, because it did seem to accommodate machine-tool setting, this "shop oriented" method of assigning tolerance became popular and prevailed. In later considerations we will see that the true position method will often provide more realistic tolerance zones. The coordinate method, however, will continue to be an important and useful dimensioning practice.

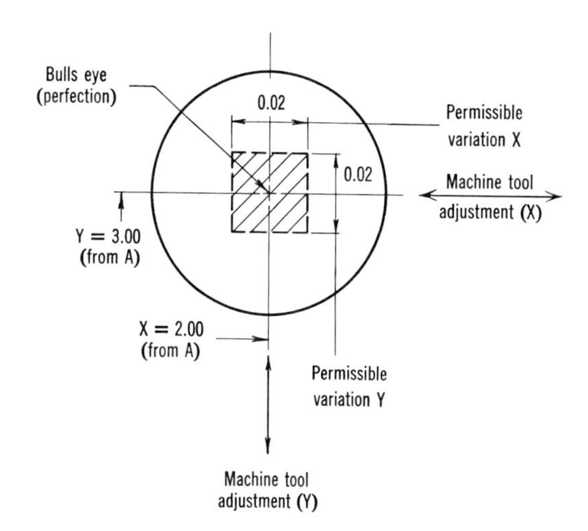

Fig. 20.7

Tolerance of Size

In the design of mating parts, e.g., a shaft and bearing, it is essential to provide "tolerances" on the parts so that production is both feasible and economical. In all cases, however, the parts must be held to the tolerances necessary for proper functioning. Dimensioning for interchangeability of parts introduces terms with which we should be familiar. The terms used and their definitions are the following:*

Nominal size is the designation which is used for general identification. For example, the depth of a 12″-wide flange steel beam weighing 40 pounds per foot is actually 12.77″.

Allowance is the *intentional* difference in the dimensions of mating parts. It is the minimum clearance (positive allowance) or maximum interference (negative allowance) between mating parts. (See Fig. 20.8.)

Basic size is that size from which the limits of size are derived by the application of allowances and tolerances.

Limits of size are applicable maximum and minimum sizes. "Limit" dimensions are the maximum and minimum sizes of the part.

Unilateral tolerance is a tolerance in which variation is permitted in only one direction from the design size. (See Fig. 20.9.)

Bilateral tolerance is a tolerance in which variation is permitted in both directions from the design size. (See Fig. 20.9.)

Clearance fit results from prescribed limits of size of mating parts so that a clearance always exists when the parts are assembled.

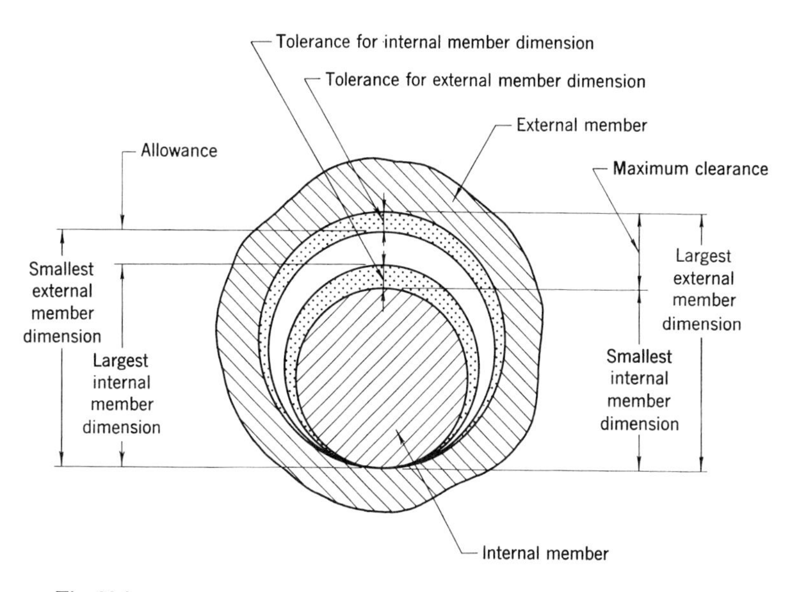

Fig. 20.8

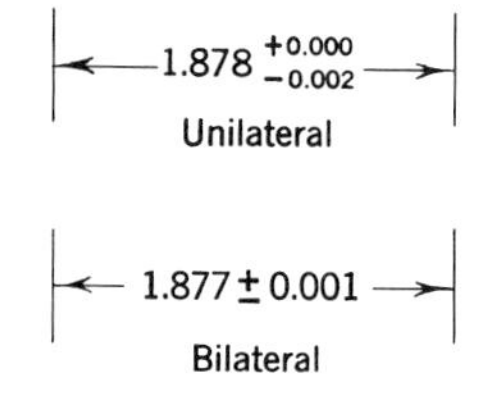

Fig. 20.9

* Courtesy ASA.

Interference fit results from prescribed limits of size of mating parts so that an interference always exists when the parts are assembled.

Transition fit results from prescribed limits of size of mating parts so that either a clearance or an interference exists when the parts are assembled.

Basic hole system is a system of fits in which the minimum size of the hole is the basic size.

Basic shaft system is a system of fits in which the maximum size of the shaft is the basic size.

The use of the two systems and the associated methods of making computations for dimensioning mating parts is discussed in the examples which follow the material on classes of fit.

Classes of Fit

In the selection of fits for a specific application, e.g., shaft and bearing design, the engineer considers such factors as bearing load, speed, length of engagement, lubrication, temperature, humidity, and materials. The engineer also recognizes that *manufacturing costs increase as closer tolerances are specified;* therefore, it is essential to prescribe the most liberal tolerances consistent with function. As yet, no single set of tables for tolerances and allowances has been developed to cover all conceivable situations. This is impossible to achieve. Nevertheless, tables are available for a fair percentage of the cases that do arise in industry. For our purpose, we will use the American Standard on *Preferred Limits and Fits for Cylindrical Parts,* ASA B4.1.

Designation of Standard Fits. Standard fits are designated by means of the following symbols to facilitate reference to classes of fit, *for educational purposes.* The symbols are *not shown* on production design drawings.

RC Running or Sliding Fit
LC Locational Clearance Fit
LT Transition Fit
LN Locational Interference Fit
FN Force or Shrink Fit

The symbols are used in conjunction with *numbers that represent the class of fit;* i.e., "RC2" represents a class 2, running or sliding fit. Each symbol, such as RC2, represents a complete fit for which the minimum and maximum clearances or interferences, and the limits of size for mating parts, are given directly in Tables 42 to 46, pp. 704–708. The classes of fit and their limits of clearance or interference are shown in these tables.

Calculations for Limit Dimensions

EXAMPLE 1 BASIC HOLE SYSTEM

The basic size of the hole is 1.500 inches and the class of fit is RC5. The portion of Table 42, p. 704, that applies to this case is shown below.

Nominal Size Range, Inches	Limits of Clearance	Standard Limits	
		Hole	Shaft
1.19–1.97	2.0 4.0	+1.0 0	−2.0 −3.0

(Limits are in thousandths of an inch)

The limit dimensions *for the hole* are:

1.500 + 0.000 = 1.500 as the minimum value, and
1.500 + 0.001 = 1.501 as the maximum value.

The limit dimensions *for the shaft* are:

1.500 − 0.002 = 1.498 as the maximum value, and
1.500 − 0.003 = 1.497 as the minimum value.

The *allowance* is 1.500 − 1.498 = 0.002, which is the difference between the smallest hole diameter and the largest shaft diameter, or the *tightest fit*. The *loosest fit* is the difference between the largest hole diameter and the smallest shaft diameter, or 1.501 − 1.497 = 0.004. Note that this value, 0.004, is equal to the allowance (0.002) plus both tolerances (each of which is 0.001). The graphic representation is shown in Fig. 20.10.

Note that the tables referred to above were designed for the basic hole system, which is the preferred practice because it allows standard hole-forming tools to be used. However, there are cases in which it is necessary to maintain constant limit dimensions for the shaft on which several components may be fitted.

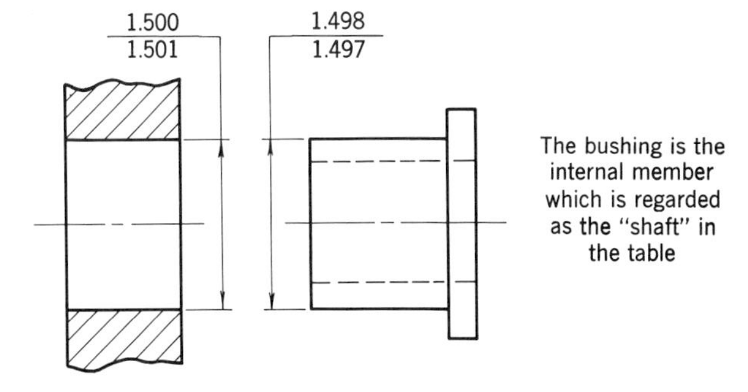

Fig. 20.10 Limit dimensions for Example 1, basic hole system.

EXAMPLE 2 BASIC SHAFT SYSTEM

The basic shaft size is 1.500 and the class of fit is RC5. To use the tabular values in Table 42, p. 704, directly the basic hole size must first be determined. This is accomplished by *adding the allowance* to the basic shaft size. The allowance is shown as 2.0 (meaning 0.002) in the "limits of clearance column"

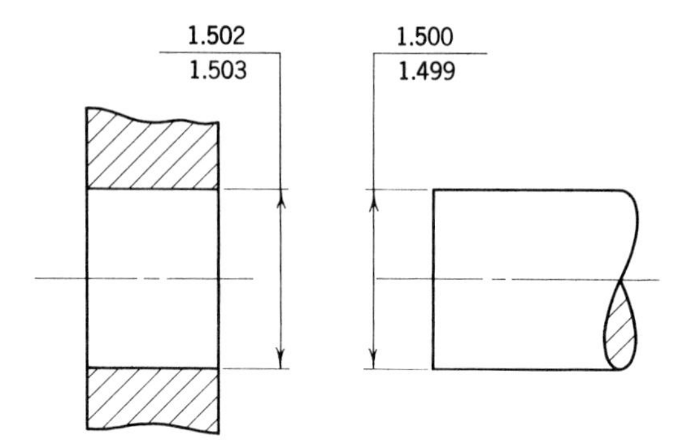

Fig. 20.11

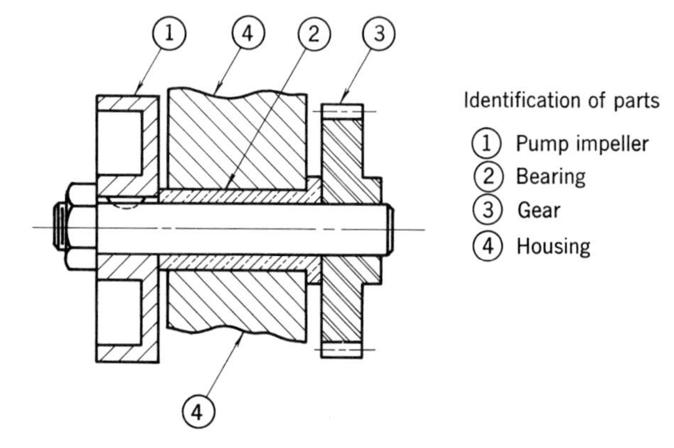

Fig. 20.12

(upper number). Therefore, the *basic hole size is* 1.500 + 0.002 = 1.502. Now we can use the table as shown in Example 1.

The limit dimensions *for the hole* are:

$$1.502 + 0.001 = 1.503, \text{ and}$$
$$1.502 + 0.000 = 1.502.$$

The limit dimensions *for the shaft* are:

$$1.502 - 0.002 = 1.500, \text{ and}$$
$$1.502 - 0.003 = 1.499.$$

The allowance is 1.502 − 1.500 = 0.002, which is the difference between the smallest hole diameter and the largest shaft diameter. The loosest fit is 1.503 − 1.499 = 0.004 which is the difference between the largest hole diameter and the smallest shaft diameter. The graphic representation is shown in Fig. 20.11.

For the following examples we will consider the assembly shown in Fig. 20.12. The gear (3) is driven by a mating gear from a power source. The gear is fixed to the shaft by a heavy force fit. The gear, shaft, and impeller will rotate as a unit. The shaft will therefore rotate in the bearing.

EXAMPLE 3 BASIC HOLE SYSTEM

Consider the case of the bearing (2) fit in the housing (4) in Fig. 20.12. For this assembly we will assume that the basic hole size is 1.5000 (as in Example 1), but the class of fit is LN2. The portion of Table 45, p. 707, that applies to this case is shown below.

Nominal Size Range, Inches	Limits of Interference	Standard Limits	
		Hole	Shaft
1.19–1.97	0 1.6	+1.0 −0	+1.6 +1.0

(Limits are in thousandths of an inch)

The limit dimensions *for the hole* are:

$$1.5000 - 0.0000 = 1.5000, \text{ and}$$
$$1.5000 + 0.0010 = 1.5010.$$

For the bearing O.D., the limit dimensions are:

$$1.5000 + 0.0016 = 1.5016, \text{ and}$$
$$1.5000 + 0.0010 = 1.5010.$$

The allowance is 1.5000 − 1.5016 = −0.0016, which is the interference in this case. The loosest fit is 1.5010 − 1.5010 = 0.0000. The graphic representation is shown in Fig. 20.13. Limit dimensions for other classes of fit are calculated in a similar manner.

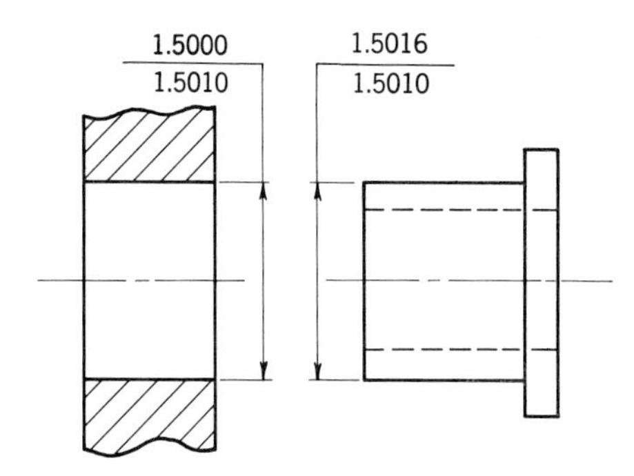

Fig. 20.13 Limit dimensions for Example 3, basic hole system.

EXAMPLE 4 BASIC SHAFT SYSTEM

For this example we again refer to the assembly, Fig. 20.12, and consider the limit dimensions for the shaft, impeller (1), bearing (2), and gear (3). The class of fit for the parts is based on the functional requirements and is as follows:

Shaft and impeller	LT1
Shaft and bearing bore	RC4
Shaft and gear	RN5

Now let us assign a nominal size of 1″ to the parts and consider the procedure for determining the limit dimensions.

When the basic hole system is used, the smallest hole limit for each part will be 1.0000. The largest shaft limit would then become 1.000 less the allowance of the fit for each component of the assembly. The required shaft configuration is shown in Fig. 20.14.

Fit	Allowance
LT1	−0.0003
RC4	+0.0008
FN5	−0.0033

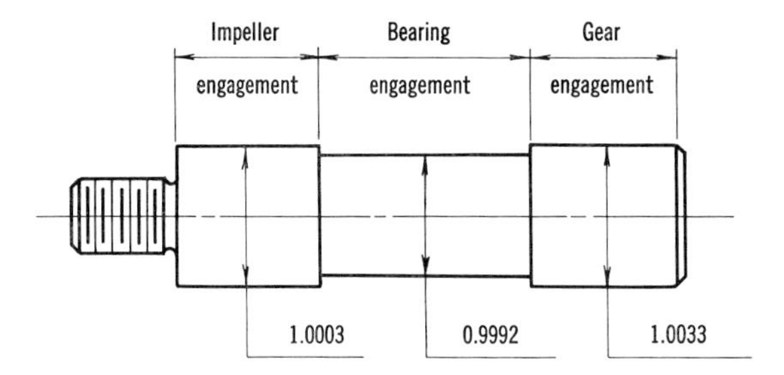

Fig. 20.14

This shaft would undoubtedly be more difficult and expensive to make than one having a constant diameter. Consider also the problem of assembly. A satisfactory solution is obtained by using the *basic shaft system* as follows:

From the Tables 42, 43, and 44 we obtain these data:

	LT1		RC4		FN5	
	Hole	Shaft	Hole	Shaft	Hole	Shaft
0.71–1.19	+0.8 −0	+0.3 −0.2	+0.8 −0	−0.8 −1.6	+0.8 0	+3.3 +2.5
	(Impeller)		(Bearing)		(Gear)	
Tolerance for each fit:	0.8	0.5	0.8	0.8	0.8	0.8
Allowances are:	−0.0003		+0.0008		−0.0033	

Now observe carefully that the least tolerance for the shaft is 0.0005 (LT1), e.g., the tolerance of the shaft may not exceed 0.0005 for a proper fit with the impeller. If the shaft diameter is specified uniform throughout its length the tolerance must also be uniform and may not exceed this value. The larger tolerance (0.0008) which could be permitted for the other components cannot be used. We therefore establish the shaft limits as 1.0000/0.9995. Remember that the largest shaft is the basic size for the basic shaft system.

To compute the minimum limits for the bores of the components we must add the allowance, after which the maximum limits are determined by adding the tolerance.

Impeller		$1.0000 + (-0.0003) = 0.9997$
	and	$0.9997 + 0.0008 = 1.0005$
Bearing		$1.0000 + 0.0008 = 1.0008$
	and	$1.0008 + 0.0008 = 1.0016$
Gear		$1.0000 + (-0.0033) = 0.9967$
	and	$0.9967 + 0.0008 = 0.9975$

A review of this important procedure will show that the requirements of each fit have been accommodated. It is, perhaps, again worthy of note that the tolerance for the shaft is better than was required for two of the fits. This is of no great significance when the over-all benefits are considered.

It is suggested that Problems 1 to 5 on pp. 530 and 532 be solved at this time in order to achieve competence in limit dimensioning.

In the preceding examples the American Standard tables were used to establish the limit dimensions for cylindrical mating parts. The engineer is frequently required to determine limit dimensions for noncylindrical mating parts and to assign tolerances and allowances determined solely by the functional requirements of the mechanism rather than use tabular values. For example, let us assume that the part used in our earlier discussion, Fig. 20.1, is our design and that it must fit with another part *M* as shown in Fig. 20.15. Suppose the functional requirements are such that the size of the tongue on *M* should not be less than 0.500 and that the minimum total clearance between the tongue and slot (allowance) must be 0.003. If we know that economical manufacturing

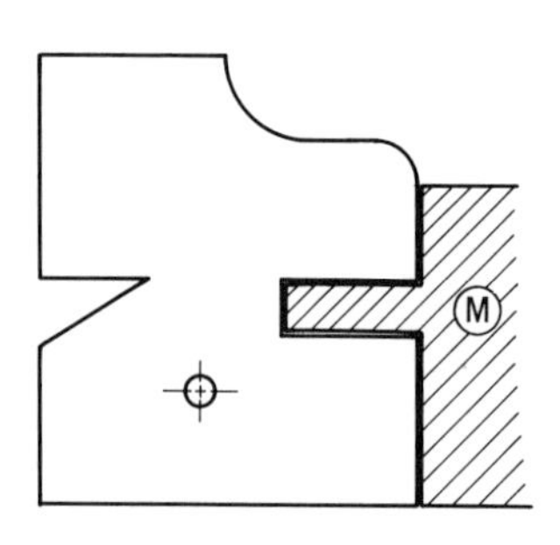

Fig. 20.15

methods require a tolerance of 0.005, and that this value will satisfy the functional requirements, we will use this knowledge to determine the limit dimensions. The limits of the tongue are immediately determined as 0.505/0.500. The limit dimensions for the slot are computed in this manner:

Allowance = (minimum dimension of external part) − (maximum dimension of internal part)

Therefore, 0.003 = minimum dimension of external part − 0.505, or minimum slot dimension = 0.505 + 0.003 = 0.508

Tolerance = (maximum limit dimension) − (minimum limit dimension)

Therefore, 0.005 = maximum limit dimension − 0.508, or maximum slot dimension = 0.508 + 0.005 = 0.513

The limit dimensions of the slot, therefore, are 0.508/0.513.

Tolerance of Position

Positional tolerance is the tolerance assigned to a dimension that locates one or more features in relation to another feature. A feature is a specific characteristic of a part such as a hole, slot, boss, etc. To illustrate this dimensioning problem we will assume that our design part is to be fastened to another as shown in Fig. 20.16. It should be noted that the tongue and slot will fit properly only if both are positioned correctly with respect to some *reference*. Further analysis will show that the logical reference is the bottom surface of our design part and the contact surface of the mating part *M* (shaded area). These reference surfaces do not lie *exactly* on the same plane. The surfaces are called datum features and establish a theoretical plane which is the datum for the dimensions. Datums are points, lines, planes, etc., assumed to be exact for computation purposes, from which the location of features of a part may be established. *Note that the same datum is used for both mating parts. This is a cardinal principle!* Datums will be treated more extensively later in this chapter.

To establish the dimensions for our design part it is imperative to know the functional requirements for the assembly. The information needed for this example is given in Fig. 20.16.

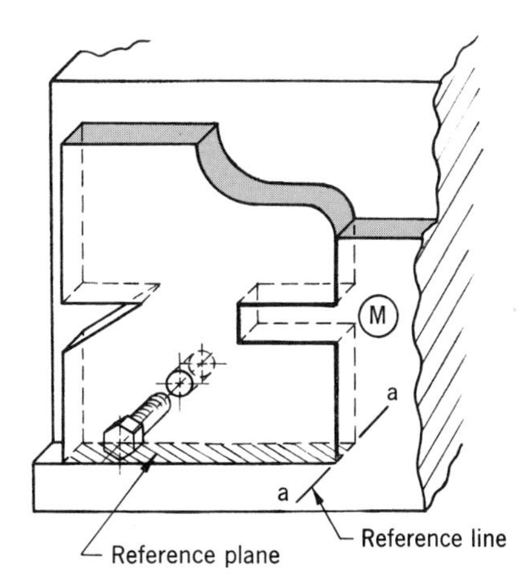

Design functional requirements
1. Minimum tongue size: 0.500
2. Minimum clearance between tongue and slot: 0.003
3. Tolerance for tongue and slot: 0.005
4. Basic dimension from reference plane to lower tongue surface: 3.000
5. Tolerance of position: 0.010
6. Fasten with $\frac{1}{4}$ cap screw.

Fig. 20.16

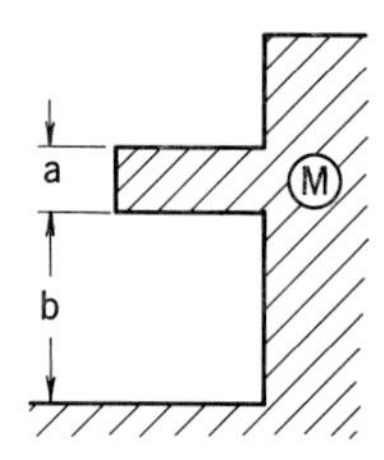

In order to determine the dimensions for our design part we must first know, or compute, the limit dimensions for the size and location of the appropriate features of the mating part. The dimensions for the tongue are determined from the functional requirements as follows:

$$a_{\text{min}} = 0.500 \text{ (given)}$$
$$a_{\text{max}} = 0.500 + 0.005 = 0.505$$

Therefore, limit dimensions for the tongue are 0.505/0.500.

The basic dimension for $b = 3.000$ (given). Total tolerance of position $= 0.010$. Using the bilateral tolerance concept the dimension for $b = 3.000^{\pm 0.005}$. Therefore, limit dimensions for locating the tongue $b = 2.995/3.005$.

Now, to compute the dimensions for our design part it is helpful to make a sketch, Fig. 20.17, to facilitate the analysis. We know the bottom of the slot must have a clearance of 0.003 with the tongue, so $e_{\text{max}} = 2.995 - 0.003 = 2.992$. The tolerance of position is 0.010; therefore, $e_{\text{min}} = 2.982$, so the limit dimensions for e are: 2.992/2.982.

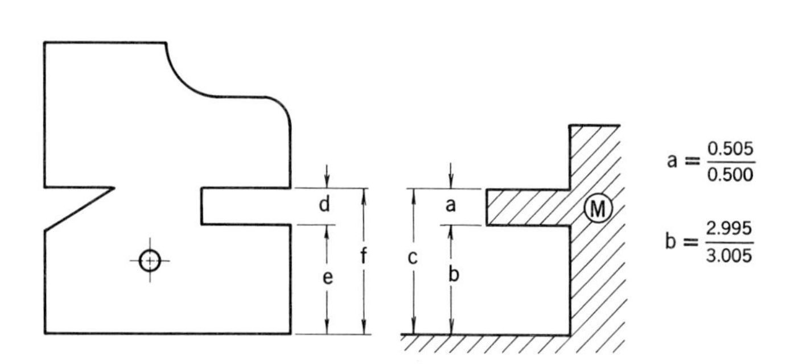

Fig. 20.17

To establish the slot dimensions we must consider the worst condition of assembly (*a cardinal rule for interchangeability*). This condition will exist when dimension c is maximum and dimension f is minimum; therefore we establish the minimum permissible value of f as follows:

$$\begin{aligned} f_{\text{min}} &= (c_{\text{max}}) + \text{clearance} \\ &= (a_{\text{max}} + b_{\text{max}}) + \text{clearance} \\ &= 0.505 + 3.005 + 0.003 = 3.513 \end{aligned}$$

The smallest permissible slot can now be determined by computing the difference between the least permissible value of f and the least value of e (again the worst condition of assembly):

$$\begin{aligned} d_{\text{min}} &= f_{\text{min}} - e_{\text{min}} \\ &= 3.513 - 2.982 = 0.531 \end{aligned}$$

then $d_{\text{max}} = 0.531 + 0.005 \text{ (tolerance)} = 0.536$

and the limit dimensions of the slot are 0.531/0.536.

It is suggested that the student make a sketch of the parts, including the dimensions for the features as calculated above. Examine the possible worst conditions. Are the limit dimensions satisfactory? It is also recommended that an analysis be made of Problem 6, page 532.

Another tolerance of position problem to be solved is that involving the location of the threaded hole in part M and the clearance hole in our design part. If these holes are not aligned properly in assembly, the cap screw will not fit into place. For this example reference line a-a is chosen, Fig. 20.16. This reference line will appear as a point in the planes of the features being located. This is a desirable datum characteristic. Again it will be noted that the datum (A) is *common to both parts* of the assembly, Fig. 20.18. If the parts were made perfectly the dimensions a and c would be equal and dimensions b and d would be equal. We will use the nominal dimensions shown in Fig. 20.4, then $a = c = 1.000$ (basic) and $b = d = 2.000$ (basic). The tolerance of position is 0.010 as cited in Fig. 20.16 design functional requirements. Again let us assign this amount as a bilateral tolerance of ± 0.005. The limit dimensions are then simply computed as:

$$a = c = \frac{0.995}{1.005} \qquad \text{and} \qquad b = d = \frac{1.995}{2.005}$$

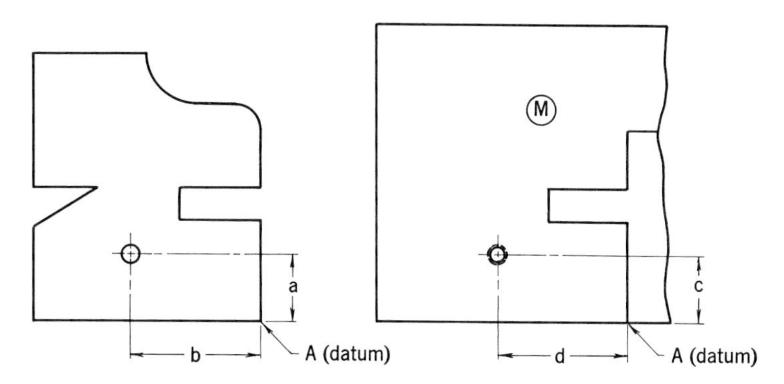

Fig. 20.18

In the case of position dimensions it is often desirable to express the basic dimension together with the tolerance. This is particularly true when automated (tape-controlled) machine tools are to be used. The dimensions for our design part are shown in Fig. 20.19. Now the question is: What size clearance hole is necessary to provide proper assembly conditions for the $\frac{1}{4}$ cap screw? This is another application of size dimensioning. Again we consider the worst possible condition of assembly that will occur when the threaded hole is located by its maximum

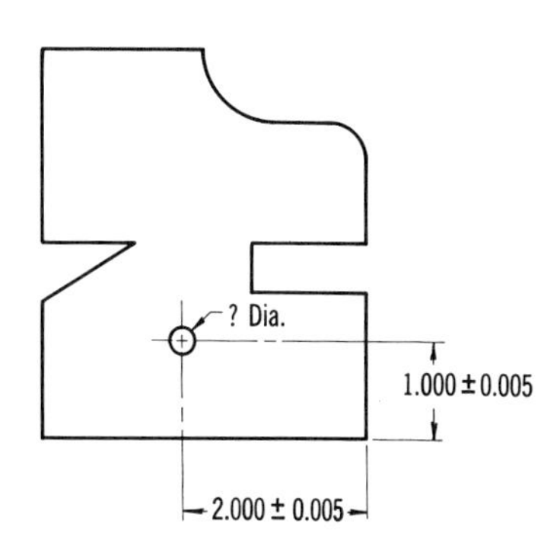

Fig. 20.19

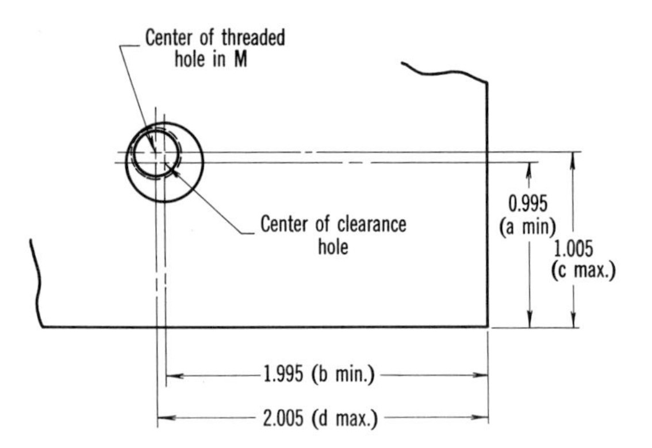

Fig. 20.20

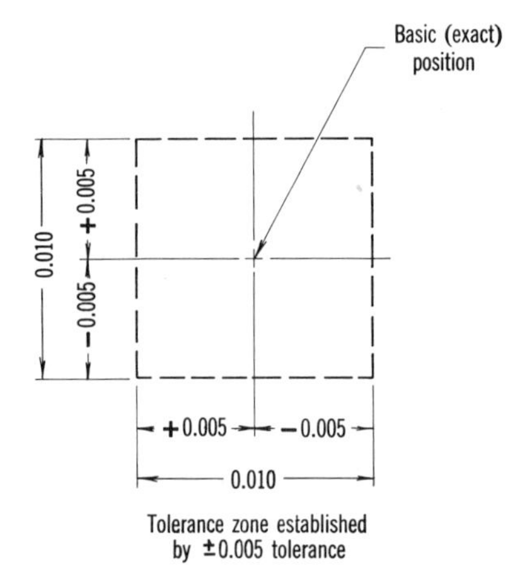

Fig. 20.21

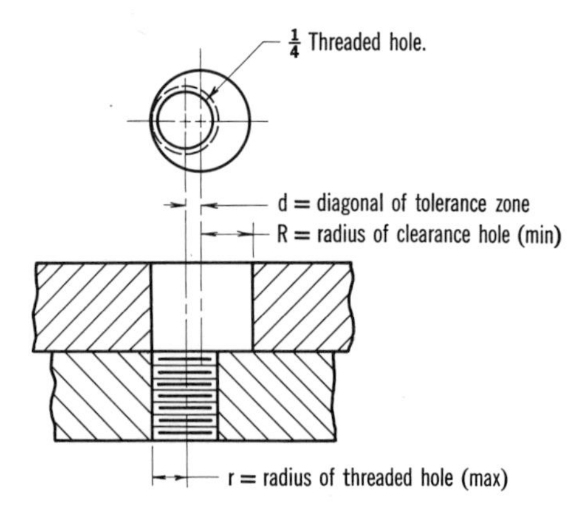

Fig. 20.22

limits and the clearance hole is located by its minimum limits, Fig. 20.20. It would also be well to emphasize again the tolerance zone which is established by the limits of the position dimensions. (See Fig. 20.21.) The significance of this zone is that the given dimensions permit the actual location of the center of each hole to fall anywhere within the zone area. It is now clearly seen that the illustrated position of the two holes, Fig. 20.20, does represent the worst condition of assembly. To determine the minimum permissible clearance hole it is desirable to first make a sketch, Fig. 20.22. Note that the threaded hole and clearance hole are in the tangent position which would exist if the clearance hole were minimum size and the threaded hole were maximum size.

To compute R:

$$\begin{aligned} r &= \frac{\text{threaded hole diameter}}{2} \\ &= 0.125 \\ d &= \text{diagonal of tolerance zone} \\ &= \sqrt{(0.010)^2 + (0.010)^2} \\ &= 0.014^{+} \\ R &= r + d \\ &= 0.125 + 0.014^{+} \\ &= 0.139^{+} \end{aligned}$$

We will use $R = 0.14$.

Then the minimum clearance hole size is 0.28.

To ascertain the maximum hole limit dimension we consult Table 33 in Appendix B for a suitable drill size to form the clearance hole. The *K* drill (0.281) would be the most acceptable standard drill size. It is an experimental fact that the hole formed by a drill will, in almost every case, be larger than the drill size. For this drilling condition the amount of oversize should not normally exceed 0.005; therefore, the largest drilled hole would be 0.281 + 0.005 = 0.286. We may now establish the limits of the clearance hole as 0.280/0.286 diameter and so note it on the drawing.

The student should be aware that a more precise design may negate the possibility of using a drill to form the hole. In all cases the functional requirements will dictate the tolerance values.

Let us go back for a moment and recall that the exact position (the bull's-eye) for the feature was established by two basic dimensions and that the square zone about the bull's-eye was the area established by the bilateral tolerances assigned to the basic dimensions. (See Fig. 20.21.) Now as we think about this, and recall the definition of tolerance, it would seem that the tolerance value (± 0.005) should identify the maximum variation (greatest miss) from the exact position. It is readily apparent that this is not true; indeed, the ± 0.005 tolerance actually allows a miss of $0.005\sqrt{2}$, or approximately 0.007 along the diagonals of the tolerance zone. This variation is 40% greater than the "apparent" value. It is essential that the design engineer be aware of this fact and consider its effect on the function of his design. The manufacturer is also perplexed by the inconsistency of the square tolerance zone. He feels that he is not given the full benefit of the functional tolerance and is thereby restricted to more precise (more expensive) manufacturing methods. The student has probably already guessed that this dilemma could be resolved quite easily by using a circular tolerance zone. This simple solution is achieved by the *true position* method. We see the results of dimensioning the clearance hole in our design part by the coordinate method, Fig. 20.23(a) and the true position method, Fig. 20.23(b).

Now let us study Fig. 20.24. We observe that the true position tolerance zone circumscribes the coordinate tolerance zone. This means that both tolerance zones permit the same maximum variation from perfection; however, the circular zone permits this variation in all directions rather than along the axes only. This provides the manufacturer with 57% more tolerance area *without any sacrifice of accuracy in functional requirements.*

Three essential differences in the two methods are immediately apparent: (1) The true position method includes the tolerance of position with the note for the feature rather than with the locating dimensions as is done in the coordinate method. (2) The true position method assigns a round tolerance zone rather than the square zone of the coordinate method. (3) The identity of the datum is specifically noted in the true position method, whereas it is "implied" in the coordinate method. By way of similarity it is noted that both methods assign a basic (exact position) dimension, a tolerance, and the limits of size for the feature.

Now let us examine a few examples that will point

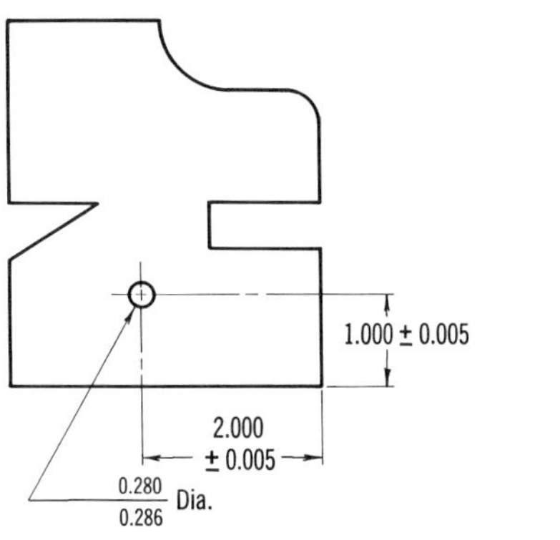

Fig. 20.23(a)

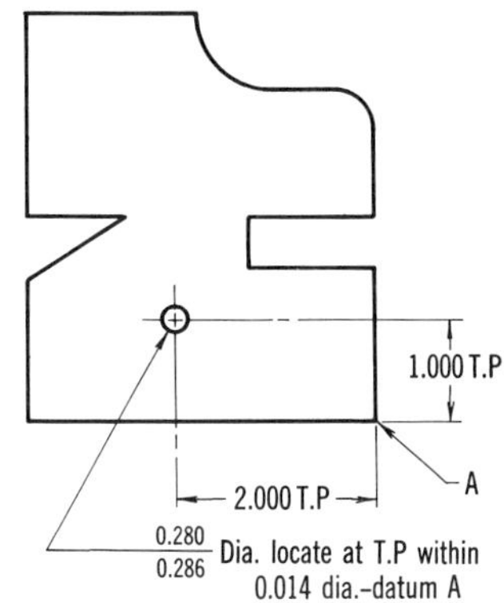

Fig. 20.23(b)

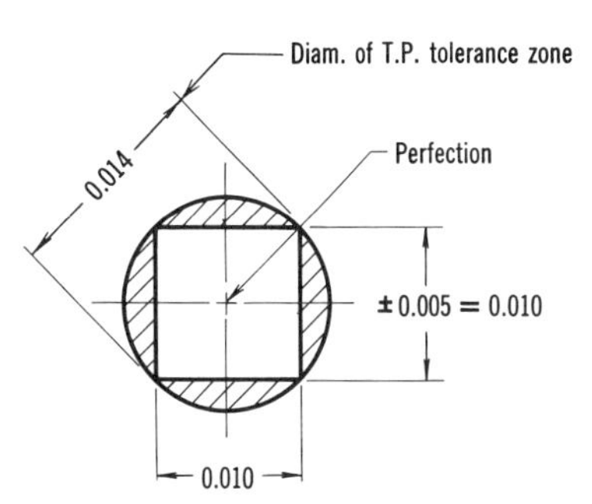

Fig. 20.24

up the difficulties arising from the use of "conventional" dimensioning practices; and then develop a rational means for overcoming them, so as to ensure a single interpretation of a design.

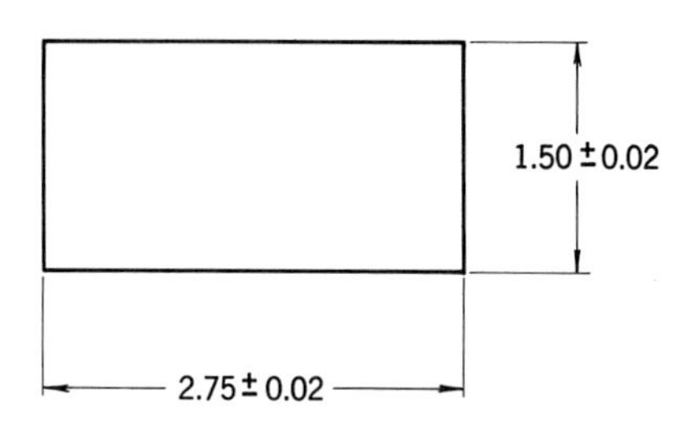

Fig. 20.25

EXAMPLE 1

Suppose we have a rectangular shape dimensioned as shown in Fig. 20.25. We observe that each dimension has a tolerance of ± 0.02. How shall we interpret this drawing? What tolerance zone is defined by the ± 0.02? What assurance do we have that the manufacturer and the inspector will interpret the specifications in the same way? Several alternative interpretations are shown in Fig. 20.26.

The student should realize that the manufactured part will *not* be perfectly straight and smooth. Then what interpretation would be meaningful? We will not attempt to identify a correct (or incorrect) interpretation at this time; instead, let us note only that the interpretation [Fig. 20.26(*c*)] might best accom-

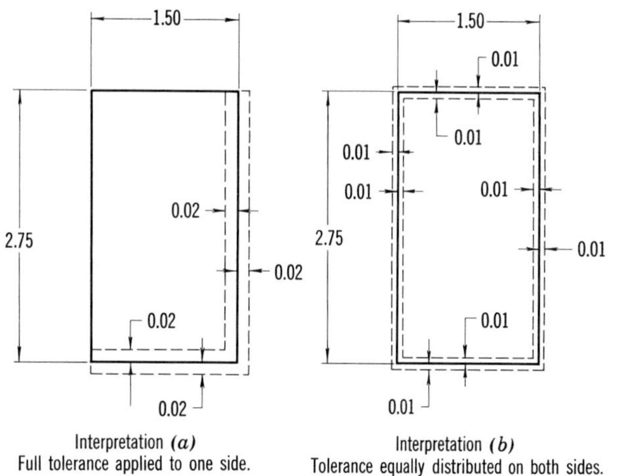

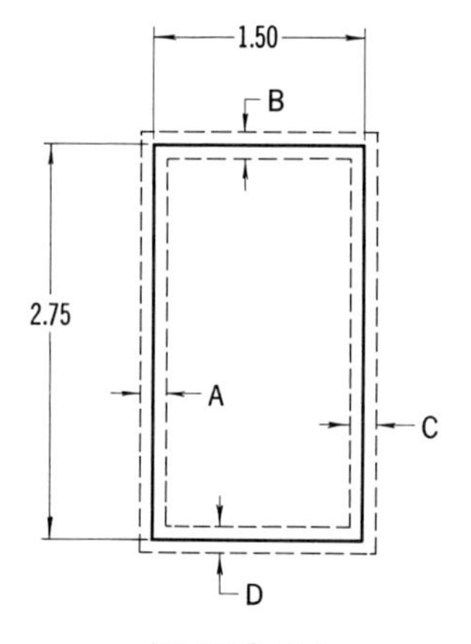

Interpretation (*a*)
Full tolerance applied to one side.

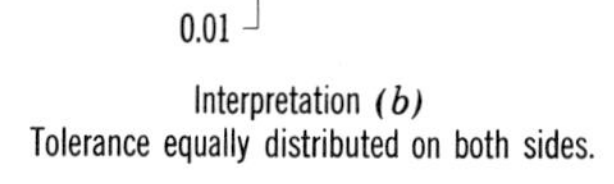

Interpretation (*b*)
Tolerance equally distributed on both sides.

Interpretation (*c*)
The tolerance may exist in any amount on either side.

Fig. 20.26

modate the shape of the part when it is manufactured. (See Fig. 20.27.) This interpretation imposed the singular requirement that the profile of the part falls within the frames established by the limit dimensions. How can the drawing be dimensioned to ensure this interpretation?

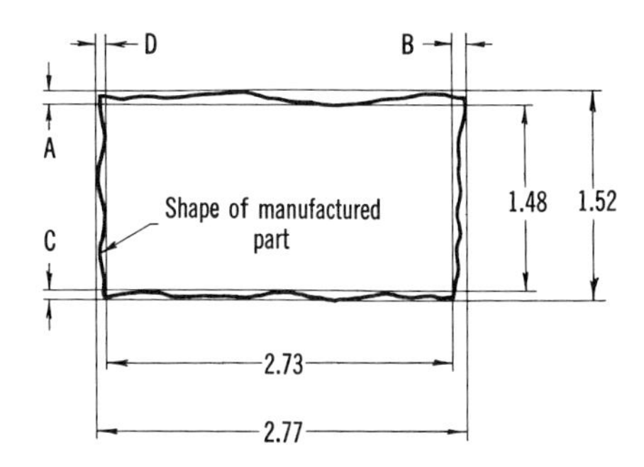

Fig. 20.27

EXAMPLE 2

Consider the cylindrical part shown in Fig. 20.28. A casual inspection of the drawing would cause no alarm. The part seems to be adequately defined by the dimensions. A more careful analysis points up a serious problem of interpretation. Observe the hole (h) which is shown to be located distances X and Y from the centerlines. Centerlines of what? As the part passes through the shop and on to inspection it does not have a pair of fixed centerlines going along with it. How then does the workman locate the hole (h)? How does the inspector verify that the hole is properly located? Should the centerlines be established by hole (H), the cylinder (d), the cylinder (D), or by some method which does not involve the features of the part? The design engineer, who knows the functional requirements of the part, is the only person qualified to answer these questions; indeed, it is *his* responsibility to provide the dimensions which will establish the single interpretation necessary to manufacture and verify a functional part. Specifically, it is his sole responsibility to establish and identify the datums properly.

Let us now study datums more thoroughly.

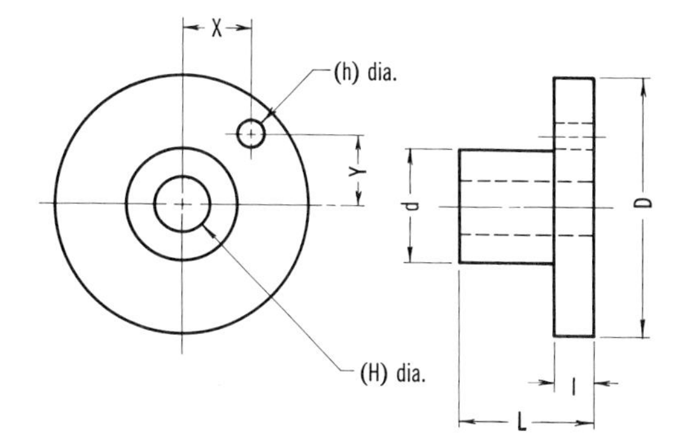

Fig. 20.28

Datums

Datums are points, lines, planes, etc., assumed to be exact for computation purposes, from which the location of features of a part may be established. This definition, given earlier, is accurate; however, it is a limited version of the total datum concept. A comprehensive exploration of the datum idea must include considerations *pertinent to the manufacture and verification* of the parts described by the drawing.

Let us return to the cylindrical part, Fig. 20.28, and note the influence of function upon the choice of a datum. Observe two possible functional requirements illustrated in Fig. 20.29. Can the same datum be used to locate the hole (h) in each case? The functional requirement for application (a) is as follows: (1) Hole H must engage the projecting cylinder

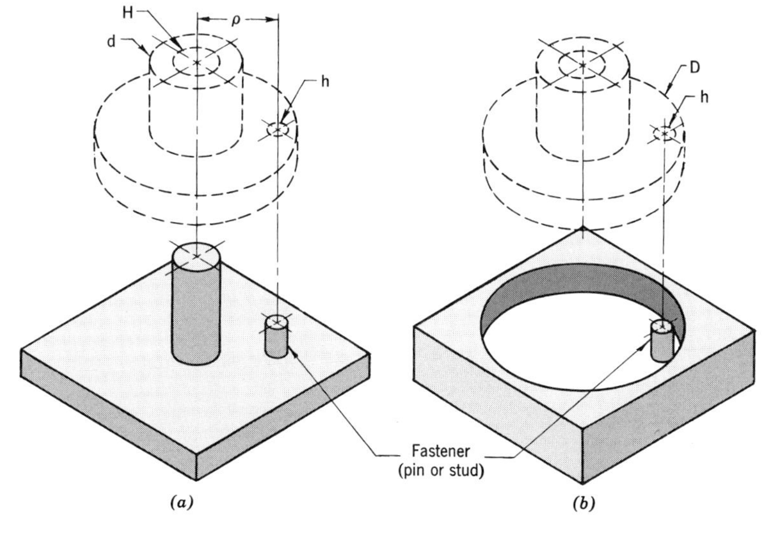

Fig. 20.29

of the mating part, and (2) the fastener must pass through hole h to permit assembly. In this case the datum for the location of hole h must be established by the feature diameter H. Using either cylinder (d or D) as the datum feature to locate h would jeopardize the probability of assembly because the parts would not have a common datum. For example, if the feature d was used to establish the datum for application (*a*), the distance ρ would be affected by the difference in position of the centerline of diameter d and the centerline of diameter H. This variation would add to the tolerance of location between diameters H and h (and the projecting cylinder and fastener on the mating part). This additional variation may well prohibit assembly. By making a similar analysis, the student should identify the proper datum feature (diameter D) for application (*b*). It is possible to imagine an application wherein the feature diameter d would properly establish the datum for locating hole h.

The above example involves a very simple part. The magnitude of the problem will increase greatly as the parts become more complex.

Let us now list some concepts which will enable the student to better understand and establish datums properly:

1. A datum is the origin of a measurement which locates, or otherwise relates, a feature to the datum feature or features. The primary function of a datum is to establish a point to serve as the origin of a measurement. In addition, the datum planes establish the *orientation* (direction) of the dimensions, Fig. 20.30.
2. The actual (physical) datum will rarely exist on the part; it is usually a feature (or features) on the processing tool and on the inspection equipment. Machine tool tables, axes of spindles, surface plates, axes and surfaces of inspection equipment, etc., are examples of features which are physical datums.
3. The features of processing and verification equipment used as physical datums are considered to be "perfect." It is necessary, of course, to check the accuracy of a physical datum carefully to justify its use as a datum (perfection). Although it is true that the use of these features will introduce an error, such equipment will usually be at least ten times more precise than the part. The

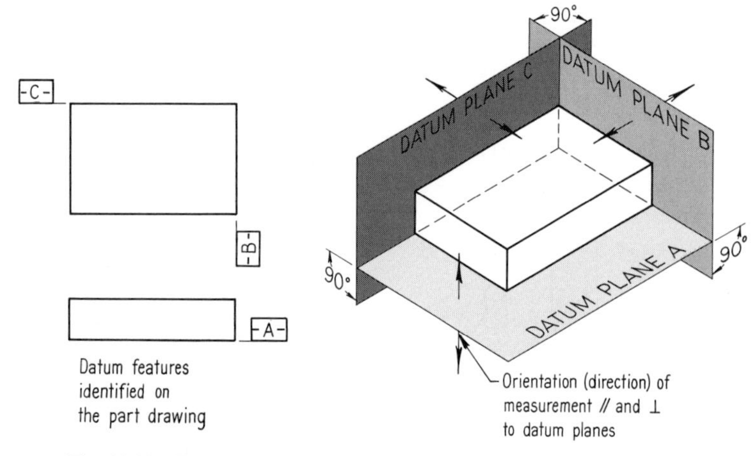

Fig. 20.30 Datum interpretation.

error introduced by this practice is therefore comparatively small.

4. The datum feature (on the part) should be carefully selected to (*a*) guarantee the proper function of the part, (*b*) allow the physical datum to be established on the processing and inspection equipment, and (*c*) ensure the use of the same theoretical datum for manufacture, verification, and assembly.

The following examples, which are given to illustrate the application of these concepts, will allow the student to observe the clarifying effect on interpretation of dimensions when datums are specified.

EXAMPLE 1

The dimensions of a rectangular shape (see Fig. 20.25) would be restricted to a single interpretation if datum planes were established, as shown in Fig. 20.31. As an additional precaution, particularly when dealing with critical and expensive parts, it may be desirable to define the datum notation. Identification and definition of datums are frequently included in Drafting Room Manuals (DRM), Process Standards, etc., which are used by the company. An example of the datum identity for our problem is shown in Fig. 20.32. The part would not generally be included, but it is shown here to illustrate the definition.

EXAMPLE 2

Let us consider the "conventional" dimensions of a simple piece and the interpretation of the dimensions (Fig. 20.33). The piece is a plate with four holes. The engineer's intent is to provide a part that may be fastened to an "identical" plate by four ½-inch diameter bolts. The edges of the plates need not match. At first glance the drawing appears to be complete except for the size dimensions of the holes. These dimensions have been omitted intentionally. What is the correct interpretation of the dimensions?

(*a*) Does the 0.75 ± 0.04 dimension apply to both holes *A* and 3 from the surface *Y*?
(*b*) Does the 0.75 ± 0.04 dimension apply to both holes *A* and 1 from surface *X*?
(*c*) Are holes 1, 2, and 3 located from hole *A* as a pattern within the tolerance ±0.01?
(*d*) Can holes 1 and 2 each vary from surface *Y* by the total amount of the tolerance ±0.04 + (±0.01)?

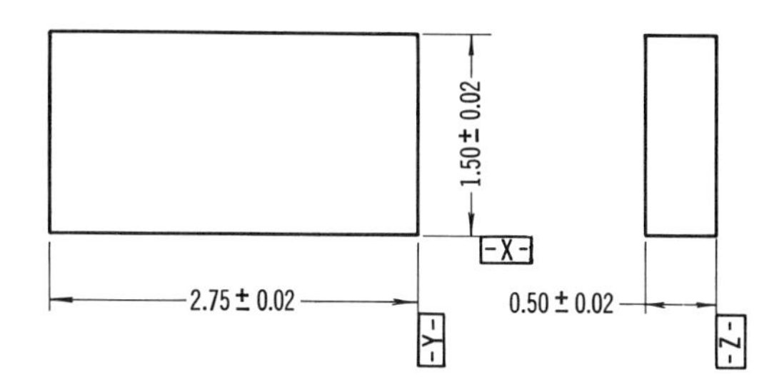

Fig. 20.31

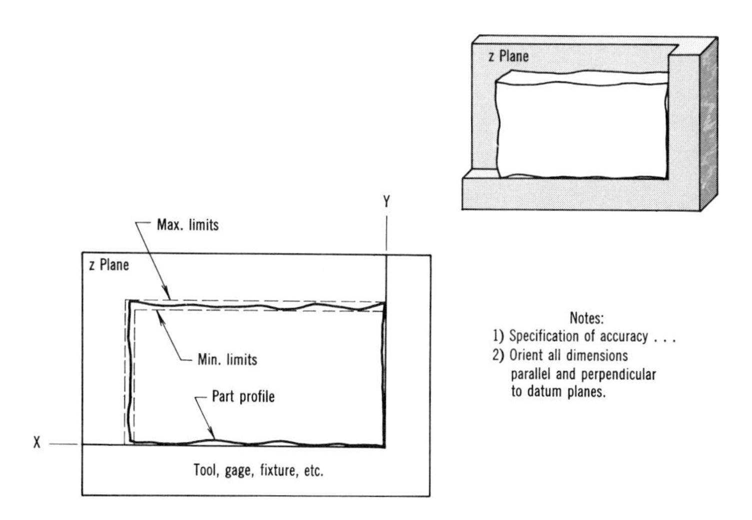

Fig. 20.32

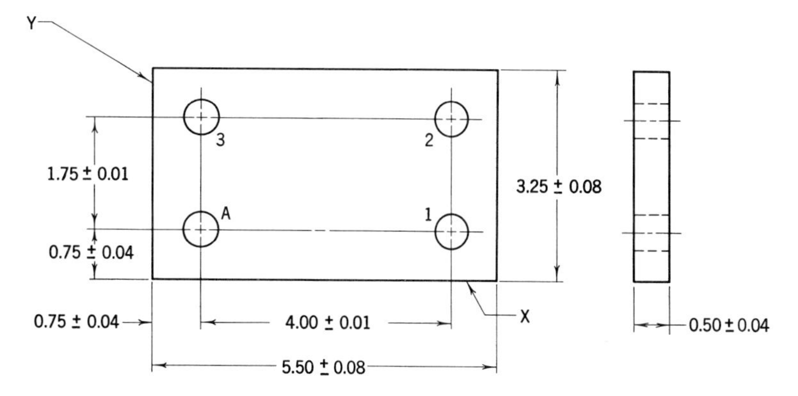

Fig. 20.33

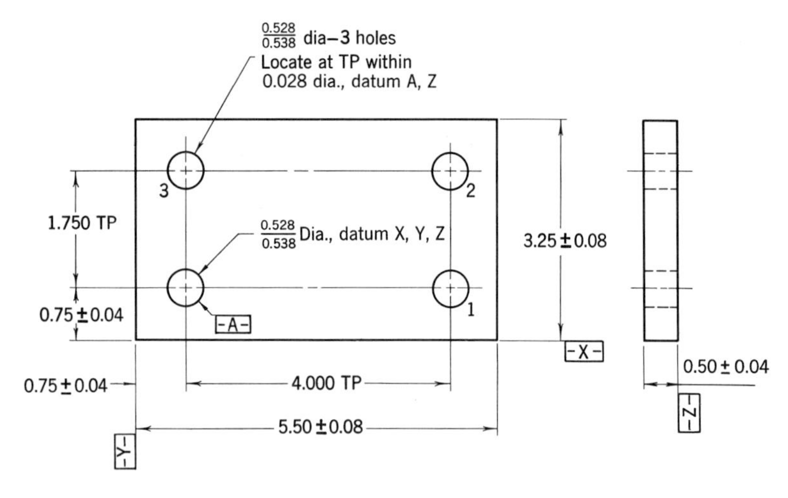

Fig. 20.34

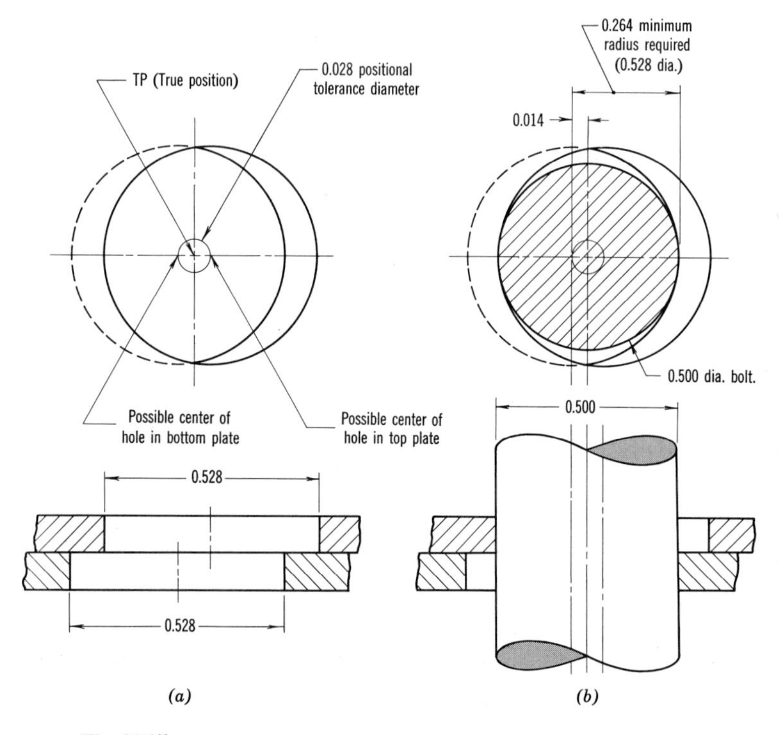

Fig. 20.35

(*e*) What is the relation of surface *X* to surface *Y*?
(*f*) From what point are measurements taken?

We now realize that interpretations of the drawing can vary, depending on the answers to the above questions and others that could be raised. Perhaps you can think of a few more. We can eliminate ambiguities by applying the true-position dimensioning method with proper datum considerations. Let us study Fig. 20.34.

The coordinate dimensions (0.75 ± 0.04) locate hole *A* from the datum established by features *X* and *Y*. Hole *A* then becomes the datum feature for the remaining holes 1, 2, and 3. This use of datums allows a large tolerance for the location of the pattern of holes relative to the edges of the plate. At the same time it retains a desirable degree of precision for the location of the holes relative to one another. This agrees with the "engineer's intent" previously stated. In cases such as this, where the part is symmetrical, the workman must mark the part to identify the datum features. This is necessary to allow the inspector to orient the plate so that he uses for verification the same datum features (*X*, *Y*, *Z*, and *A*) that were used in production.

Clearance Holes. The assembly of the plates is dependent on the size of the holes through each of the two plates. How shall we determine the size of the holes?

The datum and true position points of each plate can be aligned when placed one upon the other, since the plates are dimensioned the same.

In their extreme positions, a pair of corresponding holes (one in the upper plate and the other in the lower plate) must have a minimum diameter of 0.528 in order to accommodate a 0.500 diameter bolt. This is clearly seen in Fig. 20.35. The maximum deviation allowed from *TP* for *each plate* is the radius (0.014) of the positional tolerance zone; therefore, the total deviation with respect to both plates is 0.028. This value plus the bolt diameter establishes the minimum hole size in each plate; or $0.028 + 0.500 = 0.528$, the minimum diameter of the clearance holes.

EXAMPLE 3

Observe the drawing of the circular plate with four holes shown in Fig. 20.36. The datum *Z* is established to orient the axes of the plate (cylinder) and the holes, i.e., the axes will be perpendicular to datum *Z* if perfect. Each hole is also related to the datum feature *B* and the TP location dimensions are oriented by datum *Y*. The interpretation of the dimensions is illustrated in Fig. 20.37. It will be noted that datum feature *Z* will contact the machine work table or inspection equipment surface thus identifying the actual datum *Z* [see detail (*b*)]. Datum plane *Y* is established by the actual centers of feature diameter *B* and the hole (1) located as shown in detail (*a*). The workman must mark hole (1) so that it can be identified by the inspector, thus ensuring similar orientation. The datums *B* and planes *Z* and *Y* establish the origin and orientation for the dimensions locating the holes.

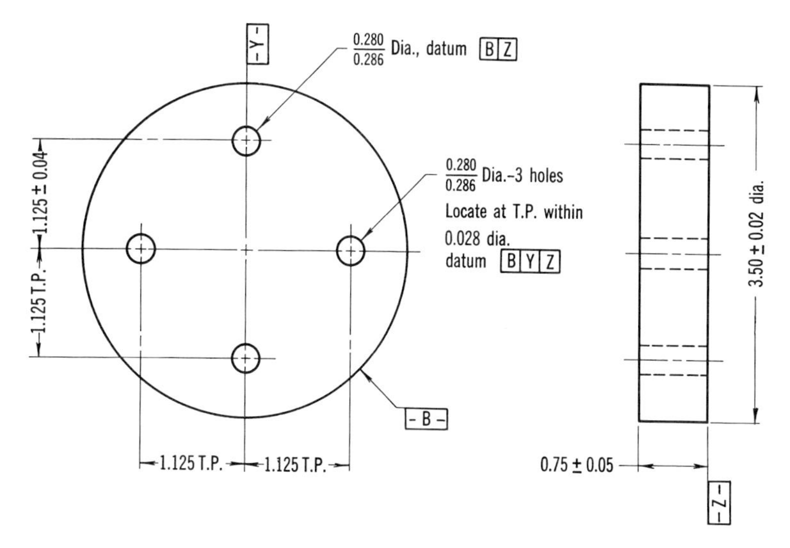

Fig. 20.36

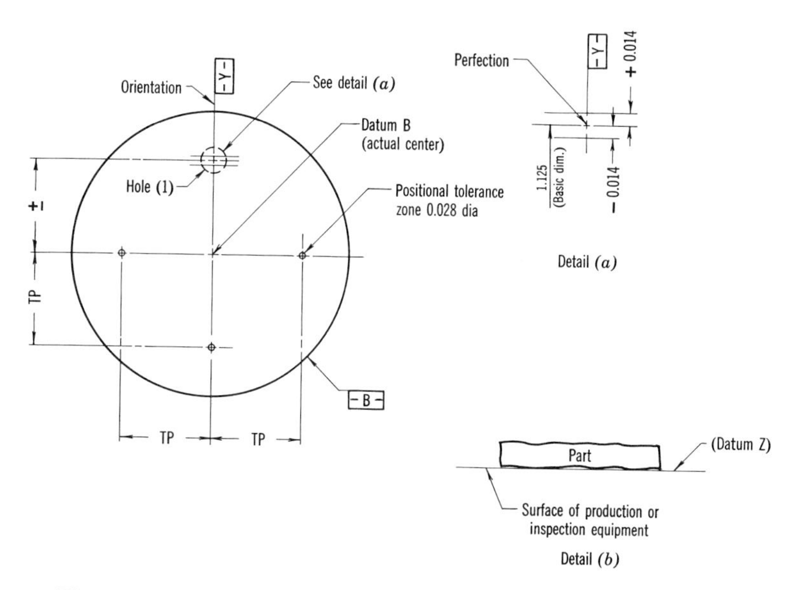

Fig. 20.37

Geometric characteristic symbols			
		Characteristic	Symbol
Form tolerance	For single feature	Flatness (1)	⏥
		Straightness (1)	—
		Roundness (circularity)	○
		Cylindricity	⌭
		Profile of any line (1)	⌒
		Profile of any surface (1)	⌓
		Parallelism	// or ||
		Perpendicularity (squareness)	⊥
		Angularity	∠
		Runout (2)	↗

(1) (∼) This symbol denotes flatness and (⌒) this symbol is also used to denote straightness by military standards.

(2) A runout tolerance control position as well as form.

Fig. 20.38 Geometric symbols. (Courtesy USASI Y14.5–1966)

Tolerances of Form

Tolerances of form control the amount of variation permitted in the geometry of a feature. Tolerances of size and position frequently control form to a certain degree; therefore, it is important to consider the influence of these tolerances before specifying tolerances of form.

Notes may be used to specify tolerances of form; however, symbols which are simple in style and easy to interpret provide a significant saving in time and effort. (See Fig. 20.38.) Complete standardization of the symbols has not as yet been achieved. Those shown in Fig. 20.38 have been approved by the International Standards Organization (ISO/TC 10) and proposed by American Standards Association (USASI Y14.5-1966). The Military Standards (Mil. Std. 8C) have adopted a symbol standard that is slightly different from those shown. It is the policy of some companies to include geometric characteristic symbols together with other drawing information in their DRM, Process Standards, or other similar publications. Recommended specifications of tolerances of form are shown in Figs. 20.39 to 20.45.

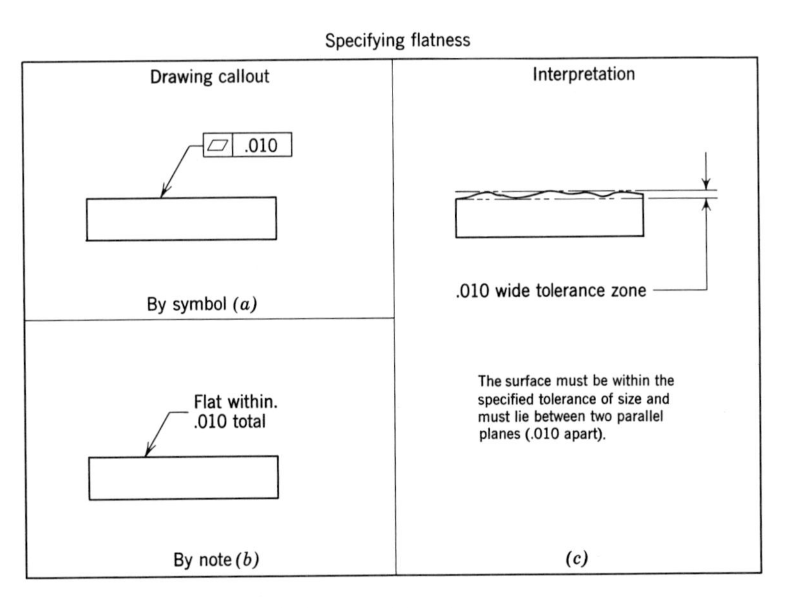

Fig. 20.39 (Courtesy USASI Y14.5–1966)

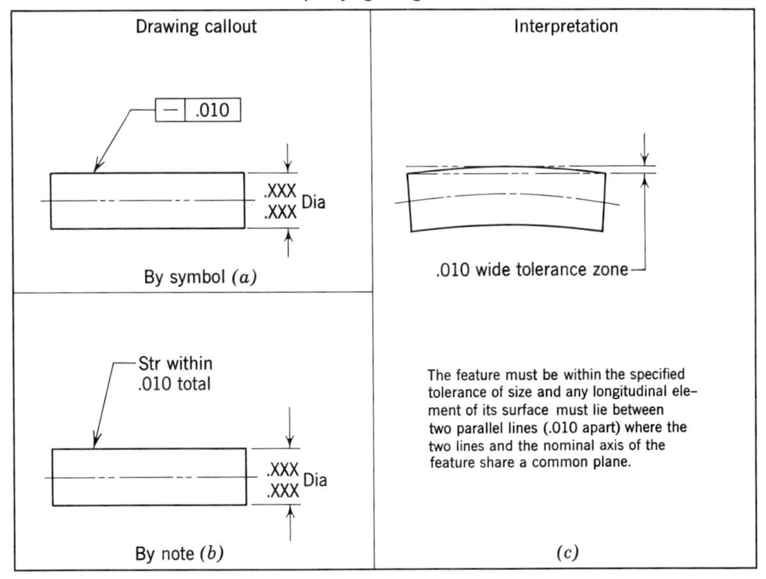

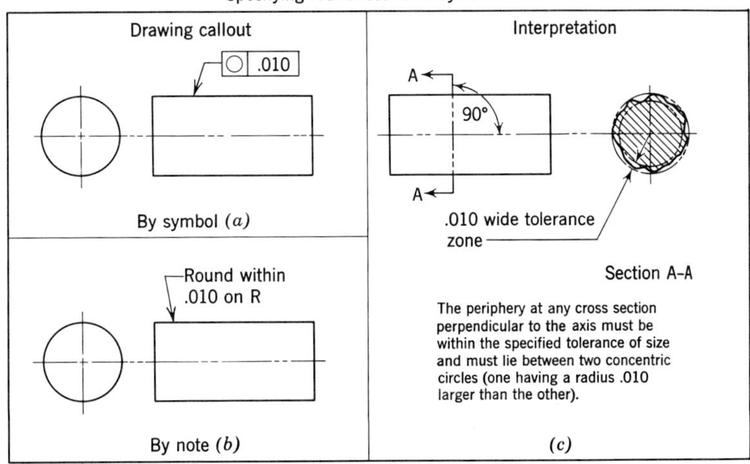

Fig. 20.40 (Courtesy USASI Y14.5–1966)

Specifying cylindricity

Drawing callout

.010

By symbol *(a)*

Cylindrical within .010 on R

By note *(b)*

Interpretation

.010 wide tolerance zone

The feature must be within the specified tolerance of size and must lie between two concentric cylinders (one having a radius .010 larger than the other).

(c)

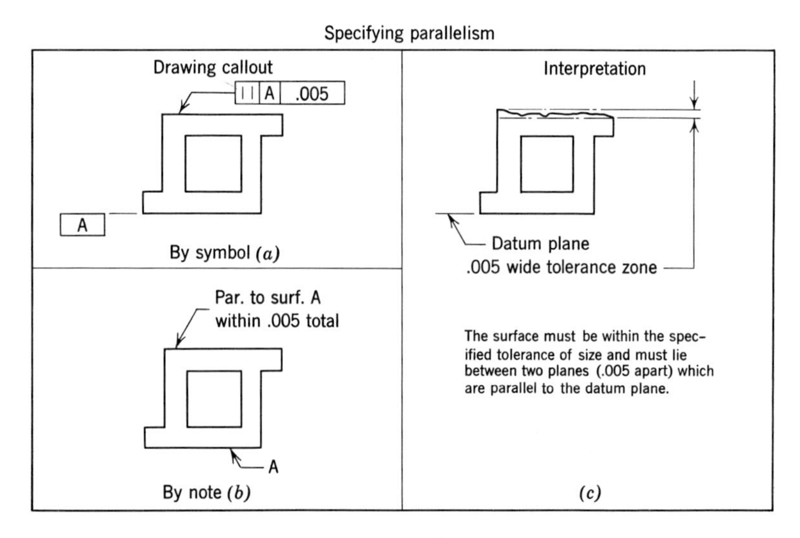

Fig. 20-41 (Courtesy USASI Y14.5–1966)

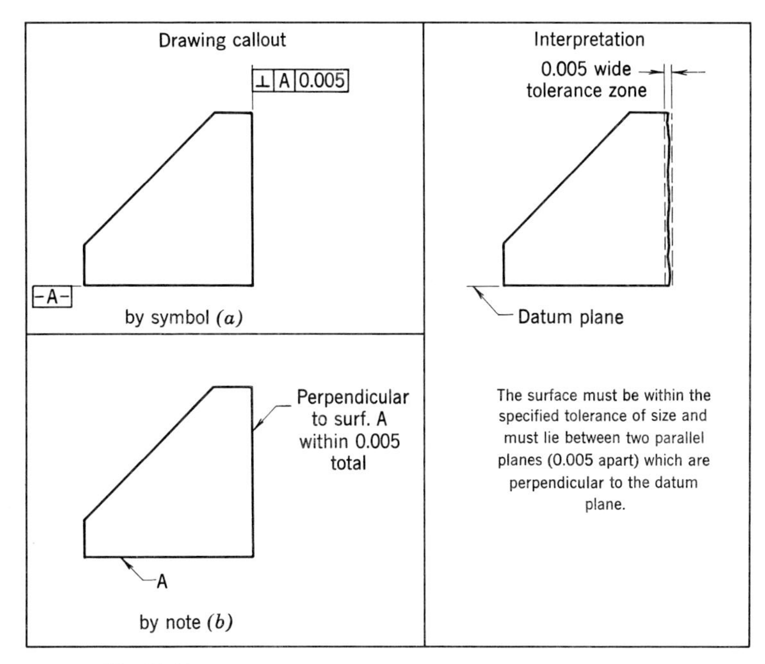

Fig. 20-42

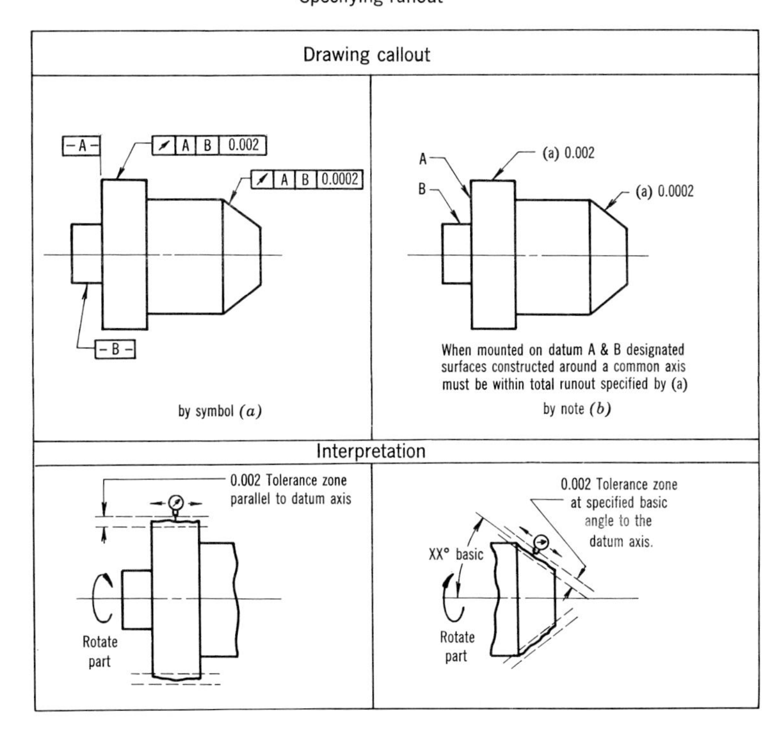

Fig. 20.43

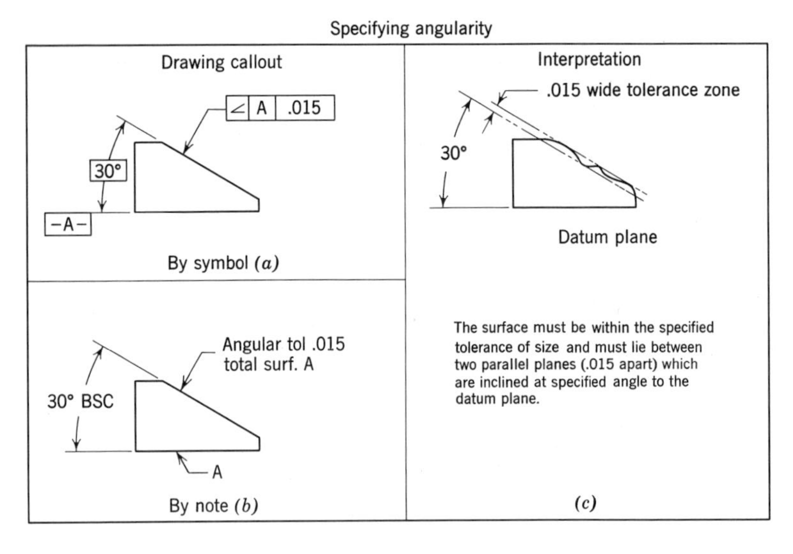

Fig. 20.44 (Courtesy USASI Y14.5–1966)

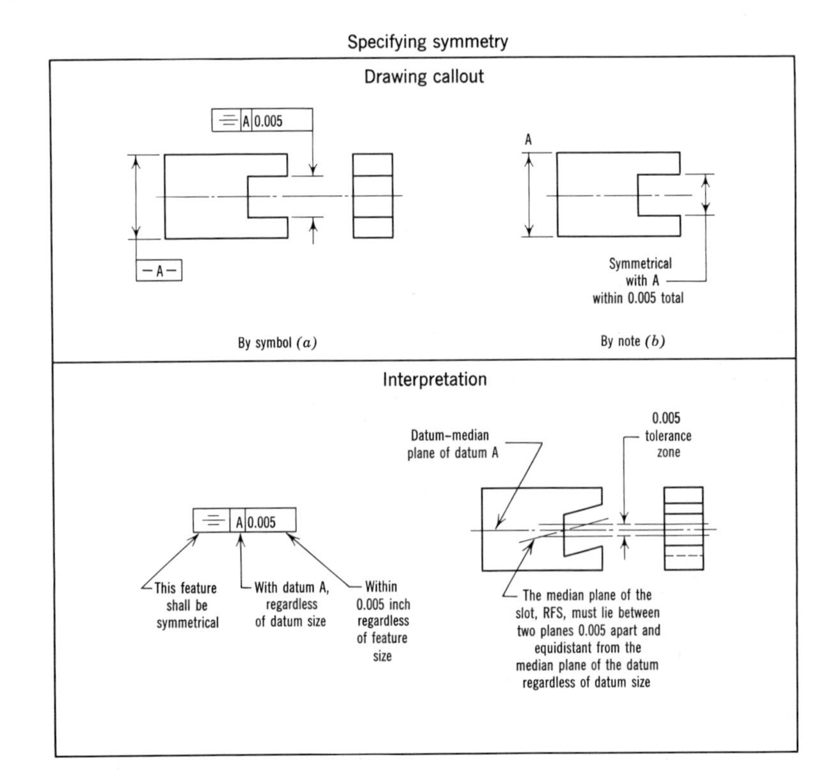

Fig. 20.45 (Courtesy Mil. Std. 8C)

Tolerances of form should be considered for all features critical to the proper function and interchangeability of the part. The extent to which these tolerances are specified will involve the following qualifications:

(*a*) Will established production practices provide the required accuracy of form?
(*b*) Do the tolerances assigned to size and location provide the necessary control of form?
(*c*) Do supplementary documents (DRM, Prod. Std., etc.) establish suitable standards of workmanship to control tolerances of form without additional specifications on the drawing?

The extent to which these qualifications provide assurance of acceptable quality of form will vary greatly, depending on company policy; therefore, it is essential that much thought be given to specifications that will provide complete information.

Profile Tolerancing

Another specification of tolerances of form is established by profile (contour) tolerances. The geometric characteristics of the profile of a part (straight line and regular or irregular curved line elements) can be controlled by specifying contour tolerances. Such tolerances establish control of size and/or location as well as form. Let us examine some examples to illustrate this practice.

EXAMPLE 1

Consider the triangular slot of our design part, Fig. 20.46. A possible interpretation of the dimension is shown in Fig. 20.47(*a*). Observe the tolerance zone established by this interpretation. Since there is no *specific* starting point for the angular dimension, accumulation of tolerances will most likely occur. On the other hand, when true position dimensions are employed, there is no accumulation of tolerances and, consequently, there is only one interpretation of the contour tolerance zone, as shown in Fig. 20.47(*b*).

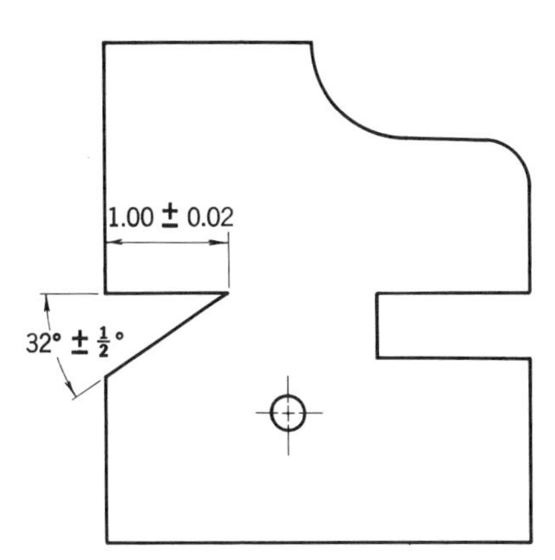

Fig. 20.46

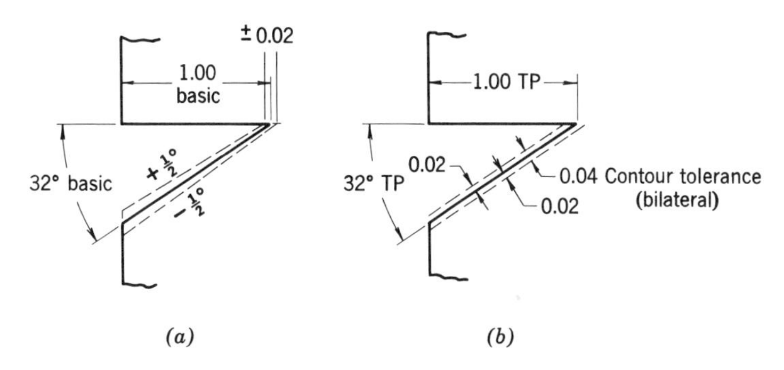

Fig. 20.47

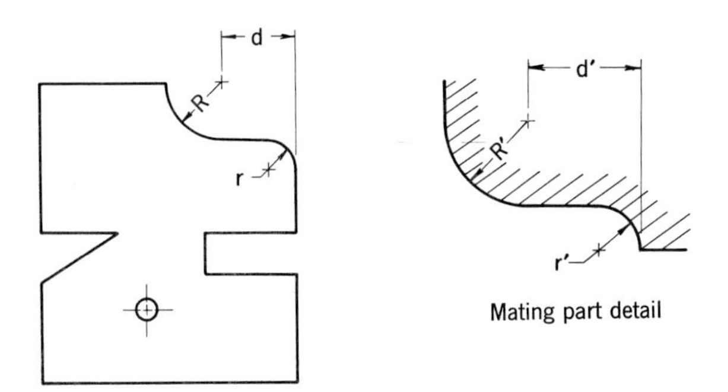

Fig. 20.48

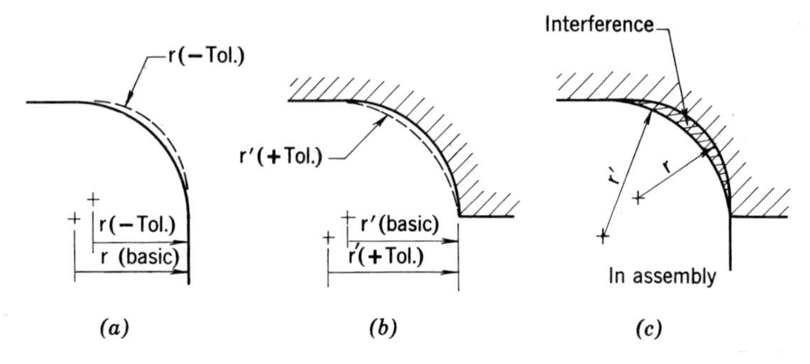

Fig. 20.49

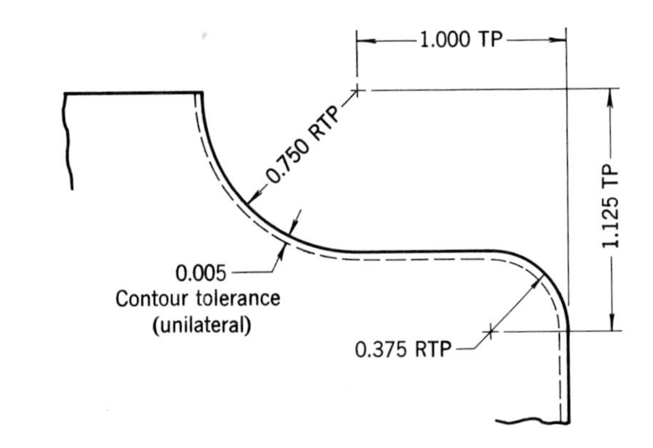

Fig. 20.50

EXAMPLE 2

Let us consider the curved features of our design part and the configuration of the mating part, Fig. 20.48. Analyze the problem of providing realistic tolerances by the coordinate method when a critical fit with limited clearances is involved. We note that the tolerances assigned to d and R are accumulative, as are R' and d' on the mating part. Also note the dilemma of establishing the tolerances of the smaller radii r and r'. These are called "free radii" in that they locate themselves. It would seem that the tolerance for r should be unilateral and negative, while the tolerance of r' should be unilateral and positive. The difficulty is that the same tolerance applies to location as well as to size. (See Fig. 20.49.) At (*a*) we see the design part with the basic location of r and the location when the minus tolerance is applied. At (*b*) we see the basic location of r' and the location when the positive tolerance is applied. At (*c*) we find that the parts will not assemble because of the radii interference. Suppose we reversed the tolerances by using a minus tolerance with r and plus tolerance with r'. Will the parts assemble properly? (It is not likely if the clearance is critical with respect to function.) A simple solution is provided by using true position dimensions and a contour tolerance, Fig. 20.50. The dimensions for the mating parts would be essentially the same, except that the basic dimensions would be adjusted to allow for the required minimum clearance. The contour tolerance would also be unilateral but opposite to that of the design part.

EXAMPLE 3

When a part contains a free form contour the shape may be defined by offset dimensions, as shown in Fig. 20.51(*a*). The interpretation of the design is shown in Fig. 20.51(*b*). Another method of specifying the true position (perfect) profile is to tabulate the coordinate distances from the datum point. (See Fig. 20.52.) This method is very convenient for numerically controlled machine-tool processes. The use of a symbol rather than a note for the profile (contour) tolerance is also illustrated in Fig. 20.52.

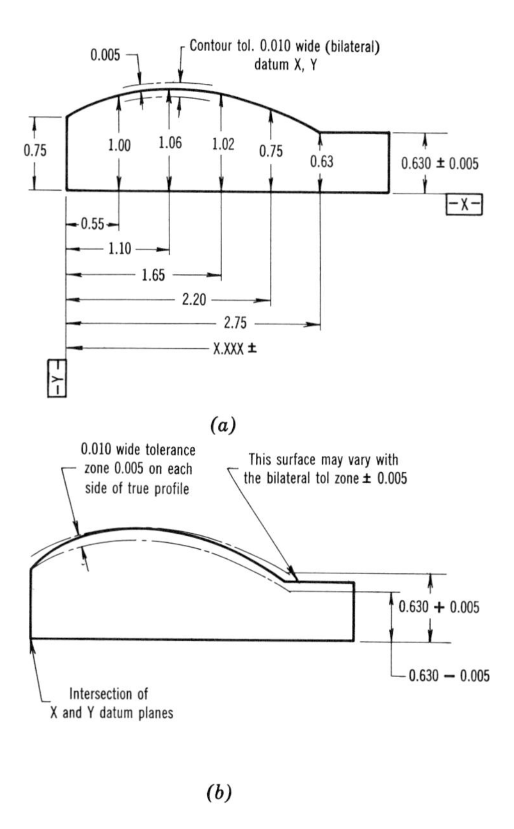

Fig. 20.51

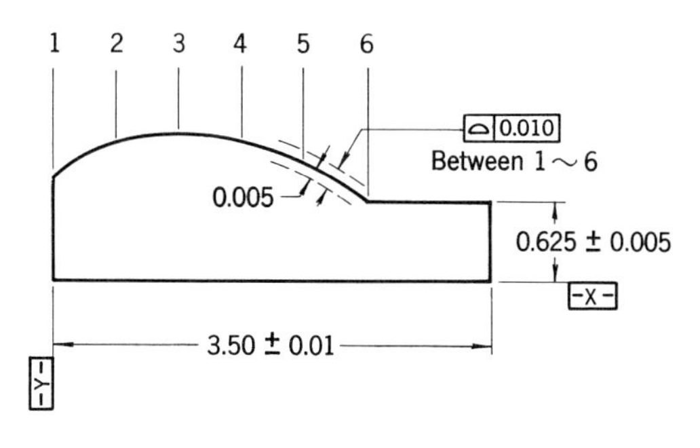

Station	1	2	3	4	5	6
X	0.000	0.500	1.000	1.500	2.000	2.500
Y	0.800	1.060	1.145	1.075	0.901	0.625

Fig. 20.52

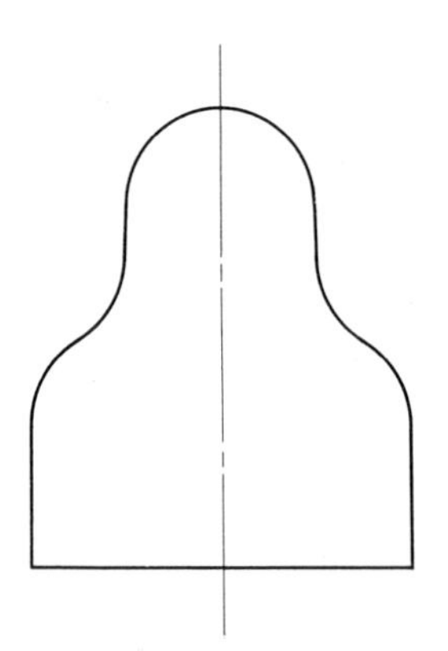

Fig. 20.53

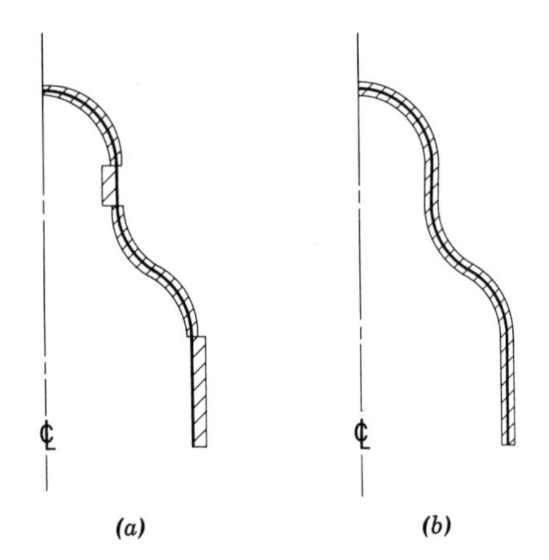

Fig. 20.54 Bilateral and true position methods applied to contour tolerance zones.

EXAMPLE 4

When the bilateral system is applied to a shell (Fig. 20.53), a possible interpretation of the tolerance zone could be similar to that shown in Fig. 20.54(*a*). When the true positioning dimensioning system is used we can obtain a uniform tolerance zone as shown in Fig. 20.54(*b*).

SPECIFICATIONS BY SYMBOLS

It was noted in the preceding section, Tolerances of Form, that the use of symbols rather than notes saves time and effort. Additional symbols which have gained wide acceptance are shown in Fig. 20.55. The student is reminded that the use of notes is acceptable and the appliance of symbols is permissive. Other symbols worthy of note are the datum identity box used in the preceding examples, | -A- |, and the use of a box, | 2.125 | or 2.125 BSC, to represent a true position (basic) dimension as an alternative to 2.125 TP. The review example, Fig. 20.66, is a good over-all illustration of the use of symbols.

Symbols for positional tolerance	
Characteristic	Symbol
True position	⌖
Concentricity (1)	◎ 2
Symmetry	⌯

Symbols for MMC and RFS	
Modifier	Symbol
(MMC) Maximum material condition	Ⓜ
(RFS) Regardless of feature size	Ⓢ

(1) Where concentricity RFS applies, it is preferred that the runout symbol be used.
(2) ◎ Recommended in Mil-Std-8C

Fig. 20.55

Maximum Material Condition

Maximum material condition (MMC) identifies the value, within the specified limits of size of a feature of a part, that will provide the maximum volume of material in the part, e.g., minimum hole diameter or maximum shaft diameter.

EXAMPLE 1

Suppose that the limit dimensions of the diameter of a shaft are 2.357/2.354. The maximum volume of the shaft occurs when the diameter is 2.357; therefore, *the MMC size of the shaft is 2.357.*

EXAMPLE 2

Suppose that the limit dimensions of the diameter of a hole are 0.528/0.538. When the formed hole is 0.528 diameter, the volume of material in the part would be maximum. The MMC size of the hole, therefore, is 0.528. Obviously, holes larger than 0.528 diameter would reduce the volume of material in the part. The effect of applying MMC is to liberalize the tolerance assigned to the form or position of a feature. Where a positional or form tolerance applies at MMC, the tolerance is limited to the stated value only if the feature is at its maximum material limit of size. The tolerance is increased, as the feature departs from MMC, by an amount equal to the departure from MMC.

Positional Tolerance (MMC)

EXAMPLE 1

Let us consider MMC as applied to tolerances of position. In Fig. 20.56 we observe the callout: "Posn Tol 0.028 Dia. (MMC)." What is the significance of MMC in this callout? The interpretation is given in the note shown in Fig. 20.57. How were the values given in the note obtained? Before reading on to learn the answer to this question, you should review the procedure for determining the clearance hole size, discussed in previous examples. (See Figs. 20.22 and 20.35.)

Let us consider the upper right-hand hole shown in Fig. 20.56 (other holes could have been selected).

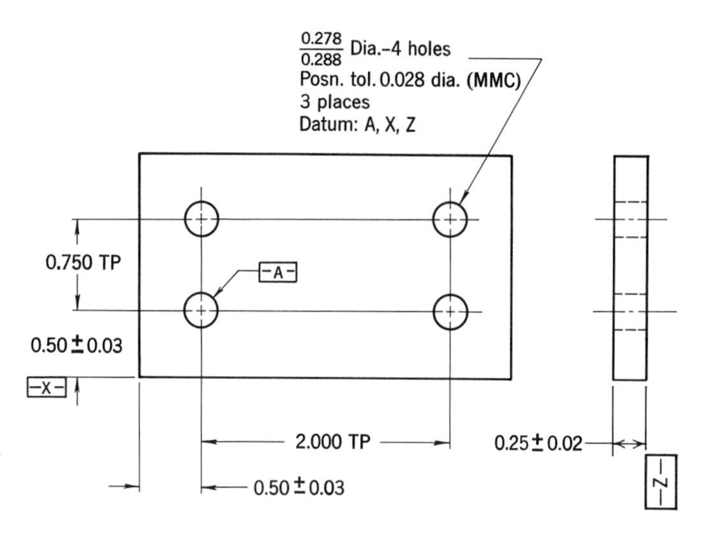

Fig. 20.56 MMC applied to tolerance of position.

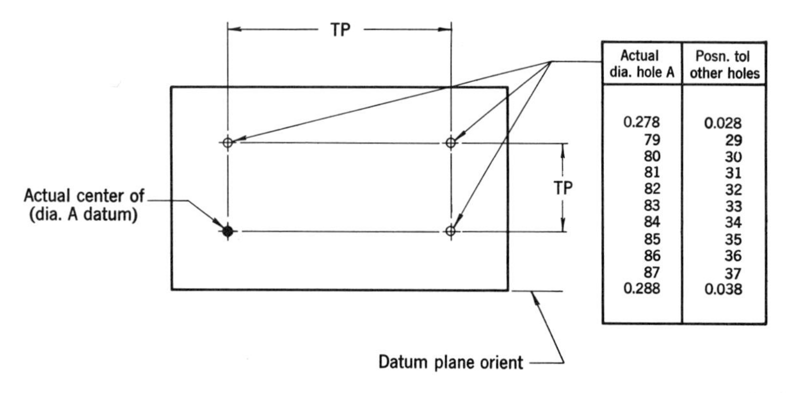

Actual dia. hole A	Posn. tol other holes
0.278	0.028
79	29
80	30
81	31
82	32
83	33
84	34
85	35
86	36
87	37
0.288	0.038

Fig. 20.57 Interpretation.

An exaggerated view of this hole is shown in Fig. 20.58(*a*). We note that the positional tolerance is 0.028 diameter when the hole is 0.278 diameter. This is the maximum material condition.

From our previous study of positional tolerancing, we know that the actual center of the hole must lie within the positional tolerance zone, 0.028 diameter. Now suppose [Fig. 20.58(*b*)] that the actual size of the hole is 0.288 diameter. We may increase the positional tolerance zone to 0.038 diameter. The *increase* is the difference between the actual diameter of the hole and the specified minimum; in this case it is 0.288 − 0.278 = 0.010. Therefore, the positional tolerance zone has a diameter equal to 0.028 + 0.010 = 0.038. Again referring back to Fig. 20.22, note that assembly is possible under the conditions established by the increased tolerance. If the actual size of the hole were any value (between the limits of size), there is an allowable increase in positional tolerance. (See Fig. 20.57.)

EXAMPLE 2

The plate shown in Fig. 20.59 has been designed to assemble with a part in which four threaded holes have been properly located by dimensions similar to those of the plate. The design intent is that four $\frac{1}{4}$-inch cap screws will pass through the clearance holes in the plate and engage the threaded holes in the mating part to fasten the two pieces. An additional requirement is that complete interchangeability shall be possible.

Now let us examine the assembly condition for the two parts. When the center-line positions of a threaded hole (and hence the cap screw) and of a corresponding clearance hole are at extreme locations within the tolerance zone, the cap-screw shank

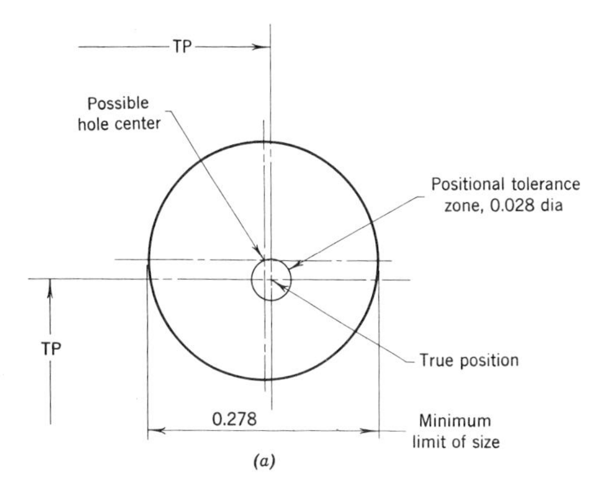

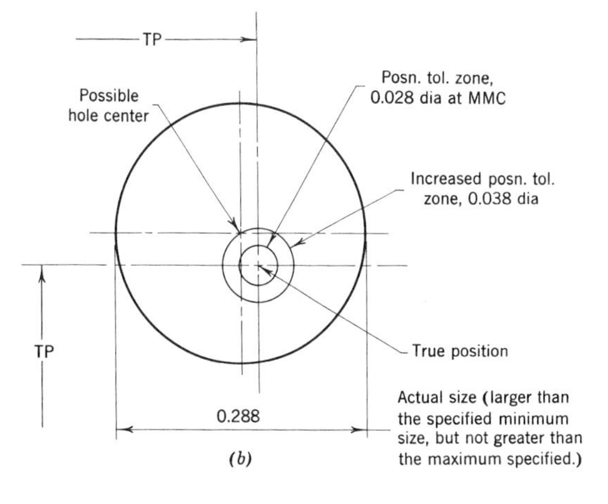

Fig. 20.58

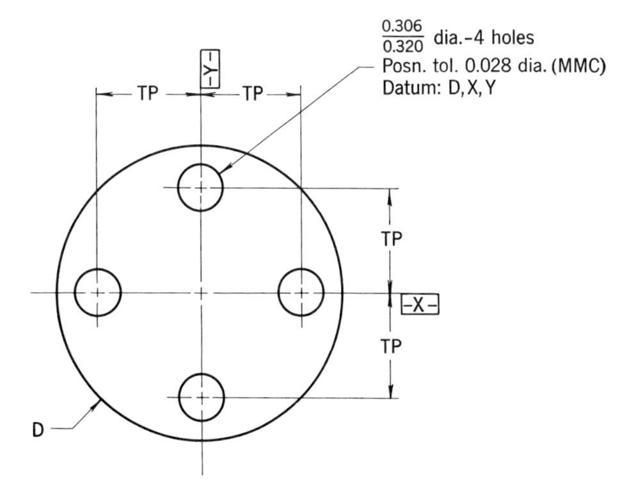

Fig. 20.59

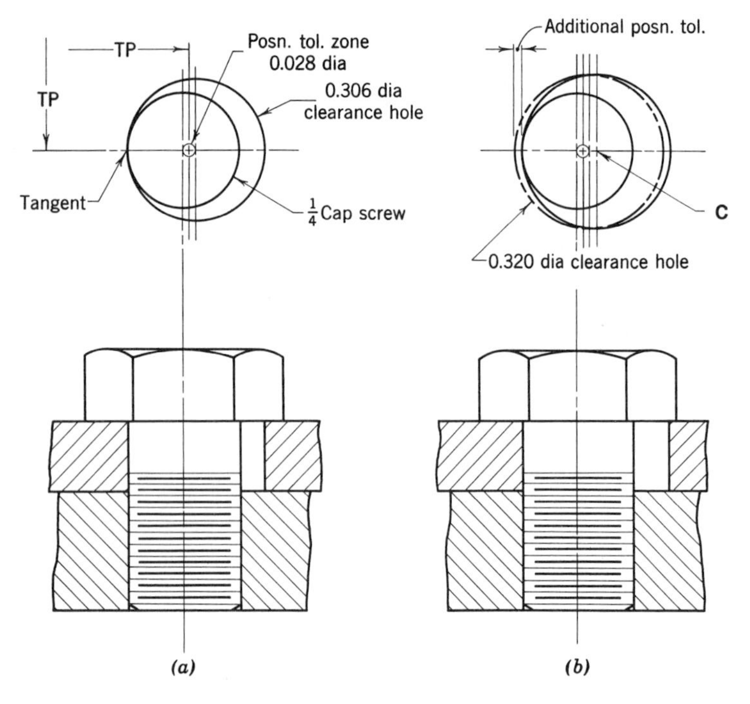

Fig. 20.60

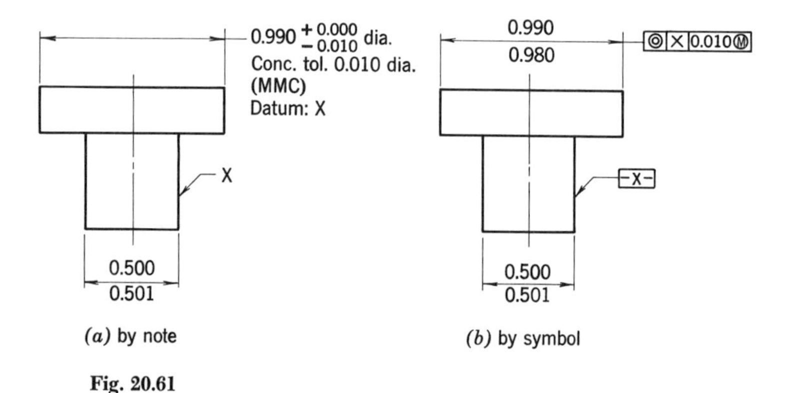

Fig. 20.61

(body) will be tangent to the clearance hole of minimum diameter 0.306. This "worst" condition is shown in Fig. 20.60(*a*).

Now suppose that the clearance hole diameter is larger than 0.306; i.e., 0.320. In this case, the center of the clearance hole could be at point *C*, as shown in Fig. 20.60(*b*), and still provide for functional assembly. Note carefully that the positional tolerance zone is increased without affecting the proper mating of the two parts.

Here you may ask: What is the importance of the callout MMC? If the engineer had not specified MMC in the callout (note), inspection personnel would reject the functional part shown in Fig. 20.60(*b*) because the clearance-hole center did not fall within the tolerance zone specified. We might be critical of the inspector for not recognizing a functional part; however, it is not his job to engineer the design. It is his duty to verify the dimensions of the part in accordance with the specification as stated on the drawing.

Tolerances of Form (MMC)

Let us consider the part shown in Fig. 20.61. When the head is at maximum material condition (diameter = 0.990) the concentricity tolerance as specified is 0.010 diameter. In this case the eccentricity is 0.005, as shown in Fig. 20.62(*a*); and at assembly, as shown in Fig. 20.62(*b*).

However, when the head is at minimum material condition (diameter 0.980) there will be an increase in the allowable eccentricity from 0.005 to 0.010, as shown in Fig. 20.63(*a*). The condition at assembly is shown in Fig. 20.63(*b*).

It may be noted that the datum feature diameter 0.500 was used in the example. What effect would result if the feature was 0.499? The student should observe that an additional amount of eccentricity could then be permitted. To take advantage of this additional form tolerance the specifications must call out datum MMC as well as feature MMC. This is accomplished by adding the modifier to the datum symbol, i.e., | ◎ | *X* Ⓜ | 0.010 Ⓜ | rather than | ◎ | *X* | 0.010 Ⓜ |, as shown in Fig. 20.61(*b*).

Although the above example dealt with MMC applied to concentricity, MMC could also be applied to straightness, squareness, roundness, etc. It should also be remarked that there are instances where the engineer would not specify MMC. It might be necessary to identify a specific fixed value of tolerance to satisfy functional requirements, for example, that vibration of the parts be held within certain limits. In such cases the use of the modifier Ⓡ, "Regardless of Feature Size," is recommended; however, at the present time the absence of the modifier Ⓜ is intended to mean (RFS)—regardless of feature size.

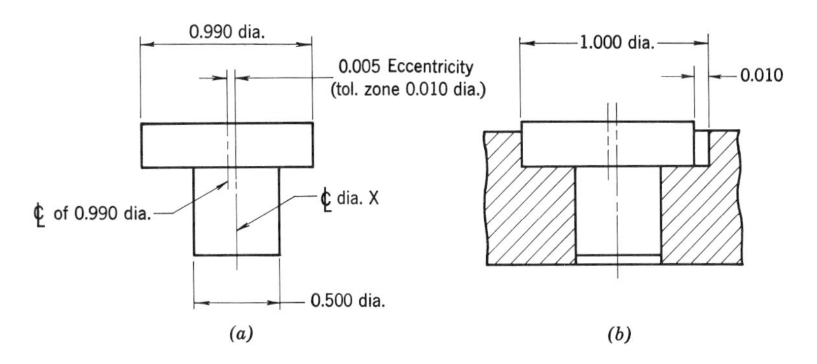

Fig. 20.62

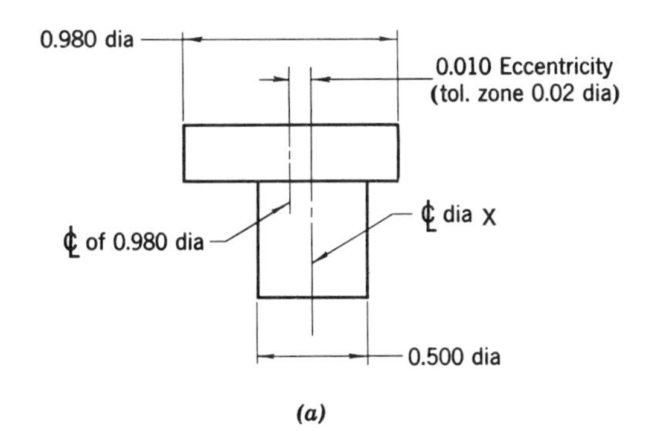

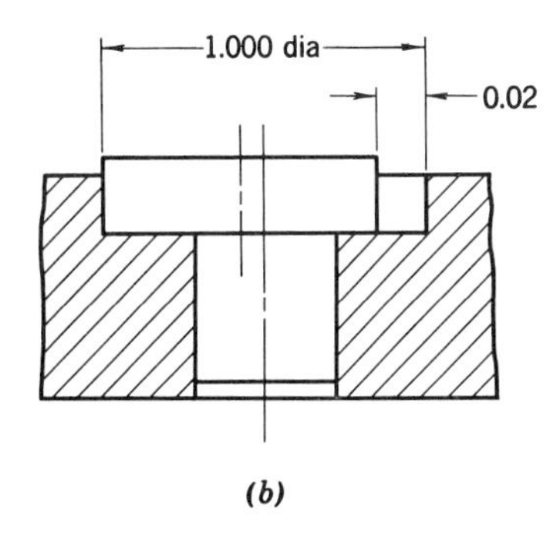

Fig. 20.63

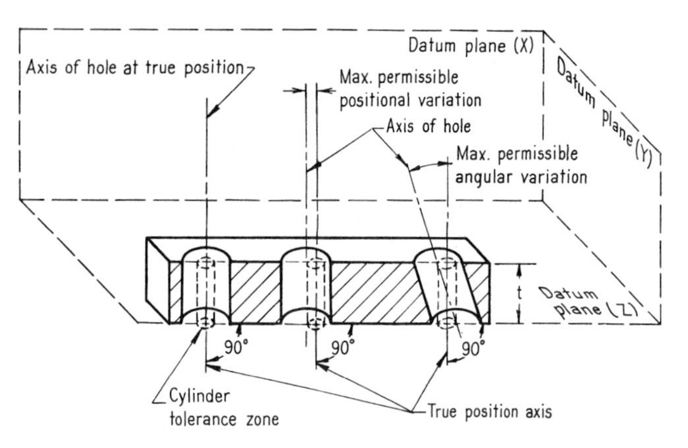

Fig. 20.64 Tolerance zones, three-dimensional.

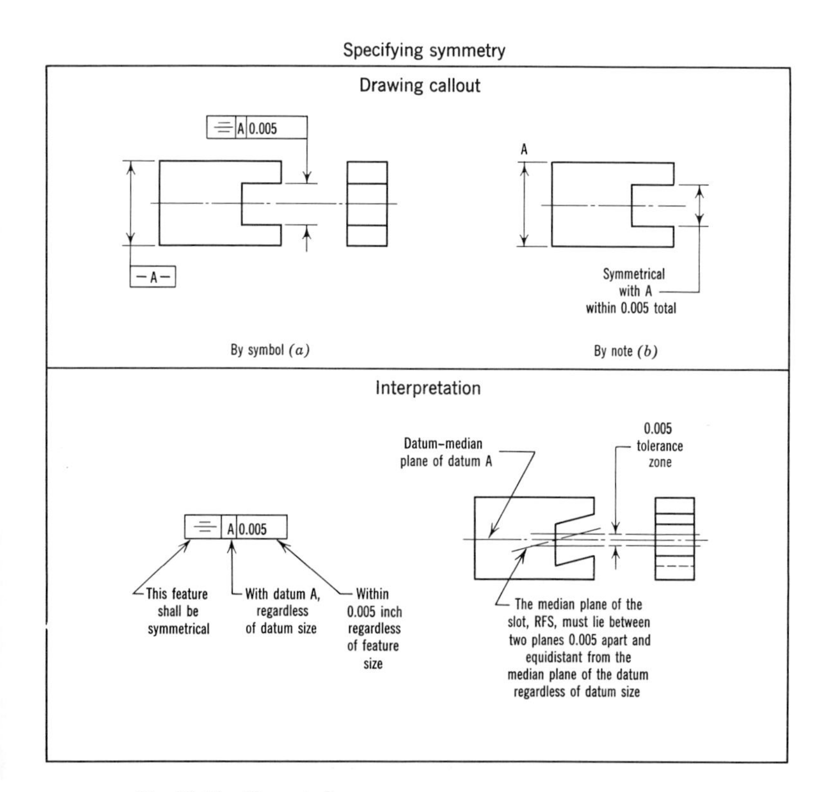

Fig. 20.45 ***(Repeated)***

Tolerance Zones—Three-Dimensional

The treatment of tolerance zones has been generally limited to a two dimensional (plane) consideration. Features such as holes, slots, etc., are three-dimensional and the tolerance zones that apply to these features must also exist in three dimensions. This is illustrated in Fig. 20.64. Observe that the true position axis extends perpendicular to the third (Z) datum plane. The tolerance zone is therefore properly interpreted as a cylinder projected from the positional tolerance circle having a height equal to the third dimension (t) of the feature. The effect of this projection is to control angularity as well as position. Note that additional restrictions of angularity may be superimposed by the callout of parallelism, perpendicularity, symmetry, etc.

It should also be observed that the third dimension projection of a tolerance zone need not be cylindrical; for example, the callout for symmetry applied to plane features will establish a right prism as the three-dimensional tolerance zone. (See Fig. 20.45.)

It is hoped that the treatment of dimensioning as presented in this chapter has given you some understanding and much appreciation of the problems of expressing the engineer's design intent with assurance that the interpretations made by the engineer, manufacturer, and inspector will be the same.

It is believed that the treatment of the true position method, although introductory, serves as a good springboard for understanding the application of this method to complex designs.

A review of the techniques and procedures discussed in this chapter is afforded by Fig. 20.65 and Fig. 20.66. The student is encouraged to "read" these drawings thoroughly to test his understanding of dimensioning.

It should be emphasized that during the past several years there has been a significant increase in the use of the true position dimensioning method because it has proved to be both reliable and economical. Inherent in the TP method is the provision for effective communication among the design engineer, manufacturer, and inspector. In the coming years we will witness the adoption of the TP method, and the use of symbols, by a much larger number of companies.

Interesting industrial examples are shown in Figs. 20.67 to 20.72.

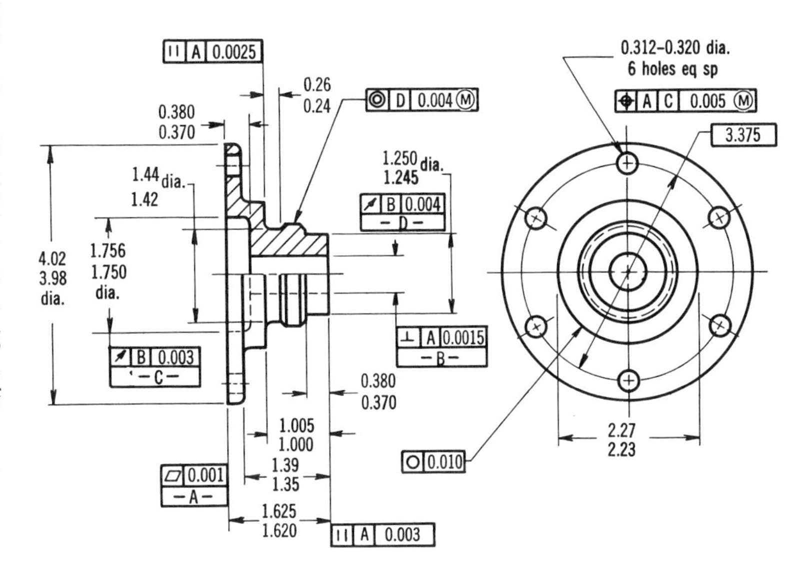

Fig. 20.65

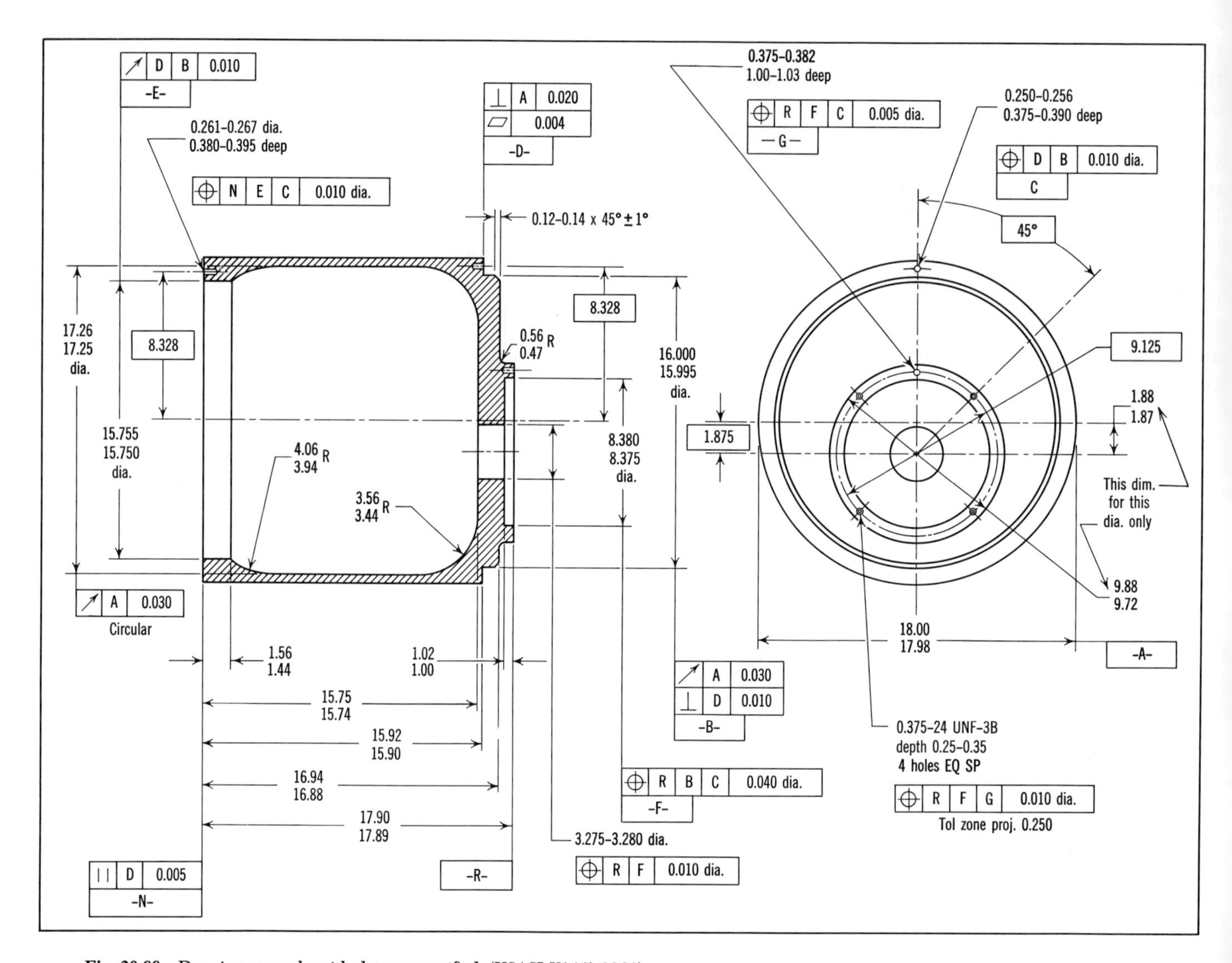

Fig. 20.66 Drawing example with datums specified. (USASI Y14.5–1966)

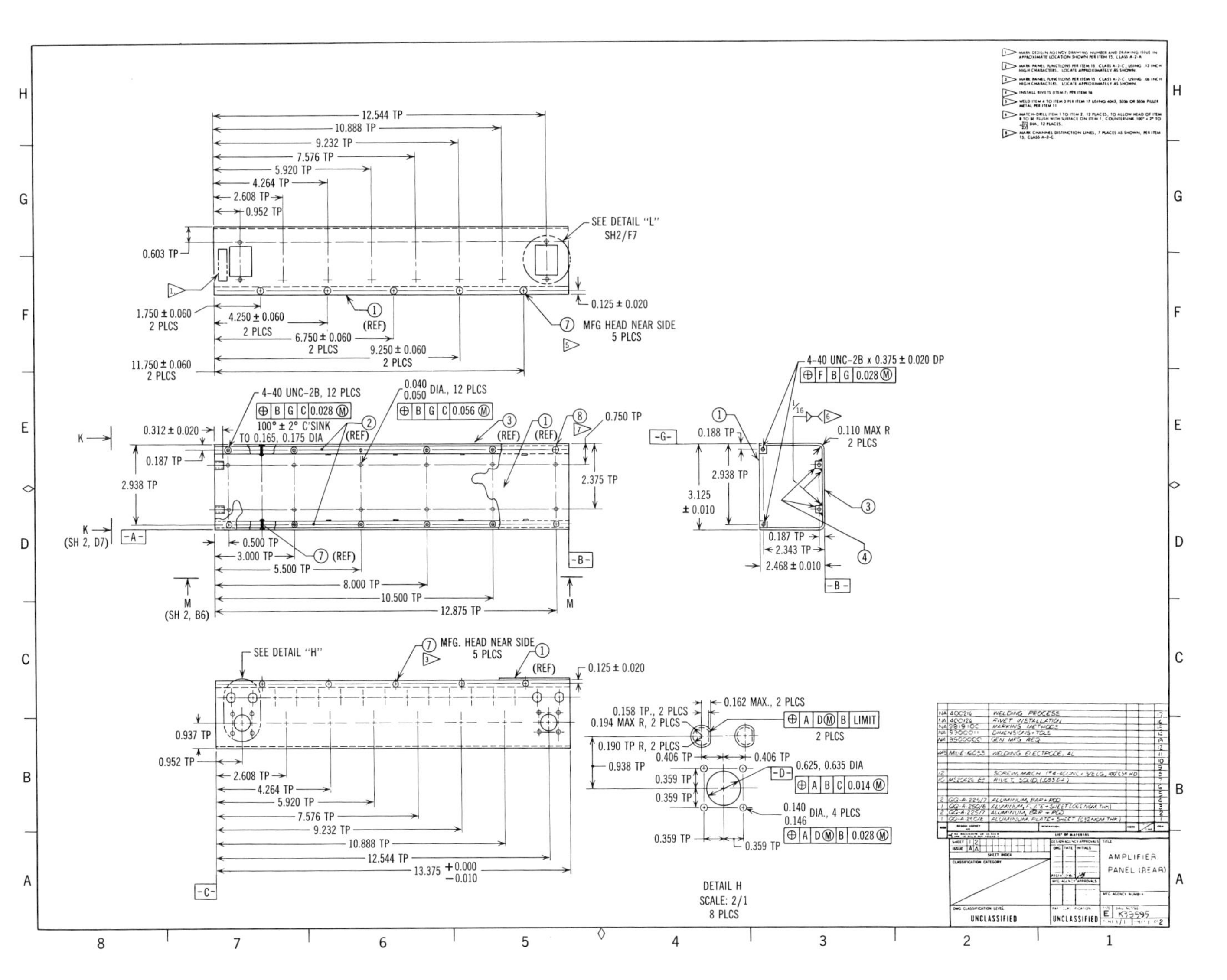

Fig. 20.67 Amplifier panel. (Courtesy Sandia Corp.)

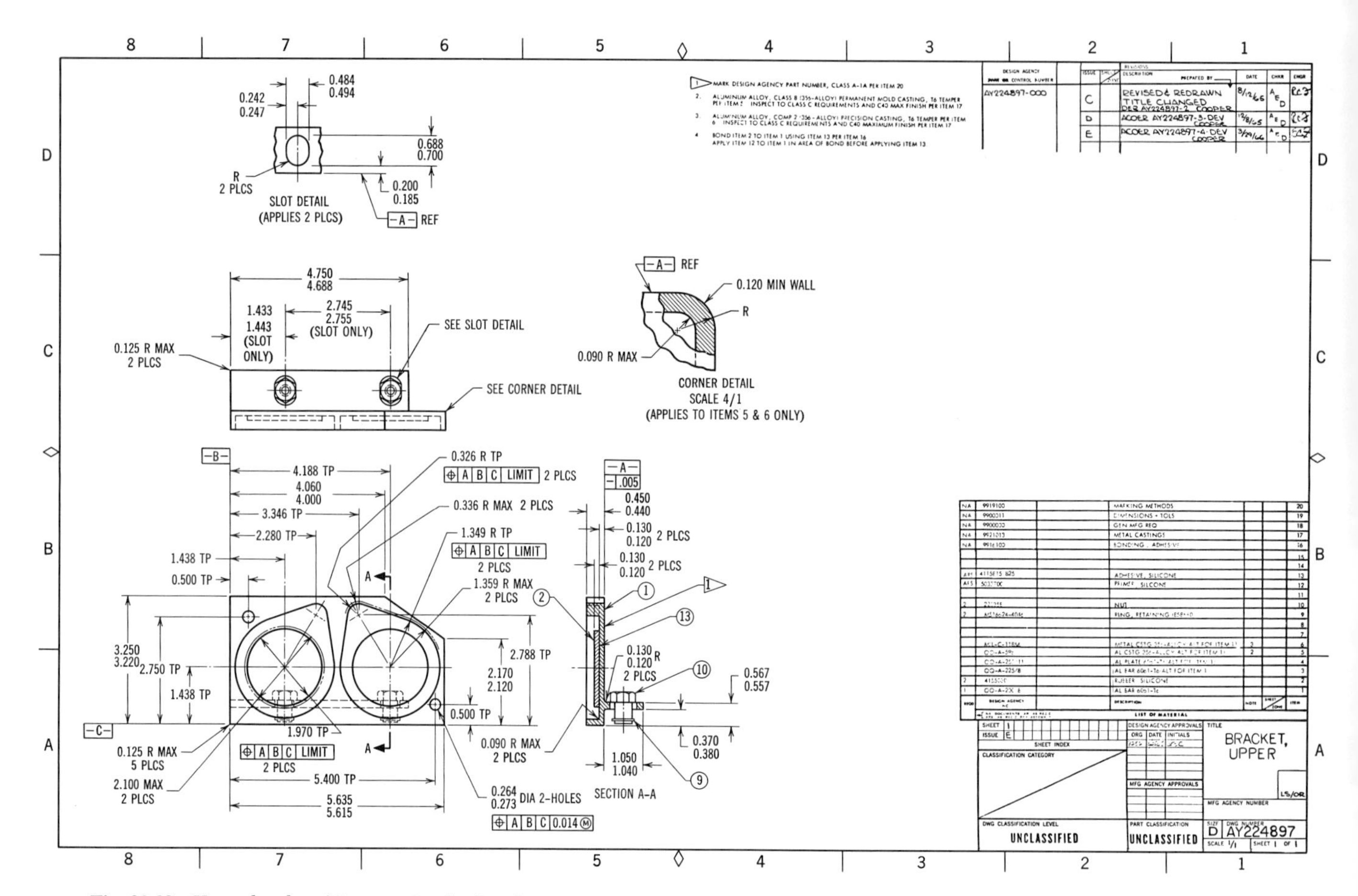

Fig. 20.68 Upper bracket. (Courtesy Sandia Corp.)

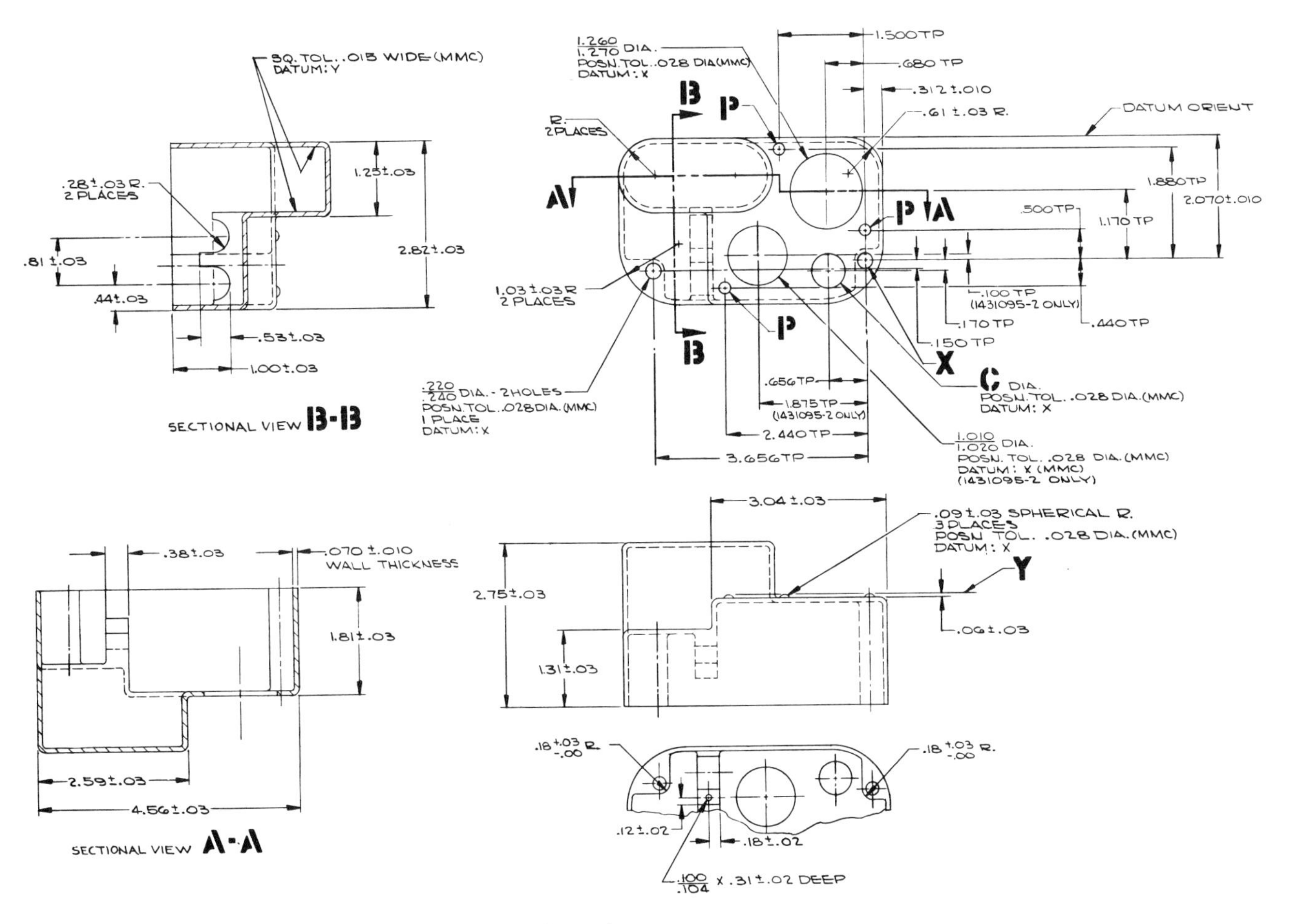

Fig. 20.69 Housing. (Courtesy Bendix Corp., Kansas City Division)

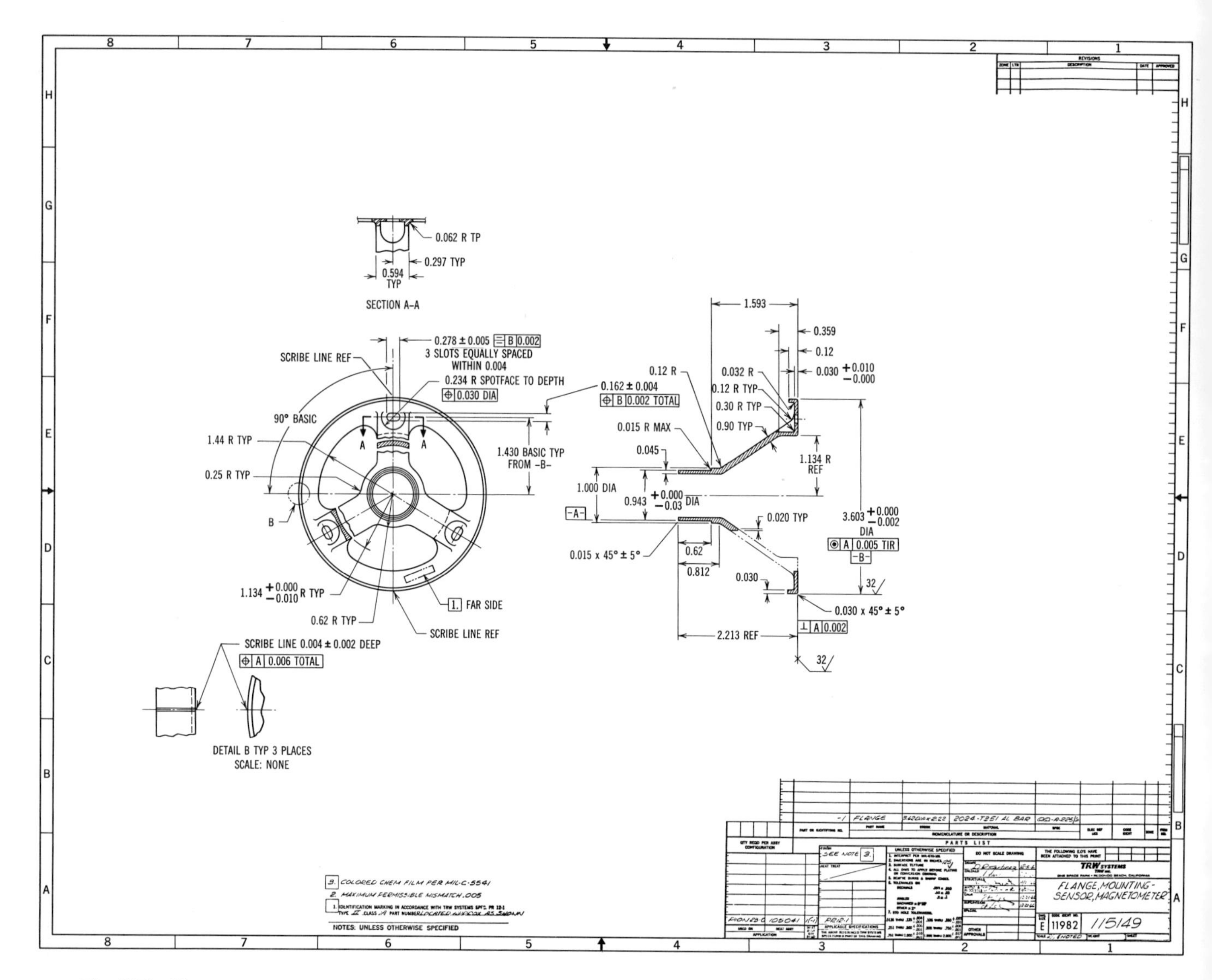

Fig. 20.70 Flange, mounting-sensor, magnetometer. (Courtesy TRW, Inc.)

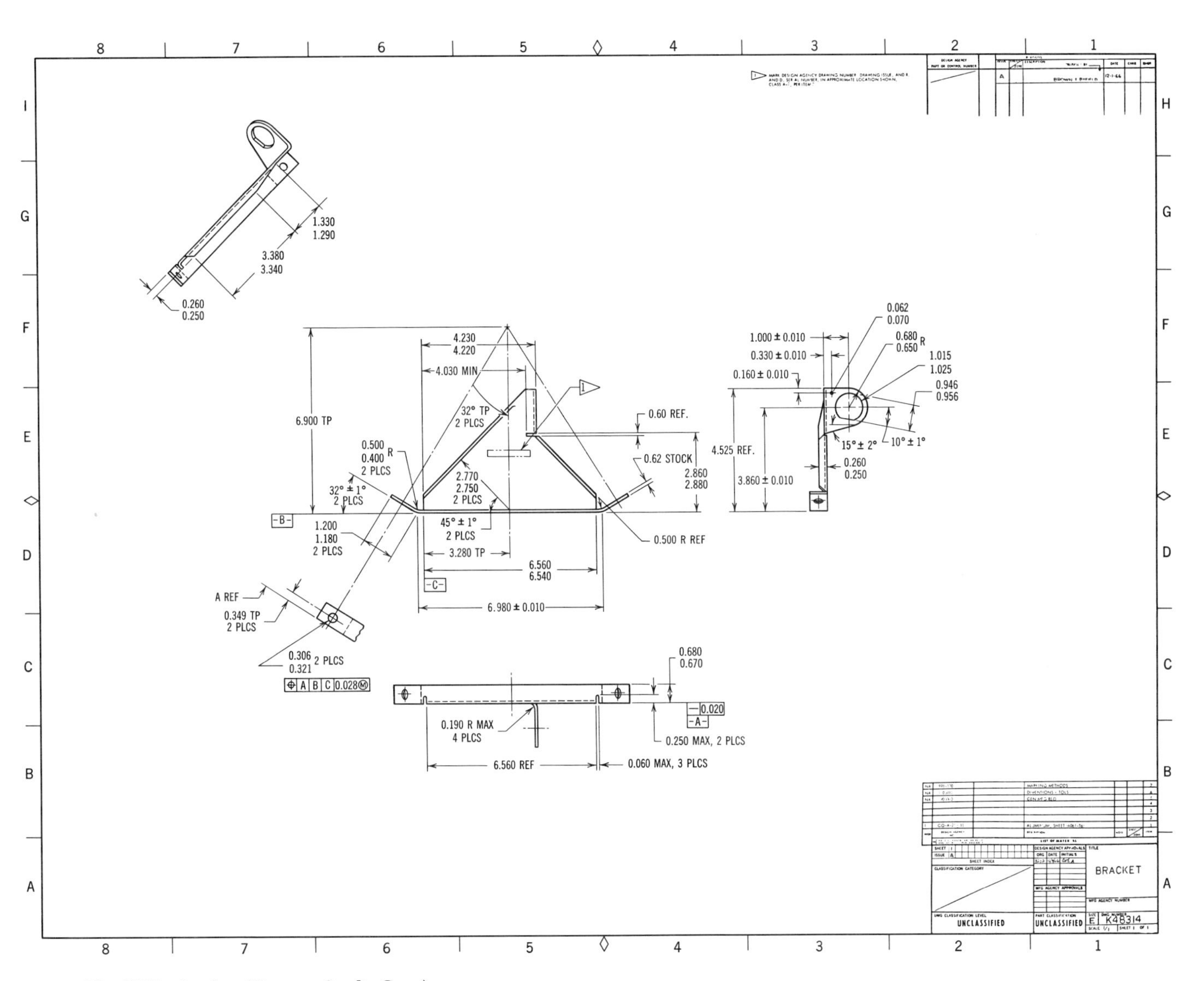

Fig. 20.71 Bracket. (Courtesy Sandia Corp.)

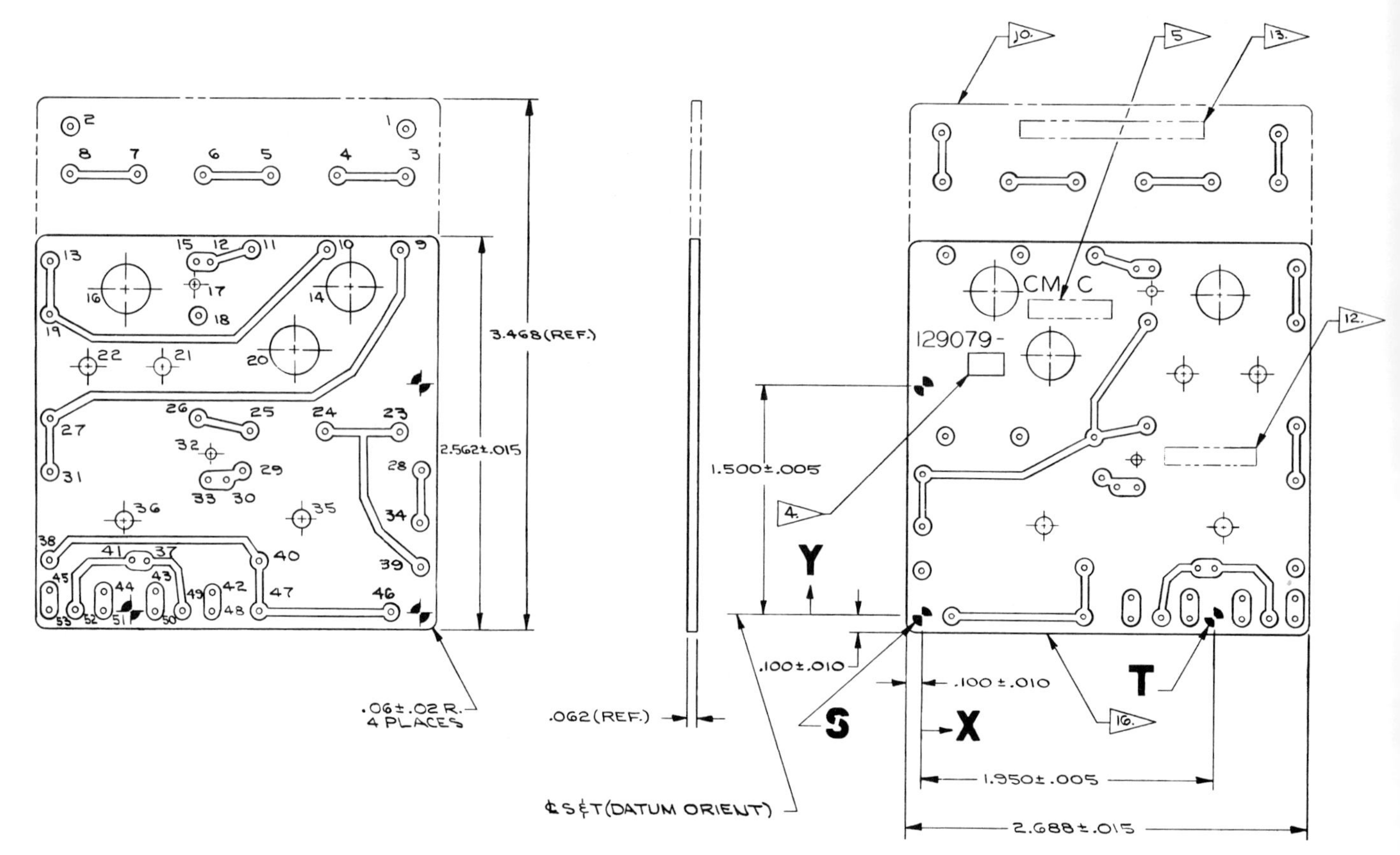

Fig. 20.72 Printed wiring board. (Courtesy Bendix Corp., Kansas City Division)

NOTES

1. FABRICATE PER BKC 28-1, WITH THE FOLLOWING EXCEPTIONS:
   A. BOARDS WILL BE GOLD PLATED PER BKC 2-14, .00010 INCH MIN. THICKNESS.
   B. THE PERIPHERY OF THE LAND SHALL, AT ALL POINTS, EXTEND A MINIMUM OF .029 BEYOND THE CIRCUMFERENCE OF THE .040/.050 HOLES AND .020 BEYOND THE CIRCUMFERENCE OF THE .047/ .057 HOLES, EXCEPT FOR THE FLATS ON TRIMMED LANDS WHICH MUST BE .014 MINIMUM FROM EDGE OF HOLE.
2. COPPER CLAD EPOXY GLASS LAMINATE PER MAT'L STD 2141514. .062 INCH NOMINAL THICKNESS.
3. HOLE NUMBERS ARE FOR MANUFACTURING PURPOSES ONLY AND ARE NOT TO APPEAR ON THE BOARD.
4. RUBBER STAMP LAST TWO DIGITS OF STOCK NUMBER PER TABLE .12 INCH CHARACTERS WITH BLACK INK PER BKC 19-13. LOCATE APPROX. AS SHOWN, NEARSIDE.
5. RUBBER STAMP MANUFACTURER'S CODE AND DATE CODE PER BKC 19-29, USING .12 INCH CHARACTERS WITH BLACK INK PER BKC 19-13. LOCATE APPROX. AS SHOWN, NEARSIDE.
6. THE TOTAL INSULATED DISTANCE BETWEEN ANY TWO CONDUCTIVE AREAS SHALL BE NOT LESS THAN .025.
7. LIMITS OF ACCEPTABLE WORKMANSHIP ARE DEFINED IN BKC 18-0.
8. INDICATES THE THREE CONTROL POINTS FOR ESTABLISHING ORIENTATION OF PATTERN.
9. .040/.050 AND .047/.057 HOLES TO BE PLATED THRU. DIAMETERS APPLY AFTER PLATING.
10. THIS PORTION OF BOARD IS 1434847-2 AND IS FOR TESTING PURPOSES ONLY. THIS PORTION OF THE BOARD SHALL BE REMOVED FROM 1434847-1 PRIOR TO ASSEMBLY.
11. FOR EXPLANTION OF TOLERANCES OF POSITION AND FORM, INCLUDING MMC, SEE BKC 25-2.
12. RUBBER STAMP SERIAL NUMBER PER BKC 19-29, USING .12 INCH CHARACTERS PER BKC 19-13. LOCATE APPROXIMATELY AS SHOWN, NEARSIDE
13. 1434847-2 PART IS TO BE MARKED WITH THE SAME MANUFACTURER'S CODE, DATE CODE AND SERIAL NUMBER AS THE 1434847-1 PART IT WAS REMOVED FROM. RUBBER STAMP .12 INCH CHARACTERS PER BKC 19-13. LOCATE APPROXIMATELY AS SHOWN, NEARSIDE.
14. CM129079 IS A REQUIREMENT OF THIS DRAWING.
15. TOTAL PLATED AREA 2.176 SQUARE INCHES.
16. THIS PORTION OF BOARD IS 1434847-1.
17. TEST REQUIREMENTS, 1434849, AND QUALITY SPEC., 1434850, ARE A REQUIREMENT OF THIS DRAWING.
18. THIS ITEM SUBJECT TO TMS APPROVAL.

HOLE	X	Y	DATUM	POSN TOL
14	.486	2.120	S	.028
16	1.988	2.100	S	.028
17	1.520	2.120	S	.028
20	.838	1.712	S	.028
21	1.738	1.590	S	.014
22	2.238	1.590	S	.014
32	1.434	1.025	S	.028
35	.818	.590	S	.014
36	2.018	.590	S	.014

HOLE	HOLE DIA 9
1	.040/ .050
2	↑
3	
4	
5	
6	
7	↓
8	.040/ .050
9	.047/ .057
10	.047/ .057
11	.047/ .057
12	.040/ .050
13	.040/ .050
14	.304/ .314
15	.040/ .050
16	.304/ .314
17	.063/ .068
18	.040/ .050
19	.040/ .050
20	.304/ .314
21	.090/ .097
22	.090/ .097
23	.047/ .057
24	.047/ .057
25	.047/ .057
26	.040/ .050
27	.040/ .050
28	.047/ .057
29	.047/ .057
30	.040/ .050
31	.040/ .050
32	.063/ .068
33	.040/ .050
34	.040/ .050
35	.090/ .097
36	.090/ .097
37	.040/ .050
38	
39	
40	
41	
42	
43	
44	
45	
46	
47	
48	
49	
50	
51	
52	
53	.040/ .050

Fig. 20.72 ***(Continued)***

EXERCISES

For problems 1 to 4 refer to Fig. E-20.1.

1. Determine the limit dimensions for the bore of the pinion (17) and the shaft (26) according to a Class LT1 fit. Use basic size = 0.7500.

2. Determine the limit dimensions for the bore of the driving flange (5) and the spindle (15) according to a Class RC5 fit. Use basic size = 1.3750.

3. Determine the limit dimensions for the spindle bushing (13) according to the following fits: (*a*) for the bore, Class RC7 (basic size = 1.5000); (*b*) for the outside diameter, Class LN3 (basic size = 1.7500).

4. Determine the limit dimensions for the following parts using the basic shaft system; consider nominal shaft size = $\frac{3}{4}''$: (*a*) shaft (26), (*b*) head (12), Class RC4; (*c*) pinion (17), Class LT1. Compare the results with the values obtained for the pinion in Problem 1—they should not be the same.

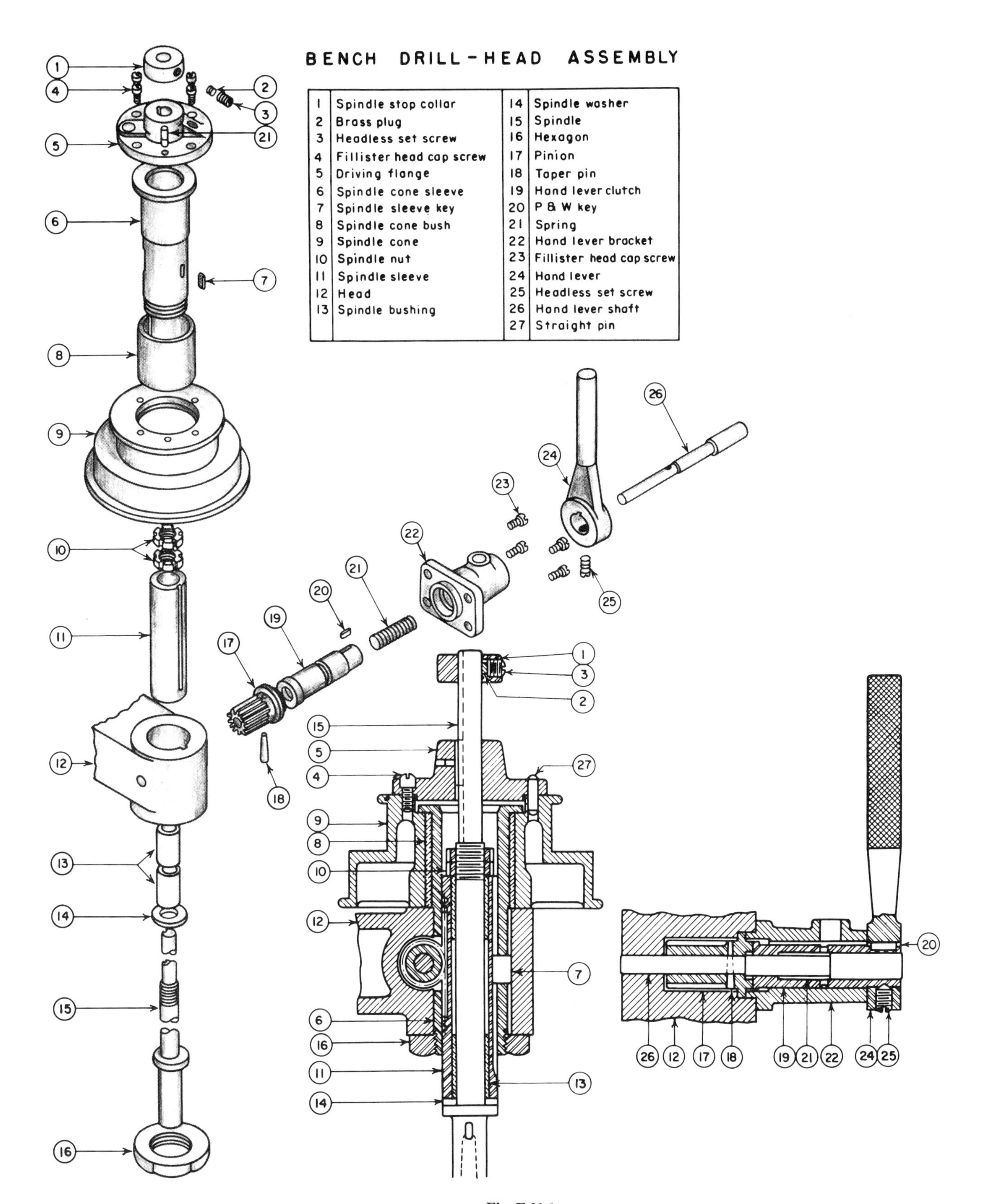

1	Spindle stop collar	14	Spindle washer
2	Brass plug	15	Spindle
3	Headless set screw	16	Hexagon
4	Fillister head cap screw	17	Pinion
5	Driving flange	18	Taper pin
6	Spindle cone sleeve	19	Hand lever clutch
7	Spindle sleeve key	20	P & W key
8	Spindle cone bush	21	Spring
9	Spindle cone	22	Hand lever bracket
10	Spindle nut	23	Fillister head cap screw
11	Spindle sleeve	24	Hand lever
12	Head	25	Headless set screw
13	Spindle bushing	26	Hand lever shaft
		27	Straight pin

Fig. E-20.1

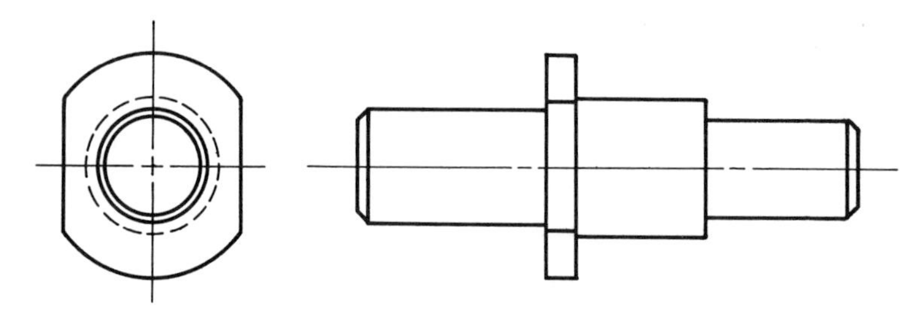

Fig. E-20.2 Shoulder shaft.

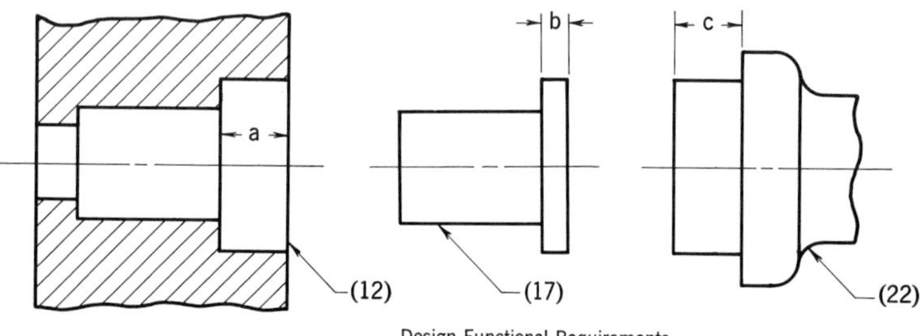

Design Functional Requirements

1. Flange limits (b) = $\frac{0.250}{0.251}$
2. Max. C'Bore depth (a) = 0.625
3. Total clearance for flange in assembly is 0.001 (min.), 0.006 (max.)

Fig. E-20.3

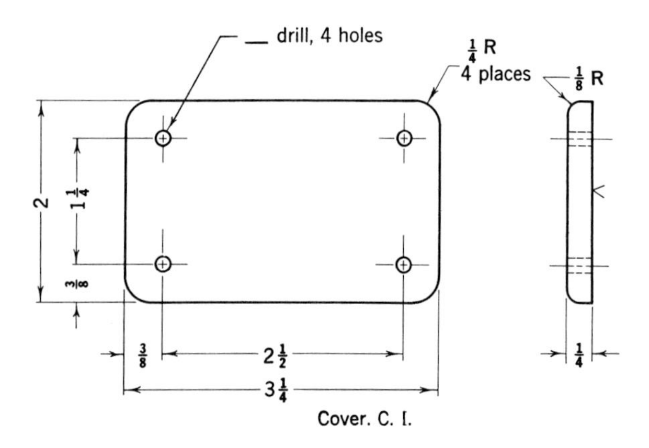

Fig. E.-20.4 Cover

5. Determine the limit dimensions for (*a*) the right end of the shaft, the mating part of which has a nominal diameter of $\frac{9}{16}''$, Class RC5; (*b*) the left end of the shaft, the mating part of which has a nominal diameter of $\frac{5}{8}''$, Class LT1; (*c*) the center section, the mating part of which has a nominal diameter of $\frac{3}{4}''$, Class FN1. (See Fig. E-20.2.)

6. Refer back to Figure 20.16. Assume the following design functional requirements:

(*a*) Minimum tongue size = 0.500.

(*b*) Minimum clearance between tongue and slot = 0.003.

(*c*) Basic dimension from reference plane to lower tongue. surface is 3.000.

(*d*) Acceptable tolerance is 0.010.

PROBLEM: Determine the limit dimensions for *e*, *f*, *c*, and *b* as shown in Figure 20.17.

7. The position and operation of the pinion (17) in the drill head, Fig. E-20.1, will depend on the proper application of tolerances to the pinion flange length and the features of the mating parts: the counterbore depth in the head (12), and the length of the boss on the hand lever bracket (22). (See Fig. E-20.3.) Determine the limit dimensions for *a* and *c* according to the given functional requirements.

8. *Design Requirements* (See Fig. E-20.4):

(*a*) Holes in cover must align with four tapped holes in mating part to allow for fastening with four $\frac{3}{16}''$ R.H. machine screws.

(*b*) Tolerance for fractional dimensions is $\pm\frac{1}{64}''$.

PROBLEM: Determine the clearance hole drill size for the dimensions shown in the figure; and the dimensions of the part using true position dimensions for the hole locations (tolerance to be assigned by instructor).

9. Prepare a full size completely dimensioned drawing of the Hand Lever Bracket shown in Fig. E-20.5. Nominal dimensions can be scaled from the figure which is half size. *Design Requirements* (See Fig. E-20.1.): $\frac{3}{8}''$ cap screws will be used to fasten the bracket (22) to the head (12). A Class LC6 fit is required for the boss diameter (A). Use datum features indicated for location of fastener holes.

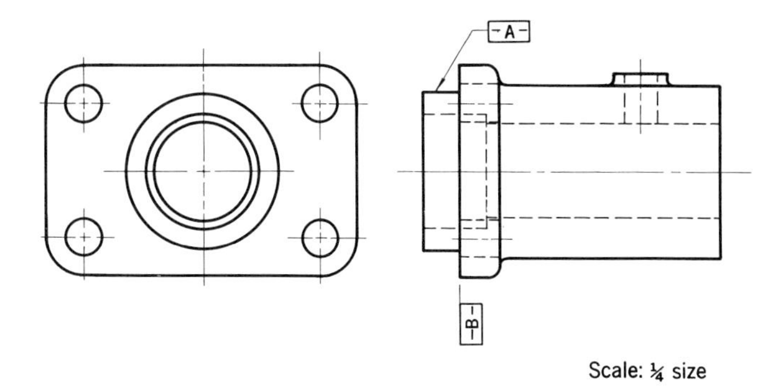

Fig. E-20.5

10. *Design Requirements* (See Fig. E-20.6):
Container: Cast iron, nominal over-all dimensions $5'' \times 5'' \times 4''$ deep; wall thickness $\frac{3}{8}''$; tap holes for $\frac{1}{4}$-20 UNC each corner.
Cover: Material $\frac{3}{8}''$ AISI 1020 steel plate; 0.005 min dia. clearance for $\frac{1}{4}''$ cap screws.
PROBLEM A: Specify dimensions for box and cover. Use ± 0.010 positional tolerance for threaded holes in the box and clearance holes in the cover.
PROBLEM B: Same as Problem A. Use true position dimensions.

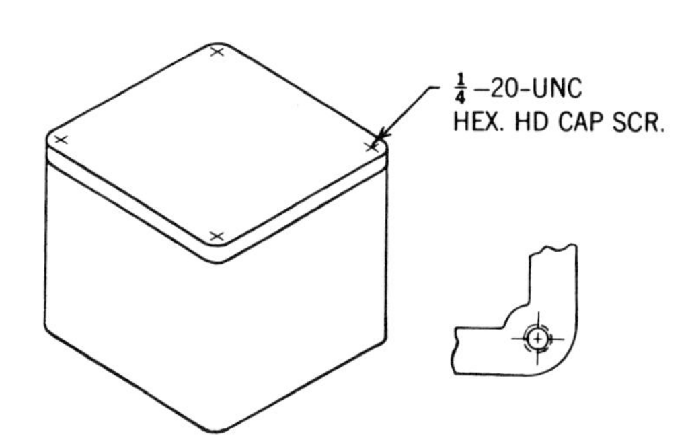

Fig. E-20.6

11. *Design Requirements* (See Fig. E-20.7):
(*a*) Dia. *A*—Press fit (hole dia. 1.2500) Class FN__*
(*b*) Dia. *B*—Clearance for $\frac{1}{4}''$ dia. drill Class RC__*
(*c*) Dia. *C*—Not critical—$\frac{3}{4}''$ nominal.
(*d*) $D = 0.500^{+0.000}_{-0.010}$
(*e*) *E*—Not critical—$\frac{3}{4}''$ nominal.
(*f*) Dia. *B* must be concentric to *A* within 0.0005 TIR.
PROBLEM A: Determine the limit dimensions for the drill jig bushing.
PROBLEM B: Design drill jig for cover in Exercise 9.

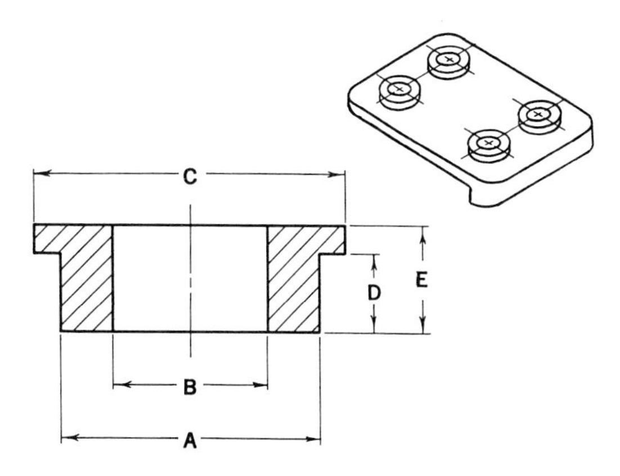

Fig. E-20.7 Drill jig bushing.

* To be assigned by the instructor.

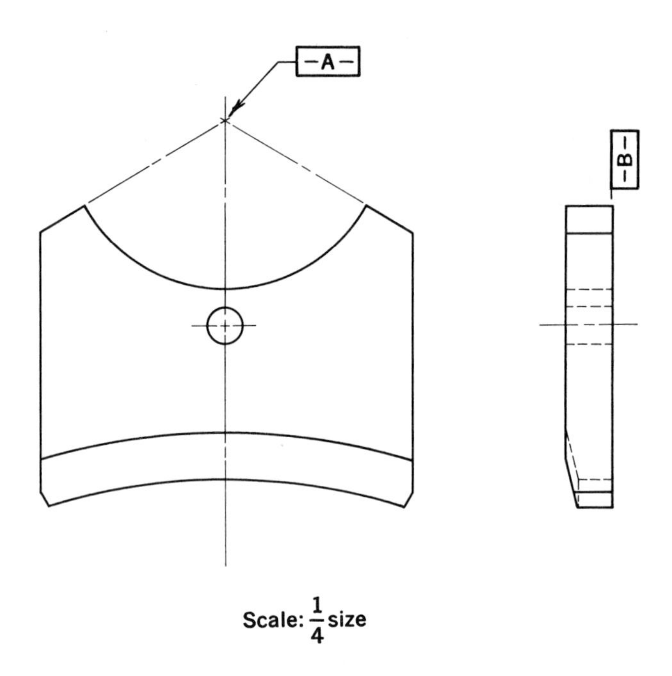

Fig. E-20.8 Lug-lifting sling. (Adapted from a drawing by Convair Division of General Dynamics Corp.)

12. Prepare a *half-size* completely dimensioned working drawing of the Lug-Lifting Sling shown in Fig. E-20.8. Measure nominal distances to the nearest 8th of an inch (on $\frac{1}{4}$-size scale) and express the dimensions as 3-place decimals. The datums have been established to conform with the functional requirements. The limit dimensions for the drilled hole should be calculated to permit a $\frac{3}{4}''$ cap screw to be used in assembly. The datums of the threaded hole on the mating part are *A* and *B*. The part is to be machined from plate stock. Scale of the drawing shown in the figure is $\frac{1}{4}$ size.

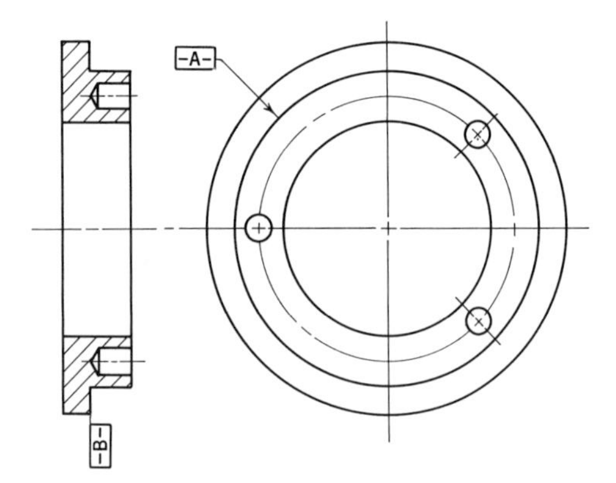

Fig. E-20.9

13. Determine the limit dimensions for the part shown in Fig. E-20.9, according to the following fits; (*a*) for the bored hole, Class LT7; (*b*) for surface *A*, Class RC6. Locate the holes using datum feature diameters *A* and *B*. Specify run-out tolerance 0.003 for bore, datum *A*. Complete the dimensions for the pieces by scaling the necessary measurements. *Scale:* $\frac{1}{2}$ size.

14. *Design Requirements* (See Fig. E-20.10):

(*a*) Allow 0.01–0.02 clearance between diameters *A* and *B*.

(*b*) Use tube O.D. as datum feature (dia. *B*).

(*c*) Use MMC for clearance holes.

PROBLEM: Dimension *CAP*. Use true position dimensions.

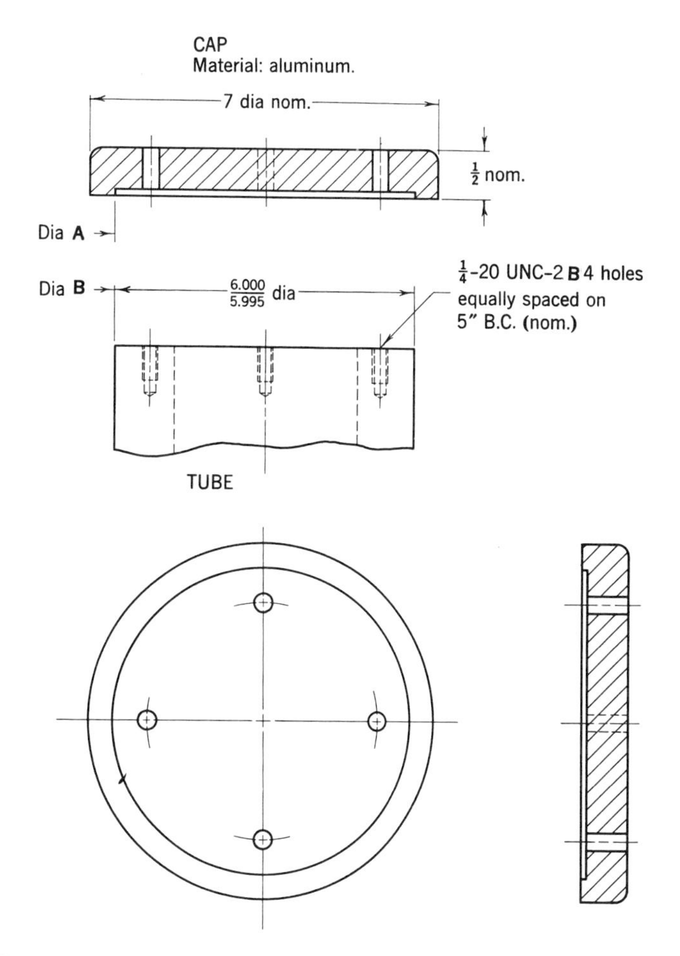

Fig. E-20.10

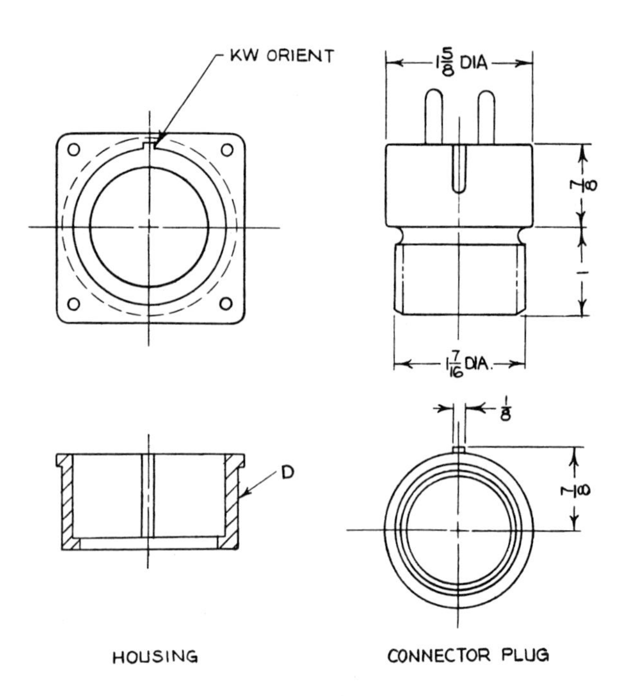

Fig. E-20.11 Housing for electrical connector.

15. *Design Requirements* (See Fig. E-20.11):
(*a*) Datum *D*, orient symmetrical tolerance on *KW*.
(*b*) Mounting holes for 10-32 machine screws.
(*c*) Allowance for housing-plug fit to be assigned by the instructor.
PROBLEM: Dimension the housing.

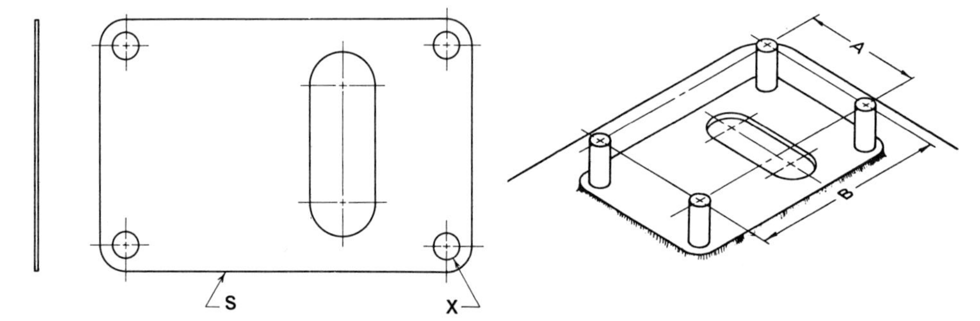

Fig. E-20.12 Gasket

16. *Design Specifications* (See Fig. E-20.12):
(*a*) $\frac{1}{32}''$ rubber, silicone gasket stock (± 0.010 tolerance on thickness).
(*b*) Gasket must fit over $4\frac{3}{8}''$ studs on part shown. Stud spacing $A = 3.00 \pm 0.015$; $B = 5.00 \pm 0.015$.
PROBLEM: Dimension the gasket. Use true position dimensions, MMC for stud holes. Use datum *X*, orientation *S* for all dimensions.

17. *Design specifications* (See Fig. E-20.13):
(*a*) 4 tapped holes for $\frac{1}{4}$ UNC, Posn. tol. 0.014 dia. (MMC), datum *X*, orient S_1.
(*b*) 2 holes (*B*) for $\frac{3}{8}''$ fillister-head cap screws, datum *X*, orient S_1. Threaded holes in mating part are spaced 1.875 ± 0.015.
(*c*) Surfaces S_1 and S_2 should be symmetrical to holes (*B*) within 0.003. Dimension of mating tongue 1.000/0.998. Use RC6 fit to the slot.
(*d*) Boss surfaces should datum to the horizontal surface of the slot.
(*e*) Material: magnesium.
Additional specifications may be assigned by instructor.
PROBLEM: Dimension the Base Mounting Plate. Use MMC dimensions for the holes (*B*).

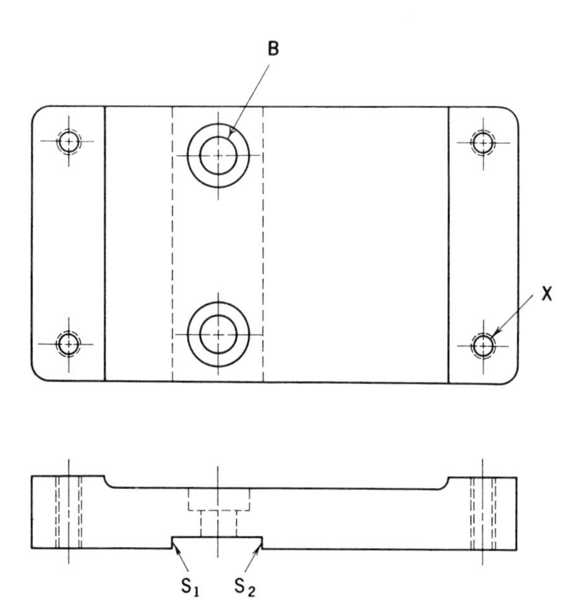

Fig. E-20.13 Base-mounting plate.

18. Prepare detail and assembly working drawings of the Butterfly Valve shown in Fig. E-20.14. Details should be drawn *freehand.* Consider concentricity requirements for shaft and housing bores. Exercise particular care in choice of datums. Calculate critical dimensions using nominal dimensions given.

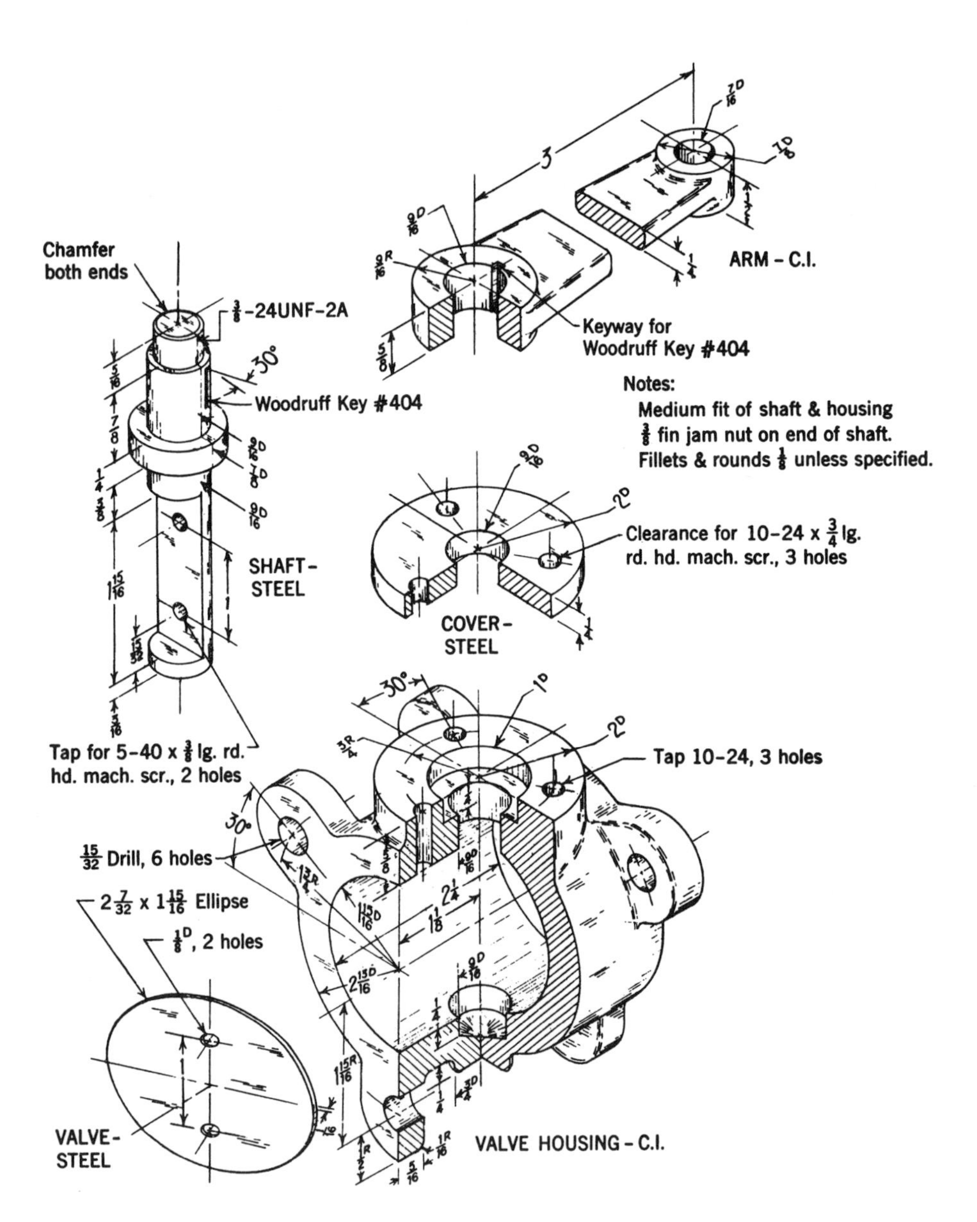

Fig. E-20.14 Detail sketches of butterfly valve assembly.

19. Prepare detail and assembly working drawings of the Shaft Bearing Pedestal shown in Fig. E-20.15. Details should be drawn *freehand*. Calculate critical dimensions from nominal dimensions given.

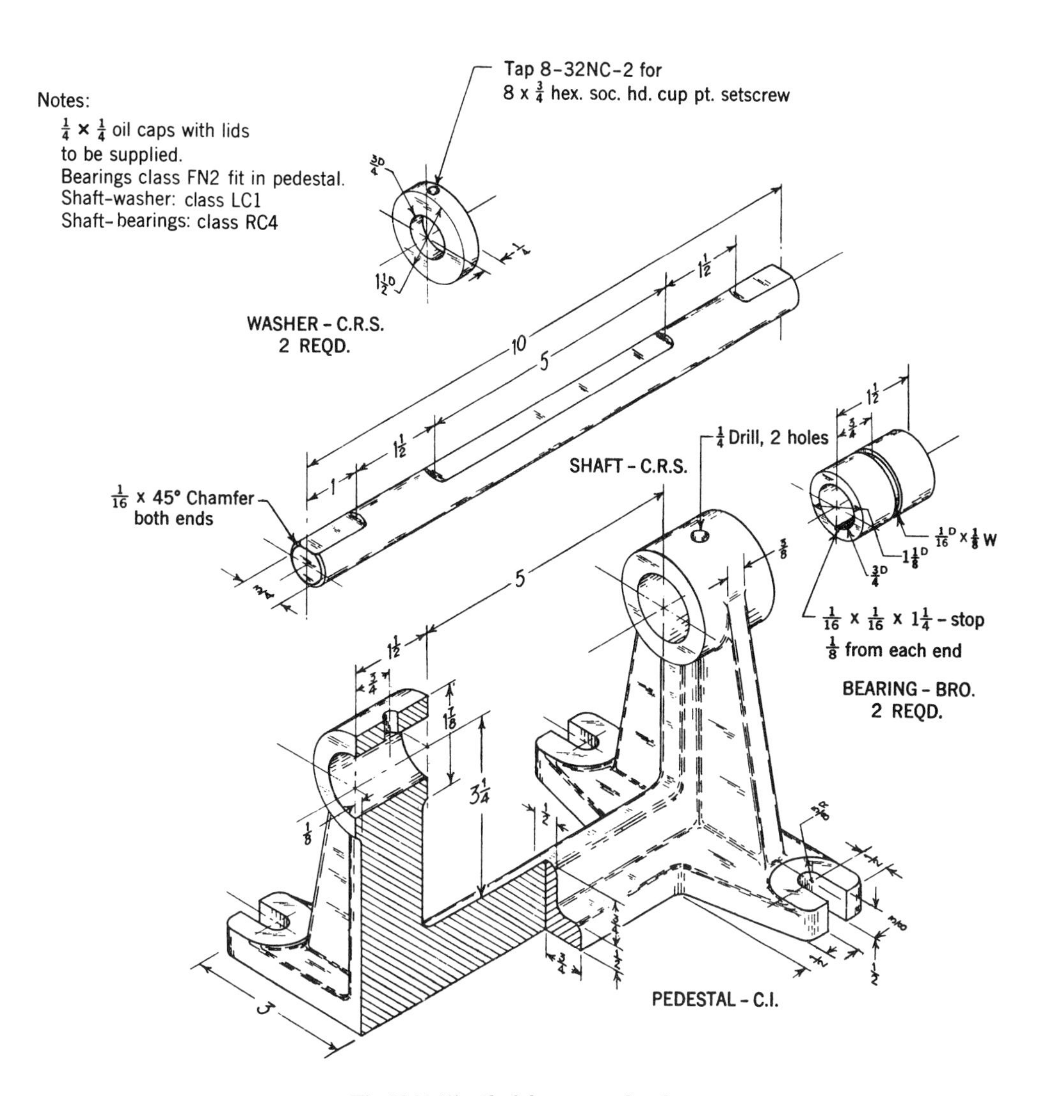

Fig. E-20.15 Shaft-bearing pedestal.

CONCEPTUAL DESIGN—DEVELOPING CREATIVITY 21

"The scientists explore what is, and the engineers create what has never been."

(Theodor von Kármán)

The scientist is concerned primarily with new knowledge of nature. The engineer is concerned with the implementation of new scientific discoveries. The design of useful products requires of the professional engineer effective knowledge of mathematics, of the physical sciences, of the engineering sciences—mechanics, graphics, properties of materials, etc.; professional experience which develops good judgment in assessing need, feasibility, economy, and reliability of proposed projects; and proficiency in the use of the English language in both its written and oral forms.

Design is the core of engineering. In a broad sense design includes circuits, machines, structures, processes, and combinations of these components into systems and plants. Moreover, the professional engineer must be capable of predicting the performances and costs of components, systems, and plants to meet specific requirements.

Designing is a conceptual process which is done largely in the mind, and the making of sketches is a recording process, a reliable memory system which the engineer uses for self-communication—talking to himself—to help him "think-through" the various aspects of his project. Graphics is an integral part of the conceptual phase because, more often than not, the making of a simple sketch to express a design conception does of itself suggest further items of a conceptual nature.

Engineers who have developed the ability to form a visual image of geometrical and physical configurations and to "think graphically" have a tremendous advantage in creating a physical means of achieving a technological objective. The "thinking-through" process is an exercise of the mental powers of judgment, conception, and reflection for the purpose of reaching a conclusion, i.e., a design which is the "best compromise" solution to a given project.

In the design process it is essential to:

1. *Establish the need* for the product and its economic justification. This step requires a *feasibility* study.
2. *Define* the "real" problem. For example, a client may request the design of a crane capable of

lifting a 5 ton load. Upon further probing the engineer discovers that the load is to be moved once every two months; that the load package is rectangular with dimensions 4 ft × 5 ft × 7 ft; that the distance the load is to be moved is within a 50-ft radius. The "real" problem is *moving the load* and not the design of a crane which is one possible solution. Other solutions could be the use of a lift fork, or perhaps the use of a helicopter. To cite another example, suppose you are asked to design a "better" saw to cut wood. Note that the *request* is the design of a saw; however, a more fundamental approach would be to think of possible methods to cut wood. This is the *real* problem. Again, the request might be to design a better hammer to drive nails. I believe you can readily see that the real problem is to design a new device to drive nails economically. You might conclude, after consideration of several alternatives, that the hammer-head design is the best.

3. *Synthesize*—to prepare a number of conceptual design sketches which disclose possible alternative solutions. There is often the tendency on the part of the engineering designer to move too rapidly from the statement of the problem to a finalized solution, only to discover later that another solution would have been much better. It is essential, considering the time constraint (i.e., next year's automobile models must be on display September 18) to remain noncommittal as long as possible; and to continue the search for new ideas and new combinations of old ones, etc.
4. *Evaluate* the possible solutions in light of the specified requirements.
5. *Select* the "best" compromise solution, then carefully re-evaluate this choice. Can two or more parts be combined? Can a complicated component be replaced by two or more simpler units? Has consideration been given to ease of manufacture? Is the choice of materials best suited to the design? And so on.
6. *Build and test* a prototype. (It is assumed that the necessary "layout" drawings and sufficient shop drawings have been prepared to construct the prototype.) This is an essential step to verify the proposed design. In a sense, this is "debugging" the design.

7. *Revise the design* as a consequence of the de-bugging step in the testing of the prototype.
8. *Prepare final design drawings* for manufacture of the components and their proper assembly.
9. *Obtain and analyze* marketing reports of consumers' reactions to the product.

   A study* by the Bank of America shows that 90% of new American products fail within four years. Why is this so? The study suggests the following:

   (*a*) Inadequate market analysis.
   (*b*) Defects in components and malfunction of some elements.
   (*c*) High costs; poor timing; competition.
   (*d*) Inadequate marketing efforts; too small sales force; poor distribution.

   Many of these failures could have been avoided had sufficient time been given to answering such questions as:

   (*a*) What *needs* does the product really satisfy?
   (*b*) Is it superior to the competition?
   (*c*) Has the product been properly de-bugged?
   (*d*) Is it safe?
   (*e*) Is its size, shape, style, and "eye appeal" satisfactory?
   (*f*) Can it be sold to a large enough market profitably?
10. *Modify and redesign*—that is, go back to step 1 and start the cycle again.

Documentation

Once a design project has been completed it is important to *document the design* so that (1) patent rights may be protected; and (2) that recall of the thinking that went into the design will be facilitated. This is most helpful in furthering new efforts.

How can we document our designs? One of the steps that can be taken is to write a report of the design history which should include:

(*a*) Statement of the problem.
(*b*) Definition of the *real* problem.
(*c*) Design parameters—operational factors, human factors, size, shape, function, environmental conditions, etc.
(*d*) Problem considered—how the parameters were satisfied; how constraints were complied with;

* "Marketing a New Consumer Product," 1965, *Small Business Reporter*, Bank of America National Trust and Savings Association.

conceptual design choices (retain all sketches); reasons for "best" choices; basis for selection of the best compromise solution; etc.
(*e*) Modifications resulting from prototype testing.
(*f*) Recommended changes in the design.
(*g*) Patentable features (be sure to retain notes and sketches).

Thus far you have devoted time and effort to (*i*) the attainment of a good grasp of the fundamental principles of orthogonal projection and their application to the solution of a variety of three-dimensional problems arising in the various fields of engineering; (*ii*) the development of facility in freehand sketching—very important to the professional engineer; (*iii*) graphical presentation of data; (*iv*) graphical mathematics; (*v*) elements of nomography; (*vi*) pictorial drawing; and (*vii*) familiarization and use of recognized practices in expressing shape and size graphically. In addition, in other courses, you have achieved some success in an understanding and use of analytic geometry and elementary calculus. You have gained knowledge of first courses in college chemistry and physics, and you have had experience with a few nontechnical courses.

This background, plus some work experience you may have had, along with hobbies, reading, etc., will be very helpful in undertaking "open-ended" projects.

An opportunity "to be on your own" to a much greater extent than heretofore presents itself in this part of our study. The experience gained from undertaking projects that have *many solutions* is invaluable. *It is the kind of experience that will enable you to approach with greater confidence real engineering situations that will confront you later.* While it is recognized that many achievements in design, research, development, and production result from "team effort," it is essential for the engineer to attain the ability to make decisions on his own.

The creative art of conceiving a physical means of achieving an objective is the first and most crucial step in an engineering project. As an engineering student you cannot begin too early to learn how to make the difficult and often arbitrary design decisions whose validity will ultimately be tested; and to recognize that *there are numerous solutions to any real problem.* Analysis of a design is the second step. The study of analytical courses is very important, because it is so difficult to predict performance of a conceptual design, still on paper, without analysis. You will appreciate and recognize the need for courses in mechanics, properties of materials, strength of materials, manufacturing, report writing, etc. Later in your engineering career you will appreciate the meaning of creative design: "The process of bringing into being, by the use of scientific principles, technical information, and imagination, the definition of a mechanical structure, machine or system, to perform pre-specified functions with maximum efficiency and economy."*

At this early stage of your engineering career your background is quite limited, and because of this the proposed problems are relatively simple and limited in scope. But *in principle* they are of the same character as some of the most advanced problems in engineering practice.

It is hoped that from your experience in solving some of the proposed problems you will get a "feel" for conceptual design and an appreciation of the need for developing a spirit of inquiry; desire for probing and searching; a recognition of the importance of meeting time schedules and of costs. The element of competition (often an impelling force) will become quite evident when you vie with your classmates as to who has the "best design." The satisfaction of having performed a given task largely through your own effort and ability, and with a reasonable degree of success, will be most gratifying.

Real engineering projects are "open-ended," that is, there are several solutions which can satisfy the specified requirements. Among the possible solutions there may be one which is the "best" compromise.

With your present engineering background we would hardly expect you to undertake projects of a complex nature; i.e., the "best design" of a space vehicle that could land two astronauts on Mars and return them to Earth safely; or the design of a supersonic transport capable of carrying 300 passengers and baggage a non-stop distance of 7000 miles at a speed of 1800 mph. A very possible solution to the latter problem is the proposed Boeing SST shown in Fig. 21.1. Moreover, would we expect you to undertake a *feasibility* study to determine the advisability of designing a completely new long-range cargo air-

* From Report No. CE 521, Joseph Lucas Ltd, Birmingham, England.

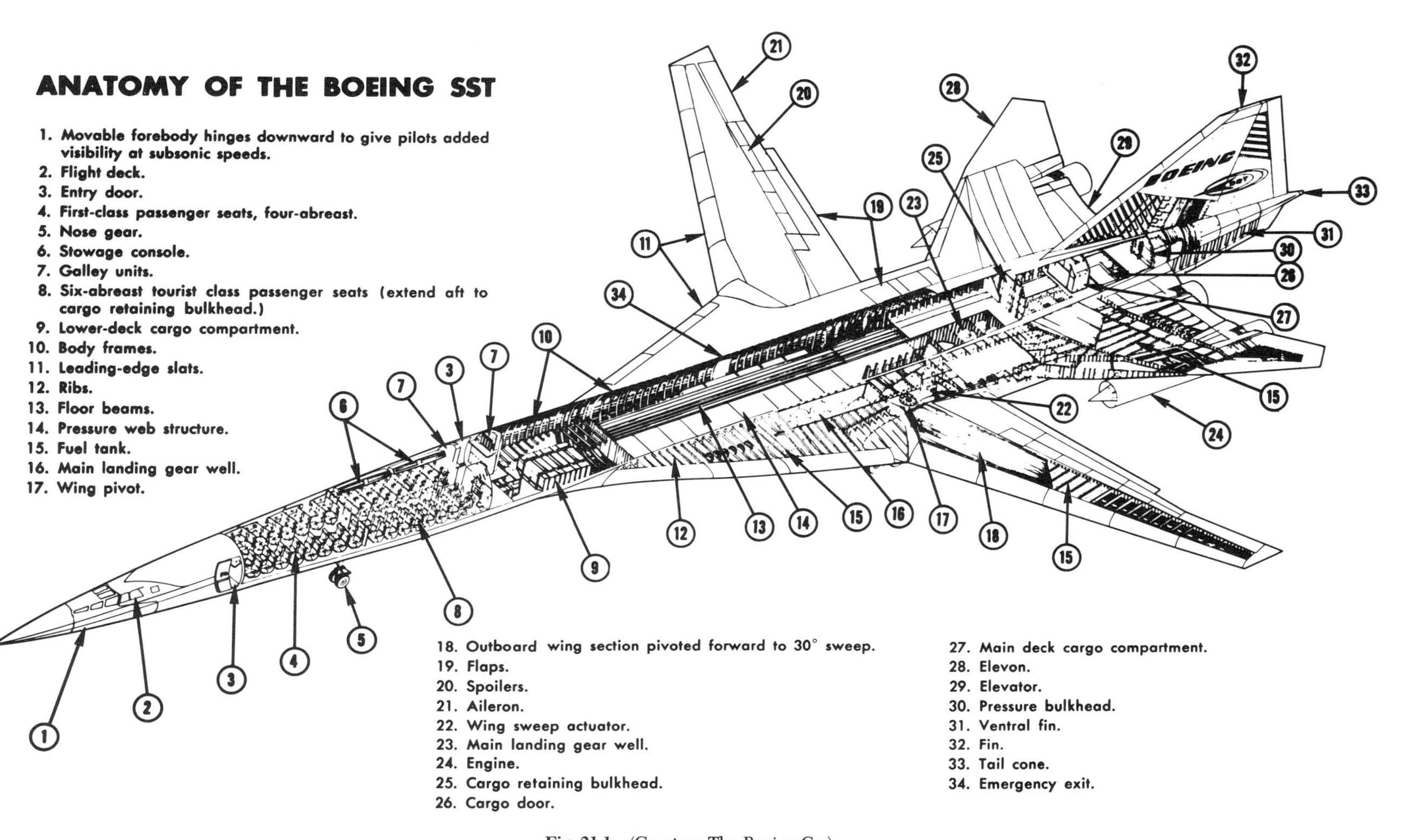
ANATOMY OF THE BOEING SST
1. Movable forebody hinges downward to give pilots added visibility at subsonic speeds.
2. Flight deck.
3. Entry door.
4. First-class passenger seats, four-abreast.
5. Nose gear.
6. Stowage console.
7. Galley units.
8. Six-abreast tourist class passenger seats (extend aft to cargo retaining bulkhead.)
9. Lower-deck cargo compartment.
10. Body frames.
11. Leading-edge slats.
12. Ribs.
13. Floor beams.
14. Pressure web structure.
15. Fuel tank.
16. Main landing gear well.
17. Wing pivot.
18. Outboard wing section pivoted forward to 30° sweep.
19. Flaps.
20. Spoilers.
21. Aileron.
22. Wing sweep actuator.
23. Main landing gear well.
24. Engine.
25. Cargo retaining bulkhead.
26. Cargo door.
27. Main deck cargo compartment.
28. Elevon.
29. Elevator.
30. Pressure bulkhead.
31. Ventral fin.
32. Fin.
33. Tail cone.
34. Emergency exit.
BOEING
SST

Fig. 21.1 (Courtesy The Boeing Co.)

plane. For example, some of the questions that must be answered before steps are taken to implement a study phase of possible designs are:

1. Is the timing right for the assumed business level?
2. How fast must the plane fly to meet competition?
3. Will a one-stop plane really sell?
4. What is the effect of reduced reliability of the one-stop versus a nonstop plane?
5. Are two 100,000-pound payload airplanes easier to schedule and sell than one 200,000-pound payload plane?
6. How does future technical development affect the optimum cargo airplane of today?
7. Is there or will there be a military market and, if so, will that change the business level, the operating cost, and price or number of airplanes produced?
8. Are there existing long-range cargo planes (derivatives of types that were designed for some other purpose) satisfactory for demands of the next fifteen years?

It becomes quite clear from these questions and others that would be raised that there are many variables which must be considered—i.e., size, speed, range, market potential, state-of-the-art, etc.

Such a study was made by the Boeing Company. The assumptions made are shown in Fig. 21.2. The intercontinental cargo market was narrowed down to the consideration of the North Atlantic problem because this route was considered to be the largest single traffic-generating route.

NORTH ATLANTIC RANGES
NON-STOP BOTH DIRECTIONS
NON-STOP EASTBOUND..... ONE STOP WESTBOUND
PAYLOAD..... 25,000 TO 200,000 POUNDS
CRUISE..... .60 TO .90 MACH
35,000 FT. CONSTANT ALTITUDE
F.A.A. TAKEOFF..... MAXIMUM 10,000 FT. ON 100°F DAY
F.A.A. LANDING..... MAXIMUM 7,000 FT.
1960 AND 1970 "STATE-OF-ART"
1960 AIRPLANE PRICES
D.O.C...... MODIFIED 1960 ATA RULES
CARGO GROWTH..... 20% PER YEAR AT 5¢ PER TON MILE D.O.C.

Fig. 21.2 Assumptions. (Courtesy The Boeing Co.)

The geographic ranges are shown in Fig. 21.3. The still air ranges assumed for study purposes were 3600 nautical miles nonstop and 2500 nautical miles for the one-stop airplane; both, however, with full payload.

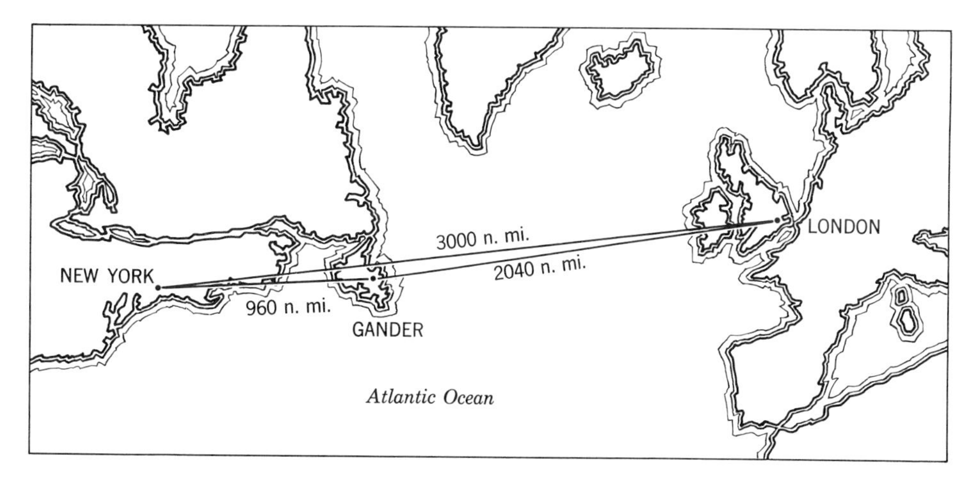

Fig. 21.3 Design route.

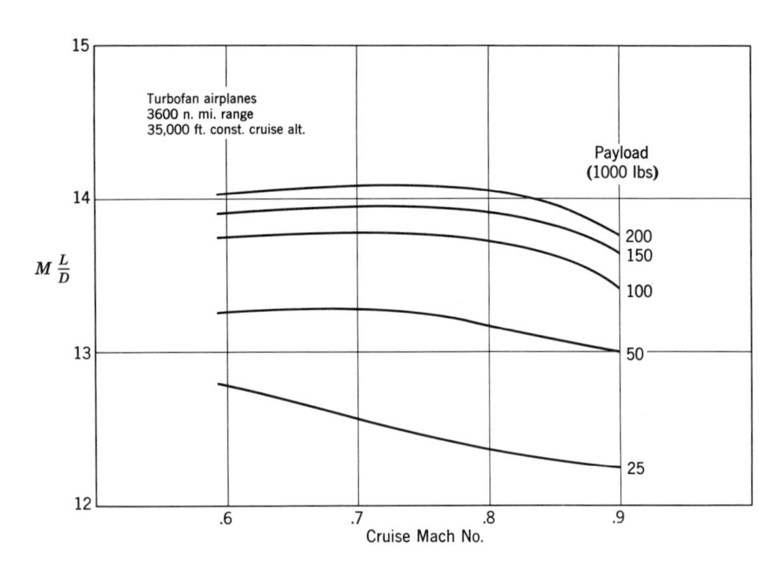

Fig. 21.4 Effect of speed on cruise efficiency. (Courtesy The Boeing Co.)

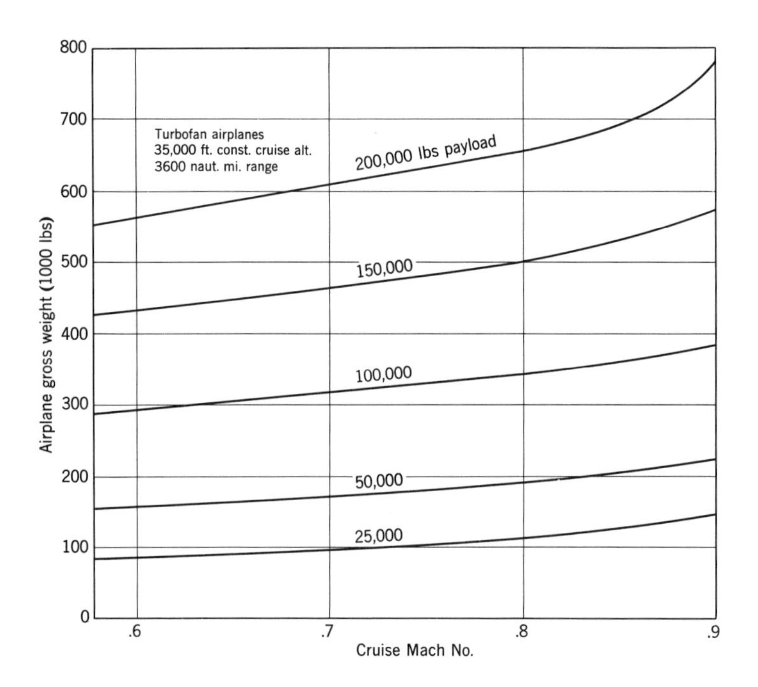

Fig. 21.5 Effect of speed and payload on cruise weight. (Courtesy The Boeing Co.)

The effect of speed on cruise efficiency as expressed by ML/D (Mach number times the lift/drag ratio) for a number of sizes of airplanes was studied and, as shown in Fig. 21.4, revealed that there was relatively little effect of cruise speed on the aerodynamic efficiency parameter. In making the study, "wing sweep and thickness were optimized for the particular speed; and power plant size was changed as necessary to provide the desired cruise speed." Consideration was given to the effect of both speed and payload on airplane cruise weight. It was noted (see Fig. 21.5) that with all payloads "there is a definite increase in airplane gross weight with speed as a result of higher structural weight, higher power required at some loss in aerodynamics efficiency at maximum speeds."

The effect of the increase in size due to speed (Fig. 21.6) was very nearly offset by the increased productivity of the airplane due to speed up to approximately 0.8 *M*. Above this speed there was some increase in cost as a result of flying faster.

Figure 21.7 shows a re-plot of the 0.8 *M* cruise speed with D.O.C. (Direct Operating Costs) varying with the design payload, and with the initial design range of 3600 nautical miles. Also shown is the lower cost of a series of airplanes of the same speed and payload but having a range of only 2500 nautical miles. A significant reduction of approximately $\frac{1}{2}$ cent per ton-mile is immediately evident. "Since these airplanes were not capable of carrying a full payload

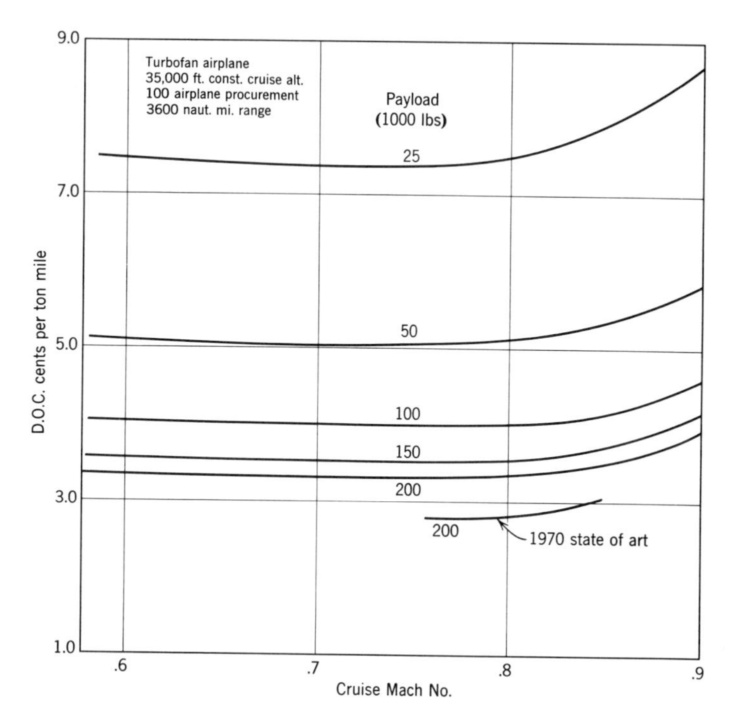

Fig. 21.6 (Courtesy The Boeing Co.)

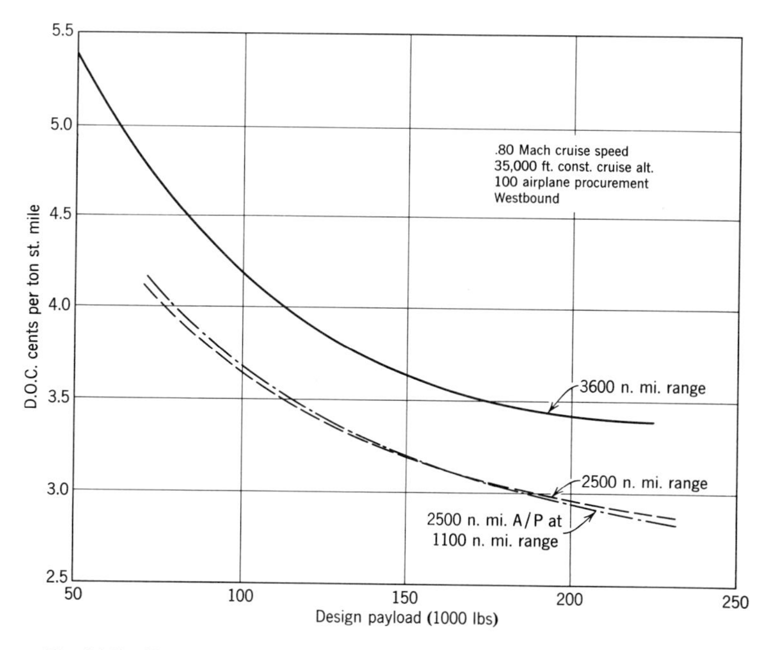

Fig. 21.7 (Courtesy The Boeing Co.)

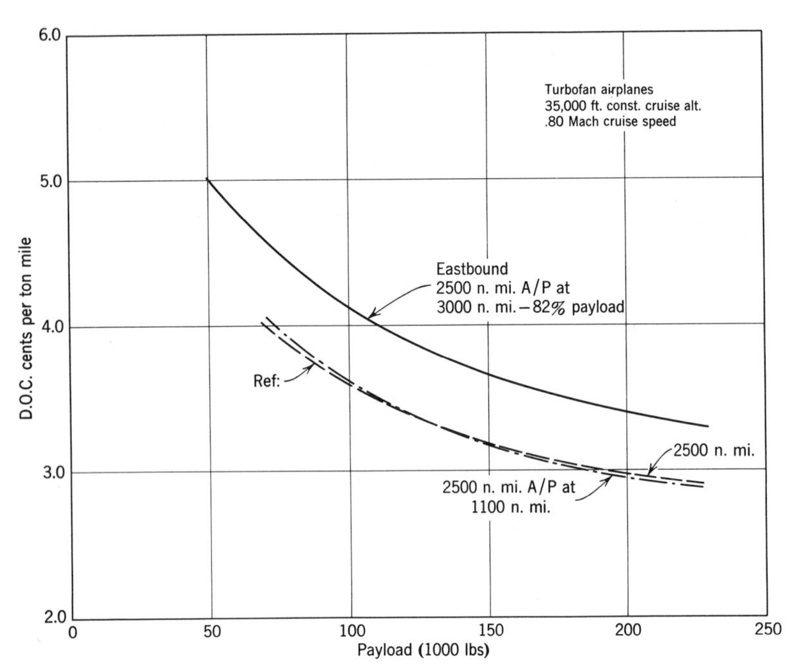

Fig. 21.8 (Courtesy The Boeing Co.)

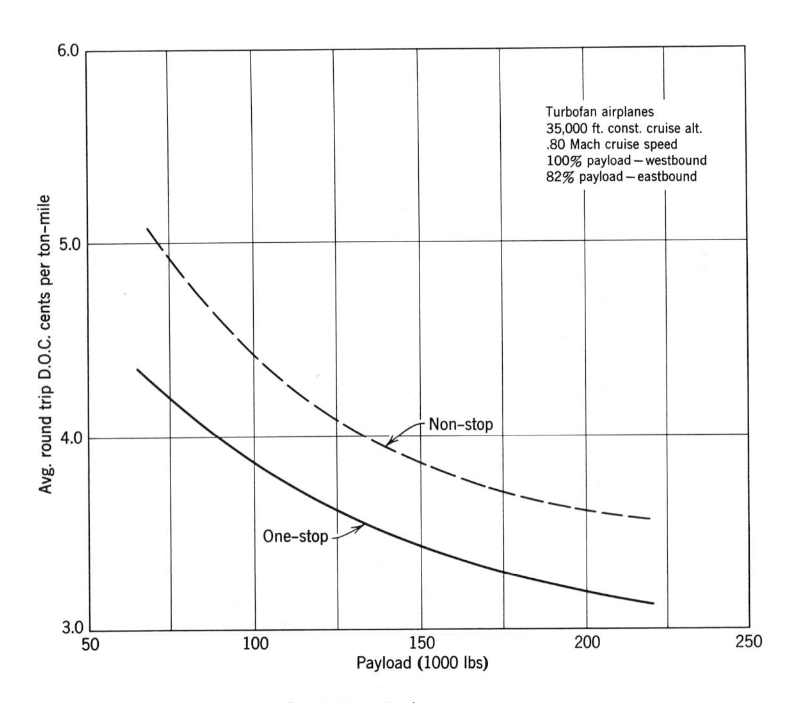

Fig. 21.9 (Courtesy The Boeing Co.)

eastbound (Fig. 21.8) non-stop (3000 nautical miles still air range) a further examination was made to determine what the operating cost effect would be when flying at a reduced payload. It was assumed, based on present statistics, that on an average only 82% of the payload was available eastbound as compared with westbound traffic. The effect is an appreciable increase in ton-mile costs. However, if a similarly reduced payload eastbound is also assumed for the non-stop airplane, the resulting average round trip cost per ton mile is as shown on Fig. 21.9 and the one-stop airplane, being smaller and less expensive, is appreciably superior to the non-stop airplane."

Other studies concerning the effect of limited market size (Fig. 21.10), the number of aircraft which one manufacturer would produce under the conditions of market limitations, the effectiveness of existing cargo airplanes derived from present passenger aircraft (Fig. 21.11), etc., were also made.

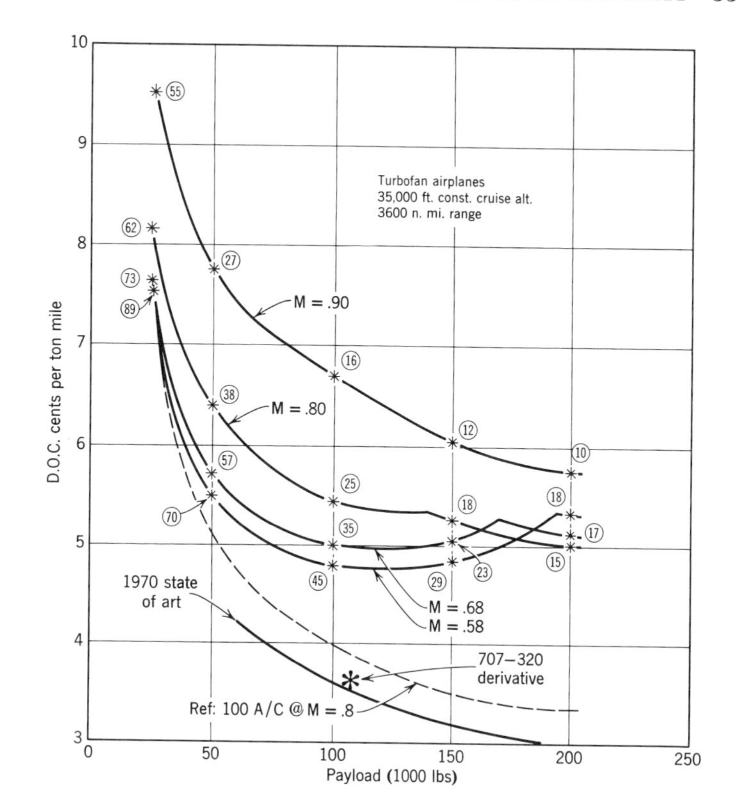

Fig. 21.10 (Courtesy The Boeing Co.)

Stimulating Creativity

At the beginner's level, let us consider the following examples which were among several projects submitted to the class.

EXAMPLE 1

Design an instant-coffee dispenser to satisfy the following requirements:

1. Attaches directly to the jar.
2. Accommodates most standard brands—small, medium and large jars.
3. Dispenses coffee in $\frac{1}{2}$- or 1-teaspoon increments.
4. Is wall mounted.
5. Can be used to dispense other instant beverages—cream, milk, tea, cocoa, etc.—packaged in standard jars.
6. The device should be light in weight (not more than 8 pounds), easily cleaned, readily mounted, and should sell for $10.

Once you have determined your "best" solution, submit a report of your project including (*a*) the statement of the problem; (*b*) a short description of the operation of the product; (*c*) an assembly drawing and parts list; (*d*) *freehand* detail drawings of each component; (*e*) conceptual design sketches; and (*f*) an index. Total time to be devoted to the project should not exceed 25 hours.

How shall we proceed to solve our problem? A no-solution approach—the lazy-mind—is quite simple: "Why bother? There are many good dispensers on the market that satisfy the specification." If this attitude were prevalent we would have witnessed little or no change in the automobile, the airplane, household appliances, etc., over the past 25 years. In fact, we would have experienced little progress in most areas of "hardware."

To know what improvements can be made over the existing state-of-the-art we must *search;* we must

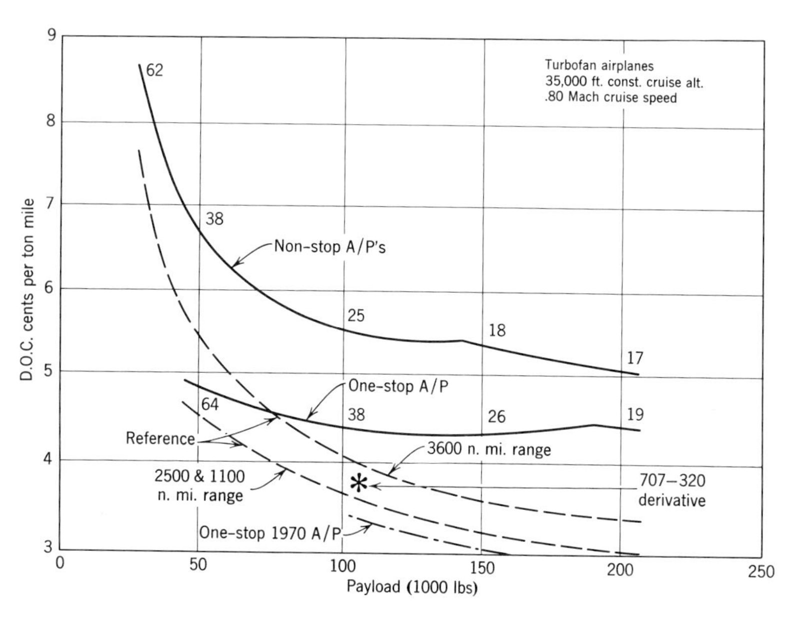

Fig. 21.11 (Courtesy The Boeing Co.)

be up-to-date with what has been developed in recent years. Before we undertake the search let us *first accept the challenge* of creating several conceptual designs that could satisfy the stated requirements. This step will help us resist the tendency to copy an existing design. To create a useful conceptual design is to achieve a *new and useful* combination of existing elements. To create we must think—at times a painful process. To think we must exercise our mental powers of conception, judgment, and reflection for the purpose of accomplishing our task.

At the outset, it is essential that you understand the specification. If necessary read it again, and perhaps again, until you grasp the situation completely. Now you have a problem. You have reasonable facility in freehand sketching. Try to record several ideas (good or bad) in sketch form without regard to sizes of components, i.e., the dispensing mechanism, the mounting unit, etc. Of these rather crude solutions, is there a possibility that a few might have merit? Your answer could be, "I don't know if any of my ideas are good." Should this be the case, and it is likely to be, what next?

Quite likely you have not actually "seen" an instant-coffee dispenser although you have "looked" at several. Our next step—let's "see" how different instant coffee dispensers work. This means we must do some *searching*. We can examine manufacturers' catalogs; we can visit the engineering library and read articles pertaining to dispensers; we can get patent information relating to various dispensers. Yes, we could visit a few coffee shops and observe the methods used for dispensing instant coffee.

Stated simply, we must collect available data that may have a bearing on our choice of the "best solution" which meets the requirements most economically. Now we may be ready to refine our first crude designs since we have information that should be quite helpful. We do not want to copy an existing design, but rather achieve a new and useful combination of existing elements. Here *evaluation* is important. Which of the finished products that you studied had the fewest number of parts? Although this may be desirable, do you believe that such designs were the best? What about the materials that were used, and their relative costs? What about function of the parts, ease of cleaning, etc.? Undoubtedly, you will think of other questions that could be raised. When do you stop the evaluation phase? Have you forgotten that the total time allocated for this project is 25 hours? Did you budget the available time when you first started? Did you plan ahead? How could you do this? In some cases past experience would serve as a guide. If your experience is quite limited, you may have to resort to an arbitrary allocation of time for each phase and make some modification as you proceed. Your past experience in making freehand sketches during the earlier part of your preparation could be helpful. Computing time could be judged by your experience in solving mathematics problems of an elementary nature or in your work that dealt with empirical equations and nomography. You may have to *make an intelligent guess!*

Your goal is to create your "best" conceptual design of an instant-coffee dispenser that will satisfy the stated requirements; and to record your creation in the form of freehand detail sketches and an assembly drawing—all within a total time of 25 hours. Your final product—the freehand detail sketches and the assembly—presents, we hope, your best effort to solve the problem. Quite likely you will do much better on subsequent assignments, as a consequence of your experience with the first problem you undertake.

A SOLUTION

One of some fifteen different solutions submitted by first-year engineering students is presented in Figs. 21.12 to 21.27.

The student realized early in his study of the problem that he needed information about the dimensions of various brands of instant coffee jars. He collected his own data by visiting a supermarket that had a supply of various brands of instant coffee. He then obtained permission from the manager to measure the various jars. The results of his efforts are shown in Figs. 21.12 to 21.14. Then he pro-

TYPE / DESCRIPTION	MJB 2 OZ.	FOLGER'S 2 OZ.	HILLS BROS. 2 OZ	YUBAN 2 OZ	NES-CAFÉ 2 OZ	SANKA 2 OZ	DECAF 2 OZ	S. I. ESPRESSO 2 OZ	LIPTON TEA $1\frac{1}{2}$ OZ	POSTUM 4 OZ	COFFE-MATE 3 OZ.
TOTAL HEIGHT	$4\frac{23}{32}$	$4\frac{22}{32}$	$4\frac{17}{32}$	$4\frac{18}{32}$	$4\frac{3}{4}$	$4\frac{3}{4}$	$4\frac{3}{4}$	$4\frac{18}{32}$	$4\frac{18}{32}$	$4\frac{3}{4}$	$4\frac{1}{4}$
MAX. WIDTH	$2\frac{13}{32}$	$2\frac{15}{32}$	$2\frac{11}{32}$	$2\frac{11}{32}$	$2\frac{3}{8}$	$2\frac{7}{32}$	$2\frac{13}{32}$	$2\frac{10}{32}$	$2\frac{13}{32}$	$2\frac{14}{32}$	$2\frac{10}{32}$
DISTANCE FROM TOP TO MAX.	$\frac{3}{8}$	$\frac{3}{16}$	$\frac{3}{8}$	$\frac{1}{8}$	$\frac{1}{8}$	$\frac{1}{4}$	$\frac{1}{4}$	$\frac{1}{4}$	$\frac{1}{4}$	$\frac{3}{8}$	$\frac{3}{8}$
MAX. LID DIA.	$2\frac{15}{64}$	$2\frac{15}{64}$	$2\frac{15}{64}$	$2\frac{15}{64}$	$2\frac{15}{64}$	$2\frac{5}{32}$	$2\frac{15}{64}$	$2\frac{15}{64}$	$2\frac{15}{64}$	$2\frac{15}{64}$	$2\frac{15}{64}$
MIN LID DIA	$2\frac{11}{64}$	$2\frac{11}{64}$	$2\frac{11}{64}$	$2\frac{11}{64}$	$2\frac{11}{64}$	$2\frac{5}{64}$	$2\frac{11}{64}$	$2\frac{11}{64}$	$2\frac{11}{64}$	$2\frac{11}{64}$	$2\frac{11}{64}$
LID HEIGHT	$\frac{19}{32}$	$\frac{22}{32}$	$\frac{20}{32}$	$\frac{20}{32}$	$\frac{21}{32}$	$\frac{19}{32}$	$\frac{22}{32}$	$\frac{22}{32}$	$\frac{21}{32}$	$\frac{21}{32}$	$\frac{21}{32}$
OUTLINE	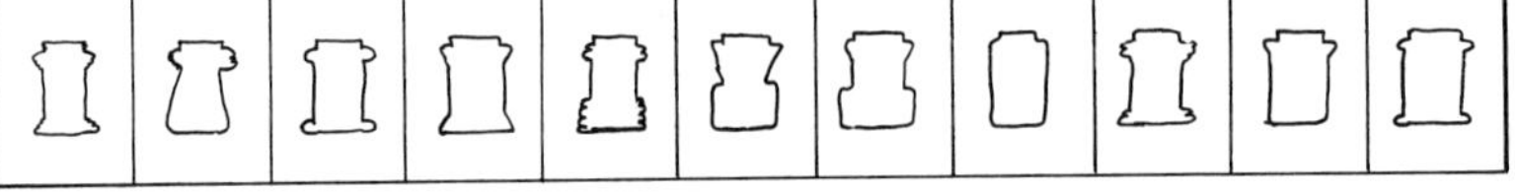										

CONTAINER DESCRIPTION—INSTANT COFFEE

SMALL JARS

Fig. 21.12

TYPE / DESCRIPTION	YUBAN 5 OZ	SANKA 5 OZ	FOLGER'S 6 OZ	M J B 6 OZ	MAXWELL HOUSE 6 OZ	HILLS BROS. 6 OZ	BONNIE HUBBARD 6 OZ	BORDEN 5 OZ	S I ANTIGUA 4 OZ.	LIPTON TEA 3 OZ	NESTEA 2 OZ
TOTAL HEIGHT	6 7/32	6 1/2	6 29/32	6 29/32	6 29/32	6 7/32	6 7/32	6 9/16	5 19/32	5 1/2	7 1/4
MAX. WIDTH	3 3/32	2 28/32	3 14/32	3 9/32	3 7/64	3 13/64	3 7/32	3 3/32	2 27/32	2 31/32	2 23/32
DISTANCE FROM TOP TO MAX.	1/4	1/2	1/2	3/8	1/2	3/4	3/4	1/4	1/4	1/4	1/4
MAX. LID DIA.	2 47/64	2 47/64	2 47/64	2 47/64	2 47/64	2 47/64	2 47/64	2 47/64	2 47/64	2 47/64	2 47/64
MIN. LID DIA.	2 39/64	2 39/64	2 39/64	2 39/64	2 39/64	2 39/64	2 39/64	2 39/64	2 39/64	2 39/64	2 39/64
LID HEIGHT	20/32	21/32	39/64	22/32	20/32	20/32	20/32	21/32	20/32	19/32	22/32
OUTLINE	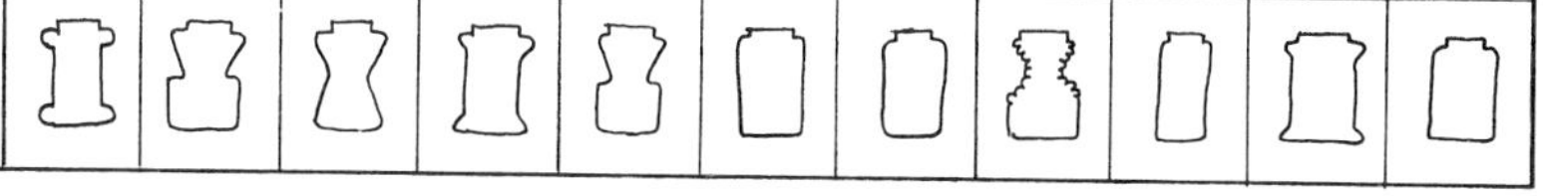										

CONTAINER DESCRIPTION — INSTANT COFFEE

MEDIUM JARS

Fig. 21.13

	MEDIUM				LARGE				
TYPE / DESCRIPTION	POSTUM 8 OZ	OVALTINE 12 OZ	COFFEE-MATE 11 OZ	PREAM 12 OZ	M J B 10 OZ	FOLGER'S 10 OZ	HILLS BROS 10 OZ	YUBAN 9 OZ	MAXWELL HOUSE 10 OZ
TOTAL HEIGHT	5 3/4	6 11/32	7 1/32	7 1/32	8	7 3/4	7 17/32	7 23/32	7 25/32
MAX. WIDTH	2 30/32	3 20/32	3 3/8	3 1/4	4	4 7/32	3 27/32	3 3/4	3 23/32
DISTANCE FROM TOP TO MAX.	1/4	1/2	1/2	1/2	1/4	1/2	3/4	1/2	1/2
MAX. LID DIA.	2 47/64	2 47/64	2 47/64	2 47/64	3 15/64	3 12/32	3 15/64	3 15/64	3 15/64
MIN LID DIA.	2 39/64	2 39/64	2 39/64	2 39/64	3 7/64	3 5/32	3 7/64	3 7/64	3 7/64
LID HEIGHT	20/32	20/32	22/32	22/32	20/32	22/32	22/32	21/32	20/32
OUTLINE	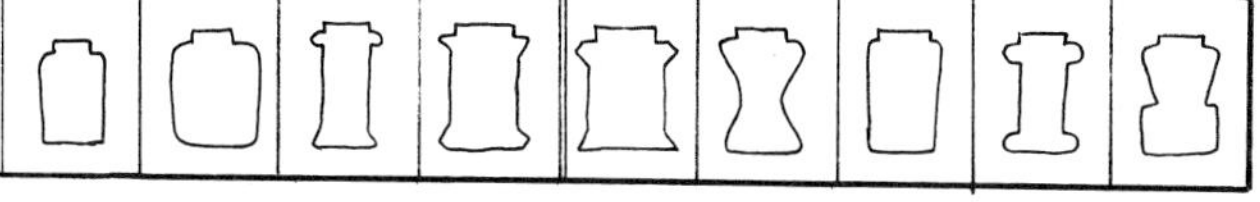								

CONTAINER DESCRIPTION — INSTANT COFFEE

MEDIUM and LARGE JARS

Fig. 21.14

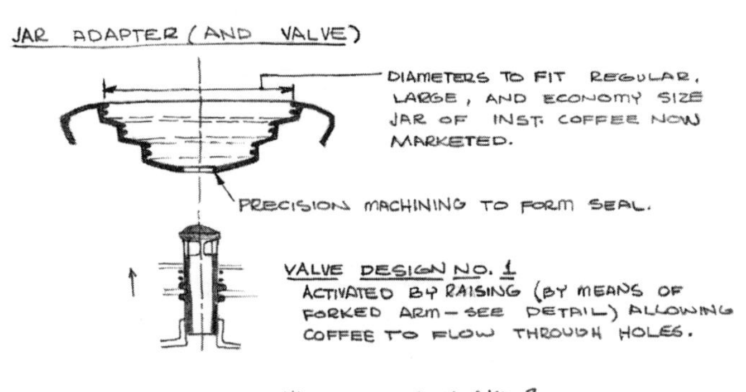

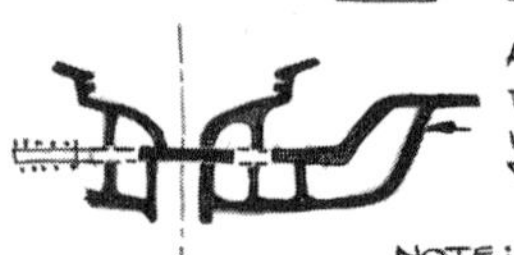

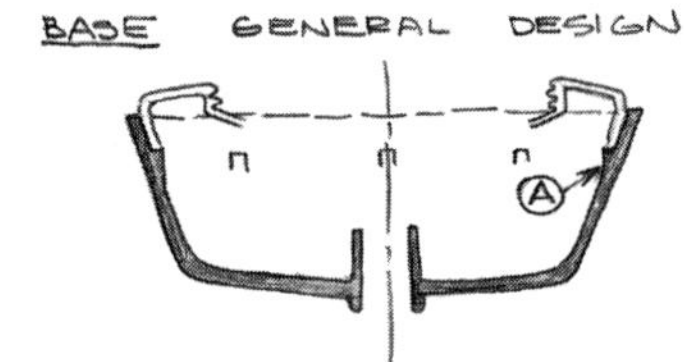

Fig. 21.15 Preliminary sketches.

ceeded to prepare some preliminary sketches—conceptual designs—of jar adapters, valves, base, etc. These are shown in Figs. 21.15 to 21.17. After an evaluation of possible alternatives he prepared a "layout design" of his "best" solution. This was fol-

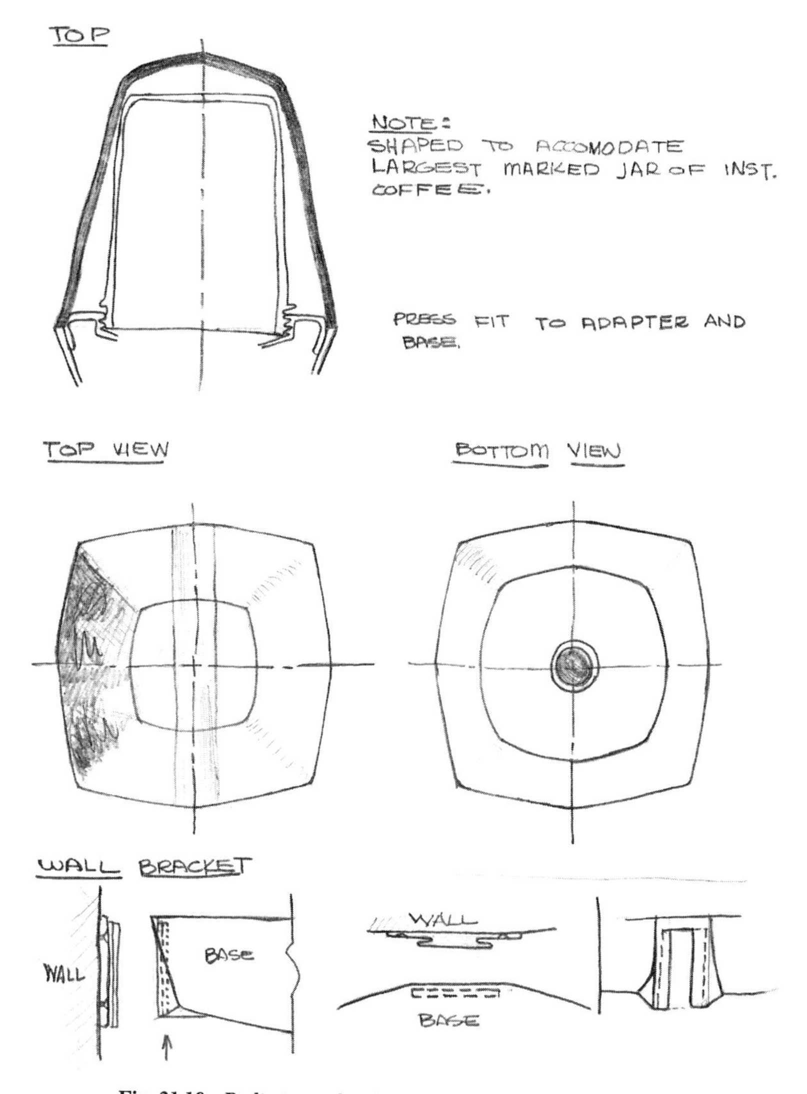

Fig. 21.16 Preliminary sketches.

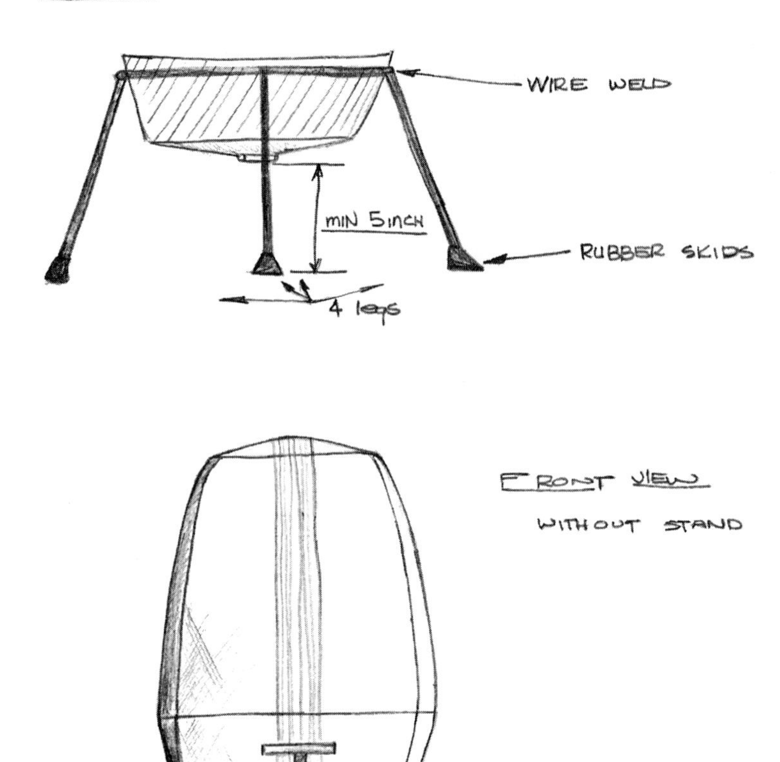

Fig. 21.17 Preliminary sketches.

lowed by *freehand* detail drawings of the components shown in Figs. 21.18 to 21.25. Finally, an assembly drawing and bill of materials were prepared. These are shown in Figs. 21.26 and 21.27, respectively.

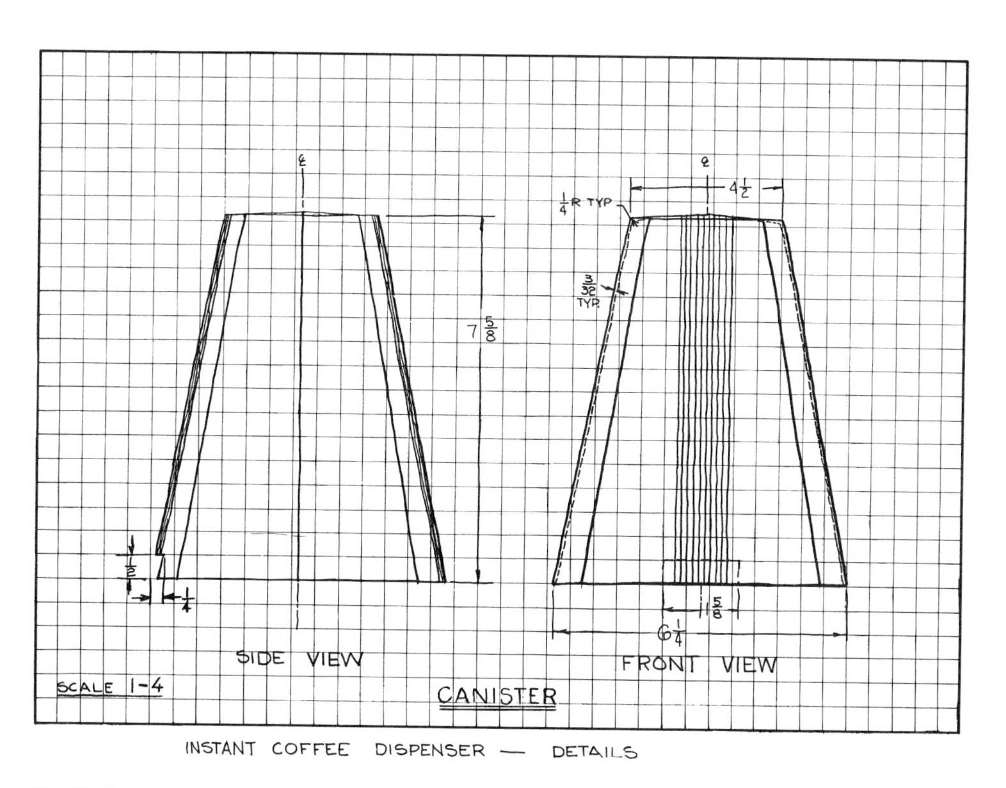

Fig. 21.18

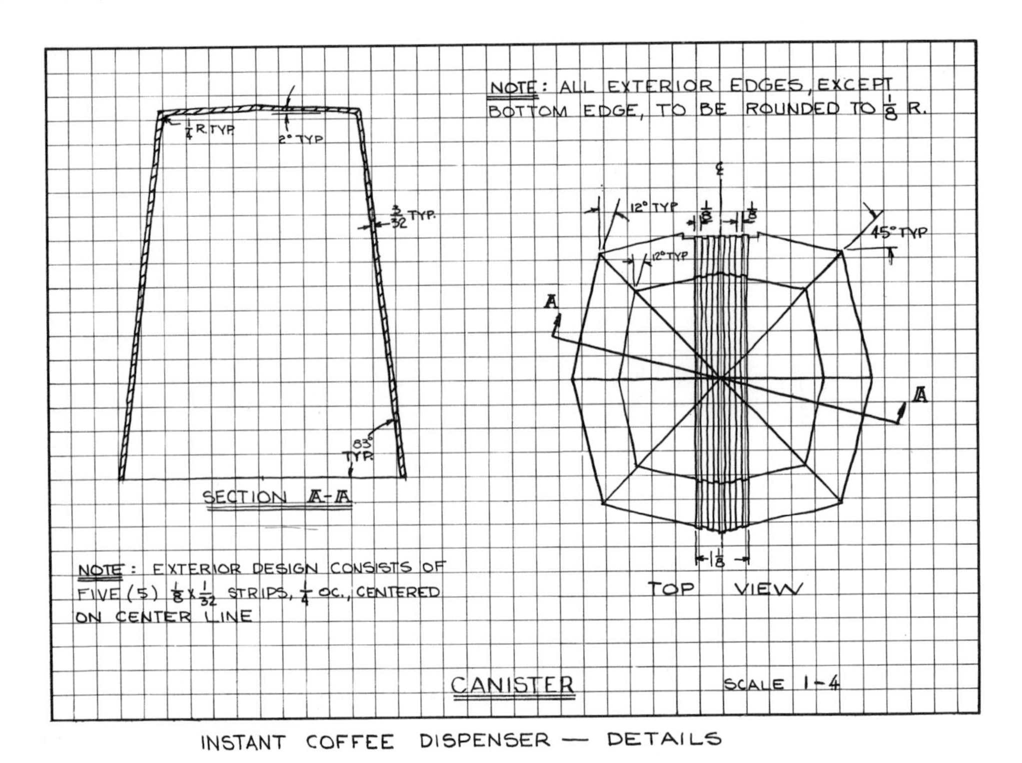
NOTE: ALL EXTERIOR EDGES, EXCEPT BOTTOM EDGE, TO BE ROUNDED TO 1/8 R.
1/4 R. TYP.
2° TYP.
3/32 TYP.
83° TYP.
SECTION A-A
NOTE: EXTERIOR DESIGN CONSISTS OF FIVE (5) 1/8 X 1/32 STRIPS, 1/4 OC., CENTERED ON CENTER LINE
12° TYP
45° TYP
A
A
1 1/8
TOP VIEW
CANISTER
SCALE 1-4
INSTANT COFFEE DISPENSER — DETAILS

Fig. 21.19

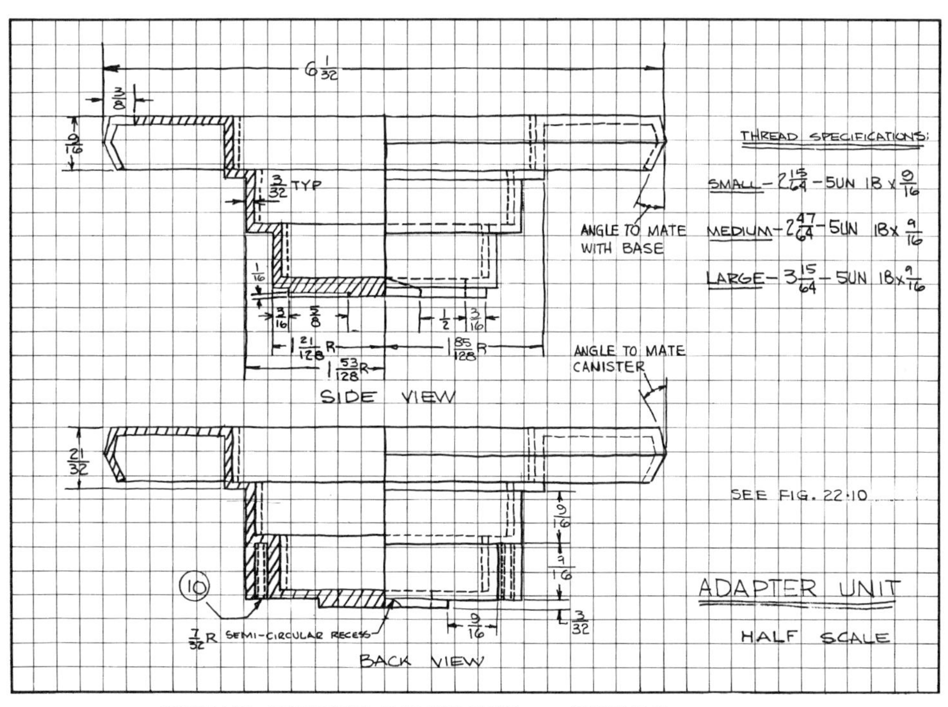
THREAD SPECIFICATIONS:
SMALL – 2 15/64 – 5UN 18 x 9/16
MEDIUM – 2 47/64 – 5UN 18x 9/16
LARGE – 3 15/64 – 5UN 18x 9/16
ANGLE TO MATE WITH BASE
ANGLE TO MATE CANISTER
SIDE VIEW
SEE FIG. 22·10
ADAPTER UNIT
HALF SCALE
7/32 R SEMI-CIRCULAR RECESS
BACK VIEW
INSTANT COFFEE DISPENSER — DETAILS

Fig. 21.20

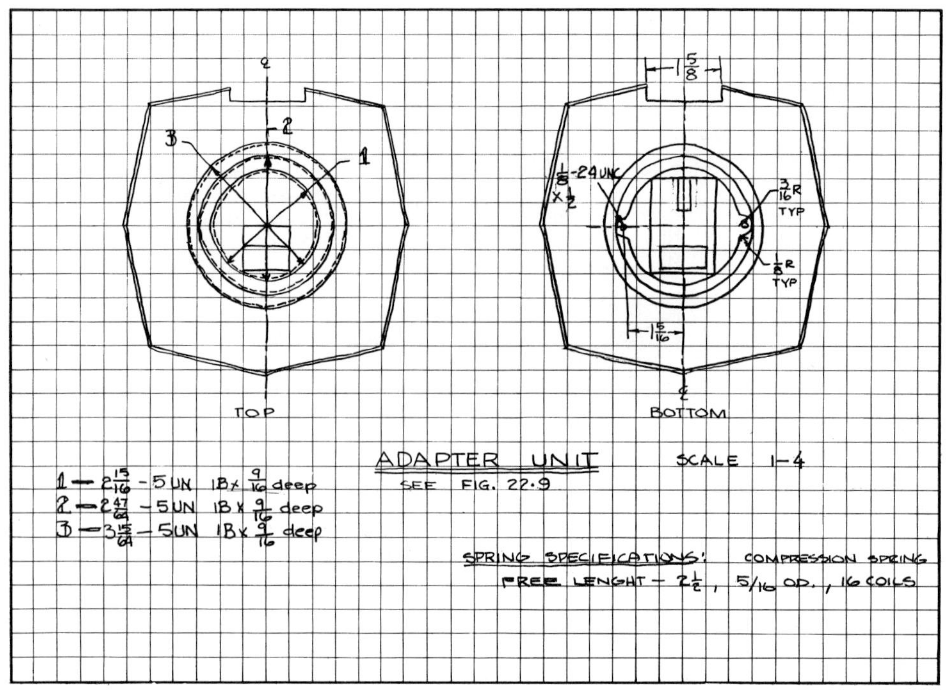

INSTANT COFFEE DISPENSER — DETAILS

Fig. 21.21

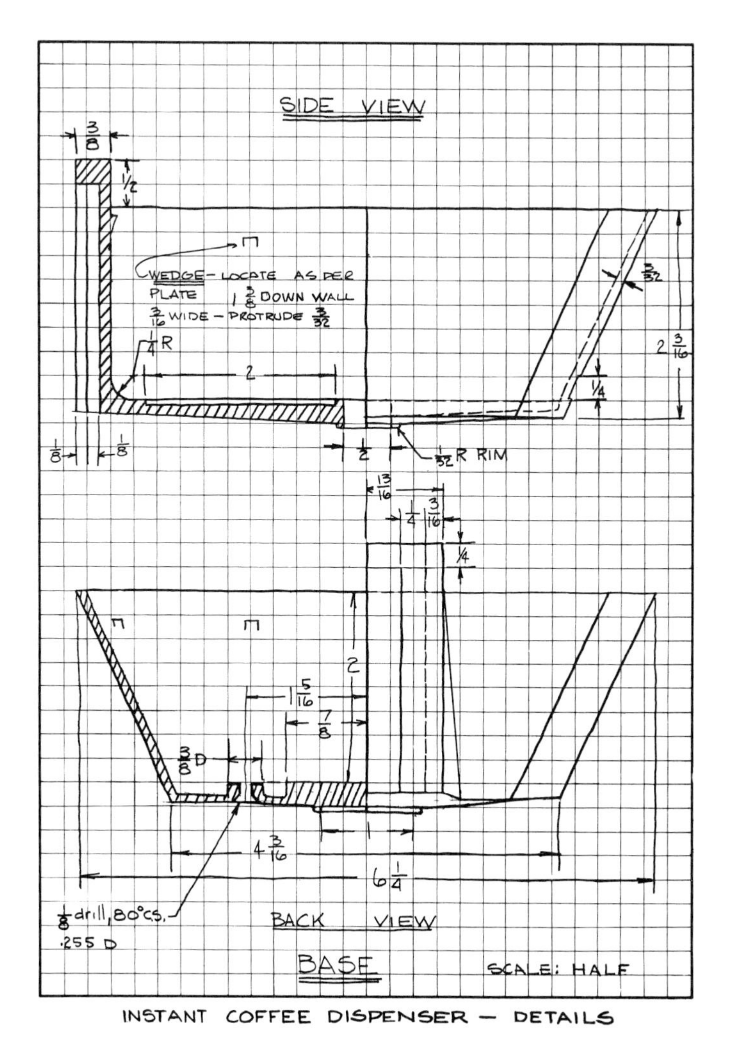
SIDE VIEW
WEDGE - LOCATE AS PER PLATE 1 3/8 DOWN WALL
3/16 WIDE - PROTRUDE 3/32
1/4 R
1/32 R RIM
1/8 drill, 80° CS.
.255 D
3/8 D
BACK VIEW
BASE
SCALE: HALF
INSTANT COFFEE DISPENSER — DETAILS

Fig. 21.22

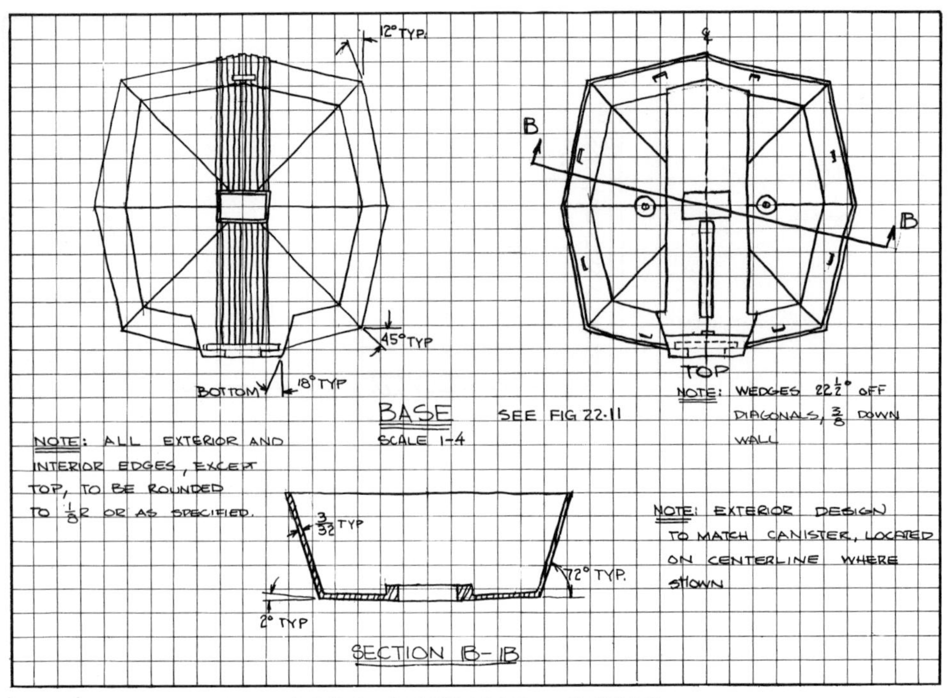
12° TYP.
45° TYP
BOTTOM
18° TYP
TOP
NOTE: WEDGES 22½° OFF DIAGONALS, 3/8 DOWN WALL
BASE
SCALE 1-4
SEE FIG 22.11
NOTE: ALL EXTERIOR AND INTERIOR EDGES, EXCEPT TOP, TO BE ROUNDED TO 1/8 R OR AS SPECIFIED.
3/32 TYP
72° TYP.
2° TYP
NOTE: EXTERIOR DESIGN TO MATCH CANISTER, LOCATED ON CENTERLINE WHERE SHOWN
SECTION B-B
INSTANT COFFEE DISPENSER — DETAILS

Fig. 21.23

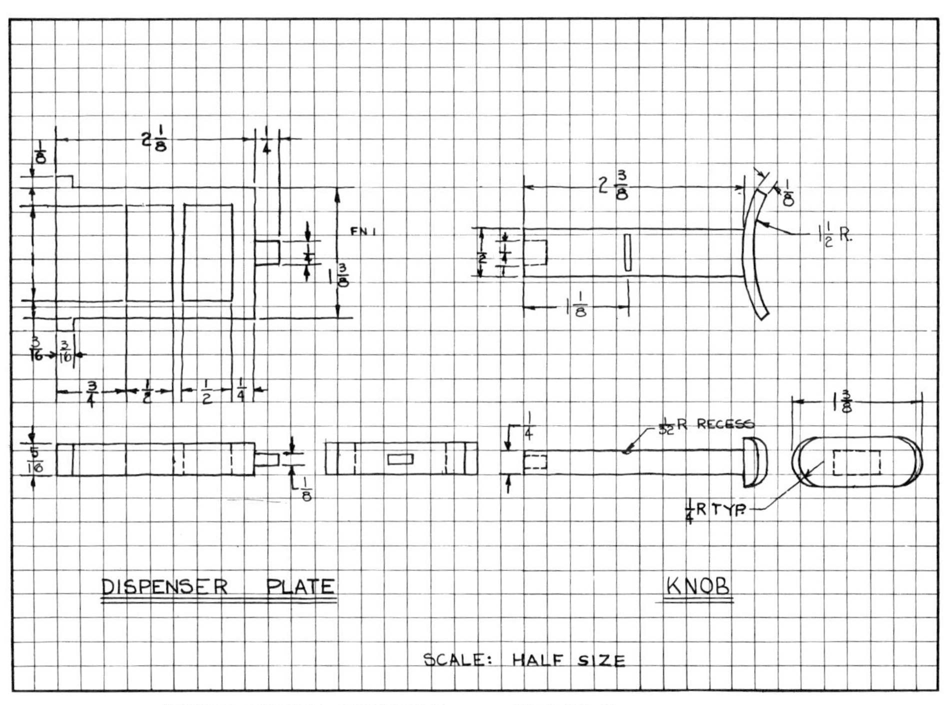
DISPENSER PLATE
KNOB
SCALE: HALF SIZE
INSTANT COFFEE DISPENSER — DETAILS

Fig. 21.24

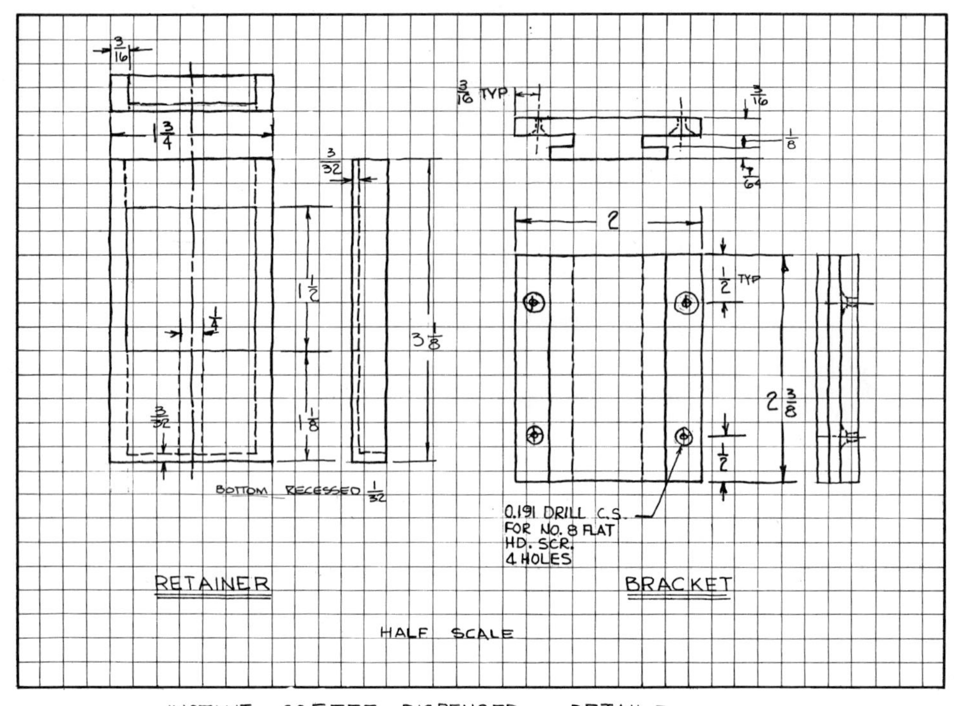

BOTTOM RECESSED 1/32
0.191 DRILL C.S.
FOR NO. 8 FLAT
HD. SCR.
4 HOLES
RETAINER
BRACKET
HALF SCALE
INSTANT COFFEE DISPENSER — DETAILS

Fig. 21.25

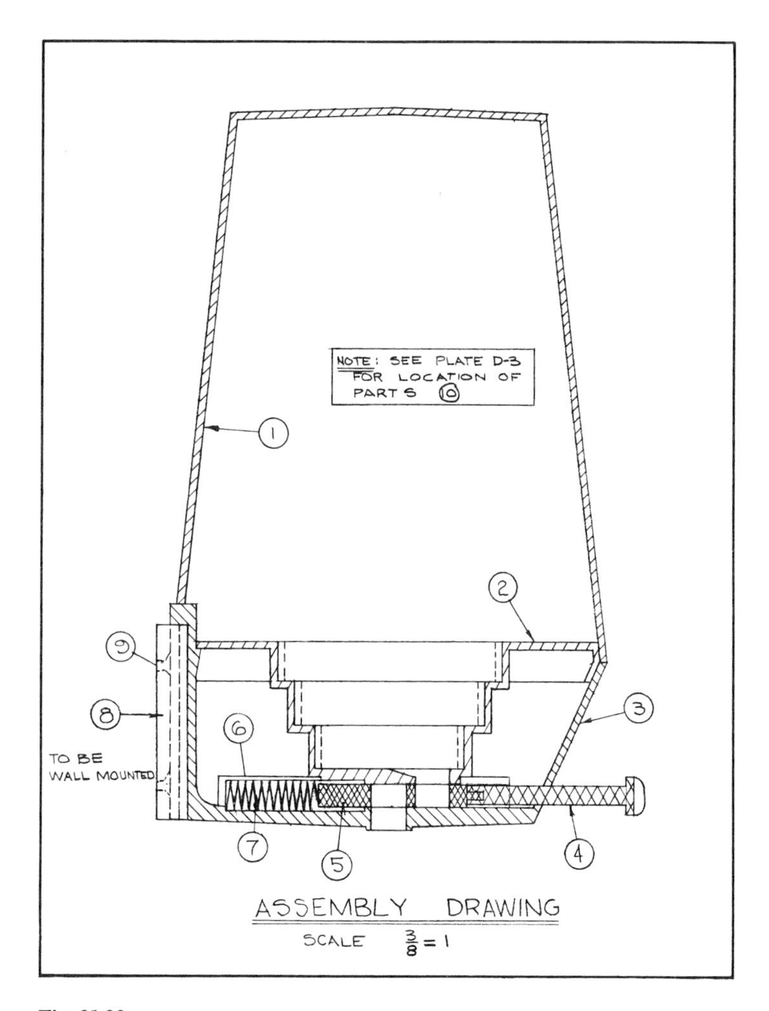
NOTE: SEE PLATE D-3
FOR LOCATION OF
PARTS 10
1
2
3
4
5
6
7
8
9
TO BE
WALL MOUNTED
ASSEMBLY DRAWING
SCALE 3/8 = 1

Fig. 21.26

BILL OF MATERIALS				
PART NO	NAME	PROCESS	NO. REQ.	MATERIAL
1	CANISTER	MOLD	1	POLYPROPYLENE
2	ADAPTER UNIT	MOLD	1	POLYPROPYLENE
3	BASE	MOLD	1	POLYPROPYLENE
4	KNOB	MOLD	1	POLYPROPYLENE
5	DISPENSER	MOLD	1	POLYPROPYLENE
6	RETAINER	MOLD	1	POLYPROPYLENE
7	SPRING	PURCHASE	1	SPRING STEEL
8	BRACKET	MACHINE	1	ALUMINUM
9	8-16-$1\frac{1}{2}$ ft. hd. w.s.	PURCHASE	4	STEEL
10	5-24-$\frac{7}{8}$ ft. hd. ms.	PURCHASE	2	STEEL

Fig. 21.27

EXAMPLE 2

Design a calligraphy pen to satisfy the following requirements:

1. Must have own ink supply.
2. Requires no dipping.
3. Easily filled from a separate dispenser.
4. Is easily cleaned when dried.
5. Has a good-grip handle.
6. Has a "moisture element" to keep pen "wet."
7. Has a good smooth form.

In the description the student points out: "The art of calligraphy is like painting, sculpture, or drawing—it demands that the tools used do not impede the artist in his attempts to create. Calligraphy deals with letter forms and their layouts; a calligrapher can neither afford to waste time unclogging his pen nor afford messy, time-consuming point or ink changes. Consequently, any improvements which can be made upon the basic tool—the pen—are a great help to his technique and performance. In short, while the visible changes are not radically new, the designer feels that his pen is superior to any pen on the market. In fact, the use of such a pen would help beginners learn easier, and would improve the more advanced person's technique."

After considerable "searching" the student realized that decisions had to be made concerning choice of materials, manufacturing methods, number of parts, etc.

Preliminary conceptual design sketches were made. These are shown in Figs. 21.28 to 21.32.

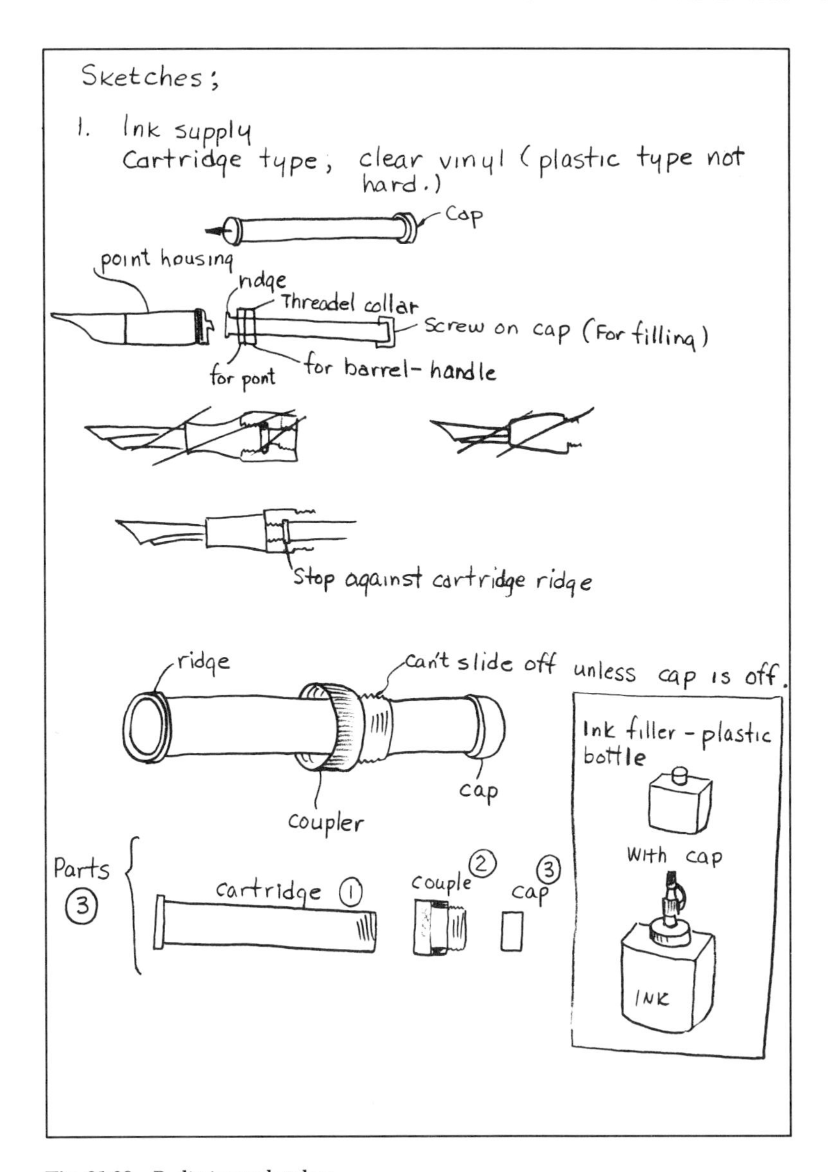

Fig. 21.28 Preliminary sketches.

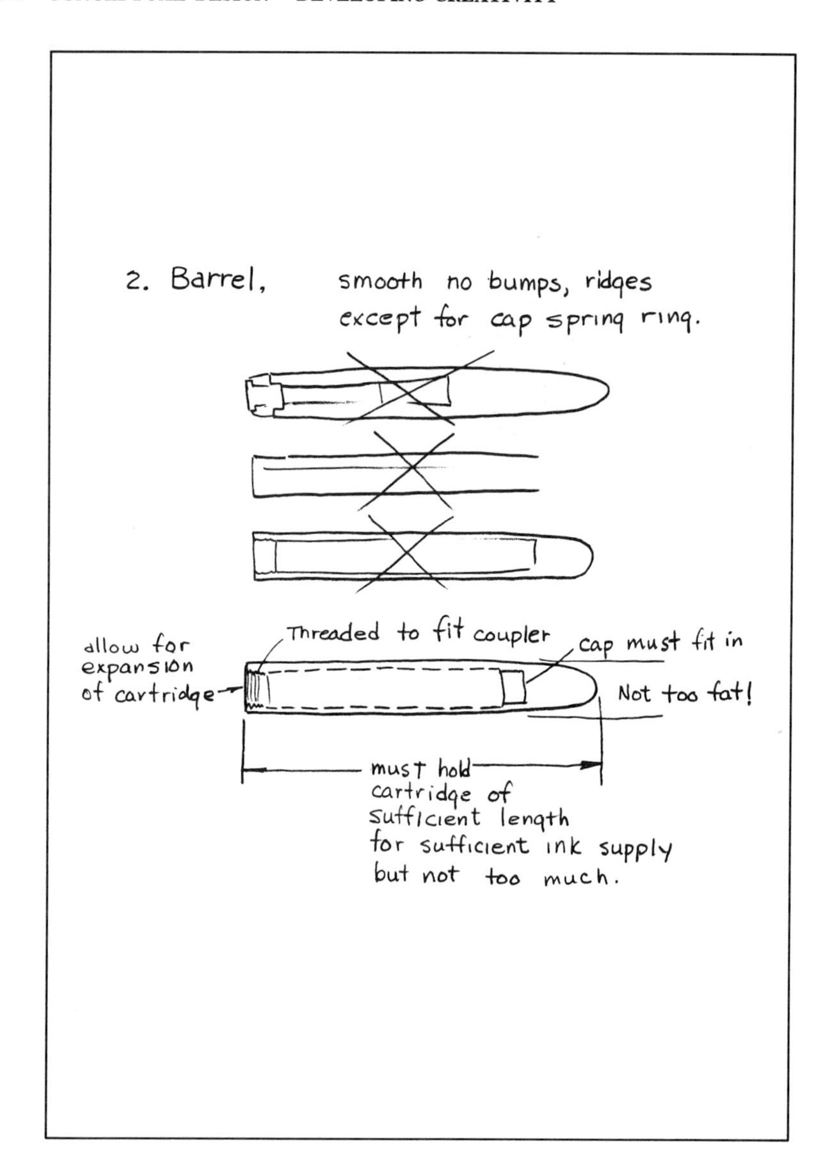

Fig. 21.29 Preliminary sketches.

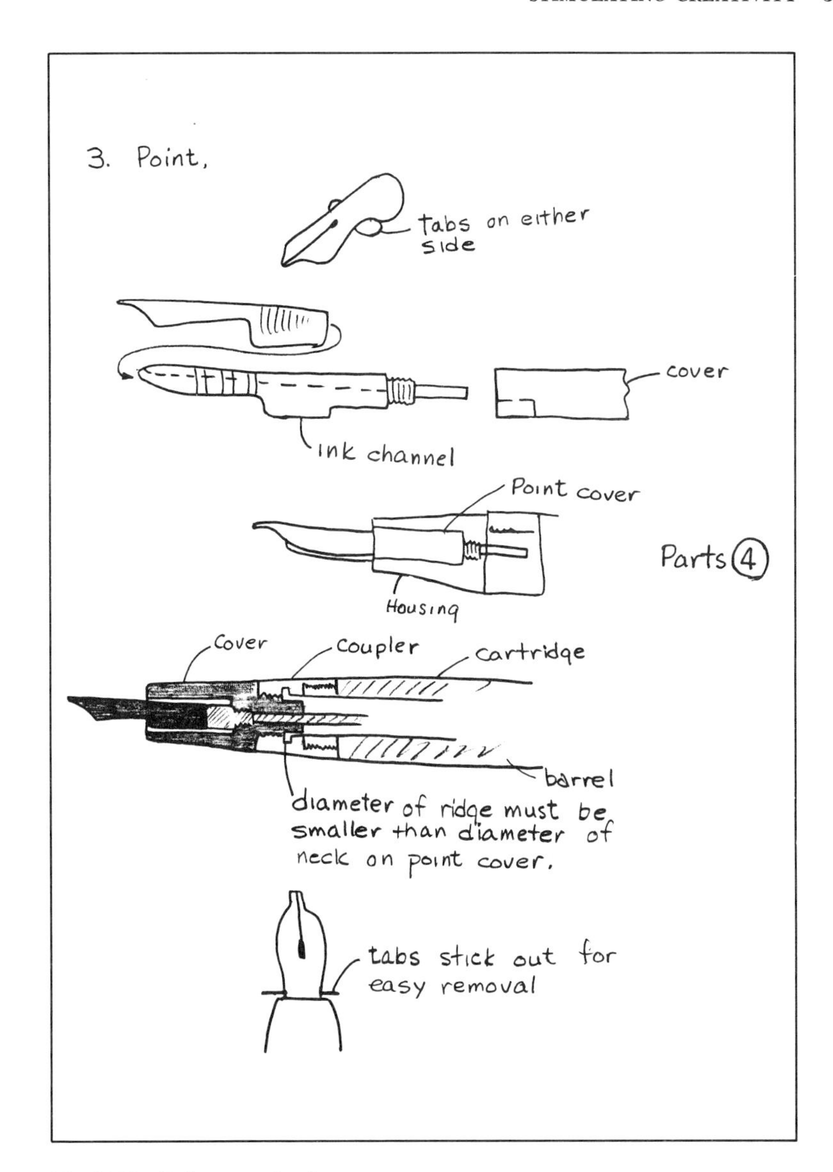

Fig. 21.30 Preliminary sketches.

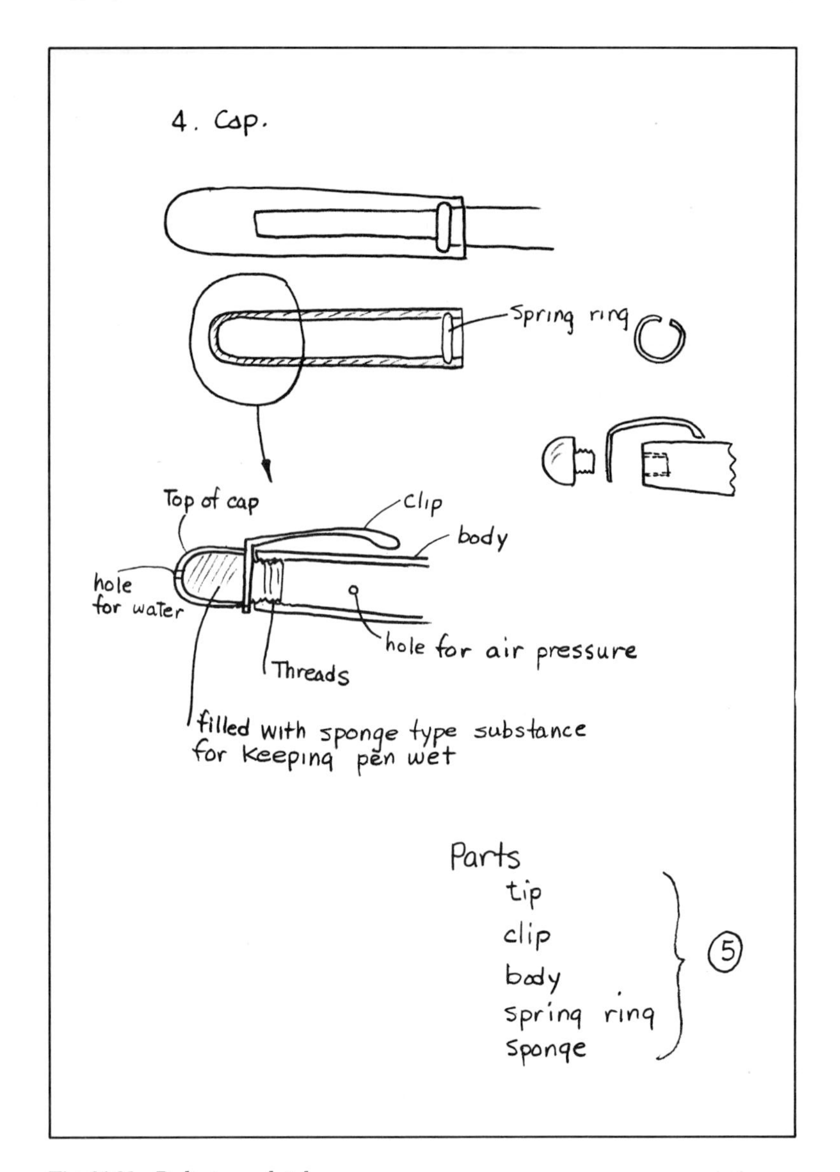

Fig. 21.31 Preliminary sketches.

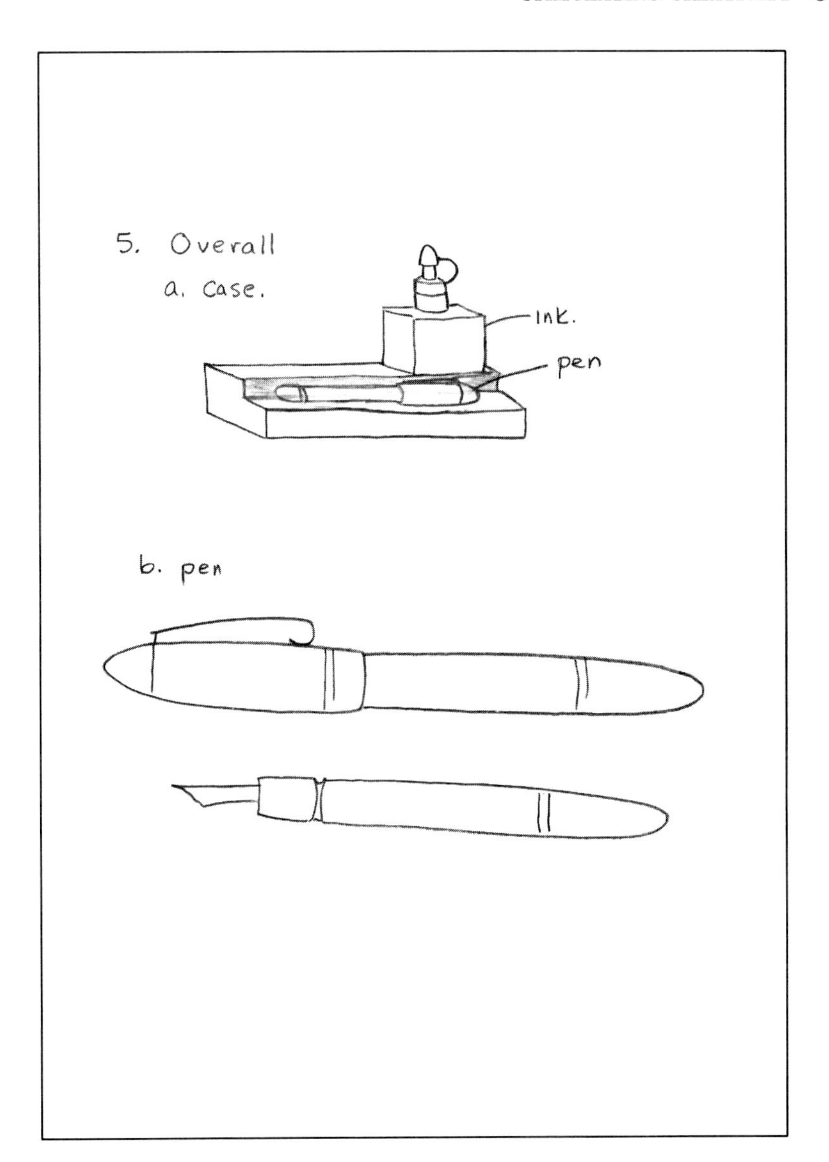

Fig. 21.32 Preliminary sketches.

After evaluating several alternative designs a "lay-out" drawing was made. Finally detail drawings, an assembly, and a parts list were prepared. These are shown in Figs. 21.33 to 21.38. A pictorial was also included. This is shown in Fig. 21.39.

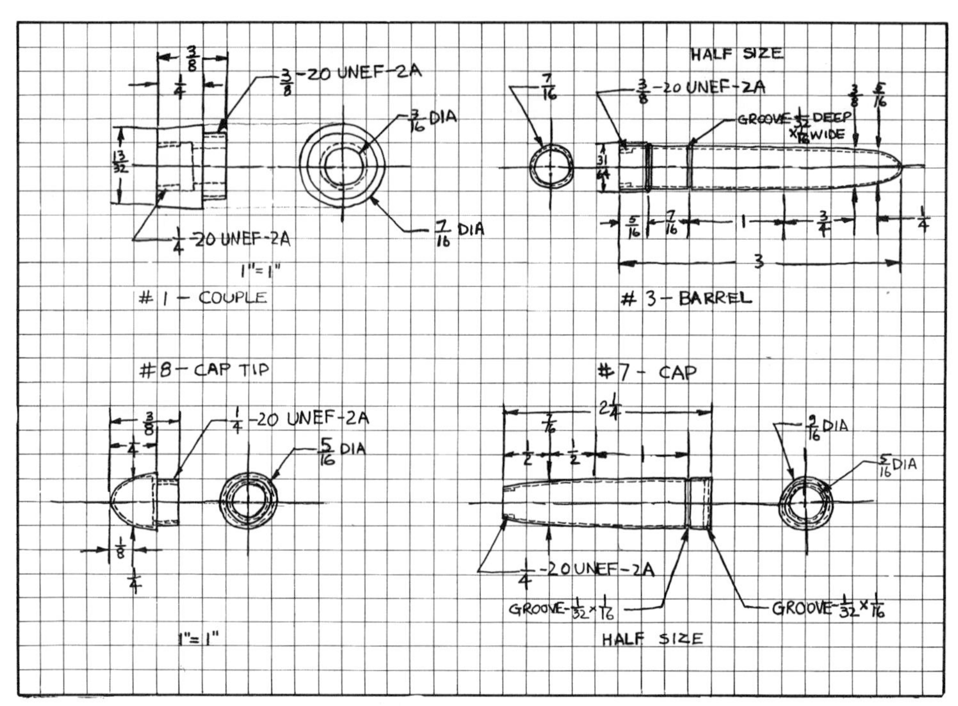

Fig. 21.33

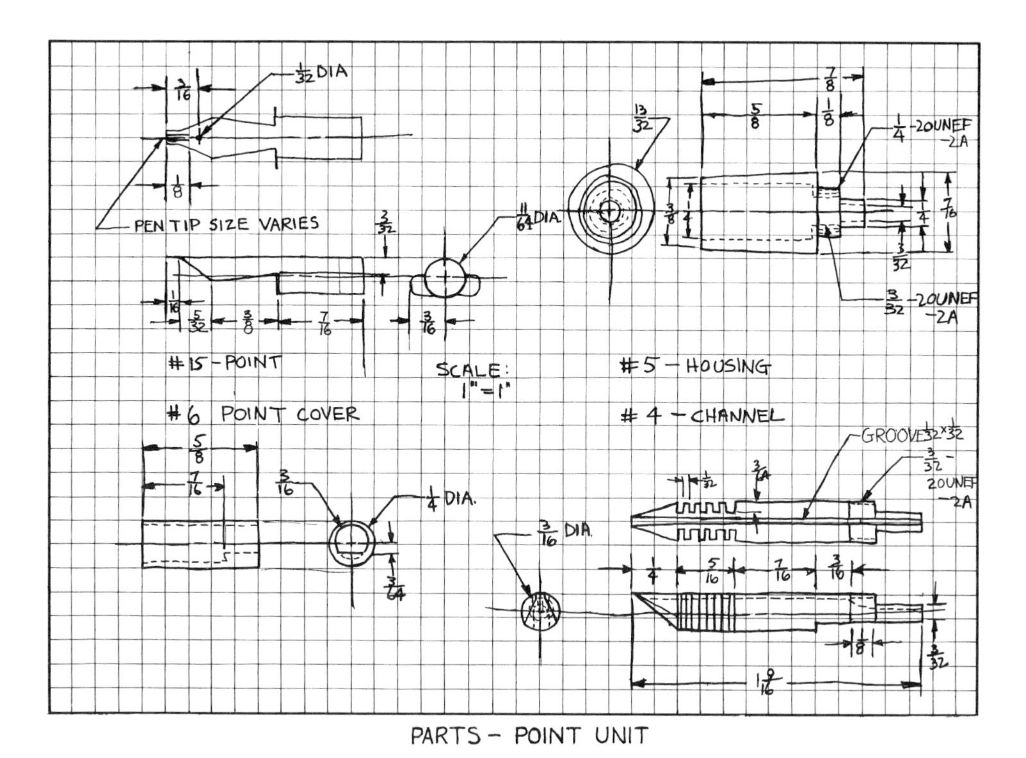

1/32 DIA
PEN TIP SIZE VARIES
11/64 DIA
1/4-20UNEF-2A
3/32-20UNEF-2A
#15 - POINT
SCALE: 1" = 1"
#5 - HOUSING
#6 POINT COVER
#4 - CHANNEL
GROOVE 1/32 x 1/32
1/4 DIA.
3/16 DIA.
PARTS - POINT UNIT

Fig. 21.34

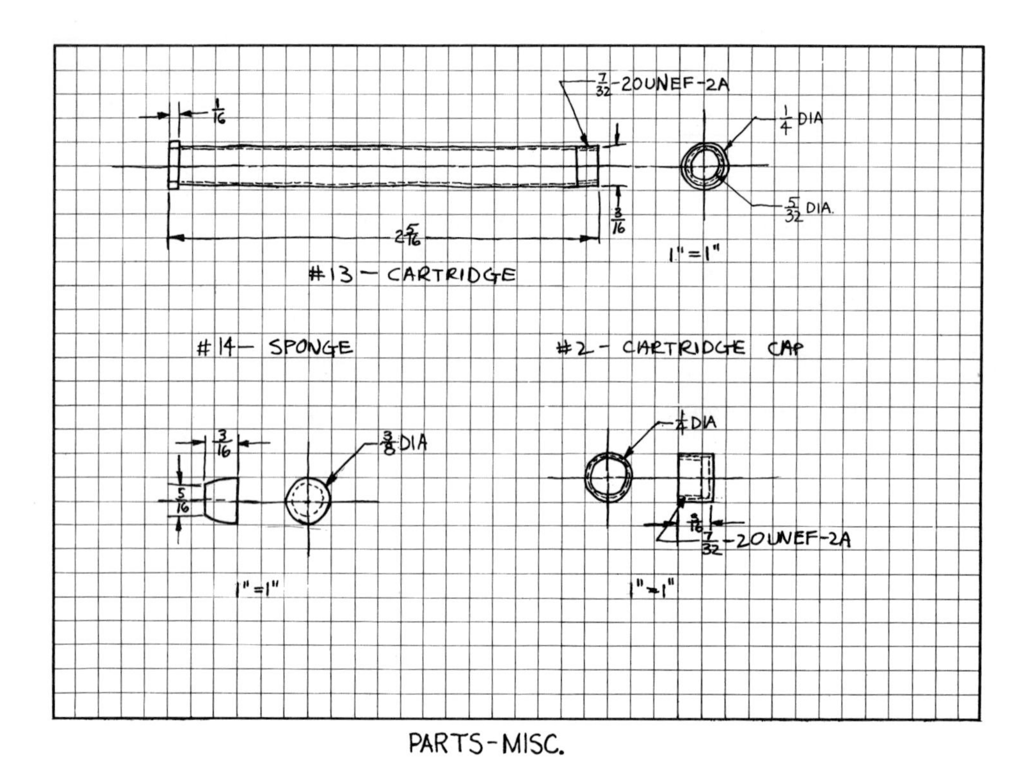
7/32-20UNEF-2A
1/16
1/4 DIA
5/32 DIA.
3/16
2 5/16
1"=1"
#13 — CARTRIDGE
#14— SPONGE
#2 - CARTRIDGE CAP
3/16
3/8 DIA
5/16
1/4 DIA
7/32-20UNEF-2A
1"=1"
1"=1"
PARTS-MISC.

Fig. 21.35

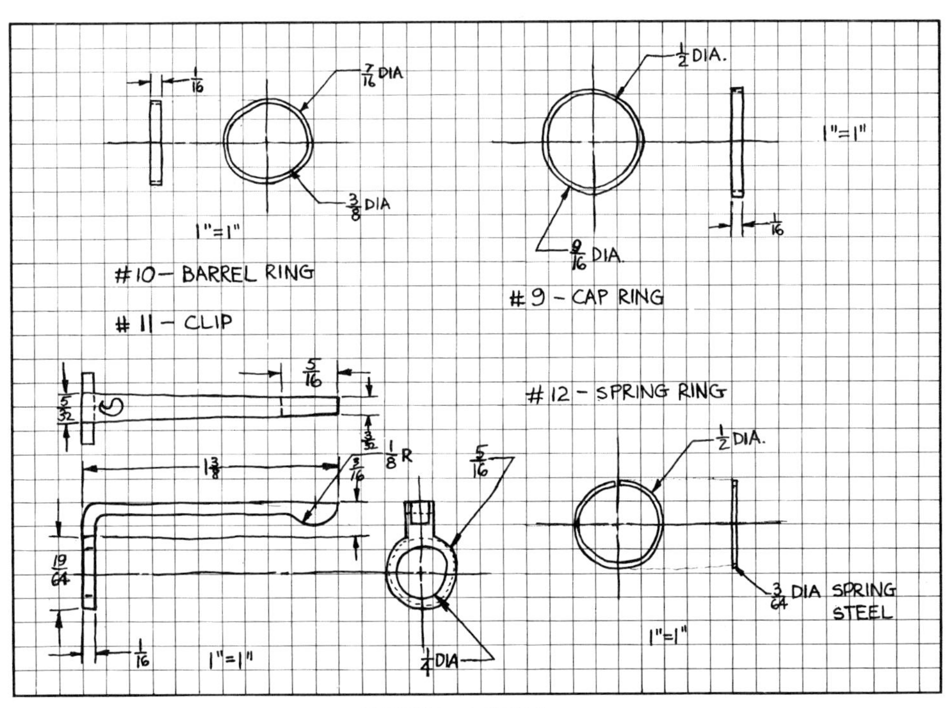

PARTS – METAL

Fig. 21.36

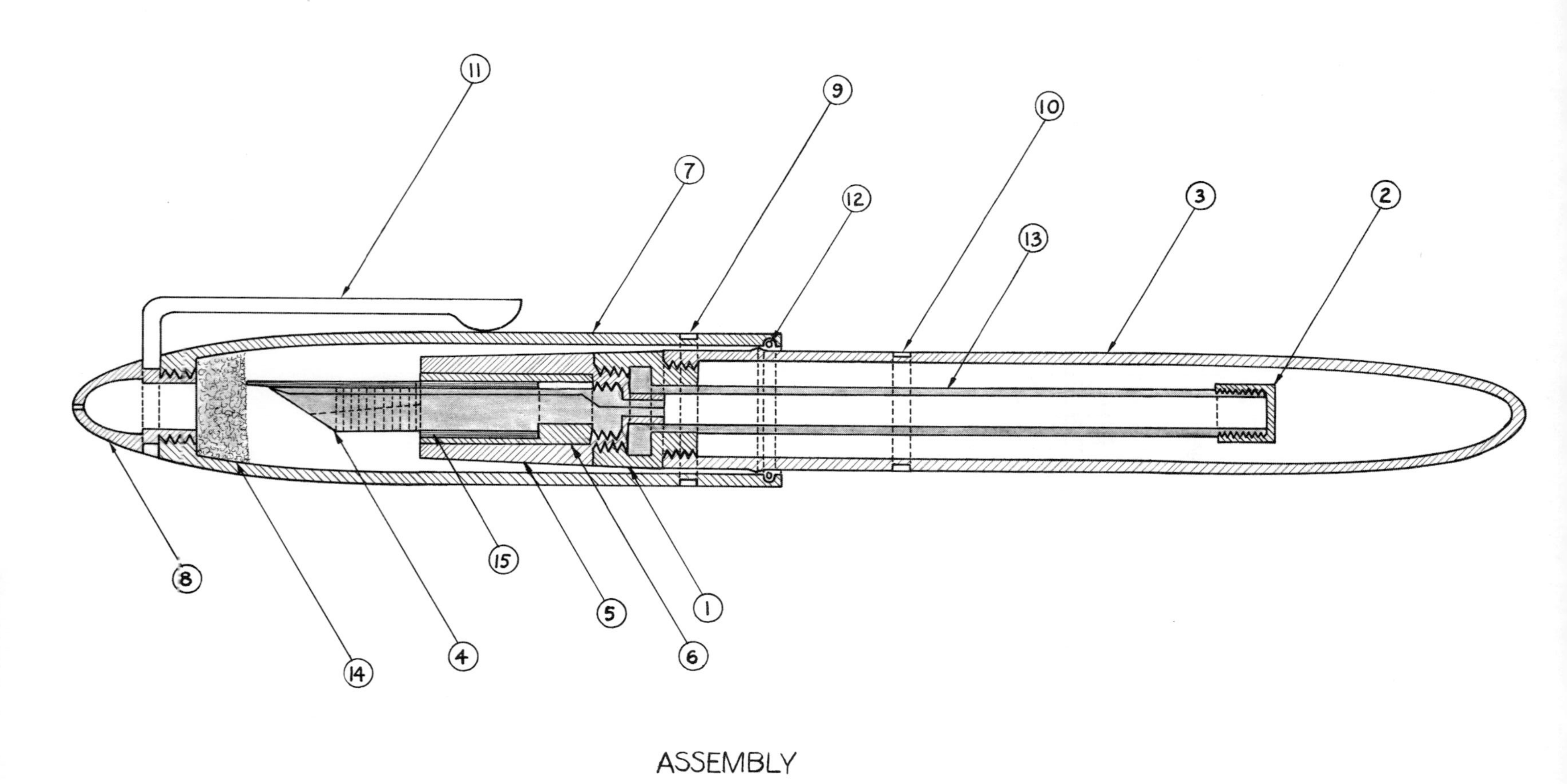

Fig. 21.37

BILL OF MATERIALS:

NO.	NAME	MFG. PROCESS	NO. REQ.	MATERIAL
1	COUPLE	PRESSURE MOLD	1	HARD PLASTIC
2	CARTRIDGE CAP	"	1	"
3	BARREL	"	1	"
4	INK CHANNEL	"	1	"
5	HOUSING	"	1	"
6	POINT COVER	"	1	"
7	CAP	"	1	"
8	CAP TIP	"	1	"
9	CAP RING	MILLED	1	ALUMINUM
10	BARREL RING	"	1	"
11	CLIP	PRESSED	1	SRING STEEL
12	SPRING RING	"	1	"
13	CARTRIDGE	PRESSURE MOLD	1	POLYETHELENE PL.
14	SPONGE	"	1	SPONGE RUBBER
15	POINT	PRESSED	1	BRASS
16				

Fig. 21.38

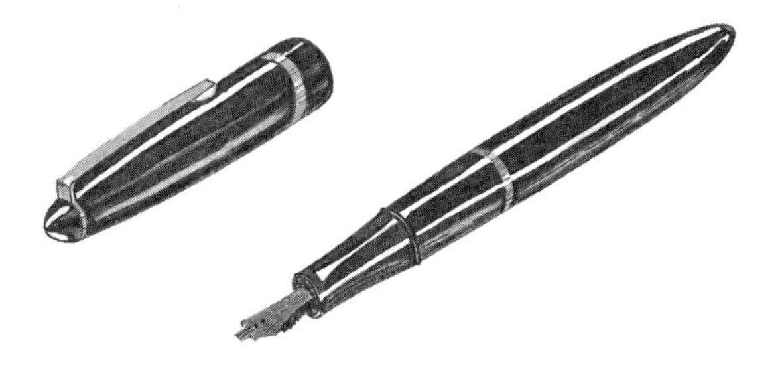

PICTORIAL

Fig. 21.39

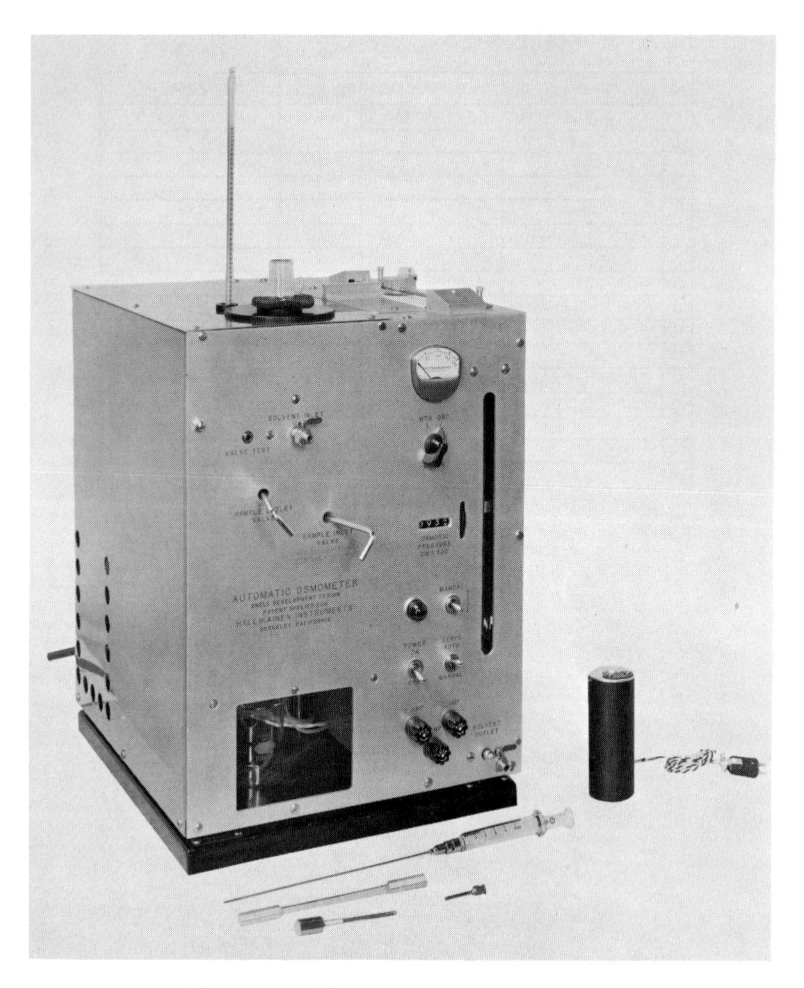

Fig. 21.40 (Courtesy Hallikainen Instrument Co.)

Two examples of product design are now presented to delineate methods used in industry.

EXAMPLE 1. HALLIKAINEN INSTRUMENT COMPANY, RICHMOND, CALIFORNIA

This task dealt with the product design of an Automatic Osmometer, an analytical instrument used for molecular weight determinations of large, chain molecules. The instrument was invented at Shell Development Company and carried as far as a "laboratory prototype," Fig. 21.40. The Hallikainen Instrument Company became attracted to the basic internal mechanism for commercial manufacture and sale. Under a licensing agreement with Shell, Hallikainen designed a new cabinet, some early conceptual sketches of which are shown in Fig. 21.41. Notice the similarity with the photograph of the

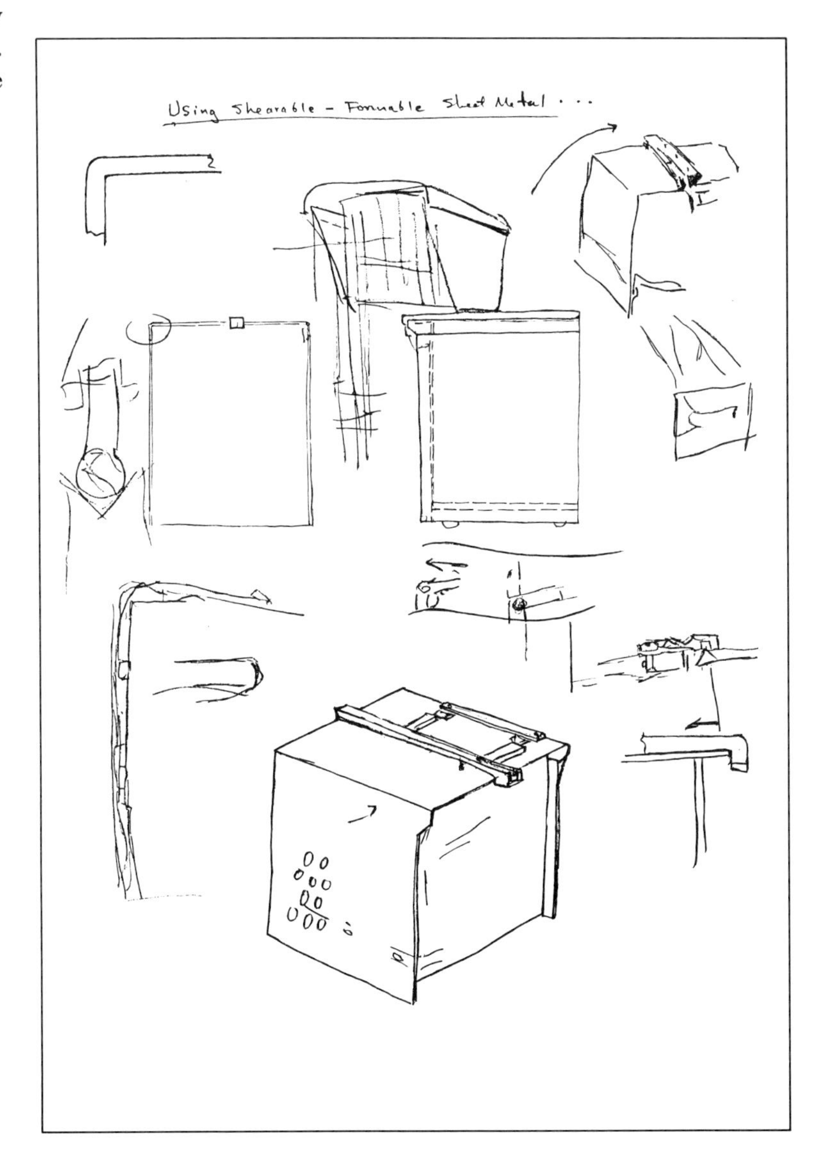

Fig. 21.41 (Courtesy Hallikainen Instrument Co.)

final product, Fig. 21.42, taken a year later. A partial interior view is shown in Fig. 21.43, which reveals the *cell-box cover*. Another partial interior

Fig. 21.42 (Courtesy Hallikainen Instrument Co.)

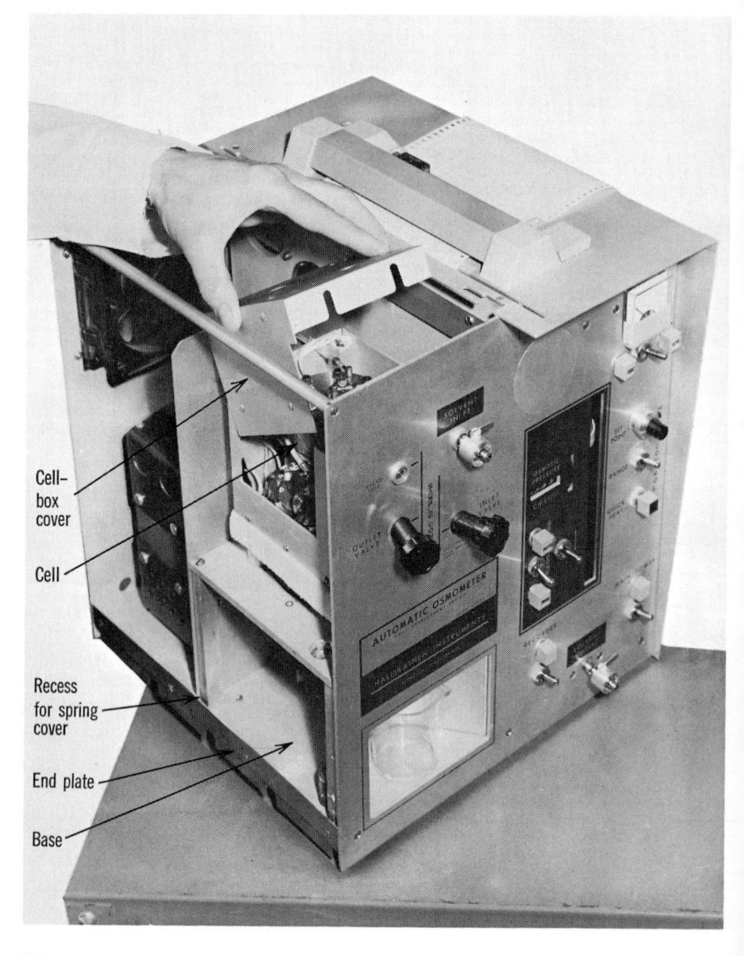

Fig. 21.43 (Courtesy Hallikainen Instrument Co.)

view, Fig. 21.44, shows the *spring cover.* Both of these covers install without tools, a primary aim of the Hallikainen redesign.

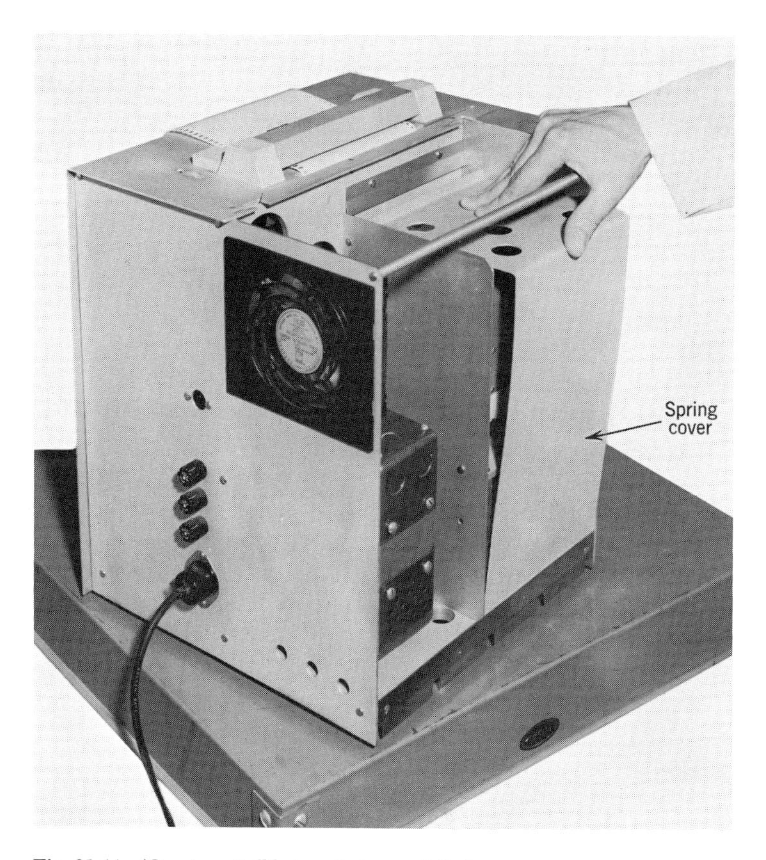

Fig. 21.44 (Courtesy Hallikainen Instrument Co.)

Fig. 21.45 (Courtesy Hallikainen Instrument Co.)

The discussion that follows deals with some of the problems that were considered in the design of the *Left Panel Assembly* shown in Fig. 21.45.

Design Objectives

(*a*) Access to the inside of the cabinet without tools.
(*b*) Left Panel Assembly as part of unified cabinet design.

Prior Decisions Governing the Design of the Left Panel Assembly

1. *Material and Finish.* Choice was limited by these criteria:
   a. Resistant to solvents
   b. Shearable and formable with relatively light shop tools
   c. Attractive appearance

   Decision:
   a. 5052-H32 Aluminum Alloy sheet
   b. 0.090 thick
   c. Straight grain and clear anodized finish
   d. Fasteners (screws and/or rivets) to be few and as inconspicuous as possible.
2. *Shape and Form.* The fundamental "square" look of the cabinet with bent-metal top corners and tapered front profile had been established, thus governing the main outline and the fore-and-aft bend of the Left Panel. (The proportions of the "hood" type design had been decided from a full size *cardboard mock-up.* The exact dimensions chosen became a matter of subjective aesthetics, with the limitation that the front over-hang could not extend so far that it blocked the view of the controls when the operator was standing slightly to the side of the instrument.)
3. *Hook.* The lower left end of the Left Panel would require a self-contained hook to catch onto the End Plate. The *details* of the hook, however, were relatively unaffected by the major decisions already made, so more will be said below about the evolution of this hook.
4. *Miscellaneous.* Several other decisions affecting the Left Panel will be mentioned but not discussed: The Latch details had already been worked out, and so were the size and positions of holes in the top surface for the glass thermometer and for the funnel through which sample fluid is admitted to the osmometry cell, Fig. 21.42.

Hook Design—Lower Left Edge of Left Panel Assembly

Theoretically, the Left Panel had to hook-on, somehow, to the Base so as to be restrained against upward and outward motion after being installed. Note the sketches (right) which illustrate the concept prior to working out the details. In approaching the Hook design, it should be mentioned that there was a certain amount of freedom remaining in the design details of the End Plate. Thus the final design of the End Plate—or whatever anchorage there was to be on the Base—was intimately tied to the design of the Hook.

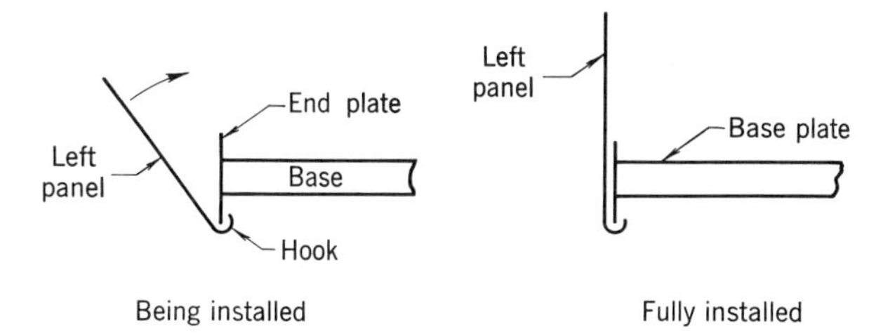

At first a number of free-hand sketches were made, one idea leading to another, as seen in Figs. 21.47 to 21.51. Then several true-size, accurately drawn sections were made on the Osmometer Design Layout, a portion of which is reproduced in Fig. 21.46. A pictorial is shown in Fig. 21.46(*a*). Some of

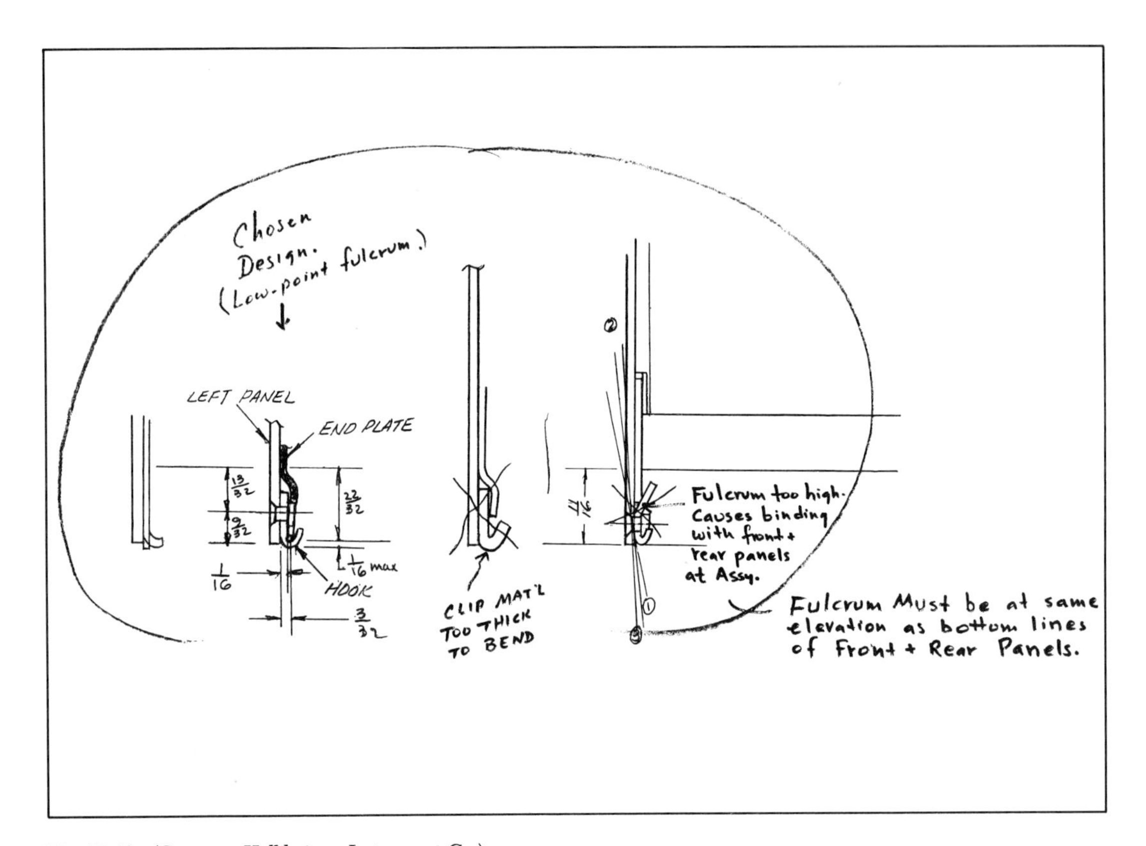

Fig. 21.46 (Courtesy Hallikainen Instrument Co.)

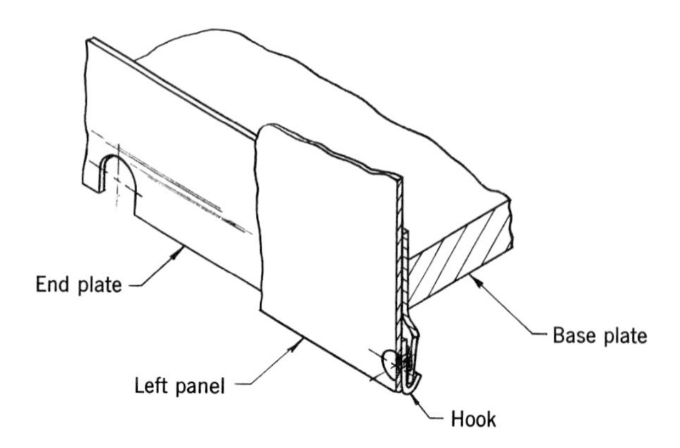

Fig. 21.46(*a*) (Courtesy Hallikainen Instrument Co.)

the earlier concepts would have been cheaper to produce, but upon reflection they were seen to have at least one clear flaw: The pivot axis was too high (Figs. 21.46, 21.47, 21.49). The panel would have to *pivot* from near its *bottom* to avoid interference with the Front and Rear Panels, so only those candidate designs having a pivot axis at the same nominal elevation as the bottom edge of the Front and Rear panels were likely to be workable (Fig. 21.45).

Many minor features, and drawbacks, were evaluated as the flow of ideas progressed. One was a "human factors" concern: The instrument would be lifted with the hands under the side panels, and so the lower edge of the Left Panel Assembly would have to be sturdy enough to take the load and smooth enough not to be uncomfortable. Another concern was over-all cabinet symmetry: An *integral* hook was considered unacceptable because it would destroy the external symmetry of the cabinet.

Eventually, the number of acceptable alternates, among the various concepts proposed, was narrowed down to the point where decisions could be made: Since it was decided the Hook was not to be integral with the Panel, it would be separate and fastened with rivets, the least conspicuous fastener. The rivet alloy was chosen for the type having the closest color shade to alloy 5052 (the Panel) when clear anodized. It was thought best to anodize the entire Panel Assembly *after* riveting. This meant, because ferrous metals are incompatible with aluminum anodizing solutions, that the clip material had to be aluminum. The exact type of aluminum chosen for the Hook was 0.063″ 5052-H32, since it was a material and size already used elsewhere in the design.

One aspect of the End Plate—the part attached to the $\frac{1}{2}$″-thick aluminum Base—took shape early: The upper edge of the End Plate would extend $\frac{5}{16}$″ above the Base, and thus provide a recess for receiving the bottom of the Spring Cover, Figs. 21.43 and 21.44. The bottom portion of the End Plate was designed to match the Hook. It would be jogged $\frac{3}{32}$″ so as to offer a horizontal spring-load when the Left Panel was installed, and it would have cutouts (Fig. 21.43) to clear the rivet heads used in assembling the Hook.

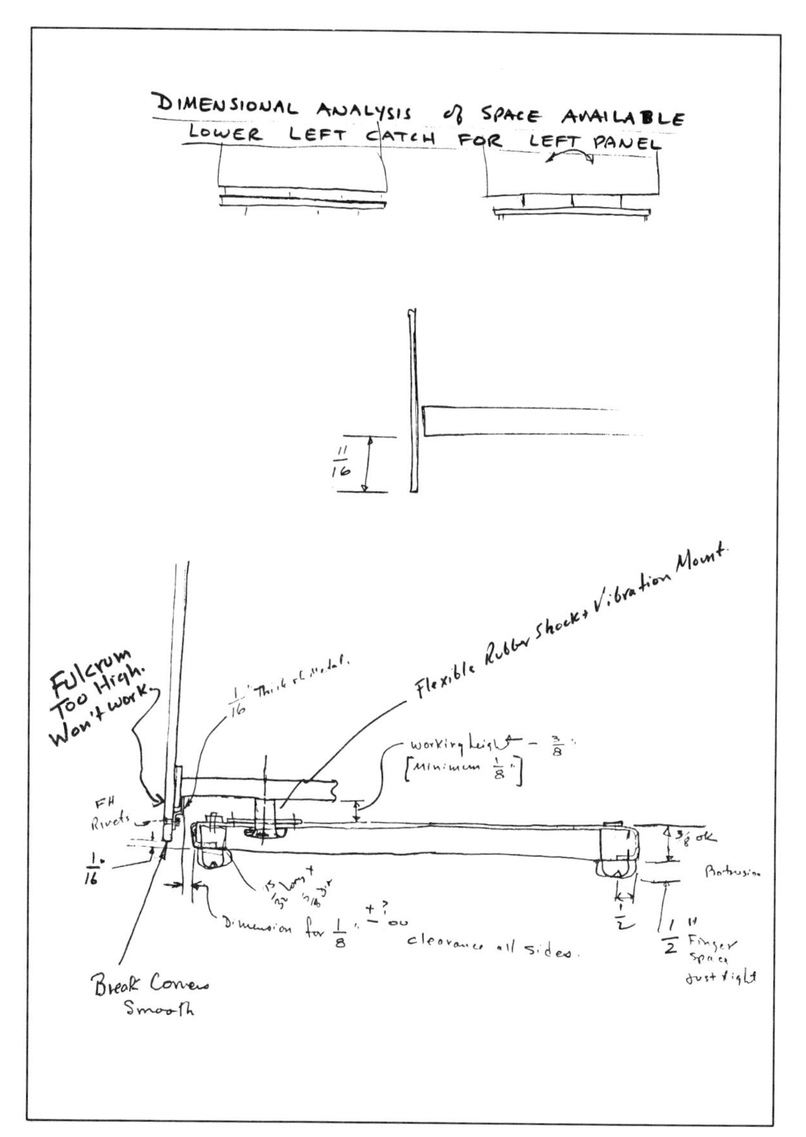

Fig. 21.47 (Courtesy Hallikainen Instrument Co.)

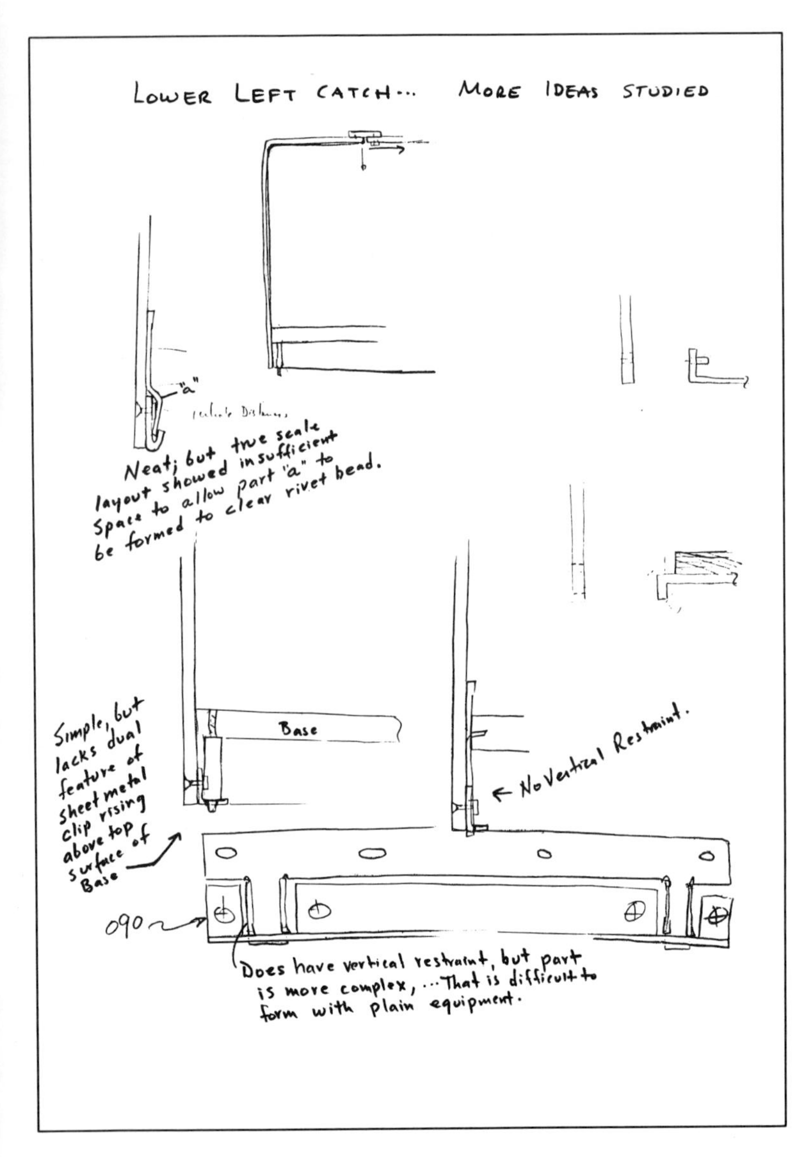

Fig. 21.48 (Courtesy Hallikainen Instrument Co.)

Production and Functional Experience

1. *General.* The catch-and-hold between the Base and the Left Panel functioned just as intended. In placing the Left Panel on the Osmometer cabinet, it catches neatly and securely at the bottom. Then it is swung up and to the right, engaging the latch. The latch bar is pushed in and the Left Panel is solidly in place, capable of supporting the Osmometer firmly when the instrument is hand-transported. Some 20 to 30 Osmometers have been manufactured, with many other design changes necessitated by experience. No fundamental change was required in the Hook-and-End Plate design, however. There were minor changes, some of which are mentioned below.
2. *Attachment of Hook to Panel.* Experience with the first Osmometer resulted in changing the material thickness of the Hook from 0.063 to 0.032. The thinner material, recommended by the Shop, was easier to bend into the U-shape. The thinner gauge was sturdy enough, but only with the addition of two more fastenings. This addition of two rivets also meant two more cut-outs on the End Plate.

   Soon after, the Shop concluded that it lacked the particular skills and/or tooling to produce quality appearing flush riveted assemblies on a routine basis. So a stainless steel oval head screw (and nut) was substituted for the rivet.
3. *Finish.* After the first three instruments were manufactured, a cost saving change was made affecting all the aluminum panels. An air-driven grit blasting was substituted for the more costly hand graining. Test panels of this blast finish were of course analyzed in advance. They demonstrated a pleasing and uniform appearance, which at the time was all that seemed necessary. So the change was authorized for a run of six Osmometers. In time, however, it was discovered that the grit-blasted surfaces of the Osmometer cabinets became impossibly smudged and dingy. Just from ordinary room dust and handling, the pores of the metal filled with dirt and could be cleaned only with a brush and vigorous scrubbing. The grit blasting was a mistake, and on the succeeding production run the grained finish was reinstated.

Unidirectional graining is easy to wipe clean, a "plus" feature of this type of surface, not previously recognized by the Engineer. This experience came at the cost of having to refinish some of the six cabinets, plus the intangible cost of the loss in product confidence by those customers (fortunately few) who received the grit-blasted cabinets.

4. *End Plate Material.* The End Plate on the Base was originally specified cold-rolled steel, nickel plated, in the belief that this was the least expensive acceptable combination of material and finish. However, since the shop was experienced in cutting and forming stainless steel, they requested a change to the latter to save the logistics and delay costs entailed in sending End Plates out to a vendor for nickel plating, even though other parts were routinely sent out for nickel plating (but not always concurrent with the day End Plates were ready for plating). Because other factors than material alone were involved, and because the part represented but a trifle in material cost, it was impractical to attempt a logical cost comparison between the two material alternatives. A change was made to stainless steel. This was a case of simply accepting the "feel" or judgment of the Shop Supervisor on which was the cheapest in the overall.
5. *Final Product.* Overall features of the Automatic Osmometer are:
   a. Speed, in comparison to manual methods.
   b. Precision, due to Shell Development's ingenious, patented, servo.
   c. Simplicity of operation, in part a contribution of Hallikainen's easy-access cabinet, and arrangement of all controls on the Front Panel.
   d. Direct readout of osmotic pressure on a numerical counter.
   e. Record of approach to equilibrium on a strip chart Recorder, completely redesigned and enlarged by Hallikainen.
   f. Easy replacement of solvent after solute permeation, an original Shell feature.
   g. Attractive appearance, due to anodized and grained aluminum with color anodized Front Panel artwork.
   h. Compact.
   i. Operates up to 135°C for analysis of polymers insoluble at room temperatures.

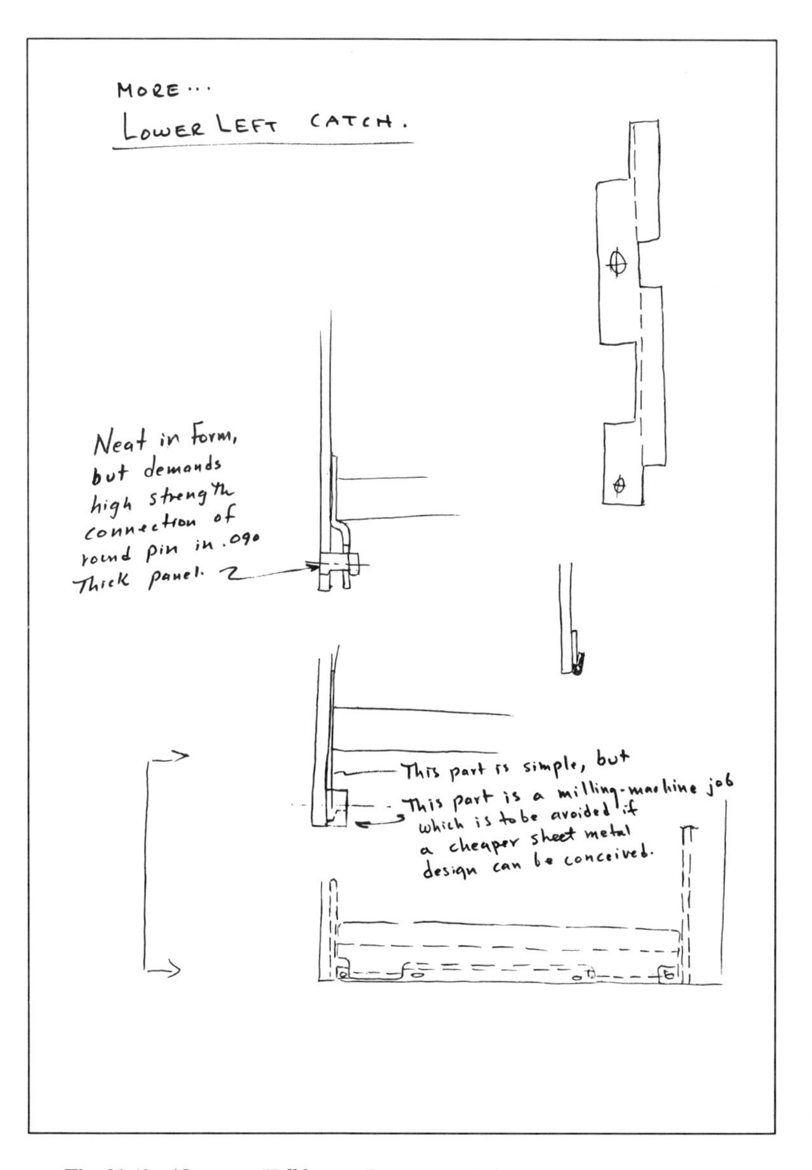

Fig. 21.49 (Courtesy Hallikainen Instrument Co.)

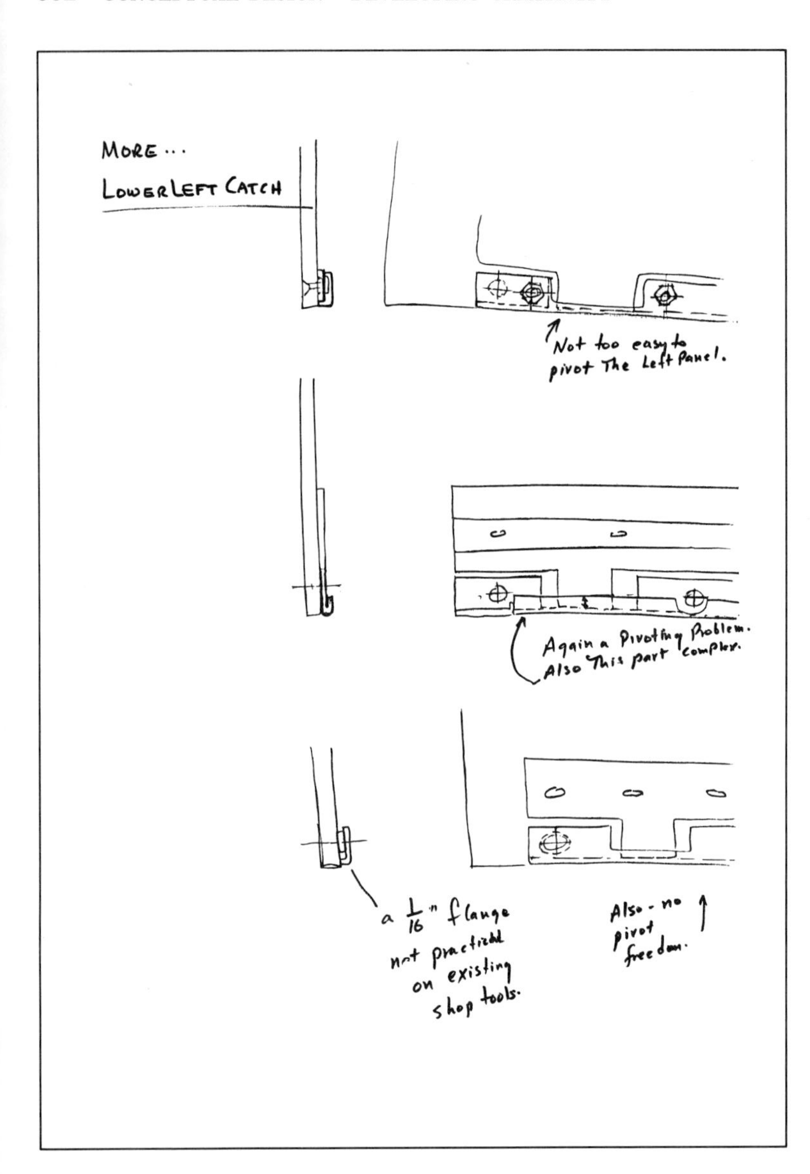

Fig. 21.50 (Courtesy Hallikainen Instrument Co.)

A brief explanation of the operational principles may be of interest to the student: The operator puts a sample solution of 5 to 10 ml into the funnel and then isolates it in the cell (Fig. 21.42) by means of the inlet and outlet valves. After a period of 5 to 10 minutes the osmotic pressure is read on a mechanical counter to one hundredth of a centimeter, over a 10-cm range. A built-in recorder, the pen of which is directly driven by the balancing servo mechanism, enables the operator to observe the balancing process and to ascertain that equilibrium has been established; solute permeation can be detected by a decrease of osmotic pressure with time. The osmometer cell consists of two cavities separated by a semipermeable membrane. On one side is the reference solvent; on the other the sample solution. The cavity containing the sample solution is bounded by a thin metal diaphragm which responds to changes of volume. Displacement of this diaphragm, due to solvent flow through the membrane, is sensed as an electrical capacity change in an oscillator circuit, causing the servo mechanism to adjust the solvent head (static pressure) for zero osmotic flow. The speed with which this instrument makes a determination results not only in improved productivity but also in increased accuracy. The reason for this is because errors caused by small solute molecules permeating the membrane increase with time. These characteristics enable osmometry to become a practical routine method for analyzing polymers (such as polyethylene) in the molecular weight range 5000 to 500,000.

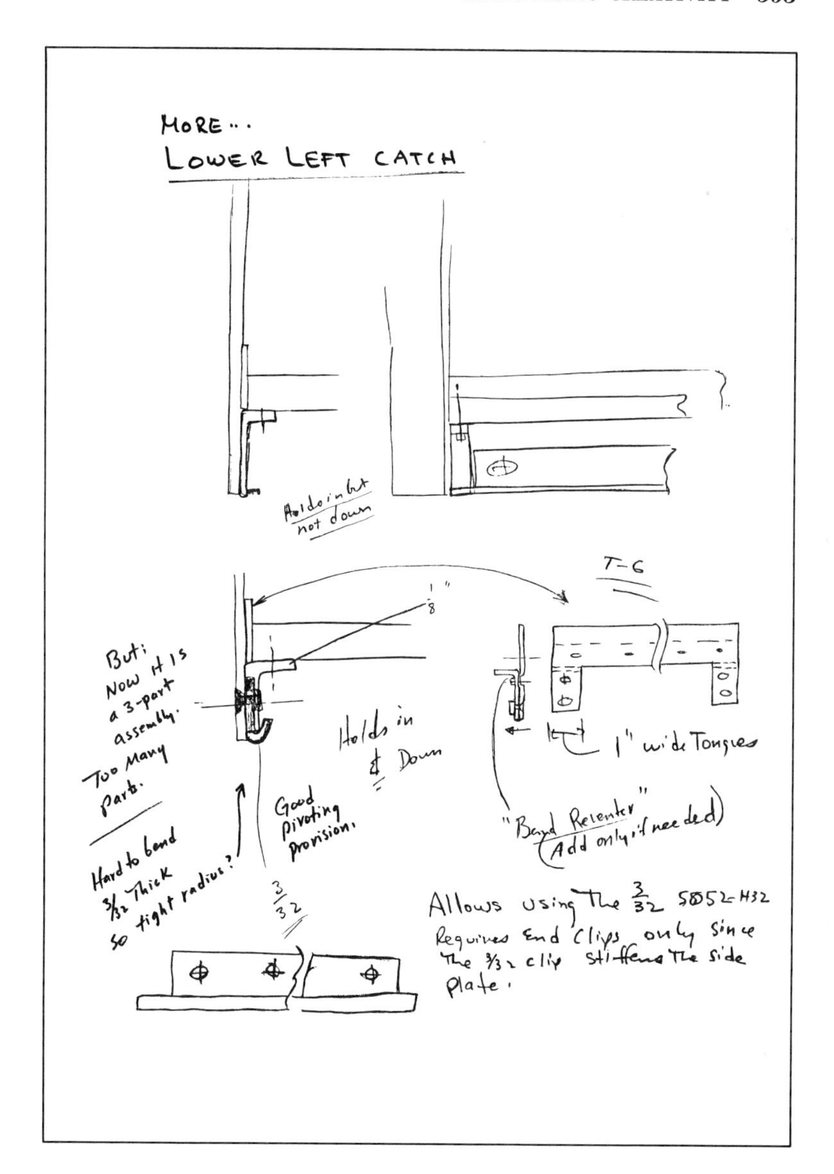

Fig. 21.51 (Courtesy Hallikainen Instrument Co.)

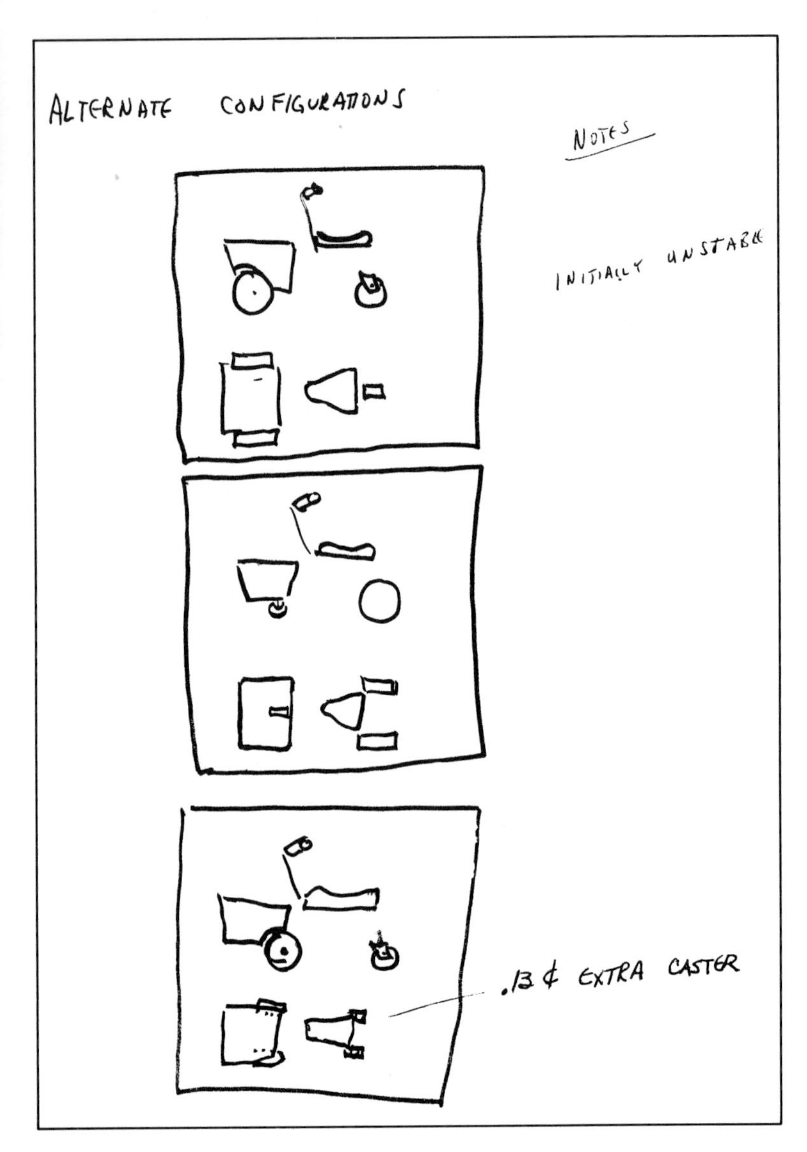

Fig. 21.52 (Courtesy Mattel, Inc.)

EXAMPLE 2. MATTEL, INC., TOYMAKERS, HAWTHORNE, CALIFORNIA

In January, 1966, the Market Research Department suggested that the basic concept of a pre-school ride-on toy would have considerable appeal to both pre-school children, ages 2 to 5 years, and their parents. Management of the corporation was receptive to the suggestion. The development of a competitive unique design was then initiated. The timing for entry into the pre-school toy market was excellent since there was only one major competitive product in the field. Successful designs of pre-school toys afforded a great opportunity to capture a fair percentage of the market.

A number of features of the ride-on toy were carefully considered and tested. Results were analyzed and re-evaluated. Special consideration was given to the following items:

1. Maneuverability
2. Stability
3. Playability
4. Appearance and styling
5. Cost
6. Production

It required a little over a year to develop the toy from the inception of basic concepts to the manufacture of the final product. The following is a brief history of the development of the "talking ride-away toy."

Management's initial thinking was to develop a 3-wheel ride-on (later called ride-away) toy having a front bucket arrangement. This idea was basically a "kiddie-kar" type of vehicle with two wheels in front and a single moveable caster in the rear. (See Figs. 21.52 to 21.55.) The 3-wheel concept was child-tested. These tests, however, were not satisfactory. The major difficulty was that the toy was not stable. Children could not apply enough body force to rotate the caster properly. This resulted in falls from the toy. Efforts by the engineers to correct this situation continued to the middle of August when

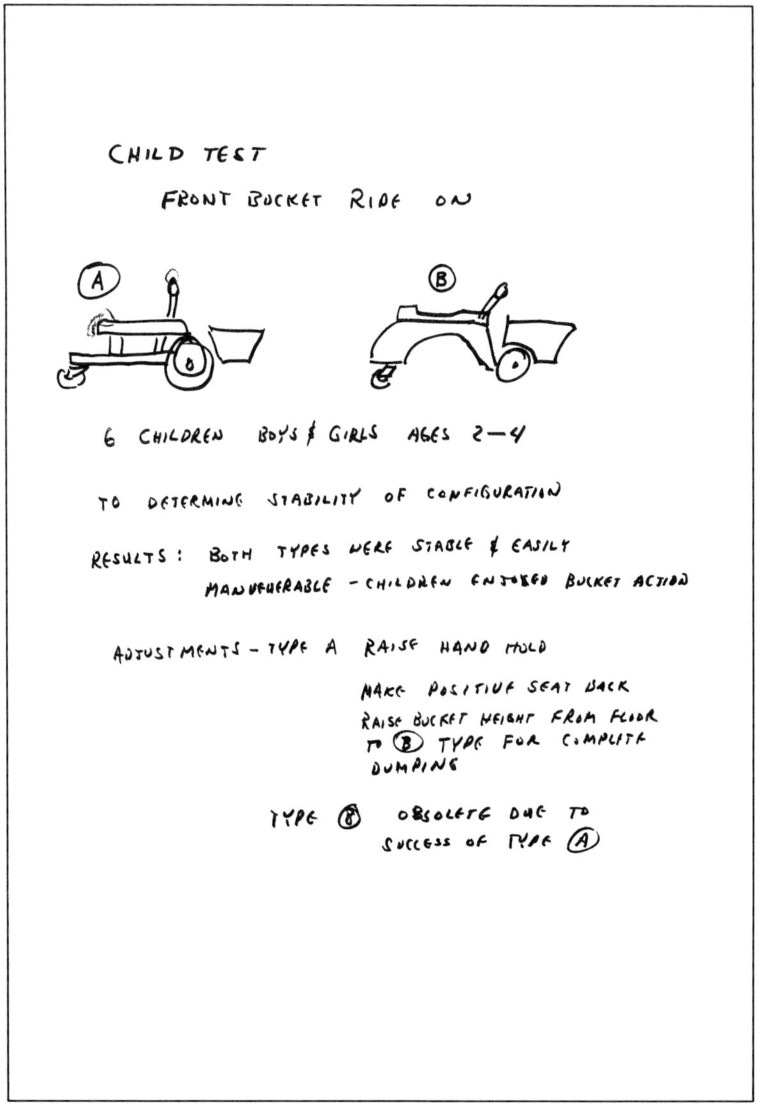

CHILD TEST

FRONT BUCKET RIDE ON

(A) (B)

6 CHILDREN BOYS & GIRLS AGES 2—4

TO DETERMINE STABILITY OF CONFIGURATION

RESULTS: BOTH TYPES WERE STABLE & EASILY MANUEVERABLE - CHILDREN ENJOYED BUCKET ACTION

ADJUSTMENTS - TYPE A RAISE HAND HOLD

MAKE POSITIVE SEAT BACK

RAISE BUCKET HEIGHT FROM FLOOR TO (B) TYPE FOR COMPLETE DUMPING

TYPE (B) OBSOLETE DUE TO SUCCESS OF TYPE (A)

Fig. 21.53 (Courtesy Mattel, Inc.)

management decided to change the design to a 4-wheel configuration. During the time that efforts were directed toward stabilizing the 3-wheel design, a number of "discoveries" were made that were

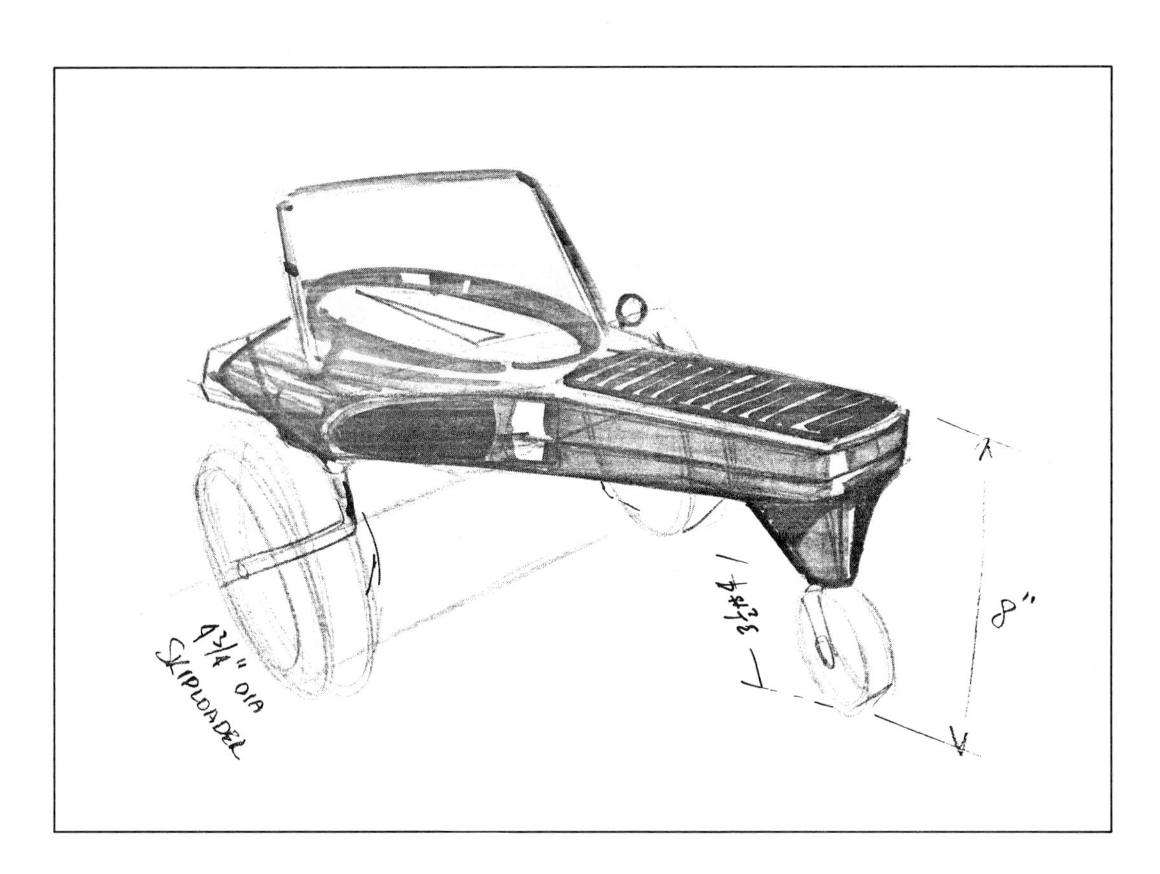

Fig. 21.54 (Courtesy Mattel, Inc.)

quite beneficial. A most important feature was added. This was the pre-selective voice unit. This unit is Mattel's own creation which is protected by patent. The voice unit is regarded as a most significant con-

Fig. 21.55 (Courtesy Mattel, Inc.)

tribution to the toy industry. Essentially it is a miniature record player which consists of a needle, a uniquely grooved nylon record, a turn table, a speaker cone, and a spring. (See Figs. 21.56 and

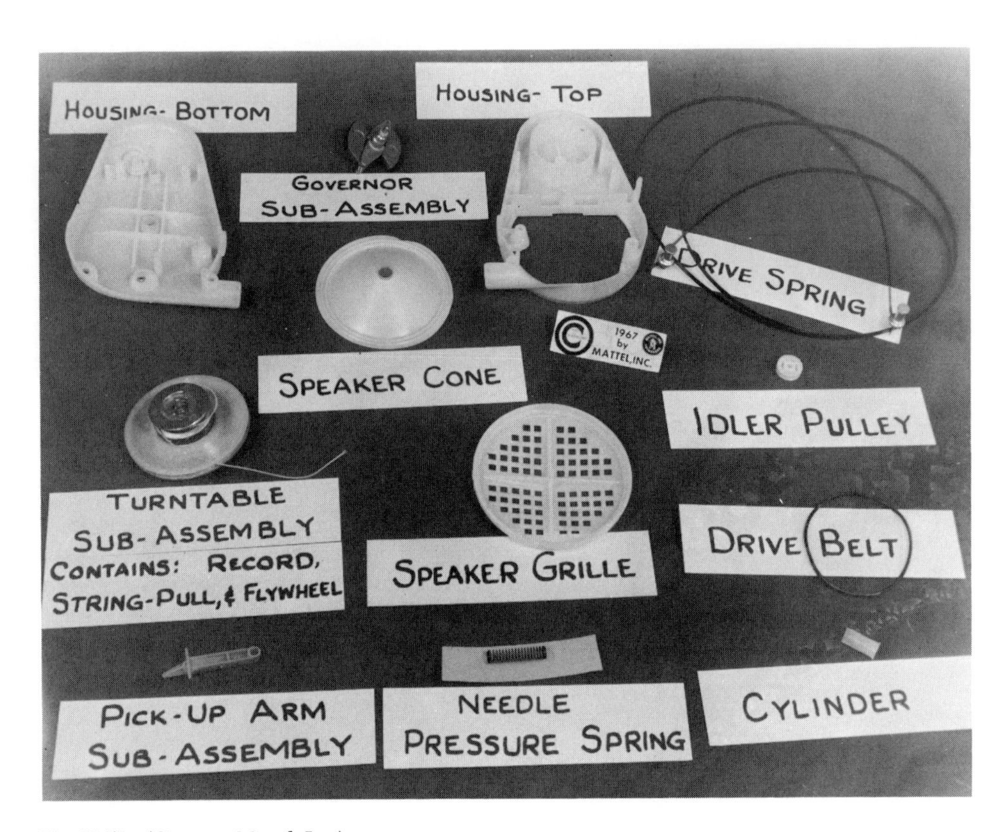

Fig. 21.56 (Courtesy Mattel, Inc.)

21.57.) By pulling the string, energy is stored in the spring which then activates the turn table. The needle drops on the record and plays music, tells a story, or makes some other meaningful sounds. The first designs of the voice unit permitted random choices of portions of the record. This prevented the child from hearing preferred parts of the record. The salient feature of the voice unit for the ride-on toy is the pre-selective feature which enables the child to choose the desired portion of the record. The selection is made prior to pulling the string by placing the pointer at the proper place.

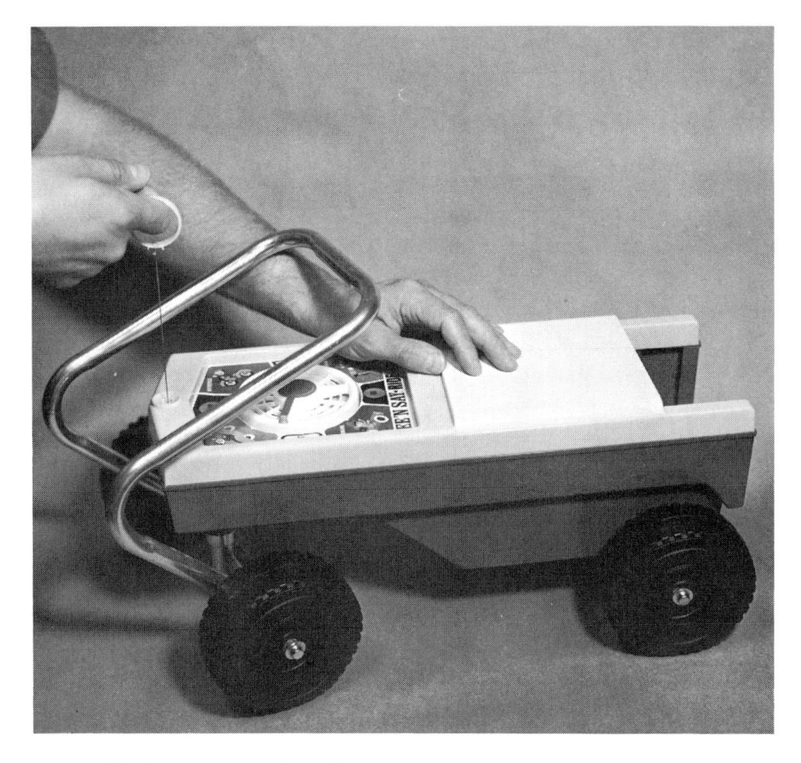

Fig. 21.57 (Courtesy Mattel, Inc.)

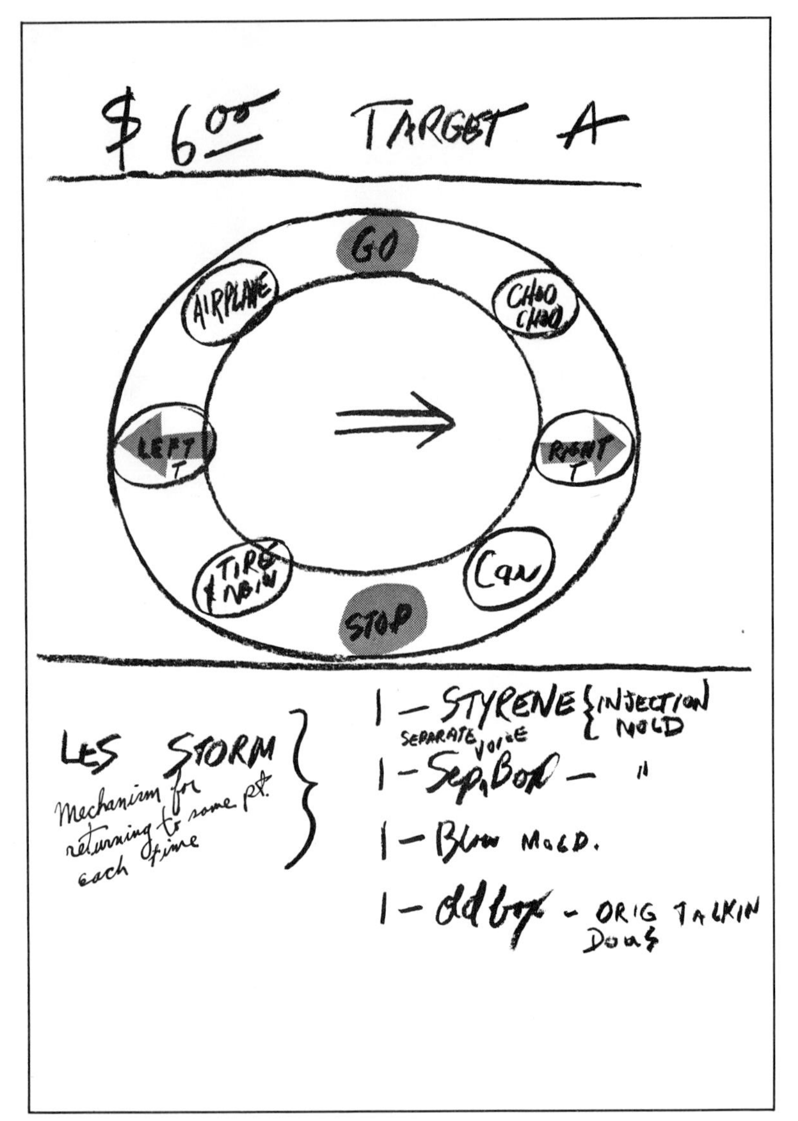

Fig. 21.58 (Courtesy Mattel, Inc.)

The pre-selective voice unit was added to the forward part of the chassis of the ride-on toy. The sound unit is a collection of vehicular noises and phrases, i.e., fire engine, locomotive, etc. Figures 21.58 and 21.59 show preliminary concepts of the labels for the voice unit.

Although the modified toy had considerably more potential appeal than the previous one in playability, because of the voice unit, the engineers were faced with the following additional problems:

1. The string on the vehicle seemed to cause more problems than it did in dolls because the string tended to get stuck on the handle or tangled in the spinner.
2. It was hard to determine the location of the string because of the existence of the U-shaped handle bar as well as the limited space.
3. The length of the string and shape of the ring which is attached to the string had to be determined to enable the child with limited arm

strength to pull the string from the sitting position. To solve these human engineering problems, considerable study and testing was done in the area of string location and spring loading.

In addition to the problems mentioned above, studies were made of the possible use of a crank mechanism to load the voice-unit spring. Increased costs precluded the use of a crank mechanism. The knowledge and know-how which was acquired through the studies during the development of the 3-wheel design enabled Mattel to complete the design of the 4-wheel toy within a relatively short period of time. These studies included:

1. The size and shape of the housing
2. Housing materials and inside space
3. Wheels, size, shape, material
4. Caster action and tube configuration
5. Seat shape

Fig. 21.59 (Courtesy Mattel, Inc.)

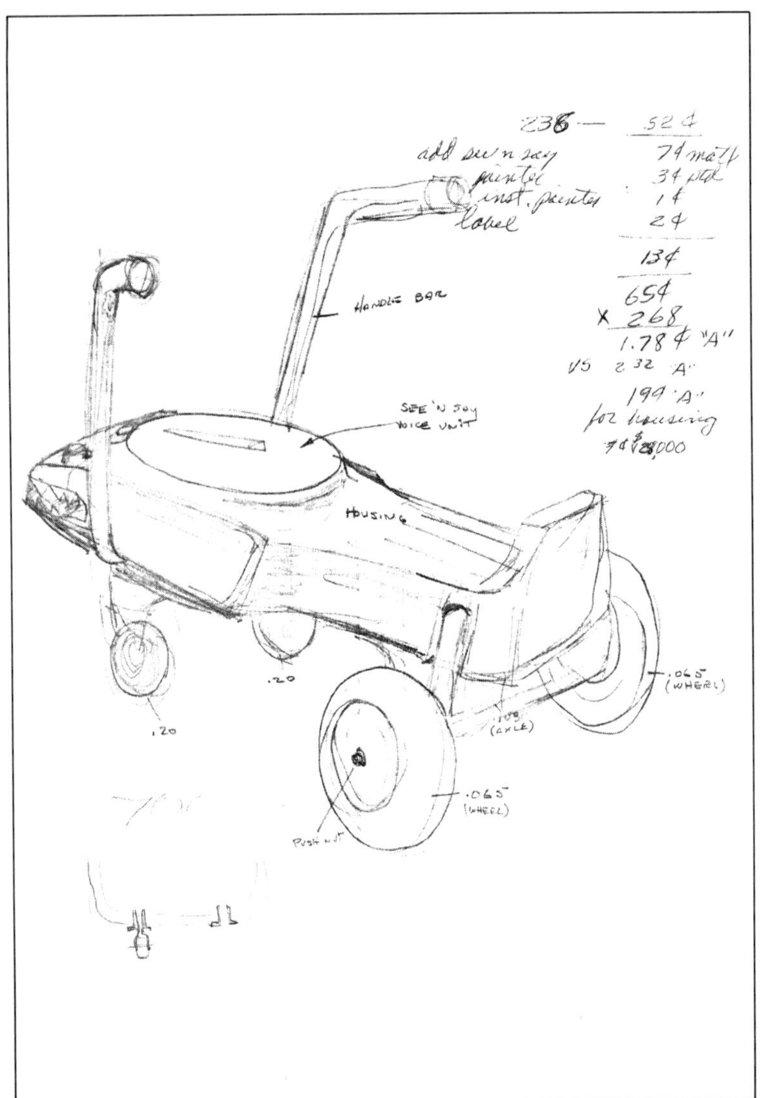

Fig. 21.60 (Courtesy Mattel, Inc.)

A sketch and picture of the 4-wheel model is shown in Figs. 21.60 to 21.62. Although minor improvements were made of the model shown in

Fig. 21.61 (Courtesy Mattel, Inc.)

Fig. 21.62 (Courtesy Mattel, Inc.)

Fig. 21.63, it was considered quite satisfactory for release to component design and for the preparation of production drawings.

PROJECTS

"The creative art of conceiving a physical means of achieving an objective is the first and most crucial step in an engineering project. Analysis of a design is the second important step." Constraints that impinge upon a project, usually lead to a final design that is the "best" compromise among the parameters of the problem. In essence this means that the engineer selects, from among several possible solutions, the one which best meets the requirements.

In undertaking the design of your project make sure that you fully understand the real problem. Then allocate (estimate) the time to be devoted to the following:

1. Gathering information—include the resources of the Engineering Library—examine journals, periodicals, patents; consult engineering and science professors who are specialists in such fields as

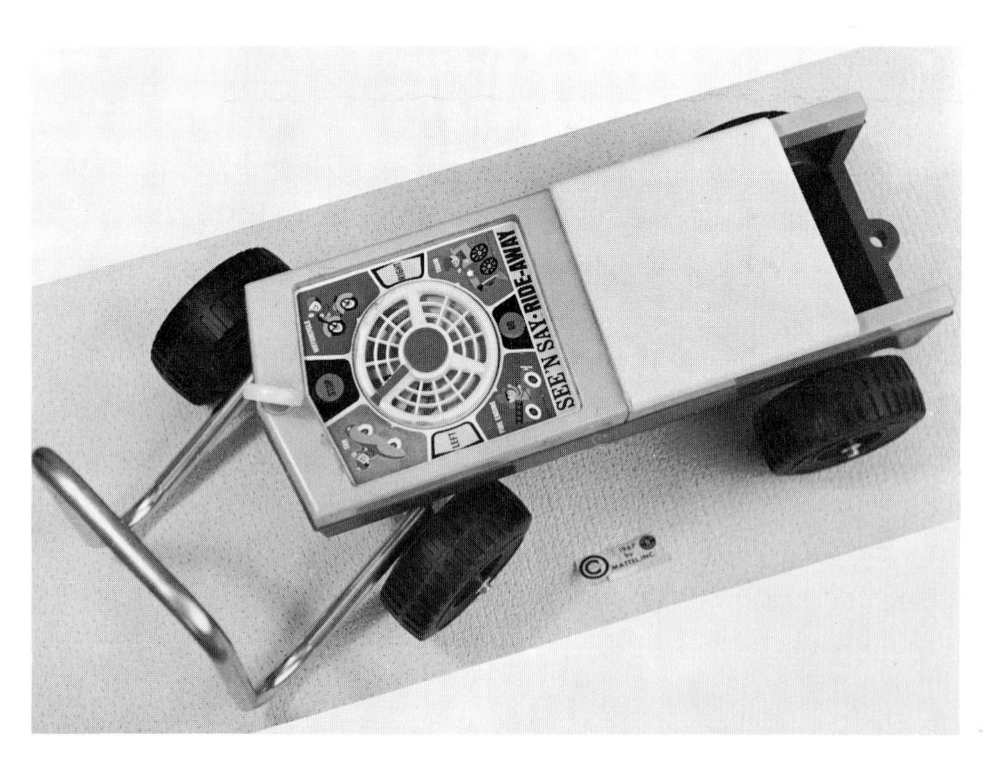

Fig. 21.63 (Courtesy Mattel, Inc.)

electronics, structures, processes, mechanisms, materials, manufacturing, etc.; local engineering and production personnel; etc.
2. Preparing freehand "idea" sketches of possible solutions.
3. Choosing the "best" ideas.
4. Modifying the best ideas.
5. Evaluating the tentative solutions.
6. Adapting and implementing the selected solution.
7. Building and testing a prototype (if time and facilities are available).
8. Revising the design and preparing final design drawings (if step 7 is accomplished).

Finally, prepare a report which should include the following:

1. Statement of the problem.
2. Brief description of the operation of the product.
3. Conceptual design sketches.
4. A "layout" design.
5. *Freehand* detailed drawings of each component.
6. An assembly drawing and parts list.
7. An index.

You may select one or more of the following projects:

1. *Indoor Coal Burning Portable Barbeque*
   a. Must have a surface area of at least 200 square inches.
   b. It should not cause smoke in the room.
   c. Should not be a fire hazard.
   d. Should be easily cleaned.
   e. Should not cost more than $12.
   f. Total time of project: 20 hours.
2. *Door Locking Mechanism.* Design a door locking mechanism which can be locked from the inside of a bath room, and which can be unlocked from the outside (but not locked), and yet protect the privacy of the occupant, a small child. The design should make possible the use of a simple tool (other than a key) to open the door from the outside. Project time: 16 hours.
3. *Salt Water Aquarium.* The problem of housing salt water animals, especially the ones usually found in tide pools, is an interesting one. Salt water can be provided by using one of a number of commercially prepared mixes. The aquarium should be about two square feet in area; a

rectangular size is not necessary. A maximum of six inches of water is needed, and it would be valuable if there were water movement approximating tidal conditions. The major problem is to keep the water rather cool, about 45 to 50°. Cost should be kept as low as possible and consideration should be given to obtaining non-shattering materials. The problem of small aquaria for use by individual students is related to this question. In this case, the price would have to be around $2. Project time: 20 hours.

4. Design a *small toy automobile* which will never fall from an edge, but will turn at least 90° as soon as the front wheels begin to drop after crossing the edge. The toy should not exceed 4 to 5 ounces, should be set in motion by a spring acting on the rear wheels, and should operate on a $1\frac{1}{2}$-ft × $1\frac{1}{2}$-ft plywood board for a minimum of 20 seconds. Project time: 18 hours.
5. Design a *teaching aid* that will graphically demonstrate the acceleration of a falling body and will make possible the computation of the speed at which the falling body accelerates. Project time: 18 hours.
6. Design an *automatic feeder for an aquarium* to meet the following requirements:
   a. Quantity of food per feeding should accommodate a normal fish population in a standard 20-gallon tank.
   b. Dry flake food will be used. Food should enter the aquarium at the same location for each feeding.
   c. A switch arrangement should be provided to turn on the aquarium lights at 6:00 A.M. and off at 6:00 P.M.
   d. The feeder must be fully automatic to operate seven days, two feedings daily (7:00 A.M. and 3:00 P.M.).

   Project time: 18 hours.
7. Design a *flour canister* to meet the following general specifications:
   a. The canister is to stand on a counter or sink and dispense the premeasured quantity into a mixing bowl.
   b. The premeasured quantity should be variable from $\frac{1}{8}$ cup to 1 cup in $\frac{1}{8}$-cup increments.
   c. The canister should store 5 pounds of flour.
   d. The device must be moistureproof and readily

cleaned. As an optional feature, it is proposed that an automatic sifter be incorporated into the design. The device may be either electrical or mechanical. The sifter should add as little as possible to the price of the canister. Project time: 20 hours.

8. Design *a toy to illustrate a scientific principle.* Specifications:
   a. Selling price under $10.00.
   b. Cannot be used as, or modified into, a lethal weapon.
   c. The scientific principle should be easily understood by a 12-year-old and his parents.
   Project time: 18 hours.
9. *Mechanical Garbage Compressor*
   a. Must accommodate the usual household size can.
   b. Must be easy to operate.
   c. Can be operated by a 10-year-old.
   d. Must be economical so that cost can be amortized over a three-year period.
   Project time: 20 hours.
10. *An Arithmetic Teaching Device*
    a. Age group 4 to 6 years.
    b. Addition and subtraction only.
    c. Lightweight.
    d. Under $8 selling price.
    Project time: 18 hours.
11. Design a simple apparatus to measure the pressure-volume relationship for gases, such as a syringe-pressure gage system where pressures below one atmosphere can be studied. Need a combination vacuum-pressure gage with negligible gas volume in the pressure actuating mechanism. Project time: 22 hours.
12. Design a simple apparatus to measure the volume change of a liquid changing to a gas. Low boiling temperature liquids must be used, e.g., Freon 11 (trichloromonofluoromethane), b.p. = 23.8°C. Project time: 22 hours.
13. Design a student apparatus (upper elementary grades) to study qualitatively the expansion and contraction of solids (metals) due to thermal interactions. Project time: 20 hours.
14. Design an apparatus for an easy way of containing and dispensing gases in the classroom. Need oxygen, nitrogen, and carbon dioxide. Need to make them as readily available as Freon in the

shaving cream type dispensing can. Want to be able to fill balloons and plastic syringes quickly and simply. Children must be able to operate apparatus. Project time: 22 hours.

15. Design an apparatus that will permit children to obtain time-temperature data for a liquid changing to a vapor and condensing to a liquid again in a closed system using low boiling temperature liquid. Project time: 22 hours.

Projects for Undeveloped Nations

The design of simple, basic pieces of equipment is not always as easy as one might believe. It means eliminating all of the non-essentials of machinery and knowing enough about materials so that products can be made locally. The end product must be inexpensive, virtually foolproof so that an unskilled person can learn to operate it in a relatively short time. Moreover, the designer must recognize the need as it exists for the group concerned. This means that the designer must know the way-of-life of the group; something about the state of its development, etc. Information about the needs of various countries can be obtained from "Development and Technical Assistance (DATA), 437 California Ave., Palo Alto, California, and VITA (Volunteers for International Technical Assistance, Inc.) at 230 State St., Schenectady, N.Y.

Some items that are in great need include the following (all must be simple and inexpensive):

(*a*) A simple rice transplanter that runs on animal power or with a small tractor. Most of the world's rice is grown in wet paddies, and transplanting is the only practical way of starting the plants if more than one crop is to be grown each year.
(*b*) A light, simple cheap tractor with no frills such as lights or an electric starter.
(*c*) Water-lifting devices.
(*d*) Gristmills.
(*e*) Mechanical rice sun-drying equipment.
(*f*) Solar cooking devices.
(*g*) Solar powered refrigerator.
(*h*) Water distillation equipment (solar stills are being used in several countries).

Other Project Areas

Another fruitful area for the design engineer is the fulfillment of the need for new hardware to harvest the sea. Design of equipments for oceanographic research is essential to the later development of means for harvesting food and energy from the sea. We need to know more about human engineering as it applies to workers in the ocean; the use of materials in the sea; remote handling devices, etc. It has been estimated that more than 500 million tons of phosphorite (an important component of fertilizers) lie from 200 to 1000 feet off the Pacific Coast. There are huge concentrations of manganese on the bottom, with an estimated value of as much as $10 million per square mile. Moreover, there is also the problem of ways and means to recover some 50 to 60 dissolved elements that are contained in sea water.

Still another area of vital interest to the design engineer is "the survival field"—smog control, desalinization of water, transportation systems and control, food supply, etc.

And still another exciting design area is in the field of "engineering medicine." The design of *heart pacers* that beat time for the hearts of thousands of Americans has extended their useful-life period considerably. Good progress is being made in the design of an artificial heart. The lives of hundreds of patients in the United States are being extended by the use of an artificial kidney. Various systems have been designed to make analyses of blood samples. Thousands of tests can be processed in a very short period of time. One such device can perform 12 different tests automatically on each sample at the rate of 27 samples per hour. In the field of prosthetic devices much can be done to improve the design of artificial legs and arms, hearing aids, braces, etc. The design of surgical instruments is another area for novel products. Dr. Robert Hall, an oral surgeon, recognized the need for bone-cutting devices that would cut bone faster and more safely than had ever been done before. He designed what he calls the Neuairtome to accomplish his purpose. He has given up his practice as an oral surgeon to devote full time to designing. With the aid of engineers on his staff he continues to develop new instruments. Engineers can make vital contributions to this important area—bio-engineering or engineering medicine.

APPENDICES

CONTENTS

APPENDIX A

LINE CONVENTIONS, GEOMETRIC CONSTRUCTIONS, DIMENSIONING PRACTICES AND TECHNIQUES

LINE CONVENTIONS

Line Symbols (See Fig. A-1)

Thickness of lines will vary according to the size and type of drawing. Where lines are close together, for example, the lines are drawn thinner. For most pencil work, *two* widths of lines are adequate—*medium thick* for visible, hidden, and cutting plane lines; and *thin* for center, extension, dimension, and section lines. In order to produce good legible prints, all pencil lines should be clean, dense black, and uniform.

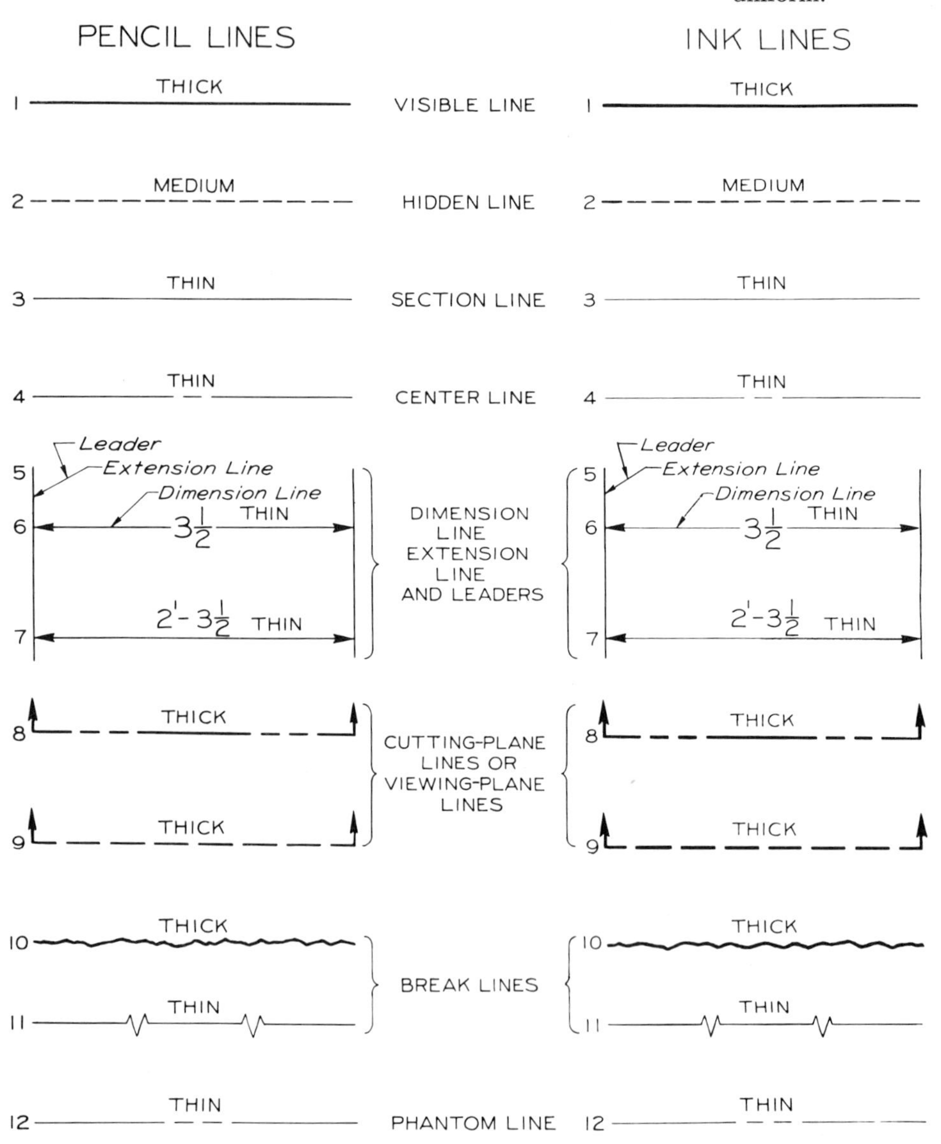

Fig. A-1 Line conventions. (Courtesy ASA)

idden Lines (See Fig. A-2)

Hidden lines should begin and end with a dash in ontact with the visible or hidden line from which ey start or end, except as shown in the figure. idden lines should not be used unless they add to e clearness of the interpretation of the design rawing.

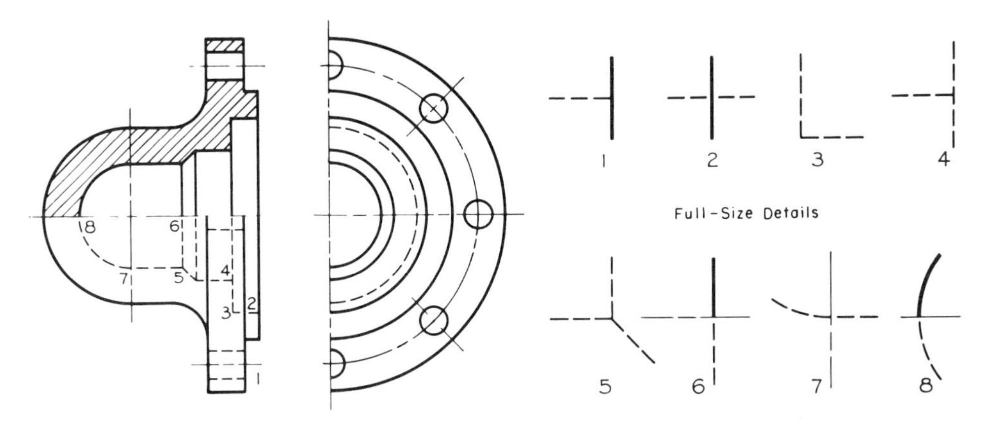

Fig. A-2

ection Lining (See Fig. A-3)

Thin solid lines should be used for section lining. Vhere more than one part is shown, the directions f the lines should be changed as illustrated in the gure for parts A, B, and C.

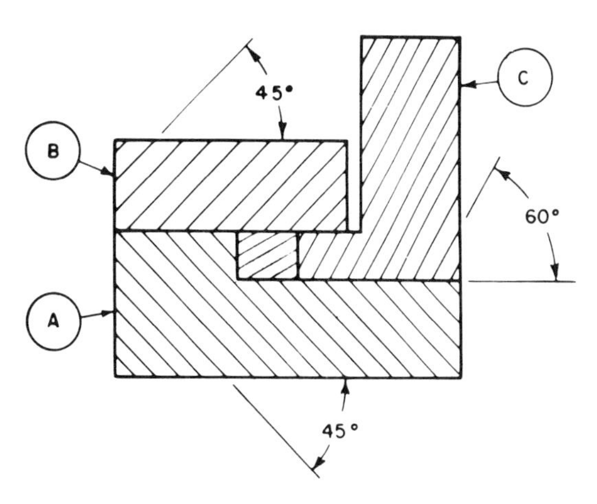

Fig. A-3

Lettering (See Figs. A-4 and A-5)

The single-stroke Gothic style shown in the figur is used on most engineering design drawings. Tl style is easy to use and provides a means for rap execution. Above all the most important requireme is legibility. Both the vertical and inclined styles a acceptable. The trend, however, is toward the u of the vertical style. In many industries notes a typed on the drawing.

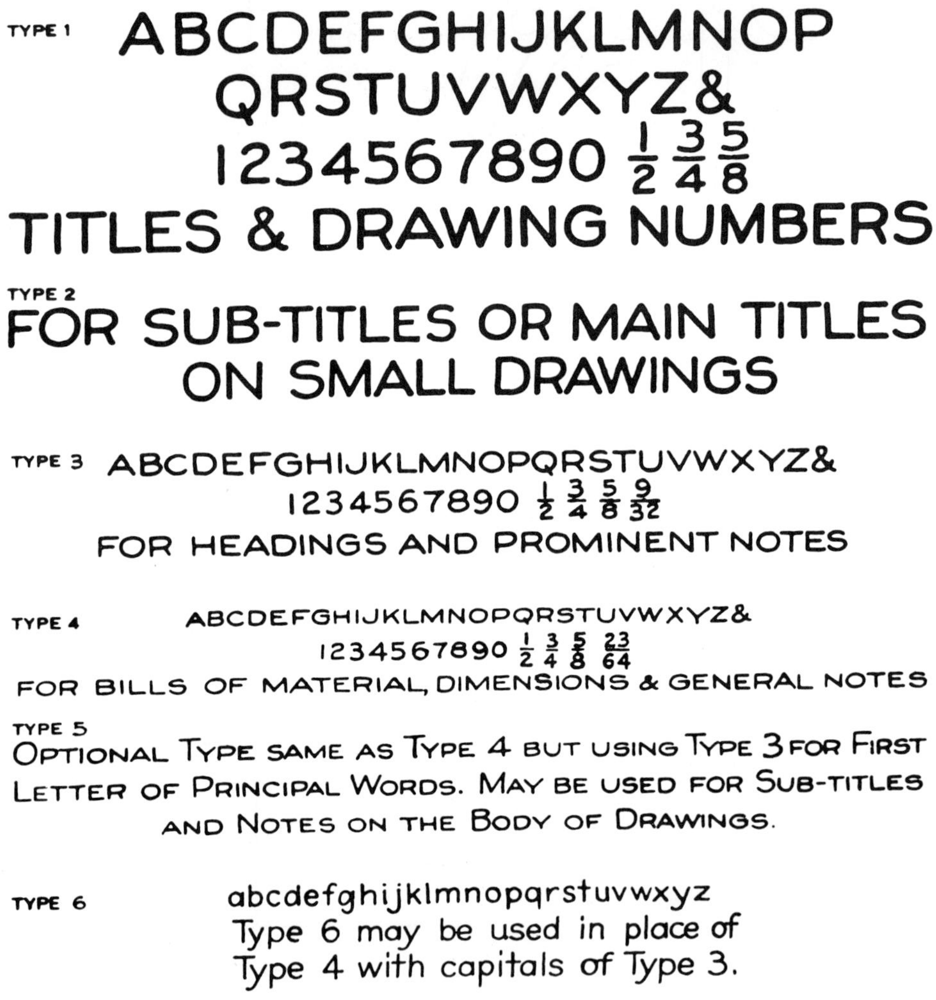

TYPE 1 ABCDEFGHIJKLMNOP
QRSTUVWXYZ&
1234567890 $\frac{1}{2}$ $\frac{3}{4}$ $\frac{5}{8}$
TITLES & DRAWING NUMBERS

TYPE 2
FOR SUB-TITLES OR MAIN TITLES
ON SMALL DRAWINGS

TYPE 3 ABCDEFGHIJKLMNOPQRSTUVWXYZ&
1234567890 $\frac{1}{2}$ $\frac{3}{4}$ $\frac{5}{8}$ $\frac{9}{32}$
FOR HEADINGS AND PROMINENT NOTES

TYPE 4 ABCDEFGHIJKLMNOPQRSTUVWXYZ&
1234567890 $\frac{1}{2}$ $\frac{3}{4}$ $\frac{5}{8}$ $\frac{23}{64}$
FOR BILLS OF MATERIAL, DIMENSIONS & GENERAL NOTES

TYPE 5
OPTIONAL TYPE SAME AS TYPE 4 BUT USING TYPE 3 FOR FIRST LETTER OF PRINCIPAL WORDS. MAY BE USED FOR SUB-TITLES AND NOTES ON THE BODY OF DRAWINGS.

TYPE 6 abcdefghijklmnopqrstuvwxyz
Type 6 may be used in place of
Type 4 with capitals of Type 3.

Fig. A-4

TYPE 1

ABCDEFGHIJKLMNOP 5
QRSTUVWXYZ& 2
1234567890 $\frac{1}{2}$ $\frac{3}{4}$ $\frac{5}{8}$ $\frac{7}{16}$
TO BE USED FOR MAIN TITLES
& DRAWING NUMBERS

TYPE 2

ABCDEFGHIJKLMNOPQR
STUVWXYZ&
1234567890 $\frac{13}{64}$ $\frac{5}{8}$ $\frac{1}{2}$
TO BE USED FOR SUB-TITLES

TYPE 3

ABCDEFGHIJKLMNOPQRSTUVWXYZ&
1234567890 $\frac{1}{2}$ $\frac{3}{4}$ $\frac{5}{8}$ $\frac{7}{16}$
FOR HEADINGS AND PROMINENT NOTES

TYPE 4

ABCDEFGHIJKLMNOPQRSTUVWXYZ&
1234567890 $\frac{1}{2}$ $\frac{1}{4}$ $\frac{3}{8}$ $\frac{5}{16}$ $\frac{7}{32}$ $\frac{1}{8}$
FOR BILLS OF MATERIAL, DIMENSIONS & GENERAL NOTES

TYPE 5

OPTIONAL TYPE SAME AS TYPE 4 BUT USING TYPE 3 FOR FIRST LETTER OF PRINCIPAL WORDS. MAY BE USED FOR SUB-TITLES & NOTES ON THE BODY OF DRAWINGS.

TYPE 6

abcdefghijklmnopqrstuvwxyz
Type 6 may be used in place of
Type 4 with capitals of Type 3

Fig. A-5

GEOMETRIC CONSTRUCTIONS

Many engineering designs make use of we known geometric shapes and geometric constru tions. In the design of components, for example, su geometric shapes as triangles, squares, trapezoic pentagons, hexagons, circles, ellipses, parabol hyperbolas and spirals are quite common. In ad tion, a number of geometric constructions are fou quite useful in engineering design, in the constru tion of scales for nomograms, and in graphical so tions of problems arising in engineering and scienc Most students are already familiar with many of t simple geometric shapes and elementary geomet constructions.

A number of these are presented *for review,* a several others are included that will be found help in effecting graphical solutions.

A. The Straight Line and Its Division

1. To Divide a Line Segment into a Specifi Number of Equal Parts (Fig. A-6). Suppose line se ment AB is given and that it is required to divide into seven equal parts.

Solution. Through point A draw line m and th lay off seven equal distances, starting at point Join points E and B and then draw parallels to through the points on m. The points of intersecti of these parallels with line segment AB divide it in the required seven equal parts. Why is this true?

Suggestion. In laying off distances on a line se ment, use needle-point dividers, alternating t rotation of the dividers as shown in Fig. A-7.

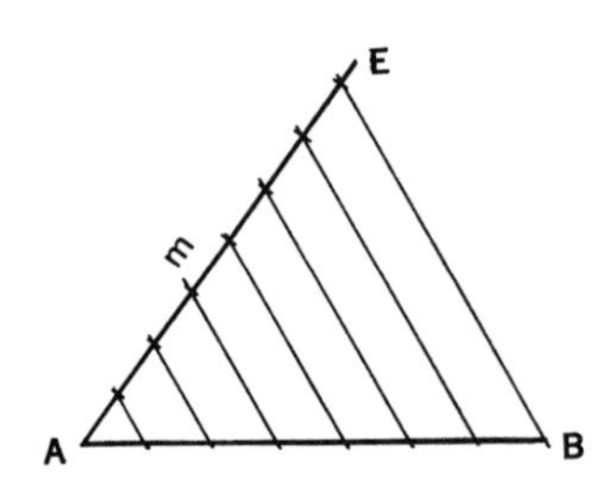

Fig. A-6

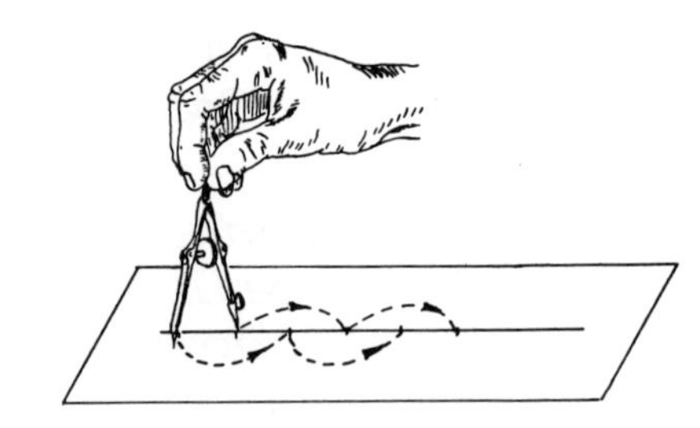

Fig. A-7

2. To Divide a Line Segment into a Given Ratio (Fig. A-8). Let AB represent the given line segment, and let the given ratio be 4:5:7. Draw line m through point A and at any convenient angle with AB. On m lay off distances $AC = 4$ units, $CD = 5$ units, $DE = 7$ units. Draw BE and then draw lines through C and D parallel to BE, cutting AB in points C' and D', which determine the required segments of AB.

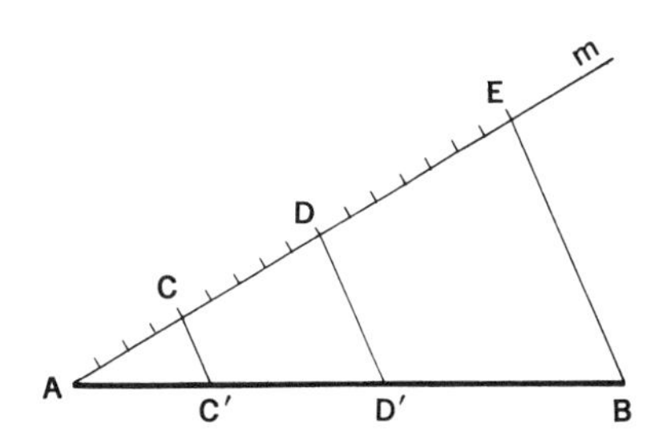

Fig. A-8

3. To Construct a Line Segment Which is the Mean Proportional (Geometric Mean) to Two Given Line Segments (Fig. A-9). Suppose the given segments are m and n. On line AB lay off consecutive segments equal to m and n. Construct a semicircle on the total length $(m + n)$ as a diameter. At the common point of the segments construct a perpendicular to the diameter. Line g is the required mean proportional. The student should prove this by showing that $g^2 = m \times n$.

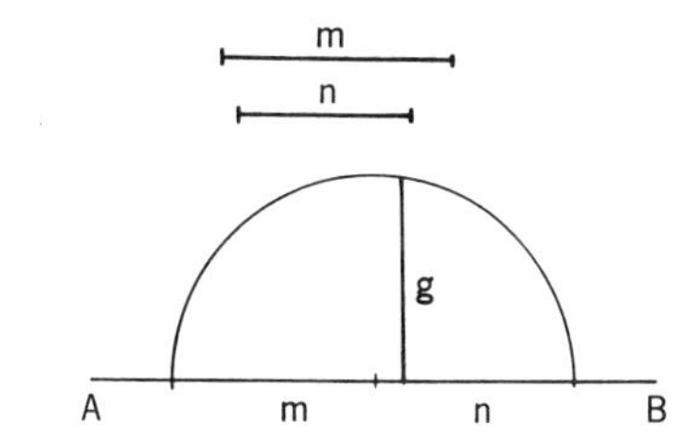

Fig. A-9

4. To Divide a Straight Line in Extreme and Mean Proportion (Fig. A-10). The given line is AB. At point B lay off line BC at 90° to AB and equal to $AB/2$. With C as center and radius CB, draw an arc cutting line AC at point D. With A as center and radius AD, draw an arc cutting line AB at point E, which divides line AB in extreme and mean proportion (i.e., the square on segment AE is equal to the rectangle having sides AB and EB).

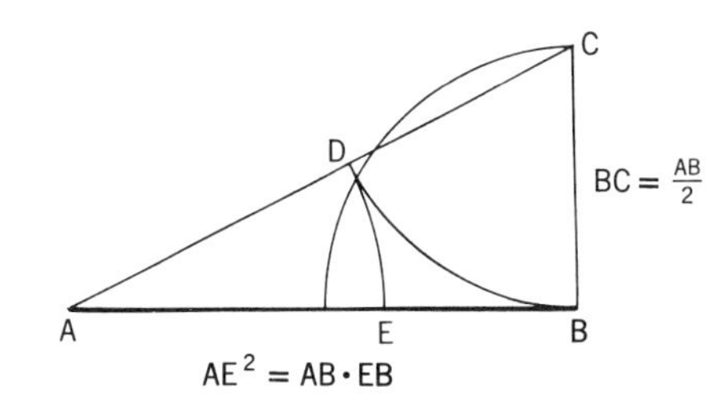

Fig. A-10

B. The Construction of Triangles and Regular Polygons

1. To Construct a Triangle with Known Lengths of the Sides (Fig. A-11). Suppose m, n, and s are the given lengths. With the end points of m as centers and radii equal to n and s, respectively, draw intersecting arcs, locating point A. Join A with the end points of m to complete the construction of the triangle.

Is it possible to construct a triangle with 7, $3\frac{1}{2}$, and 3 inches as the lengths?

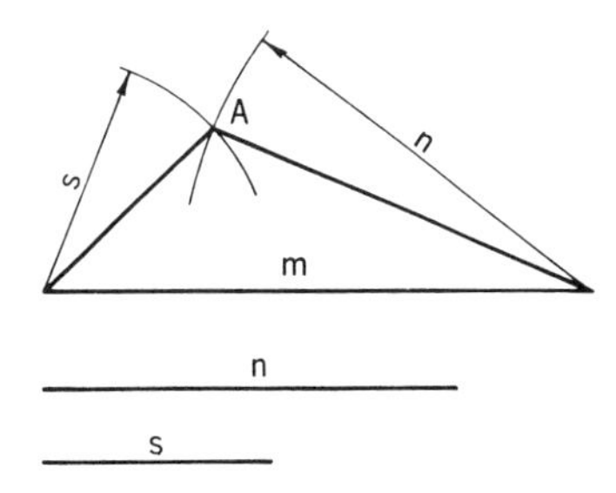

Fig. A-11

2. To Construct a Right Triangle when the Lengths of the Hypotenuse and One Side Are Known (Fig. A-12). Let line m and n represent the given lengths. Construct a semicircle with diameter AB equal to length m. With A as center (could use B) and radius equal to length n, draw an arc cutting the semicircle at point C. Triangle ABC is the required right triangle. Why is this true?

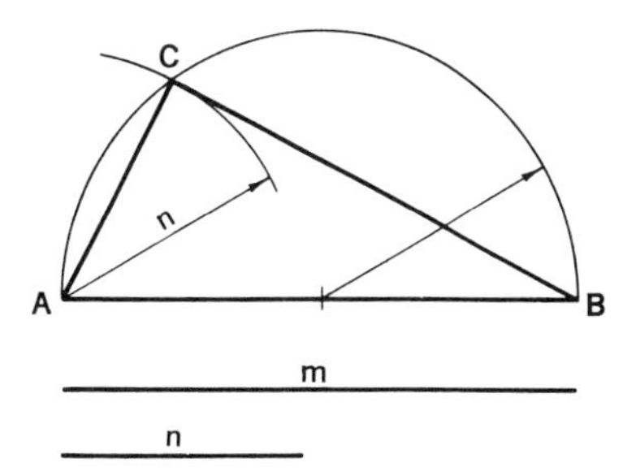

Fig. A-12

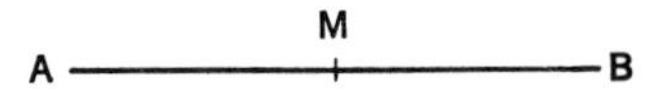

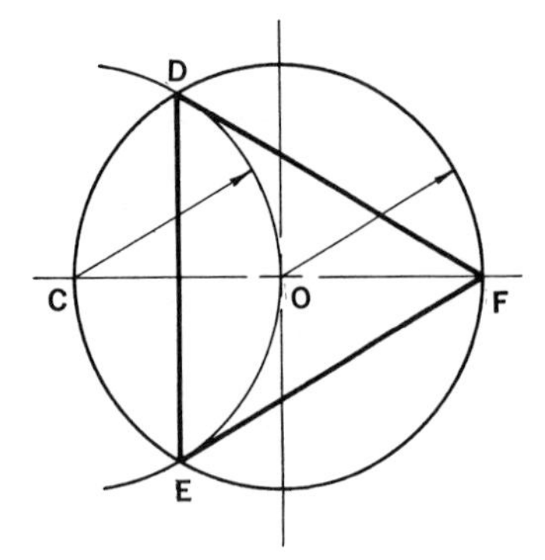

Fig. A-13

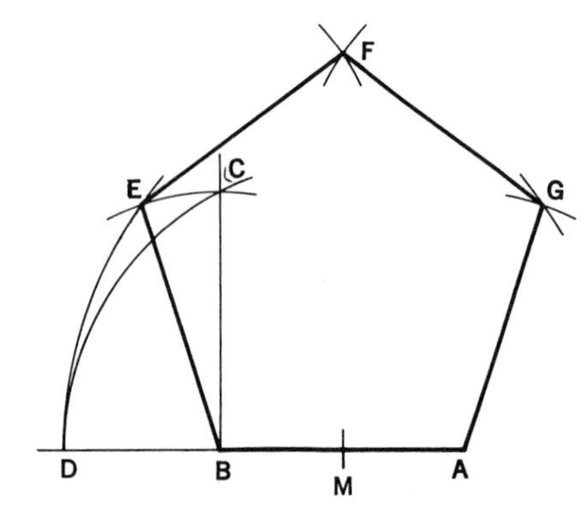

Fig. A-14

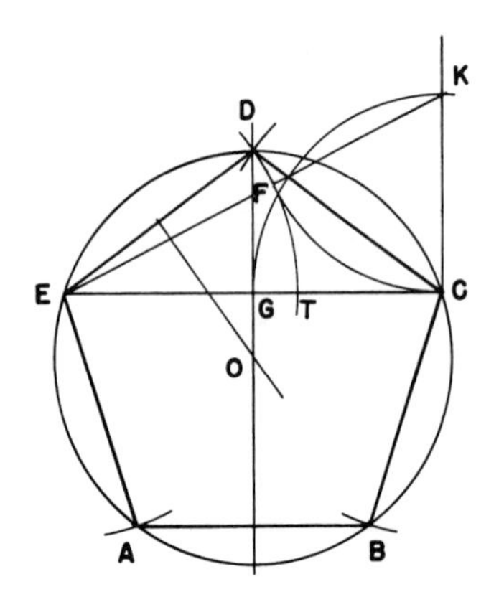

Fig. A-15

3. To Inscribe an Equilateral Triangle within a Circle Having a Given Diameter, AB (Fig. A-13). With center O and radius equal to AM (M is midpoint of AB), draw the circle. With C as center and the same length of radius, draw an arc cutting the circle in points D and E. Join points D, E, and F. The required triangle is DEF.

4. To Construct a Regular Pentagon when the Length AB of the Sides Is Known (Fig. A-14). First construct $BC = AB$, and perpendicular to AB. With M (midpoint of AB) as center and MC as radius, draw an arc cutting AB extended at point D. Now with A as center and AD as radius, draw an arc; and, with B as center and radius BA, draw an intersecting arc to locate point E. Line BE is a side of the regular pentagon. The construction for locating point F and G is fairly obvious. $ABEFG$ is the required pentagon. (The solution of this problem is based upon the fact that the larger segment of a diagonal of the pentagon, when divided in extreme and mean proportion, is the length of a side of the pentagon. Note that point B divides AD in extreme and mean proportion and that AB is the larger segment of AD.)

5. To Construct a Regular Pentagon when the Length EC of a Diagonal Is known (Fig. A-15). First divide EC in extreme and mean proportion ($ET^2 = EC \times TC$). With E and C as centers and radius ET, draw arcs that intersect at point D. Draw the perpendicular bisector of ED and locate point C on the vertical line passing through D. Point O is the center of the circle (radius OE) which circumscribes the pentagon.

With E and C as centers and a radius equal to CD (or ED), draw arcs cutting the circle at points A and B. Draw the necessary lines to form pentagon $ABCDE$.

6. To Inscribe a Hexagon within a Given Circle (Fig. A-16). With A and D as centers and a radius equal to the radius of the circle, draw arcs which intersect the given circle in points B, F, C, and E. The required hexagon is $ABCDEF$. An alternative construction is shown in Fig. A-17.

7. To Construct a Regular Polygon Having n Sides (Fig. A-18). The polygon in this example is a nonagon (nine equal sides) and AB is the given length of each side. With B as center and AB as radius, describe a semicircle, and by trial divide it into nine equal parts. Starting from point T, locate the second division mark, C. Locate point O, the center of the circumscribing circle. (This is easily done by finding the intersection of the perpendicular bisectors of AB and BC.) Draw the circle with center O and radius OA and complete the nonagon.

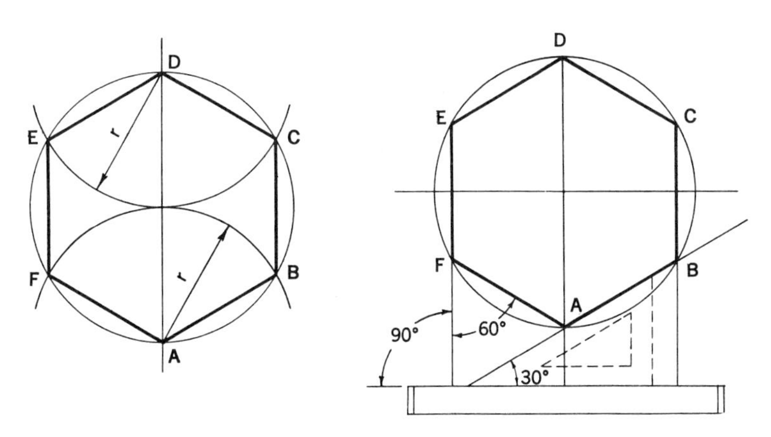

Fig. A-16

Fig. A-17

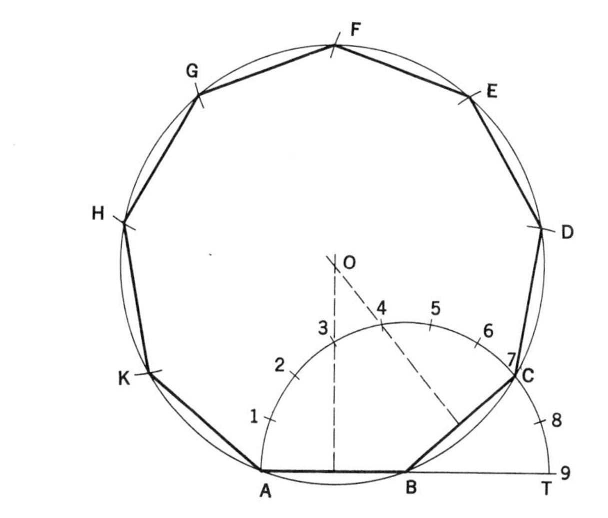

Fig. A-18

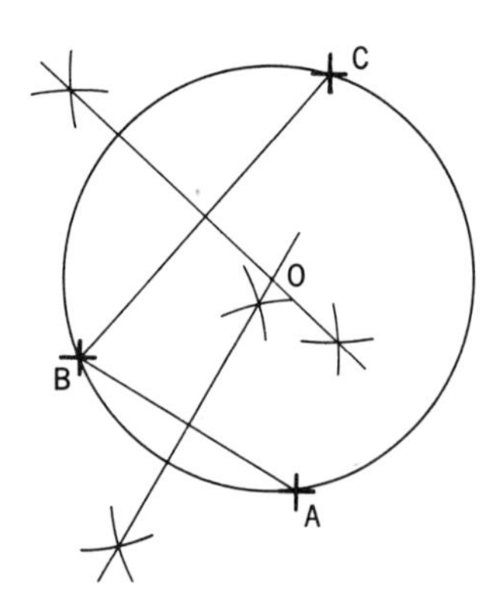

Fig. A-19

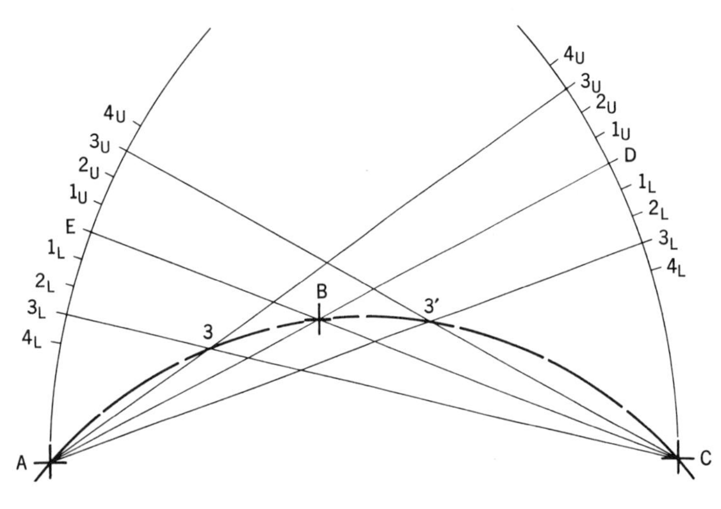

Fig. A-20

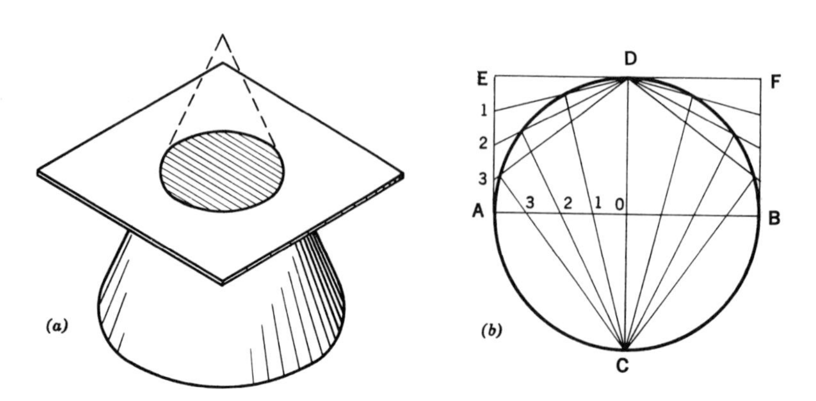

Fig. A-21

C. Circles and Their Tangents

1. To Draw a Circle Through Three Given Points (Figs. A-19 and A-20). Let us assume A, B, and C as the given points. Draw the perpendicular bisectors of lines AB and BC. The intersections of the bisectors is point O, the center of the circle which passes through the three given points.

Now suppose that the center of the circle is inaccessible (Fig. A-20). Again the given points are A, B, and C. Draw arcs with A and C as centers and radius equal to AC. Now draw lines ABD and CBE. On arc AE lay off, from point E, relatively short equal distances 1_U, 2_U, 3_U, etc., above E, and similarly, equal distances 1_L, 2_L, 3_L, etc., below E.

In like manner lay off distances 1_U, 2_U, 3_U, etc., above D and 1_L, 2_L, 3_L, etc., below D. Now draw lines A–3_U and C–3_L. Their intersection is point 3. Similarly locate point 3′ and, in like manner, additional points. The required circle will pass through points A, 3, B, 3′, and such additional points as desired. (The proof is left to the reader.)

2. To Inscribe a Circle in a Given Square, Using the Method of Intersecting Rays (Fig. A-21). If a plane parallel to the base of a right circular cone cuts the cone, the intersection is a circle [Fig. A-21(*a*)]. Points on the circle may be located by finding the intersection of rays C–1, C–2, etc., drawn through point C (Fig. A-21*b*), with the corresponding rays D–1, D–2, etc., drawn through point D. Rectangle $AEFB$ is half of the given square.

3. To Draw Lines Tangent to a Circle, and Passing Through a Given Point, A (Fig. A-22). Locate point M, the midpoint of line OA. With M as center and radius MO, draw an arc cutting the given circle at the points of tangency, T and T'. The tangent lines are AT and AT'.

4. To Draw a Line Tangent to the Arc of a Circle, and Passing Through a Given Point, P (Fig. A-23). First draw through point P secant line PAB. Extend this line to point C such that $PA = PC$. Now find the mean proportional between PA and PB. This is shown as PD. With P as center and radius PD, draw an arc cutting the given arc at point T. Line PT is the required tangent. (The construction shown in Fig. A-23 is based on the fact that $\overline{PT}^2 = PA \times PB$.)

5. To Draw Tangents to Two Given Circles (Fig. A-24). On line O–O', which joins the centers of the circles, lay off from O', distance $O'B = (R - r)$, and $O'C = (R + r)$. With O' as center and radii $O'B$ and $O'C$, draw arcs which intersect the semicircle on O–O' at points D and E, respectively. Draw $O'D$ to intersect circle O' at F, and similarly draw $O'E$ to intersect circle O' at G. Through center O draw OH parallel to $O'F$ and OK parallel to $O'G$. Lines FH and GK are tangent to both circles.

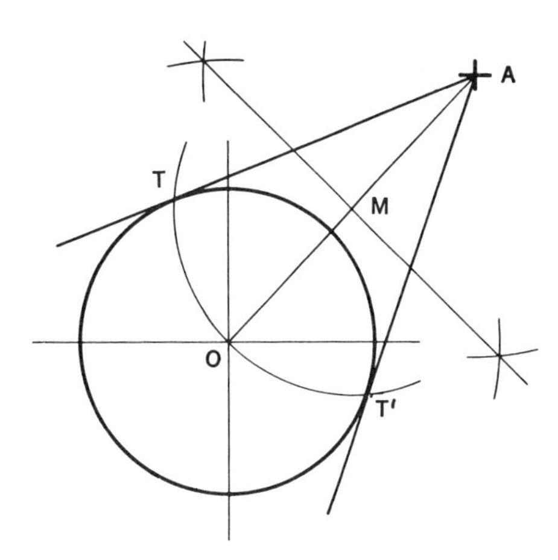

Fig. A-22

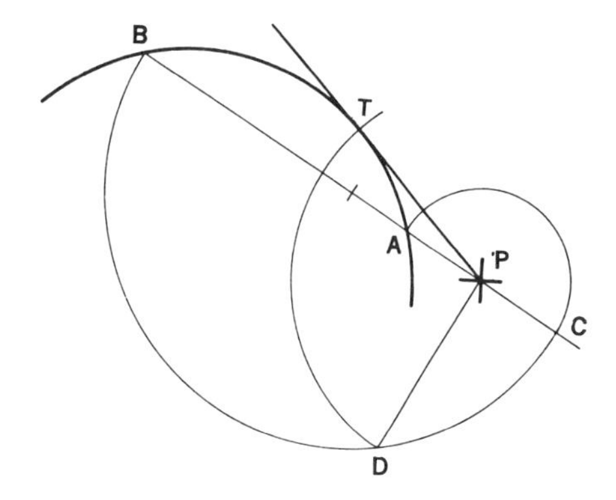

Fig. A-23

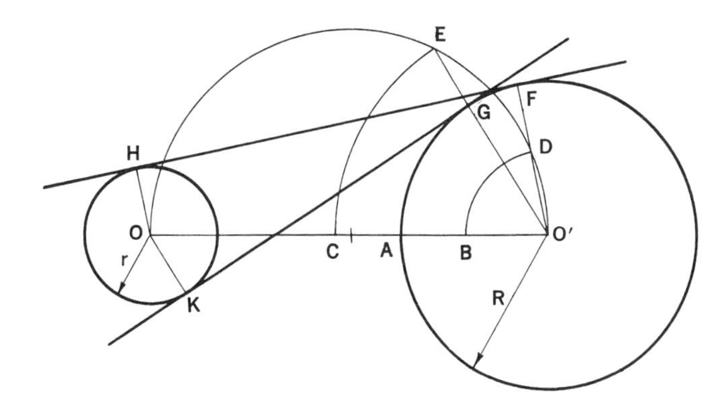

Fig. A-24

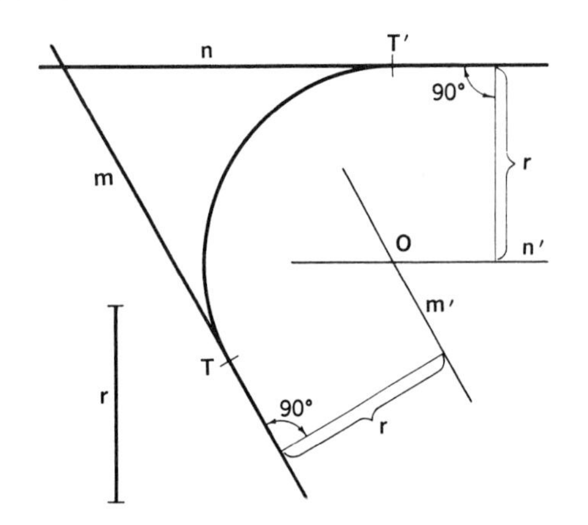

Fig. A-25

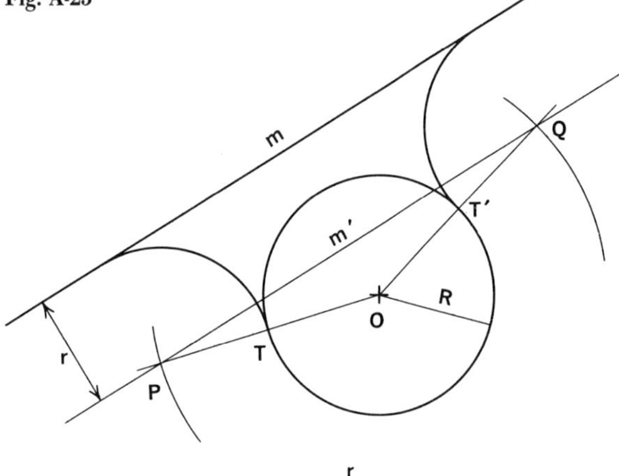

Fig. A-26

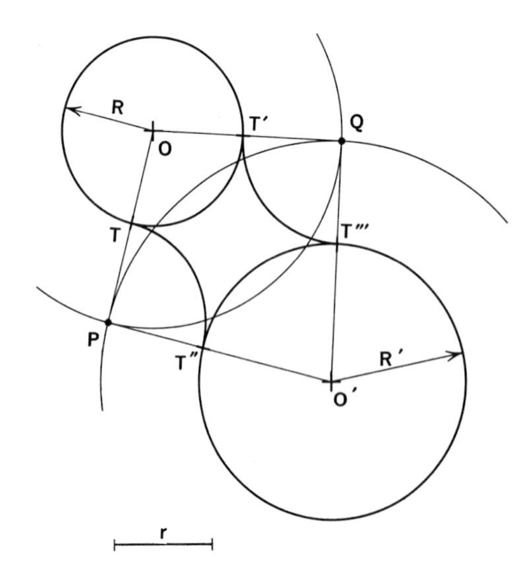

Fig. A-27

6. ***To Draw an Arc of Given Radius, r, Tangent to Two Given Lines m and n* (Fig. A-25).** The center O of the required arc is at the intersection of lines m' and n', which are parallel to lines m and n, respectively, at distance r. The construction is clearly shown in the figure.

7. ***To Draw Arcs of Radius r, Tangent to a Given Line, m, and a Given Circle, O* (Fig. A-26).** With O as center and radius $(R + r)$, draw an arc cutting line m' (line m' is parallel to m at distance r) at points P and Q, which are the centers of the required arcs.

8. ***To Draw Arcs of Radius r, Tangent to Two Given Circles* (Fig. A-27).** With center O and radius $(R + r)$, draw an arc; and, with center O' and radius $(R' + r)$, draw an arc. The intersections of the two arcs are P and Q, which are the centers of the required arcs.

***9. To Rectify a Circular Arc* (*Fig. A-28*).** Arc *Om* in Fig. A-28(*a*) is first divided into a number of short equal segments, in this case 6. These distances, such as 0–1, 1–2, etc., are laid off on tangent *OT*, as 0–1′, 1′–2′, etc. Length 0–6′ is (very nearly) the length of the arc from 0 to 6.

An alternative solution is shown in Fig. A-28(*b*). With *K* as center and *KB* as radius, an arc is drawn to intersect the tangent, *t*, at point *C*. *AC* is a close approximation to the length of the given arc *m*.

***10. To Divide a Circle into Seven Equal Parts by Concentric Circles* (*Fig. A-29*).** First, draw a semicircle on *OA* as diameter. Then divide *OA* into seven equal parts and construct verticals to intersect the semicircle in points 1, 2, etc. Finally, draw the required concentric circles with radii 0–1, 0–2, 0–3, etc.

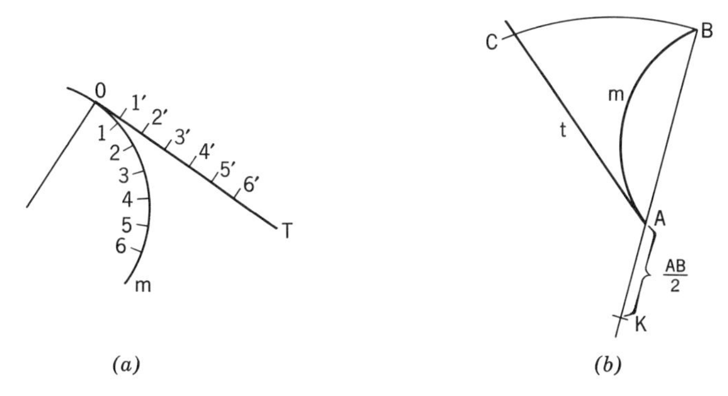

Fig. A-28

D. Conic Sections

The Ellipse. The *ellipse* may be defined as the locus of all coplanar* points, the sum of whose distances from two fixed points (foci) is a constant.

1. *To construct an ellipse when the foci F_1 and F_2 and the constant distant AB are given* (*Fig. A-30*). With F_1 as center and radius *AC* (any portion of *AB*), an arc is drawn. Now with F_2 as center and radius *CB*, an arc is drawn intersecting the first arc in points 1 and 2, which are two points on the ellipse.

This construction is repeated for the location of additional points. For example, with F_1 as center and radius *AD*, an arc is drawn, and with F_2 as center and radius *DB* an intersecting arc is drawn, thus locating two additional points 3 and 4. The smooth curve passing through these points and others (not shown) is the ellipse. The major and minor axes are *AB* and *EF*, respectively.

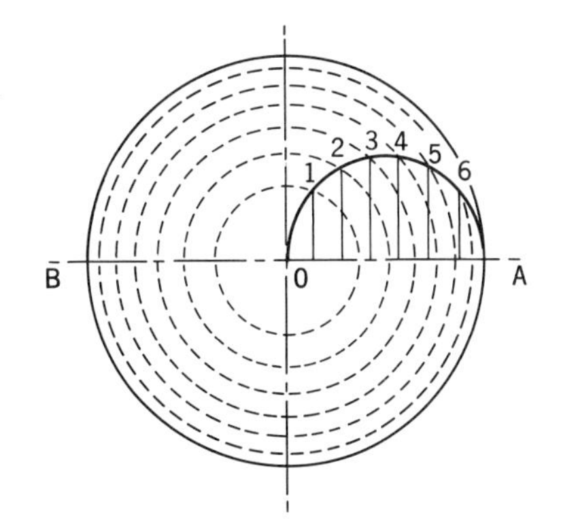

Fig. A-29

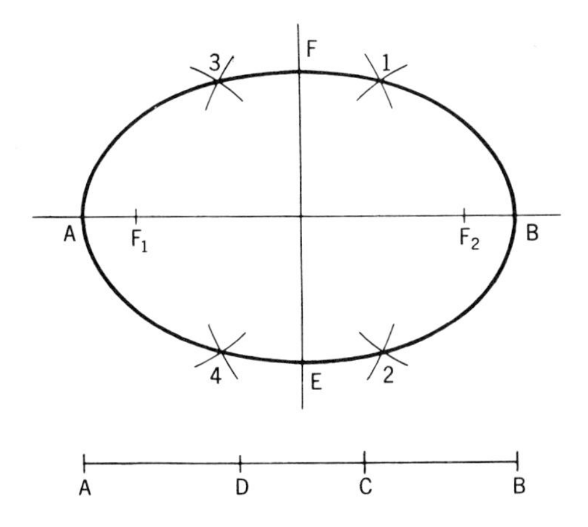

Fig. A-30

* In the same plane.

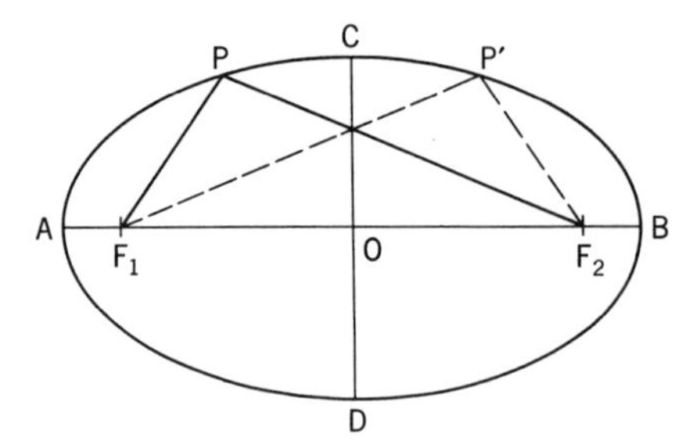

Fig. A-31

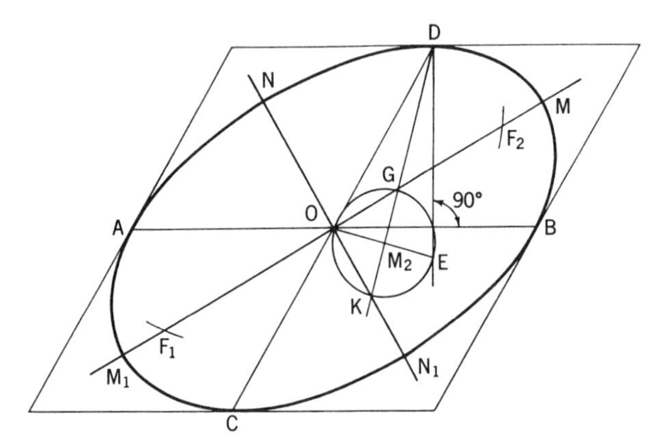

Fig. A-32

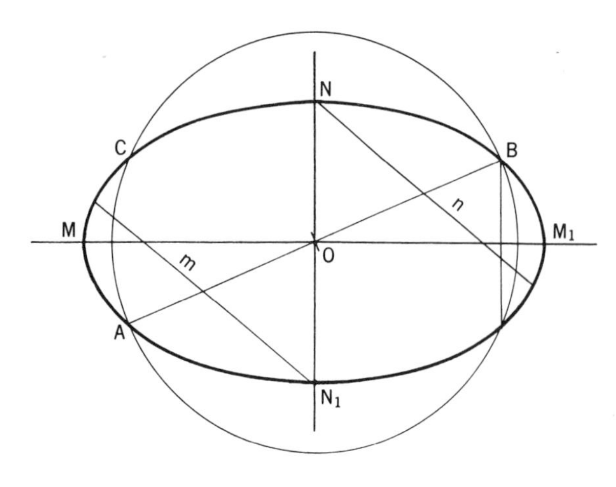

Fig. A-33

2. *To draw an ellipse by the* **pin-and-string method** *when the major axis, AB, and the minor axis, CD, are given* (*Fig. A-31*). With center C and radius equal to OA draw an arc intersecting the major axis at points F_1 and F_2, which are the foci. Now fix the ends of a string at points F_1 and F_2, such that the length of the string is equal to AB. For any point on the ellipse, such as point P or P', the sum of distances PF_1 and PF_2 (or of $P'F_1$ and $P'F_2$) remains equal to the constant length of the string. Therefore, the ellipse is easily drawn by maintaining taut segments of the string, as a pencil (or other marking device) is used to draw the curve.

3. *To draw an ellipse when two conjugate axes are given* (*Fig. A-32*). Let us assume lines AB and CD as the given conjugate axes (each axis is parallel to the tangents to the ellipse at the end points of the other axis). Lines drawn through points C and D, parallel to axis AB, are tangent to the ellipse at these points; and, similarly, lines drawn through points A and B, parallel to axis CD, are tangent to the ellipse at points A and B. The parallelogram formed by the four tangents circumscribes the required ellipse. Once the major and minor axes are located we can establish the positions of the foci, and then describe the ellipse. The following construction is used to locate the major and minor axes of the ellipse. Through D draw DE perpendicular to AB and equal to OB. Draw OE and describe a circle with radius M_2O (or M_2E). Now draw line DM_2 to intersect the circle at points G and K. The minor axis, NN_1, of the ellipse contains line KO, and the major axis, MM_1, contains line GO. Length ON (or ON_1), the semiminor axis, is equal to length DG, and OM (or OM_1), the semimajor axis, is equal to length DK. The foci are easily located. With N as center and radius OM, describe an arc to intersect the major axis at the foci F and F_1. The ellipse may now be constructed.

4. *To determine the major and minor axes, and the foci of a given ellipse* (*Fig. A-33*). First draw two parallels such as m and n. Now draw AB, which bisects lines m and n. Locate O, the midpoint of AB. With O as center and radius OA, draw a circle to intersect the ellipse at points C and B. Through O draw a line NN_1 parallel to line CA and draw line MM_1 perpendicular to CA. Lines NN_1 and MM_1 are the minor and major axes, respectively.

5. *To draw an ellipse by the* **concentric circle** *ɪethod, given the lengths of the major and minor* *ces (Fig. A-34).* Lines AB and CD are the major ɪd minor axes, respectively. Through point O, the ɜnter of the ellipse, draw radial lines such as m to ɪtersect the concentric circles having radii OB and 'C in points E, F, G, and K. Through points E ɪd F draw vertical lines to intersect the horizontals rawn through G and K, in points P and Q, which e two points on the ellipse. Repeat this construc-on for additional points and then draw a smooth ɪrve through these points to form the ellipse.

The tangent, t, at point P passes through point R, hich is the intersection of the tangent t' at point E the major circle and the major axis extended.

6. *To draw an ellipse by the* **trammel method** *hen the major and minor axes are given* (*Fig. A-35*). irst a strip is marked with distance $O'B'$ equal to B and $O'C'$ equal to OC. Now the strip is moved that point B' travels along the minor axis while C' ɪoves along the major axis. For any such position, ' will locate a point on the ellipse.

7. *To draw an ellipse by the use of circular arcs hen the axes are given.* This will result in a close pproximation to a true ellipse (Fig. A-36). Join oints A and C. Lay off distance CD equal to CE vhere $CE = OA - OC$). Now draw the perpen-icular bisector of AD and locate points G and K. Vith G as center and radius GA, describe arc TAT_1. Vith K as center and radius KT, describe arc TCT_2. enter G' and radius $G'B$ are used to draw arc $_2BT_3$; and center K' and radius $K'T_3$ are used to raw arc T_3T_1.

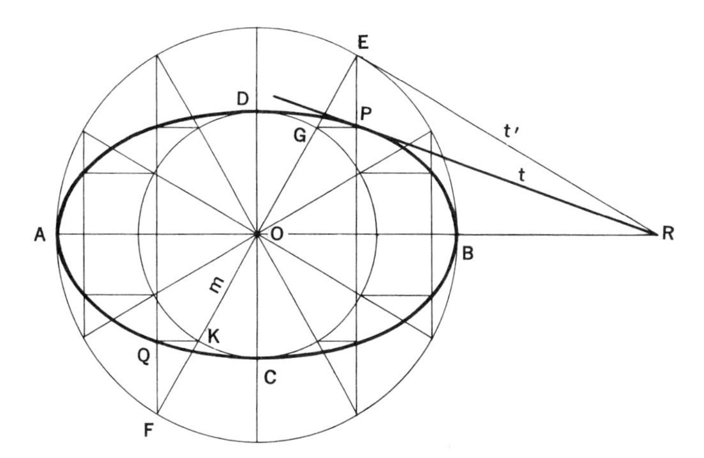

Fig. A-34

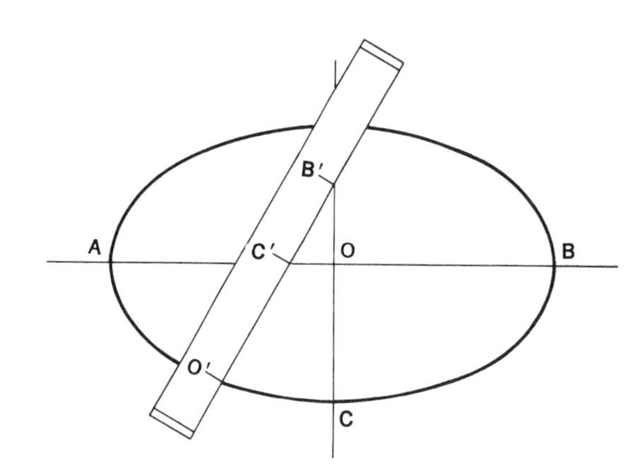

Fig. A-35

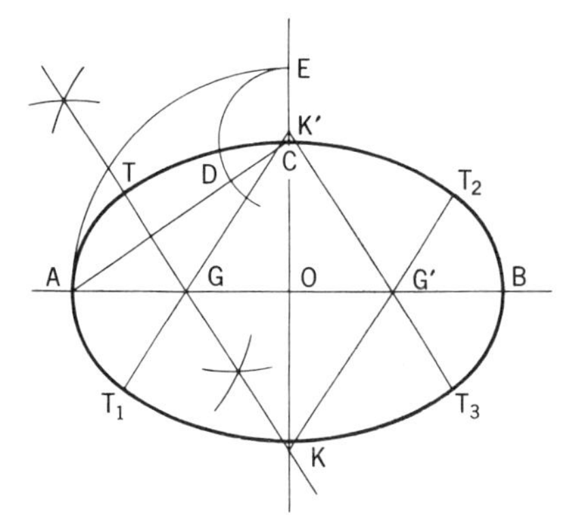

Fig. A-36

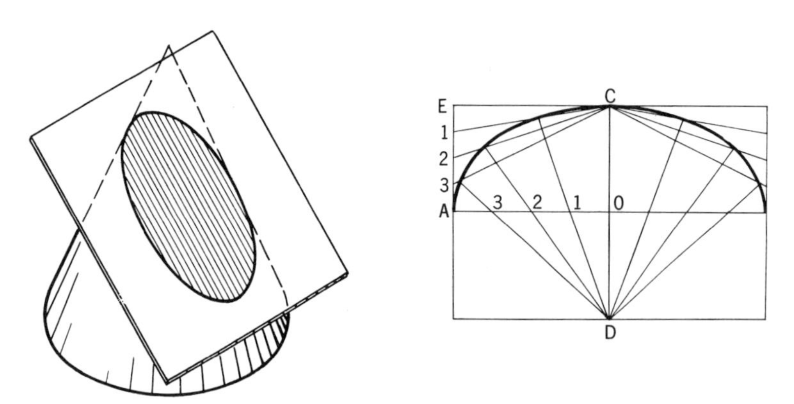

Fig. A-37

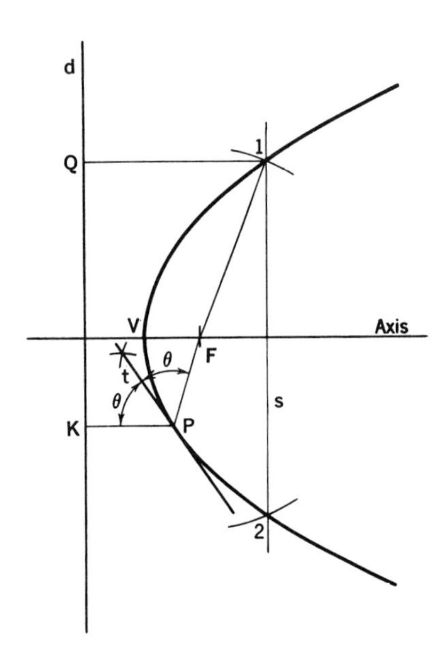

Fig. A-38

8. *To inscribe an ellipse in a given rectang* (*Fig. A-37*). Divide *OA* into a number of equal par (four are shown), and then divide *AE* into the san number of equal parts. Now draw rays *D*–1, *D*– etc., to intersect the corresponding rays *C*–1, *C*– etc., in points which lie on the ellipse. The constru tion shown may be repeated for the other quarte of the rectangle in order to obtain additional poir on the ellipse. The pictorial shows a right circul cone intersected by an inclined plane that cuts the elements of the cone. The intersection is ellipse.

The Parabola. A parabola is a plane curve ar point of which is the same distance from a poi called the focus as it is from a straight line known the directrix.

1. *To locate points on a parabola when the foc F, and the directrix, d, are given* (*Fig. A-38*). Poir such as 1 and 2 are determined by locating t intersection of line *s* (any line parallel to the dire trix) and an arc having center *F* and a radius equ to the distance between the parallel lines *d* and Now it is quite apparent that points 1 and 2 are t same distance from *F*, the focus, as they are from the directrix. The tangent, *t*, at point *P* bisects t angle *KPF*.

2. *To determine the axis, focus, and directrix of a iven parabola (Fig. A-39).* The axis is located in the ollowing manner. Draw two parallel chords such as ı and n. The line t joining the midpoints of these hords is parallel to the axis. Now introduce a line ıch as s perpendicular to line t. The required axis the perpendicular bisector of line s. The focus, F, located by making angle FTC equal to angle CTA. oint C is the intersection of the axis with the perendicular to tangent line k at point T. The direcix, d, is perpendicular to the axis and at a distance om V equal to VF, that is, $VB = VF$.

3. *To construct a parabola, given the axis, vertex V, nd a point P through which the parabola passes 'ig. A-40).* First draw rectangle $PABC$. Now divide 'P and CV into the same number of equal parts. ıtroduce lines parallel to the axis and passing ırough points 1, 2, and 3 on side VC. Draw rays '–1, V–2, and V–3. Finally locate the points in 'hich the parallels intersect the corresponding rays, e., the parallel through point 1 intersects ray V–1, tc. The curve through the points thus located is the arabola. The pictorial shows a right circular cone ıtersected by a plane parallel to an element of the one. The intersection is a parabola.

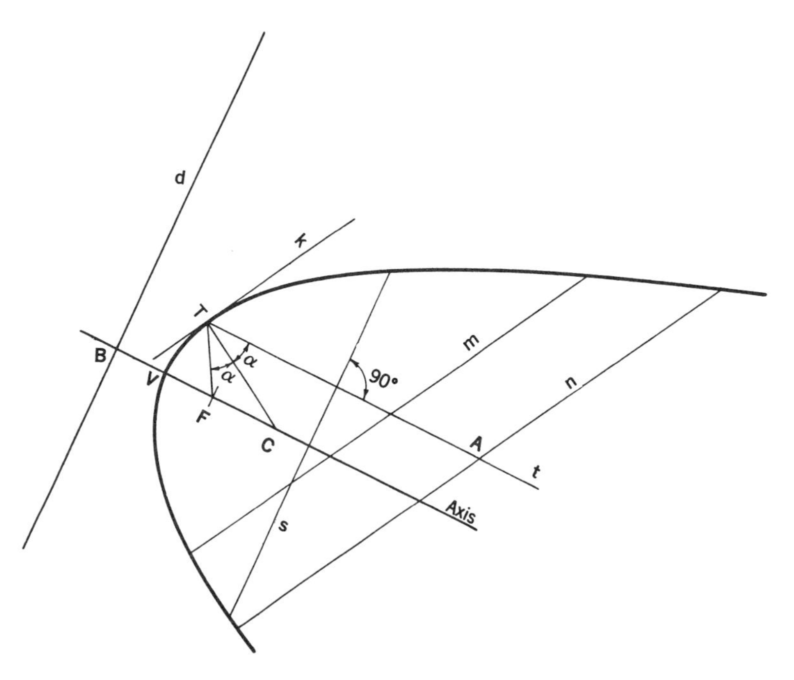

Fig. A-39

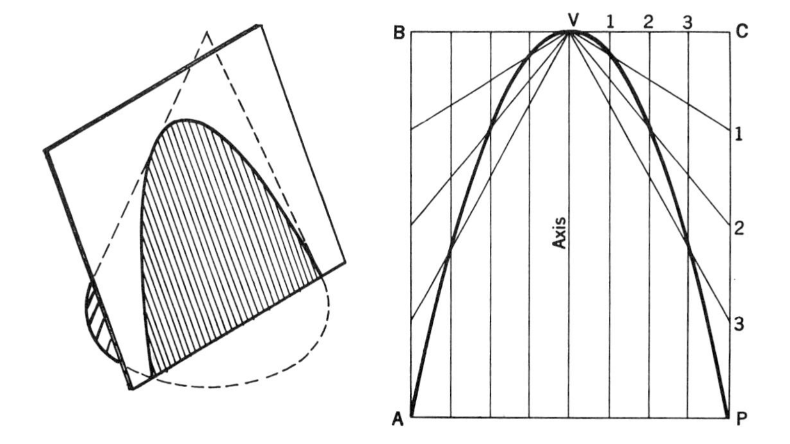

Fig. A-40

The Hyperbola. The *hyperbola* may be defined a the locus of all coplanar points the differences o whose distances from two fixed points (foci) is constant.

1. *To construct a hyperbola when the foci, F and F_2, and the constant distance, AB, are give (Fig. A-41).* With F_1 as center and a radius greate than F_1B, an arc is drawn. Now with F_2 as cente and a radius which is equal to the difference betwee the first radius and length AB, an arc is drawn t intersect the first arc in points P and Q, which ar two points on the hyperbola. It is clearly seen tha $F_1P - F_2P = AB$ and that $F_1Q - F_2Q = AB$. Add tional points may be found in a similar manner. Th smooth curve passing through the points is th hyperbola. It should be noted that the curve has tw branches which are symmetrical with respect to th axes. The asymptotes pass through the center O an are tangent to the curve at infinity. They are locate by joining point O the center of the hyperbola wit points K and K'. These points are found by locatin the intersections of the verticals through points and B with the circle of radius OF_1. The tangent, at point P bisects the angle F_2PF_1.

2. *To construct a rectangular hyperbola (asymp totes are at right angles), given the asymptotes and n, and one point P on the curve (Fig. A-42* Draw lines k and t through point P, respectivel parallel to n and m. Select any point Q on line and then draw line OQ. Locate point R, the inte section of OQ and t. Draw a horizontal line throug point R and a vertical line through point Q. Th intersection of these lines is point S, a point on th hyperbola. In a similar manner additional points ar located.

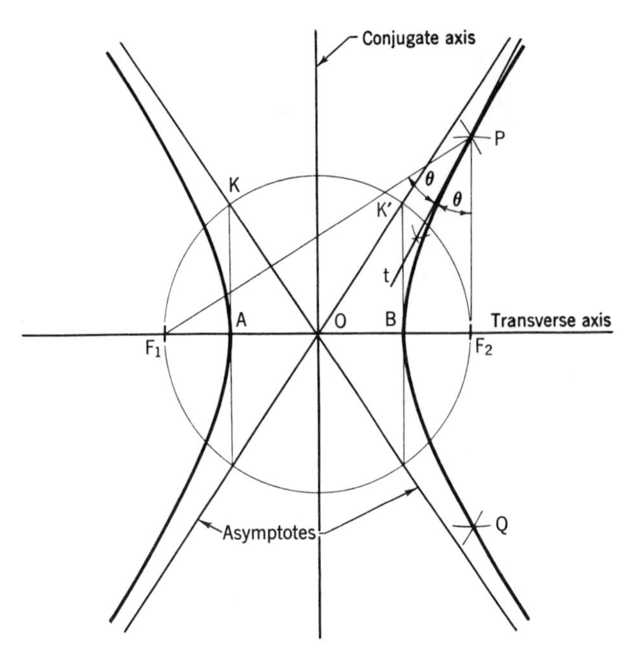

Fig. A-41

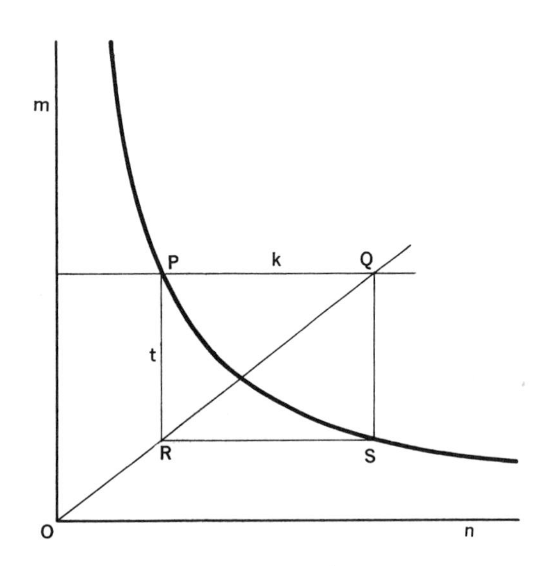

Fig. A-42

3. To construct a hyperbola, given the transverse axis AB, and a point P on the curve (Fig. A-43). First construct the rectangle *PCDE*. Now divide side *EP* into a number of equal parts (four are shown) and the right half of side *CP* into the same number of equal parts. Find the intersection of rays *A*–1, *A*–2, etc., with the corresponding rays *B*–1, *B*–2, etc. Repeat the procedure for the left half of the rectangle. The smooth curve which passes through the points thus located is one branch of the hyperbola. The other branch may be determined in a similar manner. The pictorial shows a right circular cone intersected by a plane parallel to the axis of the cone. The intersection is a hyperbola (one branch shown).

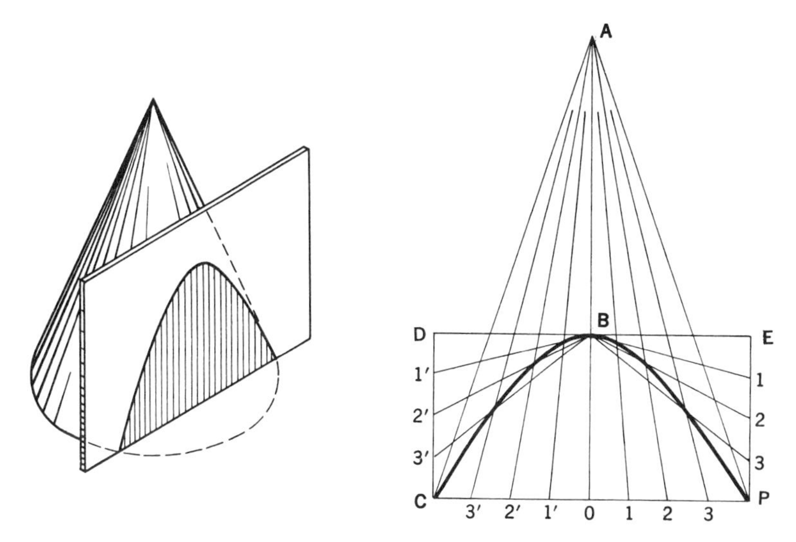

Fig. A-43

Pascal and His Theorem. In 1640, at the age of 16, Pascal discovered the relationship that "the opposite sides of a hexagon, which is inscribed in a conic, intersect in points that lie on one line." For example, in Fig. A-44, the sides of the inscribed hexagon are 1–2, 2–3, 3–4, 4–5, 5–6, and 6–1. Opposite sides 1–2 and 4–5 intersect at point *L*; opposite sides 2–3 and 5–6 intersect at point *M*; and opposite sides 3–4 and 6–1 intersect at point *N*. The line which passes through points *L*, *M*, and *N* is known as Pascal's line.

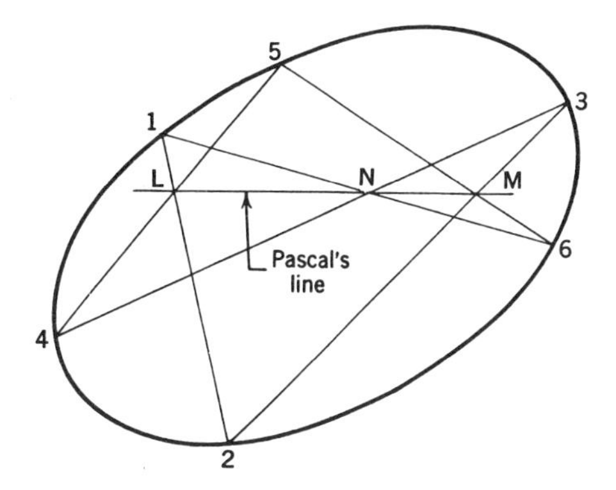

Fig. A-44

1. To locate a sixth point on a conic when five points are known (Fig. A-45). Let us suppose that points 1, 2, 3, 4, and 5 are known and that it is required to locate a sixth point, *K*. Basing our construction upon Pascal's theorem, we can first establish point *L* which is the intersection of the opposite sides 1–2 and 4–5 of the inscribed hexagon 1, 2, 3, 4, 5, *K*. Now we know that another pair of opposite sides is 2–3 and 5–*K*. Therefore, we may draw any line through point 5 to intersect side 2–3 in point *M*. We know that point *K* is somewhere on line 5–*M*. We also know that Pascal's line passes through points *L* and *M*. The third pair of opposite sides is 3–4 and *K*–1. If we join points 3 and 4, then line 3–4 must intersect Pascal's line in a point, *N*, through which side *K*–1 must pass. It is now seen that lines 1–*N* and 5–*M* must intersect at point *K*, which is a sixth point on the conic. Additional points may be located by drawing other lines through point 5 to intersect side 2–3 in a new *M*-point, and then repeating the construction described above.

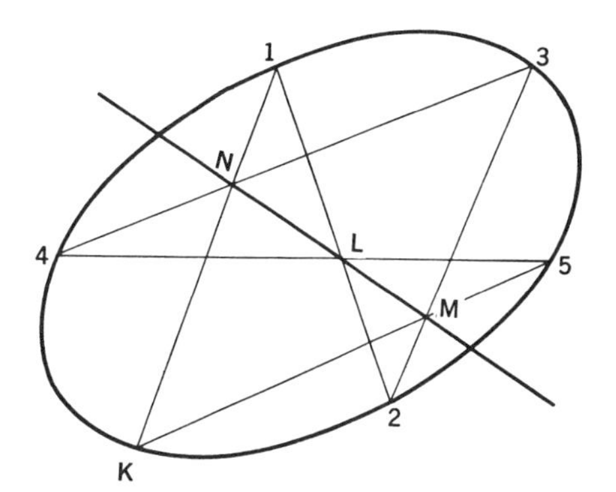

Fig. A-45

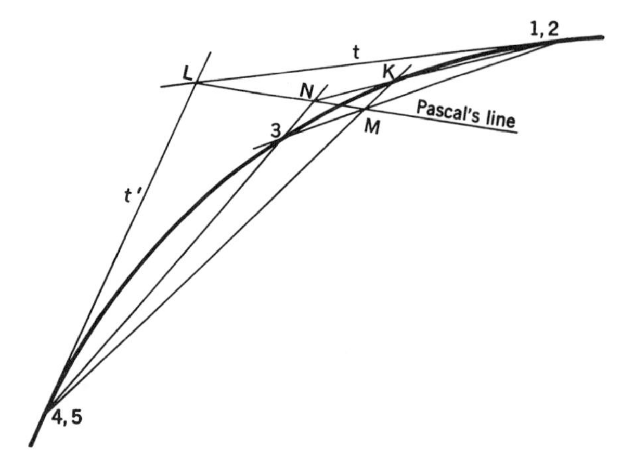

Fig. A-46

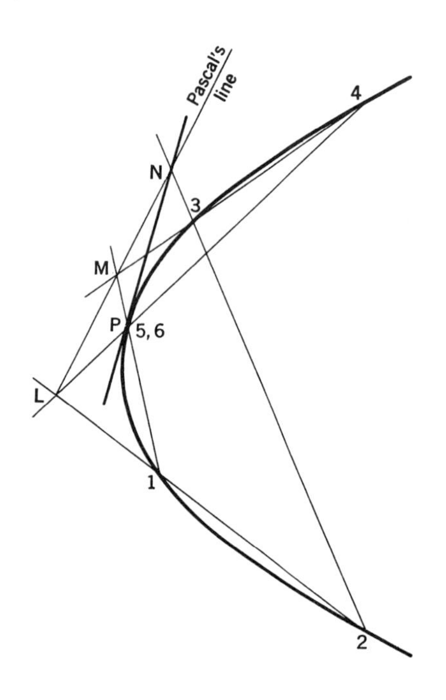

Fig. A-47

2. *To locate points on a conic that passes through a given point and is tangent to given lines at two points (Fig. A-46).* It is assumed that the conic passes through the given point 3 and tangent to lines t and t' at points 2 and 4, respectively. We recall (Fig. A-45) that when five points on a conic are known it is possible to locate a sixth point. How can we reduce the problem shown in Fig. A-46 to the previous one? How shall we establish five known points on the conic, when apparently only three are given?

If point 5, for example (Fig. A-45), were moved along the curve until it coincided with point 4, chord 4–5 would become a tangent to the curve at point 4. Therefore, we can show point 5, coincident with point 4 in Fig. A-46. Similarly, points 2 and 1 are coincident.

Now the intersection of opposite sides 1–2(t) and 4–5(t') of the inscribed hexagon 1, 2, 3, 4, 5, K (K is a sixth point on the conic) is point L. The intersection of opposite sides, 2–3 and 5–K, is point M. If we take M as any point on side 2–3, we know that point K is somewhere on line 5–M. The line joining points L and M is a Pascal line. We know that the opposite sides, 3–4 and K–1, must meet on the Pascal line; therefore, the intersection of side 3–4 with the Pascal line is point N, through which side 1–K must pass. Therefore, the intersection of lines 1–N and 5–M is point K, another point on the conic. Additional points may be located in a similar manner.

3. *To construct a tangent to a given conic at a point of the conic (Fig. A-47).* Let us construct the tangent to the conic at point P. Inscribe a hexagon 1, 2, 3, 4, 5, 6 such that side 5–6 will be the tangent. Two points L and M of the Pascal line are located by finding the intersection of sides 1–2 and 4–5, and of sides 3–4 and 6–1, respectively. Sides 2–3 and 5–6 must meet on the Pascal line, at point N, which is located by finding the intersection of side 2–3 with the Pascal line. The required tangent is line NP.

Brianchon's Theorem. This theorem, which is most useful in locating tangents to a conic, states that "the three lines joining the three pairs of opposite vertices of a hexagon circumscribed about a conic meet in a point."

In Fig. A-48, line *AB* joins the pair of opposite vertices determined by tangents 1 and 2, and 4 and 5. Similarly line *CD* joins the pair of opposite vertices determined by tangents 2 and 3, and 5 and 6. Lines *AB* and *CD* intersect in Brianchon's point, *P*. Line *EF*, which joins the remaining pair of opposite vertices, also passes through point *P*.

1. *To determine a sixth tangent to a conic when five tangents (sides of the circumscribed hexagon) are given (Fig. A-49).* Suppose that tangents 1, 2, 3, 4, and 5 are given. It is required to locate a sixth side of a circumscribed hexagon. Line *AB*, which joins one pair of opposite vertices, is easily determined. Now through point *C* (the intersection of tangents 2 and 3) a line is drawn intersecting *AB* at point *P* (Brianchon's point). Line *CP* intersects tangent 5 at point *D*. Line *FP* intersects tangent 1 at point *E*. Line *DE* is the required tangent or sixth side of the circumscribed hexagon.

2. *To determine the point of contact of a tangent to a conic (Fig. A-50).* Suppose we wish to locate the point of tangency of tangent 1. If tangent 2 approaches tangent 1 as a limiting position, the point of intersection of tangents 1 and 2 approaches the contact point of tangent 1 with the conic. Let us denote this point as point 6. Now we may make use of Brianchon's point to locate point 6. The intersection of lines *l* and *m* determines Brianchon's point, *P*. Line *n* passes through point *P* and the intersection of tangents 3 and 4. Point 6, the required point of contact of tangent 1 with the conic, is at the intersection of line *n* with the tangent line 1.

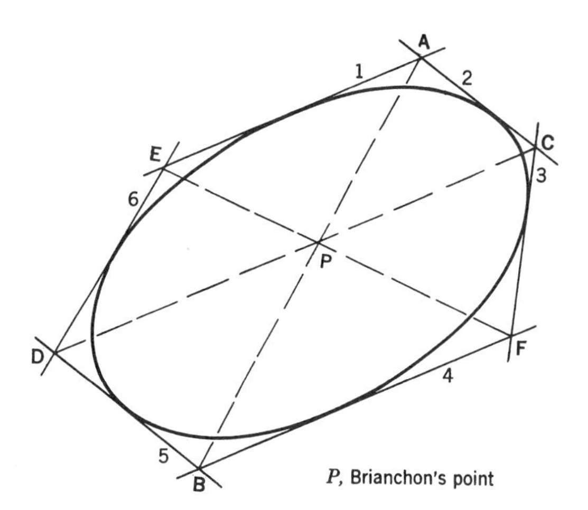

Fig. A-48

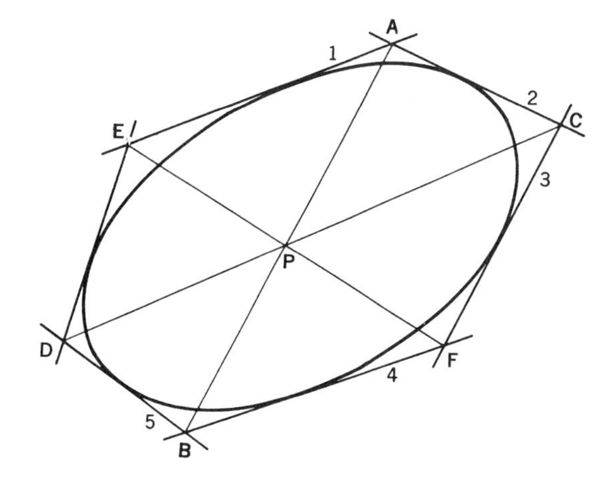

Fig. A-49

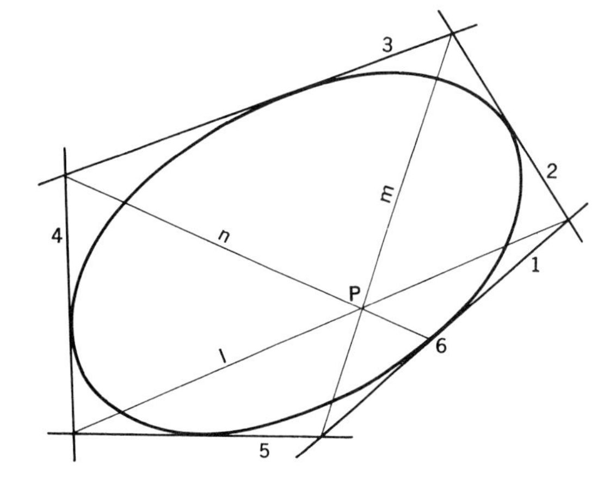

Fig. A-50

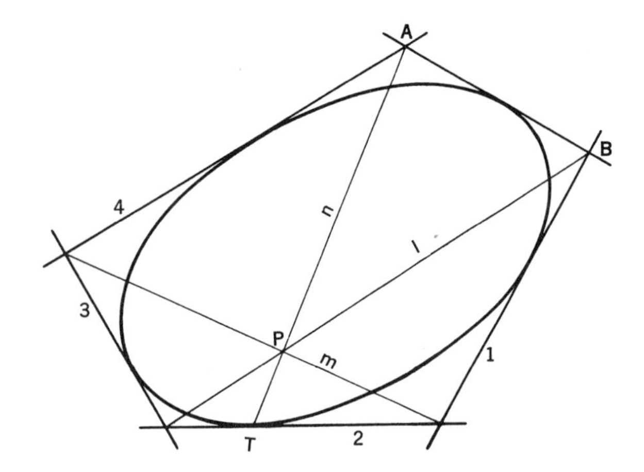

Fig. A-51

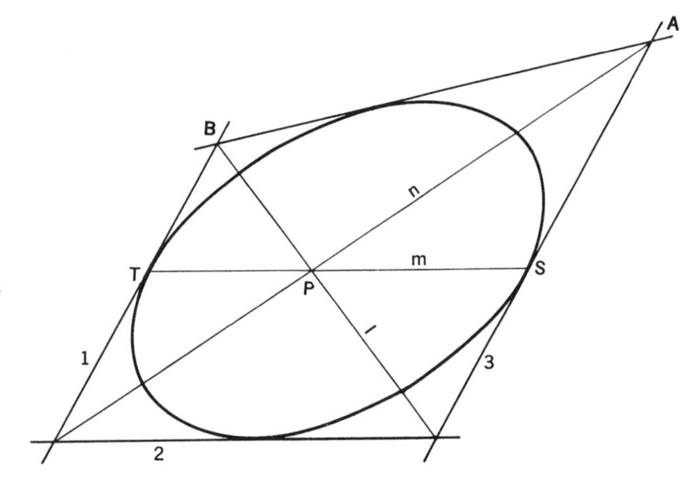

Fig. A-52

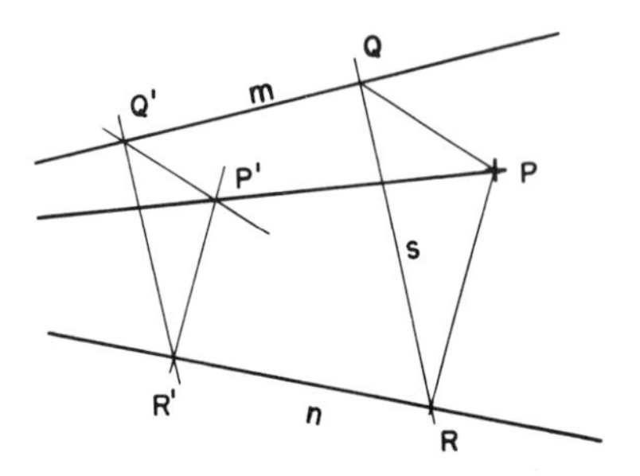

Fig. A-53

3. *To determine additional tangents to a conic when four tangents and the point of contact on one of them are known* (*Fig. A-51*). The four given tangents are 1, 2, 3, and 4, and the point of contact on tangent 2 is point T. Line m is determined by joining the pair of opposite vertices that is given. Tangent 2 is actually two tangents that intersect at point T. Now through point T a line n is drawn to intersect line m at point P (a Brianchon's point), and tangent 4 at point A. Line l is drawn through the common point of tangents 2 and 3 and point P to intersect tangent 1 at point B. Line AB is an additional tangent to the conic. In a similar manner more tangents may be determined.

4. *To determine additional tangents to a conic when three tangents and the points of contact on two of them are known* (*Fig. A-52*). The three known tangents are 1, 2, and 3, and the points of contact, points S and T. Line m joins points S and T. Now through the intersection of tangents 1 and 2 draw a line n to intersect line m at point P (a Brianchon's point) and to intersect tangent 3 at point A. Finally draw line l through the common point of tangents 2 and 3, and point P to intersect tangent 1 at point B. Line AB is an additional tangent to the conic.

Additional Useful Geometric Constructions.

1. *To draw a line through a given point, P, and the inaccessible copoint of given lines m and n* (*Fig. A-53*). Draw a line such as s to intersect m and n at points Q and R, respectively. Form triangle PQR. Now select a point such as Q' on line m and draw through Q' lines respectively parallel to QP and QR. Through R' draw a line parallel to RP and form the triangle $P'Q'R'$. The line, PP', is the solution.

2. *To draw a line perpendicular to a given line AB and through the inaccessible copoint of given lines m and n that pass through points A and B, respectively (Fig. A-54).* Through points *B* and *A* draw lines respectively perpendicular to *m* and *n*, and intersecting at point *P*. Now construct the required line through point *P* perpendicular to line *AB*.

3. *To divide a given quadrilateral ABCD into two equal areas by a line which passes through one of the corners, A (Fig. A-55).* Draw line *BD* and locate its midpoint, *M*. Draw a line through point *M* parallel to diagonal *AC* to intersect side *BC* at point *E*. Line *AE* divides the quadrilateral into equal areas *AEB* and *AECD*.

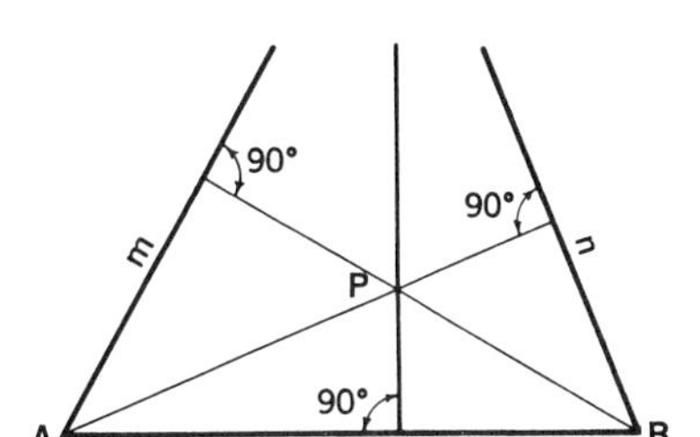

Fig. A-54

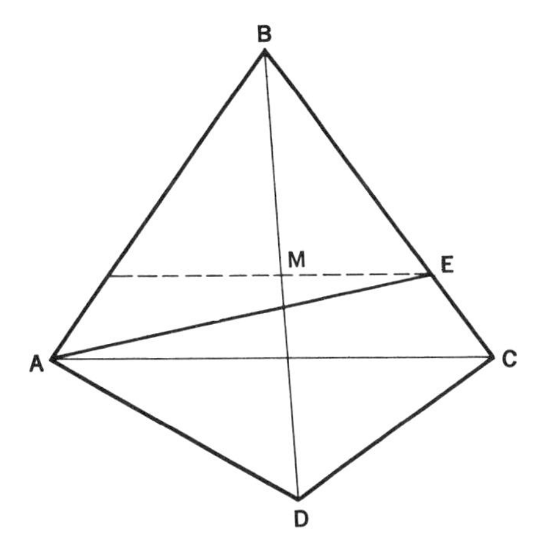

Fig. A-55

DIMENSIONING PRACTICES AND TECHNIQUES

The following practices have been strongly recommended:

(*a*) Dimensions should be so arranged that it will not be necessary to make calculations, scale the drawing, or assume any dimension in order to produce the part.

(*b*) Each dimension should be stated clearly and with no ambiguity of meaning.

(*c*) Dimension between points, lines, or surfaces that have a necessary and specific relation to each other or which control the location of components or of mating parts.

(*d*) Dimensions should not be duplicated. Only those required to produce the part should be given.

(*e*) Dimension lines should be thin full lines, broken where the dimensions are placed, except for numerals in two lines (see Fig. A-56); and for structural drawings where the figures are placed above the line. (See Fig. A-57.)

(*a*)

Break a dimension line for numerals in single line.

(*b*) 0.375 ± 0.001

(*c*)

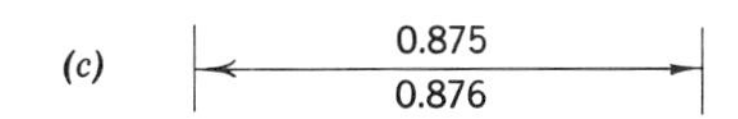

Do not break a dimension line for numerals in two lines.

Fig. A-56

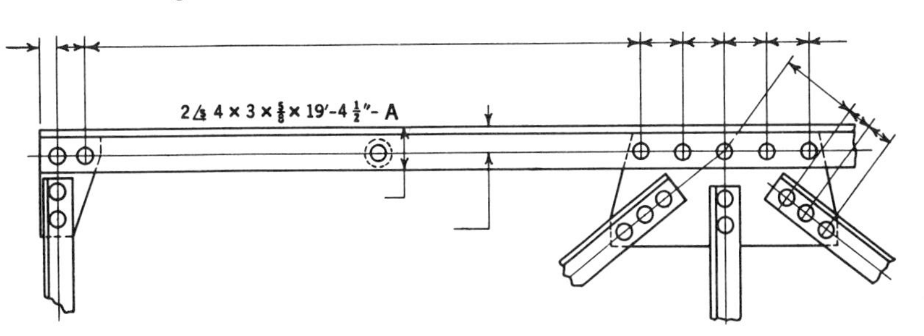

Fig. A-57

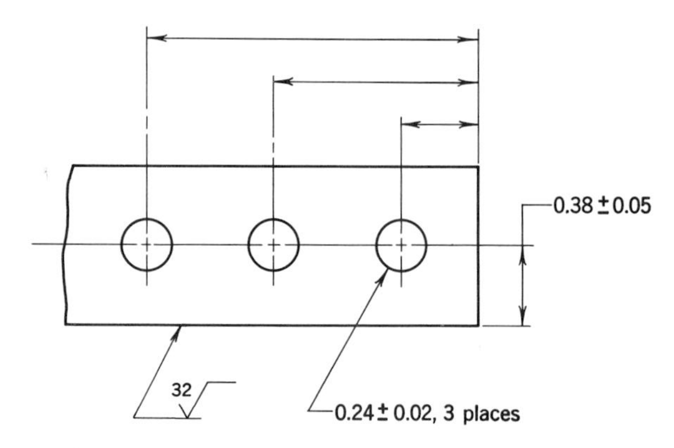

Fig. A-58

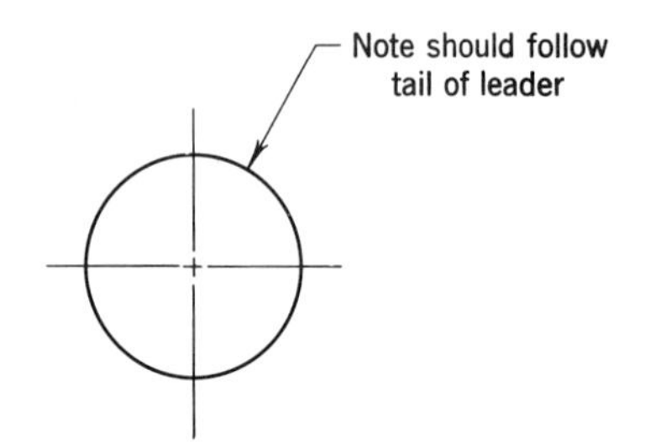

Fig. A-59

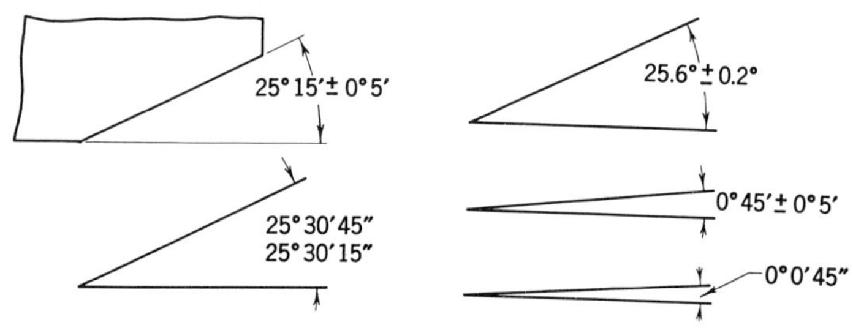

Fig. A-60

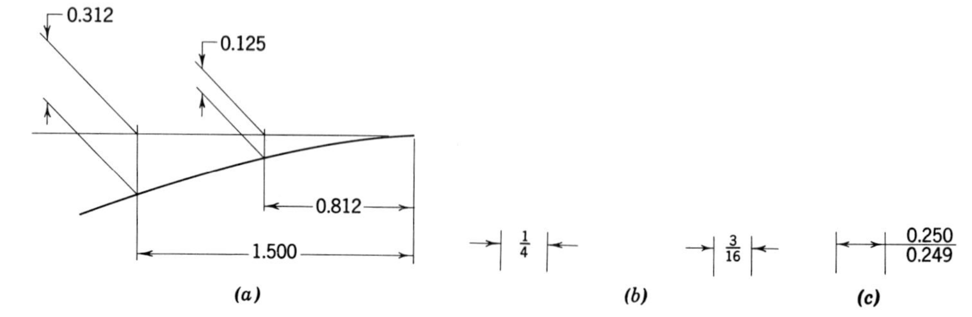

Fig. A-61

(*f*) *Dimension lines* should be terminated by arrowheads whose length is approximately three times the spread.

(*g*) *Extension lines* are used to indicate the distance measured, when the dimension is placed outside of the view. Extension lines are thin full lines extending a short distance beyond the arrowhead and not touching the feature dimensioned. (See Fig. A-58.)

(*h*) *Leaders* are thin straight lines terminated by arrowheads. The arrowhead points to an element of the part, whereas the plain end is followed or preceded by a note which describes the part. (See Figs. A-58 and A-59.)

(*i*) *Angular dimensions* are expressed in degrees (°), minutes (′), and seconds (″); i.e., 22°15′30″. Additional examples are shown in Fig. A-60.

(*j*) *Dimensioning in crowded spaces* is best illustrated in Fig. A-61.

Placing Dimensions

Dimension lines and their corresponding numbers are placed so that they may be read from the bottom or right-hand edge of the drawing sheet. The *aligned* system for orienting the dimensions is shown in Fig. A-62. In dimensioning circles, the shaded area should be avoided. (See Fig. A-63.) The *unidirectional* system is shown in Fig. A-64. Note that common fractions are written with the fraction bar parallel to the bottom of the drawing. The latter system was started in the airframe and automobile industries because it was found that, especially for large drawings, it was advantageous in marking and reading the drawing. This system, in fact, has been adopted by many companies. In both systems, dimensions and notes shown with leaders should be aligned with the bottom of the drawing.

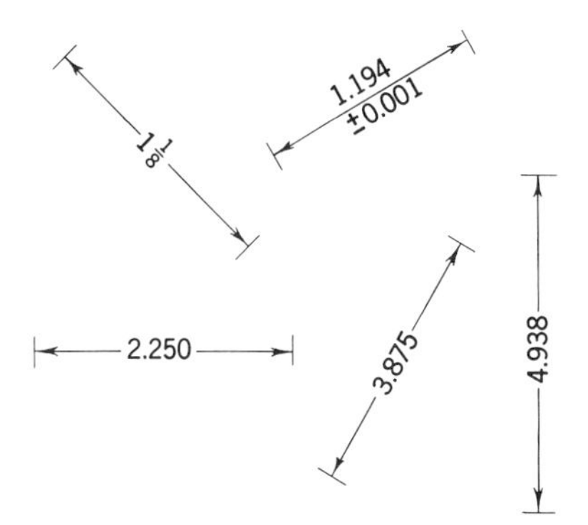

Fig. A-62

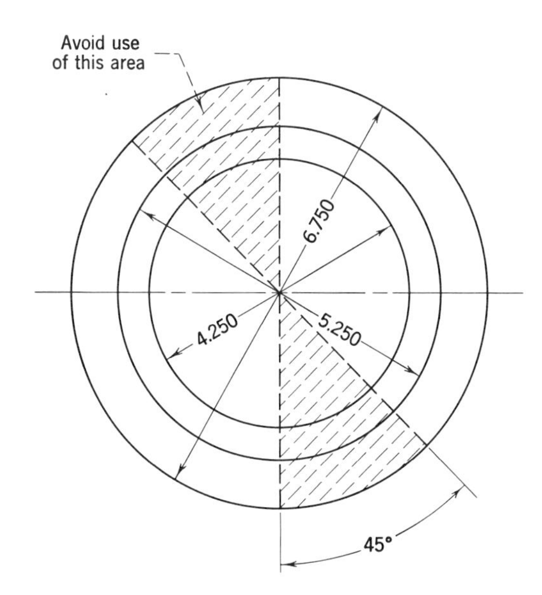

Fig. A-63

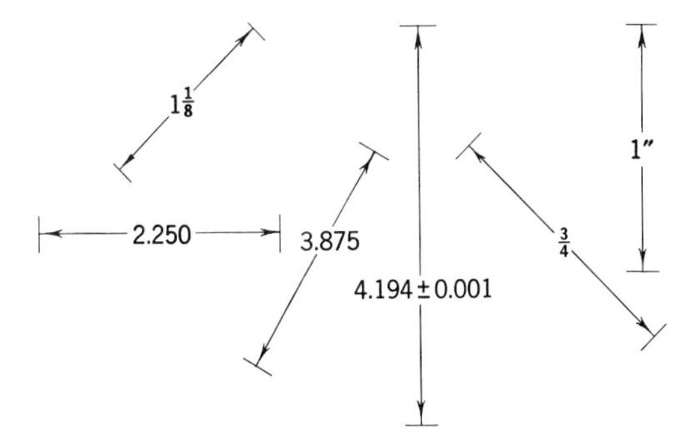

Fig. A-64

Dimensions should preferably be placed outside the views. Dimensions, however, are placed within the view when directness of application is justified. Where dimensions are placed within a sectioned area, the cross-hatching should not pass through the numerals. (See Fig. A-65.)

(*a*) *Parallel dimension lines* should be placed at least $\frac{1}{4}$-inch apart; no dimension line should be closer to the outline of the view than $\frac{3}{8}$ inch. The numerals should be staggered to facilitate reading. (See Fig. A-66.)

(*b*) *Center lines, extension lines,* and *lines of the views should not be used as dimension lines.*

(*c*) *Dimensions should generally be placed between views,* and closest to the associated view.

(*d*) *Specific features* of a part should be dimensioned in the view which most clearly describes the features. For example, in Fig. A-67 the holes are best described in the top view.

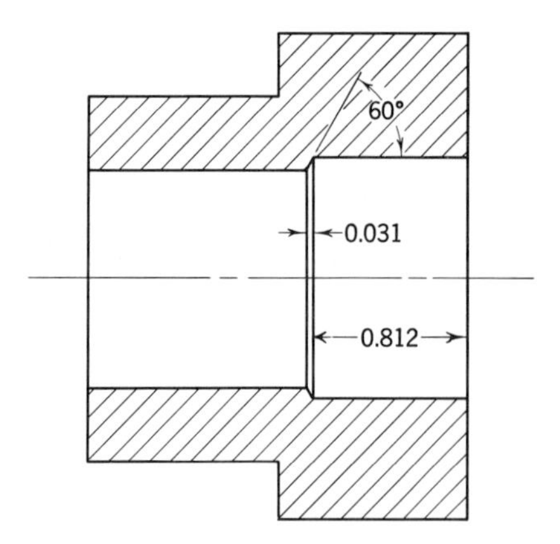

Fig. A-65

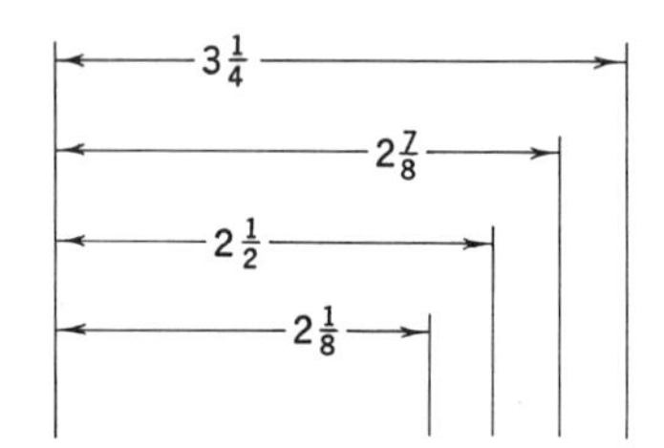

Fig. A-66

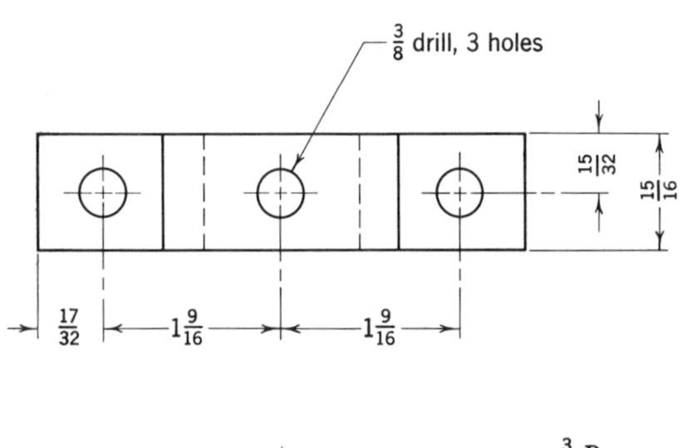

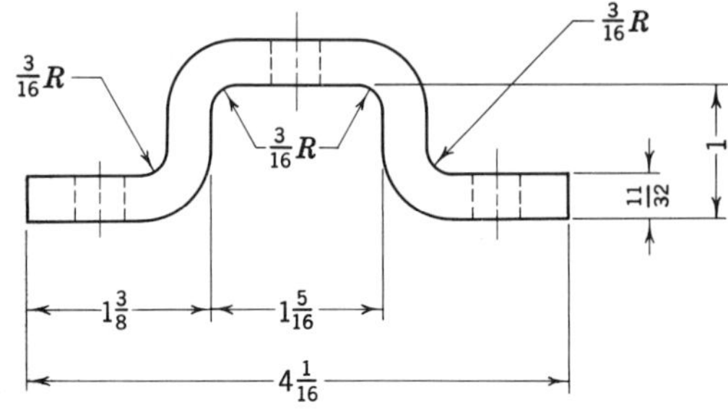

Fig. A-67

(*e*) *Over-all dimensions* should be placed outside the intermediate dimensions. If the over-all dimension is given together with all intermediate dimensions, one dimension must be marked "REF." (a reference dimension). A reference dimension is given as a convenience and does not control the size or shape of the part in any way. The selection of the dimension to be designated as REF. should be such that the selected feature will accept the tolerance accumulation. (See Fig. A-68.)

Dimensions and Notes for Standard Details

Various parts of devices, machines, and structures are produced by such common shop operations as drilling, reaming, boring, chamfering, countersinking, spotfacing, etc. These operations may be specified on the drawing by appropriate notes. Likewise, standard fasteners, structural shapes, etc., are specified by note. Graphical representations and accompanying notes that specify the more common operations are the following:

Round Holes. A hole is specified by giving its diameter and depth. When the operation for drilling the hole is noted it is shown as in Fig. A-69. When the hole does not go through the material, it is known as a *blind* hole and the depth dimension is only to the shoulder, as shown in Fig. A-70.

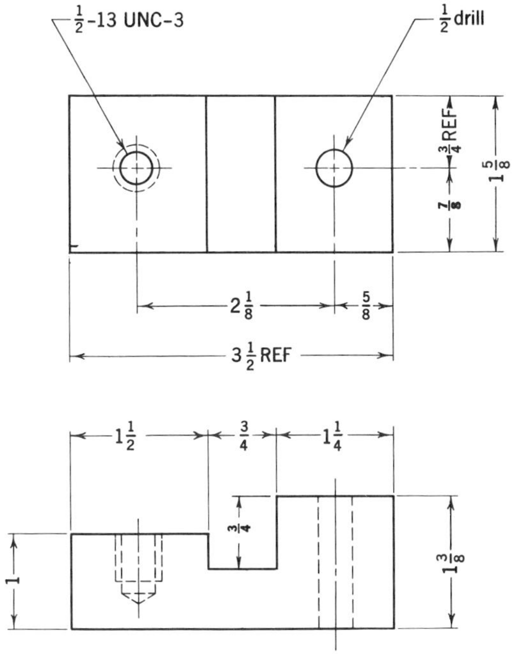

Fig. A-68

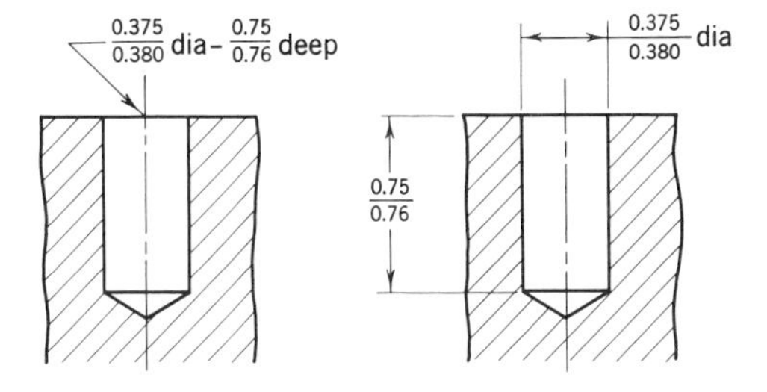

Fig. A-69

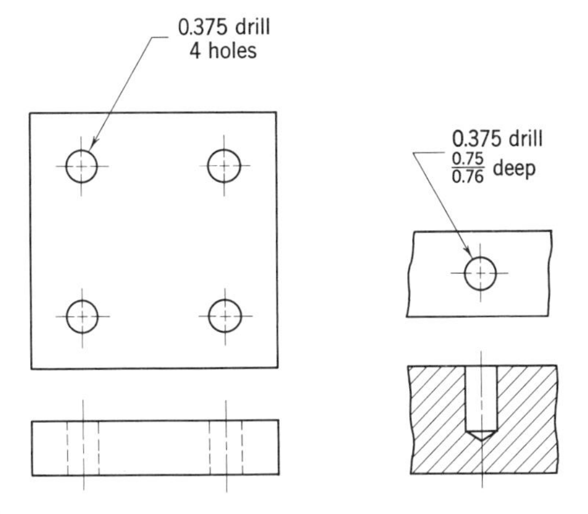

Fig. A-70

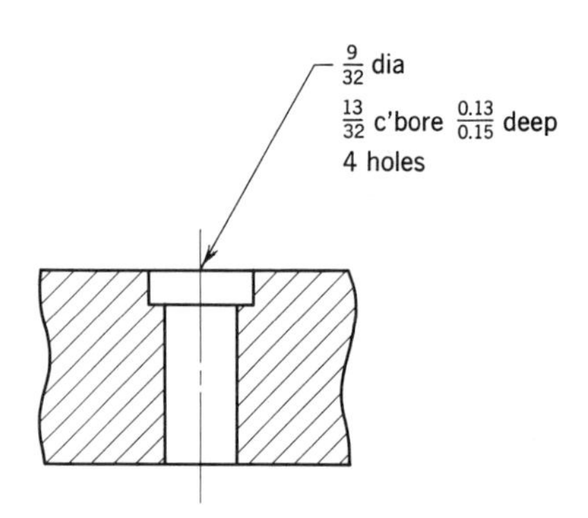

Fig. A-71

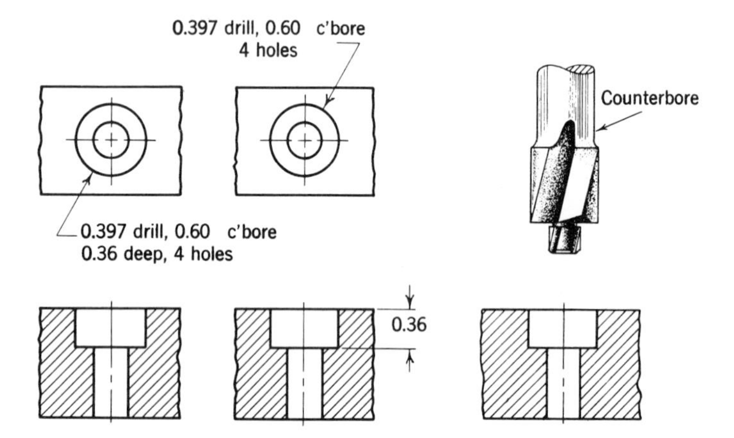

Fig. A-72

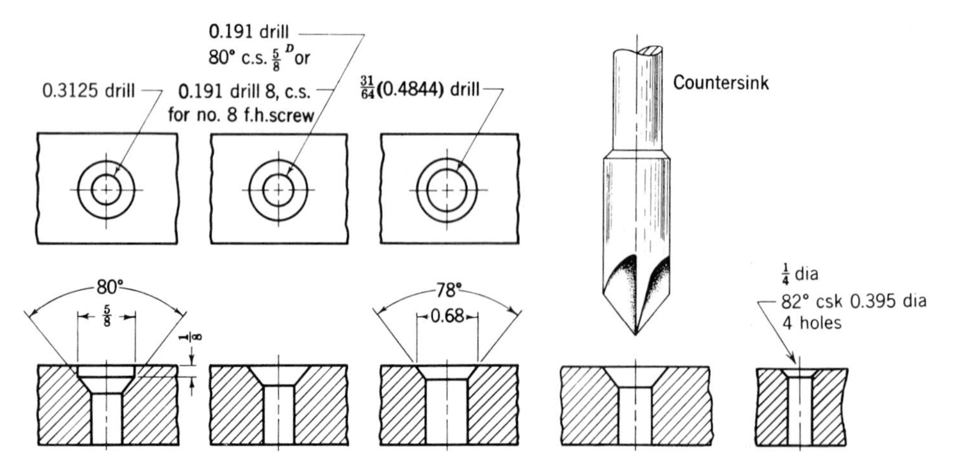

Fig. A-73

Counterbored Holes. Counterboring provides a space for recessing a bolt head or the head of a screw. The diameter and depth of the counterbore are noted, as shown in Fig. A-71. In some cases the note may include the drill operation. (See Fig. A-72.)

Countersunk Holes. A countersunk hole provides a seat for flat head screws. The diameter and the angle of the countersink are noted. (See Fig. A-73.)

Spotfaced Holes. To provide a good bearing surface for a bolt head or nut, the area around the hole, through which the bolt passes, must be made smooth. This is done by spotfacing. (See Fig. A-74.)

Chamfers. The recommended method for indicating a chamfer (beveled edge) is shown in Fig. A-75.

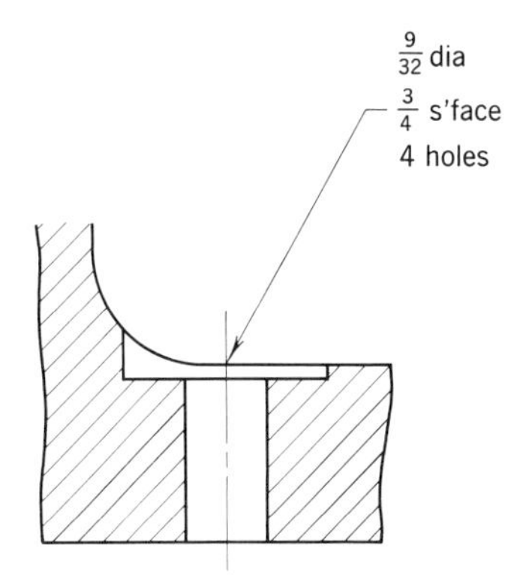

Fig. A-74

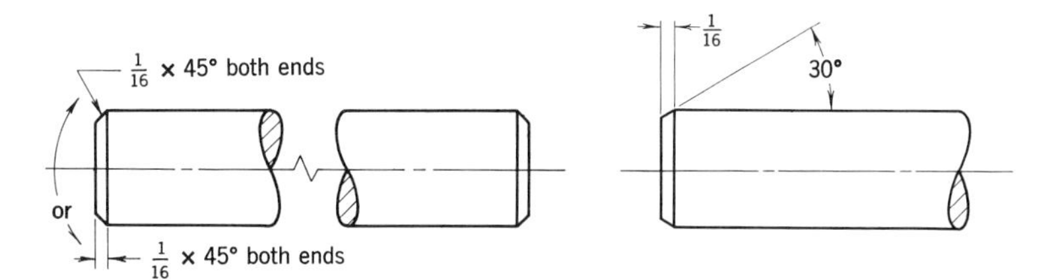

Fig. A-75 Chamfers.

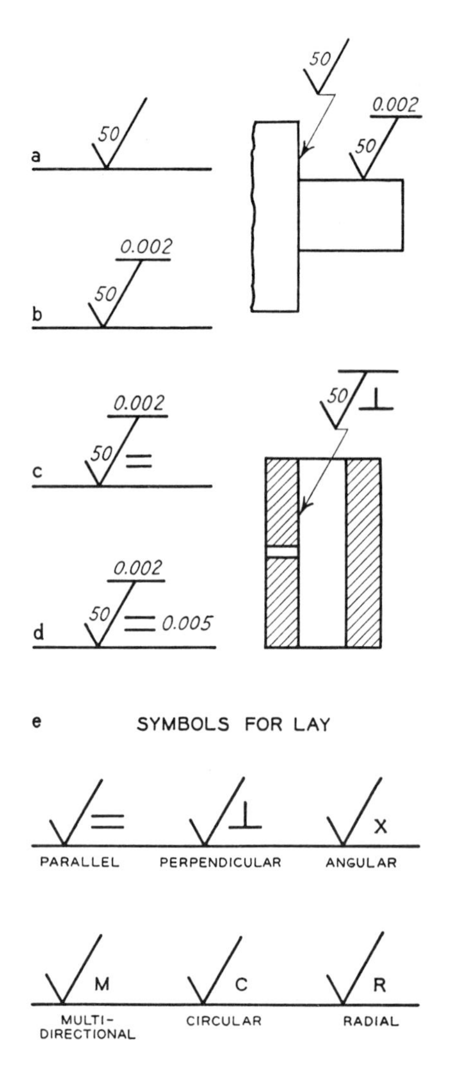

Fig. A-76 Surface quality symbols. (Courtesy ASA)

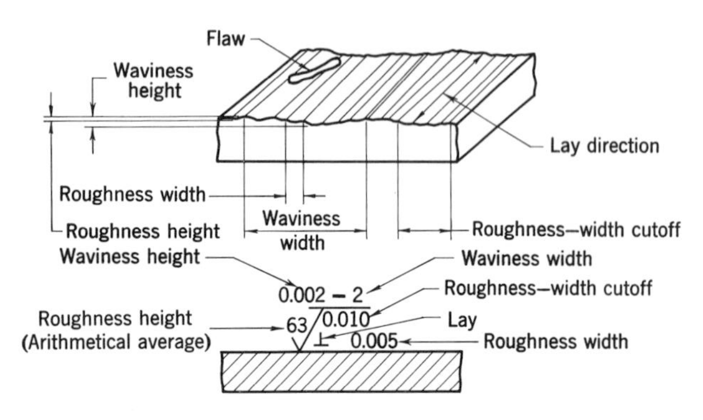

Fig. A-77 Relation of symbols to surface characteristics.

Surface Quality

The engineer can specify the desired surface quality by the use of well-recognized notes and symbols to indicate *roughness, waviness,* and *lay.* The following definitions of these terms are based on American Standard, ASA B46.1.

Roughness is defined as the predominant surface pattern resulting from the finely spaced surface irregularities produced by cutting edges of machine tools. The *height* of the irregularities is rated in microinches (one millionth of an inch).

Waviness refers to irregularities that result from such factors as machine or work deflections, vibration, heat treatment, or warping strains. These irregularities are of greater spacing than roughness. The height is expressed in inches.

Lay is the direction of the predominant surface pattern, produced by tool marks or grains of the surface ordinarily determined by the methods of production.

Surface symbols are used to specify surface quality. Where it is only necessary to specify surface roughness height, the simplest form of the symbol is used, as shown in Fig. A-76(*a*). The height, represented by the numeral 50, is in microinches, which may be the average "peak-to-valley" height, or average deviation from the mean (RMS or arithmetical).

Where it is necessary to include waviness height, in addition to roughness height, a horizontal line is added to the symbol, as shown in Fig. A-76(*b*). The numerical value of the height of waviness is placed above the horizontal line. In the example shown the value is 0.002 inch.

Now, where, in addition to roughness and waviness, it is necessary to specify *lay*, additional symbols are included as shown in Fig. A-76(*c*). The "parallel-lines" part of the symbol indicates that the dominant lines of the surface are parallel to the boundary line of the surface in contact with the symbol. The "perpendicular-lines" part of the symbol (on the right) indicates that the dominant lines of the surface are perpendicular to the boundary line of the surface in contact with the symbol. A symbol including the roughness width, placed to the right of the lay symbol, is shown in Fig. A-76(*d*).

The relation of symbols to surface characteristics is shown in Fig. A-77.

APPENDIX B
TABLES

TABLE 1

TABLE OF CHORDS

Values shown are the chord lengths of arcs subtending the given angles. Radius of arc equals one unit.

Deg	*0′*	*10′*	*20′*	*30′*	*40′*	*50′*
0	0.0000	0.0029	0.0058	0.0087	0.0116	0.0145
1	0.0175	0.0204	0.0233	0.0262	0.0291	0.0320
2	0.0349	0.0378	0.0407	0.0436	0.0465	0.0494
3	0.0524	0.0553	0.0582	0.0611	0.0640	0.0669
4	0.0698	0.0727	0.0756	0.0785	0.0814	0.0843
5	0.0872	0.0901	0.0931	0.0960	0.0989	0.1018
6	0.1047	0.1076	0.1105	0.1134	0.1163	0.1192
7	0.1221	0.1250	0.1279	0.1308	0.1337	0.1366
8	0.1395	0.1424	0.1453	0.1482	0.1511	0.1540
9	0.1569	0.1598	0.1627	0.1656	0.1685	0.1714
10	0.1743	0.1772	0.1801	0.1830	0.1859	0.1888
11	0.1917	0.1946	0.1975	0.2004	0.2033	0.2062
12	0.2091	0.2119	0.2148	0.2177	0.2206	0.2235
13	0.2264	0.2293	0.2322	0.2351	0.2380	0.2409
14	0.2437	0.2466	0.2495	0.2524	0.2553	0.2582
15	0.2611	0.2639	0.2668	0.2697	0.2726	0.2755
16	0.2783	0.2812	0.2841	0.2870	0.2899	0.2927
17	0.2956	0.2985	0.3014	0.3042	0.3071	0.3100
18	0.3129	0.3157	0.3186	0.3215	0.3244	0.3272
19	0.3301	0.3330	0.3358	0.3387	0.3416	0.3444
20	0.3473	0.3502	0.3530	0.3559	0.3587	0.3616
21	0.3645	0.3673	0.3702	0.3730	0.3759	0.3788
22	0.3816	0.3845	0.3873	0.3902	0.3930	0.3959
23	0.3987	0.4016	0.4044	0.4073	0.4101	0.4130
24	0.4158	0.4187	0.4215	0.4244	0.4272	0.4300
25	0.4329	0.4357	0.4386	0.4414	0.4442	0.4471
26	0.4499	0.4527	0.4556	0.4584	0.4612	0.4641
27	0.4669	0.4697	0.4725	0.4754	0.4782	0.4810
28	0.4838	0.4867	0.4895	0.4923	0.4951	0.4979
29	0.5008	0.5036	0.5064	0.5092	0.5120	0.5148
30	0.5176	0.5204	0.5233	0.5261	0.5289	0.5317
31	0.5345	0.5373	0.5401	0.5429	0.5457	0.5485
32	0.5513	0.5541	0.5569	0.5597	0.5625	0.5652
33	0.5680	0.5708	0.5736	0.5764	0.5792	0.5820
34	0.5847	0.5875	0.5903	0.5931	0.5959	0.5986
35	0.6014	0.6042	0.6070	0.6097	0.6125	0.6153
36	0.6180	0.6208	0.6236	0.6263	0.6291	0.6319
37	0.6346	0.6374	0.6401	0.6429	0.6456	0.6484
38	0.6511	0.6539	0.6566	0.6594	0.6621	0.6649
39	0.6676	0.6704	0.6731	0.6758	0.6786	0.6813
40	0.6840	0.6868	0.6895	0.6922	0.6950	0.6977
41	0.7004	0.7031	0.7059	0.7086	0.7113	0.7140
42	0.7167	0.7195	0.7222	0.7249	0.7276	0.7303
43	0.7330	0.7357	0.7384	0.7411	0.7438	0.7465
44	0.7492	0.7519	0.7546	0.7573	0.7600	0.7627
45	0.7654	0.7681	0.7707	0.7734	0.7761	0.7788

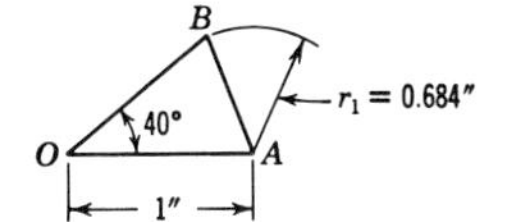

Example 1. Angle = 40°, radius = 1″ r_1 = 0.684 from the above table

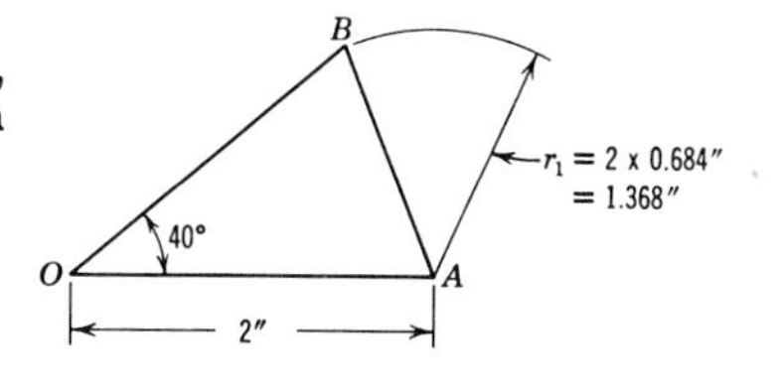

Example 2. Angle = 40°, radius = 2″
$r_1 = 2 \times 0.684 = 1.368$

TABLE 2

NATURAL TRIGONOMETRIC FUNCTIONS

Angles	Sines	Cosines	Tangents	Cotangents	Angles
0° 00′	.0000	1.0000	.0000	∞	90° 00′
10	.0029	1.0000	.0029	343.77	50
20	.0058	1.0000	.0058	171.89	40
30	.0087	1.0000	.0087	114.59	30
40	.0116	.9999	.0116	85.940	20
50	.0145	.9999	.0145	68.750	10
1° 00′	.0175	.9998	.0175	57.290	89° 00′
10	.0204	.9998	.0204	49.104	50
20	.0233	.9997	.0233	42.964	40
30	.0262	.9997	.0262	38.188	30
40	.0291	.9996	.0291	34.368	20
50	.0320	.9995	.0320	31.242	10
2° 00′	.0349	.9994	.0349	28.636	88° 00′
10	.0378	.9993	.0378	26.432	50
20	.0407	.9992	.0407	24.542	40
30	.0436	.9990	.0437	22.904	30
40	.0465	.9989	.0466	21.470	20
50	.0494	.9988	.0495	20.206	10
3° 00′	.0523	.9986	.0524	19.081	87° 00′
10	.0552	.9985	.0553	18.075	50
20	.0581	.9983	.0582	17.169	40
30	.0610	.9981	.0612	16.350	30
40	.0640	.9980	.0641	15.605	20
50	.0669	.9978	.0670	14.924	10
4° 00′	.0698	.9976	.0699	14.301	86° 00′
10	.0727	.9974	.0729	13.727	50
20	.0756	.9971	.0758	13.197	40
30	.0785	.9969	.0787	12.706	30
40	.0814	.9967	.0816	12.251	20
50	.0843	.9964	.0846	11.826	10
5° 00′	.0872	.9962	.0875	11.430	85° 00′
10	.0901	.9959	.0904	11.059	50
20	.0929	.9957	.0934	10.712	40
30	.0958	.9954	.0963	10.385	30
40	.0987	.9951	.0992	10.078	20
50	.1016	.9948	.1022	9.7882	10
6° 00′	.1045	.9945	.1051	9.5144	84° 00′
10	.1074	.9942	.1080	9.2553	50
20	.1103	.9939	.1110	9.0098	40
30	.1132	.9936	.1139	8.7769	30
40	.1161	.9932	.1169	8.5555	20
50	.1190	.9929	.1198	8.3450	10
7° 00′	.1219	.9925	.1228	8.1443	83° 00′
10	.1248	.9922	.1257	7.9530	50
20	.1276	.9918	.1287	7.7704	40
30	.1305	.9914	.1317	7.5958	30
40	.1334	.9911	.1346	7.4287	20
50	.1363	.9907	.1376	7.2687	10
Angles	Cosines	Sines	Cotangents	Tangents	Angles

TABLE 2 (*continued*)

NATURAL TRIGONOMETRIC FUNCTIONS

Angles	Sines	Cosines	Tangents	Cotangents	Angles
8° 00′	.1392	.9903	.1405	7.1154	82° 00′
10	.1421	.9899	.1435	6.9682	50
20	.1449	.9894	.1465	6.8269	40
30	.1478	.9890	.1495	6.6912	30
40	.1507	.9886	.1524	6.5606	20
50	.1536	.9881	.1554	6.4348	10
9° 00′	.1564	.9877	.1584	6.3138	81° 00′
10	.1593	.9872	.1614	6.1970	50
20	.1622	.9868	.1644	6.0844	40
30	.1650	.9863	.1673	5.9758	30
40	.1679	.9858	.1703	5.8707	20
50	.1708	.9853	.1733	5.7694	10
10° 00′	.1736	.9848	.1763	5.6713	80° 00′
10	.1765	.9843	.1793	5.5764	50
20	.1794	.9838	.1823	5.4845	40
30	.1822	.9833	.1853	5.3955	30
40	.1851	.9827	.1883	5.3093	20
50	.1880	.9822	.1914	5.2257	10
11° 00′	.1908	.9816	.1944	5.1446	79° 00′
10	.1937	.9811	.1974	5.0658	50
20	.1965	.9805	.2004	4.9894	40
30	.1994	.9799	.2035	4.9152	30
40	.2022	.9793	.2065	4.8430	20
50	.2051	.9787	.2095	4.7729	10
12° 00′	.2079	.9781	.2126	4.7046	78° 00′
10	.2108	.9775	.2156	4.6382	50
20	.2136	.9769	.2186	4.5736	40
30	.2164	.9763	.2217	4.5107	30
40	.2193	.9757	.2247	4.4494	20
50	.2221	.9750	.2278	4.3897	10
13° 00′	.2250	.9744	.2309	4.3315	77° 00′
10	.2278	.9737	.2339	4.2747	50
20	.2306	.9730	.2370	4.2193	40
30	.2334	.9724	.2401	4.1653	30
40	.2363	.9717	.2432	4.1126	20
50	.2391	.9710	.2462	4.0611	10
14° 00′	.2419	.9703	.2493	4.0108	76° 00′
10	.2447	.9696	.2524	3.9617	50
20	.2476	.9689	.2555	3.9136	40
30	.2504	.9681	.2586	3.8667	30
40	.2532	.9674	.2617	3.8208	20
50	.2560	.9667	.2648	3.7760	10
15° 00′	.2588	.9659	.2679	3.7321	75° 00′
10	.2616	.9652	.2711	3.6891	50
20	.2644	.9644	.2742	3.6470	40
30	.2672	.9636	.2773	3.6059	30
40	.2700	.9628	.2805	3.5656	20
50	.2728	.9621	.2836	3.5261	10
Angles	Cosines	Sines	Cotangents	Tangents	Angles

TABLE 2 (*continued*)

NATURAL TRIGONOMETRIC FUNCTIONS

Angles	Sines	Cosines	Tangents	Cotangents	Angles
16° 00′	.2756	.9613	.2867	3.4874	74° 00′
10	.2784	.9605	.2899	3.4495	50
20	.2812	.9596	.2931	3.4124	40
30	.2840	.9588	.2962	3.3759	30
40	.2868	.9580	.2994	3.3402	20
50	.2896	.9572	.3026	3.3052	10
17° 00′	.2924	.9563	.3057	3.2709	73° 00′
10	.2952	.9555	.3089	3.2371	50
20	.2979	.9546	.3121	3.2041	40
30	.3007	.9537	.3153	3.1716	30
40	.3035	.9528	.3185	3.1397	20
50	.3062	.9520	.3217	3.1084	10
18° 00′	.3090	.9511	.3249	3.0777	72° 00′
10	.3118	.9502	.3281	3.0475	50
20	.3145	.9492	.3314	3.0178	40
30	.3173	.9483	.3346	2.9887	30
40	.3201	.9474	.3378	2.9600	20
50	.3228	.9465	.3411	2.9319	10
19° 00′	.3256	.9455	.3443	2.9042	71° 00′
10	.3283	.9446	.3476	2.8770	50
20	.3311	.9436	.3508	2.8502	40
30	.3338	.9426	.3541	2.8239	30
40	.3365	.9417	.3574	2.7980	20
50	.3393	.9407	.3607	2.7725	10
20° 00′	.3420	.9397	.3640	2.7475	70° 00′
10	.3448	.9387	.3673	2.7228	50
20	.3475	.9377	.3706	2.6985	40
30	.3502	.9367	.3739	2.6746	30
40	.3529	.9356	.3772	2.6511	20
50	.3557	.9346	.3805	2.6279	10
21° 00′	.3584	.9336	.3839	2.6051	69° 00′
10	.3611	.9325	.3872	2.5826	50
20	.3638	.9315	.3906	2.5605	40
30	.3665	.9304	.3939	2.5386	30
40	.3692	.9293	.3973	2.5172	20
50	.3719	.9283	.4006	2.4960	10
22° 00′	.3746	.9272	.4040	2.4751	68° 00′
10	.3773	.9261	.4074	2.4545	50
20	.3800	.9250	.4108	2.4342	40
30	.3827	.9239	.4142	2.4142	30
40	.3854	.9228	.4176	2.3945	20
50	.3881	.9216	.4210	2.3750	10
23° 00′	.3907	.9205	.4245	2.3559	67° 00′
10	.3934	.9194	.4279	2.3369	50
20	.3961	.9182	.4314	2.3183	40
30	.3987	.9171	.4348	2.2998	30
40	.4014	.9159	.4383	2.2817	20
50	.4041	.9147	.4417	2.2637	10
Angles	Cosines	Sines	Cotangents	Tangents	Angles

TABLE 2 (*continued*)

NATURAL TRIGONOMETRIC FUNCTIONS

Angles	Sines	Cosines	Tangents	Cotangents	Angles
24° 00′	.4067	.9135	.4452	2.2460	66° 00′
10	.4094	.9124	.4487	2.2286	50
20	.4120	.9122	.4522	2.2113	40
30	.4147	.9100	.4557	2.1943	30
40	.4173	.9088	.4592	2.1775	20
50	.4200	.9075	.4628	2.1609	10
25° 00′	.4226	.9063	.4663	2.1445	65° 00′
10	.4253	.9051	.4699	2.1283	50
20	.4279	.9038	.4734	2.1123	40
30	.4305	.9026	.4770	2.0965	30
40	.4331	.9013	.4806	2.0809	20
50	.4358	.9001	.4841	2.0655	10
26° 00′	.4384	.8988	.4877	2.0503	64° 00′
10	.4410	.8975	.4913	2.0353	50
20	.4436	.8962	.4950	2.0204	40
30	.4462	.8949	.4986	2.0057	30
40	.4488	.8936	.5022	1.9912	20
50	.4514	.8923	.5059	1.9768	10
27° 00′	.4540	.8910	.5095	1.9626	63° 00′
10	.4566	.8897	.5132	1.9486	50
20	.4592	.8884	.5169	1.9347	40
30	.4617	.8870	.5206	1.9210	30
40	.4643	.8857	.5243	1.9074	20
50	.4669	.8843	.5280	1.8940	10
28° 00′	.4695	.8829	.5317	1.8807	62° 00′
10	.4720	.8816	.5354	1.8676	50
20	.4746	.8802	.5392	1.8546	40
30	.4772	.8788	.5430	1.8418	30
40	.4797	.8774	.5467	1.8291	20
50	.4823	.8760	.5505	1.8165	10
29° 00′	.4848	.8746	.5543	1.8040	61° 00′
10	.4874	.8732	.5581	1.7917	50
20	.4899	.8718	.5619	1.7796	40
30	.4924	.8704	.5658	1.7675	30
40	.4950	.8689	.5696	1.7556	20
50	.4975	.8675	.5735	1.7437	10
30° 00′	.5000	.8660	.5774	1.7321	60° 00′
10	.5025	.8646	.5812	1.7205	50
20	.5050	.8631	.5851	1.7090	40
30	.5075	.8616	.5890	1.6977	30
40	.5100	.8601	.5930	1.6864	20
50	.5125	.8587	.5969	1.6753	10
31° 00′	.5150	.8572	.6009	1.6643	59° 00′
10	.5175	.8557	.6048	1.6534	50
20	.5200	.8542	.6088	1.6426	40
30	.5225	.8526	.6128	1.6319	30
40	.5250	.8511	.6168	1.6212	20
50	.5275	.8496	.6208	1.6107	10
Angles	Cosines	Sines	Cotangents	Tangents	Angles

TABLE 2 (*continued*)

NATURAL TRIGONOMETRIC FUNCTIONS

Angles	Sines	Cosines	Tangents	Cotangents	Angles
32° 00′	.5299	.8480	.6249	1.6003	58° 00′
10	.5324	.8465	.6289	1.5900	50
20	.5348	.8450	.6330	1.5798	40
30	.5373	.8434	.6371	1.5697	30
40	.5398	.8418	.6412	1.5597	20
50	.5422	.8403	.6453	1.5497	10
33° 00′	.5446	.8387	.6494	1.5399	57° 00′
10	.5471	.8371	.6536	1.5301	50
20	.5495	.8355	.6577	1.5204	40
30	.5519	.8339	.6619	1.5108	30
40	.5544	.8323	.6661	1.5013	20
50	.5568	.8307	.6703	1.4919	10
34° 00′	.5592	.8290	.6745	1.4826	56° 00′
10	.5616	.8274	.6787	1.4733	50
20	.5640	.8258	.6830	1.4641	40
30	.5664	.8241	.6873	1.4550	30
40	.5688	.8225	.6916	1.4460	20
50	.5712	.8208	.6959	1.4370	10
35° 00′	.5736	.8192	.7002	1.4281	55° 00′
10	.5760	.8175	.7046	1.4193	50
20	.5783	.8158	.7089	1.4106	40
30	.5807	.8141	.7133	1.4019	30
40	.5831	.8124	.7177	1.3934	20
50	.5854	.8107	.7221	1.3848	10
36° 00′	.5878	.8090	.7265	1.3764	54° 00′
10	.5901	.8073	.7310	1.3680	50
20	.5925	.8056	.7355	1.3597	40
30	.5948	.8039	.7400	1.3514	30
40	.5972	.8021	.7445	1.3432	20
50	.5995	.8004	.7490	1.3351	10
37° 00′	.6018	.7986	.7536	1.3270	53° 00′
10	.6041	.7969	.7581	1.3190	50
20	.6065	.7951	.7627	1.3111	40
30	.6088	.7934	.7673	1.3032	30
40	.6111	.7916	.7720	1.2954	20
50	.6134	.7898	.7766	1.2876	10
38° 00′	.6157	.7880	.7813	1.2799	52° 00′
10	.6180	.7862	.7860	1.2723	50
20	.6202	.7844	.7907	1.2647	40
30	.6225	.7826	.7954	1.2572	30
40	.6248	.7808	.8002	1.2497	20
50	.6271	.7790	.8050	1.2423	10
39° 00′	.6293	.7771	.8098	1.2349	51° 00′
10	.6316	.7753	.8146	1.2276	50
20	.6338	.7735	.8195	1.2203	40
30	.6361	.7716	.8243	1.2131	30
40	.6383	.7698	.8292	1.2059	20
50	.6406	.7679	.8342	1.1988	10
Angles	Cosines	Sines	Cotangents	Tangents	Angles

TABLE 2 (*continued*)

Natural Trigonometric Functions

Angles	Sines	Cosines	Tangents	Cotangents	Angles
40° 00′	.6428	.7660	.8391	1.1918	50° 00′
10	.6450	.7642	.8441	1.1847	50
20	.6472	.7623	.8491	1.1778	40
30	.6494	.7604	.8541	1.1708	30
40	.6517	.7585	.8591	1.1640	20
50	.6539	.7566	.8642	1.1571	10
41° 00′	.6561	.7547	.8693	1.1504	49° 00′
10	.6583	.7528	.8744	1.1436	50
20	.6604	.7509	.8796	1.1369	40
30	.6626	.7490	.8847	1.1303	30
40	.6648	.7470	.8899	1.1237	20
50	.6670	.7451	.8952	1.1171	10
42° 00′	.6691	.7431	.9004	1.1106	48° 00′
10	.6713	.7412	.9057	1.1041	50
20	.6734	.7392	.9110	1.0977	40
30	.6756	.7373	.9163	1.0913	30
40	.6777	.7353	.9217	1.0850	20
50	.6799	.7333	.9271	1.0786	10
43° 00′	.6820	.7314	.9325	1.0724	47° 00′
10	.6841	.7294	.9380	1.0661	50
20	.6862	.7274	.9435	1.0599	40
30	.6884	.7254	.9490	1.0538	30
40	.6905	.7234	.9545	1.0477	20
50	.6926	.7214	.9601	1.0416	10
44° 00′	.6947	.7193	.9657	1.0355	46° 00′
10	.6967	.7173	.9713	1.0295	50
20	.6988	.7153	.9770	1.0235	40
30	.7009	.7133	.9827	1.0176	30
40	.7030	.7112	.9884	1.0117	20
50	.7050	.7092	.9942	1.0058	10
45° 00′	.7071	.7071	1.0000	1.0000	45° 00′
Angles	Cosines	Sines	Cotangents	Tangents	Angles

TABLE 3

COMMON LOGARITHMS

N	0	1	2	3	4	5	6	7	8	9
10	0000	0043	0086	0128	0170	0212	0253	0294	0334	0374
11	0414	0453	0492	0531	0569	0607	0645	0682	0719	0755
12	0792	0828	0864	0899	0934	0969	1004	1038	1072	1106
13	1139	1173	1206	1239	1271	1303	1335	1367	1399	1430
14	1461	1492	1523	1553	1584	1614	1644	1673	1703	1732
15	1761	1790	1818	1847	1875	1903	1931	1959	1987	2014
16	2041	2068	2095	2122	2148	2175	2201	2227	2253	2279
17	2304	2330	2355	2380	2405	2430	2455	2480	2504	2529
18	2553	2577	2601	2625	2648	2672	2695	2718	2742	2765
19	2788	2810	2833	2856	2878	2900	2923	2945	2967	2989
20	3010	3032	3054	3075	3096	3118	3139	3160	3181	3201
21	3222	3243	3263	3284	3304	3324	3345	3365	3385	3404
22	3424	3444	3464	3483	3502	3522	3541	3560	3579	3598
23	3617	3636	3655	3674	3692	3711	3729	3747	3766	3784
24	3802	3820	3838	3856	3874	3892	3909	3927	3945	3962
25	3979	3997	4014	4031	4048	4065	4082	4099	4116	4133
26	4150	4166	4183	4200	4216	4232	4249	4265	4281	4298
27	4314	4330	4346	4362	4378	4393	4409	4425	4440	4456
28	4472	4487	4502	4518	4533	4548	4564	4579	4594	4609
29	4624	4639	4654	4669	4683	4698	4713	4728	4742	4757
30	4771	4786	4800	4814	4829	4843	4857	4871	4886	4900
31	4914	4928	4942	4955	4969	4983	4997	5011	5024	5038
32	5051	5065	5079	5092	5105	5119	5132	5145	5159	5172
33	5185	5198	5211	5224	5237	5250	5263	5276	5289	5302
34	5315	5328	5340	5353	5366	5378	5391	5403	5416	5428
35	5441	5453	5465	5478	5490	5502	5514	5527	5539	5551
36	5563	5575	5587	5599	5611	5623	5635	5647	5658	5670
37	5682	5694	5705	5717	5729	5740	5752	5763	5775	5786
38	5798	5809	5821	5832	5843	5855	5866	5877	5888	5899
39	5911	5922	5933	5944	5955	5966	5977	5988	5999	6010
40	6021	6031	6042	6053	6064	6075	6085	6096	6107	6117
41	6128	6138	6149	6160	6170	6180	6191	6201	6212	6222
42	6232	6243	6253	6263	6274	6284	6294	6304	6314	6325
43	6335	6345	6355	6365	6375	6385	6395	6405	6415	6425
44	6435	6444	6454	6464	6474	6484	6493	6503	6513	6522
45	6532	6542	6551	6561	6571	6580	6590	6599	6609	6618
46	6628	6637	6646	6656	6665	6675	6684	6693	6702	6712
47	6721	6730	6739	6749	6758	6767	6776	6785	6794	6803
48	6812	6821	6830	6839	6848	6857	6866	6875	6884	6893
49	6902	6911	6920	6928	6937	6946	6955	6964	6972	6981
50	6990	6998	7007	7016	7024	7033	7042	7050	7059	7067
51	7076	7084	7093	7101	7110	7118	7126	7135	7143	7152
52	7160	7168	7177	7185	7193	7202	7210	7218	7226	7235
53	7243	7251	7259	7267	7275	7284	7292	7300	7308	7316
54	7324	7332	7340	7348	7356	7364	7372	7380	7388	7396

TABLE 3 (*continued*)
Common Logarithms

N	0	1	2	3	4	5	6	7	8	9
55	7404	7412	7419	7427	7435	7443	7451	7459	7466	7474
56	7482	7490	7497	7505	7513	7520	7528	7536	7543	7551
57	7559	7566	7574	7582	7589	7597	7604	7612	7619	7627
58	7634	7642	7649	7657	7664	7672	7679	7686	7694	7701
59	7709	7716	7723	7731	7738	7745	7752	7760	7767	7774
60	7782	7789	7796	7803	7810	7818	7825	7832	7839	7846
61	7853	7860	7868	7875	7882	7889	7896	7903	7910	7917
62	7924	7931	7938	7945	7952	7959	7966	7973	7980	7987
63	7993	8000	8007	8014	8021	8028	8035	8041	8048	8055
64	8062	8069	8075	8082	8089	8096	8102	8109	8116	8122
65	8129	8136	8142	8149	8156	8162	8169	8176	8182	8189
66	8195	8202	8209	8215	8222	8228	8235	8241	8248	8254
67	8261	8267	8274	8280	8287	8293	8299	8306	8312	8319
68	8325	8331	8338	8344	8351	8357	8363	8370	8376	8382
69	8388	8395	8401	8407	8414	8420	8426	8432	8439	8445
70	8451	8457	8463	8470	8476	8482	8488	8494	8500	8506
71	8513	8519	8525	8531	8537	8543	8549	8555	8561	8567
72	8573	8579	8585	8591	8597	8603	8609	8615	8621	8627
73	8633	8639	8645	8651	8657	8663	8669	8675	8681	8686
74	8692	8698	8704	8710	8716	8722	8727	8733	8739	8745
75	8751	8756	8762	8768	8774	8779	8785	8791	8797	8802
76	8808	8814	8820	8825	8831	8837	8842	8848	8854	8859
77	8865	8871	8876	8882	8887	8893	8899	8904	8910	8915
78	8921	8927	8932	8938	8943	8949	8954	8960	8965	8971
79	8976	8982	8987	8993	8998	9004	9009	9015	9020	9025
80	9031	9036	9042	9047	9053	9058	9063	9069	9074	9079
81	9085	9090	9096	9101	9106	9112	9117	9122	9128	9133
82	9138	9143	9149	9154	9159	9165	9170	9175	9180	9186
83	9191	9196	9201	9206	9212	9217	9222	9227	9232	9238
84	9243	9248	9253	9258	9263	9269	9274	9279	9284	9289
85	9294	9299	9304	9309	9315	9320	9325	9330	9335	9340
86	9345	9350	9355	9360	9365	9370	9375	9380	9385	9390
87	9395	9400	9405	9410	9415	9420	9425	9430	9435	9440
88	9445	9450	9455	9460	9465	9469	9474	9479	9484	9489
89	9494	9499	9504	9509	9513	9518	9523	9528	9533	9538
90	9542	9547	9552	9557	9562	9566	9571	9576	9581	9586
91	9590	9595	9600	9605	9609	9614	9619	9624	9628	9633
92	9638	9643	9647	9652	9657	9661	9666	9671	9675	9680
93	9685	9689	9694	9699	9703	9708	9713	9717	9722	9727
94	9731	9736	9741	9745	9750	9754	9759	9763	9768	9773
95	9777	9782	9786	9791	9795	9800	9805	9809	9814	9818
96	9823	9827	9832	9836	9841	9845	9850	9854	9859	9863
97	9868	9872	9877	9881	9886	9890	9894	9899	9903	9908
98	9912	9917	9921	9926	9930	9934	9939	9943	9948	9952
99	9956	9961	9965	9969	9974	9978	9983	9987	9991	9996

TABLE 4

Hyperbolic or Naperian Logarithms

No.	H. Log.	No.	H. Log.	No.	H. Log.	No.	H. Log.	No.	H. Log.
1.00	0.0000								
1.01	0.0099	1.51	0.4121	2.01	0.6981	2.51	0.9203	3.01	1.1019
1.02	0.0198	1.52	0.4187	2.02	0.7031	2.52	0.9243	3.02	1.1053
1.03	0.0296	1.53	0.4253	2.03	0.7080	2.53	0.9282	3.03	1.1086
1.04	0.0392	1.54	0.4318	2.04	0.7129	2.54	0.9322	3.04	1.1119
1.05	0.0488	1.55	0.4383	2.05	0.7178	2.55	0.9361	3.05	1.1151
1.06	0.0583	1.56	0.4447	2.06	0.7227	2.56	0.9400	3.06	1.1184
1.07	0.0677	1.57	0.4511	2.07	0.7275	2.57	0.9439	3.07	1.1216
1.08	0.0770	1.58	0.4574	2.08	0.7324	2.58	0.9478	3.08	1.1249
1.09	0.0862	1.59	0.4637	2.09	0.7372	2.59	0.9517	3.09	1.1282
1.10	0.0953	1.60	0.4700	2.10	0.7419	2.60	0.9555	3.10	1.1314
1.11	0.1044	1.61	0.4762	2.11	0.7467	2.61	0.9594	3.11	1.1346
1.12	0.1133	1.62	0.4824	2.12	0.7514	2.62	0.9632	3.12	1.1378
1.13	0.1222	1.63	0.4886	2.13	0.7561	2.63	0.9670	3.13	1.1410
1.14	0.1310	1.64	0.4947	2.14	0.7608	2.64	0.9708	3.14	1.1442
1.15	0.1398	1.65	0.5008	2.15	0.7655	2.65	0.9746	3.15	1.1474
1.16	0.1484	1.66	0.5068	2.16	0.7701	2.66	0.9783	3.16	1.1506
1.17	0.1570	1.67	0.5128	2.17	0.7747	2.67	0.9821	3.17	1.1537
1.18	0.1655	1.68	0.5188	2.18	0.7793	2.68	0.9858	3.18	1.1569
1.19	0.1740	1.69	0.5247	2.19	0.7839	2.69	0.9895	3.19	1.1600
1.20	0.1823	1.70	0.5306	2.20	0.7885	2.70	0.9933	3.20	1.1632
1.21	0.1906	1.71	0.5365	2.21	0.7930	2.71	0.9969	3.21	1.1663
1.22	0.1988	1.72	0.5423	2.22	0.7975	2.72	1.0006	3.22	1.1694
1.23	0.2070	1.73	0.5481	2.23	0.8020	2.73	1.0043	3.23	1.1725
1.24	0.2151	1.74	0.5539	2.24	0.8065	2.74	1.0080	3.24	1.1756
1.25	0.2231	1.75	0.5596	2.25	0.8109	2.75	1.0116	3.25	1.1787
1.26	0.2311	1.76	0.5653	2.26	0.8154	2.76	1.0152	3.26	1.1817
1.27	0.2390	1.77	0.5710	2.27	0.8198	2.77	1.0188	3.27	1.1848
1.28	0.2469	1.78	0.5766	2.28	0.8242	2.78	1.0225	3.28	1.1878
1.29	0.2546	1.79	0.5822	2.29	0.8286	2.79	1.0260	3.29	1.1909
1.30	0.2624	1.80	0.5878	2.30	0.8329	2.80	1.0296	3.30	1.1939
1.31	0.2700	1.81	0.5933	2.31	0.8372	2.81	1.0332	3.31	1.1969
1.32	0.2776	1.82	0.5988	2.32	0.8416	2.82	1.0367	3.32	1.1999
1.33	0.2852	1.83	0.6043	2.33	0.8458	2.83	1.0403	3.33	1.2030
1.34	0.2927	1.84	0.6098	2.34	0.8502	2.84	1.0438	3.34	1.2060
1.35	0.3001	1.85	0.6152	2.35	0.8544	2.85	1.0473	3.35	1.2090
1.36	0.3075	1.86	0.6206	2.36	0.8587	2.86	1.0508	3.36	1.2119
1.37	0.3148	1.87	0.6259	2.37	0.8629	2.87	1.0543	3.37	1.2149
1.38	0.3221	1.88	0.6313	2.38	0.8671	2.88	1.0578	3.38	1.2179
1.39	0.3293	1.89	0.6366	2.39	0.8713	2.89	1.0613	3.39	1.2208
1.40	0.3365	1.90	0.6419	2.40	0.8755	2.90	1.0647	3.40	1.2238
1.41	0.3436	1.91	0.6471	2.41	0.8796	2.91	1.0682	3.41	1.2267
1.42	0.3507	1.92	0.6523	2.42	0.8838	2.92	1.0716	3.42	1.2296
1.43	0.3577	1.93	0.6575	2.43	0.8879	2.93	1.0750	3.43	1.2326
1.44	0.3646	1.94	0.6627	2.44	0.8920	2.94	1.0784	3.44	1.2355
1.45	0.3716	1.95	0.6678	2.45	0.8961	2.95	1.0818	3.45	1.2384
1.46	0.3784	1.96	0.6729	2.46	0.9002	2.96	1.0852	3.46	1.2413
1.47	0.3853	1.97	0.6780	2.47	0.9042	2.97	1.0886	3.47	1.2442
1.48	0.3920	1.98	0.6831	2.48	0.9083	2.98	1.0919	3.48	1.2470
1.49	0.3988	1.99	0.6881	2.49	0.9123	2.99	1.0953	3.49	1.2499
1.50	0.4055	2.00	0.6931	2.50	0.9163	3.00	1.0986	3.50	1.2528

TABLE 4 (*continued*)

HYPERBOLIC OR NAPERIAN LOGARITHMS

No.	H. Log.	No.	H. Log.	No.	H. Log.	No.	H. Log.	No.	H. Log.
3.51	1.2556	4.01	1.3888	4.51	1.5063	5.01	1.6114	5.51	1.7066
3.52	1.2585	4.02	1.3913	4.52	1.5085	5.02	1.6134	5.52	1.7084
3.53	1.2613	4.03	1.3938	4.53	1.5107	5.03	1.6154	5.53	1.7102
3.54	1.2641	4.04	1.3962	4.54	1.5129	5.04	1.6174	5.54	1.7120
3.55	1.2669	4.05	1.3987	4.55	1.5151	5.05	1.6194	5.55	1.7138
3.56	1.2698	4.06	1.4012	4.56	1.5173	5.06	1.6214	5.56	1.7156
3.57	1.2726	4.07	1.4036	4.57	1.5195	5.07	1.6233	5.57	1.7174
3.58	1.2754	4.08	1.4061	4.58	1.5217	5.08	1.6253	5.58	1.7192
3.59	1.2782	4.09	1.4085	4.59	1.5239	5.09	1.6273	5.59	1.7210
3.60	1.2809	4.10	1.4110	4.60	1.5261	5.10	1.6292	5.60	1.7228
3.61	1.2837	4.11	1.4134	4.61	1.5282	5.11	1.6312	5.61	1.7246
3.62	1.2865	4.12	1.4159	4.62	1.5304	5.12	1.6332	5.62	1.7263
3.63	1.2892	4.13	1.4183	4.63	1.5326	5.13	1.6351	5.63	1.7281
3.64	1.2920	4.14	1.4207	4.64	1.5347	5.14	1.6371	5.64	1.7299
3.65	1.2947	4.15	1.4231	4.65	1.5369	5.15	1.6390	5.65	1.7317
3.66	1.2975	4.16	1.4255	4.66	1.5390	5.16	1.6409	5.66	1.7334
3.67	1.3002	4.17	1.4279	4.67	1.5412	5.17	1.6429	5.67	1.7352
3.68	1.3029	4.18	1.4303	4.68	1.5433	5.18	1.6448	5.68	1.7370
3.69	1.3056	4.19	1.4327	4.69	1.5454	5.19	1.6467	5.69	1.7387
3.70	1.3083	4.20	1.4351	4.70	1.5476	5.20	1.6487	5.70	1.7405
3.71	1.3110	4.21	1.4375	4.71	1.5497	5.21	1.6506	5.71	1.7422
3.72	1.3137	4.22	1.4398	4.72	1.5518	5.22	1.6525	5.72	1.7440
3.73	1.3164	4.23	1.4422	4.73	1.5539	5.23	1.6544	5.73	1.7457
3.74	1.3191	4.24	1.4446	4.74	1.5560	5.24	1.6563	5.74	1.7475
3.75	1.3218	4.25	1.4469	4.75	1.5581	5.25	1.6582	5.75	1.7492
3.76	1.3244	4.26	1.4493	4.76	1.5602	5.26	1.6601	5.76	1.7509
3.77	1.3271	4.27	1.4516	4.77	1.5623	5.27	1.6620	5.77	1.7527
3.78	1.3297	4.28	1.4540	4.78	1.5644	5.28	1.6639	5.78	1.7544
3.79	1.3324	4.29	1.4563	4.79	1.5665	5.29	1.6658	5.79	1.7561
3.80	1.3350	4.30	1.4586	4.80	1.5686	5.30	1.6677	5.80	1.7579
3.81	1.3376	4.31	1.4609	4.81	1.5707	5.31	1.6696	5.81	1.7596
3.82	1.3403	4.32	1.4633	4.82	1.5728	5.32	1.6715	5.82	1.7613
3.83	1.3429	4.33	1.4656	4.83	1.5748	5.33	1.6734	5.83	1.7630
3.84	1.3455	4.34	1.4679	4.84	1.5769	5.34	1.6752	5.84	1.7647
3.85	1.3481	4.35	1.4702	4.85	1.5790	5.35	1.6771	5.85	1.7664
3.86	1.3507	4.36	1.4725	4.86	1.5810	5.36	1.6790	5.86	1.7681
3.87	1.3533	4.37	1.4748	4.87	1.5831	5.37	1.6808	5.87	1.7699
3.88	1.3558	4.38	1.4770	4.88	1.5851	5.38	1.6827	5.88	1.7716
3.89	1.3584	4.39	1.4793	4.89	1.5872	5.39	1.6845	5.89	1.7733
3.90	1.3610	4.40	1.4816	4.90	1.5892	5.40	1.6864	5.90	1.7750
3.91	1.3635	4.41	1.4839	4.91	1.5913	5.41	1.6882	5.91	1.7766
3.92	1.3661	4.42	1.4861	4.92	1.5933	5.42	1.6901	5.92	1.7783
3.93	1.3686	4.43	1.4884	4.93	1.5953	5.43	1.6919	5.93	1.7800
3.94	1.3712	4.44	1.4907	4.94	1.5974	5.44	1.6938	5.94	1.7817
3.95	1.3737	4.45	1.4929	4.95	1.5994	5.45	1.6956	5.95	1.7834
3.96	1.3762	4.46	1.4951	4.96	1.6014	5.46	1.6974	5.96	1.7851
3.97	1.3788	4.47	1.4974	4.97	1.6034	5.47	1.6993	5.97	1.7867
3.98	1.3813	4.48	1.4996	4.98	1.6054	5.48	1.7011	5.98	1.7884
3.99	1.3838	4.49	1.5019	4.99	1.6074	5.49	1.7029	5.99	1.7901
4.00	1.3863	4.50	1.5041	5.00	1.6094	5.50	1.7047	6.00	1.7918

TABLE 4 (*continued*)

Hyperbolic or Naperian Logarithms

No.	H. Log.	No.	H. Log.	No.	H. Log.	No.	H. Log.	No.	H. Log.
6.01	1.7934	6.51	1.8733	7.01	1.9473	7.51	2.0162	8.01	2.0807
6.02	1.7951	6.52	1.8749	7.02	1.9488	7.52	2.0176	8.02	2.0819
6.03	1.7967	6.53	1.8764	7.03	1.9502	7.53	2.0189	8.03	2.0832
6.04	1.7984	6.54	1.8779	7.04	1.9516	7.54	2.0202	8.04	2.0844
6.05	1.8001	6.55	1.8795	7.05	1.9530	7.55	2.0215	8.05	2.0857
6.06	1.8017	6.56	1.8810	7.06	1.9544	7.56	2.0229	8.06	2.0869
6.07	1.8034	6.57	1.8825	7.07	1.9559	7.57	2.0242	8.07	2.0882
6.08	1.8050	6.58	1.8840	7.08	1.9573	7.58	2.0255	8.08	2.0894
6.09	1.8066	6.59	1.8856	7.09	1.9587	7.59	2.0268	8.09	2.0906
6.10	1.8083	6.60	1.8871	7.10	1.9601	7.60	2.0281	8.10	2.0919
6.11	1.8099	6.61	1.8886	7.11	1.9615	7.61	2.0295	8.11	2.0931
6.12	1.8116	6.62	1.8901	7.12	1.9629	7.62	2.0308	8.12	2.0943
6.13	1.8132	6.63	1.8916	7.13	1.9643	7.63	2.0321	8.13	2.0956
6.14	1.8148	6.64	1.8931	7.14	1.9657	7.64	2.0334	8.14	2.0968
6.15	1.8165	6.65	1.8946	7.15	1.9671	7.65	2.0347	8.15	2.0980
6.16	1.8181	6.66	1.8961	7.16	1.9685	7.66	2.0360	8.16	2.0992
6.17	1.8197	6.67	1.8976	7.17	1.9699	7.67	2.0373	8.17	2.1005
6.18	1.8213	6.68	1.8991	7.18	1.9713	7.68	2.0386	8.18	2.1017
6.19	1.8229	6.69	1.9006	7.19	1.9727	7.69	2.0399	8.19	2.1029
6.20	1.8245	6.70	1.9021	7.20	1.9741	7.70	2.0412	8.20	2.1041
6.21	1.8262	6.71	1.9036	7.21	1.9755	7.71	2.0425	8.21	2.1054
6.22	1.8278	6.72	1.9051	7.22	1.9769	7.72	2.0438	8.22	2.1066
6.23	1.8294	6.73	1.9066	7.23	1.9782	7.73	2.0451	8.23	2.1078
6.24	1.8310	6.74	1.9081	7.24	1.9796	7.74	2.0464	8.24	2.1090
6.25	1.8326	6.75	1.9095	7.25	1.9810	7.75	2.0477	8.25	2.1102
6.26	1.8342	6.76	1.9110	7.26	1.9824	7.76	2.0490	8.26	2.1114
6.27	1.8358	6.77	1.9125	7.27	1.9838	7.77	2.0503	8.27	2.1126
6.28	1.8374	6.78	1.9140	7.28	1.9851	7.78	2.0516	8.28	2.1138
6.29	1.8390	6.79	1.9155	7.29	1.9865	7.79	2.0528	8.29	2.1150
6.30	1.8405	6.80	1.9169	7.30	1.9879	7.80	2.0541	8.30	2.1163
6.31	1.8421	6.81	1.9184	7.31	1.9892	7.81	2.0554	8.31	2.1175
6.32	1.8437	6.82	1.9199	7.32	1.9906	7.82	2.0567	8.32	2.1187
6.33	1.8453	6.83	1.9213	7.33	1.9920	7.83	2.0580	8.33	2.1199
6.34	1.8469	6.84	1.9228	7.34	1.9933	7.84	2.0592	8.34	2.1211
6.35	1.8485	6.85	1.9242	7.35	1.9947	7.85	2.0605	8.35	2.1223
6.36	1.8500	6.86	1.9257	7.36	1.9961	7.86	2.0618	8.36	2.1235
6.37	1.8516	6.87	1.9272	7.37	1.9974	7.87	2.0631	8.37	2.1247
6.38	1.8532	6.88	1.9286	7.38	1.9988	7.88	2.0643	8.38	2.1258
6.39	1.8547	6.89	1.9301	7.39	2.0001	7.89	2.0656	8.39	2.1270
6.40	1.8563	6.90	1.9315	7.40	2.0015	7.90	2.0669	8.40	2.1282
6.41	1.8579	6.91	1.9330	7.41	2.0028	7.91	2.0681	8.41	2.1294
6.42	1.8594	6.92	1.9344	7.42	2.0041	7.92	2.0694	8.42	2.1306
6.43	1.8610	6.93	1.9359	7.43	2.0055	7.93	2.0707	8.43	2.1318
6.44	1.8625	6.94	1.9373	7.44	2.0069	7.94	2.0719	8.44	2.1330
6.45	1.8641	6.95	1.9387	7.45	2.0082	7.95	2.0732	8.45	2.1342
6.46	1.8656	6.96	1.9402	7.46	2.0096	7.96	2.0744	8.46	2.1353
6.47	1.8672	6.97	1.9416	7.47	2.0109	7.97	2.0757	8.47	2.1365
6.48	1.8687	6.98	1.9430	7.48	2.0122	7.98	2.0769	8.48	2.1377
6.49	1.8703	6.99	1.9445	7.49	2.0136	7.99	2.0782	8.49	2.1389
6.50	1.8718	7.00	1.9459	7.50	2.0149	8.00	2.0794	8.50	2.1401

TABLE 4 (*continued*)

Hyperbolic or Naperian Logarithms

No.	H. Log.	No.	H. Log.	No.	H. Log.	No.	H. Log.	No.	H. Log.
8.51	2.1412	9.01	2.1983	9.51	2.2523	10.25	2.3273	41	3.7136
8.52	2.1424	9.02	2.1994	9.52	2.2534	10.50	2.3514	42	3.7377
8.53	2.1436	9.03	2.2006	9.53	2.2544	10.75	2.3749	43	3.7612
8.54	2.1448	9.04	2.2017	9.54	2.2555	11.00	2.3979	44	3.7842
8.55	2.1459	9.05	2.2028	9.55	2.2565	11.25	2.4204	45	3.8067
8.56	2.1471	9.06	2.2039	9.56	2.2576	11.50	2.4423	46	3.8286
8.57	2.1483	9.07	2.2050	9.57	2.2586	11.75	2.4638	47	3.8501
8.58	2.1494	9.08	2.2061	9.58	2.2597	12.00	2.4849	48	3.8712
8.59	2.1506	9.09	2.2072	9.59	2.2607	12.25	2.5055	49	3.8918
8.60	2.1518	9.10	2.2083	9.60	2.2618	12.50	2.5257	50	3.9120
8.61	2.1529	9.11	2.2094	9.61	2.2628	12.75	2.5455	51	3.9318
8.62	2.1541	9.12	2.2105	9.62	2.2638	13.00	2.5649	52	3.9512
8.63	2.1552	9.13	2.2116	9.63	2.2649	13.25	2.5840	53	3.9703
8.64	2.1564	9.14	2.2127	9.64	2.2659	13.50	2.6027	54	3.9890
8.65	2.1576	9.15	2.2138	9.65	2.2670	13.75	2.6210	55	4.0073
8.66	2.1587	9.16	2.2148	9.66	2.2680	14.00	2.6391	56	4.0254
8.67	2.1599	9.17	2.2159	9.67	2.2690	14.25	2.6568	57	4.0431
8.68	2.1610	9.18	2.2170	9.68	2.2701	14.50	2.6741	58	4.0604
8.69	2.1622	9.19	2.2181	9.69	2.2711	14.75	2.6912	59	4.0775
8.70	2.1633	9.20	2.2192	9.70	2.2721	15.00	2.7081	60	4.0943
8.71	2.1645	9.21	2.2203	9.71	2.2732	15.50	2.7408	61	4.1109
8.72	2.1656	9.22	2.2214	9.72	2.2742	16.00	2.7726	62	4.1271
8.73	2.1668	9.23	2.2225	9.73	2.2752	16.50	2.8034	63	4.1431
8.74	2.1679	9.24	2.2235	9.74	2.2762	17.00	2.8332	64	4.1589
8.75	2.1691	9.25	2.2246	9.75	2.2773	17.50	2.8622	65	4.1744
8.76	2.1702	9.26	2.2257	9.76	2.2783	18.00	2.8904	66	4.1897
8.77	2.1713	9.27	2.2268	9.77	2.2793	18.50	2.9178	67	4.2047
8.78	2.1725	9.28	2.2279	9.78	2.2803	19.00	2.9444	68	4.2195
8.79	2.1736	9.29	2.2289	9.79	2.2814	19.50	2.9704	69	4.2341
8.80	2.1748	9.30	2.2300	9.80	2.2824	20.00	2.9957	70	4.2485
8.81	2.1759	9.31	2.2311	9.81	2.2834	21	3.0445	71	4.2627
8.82	2.1770	9.32	2.2322	9.82	2.2844	22	3.0910	72	4.2767
8.83	2.1782	9.33	2.2332	9.83	2.2854	23	3.1355	73	4.2905
8.84	2.1793	9.34	2.2343	9.84	2.2865	24	3.1781	74	4.3041
8.85	2.1804	9.35	2.2354	9.85	2.2875	25	3.2189	75	4.3175
8.86	2.1815	9.36	2.2364	9.86	2.2885	26	3.2581	76	4.3307
8.87	2.1827	9.37	2.2375	9.87	2.2895	27	3.2958	77	4.3438
8.88	2.1838	9.38	2.2386	9.88	2.2905	28	3.3322	78	4.3567
8.89	2.1849	9.39	2.2396	9.89	2.2915	29	3.3673	79	4.3694
8.90	2.1861	9.40	2.2407	9.90	2.2925	30	3.4012	80	4.3820
8.91	2.1872	9.41	2.2418	9.91	2.2935	31	3.4340	82	4.4067
8.92	2.1883	9.42	2.2428	9.92	2.2946	32	3.4657	84	4.4308
8.93	2.1894	9.43	2.2439	9.93	2.2956	33	3.4965	86	4.4543
8.94	2.1905	9.44	2.2450	9.94	2.2966	34	3.5264	88	4.4773
8.95	2.1917	9.45	2.2460	9.95	2.2976	35	3.5553	90	4.4998
8.96	2.1928	9.46	2.2471	9.96	2.2986	36	3.5835	92	4.5218
8.97	2.1939	9.47	2.2481	9.97	2.2996	37	3.6109	94	4.5433
8.98	2.1950	9.48	2.2492	9.98	2.3006	38	3.6376	96	4.5643
8.99	2.1961	9.49	2.2502	9.99	2.3016	39	3.6636	98	4.5850
9.00	2.1972	9.50	2.2513	10.00	2.3026	40	3.6889	100	4.6052

TABLE 5

DECIMAL AND METRIC EQUIVALENTS OF FRACTIONS OF ONE INCH

Fraction	Decimal Equivalents	Metric Equivalents, mm	Fraction	Decimal Equivalents	Metric Equivalents, mm
1/64	0.015625	0.397	33/64	0.515625	13.096
1/32	0.03125	0.794	17/32	0.53125	13.493
3/64	0.046875	1.191	35/64	0.546875	13.890
1/16	0.0625	1.587	9/16	0.5625	14.287
5/64	0.078125	1.984	37/64	0.578125	14.684
3/32	0.09375	2.381	19/32	0.59375	15.081
7/64	0.109375	2.778	39/64	0.609375	15.478
1/8	0.1250	3.175	5/8	0.6250	15.875
9/64	0.140625	3.572	41/64	0.640625	16.272
5/32	0.15625	3.968	21/32	0.65625	16.668
11/64	0.171875	4.365	43/64	0.671875	17.065
3/16	0.1875	4.762	11/16	0.6875	17.462
13/64	0.203125	5.159	45/64	0.703125	17.859
7/32	0.21875	5.556	23/32	0.71875	18.256
15/64	0.234375	5.953	47/64	0.734375	18.653
1/4	0.2500	6.349	3/4	0.7500	19.050
17/64	0.265625	6.746	49/64	0.765625	19.447
9/32	0.28125	7.144	25/32	0.78125	19.843
19/64	0.296875	7.541	51/64	0.796875	20.240
5/16	0.3125	7.937	13/16	0.8125	20.637
21/64	0.328125	8.334	53/64	0.828125	21.034
11/32	0.34375	8.731	27/32	0.84375	21.431
23/64	0.359375	9.128	55/64	0.859375	21.828
3/8	0.3750	9.525	7/8	0.8750	22.225
25/64	0.390625	9.922	57/64	0.890625	22.622
13/32	0.40625	10.319	29/32	0.90625	23.018
27/64	0.421875	10.716	59/64	0.921875	23.415
7/16	0.4375	11.112	15/16	0.9375	23.812
29/64	0.453125	11.509	61/64	0.953125	24.209
15/32	0.46875	11.906	31/32	0.96875	24.606
31/64	0.484375	12.303	63/64	0.984375	25.003
1/2	0.5000	12.699	1	1.000	25.400

TABLE 6

Squares and Square Roots

No.	Square	Square Root	No.	Square	Square Root
1	1	1.000	24	576	4.899
2	4	1.414	25	625	5.000
3	9	1.732	26	676	5.099
4	16	2.000	27	729	5.196
5	25	2.236	28	784	5.292
6	36	2.450	29	841	5.385
7	49	2.646	30	900	5.477
8	64	2.828	31	961	5.568
9	81	3.000	32	1024	5.657
10	100	3.162	33	1089	5.745
11	121	3.317	34	1156	5.831
12	144	3.464	35	1225	5.916
13	169	3.606	36	1296	6.000
14	196	3.742	37	1369	6.083
15	225	3.873	38	1444	6.164
16	256	4.000	39	1521	6.245
17	289	4.123	40	1600	6.325
18	324	4.243	50	2500	7.071
19	361	4.359	60	3600	7.746
20	400	4.472	70	4900	8.367
21	441	4.583	80	6400	8.944
22	484	4.690	90	8100	9.487
23	529	4.796	100	10,000	10.000

TABLE 7

Unified and American Screw Threads

Coarse, Fine, Extra-Fine, 8-, 12-, and 16-Thread Series

Sizes	Basic Major Diameter	Series with graded pitches						Series with constant pitches					
		Coarse UNC; NC		Fine UNF; NF		Extra-Fine UNEF; NEF		8-Thread 8 UN; 8N		12-Thread 12 UN; 12 N		16-Thread 16 UN; 16 N	
		Thds. per in.	Tap Drill	Thds. per in.	Tap Drill	Thds. per in.	Tap Drill	Thds. per in.	Tap Drill	Thds. per in.	Tap Drill	Thds. per in.	Tap Drill
0	0.0600	...	...	80	3/64	...	...	...	...	...	...	...	...
1	0.0730	64	53	72	53	...	...	...	...	...	...	...	...
2	0.0860	56	50	64	50	...	...	...	...	...	...	...	...
3	0.0990	48	47	56	45	...	...	...	...	...	...	...	...
4	0.1120	40	43	48	42	...	...	...	...	...	...	...	...
5	0.1250	40	38	44	37	...	...	...	...	...	...	...	...
6	0.1380	32	36	40	33	...	...	...	...	...	...	...	...
8	0.1640	32	29	36	29	...	...	...	...	...	...	...	...
10	0.1900	24	25	32	21	...	...	...	...	...	...	...	...
12	0.2160	24	16	28	14	32	13	...	...	...	...	...	...
1/4	0.2500	20	7	28	3	32	7/32	...	...	...	...	...	...
5/16	0.3125	18	F	24	I	32	9/32	...	...	...	...	...	...
3/8	0.3750	16	5/16	24	Q	32	11/32	...	...	...	...	UNC	5/16
7/16	0.4375	14	U	20	25/64	28	13/32	...	...	...	...	16	3/8
1/2	0.5000	13	27/64	20	29/64	28	15/32	...	...	...	...	16	7/16
9/16	0.5625	12	31/64	18	33/64	24	33/64	...	...	UNC	31/64	16	1/2
5/8	0.6250	11	17/32	18	37/64	24	37/64	...	...	12	35/64	16	9/16
11/16	0.6875	...	...	...	...	24	41/64	...	...	12	39/64	16	5/8
3/4	0.7500	10	21/32	16	11/16	20	45/64	...	...	12	43/64	UNF	11/16
13/16	0.8125	...	...	...	...	20	49/64	...	...	12	47/64	16	3/4
7/8	0.8750	9	49/64	14	13/16	20	53/64	...	...	12	51/64	16	13/16
15/16	0.9375	...	...	...	...	20	57/64	...	...	12	55/64	16	7/8
1	1.0000	8	7/8	12	59/64	20	61/64	UNC	7/8	UNF	59/64	16	15/16
1 1/16	1.0625	...	...	...	...	18	1	8	15/16	12	63/64	16	1
1 1/8	1.1250	7	63/64	12	1 3/64	18	1 5/64	8	1	UNF	1 3/64	16	1 1/16
1 3/16	1.1875	...	...	...	...	18	1 9/64	8	1 1/16	12	1 7/64	16	1 1/8
1 1/4	1.2500	7	1 7/64	12	1 11/64	18	1 3/16	8	1 1/8	UNF	1 11/64	16	1 3/16
1 5/16	1.3125	...	...	...	...	18	1 17/64	8	1 3/16	12	1 15/64	16	1 1/4

1 3/8	1.3750	6	1 7/32	12	1 19/64	18	1 5/16	8	1 1/4	UNF	1 19/64	16	1 5/16
1 7/16	1.4375	...	...	...	...	18	1 3/8	8	1 5/16	12	1 23/64	16	1 3/8
1 1/2	1.5000	6	1 11/32	12	1 27/64	18	1 7/16	8	1 3/8	UNF	1 27/64	16	1 7/16
1 9/16	1.5625	...	...	...	...	18	1 1/2	8	1 7/16	12	1 31/64	16	1 1/2
1 5/8	1.6250	...	...	...	...	18	1 9/16	8	1 1/2	12	1 35/64	16	1 9/16
1 11/16	1.6875	...	...	...	...	18	1 5/8	8	1 9/16	12	1 39/64	16	1 5/8
1 3/4	1.7500	5	1 9/16	...	...	16	1 11/16	8	1 5/8	12	1 43/64	16	1 11/16
1 13/16	1.8125	...	...	...	...	...	...	8	1 11/16	12	1 47/64	16	1 3/4
1 7/8	1.8750	...	...	...	...	...	...	8	1 3/4	12	1 51/64	16	1 13/16
1 15/16	1.9375	...	...	...	...	...	...	8	1 13/16	12	1 55/64	16	1 7/8
2	2.0000	4 1/2	1 25/32	...	...	16	1 15/16	8	1 7/8	12	1 59/64	16	1 15/16
2 1/8	2.1250	...	...	...	...	...	...	8	2	12	2 3/64	16	2 1/16
2 1/4	2.2500	4 1/2	2 1/32	...	...	...	...	8	2 1/8	12	2 11/64	16	2 3/16
2 3/8	2.3750	...	...	...	...	...	...	8	2 1/4	12	2 19/64	16	2 5/16
2 1/2	2.5000	4	2 1/4	...	...	...	...	8	2 3/8	12	2 27/64	16	2 7/16
2 5/8	2.6250	...	...	...	...	...	...	8	2 1/2	12	2 35/64	16	2 9/16
2 3/4	2.7500	4	2 1/2	...	...	...	...	8	2 5/8	12	2 43/64	16	2 11/16
2 7/8	2.8750	...	...	...	...	...	...	8	2 3/4	12	2 51/64	16	2 13/16
3	3.0000	4	2 3/4	...	...	...	...	8	2 7/8	12	2 59/64	16	2 15/16
3 1/8	3.1250	...	...	...	...	...	...	8	3	12	3 3/64	16	3 1/16
3 1/4	3.2500	4	3	...	...	...	...	8	3 1/8	12	3 11/64	16	3 3/16
3 3/8	3.3750	...	...	...	...	...	...	8	3 1/4	12	3 19/64	16	3 5/16
3 1/2	3.5000	4	3 1/4	...	...	...	...	8	3 3/8	12	3 27/64	16	3 7/16
3 5/8	3.6250	...	...	...	...	...	...	8	3 1/2	12	3 35/64	16	3 9/16
3 3/4	3.7500	4	3 1/2	...	...	...	...	8	3 5/8	12	3 43/64	16	3 11/16
3 7/8	3.8750	...	...	...	...	...	...	8	3 3/4	12	3 51/64	16	3 13/16
4	4.0000	4	3 3/4	...	...	...	...	8	3 7/8	12	3 59/64	16	3 15/16
4 1/8	4.1250	...	...	...	...	...	...	8	4	12	4 3/64	16	4 1/16
4 1/4	4.2500	...	...	...	...	...	...	8	4 1/8	12	4 11/64	16	4 3/16
4 3/8	4.3750	...	...	...	...	...	...	8	4 1/4	12	4 19/64	16	4 5/16
4 1/2	4.5000	...	...	...	...	...	...	8	4 3/8	12	4 27/64	16	4 7/16
4 5/8	4.6250	...	...	...	...	...	...	8	4 1/2	12	4 35/64	16	4 9/16
4 3/4	4.7500	...	...	...	...	...	...	8	4 5/8	12	4 43/64	16	4 11/16
4 7/8	4.8750	...	...	...	...	...	...	8	4 3/4	12	4 51/64	16	4 13/16
5	5.0000	...	...	...	...	...	...	8	4 7/8	12	4 59/64	16	4 15/16
5 1/8	5.1250	...	...	...	...	...	...	8	5	12	5 3/64	16	5 1/16
5 1/4	5.2500	...	...	...	...	...	...	8	5 1/8	12	5 11/64	16	5 3/16

(Compiled from ASA B1.1-1960)

TABLE 8. CAP SCREWS

Hexagon Head Cap Screws

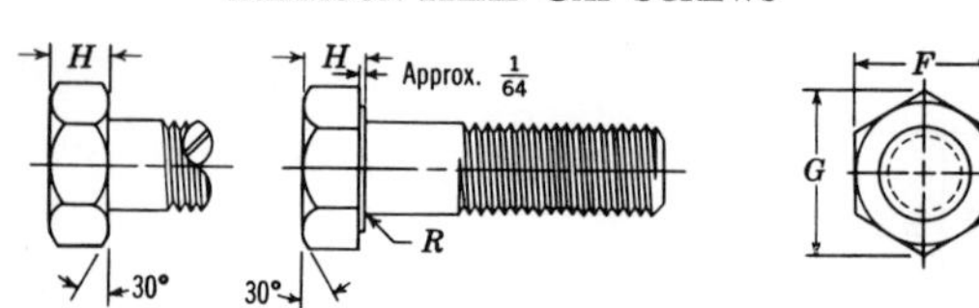

Nominal Size	Width across Flats, F (Max)	Width across Corners, G (Max)	Height, H (Nom)
1/4	**7/16**	**0.505**	**5/32**
5/16	**1/2**	**0.577**	**13/64**
3/8	**9/16**	**0.650**	**15/64**
7/16	**5/8**	**0.722**	**9/32**
1/2	**3/4**	**0.866**	**5/16**
9/16	**13/16**	**0.938**	**23/64**
5/8	**15/16**	**1.083**	**25/64**
3/4	**1 1/8**	**1.299**	**15/32**
7/8	**1 5/16**	**1.516**	**35/64**
1	**1 1/2**	**1.732**	**39/64**
1 1/8	1 11/16	1.949	11/16
1 1/4	1 7/8	2.165	25/32
1 3/8	2 1/16	2.382	27/32
1 1/2	2 1/4	2.598	15/16

Threads shall be coarse-, fine-, or 8-thread series, class 2A for plain (unplated) cap screws. For plated cap screws, the diameters may be increased by the amount of class 2A allowance. Thickness or quality of plating shall be measured or tested on the side of the head.

All dimensions given in inches.

BOLD TYPE INDICATES PRODUCTS UNIFIED DIMENSIONALLY WITH BRITISH AND CANADIAN STANDARDS.

Minimum thread length shall be twice the diameter plus 1/4 in. for lengths up to and including 6 in.; twice the diameter plus 1/2 in. for lengths over 6 in. The tolerance shall be plus 3/16 in. or 2 1/2 threads, whichever is greater. On products that are too short for minimum thread lengths the distance from the bearing surface of the head to the first complete thread shall not exceed the length of 2 1/2 threads, as measured with a ring thread gage, for sizes up to and including 1 in., and 3 1/2 threads for sizes larger than 1 in.

Dimensions of Fillister Head Cap Screws

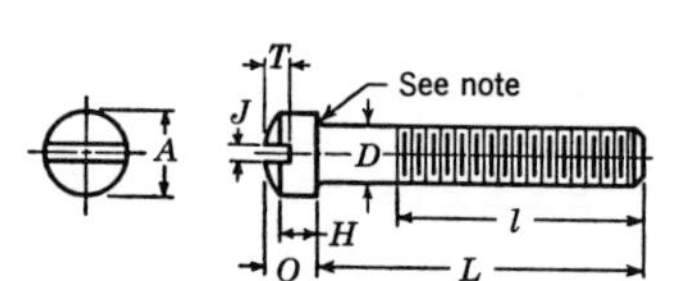

Nominal Size, D	Head Diameter, A (Max)	Height of Head, H (Max)	Total Height of Head, O (Max)	Width of Slot, J (Max)	Depth of Slot, T (Max)
1/4	0.375	0.172	0.216	0.075	0.097
5/16	0.437	0.203	0.253	0.084	0.115
3/8	0.562	0.250	0.314	0.094	0.143
7/16	0.625	0.297	0.368	0.094	0.168
1/2	0.750	0.328	0.412	0.106	0.188
9/16	0.812	0.375	0.466	0.118	0.214
5/8	0.875	0.422	0.521	0.133	0.240
3/4	1.000	0.500	0.612	0.149	0.283
7/8	1.125	0.594	0.720	0.167	0.334
1	1.312	0.656	0.802	0.188	0.372

All dimensions are given in inches.
The radius of the fillet at the base of the head:
For sizes 1/4 to 3/8 in. incl. is 0.016 min and 0.031 max.
7/16 to 9/16 in. incl. is 0.016 min and 0.047 max.
5/8 to 1 in. incl. is 0.031 min and 0.062 max.

Dimensions of Round Head Cap Screws

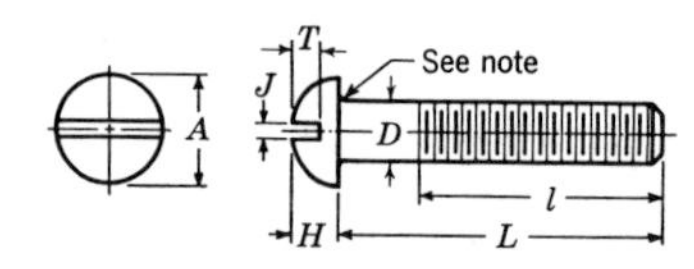

Nominal Size, D	Head Diameter, A (Max)	Height of Head, H (Max)	Width of Slot, J (Max)	Depth of Slot, T (Max)
1/4	0.437	0.191	0.075	0.117
5/16	0.562	0.246	0.084	0.151
3/8	0.625	0.273	0.094	0.168
7/16	0.750	0.328	0.094	0.202
1/2	0.812	0.355	0.106	0.219
9/16	0.937	0.410	0.118	0.253
5/8	1.000	0.438	0.133	0.270
3/4	1.250	0.547	0.149	0.337

All dimensions are given in inches.
Radius of the fillet at the base of the head:
For sizes 1/4 to 3/8 in. incl. is 0.016 min and 0.031 max.
7/16 to 9/16 in. incl. is 0.016 min to 0.047 max.
5/8 to 1 in. incl. is 0.031 min and 0.062 max.

Dimensions of Flat Head Cap Screws

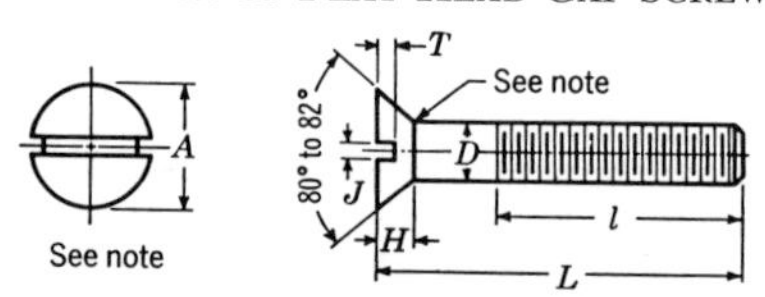

Nominal Size, D	Head Diameter, A (Max)	Height of Head, H (Average)	Width of Slot, J (Max)	Depth of Slot, T (Max)
1/4	0.500	0.140	0.075	0.069
5/16	0.625	0.176	0.084	0.086
3/8	0.750	0.210	0.094	0.103
7/16	0.8125	0.210	0.094	0.103
1/2	0.875	0.210	0.106	0.103
9/16	1.000	0.245	0.118	0.120
5/8	1.125	0.281	0.133	0.137
3/4	1.375	0.352	0.149	0.171
7/8	1.625	0.423	0.167	0.206
1	1.875	0.494	0.188	0.240

All dimensions are given in inches.
The maximum head diameters, A, are extended to the theoretical sharp corners.
The radius of the fillet at the base of the head shall not exceed twice the pitch of the screw thread.

(Courtesy ASA)

TABLE 9. MACHINE SCREWS

Dimensions for Hexagon-Head Machine Screws, Plain or Slotted

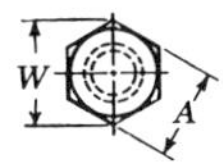

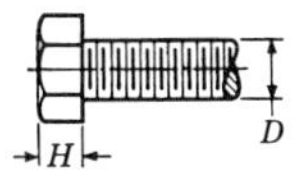

Trimmed Head

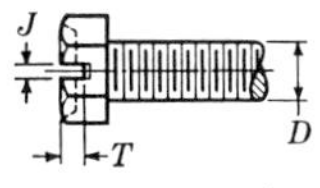

Upset Head

Nominal Size	Basic Diameter, D	Standard Trimmed or Upset Head: Head Diameter, A (Max)	Standard Trimmed or Upset Head: Across Corners, W (Min)	Optional Upset Type Head for Special Requirements: Head Diameter, A (Max)	Optional Upset Type Head for Special Requirements: Across Corners, W (Min)	Height of Head, H (Max)	Width of Slot, J (Max)	Depth of Slot, T (Max)
2	0.0860	0.125	0.134	.....	.....	0.050	.....	.....
3	0.0990	0.187	0.202	.....	.....	0.055	.....	.....
4	0.1120	0.187	0.202	0.219	0.238	0.060	0.039	0.036
5	0.1250	0.187	0.202	0.250	0.272	0.070	0.043	0.042
6	0.1380	0.250	0.272	.....	.....	0.080	0.048	0.046
8	0.1640	0.250	0.272	0.312	0.340	0.110	0.054	0.066
10	0.1900	0.312	0.340	.....	.....	0.120	0.060	0.072
12	0.2160	0.312	0.340	0.375	0.409	0.155	0.067	0.093
1/4	0.2500	0.375	0.409	0.437	0.477	0.190	0.075	0.101
5/16	0.3125	0.500	0.548	.....	.....	0.230	0.084	0.122
3/8	0.3750	0.562	0.616	.....	.....	0.295	0.094	0.156

All dimensions are given in inches.
Hexagon-head machine screws are usually not slotted. The slot is optional.

Dimensions of Round Head Machine Screws

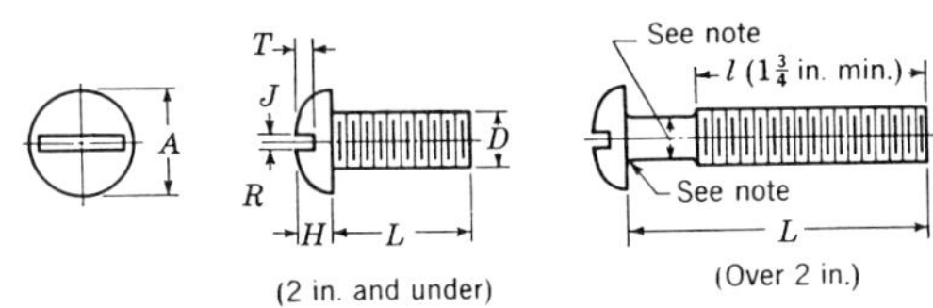

Nominal Size	Diameter of Screw, D (Max)	Head Diameter, A (Max)	Height of Head, H (Max)	Width of Slot, J (Max)	Depth of Slot, T (Max)
0	0.060	0.113	0.053	0.023	0.039
1	0.073	0.138	0.061	0.026	0.044
2	0.086	0.162	0.069	0.031	0.048
3	0.099	0.187	0.078	0.035	0.053
4	0.112	0.211	0.086	0.039	0.058
5	0.125	0.236	0.095	0.043	0.063
6	0.138	0.260	0.103	0.048	0.068
8	0.164	0.309	0.120	0.054	0.077
10	0.190	0.359	0.137	0.060	0.087
12	0.216	0.408	0.153	0.067	0.096
1/4	0.250	0.472	0.175	0.075	0.109
5/16	0.3125	0.590	0.216	0.084	0.132
3/8	0.375	0.708	0.256	0.094	0.155
7/16	0.4375	0.750	0.328	0.094	0.196
1/2	0.500	0.813	0.355	0.106	0.211
9/16	0.5625	0.938	0.410	0.118	0.242
5/8	0.625	1.000	0.438	0.133	0.258
3/4	0.750	1.250	0.547	0.149	0.320

All dimensions are given in inches.
Head dimensions for sizes 7/16 in. and larger are in agreement with round head cap screw dimensions.
The diameter of the unthreaded portion of machine screws shall not be less than the minimum pitch diameter nor more than the maximum major diameter of the thread.
The radius of the fillet at the base of the head shall not exceed one-half the pitch of the screw thread.

(Courtesy ASA)

TABLE 9 (*continued*)

DIMENSIONS OF OVAL HEAD AND FLAT HEAD MACHINE SCREWS

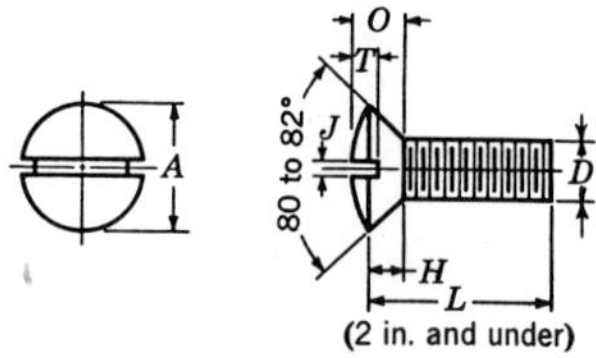

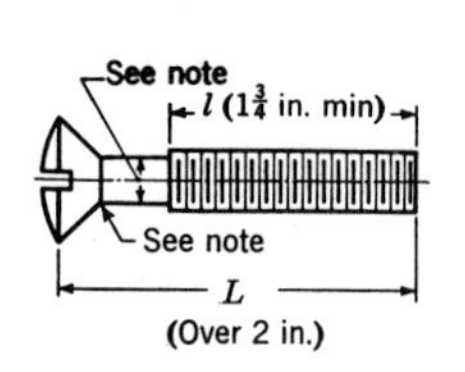

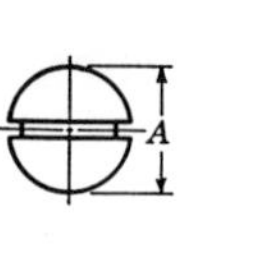

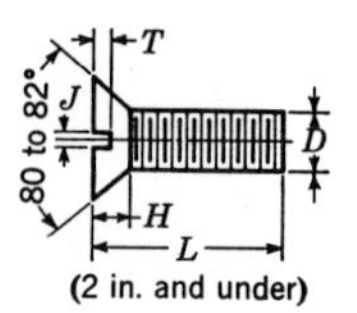

Nominal Size	*Max Diameter of Screw, D*	*Head Diameter, A (Max Sharp)*	*Height of Head, H (Max)*	*Total Height of Head, O (Max)*	*Width of Slot, J (Max)*	*Depth of Slot, T*	
						Oval (Max)	*Flat (Max)*
0	0.060	0.119	0.035	0.056	0.023	0.030	0.015
1	0.073	0.146	0.043	0.068	0.026	0.038	0.019
2	0.086	0.172	0.051	0.080	0.031	0.045	0.023
3	0.099	0.199	0.059	0.092	0.035	0.052	0.027
4	0.112	0.225	0.067	0.104	0.039	0.059	0.030
5	0.125	0.252	0.075	0.116	0.043	0.067	0.034
6	0.138	0.279	0.083	0.128	0.048	0.074	0.038
8	0.164	0.332	0.100	0.152	0.054	0.088	0.045
10	0.190	0.385	0.116	0.176	0.060	0.103	0.053
12	0.216	0.438	0.132	0.200	0.067	0.117	0.060
1/4	0.250	0.507	0.153	0.232	0.075	0.136	0.070
5/16	0.3125	0.635	0.191	0.290	0.084	0.171	0.088
3/8	0.375	0.762	0.230	0.347	0.094	0.206	0.106
7/16	0.4375	0.812	0.223	0.345	0.094	0.210	0.103
1/2	0.500	0.875	0.223	0.354	0.106	0.216	0.103
9/16	0.5625	1.000	0.260	0.410	0.118	0.250	0.120
5/8	0.625	1.125	0.298	0.467	0.133	0.285	0.137
3/4	0.750	1.375	0.372	0.578	0.149	0.353	0.171

TABLE 9 (*continued*)

DIMENSIONS OF FILLISTER HEAD MACHINE SCREWS

(2 in. and under)

See note
l ($1\frac{3}{4}$ in. min.)
See note
(Over 2 in.)

minal Size	Diameter of Screw, *D* (*Max*)	Head Diameter, *A* (*Max*)	Height of Head, *H* (*Max*)	Total Height of Head, *O* (*Max*)	Width of Slot, *J* (*Max*)	Depth of Slot, *T* (*Max*)
0	0.060	0.096	0.045	0.059	0.023	0.025
1	0.073	0.118	0.053	0.071	0.026	0.031
2	0.086	0.140	0.062	0.083	0.031	0.037
3	0.099	0.161	0.070	0.095	0.035	0.043
4	0.112	0.183	0.079	0.107	0.039	0.048
5	0.125	0.205	0.088	0.120	0.043	0.054
6	0.138	0.226	0.096	0.132	0.048	0.060
8	0.164	0.270	0.113	0.156	0.054	0.071
10	0.190	0.313	0.130	0.180	0.060	0.083
12	0.216	0.357	0.148	0.205	0.067	0.094
1/4	0.250	0.414	0.170	0.237	0.075	0.109
5/16	0.3125	0.518	0.211	0.295	0.084	0.137
3/8	0.375	0.622	0.253	0.355	0.094	0.164
7/16	0.4375	0.625	0.265	0.368	0.094	0.170
1/2	0.500	0.750	0.297	0.412	0.106	0.190
9/16	0.5625	0.812	0.336	0.466	0.118	0.214
5/8	0.625	0.875	0.375	0.521	0.133	0.240
3/4	0.750	1.000	0.441	0.612	0.149	0.281

dimensions are given in inches.

diameter of the unthreaded portion of machine screws shall not be less than the minimum pitch diameter nor more than the maximum major neter of the thread. The radius of the fillet at the base of the head shall not exceed one-half the pitch of the screw thread.

(Courtesy ASA)

TABLE 10. SET SCREWS

Square Head Set Screws

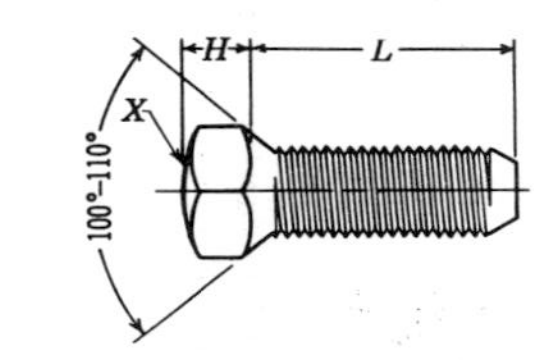

Nominal Size		Width across Flats, F (Max)	Width across Corners, G (Min)	Height of Head, H (Nom)	Diameter of Neck Relief, K (Max)	Radius of Head, X (Nom)	Rad of Neck Relief, R (Max)	Width of Neck Relief, U (Max)
10	0.190	0.1875	0.247	9/64	0.145	15/32	0.027	0.083
12	0.216	0.216	0.292	5/32	0.162	35/64	0.029	0.091
1/4	0.250	0.250	0.331	3/16	0.185	5/8	0.032	0.100
5/16	0.3125	0.3125	0.415	15/64	0.240	25/32	0.036	0.111
3/8	0.3750	0.375	0.497	9/32	0.294	15/16	0.041	0.125
7/16	0.4375	0.4375	0.581	21/64	0.345	1 3/32	0.046	0.143
1/2	0.500	0.500	0.665	3/8	0.400	1 1/4	0.050	0.154
9/16	0.5625	0.5625	0.748	27/64	0.454	1 13/32	0.054	0.167
5/8	0.6250	0.625	0.833	15/32	0.507	1 9/16	0.059	0.182
3/4	0.750	0.750	1.001	9/16	0.620	1 7/8	0.065	0.200
7/8	0.875	0.875	1.170	21/32	0.731	2 3/16	0.072	0.222
1	1.000	1.000	1.337	3/4	0.838	2 1/2	0.081	0.250
1 1/8	1.125	1.125	1.505	27/32	0.939	2 13/16	0.092	0.283
1 1/4	1.250	1.250	1.674	15/16	1.064	3 1/8	0.092	0.283
1 3/8	1.375	1.375	1.843	1 1/32	1.159	3 7/16	0.109	0.333
1 1/2	1.500	1.500	2.010	1 1/8	1.284	3 3/4	0.109	0.333

All dimensions given in inches.
Threads shall be coarse-, fine-, or 8-thread series, class 2A. Square head set screws 1/4 in. size and larger are normally stocked in coarse thread series only.
Square head set screws shall be made from alloy or carbon steel suitably hardened. Screws made from nonferrous material or corrosion-resisting steel shall be made from a material

TABLE 10 (*continued*)

SQUARE HEAD SET SCREW POINTS

Cone Point

Full Dog Point

Half Dog Point

Cup Point

Flat Point

Oval (Round) Point

80° to 90° included angle

$35^{\circ}\,{}^{+5}_{-0}$

$118^{\circ} \pm 5^{\circ}$

Nominal Size	*Cup and Flat Point Diameter, C*		*Oval (Round) Point Radius, J*	*Full Dog, Half Dog, Pivot Point*		
	(Nom)	*(Max)*	*(Nom)*	*Diameter, P (Max)*	*Full Dog Pivot, Q*	*Half Dog Pivot, q*
10	3/32	0.102	0.141	0.127	0.090	0.045
12	7/64	0.115	0.156	0.144	0.110	0.055
1/4	1/8	0.132	0.188	0.156	0.125	0.063
5/16	11/64	0.172	0.234	0.203	0.156	0.078
3/8	13/64	0.212	0.281	0.250	0.188	0.094
7/16	15/64	0.252	0.328	0.297	0.219	0.109
1/2	9/32	0.291	0.375	0.344	0.250	0.125
9/16	5/16	0.332	0.422	0.391	0.281	0.140
5/8	23/64	0.371	0.469	0.469	0.313	0.156
3/4	7/16	0.450	0.563	0.563	0.375	0.188
7/8	33/64	0.530	0.656	0.656	0.438	0.219
1	19/32	0.609	0.750	0.750	0.500	0.250
1 1/8	43/64	0.689	0.844	0.844	0.562	0.281
1 1/4	3/4	0.767	0.938	0.938	0.625	0.312
1 3/8	53/64	0.848	1.031	1.031	0.688	0.344
1 1/2	29/32	0.926	1.125	1.125	0.750	0.375

All dimensions given in inches.
Pivot points are similar to full dog point except that the point is rounded by a radius equal to J.
Where usable length of thread is less than the nominal diameter, half-dog point shall be used.
When length equals nominal diameter or less, $Y = 118$ deg ± 2 deg; when length exceeds nominal diameter, $Y = 90$ deg ± 2 deg.

(Courtesy ASA)

TABLE 10 (*continued*)

DIMENSIONS OF FLUTED AND HEXAGONAL SOCKET HEADLESS SET SCREWS

Section through Hexagonal Socket | Section through Fluted Socket | Cup Point | Flat Point | Oval Point | Cone Point | Full Dog Point | Half Dog Point

D	C	R	Y		P	Q	q		J	M	N	J (Hex)
Nominal Diameter	Cup and Flat Point Diameter (Mean)	Oval Point Radius	Cone Point Angle		Full Dog Point and Half Dog Point			Number of Flutes	Socket Diameter, Minor (Max)	Socket Diameter, Major (Max)	Socket Land Width (Max)	Socket Width across Flats (Max)
			118° ± 2° for these Lengths and Under	90° ± 2° for these Lengths and Over	Diameter (Max)	Full	Half					
5	1/16	3/32	1/8	3/16	0.083	0.06	0.03	4	0.053	0.071	0.022	0.0635
6	.069	7/64	1/8	3/16	0.092	0.07	0.03	4	0.056	0.079	0.023	0.0635
8	5/64	1/8	3/16	1/4	0.109	0.08	0.04	6	0.082	0.098	0.022	0.0791
10	3/32	9/64	3/16	1/4	0.127	0.09	0.04	6	0.098	0.115	0.025	0.0947
12	7/64	5/32	3/16	1/4	0.144	0.11	0.06	6	0.098	0.115	0.025	0.0947
1/4	1/8	3/16	1/4	5/16	5/32	1/8	1/16	6	0.128	0.149	0.032	0.1270
5/16	11/64	15/64	5/16	3/8	13/64	5/32	5/64	6	0.163	0.188	0.039	0.1582
3/8	13/64	9/32	3/8	7/16	1/4	3/16	3/32	6	0.190	0.221	0.050	0.1895

7/16	15/64	21/64	7/16	1/2	19/64	7/32	7/64	6	0.221	0.256	0.060	0.2207
1/2	9/32	3/8	1/2	9/16	11/32	1/4	1/8	6	0.254	0.298	0.068	0.2520
9/16	5/16	27/64	9/16	5/8	25/64	9/32	9/64	6	0.254	0.298	0.068	0.2520
5/8	23/64	15/32	5/8	3/4	15/32	5/16	5/32	6	0.319	0.380	0.092	0.3155
3/4	7/16	9/16	3/4	7/8	9/16	3/8	3/16	6	0.386	0.463	0.112	0.3780
7/8	33/64	21/32	7/8	1	21/32	7/16	7/32	6	0.509	0.604	0.138	0.5030
1	19/32	3/4	1	1 1/8	3/4	1/2	1/4	6	0.535	0.631	0.149	0.5655
1 1/8	43/64	27/32	1 1/8	1 1/4	27/32	9/16	9/32	6	0.604	0.709	0.168	0.5655
1 1/4	3/4	15/16	1 1/4	1 1/2	15/16	5/8	5/16	6	0.685	0.801	0.189	0.6290
1 3/8	53/64	1 1/32	1 3/8	1 5/8	1 1/32	11/16	11/32	6	0.744	0.869	0.207	0.6290
1 1/2	29/32	1 1/8	1 1/2	1 3/4	1 1/8	3/4	3/8	6	0.828	0.970	0.231	0.7540
1 3/4	1 1/16	1 5/16	1 3/4	2	1 5/16	7/8	7/16	6	1.007	1.275	0.298	1.0040
2	1 7/32	1 1/2	2	2 1/4	1 1/2	1	1/2	6	1.007	1.275	0.298	1.0040

All dimensions in inches. (Courtesy ASA)

Where usable length of thread is less than nominal diameter, half dog point shall be used.

Length (L). The length of the screw shall be measured overall on a line parallel to the axis. The difference between consecutive lengths shall be as follows:

(*a*) for screw lengths 1/4 to 5/8 in., difference = 1/16 in.
(*b*) for screw lengths 5/8 to 1 in., difference = 1/8 in.
(*c*) for screw lengths 1 to 4 in., difference = 1/4 in.
(*d*) for screw lengths 4 to 6 in., difference = 1/2 in.

TABLE 11. SQUARE BOLTS

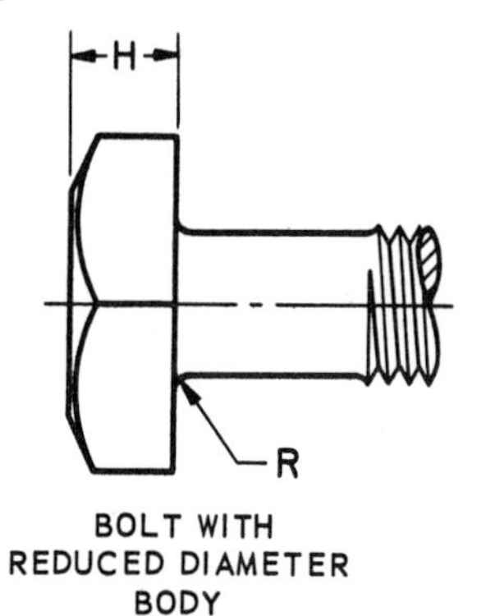

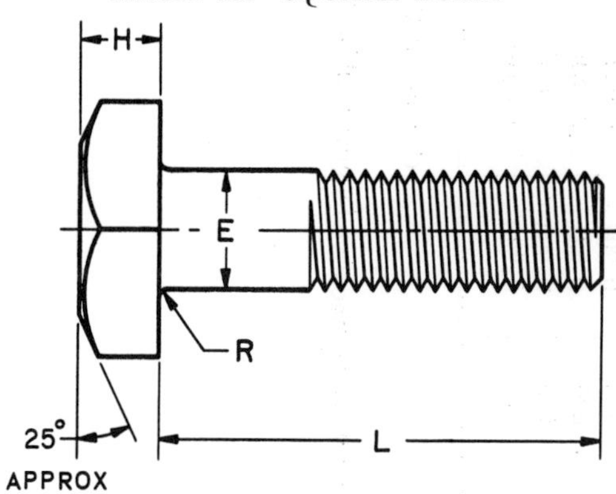

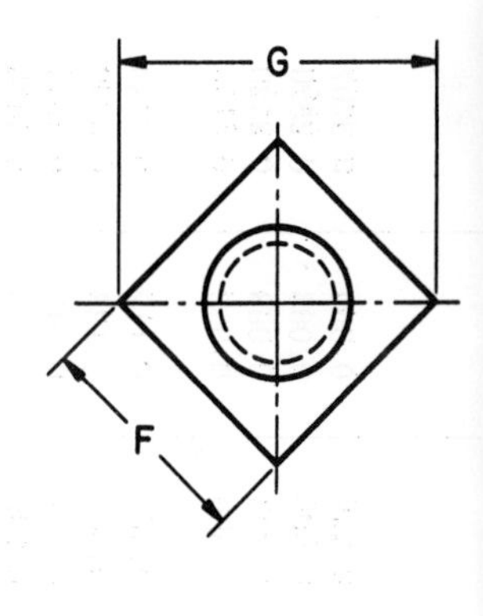

DIMENSIONS OF SQUARE BOLTS

Nominal Size[1] or Basic Product Dia		Body Dia[2] E	Width Across Flats F			Width Across Corners G		Height H			Radius of Fillet R
		Max	Basic	Max	Min	Max	Min	Basic	Max	Min	Max
1/4	0.2500	0.260	3/8	0.3750	0.362	0.530	0.498	11/64	0.188	0.156	0.031
5/16	0.3125	0.324	1/2	0.5000	0.484	0.707	0.665	13/64	0.220	0.186	0.031
3/8	0.3750	0.388	9/16	0.5625	0.544	0.795	0.747	1/4	0.268	0.232	0.031
7/16	0.4375	0.452	5/8	0.6250	0.603	0.884	0.828	19/64	0.316	0.278	0.031
1/2	0.5000	0.515	3/4	0.7500	0.725	1.061	0.995	21/64	0.348	0.308	0.031
5/8	0.6250	0.642	15/16	0.9375	0.906	1.326	1.244	27/64	0.444	0.400	0.062
3/4	0.7500	0.768	1 1/8	1.1250	1.088	1.591	1.494	1/2	0.524	0.476	0.062
7/8	0.8750	0.895	1 5/16	1.3125	1.269	1.856	1.742	19/32	0.620	0.568	0.062
1	1.0000	1.022	1 1/2	1.5000	1.450	2.121	1.991	21/32	0.684	0.628	0.093
1 1/8	1.1250	1.149	1 11/16	1.6875	1.631	2.386	2.239	3/4	0.780	0.720	0.093
1 1/4	1.2500	1.277	1 7/8	1.8750	1.812	2.652	2.489	27/32	0.876	0.812	0.093
1 3/8	1.3750	1.404	2 1/16	2.0625	1.994	2.917	2.738	29/32	0.940	0.872	0.093
1 1/2	1.5000	1.531	2 1/4	2.2500	2.175	3.182	2.986	1	1.036	0.964	0.093

(Courtesy ASA B18.2.1-1965)

All dimensions given in inches.

BOLD TYPE INDICATES PRODUCTS UNIFIED DIMENSIONALLY WITH BRITISH AND CANADIAN STANDARDS.

Bolt need not be finished on any surface except threads.

Minimum thread length shall be twice the basic bolt diameter plus 0.25 in. for lengths up to and including 6 in., and twice the basic diameter plus 0.50 in. for lengths over 6 in. Bolts too short for the formula thread length shall be threaded as close to the head as practical.

Threads shall be in the Unified coarse thread series (UNC Series), Class 2A.

TABLE 12. SQUARE NUTS

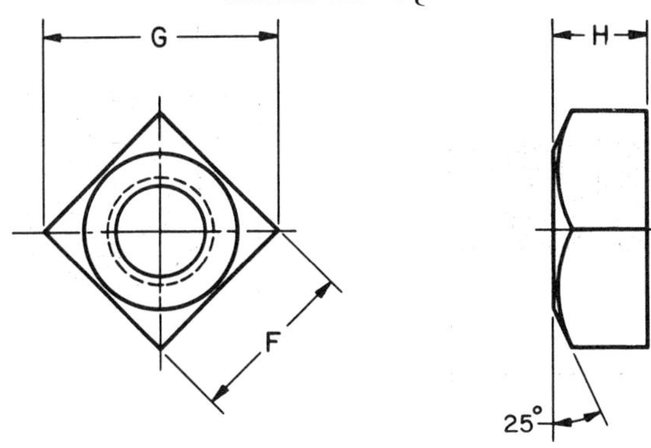

DIMENSIONS OF SQUARE NUTS

Nominal Size[1] or Basic Major Dia of Thread		Width Across Flats F			Width Across Corners G		Thickness H		
		Basic	Max	Min	Max	Min	Basic	Max	Min
1/4	0.2500	7/16	0.4375	0.425	0.619	0.584	7/32	0.235	0.203
5/16	0.3125	9/16	0.5625	0.547	0.795	0.751	17/64	0.283	0.249
3/8	0.3750	5/8	0.6250	0.606	0.884	0.832	21/64	0.346	0.310
7/16	0.4375	3/4	0.7500	0.728	1.061	1.000	3/8	0.394	0.356
1/2	0.5000	13/16	0.8125	0.788	1.149	1.082	7/16	0.458	0.418
5/8	0.6250	1	1.0000	0.969	1.414	1.330	35/64	0.569	0.525
3/4	0.7500	1 1/8	1.1250	1.088	1.591	1.494	21/32	0.680	0.632
7/8	0.8750	1 5/16	1.3125	1.269	1.856	1.742	49/64	0.792	0.740
1	1.0000	1 1/2	1.5000	1.450	2.121	1.991	7/8	0.903	0.847
1 1/8	1.1250	1 11/16	1.6875	1.631	2.386	2.239	1	1.030	0.970
1 1/4	1.2500	1 7/8	1.8750	1.812	2.652	2.489	1 3/32	1.126	1.062
1 3/8	1.3750	2 1/16	2.0625	1.994	2.917	2.738	1 13/64	1.237	1.169
1 1/2	1.5000	2 1/4	2.2500	2.175	3.182	2.986	1 5/16	1.348	1.276

(Courtesy ASA B18.2.2-1965)

All dimensions given in inches.
Threads shall be coarse-threaded series, Class 2B.

TABLE 13. HEXAGON BOLTS

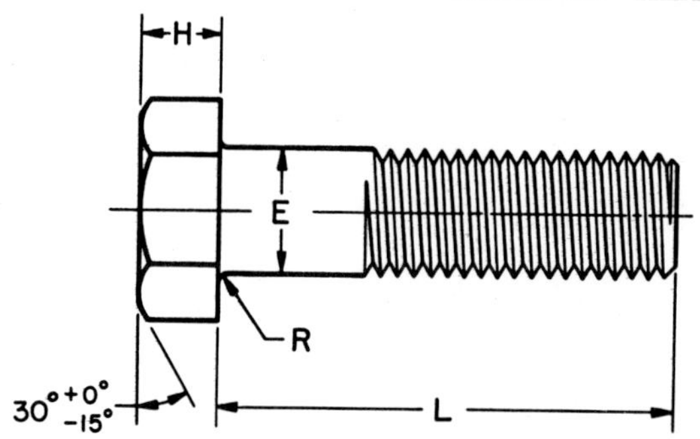

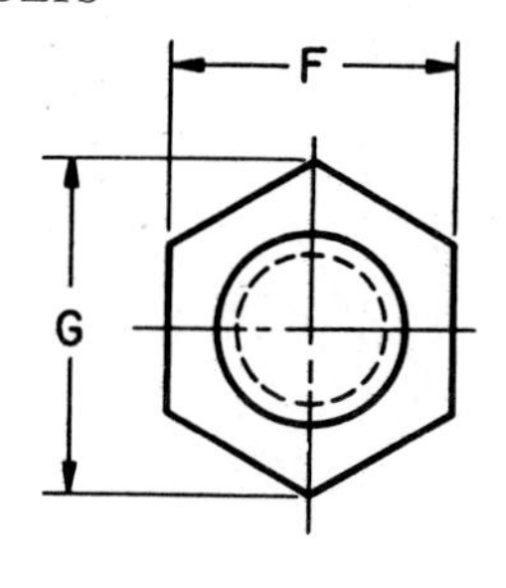

DIMENSIONS OF HEX BOLTS

Nominal Size[1] or Basic Product Dia		Body Dia[2] E	Width Across Flats F			Width Across Corners G		Height H			Radius of Fillet R
		Max	Basic	Max	Min	Max	Min	Basic	Max	Min	Max
1/4	**0.2500**	**0.260**	**7/16**	**0.4375**	**0.425**	**0.505**	**0.484**	**11/64**	**0.188**	**0.150**	**0.031**
5/16	**0.3125**	**0.324**	**1/2**	**0.5000**	**0.484**	**0.577**	**0.552**	**7/32**	**0.235**	**0.195**	**0.031**
3/8	**0.3750**	**0.388**	**9/16**	**0.5625**	**0.544**	**0.650**	**0.620**	**1/4**	**0.268**	**0.226**	**0.031**
7/16	**0.4375**	**0.452**	**5/8**	**0.6250**	**0.603**	**0.722**	**0.687**	**19/64**	**0.316**	**0.272**	**0.031**
1/2	**0.5000**	**0.515**	**3/4**	**0.7500**	**0.725**	**0.866**	**0.826**	**11/32**	**0.364**	**0.302**	**0.031**
5/8	**0.6250**	**0.642**	**15/16**	**0.9375**	**0.906**	**1.083**	**1.033**	**27/64**	**0.444**	**0.378**	**0.062**
3/4	**0.7500**	**0.768**	**1 1/8**	**1.1250**	**1.088**	**1.299**	**1.240**	**1/2**	**0.524**	**0.455**	**0.062**
7/8	**0.8750**	**0.895**	**1 5/16**	**1.3125**	**1.269**	**1.516**	**1.447**	**37/64**	**0.604**	**0.531**	**0.062**
1	**1.0000**	**1.022**	**1 1/2**	**1.5000**	**1.450**	**1.732**	**1.653**	**43/64**	**0.700**	**0.591**	**0.093**
1 1/8	**1.1250**	**1.149**	**1 11/16**	**1.6875**	**1.631**	**1.949**	**1.859**	**3/4**	**0.780**	**0.658**	**0.093**
1 1/4	**1.2500**	**1.277**	**1 7/8**	**1.8750**	**1.812**	**2.165**	**2.066**	**27/32**	**0.876**	**0.749**	**0.093**
1 3/8	**1.3750**	**1.404**	**2 1/16**	**2.0625**	**1.994**	**2.382**	**2.273**	**29/32**	**0.940**	**0.810**	**0.093**
1 1/2	**1.5000**	**1.531**	**2 1/4**	**2.2500**	**2.175**	**2.598**	**2.480**	**1**	**1.036**	**0.902**	**0.093**
1 3/4	**1.7500**	**1.785**	**2 5/8**	**2.6250**	**2.538**	**3.031**	**2.893**	**1 5/32**	**1.196**	**1.054**	**0.125**
2	**2.0000**	**2.039**	**3**	**3.0000**	**2.900**	**3.464**	**3.306**	**1 11/32**	**1.388**	**1.175**	**0.125**
2 1/4	2.2500	2.305	3 3/8	3.3750	3.262	3.897	3.719	1 1/2	1.548	1.327	0.188
2 1/2	2.5000	2.559	3 3/4	3.7500	3.625	4.330	4.133	1 21/32	1.708	1.479	0.188
2 3/4	2.7500	2.827	4 1/8	4.1250	3.988	4.763	4.546	1 13/16	1.869	1.632	0.188
3	3.0000	3.081	4 1/2	4.5000	4.350	5.196	4.959	2	2.060	1.815	0.188
3 1/4	3.2500	3.335	4 7/8	4.8750	4.712	5.629	5.372	2 3/16	2.251	1.936	0.188
3 1/2	3.5000	3.589	5 1/4	5.2500	5.075	6.062	5.786	2 5/16	2.380	2.057	0.188
3 3/4	3.7500	3.858	5 5/8	5.6250	5.437	6.495	6.198	2 1/2	2.572	2.241	0.188
4	4.0000	4.111	6	6.0000	5.800	6.928	6.612	2 11/16	2.764	2.424	0.188

(Courtesy ASA B18.2.1-1965)

All dimensions given in inches.

BOLD TYPE INDICATES PRODUCT FEATURES UNIFIED DIMENSIONALLY WITH BRITISH AND CANADIAN STANDARDS.

Bolt need not be finished on any surface except threads.

Threads shall be in the Unified coarse thread series (UNC Series), Class 2A.

TABLE 14. HEX FLAT NUTS AND HEX FLAT JAM NUTS

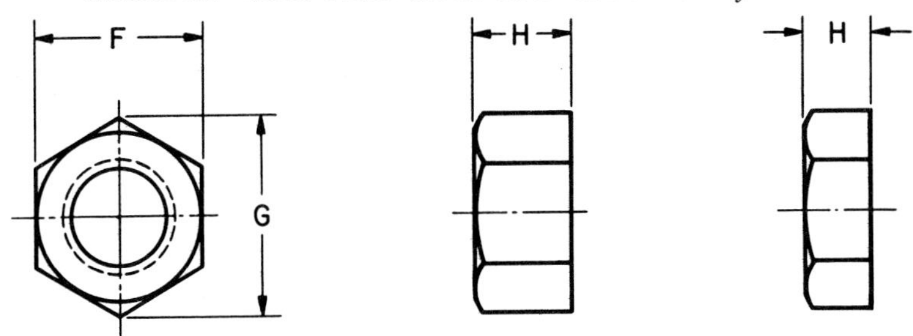

Dimensions of Hex Flat Nuts and Hex Flat Jam Nuts

Nominal Size[1] or Basic Major Dia of Thread		Width Across Flats F			Width Across Corners G		Thickness Hex Flat Nuts H			Thickness Hex Flat Jam Nuts H		
		Basic	Max	Min	Max	Min	Basic	Max	Min	Basic	Max	Min
1 1/8	**1.1250**	**1 11/16**	**1.6875**	**1.631**	**1.949**	**1.859**	**1**	**1.030**	**0.970**	5/8	**0.655**	**0.595**
1 1/4	**1.2500**	**1 7/8**	**1.8750**	**1.812**	**2.165**	**2.066**	**1 3/32**	**1.126**	**1.062**	3/4	**0.782**	**0.718**
1 3/8	**1.3750**	**2 1/16**	**2.0625**	**1.994**	**2.382**	**2.273**	**1 13/64**	**1.237**	**1.169**	13/16	**0.846**	**0.778**
1 1/2	**1.5000**	**2 1/4**	**2.2500**	**2.175**	**2.598**	**2.480**	**1 5/16**	**1.348**	**1.276**	7/8	**0.911**	**0.839**

(Courtesy ASA B18.2.2-1965)

All dimensions are in inches.
BOLD TYPE INDICATES PRODUCTS UNIFIED DIMENSIONALLY WITH BRITISH AND CANADIAN STANDARDS.
Threads shall be in the Unified coarse thread series, Class 2B.

TABLE 15. REGULAR SEMIFINISHED HEXAGON BOLTS

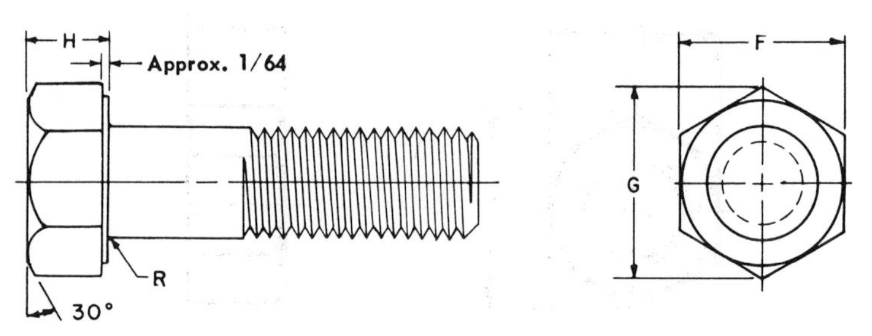

Dimensions of Regular Semifinished Hexagon Bolts

Nominal Size or Basic Major Diameter of Thread	Body Diam*	Width Across Flats F		Width Across Corners G		Height H			Radius of Fillet R	
	Max	Max (Basic)	Min	Max	Min	Nom	Max	Min	Min	Max
1/4 0.2500	0.260	7/16 0.4375	0.425	0.505	0.484	5/32	0.163	0.150	0.009	0.031
5/16 0.3125	0.324	1/2 0.5000	0.484	0.577	0.552	13/64	0.211	0.195	0.009	0.031
3/8 0.3750	0.388	9/16 0.5625	0.544	0.650	0.620	15/64	0.243	0.226	0.009	0.031
7/16 0.4375	0.452	5/8 0.6250	0.603	0.722	0.687	9/32	0.291	0.272	0.009	0.031
1/2 0.5000	0.515	3/4 0.7500	0.725	0.866	0.826	5/16	0.323	0.302	0.009	0.031
5/8 0.6250	0.642	15/16 0.9375	0.906	1.083	1.033	25/64	0.403	0.378	0.021	0.062
3/4 0.7500	0.768	1 1/8 1.1250	1.088	1.299	1.240	15/32	0.483	0.455	0.021	0.062
7/8 0.8750	0.895	1 5/16 1.3125	1.269	1.516	1.447	35/64	0.563	0.531	0.031	0.062
1 1.0000	1.022	1 1/2 1.5000	1.450	1.732	1.653	39/64	0.627	0.591	0.062	0.093
1 1/8 1.1250	1.149	1 11/16 1.6875	1.631	1.949	1.859	11/16	0.718	0.658	0.062	0.093
1 1/4 1.2500	1.277	1 7/8 1.8750	1.812	2.165	2.066	25/32	0.813	0.749	0.062	0.093
1 3/8 1.3750	1.404	2 1/16 2.0625	1.994	2.382	2.273	27/32	0.878	0.810	0.062	0.093
1 1/2 1.5000	1.531	2 1/4 2.2500	2.175	2.598	2.480	15/16	0.974	0.902	0.062	0.093

(Courtesy ASA

All dimensions given in inches. Thread shall be coarse-thread series, class 2A. Semifinished bolt is processed to produce a flat bearing surface under head only.

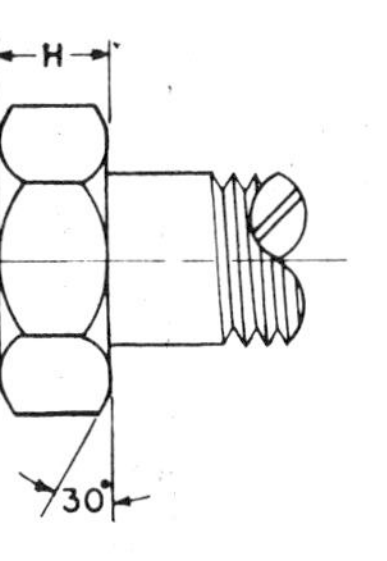

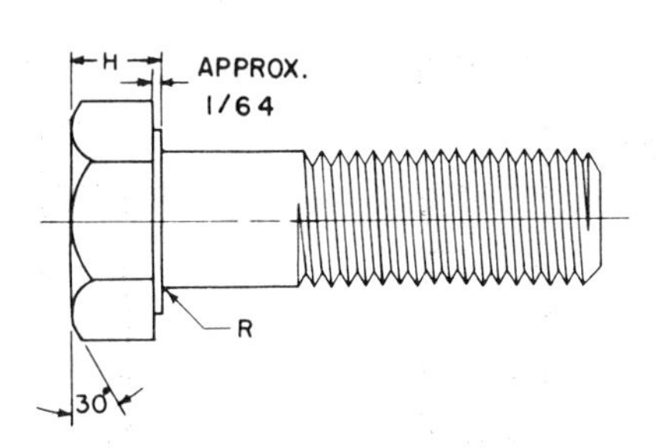

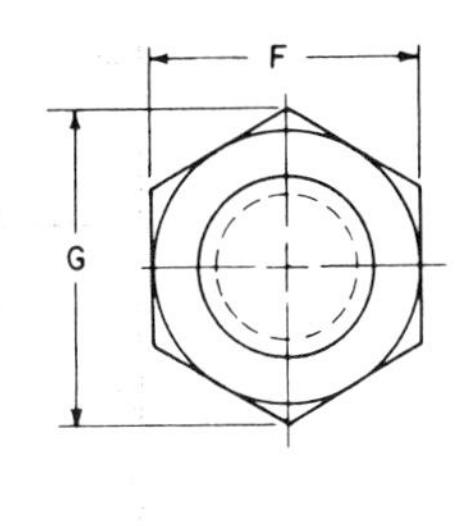

DIMENSIONS OF FINISHED HEXAGON BOLTS

Nominal Size or Basic Major Diameter of Thread	Body Diameter Min (Maximum Equal to Nominal Size)	Width Across Flats F		Width Across Corners G		Height H			Radius of Fillet R	
		Max (Basic)	Min°	Max	Min	Nom	Max	Min	Max	Min
¼ 0.2500	**0.2450**	**7/16 0.4375**	**0.428**	**0.505**	**0.488**	**5/32**	**0.163**	**0.150**	**0.023**	**0.009**
5/16 0.3125	**0.3065**	**½ 0.5000**	**0.489**	**0.577**	**0.557**	**13/64**	**0.211**	**0.195**	**0.023**	**0.009**
⅜ 0.3750	**0.3690**	**9/16 0.5625**	**0.551**	**0.650**	**0.628**	**15/64**	**0.243**	**0.226**	**0.023**	**0.009**
7/16 0.4375	**0.4305**	**⅝ 0.6250**	**0.612**	**0.722**	**0.698**	**9/32**	**0.291**	**0.272**	**0.023**	**0.009**
½ 0.5000	**0.4930**	**¾ 0.7500**	**0.736**	**0.866**	**0.840**	**5/16**	**0.323**	**0.302**	**0.023**	**0.009**
9/16 0.5625	**0.5545**	**13/16 0.8125**	**0.798**	**0.938**	**0.910**	**23/64**	**0.371**	**0.348**	**0.041**	**0.021**
⅝ 0.6250	**0.6170**	**15/16 0.9375**	**0.922**	**1.083**	**1.051**	**25/64**	**0.403**	**0.378**	**0.041**	**0.021**
¾ 0.7500	**0.7410**	**1⅛ 1.1250**	**1.100**	**1.299**	**1.254**	**15/32**	**0.483**	**0.455**	**0.041**	**0.021**
⅞ 0.8750	**0.8660**	**1 5/16 1.3125**	**1.285**	**1.516**	**1.465**	**35/64**	**0.563**	**0.531**	**0.062**	**0.041**
1 1.0000	**0.9900**	**1½ 1.5000**	**1.469**	**1.732**	**1.675**	**39/64**	**0.627**	**0.591**	**0.093**	**0.062**
1⅛ 1.1250	1.1140	1 11/16 1.6875	1.631	1.949	1.859	11/16	0.718	0.658	0.093	0.062
1¼ 1.2500	1.2390	1⅞ 1.8750	1.812	2.165	2.066	25/32	0.813	0.749	0.093	0.062
1⅜ 1.3750	1.3630	2 1/16 2.0625	1.994	2.382	2.273	27/32	0.878	0.810	0.093	0.062
1½ 1.5000	1.4880	2¼ 2.2500	2.175	2.598	2.480	15/16	0.974	0.902	0.093	0.062

(Courtesy ASA)

All dimensions given in inches.

Bold type indicates products unified dimensionally with British and Canadian standards.

"Finished" in the title refers to the quality of manufacture and the closeness of tolerance and does not indicate that surfaces are completely machined.

Threads shall be coarse-, fine-, or 8-thread series, class 2A for plain (unplated) bolts. For plated bolts, the diameters may be increased by the amount of class 2A allowance.

TABLE 17. HEX NUTS AND HEX JAM NUTS

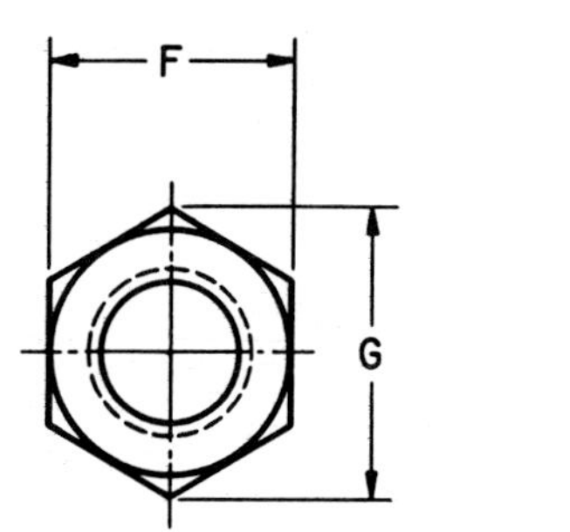

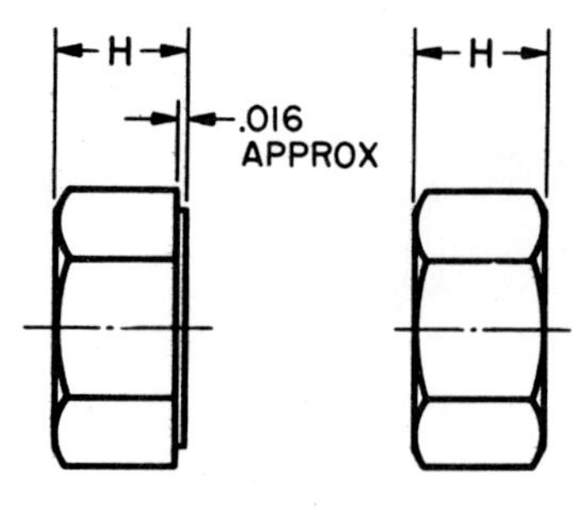

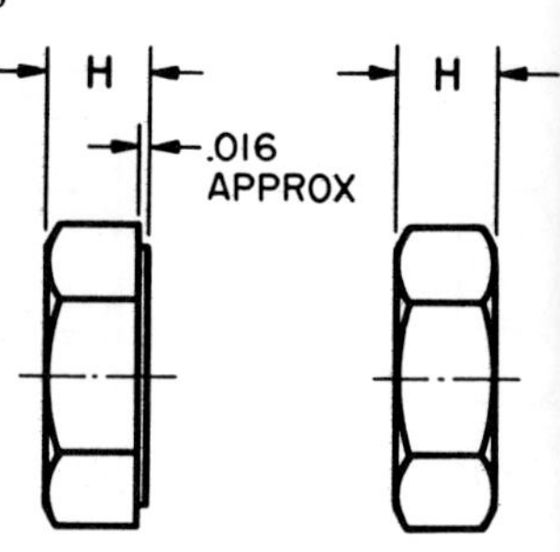

DIMENSIONS OF HEX NUTS AND HEX JAM NUTS

Nominal Size or Basic Major Dia of Thread		Width Across Flats F			Width Across Corners G		Thickness Hex Nuts H			Thickness Hex Jam Nuts H		
		Basic	Max	Min	Max	Min	Basic	Max	Min	Basic	Max	Min
1/4	**0.2500**	**7/16**	**0.4375**	**0.428**	**0.505**	**0.488**	**7/32**	**0.226**	**0.212**	**5/32**	**0.163**	**0.15**
5/16	**0.3125**	**1/2**	**0.5000**	**0.489**	**0.577**	**0.557**	**17/64**	**0.273**	**0.258**	**3/16**	**0.195**	**0.18**
3/8	**0.3750**	**9/16**	**0.5625**	**0.551**	**0.650**	**0.628**	**21/64**	**0.337**	**0.320**	**7/32**	**0.227**	**0.21**
7/16	**0.4375**	**11/16**	**0.6875**	**0.675**	**0.794**	**0.768**	**3/8**	**0.385**	**0.365**	**1/4**	**0.260**	**0.24**
1/2	**0.5000**	**3/4**	**0.7500**	**0.736**	**0.866**	**0.840**	**7/16**	**0.448**	**0.427**	**5/16**	**0.323**	**0.30**
9/16	**0.5625**	**7/8**	**0.8750**	**0.861**	**1.010**	**0.982**	**31/64**	**0.496**	**0.473**	**5/16**	**0.324**	**0.30**
5/8	**0.6250**	**15/16**	**0.9375**	**0.922**	**1.083**	**1.051**	**35/64**	**0.559**	**0.535**	**3/8**	**0.387**	**0.363**
3/4	**0.7500**	**1 1/8**	**1.1250**	**1.088**	**1.299**	**1.240**	**41/64**	**0.665**	**0.617**	**27/64**	**0.446**	**0.398**
7/8	**0.8750**	**1 5/16**	**1.3125**	**1.269**	**1.516**	**1.447**	**3/4**	**0.776**	**0.724**	**31/64**	**0.510**	**0.458**
1	**1.0000**	**1 1/2**	**1.5000**	**1.450**	**1.732**	**1.653**	**55/64**	**0.887**	**0.831**	**35/64**	**0.575**	**0.519**
1 1/8	**1.1250**	**1 11/16**	**1.6875**	**1.631**	**1.949**	**1.859**	**31/32**	**0.999**	**0.939**	**39/64**	**0.639**	**0.579**
1 1/4	**1.2500**	**1 7/8**	**1.8750**	**1.812**	**2.165**	**2.066**	**1 1/16**	**1.094**	**1.030**	**23/32**	**0.751**	**0.687**
1 3/8	**1.3750**	**2 1/16**	**2.0625**	**1.994**	**2.382**	**2.273**	**1 11/64**	**1.206**	**1.138**	**25/32**	**0.815**	**0.747**
1 1/2	1.5000	2 1/4	2.2500	2.175	2.598	2.480	1 9/32	1.317	1.245	27/32	0.880	0.808

(Courtesy ASA B18.2.2-196

All dimensions are in inches.
BOLD TYPE INDICATES PRODUCTS UNIFIED DIMENSIONALLY WITH BRITISH AND CANADIAN STANDARD
Threads shall be in the Unified coarse, fine, or 8-thread series, Class 2B.

TABLE 18. HEX SLOTTED NUTS

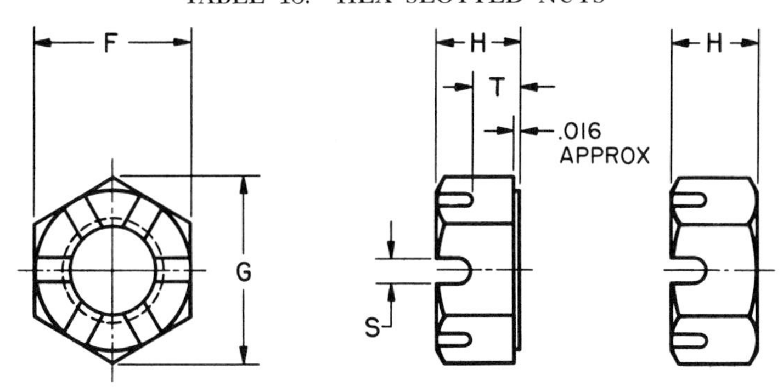

Dimensions of Hex Slotted Nuts

Nominal Size or Basic Major Dia of Thread		Width Across Flats F			Width Across Corners G		Thickness H			Unslotted Thickness T		Width of Slot S	
		Basic	Max	Min	Max	Min	Basic	Max	Min	Max	Min	Max	Min
1/4	0.2500	7/16	0.4375	0.428	0.505	0.488	7/32	0.226	0.212	0.14	0.12	0.10	0.07
5/16	0.3125	1/2	0.5000	0.489	0.577	0.557	17/64	0.273	0.258	0.18	0.16	0.12	0.09
3/8	0.3750	9/16	0.5625	0.551	0.650	0.628	21/64	0.337	0.320	0.21	0.19	0.15	0.12
7/16	0.4375	11/16	0.6875	0.675	0.794	0.768	3/8	0.385	0.365	0.23	0.21	0.15	0.12
1/2	0.5000	3/4	0.7500	0.736	0.866	0.840	7/16	0.448	0.427	0.29	0.27	0.18	0.15
9/16	0.5625	7/8	0.8750	0.861	1.010	0.982	31/64	0.496	0.473	0.31	0.29	0.18	0.15
5/8	0.6250	15/16	0.9375	0.922	1.083	1.051	35/64	0.559	0.535	0.34	0.32	0.24	0.18
3/4	0.7500	1 1/8	1.1250	1.088	1.299	1.240	41/64	0.665	0.617	0.40	0.38	0.24	0.18
7/8	0.8750	1 5/16	1.3125	1.269	1.516	1.447	3/4	0.776	0.724	0.52	0.49	0.24	0.18
1	1.0000	1 1/2	1.5000	1.450	1.732	1.653	55/64	0.887	0.831	0.59	0.56	0.30	0.24
1 1/8	1.1250	1 11/16	1.6875	1.631	1.949	1.859	31/32	0.999	0.939	0.64	0.61	0.33	0.24
1 1/4	1.2500	1 7/8	1.8750	1.812	2.165	2.066	1 1/16	1.094	1.030	0.70	0.67	0.40	0.31
1 3/8	1.3750	2 1/16	2.0625	1.994	2.382	2.273	1 11/64	1.206	1.138	0.82	0.78	0.40	0.31
1 1/2	1.5000	2 1/4	2.2500	2.175	2.598	2.480	1 9/32	1.317	1.245	0.86	0.82	0.46	0.37

(Courtesy ASA B18.2.2-1965)

All dimensions are in inches.

BOLD TYPE INDICATES PRODUCT FEATURES UNIFIED DIMENSIONALLY WITH BRITISH AND CANADIAN STANDARDS.

Threads shall be in the Unified coarse, fine, or 8-thread series, Class 2B.

TABLE 19. HEX THICK NUTS

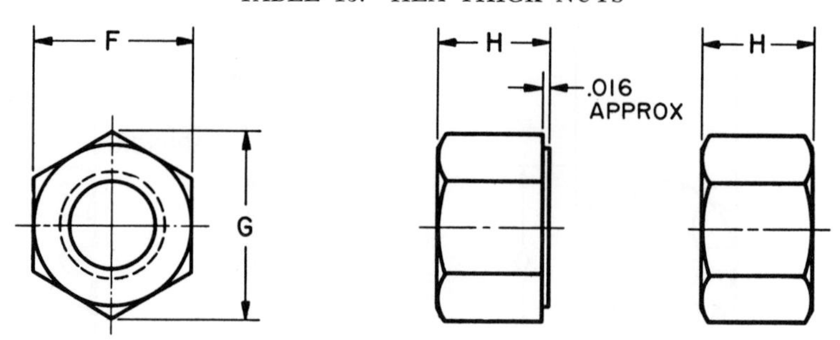

DIMENSIONS OF HEX THICK NUTS

Nominal Size or Basic Major Dia of Thread		Width Across Flats F			Width Across Corners G		Thickness H		
		Basic	Max	Min	Max	Min	Basic	Max	Min
1/4	0.2500	7/16	0.4375	0.428	0.505	0.488	9/32	0.288	0.274
5/16	0.3125	1/2	0.5000	0.489	0.577	0.557	21/64	0.336	0.320
3/8	0.3750	9/16	0.5625	0.551	0.650	0.628	13/32	0.415	0.398
7/16	0.4375	11/16	0.6875	0.675	0.794	0.768	29/64	0.463	0.444
1/2	0.5000	3/4	0.7500	0.736	0.866	0.840	9/16	0.573	0.552
9/16	0.5625	7/8	0.8750	0.861	1.010	0.982	39/64	0.621	0.598
5/8	0.6250	15/16	0.9375	0.922	1.083	1.051	23/32	0.731	0.706
3/4	0.7500	1 1/8	1.1250	1.088	1.299	1.240	13/16	0.827	0.798
7/8	0.8750	1 5/16	1.3125	1.269	1.516	1.447	29/32	0.922	0.890
1	1.0000	1 1/2	1.5000	1.450	1.732	1.653	1	1.018	0.982
1 1/8	1.1250	1 11/16	1.6875	1.631	1.949	1.859	1 5/32	1.176	1.136
1 1/4	1.2500	1 7/8	1.8750	1.812	2.165	2.066	1 1/4	1.272	1.228
1 3/8	1.3750	2 1/16	2.0625	1.994	2.382	2.273	1 3/8	1.399	1.351
1 1/2	1.5000	2 1/4	2.2500	2.175	2.598	2.480	1 1/2	1.526	1.474

(Courtesy ASA B18.2.2-1965)

All dimensions are in inches.

Threads shall be in the Unified coarse, fine, or 8-thread series (UNC, UNF, or 8UN), Class 2B.

TABLE 20. HEX THICK SLOTTED NUTS

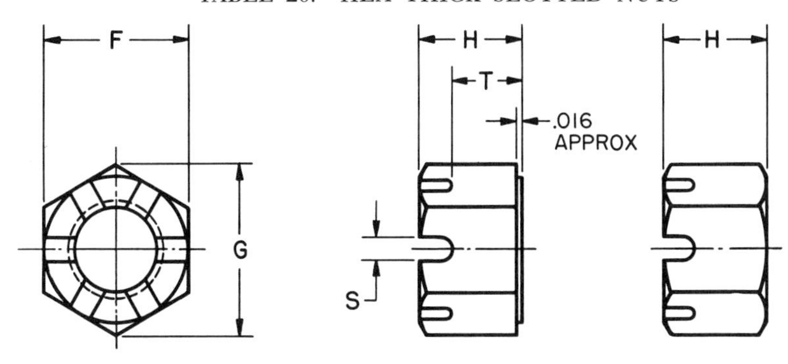

Dimensions of Hex Thick Slotted Nuts

Nominal Size or Basic Major Dia of Thread		Width Across Flats F			Width Across Corners G		Thickness H			Unslotted Thickness T		Width of Slot S	
		Basic	Max	Min	Max	Min	Basic	Max	Min	Max	Min	Max	Min
1/4	**0.2500**	**7/16**	**0.4375**	**0.428**	**0.505**	**0.488**	**9/32**	**0.288**	**0.274**	0.20	0.18	0.10	0.07
5/16	**0.3125**	**1/2**	**0.5000**	**0.489**	**0.577**	**0.557**	**21/64**	**0.336**	**0.320**	0.24	0.22	0.12	0.09
3/8	**0.3750**	**9/16**	**0.5625**	**0.551**	**0.650**	**0.628**	**13/32**	**0.415**	**0.398**	0.29	0.27	0.15	0.12
7/16	**0.4375**	**11/16**	**0.6875**	**0.675**	**0.794**	**0.768**	**29/64**	**0.463**	**0.444**	0.31	0.29	0.15	0.12
1/2	**0.5000**	**3/4**	**0.7500**	**0.736**	**0.866**	**0.840**	**9/16**	**0.573**	**0.552**	0.42	0.40	0.18	0.15
9/16	**0.5625**	**7/8**	**0.8750**	**0.861**	**1.010**	**0.982**	**39/64**	**0.621**	**0.598**	0.43	0.41	0.18	0.15
5/8	**0.6250**	**15/16**	**0.9375**	**0.922**	**1.083**	**1.051**	**23/32**	**0.731**	**0.706**	0.51	0.49	0.24	0.18
3/4	**0.7500**	**1 1/8**	**1.1250**	**1.088**	**1.299**	**1.240**	**13/16**	**0.827**	**0.798**	0.57	0.55	0.24	0.18
7/8	**0.8750**	**1 5/16**	**1.3125**	**1.269**	**1.516**	**1.447**	**29/32**	**0.922**	**0.890**	0.67	0.64	0.24	0.18
1	**1.0000**	**1 1/2**	**1.5000**	**1.450**	**1.732**	**1.653**	**1**	**1.018**	**0.982**	0.73	0.70	0.30	0.24
1 1/8	**1.1250**	**1 11/16**	**1.6875**	**1.631**	**1.949**	**1.859**	**1 5/32**	**1.176**	**1.136**	0.83	0.80	0.33	0.24
1 1/4	**1.2500**	**1 7/8**	**1.8750**	**1.812**	**2.165**	**2.066**	**1 1/4**	**1.272**	**1.228**	0.89	0.86	0.40	0.31
1 3/8	**1.3750**	**2 1/16**	**2.0625**	**1.994**	**2.382**	**2.273**	**1 3/8**	**1.399**	**1.351**	1.02	0.98	0.40	0.31
1 1/2	**1.5000**	**2 1/4**	**2.2500**	**2.175**	**2.598**	**2.480**	**1 1/2**	**1.526**	**1.474**	1.08	1.04	0.46	0.37

(Courtesy ASA B18.2.2-1965)

All dimensions are in inches.

BOLD TYPE INDICATES PRODUCT FEATURES UNIFIED DIMENSIONALLY WITH BRITISH AND CANADIAN STANDARDS

Threads shall be in the Unified coarse, fine, or 8-thread series (UNC, UNF or 8UN series), Class 2B, Unification of fine thread products is limited to sizes 1 in. and under.

TABLE 21. HEAVY HEX SCREWS

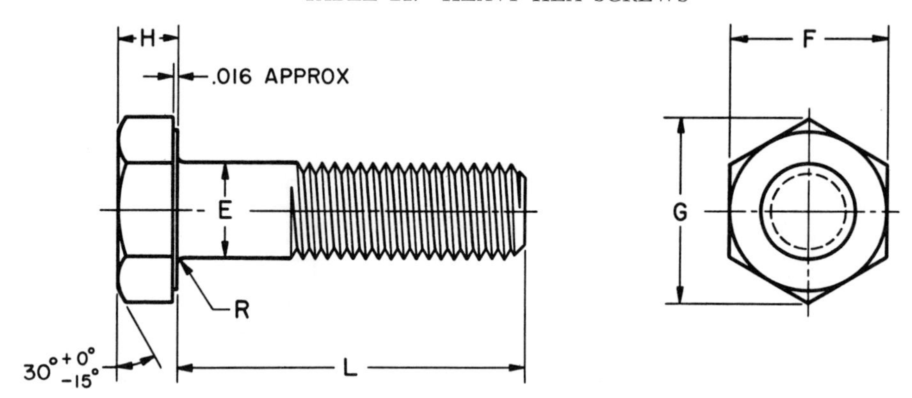

Dimensions of Heavy Hex Screws

Nominal Size or Basic Product Dia		Body Dia[2] E		Width Across Flats F			Width Across Corners G		Height H			Radius of Fillet R	
		Mix	Min	Basic	Max	Min	Max	Min	Basic	Max	Min	Max	Min
1/2	**0.5000**	**0.5000**	0.482	**7/8**	**0.8750**	**0.850**	**1.010**	**0.969**	5/16	0.323	0.302	**0.031**	**0.009**
5/8	**0.6250**	**0.6250**	0.605	**1 1/16**	**1.0625**	**1.031**	**1.227**	**1.175**	25/64	0.403	0.378	**0.062**	**0.021**
3/4	**0.7500**	**0.7500**	0.729	**1 1/4**	**1.2500**	**1.212**	**1.443**	**1.383**	15/32	0.483	0.455	**0.062**	**0.021**
7/8	**0.8750**	**0.8750**	0.852	**1 7/16**	**1.4375**	**1.394**	**1.660**	**1.589**	35/64	0.563	0.531	**0.062**	**0.031**
1	**1.0000**	**1.0000**	0.976	**1 5/8**	**1.6250**	**1.575**	**1.876**	**1.796**	39/64	0.627	0.591	**0.093**	**0.062**
1 1/8	**1.1250**	**1.1250**	1.098	**1 13/16**	**1.8125**	**1.756**	**2.093**	**2.002**	11/16	0.718	0.658	**0.093**	**0.062**
1 1/4	**1.2500**	**1.2500**	1.223	**2**	**2.0000**	**1.938**	**2.309**	**2.209**	25/32	0.813	0.749	**0.093**	**0.062**
1 3/8	**1.3750**	**1.3750**	1.345	**2 3/16**	**2.1875**	**2.119**	**2.526**	**2.416**	27/32	0.878	0.810	**0.093**	**0.062**
1 1/2	**1.5000**	**1.5000**	1.470	**2 3/8**	**2.3750**	**2.300**	**2.742**	**2.622**	15/16	0.974	0.902	**0.093**	**0.062**
1 3/4	**1.7500**	**1.7500**	1.716	**2 3/4**	**2.7500**	**2.662**	**3.175**	**3.035**	1 3/32	1.134	1.054	**0.125**	**0.078**
2	**2.0000**	**2.0000**	1.964	**3 1/8**	**3.1250**	**3.025**	**3.608**	**3.449**	1 7/32	1.263	1.175	**0.125**	**0.078**
2 1/4	2.2500	2.2500	2.214	3 1/2	3.5000	3.388	4.041	3.862	1 3/8	1.423	1.327	0.188	0.125
2 1/2	2.5000	2.5000	2.461	3 7/8	3.8750	3.750	4.474	4.275	1 17/32	1.583	1.479	0.188	0.125
2 3/4	2.7500	2.7500	2.711	4 1/4	4.2500	4.112	4.907	4.688	1 11/16	1.744	1.632	0.188	0.125
3	3.0000	3.0000	2.961	4 5/8	4.6250	4.475	5.340	5.102	1 7/8	1.935	1.815	0.188	0.125

(Courtesy ASA B18.2.1-1965)

All dimensions given in inches.

BOLD TYPE INDICATES PRODUCT FEATURES UNIFIED DIMENSIONALLY WITH BRITISH AND CANADIAN STANDARDS. Unification of fine thread products is limited to sizes 1 in. and under.

Threads shall be in the Unified coarse, fine, or 8-thread series (UNC, UNF or 8UN Series), Class 2A.

TABLE 22. HEXAGON SOCKET HEAD SHOULDER SCREWS

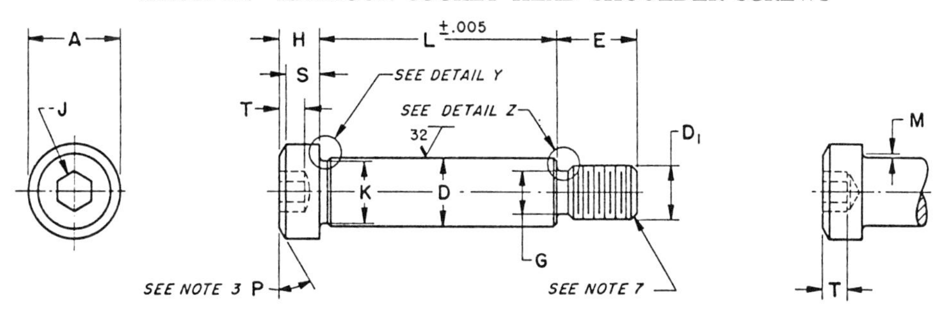

DIMENSIONS OF HEXAGON SOCKET HEAD SHOULDER SCREWS

	D		A		H		S	D_1	E
Nominal Size	Shoulder Diameter		Head Diameter		Head Height		Head Side Height	Nominal Thread Size	Thread Length
	Max	Min	Max	Min	Max	Min	Min		
1/4	0.2480	0.2460	3/8	0.357	3/16	0.182	0.157	10–24	0.375
5/16	0.3105	0.3085	7/16	0.419	7/32	0.213	0.183	1/4–20	0.438
3/8	0.3730	0.3710	9/16	0.543	1/4	0.244	0.209	5/16–18	0.500
1/2	0.4980	0.4960	3/4	0.729	5/16	0.306	0.262	3/8–16	0.625
5/8	0.6230	0.6210	7/8	0.853	3/8	0.368	0.315	1/2–13	0.750
3/4	0.7480	0.7460	1	0.977	1/2	0.492	0.421	5/8–11	0.875
1	0.9980	0.9960	1 5/16	1.287	5/8	0.616	0.527	3/4–10	1.000
1 1/4	1.2480	1.2460	1 3/4	1.723	3/4	0.741	0.633	7/8–9	1.125
See Notes	1		2					13	6

(Courtesy ASA B18.3-1965)

NOTES:

(1) *Shoulder.* The shoulder refers to the unthreaded portion of the screw and the maximum diameter shall conform to the nominal screw diameter less 0.002 in.

(2) *Head diameter.* The head may be plain or knurled at the option of the manufacturer.

(3) *Head chamfer.* The head shall be flat and chamfered. The flat shall be normal to the axis of the screw and the chamfer P shall be at an angle of 30 deg to 45 deg with the surface of the flat. The edge between the flat and the chamfer may be slightly rounded.

(4) *Length.* The length of the screw shall be measured, on a line parallel to the axis, from the plane of the bearing surface under the head to the plane of the shoulder at the threaded end.

(5) *Standard lengths.* The difference between consecutive lengths of standard screws shall be:

Nominal Screw Length	Standard Length Increment
1/4 to 3/4	1/8
3/4 to 5	1/4
Over 5	1/2

(6) *Thread length tolerance.* The tolerance on thread length E shall be unilateral. For screw sizes 1/4 through 3/8 in., inclusive, the tolerance shall be minus 0.020 in.; and for sizes larger than 3/8 in., the tolerance shall be minus 0.030 in.

(7) *Screw point chamfer.* The point shall be flat and chamfered. The flat shall be normal to the axis of the screw. The chamfer shall extend slightly below the root of the thread and the edge between the flat and chamfer may be slightly rounded. The included angle of the point should be approximately 90 deg.

TABLE 23

WRENCH OPENINGS FOR SQUARE AND HEX BOLTS AND SCREWS

Nominal Size of Wrench also Basic (Maximum) Width Across Flats of Bolt and Screw Heads	Allowance between Bolt or Screw Head and Jaws of Wrench	Wrench Openings			Square Bolt / Hex Bolt / Hex Cap Screw (Finished Hex Bolt) / Lag Screw	Heavy Hex Bolt / Heavy Hex Screw / Heavy Hex Structural Bolt
		Min	Tol	Max		
9/32 0.2812	0.002	0.283	0.005	0.288	No. 10	
5/16 0.3125	0.003	0.316	0.006	0.322		
11/32 0.3438	0.003	0.347	0.006	0.353		
3/8 0.3750	0.003	0.378	0.006	0.384	1/4*	
7/16 0.4375	0.003	0.440	0.006	0.446	1/4	
1/2 0.5000	0.004	0.504	0.006	0.510	5/16	
9/16 0.5625	0.004	0.566	0.007	0.573	3/8	
5/8 0.6250	0.004	0.629	0.007	0.636	7/16	
11/16 0.6875	0.004	0.692	0.007	0.699		
3/4 0.7500	0.005	0.755	0.008	0.763	1/2	
13/16 0.8125	0.005	0.818	0.008	0.826	9/16	
7/8 0.8750	0.005	0.880	0.008	0.888		1/2
15/16 0.9375	0.006	0.944	0.009	0.953	5/8	
1 1.0000	0.006	1.006	0.009	1.015		
1 1/16 1.0625	0.006	1.068	0.009	1.077		5/8
1 1/8 1.1250	0.007	1.132	0.010	1.142	3/4	
1 1/4 1.2500	0.007	1.257	0.010	1.267		3/4
1 5/16 1.3125	0.008	1.320	0.011	1.331	7/8	
1 3/8 1.3750	0.008	1.383	0.011	1.394		
1 7/16 1.4375	0.008	1.446	0.011	1.457		7/8
1 1/2 1.5000	0.008	1.508	0.012	1.520	1	
1 5/8 1.6250	0.009	1.634	0.012	1.646		1
1 11/16 1.6875	0.009	1.696	0.012	1.708	1 1/8	
1 13/16 1.8125	0.010	1.822	0.013	1.835		1 1/8
1 7/8 1.8750	0.010	1.885	0.013	1.898	1 1/4	
2 2.0000	0.011	2.011	0.014	2.025		1 1/4
2 1/16 2.0625	0.011	2.074	0.014	2.088	1 3/8	
2 3/16 2.1875	0.012	2.200	0.015	2.215		1 3/8
2 1/4 2.2500	0.012	2.262	0.015	2.277	1 1/2	
2 3/8 2.3750	0.013	2.388	0.016	2.404		1 1/2
2 7/16 2.4375	0.013	2.450	0.016	2.466	1 5/8	
2 9/16 2.5625	0.014	2.576	0.017	2.593		1 5/8
2 5/8 2.6250	0.014	2.639	0.017	2.656	1 3/4	
2 3/4 2.7500	0.014	2.766	0.017	2.783		1 3/4
2 13/16 2.8125	0.015	2.827	0.018	2.845	1 7/8	
2 15/16 2.9375	0.016	2.954	0.019	2.973		1 7/8
3 3.0000	0.016	3.016	0.019	3.035	2	
3 1/8 3.1250	0.017	3.142	0.020	3.162		2
3 3/8 3.3750	0.018	3.393	0.021	3.414	2 1/4	
3 1/2 3.5000	0.019	3.518	0.022	3.540		2 1/4
3 3/4 3.7500	0.020	3.770	0.023	3.793	2 1/2	
3 7/8 3.8750	0.020	3.895	0.023	3.918		2 1/2
4 1/8 4.1250	0.022	4.147	0.025	4.172	2 3/4	
4 1/4 4.2500	0.022	4.272	0.025	4.297		2 3/4
4 1/2 4.5000	0.024	4.524	0.026	4.550	3	
4 5/8 4.6250	0.024	4.649	0.027	4.676		3
4 7/8 4.8750	0.025	4.900	0.028	4.928	3 1/4	
5 5.0000	0.026	5.026	0.029	5.055		
5 1/4 5.2500	0.027	5.277	0.030	5.307	3 1/2	
5 3/8 5.3750	0.028	5.403	0.031	5.434		
5 5/8 5.6250	0.029	5.654	0.032	5.686	3 3/4	
5 3/4 5.7500	0.030	5.780	0.033	5.813		
6 6.0000	0.031	6.031	0.034	6.065	4	

(Courtesy ASA)

All dimensions given in inches.

*Square bolt and lag screw only.

Wrenches shall be marked with the "Nominal Size of Wrench" which is equal to the basic (maximum) width across flats of the corresponding bolt or screw head.

Allowance (minimum clearance) between maximum width across flats of bolt or screw head and jaws of wrench equals $(0.005W + 0.001)$. Tolerance on wrench opening equals plus $(0.005W + 0.004$ from minimum). W equals nominal size of wrench.

TABLE 24. FORMULAS FOR BOLT AND SCREW HEADS

Product	Width Across Flats		Height of Head	
	Basic[1]	Tolerance (Minus)	Basic[2]	Tolerance (Plus or Minus)
Square Bolt Lag Screw	$1\frac{1}{2}D$	0.050D	$\frac{2}{3}D$	0.016D + 0.012
Hex Bolt	Size — Width $\frac{1}{4}$ — $1\frac{1}{2}D + \frac{1}{16}$ $\frac{5}{16}$ to 4 — $1\frac{1}{2}D$	0.050D	Size — Height $\frac{1}{4}$ to $\frac{7}{16}$ — $\frac{5}{8}D + \frac{1}{64}$ $\frac{1}{2}$ to $\frac{7}{8}$ — $\frac{5}{8}D + \frac{1}{32}$ 1 to $1\frac{7}{8}$ — $\frac{5}{8}D + \frac{1}{16}$ 2 to $3\frac{3}{4}$ — $\frac{5}{8}D + \frac{1}{8}$ 4 — $\frac{5}{8}D + \frac{3}{16}$	Plus tolerance only 0.016D + 0.012 Minus tolerance adjusted so that minimum head height is same as minimum head height of hex cap screw (finished hex bolt).
Hex Cap Screw (Finished Hex Bolt)	$\frac{1}{4}$ — $1\frac{1}{2}D + \frac{1}{16}$ $\frac{5}{16}$ to 3 — $1\frac{1}{2}D$	Size — Tolerance $\frac{1}{4}$ to $\frac{5}{8}$ — 0.015D + 0.006 $\frac{3}{4}$ to 1 — 0.025D + 0.006 $1\frac{1}{8}$ to 3 — 0.050D	$\frac{1}{4}$ to $\frac{7}{8}$ — $\frac{5}{8}D$ 1 to $1\frac{7}{8}$ — $\frac{5}{8}D - \frac{1}{64}$ 2 to $2\frac{3}{4}$ — $\frac{5}{8}D - \frac{1}{32}$ 3 — $\frac{5}{8}D$	Size — Tolerance $\frac{1}{4}$ to 1 — 0.015D + 0.003 $1\frac{1}{8}$ to 4 — 0.016D + 0.012
Heavy Hex Bolt	$1\frac{1}{2}D + \frac{1}{8}$	0.050D	Same as for * Hex Bolt	Same as for* Hex Bolt
Heavy Hex Screw Heavy Hex Structural Bolt	$1\frac{1}{2}D + \frac{1}{8}$	0.050D	Same as for** Hex Cap Screw-Finished Hex Bolt	Same as for** Hex Cap Screw (Finished Hex Bolt)

All dimensions given in inches.

*In 1960 head heights for heavy hex bolts were reduced. Prior to 1960 head heights were 3D/4 + 1/16 in. Plus tolerance was 0.016D + 0.012 in. Minus tolerance was adjusted so that minimum head height would be the same as minimum head height of heavy hex screw.

**In 1960 head heights for heavy hex screws were reduced. Prior to 1960 head heights were 3D/4 + 1/32 in. for sizes 1/2 to 7/8 in.; 3D/4 for sizes 1 to 1-7/8 in.; and 3D/4–1/16 in. for sizes 2 to 3 in. Tolerances on head height for all sizes were plus and minus 0.016D + 0.012 in.

[1] Adjusted to sixteenths.

[2] 1/4 to 1 in. sizes adjusted to sixty-fourths. 1-1/8 to 2-1/2 in. sizes adjusted upward to thirty-seconds. 2-3/4 to 4 in. sizes adjusted upward to sixteenths.

For all square bolt heads, maximum width across corners equals 1.4142 × F (max) and minimum width across corners equals 1.373 × F (min).

For all hexagon bolt or screw heads, maximum width across corners equals 1.1547 × F (max) and minimum width across corners equals 1.14 × F (min).

D = Basic bolt or screw size.

F = Width across flats.

TABLE 25. WRENCH OPENINGS FOR NUTS

Nominal Size of Wrench also Basic (Maximum) Width Across Flats of Nuts		Allowance between Nut Flats and Jaws of Wrench	Wrench Openings			Square Nut	Hex Flat Hex Flat Jam Hex Hex Jam Hex Slotted Hex Thick Hex Thick Slotted Hex Castle	Heavy Square Heavy Hex Flat Heavy Hex Flat Jam Heavy Hex Heavy Hex Jam Heavy Hex Slotted
			Min	Tol	Max			
7/16	0.4375	0.003	0.440	0.006	0.446	1/4	1/4	
1/2	0.5000	0.004	0.504	0.006	0.510		5/16	1/4
9/16	0.5625	0.004	0.566	0.007	0.573	5/16	3/8	5/16
5/8	0.6250	0.004	0.629	0.007	0.636	3/8		
11/16	0.6875	0.004	0.692	0.007	0.699		7/16	3/8
3/4	0.7500	0.005	0.755	0.008	0.763	7/16	1/2	7/16
13/16	0.8125	0.005	0.818	0.008	0.826	1/2		
7/8	0.8750	0.005	0.880	0.008	0.888		9/16	1/2
15/16	0.9375	0.006	0.944	0.009	0.953		5/8	9/16
1	1.0000	0.006	1.006	0.009	1.015	5/8		
1 1/16	1.0625	0.006	1.068	0.009	1.077			5/8
1 1/8	1.1250	0.007	1.132	0.010	1.142	3/4	3/4	
1 1/4	1.2500	0.007	1.257	0.010	1.267			3/4
1 5/16	1.3125	0.008	1.320	0.011	1.331	7/8	7/8	
1 3/8	1.3750	0.008	1.383	0.011	1.394			
1 7/16	1.4375	0.008	1.446	0.011	1.457			7/8
1 1/2	1.5000	0.008	1.508	0.012	1.520	1	1	
1 5/8	1.6250	0.009	1.634	0.012	1.646			1
1 11/16	1.6875	0.009	1.696	0.012	1.708	1 1/8	1 1/8	
1 13/16	1.8125	0.010	1.822	0.013	1.835			1 1/8
1 7/8	1.8750	0.010	1.885	0.013	1.898	1 1/4	1 1/4	
2	2.0000	0.011	2.011	0.014	2.025			1 1/4
2 1/16	2.0625	0.011	2.074	0.014	2.088	1 3/8	1 3/8	
2 3/16	2.1875	0.012	2.200	0.015	2.215			1 3/8
2 1/4	2.2500	0.012	2.262	0.015	2.277	1 1/2	1 1/2	
2 3/8	2.3750	0.013	2.388	0.016	2.404			1 1/2
2 7/16	2.4375	0.013	2.450	0.016	2.466			
2 9/16	2.5625	0.014	2.576	0.017	2.593			1 5/8
2 5/8	2.6250	0.014	2.639	0.017	2.656			
2 3/4	2.7500	0.014	2.766	0.017	2.783			1 3/4
2 13/16	2.8125	0.015	2.827	0.018	2.845			
2 15/16	2.9375	0.016	2.954	0.019	2.973			1 7/8
3	3.0000	0.016	3.016	0.019	3.035			
3 1/8	3.1250	0.017	3.142	0.020	3.162			2
3 3/8	3.3750	0.018	3.393	0.021	3.414			
3 1/2	3.5000	0.019	3.518	0.022	3.540			2 1/4
3 3/4	3.7500	0.020	3.770	0.023	3.793			
3 7/8	3.8750	0.020	3.895	0.023	3.918			2 1/2
4 1/8	4.1250	0.022	4.147	0.025	4.172			
4 1/4	4.2500	0.022	4.272	0.025	4.297			2 3/4
4 1/2	4.5000	0.024	4.524	0.026	4.550			
4 5/8	4.6250	0.024	4.649	0.027	4.676			3
4 7/8	4.8750	0.025	4.900	0.028	4.928			
5	5.0000	0.026	5.026	0.029	5.055			3 1/4
5 1/4	5.2500	0.027	5.277	0.030	5.307			3 1/2
5 3/8	5.3750	0.028	5.403	0.031	5.434			3 1/2
5 5/8	5.6250	0.029	5.654	0.032	5.686			3 3/4
5 3/4	5.7500	0.030	5.780	0.033	5.813			3 3/4
6	6.0000	0.031	6.031	0.034	6.065			4
6 1/8	6.1250	0.032	6.157	0.035	6.192			4

(Courtesy ASA)

All dimensions given in inches.

Wrenches shall be marked with the "Nominal Size of Wrench" which is equal to the basic (maximum) width across flats of the corresponding nut.

Allowance (minimum clearance) between maximum width across flats of the nut and jaws of wrench equals $(0.005W + 0.001)$. Tolerance on wrench opening equals plus $(0.005W + 0.004$ from minimum$)$. W equals nominal size of wrench.

TABLE 26. FORMULAS FOR NUTS

Type of Nut	Width Across Flats				Thickness of Nut			
	Basic[1]		Tolerance (Minus)		Basic[2]		Tolerance (Plus or Minus)	
	Size	Width	Size	Tolerance	Size	Thickness	Size	Tolerance
Hex Hex Slotted	$\frac{1}{4}$ $\frac{5}{16}$ to $1\frac{1}{2}$	$1\frac{1}{2}D + \frac{1}{16}$ $1\frac{1}{2}D$	$\frac{1}{4}$ to $\frac{5}{8}$ $\frac{3}{4}$ to $1\frac{1}{2}$	$0.015D + 0.006$ $0.050D$	$\frac{1}{4}$ to $\frac{5}{8}$ $\frac{3}{4}$ to $1\frac{1}{8}$ $1\frac{1}{4}$ to $1\frac{1}{2}$	$\frac{7}{8}D$ $\frac{7}{8}D - \frac{1}{64}$ $\frac{7}{8}D - \frac{1}{32}$	$\frac{1}{4}$ to $\frac{5}{8}$ $\frac{3}{4}$ to $1\frac{1}{2}$	$0.015D + 0.003$ $0.016D + 0.012$
Hex Jam	$\frac{1}{4}$ $\frac{5}{16}$ to $1\frac{1}{2}$	$1\frac{1}{2}D + \frac{1}{16}$ $1\frac{1}{2}D$	$\frac{1}{4}$ to $\frac{5}{8}$ $\frac{3}{4}$ to $1\frac{1}{2}$	$0.015D + 0.006$ $0.050D$	$\frac{1}{4}$ to $\frac{5}{8}$ $\frac{3}{4}$ to $1\frac{1}{8}$ $1\frac{1}{4}$ to $1\frac{1}{2}$	*(See Table)* $\frac{1}{2}D + \frac{3}{64}$ $\frac{1}{2}D + \frac{3}{32}$	$\frac{1}{4}$ to $\frac{5}{8}$ $\frac{3}{4}$ to $1\frac{1}{2}$	$0.015D + 0.003$ $0.016D + 0.012$
Hex Thick Hex Thick Slotted Hex Castle	$\frac{1}{4}$ $\frac{5}{16}$ to $1\frac{1}{2}$	$1\frac{1}{2}D + \frac{1}{16}$ $1\frac{1}{2}D$	$\frac{1}{4}$ to $\frac{5}{8}$ $\frac{3}{4}$ to $1\frac{1}{2}$	$0.015D + 0.006$ $0.050D$	*(See Table)*		$0.015D + 0.003$	
Square Hex Flat	$\frac{1}{4}$ to $\frac{5}{8}$ $\frac{3}{4}$ to $1\frac{1}{2}$	$1\frac{1}{2}D + \frac{1}{16}$ $1\frac{1}{2}D$	$0.050D$		$\frac{7}{8}D$		$0.016D + 0.012$	
Hex Flat Jam	$1\frac{1}{8}$ to $1\frac{1}{2}$	$1\frac{1}{2}$	$0.050D$		$1\frac{1}{8}$ $1\frac{1}{4}$ to $1\frac{1}{2}$	$\frac{1}{2}D + \frac{1}{16}$ $\frac{1}{2}D + \frac{1}{8}$	$0.016D + 0.012$	
Heavy Square Heavy Hex Flat	$1\frac{1}{2}D + \frac{1}{8}$		$0.050D$		D		Plus tolerance only $0.016D + 0.012$ Minus tolerance adjusted so that minimum thickness is equal to minimum thickness of heavy hex nut.	
Heavy Hex Flat Jam	$1\frac{1}{2}D + \frac{1}{8}$		$0.050D$		$\frac{1}{4}$ to $1\frac{1}{8}$ $1\frac{1}{4}$ to $2\frac{1}{4}$ $2\frac{1}{2}$ to 4	$\frac{1}{2}D + \frac{1}{16}$ $\frac{1}{2}D + \frac{1}{8}$ $\frac{1}{2}D + \frac{1}{4}$	Plus tolerance only $0.016D + 0.012$ Minus tolerance adjusted so that minimum thickness is equal to minimum thickness of heavy hex jam nut.	
Heavy Hex Heavy Hex Slotted	$1\frac{1}{2}D + \frac{1}{8}$		$0.050D$		$\frac{1}{4}$ to $1\frac{1}{8}$ $1\frac{1}{4}$ to 2 $2\frac{1}{4}$ to 3 $3\frac{1}{4}$ to 4	$D - \frac{1}{64}$ $D - \frac{1}{32}$ $D - \frac{3}{64}$ $D - \frac{1}{16}$	$0.016D + 0.012$	
Heavy Hex Jam	$1\frac{1}{2}D + \frac{1}{8}$		$0.050D$		$\frac{1}{4}$ to $1\frac{1}{8}$ $1\frac{1}{4}$ to 2 $2\frac{1}{4}$ $2\frac{1}{2}$ to 3 $3\frac{1}{4}$ to 4	$\frac{1}{2}D + \frac{3}{64}$ $\frac{1}{2}D + \frac{3}{32}$ $\frac{1}{2}D + \frac{5}{64}$ $\frac{1}{2}D + \frac{13}{64}$ $\frac{1}{2}D + \frac{3}{16}$	$0.016D + 0.012$	

(Courtesy ASA)

All dimensions given in inches.

[1] Adjusted to sixteenths.

[2] $\frac{1}{4}$ to 1 in. sizes adjusted to sixty-fourths. $1\frac{1}{8}$ to $2\frac{1}{2}$ in. sizes adjusted upward to thirty-seconds. $2\frac{3}{4}$ to 4 in. sizes adjusted upward to sixteenths.

For all square nuts, maximum width across corners equals $1.4142 \times F$ (max) and minimum width across corners equals $1.373 \times F$ (min).

For all hex nuts, maximum width across corners equals $1.1547 \times F$ (max) and minimum width across corners equals $1.14 \times F$ (min).

D = Nominal nut size.

F = Width across flats.

TABLE 27. PLAIN WASHERS—TYPE A

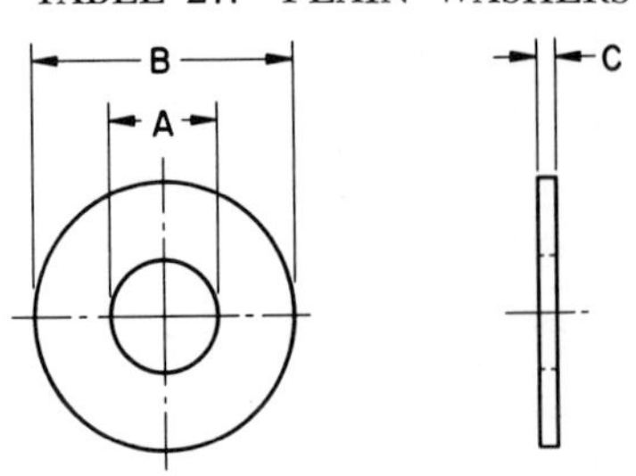

Dimensions of Preferred Sizes of Type A Plain Washers**

Nominal Washer Size***			Inside Diameter A			Outside Diameter B			Thickness C		
			Basic	Tolerance Plus	Tolerance Minus	Basic	Tolerance Plus	Tolerance Minus	Basic	Max	Min
—	—		0.078	0.000	0.005	0.188	0.000	0.005	0.020	0.025	0.016
—	—		0.094	0.000	0.005	0.250	0.000	0.005	0.020	0.025	0.016
—	—		0.125	0.008	0.005	0.312	0.008	0.005	0.032	0.040	0.025
No. 6	0.138		0.156	0.008	0.005	0.375	0.015	0.005	0.049	0.065	0.036
No. 8	0.164		0.188	0.008	0.005	0.438	0.015	0.005	0.049	0.065	0.036
No. 10	0.190		0.219	0.008	0.005	0.500	0.015	0.005	0.049	0.065	0.036
3/16	0.188		0.250	0.015	0.005	0.562	0.015	0.005	0.049	0.065	0.036
No. 12	0.216		0.250	0.015	0.005	0.562	0.015	0.005	0.065	0.080	0.051
1/4	0.250	N	0.281	0.015	0.005	0.625	0.015	0.005	0.065	0.080	0.051
1/4	0.250	W	0.312	0.015	0.005	0.734*	0.015	0.007	0.065	0.080	0.051
5/16	0.312	N	0.344	0.015	0.005	0.688	0.015	0.007	0.065	0.080	0.051
5/16	0.312	W	0.375	0.015	0.005	0.875	0.030	0.007	0.083	0.104	0.064
3/8	0.375	N	0.406	0.015	0.005	0.812	0.015	0.007	0.065	0.080	0.051
3/8	0.375	W	0.438	0.015	0.005	1.000	0.030	0.007	0.083	0.104	0.064
7/16	0.438	N	0.469	0.015	0.005	0.922	0.015	0.007	0.065	0.080	0.051
7/16	0.438	W	0.500	0.015	0.005	1.250	0.030	0.007	0.083	0.104	0.064
1/2	0.500	N	0.531	0.015	0.005	1.062	0.030	0.007	0.095	0.121	0.074
1/2	0.500	W	0.562	0.015	0.005	1.375	0.030	0.007	0.109	0.132	0.086
9/16	0.562	N	0.594	0.015	0.005	1.156*	0.030	0.007	0.095	0.121	0.074
9/16	0.562	W	0.625	0.015	0.005	1.469*	0.030	0.007	0.109	0.132	0.086
5/8	0.625	N	0.656	0.030	0.007	1.312	0.030	0.007	0.095	0.121	0.074
5/8	0.625	W	0.688	0.030	0.007	1.750	0.030	0.007	0.134	0.160	0.108
3/4	0.750	N	0.812	0.030	0.007	1.469	0.030	0.007	0.134	0.160	0.108
3/4	0.750	W	0.812	0.030	0.007	2.000	0.030	0.007	0.148	0.177	0.122
7/8	0.875	N	0.938	0.030	0.007	1.750	0.030	0.007	0.134	0.160	0.108
7/8	0.875	W	0.938	0.030	0.007	2.250	0.030	0.007	0.165	0.192	0.136
1	1.000	N	1.062	0.030	0.007	2.000	0.030	0.007	0.134	0.160	0.108
1	1.000	W	1.062	0.030	0.007	2.500	0.030	0.007	0.165	0.192	0.136
1 1/8	1.125	N	1.250	0.030	0.007	2.250	0.030	0.007	0.134	0.160	0.108
1 1/8	1.125	W	1.250	0.030	0.007	2.750	0.030	0.007	0.165	0.192	0.136
1 1/4	1.250	N	1.375	0.030	0.007	2.500	0.030	0.007	0.165	0.192	0.136
1 1/4	1.250	W	1.375	0.030	0.007	3.000	0.030	0.007	0.165	0.192	0.136
1 3/8	1.375	N	1.500	0.030	0.007	2.750	0.030	0.007	0.165	0.192	0.136
1 3/8	1.375	W	1.500	0.045	0.010	3.250	0.045	0.010	0.180	0.213	0.153
1 1/2	1.500	N	1.625	0.030	0.007	3.000	0.030	0.007	0.165	0.192	0.136
1 1/2	1.500	W	1.625	0.045	0.010	3.500	0.045	0.010	0.180	0.213	0.153
1 5/8	1.625		1.750	0.045	0.010	3.750	0.045	0.010	0.180	0.213	0.153
1 3/4	1.750		1.875	0.045	0.010	4.000	0.045	0.010	0.180	0.213	0.153
1 7/8	1.875		2.000	0.045	0.010	4.250	0.045	0.010	0.180	0.213	0.153
2	2.000		2.125	0.045	0.010	4.500	0.045	0.010	0.180	0.213	0.153
2 1/4	2.250		2.375	0.045	0.010	4.750	0.045	0.010	0.220	0.248	0.193
2 1/2	2.500		2.625	0.045	0.010	5.000	0.045	0.010	0.238	0.280	0.210
2 3/4	2.750		2.875	0.065	0.010	5.250	0.065	0.010	0.259	0.310	0.228
3	3.000		3.125	0.065	0.010	5.500	0.065	0.010	0.284	0.327	0.249

(Courtesy ASA B27.2-1965)

* The 0.734 in., 1.156 in., and 1.469 in. outside diameters avoid washers which could be used in coin operated devices.

** Preferred sizes are for the most part from series previously designated "Standard Plate" and "SAE." Where common sizes existed in the two series, the SAE size is designated "N" (narrow) and the Standard Plate "W" (wide). These sizes as well as all other sizes of Type A Plain Washers are to be ordered by ID, OD, and thickness dimensions.

*** Nominal washer sizes are intended for use with comparable nominal screw or bolt sizes.

TABLE 28. PLAIN WASHERS—TYPE B

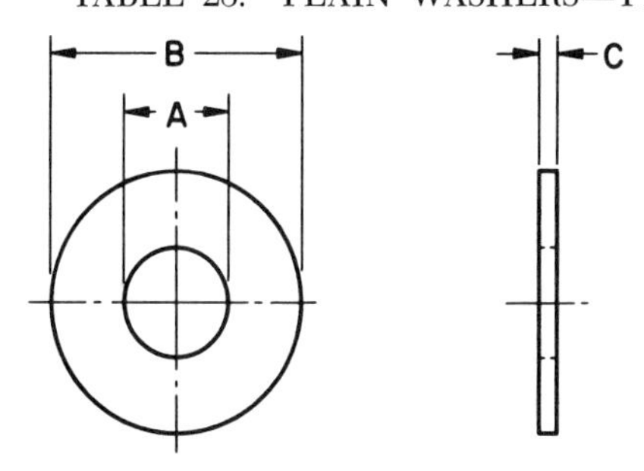

DIMENSIONS OF TYPE B PLAIN WASHERS

Nominal Washer Size**		Series	Inside Diameter A			Outside Diameter B			Thickness C		
			Basic	Tolerance Plus	Tolerance Minus	Basic	Tolerance Plus	Tolerance Minus	Basic	Max	Min
No. 0	0.060	Narrow	0.068	0.000	0.005	0.125	0.000	0.005	0.025	0.028	0.022
		Regular	0.068	0.000	0.005	0.188	0.000	0.005	0.025	0.028	0.022
		Wide	0.068	0.000	0.005	0.250	0.000	0.005	0.025	0.028	0.022
No. 1	0.073	Narrow	0.084	0.000	0.005	0.156	0.000	0.005	0.025	0.028	0.022
		Regular	0.084	0.000	0.005	0.219	0.000	0.005	0.025	0.028	0.022
		Wide	0.084	0.000	0.005	0.281	0.000	0.005	0.032	0.036	0.028
No. 2	0.086	Narrow	0.094	0.000	0.005	0.188	0.000	0.005	0.025	0.028	0.022
		Regular	0.094	0.000	0.005	0.250	0.000	0.005	0.032	0.036	0.028
		Wide	0.094	0.000	0.005	0.344	0.000	0.005	0.032	0.036	0.028
No. 3	0.099	Narrow	0.109	0.000	0.005	0.219	0.000	0.005	0.025	0.028	0.022
		Regular	0.109	0.000	0.005	0.312	0.000	0.005	0.032	0.036	0.028
		Wide	0.109	0.008	0.005	0.406	0.008	0.005	0.040	0.045	0.036
No. 4	0.112	Narrow	0.125	0.000	0.005	0.250	0.000	0.005	0.032	0.036	0.028
		Regular	0.125	0.008	0.005	0.375	0.008	0.005	0.040	0.045	0.036
		Wide	0.125	0.008	0.005	0.438	0.008	0.005	0.040	0.045	0.036
No. 5	0.125	Narrow	0.141	0.000	0.005	0.281	0.000	0.005	0.032	0.036	0.028
		Regular	0.141	0.008	0.005	0.406	0.008	0.005	0.040	0.045	0.036
		Wide	0.141	0.008	0.005	0.500	0.008	0.005	0.040	0.045	0.036
No. 6	0.138	Narrow	0.156	0.000	0.005	0.312	0.000	0.005	0.032	0.036	0.028
		Regular	0.156	0.008	0.005	0.438	0.008	0.005	0.040	0.045	0.036
		Wide	0.156	0.008	0.005	0.562	0.008	0.005	0.040	0.045	0.036
No. 8	0.164	Narrow	0.188	0.008	0.005	0.375	0.008	0.005	0.040	0.045	0.036
		Regular	0.188	0.008	0.005	0.500	0.008	0.005	0.040	0.045	0.036
		Wide	0.188	0.008	0.005	0.625	0.015	0.005	0.063	0.071	0.056
No. 10	0.190	Narrow	0.203	0.008	0.005	0.406	0.008	0.005	0.040	0.045	0.036
		Regular	0.203	0.008	0.005	0.562	0.008	0.005	0.040	0.045	0.036
		Wide	0.203	0.008	0.005	0.734*	0.015	0.007	0.063	0.071	0.056
No. 12	0.216	Narrow	0.234	0.008	0.005	0.438	0.008	0.005	0.040	0.045	0.036
		Regular	0.234	0.008	0.005	0.625	0.015	0.005	0.063	0.071	0.056
		Wide	0.234	0.008	0.005	0.875	0.015	0.007	0.063	0.071	0.056
¼	0.250	Narrow	0.281	0.015	0.005	0.500	0.015	0.005	0.063	0.071	0.056
		Regular	0.281	0.015	0.005	0.734*	0.015	0.007	0.063	0.071	0.056
		Wide	0.281	0.015	0.005	1.000	0.015	0.007	0.063	0.071	0.056

*The 0.734 in. outside diameter avoids washers which could be used in coin operated devices.

**Nominal washer sizes are intended for use with comparable nominal screw or bolt sizes.

Inside and outside diameters shall be concentric within at least the inside diameter tolerance.

Washers shall be flat within 0.005 in. for basic outside diameters up to and including 0.875 in. and within 0.010 in. for larger outside diameters.

TABLE 28. *(continued)*

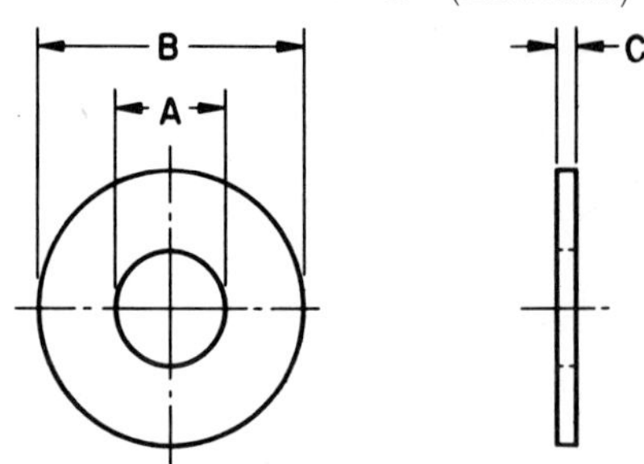

DIMENSIONS OF TYPE B PLAIN WASHERS *(continued)*

Nominal Washer Size**		Series	Inside Diameter A			Outside Diameter B			Thickness C		
			Basic	Tolerance Plus	Tolerance Minus	Basic	Tolerance Plus	Tolerance Minus	Basic	Max	Min
5/16	0.312	Narrow	0.344	0.015	0.005	0.625	0.015	0.005	0.063	0.071	0.056
		Regular	0.344	0.015	0.005	0.875	0.015	0.007	0.063	0.071	0.056
		Wide	0.344	0.015	0.005	1.125	0.015	0.007	0.063	0.071	0.056
3/8	0.375	Narrow	0.406	0.015	0.005	0.734*	0.015	0.007	0.063	0.071	0.056
		Regular	0.406	0.015	0.005	1.000	0.015	0.007	0.063	0.071	0.056
		Wide	0.406	0.015	0.005	1.250	0.030	0.007	0.100	0.112	0.090
7/16	0.438	Narrow	0.469	0.015	0.005	0.875	0.015	0.007	0.063	0.071	0.056
		Regular	0.469	0.015	0.005	1.125	0.015	0.007	0.063	0.071	0.056
		Wide	0.469	0.015	0.005	1.469*	0.030	0.007	0.100	0.112	0.090
1/2	0.500	Narrow	0.531	0.015	0.005	1.000	0.015	0.007	0.063	0.071	0.056
		Regular	0.531	0.015	0.005	1.250	0.030	0.007	0.100	0.112	0.090
		Wide	0.531	0.015	0.005	1.750	0.030	0.007	0.100	0.112	0.090
9/16	0.562	Narrow	0.594	0.015	0.005	1.125	0.015	0.007	0.063	0.071	0.056
		Regular	0.594	0.015	0.005	1.469*	0.030	0.007	0.100	0.112	0.090
		Wide	0.594	0.015	0.005	2.000	0.030	0.007	0.100	0.112	0.090
5/8	0.625	Narrow	0.656	0.030	0.007	1.250	0.030	0.007	0.100	0.112	0.090
		Regular	0.656	0.030	0.007	1.750	0.030	0.007	0.100	0.112	0.090
		Wide	0.656	0.030	0.007	2.250	0.030	0.007	0.160	0.174	0.146
3/4	0.750	Narrow	0.812	0.030	0.007	1.375	0.030	0.007	0.100	0.112	0.090
		Regular	0.812	0.030	0.007	2.000	0.030	0.007	0.100	0.112	0.090
		Wide	0.812	0.030	0.007	2.500	0.030	0.007	0.160	0.174	0.146
7/8	0.875	Narrow	0.938	0.030	0.007	1.469*	0.030	0.007	0.100	0.112	0.090
		Regular	0.938	0.030	0.007	2.250	0.030	0.007	0.160	0.174	0.146
		Wide	0.938	0.030	0.007	2.750	0.030	0.007	0.160	0.174	0.146
1	1.000	Narrow	1.062	0.030	0.007	1.750	0.030	0.007	0.100	0.112	0.090
		Regular	1.062	0.030	0.007	2.500	0.030	0.007	0.160	0.174	0.146
		Wide	1.062	0.030	0.007	3.000	0.030	0.007	0.160	0.174	0.146
1 1/8	1.125	Narrow	1.188	0.030	0.007	2.000	0.030	0.007	0.100	0.112	0.090
		Regular	1.188	0.030	0.007	2.750	0.030	0.007	0.160	0.174	0.146
		Wide	1.188	0.030	0.007	3.250	0.030	0.007	0.160	0.174	0.146
1 1/4	1.250	Narrow	1.312	0.030	0.007	2.250	0.030	0.007	0.160	0.174	0.146
		Regular	1.312	0.030	0.007	3.000	0.030	0.007	0.160	0.174	0.146
		Wide	1.312	0.045	0.010	3.500	0.045	0.010	0.250	0.266	0.234

*The 0.734 in. and 1.469 in. outside diameters avoid washers which could be used in coin operated devices.

**Nominal washer sizes are intended for use with comparable nominal screw or bolt sizes.

Inside and outside diameters shall be concentric within at least the inside diameter tolerance.

Washers shall be flat within 0.005 in. for basic outside diameters up to and including 0.875 in., and within 0.010 in. for larger outside diameters.

TABLE 28. *(concluded)*

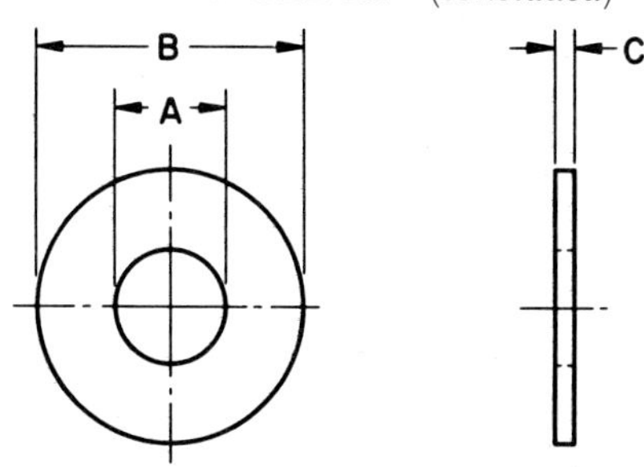

DIMENSIONS OF TYPE B PLAIN WASHERS *(concluded)*

Nominal Washer Size*		Series	Inside Diameter A			Outside Diameter B			Thickness C		
			Basic	Tolerance Plus	Tolerance Minus	Basic	Tolerance Plus	Tolerance Minus	Basic	Max	Min
1⅜	1.375	Narrow	1.438	0.030	0.007	2.500	0.030	0.007	0.160	0.174	0.146
		Regular	1.438	0.030	0.007	3.250	0.030	0.007	0.160	0.174	0.146
		Wide	1.438	0.045	0.010	3.750	0.045	0.010	0.250	0.266	0.234
1½	1.500	Narrow	1.562	0.030	0.007	2.750	0.030	0.007	0.160	0.174	0.146
		Regular	1.562	0.045	0.010	3.500	0.045	0.010	0.250	0.266	0.234
		Wide	1.562	0.045	0.010	4.000	0.045	0.010	0.250	0.266	0.234
1⅝	1.625	Narrow	1.750	0.030	0.007	3.000	0.030	0.007	0.160	0.174	0.146
		Regular	1.750	0.045	0.010	3.750	0.045	0.010	0.250	0.266	0.234
		Wide	1.750	0.045	0.010	4.250	0.045	0.010	0.250	0.266	0.234
1¾	1.750	Narrow	1.875	0.030	0.007	3.250	0.030	0.007	0.160	0.174	0.146
		Regular	1.875	0.045	0.010	4.000	0.045	0.010	0.250	0.266	0.234
		Wide	1.875	0.045	0.010	4.500	0.045	0.010	0.250	0.266	0.234
1⅞	1.875	Narrow	2.000	0.045	0.010	3.500	0.045	0.010	0.250	0.266	0.234
		Regular	2.000	0.045	0.010	4.250	0.045	0.010	0.250	0.266	0.234
		Wide	2.000	0.045	0.010	4.750	0.045	0.010	0.250	0.266	0.234
2	2.000	Narrow	2.125	0.045	0.010	3.750	0.045	0.010	0.250	0.266	0.234
		Regular	2.125	0.045	0.010	4.500	0.045	0.010	0.250	0.266	0.234
		Wide	2.125	0.045	0.010	5.000	0.045	0.010	0.250	0.266	0.234
2¼	2.250	Narrow	2.375	0.045	0.010	4.000	0.045	0.010	0.250	0.266	0.234
		Regular	2.375	0.045	0.010	5.000	0.045	0.010	0.250	0.266	0.234
		Wide	2.375	0.065	0.010	5.500	0.065	0.010	0.375	0.393	0.357
2½	2.500	Narrow	2.625	0.045	0.010	4.500	0.045	0.010	0.250	0.266	0.234
		Regular	2.625	0.065	0.010	5.500	0.065	0.010	0.375	0.393	0.357
		Wide	2.625	0.065	0.010	6.000	0.065	0.010	0.375	0.393	0.357
2¾	2.750	Narrow	2.875	0.045	0.010	5.000	0.045	0.010	0.250	0.266	0.234
		Regular	2.875	0.065	0.010	6.000	0.065	0.010	0.375	0.393	0.357
		Wide	2.875	0.065	0.010	6.500	0.065	0.010	0.375	0.393	0.357
3	3.000	Narrow	3.125	0.065	0.010	5.500	0.065	0.010	0.375	0.393	0.357
		Regular	3.125	0.065	0.010	6.500	0.065	0.010	0.375	0.393	0.357
		Wide	3.125	0.065	0.010	7.000	0.065	0.010	0.375	0.393	0.357

*Nominal washer sizes are intended for use with comparable nominal screw or bolt sizes.
Inside and outside diameters shall be concentric within at least the inside diameter tolerance.
Washers of sizes shown above shall be flat within 0.010 in.

TABLE 29. MEDIUM LOCK WASHERS

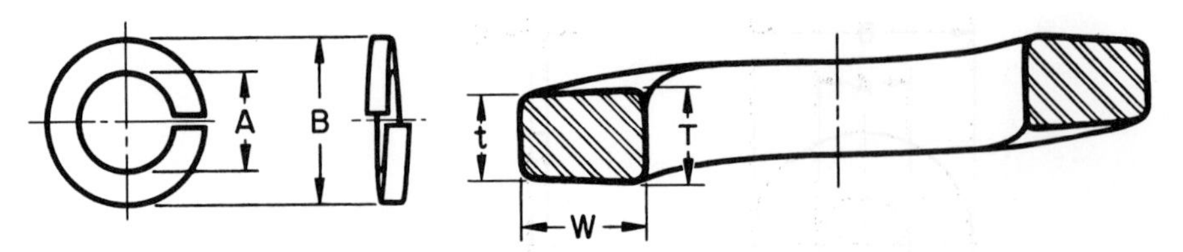

DIMENSIONS OF REGULAR* HELICAL SPRING LOCK WASHERS

Nominal Washer Size		Inside Diameter A		Outside Diameter B	Washer Section	
					Width W	Thickness $\frac{T+t}{2}$
		Min	Max	Max**	Min	Min
No. 2	0.086	0.088	0.094	0.172	0.035	0.020
No. 3	0.099	0.101	0.107	0.195	0.040	0.025
No. 4	0.112	0.115	0.121	0.209	0.040	0.025
No. 5	0.125	0.128	0.134	0.236	0.047	0.031
No. 6	0.138	0.141	0.148	0.250	0.047	0.031
No. 8	0.164	0.168	0.175	0.293	0.055	0.040
No. 10	0.190	0.194	0.202	0.334	0.062	0.047
No. 12	0.216	0.221	0.229	0.377	0.070	0.056
1/4	0.250	0.255	0.263	0.489	0.109	0.062
5/16	0.312	0.318	0.328	0.586	0.125	0.078
3/8	0.375	0.382	0.393	0.683	0.141	0.094
7/16	0.438	0.446	0.459	0.779	0.156	0.109
1/2	0.500	0.509	0.523	0.873	0.171	0.125
9/16	0.562	0.572	0.587	0.971	0.188	0.141
5/8	0.625	0.636	0.653	1.079	0.203	0.156
11/16	0.688	0.700	0.718	1.176	0.219	0.172
3/4	0.750	0.763	0.783	1.271	0.234	0.188
13/16	0.812	0.826	0.847	1.367	0.250	0.203
7/8	0.875	0.890	0.912	1.464	0.266	0.219
15/16	0.938	0.954	0.978	1.560	0.281	0.234
1	1.000	1.017	1.042	1.661	0.297	0.250
1 1/16	1.062	1.080	1.107	1.756	0.312	0.266
1 1/8	1.125	1.144	1.172	1.853	0.328	0.281
1 3/16	1.188	1.208	1.237	1.950	0.344	0.297
1 1/4	1.250	1.271	1.302	2.045	0.359	0.312
1 5/16	1.312	1.334	1.366	2.141	0.375	0.328
1 3/8	1.375	1.398	1.432	2.239	0.391	0.344
1 7/16	1.438	1.462	1.497	2.334	0.406	0.359
1 1/2	1.500	1.525	1.561	2.430	0.422	0.375

(Courtesy ASA B27.1-1965)

* Formerly designated Medium Helical Spring Lock Washers.
** The maximum outside diameters specified allow for the commercial tolerances on cold drawn wire.

TABLE 30. HEAVY LOCK WASHERS

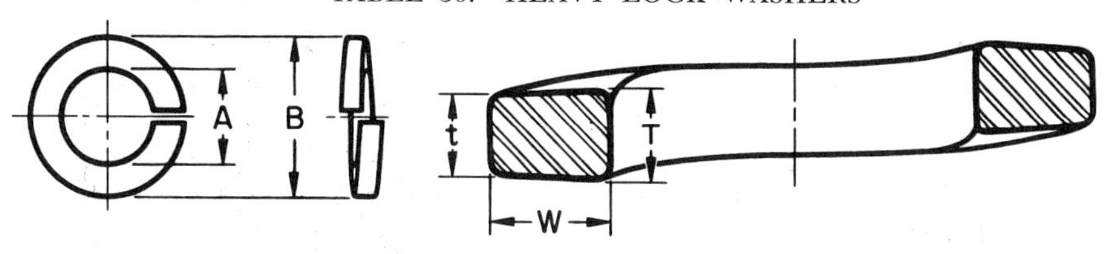

Dimensions of Heavy Helical Spring Lock Washers

Nominal Washer Size		Inside Diameter A		Outside Diameter B	Washer Section	
					Width W	Thickness $\frac{T+t}{2}$
		Min	Max	Max*	Min	Min
No. 2	0.086	0.088	0.094	0.182	0.040	0.025
No. 3	0.099	0.101	0.107	0.209	0.047	0.031
No. 4	0.112	0.115	0.121	0.223	0.047	0.031
No. 5	0.125	0.128	0.134	0.252	0.055	0.040
No. 6	0.138	0.141	0.148	0.266	0.055	0.040
No. 8	0.164	0.168	0.175	0.307	0.062	0.047
No. 10	0.190	0.194	0.202	0.350	0.070	0.056
No. 12	0.216	0.221	0.229	0.391	0.077	0.063
1/4	0.250	0.255	0.263	0.491	0.110	0.077
5/16	0.312	0.318	0.328	0.596	0.130	0.097
3/8	0.375	0.382	0.393	0.691	0.145	0.115
7/16	0.438	0.446	0.459	0.787	0.160	0.133
1/2	0.500	0.509	0.523	0.883	0.176	0.151
9/16	0.562	0.572	0.587	0.981	0.193	0.170
5/8	0.625	0.636	0.653	1.093	0.210	0.189
11/16	0.688	0.700	0.718	1.192	0.227	0.207
3/4	0.750	0.763	0.783	1.291	0.244	0.226
13/16	0.812	0.826	0.847	1.391	0.262	0.246
7/8	0.875	0.890	0.912	1.494	0.281	0.266
15/16	0.938	0.954	0.978	1.594	0.298	0.284
1	1.000	1.017	1.042	1.705	0.319	0.306
1 1/16	1.062	1.080	1.107	1.808	0.338	0.326
1 1/8	1.125	1.144	1.172	1.909	0.356	0.345
1 3/16	1.188	1.208	1.237	2.008	0.373	0.364
1 1/4	1.250	1.271	1.302	2.113	0.393	0.384
1 5/16	1.312	1.334	1.366	2.211	0.410	0.403
1 3/8	1.375	1.398	1.432	2.311	0.427	0.422
1 7/16	1.438	1.462	1.497	2.406	0.442	0.440
1 1/2	1.500	1.525	1.561	2.502	0.458	0.458
1 5/8	1.625	1.650	1.686	2.693	0.491	0.458
1 3/4	1.750	1.775	1.811	2.818	0.491	0.458
1 7/8	1.875	1.900	1.936	2.943	0.491	0.458
2	2.000	2.025	2.061	3.068	0.491	0.458
2 1/4	2.250	2.275	2.311	3.388	0.526	0.496
2 1/2	2.500	2.525	2.561	3.638	0.526	0.496
2 3/4	2.750	2.775	2.811	3.888	0.526	0.496
3	3.000	3.025	3.061	4.138	0.526	0.496

(Courtesy ASA B27.1-1965)

*The maximum outside diameters specified allow for the commercial tolerances on cold drawn wire.

TABLE 31. INTERNAL TOOTH LOCK WASHERS

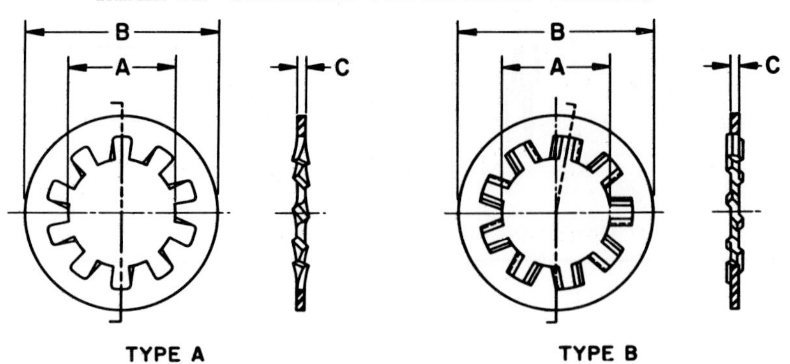

Dimensions of Internal Tooth Lock Washers

Nominal Washer Size		A Inside Diameter		B Outside Diameter		C Thickness	
		Min	Max	Max	Min	Max	Min
No. 2	0.086	0.089	0.095	0.200	0.175	0.015	0.010
No. 3	0.099	0.102	0.109	0.232	0.215	0.019	0.012
No. 4	0.112	0.115	0.123	0.270	0.255	0.019	0.015
No. 5	0.125	0.129	0.136	0.280	0.245	0.021	0.017
No. 6	0.138	0.141	0.150	0.295	0.275	0.021	0.017
No. 8	0.164	0.168	0.176	0.340	0.325	0.023	0.018
No. 10	0.190	0.195	0.204	0.381	0.365	0.025	0.020
No. 12	0.216	0.221	0.231	0.410	0.394	0.025	0.020
1/4	0.250	0.256	0.267	0.478	0.460	0.028	0.023
5/16	0.312	0.320	0.332	0.610	0.594	0.034	0.028
3/8	0.375	0.384	0.398	0.692	0.670	0.040	0.032
7/16	0.438	0.448	0.464	0.789	0.740	0.040	0.032
1/2	0.500	0.512	0.530	0.900	0.867	0.045	0.037
9/16	0.562	0.576	0.596	0.985	0.957	0.045	0.037
5/8	0.625	0.640	0.663	1.071	1.045	0.050	0.042
11/16	0.688	0.704	0.728	1.166	1.130	0.050	0.042
3/4	0.750	0.769	0.795	1.245	1.220	0.055	0.047
13/16	0.812	0.832	0.861	1.315	1.290	0.055	0.047
7/8	0.875	0.894	0.927	1.410	1.364	0.060	0.052
1	1.000	1.019	1.060	1.637	1.590	0.067	0.059
1 1/8	1.125	1.144	1.192	1.830	1.799	0.067	0.059
1 1/4	1.250	1.275	1.325	1.975	1.921	0.067	0.059

Dimensions of Heavy Internal Tooth Lock Washers

Nominal Washer Size		A Inside Diameter		B Outside Diameter		C Thickness	
		Min	Max	Max	Min	Max	Min
1/4	0.250	0.256	0.267	0.536	0.500	0.045	0.035
5/16	0.312	0.320	0.332	0.607	0.590	0.050	0.040
3/8	0.375	0.384	0.398	0.748	0.700	0.050	0.042
7/16	0.438	0.448	0.464	0.858	0.800	0.067	0.050
1/2	0.500	0.512	0.530	0.924	0.880	0.067	0.055
9/16	0.562	0.576	0.596	1.034	0.990	0.067	0.055
5/8	0.625	0.640	0.663	1.135	1.100	0.067	0.059
3/4	0.750	0.768	0.795	1.265	1.240	0.084	0.070
7/8	0.875	0.894	0.927	1.447	1.400	0.084	0.075

(Courtesy ASA B27.1-1965)

TABLE 32. EXTERNAL TOOTH LOCK WASHERS

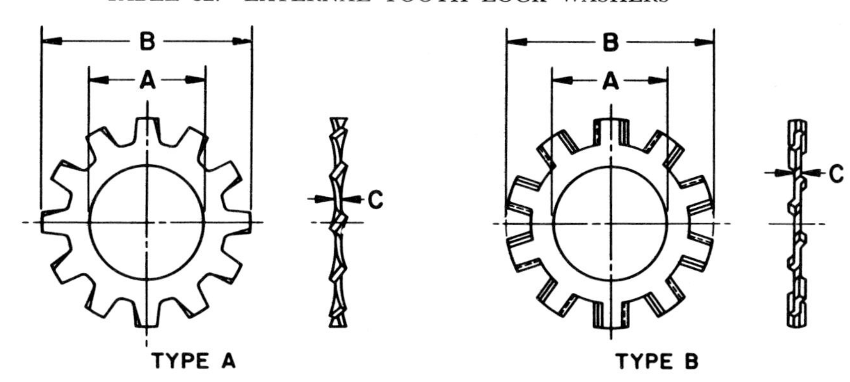

Dimensions of External Tooth Lock Washers

Nominal Washer Size		*A* Inside Diameter		*B* Outside Diameter		*C* Thickness	
		Min	Max	Max	Min	Max	Min
No. 4	0.112	0.115	0.123	0.260	0.245	0.019	0.015
No. 6	0.138	0.141	0.150	0.320	0.305	0.022	0.016
No. 8	0.164	0.168	0.176	0.381	0.365	0.023	0.018
No. 10	0.190	0.195	0.204	0.410	0.395	0.025	0.020
No. 12	0.216	0.221	0.231	0.475	0.460	0.028	0.023
1/4	0.250	0.256	0.267	0.510	0.494	0.028	0.023
5/16	0.312	0.320	0.332	0.610	0.588	0.034	0.028
3/8	0.375	0.384	0.398	0.694	0.670	0.040	0.032
7/16	0.438	0.448	0.464	0.760	0.740	0.040	0.032
1/2	0.500	0.513	0.530	0.900	0.880	0.045	0.037
9/16	0.562	0.576	0.596	0.985	0.960	0.045	0.037
5/8	0.625	0.641	0.663	1.070	1.045	0.050	0.042
11/16	0.688	0.704	0.728	1.155	1.130	0.050	0.042
3/4	0.750	0.768	0.795	1.260	1.220	0.055	0.047
13/16	0.812	0.833	0.861	1.315	1.290	0.055	0.047
7/8	0.875	0.897	0.927	1.410	1.380	0.060	0.052
1	1.000	1.025	1.060	1.620	1.590	0.067	0.059

(Courtesy ASA B27.1-1965)

TABLE 33

STRAIGHT SHANK TWIST DRILLS

Drill Size	*Decimal Equivalent Diameter*	*Drill Size*	*Decimal Equivalent Diameter*	*Drill Size*	*Decimal Equivalent Diameter*	*Drill Size*	*Decimal Equivalent Diameter*
80	0.0135	7/64	0.1094	*G*	0.261	47/64	0.7344
79	0.0145	35	0.110	17/64	0.2656	3/4	0.7500
1/64	0.0156	34	0.111	*H*	0.266	49/64	0.7656
78	0.016	33	0.113	*I*	0.272	25/32	0.7812
77	0.018	32	0.116	*J*	0.277	51/64	0.7969
76	0.020	31	0.120	*K*	0.281	13/16	0.8125
75	0.021	1/8	0.1250	9/32	0.2812	53/64	0.8281
74	0.0225	30	0.1285	*L*	0.290	27/32	0.8437
73	0.024	29	0.136	*M*	0.295	55/64	0.8594
72	0.025	28	0.1405	19/64	0.2969	7/8	0.8750
71	0.026	9/64	0.1406	*N*	0.302	57/64	0.8906
70	0.028	27	0.144	5/16	0.3125	29/32	0.9062
69	0.0292	26	0.147	*O*	0.316	59/64	0.9219
68	0.031	25	0.1495	*P*	0.323	15/16	0.9375
1/32	0.0312	24	0.152	21/64	0.3281	61/64	0.9531
67	0.032	23	0.154	*Q*	0.332	31/32	0.9687
66	0.033	5/32	0.1562	*R*	0.339	63/64	0.9844
65	0.035	22	0.157	11/32	0.3437	1	1.0000
64	0.036	21	0.159	*S*	0.348	1 1/64	1.0156
63	0.037	20	0.161	*T*	0.358	1 1/32	1.0312
62	0.038	19	0.166	23/64	0.3594	1 3/64	1.0469
61	0.039	18	0.1695	*U*	0.368	1 1/16	1.0625
60	0.040	11/64	0.1719	3/8	0.375	1 5/64	1.0781
59	0.041	17	0.173	*V*	0.377	1 3/32	1.0937
58	0.042	16	0.177	*W*	0.386	1 7/64	1.1094
57	0.043	15	0.180	25/64	0.3906	1 1/8	1.1250
56	0.0465	14	0.182	*X*	0.397	1 9/64	1.1406
3/64	0.0468	13	0.185	*Y*	0.404	1 5/32	1.1562
55	0.052	3/16	0.1875	13/32	0.4062	1 11/64	1.1719
54	0.055	12	0.189	*Z*	0.413	1 3/16	1.1875
53	0.0595	11	0.191	27/64	0.4219	1 13/64	1.2031
1/16	0.0625	10	0.1935	7/16	0.4375	1 7/32	1.2187
52	0.0635	9	0.196	29/64	0.4531	1 15/64	1.2344
51	0.067	8	0.199	15/32	0.4687	1 1/4	1.2500
50	0.070	7	0.201	31/64	0.4844	1 9/32	1.2812
49	0.073	13/64	0.2031	1/2	0.5000	1 5/16	1.3125
48	0.076	6	0.204	33/64	0.5156	1 11/32	1.3437
5/64	0.0781	5	0.2055	17/32	0.5312	1 3/8	1.3750
47	0.0785	4	0.209	35/64	0.5469	1 13/32	1.4062
46	0.081	3	0.213	9/16	0.5625	1 7/16	1.4375
45	0.082	7/32	0.2187	37/64	0.5781	1 15/32	1.4687
44	0.086	2	0.221	19/32	0.5937	1 1/2	1.5000
43	0.089	1	0.228	39/64	0.6094	1 9/16	1.5625
42	0.0935	*A*	0.234	5/8	0.6250	1 5/8	1.6250
3/32	0.0937	15/64	0.2344	41/64	0.6406	1 11/16	1.6875
41	0.096	*B*	0.238	21/32	0.6562	1 3/4	1.7500
40	0.098	*C*	0.242	43/64	0.6719	1 13/16	1.8125
39	0.0995	*D*	0.246	11/16	0.6875	1 7/8	1.8750
38	0.1015	*E* & 1/4	0.250	45/64	0.7031	1 15/16	1.9375
37	0.104	*F*	0.257	23/32	0.7187	2	2.0000
36	0.1065						

(Courtesy ASA)

TABLE 34

DIMENSIONS OF SUNK KEYS*

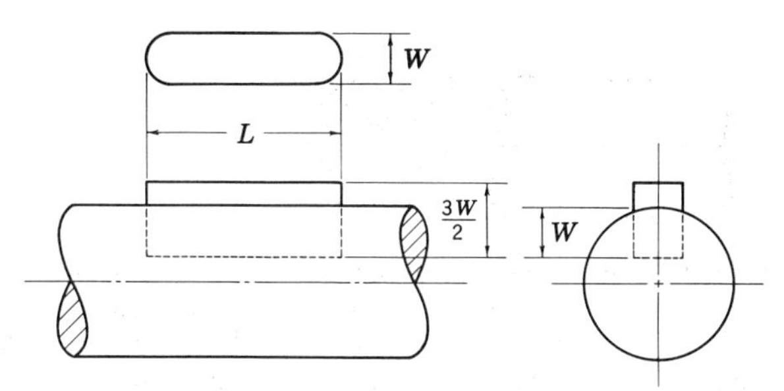

(Dimensions in Inches)

Key No.	*L*	*W*	*Key No.*	*L*	*W*
1	1/2	1/16	22	1 3/8	1/4
2	1/2	3/32	23	1 3/8	5/16
3	1/2	1/8	*F*	1 3/8	3/8
4	5/8	3/32	24	1 1/2	1/4
5	5/8	1/8	25	1 1/2	5/16
6	5/8	5/32	*G*	1 1/2	3/8
7	3/4	1/8	51	1 3/4	1/4
8	3/4	5/32	52	1 3/4	5/16
9	3/4	3/16	53	1 3/4	3/8
10	7/8	5/32	26	2	3/16
11	7/8	3/16	27	2	1/4
12	7/8	7/32	28	2	5/16
A	7/8	1/4	29	2	3/8
13	1	3/16	54	2 1/4	1/4
14	1	7/32	55	2 1/4	5/16
15	1	1/4	56	2 1/4	3/8
B	1	5/16	57	2 1/4	7/16
16	1 1/8	3/16	58	2 1/2	5/16
17	1 1/8	7/32	59	2 1/2	3/8
18	1 1/8	1/4	60	2 1/2	7/16
C	1 1/8	5/16	61	2 1/2	1/2
19	1 1/4	3/16	30	3	3/8
20	1 1/4	7/32	31	3	7/16
21	1 1/4	1/4	32	3	1/2
D	1 1/4	5/16	33	3	9/16
E	1 1/4	3/8	34	3	5/8

* Manufactured by Pratt and Whitney, Hartford, Conn.

TABLE 35

WOODRUFF KEYS AND KEYSLOTS

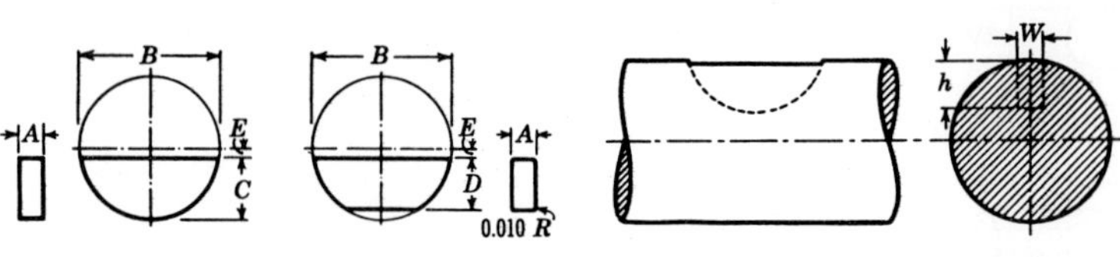

*Key * Number*	*Nominal Size, A × B*	*Maximum Width of Key, A*	*Maximum Diameter of Key, B*	*Maximum Height of Key*		*Distance below Center, E*	*Keyslot*	
				C	D		*Maximum Width, W*	*Maximum Depth, h*
204	$\frac{1}{16} \times \frac{1}{2}$	0.0635	0.500	0.203	0.194	$\frac{3}{64}$	0.0630	0.1718
304	$\frac{3}{32} \times \frac{1}{2}$	0.0948	0.500	0.203	0.194	$\frac{3}{64}$	0.0943	0.1561
305	$\frac{3}{32} \times \frac{5}{8}$	0.0948	0.625	0.250	0.240	$\frac{1}{16}$	0.0943	0.2031
404	$\frac{1}{8} \times \frac{1}{2}$	0.1260	0.500	0.203	0.194	$\frac{3}{64}$	0.1255	0.1405
405	$\frac{1}{8} \times \frac{5}{8}$	0.1260	0.625	0.250	0.240	$\frac{1}{16}$	0.1255	0.1875
406	$\frac{1}{8} \times \frac{3}{4}$	0.1260	0.750	0.313	0.303	$\frac{1}{16}$	0.1255	0.2505
505	$\frac{5}{32} \times \frac{5}{8}$	0.1573	0.625	0.250	0.240	$\frac{1}{16}$	0.1568	0.1719
506	$\frac{5}{32} \times \frac{3}{4}$	0.1573	0.750	0.313	0.303	$\frac{1}{16}$	0.1568	0.2349
507	$\frac{5}{32} \times \frac{7}{8}$	0.1573	0.875	0.375	0.365	$\frac{1}{16}$	0.1568	0.2969
606	$\frac{3}{16} \times \frac{3}{4}$	0.1885	0.750	0.313	0.303	$\frac{1}{16}$	0.1880	0.2193
607	$\frac{3}{16} \times \frac{7}{8}$	0.1885	0.875	0.375	0.365	$\frac{1}{16}$	0.1880	0.2813
608	$\frac{3}{16} \times 1$	0.1885	1.000	0.438	0.428	$\frac{1}{16}$	0.1880	0.3443
609	$\frac{3}{16} \times 1\frac{1}{8}$	0.1885	1.125	0.484	0.475	$\frac{5}{64}$	0.1880	0.3903
807	$\frac{1}{4} \times \frac{7}{8}$	0.2510	0.875	0.375	0.365	$\frac{1}{16}$	0.2505	0.2500
808	$\frac{1}{4} \times 1$	0.2510	1.000	0.438	0.428	$\frac{1}{16}$	0.2505	0.3130
809	$\frac{1}{4} \times 1\frac{1}{8}$	0.2510	1.125	0.484	0.475	$\frac{5}{64}$	0.2505	0.3590
810	$\frac{1}{4} \times 1\frac{1}{4}$	0.2510	1.250	0.547	0.537	$\frac{5}{64}$	0.2505	0.4220
811	$\frac{1}{4} \times 1\frac{3}{8}$	0.2510	1.375	0.594	0.584	$\frac{3}{32}$	0.2505	0.4690
812	$\frac{1}{4} \times 1\frac{1}{2}$	0.2510	1.500	0.641	0.631	$\frac{7}{64}$	0.2505	0.5160
1008	$\frac{5}{16} \times 1$	0.3135	1.000	0.438	0.428	$\frac{1}{16}$	0.3130	0.2818
1009	$\frac{5}{16} \times 1\frac{1}{8}$	0.3135	1.125	0.484	0.475	$\frac{5}{64}$	0.3130	0.3278
1010	$\frac{5}{16} \times 1\frac{1}{4}$	0.3135	1.250	0.547	0.537	$\frac{5}{64}$	0.3130	0.3908
1011	$\frac{5}{16} \times 1\frac{3}{8}$	0.3135	1.375	0.594	0.584	$\frac{3}{32}$	0.3130	0.4378
1012	$\frac{5}{16} \times 1\frac{1}{2}$	0.3135	1.500	0.641	0.631	$\frac{7}{64}$	0.3130	0.4848
1210	$\frac{3}{8} \times 1\frac{1}{4}$	0.3760	1.250	0.547	0.537	$\frac{5}{64}$	0.3755	0.3595
1211	$\frac{3}{8} \times 1\frac{3}{8}$	0.3760	1.375	0.594	0.584	$\frac{3}{32}$	0.3755	0.4065
1212	$\frac{3}{8} \times 1\frac{1}{2}$	0.3760	1.500	0.641	0.631	$\frac{7}{64}$	0.3755	0.4535

All dimensions given in inches.

* Note: Key numbers indicate the nominal key dimensions. The last two digits give the nominal diameter (B) in eighths of an inch and the digits preceding the last two give the nominal width (A) in thirty-seconds of an inch. Thus, 204 indicates a key $\frac{2}{32} \times \frac{4}{8}$ or $\frac{1}{16} \times \frac{1}{2}$ inches.

(Courtesy ASA)

TABLE 36

DIMENSIONS OF SQUARE AND FLAT PLAIN PARALLEL STOCK KEYS

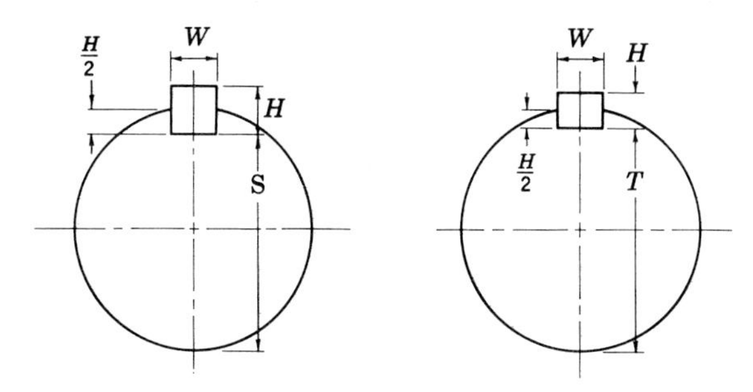

(Dimensions in Inches)

Shaft Diameter Range	*Square Key, W x H*	*Flat Key, W x H*	*Shaft Diameter Range*	*Square Key, W x H*	*Flat Key, W x H*
1/2 – 9/16	1/8 x 1/8	1/8 x 3/32	2 5/16–2 3/4	5/8 x 5/8	5/8 x 7/16
5/8 – 7/8	3/16 x 3/16	3/16 x 1/8	2 7/8 –3 1/4	3/4 x 3/4	3/4 x 1/2
15/16–1 1/4	1/4 x 1/4	1/4 x 3/16	3 3/8 –3 3/4	7/8 x 7/8	7/8 x 5/8
1 5/16 –1 3/8	5/16 x 5/16	5/16 x 1/4	3 7/8 –4 1/2	1 x 1	1 x 3/4
1 7/16 –1 3/4	3/8 x 3/8	3/8 x 1/4	4 3/4 –5 1/2	1 1/4 x 1 1/4	1 1/4 x 7/8
1 13/16–2 1/4	1/2 x 1/2	1/2 x 3/8	5 3/4 –6	1 1/2 x 1 1/2	1 1/2 x 1

Stock keys are applicable to the general run of work.

TABLE 37

SQUARE AND FLAT PLAIN TAPER STOCK KEYS

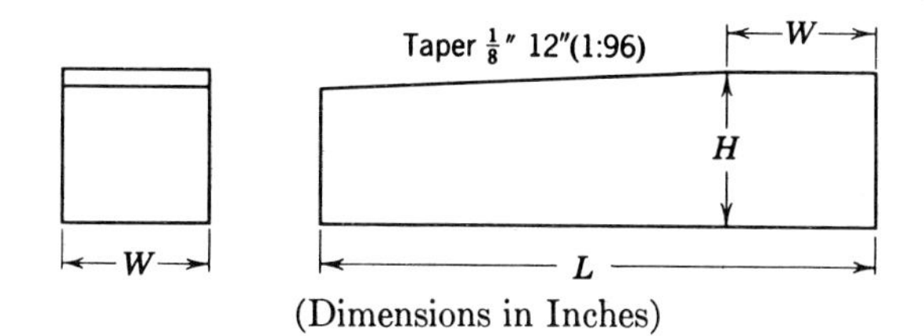

(Dimensions in Inches)

	Square Type		*Flat Type*	
Shaft Diameter Range	*Width Maximum, W*	*Height Minimum, H*	*Width Maximum, W*	*Height Minimum, H*
1/2 – 9/16	1/8	1/8	1/8	3/32
5/8 – 7/8	3/16	3/16	3/16	1/8
15/16–1 1/4	1/4	1/4	1/4	3/16
1 5/16 –1 3/8	5/16	5/16	5/16	1/4
1 7/16 –1 3/4	3/8	3/8	3/8	1/4
1 13/16–2 1/4	1/2	1/2	1/2	3/8
2 5/16 –2 3/4	5/8	5/8	5/8	7/16
2 7/8 –3 1/4	3/4	3/4	3/4	1/2
3 3/8 –3 3/4	7/8	7/8	7/8	5/8
3 7/8 –4 1/2	1	1	1	3/4
4 3/4 –5 1/2	1 1/4	1 1/4	1 1/4	7/8
5 3/4 –6	1 1/2	1 1/2	1 1/2	1

The minimum stock length of keys is $4W$, and the maximum stock length is $16W$.

(Courtesy ASA)

TABLE 38

DIMENSIONS OF SQUARE AND FLAT GIB-HEAD TAPER STOCK KEYS

(Dimensions in Inches)

Shaft Diameter Range	Square Type of Key					Flat Type of Key				
	Key		Gib-head			Key		Gib-head		
	Maximum Width, W	Height, H	Height, C	Length, D	Height to Edge Chamfer, E	Maximum Width, W	Height, H	Height, C	Length, D	Height to Edge Chamfer, E
1/2 – 9/16	1/8	1/8	1/4	7/32	5/32	1/8	3/32	3/16	1/8	1/8
5/8 – 7/8	3/16	3/16	5/16	9/32	7/32	3/16	1/8	1/4	3/16	5/32
15/16–1 1/4	1/4	1/4	7/16	11/32	11/32	1/4	3/16	5/16	1/4	3/16
1 5/16 –1 3/8	5/16	5/16	9/16	13/32	13/32	5/16	1/4	3/8	5/16	1/4
1 7/16 –1 3/4	3/8	3/8	11/16	15/32	15/32	3/8	1/4	7/16	3/8	5/16
1 13/16–2 1/4	1/2	1/2	7/8	19/32	5/8	1/2	3/8	5/8	1/2	7/16
2 5/16 –2 3/4	5/8	5/8	1 1/16	23/32	3/4	5/8	7/16	3/4	5/8	1/2
2 7/8 –3 1/4	3/4	3/4	1 1/4	7/8	7/8	3/4	1/2	7/8	3/4	5/8
3 3/8 –3 3/4	7/8	7/8	1 1/2	1	1	7/8	5/8	1 1/16	7/8	3/4
3 7/8 –4 1/2	1	1	1 3/4	1 3/16	1 3/16	1	3/4	1 1/4	1	1 3/16
4 3/4 –5 1/2	1 1/4	1 1/4	2	1 7/16	1 7/16	1 1/4	7/8	1 1/2	1 1/4	1
5 3/4 –6	1 1/2	1 1/2	2 1/2	1 3/4	1 3/4	1 1/2	1	1 3/4	1 1/2	1 1/4

Stock keys are applicable to the general run of work. They are not intended to cover the finer applications where a close fit may be required. The minimum stock length of keys is $4W$ and the maximum stock length is $16W$.

(Courtesy ASA)

TABLE 39
Taper Pins
Dimensions of Taper Pins

Number	7/0	6/0	5/0	4/0	3/0	2/0	0	1	2	3	4	5	6	7	8	9	10
Size (Large End)	0.0625	0.0780	0.0940	0.1090	0.1250	0.1410	0.1560	0.1720	0.1930	0.2190	0.2500	0.2890	0.3410	0.4090	0.4920	0.5910	0.7060
Length, L																	
0.375	X	X															
0.500	X	X	X	X	X	X	X										
0.625	X	X	X	X	X	X	X										
0.750		X	X	X	X	X	X	X	X	X							
0.875					X	X	X	X	X	X							
1.000			X	X	X	X	X	X	X	X	X	X					
1.250						X	X	X	X	X	X	X	X				
1.500							X	X	X	X	X	X	X				
1.750								X	X	X	X	X	X				
2.000								X	X	X	X	X	X	X	X		
2.250									X	X	X	X	X	X	X		
2.500									X	X	X	X	X	X	X		
2.750										X	X	X	X	X	X	X	
3.000										X	X	X	X	X	X	X	
3.250													X	X	X	X	
3.500													X	X	X	X	X
3.750													X	X	X	X	X
4.000													X	X	X	X	X
4.250															X	X	X
4.500															X	X	X
4.750															X	X	X
5.000															X	X	X
5.250																X	X
5.500																X	X
5.750																X	X
6.000																X	X

(Courtesy ASA)

All dimensions are given in inches. Standard reamers are available for pins given above the line.
Pins Nos. 11 (size 0.8600), 12 (size 1.032), 13 (size 1.241), and 14 (1.523) are special sizes—hence their lengths are special.
To find small diameter of pin, multiply the length by 0.02083 and subtract the result from the large diameter.

TYPES	COMMERCIAL TYPE	PRECISION TYPE	
Sizes	7/0 to 14	7/0 to 10	0.0005 up to 1 in. long
Tolerance on Diameter	(+0.0013, −0.0007)	(+0.0013, −0.0007)	0.001 1 1/16 to 2 in. long
Taper	1/4 In. per Ft	1/4 In. per Ft	0.002 2 1/16 and longer
Length Tolerance	(±0.030)	(±0.030)	
Concavity Tolerance	None		

TABLE 40. DIMENSIONS OF COTTER PINS

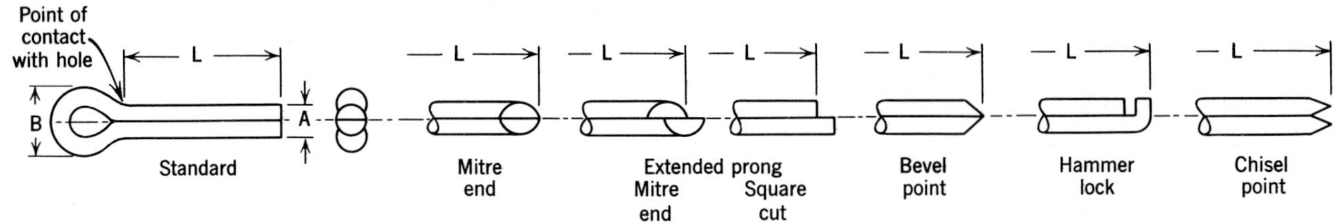

Diameter Nominal	Diameter A		Outside Eye Diameter B Min	Hole Sizes Recom-mended
	Max	Min		
0.031	0.032	0.028	1/16	3/64
0.047	0.048	0.044	3/32	1/16
0.062	0.060	0.056	1/8	3/64
0.078	0.076	0.072	5/32	3/32
0.094	0.090	0.086	3/16	7/64
0.109	0.104	0.100	7/32	1/8
0.125	0.120	0.116	1/4	9/64
0.141	0.134	0.130	9/32	3/32
0.156	0.150	0.146	3/16	11/64
0.188	0.176	0.172	3/8	13/64
0.219	0.207	0.202	7/16	15/64
0.250	0.225	0.220	1/2	17/64
0.312	0.280	0.275	5/6	5/16
0.375	0.335	0.329	3/4	3/8
0.438	0.406	0.400	7/8	7/16
0.500	0.473	0.467	1	1/2
0.625	0.598	0.590	1 3/4	5/8
0.750	0.723	0.715	1 1/2	3/4

(Courtesy ASA B5.20-1958)

All dimensions are given in inches.

A certain amount of leeway is permitted in the design of the head, however the outside diameters given should be adhered to.

Prongs are to be parallel, ends shall not be open.

Points may be blunt, bevel, extended prong, mitre, etc., and purchaser may specify type required.

Lengths shall be measured as shown on the above illustration. (L-Dimension)

Cotter pins shall be free from burrs or any defects that will affect their serviceability.

TABLE 41

Drilling Specifications for Taper Pins

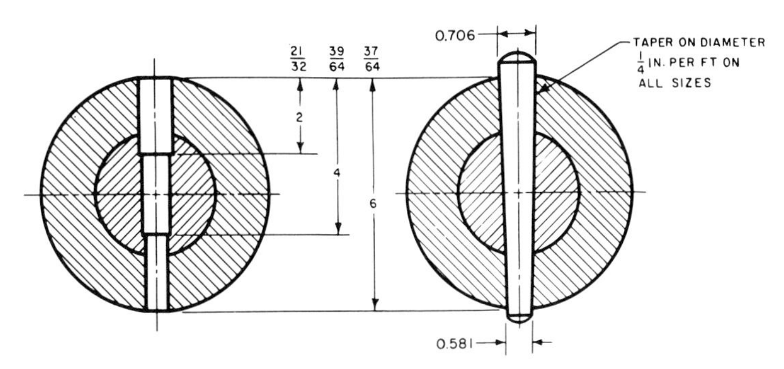

EXAMPLE— NO. 10 X 6" TAPER PIN USING 3 DRILLS FOR STEP DRILLING AND STRAIGHT FLUTED REAMERS

NOTE— SEE DRILL CHART AT BOTTOM OF PAGE FOR SIZE OF DRILL AND NUMBER REQUIRED

DRILL DIA | PIN DIA | LENGTH OF PIN IN INCHES 1 2 3 4 5 6

DRILL DIA: 0.0156, 0.0312, 0.0469, 0.0625, 0.0781, 0.0938, 0.1094, 0.1250, 0.1406, 0.1562, 0.1719, 0.1875, 0.2031, 0.2188, 0.2344, 0.2500, 0.2656, 0.2812, 0.2969, 0.3125, 0.3281, 0.3438, 0.3594, 0.3750, 0.3906, 0.4062, 0.4219, 0.4375, 0.4531, 0.4688, 0.4844, 0.5000, 0.5156, 0.5312, 0.5469, 0.5625, 0.5781, 0.5938, 0.6094, 0.6250, 0.6406, 0.6562, 0.6719, 0.6875, 0.7031

PIN DIA: 0.0625 (7/0), 0.0780 (6/0), 0.0940 (5/0), 0.1090 (4/0), 0.1250 (3/0), 0.1410 (2/0), 0.1560 (0), 0.1720 (1), 0.1930 (2), 0.2190 (3), 0.2500 (4), 0.2890 (5), 0.3410 (6), 0.4090 (7), 0.4920 (8), 0.5910 (9), 0.7060 (10)

TO OBTAIN DRILL SIZES

DETERMINE DEPTH OF HOLE (PIN LENGTH)

DETERMINE INTERSECTION OF DEPTH LINE WITH TAPER LINE

DRILL DIAMETER WILL BE THE NEXT SMALLER DIAMETER ON THE HORIZONTAL LINE ABOVE THE INTERSECTION.

NOTE THE NUMBER OF DRILLS RECOMMENDED FOR MAXIMUM LENGTH

IF CHART CALLS FOR 3 DRILLS, DIVIDE THE DRILLING DEPTH INTO 3 EQUAL SPACES (NEAREST 1/4 INCH). IF PIN INDICATES 2 DRILLS, DIVIDE THE DRILLING DEPTH INTO 2 SPACES.

EXAMPLE A
#10 PIN – 6" LONG
USE 3 DRILLS
0.5781 DRILL THROUGH
0.6094 DRILL 4" DEEP
0.6562 DRILL 2" DEEP

EXAMPLE B
#10 PIN – 4" LONG
USE 2 DRILLS
0.6094 DRILL THROUGH
0.6562 DRILL 2" DEEP

ALL DETAIL DRAWINGS USING TAPER PINS SHOULD CARRY THE ABOVE TYPE OF NOTE AS INDICATED IN (A) OR (B).

TO OBTAIN THE DIAMETER AT THE SMALL END MULTIPLY THE LENGTH BY 0.02093 AND SUBTRACT FROM THE LARGE DIAMETER.

DRILL CHART

SIZE PIN	1ST DRILL THROUGH SIZE	SECOND DRILL		THIRD DRILL	
		SIZE	DEPTH	SIZE	DEPTH
7/0	0.0469				
6/0	0.0469				
5/0	0.0625				
4/0	0.0781				
3/0	0.0938				
2/0	0.0938	0.1094	1-1/4		
0	0.0938	0.1250	1-1/2		
1	0.1094	0.1406	1-1/2		
2	0.1094	0.1406	1-3/4		
3	0.1406	0.1719	1-3/4		
4	0.1719	0.2031	1-3/4		
5	0.1875	0.2344	2-		
6	0.2344	0.2656	3-1/4	0.2969	1-3/4
7	0.2969	0.3281	3-1/4	0.3750	1-5/8
8	0.3906	0.4219	3-1/4	0.4531	1-5/8
9	0.4688	0.5000	4-	0.5469	2-
10	0.5781	0.6094	4-	0.6562	2-

IF HELICALLY FLUTED TAPER REAMERS ARE USED INSTEAD OF STEP DRILLING AND STRAIGHT FLUTED REAMERS, THE DIAMETER AT THE SMALL END OF THE PIN IS THE SIZE FOR THE THROUGH DRILL

(Courtesy ASA)

DESCRIPTIONS OF FITS

FOR TABLES 42 to 46

Running and Sliding Fits. Running and sliding fits, for which limits of clearance are given in Table 42, are intended to provide a similar running performance, with suitable lubrication allowance, throughout the range of sizes. The clearance for the first two classes, used chiefly as slide fits, increase more slowly with diameter than the other classes, so that accurate location is maintained even at the expense of free relative motion.

These fits may be described briefly as follows:

RC 1 *Close sliding fits* are intended for the accurate location of parts which must assemble without perceptible play.

RC 2 *Sliding fits* are intended for accurate location, but with greater maximum clearance than class RC 1. Parts made to fit move and turn easily, but are not intended to run freely, and in the larger sizes may seize with small temperature changes.

RC 3 *Precision running fits* are about the closest fits which can be expected to run freely, and are intended for precision work at slow speeds and light journal pressures, but are not suitable where appreciable temperature differences are likely to be encountered.

RC 4 *Close running fits* are intended chiefly for running fits on accurate machinery with moderate surface speeds and journal pressures, where accurate location and minimum play is desired.

RC 5, RC 6 *Medium running fits* are intended for higher running speeds, or heavy journal pressures, or both.

RC 7 *Free running fits* are intended for use where accuracy is not essential, or where large temperature variations are likely to be encountered, or under both these conditions.

RC 8, RC 9 *Loose running fits* are intended for use where materials, such as cold-rolled shafting and tubing, made to commercial tolerances are involved.

Locational Fits. Locational fits are intended to determine only the location of the mating parts; they may provide rigid or accurate location, as with interference fits, or provide some freedom of location, as with clearance fits. Accordingly, they are divided into three groups: clearance fits, transition fits, and interference fits.

These fits are more fully described as follows:

LC *Locational clearance fits* are intended for parts which are normally stationary, but which can be freely assembled or disassembled. They run from snug fits for parts requiring accuracy of location, through the medium clearance fits for parts such as spigots, to the looser fastener fits where freedom of assembly is of prime importance.

LT *Transition fits* are a compromise between clearance and interference fits, for application where accuracy of location is important, but either a small amount of clearance or interference is permissible.

LN *Locational interference fits* are used where accuracy of location is of prime importance, and for parts requiring rigidity and alignment with special requirements for bore pressure. Such fits are not intended for parts designed to transmit frictional loads from one part to another by virtue of the tightness of fit, as these conditions are covered by force fits.

Force Fits. Force or shrink fits constitute a special type of interference fit, normally characterized by maintenance of constant bore pressures through the range of sizes. The interference therefore varies almost directly with diameter, and the difference between its minimum and maximum value is small, to maintain the resulting pressures within reasonable limits.

These fits may be described briefly as follows:

FN 1 *Light drive fits* are those requiring light assembly pressures, and produce more or less permanent assemblies. They are suitable for thin sections or long fits, or in cast-iron external members.

FN 2 *Medium drive fits* are suitable for ordinary steel parts, or for shrink fits on light sections. They are about the tightest fits that can be used with high-grade cast-iron external members.

FN 3 Heavy drive fits are suitable for heavier steel parts or for shrink fits in medium sections.

FN 4, FN 5 Force fits are suitable for parts which can be highly stressed, or for shrink fits where the heavy pressing forces required are impractical.

TABLE 42. RUNNING AND SLIDING FITS, RC

Limits are in thousandths of an inch. Limits for hole and shaft are applied algebraically to the basic size to obtain the limits of size for the parts. Data in bold face are in accordance with ABC agreements. Symbols H5, g5, etc., are Hole and Shaft designations used in ABC System.

Nominal Size Range Inches		Class RC 1			Class RC 2			Class RC 3			Class RC 4		
		Limits of Clearance	Standard Limits		Limits of Clearance	Standard Limits		Limits of Clearance	Standard Limits		Limits of Clearance	Standard Limits	
Over	To		Hole H5	Shaft g4		Hole H6	Shaft g5		Hole H6	Shaft f6		Hole H7	Shaft f7
0.04	0.12	0.1 0.45	+0.2 0	− 0.1 − 0.25	0.1 0.55	+ 0.25 0	− 0.1 − 0.3	0.3 0.8	+ 0.25 0	− 0.3 − 0.55	0.3 1.1	+ 0.4 0	− 0.3 − 0.7
0.12	0.24	0.15 0.5	+0.2 0	− 0.15 − 0.3	0.15 0.65	+ 0.3 0	− 0.15 − 0.35	0.4 1.0	+ 0.3 0	− 0.4 − 0.7	0.4 1.4	+ 0.5 0	− 0.4 − 0.9
0.24	0.40	0.2 0.6	+0.25 0	− 0.2 − 0.35	0.2 0.85	+ 0.4 0	− 0.2 − 0.45	0.5 1.3	+ 0.4 0	− 0.5 − 0.9	0.5 1.7	+ 0.6 0	− 0.5 − 1.1
0.40	0.71	0.25 0.75	+0.3 0	− 0.25 − 0.45	0.25 0.95	+ 0.4 0	− 0.25 − 0.55	0.6 1.4	+ 0.4 0	− 0.6 − 1.0	0.6 2.0	+ 0.7 0	− 0.6 − 1.3
0.71	1.19	0.3 0.95	+0.4 0	− 0.3 − 0.55	0.3 1.2	+ 0.5 0	− 0.3 − 0.7	0.8 1.8	+ 0.5 0	− 0.8 − 1.3	0.8 2.4	+ 0.8 0	− 0.8 − 1.6
1.19	1.97	0.4 1.1	+0.4 0	− 0.4 − 0.7	0.4 1.4	+ 0.6 0	− 0.4 − 0.8	1.0 2.2	+ 0.6 0	− 1.0 − 1.6	1.0 3.0	+ 1.0 0	− 1.0 − 2.0
1.97	3.15	0.4 1.2	+0.5 0	− 0.4 − 0.7	0.4 1.6	+ 0.7 0	− 0.4 − 0.9	1.2 2.6	+ 0.7 0	− 1.2 − 1.9	1.2 3.6	+ 1.2 0	− 1.2 − 2.4
3.15	4.73	0.5 1.5	+0.6 0	− 0.5 − 0.9	0.5 2.0	+ 0.9 0	− 0.5 − 1.1	1.4 3.2	+ 0.9 0	− 1.4 − 2.3	1.4 4.2	+ 1.4 0	− 1.4 − 2.8

Nominal Size Range Inches		Class RC 5			Class RC 6			Class RC 7			Class RC 8			Class RC 9		
		Limits of Clearance	Standard Limits		Limits of Clearance	Standard Limits		Limits of Clearance	Standard Limits		Limits of Clearance	Standard Limits		Limits of Clearance	Standard Limits	
Over	To		Hole H7	Shaft e7		Hole H8	Shaft e8		Hole H9	Shaft d8		Hole H10	Shaft c9		Hole H11	Shaft
0.04	0.12	0.6 1.4	+ 0.4 0	− 0.6 − 1.0	0.6 1.8	+ 0.6 0	− 0.6 − 1.2	1.0 2.6	+ 1.0 0	− 1.0 − 1.6	2.5 5.1	+ 1.6 0	− 2.5 − 3.5	4.0 8.1	+ 2.5 0	− 4.0 − 5.6
0.12	0.24	0.8 1.8	+ 0.5 0	− 0.8 − 1.3	0.8 2.2	+ 0.7 0	− 0.8 − 1.5	1.2 3.1	+ 1.2 0	− 1.2 − 1.9	2.8 5.8	+ 1.8 0	− 2.8 − 4.0	4.5 9.0	+ 3.0 0	− 4.5 − 6.0
0.24	0.40	1.0 2.2	+ 0.6 0	− 1.0 − 1.6	1.0 2.8	+ 0.9 0	− 1.0 − 1.9	1.6 3.9	+ 1.4 0	− 1.6 − 2.5	3.0 6.6	+ 2.2 0	− 3.0 − 4.4	5.0 10.7	+ 3.5 0	− 5.0 − 7.2
0.40	0.71	1.2 2.6	+ 0.7 0	− 1.2 − 1.9	1.2 3.2	+ 1.0 0	− 1.2 − 2.2	2.0 4.6	+ 1.6 0	− 2.0 − 3.0	3.5 7.9	+ 2.8 0	− 3.5 − 5.1	6.0 12.8	+ 4.0 0	− 6.0 − 8.8
0.71	1.19	1.6 3.2	+ 0.8 0	− 1.6 − 2.4	1.6 4.0	+ 1.2 0	− 1.6 − 2.8	2.5 5.7	+ 2.0 0	− 2.5 − 3.7	4.5 10.0	+ 3.5 0	− 4.5 − 6.5	7.0 15.5	+ 5.0 0	− 7.0 − 10.5
1.19	1.97	2.0 4.0	+ 1.0 0	− 2.0 − 3.0	2.0 5.2	+ 1.6 0	− 2.0 − 3.6	3.0 7.1	+ 2.5 0	− 3.0 − 4.6	5.0 11.5	+ 4.0 0	− 5.0 − 7.5	8.0 18.0	+ 6.0 0	− 8.0 − 12.0
1.97	3.15	2.5 4.9	+ 1.2 0	− 2.5 − 3.7	2.5 6.1	+ 1.8 0	− 2.5 − 4.3	4.0 8.8	+ 3.0 0	− 4.0 − 5.8	6.0 13.5	+ 4.5 0	− 6.0 − 9.0	9.0 20.5	+ 7.0 0	− 9.0 − 13.5
3.15	4.73	3.0 5.8	+ 1.4 0	− 3.0 − 4.4	3.0 7.4	+ 2.2 0	− 3.0 − 5.2	5.0 10.7	+ 3.5 0	− 5.0 − 7.2	7.0 15.5	+ 5.0 0	− 7.0 − 10.5	10.0 24.0	+ 9.0 0	− 10.0 − 15.0

TABLE 43. CLEARANCE LOCATIONAL FITS

Limits are in thousandths of an inch. Limits for hole and shaft are applied algebraically to the basic size to obtain the limits of size for the parts. Data in bold face are in accordance with ABC agreements. Symbols H6, h5, etc., are Hole and Shaft designations used in ABC System.

Nominal Size Range Inches Over	To	Class LC 1 Limits of Clearance	Class LC 1 Standard Limits Hole H6	Class LC 1 Standard Limits Shaft h5	Class LC 2 Limits of Clearance	Class LC 2 Standard Limits Hole H7	Class LC 2 Standard Limits Shaft h6	Class LC 3 Limits of Clearance	Class LC 3 Standard Limits Hole H8	Class LC 3 Standard Limits Shaft h7	Class LC 4 Limits of Clearance	Class LC 4 Standard Limits Hole H9	Class LC 4 Standard Limits Shaft h9	Class LC 5 Limits of Clearance	Class LC 5 Standard Limits Hole H7	Class LC 5 Standard Limits Shaft g6
0.04–	0.12	0 0.45	+ 0.25 − 0	+ 0 −0.2	0 0.65	+ 0.4 − 0	+ 0 − 0.25	0 1	+ 0.6 − 0	+ 0 − 0.4	0 2.0	+ 1.0 − 0	+ 0 − 1.0	0.1 0.75	+ 0.4 − 0	− 0.1 − 0.35
0.12–	0.24	0 0.5	+ 0.3 − 0	+ 0 −0.2	0 0.8	+ 0.5 − 0	+ 0 − 0.3	0 1.2	+ 0.7 − 0	+ 0 − 0.5	0 2.4	+ 1.2 − 0	+ 0 − 1.2	0.15 0.95	+ 0.5 − 0	− 0.15 − 0.45
0.24–	0.40	0 0.65	+ 0.4 − 0	+ 0 −0.25	0 1.0	+ 0.6 − 0	+ 0 − 0.4	0 1.5	+ 0.9 − 0	+ 0 − 0.6	0 2.8	+ 1.4 − 0	+ 0 − 1.4	0.2 1.2	+ 0.6 − 0	− 0.2 − 0.6
0.40–	0.71	0 0.7	+ 0.4 − 0	+ 0 −0.3	0 1.1	+ 0.7 − 0	+ 0 − 0.4	0 1.7	+ 1.0 − 0	+ 0 − 0.7	0 3.2	+ 1.6 − 0	+ 0 − 1.6	0.25 1.35	+ 0.7 − 0	− 0.25 − 0.65
0.71–	1.19	0 0.9	+ 0.5 − 0	+ 0 −0.4	0 1.3	+ 0.8 − 0	+ 0 − 0.5	0 2	+ 1.2 − 0	+ 0 − 0.8	0 4	+ 2.0 − 0	+ 0 − 2.0	0.3 1.6	+ 0.8 − 0	− 0.3 − 0.8
1.19–	1.97	0 1.0	+ 0.6 − 0	+ 0 −0.4	0 1.6	+ 1.0 − 0	+ 0 − 0.6	0 2.6	+ 1.6 − 0	+ 0 − 1	0 5	+ 2.5 − 0	+ 0 − 2.5	0.4 2.0	+ 1.0 − 0	− 0.4 − 1.0
1.97–	3.15	0 1.2	+ 0.7 − 0	+ 0 −0.5	0 1.9	+ 1.2 − 0	+ 0 − 0.7	0 3	+ 1.8 − 0	+ 0 − 1.2	0 6	+ 3 − 0	+ 0 − 3	0.4 2.3	+ 1.2 − 0	− 0.4 − 1.1
3.15–	4.73	0 1.5	+ 0.9 − 0	+ 0 −0.6	0 2.3	+ 1.4 − 0	+ 0 − 0.9	0 3.6	+ 2.2 − 0	+ 0 − 1.4	0 7	+ 3.5 − 0	+ 0 − 3.5	0.5 2.8	+ 1.4 − 0	− 0.5 − 1.4

Nominal Size Range Inches Over	To	Class LC 6 Limits of Clearance	Class LC 6 Standard Limits Hole H8	Class LC 6 Standard Limits Shaft f8	Class LC 7 Limits of Clearance	Class LC 7 Standard Limits Hole H9	Class LC 7 Standard Limits Shaft e9	Class LC 8 Limits of Clearance	Class LC 8 Standard Limits Hole H10	Class LC 8 Standard Limits Shaft d9	Class LC 9 Limits of Clearance	Class LC 9 Standard Limits Hole H11	Class LC 9 Standard Limits Shaft c11	Class LC 10 Limits of Clearance	Class LC 10 Standard Limits Hole H12	Class LC 10 Standard Limits Shaft	Class LC 11 Limits of Clearance	Class LC 11 Standard Limits Hole H13	Class LC 11 Standard Limits Shaft
0.04–	0.12	0.3 1.5	+ 0.6 − 0	− 0.3 − 0.9	0.6 2.6	+1.0 − 0	−0.6 −1.6	1.0 3.6	+1.6 − 0	− 1.0 − 2.0	2.5 7.5	+2.5 − 0	−2.5 −5.0	4 12	+ 4 − 0	− 4 − 8	5 17	+ 6 − 0	− 5 − 11
0.12–	0.24	0.4 1.8	+ 0.7 − 0	− 0.4 − 1.1	0.8 3.2	+1.2 − 0	−0.8 −2.0	1.2 4.2	+1.8 − 0	− 1.2 − 2.4	2.8 8.8	+3.0 − 0	−2.8 −5.8	4.5 14.5	+ 5 − 0	−4.5 −9.5	6 20	+ 7 − 0	− 6 − 13
0.24–	0.40	0.5 2.3	+ 0.9 − 0	− 0.5 − 1.4	1.0 3.8	+1.4 − 0	−1.0 −2.4	1.6 5.2	+2.2 − 0	− 1.6 − 3.0	3.0 10.0	+3.5 − 0	−3.0 −6.5	5 17	+ 6 − 0	− 5 − 11	7 25	+ 9 − 0	− 7 − 16
0.40–	0.71	0.6 2.6	+ 1.0 − 0	− 0.6 − 1.6	1.2 4.4	+1.6 − 0	−1.2 −2.8	2.0 6.4	+2.8 − 0	− 2.0 − 3.6	3.5 11.5	+4.0 − 0	−3.5 −7.5	6 20	+ 7 − 0	− 6 − 13	8 28	+ 10 − 0	− 8 − 18
0.71–	1.19	0.8 3.2	+ 1.2 − 0	− 0.8 − 2.0	1.6 5.6	+2.0 − 0	−1.6 −3.6	2.5 8.0	+3.5 − 0	− 2.5 − 4.5	4.5 14.5	+5.0 − 0	−4.5 −9.5	7 23	+ 8 − 0	− 7 − 15	10 34	+ 12 − 0	− 10 − 22
1.19–	1.97	1.0 4.2	+ 1.6 − 0	− 1.0 − 2.6	2.0 7.0	+2.5 − 0	−2.0 −4.5	3.0 9.5	+4.0 − 0	− 3.0 − 5.5	5 17	+6 − 0	− 5 − 11	8 28	+ 10 − 0	− 8 − 18	12 44	+ 16 − 0	− 12 − 28
1.97–	3.15	1.2 4.8	+ 1.8 − 0	− 1.2 − 3.0	2.5 8.5	+3.0 − 0	−2.5 −5.5	4.0 11.5	+4.5 − 0	− 4.0 − 7.0	6 20	+7 − 0	− 6 − 13	10 34	+ 12 − 0	− 10 − 22	14 50	+ 18 − 0	− 14 − 32
3.15–	4.73	1.4 5.8	+ 2.2 − 0	− 1.4 − 3.6	3.0 10.0	+3.5 − 0	−3.0 −6.5	5.0 13.5	+5.0 − 0	− 5.0 − 8.5	7 25	+9 − 0	− 7 − 16	11 39	+ 14 − 0	− 11 − 25	16 60	+ 22 − 0	− 16 − 38

TABLE 44. TRANSITION LOCATIONAL FITS

Limits are in thousandths of an inch. Limits for hole and shaft are applied algebraically to the basic size to obtain the limits of size for the mating parts. Data in bold face are in accordance with ABC agreements. "Fit" represents the maximum interference (minus values) and the maximum clearance (plus values). Symbols H8, j6, etc., are Hole and Shaft designations used in ABC System.

Nominal Size Range Inches	Class LT 1			Class LT 2			Class LT 3			Class LT 4			Class LT 6			Class LT 7		
	Fit	Standard Limits		Fit	Standard Limits		Fit	Standard Limits		Fit	Standard Limits		Fit	Standard Limits		Fit	Standard Limits	
Over To		Hole H7	Shaft j6		Hole H8	Shaft j7		Hole H7	Shaft k6		Hole H8	Shaft k7		Hole H8	Shaft m7		Hole H7	Shaft n6
0.04– 0.12	−0.15 +0.5	+0.4 − 0	+0.15 −0.1	−0.3 +0.7	+0.6 − 0	+0.3 −0.1							−0.55 +0.45	+0.6 − 0	+0.55 +0.15	−0.5 +0.15	+0.4 − 0	+0.5 +0.25
0.12– 0.24	−0.2 +0.6	+0.5 − 0	+0.2 −0.1	−0.4 +0.8	+0.7 − 0	+0.4 −0.1							−0.7 +0.5	+0.7 − 0	+0.7 +0.2	−0.6 +0.2	+0.5 − 0	+0.6 +0.3
0.24– 0.40	−0.3 +0.7	+0.6 − 0	+0.3 −0.1	−0.4 +1.1	+0.9 − 0	+0.4 −0.2	−0.5 +0.5	+0.6 − 0	+0.5 +0.1	−0.7 +0.8	+0.9 − 0	+0.7 +0.1	−0.8 +0.7	+0.9 − 0	+0.8 +0.2	−0.8 +0.2	+0.6 − 0	+0.8 +0.4
0.40– 0.71	−0.3 +0.8	+0.7 − 0	+0.3 −0.1	−0.5 +1.2	+1.0 − 0	+0.5 −0.2	−0.5 +0.6	+0.7 − 0	+0.5 +0.1	−0.8 +0.9	+1.0 − 0	+0.8 +0.1	−1.0 +0.7	+1.0 − 0	+1.0 +0.3	−0.9 +0.2	+0.7 − 0	+0.9 +0.5
0.71– 1.19	−0.3 +1.0	+0.8 − 0	+0.3 −0.2	−0.5 +1.5	+1.2 − 0	+0.5 −0.3	−0.6 +0.7	+0.8 − 0	+0.6 +0.1	−0.9 +1.1	+1.2 − 0	+0.9 +0.1	−1.1 +0.9	+1.2 − 0	+1.1 +0.3	−1.1 +0.2	+0.8 − 0	+1.1 +0.6
1.19– 1.97	−0.4 +1.2	+1.0 − 0	+0.4 −0.2	−0.6 +2.0	+1.6 − 0	+0.6 −0.4	−0.7 +0.9	+1.0 − 0	+0.7 +0.1	−1.1 +1.5	+1.6 − 0	+1.1 +0.1	−1.4 +1.2	+1.6 − 0	+1.4 +0.4	−1.3 +0.3	+1.0 − 0	+1.3 +0.7
1.97– 3.15	−0.4 +1.5	+1.2 − 0	+0.4 −0.3	−0.7 +2.3	+1.8 − 0	+0.7 −0.5	−0.8 +1.1	+1.2 − 0	+0.8 +0.1	−1.3 +1.7	+1.8 − 0	+1.3 +0.1	−1.7 +1.3	+1.8 − 0	+1.7 +0.5	−1.5 +0.4	+1.2 − 0	+1.5 +0.8
3.15– 4.73	−0.5 +1.8	+1.4 − 0	+0.5 −0.4	−0.8 +2.8	+2.2 − 0	+0.8 −0.6	−1.0 +1.3	+1.4 − 0	+1.0 +0.1	−1.5 +2.1	+2.2 − 0	+1.5 +0.1	−1.9 +1.7	+2.2 − 0	+1.9 +0.5	−1.9 +0.4	+1.4 − 0	+1.9 +1.0

TABLE 45. INTERFERENCE LOCATIONAL FITS

Limits are in thousandths of an inch. Limits for hole and shaft are applied algebraically to the basic size to obtain the limits of size for the parts. Data in bold face are in accordance with ABC agreements. Symbols H7, p6, etc., are Hole and Shaft designations used in ABC System.

Nominal Size Range Inches		Class LN 2			Class LN 3		
		Limits of Interference	Standard Limits		Limits of Interference	Standard Limits	
Over	To		Hole H7	Shaft p6		Hole H7	Shaft r6
0.04–	0.12	0 0.65	+ 0.4 − 0	+ 0.65 + 0.4	0.1 0.75	+ 0.4 − 0	+ 0.75 + 0.5
0.12–	0.24	0 0.8	+ 0.5 − 0	+ 0.8 + 0.5	0.1 0.9	+ 0.5 − 0	+ 0.9 + 0.6
0.24–	0.40	0 1.0	+ 0.6 − 0	+ 1.0 + 0.6	0.2 1.2	+ 0.6 − 0	+ 1.2 + 0.8
0.40–	0.71	0 1.1	+ 0.7 − 0	+ 1.1 + 0.7	0.3 1.4	+ 0.7 − 0	+ 1.4 + 1.0
0.71–	1.19	0 1.3	+ 0.8 − 0	+ 1.3 + 0.8	0.4 1.7	+ 0.8 − 0	+ 1.7 + 1.2
1.19–	1.97	0 1.6	+ 1.0 − 0	+ 1.6 + 1.0	0.4 2.0	+ 1.0 − 0	+ 2.0 + 1.4
1.97–	3.15	0.2 2.1	+ 1.2 − 0	+ 2.1 + 1.4	0.4 2.3	+ 1.2 − 0	+ 2.3 + 1.6
3.15–	4.73	0.2 2.5	+ 1.4 − 0	+ 2.5 + 1.6	0.6 2.9	+ 1.4 − 0	+ 2.9 + 2.0

TABLE 46. FORCE AND SHRINK FITS

Limits are in thousandths of an inch. Limits for hole and shaft are applied algebraically to the basic size to obtain the limits of size for the parts. Data in bold face are in accordance with ABC agreements. Symbols H7, s6, etc., are Hole and Shaft designations used in ABC System.

Nominal Size Range Inches		Class FN 1			Class FN 2			Class FN 3			Class FN 4			Class FN 5		
		Limits of Interference	Standard Limits		Limits of Interference	Standard Limits		Limits of Interference	Standard Limits		Limits of Interference	Standard Limits		Limits of Interference	Standard Limits	
Over	To		Hole H6	Shaft		Hole H7	Shaft s6		Hole H7	Shaft t6		Hole H7	Shaft u6		Hole H7	Shaft x7
0.04	0.12	0.05	+0.25	+ 0.5	0.2	+ 0.4	+ 0.85				0.3	+ 0.4	+ 0.95	0.5	+ 0.4	+ 1.3
		0.5	− 0	+ 0.3	0.85	− 0	+ 0.6				0.95	− 0	+ 0.7	1.3	− 0	+ 0.9
0.12	0.24	0.1	+0.3	+ 0.6	0.2	+ 0.5	+ 1.0				0.4	+ 0.5	+ 1.2	0.7	+ 0.5	+ 1.7
		0.6	− 0	+ 0.4	1.0	− 0	+ 0.7				1.2	− 0	+ 0.9	1.7	− 0	+ 1.2
0.24	0.40	0.1	+0.4	+ 0.75	0.4	+ 0.6	+ 1.4				0.6	+ 0.6	+ 1.6	0.8	+ 0.6	+ 2.0
		0.75	− 0	+ 0.5	1.4	− 0	+ 1.0				1.6	− 0	+ 1.2	2.0	− 0	+ 1.4
0.40	0.56	0.1	+0.4	+ 0.8	0.5	+ 0.7	+ 1.6				0.7	+ 0.7	+ 1.8	0.9	+ 0.7	+ 2.3
		0.8	− 0	+ 0.5	1.6	− 0	+ 1.2				1.8	− 0	+ 1.4	2.3	− 0	+ 1.6
0.56	0.71	0.2	+0.4	+ 0.9	0.5	+ 0.7	+ 1.6				0.7	+ 0.7	+ 1.8	1.1	+ 0.7	+ 2.5
		0.9	− 0	+ 0.6	1.6	− 0	+ 1.2				1.8	− 0	+ 1.4	2.5	− 0	+ 1.8
0.71	0.95	0.2	+0.5	+ 1.1	0.6	+ 0.8	+ 1.9				0.8	+ 0.8	+ 2.1	1.4	+ 0.8	+ 3.0
		1.1	− 0	+ 0.7	1.9	− 0	+ 1.4				2.1	− 0	+ 1.6	3.0	− 0	+ 2.2
0.95	1.19	0.3	+0.5	+ 1.2	0.6	+ 0.8	+ 1.9	0.8	+ 0.8	+ 2.1	1.0	+ 0.8	+ 2.3	1.7	+ 0.8	+ 3.3
		1.2	− 0	+ 0.8	1.9	− 0	+ 1.4	2.1	− 0	+ 1.6	2.3	− 0	+ 1.8	3.3	− 0	+ 2.5
1.19	1.58	0.3	+0.6	+ 1.3	0.8	+ 1.0	+ 2.4	1.0	+ 1.0	+ 2.6	1.5	+ 1.0	+ 3.1	2.0	+ 1.0	+ 4.0
		1.3	− 0	+ 0.9	2.4	− 0	+ 1.8	2.6	− 0	+ 2.0	3.1	− 0	+ 2.5	4.0	− 0	+ 3.0
1.58	1.97	0.4	+0.6	+ 1.4	0.8	+ 1.0	+ 2.4	1.2	+ 1.0	+ 2.8	1.8	+ 1.0	+ 3.4	3.0	+ 1.0	+ 5.0
		1.4	− 0	+ 1.0	2.4	− 0	+ 1.8	2.8	− 0	+ 2.2	3.4	− 0	+ 2.8	5.0	− 0	+ 4.0
1.97	2.56	0.6	+0.7	+ 1.8	0.8	+ 1.2	+ 2.7	1.3	+ 1.2	+ 3.2	2.3	+ 1.2	+ 4.2	3.8	+ 1.2	+ 6.2
		1.8	− 0	+ 1.3	2.7	− 0	+ 2.0	3.2	− 0	+ 2.5	4.2	− 0	+ 3.5	6.2	− 0	+ 5.0
2.56	3.15	0.7	+0.7	+ 1.9	1.0	+ 1.2	+ 2.9	1.8	+ 1.2	+ 3.7	2.8	+ 1.2	+ 4.7	4.8	+ 1.2	+ 7.2
		1.9	− 0	+ 1.4	2.9	− 0	+ 2.2	3.7	− 0	+ 3.0	4.7	− 0	+ 4.0	7.2	− 0	+ 6.0
3.15	3.94	0.9	+0.9	+ 2.4	1.4	+ 1.4	+ 3.7	2.1	+ 1.4	+ 4.4	3.6	+ 1.4	+ 5.9	5.6	+ 1.4	+ 8.4
		2.4	− 0	+ 1.8	3.7	− 0	+ 2.8	4.4	− 0	+ 3.5	5.9	− 0	+ 5.0	8.4	− 0	+ 7.0
3.94	4.73	1.1	+0.9	+ 2.6	1.6	+ 1.4	+ 3.9	2.6	+ 1.4	+ 4.9	4.6	+ 1.4	+ 6.9	6.6	+ 1.4	+ 9.4
		2.6	− 0	+ 2.0	3.9	− 0	+ 3.0	4.9	− 0	+ 4.0	6.9	− 0	+ 6.0	9.4	− 0	+ 8.0

TABLE 47

WIRE AND SHEET-METAL GAGES

No. of gage	American copper or B. & S. wire gage	British imperial wire gage	U. S. St'd. gage for plate	No. of gage	American copper or B. & S. wire gage	British imperial wire gage	U. S. St'd. gage for plate
0000000	. . .	0.5000	0.5000	20	0.0320	0.0360	0.0375
000000	0.5800	0.4640	0.4688	21	0.0285	0.0320	0.0344
00000	0.5165	0.4320	0.4375	22	0.0253	0.0280	0.0313
0000	0.4600	0.4000	0.4063	23	0.0226	0.0240	0.0281
000	0.4096	0.3720	0.3750	24	0.0201	0.0220	0.0250
00	0.3648	0.3480	0.3438	25	0.0179	0.0200	0.0219
0	0.3249	0.3240	0.3125	26	0.0159	0.0180	0.0188
1	0.2893	0.3000	0.2813	27	0.0142	0.0164	0.0172
2	0.2576	0.2760	0.2656	28	0.0126	0.0148	0.0156
3	0.2294	0.2520	0.2500	29	0.0113	0.0136	0.0141
4	0.2043	0.2320	0.2344	30	0.0100	0.0124	0.0125
5	0.1819	0.2120	0.2188	31	0.0089	0.0116	0.0109
6	0.1620	0.1920	0.2031	32	0.0080	0.0108	0.0102
7	0.1443	0.1760	0.1875	33	0.0071	0.0100	0.0094
8	0.1285	0.1600	0.1719	34	0.0063	0.0092	0.0086
9	0.1144	0.1440	0.1563	35	0.0056	0.0084	0.0078
10	0.1019	0.1280	0.1406	36	0.0050	0.0076	0.0070
11	0.0907	0.1160	0.1250	37	0.0045	0.0068	0.0066
12	0.0808	0.1040	0.1094	38	0.0040	0.0060	0.0063
13	0.0720	0.0920	0.0938	39	0.0035	0.0052	. . .
14	0.0641	0.0800	0.0781	40	0.0031	0.0048	. . .
15	0.0571	0.0720	0.0703	41	0.0028	0.0044	. . .
16	0.0508	0.0640	0.0625	42	0.0025	0.0040	. . .
17	0.0453	0.0560	0.0563	43	0.0022	0.0036	. . .
18	0.0403	0.0480	0.0500	44	0.0020	0.0032	. . .
19	0.0359	0.0400	0.0438	45	0.00176	0.0028	. . .

TABLE 48

Degrees, Minutes, and Seconds to Radians

Deg	Rad	Min	Rad	Sec	Rad
1	0.0174533	1	0.0002909	1	0.0000048
2	0.0349066	2	0.0005818	2	0.0000097
3	0.0523599	3	0.0008727	3	0.0000145
4	0.0698132	4	0.0011636	4	0.0000194
5	0.0872665	5	0.0014544	5	0.0000242
6	0.1047198	6	0.0017453	6	0.0000291
7	0.1221730	7	0.0020362	7	0.0000339
8	0.1396263	8	0.0023271	8	0.0000388
9	0.1570796	9	0.0026180	9	0.0000436
10	0.1745329	10	0.0029089	10	0.0000485
20	0.3490659	20	0.0058178	20	0.0000970
30	0.5235988	30	0.0087266	30	0.0001454
40	0.6981317	40	0.0116355	40	0.0001939
50	0.8726646	50	0.0145444	50	0.0002424
60	1.0471976	60	0.0174533	60	0.0002909
70	1.2217305				
80	1.3962634				
90	1.5707963				

Radians to Degrees, Minutes, and Seconds

Rad		Rad		Rad		Rad	
1	57°17′44″.8	.1	5°43′46″.5	.01	0°34′22″.6	.001	0° 3′26″.3
2	114°35′29″.6	.2	11°27′33″.0	.02	1° 8′45″.3	.002	0° 6′52″.5
3	171°53′14″.4	.3	17°11′19″.4	.03	1°43′07″.9	.003	0°10′18″.8
4	229°10′59″.2	.4	22°55′05″.9	.04	2°17′30″.6	.004	0°13′45″.1
5	286°28′44″.0	.5	28°38′52″.4	.05	2°51′53″.2	.005	0°17′11″.3
6	343°46′28″.8	.6	34°22′38″.9	.06	3°26′15″.9	.006	0°20′37″.6
7	401° 4′13″.6	.7	40° 6′25″.4	.07	4° 0′38″.5	.007	0°24′03″.9
8	458°21′58″.4	.8	45°50′11″.8	.08	4°35′01″.2	.008	0°27′30″.1
9	515°39′43″.3	.9	51°33′58″.3	.09	5° 9′23″.8	.009	0°30′56″.4

TABLE 49

CONVERSION VALUES FOR 2.54 I = C, WHERE
I = NUMBER OF INCHES, AND C = NUMBER OF CENTIMETERS

Inches	Centimeters	Centimeters	Inches
0	0	0	0
1	2.54	1	0.3937
2	5.08	2	0.7874
3	7.62	3	0.1811
4	10.16	4	1.5748
5	12.70	5	1.9685
6	15.24	6	2.3622
7	17.78	7	2.7559
8	20.32	8	3.1496
9	22.86	9	3.5433
10	25.40	10	3.9370
11	27.94	11	4.3307
12	30.48	12	4.7244
13	33.02	13	5.1181
14	35.56	14	5.5118
15	38.10	15	5.9055
16	40.64	16	6.2492
17	43.18	17	6.6929
18	45.72	18	7.0866
19	48.26	19	7.4803
20	50.80	20	7.8740
21	53.34	21	8.2677
22	55.88	22	8.6614
23	58.42	23	9.0551
24	60.96	24	9.4488
25	63.50	25	9.8425
26	66.04	26	10.2362
27	68.58	27	10.6299
28	71.12	28	11.0236
29	73.66	29	11.4173
30	76.20	30	11.8110
31	78.74	31	12.2047
32	81.28	32	12.5989
33	83.82	33	12.9921
34	86.36	34	13.3858
35	88.90	35	13.7795
36	91.44	36	14.1732
37	93.98	37	14.5669
38	96.52	38	14.9606
39	99.06	39	15.3543
40	101.60	40	15.7480
50	127.00	50	19.6850
100	254.00	100	39.3701

APPENDIX C
ABBREVIATIONS AND SYMBOLS

ABBREVIATIONS FOR USE ON DRAWINGS

Word	*Abbr.*
Abampere, absolute ampere	ABAMP
Abrasive Resistant	ABRSV RES
Absolute	ABS
Accelerate	ACCEL
Acetylene	ACET
Acid Resisting	AR
Acoustic	ACST
Acre Foot	AC FT
Adapter	ADPT
Addendum	ADD.
Aerodynamic	AERODYN
Aeronautic	AERO
Aeronautical Material Specifications	AMS
Aeronautical Recommended Practice	ARP
Afterburner	AB
Aileron	AIL.
Air Blast Circuit Breaker	ABCB
Air Break Switch	ABS
Air Circuit Breaker	ACB
Air Force—Navy	AN.
Air Force—Navy Aeronautical	ANA
Air Force—Navy Civil	ANC
Air Force—Navy Design	AND.
Air Horsepower	AHP
Airborne	ABN
Aircooled	ACLD
Aircraft	ACFT
Airplane	APL
Airport	AP
Airscoop	AS.
Airspeed	A/S
Air-to-air	A-A
Air-to-ground	A-G
Alarm	ALM
Alignment	ALIGN.
Allowance	ALLOW.
Alternating Current	AC

Word	*Abbr.*
Alternating Current Synchronous	
Alternator	
Altimeter	A
Altitude	
Ambient	
American	A
American War Standard	
American Wire Gage	A
Ammeter	
Ampere	
Ampere Turn	
Ampere-hour	AM
Ampere-hour Meter	
Amphibian-Amphibious	A
Amplitude	
Amplifier	A
Angstrom Unit	
Anneal	
Anode	
Anodize	A
Antenna	
Anti-friction Bearing	
Anti-icing	
Antilogarithm	ANTI
Apparatus	
Apparent Watts	
Arc Weld	AR
Area	
Armature	
Armature Shunt	
Article	
As Required	
Assemble	AS
Assembly	
Astronomical Time Switch	
Atmosphere	
Atomic	

Word	Abbr.
Attenuator	ATTEN
Audio Frequency	AF
Automatic Direction Finder	ADF
Automatic Mixture Control	AMC
Automatic Phase Control	APC
Automatic Volume Control	AVC
Automotive	AUTOM
Auxiliary	AUX
Auxiliary Power Unit	APU
Auxiliary Switch (breaker) Normally Closed	ASC
Auxiliary Switch (breaker) Normally Open	ASO
Average	AVG
Aviation	AVI
Aviation Gas Turbine	AGT
Axial Flow	AX FL
Azimuth	AZ
Babbitt	BAB
Back Pressure	BP
Back to Back	B to B
Bacteriological	BACT
Balanced Voltage	BV
Ball Bearing	BB
Barometer	BAR
Barrel	BBL
Baume	BE
Bearing	BRG
Bell and Bell	B&B
Bell and Flange	B&F
Bell and Spigot	B&S
Bell Crank	BELCRK
Bench Mark	BM
Bending Moment	M
Between Centers	BC
Between Perpendiculars	BP
Bill of Material	B/M
Billion Electron Volts	BEV
Biochemical Oxygen Demand	BOD
Birmingham Wire Gage	BWG
Blower	BLO
Board Foot	FBM
Boiler Feed Pump	BFP
Boiler Feed Water	BFW
Boiler Horsepower	BHP
Bolt Circle	BC
Both Sides	BS
Bottoming	BOTMG
Boundary	BDY
Bracket	BRKT
Brake Horsepower	BHP
Brake Mean Effective Pressure	BMEP
Brazing	BRZG
Bridge	BRDG
Brinnell Hardness	BH
Brinnell Hardness Number	BHN
British Thermal Units	BTU
Broach	BRO

Word	Abbr.
Bronze	BRZ
Brown & Sharp (Wire Gage, same as AWG)	B&S
Bulkhead	BHD
Bureau of Standards	BU STD
Burnish	BNH
Bushing	BUSH.
Bushing Current Transformer	BCT
Buttock Line	BL
Cadmium Plate	CD PL
Calculate	CALC
Calibrate	CAL
Caliper	CLPR
Calked Joint	CAJ
Calking	CLKG
Calorie	CAL
Cap Screw	CAP. SCR
Capacitor	CAP.
Capitance	C
Carload	CL
Case Harden	CH
Casing	CSG
Cast Iron	CI
Cast Steel	CS
Casting	CSTG
Castle Nut	CAS NUT
Cathode-ray	CR
Cathode-ray Oscilloscope or Oscillograph	CRO
Cathode-ray Tube	CRT
Center	CTR
Center Line	℄ or CL
Center of Gravity	CG
Center of Pressure	CP
Center to Center	C to C
Centigrade	C
Centimeter	CM
Centimeter-Gram-Second System	CGS
Centimeters per Second	CMPS
Centipoises	CP
Centrifugal	CENT.
Centrifugal Force	CF
Ceramic	CER
Chamfer	CHAM
Chemically Pure	CP
Chromium Plate	CR PL
Chrome Vanadium	CR VAN
Cinematographic	CINE
Circle	CIR
Circuit	CKT
Circuit Breaker	CB
Circular Mil	CM
Circular Mils, Thousands	MCM
Circumference	CIRC
Clear	CLR
Clearance	CL
Clevis	CLV

Word	*Abbr.*
Clockwise	CW
Coaxial	COAX
Cockpit	CKPT
Coefficient	COEF
Coils per Slot	CPS
Cold Drawn Steel	CDS
Column	COL
Combustion	COMB
Commercial	COML
Communication	COMM
Commutator	COMM
Compressor	COMPR
Concentric	CONC
Condensate	CNDS
Condenser	COND
Conductor	COND
Conduit	CND
Cone Point	CP
Constant	CONST
Constant Current Transformer	CCT
Construction	CONST
Contact-making Voltmeter	CMVM
Contact Potential Difference	CPD
Control Switch	CS
Coolant	COOL.
Corrosion Resistant	CRE
Counter Clockwise	CCW
Counter Electromotive Force	CEMF
Counterbore	CBORE
Counter-radar Measures	CRM
Countersink	CSK
Cowling	COWL.
Cubic Feet per Minute	CFM
Cubic Foot	CU FT
Cubic Inch	CU IN.
Cubic Meter	CU M
Cubic Micron	CU MU
Cubic Yard	CU YD
Current Directional Relay	CDR
Current Transformer	CT
Current-limiting Resistor	CLR
Cycles per Minute	CPM
Cycles per Second	CPS
Cylinder	CYL
Damage Control	DC
Dash Pot	DP
Datum	DAT
Decibel	DB
Decontamination	DECONTN
Dedendum	DED
Deep Drawn	DD
Deflect	DEFL
Demand Meter	DM
Demodulator	DEM
Density	D
Design	DSGN

Word	*Abbr.*
Detail	DET
Diagonal	DIAG
Diagram	DIAG
Diameter	DIA
Diametrical Pitch	DP
Differential Time Relay	DIFF TR
Dimension	DIM.
Diode	DIO
Direct Current	DC
Direction Finder	DF
Displacement	DISPL
Distance	DIST
Double Extra Strong	XXSTR
Double Pole Both Connected	DPBC
Double Pole, Single Throw	DPST
Dowel	DWL
Drafting Room Manual	DRM
Drawbar Horsepower	DBHP
Drill	DR
Drop Forge	DF
Dynamic	DYN
Dynamotor	DYNM
Eccentric	ECC
Effective	EFF
Effective Horsepower	EHP
Efficiency	EFF
Ejector	EJECT.
Electric	ELEC
Electric Horsepower	EHP
Electromotive Force	EMF
Elevation	EL
Elevator	ELEV
Elongation	ELONG
Elongation in 2 Inches	EL2
Emergency	EMER
Empennage	EMP
Engineer	ENGR
Engineering	ENGRG
Engineering Change Order	ECO
Engineering Order	EO
Engineering Work Order	EWO
Equation	EQ
Equivalent	EQUIV
Estimate	EST
Exhaust	EXH
Exhaust Gas Temperature	EGT
Experiment	EXP
Explosive	XPL
Extension	EXT
Extra	EXT
Extra Heavy	X HVY
Extra Strong	X STR
Fabricate	FAB
Facsimile	FAX
Fahrenheit	F

Word	Abbr.
Fast Operating (Relay)	FO
Fast Release (Relay)	FR
Feet Board Measure	FBM
Feet per Minute	FPM
Feet per Second	FPS
Field Accelerator	FAC
Field Decelerator	FDE
Field Forcing (Decreasing)	FFD
Field Forcing (Increasing)	FFI
Figure	FIG.
Filament	FIL
Fillet	FIL
Fillister	FIL
Fillister Head	FILH
Filter	FLT
Finish	FIN.
Finish All Over	FAO
Fireproof	FPRF
Fitting	FTG
Flange	FLG
Flat Fillister Head	FFILH
Flat Head	FH
Flat Oval	FO
Fluorescent	FLUOR
Focus	FOC
Foot Candle	FC
Foot Pounds	FT LB
Force	F
Forged Steel	FST
Forging	FORG
Freeboard	FREEBD
Frequency	FREQ
Frequency, Extremely High	EHF
Frequency, High	HF
Frequency, Low	LF
Frequency, Medium	MF
Frequency Meter	FRM
Frequency Modulation	FM
Fuselage	FUS
Gage or Gauge	GA
Gallons per Minute	GPM
Gallons per Second	GPS
Galvanize	GALV
Gas	G
Gasket	GSKT
Gasoline	GASO
Glass Block	GLB
Glaze	GL
Government	GOVT
Gram	G
Gram-calorie	G-CAL
Graphic	GRAPH.
Gravity	G
Grommet	GROM
Ground-controlled Approach	GCA
Ground-position Indicator	GPI

Word	Abbr.
Ground-to-ground	G-G
Gyroscope	GYRO
Half Dog Point	½DP
Hanger	HGR
Hard Chromium	HD CR
Hardware	HDW
Head	HD
Headless	HDLS
Heat Resisting	HR
Heat Treat	HT TR
Heavy	HVY
Hexagon	HEX
Hexagonal Head	HXH
Hexagonal Socket	HXSOC
High Frequency	HF
High-speed	HS
High-speed Steel	HSS
High Tension	HT
High Voltage	HV
Highway	HWY
Horizontal	HOR
Horizontal Center Line	HCL
Horizontal Reference Line	HRL
Hundredweight	CWT
Hydraulic	HYD
Hydrostatic	HYDRO
Identify	IDENT
Illustrate	ILLUS
Impact	IMP
Inboard	INBD
Indicated Horsepower Hour	IHPH
Inductance or Induction	IND
Inductance-capacitance	LC
Inductance-capacitance Resistance	LCR
Inductance Coil	L
Inside Diameter	ID
Instrument	INST
Instrument Landing System	ILS
Interchangeable	INTCHG
Intercommunication	INTERCOM
Intercooler	INCLR
Iron Pipe	IP
Isometric	ISO
Job Order	JO
Joint	JT
Joule	J
Junction	JCT
Junction Box	JB
Kelvin	K
Keyseat	KST
Kilo	K
Kilocycle	KC
Kilocycles per Second	KCPS
Kilogram	KG

Word	Abbr.
Kilograms per Second	KGPS
Kilohm	K
Kiloliter	KL
Kilovolt-ampere	KVA
Kilowatt Hour	KWH
Kip (1000 lb)	K
Kips per Square Inch	KSI
Landing Gear	LG
Lateral	LAT
Length	LG
Length Over All	LOA
Linear	LIN
Liquid	LIQ
Liter	L
Logarithm	LOG.
Longeron	LONGN
Longitude	LONG.
Longitudinal Expansion Joint	LEJ
Lubricate	LUB
Machine Screw	MS
Machine Steel	MS
Magnet	MAG
Male & Female	M&F
Manual	MAN.
Manufacture	MFR
Manufactured	MFD
Manufacturing	MFG
Material	MATL
Material List	ML
Mechanical	MECH
Membrane	MEMB
Metal	MET.
Micro	μ or U
Microampere	μA or UA
Microangstrom	μA
Microfarad	μF or UF
Micrometer	MIC
Micro-micro	μ-μ or U-U
Micron	μ or U
Microphone	MIKE
Milliampere	MA
Million Gallons per Day	MGD
Milliwatt	MW
Minimum	MIN
Miscellaneous	MISC
Molecular Weight	MOL WT
Nacelle	NAC
National	NATL
National Aircraft Standards	NAS
National Electrical Code	NEC
No Good	NG
Nomenclature	NOM
Noon	M
Normalize	NORM
Not to Scale	NTS

Word	Abbr.
Oblique	OBL
Observe	OBS
Obsolete	OBS
Ohm	Ω
Ohmmeter	OHM
Oil Circuit Breaker	OCB
Oil Ring	OR.
Oil Seal	OSL
On Center	OC
Open-close-open	OCO
Operate	OPR
Optical	OPT
Orifice	ORF
Outboard	OUTBD
Outside Radius	OR
Out to Out	O to O
Oval Head	OVH
Oval Point	OVP
Overhead	OVHD
Overload	OVLD
Oxidized	OXD
Painted	PTD
Panel	PNL
Pantograph	PANT.
Pantry	PAN.
Parabola	PRB
Paraboloid	PRBD
Parallel	PAR.
Patent.	PAT.
Penny (Nails, etc)	d
Pennyweight	dWT
Perpendicular	PERP
Perspective	PERS
Phase	PH
Phase Meter	PHM
Photograph	PHOTO
Piece	PC
Pipe Tap	PT
Pitch	P
Pitch Diameter	PD
Plain Washer	PW
Plastic	PLSTC
Plate	PL
Plotting	PLOT.
Pneumatic	PNEU
Point of Curve	PC
Point of Tangent	PT
Polar	POL
Polyphase	PYPH
Potential	POT.
Potentiometer	POT.
Pound	LB
Pounder	PDR
Pounds per Square Foot	PSF
Power	PWR
Power Circuit Breaker	PCB
Power Factor	PF

Word	Abbr.
Preheater	PHR
Printed Circuit	PCKT
Process	PROC
Procurement	PROC
Production	PROD
Profile	PF
Project	PROJ
Propeller	PROP
Pulley	PUL
Pulse-frequency	PF
Pulse-position Modulation	PPM
Pulses per Second	PPS
Punch	PCH
Purchase	PUR
Push-pull	P-P
Pyrometer	PYR
Quadrangle	QUAD
Quality	QUAL
Quantity	QTY
Quick-opening Device	QOD
Rabbet	RAB
Radar	RDR
Radar Counter Measure	RCM
Radial	RAD
Radius	R
Reactor	REAC
Ream	RM
Rankine	R
Recommend	RECM
Recovery	RECY
Reduce	RED.
Reference	RF
Reference Line	REF L
Reinforce	REINF
Relative Humidity	RH
Relief Valve	RV
Remote Control	RC
Remove	REM
Request	REQ
Required	REQD
Requisition	REQ
Residual	RESID
Resistance	RES
Revolutions per Minute	RPM
Right of Way	R/W
Rivet	RIV
Rocket	RKT
Rocket Launcher	RL
Rockwell Hardness	RH
Roentgen	R
Roller Bearing	RB
Root Diameter	RD
Root Mean Square	RMS
Root Sum Square	RSS
Rough	RGH
Round	RD

Word	Abbr.
Rubber	RUB.
Runout	RO
Safe Working Pressure	SWP
Sand Blast	SD BL
Saturate	SAT.
Saybolt Seconds Furol (Oil Viscosity)	SSF
Saybolt Seconds Universal (Oil Viscosity)	SSU
Schedule	SCH
Schematic	SCHEM
Scleroscope Hardness	SH
Screw	SCR
Sea Level	SL
Section	SECT
Selsyn	SELS
Semi-finished	SF
Set Screw	SS
Shaft	SFT
Sheathing	SHTHG
Shield	SHLD
Shop Order	SO
Shot Blast	SH BL
Shoulder	SHLD
Signal	SIG
Signal-to-noise Ratio	SNR
Silver Solder	SILS
Simplex	SX
Sink	SK
Sketch	SK
Sleeve	SLV
Sleeve Bearing	SB
Sliding Expansion Joint	SEJ
Slope	S
Socket	SOC
Solder	SLD
Sound	SND
Space	SP
Space Heater	SPH
Speaker	SPKR
Specific Gravity	SP GR
Specific Heat	SP HT
Specification	SPEC
Speedometer	SPEEDO
Spherical	SPHER
Spot Face	SF
Spot Face Other Side	SF-O
Spot Weld	SW
Spring	SPG
Square	SQ
Square Head	SQH
Stabilize	STAB
Stainless	STN
Stanchion	STAN
Starboard	STBD
Static Pressure	SP
Steam	ST
Steel	STL
Stock	STK

Word	Abbr.
Strength	STR
Structural	STR
Structural Carbon Steel Hard	SCSH
Substation	SUBSTA
Substitute	SUB
Substructure	SUBSTR
Supercharge	S-CHG
Superheater	SUPHTR
Supersede	SUPSD
Switch and Relay Types	
Single Pole Switch	SP SW
Single Pole Single Throw Switch	SPST SW
Single Pole Double Throw Switch	SPDT SW
Double Pole Switch	DP SW
Double Pole Single Throw Switch	DPST SW
Double Pole Double Throw Switch	DPDT SW
Triple Pole Switch	3P SW
Triple Pole Single Throw Switch	3PST SW
Triple Pole Double Throw Switch	3PDT SW
4 Pole Switch	4P SW
4 Pole Single Throw Switch	4PST SW
4 Pole Double Throw Switch	4PDT SW
etc	
Switchboard	SWBD
Switchgear	SWGR
Symmetrical	SYM
System	SYS
Tabulate	TAB.
Tachometer	TACH
Tail Landing Gear	TLG
Tank	TK
Taper	TPR
Technical	TECH
Telemeter	TLM
Temperature	TEMP
Tensile Strength	TS
Terminal	TERM.
Thermocouple	TC
Thermostat	THERMO
Thousand Pound	KIP
Tinned	TD
Tolerance	TOL
Tongue & Groove	T&G
Torque	TOR
Total Indicator Reading	TIR
Transistor	TSTR
Transmission	XMSN
Transmitter	XMTR
Transportation	TRANS
Transverse	TRANSV
Treatment	TREAT.
Tubing	TUB.
Turbine	TURB
Turbine Drive	TD
Turbine Generator	TURBO GEN
Ultra-high Frequency	UHF
Unfinished	UNFIN
United States Standard	USS
Unless Otherwise Specified	UOS
Vacuum	VAC
Vacuum Tube	VT
Valve	V
Valve Box	VB
Velocity	V
Vertical	VERT
Vertical Center Line	VCL
Vertical Reference Line	VRL
Very-high Frequency	VHF
Viscosity	VISC
Volt	V
Voltammeter	VAM
Volume	VOL
Washer	WASH.
Water Cooled	WCLD
Watt	W
Watthour	WHR
Wavelength	WL
Withdrawn	W/D
Without	W/O
Without Equipment and Spare Parts	W/O E&SP
Woodruff	WDF
Working Point	WP
Working Pressure	WP
Wrought	WRT
Wrought Brass	W BRS
Wrought Iron	WI

Abbreviations for Chemical Symbols

Word	Abbr.
Actinium	AC
Aluminum	AL
Antimony (stibium)	SB
Argon	A
Arsenic	AS
Barium	BA
Beryllium (glucinum)	BE
Bismuth	BI
Boron	B
Bromine	BR
Cadmium	CD
Caesium	CS
Calcium	CA
Carbon	C
Cerium	CE
Chlorine	CL
Chromium	CR
Cobalt	CO
Columbium (niobium)	CB
Copper	CU
Dysprosium	DY
Erbium	ER
Europium	EU
Fluorine	F
Gadolinium	GD
Gallium	GA
Germanium	GE
Gold (aurum)	AU
Hafnium	HF
Helium	HE
Holmium	HO
Hydrogen	H
Illinium	IL
Indium	IN
Iodine	I
Ionium	IO
Iridium	IR
Iron (ferrum)	FE
Krypton	KR
Lanthanum	LA
Lead (plumbum)	PB
Lithium	LI
Lutecium	LU
Magnesium	MG
Manganese	MN

Word	Abbr.
Masurium	MA
Mercury (hydrargyrum)	HG
Molybdenum	MO
Neodymium	ND
Neon	NE
Nickel	NI
Nitrogen	N
Osmium	OS
Oxygen	O
Palladium	PD
Phosphorus	P
Platinum	PT
Polonium	PO
Potassium (kalium)	K
Praseodymium	PR
Proactinium	PA
Radium	RA
Radon (niton)	RN
Rhenium	RE
Rhodium	RH
Rubidium	RB
Ruthenium	RU
Samarium	SM
Scandium	SC
Selenium	SE
Silicon	SI
Silver (argentum)	AG
Sodium (natrium)	NA
Strontium	SR
Sulfur	S
Tantalum	TA
Tellurium	TE
Terbium	TB
Thallium	TL
Thorium	TH
Thulium	TM
Tin (stannum)	SN
Titanium	TI
Tungsten (wolfranium)	W
Uranium	U
Vanadium	V
Xenon	XE
Ytterbium	YB
Yttrium	YT
Zinc	ZN
Zirconium	ZR

Abbreviations for Engineering Societies

American Association of Engineers . . . AAE
American Boiler Manufacturers' Association & Affiliated Industries . . . . . ABMA
American Bureau of Shipping . . . . . ABS
Air Conditioning & Refrigerating Machinery Association . . . . . ACRMA
American Chemical Society . . . . . . ACS
American Concrete Institute . . . . . ACI
American Electrochemical Society . . . AES
American Electroplaters Society . . . . AES
American Engineering Council . . . . . AEC
American Foundrymen's Association . . . AFA
American Gas Association . . . . AGA
American Gear Manufacturers' Association. . . . . AGMA
American Institute of Architects . . . . AIA
American Institute of Chemical Engineers . . . . AIChE
American Institute of Electrical Engineers . . . . AIEE
American Institute of Mining & Metallurgical Engineers. . . . . AIMME
American Institute of Steel Construction. AISC
American Iron & Steel Institute . . . . AISI
American Petroleum Institute . . . . . API
American Railway Engineering Association . . . . AREA
American Railway Bridge & Building Association . . . . ARBBA
American Society of Aeronautical Engineers . . . . ASAE
American Society of Body Engineers . . ASBE
American Society of Civil Engineers . . ASCE
American Society of Engineers and Architects . . . . ASEA
American Society of Heating & Ventilating Engineers . . . . . ASHVE
American Society of Lubricating Engineers . . . . ASLE
American Society of Mechanical Engineers . . . . ASME
American Society of Metals . . . . . ASM
American Society of Refrigerating Engineers . . . . ASRE
American Society of Safety Engineers . . ASSE
American Society of Sanitary Engineering . ASSE
American Society for Steel Treating. . . ASST
American Society for Testing Materials . ASTM
American Society of Tool Engineers . . ASTE
American Standards Association . . . . ASA
American Steel Foundrymen's Association . . . . ASFA
American Transit Association. . . . . ATA
American Water Works Association . . . AWWA
American Welding Society . . . . . . AWS
American Wood Preservers' Association. . AWPA
Anti-friction Bearing Manufacturers' Association . . . . AFBMA
Association of American Railroads . . . AAR
Association of American Steel Manufacturers . . . . AASM
Association of Iron & Steel Engineers . . AISE
Automobile Manufacturers' Association . AMA
Canadian Lumbermen's Association . . CLA
Canadian Standards Association . . . . CSA
Compressed Air Institute . . . . . . CAI
Edison Electric Institute . . . . . . EEI
Electrochemical Society . . . . ES
Gas Appliances Manufacturers' Association . . . . GAMA
Hydraulic Institute . . . . HI
Illuminating Engineering Society. . . . IES
Institute of Radio Engineers . . . . . .IRE
Institute of Traffic Engineers. . . . . .ITE
Insulated Power Cable Engineers' Association . . . . .IPCEA
Joint Electron Tube Engineering Council . . . . JETEC
Manufacturers,Standardization Society of the Valve and Fittings Industry . . . .MSS
National Advisory Committee for Aeronautics . . . . NACA
National Aircraft Standards . . . . . NAS
National Bureau of Standards . . . . . NBS
National Association of Manufacturers . . NAM
National Conservation Bureau . . . . NCB
National Electrical Manufacturers' Association . . . . NEMA
National Hardwood Lumber Association . NHLA
National Housing Agency . . . . . . NHA
National Lumber Manufacturers' Association . . . . NLMA
National Machine Tool Builders' Association . . . . NMTBA
National Petroleum Association . . . . NPA
National Safety Council . . . . . . NSC
Oil Heat Institute of America . . . . . OHIA
Radio Manufacturers' Association . . . RMA
Refrigeration Equipment Manufacturers' Association . . . . REMA
Society for the Advancement of Management . . . . SAM
Society of Automotive Engineers . . . . SAE
Society of Fire Engineers . . . . . . SFE
Society of Industrial Engineers . . . . SIE
Society of Military Engineers . . . . . SME
Society of Naval Architects and Marine Engineers . . . . .SNA&ME
Society of Tractor Engineers . . . . . STE
Standards Engineers' Society . . . . . SES
Underwriters' Laboratories, Inc . . . . .UL

PIPE FITTINGS AND VALVES

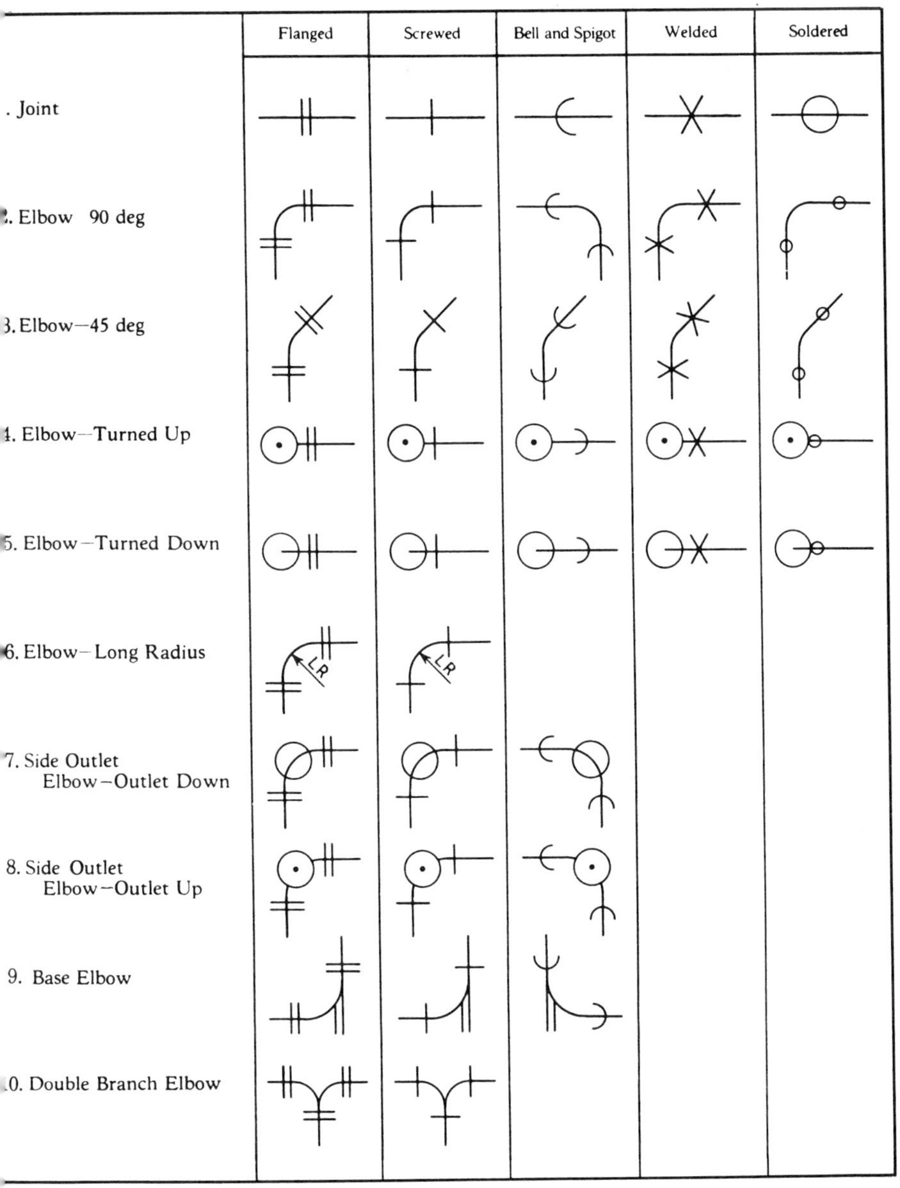
Flanged
Screwed
Bell and Spigot
Welded
Soldered
. Joint
2. Elbow 90 deg
3. Elbow—45 deg
4. Elbow—Turned Up
5. Elbow—Turned Down
6. Elbow—Long Radius
LR
LR
7. Side Outlet Elbow—Outlet Down
8. Side Outlet Elbow—Outlet Up
9. Base Elbow
10. Double Branch Elbow

(Courtesy ASA)

Pipe Fittings and Valves (*continued*)

	Flanged	Screwed	Bell and Spigot	Welded	Soldered
11. Single Sweep Tee					
12. Double Sweep Tee					
13. Reducing Elbow					
14. Tee					
15. Tee—Outlet Up					
16. Tee—Outlet Down					
17. Side Outlet Tee Outlet Up					
18. Side Outlet Tee Outlet Down					
19. Cross					
20. Reducer, Concentric					

(Courtesy AS

PIPE FITTINGS AND VALVES (*continued*)

	Flanged	Screwed	Bell and Spigot	Welded	Soldered
21. Reducer, Eccentric					
22. **Lateral**					
23. **Gate Valve** Elevation (See 169)					
24. **Globe Valve** Elevation (See 170)					
25. Angle Gate Valve Elevation (See 171)					
26. Angle Globe Valve Elevation (See 172)					
27. **Check Valve**					
28. **Angle Check Valve**					
29. **Stop Cock**					
30. **Safety Valve**					

(Courtesy ASA)

WELDING

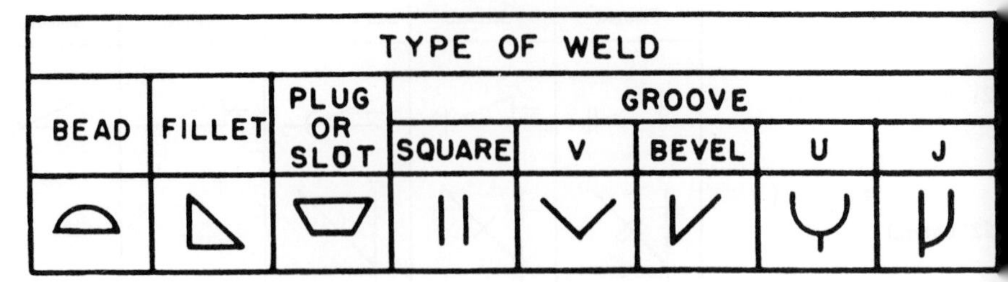

TYPE OF WELD							
BEAD	FILLET	PLUG OR SLOT	GROOVE				
			SQUARE	V	BEVEL	U	J

Fig. C-1 Basic arc and gas weld symbols.

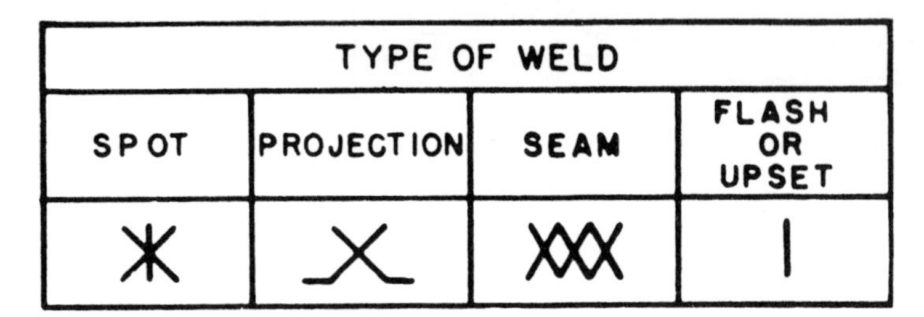

TYPE OF WELD			
SPOT	PROJECTION	SEAM	FLASH OR UPSET

Fig. C-2 Basic resistance weld symbols.

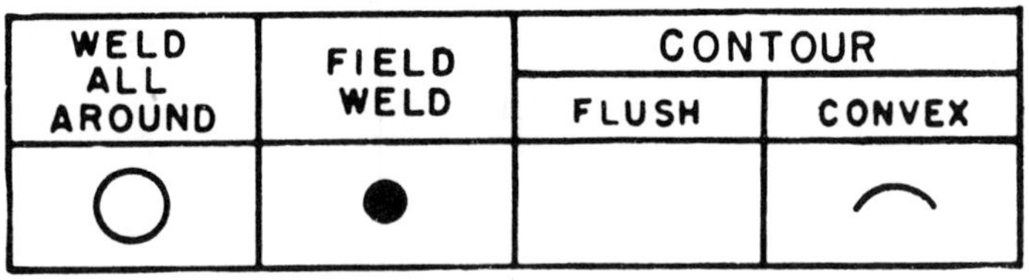

WELD ALL AROUND	FIELD WELD	CONTOUR	
		FLUSH	CONVEX
○	●		⌒

Fig. C-3 Supplementary symbols.

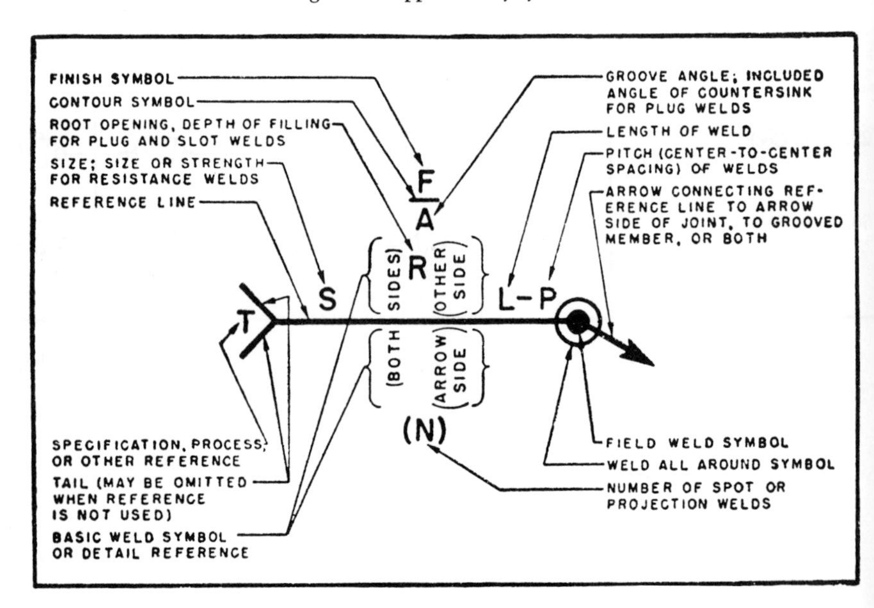

Fig. C-4 Standard location of elements of a welding symbol.

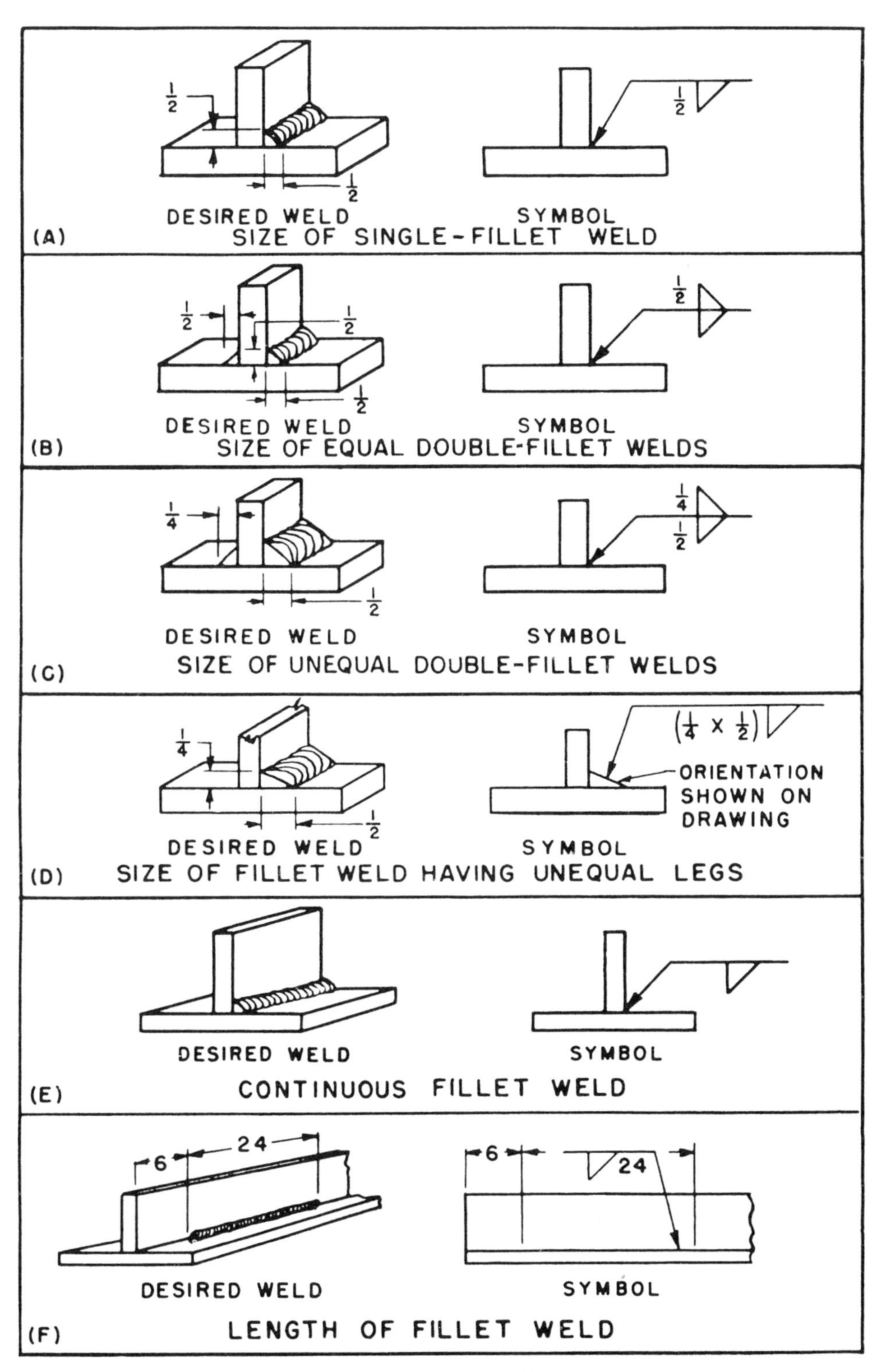

Fig. C-5 Application of dimensions to fillet welding symbols.

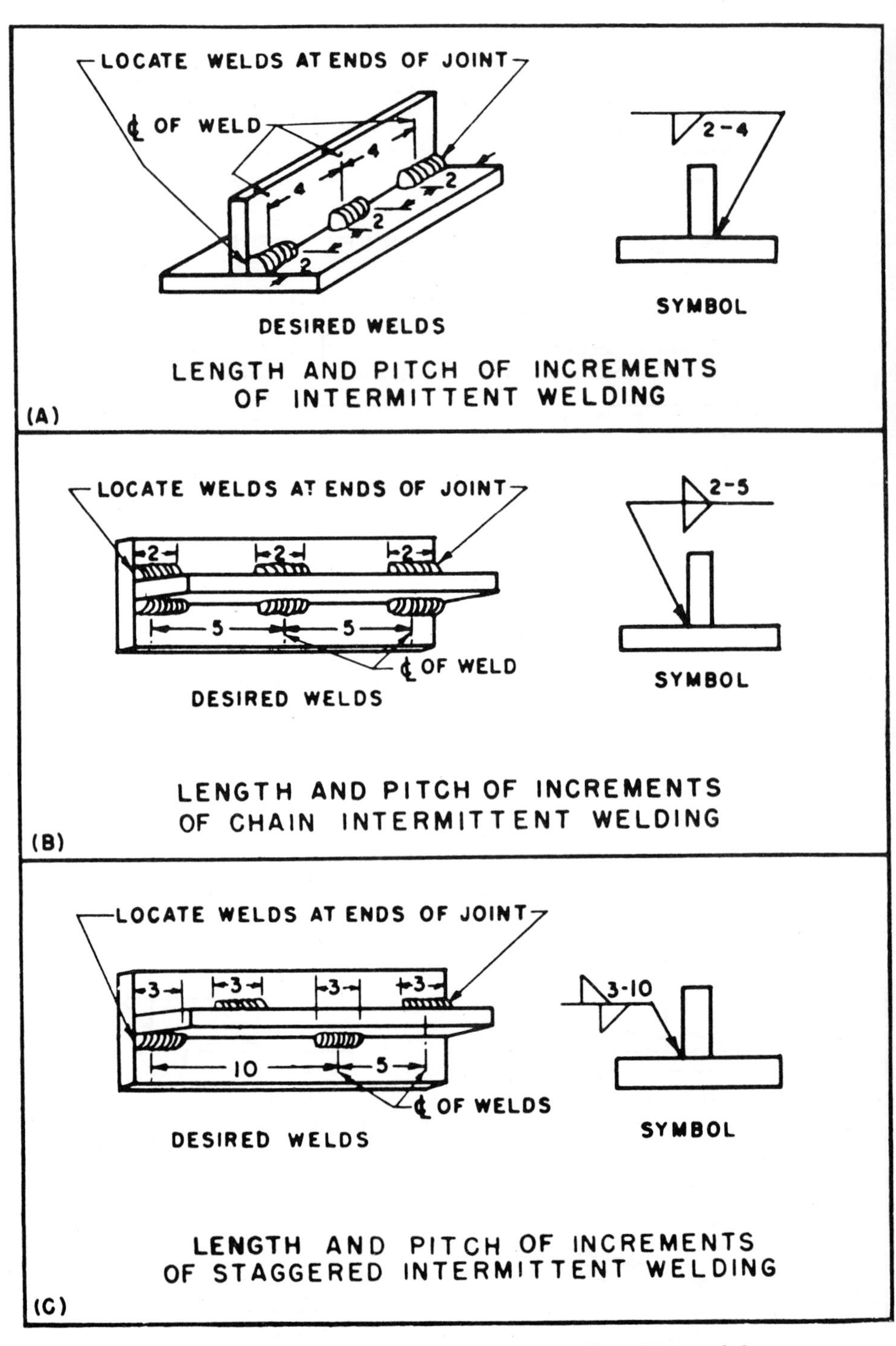

Fig. C-6 Application of dimensions to intermittent fillet welding symbols.

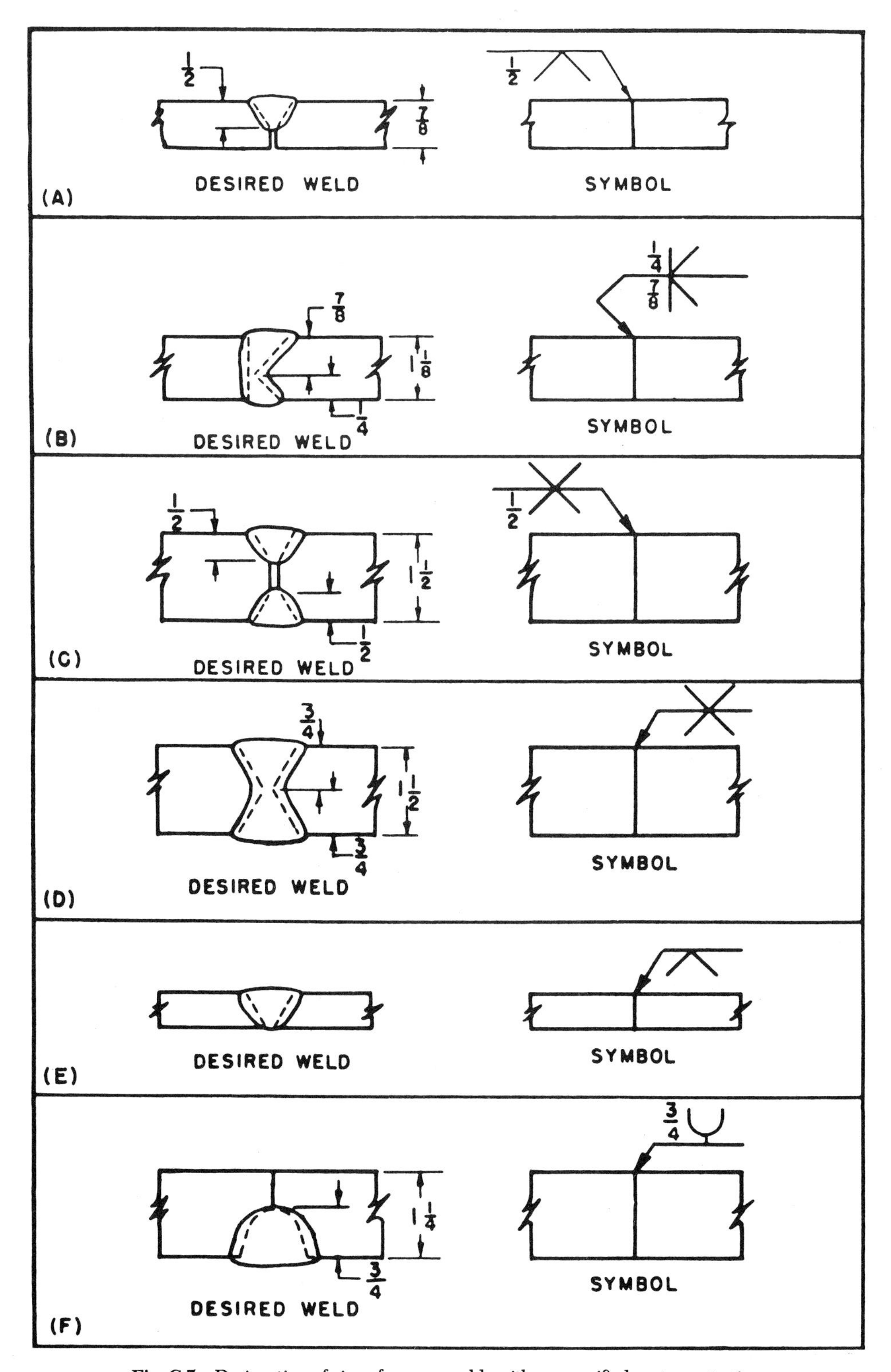

Fig. C-7 Designation of size of groove welds with no specified root penetration.

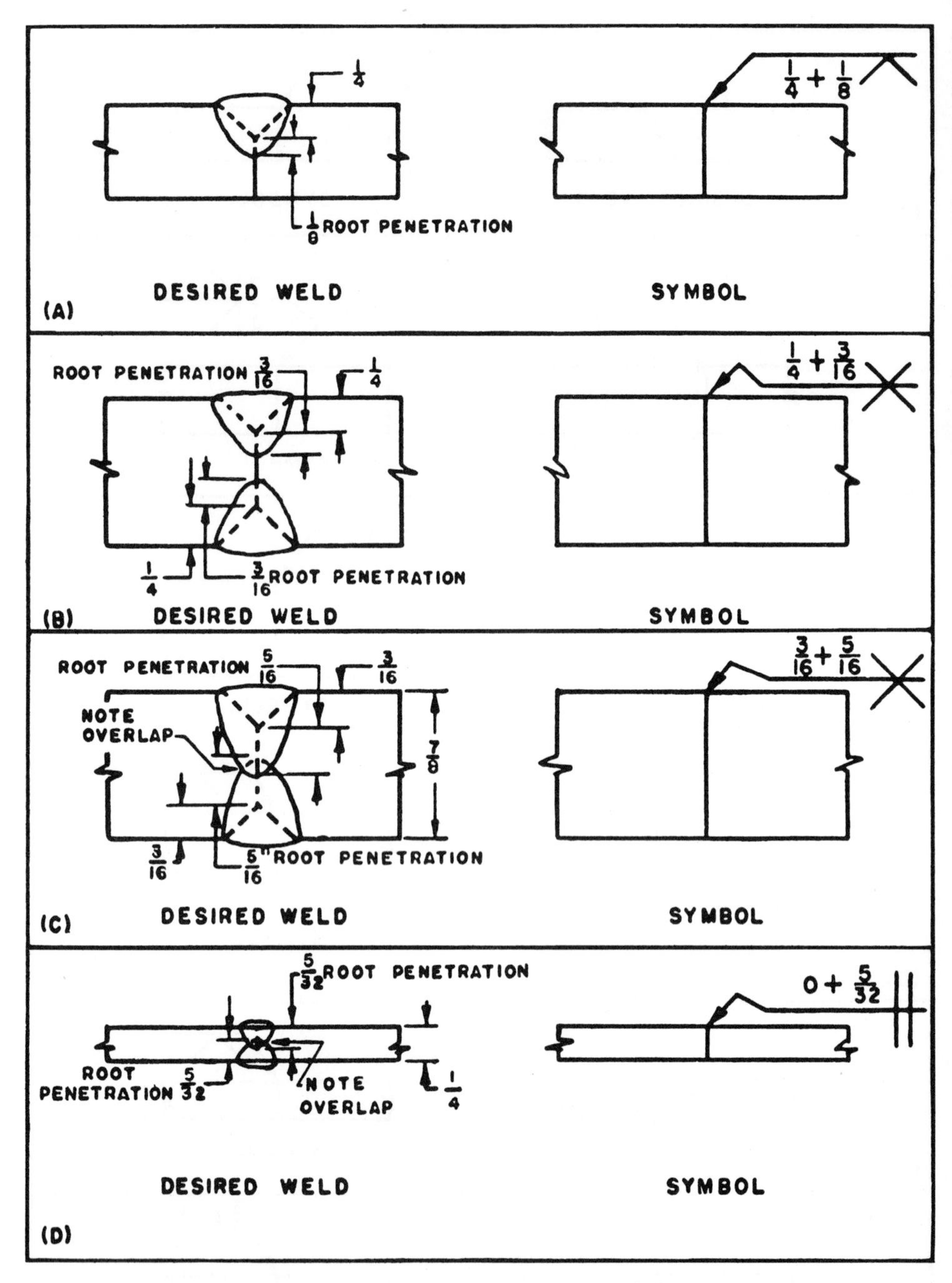

Fig. C-8 Designation of size of groove welds with specified root penetration.

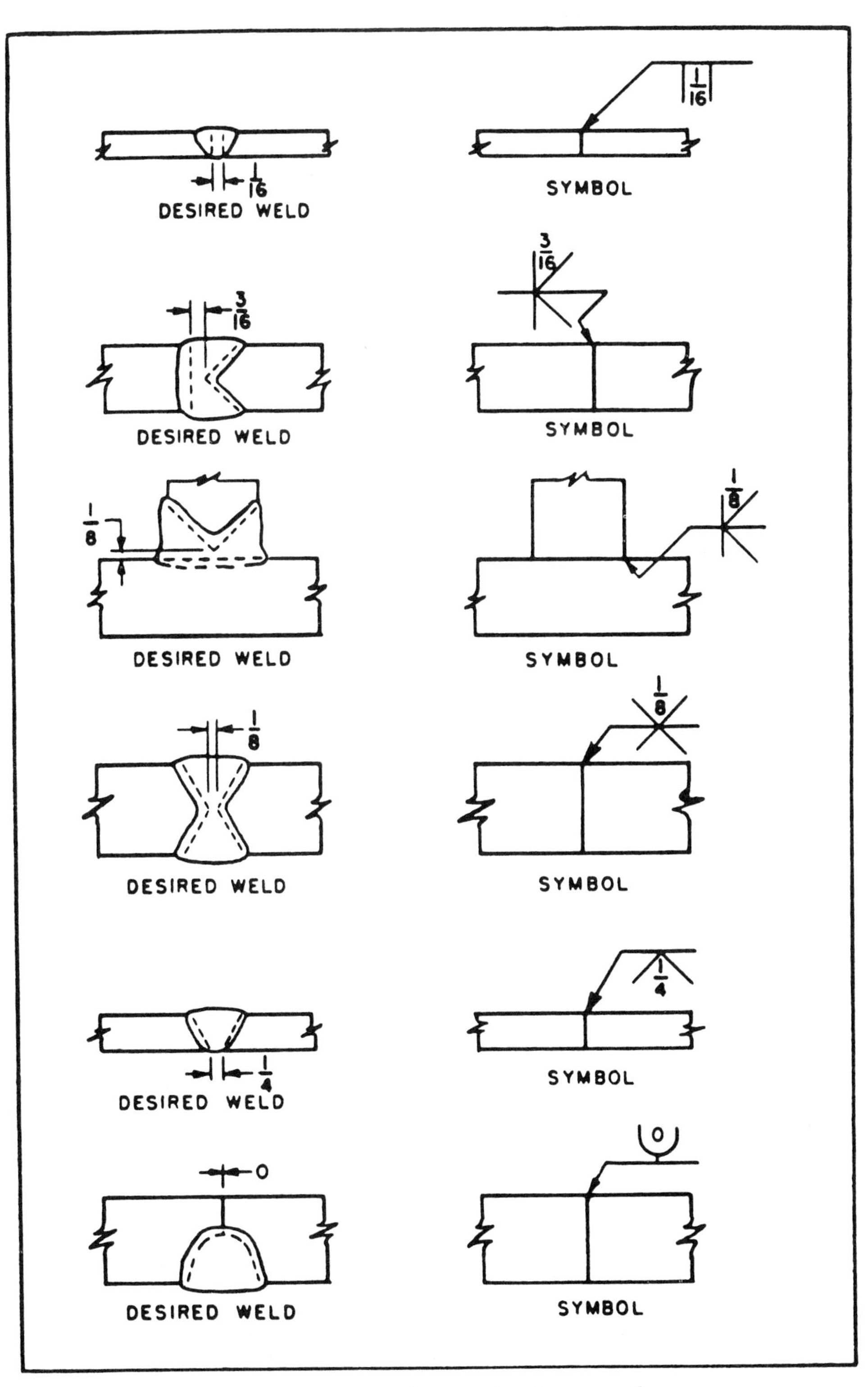

Fig. C-9 Designation of root opening of groove welds.

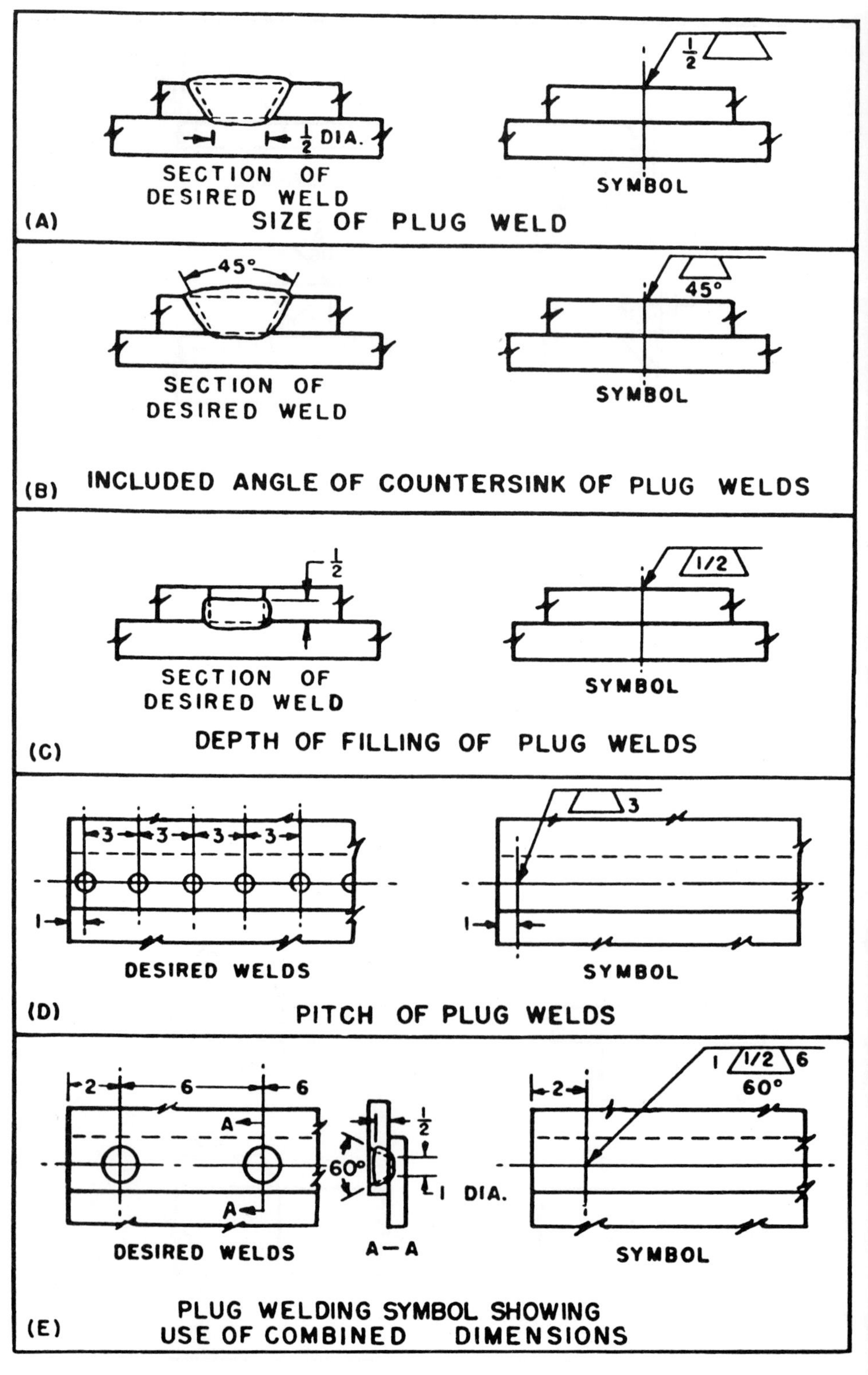

Fig. C-10 Application of dimensions to plug welding symbols.

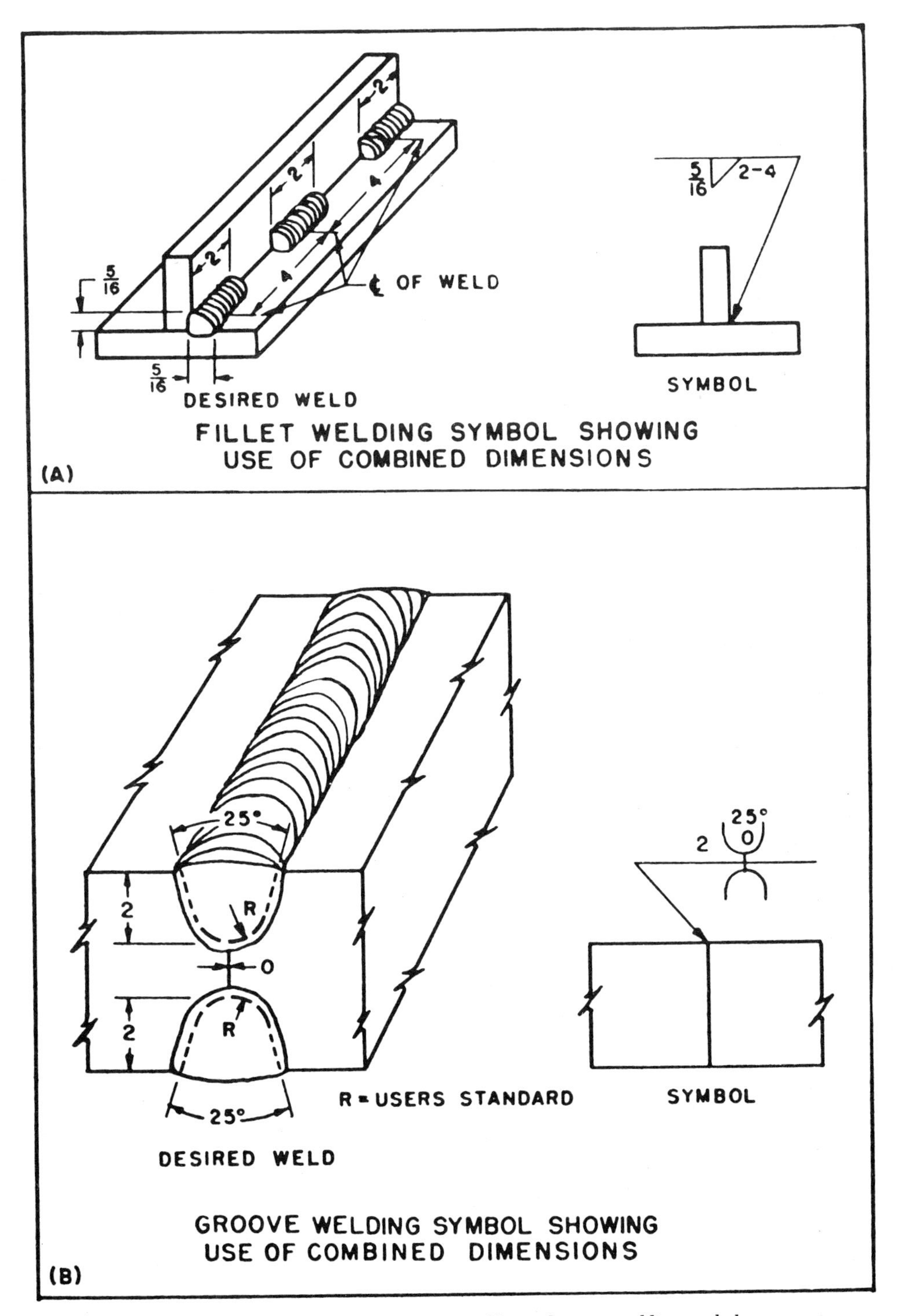

Fig. C-11 Application of dimensions to fillet and groove welding symbols.

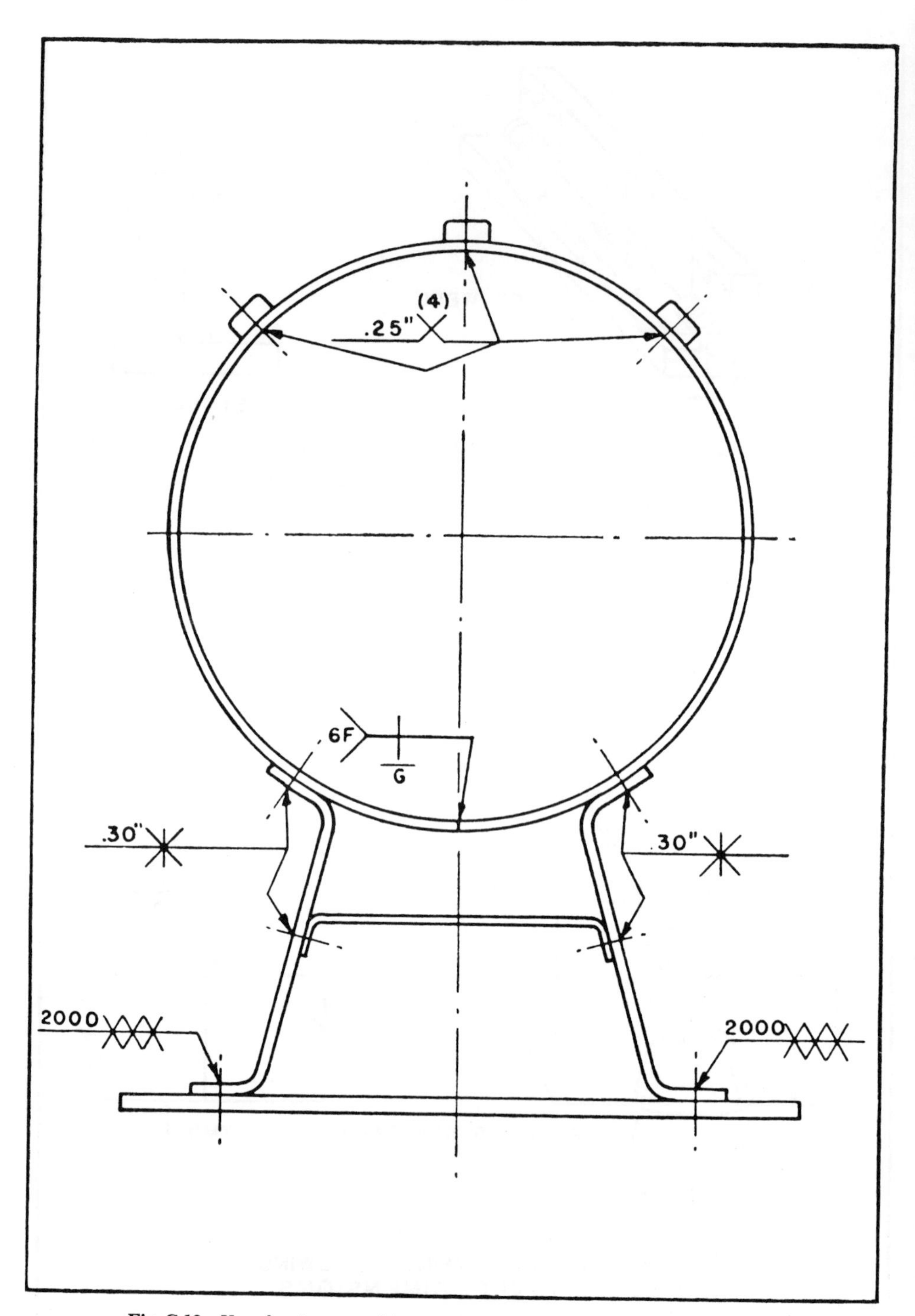

Fig. C-12 Use of resistance welding symbols on sheet-metal fabrication drawing.

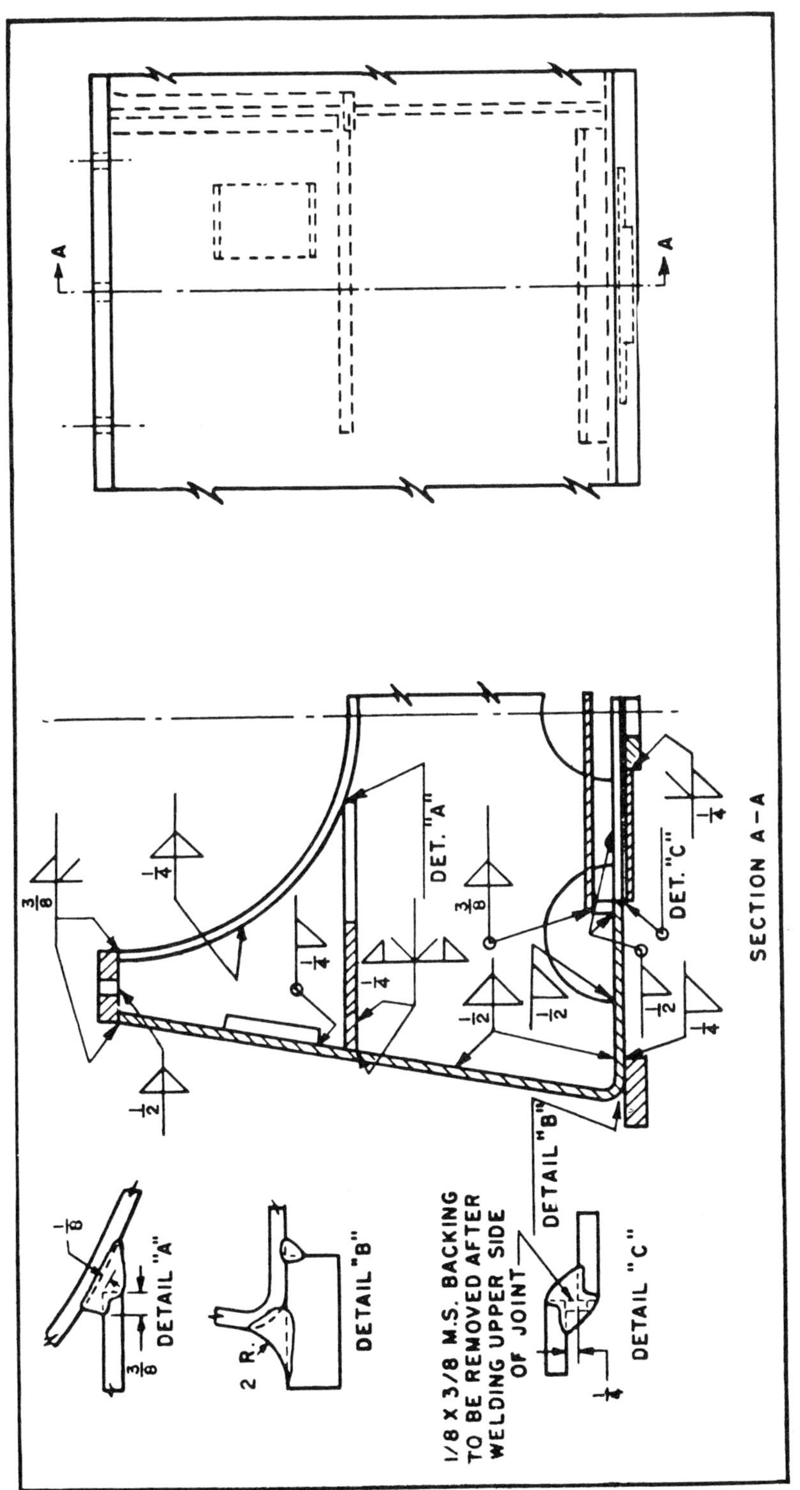

Fig. C-13 Use of arc and gas welding symbols on machinery drawing.

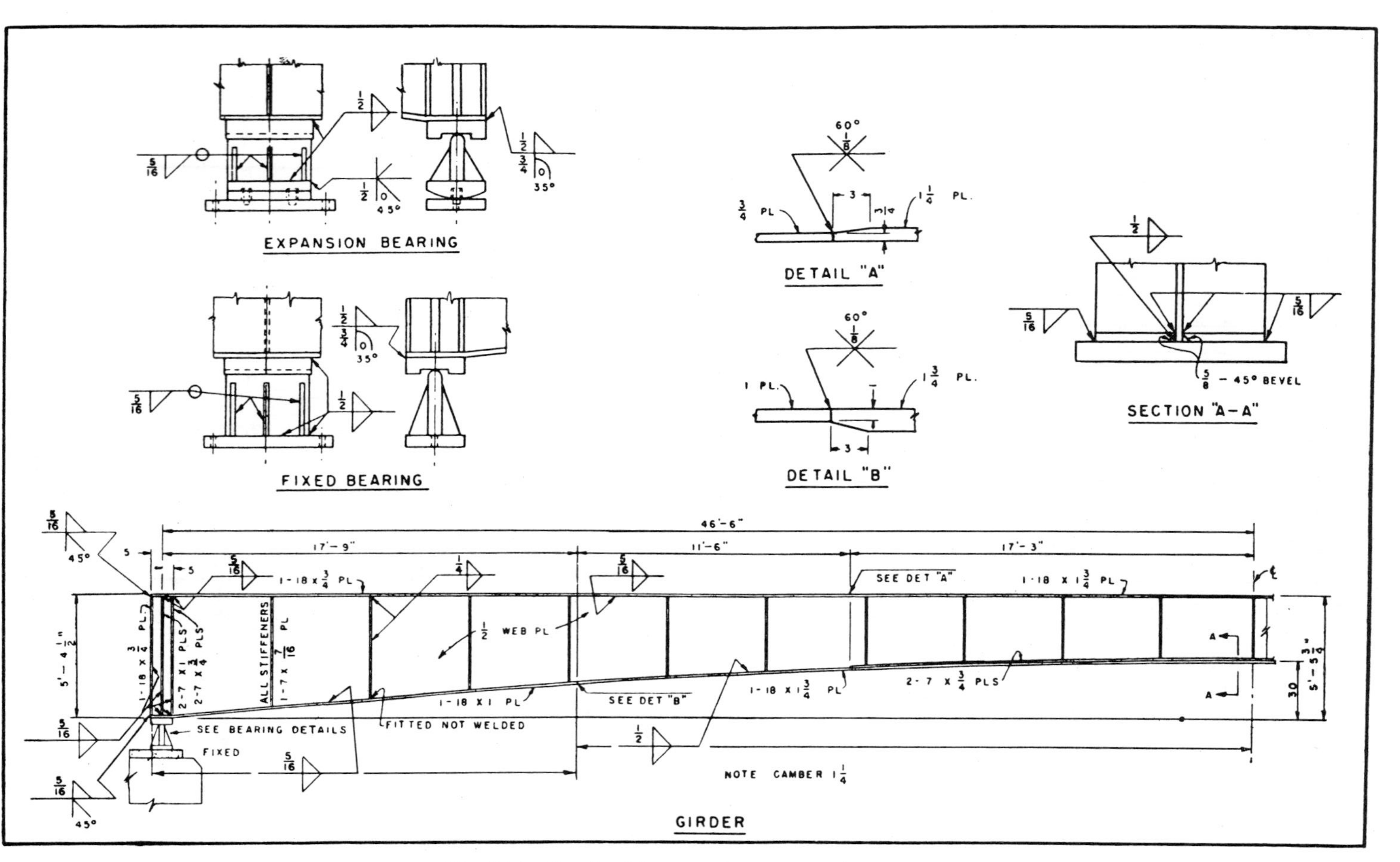

Fig. C-14 Use of welding symbols on structural drawing.

APPENDIX D
MATHEMATICAL CALCULATIONS FOR ANGLES AND SCALE RATIOS IN PICTORIAL ORTHOGRAPHIC VIEWS

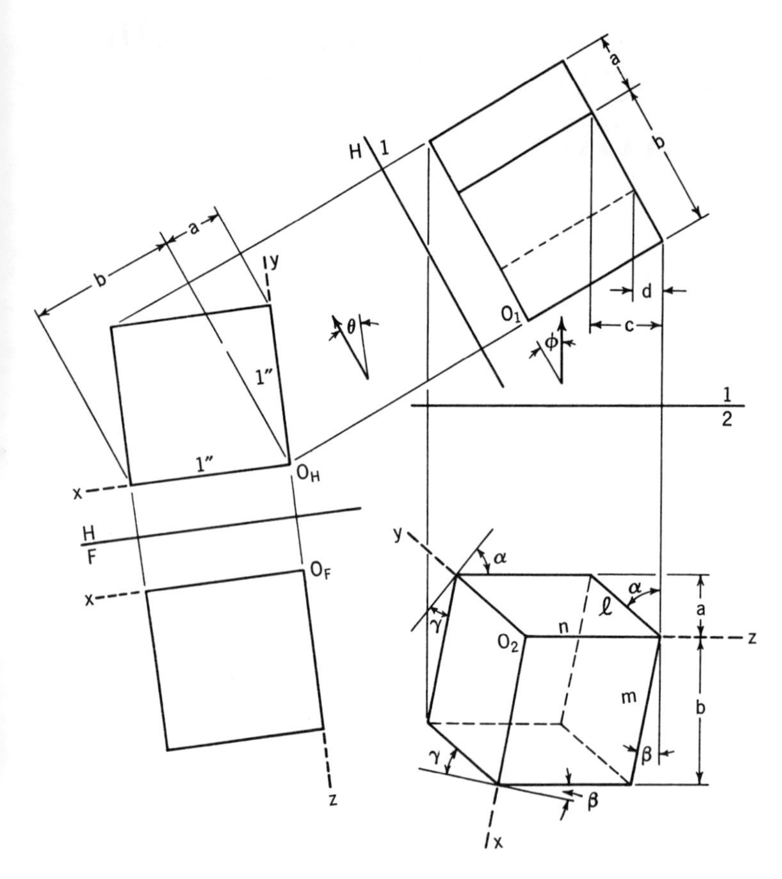

Fig. D-1 Angles and foreshortening of axes in pictorial orthographic views.

In the supplementary view 2 (plane 2 of Fig. D-1),

$$l = \sqrt{a^2 + c^2}$$

But in the H-view, $a = \sin\theta$, and in the supplementary view, 1,

$$\begin{aligned} c &= b \sin\phi \\ &= \cos\theta \sin\phi \end{aligned}$$

from the H-view. Hence

$$l = \sqrt{\sin^2\theta + \cos^2\theta \sin^2\phi}$$

Similarly,

$$m = \sqrt{b^2 + d^2}$$

$$d = a \sin\phi;$$

thus

$$m = \sqrt{\cos^2\theta + \sin^2\theta \sin^2\phi}$$

The ratios of foreshortening l, m, n of lines in the directions of axes x, y, z, respectively, are:

$$l = \sqrt{\sin^2\theta + \cos^2\theta \sin^2\phi}$$

$$m = \sqrt{\cos^2\theta + \sin^2\theta \sin^2\phi}$$

$$n = \cos\phi$$

The angles α, β, γ as defined in the supplementary view 2 are given by

$$\tan\alpha = \frac{\sin\phi}{\tan\theta}$$

$$\tan\beta = \sin\phi \tan\theta$$

$$\gamma = 90° - (\alpha + \beta)$$

Application of Above Results: The following relations result from combining those above.

$$l^2 + m^2 + n^2 = 2$$

Also, from

$$\begin{aligned} l &= \sqrt{\sin^2\theta + \cos^2\theta \sin^2\phi} \\ &= \sqrt{(1 - \cos^2\theta) + \cos^2\theta \sin^2\phi} \\ &= \sqrt{1 - \cos^2\theta\,(1 - \sin^2\phi)} \\ &= \sqrt{1 - \cos^2\theta \cos^2\phi} \end{aligned}$$

there results

$$\cos\theta = \frac{1}{\cos\phi}\sqrt{1 - l^2} = \frac{\sqrt{1 - l^2}}{n};$$

similarly

$$\sin\theta = \frac{\sqrt{1 - m^2}}{n}$$

Sample Solution. Given: $l:m:n = \frac{3}{4}:\frac{7}{8}:1$
Required: l, m, n; θ, ϕ; α, β, γ
Solution:

	Results
$l^2 + m^2 + n^2 = 2$	$l = 0.695$
$[(\frac{3}{4})^2 + (\frac{7}{8})^2 + 1^2]n^2 = 2;\ n = 0.927$	$m = 0.811$

Therefore,

$m = \frac{7}{8}n = 0.811;\ l = \frac{3}{4}n = 0.695$	$n = 0.927$
$n = \cos\phi;\ \phi = 22°3'$	$\phi = 22°3'$
$\cos\theta = \dfrac{\sqrt{1 - l^2}}{n};\ \theta = 39°8'$	$\theta = 39°8'$
$\tan\alpha = \dfrac{\sin\phi}{\tan\theta};\ \alpha = 24°46'$	$\alpha = 24°46'$
$\tan\beta = \sin\phi\tan\theta;\ \beta = 16°59'$	$\beta = 16°59'$
$\gamma = 90° - (\alpha + \beta);\ \gamma = 48°15'$	$\gamma = 48°15'$

Application to Dimetric Views: When $l = m$, there results $\theta = 45°$; thus $\alpha = \beta$. The calculations are considerably shorter than in the example above.

When $l \neq m$, but $l = n$ or $n = m$, direct substitution into the equations above results in a solution as lengthy as that for the trimetric case. This is unnecessary if use is made of the angle γ found for $\theta = 45°$, and of the same relative amounts of foreshortening. Consider the case of relative foreshortening of $1:\frac{3}{4}:\frac{3}{4}$, in any order:

$$l:m:n = \tfrac{3}{4}:\tfrac{3}{4}:1 \text{ gives } \theta = 45°$$

and results in Fig. D-2.

If this figure is rotated until axis b–b is horizontal, Fig. D-3 results. Making axis c–c horizontal gives the orientation shown in Fig. D-4. Although Fig. D-3 corresponds to the mathematical case $l:m:n = \frac{3}{4}:1:\frac{3}{4}$, for which $\theta = 19°28'$ and $\phi = 43°19'$, preparation of the dimetric view requires no further calculation beyond those for the mathematically simpler case $l:m:n = \frac{3}{4}:\frac{3}{4}:1$, $\theta = 45°$. The same figure has merely been reoriented.

It should be noted that, since isometric, dimetric, and trimetric *drawings* are geometrically similar to the corresponding *views*, it is only necessary to maintain the *relative* foreshortening of the axes. For example, the dimetric drawing corresponding to Fig. D-2 could be made by laying off vertical (z) distances full size and the x and y distances three-quarter size.

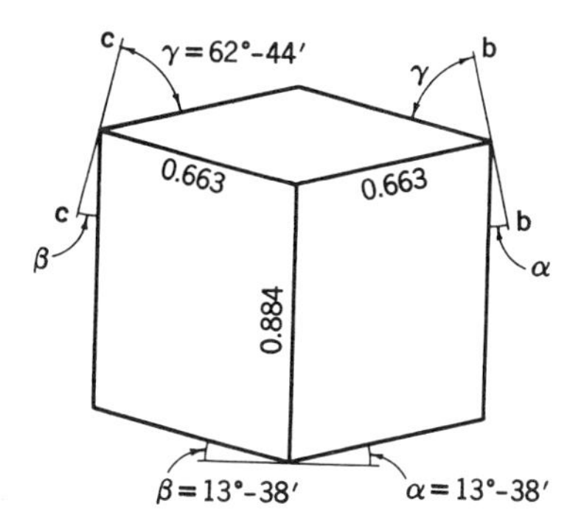

Fig. D-2

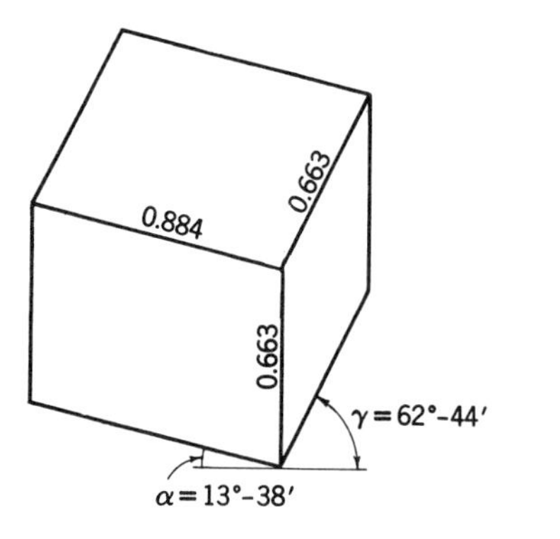

Fig. D-3

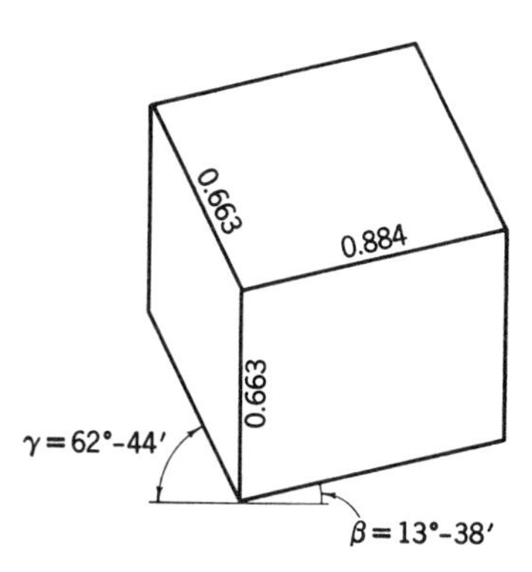

Fig. D-4

APPENDIX E
MATHEMATICAL (ALGEBRAIC) SOLUTIONS OF SPACE PROBLEMS

1. Length of a Line Segment (Fig. E-1). When the coordinates of the end points of the line segment are known or can be determined, the length of the segment can be computed.

In Fig. E-1 it is clearly seen that the line segment AB can be regarded as the hypotenuse of a right triangle, the vertical side of which is equal to $(Z_B - Z_A)$ and the horizontal side equal to

$$\sqrt{(X_B - X_A)^2 + (Y_B - Y_A)^2}.$$

From these values, d or

$$\overline{AB} = \sqrt{(X_B - X_A)^2 + (Y_B - Y_A)^2 + (Z_B - Z_A)^2}.$$

If the coordinates of the points shown in the figure are $A(10, 6, 9)$ and $B(2, 16, 3)$, the true length of the line is

$$d = \sqrt{(2 - 10)^2 + (16 - 6)^2 + (3 - 9)^2}$$
$$= \sqrt{64 + 100 + 36} = \sqrt{200} = 14.1 \text{ units}.$$

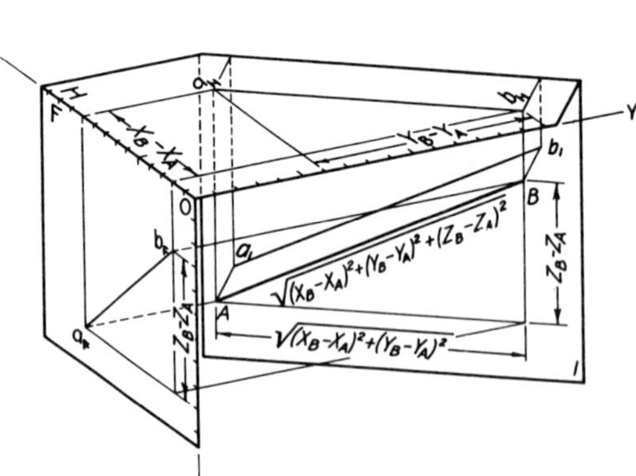

Fig. E-1

2. Perpendicular Distance from a Point to a Line (Fig. E-2). Let us assume that the given point is $P(x_2 y_2 z_2)$ and the line is given by the equations,

$$\frac{x - x_1}{\lambda} = \frac{y - y_1}{\mu} = \frac{z - z_1}{\upsilon}$$

where the denominator values are the direction cosines of the line; i.e., $\lambda = \cos \alpha$; $\mu = \cos \beta$; $\upsilon = \cos \gamma$ (α, β, and γ are the angles that the line makes with the X, Y, and Z axes, respectively).

From the figure it is evident that

$$d = PP_1 \sin \phi = PP_1 \sqrt{1 - \cos^2 \phi}$$
$$= PP_1 \sqrt{1 - \left[\lambda \cdot \frac{x_1 - x_2}{PP_1} + \mu \cdot \frac{y_1 - y_2}{PP_1} + \nu \cdot \frac{z_1 - z_2}{PP_1}\right]^2}$$
$$= \sqrt{PP_1^2 - [\lambda(x_1 - x_2) + \mu(y_1 - y_2) + \nu(z_1 - z_2)]^2}$$
$$= \sqrt{(x_1 - x_2)^2 + (y_1 - y_2)^2 + (z_1 - z_2)^2 - [\lambda(x_1 - x_2) + \mu(y_1 - y_2) + \nu(z_1 - z_2)]^2}$$

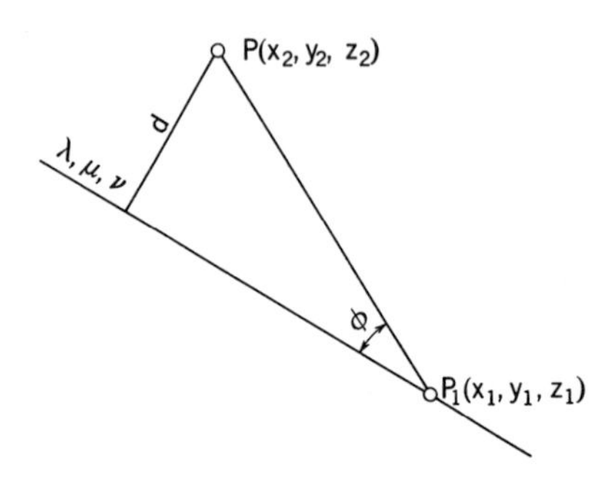

Fig. E-2

3. Distance from a Point to a Plane. The given plane is $Ax + By + Cz + D = 0$ and the point is $P(x_1, y_1, z_1)$.

The plane through point P and parallel to the given plane is

$$Ax + By + Cz - Ax_1 - By_1 - Cz_1 = 0$$

The distance from the point to the given plane is the same as the distance between the parallel planes, or,

$$\text{Distance} = \frac{Ax_1 + By_1 + Cz_1 + D}{\sqrt{A^2 + B^2 + C^2}}$$

4. Distance between two skew lines (Fig. E-3). The distance between two skew lines is the distance between the two parallel planes, each of which contains one of the lines and is perpendicular to the common perpendicular.

The equations of the two planes are

$$\text{I. } (\lambda x + \mu y + \nu z = \lambda x_1 + \mu y_1 + \nu z_1)$$
$$\text{II. } (\lambda x + \mu y + \nu z = \lambda x_2 + \mu y_2 + \nu z_2)$$

The distance between the two planes is the difference between their constant terms (the equations are in the normal form)

$$\therefore \text{Distance} = \lambda(x_1 - x_2) + \mu(y_1 - y_2) + \nu(z_1 - z_2)$$

In terms of the given elements the above equation is

$$\text{Distance} = \frac{(\mu_1\nu_2 - \mu_2\nu_1)(x_1 - x_2) + (\nu_1\lambda_2 - \nu_2\lambda_1)(y_1 - y_2) + (\lambda_1\mu_2 - \lambda_2\mu_1)(z_1 - z_2)}{[(\mu_1\nu_2 - \mu_2\nu_1)^2 + (\nu_1\lambda_2 - \nu_2\lambda_1)^2 + (\lambda_1\mu_2 - \lambda_2\mu_1)^2]^{1/2}}$$

which may be written in the form,

$$\text{Distance} = \frac{\begin{vmatrix} x_1 - x_2 & y_1 - y_2 & z_1 - z_2 \\ \lambda_1 & \mu_1 & \nu_1 \\ \lambda_2 & \mu_2 & \nu_2 \end{vmatrix}}{[(\mu_1\nu_2 - \mu_2\nu_1)^2 + (\nu_1\lambda_2 - \nu_2\lambda_1)^2 + (\lambda_1\mu_2 - \lambda_2\mu_1)^2]^{1/2}}$$

which in turn may be expressed as

$$\text{Distance} = \frac{\begin{vmatrix} x_1 - x_2 & y_1 - y_2 & z_1 - z_2 \\ L_1 & M_1 & N_1 \\ L_2 & M_2 & N_2 \end{vmatrix}}{\left[\begin{vmatrix} M_1 N_1 \\ M_2 N_2 \end{vmatrix}^2 + \begin{vmatrix} N_1 L_1 \\ N_2 L_2 \end{vmatrix}^2 + \begin{vmatrix} L_1 M_1 \\ L_2 M_2 \end{vmatrix}^2\right]^{1/2}}$$

(L_1, M_1, N_1, L_2, M_2, and N_2 are direction numbers which are proportional, respectively, to the direction cosines.)

5. Line through a Point and Perpendicular to a Given Plane. It is required to determine the equa-

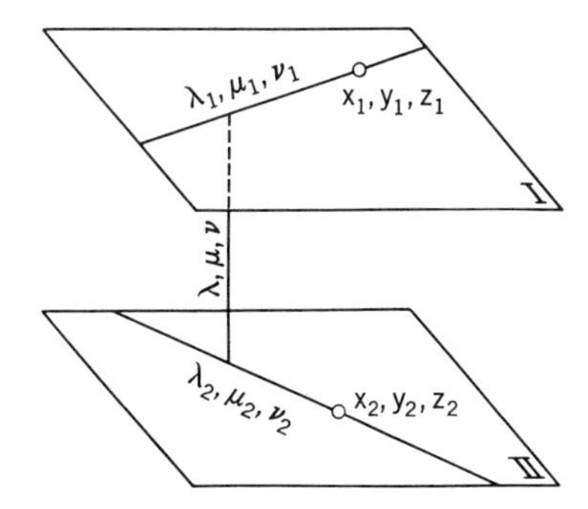

Fig. E-3

tions of the line that passes through point $P(X_1, Y_1, Z_1)$ and perpendicular to the plane,

$$Ax + By + Cz + D = 0$$

A set of direction numbers of the required line is A, B, and C.

The equations of the line are

$$\frac{X - X_1}{A} = \frac{Y - Y_1}{B} = \frac{Z - Z_1}{C}$$

Example. Suppose the point is $P(-8, 4, 1)$ and the plane is $(2x + y - 6z - 32 = 0)$. The equations of the line through P and perpendicular to the plane are,

$$\frac{X + 8}{2} = \frac{Y - 4}{1} = \frac{Z - 1}{-6}$$

$$\text{or } \begin{cases} X - 2Y + 16 = 0 \\ -6X - 2Z - 46 = 0 \end{cases}$$

6. Plane through a Point and Perpendicular to a Given Line. Let us suppose that the given point is $P(X_1, Y_1, Z_1)$ and that the direction of the given line is given by its direction cosines (λ, μ, ν) or its direction numbers (L, M, N). Now any plane that is perpendicular to the given line has an equation of the form,

$$\lambda X + \mu Y + \nu Z = p \tag{1}$$

Since the plane must pass through point P, then,

$$\lambda X_1 + \mu Y_1 + \nu Z_1 = p \tag{2}$$

From equations (1) and (2) we obtain

$$\lambda X + \mu Y + \nu Z = \lambda X_1 + \mu Y_1 + \nu Z_1$$

which is the required plane.

The equation can be written as

$$\lambda(X - X_1) + \mu(Y - Y_1) + \nu(Z - Z_1) = 0$$

or

$$L(X - X_1) + M(Y - Y_1) + N(Z - Z_1) = 0$$

7. The Angle between Two Lines (Fig. E-4). The angle between two non-intersecting lines is defined as the angle between their directions, or the angle between two lines parallel respectively to the two given lines and passing through any selected point.

Let the two given lines have direction cosines λ_1, μ_1, ν_1, and λ_2, μ_2, ν_2 respectively, and let ϕ represent the angle between them. Through the origin let us pass lines parallel to the two given lines. Let $P_1(x_1, y_1, z_1)$ be any point on one of these lines, and $P_2(x_2, y_2, z_2)$ be any point on the other. Connect P_1 and P_2.

In triangle OP_1P_2, by virtue of the cosine law of trigonometry, we have

$$\overline{P_1P_2^2} = \overline{OP_1^2} + \overline{OP_2^2} - 2OP_1 \cdot OP_2 \cos \phi$$

or

$$\cos \phi = \frac{\overline{OP_1^2} + \overline{OP_2^2} - \overline{P_1P_2^2}}{2 \cdot OP_1 \cdot OP_2} \tag{1}$$

Hence, by the distance formula

$$\cos \phi = \frac{[x_1^2 + y_1^2 + z_1^2] + [x_2^2 + y_2^2 + z_2^2]}{2OP_1 \cdot OP_2} - \frac{[(x_2 - x_1)^2 + (y_2 - y_1)^2 + (z_2 - z_1)^2]}{2OP_1 \cdot OP_2}$$

$$= \frac{x_1^2 + y_1^2 + z_1^2 + x_2^2 + y_2^2 + z_2^2 - x_2^2}{2OP_1 \cdot OP_2} + \frac{2x_2x_1 - x_1^2 - y_2^2 + 2y_1y_2 - y_1^2 - z_2^2 + 2z_2z_1 - z_1^2}{2OP_1 \cdot OP_2}$$

$$= \frac{2x_1x_2 + 2y_1y_2 + 2z_1z_2}{2OP_1 \cdot OP_2} = \frac{x_1x_2 + y_1y_2 + z_1z_2}{OP_1 \cdot OP_2}$$

$$= \frac{x_1}{OP_1} \cdot \frac{x_2}{OP_2} + \frac{y_1}{OP_1} \cdot \frac{y_2}{OP_2} + \frac{z_1}{OP_1} \cdot \frac{z_2}{OP_2}$$

But

$$\frac{x_1}{OP_1} = \frac{\text{Difference in the } x \text{ coordinates of points } O \text{ and } P_1}{\text{Distance } OP_1} = \lambda_1$$

And similarly

$$\frac{x_2}{OP_2} = \lambda_2, \frac{y_1}{OP_1} = \mu_1, \frac{y_2}{OP_2} = \mu_2, \text{ etc.}$$

$$\therefore \cos \phi = \lambda_1\lambda_2 + \mu_1\mu_2 + \nu_1\nu_2.$$

8. The Angle between Two Planes. The angle between two planes may be defined as the angle between normals to the two planes respectively.

Let the two planes be

$$A_1x + B_1y + C_1z + D_1 = 0 \tag{1}$$

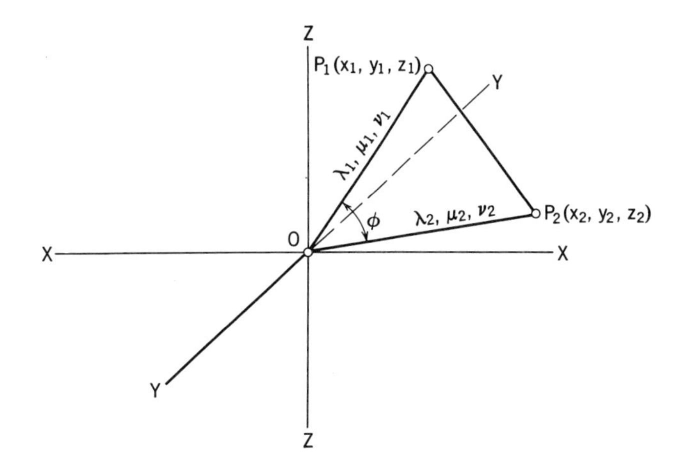

Fig. E-4

and

$$A_2x + B_2y + C_2z + D_2 = 0 \qquad (2)$$

Let ϕ represent the angle between them.

The direction cosines of the normal to the first plane are

$$\lambda_1 = \frac{A_1}{\sqrt{A_1^2 + B_1^2 + C_1^2}}$$

$$\mu_1 = \frac{B_1}{\sqrt{A_1^2 + B_1^2 + C_1^2}}$$

$$\nu_1 = \frac{C_1}{\sqrt{A_1^2 + B_1^2 + C_1^2}}$$

The direction cosines of the normal to the second plane are

$$\lambda_2 = \frac{A_2}{\sqrt{A_2^2 + B_2^2 + C_2^2}}$$

$$\mu_2 = \frac{B_2}{\sqrt{A_2^2 + B_2^2 + C_2^2}}$$

$$\nu_2 = \frac{C_2}{\sqrt{A_2^2 + B_2^2 + C_2^2}}$$

The formula for the angle between the planes then is

$$\cos\phi = \lambda_1\lambda_2 + \mu_1\mu_2 + \nu_1\nu_2$$

$$= \frac{A_1A_2 + B_1B_2 + C_1C_2}{\sqrt{A_1^2 + B_1^2 + C_1^2} \cdot \sqrt{A_2^2 + B_2^2 + C_2^2}}$$

9. *The Angle between a Line and a Plane.* The angle between a line and a plane is *defined as the angle between the line and its projection on the plane.*

If α represents the angle between the line and its projection on the plane, and if β represents the angle between the line and a normal to the plane, then

$$\alpha + \beta = 90°$$

If the given line has direction cosines λ_1, μ_1, ν_1, and if the given plane is

$$A_1x + B_1y + C_1z + D_1 = 0$$

then

$$\cos\beta = \frac{A_1\lambda_1 + B_1\mu_1 + C_1\nu_1}{\sqrt{A_1^2 + B_1^2 + C_1^2}}$$

Hence the formula for the required angle α is

$$\sin\alpha = \frac{A_1\lambda_1 + B_1\mu_1 + C_1\nu_1}{\sqrt{A_1^2 + B_1^2 + C_1^2}}$$

10. The Angles between a Given Plane and the Three Reference Planes. Let the given plane be

$$Ax + By + Cz + D = 0 \quad (1)$$

The equation of the horizontal plane ($z = 0$) may be written

$$Ox + Oy + Kz = 0 \quad (2)$$

where K is any finite constant different from zero.

Applying the formula for the angle between the two planes, namely,

$$\cos \phi = \frac{A_1A_2 + B_1B_2 + C_1C_2}{\sqrt{A_1^2 + B_1^2 + C_1^2} \cdot \sqrt{A_2^2 + B_2^2 + C_2^2}}$$

to the two planes under consideration, we have

$$\cos H = \frac{A \cdot O + B \cdot O + C \cdot K}{\sqrt{A^2 + B^2 + C^2} \cdot \sqrt{O^2 + O^2 + K^2}} = \frac{C}{\sqrt{A^2 + B^2 + C^2}}$$

Similarly,

$$\cos F = \frac{B}{\sqrt{A^2 + B^2 + C^2}}$$

$$\cos P = \frac{A}{\sqrt{A^2 + B^2 + C^2}}$$

The letters H, F and P are used to designate the angles which the given plane makes with the H, F and P planes respectively.

11. The Angles between a Given Plane and the Reference Axes. Let the given plane be

$$Ax + By + Cz + D = 0 \quad (1)$$

The formula for the angle between a plane and a line is

$$\sin \phi = \frac{A_1\lambda_1 + B_1\mu_1 + C_1\nu_1}{\sqrt{A_1^2 + B_1^2 + C_1^2}} \quad (2)$$

The direction cosines of the x-axis are $\lambda_1 = 1$, $\mu_1 = 0$, $\nu_1 = 0$.

Letting α, β and ζ represent the angles between the given plane and the x, y and z-axes respectively, we have

$$\sin \alpha = \frac{A \cdot 1 + B \cdot 0 + C \cdot 0}{\sqrt{A^2 + B^2 + C^2}} = \frac{A}{\sqrt{A^2 + B^2 + C^2}}$$

$$\sin \beta = \frac{B}{\sqrt{A^2 + B^2 + C^2}}$$

$$\sin \zeta = \frac{C}{\sqrt{A^2 + B^2 + C^2}}$$

Comparing these formulae with those for the angles which the plane makes with the reference planes, we observe

$$\sin \alpha = \cos P$$
$$\sin \beta = \cos F$$
$$\sin \zeta = \cos H$$

or

$$\alpha + P = \beta + F = \zeta + H = 90^\circ$$

In other words, if a plane makes a certain angle with a reference plane it makes the complement of that angle with the reference axis that is the normal to the reference plane under consideration.

APPENDIX F

GRAPHICAL SOLUTIONS OF DIFFERENTIAL EQUATIONS

1. *The Slope-Field Method.* Most engineering and science students are familiar with differential equations. The brief treatment presented here deals with the use of the "slope-field method" in solving ordinary differential equations of the first order.

Example 1. Let us consider the differential equation

$$\frac{dy}{dx} = 2x + 2 \tag{1}$$

It is recognized that Eq. (1) may be integrated directly to give $y = x^2 + 2x + C$; however, it will be instructive to solve this equation by the slope-field method before dealing with equations which are not so easily solved by algebraic methods.

Suppose we let $x = 1$ in Eq. (1); then

$$\frac{dy}{dx} = 4$$

On the line $x = 1$ a number of short lines having a slope of 4 are drawn. (See Fig. F-1.) Additional slope lines are drawn for $x = 0, \frac{1}{2}, 1\frac{1}{2}, 2$, etc. The totality of slope lines is the slope field.

Let us assume that a particular solution of the equation is the integral curve which passes through

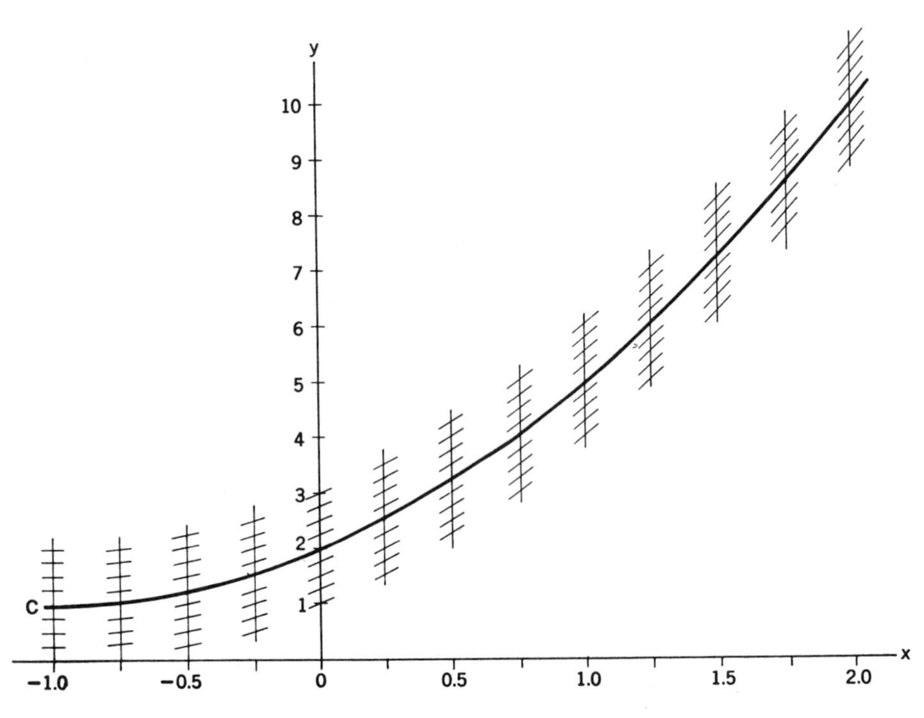

Fig. F-1

point $C(-1, 1)$. The curve is started at point C and sketched in, by eye, so that the slope of the curve at each value of x is the same as that of the slope lines drawn previously.

Other solutions may be sketched in similarly.

Example 2. Let us consider the differential equation

$$\frac{dy}{dx} = y - x^2 \qquad (2)$$

Again, let $x = 0, \frac{1}{2}, 1, 1\frac{1}{2}$, etc., and then let us draw corresponding slope lines, as shown in Fig. F-2. For example, when $x = 0$

$$\frac{dy}{dx} = y$$

The slopes of the short lines crossing $x = 0$ are equal in magnitude to the corresponding values of y. It should be observed that these slope lines pass through the point $(-1, 0)$. Once this is recognized it becomes a very simple process to draw as many slope lines as one finds necessary. Again, when $x = 1$,

$$\frac{dy}{dx} = y - 1$$

The slope lines on $x = 1$ pass through the point $(0, 1)$.

Two solutions are shown in Fig. F-2, one the integral curve passing through point $(0, 2.5)$, and the other the curve passing through point $(0, 1)$.

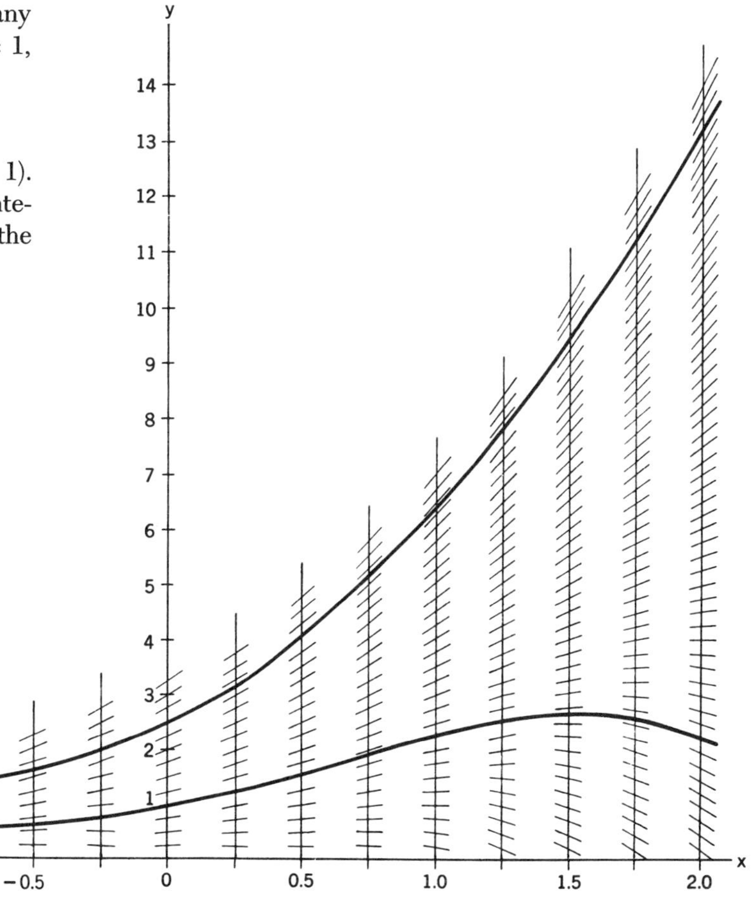

Fig. F-2

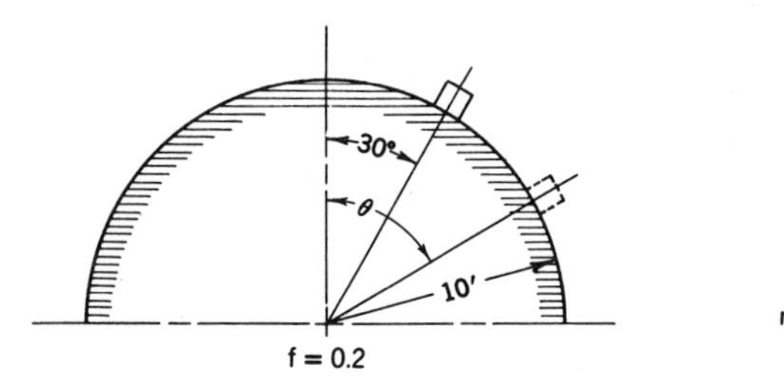

Fig. F-3

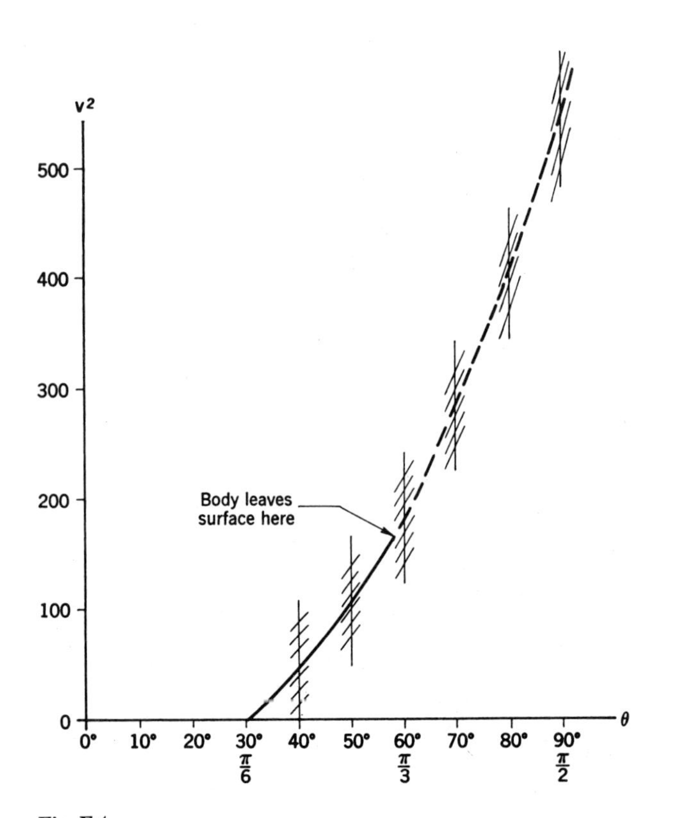

Fig. F-4

Example 3. Figure F-3 shows a small object which slides down the surface of a rough cylinder, starting from rest in the position defined by $\theta = 30°$. The free-body diagram shows the forces acting on the block for a general position θ. From the equations of motion, we obtain

$$W \sin \theta - fN = \frac{W}{g} a_t$$

and

$$W \cos \theta - N = \frac{W}{g} a_n = \frac{W}{g} \frac{v^2}{r}$$

Eliminating N between the two equations gives

$$a_t = g(\sin \theta - f \cos \theta) + \frac{fv^2}{r}$$

If the definitions $a_t = d^2s/dt^2 = r\, d^2\theta/dt^2$ and $v = ds/dt = r\, d\theta/dt$ are substituted, the equation becomes a second-order nonlinear equation. If, on the other hand, a_t is replaced by its equivalent, $a_t = v\, dv/ds = d(v^2)/2\, ds = d(v^2)/2r\, d\theta$, the equation becomes

$$\frac{d(v^2)}{d\theta} = 2gr(\sin \theta - f \cos \theta) + 2fv^2 \qquad (3)$$

which is a linear first-order equation in the variables v^2 and θ. The initial condition for Eq. (3) is

$$v^2 = 0 \qquad \text{when} \qquad \theta = 30° \qquad (4)$$

The slope-field solution of the differential Eq. (3), with its initial condition (4), is accomplished just as in the first two examples. (See Fig. F-4.)

While application of the slope-field method herein illustrated is restricted to first-order equations—or to higher order equations reducible to the first order (as in Example 3)—many practical problems fall within this group, so that the method has, in a number of instances, proved itself a useful tool in research.

2. ***The Pole-and-Ray Method.*** In Chapter 13 we employed the pole-and-ray method for both integration and differentiation. Now let us see what the relationship is between graphical integration and differentiation. Let us assume (Fig. F-5) that the given curve has been integrated by the pole-and-ray method. From the figure we obtain the relation,

$$\frac{BC}{AC} = \frac{OQ}{OP} \qquad (1)$$

from which,

$$BC = \frac{AC \times OQ}{OP} = \frac{\text{area of strip (shown shaded)}}{\text{pole distance}}$$

We also note from equation (1) that

$$OQ = OP \times \frac{BC}{AC} = \begin{array}{l}\text{pole distance times the average}\\ \text{slope of the integral curve for}\\ \text{interval } AC\end{array}$$

If the width AC approaches zero (A considered fixed), chord AB approaches the tangent to the integral curve at point A; and the ordinate of the given curve approaches the value of this slope, multiplied by the pole distance. *This shows that the given curve is the derived curve* (curve of slopes) *of the integral curve.*

When the pole distance is unity, OQ represents the value of the average slope for interval AC. Now we will use the above relation in solving the following problems.

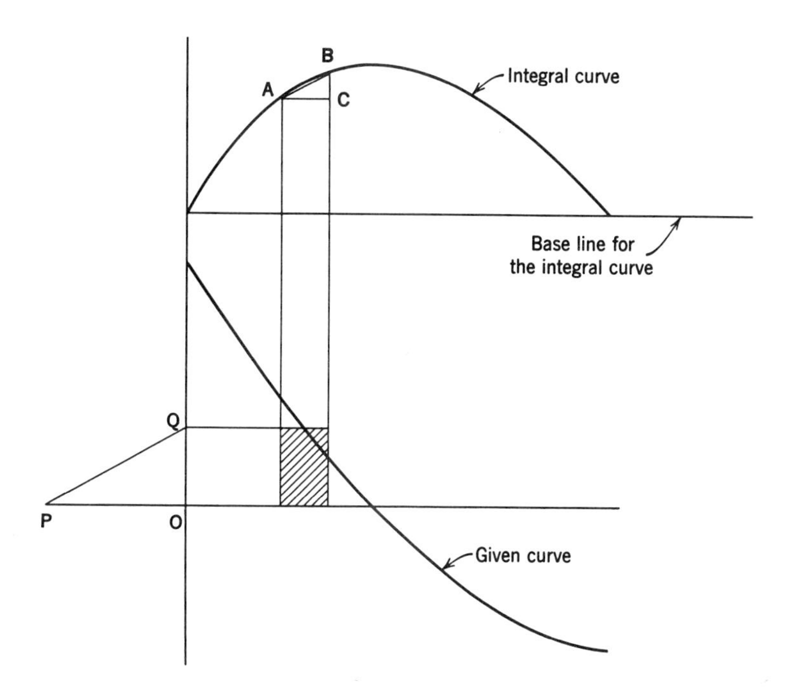

Fig. F-5 Relationship between graphical integration and differentiation.

Example 1. Let us consider the differential equation, $dy/dx = 1 - X$. We wish to determine the integral curve which when differentiated will yield $dy/dx = 1 - X$. We will assume that X varies from 0 to 1, and that $\Delta X = 0.1$. First we plot the "curve of slopes" from the expression $dy/dx = 1 - X$. Then this curve is integrated. *The integral curve is the graphical solution of the differential equation.* Figure F-6 shows the solution for the condition that the curve passes through the point (0, 0). In this elementary example, we can check the accuracy of the integral curve algebraically since we can easily find the solution, $Y = X - X^2/2 + C$. The advantages in using the pole-and-ray method, however, should be quite evident for such cases where an algebraic solution is not always feasible nor necessary.

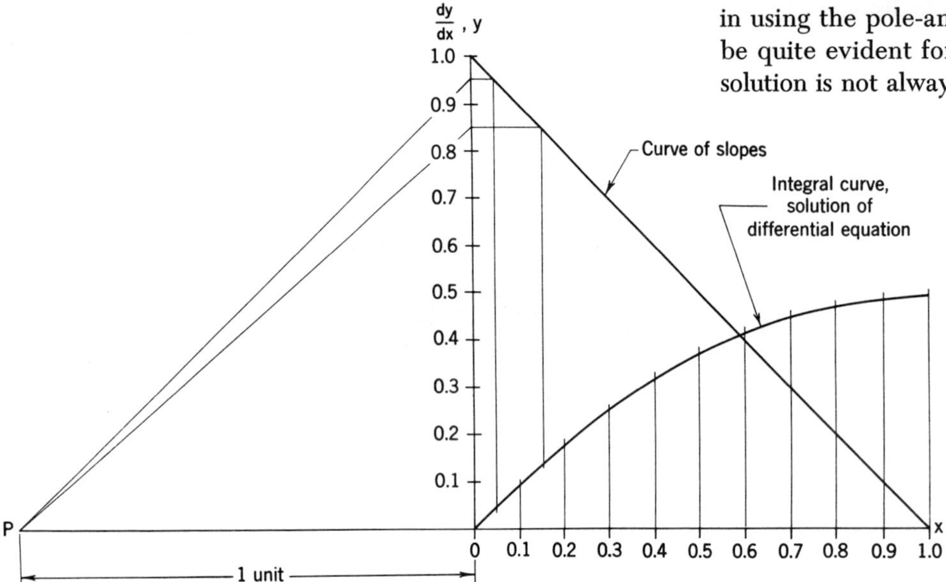

Fig. F-6 Graphical solution of differential equation Example 1. Pole-and-ray method.

Example 2. Let us apply the pole-and-ray method to equation (1) which was solved by the slope-field method.

The differential equation is

$$\frac{dy}{dx} = 2X + 2$$

We will assume that X varies from 0 to 2; that $\Delta X = 0.2$; and that the integral curve contains the point (0, 2). As in the previous example, a plot of the "curve-of-slopes" is made. This curve is then integrated to establish the integral curve which is the graphical solution of the differential equation. (See Fig. F-7.)

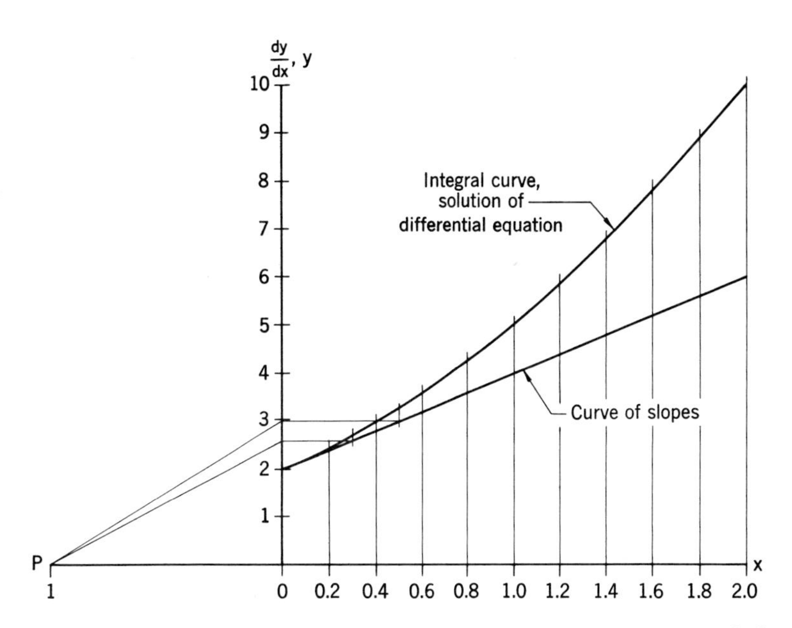

Fig. F-7 Graphical solution of differential equation Example 2. Pole-and-ray method.

Example 3. Let us consider the differential equation,

$$\frac{dy}{dx} = y - x^2$$

We will assume that X varies from 0 to 2; that $\Delta X = 0.2$; and that the integral curve passes through point (0, 2). The graphical solution, a combination of "curves-of-slope" and "pole-and-ray," is shown in Fig. F-8.

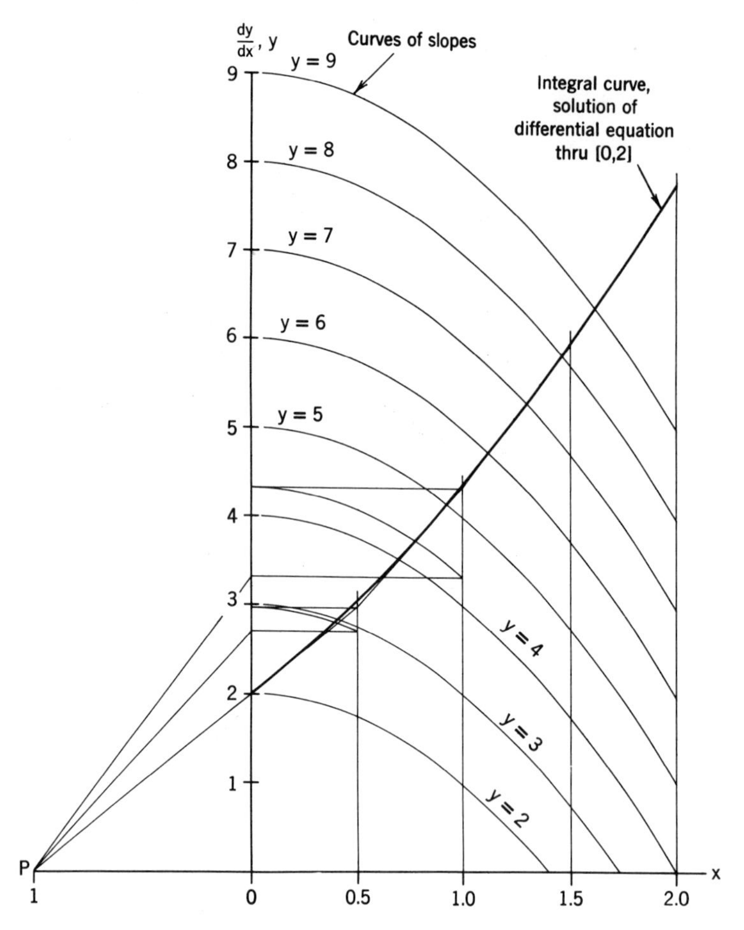

Fig. F-8 Graphical solution of differential equation Example 3. Curves of slopes and pole-and-ray method.

APPENDIX G
USEFUL TECHNICAL TERMS

Allen Screws—Socket-type screws with a hexagonal socket in the head. (See Fig. G-1.)

Alloy—Two or more metals mixed or combined to make a substance which is different from the pure metal, e.g., copper + tin = bronze.

Alternating Current—An electric current which first increases to a maximum in one direction, decreases to zero, then increases in the opposite direction, and so on.

Ampere—Unit for measuring the rate of flow of an electric current.

Amplifier—Radio valve used to increase the amplitude of an alternating current, and so strengthen the sound.

Anneal—When glass or metal is shaped stresses are caused. Annealing reduces the stresses; e.g., the glass (or metal) is heated, then cooled slowly to relieve the forces. The heat treatment may be to remove stresses; induce softness; refine the structure; or to alter toughness, electrical, magnetic, or other properties of the material.

Backlash—Looseness in a joint of a machine so that there is some free movement in one part before another part (which is joined to it) begins to move; e.g., between gear teeth of mating gears.

Bascule—A bridge of which one end lifts up so that ships may pass; or both ends, thus opening in the middle.

Bearing—The part of a machine in which a revolving rod (or shaft) is held, or which turns on a fixed rod (or shaft).

Bearing, Ball—One in which steel balls are used between the bearing surfaces to permit rolling action. (See Fig. G-2.)

Bearing, Roller—One in which cylindrical rolls are employed in place of steel balls. When rolls are quite small the bearing is referred to as a *needle bearing*. (See Fig. G-3.)

Bell Crank—A solid lever having two fixed arms, and pivoted where they join. (See Fig. G-4.)

Blocks, Gage—Hardened steel blocks accurately ground and finished to a specified dimension. Used in precision measurement.

Blocks, Lapping—Flat cast iron blocks with slots that criss-cross the surface which, when impregnated with fine abrasives, is used to produce a very smooth (lapped) surface finish.

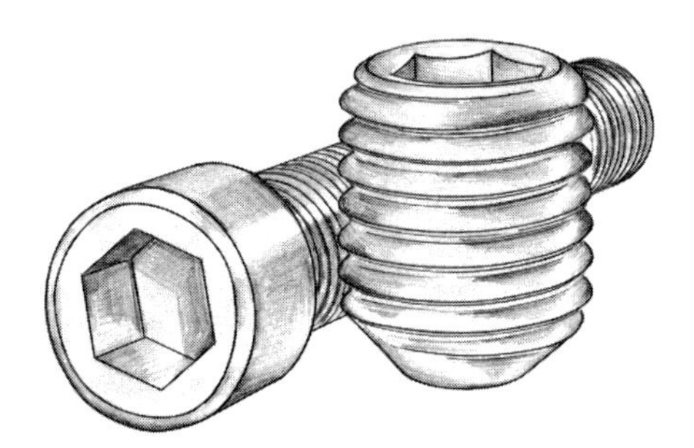

Fig. G-1 Allen screws.

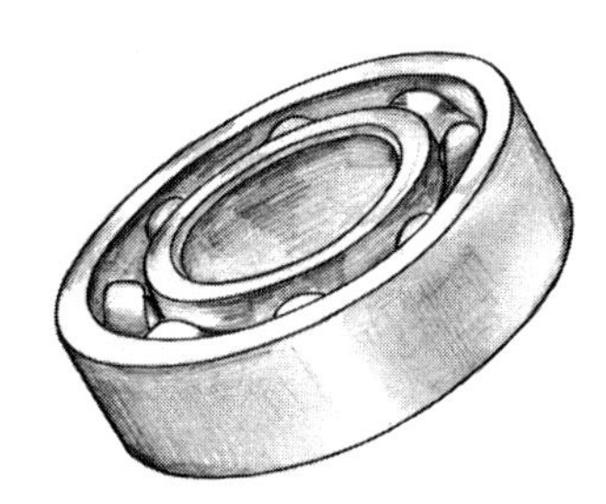

Fig. G-2 Ball bearing.

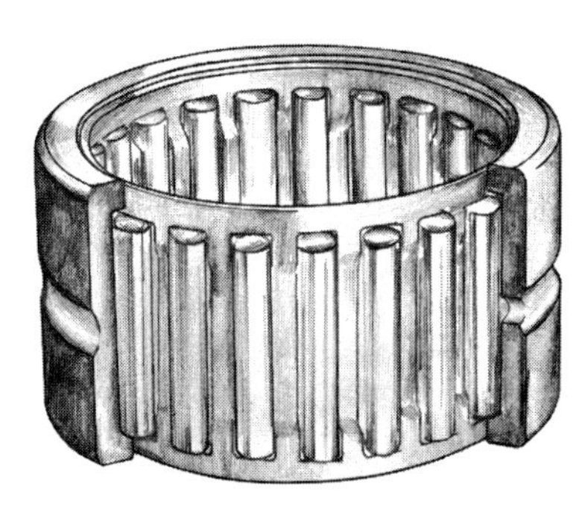

Fig. G-3 Roller bearing.

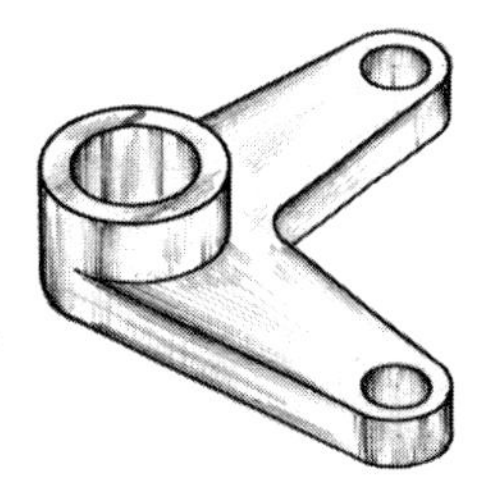

Fig. G-4 Bell crank.

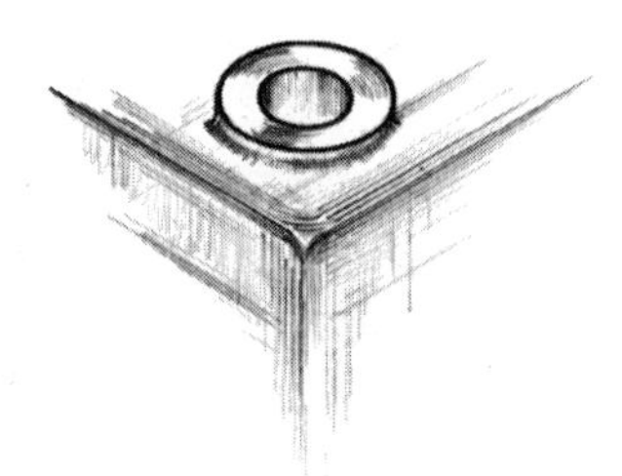

Fig. G-5 Boss.

Fig. G-6 Pad.

Fig. G-7 Broaches.

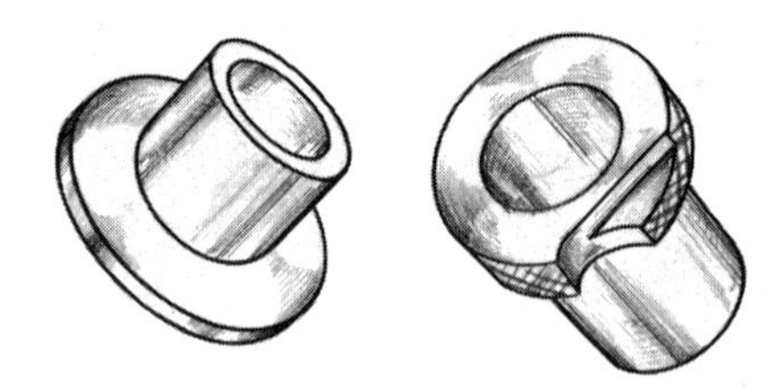

Fig. G-8 Bushings.

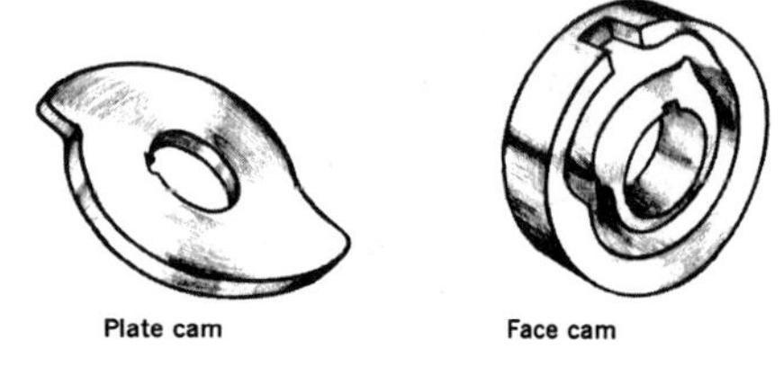

Fig. G-9 Cams.

BLOOM—Mass of melted metal or glass.

BLOOMING MILLS—Machines for rolling out iron and steel into sheets.

BOARD FOOT—Unit-measure of wood, 1 × 12 × 12 inches.

BOSS—A circular raised portion above the surface of a part. Usually provides a bearing surface around a hole. (See Fig. G-5.) Note the difference between a boss and a *pad.* (See Fig. G-6.)

BRAZING—A process for joining metals with alloys of copper and zinc. The alloys have melting temperatures quite below the melting temperatures of the metals.

BROACH—A tool with a square or six-sided or special-shaped blade, smaller toward the point, pushed or pulled to make a hole a certain shape. (See Fig. G-7.)

BULLDOZER—Machine like an army "tank" with a large plate in front used to push masses of earth, etc., in road-making and in clearing.

BUSHING—A cylindrical lining used as a bearing for a shaft or similar parts. Also a removable liner, inserted into the guide holes of a drill jig, to position the drill. (See Fig. G-8.)

CAM—A device used on a rotating shaft to transform rotary motion to lateral motion. There are several types of cams; e.g., disc (or plate), face, and cylindrical. The follower travels along the periphery of a disc cam; travels in a groove of a face cam; or in a groove cut in the outer surface of a cylindrical part to provide motion parallel to the axis of the cam. (See Fig. G-9.)

CAPACITANCE—Measure of the power of a capacitor (electric condenser) to hold electricity. The measure (farads) = amount of electricity (coulombs) held, divided by the voltage between the plates. Capacitance depends upon the distance between the plates, the substance (air, paper, plastic) between the plates and the area of the plates.

CARBURIZE—To cause the surface layer of steel to be combined with carbon, e.g., by heating the steel in a box with carbon (charcoal) packed around it (= case hardening).

CASTING—Any object made by pouring molten metal into a mold.

CATHODE—A negative electrical plate. In an X-ray tube or radio valve, electrons pass from the cathode through the gas (or vacuum) to the anode (the positive plate).

COLLET—A device to hold a tool or piece of stock during a machining operation. (See Fig. G-10.)

CONDENSER—(Electrical) also called a capacitor.

CONDUCTION—Passing electricity along a substance (e.g., wire), passing heat along a substance (e.g., metal bar).

CONDUIT—A large pipe (e.g., underground pipe) to carry water, or to contain electrical supply wires.

CURRENT—Flow of water in one direction; flow of electricity along a wire. Direct current: electric flow in one direction. Alternating current: flow first in one and then in the other direction.

COUNTERBORE—A tool to enlarge, to a specified depth, a previously drilled hole so that the bottom of the enlarged hole has a square shoulder. (See Fig. G-11.)

COUNTERSINK—A cone-shaped tool to provide a seat for a flat-head screw or rivet. (See Fig. G-12.)

DECIBEL—One tenth of a bel, a unit used to describe how many times one sound is more powerful than another.

DIELECTRIC (substance)—A substance which prevents the flow of electricity, (e.g., the air, paper).

DIE CASTING—A method for producing castings by forcing metal into a metallic mold or die under mechanical or pneumatic pressure.

DIODE—A radio valve containing only two electrodes; cathode (negative) which is usually a hot wire, anode (positive) plate.

ELASTIC LIMIT—The largest load per unit area which will not produce a measureable permanent deformation after the load is released.

ELECTRON—Negative electric charge which forms part of an atom.

ELECTRONICS—The study of instruments in which free electrons move, e.g., radio valves, cathode ray tubes.

ELECTRON-VOLT—Unit for describing the energy of a moving electron.

ENERGY—The capacity of a body to do work.

ENGINEERING—Engineering as a profession is concerned primarily with the design of circuits, machines, processes, and structures, or with combinations of these components into plants and systems, and the characteristic activity of full-fledged professional engineers is the predic-

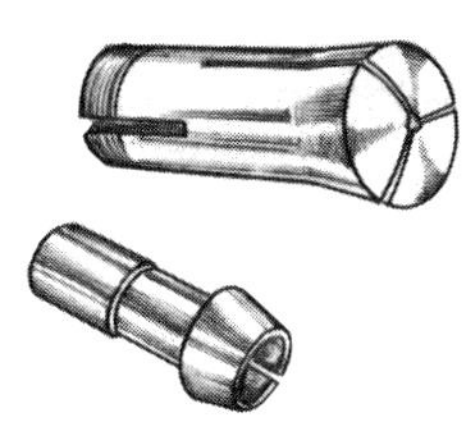

Fig. G-10 Collets.

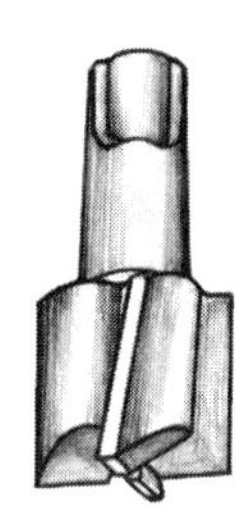

Fig. G-11 Counterbore.

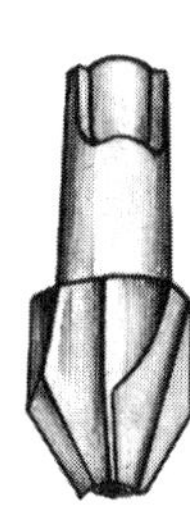

Fig. G-12 Countersink.

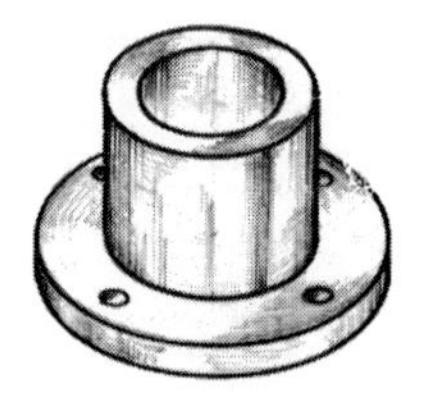

Fig. G-13 Flange.

tion of performance and cost under specified conditions.

FATIGUE—Change in structure of a material when subjected to repeated strain.

FIXTURE—A special tool used for holding work while performing operations upon the work.

FLANGE—A rim which extends from the main part, e.g., the edge which stands out from a railway wheel so the wheel will remain on the track. (See Fig. G-13.)

FORCE—The action of one body on another; it changes, or tends to change, the state of rest or motion of the body acted upon.

FLUX—A substance used in soldering or brazing to promote the fusion of the metals.

FLUX (magnetic)—The total amount of magnetic power passing through an area, e.g., through the coil of an electric generator.

FORGING—Metal that is shaped or formed, while hot or cold, by a hammer, press, or drop hammer.

GAGE—An instrument for determining the accuracy of specified manufactured parts.

GALVANIZING—A process of coating metal parts by dipping the parts in a molten bath containing another metal, (e.g., coating iron with zinc). Other methods can be used.

GUSSET PLATE—Metal plate fixed over two or more members meeting at an angle to strengthen the joint, (e.g., in fabricating a roof truss).

HOPPER—A container with sides sloping in toward an opening in the bottom.

HYDRAULIC—Having to do with water in movement.

INDEX HEAD—An attachment to milling machines that divides the circumference of a cylindrical part into a number of equal spaces.

INDICATOR, DIAL—A measuring instrument with a graduated dial face upon which the movement of an attached spindle that contacts the work is registered.

INDUCTANCE—When a changing current flows through a coil, pressure of electricity is caused in the coil.

ION—An ion is an atom which lacks its full number of electrons and so is positive; or has too many electrons and so is negative.

ISOTOPES—Atoms of different atomic weight, but chemically the same substance.

JIG—A special device used to guide the tool or to

hold the material in the right position for cutting, shaping, or fitting together.

JOURNAL—That part of a shaft that rotates in a bearing.

KNURLING—Forming of fine ridges on the surface of a part to increase gripping power. (See Fig. G-14.)

LATHE—A machine tool with a horizontal spindle that supports and rotates a part while it is machined by a cutting tool that is supported on a cross-slide. The lathe is used primarily for turning cylindrical pieces.

MACH NUMBER—A number that expresses the relation between speed of air-flow and the speed of sound. A number less than one indicates speed less than the speed of sound.

MACHINABILITY—A term used to denote the relative ease with which parts can be machined.

MAGNETRON—A kind of radio valve used for producing high frequency oscillations (swings of current backward and forward).

MICRO-WAVES—Very short radio waves (less than 20 centimeters).

MILLING—An operation for removing metal from a piece by means of a revolving cutter which is a multiple-toothed cutter.

MILLING, STRADDLE—An operation for milling opposite sides of one part at one time so that the surfaces will be parallel. (See Fig. G-15.)

MOLD—A form into which molten metal is poured to produce a casting of a desired shape.

NITRIDING—Hardening the surface of steel by passing ammonia gas over it.

NORMALIZING—Heating iron-base alloys above the critical temperature range and cooling in air at room temperature to reduce internal forces.

OHM—The unit for measuring electrical resistance. A conductor has a resistance of one ohm when one ampere flows through it at a pressure of one volt.

PAWL—A device used to prevent a toothed wheel from turning backwards; or a device that stops, locks, or releases a mechanism. (See Fig. G-16.)

PEEN—A process to expand or stretch metal by hammering with a peen hammer or by a shot-peening method.

PICKLE—A process for removing scale from castings or forgings by immersion in a water solution of

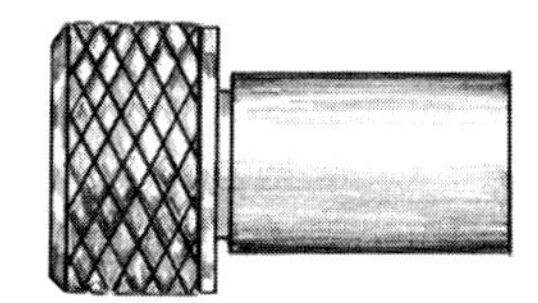

Fig. G-14 Knurling.

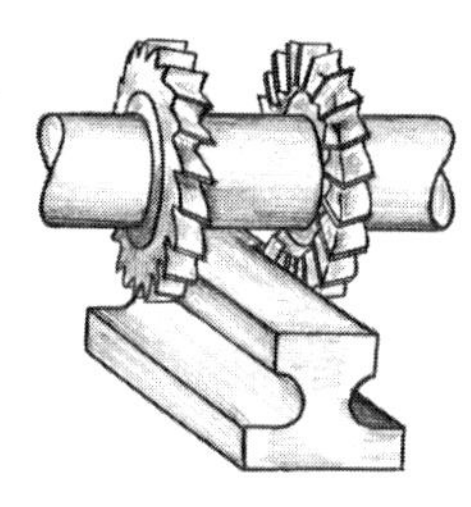

Fig. G-15 Straddle milling.

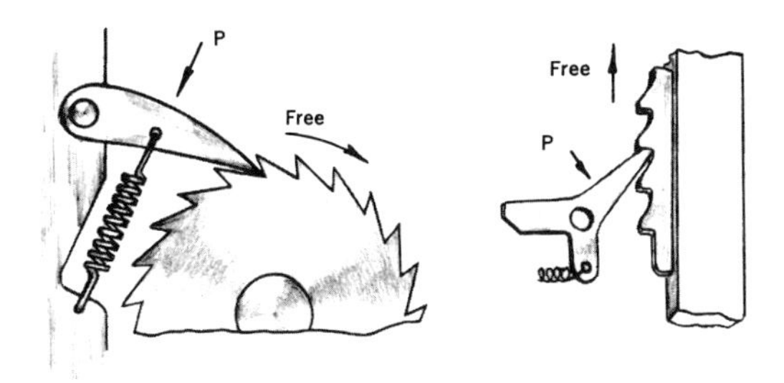

Fig. G-16 Pawls.

sulphuric or nitric acids. Pickling conditions the surfaces for plating.

PLANER—Machine tool for producing flat surfaces on metal parts.

POLISH—To produce a smooth surface, usually by a polishing wheel of leather, felt, or wool to which a fine abrasive is glued.

PUNCH—A tool that pierces holes in metal. Also a small hand tool used to set nails or to mark centers.

QUENCH—Cooling hot metal quickly by immersion in liquids or gases.

RACK—A toothed bar, acting on (or acted upon by) a gear-wheel. (See Fig. G-17.)

Fig. G-17 Rack.

RADAR—(Radio-Directing and Ranging), a way of finding the distance and direction of objects. Radio pulses are sent out and reflected back from the object. The return pulses are seen on a screen like that of television.

REAMER—A tool used to finish drilled holes. The fluted tool produces the finish by rotation through the drilled hole. Commonly used reamers are straight or tapered. (See Fig. G-18.)

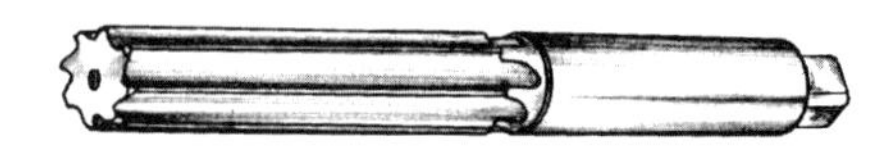

Fig. G-18 Reamer.

RELIEF—A goove (e.g., a cut next to a shoulder) on a part to facilitate machining operations (e.g., grinding). (See Fig. G-19.)

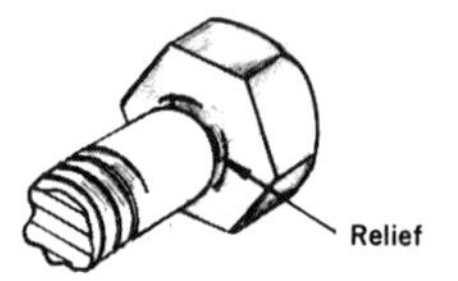

Fig. G-19

RESISTOR—Piece of wire (e.g., coil of high-resistance wire) used in an electric circuit to offer resistance to the flow of current.

SANDBLASTING—A process of cleaning castings by driving sand, under pressure, against the surfaces of the pieces. Also used to give a rough surface to glass.

SHAPER—A machine tool that has a sliding ram; used to produce flat surfaces. The work is stationary. The cutting tool is carried on the ram which has a reciprocating motion.

SHEAR—A tool for cutting metal. The cutting can be accomplished by two blades, one fixed and the other movable.

SHIM—A thin strip of metal (often laminated) placed between surfaces to permit adjustment for fit.

SPLINE—A key for a relatively long slot. (See Fig. G-20.)

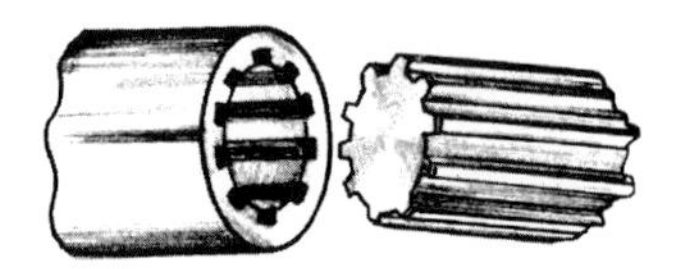

Fig. G-20 Spline.

SPOT-FACING—A drilling operation, using a counterbore tool, to smooth the surface around a hole to provide a good bearing surface for a bolt or screw head.

STOCHASTIC—Implies the presence of a random variable.

SWAGING—A method for forming a piece of cold metal by drawing, sqeezing or by hammering with a pair of dies shaped to the desired form. Swage blocks can be used to shape the part.

TAP—A tool for cutting internal threads. The tool has a thread cut on it, and is fluted to produce cutting edges and to remove chips. See Fig. G-21.

THERMIONICS—The science dealing with the emission of electrons from hot bodies.

TRANSDUCER—A power-transforming device for insertion between electrical, mechanical, or acoustic parts of systems of communication.

TRANSISTOR—An electronic device for rectification and/or amplification, consisting of a semiconducting material to which contact is made by three or more electrodes, which are usually metal points, or by soldered junctions.

UPSETTING—A process for enlarging the diameter of a metal rod or the end of a tenon or stud. Rivet shanks are upset to provide material to form rivet heads in fastening parts.

WELDING—Method for joining two metal pieces by heating the joint to fusion temperature.

Note: For additional terms that relate to the manufacture of fasteners, see "Glossary of Terms for Mechanical Fasteners," ASA B18.12-1962.

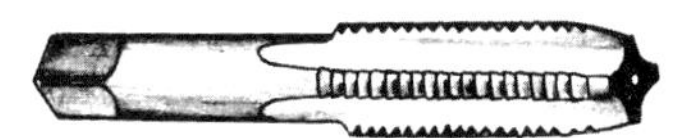

Fig. G-21 Tap.

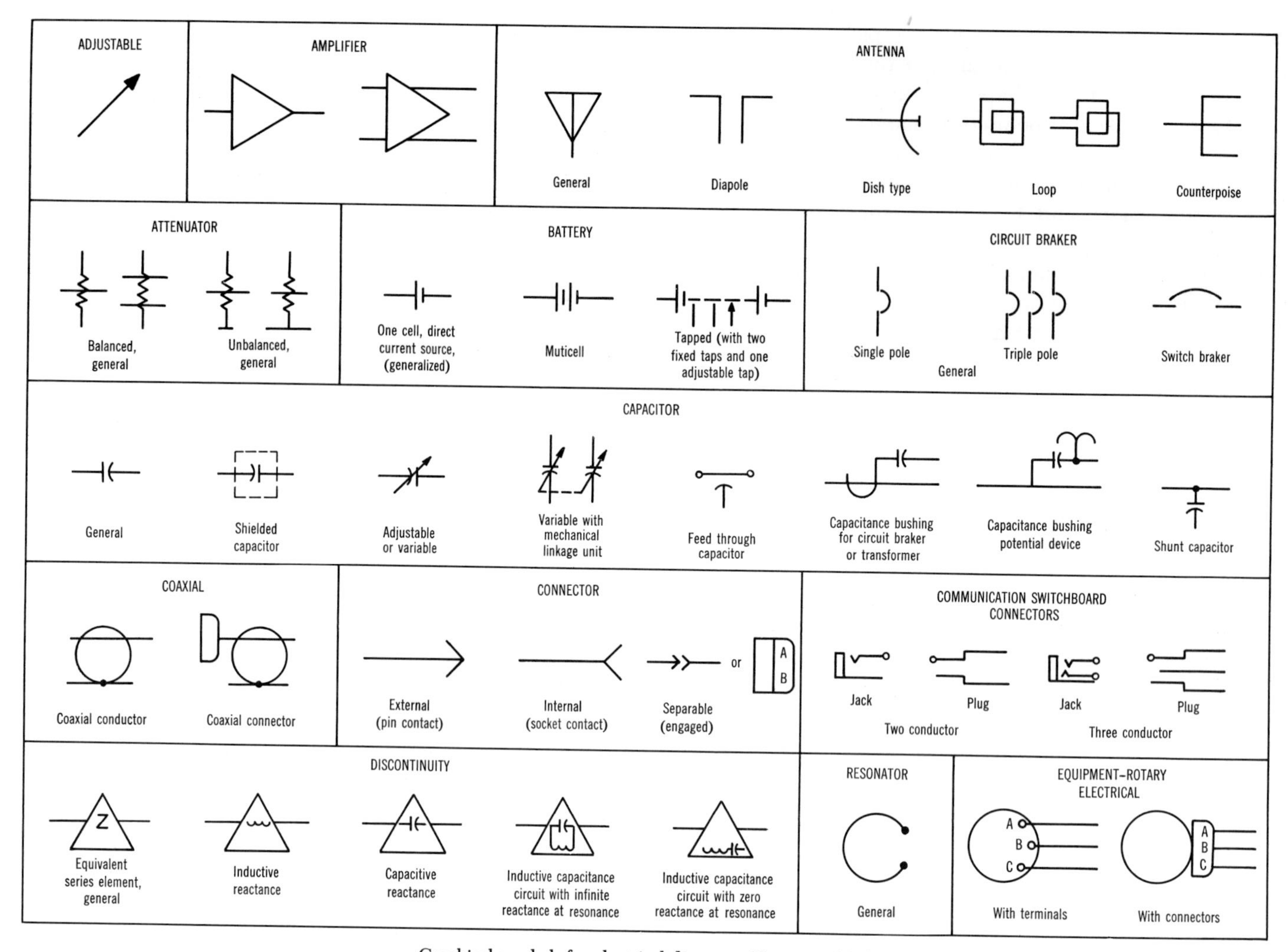

Graphical symbols for electrical diagrams (Courtesy ASA)

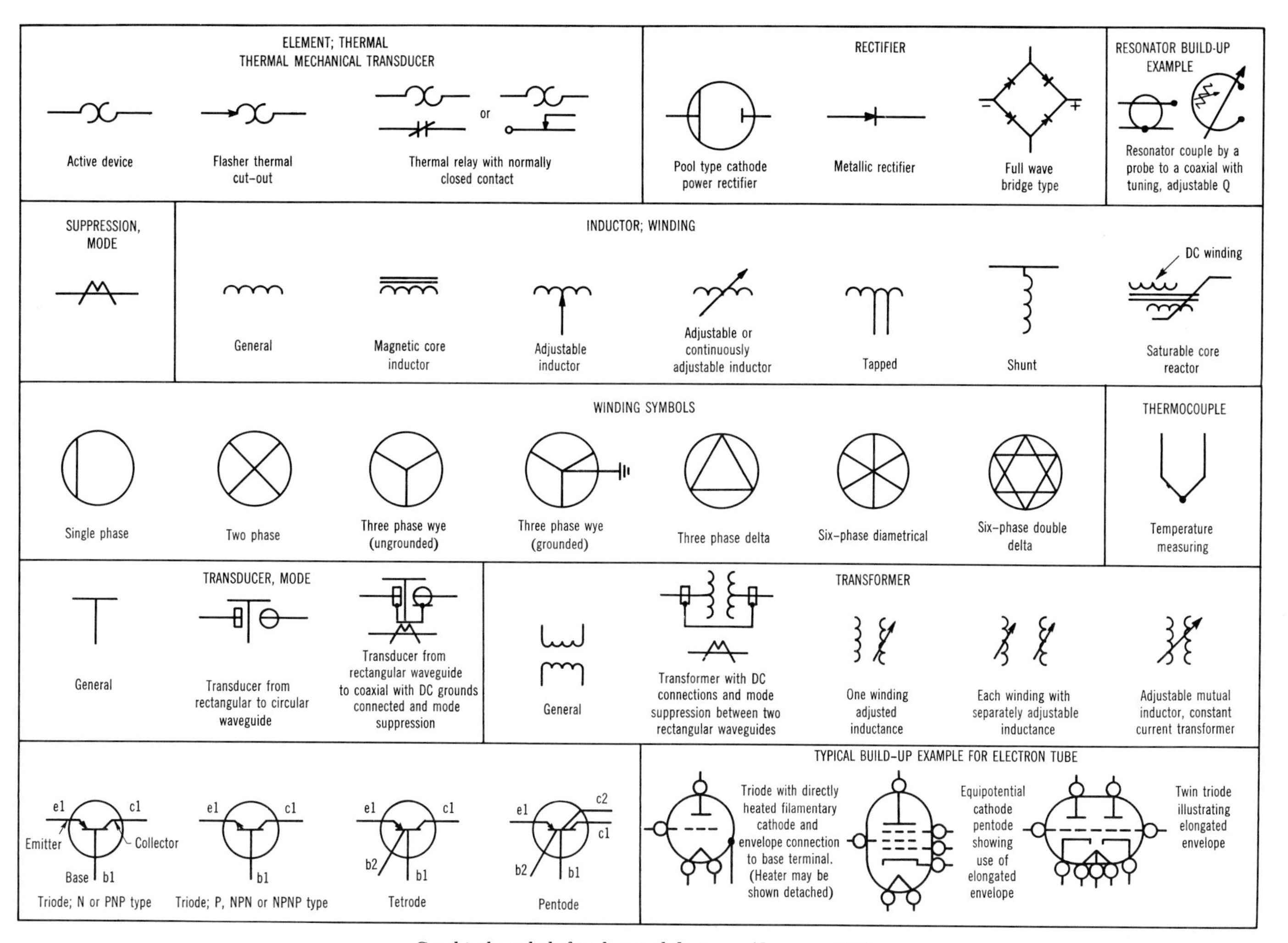

Graphical symbols for electrical diagrams (Courtesy ASA)

INDEX